STALLCUP'S®

Electrical Design, 2008 Edition

Based on the NEC® and Related Standards

James G. Stallcup
with
James W. Stallcup

JONES AND BARTLETT PUBLISHERS

Sudbury, Massachusetts

BOSTON   TORONTO   LONDON   SINGAPORE

Jones and Bartlett Publishers
40 Tall Pine Drive
Sudbury, MA 01776
978-443-5000
www.jbpub.com

Jones and Bartlett Publishers Canada
6339 Ormindale Way
Mississauga, Ontario L5V 1J2
Canada

Jones and Bartlett Publishers International
Barb House, Barb Mews
London W6 7PA
United Kingdom

National Fire Protection Association
1 Batterymarch Park
Quincy, MA 02169
www.NFPA.org

Jones and Bartlett's books and products are available through most bookstores and online booksellers. To contact Jones and Bartlett Publishers directly, call 800-832-0034, fax 978-443-8000, or visit our website www.jbpub.com.

Substantial discounts on bulk quantities of Jones and Bartlett's publications are available to corporations, professional associations, and other qualified organizations. For details and specific discount information, contact the special sales department at Jones and Bartlett Publishers via the above contact information or send an email to specialsales@jbpub.com.

Production Credits

Chief Executive Officer: Clayton E. Jones
Chief Operating Officer: Donald W. Jones, Jr.
President, Higher Education and Professional Publishing: Robert W. Holland, Jr.
Sr. V.P., Sales and Marketing: James Homer
V.P., Production and Design: Anne Spencer
V.P., Manufacturing and Inventory Control: Therese Connell
Publisher, Public Safety Group: Kimberly Brophy

Acquisitions Editor: Martin Schumacher
Reprints Coordinator/Associate Production Editor: Amy Browning
Director of Marketing: Alisha Weisman
Marketing Associate: Meagan Norlund
Text Design, Graphics, and Layout: Billy G. Stallcup
Cover Design: Anne Spencer
Text Printing and Binding: Courier Corporation
Cover Printing: Courier Corporation

Copyright © 2009 by Grayboy, Inc.

All rights reserved. No part of the material protected by this copyright notice may be reproduced or utilized in any form, electronic or mechanical, including photocopying, recording, or by any information storage and retrieval system, without written permission from the copyright owner and publisher.

Publication of this work is for the purpose of circulating information and opinion among those concerned for fire and electrical safety and related subjects. While every effort has been made to achieve a work of high quality, neither the publisher, the NFPA, the authors, or the contributors to this work guarantee the accuracy or completeness of or assume any liability in connection with the information and opinions contained in this work. The publisher, NFPA, authors, and contributors shall in no event be liable for any personal injury, property, or other damages of any nature whatsoever, whether special, indirect, or consequential, or compensatory, directly or indirectly resulting from the publication, use of or reliance upon this work. This work is published with the understanding that the publisher, NFPA, authors, and contributors to this work are supplying information and opinion but are not attempting to render engineering or other professional services. If such services are required, the assistance of an appropriate professional should be sought.

National Electrical Code® and NEC® are registered trademarks of the National Fire Protection Association, Inc.

Library of Congress Cataloging-in-Publication Data
Stallcup, James G.
  Stallcup's electrical design 2008 / James G. Stallcup.
     p. cm.
  Includes index.
  ISBN-13: 978-0-7637-5253-8
  ISBN-10: 0-7637-5253-3
  1. Electric wiring. 2. Electric wiring—Problems, exercises, etc. 3. Electric apparatus and appliances—Installation—Problems, exercises, etc. I. Title.
  TK3201.S7324 2008
  621.319'24—dc22
                          2007048634

6048

Printed in the United States of America
12 11 10 09 08   10 9 8 7 6 5 4 3 2 1

Introduction

James Stallcup, Sr. has reinvented the electrical design book that catapulted him to the forefront of the electrical industry as a **prominent author**, **effective instructor** and **respected authority** on the application of the *National Electrical Code*®.

Everyone from **engineers**, **electrical contractors**, **inspectors**, **electricians,** and **instructors** of the Code have anticipated the arrival of this book. The large workbook format allows a masterful blending of valuable Design Tips, NEC Loops, Examples, Quick Calcs and effective illustrations with authoritative Code references. Because of the abundant amount of detailed information included, it is the most comprehensive design book of its kind.

Stallcup's® *Electrical Design* book explains the purpose of the *National Electrical Code* (NEC) and more particularly, its use as it applies to the design and installation of electrical wiring systems and equipment.

While the substance of design is found in the NEC, the art of the design is found in the applicability of that same NEC. With the advancement of today's techology and ever-increasing liabilities, effective electrical design must now, more than ever, consider the use of certified products, energy conservation, economy vs. quality, anticipated load growth, local codes, special applications of electrical equipment, and the use and interpretation of NFPA and IEEE standards that relate to special areas, etc. For better understanding and interpretation of these advancements, considerable effort has been made by the author to condense the more complicated rules pertaining to the design, installation and selection of wiring methods and equipment.

For the convenience of the reader, the Electrical Design not only contains discussions and explanations of code rules, but also includes detailed illustrations and sample calculations that will help tremendously in understanding and becoming proficient in the application of the NEC. The Electrical Design also points out common industry problems and shows in detail the proper procedures and techniques to use in order to ensure proper code compliance. Design Tips, Calculation Tips and guidelines for "rule of thumb" methods for instances where a fast and approximate design answer is needed are also provided.

In order for the reader to measure his or her design skills or to prepare for a maintenance, journeyman or master electrician examination or for inspector certification, an ample number of quizzes, tests and final examinations have been included.

With the wealth of information found in this book, it will make a valuable addition to the library of any consultant, serve as an excellent reference to the seasoned professional, and also provide valuable introductory material to the beginner.

Table of Contents

Electrical Codes
and Standards

The *National Electrical Code®* (NEC®) shall be used in conjunction with other codes and standards when designing and installing electrical systems if they are to be considered safe, dependable and reliable. This chapter discusses such codes and standards and outlines how they are utilized with the rules and regulations of the NEC.

HISTORY OF THE *NATIONAL ELECTRICAL CODE*

The *National Electrical Code* (NEC) was first published in 1897 under the sponsorship of the National Conference of Electrical Rules, the membership of which consisted of delegates from interested national associations. The 1897 NEC was the forerunner of a standard that has been responsible in saving many lives and many dollars in property. This organization was disbanded in 1911, at which time the National Fire Protection Association (NFPA) took over the great responsibility of writing and publishing the NEC, which has been continued since that time. The NEC has been adopted by the American National Standards Institute (ANSI) as a National Consensus Document.

NFPA set up a National Electrical Code Committee, which is headed by the Technical Correlating Committee, and its members work with the various code-making panels. Each code-making panel has the responsibility of certain Articles during the proposal and comment stage of upgrading the NEC. The Technical Correlating Committee and code-making panels include dedicated personnel from all phases of the electrical industry. Members are assigned to the particular code-making panel where it is felt they have the most expertise. The NEC is a minimum standard used for electrical installations and for the protection of life and property.

PURPOSE OF THE NEC

Electricity can be very hazardous if proper precautions are not adhered to. The NEC is intended to provde guidelines for proper installations, in order to safeguard personnel and property from the dangers of electricity.

NEC requirements are essential for safety and compliance. However, such requirements may not necessarily result in efficient, convenient or adequate service if not used as designed.

> **For example,** provisions for future expansion of the electrical system may not be provided. However, the electrical system is essentially free from hazards that may be encountered during the use of its elements. If future needs are taken into consideration at the time of the original design, and installation and adequate measures taken to provide for the increased usage of electricity, such hazards and overloading should be greatly eliminated.

The NEC is not intended to be used for design specifications, nor as an instruction manual for untrained persons. The requirements of the NEC will, however, serve as rules that may be used to help design electrical systems properly.

The NEC is usually adopted by government agencies and becomes effective January 1 of the year that it is published. Its rules and regulations govern wiring methods and equipment used in residential, commercial and industrial facilities.

> **Design Tip:** There may be additional requirements by local agencies, and such amendments must be checked, for they are usually more restrictive and require in some cases larger components to be used in completing electrical systems.

APPLYING THE NEC

The NEC is intended to be used as a mandatory standard by governmental bodies and inspection authorities. All fields of the electrical industry apply the NEC as a minimum standard for the installation of electrical systems.

LAYOUT OF THE NEC

The rules and regulations of the NEC are laid out by Chapters, Articles, Parts, and Sections, which contain the requirements for design, installation, use, and maintenance of electrical systems. The arrangement of the Chapters and what they pertain to are as follows:

- Introduction
- Chapters 1 through 4 - General Requirements
  Chapter 1 - General
  Chapter 2 - Wiring and Protection
  Chapter 3 - Wiring Methods and Materials
  Chapter 4 - Equipment for General Use
- Chapter 5 - Special Occupancies
- Chapter 6 - Special Equipment
- Chapter 7 - Special Conditions
- Chapter 8 - Communications Systems
- Chapter 9 - Tables
  Chapters 1 through 4 apply generally.
  Chapters 5 through 7 supplement or modify the general rules in Chapters 1 through 4.
  Chapter 8 is independent of other chapters except as referenced.
- Annex H - Administration and Enforcement

See Figure 1-1 for a detailed illustration of how the NEC is laid out.

ADMINISTRATION AND ENFORCEMENT ANNEX H

Annex H covers code administration and enforcement rules. Included in this Annex, located at the back of the NEC, are provisions for existing nonconforming installations and appointment requirements for the authority of electrical inspectors and inspection procedures. Another provision is the creation of an electrical board having the power to appoint, regulate and discipline inspectors. This Annex also has requirements for plan review, permits and record keeping. For this Annex to be enforceable, its provisions must be adopted by a jurisdiction. Until it is adopted, its rules may only function in an advisory capacity. **(See Figure 1-2)**

INTRODUCTION AND CHAPTER 1 ARTICLES 90, 100, AND 110

For many years in the electrical industry, **Chapter 1** has been called the "get acquainted" chapter. Everyone in the electrical industry who has the responsibility of applying the rules of the NEC must know the information and requirements of **Chapter 1**. When using the definitions and rules of **Chapter 1**, the "get acquainted chapter," the user of the NEC must use either **Article 90** or one of the Articles of the 100 series.

THE CODE IS DIVIDED INTO THE INTRODUCTION AND NINE CHAPTERS. CHAPTERS 1, 2, 3, AND 4 APPLY GENERALLY; CHAPTERS 5, 6 AND 7 APPLY TO SPECIAL OCCUPANCIES, SPECIAL EQUIPMENT OR OTHER SPECIAL CONDITIONS. THE LATTER CHAPTERS SUPPLEMENT OR MODIFY THE GENERAL RULES. CHAPTERS 1 THROUGH 4 APPLY EXCEPT AS AMENDED BY CHAPTERS 5, 6 AND 7 FOR THE PARTICULAR CONDITIONS.

CHAPTER 8 COVERS COMMUNICATIONS SYSTEMS AND IS INDEPENDENT OF THE OTHER CHAPTERS EXCEPT WHERE THEY ARE SPECIFICALLY REFERENCED THEREIN.

CHAPTER 9 CONSISTS OF TABLES .

VERTICAL LINE BESIDE SECTION INDICATES A CHANGE IN RULES

Figure 1-1. Electrical workers and maintenance personnel shall be trained on applying the rules of the NEC as a means of providing safety for electrical or nonelectrical workers that work around or near electrical circuits and components.

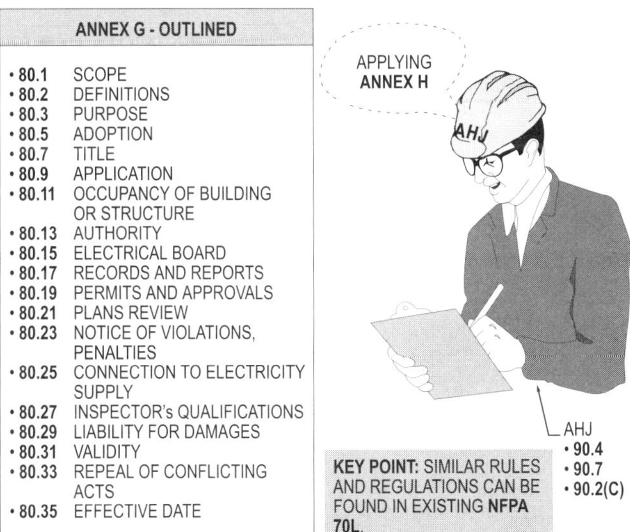

Figure 1-2. Annex H contains a standarized set of administrative requirements, which may be utilized by adopting jurisdictions.

INTRODUCTION
ARTICLE 90

Article 90 contains requirements that apply to the purpose of the NEC and other pertinent information that shall be utilized with such purpose.

For example, what is "covered" and what is "not covered" by the NEC is found in **Article 90**. The word "purpose" simply means the intent, aim or goal of the requirements of the NEC. **(See Figure 1-3)**

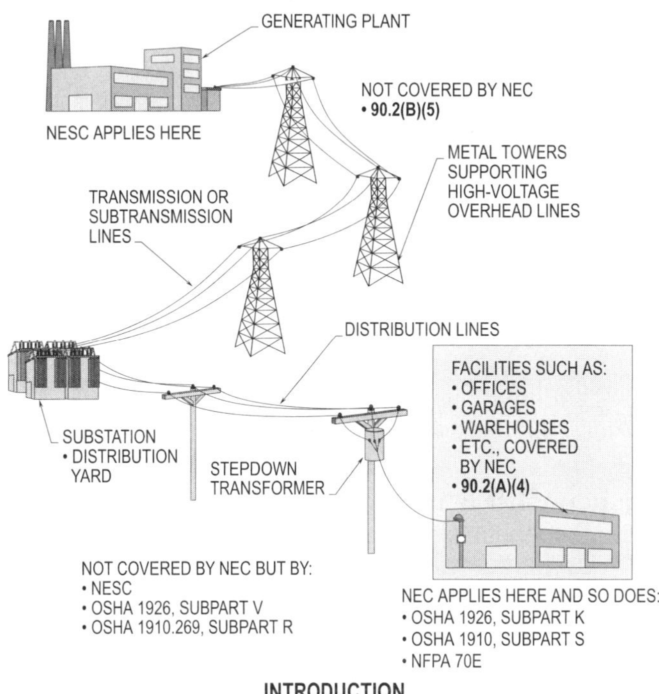

Figure 1-3. Article 90 contains requirements that give the purpose of the NEC and other pertinent information that shall be utilized with such purpose.

CHAPTER 1
ARTICLE 100

Article 100 breaks down the language barrier for the user of the NEC. When users of the NEC fully understand the meanings of the words, terms or phrases that are used, they have a greater chance to comprehend and properly interpret the rules of the NEC.

For example, if the designer wanted to run two services to a building and if he or she was to obtain the definition of a building from *Webster's Dictionary* instead of **Article 100**, the designer would be very confused as to whether an individual service could be provided for such a building because of different definitions. The definition in **Article 100** is defined specifically for the installation of wiring methods and equipment in a building.

Commonly defined terms are not included in **Article 100**. Only the terms essential to the proper application or enforcement of the NEC rules are listed. **(See Figure 1-4)**

> **Design Tip:** When determining clearances in front of electrical installations for 600 volts or less or over 600 volts, **Article 110** of the 100 series in **Chapter 1** shall be used.

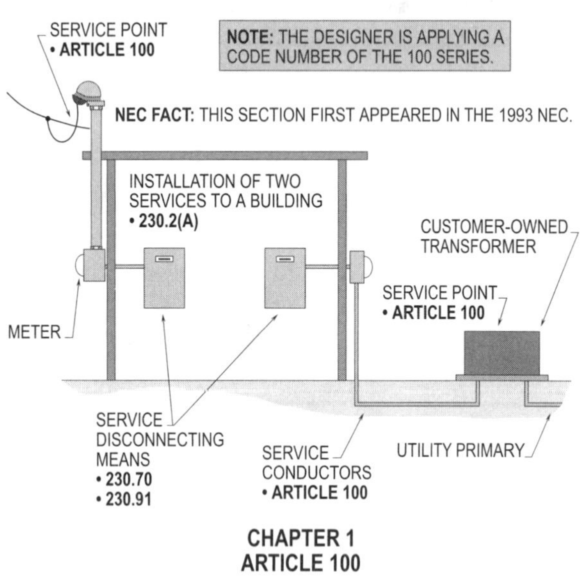

Figure 1-4. Article 100 contains definitions that are essential to the proper application and enforcement of the NEC.

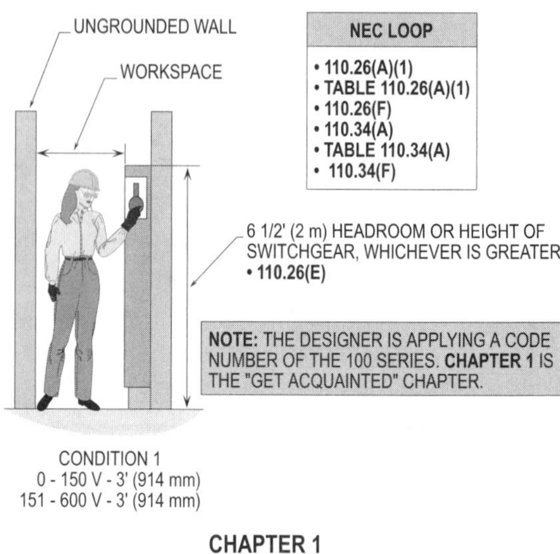

Figure 1-5. Article 110 contains requirements for electrical installations except those granted, supplemented or modified by a rule in **Chapters 5, 6,** or **7** in the NEC.

The definitions for systems of 600 volts or less are listed in **Part I,** and those for over 600 volts are listed in **Part II** of **Article 100**.

> **Design Tip:** For a definition to be eligible to be included in **Article 100**, such definition shall appear in at least two or more Articles of the NEC. Otherwise, definitions appear in front of an Article, such as **240.2** for definitions used in **Article 240**.

CHAPTER 1
ARTICLE 110

Article 110 of the NEC covers the requirements for electrical installations. The rules of **Article 110** apply to all types of electrical installations after **Chapter 1** unless there is an exception granted or it is supplemented or modified by a rule in **Chapters 5, 6,** or **7** in the NEC.

For example, the clearance in front of a 277/480 volt panelboard is 3 ft (914 mm) per **Table 110.26(A)(1),** if the electrical installation falls under the provisions of **Condition 1** to such Table. **(See Figure 1-5)**

CHAPTER 2
ARTICLES 200 - 285

Chapter 2 with its accompanying Articles is used by designers to design the elements and components of branch-circuits, feeder-circuits, and service equipment that make up the premises electrical system. The title of **Chapter 2** is "Wiring and Protection," which pertains to designing loads for selecting wiring methods and equipment and sizing overcurrent protection devices to protect such elements as conductors, parts, and equipment.

When designers are in the planning stage and calculations have to be made by which the elements are determined and equipment is selected, the designer shall apply one of the Articles of **Chapter 2**. In other words, one of the Articles in the 200 series shall be chosen.

For example, if the designer wants to know how many disconnects are allowed for each service, the subject is services and disconnects, the Article is **230,** and **230.71,** and **230.72** are applied.

When designing the electrical system, the designer must think of **Chapter 2** and use one of the 200 series Articles that deals with his or her particular design.

If one of the Articles in **Chapter 2** can't be used to design a certain installation, the designer will be referred to another Chapter that covers such installation.

For example, 220.50 in the NEC refers the designer to Article **430 (430.24, 430.25,** and **430.26)** for calculating motor loads. **(See Figure 1-6)**

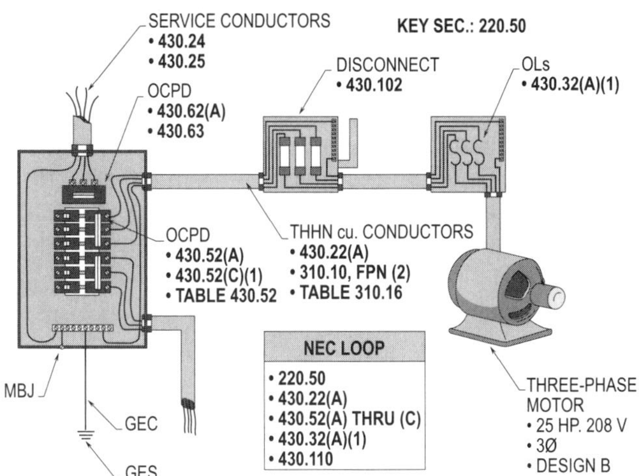

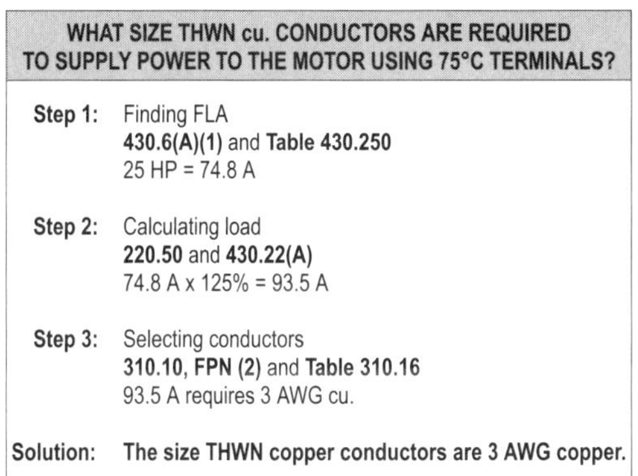
WHAT SIZE THWN cu. CONDUCTORS ARE REQUIRED
TO SUPPLY POWER TO THE MOTOR USING 75°C TERMINALS?

Step 1: Finding FLA
430.6(A)(1) and **Table 430.250**
25 HP = 74.8 A

Step 2: Calculating load
220.50 and **430.22(A)**
74.8 A x 125% = 93.5 A

Step 3: Selecting conductors
310.10, FPN (2) and **Table 310.16**
93.5 A requires 3 AWG cu.

Solution: The size THWN copper conductors are 3 AWG copper.

CHAPTER 2
ARTICLES 200 - 285

Figure 1-6. Chapter 2 is used by designers to design the elements and components of branch circuits, feeders, and service equipment. (See **220.50**)

Design Tip: The first six Articles of **Chapter 2** are used for designing the electrical system and the last three are used for protecting people, wiring, and equipment.

CHAPTER 3
ARTICLES 300 - 398

Chapter 3 with its accompanying Articles is used by installers to install the electrical wiring methods and equipment after they have been designed by the designer

per **Chapter 2** in the NEC. Electricians cannot install electrical wiring methods and equipment that comply with NEC rules without reading, studying and applying the requirements of **Chapter 3**.

The rules that electricians must use for routing raceways and cable systems, mounting and securing boxes, outlet boxes, junction boxes, panelboards, switchgear, etc. are outlined.

For example, how often does a run of EMT between a process machine and panelboard have to be supported?

Section **358.30(A)** requires supports to be provided within 3 ft (900 mm) of the machine and panelboard and at intervals of 10 ft (3 m) for EMT installed between such equipment. **(See Figure 1-7)**

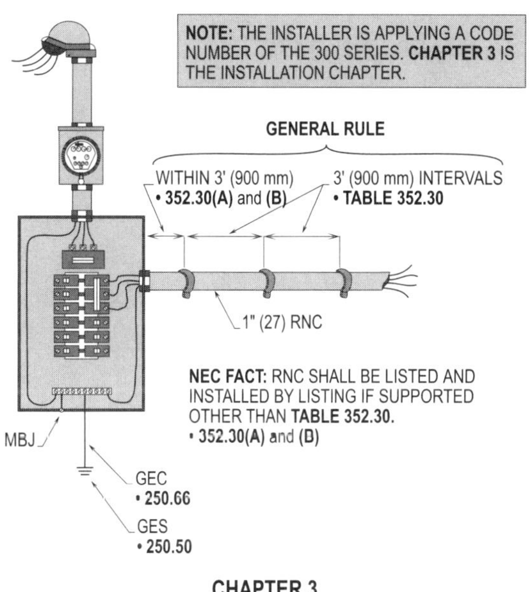

CHAPTER 3
ARTICLES 300 - 398

Figure 1-7. Chapter 3 is used by installers to install the electrical wiring methods and equipment after they have been designed by **Chapter 2**.

CHAPTER 4
ARTICLES 400 - 490

Chapter 4 with its accompanying Articles is used by users and maintainers. The electrical equipment in **Chapter 4** is current consuming and requires electricity to operate. Therefore this chapter is usually applied after the design and installation of such systems and equipment have been completed.

For example, what Article and Section in the 400 series is used to replace a 120 volt ballast (1000 volts or less output) in an electrical discharge luminaire?

Section **410.130(E)** of **Article 410** in **Chapter 4** is used to replace the ballast of 1000 volts or less. The maintainer may use the rules of **410.130** when dealing with replacement parts in electric-discharge luminaires. **(See Figure 1-8)**

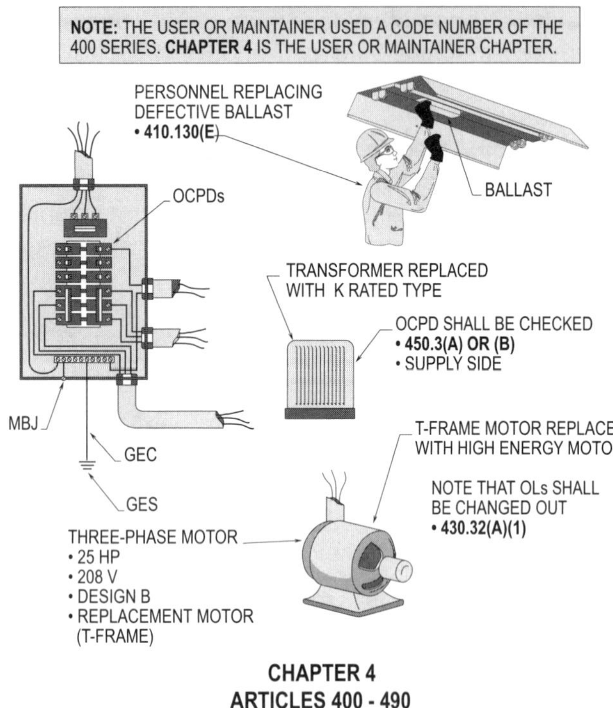

Figure 1-8. Chapter 4 with its accompanying Articles is used by users and maintainers.

CHAPTER 5
ARTICLES 500 - 590

Chapter 5 deals with special occupancies, and the first four Articles deal with identifying specific classifications and general provisions. Rules and regulations pertaining to classifying these areas are discussed. The remaining Articles cover specific provisions for specific locations such as commercial garages, aircraft hangars, service stations, bulk storage plants, spray applications, etc., as well as health care facilities.

Design Tip: When special occupancies are involved, one of the Articles of the 500 series shall be used based upon the type of occupancy listed in **Chapter 5**. **(See Figure 1-9)**

CHAPTER 6
ARTICLES 600 - 695

Chapter 6 is completely devoted to special equipment that requires special requirements for design and installation. In other words, the needed rules cannot be found in **Chapter 4,** which covers "equipment for general use."

If an engineer is designing the electrical system for an elevator, he or she shall apply the rules and regulations of **Article 620** in **Chapter 6**.

For example, to design the disconnecting means to disconnect power to the machine room for elevator equipment, the designer shall reference **620.51** in the NEC for requirements related to sizing, selecting, and locating this disconnect.

Design Tip: When the requirements for special equipment are not found in **Chapter 4**, the designer shall use one of the Articles found in the 600 series of **Chapter 6**. **(See Figure 1-10)**

CHAPTER 7
ARTICLES 700 - 770

Chapter 7 contains requirements for special conditions that arise from the particular type, use, and application of the power source used to supply circuits and equipment. The number used to identify Articles and Sections of **Chapter 7** is 700. When special conditions pertaining to the power source, circuits, and equipment are addressed, one of the Articles of the 700 series shall be used.

For example, the designer wants to know if the wiring of an emergency system may be routed in the same raceway as the wiring of the normal power source.

Special conditions are needed to validate this requirement. **Article 700** of **Chapter 7** is used, and **700.9(B)** in the NEC does not permit the emergency wiring and normal wiring to be mixed in a raceway system. **(See Figure 1-11)**

Design Tip: Designers must take notice and carefully review how predominantly the **source**, the **supply**, and the **circuit(s)** are applied in each Article of **Chapter 7**, to ensure the reliability of such wiring methods and equipment. Remember these "three words" are main "key words" in applying the rules and regulations of **Chapter 7**.

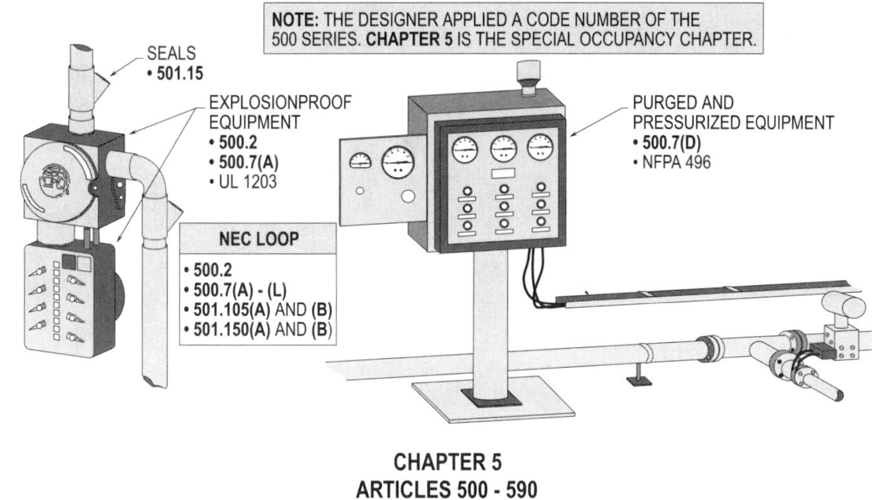

Figure 1-9. Chapter 5 deals with special occupancies and the first four Articles deal with identifying specific classifications and general provisions.

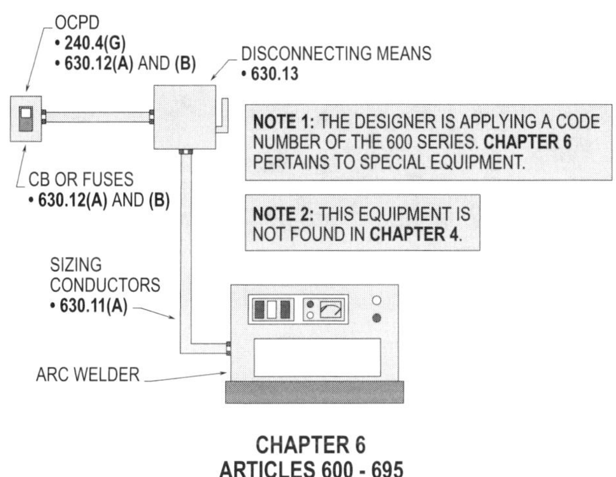

Figure 1-10. Chapter 6 deals with special equipment that requires special requirements for design and installation.

CHAPTER 8
ARTICLES 800 - 830

Chapter 8 covers communications circuits and equipment and is independent of the other Chapters except where they are specifically referenced.

Communications systems regulated by the Articles in **Chapter 8** are telephone, telegraph, radio, and alarm systems. The requirements basically cover systems that connect to a central alarm station to sound some kind of alarm.

For example, an alarm could communicate to a central control room that the ventilation system has failed in a hazardous location. **(See Figure 1-12)**

Design Tip: When requirements for communications systems are needed, one of the Articles in the 800 series in **Chapter 8** shall be utilized by the designer.

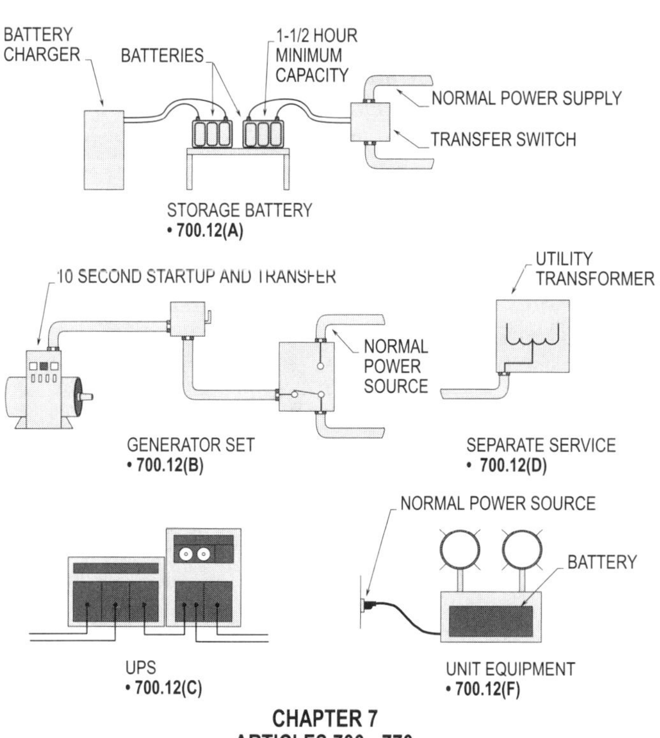

Figure 1-11. Chapter 7 contains requirements for special conditions that arise from the particular type, use, and application of the power source used to supply circuits and equipment.

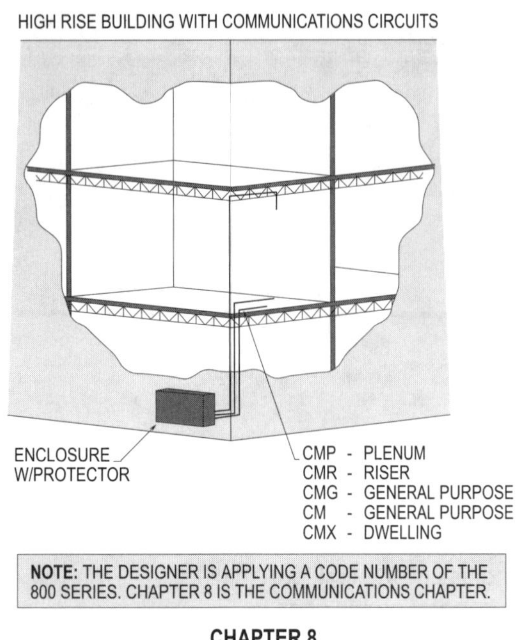

Figure 1-12. Chapter 8 covers communications circuits and equipment.

CHAPTER 9
TABLES 1 - 12(B)

Chapter 9 contains Tables that are mainly used to size raceways, based upon the conductor arrangement. **Annex D** contains examples to illustrate how calculations are made for various occupancies, motor circuits, elevator systems, etc. Designers may use these examples to give them basic procedures for calculating the load for a simple occupancy, branch circuit, or feeder.

For example: What size EMT is required for 3 - 4 AWG THHN and 1 - 6 AWG THHN copper conductors?

Step 1: Finding sq. in. area
Table 5, Ch. 9
4 AWG = .0824 sq. in.
6 AWG = .0507 sq. in.

Step 2: Calculating sq. in. area
Table 5, Ch. 9
3 x .0824 sq. in.   = .2472 sq. in.
1 x .0507 sq. in.   = .0507 sq. in.
Total   = .2979 sq. in.

Step 3: Selecting EMT
Table 4, Ch. 9
.2979 sq. in. requires 1"(27)

Solution: A 1 in. (27) EMT is required.

Annex C shall be permitted to be used per AHJ to size the raceway where the conductors are the same size. There are approximately 26 numbered Tables in **Annex C** that shall be permitted to be used for the purpose of selecting raceways based upon the same size conductors. See **Note (1)** to the **Tables** in **Chapter 9**.

For example: What size EMT is required for 4 - 4 AWG THHN copper conductors?

Step 1: Finding number
Table C1, Annex C
4 - 4 AWG THHN requires 1" (27)

Solution: A 1 in. (27) EMT is required.

See Figure 1-13 for a step-by-step procedure for sizing a raceway system.

Design Tip: All the Tables and Examples shall be reviewed carefully by designers, installers, maintainers, and inspectors in order to be familiar with such information that may be pertinent with the duty and performance of one's job. The "new" Tables may require different size conduits based on the type of conductors and conduit used.

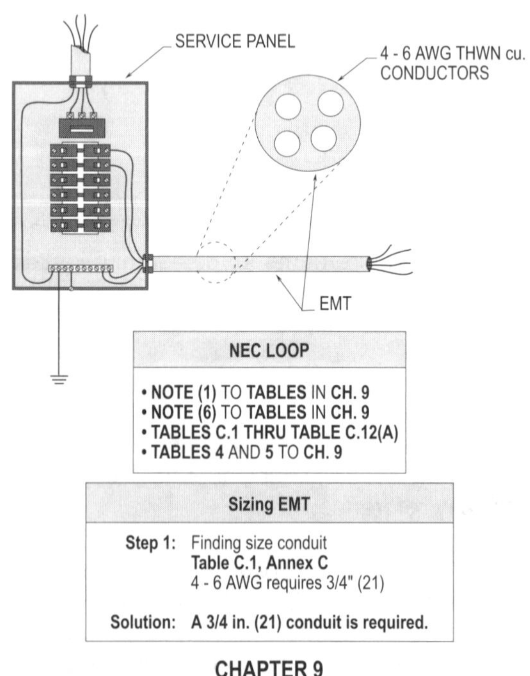

Figure 1-13. Chapter 9 contains Tables that shall be permitted to be used to size raceways based upon the conductor arrangement; in **Annex D** examples are available to illustrate how calculations are made.

PURPOSE OF OSHA

The OSHA Safety Standards, **29 CFR 1910, Subpart S** and **1926, Subpart K (Electrical)** is designed to help protect employees from electrical hazards during the construction, repair, and maintenance of such systems. These Standards are the industrial standards utilized by employers to ensure that their employees have a safe workplace.

APPLYING OSHA 1910, SUBPART S

OSHA defines a workplace as a location where one or more persons performing a work task are employed. OSHA puts the sole responsibility on the employer to see that employees are trained to use safety related work practices. Employees shall adhere to these procedures and implement the practices in their respective workplace. **OSHA 1910, Subpart S** requires these safety-related work practices to be applied so that employees are safeguarded from injury due to electrical hazards while they are working around or near electrical circuits and components. This required training shall be permitted to be classroom training, on the job training, or both. OSHA requires electrical circuits and components to be maintained so that they continue to comply with the provisions listed in **Subpart S, Electrical**. These rules, if applied, prevent employees from being exposed to electrical shock and other dangerous hazards due to energized electrical circuits and components. **(See Figure 1-14)**

> **Design Tip: OSHA 1910, Subpart S** covers four main categories with requirements designed to protect employees in his or her workplace, and they are as follows:
>
> • Design and installation safety,
> • Safety-related work practices,
> • Safety-related maintenance requirements, and
> • Special equipment.

APPLYING OSHA 1926, SUBPART K

OSHA 1926, Subpart K contains rules and regulations that are necessary to protect employees from electrical hazards posing significant risks in construction site workplaces.

Subpart K has incorporated, by reference, all relevant requirements pertaining to the NEC that directly affect employee safety in construction workplaces. By placing such NEC requirements in this OSHA Standard, there is no need to reference the NEC to obtain such requirements.

The rules and regulations in **Subpart K** of **OSHA 1926** have been designed to protect electrical personnel performing work in workplaces of construction sites while the requirements in the NEC are used to design, install, and inspect the wiring methods and equipment used to build the electrical system. **(See Figure 1-15)**

> **Design Tip: OSHA 1926, Subpart K** covers four main categories with requirements designed to protect construction workers in his or her workplace, and they are as follows:
>
> • Installation safety requirements,
> • Safety-related work practices,
> • Safety-related maintenance and environmental consideration (construction), and
> • Safety requirements for special equipment (construction).

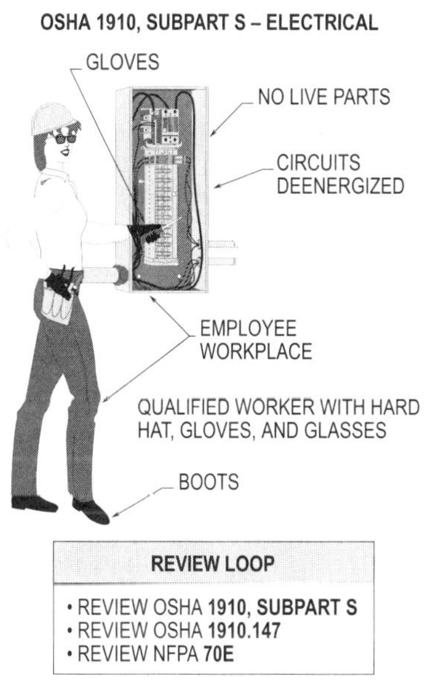

Figure 1-14. OSHA 1910, Subpart S deals with design and installation safety, safety-related work practices, safety-related maintenance requirements, and special equipment.

APPLYING OSHA 1926, SUBPART V

Subpart V of OSHA 1926 deals with work practices to be used during the construction, operation, and maintenance of electric power generation, transmission, and distribution facilities. Work clearances, deenergizing methods, and emergency procedures are covered and related in such a manner as to provide safe and reliable working conditions for personnel serving high-voltage systems.

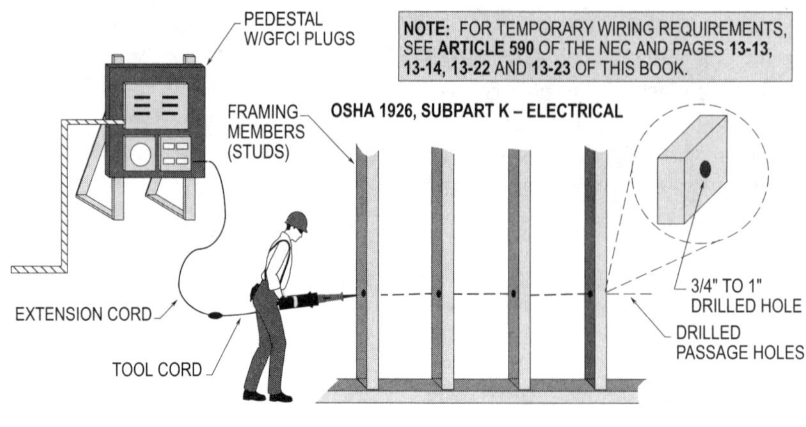

Figure 1-15. OSHA 1926, Subpart K deals with installation safety requirements, safety-related work practices, safety-related maintenance, and environmental considerations (construction) and safety requirements for special equipment (construction).

As used in **Subpart V**, the term "construction" includes the erection of new electric transmission and distribution lines and equipment, and the alteration, conversion, and improvement of existing electric transmission and distribution lines and equipment.

Existing electric transmission and distribution lines and electrical equipment need not be modified to conform to the requirements of applicable standards until such work as described above is performed on such lines or equipment.

The standards set forth are minimum requirements for safety and health. Employers may require adherence to additional standards not in conflict with the requirements contained in **OSHA Subpart V**. **(See Figure 1-16)**

> **Safety Tip:** Working clearances, deenergizing methods, and emergency procedures are covered in **Subpart V**, including recommended practices. Special tools that are used for climbing are discussed, and special in-service caring techniques are outlined. Safe operating rules to be followed before operating mechanically related equipment are highlighted. Proper ways to tie down loads in transit to prevent shifting are covered in detail, and unloading of such materials with safety tips are presented. The correct procedures of grounding and bonding overhead lines and equipment to protect workers are shown. Overhead and underground wiring techniques are outlined. Safety-related work practices involved with such systems are discussed.

APPLYING OSHA 1910, SUBPART R

OSHA 1910, Subpart R covers the work practices to be used during the operation and maintenance of electric power generation, transmission, and distribution facilities. Employees engaged in the construction of electric power transmission or distribution systems are protected by the provisions of **OSHA 1926, Subpart V**. Employee training techniques on existing facilities are covered and related to in such a manner as to provide safe and reliable working conditions for personnel maintaining high-voltage electrical systems.

Figure 1-16. OSHA 1926, Subpart V deals with work practices to be used during the construction, operation, and maintenance of electric power generation, transmission, and distribution facilities.

OSHA 1910, Subpart R covers five main areas of operation and maintenance, which are electric power generation, control, transformation, transmission, and distribution lines and equipment. These provisions apply to:

- Power generation,

- Transmission and distribution installations,

- Including related equipment for the purpose of communications or metering, and

- That are accessible only to qualified employees.

> **Design Tip:** The types of installations covered by this paragraph include the generation, transmission, and distribution installations of electric utilities, as well as equivalent installations of industrial establishments. Supplementary electric generating equipment that is used to supply a workplace for emergency, standby, or similar purposes only is covered under **Subpart S** of this Part. (See **paragraph (a)(1)(ii)(B) in OSHA 1910.269.**)

OSHA 1910.303 through 1910.308 in **Subpart S** has never actually covered the operation and maintenance of electric power generation, transmission, and distribution installations under the exclusive control of electric utilities or industrial facilities. **Subpart S** specifically exempts work performed by qualified persons on or directly associated with such systems and lines regardless who owns or controls them, per **OSHA 1910.331(c)(1)** in **Subpart S**.

In contrast, telecommunications workers facing similar hazards have always been covered under the rules and regulations of the telecommunications standard **OSHA 1910.268**. This regulation protects employees performing communications work from two major hazards of falling and electric shock. (**See Figure 1-17**)

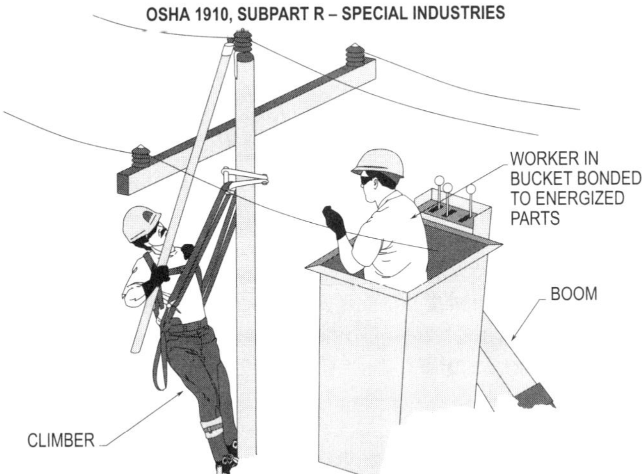

Figure 1-17. OSHA 1910, Subpart R deals with work practices to be used during the operation and maintenance of electric power generation, transmission, and distribution facilities.

PURPOSE OF NFPA 70E

NFPA 70E, *Standard for Electrical Safety in the Workplace,* was developed by NFPA at the request of OSHA.

NFPA 70E covers rules and regulation procedures on installing equipment, providing working clearances, guarding live parts, wiring design and protection, grounding connections, wiring methods, damp and wet locations, specific-purpose equipment, electric signs, cranes and hoists, electric welders, portable electric equipment, hazardous locations, and special systems. **NFPA 70E** seems to apply directly to electrical personnel.

APPLYING NFPA 70E

NFPA 70E is a companion document to the NEC. It basically covers the same industries as the NEC. **NFPA 70E** addresses electrical safety rules and regulations that are needed for safeguarding employees in the workplace. Three major installations are covered as follows:

- Electrical utilization installations including carnival and parking lots, mobile homes and recreational vehicles, and industrial substations.

- Conductors and wiring methods that connect installations to a supply of electricity.

- Other outside premises wiring.

These three major installations are in addition to **OSHA 1910, Subpart S** and are intended to be utilized to provide a safe work area for employees in his or her workplace. They are not to be used as requirements for design or installation. (**See Figure 1-18**)

> **Safety Tip:** OSHA covers similar requirements in **OSHA 1910.302 through 1910.308** in **Subpart S**. OSHA uses **NFPA 70E** to help keep **Subpart S** rules more current.

OSHA's main thrust in using **NFPA 70E** is to provide directions to employees on how to perform their work so that electrical hazards may be avoided. Information on working on or near energized parts, working on or near deenergized parts, the safe deenergizing of systems, lockout-tagout procedures, the safe use of tools and equipment, personal and other protective equipment, safe approach distances for overhead lines, switching of protective devices after operation, safety training, and safe voltage-measurement practices are covered in detail.

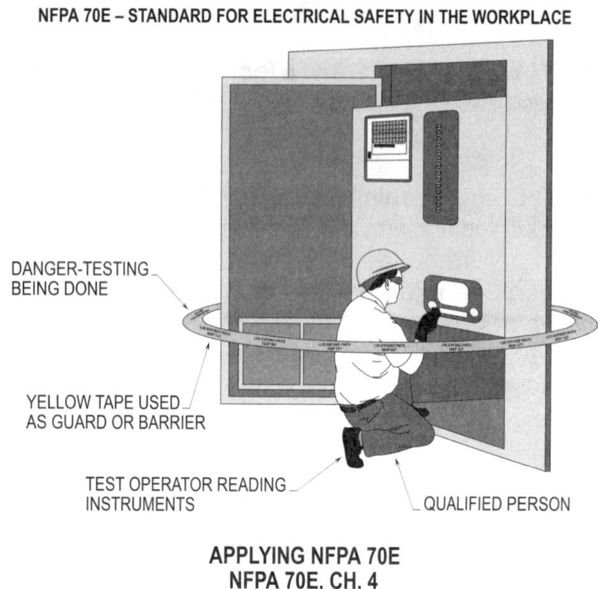

Figure 1-18. NFPA 70E deals with electrical safety rules and regulations that are needed for safeguarding employees in the workplace.

PURPOSE OF NFPA 70B

Like **NFPA 70E, NFPA 70B,** *Recommended Practice for Electrical Equipment Maintenance,* is a companion standard to the NEC. It contains recommended practice information for the proper maintenance of electric equipment.

NFPA 70B covers maintenance of industrial-type electric systems and equipment. Such systems and equipment are typical of those installed in commercial buildings, industrial plants, and multifamily dwellings. The standard makes it clear that it does not cover consumer appliances or home-type equipment, and its rules are not intended to supersede manufacturer-recommended maintenance procedures.

APPLYING NFPA 70B

NFPA 70B includes 31 chapters and 13 annexes which cover proper preventive maintenance procedures. The standard covers important items such as detailed maintenance techniques for switchgear, circuit breakers, cables, motor control centers, rotating equipment, wiring devices, and other related electric equipment.

Specific details of electrical testing in areas such as insulation resistance, protective device testing, fault-gas analysis, and many other such modern testing procedures are discussed. The maintenance of equipment subject to long intervals between shutdowns and the methods that are to be used for deenergizing equipment are highlighted and discussed.

Maintenance Tip: The annexes of **NFPA 70B** are equipped with information for walk-through inspections, instruction techniques, symbols, diagrams, and recommended test sheets and forms that are designed to be used for maintenance programs. **(See Figure 1-19)**

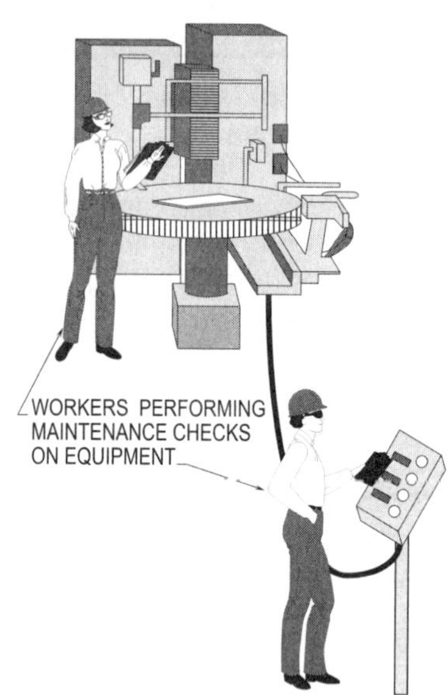

Figure 1-19. NFPA 70B covers important items such as detailed maintenance techniques for switchgear, circuit breakers, cables, motor control centers, rotating equipment, wiring devices, and other related equipment.

PURPOSE OF NFPA 79

The purpose of **NFPA 79,** *Electrical Standard for Industrial Machinery,* is to provide detailed information for the application of electrical/electronic equipment, apparatus or systems supplied as part of industrial machinery that will promote safety to life and property.

Information in this standard is not intended to limit or inhibit the advancement of the state-of-the-art designing techniques of such equipment. Each type of machine has unique requirements that shall be accommodated to provide adequate safety for the operator.

APPLYING NFPA 79

The rules of **NFPA 79** apply to the electrical/electronic equipment, apparatus, or systems of industrial machinery operating from a nominal voltage of 600 volts or less and commencing at the place of connection of the supply to the electrical equipment of the machines.

Modern machine tool electrical equipment may range from single-motor machines, such as drill presses, that perform simple, repetitive operations, to very large, multimotored automatic machines that involve highly complex electrical control systems, including electronic and solid-state devices and equipment. Generally, these machines are especially designed, factory-wired, and tested by the builder and then erected in the plant in which they will be used. Because of their importance to plant production and their usually high cost, they are customarily provided with safeguards and other devices not often incorporated in the usual motor and control application as contemplated by the NEC.

Although these machines may be completely automatic, they are constantly attended, when operating, by highly skilled operators. The machine usually incorporates many special devices to protect the operator, protect the machine and building against fires of electrical origin, protect the machine and work in process against damage due to electrical failures, and protect against loss of production due to failure of a machine component. To provide these safeguards, it may be preferable to deliberately sacrifice a motor or some other component, rather than to chance injury to the operator, the work, or the machine. **(See Figure 1-20)** It is due to proper supervision and qualified personnel, that **NFPA 79** allows such procedures when operating plant machine tool electrical equipment.

Design Tip: It is because of the above consideration that **NFPA 79** differs from the basic concepts of motor protection and other protection techniques as required by the provisions in the NEC.

NFPA 79 – ELECTRICAL STANDARD FOR INDUSTRIAL MACHINERY

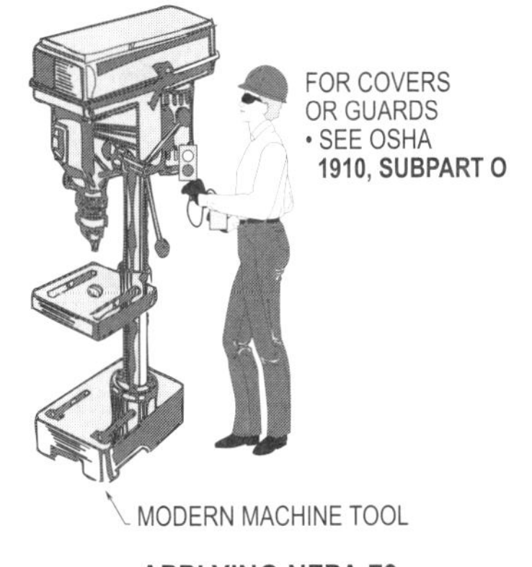

APPLYING NFPA 79
NFPA 79, CH. 4

Figure 1-20. NFPA 79 covers electrical/electronic equipment, apparatus, or systems of industrial machinery operating from a nominal voltage of 600 volts or less.

PURPOSE OF THE NESC

The **NESC,** *National Electrical Safety Code,* was developed in 1913 by the National Bureau of Standards, became a consensus standard and is used in the industry as such today. The **NESC** is an American National Standard published by IEEE. It provides practical rules for safeguarding personnel during the installation, operation, or maintenance of electric supply and communications lines and associated apparatus and equipment. The NESC contains three main general rules, and they are as follows:

- All electric supply and communications lines and equipment shall be designed, constructed, operated, and maintained to comply with its requirements.

- The utilities, authorized contractors, or other entities performing design, construction, operation, or maintenance procedures for electric supply or communications lines or equipment by the **NESC** are responsible for complying with specific requirements.

- For all conditions not specified in the **NESC**, construction and maintenance procedures should be performed and accepted with good practices based upon given local conditions. **(See Figure 1-21)**

Design Tip: The NESC covers supply and communications lines, equipment, and associated work practices used by both public and private electric supply, communications, railway, or similar utilities. Similar systems under the control of qualified persons are also covered.

PURPOSE OF THIRD-PARTY CERTIFICATION

Underwriters laboratories (UL) and other such laboratories are nonprofit organizations supported by the manufacturers who submit merchandise, that is, for the testing of merchandise for safety when being used.

The **UL** or any other approved testing laboratory label assures the user that the manufacturer of the item has submitted samples of the item to the laboratory for testing and that they were found to meet the required minimum safety standards.

Manufacturers submit samples of their products to a laboratory such as **UL** for testing before they are manufactured on a large scale. It is then presumed that the manufacturer will maintain the same quality in the future production of the same item. Such testing facilities also perform on-site inspections to ensure that the manufacturer continues to manufacture the product by their standards.

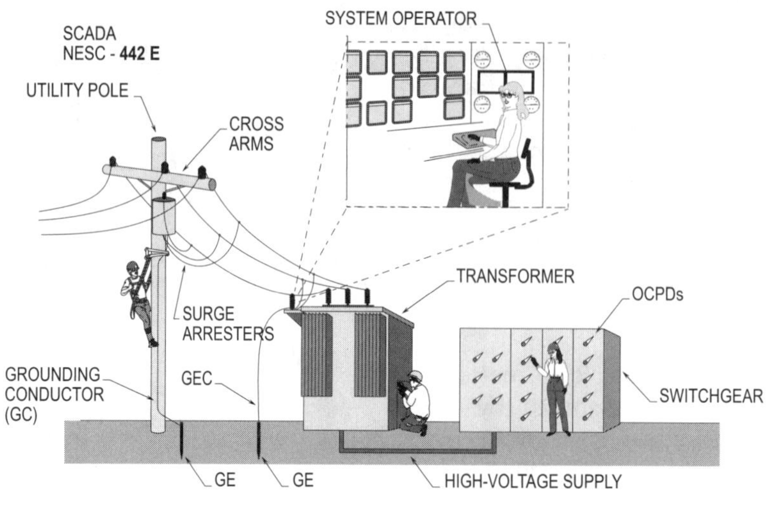

Figure 1-21. The **NESC** provides practical rules for safeguarding personnel during the installation, operation, or maintenance of electric supply and communications lines and associated apparatus equipment.

For example, if the product passes the exhaustive tests of **UL** in accordance with established standards, it is added to the official **UL** published list and the product is then known as "**Listed by Underwriters Laboratories, Inc.**"

Such **UL** items usually have a **UL** label attached directly to the product. In some cases, this **UL** "label" is molded or stamped into the merchandise.

Note that this same procedure is used by other testing laboratories that are approved by OSHA.

An item approved by Testing Laboratories, however, does not mean that the item is approved for all uses. Rather, a testing label means that the item or device, as labeled, is safe only for the purpose for which it was intended. **(See Figure 1-22)**

Design Tip: Once equipment has been approved, it is not to be modified in any way, but must be used as instructed by the manufacturer.

For example, the wiring or any of the other elements are not to be changed out or rearranged in a manner that does not comply with the listing and labeling of the nameplate on the equipment or accompanying literature.

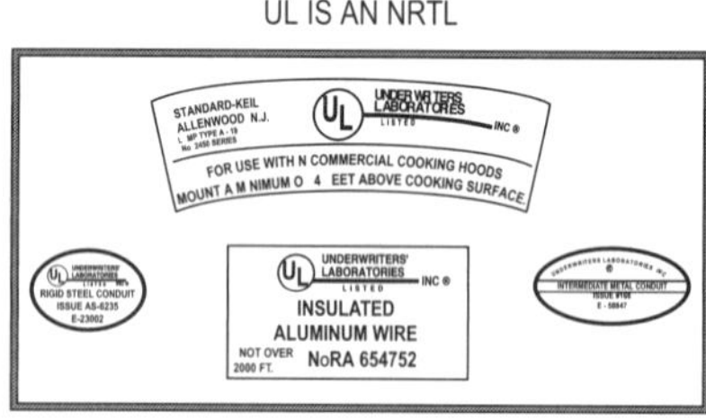

THIRD PARTY CERTIFICATION
- UL
- FM
- ETL
- CSA
- MET
- D,S,G
- NOTE THAT THERE ARE ABOUT 15 TESTING LABS NOW APPROVED BY OSHA.

PURPOSE OF THIRD PARTY CERTIFICATION

Figure 1-22. UL or any other approved testing laboratory label assures the user that the product being installed is listed and labeled to perform the function intended.

UL WHITE BOOK

The **UL White Book**, known in the electrical industry as the "**General Information Electrical Construction, Hazardous Locations, and Electrical Heating and Air Conditioning Equipment Directory,**" provides a service for the classification of products that have been determined to meet the appropriate requirements of the applicable international publication(s). For those products which comply with the requirements of an international publication(s), the Classification Marking may appear in various forms as authorized by **UL**.

For example, a form may include and illustrate the word "Classified" with a control number assigned by **UL**. The product name as indicated in this *Directory* under each of the product categories is generally included as part of the Classification Marking text but may be omitted when, in **UL's** opinion, the use of the name is superfluous and the Classification Marking is directly and permanently applied to the product by stamping, molding, ink stamping, silk screening, or similar processes.

Separable Classification Markings (not part of a nameplate and in the form of decals, stickers, or labels) will always include four elements: **UL's** name and/or symbol, the word Classified, the product category name, and a control number.

The complete Classification Marking will appear on the smallest unit container in which the product is packaged when the product is of such a size that the complete Classification Marking cannot be applied to the product or when the product size, shape, material, or surface texture makes it impossible to apply any legible marking to the product. When the complete Classification Marking cannot be applied to the product, no reference to **UL** on the product is permitted. **(See Figure 1-23)**

APPROVAL

OSHA requires conductors and equipment to be accepted only when approved. The definition of *approved* is to be acceptable to the authority having jurisdiction (AHJ) per **Article 100** in the **NEC**. **OSHA 1910.303(a)** and **NEC 110.2** now have the same requirements considering the acceptability of components, wiring methods, and equipment. The AHJ will usually require third-party certification of all equipment. For third-party certification, a nationally recognized testing laboratory performs tests to verify if the equipment will do what it is designed to do and still be safe to the user. The only problem with accepting equipment that is not certified by a third party is that the authority having jurisdiction takes the sole responsibility that the equipment is safe.

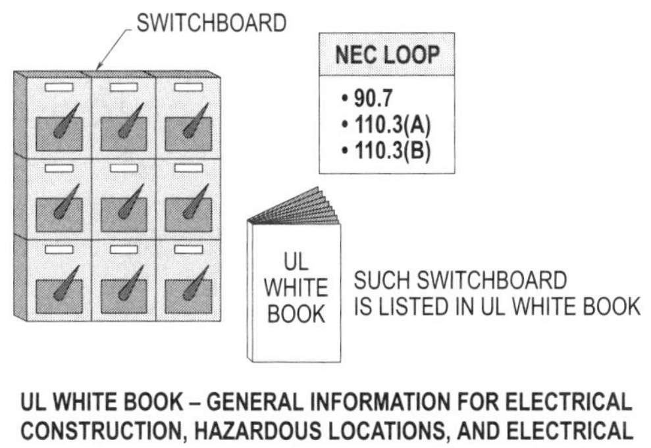

Figure 1-23. The **UL White Book** provides a service for the classification of products that have been determined to meet the appropriate requirements of the applicable international publication(s).

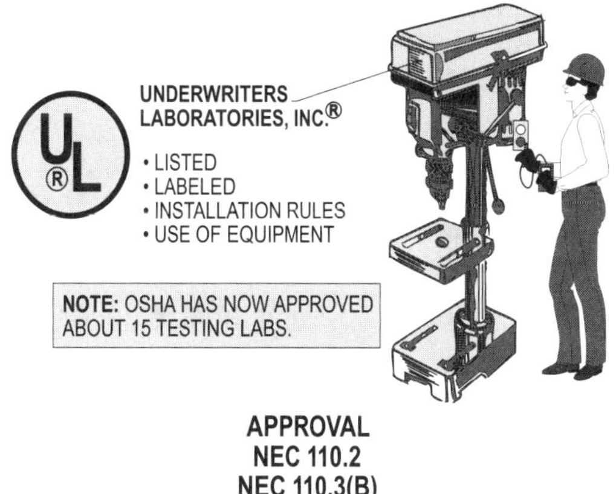

Figure 1-24. OSHA requires the conductors and equipment to be accepted only when approved.

<table>
<tr><td>Codes and Standards</td></tr>
<tr><td>

• **NFPA 70E - 400.1**
• **NEC 90.7**
• **NEC 110.2** and **NEC Art. 100**
• **NEC 110.3(B)**
• **OSHA 1910.303(a)** and **1910.399**
• **OSHA 1926.403(a)**

</td></tr>
</table>

EXAMINATION, INSTALLATION AND USE OF EQUIPMENT

All electrical equipment and work must be related to the Occupational Safety and Health Administration (OSHA) regulations and comply with the provisions of standards and installation instructions of qualified testing laboratories that have tested and certified the equipment. Such Nationally Recognized Testing Laboratories (NRTLs) and certifying organizations are:

- Underwriters Laboratories (UL)
- Factory Mutual Engineering Corp. (FM)
- Electrical Testing Laboratories (ETL)
- MET Testing Company, Inc.
- Dash, Straus, and Goodhue, Inc.
- Canadian Standards Association (CSA)
- (Limited Certification) Note, there are others

EXAMINATION

The NEC, OSHA, and **NFPA 70E** do not include detailed requirements for wiring inside of electrical equipment. Information pertaining to the internal wiring of equipment is usually found in individual standards that contain special requirements for installing the equipment concerned. It is not the intent of the NEC and OSHA to take away the authority of inspectors to examine and approve equipment. However, the NEC, OSHA, and **NFPA 70E** indicate that the rules pertaining to the internal construction of equipment are usually investigated and approved by a qualified testing laboratory. The NEC does not make a flat rule to the effect that all electrical equipment shall have third-party certification of equipment by independent testing laboratories per **90.7, 110.2,** and **110.3**. OSHA, on the other hand, does require electrical equipment to have third-party certification for the safety of personnel using such equipment per **Subpart S** to **1910.303(a)**. **NFPA 70E** recognizes the same rule in **400.3(A),** and OSHA recommends that the requirements in **NFPA 70E** be enforced to provide safety for employees in employee workplaces.

In judging and evaluating electrical equipment that is installed in the employee workplace, the following steps shall be followed:

- Judge the wiring installation to ensure that it complies with the provisions of **NFPA 70E**. Evaluate the equipment for suitability for installation and use according to nameplate ratings and instructions. List the appropriate standard for such equipment.

For example, every product listed in the **UL Electrical Construction Materials Directory (Green Book)** is required to be installed as described in the application data given with the listing in the book. It becomes the responsibility of the AHJ to decide and determine the suitability of electrical equipment that has not been tested and certified by a qualified testing laboratory.

- Mechanical strength and durability of enclosures, raceways, etc. enclosing electrical parts and wiring shall be adequate to provide proper protection of such elements.

- Electrical insulation shall be installed by **Table 310.13** in the **NEC** and loaded according to the ampacities of **Table 310.16** for systems rated at 2000 volts or less. Insulation of conductors may be checked with any good insulation resistance tester.

- Equipment and insulation shall be sized and selected to operate safely under normal conditions of use and also under abnormal conditions without adverse heating effects. (See **310.10** in the NEC.)

- Electrical equipment shall be equipped with shields, etc. to disguise the arcing effect and limit the burning effect of the arc. (See **110.18** in the NEC.)

- Electrical equipment with components shall be classified by its type, size, voltage, current capacity, and specific use. An evaluation shall be made to verify that the type and size of equipment is being supplied with the proper voltage and is loaded to a value no greater than it is designed to carry. (See **90.7** and **110.3(B)** in the NEC and **OSHA 1910.399.**)

<table>
<tr><td>Codes and Standards</td></tr>
<tr><td>

• **NFPA 70E - 400.3(A)**
• **NEC 110.3(A)**
• **OSHA 1910.399**
• **OSHA 1910.303(b)(1)**
• **OSHA 1926.403(b)(1)**

</td></tr>
</table>

INSTALLATION AND USE

OSHA'S relationship to electrical equipment is very clear in **1910, Subpart S, Electrical.** Electrical equipment shall have third-party certification for the essential safety of the equipment and components used to assemble electrical installations. For the safety of personnel utilizing such equipment, OSHA requires the following:

Utilization equipment. Utilization equipment is equipment that utilizes electric energy for mechanical, chemical, heating, lighting, or similar useful purposes.

Approval. The conductors and equipment required or permitted by **Subpart S** shall be acceptable only if approved.

Approved. Acceptable to the authority enforcing **Subpart S**. The authority enforcing **Subpart S** is the Assistant Secretary of Labor for Occupational Safety and Health. The definition of "acceptable" indicates what is acceptable to the Assistant Secretary of Labor, and therefore approved within the meaning of **Subpart S**.

Acceptable. An installation or equipment is acceptable to the Assistant Secretary of Labor, and approved within the meaning of **Subpart S**: if it is accepted, certified, listed, labeled, or otherwise determined to be safe by a qualified testing laboratory, or with respect to an installation or equipment of a kind which no qualified testing laboratory accepts, certifies, lists, labels, or determines to be safe, if it is inspected or tested by another Federal Agency, or by a state, municipality, or local authority responsible for enforcing the safety provisions of the **National Electrical Code (NEC)** and found in compliance with the provisions of the NEC as applied in **Subpart S**, or with respect to custom-made equipment or related installations which are designed, fabricated for, and intended for use by a customer if it is determined to be safe for its intended use by its manufacturer on the basis of test data which the employer keeps and makes available for inspection to the Assistant Secretary and his authorized representatives in the field (OSHA inspectors). Custom-made equipment should be designed and built by the provisions listed in **UL 508** or one of the C series published by ANSI, based on the type equipment involved. The customer specifies to the manufacturer the function that their equipment must perform. **(See Figure 1-25)**

Labeled. Equipment is labeled if there is attached to it a label, symbol, or other identifying mark of a nationally recognized testing laboratory, which (a) makes periodic inspection of the production of such equipment and by (b) whose labeling indicates compliance with nationally recognized standards or tests to determine safe use in a specified manner.

Listed. Equipment is listed if it is of a kind mentioned in a list which (a) is published by a nationally recognized laboratory which makes periodic inspection of the production of such equipment and (b) states that such equipment, meets the nationally recognized standards or has been tested and found safe for use in specified manner.

Accepted. An installation is "accepted" if it has been inspected and found by a nationally recognized testing laboratory to conform to specified plans or to procedures of applicable codes.

Certified. Equipment is "certified" if it:

(a) has been tested and found by a nationally recognized testing laboratory to meet nationally recognized standards or be safe for use in a specified manner, or

(b) is of a kind whose production is periodically inspected by a nationally recognized testing laboratory, and

(c) it bears a label, tag, or other record of certification.

> **Design Tip:** OSHA requires all electrical equipment, where available, to be listed and labeled by a Nationally Recognized Testing Laboratory. UL and FM are laboratories without limited certification that are accepted for industrial locations. However, if such needed electrical equipment is not listed, labeled, and available, then custom-made equipment that complies with the requirements of the Assistant Secretary is acceptable. OSHA recognizes a NRTL while the NEC recognizes a qualified testing laboratory (QTL).

There is a conflict in the standards due to the change in the 1981 NEC, which changed the wording from NRTL to QTL.

Codes and Standards
• NFPA 70E - 400.2
• NEC 110.3(B)
• OSHA 1910.303(b)(2)
• OSHA 1926.403(b)(2)

PURPOSE OF CUSTOM-MADE ELECTRICAL EQUIPMENT

OSHA requires all electrical equipment, where available, to be listed and labeled by a nationally recognized testing laboratory (NRTL). However, if such electrical equipment is not listed, labeled, and available, then custom-made equipment that complies with the requirements of the Assistant Secretary of Labor is acceptable. (See **OSHA 1910.399**)

Custom-made equipment or related installations is equipment that is designed, fabricated for, and intended for use by a customer, where it is determined to be safe for its intended use by its manufacturer on the basis of test data which the employer keeps and makes available for inspection to the Assistant Secretary and his or her authorized representatives in the field (OSHA inspectors). Custom-made equipment should be designed and built by the provisions listed in the **NEC, UL 508, NFPA 79**, or one of the C series published by ANSI, based on the type or equipment involved. The customer specifies to the manufacturer the function that their equipment shall perform. **(See Figure 1-25)**

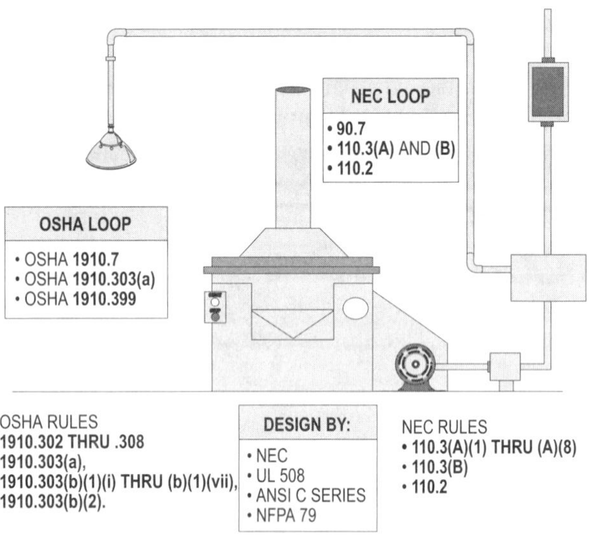

Figure 1-25. An installation or equipment is acceptable to the Assistant Secretary of Labor, and approved within the meaning of **Subpart S**, where it is accepted, certified, listed, labeled, or otherwise determined to be safe by a nationally recognized testing laboratory.

INSTALLING AND MAINTAINING ELECTRICAL EQUIPMENT

It is the sole responsibility of personnel who have been designated to install and maintain electrical equipment, including all electrical apparatus, to do so in a safe and reliable manner. Such personnel do not have the authority to wire in and modify the electrical element, where they do not comply with the **NEC**, OSHA and various testing laboratory's related standards. This rule also applies to custom-made equipment.

OSHA'S RELATIONSHIP TO ELECTRICAL EQUIPMENT OSHA 1910.7, 1910.303(a), 1910.399(a), NEC 110.2, 110.3(A) AND (B), AND ARTICLE 100 (DEFINITIONS OF APPROVAL)

The requirements set forth in **OSHA 29 CFR, 1910.302 through 1910.308** pertain to electrical equipment installed before and after certain mandatory dates listed in **Subpart S**. All electrical utilization systems and pieces of equipment that were designed, installed, and inspected after March 15, 1972 shall comply with all the provisions according to **OSHA 1910.302 through 1910.308**. All major replacement, modification, repair, or rehabilitation performed after March 15, 1972 on such systems and equipment shall comply with these requirements. These requirements also include equipment installed before March 15, 1972. The above rules are retroactive to such electrical systems and equipment.

There are certain provisions in **OSHA 1910.302 through 1910.308** that are intended to apply only to electrical systems and equipment installed after April 16, 1981, which are not meant to be retroactive. These nonretroactive requirements and the electrical installations and equipment they apply to are listed in **OSHA 1910.302(b)**.

> **Design Tip:** Major replacements, modifications, or rehabilitation (as outlined in OSHA regulations) include work similar to that involved where a new building or facility is built, a new wing is added or an entire floor is renovated.

PURPOSE OF UL 508

UL 508 contains basic requirements for products covered by Underwriters Laboratories, Inc. under its Follow-Up Service for this category within the rules and regulations of the Standard. These requirements are based upon sound engineering principles, research, records of tests, field experience, and an appreciation of the problems of manufacture, installation, and use derived from consultation with and information obtained from manufacturers, users, inspection authorities, and others having specialized experience. They are subject to revision as further experience and investigation may show is necessary or desirable.

A product that complies with the text of **UL 508** will not necessarily be judged to comply with the Standard if, when examined and tested, it is found to have other features that impair the level of safety contemplated by these requirements.

A product employing materials or having forms of construction differing from those detailed in the requirements of this Standard may be examined and tested according to the intent of the requirements and, if found to be substantially equivalent, may be judged to comply with the Standard.

APPLYING UL 508

The requirements of **UL 508** cover industrial control devices, and device accessories thereto, for starting, stopping, regulating, controlling, or protecting electric motors. Such requirements also cover industrial control devices or systems that store or process information and are provided with an output motor control function(s). This equipment is for use in ordinary locations in accordance with the NEC.

Requirements also cover industrial control panels that are assemblies of industrial control devices and other devices associated with the control of motor-operated and related industrial equipment. Examples of devices in an industrial control panel are disconnecting means, motor branch circuit protective devices, temperature control devices, electrical instruments, etc. **(See Figure 1-26)**

For example, an industrial control panel for the control of metal-working machine tools or plastics machinery is judged on the basis of compliance with the applicable requirements of this Standard as well as the requirements in the **Standard for Industrial Machinery, NFPA 79**.

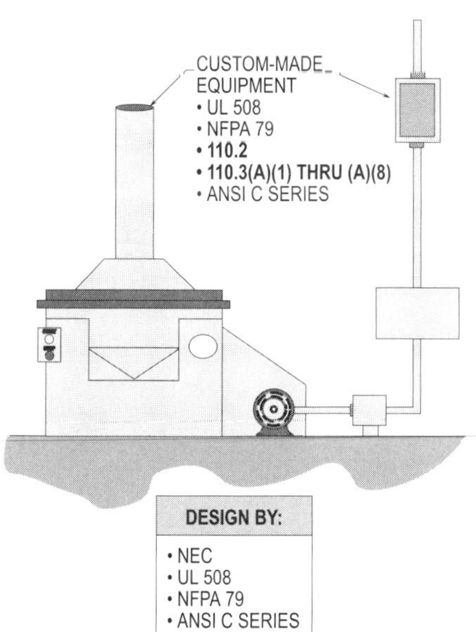

Figure 1-26. UL 508 covers industrial control devices, and device accessories thereto, for starting, stopping, regulating, controlling, or protecting electric motors.

> **Design Tip:** Electrical personnel, due to the type of apparatus or equipment involved, may need to use other Codes and Standards in addition to **UL 508**.

UL 508 covers five main categories with requirements designed to judge and evaluate industrial control devices and accessories and they are as follows:
- Construction,
- Performance,
- Rating,
- Marking,
- Instruction, and use.

> **Design Tip:** The rules and regulations in **UL 508** cover devices rated at 1500 volts or less. Such rules and regulations must be reviewed very carefully before attempting to apply them to the design of such apparatus.

AUTHORITY HAVING JURISDICTION

The authority having jurisdiction (AHJ) is the organization, office, or individual responsible for "approving" equipment, installations, or procedures.

> **Design Tip:** The phrase "AHJ" is used in NFPA documents in a broad manner since jurisdictions and "approval" agencies vary, as do their responsibilities. Where public safety is primary, the AHJ may be a federal, state, local, or other regional department or individual such as a fire chief, fire marshal, chief of a fire prevention bureau, labor department, health department, building official, electrical inspector, or others having such authority. For insurance purposes, an insurance inspection department, rating bureau, or other insurance company representative may be the "AHJ". In many circumstances, the property owner or his designated agent assumes the role of the "AHJ" at government installations or county areas not requiring a permit and inspection. (See definition of AHJ in **Article 100**)

The AHJ has the option of permitting alternative methods where specific rules are not established in the NEC. This allows the local authority to waive specific requirements in industrial occupancies, research and testing laboratories, and other occupancies where the specific type of installation is not contemplated in the NEC requirements.

The AHJ may waive a new NEC requirement during the interim period between the acceptance of a new edition of the NEC and the availability of a new product, construction, or material redesigned to comply with the increased safety required by the new NEC edition. It is difficult to establish a viable future effective date in the NEC because the time needed to change existing products and standards and to develop new materials and test methods is not usually known at the time of adoption of the new requirement in the NEC.

It is also the responsibility of the local authority (AHJ) enforcing the NEC to interpret the specific rules and regulations of the NEC. **(See Figure 1-27)**

By special permission, the AHJ has the responsibility of giving the engineer, contractor, electrician, etc. written consent of his or her decision to waive a requirement in the NEC pertaining to a specific installation.

Figure 1-27. The AHJ is the organization, office, or individual responsible for "approving" equipment, installations or procedures.

Chapter 1. Electrical Codes and Standards

Section   Answer

_________   _________   **1.** The NEC was first published in
         (a) 1897   (b) 1900
         (c) 1911   (d) 1928

_________   _________   **2.** The NEC is usually adopted by government agencies and becomes effective
      _____ of the year that it is published.
         (a) August 1   (b) September 1
         (c) December 1   (d) January 1

_________   _________   **3.** Chapter 1 of the NEC covers
         (a) wiring and protection   (b) equipment for general use
         (c) general   (d) communications systems

_________   _________   **4.** Chapter 2 of the NEC covers
         (a) wiring and protection   (b) general
         (c) wiring methods and materials   (d) special equipment

_________   _________   **5.** Chapter 3 of the NEC covers
         (a) general   (b) equipment for general use
         (c) wiring methods and materials   (d) special equipment

_________   _________   **6.** Chapter 4 of the NEC covers
         (a) special occupancies   (b) wiring and protection
         (c) communications systems   (d) equipment for general use

_________   _________   **7.** Chapter 5 of the NEC covers
         (a) special eqiupment   (b) special occupancies
         (c) Tables   (d) wiring and protection

_________   _________   **8.** Chapter 6 of the NEC covers
         (a) equipment for general use   (b) special equipment
         (c) general   (d) Tables

_________   _________   **9.** Chapter 7 of the NEC covers
         (a) special occupancies   (b) special equipment
         (c) special conditions   (d) communications systems

_________   _________   **10.** Chapter 8 of the NEC covers
         (a) communications systems   (b) Tables
         (c) special conditions   (d) wiring and protection

_________   _________   **11.** Chapter 9 of the NEC covers
         (a) equipment for general use   (b) communications systems
         (c) wiring methods and materials   (d) Tables

_________   _________   **12.** Annex H of the NEC covers
         (a) examples   (b) types of construction
         (c) supervisory control and data   (d) administration and
             acquisitions   enforcement

13. Annex D of the NEC covers
    (a) examples   (b) Tables
    (c) types of construction   (d) administration and enforcement

14. OSHA defines a workplace as a location where _____ or more persons performing a work task are employed.
    (a) one   (b) two
    (c) three   (d) four

15. Safety training, per OSHA, shall be permitted to be
    (a) classroom   (b) on the job
    (c) all of the above   (d) none of the above

16. NFPA 70B contains recommended practice information for the proper _____ of electrical equipment
    (a) design   (b) installation
    (c) safety   (d) maintenance

17. NFPA 79 applies to electrical/electronic equipment, apparatus, or systems of industrial machinery operating from a nominal voltage of _____ volts or less.
    (a) 150   (b) 277
    (c) 480   (d) 600

18. All electrical utilization systems and pieces of equipment that were designed, installed, and inspected after March 15, _____ shall comply with all of the provisions found in 1910.302 through 1910.308.
    (a) 1968   (b) 1972
    (c) 1975   (d) 1980

19. The rules and regulations found in UL 509 cover devices rated at _____ volts or less.
    (a) 150   (b) 480
    (c) 600   (d) 1500

20. The AHJ is the organization, office, or individual responsible for _____ equipment, installations, or procedures.
    (a) approving   (b) labeling
    (c) listing   (d) identifying

2

Electrical Safety and First Aid

Because of the hazard of electric shock or burns from an arc or blast when employees are working near or performing work on energized lines and equipment, workers may suffer electrocution or severe injury on the job.

The employer shall provide training or require that his or her employees are knowledgeable and proficient in the procedures involving emergency situations and first aid fundamentals. Medical services, first aid, and job briefing procedures are covered and related in such a manner that, if followed, will provide a basic concept of first aid for personnel servicing electrical wiring and equipment. Note that only trained qualified personnel shall be permitted to administer first aid to hurt employees.

THE FLOW OF ELECTRICITY

When an electrical current passes through the body, it takes a path drawn from the two points of contact where it enters and exits. Along this path appears the greatest magnitude of electrical current. Damage to tissues or organs along this current pathway dictates the severity of the injury.

For example, a common path for current to pass through the body is where it enters on one of the arms and exits through the other arm or a leg. This type of shock is very dangerous because it can cause severe burning, respiratory arrest, or ventricular fibrillation when it crosses the chest area.

ELECTRIC SHOCK

Electric shock is usually associated with alternating current (AC). However, there is still the danger of shock with direct current (DC). Other factors that enhance electric shock potential are as follows:

- Wet and/or damp locations,
- Ground/grounded objects,
- Current loop from source back to source,
- Path of current through body/duration of contact,
- Area of body contact and pressure of contact,
- Physical size/condition/age of person,
- Type and/or amount of voltage,
- Personal protective equipment/gloves/shoes,
- Metal objects such as watches, necklaces, rings, etc., and
- Miscellaneous
    - **(a)** Poor workplace illumination,
    - **(b)** Color blindness,
    - **(c)** Lack of training, knowledge, and
    - **(d)** No safe work procedures.

> **Safety Tip:** The above conditions may have a direct effect on how much current the voltage can push through a body from phase-to-phase conditions or phase-to-ground conditions.

DANGER OF ELECTRIC SHOCK

Mild exposures to electric shock can be very painful and may cause muscular contractions, respiratory arrest, severe tissue burning, or ventricular fibrillation, all of which can be fatal.

MUSCULAR CONTRACTIONS

The electrical stimulation of nerves along the current path in the body can cause the muscles in that area to involuntarily contract.

For example, if a person held on to a de-energized conductor and had someone gradually increase the current flow, a sensation would be detected when the current reaches 1 mA. As the current is increased, pain and heat will increase, until at about 16 mA, the person would not be able to let go of the conductor because the forearm muscles would involuntarily contract due to current flow. If the current path went across the chest, the diaphragm muscles would contract at about 20-40 mA, preventing the person from breathing. This magnitude of current is fatal unless it is interrupted. It is often referred to as "let-go" current.

> **Safety Tip:** If current is interrupted in time, artificial respiration is usually suggested as the correct treatment and may restore breathing and prevent death or brain damage.

RESPIRATORY CONTROLS

Respiratory arrest may also be caused by a shock passing through the spinal cord to the lower portion of the brain, causing the respiratory controls of the central nervous system to shut down. In some cases, respiratory arrest may continue even after interrupting the current flow, which is called persistent respiratory arrest. Unless ventricular fibrillation has taken place (there is no pulse), artificial respiration is the correct treatment until medical help arrives.

TISSUE BURNING

As indicated above, the areas of major damage from an electrical shock in the body lie along the current path. The current does not follow the blood vessels or other parts of the circulatory system, it strikes a path through the body and burns any tissue or organ along that path. This can lead to internal bleeding or severe organ damage and requires prompt medical attention as soon as possible.

VENTRICULAR FIBRILLATION

Another result of electric current passing through the chest is a disruption of the natural rhythm of the heart, or ventricular fibrillation. When this happens, regular heart action ceases, there is no pulse and blood stops circulating to the organs. This condition is nearly always fatal unless it is treated promptly.

Normal heart function can sometimes be restored through a defibrillator, an apparatus that applies electric impulses to the heart muscles in an attempt to restore regular pumping action of the heart.

> **Safety Tip:** While waiting for medical assistance, it is appropriate to administer cardiopulmonary resuscitation (artificial respiration and chest compression). In some cases, the fibrillation may reverse itself spontaneously, and it is important that oxygen is supplied to vital organs during this time.

FREEING VICTIM

An electric shock generally occurs when a person touches a live wire with his hand, a noninsulated foot, or any other part of his body. When freeing the victim, the rescuer should separate the victim from the contact by utilizing a long dry stick, a long dry rope, or a long length of dry cloth. The rescuer needs to be especially careful and not attempt to free the victim with bare hands alone; the rescuer might become another victim if proper protection is not used.

ARTIFICIAL RESPIRATION

When a victim's natural breathing is inadequate or ceases from electric shock, artificial respiration should be applied immediately to maintain an open airway through the mouth and nose (or through the stoma) and restore breathing by maintaining an alternating increase and decrease in the expansion of the chest. Artificial respiration should always be continued by the following recommended practices until the victim:

- Begins to breathe spontaneously,
- Is pronounced dead by a doctor, or
- Is dead beyond any doubt.

MOUTH-TO-MOUTH RESUSCITATION

Mouth-to-mouth resuscitation is a technique used to ensure the victim of an air passage and increase the volume of air exchange in his or her lungs. There are nine techniques suggested when applying mouth-to-mouth resuscitation:

- Determine the consciousness of victim.
- Tilt the victim's head back so that his/her chin is pointing upward.
- For about 5 seconds, listen and feel for air to be exhaled and look at the victim's chest to see if it rises and falls.
- If victim is not breathing, use the thumb and index finger to pinch the victim's nostrils shut.
- Use the following techniques to blow air into the victim's mouth:
  - (a) Open your mouth wide,
  - (b) Take a deep breath,
  - (c) Blow air into the victim's mouth with your mouth sealed tightly around the victim's mouth,
  - (d) Initially, give four quick, full breaths without allowing the lungs to fully deflate (empty) between each breath,
  - (e) For about 5 to 10 seconds, listen and feel for air to be exhaled and look at the victim's chest to see if it rises and falls,
  - (f) Provide at least one breath every 5 seconds, if there is a pulse and no breathing, and
  - (g) Check for moderate resistance to blowing, which will be felt if the airway is clear.
- Watch victim's chest to see when it rises.
- Stop blowing when the victim's chest has expanded, and look and listen for air to be exhaled.
- Watch victim's chest to see when it falls.
- Repeat recommended technique.

HEART COMPRESSION

The best technique of performing emergency first aid is a combination of artificial respiration and artificial circulation by external heart or cardiac compression. Only properly trained personnel should perform this technique.

Inflate the victim's lungs about 5 times by artificial respiration, then check the pulse of the victim in the neck or on the side of the Adam's apple. If the victim's pulse did not start, the rescuer must start cardiac compression in addition to artificial respiration in order to restore the circulation. Two rescuers should perform this technique. One rescuer should perform the artificial respiration, and the other rescuer should rhythmically compress the victim's heart. **(See Figure 2-1)**

INHALATION TREATMENT

Inhalation treatment should be used to make breathing failures less likely. The victim should breathe pure oxygen from a tank when assistance is obtained from police or fire department or the emergency crew of the local electric or gas company. The artificial breathing method and inhalation treatment should be used together for best results.

KEEPING THE VICTIM WARM

Victims of electric shock should be kept warm by using a blanket or some loose clothing to prevent a sudden chill or pneumonia.

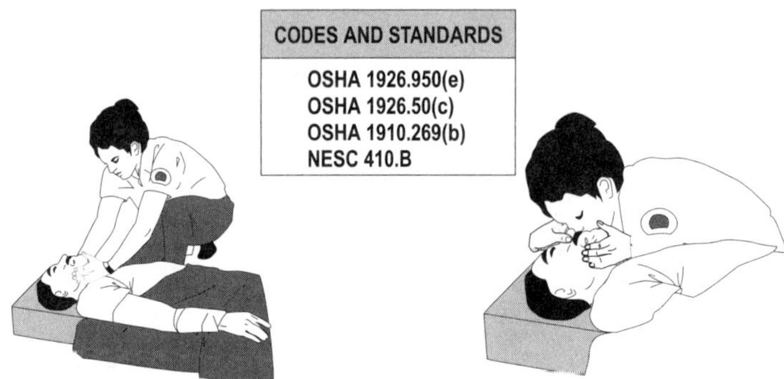

To open the airway:
The tongue is attached to the lower jaw and floor of the mouth. In an unconscious victim, the head flexes, the jaw relaxes and the tongue drops back to obstruct the air passage. To restore breathing, the first thing to do is open the airway. The simplest and quickest method is:

• Lift the neck with one hand.
• Tilt the head backward into maximum extension with the other hand; this draws the tongue forward and opens the airway. This should always be the first maneuver performed. Sometimes this alone will allow the victim to resume spontaneous breathing.

Note 1: Only qualified trained personnel should perform artificial respiration on a hurt victim.

To restore breathing:
If the victim does not resume spontaneous breathing, start mouth-to-mouth respiration.

• First pinch the nose shut with your thumb and index finger to prevent the escape of air. Your cheek can also be used to seal the victim's nose.
• Then open your mouth widely, take a deep breath, place your mouth over the victim's mouth to get a good seal and inflate their lungs with about twice the amount of a normal breath.
• Remove your mouth and allow them to exhale.

Note 2: Employers must have their staff who have the responsibility of performing first aid to hurt employees trained by Red Cross instructors or other qualified agencies.

Figure 2-1. The best technique of performing emergency first aid is a combination of artificial respiration and artificial circulation by external heart or cardiac compression. Only properly trained personnel should perform this technique.

Chapter 2. Electrical Safety and First Aid

Section   Answer

1. When an electrical current passes through the body, there is a path drawn from the _______ point(s) of contact where it enters and exits.
   (a) one							(b) two
   (c) three						(d) four

2. If a person holds on to a deenergized conductor and has someone gradually increase the current flow, a sensation would be detected when the current reached _______ mA.
   (a) 1							(b) 5
   (c) 10							(d) 16

3. When freeing a victim from an electrical shock, the rescuer should separate the victim from the contact by _______.
   (a) a long dry stick				(b) a long dry rope
   (c) a long length of dry cloth		(d) any of the above

4. There are _______ techniques suggested when applying mouth-to-mouth resuscitation to a victim
   (a) 4							(b) 6
   (c) 9							(d) 12

5. When performing mouth-to-mouth resuscitation, the rescuer should provide at least one breath every _______ seconds, if there is a pulse and no breathing.
   (a) 3							(b) 5
   (c) 10							(d) 15

Dress Standards

Employees working in locations where electrical hazards exist shall be provided with and use electrical protective equipment that is suitable for the parts of the body to be protected and for the specific work to be performed. All protective equipment shall be maintained in a safe reliable condition, and periodic checks shall be made to ensure such equipment remains in a safe and usable condition. Note that when the requirements of **NFPA 70E®** are being applied to specific job functions, the latest edition published by NFPA should be used.

HEAD PROTECTION

Nonconductive head protection shall be worn if there is danger of head injury from electric shock, burns, or flying or falling objects resulting from electrical explosions. OSHA requires all employees to wear hard hats or other suitable means to protect the head, which is always vulnerable to injury. Helmets used for the protection of employees against impact and penetration of falling or flying objects shall comply with **ANSI Z89.1**, *Requirements for Protective Headwear for Industrial Workers, 1997*. **(See Figure 3-1)**

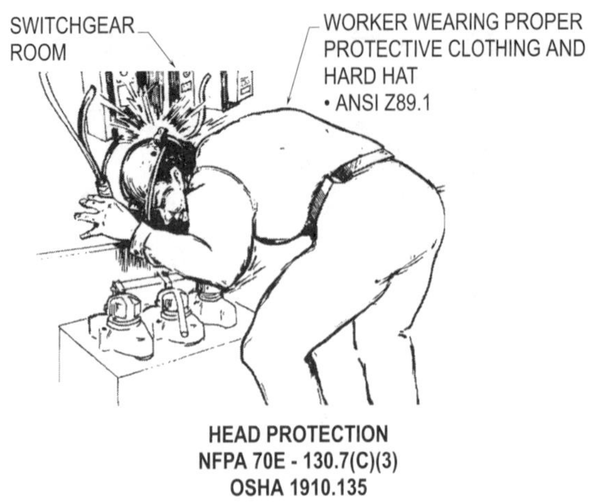

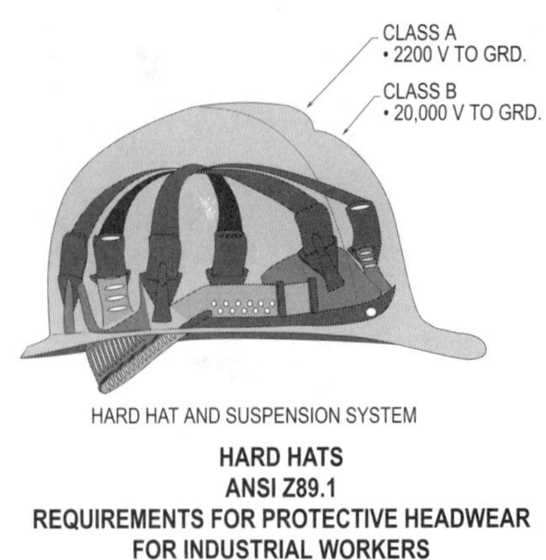

Figure 3-1. Nonconductive head protection shall be worn if there is danger of head injury from electric shock, burns, or flying or falling objects resulting from electrical explosion.

Figure 3-2. Hard hats shall be approved for the purpose of electrical work being performed.

HARD HATS

Hard hats shall be approved for the purpose of electrical work. Metal hard hats are not acceptable. Hard hats are required in the following cases:

- Work is being performed above the head,

- Elevated work is being performed,

- Underground work or work in confined spaces is being performed,

- Work is being performed in electrical substations or switchgear,

- Work is being performed on or near energized electrical distribution equipment, and

- In all construction areas. **(See Figure 3-2)**

EAR PROTECTION

The appropriate type of ear protection shall be worn when working in areas where such protection is needed.

EYES AND FACE PROTECTION

All employees shall wear protective equipment for the eyes or face whenever there is danger of injury due to electrical arcs or flashes or from flying or falling objects resulting from electrical explosion. Eye and face protection equipment shall comply with **ANSI Standard Z87.1,** *Practice for Occupational and Educational Eye and Face Protection, 1998.* (See Figure 3-3)

Figure 3-3. All employees shall wear protective equipment for the eyes or face whenever there is danger of injury due to electrical arcs or flashes or from flying or falling objects resulting from electrical explosion.

FACTOR IN PROTECTION

To protect the eyes from potentially flying or falling objects, the following types of personal protective equipment shall be used:

- Face shields,
- Windows, and
- Safety goggles.

Where there is potential exposure to arc flash conditions, they shall protect the eyes from the resulting thermal and luminous energy.

PROTECTIVE GOGGLES AND GLASSES

There are two main types of protective glasses and goggles that are available in a variety of styles to protect against a wide range of hazards. Spectacle-type safety glasses are designed for protection from the front only with some models having side shields for the protection of objects that fly or bounce. Eye-cup goggle safety glasses are designed to provide protection from all directions. **(See Figure 3-4)**

Figure 3-4. Types of safety glasses and goggles used to protect the employee against a wide range of hazards.

HAND AND ARM PROTECTION

Rubber gloves shall be tested each time they are used and shall be worn under leather gloves to protect them against mechanical damage, oil, and grease.

Employees shall wear outer clothing with full-length sleeves, with the sleeves rolled down for better protection. The main function of rubber sleeves is to protect employees from electrical hazards when they are working on or in proximity to live circuits and parts in or near electrical equipment. **(See Figure 3-5)**

FOOT AND LEG PROTECTION

Dielectric overshoes shall be required where insulated footwear must be depended on as the primary personal protection against step and touch potentials. Appropriate foot and leg protection shall be worn. Rubber insulated mats shall not be permitted as an alternate method to the use of insulated shoes. **(See Figure 3-6)**

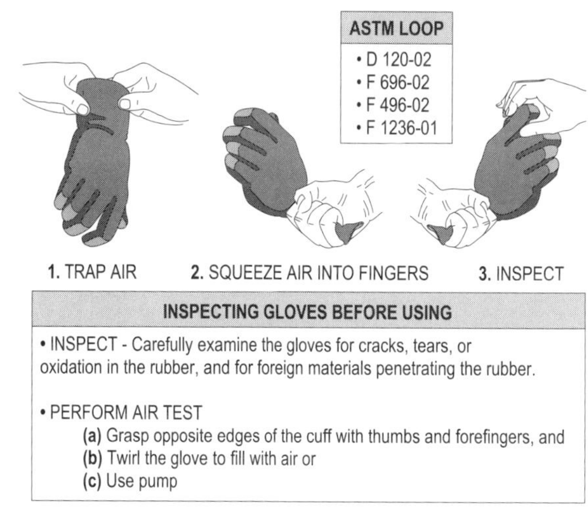

Figure 3-5. Employees shall test gloves before they are used.

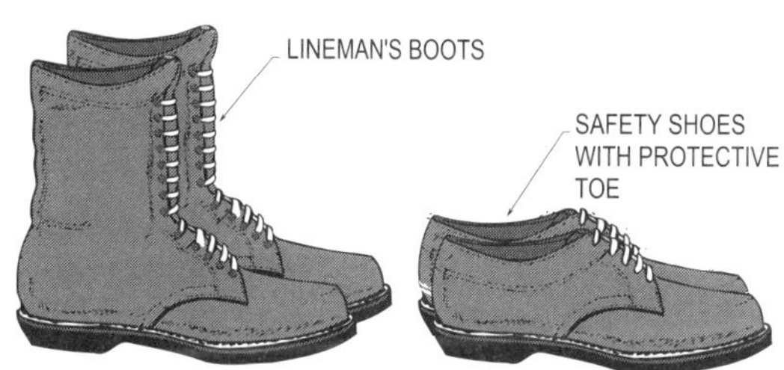

Figure 3-6. Appropriate foot and leg protection shall be worn to protect personnel from electrical hazards.

OVERALL PROTECTION

To perform a job safely, the proper protective clothing and tools shall be used in a safe manner. Safety should always be practiced by electricians when working with electrical circuits.

Electricians should be familiar with the standards pertaining to the type of protective equipment that they shall wear and how such equipment shall be cared for. **[See Figures 3-7(a) and (b)]**

STANDARDS ON PROTECTIVE EQUIPMENT

Subject	Number and Title
Head Protection	ANSI Z89.1 - 1997, Requirements for Protective Headwear for Industrial Workers
Eye and Face Protection	ANSI Z87.1 - 1998, Practice for Occupational and Educational Eye and Face Protection
Blankets	ASTM D 1048 - 1998, Standard Specification for Rubber Insulating Blankets
Covers	ASTM D 1049 - 1983, Standard Specification for Rubber Covers
Line Hoses	ASTM D 1050 - 1990, Standard Specification for Rubber Insulating Line Hoses
Sleeves	ASTM D 1051 - 2002, Standard Specification for Rubber Insulating Sleeves
Gloves	ASTM D 120 - 2002, Standard Specification for Rubber Insulating Gloves

(a)

**SPECIFICATIONS FOR DETERMINING IF PROTECTIVE EQUIPMENT
IS SAFE TO USE FOR A PARTICULAR JOB FUNCTION**

STANDARDS ON PROTECTIVE EQUIPMENT

Subject	Number and Title
Head Protection	ANSI Z89.2 - 1997, Requirements for Protective Headwear for Industrial Workers
Eye and Face Protection	ANSI Z87.1 - 1998, Practice for Occupational and Educational Eye and Face Protection
Gloves and Sleeves	ASTM F 496-02 - 2002, Standard Specfication for In-Service Care of Insulating Gloves and Sleeves
Line Hoses and Covers	ASTM F 478-99, Standard Specification for In-Service Care of Insulating Line Hose and Covers
Blankets	ASTM F 479-95, Standard Specification for In-Service Care of Insulating Blankets
Leather Protectors	ASTM F 696-02, Standard Specification for Leather Protectors for Rubber Insulating Gloves and Mittens

(b)

**SPECIFICATIONS TO BE UTILIZED TO PROVIDE IN-HOUSE
SERVICE CARE OF PROTECTIVE EQUIPMENT USED BY PERSONNEL**

OVERALL PROTECTION

Figures 3-7(a) and (b). If a job is to be performed safely, the proper protective clothing and tools shall be used in a safe manner and cared for, so they will be safe to use again.

Chapter 3. Dress Standards

Section   **Answer**

1. To protect the eyes from potential flying and falling objects, the following type(s) of personal protective equipment shall be used.
   (a) safety goggles (b) face shield
   (c) windows (d) all of the above

2. Rubber gloves shall be tested ______.
   (a) each time they are used (b) every day
   (c) one a week (d) once a month

3. Rubber gloves shall be worn under leather goves to protect them against ______.
   (a) mechanical protection (b) oil
   (c) none of the above (d) all of the above

4. Hard hats shall be ______ for the purpose of electrical work.
   (a) labeled (b) approved
   (c) listed (d) identified

5. When working on energized parts, the following hard hat shall not be permitted to be used.
   (a) metal (b) nonconductive
   (c) all of the above (d) none of the above

4

Electrical Systems

The distribution of electrical power for electrical systems to a premises consists of three-phase transmission lines with a typical voltage of 2400 or 13,800 volts across each phase.

For example, the voltage from phase to phase is 4160 volts for a three-phase wye configuration transformed at the substation output. This distribution power of 2400/4160 volts is stepped down by transformers to voltages suitable for each premises served. The electrical wiring system loads consist of three parts known as the service, the feeders (subfeeders), and the branch circuits. The calculated load of the service conductors is found from the total volt-amperage of all the branch circuits and feeders. Covered in this chapter are the types of electrical systems and voltages used to supply premises wiring.

Electric power is produced and distributed to the premises by the public utility companies almost exclusively as alternating current. Private generating systems in industries that require close control of motor speeds continue the use of direct current even thought it has largely been discontinued by supplying utilities. In recent years, the use of three-phase (alternating current) services has increased rapidly. Single-phase service is used mainly for power systems supplying facilities requiring smaller loads.

SYSTEMS OF 600 VOLTS OR LESS

Systems of 600 volts or less are usually called utilization systems. Such systems are used within a facility to supply power to lighting, general purpose outlets, and utilization equipment.

SINGLE-PHASE SYSTEMS

Single-phase systems are available in two- and three-wire services. Dwelling units, small business establishments, and stores use three-wire services for most single-phase applicants.

120 V, SINGLE-PHASE, TWO-WIRE SYSTEMS

Two-wire circuits consist of 120 volts between the ungrounded (phase) conductor and a grounded (neutral) conductor. 120 volt, two-wire, single-phase systems are used to supply equipment such as signboards and television boosters and other related loads. **(See Figure 4-1)**

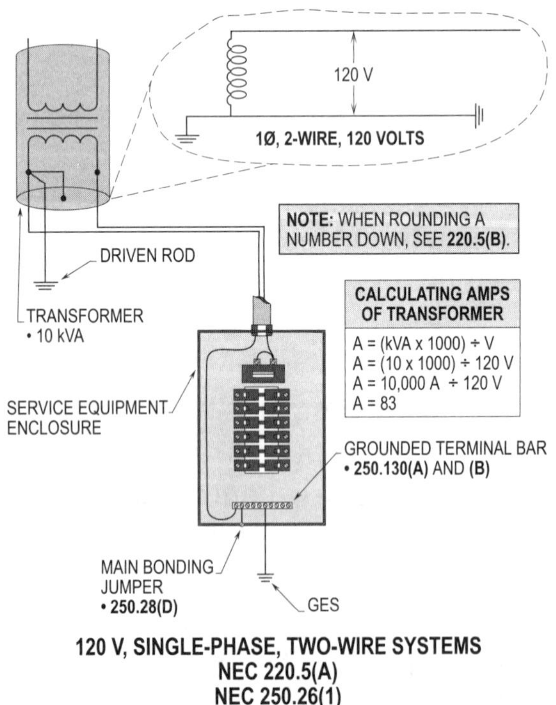

Figure 4-1. 120 volt, single-phase, two-wire systems consist of 120 volts between the ungrounded (phase) conductors and a grounded (neutral) conductor.

208/120 V, SINGLE-PHASE, THREE-WIRE SYSTEMS

Electrical supply systems of 208/120 volts are usually used for individual dwelling units in apartment complexes that are supplied by 208/120 volt, four-wire, three-phase systems. At a tap can, individual 208/120 volt, three-wire, single-phase systems are routed to the service equipment in each dwelling unit. This same arrangement of utility and service supply may be used for small commercial buildings. **(See Figure 4-2)**

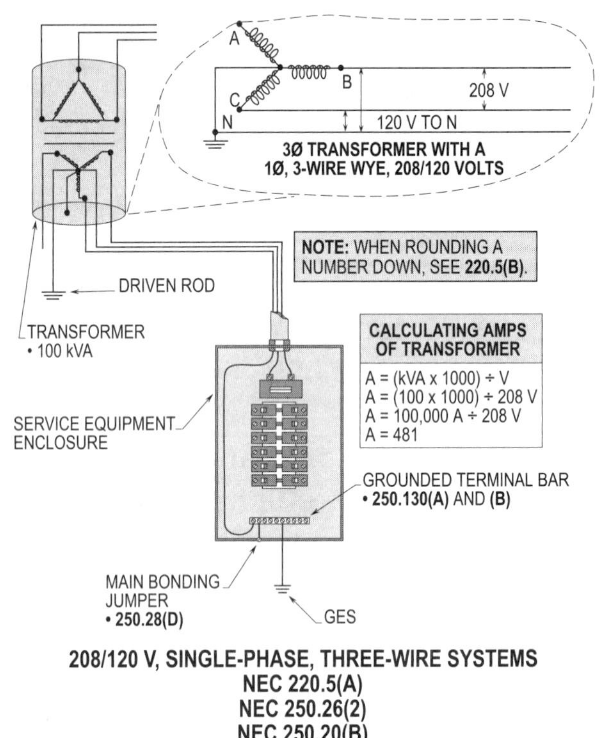

Figure 4-2. 208/120 volt, single-phase, three-wire systems are usually used for individual dwellings units in apartment complexes or small commercial buildings that are supplied by 208/120 volt, four-wire, three-phase systems.

240/120 V, SINGLE-PHASE, THREE-WIRE SYSTEMS

Electrical supply systems of 240/120 volt, single-phase, three-wire are also used in residential and small commercial occupancies to provide 120 volts for lighting and small appliance loads, and 240 volts, single-phase for heavy appliances and small motor loads. The third wire gives additional capacity over a 120 volt, two-wire system, which has been largely superseded in new installations. Systems of 240/120 volts will have two 120 volt conductors to the grounded (neutral) conductor and 240 volts between the two ungrounded (phase) conductors. **(See Figure 4-3)**

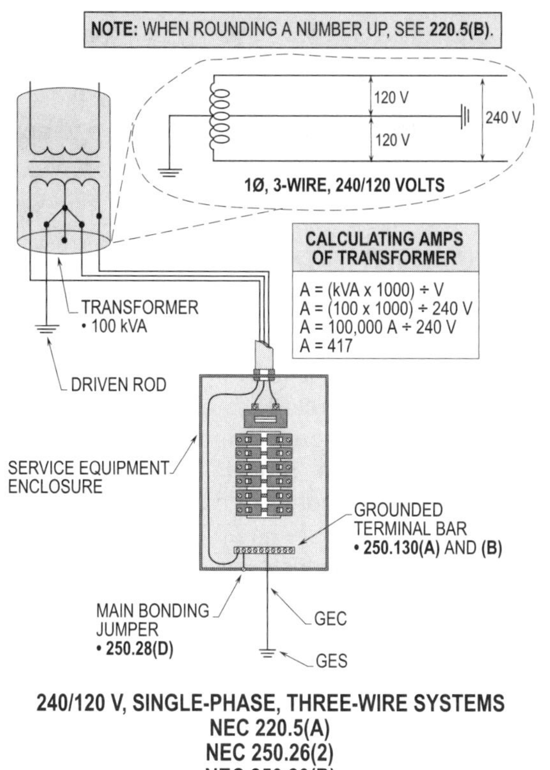

240/120 V, SINGLE-PHASE, THREE-WIRE SYSTEMS
NEC 220.5(A)
NEC 250.26(2)
NEC 250.20(B)

Figure 4-3. 240/120 volt, single-phase, three-wire systems consist of two 120 volt conductors to the grounded (neutral) conductor and 240 volts between the two ungrounded (phase) conductors.

208/120 V, THREE-PHASE, FOUR-WIRE SYSTEMS

Electrical supply systems of 208/120 volt, three-phase, four-wire are generally used for small industrial plants, office buildings, stores, and schools. They provide 120 volts, single-phase, for lighting and appliances, and 208 volts, three-phase, for various types of power loads. These systems are known in the electrical industrial as wye-connected systems. Systems of 208/120 volts have three 120 volt ungrounded (phase) conductors to the grounded (neutral) conductor and 208 volts between any two of the ungrounded (phase) conductors. **(See Figure 4-4)**

Design Tip: The fourth conductor of this system provides almost 50 percent greater capacity than a single-phase, three-wire electrical system.

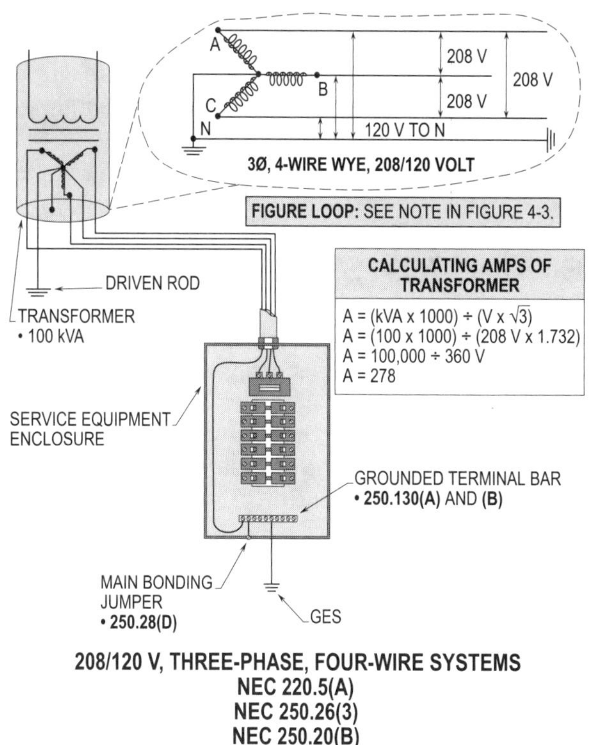

208/120 V, THREE-PHASE, FOUR-WIRE SYSTEMS
NEC 220.5(A)
NEC 250.26(3)
NEC 250.20(B)

Figure 4-4. 208/120 volt, three-phase, four-wire systems consist of three 120 volt ungrounded (phase) conductors to the grounded (neutral) conductor and 208 volts between any two of the ungrounded (phase) conductors.

240/120 V, THREE-PHASE, FOUR-WIRE SYSTEMS

Electrical supply systems of 240/120 volt, three-phase, four-wire are available using three transformers (closed-delta) or two transformers (open-delta). Closed-delta connected transformers offer a greater range of capacity than open delta connected units and provide additional advantages in servicing 120/240 volt, single-phase and 240 volt, three-phase loads.

CLOSED-DELTA SYSTEMS

Systems of 240/120 volt, three-phase, four-wire delta provide single-phase or three-phase, 240 volt supply for power between ungrounded (phase) conductors, and 120 volt, single-phase between two of the ungrounded (phases) and grounded (neutral) conductors. Voltage between the middle phase and the grounded (neutral) is 208 volts (high leg). The total load is not equally balanced over the three phases. Such systems are usually used where motor loads are large as compared with lighting, receptacle, and appliance loads. **(See Figure 4-5)**

> **Design Tip:** Where there are three individual transformers and one of the single-phase supply transformers fails, the system may operate open-delta at approximately 58 percent capacity.

> **Design Tip:** The system capacity is nearly doubled just by adding one single-phase transformer to the bank of two single-phase transformers. However, there must be three individual transformers for this rule to apply.

OPEN-DELTA SYSTEMS

The 240/120 volt, three-phase, four-wire, open-delta system is the same type of system as a closed-delta system.

The main difference is that only two single-phase transformers are used instead of three, and the total capacity is approximately 58 percent of a three-transformer unit system.

Open-delta systems are usually used where the initial load may be expected to be increased some date after the original installation has been completed. **(See Figure 4-6)**

480/277 V, THREE-PHASE, FOUR-WIRE SYSTEMS

Electrical supply systems of 480/277 volt, three-phase, four-wire are called wye connected systems. Such systems are used for circuits in large industrial plants and large commercial facilities. They provide 480 volts, three-phase, for motor and machinery loads and 277 volts, single-phase for fluorescent lighting systems and other 277 volt related loads. Equipment and conductor sizing is one big advantage that such a system provides. **(See Figure 4-7)**

For example, electric power can be delivered at 480/277 volts with less than half the conductor size (cu. or alu.) that is needed at 208/120 volts.

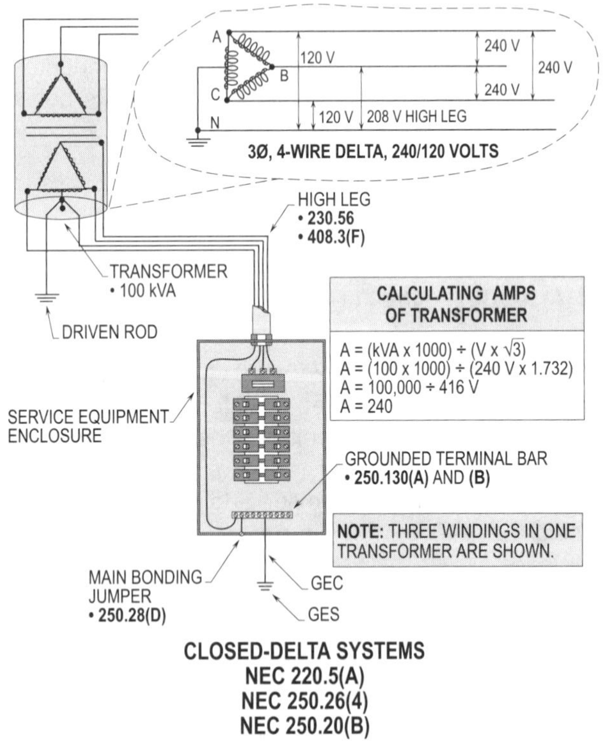

Figure 4-5. Systems of 240/120 volt, three-phase, four-wire delta provide single-phase or three-phase, 240 volt supply for power between ungrounded (phase) conductors, and 120 volt, single-phase between the ungrounded (phases) and grounded (neutral) conductors.

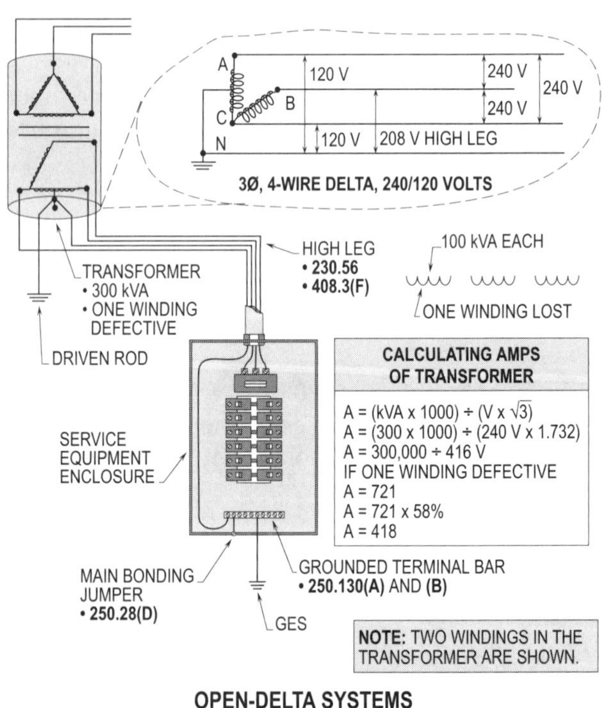

Figure 4-6. Systems that are 240/120 volt, three-phase, four-wire, open-delta are the same type of system as a closed-delta system. The main difference is that only two single-phase transformers are used instead of three. The total capacity is approximately 58 percent of a three-transformer unit system.

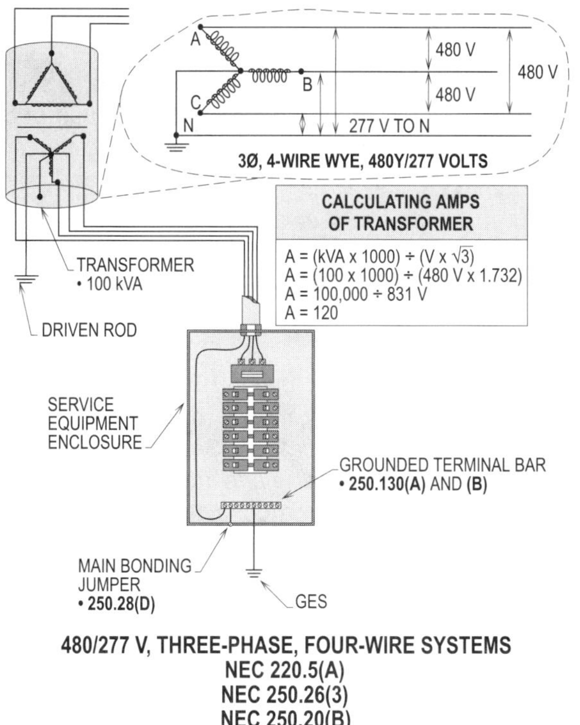

Figure 4-7. Electrical supply systems of 480/277 volt, three-phase, four-wire are called wye connected systems.

SYSTEMS OVER 600 VOLTS

Voltages of over 600 volts are usually referred to as distribution systems. They are used in large industrial facilities to supply large motors and other related type equipment and apparatus. Depending on the voltage, larger pieces of equipment may be supplied by the higher voltage, while lighting, general purpose outlets, and appliances may be supplied by lower voltage systems that are stepped down by transformers to the desired load.

SYSTEMS OF 2.4 TO 15 kV

Electrical supplying systems of 2.4 to 15 kV are usually used for primary power distribution in industrial plants. All plants use a primary voltage of this rating except some of the very large chemical plants, steel mills, etc.

When the utility voltage is below 15,000 volts, selecting the primary voltage is no problem because the *National Electrical Code* and the *National Electrical Safety Code* allow 15,000 volts in buildings. The higher voltage (2400 to 15,000 volts) may be routed in cables and conduits or in interlocked-armored cables to the load-center substations and there transformed down to utilization voltage.

SYSTEMS OF 4160 VOLTS

A problem that often arises is when to select 4160 volts over a lower voltage. There are two major reasons. The first is that 4160 volt systems have lower cost, and the second, that they provide greater allowances for expansion and growth. As for cost, 4160 volt switchgear for a given interrupting rating costs less than lower voltage switchgear, such as 2400 volt systems.

Systems of 4160 volts are three-phase and are equipped with three-phase conductors. **(See Figure 4-8)**

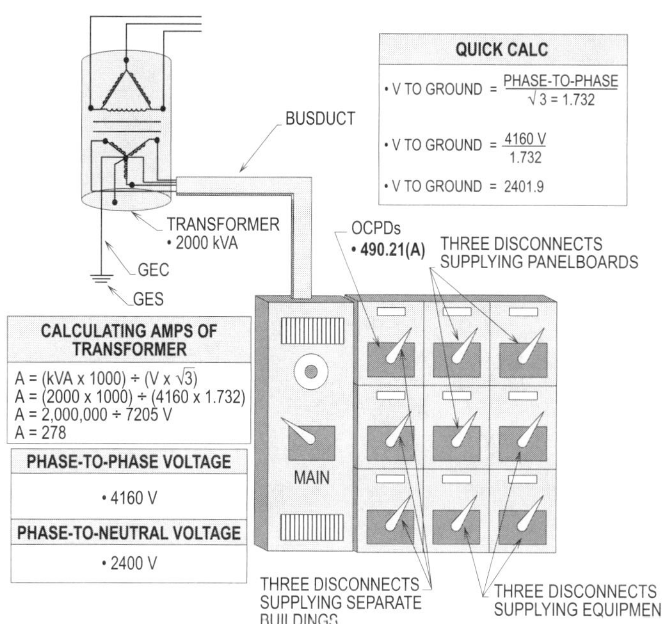

Figure 4-8. Systems of 4160 volts are three-phase and are equipped with three-phase conductors.

Design Tip: Generally, more kVA per circuit may be carried at 4160 volts than at 2400 volts, which results in fewer circuits, therefore reducing the cost of the switchgear. Conductor costs are usually less at 4160 volts than at 2400 volts because less copper is required due to the smaller load amps supplied.

SYSTEMS OF 13,800 VOLTS

In large industrial plants, much of the power is usually used by large motors. Therefore, the motor cost as well as the system cost must be considered to really get a true overall system cost when choosing the primary voltage. There are certain levels of voltage suitable for large industrial plant motors. Such voltage levels are 69,000 volts through 12,470 volts, respectively. The voltage level selected depends upon the size and ratings of the plant's service equipment, motors, and the possibility of future expansion. The electrical supply of these voltages are usually three-phase, four-wire systems.

The larger the electrical system becomes, the more factors there are that cause the cost of a 69,000 volt system to increase faster than that of a 13,800 system. This consideration must be evaluated before such system voltage is selected.

Designers usually select 12,470, 13,200, or 13,800 volt systems and route this voltage to transformers that step the voltage down to desired levels. **(See Figure 4-9)**

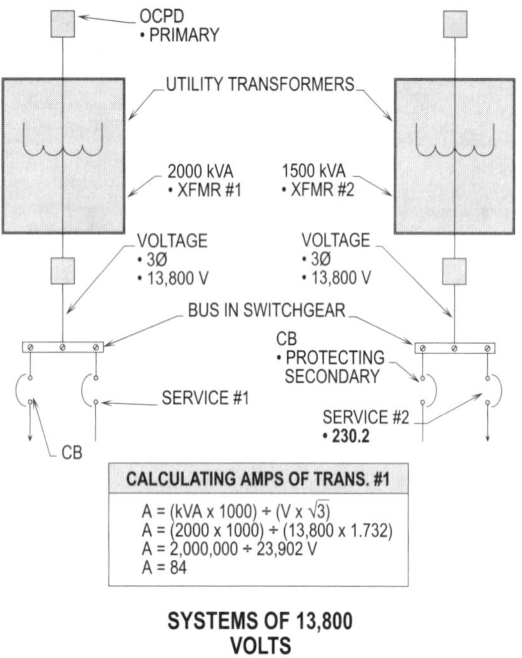

Figure 4-9. Designers usually select 12,470, 13,200, or 13,800 volt systems and route this voltage to transformers that step the voltage down to desired levels.

SYSTEMS OF LESS THAN 50 VOLTS

For a designer or installer to fully understand the requirements pertaining to designing and installing low-voltage electrical systems, he or she must review different Articles and Sections in the NEC. **Article 720** covers circuits and equipment operating at less than 50 volts. **Article 725** deals with Class 1, Class 2, and Class 3 circuits, which may fall under the rules of low-voltage systems of less than 50 volts. **Article 411** contains requirements for low-voltage lighting systems of 30 volts or less, which regulate such systems whether installed indoors or outdoors.

SYSTEMS OF LESS THAN 50 VOLTS
ARTICLE 720

The rules and regulations of low-voltage electrical systems rated at less than 50 volts are stricter than those of other low-volt systems. Conductors, lampholders, receptacles, and protection devices shall be a certain size and rating.

HAZARDOUS (CLASSIFIED) LOCATIONS
720.3

Low-voltage systems of less than 50 volts shall comply with all the requirements for hazardous (classified) locations if they are installed in such locations. **Section 500.1** makes it very clear that all voltage levels utilized in hazardous (classified) locations shall be considered dangerous and shall comply with certain installation rules to ensure that such systems are not capable of igniting an explosive mixture of gases and air.

Design Tip: It must be understood that low-voltage systems alone do not render a circuit incapable of igniting flammable atmospheres. Under some conditions, an ordinary flashlight using two 1-1/2 volt "D" cells can be a source of ignition in hazardous (classified) locations. Even nonincendive circuits can produce a dangerous ignitible source of energy if components are damaged.

CONDUCTORS
720.4

Conductors used for low-voltage systems of less than 50 volts shall not be smaller than 12 AWG copper or equivalent. Conductors for appliance branch circuits supplying more than one appliance or appliance receptacle shall not be smaller than 10 AWG copper or equivalent. Due to the energy output of such systems, conductors shall be large enough to carry normal currents and fault-grounds or short-circuit currents if any should occur. **(See Figure 4-10)**

LAMPHOLDERS
720.5

Standard lampholders used in systems of less than 50 volts shall have a rating of not less than 660 watts. Note that a similar requirement appears in **210.21(A)** for heavy-duty lampholders. In both cases, there are stricter rules for such lampholders based upon voltage and ratings. **(See Figure 4-11)**

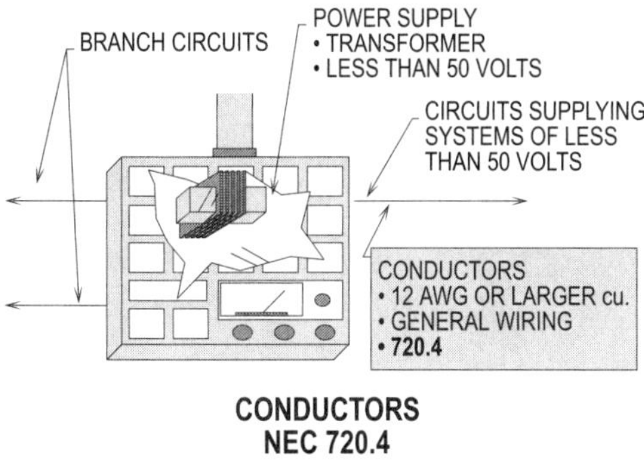

Figure 4-10. Conductors used for low-voltage systems of less than 50 volts shall not be smaller than 12 AWG copper or equivalent.

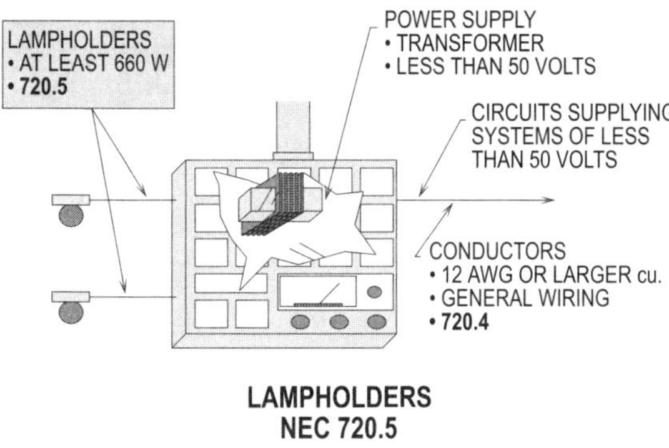

Figure 4-11. Standard lampholders used in systems of less than 50 volts shall have a rating of not less than 660 watts.

RECEPTACLES
720.6 AND 720.7

Receptacles shall have a rating of not less than 15 amperes when used for general purpose. Receptacles of not less than a 20 ampere rating shall be utilized in kitchens, laundries, and other locations where portable appliances are likely to be cord-and-plug connected. **(See Figure 4-12)**

Design Tip: Receptacles shall be sized with capacity ratings for carrying normal currents and abnormal currents due to short-circuit conditions.

BATTERIES
720.9

When storage batteries are used as the power source for systems operating at less than 50 volts, they shall be designed and installed in accordance with **480.1 through 480.4** and **480.8 through 480.10** in the NEC.

For example, ventilation systems shall meet all the rules of **480.9(A)** and **NFPA 30** to determine if the location is hazardous or nonhazardous; **480.9** must be carefully reviewed and adhered to. **(See Figure 4-13)**

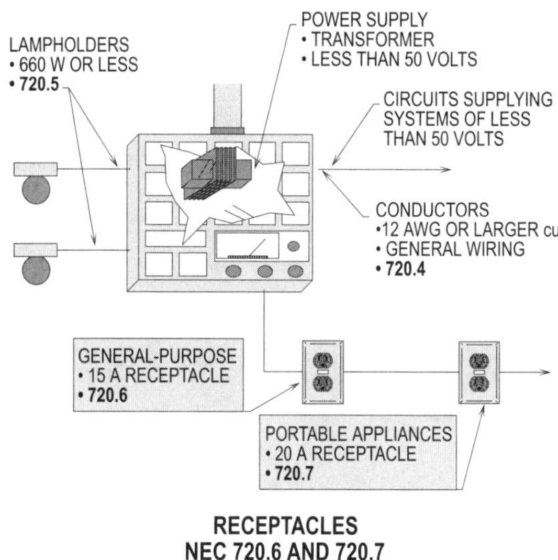

Figure 4-12. Receptacles shall have a rating of not less than 15 amperes when used for general purpose.

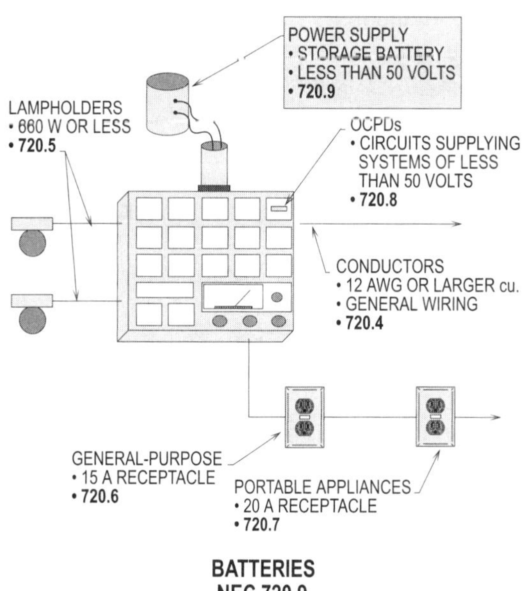

Figure 4-13. When storage batteries are used as the power source for systems operating at less than 50 volts, they shall be designed and installed in accordance with **480.1 through 480.4** and **480.8 through 480.10**.

CLASS 1, 2, AND 3 CIRCUITS
725.2

Class 1, 2, and 3 circuits are classifed by the NEC as remote control, signaling, and power-limited circuits. The NEC defines such circuits as that portion of the wiring system between the load side of the overcurrent protection device or the power-limited supply and all connected equipment. Class 1, Class 2, or Class 3 remote control, signaling, or power-limited circuits are characterized by their usage and electrical power limitation, which differentiates them from light and power circuits. These circuits are classified in accordance with their respective voltage and power limitations.

CLASS 1 CIRCUIT CLASSIFICATIONS AND POWER SOURCE REQUIREMENTS
725.41(A) AND (B)

Class 1 circuits are divided into two types, power-limited and remote-control and signaling circuits. Power-limited Class 1 circuits are limited to 30 volts and 1000 volt-amperes. Class 1 remote-control and signaling circuits are limited to 600 volts, but there aren't any limitations on the power output of the source. Note that the rules pertaining to circuits of less than 50 volts only are reviewed in this chapter.

Class 1 power-limited circuits are supplied from a power source that has a rated output of not more than 30 volts and a power limitation of 1000 volt-amps. Class 1 power-limited circuits have a current limiter on the power source that supplies them. This limiter is an overcurrent protection device that restricts the amount of supply current to the circuit in the event of an overload, short-circuit, or ground-fault. These Class 1 circuits shall be permitted to be supplied from a transformer or other type of power supply such as generators or batteries.

Class 1 remote control or signaling circuits shall be permitted to operate at up to 600 volts and have no limitation on the power rating of the source. Class 1 systems, generally, shall meet most wiring requirements for power and light circuits. Class 1 remote control circuits are commonly used in motor controllers that operate mechanical processes, elevators, conveyors, and equipment that is controlled from one or more remote locations. Class 1 signaling circuits are used in nurses' call systems in hospitals, electric clocks, bank alarm systems, and factory call systems. **(See Figure 4-14)**

Power limitations of power sources used for Class 1 circuits can be reviewed in **Tables 11(A)** and **Table 11(B)** in **Chapter 9** of Tables.

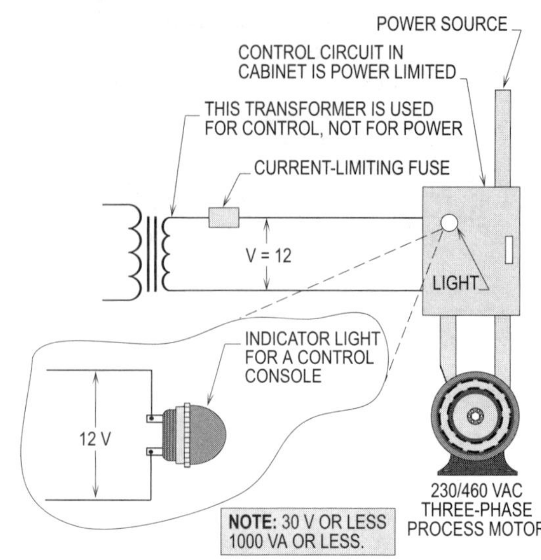

Figure 4-14. Class 1 remote control or signaling circuits shall be permitted to operate up to 600 volts with or without power limitations.

CONDUCTORS OF DIFFERENT CIRCUITS IN THE SAME CABLE, CABLE TRAY, ENCLOSURE, OR RACEWAY
725.48(A) AND (B) AND 300.3(C)(1)

Class 1 circuits shall be permitted to occupy the same cable, enclosure, or raceway without regard to whether the individual circuits are AC or DC current, provided all conductors are insulated for the maximum voltage of any conductor in the cable, enclosure, or raceway. Class 1 circuits and power supply circuits shall be permitted to occupy the same cable, enclosure, or raceway only in situations where the equipment power system is functionally associated. **(See Figure 4-15)**

Section 725.48(B)(2) clarifies that they shall be permitted to be mixed where installed in factory- or field-assembled control centers.

Section 725.48(B)(3) allows mixing of underground conductors in a manhole if all of the following conditions are complied with:

- The power-supply or Class 1 circuit conductors are in a metal-enclosed cable or Type UF cable,

- The conductors are permanently separated from the power-supply conductors by a continuous, firmly fixed nonconductor, such as flexible tubing, in addition to the installation on the wire, and

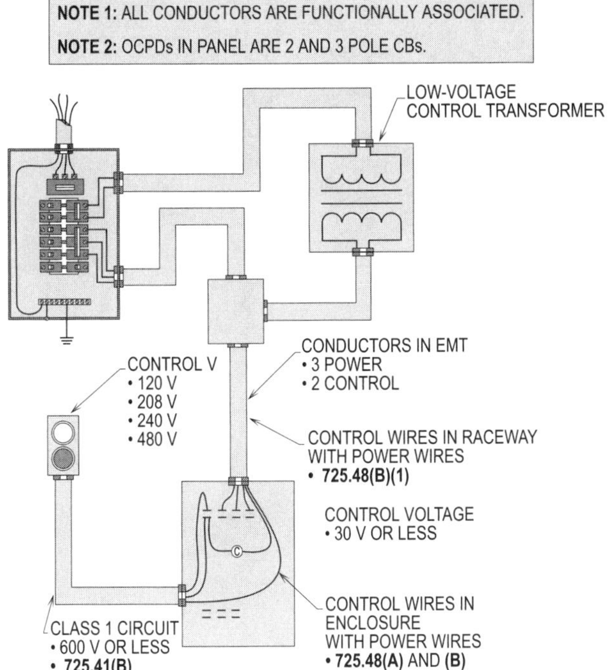

CONDUCTORS OF DIFFERENT CIRCUITS IN THE SAME CABLE, CABLE TRAY, ENCLOSURE, OR RACEWAY NEC 725.48(A) AND (B) AND 300.3(C)(1)

Figure 4-15. Class 1 circuits shall be permitted to occupy the same cable, enclosure, or raceway without regard to whether the individual circuits are AC or DC current, provided all conductors are insulated for the maximum voltage of any conductor in the cable, enclosure, or raceway.

• The conductors are permanently and effectively separated from the power supply conductors and securely fastened to racks, insulators, or other approved supporting means.

CLASS 1 CIRCUIT CONDUCTORS
725.49(A) AND (B)

Conductors of 18 AWG and 16 AWG shall be permitted to be used if they supply loads that do not exceed the ampacities given in **402.5** and are installed in a raceway, an approved enclosure, or a listed cable. Conductors larger than 16 AWG shall not be permitted to supply loads greater than the ampacities given in **310.15**. Flexible cords shall comply with the design and installation requirements of **Article 400**.

Insulation on conductors shall be suitable for 600 volts. Conductors larger than 16 AWG shall comply with the requirements of **Article 310**.

Conductors in sizes 18 AWG and 16 AWG shall be Type FFH-2, KF-2, KFF-2, PAF, PAFF, PF, PFF, PGF, PGFF, PTF, PTFF, RFH-2, RFHH-2, RFHH-3, SF-2, SFF-2, TF, TFF, TFFN, TFN, ZF or ZFF. However, conductors with other types and thicknesses of insulation shall be permitted to be used if listed for Class 1 circuit use.

NUMBER OF CONDUCTORS IN CABLE TRAYS AND RACEWAY AND DERATING
725.51(A), (B), AND (C)

Where only Class 1 circuit conductors are in a raceway, the number of conductors shall be permitted to be determined by the provisions of **300.17**. The derating factors given in **Article 310, Table 310.15(B)(2)(a)** to Ampacity Tables of 0 to 2000 volts apply only if such conductors carry continuous loads in excess of 10 percent of the ampacity of each control conductor routed through the raceway system.

The number of power-supply conductors and Class 1 circuit conductors pulled through a raceway based upon the rules of **725.48** shall be determined per **300.17** in the NEC. The derating factors given in **Article 310, Table 310.15(B)(2)(a)** to Ampacity Tables of 0 to 2000 volts apply to the following conditions:

• To all conductors where the Class 1 circuit conductors carry continuous loads in excess of 10 percent of the ampacity of each conductor and where the total number of conductors is four or more.

• To the power-supply conductors only, where the Class 1 circuit conductors do not carry continuous loads in excess of 10 percent of the ampacity of each conductor and where the number of power-supply conductors is four or more. **(See Figure 4-16)**

Design Tip: Class 1 circuit conductors installed in cable tray systems shall comply with the rules and regulations of **392.9 through 392.11** of **Article 392** in the NEC.

POWER SOURCES FOR CLASS 2 AND 3 CIRCUITS
725.121(A) AND (B)

Class 2 and Class 3 circuits are defined by two Tables, one for AC current and one for DC current. In general, a Class 2 circuit operating at 24 volts with a power supply durably marked "Class 2" and not exceeding 100 volt-amperes is the type most commonly used.

A Class 2 circuit is defined as that portion of the wiring system between the load side of a Class 2 power source and the connected equipment. Because of its power limitations, a Class 2 circuit is considered safe from a fire initiation standpoint and provides acceptable protection from electric shock.

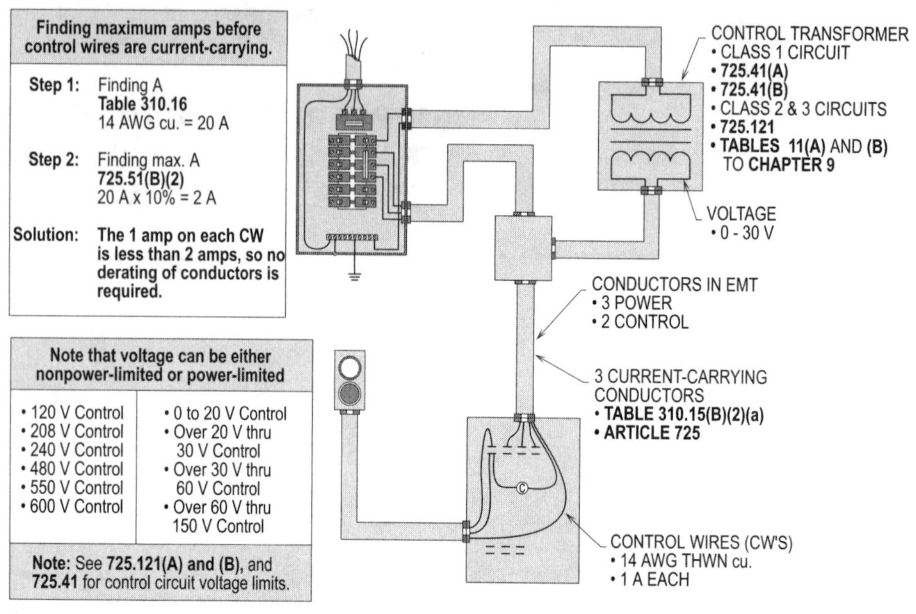

**NUMBER OF CONDUCTORS IN CABLE TRAYS
AND RACEWAY, AND DERATING
NEC 725.51(A), (B), AND (C)**

Figure 4-16. The derating factors given in **Table 310.15(B)(2)(a)** to Ampacity Tables of 0 to 2000 volts apply only if such control conductors carry continuous loads in excess of 10 percent of the ampacity of each conductor routed through the raceway system. For determining if grounded (neutral) conductors are current-carrying, see **Figure 8-8**.

A Class 3 circuit is defined as that portion of the wiring system between the load side of a Class 3 power source and the connected equipment. Due to its power limitations, a Class 3 circuit is not considered safe from a fire-initiation standpoint. Since higher levels of voltage and current are permitted than for Class 2 circuits, additional safeguards are specified to provide protection from an electric shock hazard that might be encountered.

Power for Class 2 and Class 3 circuits is limited either inherently (in which no overcurrent protection is required) or by a combination of a power source and overcurrent protection.

The maximum circuit voltage is 150 volts AC or DC for a Class 2 inherently limited power source, and 100 volts AC or DC for a Class 3 inherently limited power source. The maximum circuit voltage is 30 volts AC and 60 volts DC for a Class 2 power source limited by overcurrent protection, and 150 volts AC or DC for a Class 3 power source limited by overcurrent protection. **(See Figure 4-17)**

For example, heating system thermostats are commonly Class 2 systems, and the majority of small bell, buzzer, and annunciator systems are Class 2 circuits. Class 2 also includes small intercommunicating telephone systems in which the voice circuit is supplied by a battery and the ringing circuit by a transformer.

Class 2 and 3 systems do not require the same wiring methods as power, light, and Class 1 systems. There are cases that require a 2 in. (50 mm) separation is required between these systems.

See **Figure 725.121** in the NEC for a comparison of Class 2 and 3 remote control, signaling, and power-limited circuits.

FIRE ALARM CIRCUIT
760.2

A fire alarm circuit is defined as that portion of the wiring system between the load side of the overcurrent device or the power-limited supply and the connected equipment of all circuits powered and controlled by the fire alarm system. Fire alarm circuits are classified as either nonpower-limited or power-limited.

NONPOWER-LIMITED FIRE ALARM (NPFLA) CIRCUITS
760.2

Nonpower-limited fire-protective signaling (NPLFA) circuits may include circuits that are part of a central station signaling system, a sprinkler water flow alarm, or a local fire alarm in a building. Voltages for these circuits range up to 600 volts. These circuits shall be permitted to be located in the same enclosure, cable, or raceway as Class 1 circuits if the insulation on all of the wires within that enclosure are rated for the highest voltage of any conductor therein.

Power supply conductors are not usually permitted in the same enclosure, cable, or raceway as fire-protective conductors because a fault or overcurrent condition in the power supply conductor could damage the fire protective circuits. This damage would cause the fire protective signal circuit to malfunction and perhaps not transmit a needed alarm or fire signal. However, power supply conductors and fire-protective signaling circuits shall be permitted to occupy the same enclosure if they are connected to the same equipment. **(See Figure 4-18)**

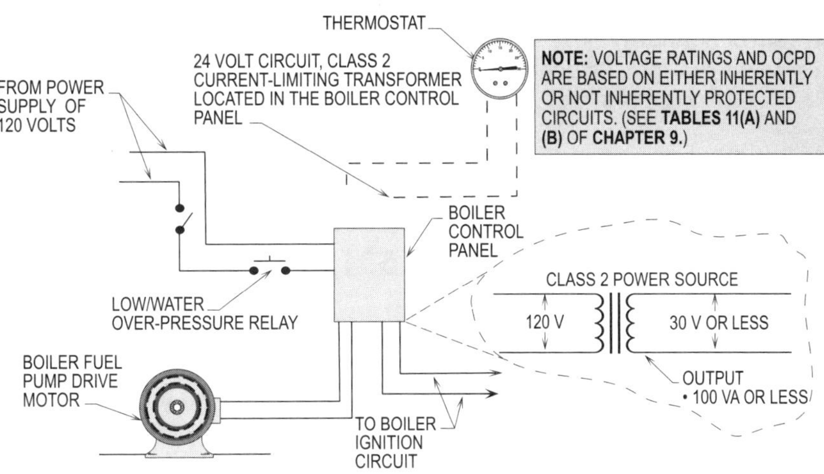

POWER SOURCES FOR CLASS 2 AND 3 CIRCUITS
NEC 725.121(A) AND (B)

Figure 4-17. The maximum voltage is usually 100 volts AC or DC for a Class 2 inherently limited power source, and 150 volts AC or DC for a Class 3 inherently limited power source.

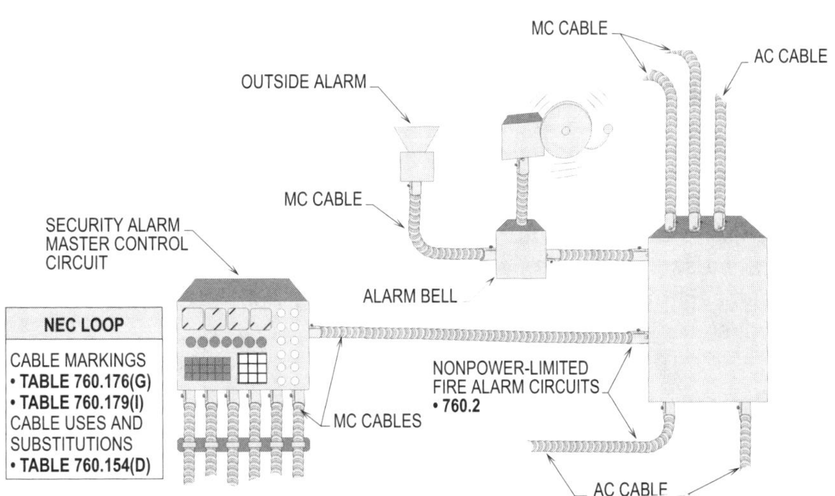

NONPOWER-LIMITED FIRE ALARM (NPLFA) CIRCUITS
NEC 760.2

Figure 4-18. Nonpower-limited fire alarm circuits may include circuits that are part of a central station signaling system, a sprinkler water flow alarm, or a local fire alarm in a building.

POWER-LIMITED FIRE ALARM (PLFA) CIRCUIT
760.2

Since power-limited fire alarm circuit conductors are usually light gauge wire with low-voltage rating and operate at lower voltages and power ratings than Class 1 circuits, power circuits, and nonpower-limited circuits, special measures shall be taken to keep these conductors physically separate. Generally, the power-limited fire alarm circuit conductors shall be separated from these other circuits by at least 2 in. (50 mm). However, the different circuits shall be permitted to be closer:

- If the light, power, Class 1, or nonpower-limited circuit is in a raceway or in a sheathed, metal-clad, or Type UF cable, or

- If the power-limited fire alarm circuit conductors are separated from the other circuits by a non-conductor, such as porcelain tubes or flexible tubing, in addition to the conductor insulation.

Because of the differing operating voltages and insulation levels of power-limited fire alarm circuits, the conductors shall only be permitted to be located where not subject to damage or interference from other types of circuits. Class 3 circuits and power-limited fire alarm circuits operate at similar voltages and power levels. Therefore, the conductors and cables of two or more power-limited fire alarm circuits or Class 3 circuits shall be permitted to occupy the same enclosure. However, power-limited fire alarm circuits and Class 2 circuits differ in operating voltages and insulation and shall be permitted in the same enclosure only when the insulation of the Class 2 conductors is comparable to that of the fire-protective circuit. **(See Figure 4-19)**

FIRE ALARM CIRCUIT IDENTIFICATION
760.30

Identification of fire alarm circuits is required so that these systems will not be interfered with during maintenance operations. Because these are essential systems, this requirement is intended to protect alarm circuits while work is being performed on other systems or while the alarm circuit itself is being serviced, avoiding damage to the circuit and creating false alarms. Therefore, fire alarm circuits shall be properly identified.

For power-limited fire alarm circuits, the marking shall indicate that the circuit is a power-limited fire alarm circuit. This rule is intended to ensure that power-limited fire alarm circuits, which operate at lower power and voltage levels, are not confused with other circuits operating at higher voltages.

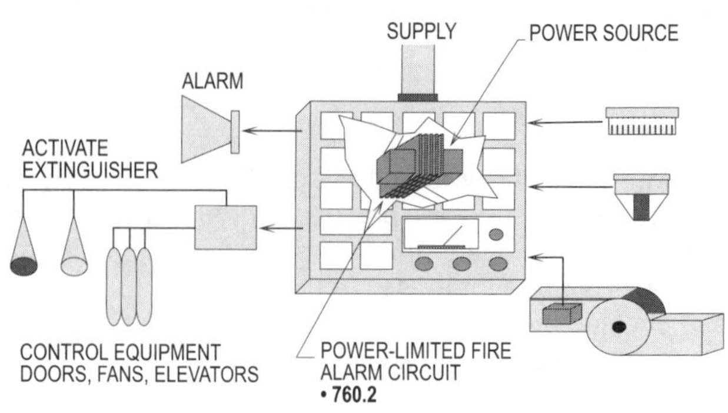

**POWER-LIMITED FIRE ALARM (PLFA) CIRCUIT
NEC 760.2**

Figure 4-19. Because of the differing operating voltage and insulation levels of power-limited signaling circuits, their conductors shall only be permitted to be located where not subject to damage or interference from other types of circuits. (For installation rules, see **760.133.**)

INTRINSICALLY SAFE SYSTEM
504.2

An intrinsically safe system is defined as an assembly of interconnected intrinsically safe apparatus, associated apparatus, and interconnecting cables, in that those parts of the system that may be used in hazardous (classified) locations are intrinsically safe circuits. A circuit in which any spark or thermal effect is incapable of causing ignition of a mixture of flammable or combustible material in air under prescribed test conditions is considered an intrinsically safe circuit (ITSC). **(See Figure 4-20)**

Associated apparatus is an apparatus in which the circuits are not necessarily intrinsically safe themselves but affect the energy in the ITSC's and are relied on to maintain intrinsic safety. Associated apparatus may be either:

- Electrical apparatus that has an alternative-type protection for use in the appropriate hazardous (classified) location or

- Electrical apparatus not so protected that shall not be used within a hazardous (classified) location.

SIMPLE APPARATUS

An electrical component or combination of components of simple construction with well-defined electrical parameters that does not generate more than 1.5 volts, 100 milliamps, 25 milliwatts, or a passive component that does not dissipate more than 1.3 watts and is compatible with the intrinsic safety of the circuit in which it is used. Examples of such apparatus are switches, thermocouples, light-emitting diodes (LED's), connectors and resistance temperature devices (RTD's).

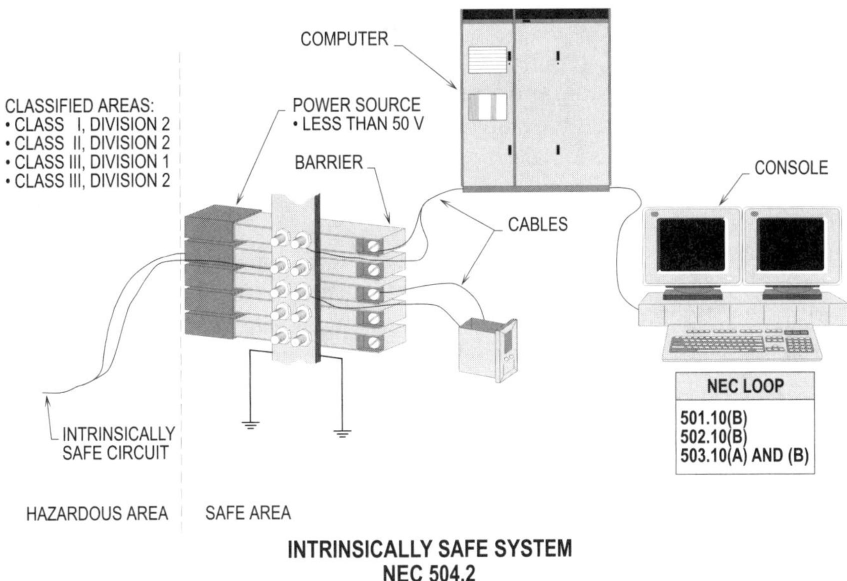

Figure 4-20. An intrinsically safe system is defined as an assembly of interconnected intrinsically safe apparatus, associated apparatus, and interconnecting cables, in that those parts of the system that may be used in hazardous (classified) locations are intrinsically safe circuits.

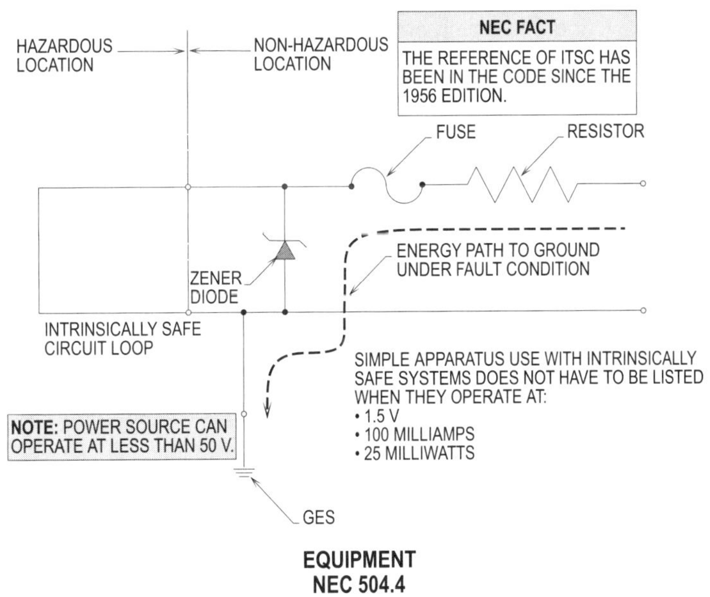

Figure 4-21. All intrinsically safe apparatus and associated apparatus shall be listed except simple apparatus, as described on control drawings.

EQUIPMENT
504.4

All intrinsically safe apparatus and associated apparatus shall be listed except simple apparatus, as described on control drawings. **(See Figure 4-21)**

CONTROL DRAWING
504.10(A) AND 500.2

Intrinsically safe apparatus, associated apparatus, and other equipment shall be installed in accordance with the control drawing(s).

LOCATION
504.10(B)

Intrinsically safe apparatus, associated apparatus, and other equipment shall be permitted to be installed if they are in accordance with the control drawing(s) except a simple apparatus that does not interconnect intrinsically safe circuits.

Intrinsically safe and associated apparatus shall be permitted to be installed in any hazardous (classified) location for which it has been identified. **(See Figure 4-22)**

Design Tip: In most installations, general-purpose enclosures shall be permitted for intrinsically safe apparatus. However, they shall be installed in either Class I, Division 2 or nonhazardous locations.

WIRING METHODS
504.20

Intrinsically safe apparatus and wiring can be installed using any of the wiring methods suitable for unclassified locations, including those covered in **Chapters 7** and **8**. Sealing shall comply with the provisions of **504.70**, and separation shall meet the requirements of **504.30**.

IN RACEWAYS, CABLE TRAYS, AND CABLES
504.30(A)(1)

Conductors of intrinsically safe circuits shall not be permitted to be placed in any raceway, cable tray, or cable with conductors of any nonintrinsically safe circuit.

Ex. 1 to **504.30(A)(1)** recognizes that where conductors of intrinsically safe circuits are separated from conductors of nonintrinsically safe circuits by a distance of at least 2 in. (50 mm) and secured, or by a grounded metal partition or an approved insulating partition.

Ex. 2 to **504.30(A)(1)** recognizes that where either (1) all of the intrinsically safe circuit conductors or (2) all of the nonintrinsically safe circuit conductors are in grounded metal-sheathed or metal-clad cables where the sheathing or cladding is capable of carrying fault current to ground. **(See Figure 4-23)**

Ex. 3 to **504.30(A)(1)** recognizes that intrinsically safe circuits in a Division 2 or Zone 2 location shall be permitted to be installed in a raceway, cable tray, or cable along with nonincendive field wiring circuits when installed in accordance with **504.30(B)**.

Ex. 4 to **504.30(A)(1)** recognizes that intrinsically safe circuits passing through a Division 2 or Zone 2 location to

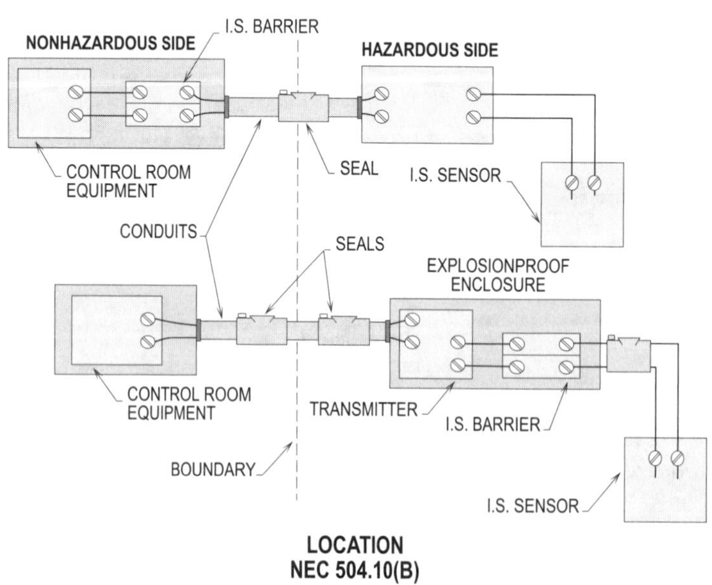

Figure 4-22. Intrinsically safe and associated apparatus shall be permitted in any hazardous (classified) location for which it has been approved.

supply an apparatus that is located in a Division 1, Zone 0 or Zone 1 location shall be permitted to be installed in a raceway, cable tray or cable along with nonincendive field wiring circuits when installed in accordance with **504.30(B)**.

WITHIN ENCLOSURES
504.30(A)(2)

Conductors of intrinsically safe circuits shall be separated at least 2 in. (50 mm) from conductors of any nonintrinsically safe circuits.

Conductors shall be secured so that any conductor that might come loose from a terminal cannot come in contact with another terminal. **(See Figure 4-24)**

Design Tip: The use of separate wiring compartments for the intrinsically safe and nonintrinsically safe terminals is the preferred method of complying with this requirement. Physical barriers shall be permitted, such as grounded metal partitions, approved insulating partitions, or where either (1) all of the intrinsically safe circuit conductors or (2) all of the nonintrinsically safe circuit conductors are in grounded metal-sheathed or metal-clad cables where the sheathing or cladding is capable of carrying fault current to ground.

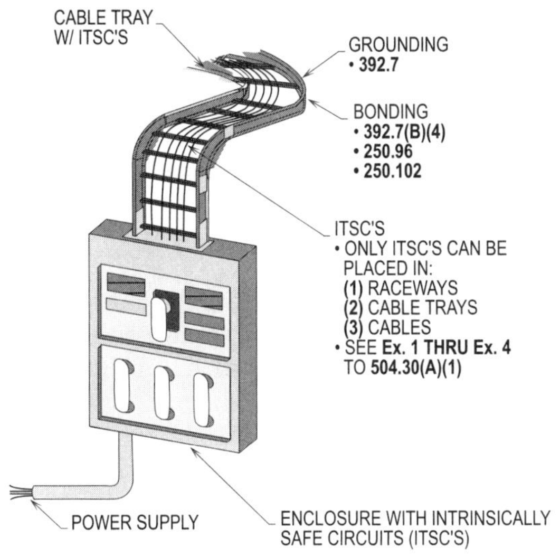

Figure 4-23. Conductors of intrinsically safe circuits shall not be permitted to be placed in any raceway, cable tray or cable with conductors of nonintrinsically safe circuits.

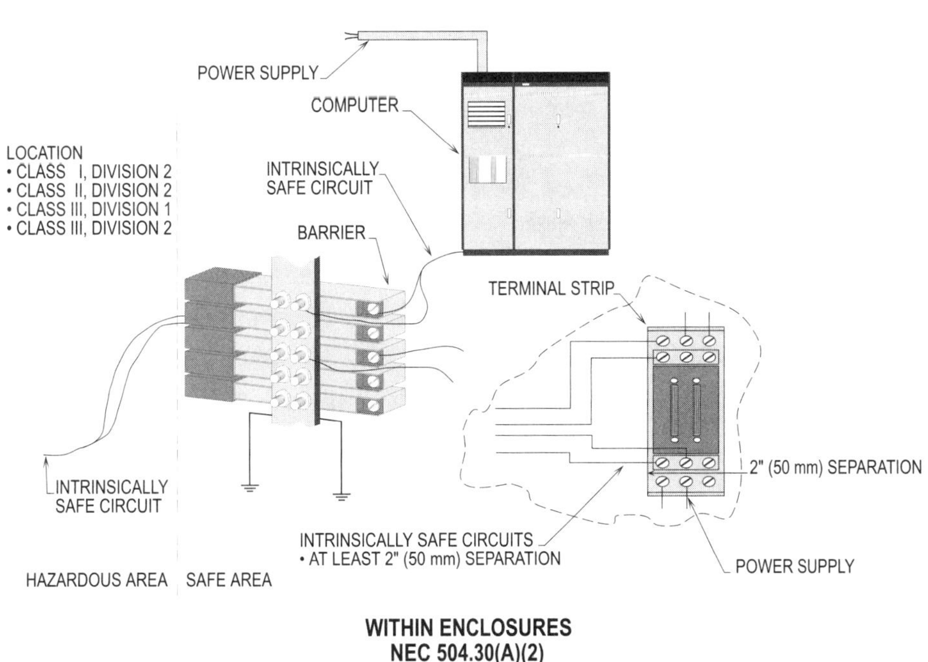

Figure 4-24. Conductors of intrinsically safe circuits shall be separated at least 2 in. (50 mm) from conductors of any nonintrinsically safe circuits when installed within enclosures.

OTHER (NOT IN RACEWAY OR CABLE TRAY SYSTEM)
504.30(A)(3)

Conductors and cables of intrinsically safe circuits that are not in raceways or cable trays shall be separated at least 2 in. (50 mm) and secured from conductors and cables of any nonintrinsically safe circuits.

The **Exception 2** to **504.30(A)(3)** does not require such separation where either (1) all of the intrinsically safe circuit conductors are in Type MI or MC cables or (2) all of the nonintrinsically safe circuit conductors are in raceways or Type MI or MC cables where the sheathing or cladding is capable of carrying fault current to ground. **(See Figure 4-25)**

FROM DIFFERENT INTRINSICALLY SAFE CIRCUIT CONDUCTORS
504.30(B)(1) AND (B)(2)

Different intrinsically safe circuits shall be in separate cables or be separated from each other by one of the following means:

- The conductors of each circuit are within a grounded metal shield.

- The conductors of each circuit have an insulation with a minimum thickness of 0.01 in. (0.25 mm).

- The clearance between two terminals for connection of different intrinsically safe circuits shall be at least 0.25 in. (6 mm), unless this clearance is permitted to be reduced by the control drawing.

IDENTIFICATION
504.80(A) AND (B)

Labels shall be suitable for the environment where they are installed, and consideration given where they are exposed to chemicals and sunlight.

Intrinsically safe circuits shall be identified at terminal and junction locations in such a manner that will prevent unintentional interference with the circuits during testing and servicing.

Raceways, cable trays, and other wiring for intrinsically safe system wiring shall be identified with permanently affixed labels with the wording "Intrinsic Safety Wiring" or equivalent. The labels shall be so located as to be visible after installation and placed so that they are readily traced through the entire length of the installation. Spacing between labels shall be provided at 25 ft (7.5 m) intervals. **(See Figure 4-26)**

The **Ex.** to **504.80(B)** permits circuits run underground, to be identified where they become accessible after emergence from the ground.

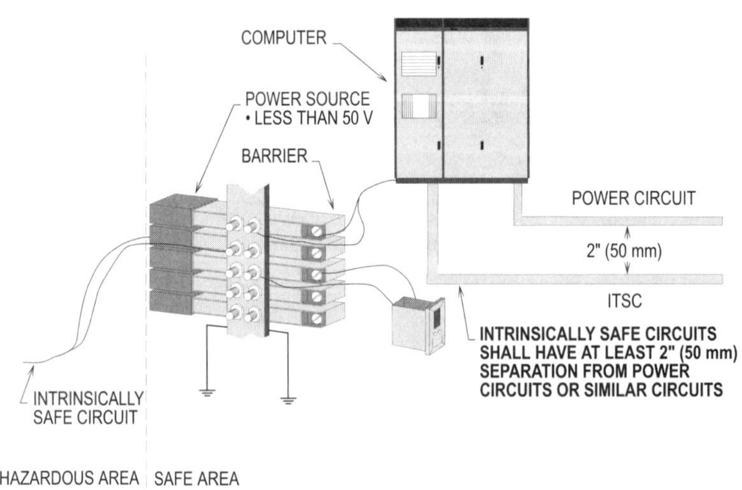

Figure 4-25. Conductors and cables of intrinsically safe circuits that are not in raceways or cable trays shall be separated at least 2 in. (50 mm) and secured from conductors and cables of any nonintrinsically safe circuits.

COLOR CODING
504.80(C)

Color coding shall be provided to identify intrinsically safe conductors, where they are colored light blue and no other conductors are colored light blue. Likewise, color coding shall be permitted in identifying raceways, cable trays, and junction boxes where they are colored light blue and contain only intrinsically safe wiring systems. **(See Figure 4-27)**

GROUNDING
504.50(A) AND (B)

Intrinsically safe apparatus, enclosures, and raceways, if of metal, shall be connected to the equipment grounding conductor.

In addition to an equipment grounding conductor connection, a connection to a grounding electrode may be required. The grounding electrode shall be as specified in **250.52(A)(1) through (A)(4)** and shall comply with **250.30(A)(7)**. Sections **250.52(A)(5)**, **(A)(6)** and **(A)(7)** shall not be permitted if electrodes specified in **250.52(A)(1) through (A)(4)** are available. **(See Figure 4-28)** Associated apparatus and shielded cables, if used, shall be grounded in accordance with the required control drawings.

BONDING
504.60(A) AND (B)

In hazardous (classified) locations, intrinsically safe apparatus shall be bonded in the hazardous (classified) location according to the provisions of **250.100**.

In unclassified locations, metal raceways used for intrinsically safe system wiring in hazardous (classified) locations shall have associated apparatus bonded in accordance with **501.30(A)**, **502.30(A)**, **503.30(A)**, **505.25**, or **506.25** where applicable. **(See Figure 4-29)**

NONINCENDIVE CIRCUITS

Nonincendive circuits (NICCs) are defined as a circuit in which any arc or thermal effect produced, under intended operating conditions of the equipment or due to opening, shorting, or grounding of field wiring, is not capable, under specified test conditions, of igniting the flammable gas, vapor, or dust-air mixture.

Simply put, NICCs are not capable of causing ignition under normal conditions of operation. The concept is the same as with other equipment operating in Class I, Division 2 locations.

> **For example,** a fault in the equipment that makes it capable of causing ignition is very unlikely to occur at the same time that a flammable atmosphere in a Division 2 location is present. (See **500.2**)

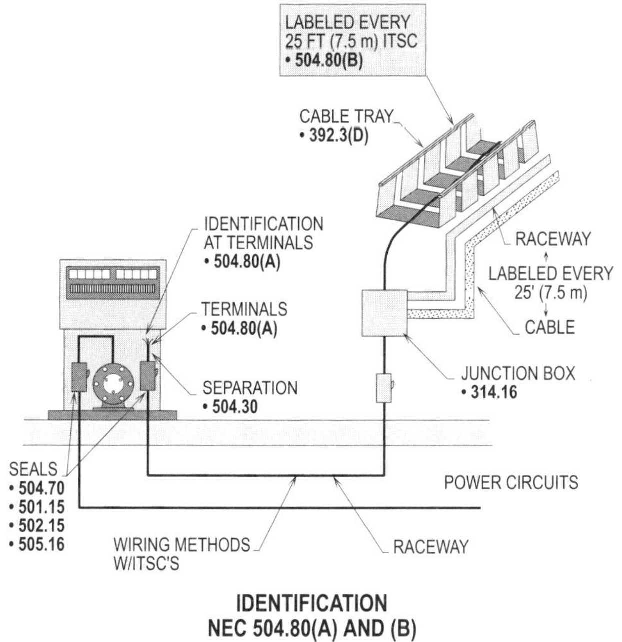

Figure 4-26. Spacing between labels on raceways and cables shall be provided at 25 ft (7.5 m) intervals.

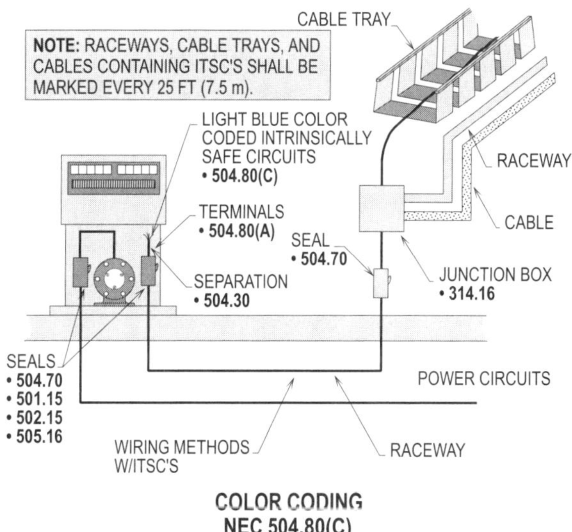

Figure 4-27. Light blue color coding shall be provided to identify intrinsically safe conductors.

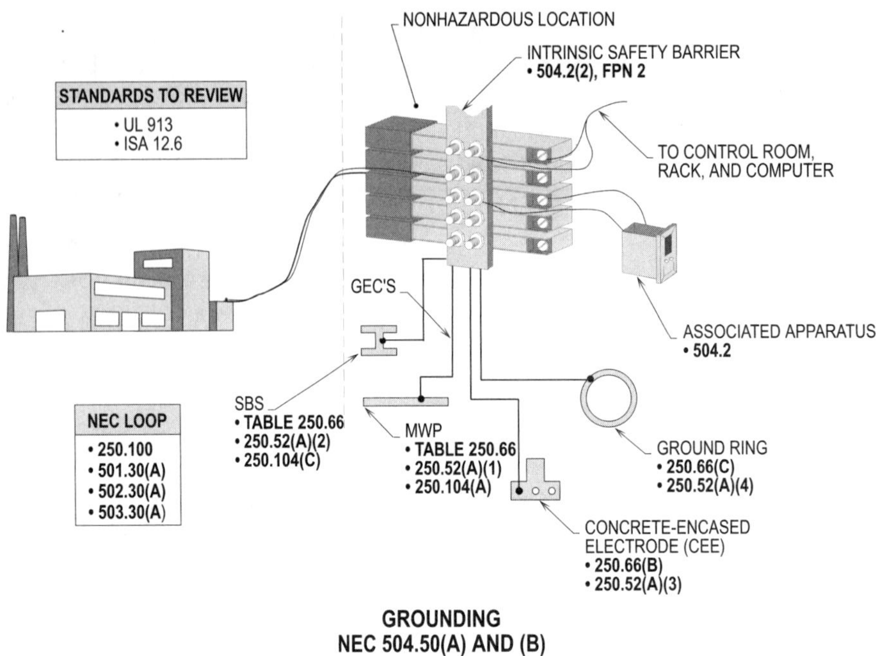

Figure 4-28. Intrinsically safe apparatus, enclosures, and raceways, if of metal, shall be grounded.

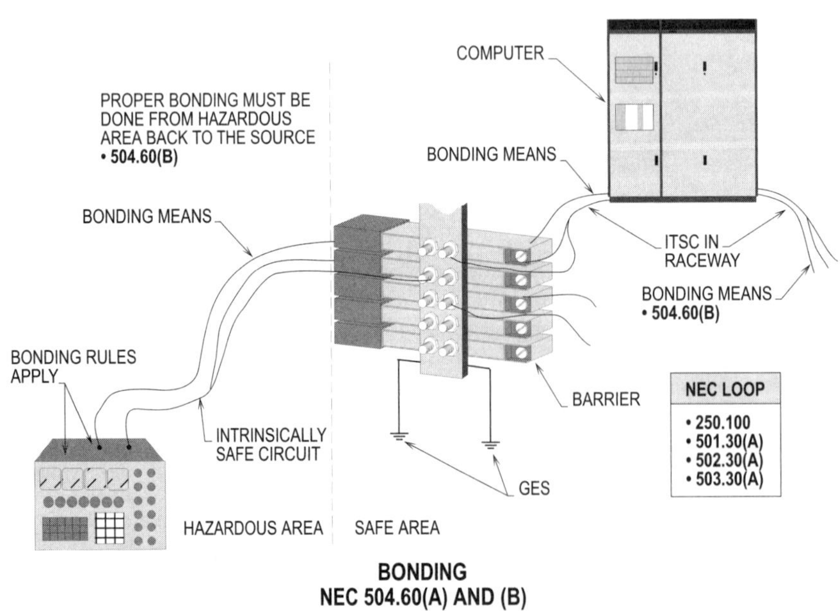

Figure 4-29. In hazardous (classified) locations, intrinsically safe apparatus shall be bonded.

FOUR TYPES OF NONINCENDIVE CIRCUITS

There are four types of systems that may be used for designing and installing a nonincendive system. Such systems are as follows:

- Nonincendive circuits
- Nonincendive contacts
- Nonincendive components
- Nonincendive equipment

NONINCENDIVE CIRCUITS

A nonincendive circuit is a low-energy circuit, and the opening, shorting, or grounding of one or more of the circuit conductors will not result in release of energy that is capable of causing ignition.

> **Design Tip:** Such circuits allow wiring methods to be used that under normal conditions do not release sufficient energy to ignite a specific ignitible atmospheric mixture by opening, shorting, or grounding. Any wiring method permitted in Class I, Division 2 locations shall be permitted to be utilized for these circuits.

NONINCENDIVE CONTACTS

Nonincendive contacts are not always utilized in a nonincendive circuit. However, they are used in a circuit that is considered a low-energy circuit.

An exception in the NEC includes a definition of nonincendive contacts that pertains to contacts used in meters, instruments, and relays, with a separate set of rules for signaling, alarm, remote-control, and communications systems. This requirement permits general-purpose enclosures to be used if current interrupting contacts are in circuits that under normal conditions do not release sufficient energy to ignite a specific ignitible atmospheric mixture. See **501.105(B)(1), Ex.** in the NEC for more information concerning this rule.

NONINCENDIVE COMPONENTS

Nonincendive components are contacts installed in such a manner that ignition is precluded, even though in the open the contacts may be incendive.

Nonincendive components are designed in such a manner to absorb energy and prevent its release, which could cause arcing and heating.

Nonincendive components are components having contacts for making and breaking an incendive circuit. They are installed within an enclosure that is so constructed that the contacts are not capable of causing ignition. Danger only occurs if these components are hit and damaged by a forklift, etc. Then enough energy could be released to ignite an atmospheric mixture.

NONINCENDIVE EQUIPMENT

Nonincendive equipment usually will not comply with all the requirements needed to be classified as intrinsic safety. This is due to the absence of protective barriers. Care must be exercised when designing and using nonincendive equipment in Class I, Division 2 locations. Such locations are usually adjacent to or above Division 1 locations.

An accident might occur that could damage the enclosure housing nonincendive contacts, and such accident may result in a fault releasing (ignition-capable) energy at the same time an explosive mixture of gas and air is present around the equipment.

> **Design Tip:** It is due to the strange fact that, an explosive mixture of gas and air being present at the time a fault capable of releasing ignitable energy occurs, nonincendive equipment is permitted to be used in Class I, Division 2 locations.

WIRING METHODS AND ENCLOSURES (RECOMMENDED REQUIREMENTS)

Nonincendive circuits and wiring shall be permitted to be installed using any of the wiring methods suitable for unclassified locations, including **Chapters 7** and **8**. Sealing shall comply with the provisions of **501.15(A)**, and separation shall comply with **UL 1604, ANSI S12.12** or manufacturer recommendations.

> **Design Tip:** General-purpose enclosures are usually used for such installations per **501.105(B)(1), Ex.'s (3)** and **(4)**.

See Figure 4-30 for a detailed illustration on requirements pertaining to nonincendive circuits and enclosures.

IDENTIFICATION OF TERMINALS AND WIRING (RECOMMENDED REQUIREMENTS)

It is recommended that labels be suitable for the environment in which they are installed, and consideration must be given when they are exposed to chemicals and sunlight.

Nonincendive circuits shall be identified at terminal and junction locations in such a manner that will prevent unintentional interference with the circuits during testing and servicing.

Raceways, cable trays, and other wiring for nonincendive system wiring shall be identified with permanently affixed labels with the wording "Nonincendive Wiring" or equivalent. The labels shall be so located as to be visible after installation and placed so that they are readily traced through the entire length of the installation. It is recommended that spacing between labels be at 25 ft (7.5 m) intervals. **(See Figure 4-31)**

Nonincendive circuits run underground should be identified where they become accessible after emergence from the ground.

COLOR CODING (RECOMMENDED REQUIREMENTS)

It is recommended that color coding (yellow) should be provided to identify nonincendive conductors, and that where they are colored yellow no other conductors shall be permitted to be colored yellow. Likewise, color coding is recommended to be used to identify raceways, cable trays, and junction boxes where they are colored yellow and contain only nonincendive wiring systems; this color code scheme distinguishes them from other wiring methods. **(See Figure 4-32)**

LIGHTING SYSTEMS OPERATING AT 30 VOLTS OR LESS ARTICLE 411

The rules and regulations of this article cover lighting systems operating at 30 volts or less and their associated components. Designers and installers using these requirements will design and install safe, dependable, and reliable lighting systems operating at 30 volts or less.

LIGHTING SYSTEMS OPERATING AT 30 VOLTS OR LESS
411.2

A lighting system operating at 30 volts or less is a lighting system consisting of an isolating power supply, the low-voltage luminaire, and associated equipment that are all identified for the use. The output circuits of the power supply are rated for not more than 25 amperes and operate at 30 volts (42.4 volt peak) or less under all load conditions. **(See Figure 4-33)**

LISTING REQUIRED
411.3

Lighting systems operating at 30 volts or less shall be listed as a complete system. The luminaires, power supply, and luminaire fittings (including the exposed bare conditions) of an exposed bare conductor ligting system shall be listed

for the use as part of the same identified lighting system. **(See Figure 4-34)**

SPECIFIC LOCATION REQUIREMENTS
411.4

Conductors concealed or extended through a wall, floor, or ceiling shall be installed using any of the wiring methods permitted in **Chapter 3** or using wiring supplied by a listed Class 2 power service and installed per **725.130**. Such lighting systems shall not be permitted within 10 ft (3 m) horizontally of pools, spas, fountains, or similar locations, except as permitted by **Article 680**. **(See Figure 4-35)**

SECONDARY CIRCUITS
411.5

Secondary circuits of lighting systems rated at 30 volts or less shall not be grounded per **250.20(A)** and **411.5** in the NEC.

All secondary circuits shall be insulated from the branch circuit by the installation of an isolation transformer.

Exposed bare conductors and current-carrying parts of such systems shall be permitted for indoor installations only. Bare conductors shall not be installed less than 7 ft (2.1 m) above the finished floor, unless specifically listed for a lower installation height and equipped with proper safeguards. **(See Figure 4-36)**

BRANCH CIRCUIT
411.6

Lighting systems operating at 30 volts or less shall be supplied from a maximum 20 ampere branch circuit. This requirement limits the number of outlets and prevents greater sized branch circuits with a greater number of outlets installed throughout a building. **(See Figure 4-37)**

> **Design Tip:** See **210.3** for branch circuit rating rules and **210.19(A)(1)** for ampacity requirements for conductors based upon load. Review **210.20(A)** for regulations concerning the selection of the overcurrent protection devices.

HAZARDOUS (CLASSIFIED) LOCATIONS
411.7

Lighting systems of 30 volts or less installed in hazardous (classified) locations shall comply with the provisions of **Articles 500 through 517** that are in addition to this Article.

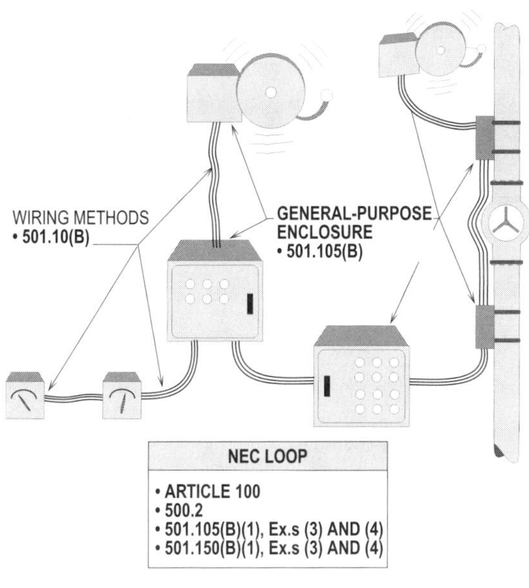

Figure 4-30. General-purpose enclosures are usually used when installing nonincendive circuits and wiring.

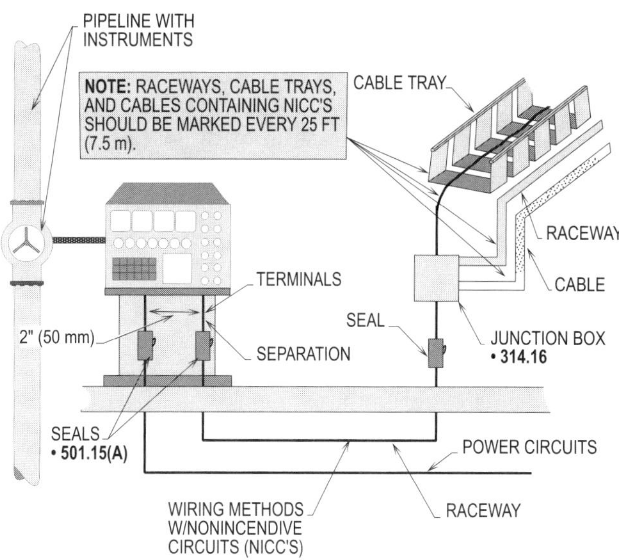

Figure 4-31. Raceways, cable trays, and other wiring for nonincendive system wiring should be identified with permanently affixed labels with the wording "Nonincendive Wiring" or equivalent at 25 ft (7.5 m) intervals.

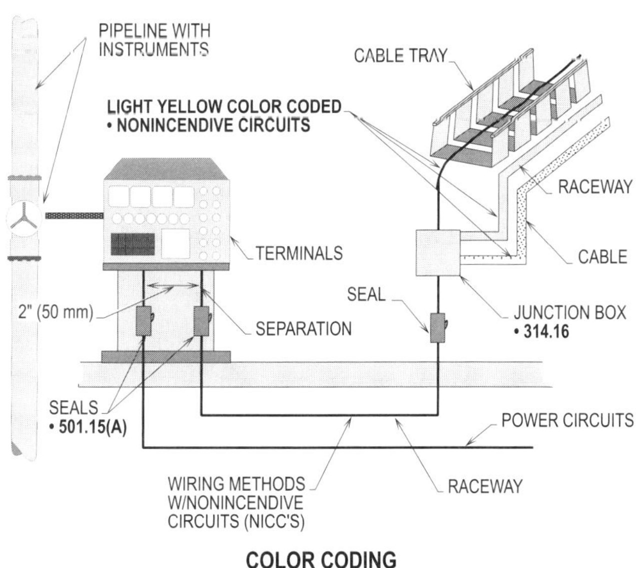

Figure 4-32. It is recommended that color coding (yellow) should be provided to identify nonincendive conductors.

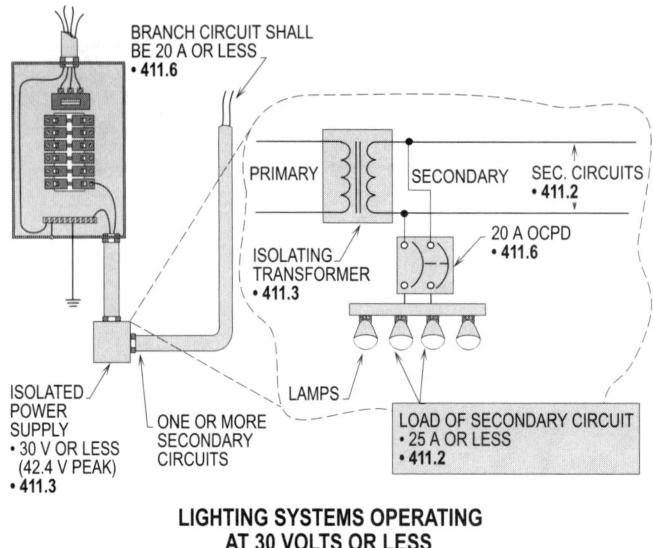

Figure 4-33. A lighting system operating at 30 volts or less is a lighting system consisting of an isolating power supply, the low-voltage luminaires, and associated equipment that are all identified for the use.

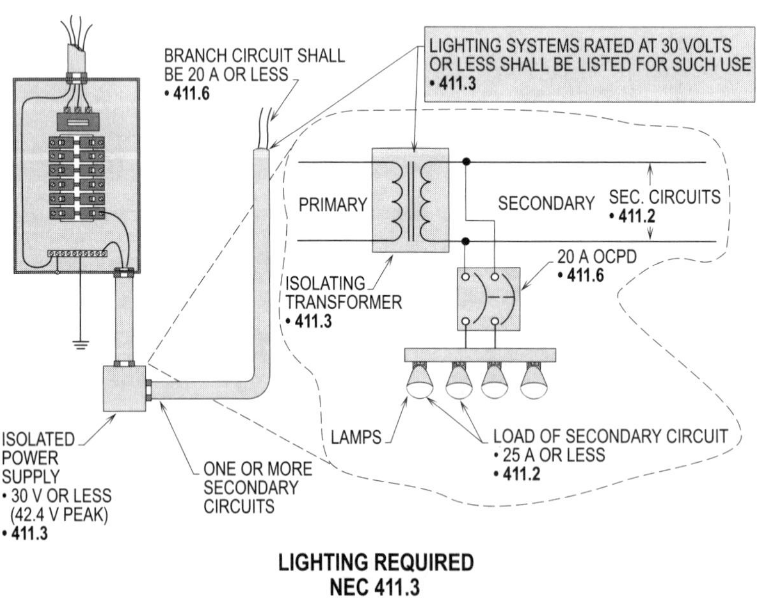

Figure 4-34. Lighting systems operating at 30 volts or less shall be listed as a complete system.

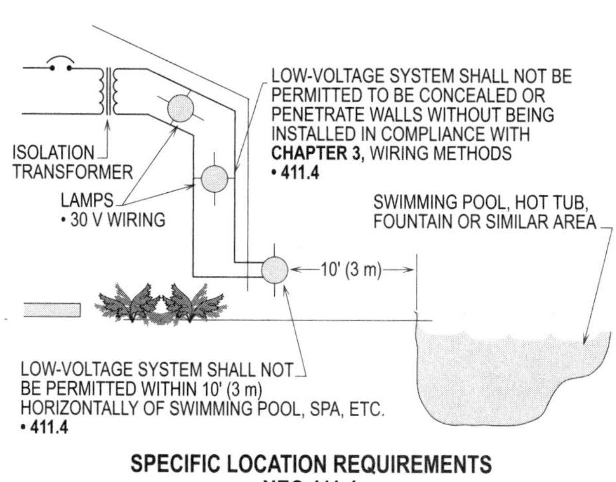

Figure 4-35. Lighting systems operating at 30 volts or less shall not be permitted within 10 ft (3 m) horizontally of pools, spas, fountains, or similar locations, except as permitted by **Article 680**.

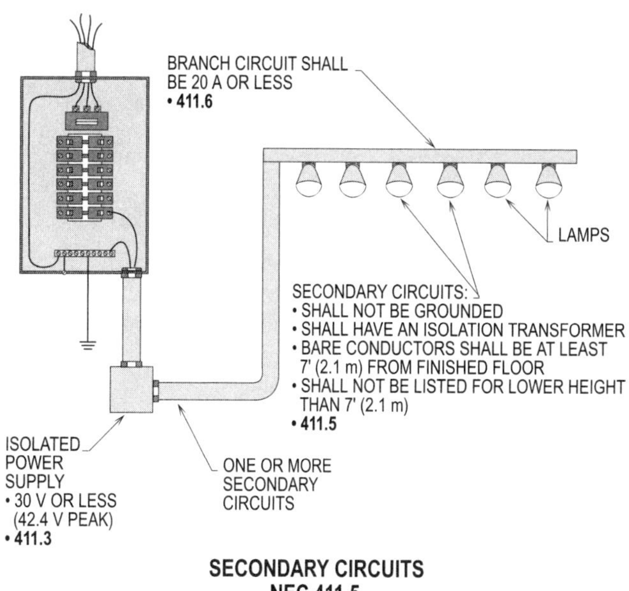

Figure 4-36. Secondary circuits of lighting systems rated at 30 volts or less shall not be permitted to be grounded.

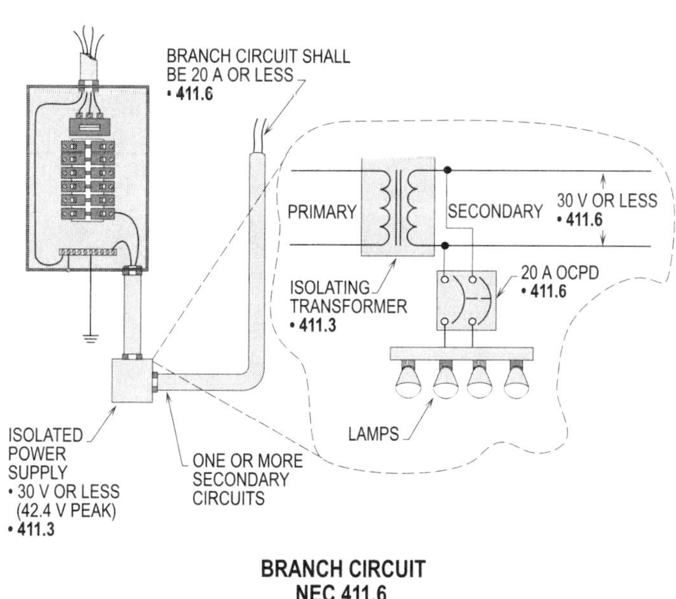

Figure 4-37. Lighting systems operating at 30 volts or less shall be supplied from a maximum 20 ampere branch circuit.

Chapter 4. Electrical Systems

Section   **Answer**

________ ________   1. It is recommended that nonincendive conductors be color coded _____.
  (a) blue   (b) yellow
  (c) red   (d) orange

________ ________   2. Electrical supply systems of _____ volts are usually wired for individual dwelling units in apartment complexes.
  (a) 120/208   (b) 240/120
  (c) 208/120   (d) 480/277

________ ________   3. Article 720 covers circuits and equipment operating at less than _____ volts.
  (a) 12   (b) 15
  (c) 50   (d) 120

________ ________   4. Voltages of over _____ volts are usually referred to as distribution systems.
  (a) 120   (b) 240
  (c) 480   (d) 600

________ ________   5. Conductors used for low-voltage systems of less than 50 volts (for appliance branch circuits) supplying more than one appliance or appliance receptacle shall not be permitted to be smaller than _____ AWG copper or equivalent.
  (a) 18   (b) 14
  (c) 12   (d) 10

________ ________   6. Conductors used for low-voltage systems of less than 50 volts for receptacles shall have a rating of not less than _____ amperes when used for general purpose.
  (a) 15   (b) 20
  (c) 25   (d) 30

________ ________   7. Class 1 power-limited circuits are limited to 30 volts and _____ volt-amperes.
  (a) 600   (b) 1000
  (c) 1200   (d) 1500

________ ________   8. The maximum circuit voltage is _____ volts AC or DC for a Class 2 inherently limited power source.
  (a) 100   (b) 120
  (c) 150   (d) 200

________ ________   9. Conductors of intrinsically safe circuits shall be separated at least _____ in. from conductors of any nonintrinsically safe circuits.
  (a) 2   (b) 4
  (c) 6   (d) 12

________ ________   10. Raceways, cable trays, and other wiring for nonincendive system wiring shall be identified with permanently affixed labels with the wording _____ wiring or equivalent.
  (a) nonincendive   (b) intrinsically safe
  (c) pressurized   (d) incendive

_________ _________

11. Lighting systems operating at 30 volts or less shall not be permitted within _____ ft of pools, spas, fountains, or similar locations.
    (a) 5   (b) 6
    (c) 10   (d) 12

12. Two-wire circuits consist of _____ volts between the ungrounded (phase) conductor and a grounded (neutral) conductor.
    (a) 120   (b) 240
    (c) 208   (d) 300

13. Standard lampholders for system of less than 50 volts shall have a rating of not less than _____ watts.
    (a) 550   (b) 660
    (c) 750   (d) 1000

14. Receptacles of not less than _____ ampere rating shall be utilized in kitchens, laundries, and other locations where portable appliances are likely to be cord-and-plug connected for systems of less than 50 volts.
    (a) 15   (b) 20
    (c) 30   (d) 40

15. Class 1 remote control or signaling circuits shall be permitted to operate up to _____ volts and with no limitation on the power rating of the source.
    (a) 120   (b) 240
    (c) 480   (d) 600

16. The maximum circuit voltage is _____ volts AC or DC for a Class 3 inherently limited power source.
    (a) 30   (b) 50
    (c) 100   (d) 150

17. Intrinsically safe circuits shall be separated from conductors of nonintrinsically safe circuits by a distance of at least _____ in. and secured.
    (a) 1.97   (b) 2
    (c) 12   (d) 24

18. Intrinsically safe circuits shall be identified by color coding the conductors _____.
    (a) yellow   (b) light blue
    (c) red   (d) purple

19. Lighting systems operating at 30 volts or less shall be permitted to be installed with bare conductors that are not less than _____ ft above the finished floor.
    (a) 6   (b) 6-1/2
    (c) 7   (d) 10

20. The maximum circuit voltage for a Class 3 power souce limited by overcurrent protection is _____ volts AC or DC.
    (a) 100   (b) 120
    (c) 150   (d) 600

21. It is recommended that spacing between labels for nonincendive circuits be provided at _____ ft intervals.
    (a) 10   (b) 15
    (c) 20   (d) 25

22. What is the amperage for a 120 volt, single-phase, two-wire, 20 kVA transformer?

    (a) 48   (b) 56

    (c) 96   (d) 167

23. What is the amperage for a 208/120 volt, single-phase, three-wire, 150 kVA transformer?

    (a) 313   (b) 605

    (c) 721   (d) 1250

24. What is the amperage for a 240/120 volt, single-phase, three-wire, 150 kVA transformer?

    (a) 313   (b) 605

    (c) 721   (d) 1250

25. What is the amperage for a 208/120 volt, three-phase, four-wire, 150 kVA transformer?

    (a) 181   (b) 361

    (c) 417   (d) 628

26. What is the amperage for a 240/120 volt, three-phase, four-wire, 150 kVA closed-delta transformer?

    (a) 181   (b) 361

    (c) 417   (d) 628

27. What is the amperage for three 240/120 volt, three-phase, four-wire, 150 kVA closed-delta transformers? (One transformer was lost on closed-delta and the system recommended open-delta.)

    (a) 181   (b) 361

    (c) 417   (d) 628

28. What is the amperage for a 480/277 volt, three-phase, four-wire, 150 kVA transformer?

    (a) 181   (b) 361

    (c) 417   (d) 628

29. What is the amperage for a 4160 volt, three-phase, four-wire, 1500 kVA transformer?

    (a) 63   (b) 109

    (c) 208   (d) 361

30. What is the amperage for a 13,800 volt, three-phase, four-wire, 1500 kVA transformer?

    (a) 63   (b) 109

    (c) 208   (d) 361

Working Clearances

All electrical equipment shall be provided with sufficient working space to allow safe access for servicing the equipment without exposing electrical workers to shock hazards. Workspace shall be maintained around and about all electrical equipment where parts may be serviced while energized.

There are three main clearances required for electrical equipment to ensure protection for personnel from electric shocks and burns. The first clearance is to maintain a measurement of 30 in. (762 mm) wide in front of the equipment; the second clearance of at least 3 ft (914 mm) is required outwardly in front of all electrical equipment; the third clearance is the minimum headroom clearance of 6 ft 6 in. (2 m) or the equipment height, whichever is greater.

Design Tip: For electrical equipment installed before the 1978 NEC, the minimum clearance is only 2 ft 6 in. in front of such electrical equipment. The 30 in. (762 mm) wide rule has been in the NEC since the 1971 edition. Headroom clearance has been required since the 1965 NEC.

WORKING SPACE (600 VOLTS OR LESS)
110.26

All electrical equipment shall be provided with sufficient working space to allow safe access for servicing the equipment without exposing electrical workers to shock hazards. Workspace shall be maintained around and about all electrical equipment where parts may be serviced while energized.

> **Design Tip:** Equipment not extending more than 6 in. (150 mm) shall be permitted to be mounted below or above another piece of equipment, if it complies with the height requirements per **110.26(A)(3)**. **[See Figure 5-3(a)]**

SPACES ABOUT ELECTRICAL EQUIPMENT
110.26(A)(1) THRU (A)(3)

A minimum working space of 30 in. (762 mm) width shall be required in front of electrical equipment operating at 600 volts or less. This space permits sufficient room to avoid contact of elbows from contacting live parts and metal parts at the same time while working on the equipment. Equipment doors and hinged panels shall have at least a 90° opening provided in the workspace. This opening allows the electrical worker to have adequate room to repair, adjust, or reset overcurrent protection devices without placing his or her body between the panel door and panelboard. **(See Figure 5-1)**

> **Design Tip:** The 90° opening of the door has been required since the 1987 NEC.

Working space shall not be required in the back of electrical equipment where there are not any removable or adjustable parts such as circuit breakers, fuses, or switches mounted on the back of the equipment. All connections and service areas for maintenance shall be accessible from locations other than the back of the equipment. If rear access is necessary to work on nonelectrical parts on the back of enclosed equipment, a minimum working space of 30 in. (762 mm) horizontally shall be provided.

The inspection authority has the power to make exceptions for smaller workspaces, where in his or her judgment there is no hazard involved. For less working space to be allowed, the installation of the equipment shall provide adequate accessibility to electrical workers. Such space shall be approved by the AHJ per **110.26(A)(1)(b)** and **90.4**.

> **Design Tip: Section 110.26(A)(1)(b)** in the *National Electrical Code* only applies where there are uninsulated parts operating at a voltage no greater than 30 volts RMS, 42 volts peak, or 60 volts DC.

Note: If working on live parts, **110.16** requires a marking to be placed on equipment that warns personnel of possible arc blast.

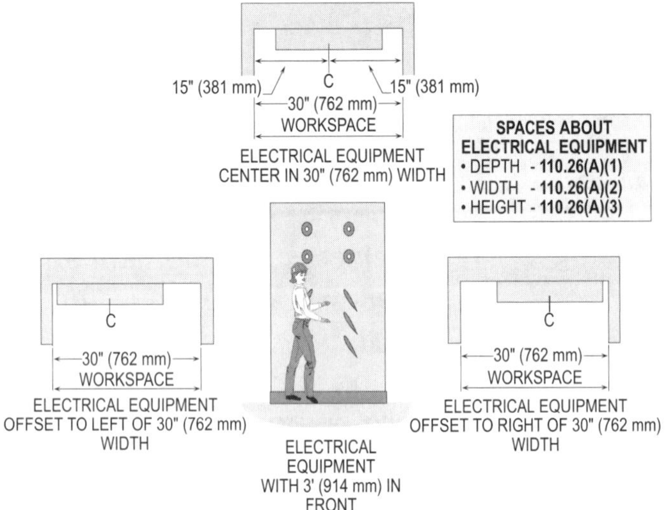

Figure 5-1. A minimum workspace (width) of 30 in. (762 mm) shall required in front of electrical equipment. The workspace may be centered or offset if water pipe or other types of piping are present.

DEPTH OF WORKING SPACE
CONDITIONS 1, 2, AND 3
110.26(A)(1) AND TABLE 110.26(A)(1)

There are different clearances for workspace required in front of electrical equipment based on the type of material that the wall directly in front of the equipment is made of or if there are any live parts opposite the equipment. The clearances are based on two voltage levels to ground:

- 150 volts or less to ground
- over 150 volts to 600 volts to ground

For example, the clearances of a 120/208 volt, three-phase, four-wire system fall under the 150 volts or less to ground, while a 277/480 volt, three-phase, four-wire system is guarded by the 151 to 600 volts to ground.

DEPTH OF WORKING SPACE
CONDITION 1
TABLE 110.26(A)(1)

Condition 1 is where the electrical equipment is installed in or on one wall, with the wall on the opposite side being an insulated wall. An insulated wall is constructed of wood or metal studs, with the wallboard consisting of sheetrock, wood panels, etc. An electrical worker making contact with the insulated wall while touching live parts is isolated from the grounded slab or earth, therefore, **Condition 1** allows less working space. **(See Figure 5-2)**

> **Design Tip:** Until the 1965 NEC was published, Condition 1 only required a 2-1/2 ft workspace in front of electrical equipment.

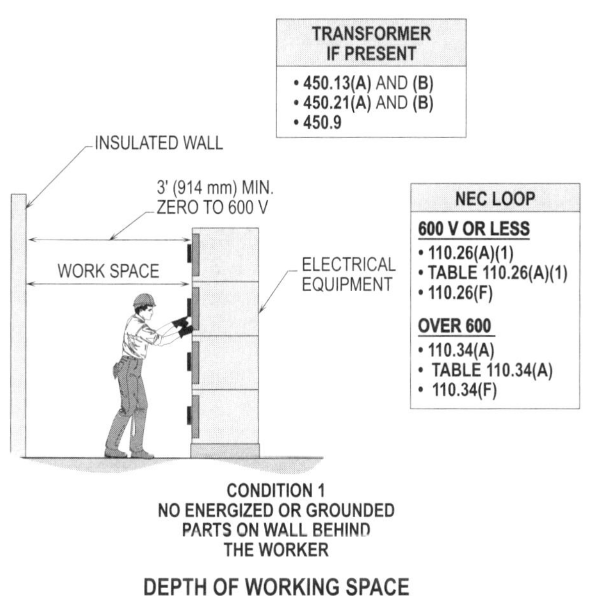

Figure 5-2. Clearances in front of electrical equipment with an insulated wall opposite equipment.

DEPTH OF WORKING SPACE
CONDITION 2
TABLE 110.26(A)(1)

Condition 2 is where the electrical equipment is installed on one wall, with the wall on the opposite side being a conductive wall. A conductive wall (grounded) is constructed of concrete, brick, or tile that when touched may connect the body to earth ground. If an electrical worker accidentally makes contact with the wall while touching a live part or conductor, a circuit is made and he or she could easily be electrocuted. Because this danger is present, a greater workspace is required for the voltages ranging from 151 to 600 volts. **[See Figure 5-3(a) and (b)]**

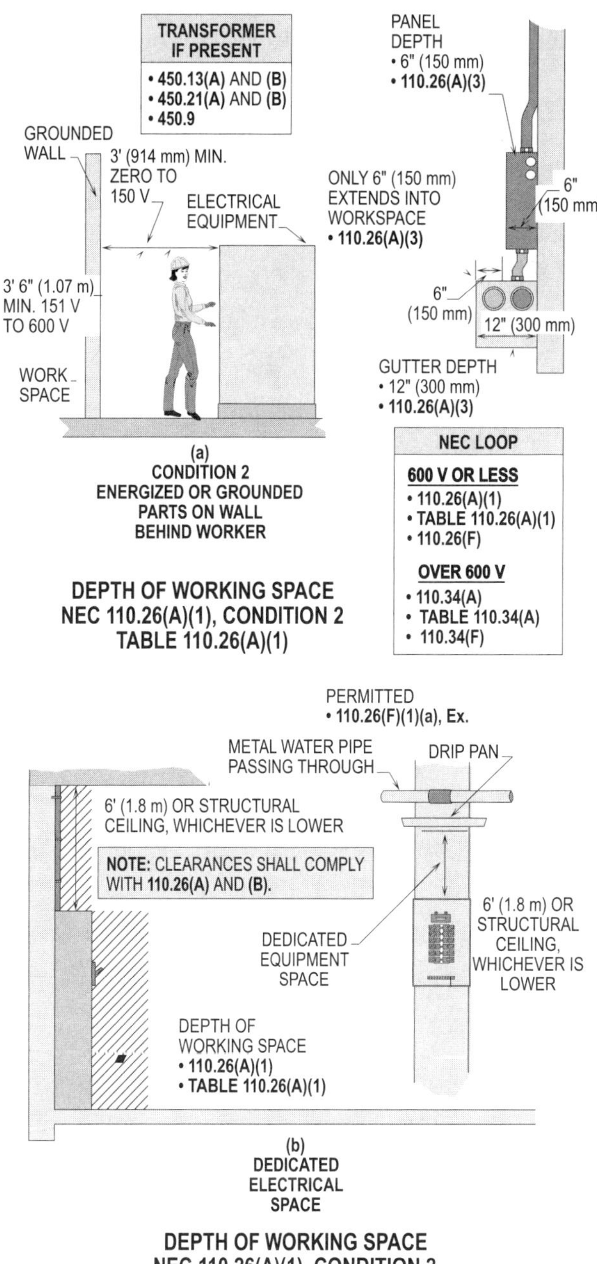

Figures 5-3(a) and (b). Figure 5-3(a) outlines clearances in front of electrical equipment with a grounded wall opposite equipment. Figure 5-3(b) describes clearance rules above electrical equipment.

DEPTH OF WORKING SPACE
CONDITION 3
TABLE 110.26(A)(1)

Condition 3 is where the electrical equipment is installed in or on one wall, with the wall on the opposite side having electrical equipment mounted or set on it. With the electrical equipment installed in this manner, there are live parts on both sides of the room. Electricians and maintenance workers are subjected to phase-to-phase voltage or phase-to-ground voltage when servicing the equipment. Where electrical equipment is mounted or set on opposite walls directly across and in front of each other, electrical workers may be exposed to live parts with panelboard covers removed. Because electrical workers could be exposed to a fatal shock from live parts on both sides of the working space, a greater clearance shall be required for their safety. **[See Figure 5-4(a)]**

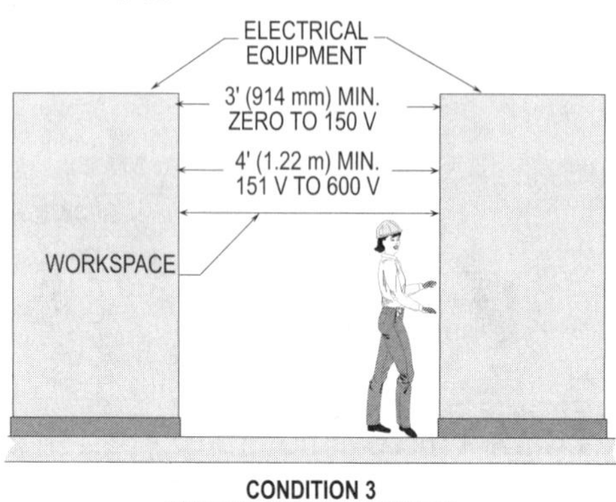

Figure 5-4(a). Clearances in front of electrical equipment opposite other electrical equipment.

EXISTING BUILDINGS
110.26(A)(1)(c)

In existing buildings where electrical equipment is being replaced, **Condition 2** working clearances shall be applied between dead-front switchboards, panelboards, or motor control centers, which are located across the aisle from each other. However, for this rule to apply, conditions of maintenance and supervision shall be ensured by written procedures adopted to prohibit equipment on both sides of the aisle from being open at the same time. In addition, only authorized, qualified and trained personnel shall be permitted to service such installations. **[See Figure 5-4(b)]**

CLEAR SPACES
110.26(B)

Electrical equipment shall be located where adequate working space is accessible to take voltage measurements, check continuity of circuits, adjust, or replace defective overcurrent protection devices and tighten loose connections. Working space in front of electrical equipment shall be free from storage of materials, etc. Mains and overcurrent protection devices shall be accessible to the users in case of emergencies as well as disconnecting

means of equipment. Any and all exposed live parts located in a passageway or open space shall be suitably guarded. **(See Figure 5-5)**

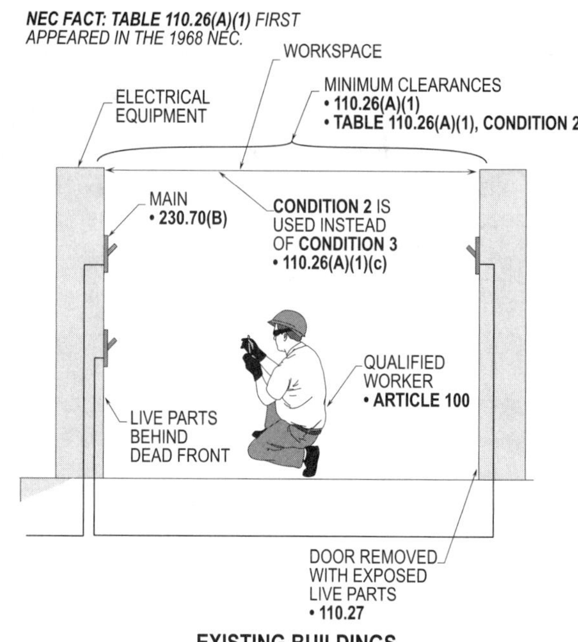

Figure 5-4(b). Section 110.26(A)(1)(c) allows the use of **Condition 2** instead of **Condition 3** when working between existing pieces of electrical equipment with dead fronts that have been removed.

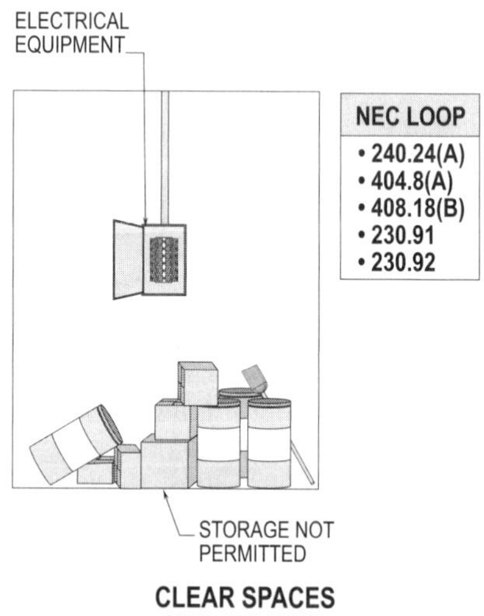

Figure 5-5. Storage shall not be permitted in front or below electrical equipment.

ENTRANCE TO AND
EGRESS FROM WORKING SPACE
110.26(C)

Electrical equipment shall be designed and installed to have at least one entrance to provide access and egress from the workspace. Easy and fast access to electrical devices is essential. Special considerations shall be given to electrical equipment that is 1200 amps or more and over 6 ft (1.8 m) wide, containing overcurrent devices, switching devices, or control devices. Such equipment shall have a clearance of 24 in. (610 mm) wide and 6-1/2 ft (2 m) high at each end for safe exit in case of a ground-fault. **(See Figure 5-6)**

> **Design Tip:** This requirement has been in the NEC since the 1978 edition.

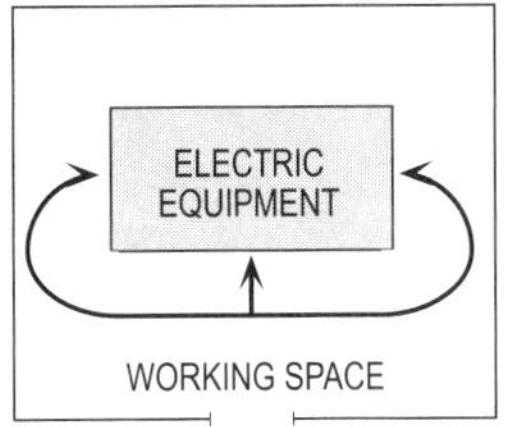
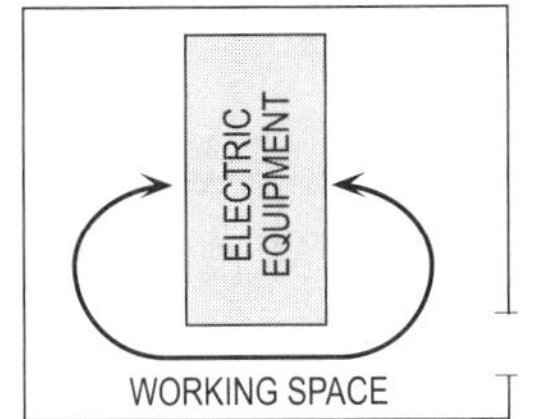

AT LEAST ONE ENTRANCE SHALL BE REQUIRED TO PROVIDE ACCESS TO THE WORKING SPACE AROUND ELECTRIC EQUIPMENT.

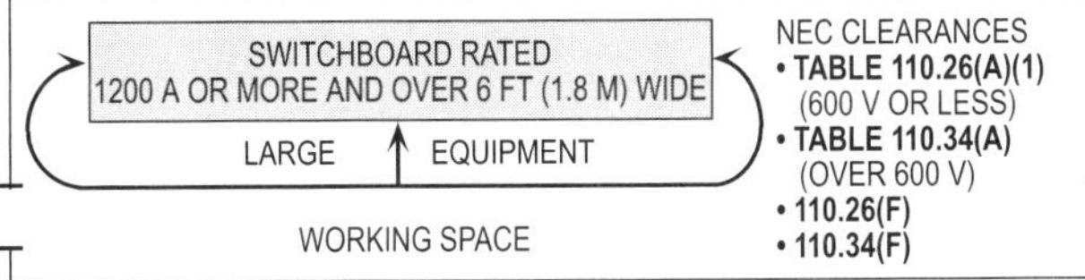

FOR EQUIPMENT RATED 1200 A OR MORE AND OVER 6 FT (1.8 m) WIDE, THERE SHALL BE ONE ENTRANCE (OR TWO) AND NOT LESS THAN 24 IN. (610 mm) WIDE AND 6 FT 6 IN.(2 m) HIGH AT EACH END, OR WORKSPACE SHALL BE DOUBLED.

**ENTRANCE TO AND EGRESS FROM WORKING SPACE
NEC 110.26(C)**

Figure 5-6. Access and entrance to working space for electrical equipment shall be provided.

LARGE EQUIPMENT
110.26(C)(2)

A single entrance shall be permitted where the location permits a continuous and unobstructed way of egress travel per **110.26(C)(2)(a)**. The workspace shall be permitted to be doubled per **110.26(C)(2)(b)** and **Table 110.26(A)(1)** based on **Conditions 1, 2, or 3,** and only one entrance is required. With the workspace doubled in front of the equipment, a worker may move out of the endangered workspace and exit along the length of the equipment.

The deeper workspace in front of the equipment provides a safe route to exit [unobstructed per **110.26(C)(2)(a),** without providing two entrances].

The ability to exit these areas using swing-out doors with panic bars, pressure plates, etc. is very beneficial to an electrical worker, if he or she is hurt from an arc blast.

PERSONNEL DOORS
110.26(C)(3)

Personnel door(s) shall be equipped with panic bars, pressure plates, or other devices (that are not normally latched but open under simple pressure) and shall open in the direction of egress where equipment is rated 1200 amps or more. The personnel door(s) shall not be less than 25 ft (7.6 m) from the nearest edge of the working space for entrance to and egress from the working space. **(See Figure 5-7)**

ILLUMINATION
110.26(D)

Service equipment, switchboards, panelboards, and motor control centers installed indoors shall be provided with adequate lighting for the safety of electrical workers servicing such equipment from the front or rear when live parts are accessible. Luminaires shall be permitted to be incandescent or fluorescent as long as they provide the proper lighting for parts to be serviced.

Luminaires shall have a headroom clearance of at least 6-1/2 ft (2 m) to give personnel sufficient room to stand in front of electrical equipment without the threat of their head or head gear contacting metal, etc. **(See Figure 5-8)**

> **Design Tip:** Additional luminaires shall not be required where the workspace is illuminated by an adjacent light source. An automatic means, by itself, shall not be permitted to be used to control the illumination in electrical equipment rooms or can be as permitted by **210.70(A)(1), Ex. 2.** Illumination was not required in the workspace until the 1965 NEC.

HEADROOM
110.26(E)

A minimum headroom clearance of 6-1/2 ft (2 m) shall be maintained from the floor or platform up to the luminaire or any overhead obstruction. This overhead workspace is mandatory and applies especially to service equipment, switchboards, panelboards, and motor control centers. The purpose of the overhead workspace is to protect employees from accidentally contacting grounded objects with their bodies (head, hands, etc.) while touching live parts and completing a circuit to ground which could cause a fatal electric shock. Electricians or maintenance workers should never have to stoop or bend down to gain access to service, repair, replace, or modify components inside electrical equipment as previously mentioned. See **110.26(A)(2)** and **90.4** for application of the 90° rule. **(See Figure 5-9)**

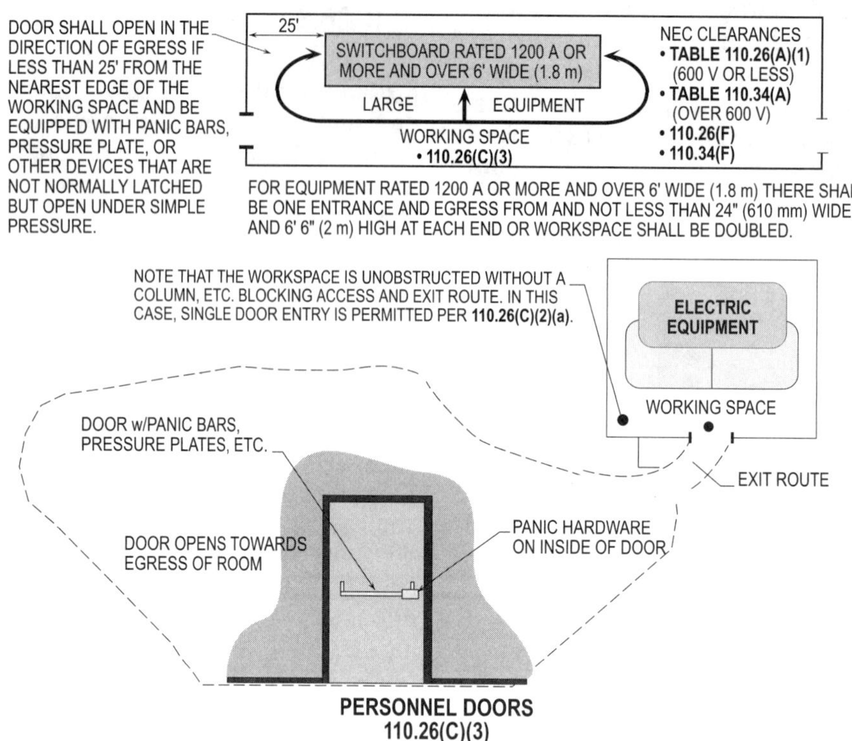

Figure 5-7. Personnel door(s) shall have panic bars and similar equipment and open in the direction of egress if located less than 25 ft (7.6 m) from the nearest edge of the working space.

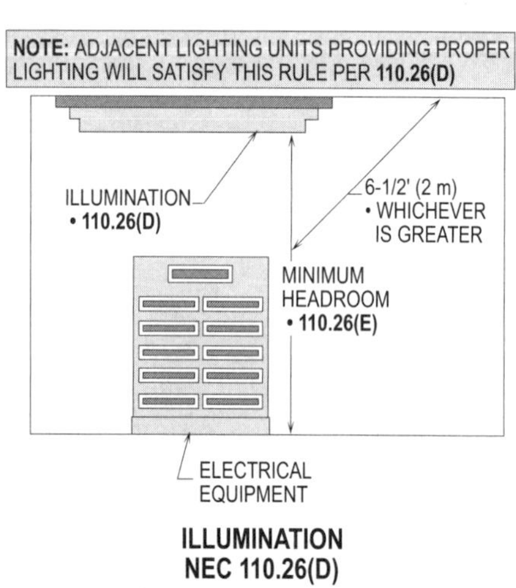

Figure 5-8. Proper lighting shall be provided for the servicing of electrical equipment.

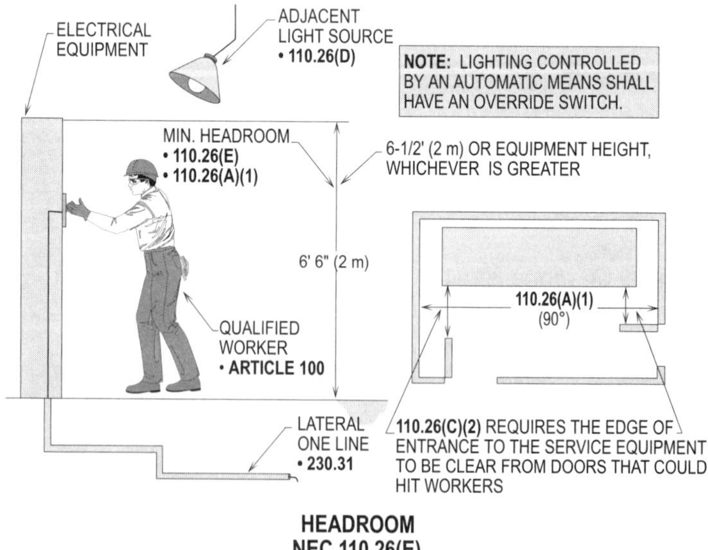

Figure 5-9. Adequate headroom space shall be provided based upon the electrical equipment height or the 6-1/2 ft (2 m) rule, whichever is greater. Also, the 90° rule shall be applied properly.

> **Design Tip:** The minimum headroom or working space shall not be permitted to be less than 6-1/2 ft. (2 m) or the height of the equipment, whichever is greater. Headroom in the workspace was not required until the 1965 NEC.

DEDICATED EQUIPMENT SPACE
110.26(F)

In the following locations, switchboards, panelboards, and distribution boards that are installed for the control of light and power circuits, battery charging supplied from light and power circuits, and motor control centers shall be located in dedicated spaces and protected from damage:
  - Indoor
  - Outdoor

INDOOR
110.26(F)(1)

The following conditions of use shall be considered when equipment is installed indoors in a dedicated equipment space:

  - Dedicated electrical space
  - Foreign systems
  - Sprinkler protection
  - Suspended ceilings

DEDICATED ELECTRICAL SPACE
110.26(F)(1)(a)

Switchboards shall be installed in a dedicated space equal to the width and depth of the equipment and extending from the floor to a height of 6 ft (1.8 m) or to the structural ceiling (roof), whichever is lower. No piping, ducts, or equipment foreign to the electrical installation shall be permitted to be installed in this dedicated space since leakage of water or condensation could damage the equipment enclosure. Any space extending 6 ft (1.8 m) above the dedicated space shall not be considered dedicated space. **(See Figure 5-10)**

Suspended ceilings with removable panels shall be permitted within the 6 ft (1.8 m) zone per **110.26(F)(1)(a), Ex.** This dedicated space is considered by electricians to be an area where they may run and connect cables and raceways without foreign items (other than electrical) blocking their access directly above equipment. **(See Figure 5-11)**

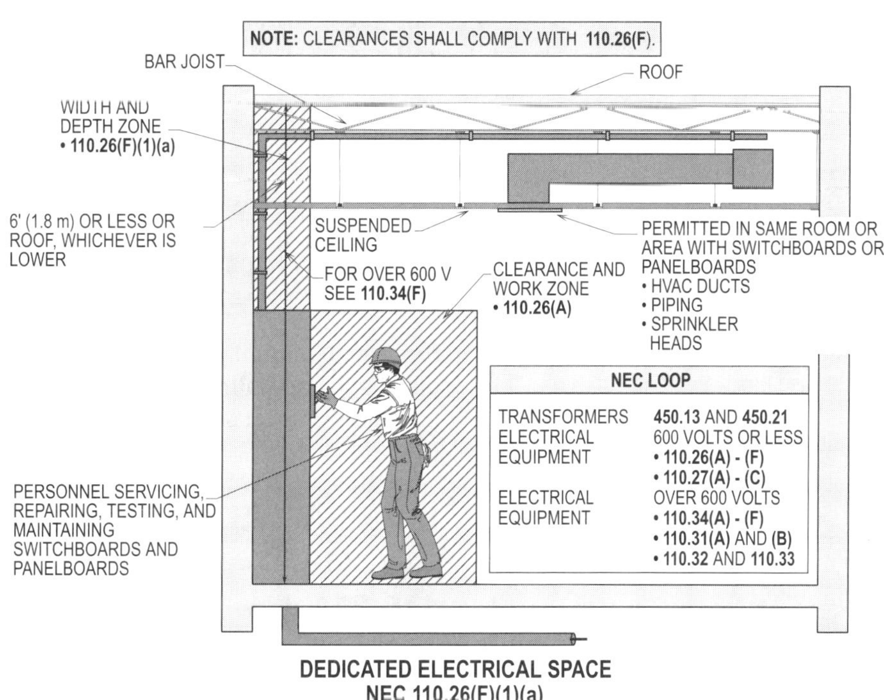

Figure 5-10. Switchboards shall be installed in a dedicated space extending 6 ft (1.8 m) above the equipment from the floor or to the structural ceiling (roof), whichever is lower.

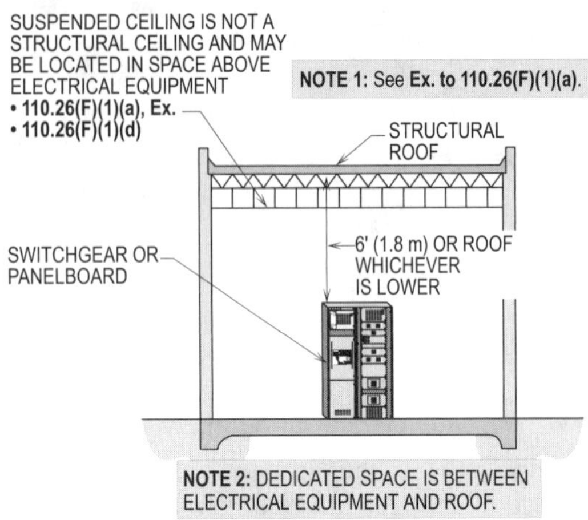

Figure 5-11. A suspended ceiling shall not be considered a structural ceiling and shall be permitted to be located in the dedicated space.

FOREIGN SYSTEMS
110.26(F)(1)(b)

The area above the dedicated space shall be permitted to contain foreign systems, provided protection is installed to avoid damage to the electrical equipment from condensation, leaks, or breaks in such foreign systems. Otherwise, equipment not associated with the electrical equipment shall not be permitted to be installed in the 6 ft (1.8 m) area above the electrical equipment. **(See Figure 5-12)**

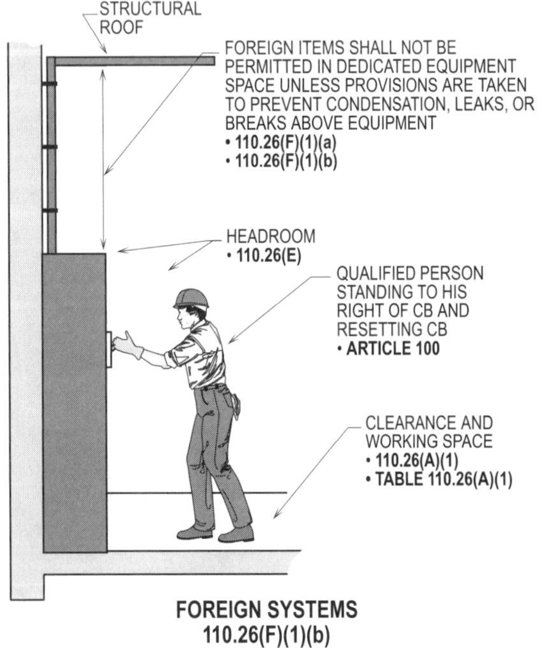

Figure 5-12. This illustration shows the foreign items permitted in the dedicated electrical space.

SPRINKLER PROTECTION
110.26(F)(1)(c)

Sprinkler protection shall be permitted to be installed in this dedicated space to protect the equipment from fire hazards.

> **Design Tip:** For further information on sprinkler systems, see **NFPA 13, 4-4.14.**

SUSPENDED CEILINGS
110.26(F)(1)(d)

A suspended ceiling such as a dropped, suspended, or similar ceiling shall be permitted to be installed in this dedicated space. The suspended ceiling does not add strength to the building structure, and therefore shall not be considered a structural ceiling. **(See Figure 5-11)**

OUTDOOR
110.26(F)(2)

Electrical equipment that is installed outdoors shall be in adequate enclosures or covers that will afford adequate mechanical protection from vehicular traffic, accidental contact by unauthorized personnel, or accidental spillage or leakage from piping systems. The working space clearance around such equipment shall comply with **110.26(A)**. No architectural appurtenance or other equipment shall be permitted to be located in this working space clearance around equipment.

LOCKED ELECTRICAL EQUIPMENT ROOMS OR ENCLOSURES
110.26(G)

Electrical equipment rooms or enclosures that house electrical apparatus and controlled by a lock(s) shall be considered accessible to qualified persons.

LIVE PARTS GUARDED AGAINST ACCIDENTAL CONTACT
110.27(A)

The general rule of protecting live parts (energized) from accidental contact is by installing them in a complete enclosure, which provides a dead front. Sometimes it is not practical to construct enclosures to house large control panels, etc., and in such cases if the apparatus is rated at 50 volts or more, suitable guards or isolation shall be provided according to one of the following rules:

- Section **110.27(A)(1)** permits the live parts in electrical equipment that are not mounted in a completely enclosed enclosure to be installed in a room, vault, or similar enclosure that is accessible only to qualified personnel. **(See Figure 5-13)**

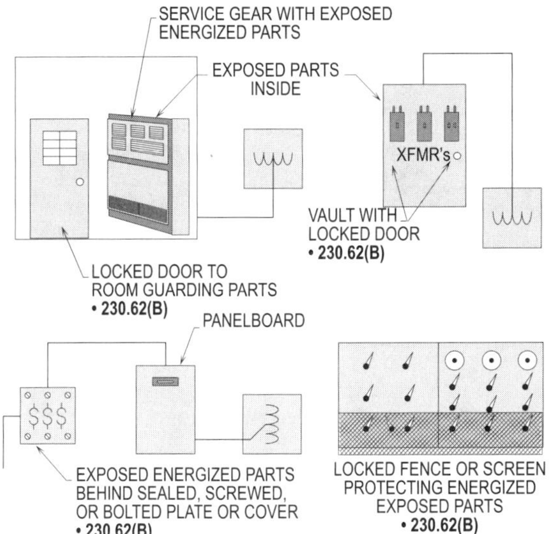

Figure 5-13. Live parts of electrical equipment shall be protected and guarded from unqualified persons.

- Section **110.27(A)(2)** permits live parts to be separated by permanent partitions or screens so located that only qualified persons have access and reach of the live parts. Openings in partitions or screens shall be designed to prevent accidental contact with live parts or bring conductive objects in contact with live parts.

> **Design Tip:** The safety thought is making the live parts in the equipment accessible without obstruction and giving special attention to the prevention of contacting conductive materials such as metal conduit, pipes, etc. carried by a passing worker. **(See Figure 5-14)**

- Section **110.27(A)(3)** permits exposed live parts in equipment to be located on a suitable balcony, gallery, or platform that is high enough or designed in such a manner to keep unqualified personnel out.

For example, a well-designed location of a balcony accessible only to qualified personnel who have been given the responsibility to maintain such equipment fully complies as a suitable means of guarding live parts. **(See Figure 5-15)**

- Section **110.27(A)(4)** permits the live electrical parts to be elevated at least 8 ft (2.5 m) above the floor.

For example, the live parts may be mounted or set on a platform attached to a utility pole or on the side of a building, etc., and this type of installation shall be considered safely guarding live parts. **(See Figure 5-15)**

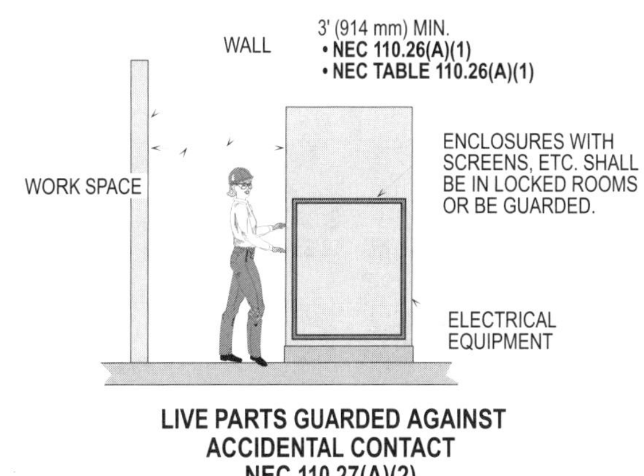

Figure 5-14. Live parts in electrical equipment shall be guarded or enclosed.

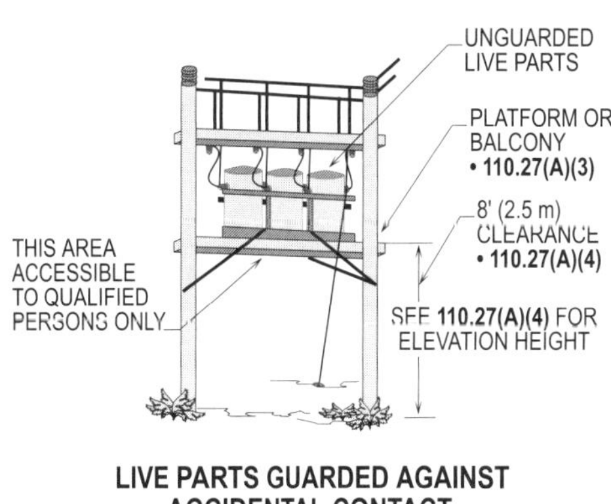

Figure 5-15. Live parts shall be considered protected where they are elevated above grade and are out of reach of the general public.

PREVENT PHYSICAL DAMAGE 110.27(B)

Section **110.27(B)** recognizes that many times electrical equipment is located in work areas where the work activity around it might damage the equipment. In such cases, the equipment shall be properly protected with enclosures or guards that provide the necessary strength to prevent any damage to the electrical equipment. **(See Figure 5-16)**

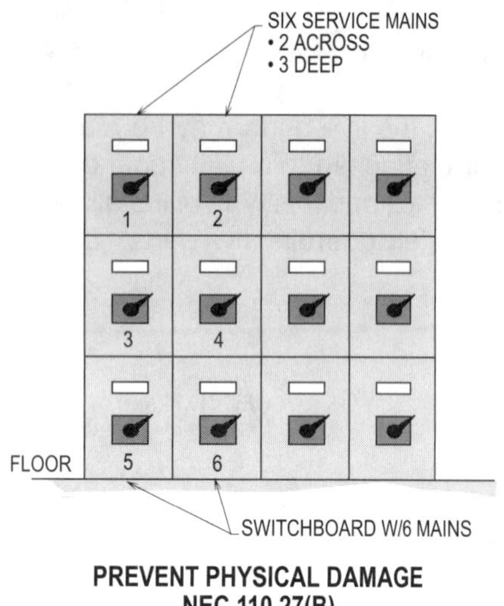

Figure 5-16. Totally enclosed gear protects live parts and equipment from physical damage.

WARNING SIGNS
110.27(C)

Section **110.27(C)** requires guarded locations to be posted with warning signs, giving warning that only qualified personnel are allowed to enter and service the electrical components. Any dangers that might exist shall be posted to warn employees to be careful. **(See Figure 5-17)**

ENTRANCE TO ENCLOSURES AND ACCESS TO WORKING SPACE (OVER 600 V)
110.33

There shall be at least one entrance to enclosures for electrical installations as described in **110.31**, that provides a minimum of 24 in. (610 mm) wide by 6-1/2 ft (2 m) high to gain access to the working space of high-voltage equipment (over 600 volts). This is a minimum, and more should be provided when possible. Where the equipment is over 6 ft (1.8 m) in width, there shall be one entrance at each end of the equipment. **(See Figure 5-18)**

> **Design Tip:** The 1984 NEC required at least one entrance of 24 in. wide by 6-1/2 ft high to be provided for electrical equipment that was over 4-1/2 ft in width.

If bare or insulated parts of more than 600 volts, nominal, are located adjacent to such entrances, there shall be suitable means designed to guard them.

When electric equipment is installed on platforms, balconies, mezzanine floors, or in attic or roof rooms or spaces, there shall be permanent ladders or stairways installed for access. There is an OSHA regulation that requires ladders to extend 3 ft (900 mm) above the platform, etc. to which they give access to service equipment. **(See Figure 5-19)**

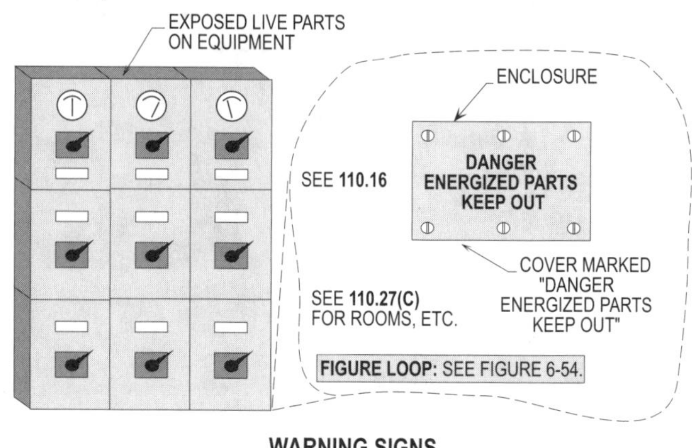

Figure 5-17. Electrical rooms with exposed live parts supplied by 600 volts or less systems shall be posted with warning signs that only qualified persons are allowed to enter.

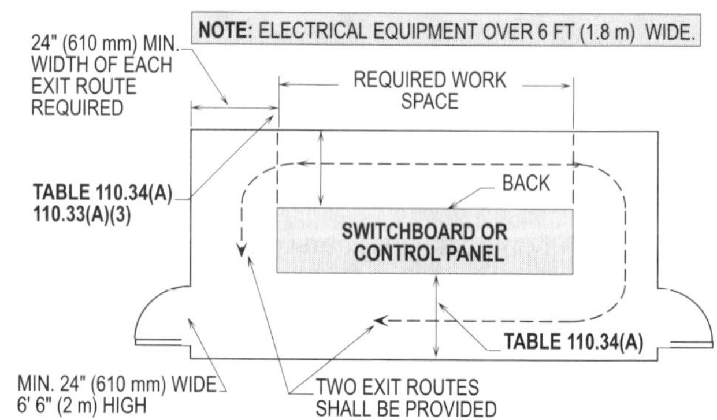

Figure 5-18. Electrical equipment shall be provided with proper exits for personnel servicing such equipment.

OVER 600 VOLTS NOMINAL
110.34(A)

The installation requirements of conductors, overcurrent protection devices, and equipment likely to require examination, adjustment, servicing, or maintenance while energized are usually more stringent for systems rated over 600 volts. Conductors are normally shielded, and special terminations are required where they terminate to the equipment, etc. Greater clearances around electrical equipment shall be required due to the threat of flashover voltage from live parts to grounded metal. High-voltage switches and overcurrent protection devices shall be marked to indicate the circuit or equipment that they supply and control. The rules in the NEC make it very clear that **110.22** applies to systems rated over 600 volts as well as systems rated at 600 volts or less. **Article 490** in the NEC is applied to design, install, and inspect systems over 600 volts.

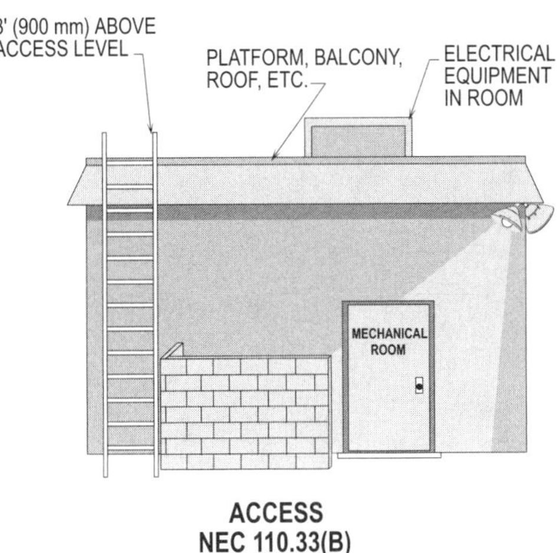

Figure 5-19. Where electrical equipment shall have a ladder to gain access to a platform, balcony, etc., the side rails of the ladder shall extend 3 ft (900 mm) above the access area.

ENCLOSURE FOR ELECTRICAL INSTALLATIONS
110.31

In locations where entrance and access is possible only for qualified personnel, doors and covers of enclosures shall meet the following location requirements:

- Electrical installations in vaults
- Electrical installations in rooms
- Electrical installations in closets
- Locations surrounded by a:
  - **(a)** Wall,
  - **(b)** Screen, or
  - **(c)** Fence.

The design and construction of enclosures shall be suitable to the nature and degree of the hazards involved.

For example, an enclosure installed in an area exposed to gas is normally required to be classified for Class I, Division 1 locations with a Group D rating.

> **Design Tip:** A wall, screen, or fence that is less than 6 ft (1.8 m) high shall not be considered as preventing access to equipment unless other features are provided with a degree of isolation equivalent to a 7 ft (2.1 m) fence. Consider a fence of 6 ft (1.8 m) in height and equipped with 1 ft (300 mm) of barbed wire at the top of the fence. This addition of the barbed wire is the extra feature needed to prevent easy access of unqualified personnel. **(See Figure 5-20)**

FIRE RESISTANCE OF ELECTRICAL VAULTS
110.31(A)

The walls, roof, floors, and doorways of vaults containing conductors and equipment over 600 volts, nominal, shall be constructed of materials that have adequate structural strength for the conditions, with a minimum fire rating of 3 hours. The floors of vaults in contact with the earth shall be of concrete that is not less than 4 in. (102 mm) thick, but where the vault is constructed with a vacant space or other stories below it, the floor shall have adequate structural strength for the load imposed on it and a minimum fire resistance of 3 hours. For the purpose of the section, studs and wallboards shall not be considered acceptable. **(See Figure 5-21)**

INDOOR INSTALLATIONS
110.31(B)

Electrical equipment installed indoors shall meet certain requirements where places are accessible to unqualified persons and where places are accessible to qualified persons only.

Indoor electrical installations shall be made with metal-enclosed equipment that is accessible to unqualified persons. Metal-enclosed switchgear, unit substations, transformers, pull boxes, connection boxes, and other similar associated equipment shall be marked with appropriate caution signs where unqualified persons have access. Ventilating or similar openings shall be so designed that foreign objects are deflected from energized parts. **(See Figure 5-22)**

Electrical installations installed indoors containing exposed live parts shall be accessible to qualified persons only and comply with **110.34**, **110.36**, and **490.24**.

OUTDOOR INSTALLATIONS
110.31(C)

Outdoor electrical installations shall comply with **Parts I, II, and III** of **Article 225** where accessible to unqualified persons. Electrical installations installed outdoors that contain exposed live parts shall be accessible to qualified personnel only, and shall comply with the provisions of **110.34, 110.36,** and **490.24**. It is essential that the live parts of high-voltage systems be totally enclosed in the equipment or be isolated to prevent access by the general public or unqualified employees. High-voltage current traveling through the body to ground can do great damage. If it doesn't kill immediately, the internal damage usually causes death later. Only qualified personnel shall be permitted to service high-voltage systems. **(See Figure 5-23)**

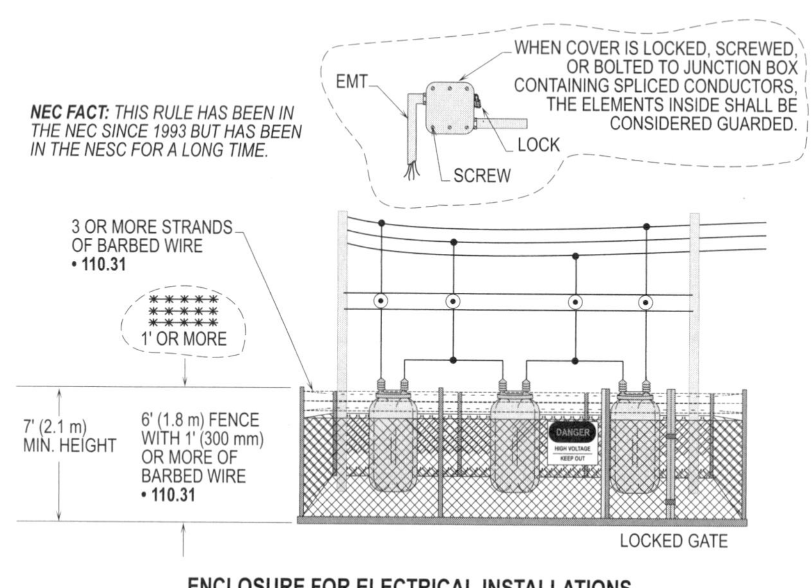

Figure 5-20. A wall, screen, or fence installed properly prevents easy access of unqualified personnel.

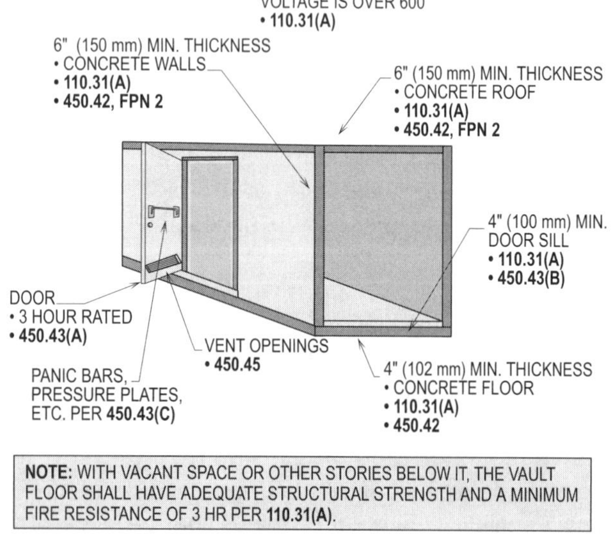

Figure 5-21. The walls, roof, floors, and doorways of vaults containing conductors and equipment over 600 volts, nominal, shall be constructed of materials that have adequate structural strength for the conditions, with a minimum fire rating of 3 hours.

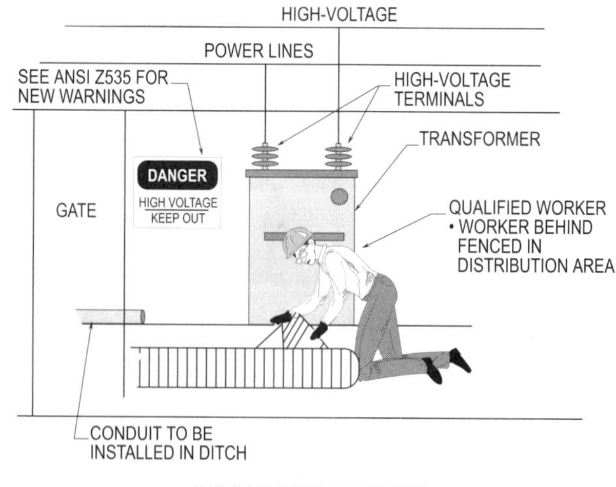

Figure 5-22. Only qualified personnel shall be permitted to service high-voltage systems.

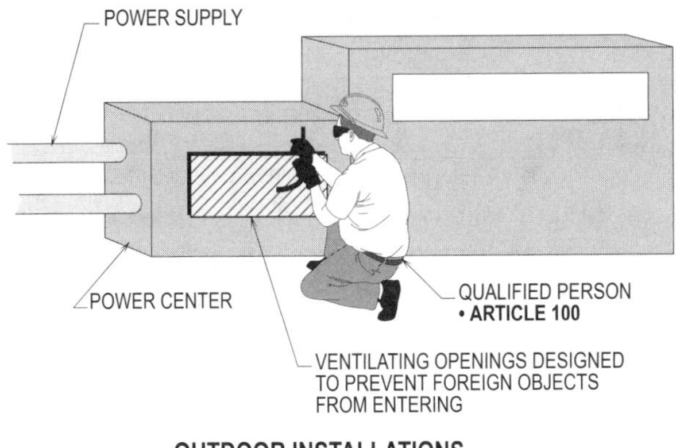

Figure 5-23. Openings in electrical equipment shall be designed to deflect foreign objects from live parts.

ENCLOSED EQUIPMENT ACCESSIBLE TO UNQUALIFIED PERSONS
110.31(D)

Ventilating or similar openings in equipment shall be so designed that foreign objects inserted through these openings are deflected from energized parts. When exposed to physical damage from vehicular traffic, suitable guards and caution signs shall be provided. Metal-enclosed equipment located outdoors that is accessible to the general public shall be designed so that exposed nuts or bolts are not readily removed, permitting access to live parts. Where metal-enclosed equipment is accessible to the general public and the bottom of the enclosure is less than 8 ft (2.5 m) above the finished floor or grade level, the enclosure door or hinged cover shall be kept locked. Doors and covers for electrical enclosures such as pull boxes, splice boxes, or junction boxes shall be locked, bolted, or screwed on. Underground box covers that weigh over 100 lb (45.4 kg) shall be considered to be complying with this requirement. **(See Figure 5-24)**

WORKSPACE ABOUT EQUIPMENT
110.32

There shall be sufficient clear working space about high-voltage equipment to permit ready and safe operation of such equipment. There have been minimum clearances set for workspace in and around electrical equipment, and they are as follows:

• Where energized parts are exposed, the minimum clear working space shall not be permitted to be less than 6-1/2 ft (2 m) measured vertically from the floor or platform or less than 3 ft (914 mm) wide measured parallel to the equipment.

See Figure 5-25 for a detailed procedure on clearances about electrical equipment supplied with voltage over 600 volts.

• Space shall be adequate to permit doors and hinged panels to open a minimum of 90° so personnel can stand in front of the equipment without making contact with their body to the doors or hinged panels while touching live parts.

• See **Table 110.34(A)** for the minimum working clearances (depth) of equipment based upon on supply voltage.

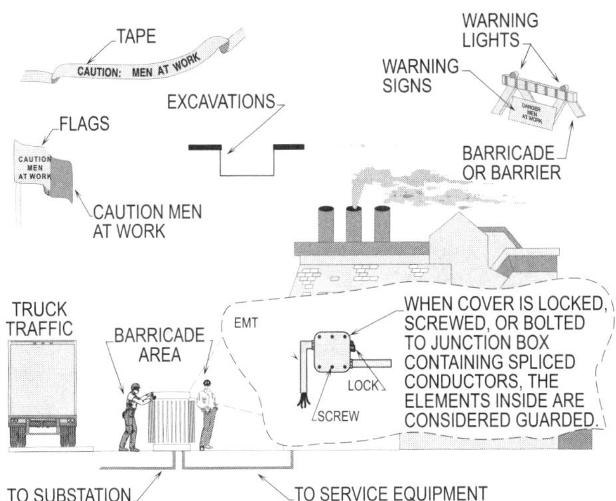

Figure 5-24. Covers for totally enclosed enclosures, with or without ventilating openings, shall be secured on or locked when accessible to the general public. Suitable guards or caution signs shall be used when necessary.

TABLE 110.34(A)
MINIMUM DEPTH OF CLEAR WORKING SPACE
AT ELECTRICAL EQUIPMENT

NOMINAL VOLTAGE-TO-GROUND	MINIMUM CLEAR DISTANCE		
	CONDITION 1	CONDITION 2	CONDITION 3
601 - 2500 V	3 ft (900 mm)	4 ft (1.2 m)	5 ft (1.5 m)
2501 - 9000 V	4 ft (1.2 m)	5 ft (1.5 m)	6 ft (1.8 m)
9001 - 25,000 V	5 ft (1.5 m)	6 ft (1.8 m)	9 ft (2.8 m)
25,001 - 75 kV	6 ft (1.8 m)	8 ft (2.5 m)	10 ft (3.0 m)
ABOVE 75 kV	8 ft (2.5 m)	10 ft (3.0 m)	12 ft (3.7 m)

WORKSPACE ABOUT EQUIPMENT
NEC 110.32

Figure 5-25. Minimum clearances (depth) of equipment based upon the supply voltage.

ENTRANCE TO ENCLOSURES AND ACCESS TO WORKING SPACE
110.33

The entrance and access to workspace shall be considered for high-voltage electrical equipment. The entrance and access to workspace may require at least one entrance at each end or allow one entrance to the workspace.

ENTRANCE
110.33(A)

There shall be at least one entrance a minimum of 24 in. (610 mm) x 6-1/2 ft (2 m) to provide access to the working space of high-voltage equipment (over 600 volts). This is a minimum, and more should be provided when possible.

LARGE EQUIPMENT
110.33(A)(1)

There shall be one entrance at each end of switchboards and control panels where the equipment is over 6 ft (1.8 m) in width. A single entrance to the required working space shall be permitted if either one of the following conditions is applied:

- The workspace permits a continuous and obstructed way of exit travel.

- The workspace is doubled per **110.34(A)**. This entrance shall be located so that the edge of the entrance nearest the electrical equipment is the minimum clear distance given in **Table 110.34(A)** away from such equipment. **(See Figure 5-26)**

GUARDING
110.33(A)(2)

Where bare or insulated parts or more than 600 volts, nominal, are located adjacent to such entrances, they shall be suitably guarded.

PERSONNEL DOORS
110.33(A)(3)

Personnel door(s) shall be equipped with panic bars, pressure plates, or other devices that are not normally latched but open under simple pressure. The personnel door(s) shall not be less than 25 ft (7.6 m) from the nearest edge of the working space for entrance to and egress from the working space. **(See Figure 5-27)**

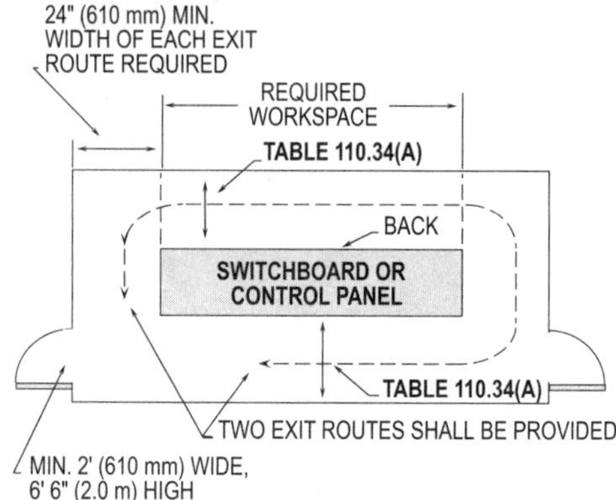

Figure 5-26. This illustration shows that electrical equipment shall be provided with proper exits for personnel servicing such equipment.

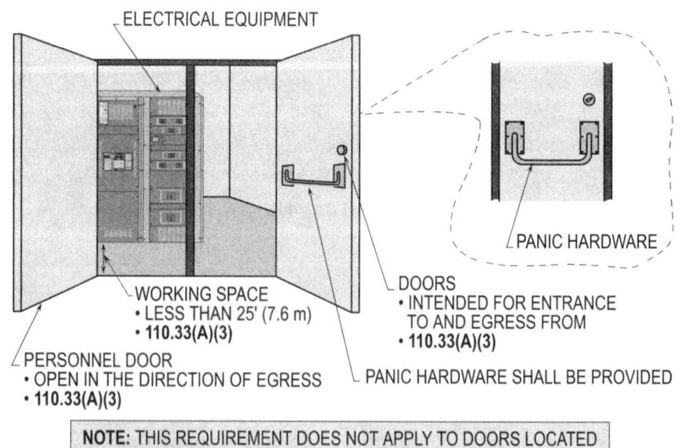

Figure 5-27. Personnel door(s) shall have panic bars and similar equipment and open in the direction of egress if located less than 25 ft (7.6 m) from the nearest edge of the working space.

ACCESS
110.33(B)

When electrical equipment is installed on platforms, balconies, mezzanine floors, or in attic roof rooms or spaces, permanent ladders or stairways shall be installed for access. **(See Figure 5-28)**

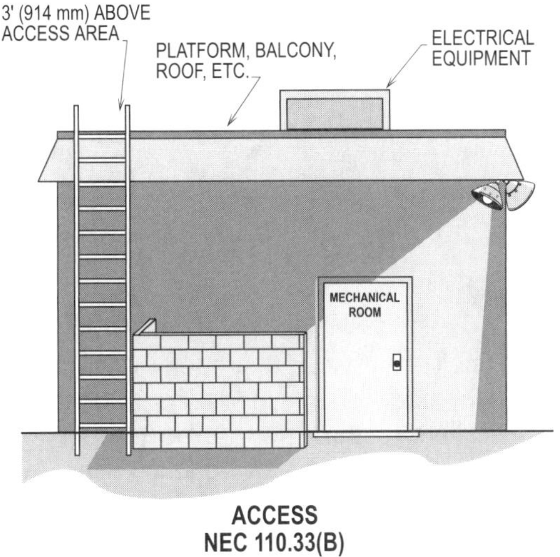

Figure 5-28. This illustration shows that electrical equipment shall have a ladder to gain access to a platform, balcony, etc. The side rails of the ladder shall extend 3 ft (900 mm) above the access area.

WORKING SPACE AND GUARDING
110.34(A)

Table 110.34(A) in the NEC lists the minimum clear working space in front of electrical equipment. This can be such equipment as switchboards, control panels, switches, circuit breakers, motor control centers, relays, etc. If other distances are spelled out elsewhere in **NFPA 70E**, those specified distances are to be applied.

In measuring the distances given in **Table 110.34(A)**, they shall be measured from exposed live parts or if in enclosures, the distance is measured from the enclosure's front or opening.

The **Ex.** to **110.34(A)** does not require working space to be provided in back of electrical equipment where there are not any renewable or adjustable parts such as fuses and circuit breakers on the back and all connections are accessible from locations other than the back. If rear access is required to work on deenergized parts on the back of enclosed equipment, a minimum distance of 30 in. (762 mm) horizontally shall be required for the working space. **(See Figure 5-25)**

WORKING SPACE
CONDITION 1
TABLE 110.34(A)

In **Condition 1,** there are exposed live parts on one side of the equipment and there are no live or grounded parts on the opposite side of the working space. Such condition is where there are exposed parts on one side that is effectively guarded by suitable wood or other insulating materials on the opposite side.

For example: What is the minimum clearance required for a switchgear supplied with 4160 volts and installed opposite an insulated wall?

Step 1: Finding clearance
Table 110.34(A), Condition 1
4160 volts requires 4 ft (1.2 m)

Solution: The minimum clearance is 4 ft (1.2 m).

WORKING SPACE
CONDITION 2
TABLE 110.34(A)

In **Condition 2,** there are exposed live parts on the equipment with the other side (directly in front) having grounded parts such as concrete, brick, or tile walls.

For example: What is the minimum clearance for a motor control center supplied with 12,470 volts and installed opposite a concrete wall?

Step 1: Finding clearance
Table 110.34(A), Condition 2
12,470 volts requires 6 ft (1.8 m)

Solution: The minimum clearance is 6 ft (1.8 m).

WORKING SPACE CONDITION 3 TABLE 110.34(A)

Condition 3 consists of exposed live parts on both sides (not guarded as in Condition 1), with the worker or operator between the equipment and live parts. Such a condition would be where two covers are removed from the equipment opposite each other, exposing live parts to troubleshoot an electrical problem. The electrician or maintenance worker could accidentally make contact with live parts and be in series with phase-to-phase voltage or be in series with a phase and the grounded metal of the equipment. Either condition would most likely be fatal.

For example: What is the minimum clearance for equipment enclosures supplied with 13,800 volts and mounted opposite one another?

> **Step 1:** Finding clearance
> **Table 110.34(A), Condition 3**
> 13,800 volts requires 9 ft (2.8 m)
>
> **Solution: The minimum clearance is 9 ft (2.8 m).**

SEPARATION FROM LOW-VOLTAGE EQUIPMENT 110.34(B)

When low-voltage (600 volts or less) equipment is installed in a room or enclosure where there are exposed high-voltage live parts, the high-voltage equipment shall be separated from the low-voltage equipment by a suitable partition, fence, or screen.

If the room or enclosure is accessible to qualified persons (only), a partition, etc. shall not be required, provided the low-voltage equipment serves only equipment within the room or enclosure per **110.34(B), Ex.**

Design Tip: Many companies will not allow low voltage in transformer vaults with high voltage, with the exception of low-voltage buses. However, this does not include lighting and other low-voltage systems that might be required in the operation of the high-voltage equipment.

LOCKED ROOMS OR ENCLOSURES 110.34(C)

The entrances to all buildings, rooms, or enclosures containing high-voltage exposed live parts shall be kept locked, except where such entrances are under the constant observation of a qualified person. **(See Figures 5-22 and 6-53)** For voltages over 600 volts, warning signs shall be posted at the entrance which reads as follows:

DANGER - HIGH VOLTAGE - KEEP OUT!

ILLUMINATION 110.34(D)

Illumination shall be provided to properly illuminate the high-voltage area for safe working conditions, and the luminaires shall be installed so that there is no danger to anyone changing bulbs, ballasts, etc., or working on the illumination system. The switching point for this illumination shall be readily accessible and in such a place that persons operating the controls for the illuminations are in no danger of coming in contact with any live parts.

Design Tip: There is not a footcandle level given. Section **110.34(D)** requires sufficient lighting to be provided when needed. **(See Figure 5-29)**

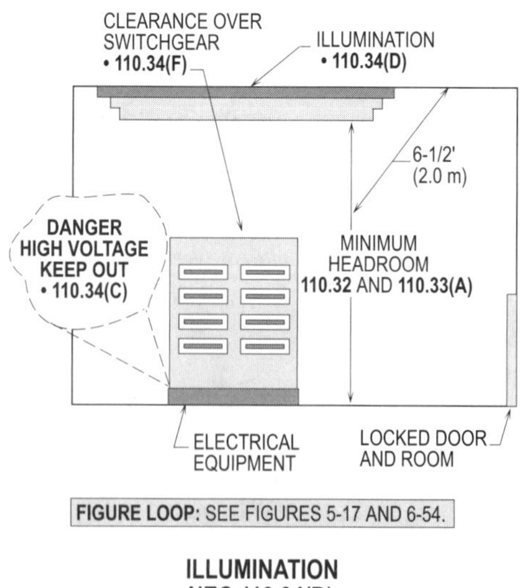

Figure 5-29. Electrical equipment shall have the proper lighting around the workspace for safe servicing.

ELEVATION OF UNGUARDED LIVE PARTS 110.34(E) AND TABLE 110.34(E)

It is permissible to install unguarded live parts at elevated heights of at least 9 ft (2.8 m) from any and all finished grade areas. For installations prior to April 16, 1981, the minimum clearance height was 8 ft (2.5 m), where the voltage between phases ranged from 601 to 6600 volts. The elevated height is based and selected from the supply voltage supplying the live parts.

For example: What is the height required for unguarded live parts operating at 35,000 volts?

Step 1:   Finding height
Table 110.34(E)
35,000 volts require 9-1/2 ft (2.9 m)

Solution: **The minimum clearance is 9-1/2 ft (2.9 m).**

See Figure 5-30 for a detailed procedure for installing live parts at elevated heights.

Design Tip: For equipment installed before April 16, 1981 that operates above 25 kV, the required working clearance shall not be required to be increased beyond the value listed for 25 kV systems.

IDENTIFICATION OF DISCONNECTING MEANS 110.22

Panelboard circuit directories shall be fully and clearly filled out to indicate the load served and its location. Marking on equipment shall be painted lettering or another suitable identification. The circuit breakers in panelboards and fuses in fusible disconnects shall be marked to match the markings on equipment they serve. This is a mandatory requirement, and it is to be applied to both new and old electrical systems. Adequate circuit mapping and proper identification shall be provided for expanding or alteration of all electrical systems. It is extremely important that the right disconnecting means be opened to deenergize a circuit quickly and safely when there is a threat of injury to personnel working, repairing, or pulling maintenance on the equipment supplied.

Painted labeling or embossed plates fixed to the equipment complies with the requirements that disconnects be "legibly marked" and the "marking shall be durable." Glue-on paper labels or marking with crayon, ink, or chalk most likely will be rejected as not being suitable permanent markings.

For example, a circuit breaker in a panelboard is marked No. 1 and supplies power to a disconnecting means in sight of a motor controller and motor.

The disconnecting means, motor controller, and motor shall be legibly marked No. 1 to correspond to the marking on the circuit directory on the panelboard enclosure. Effective identification of disconnecting means and protective devices is a must by the NEC, NFPA 70E, and OSHA to ensure the safety necessary for employees in the workplace. **(See Figure 5-31)**

FLASH PROTECTION 110.16

Electrical equipment, such as switchboards, panelboards, industrial control panels, meter socket enclosures, and motor control centers, in other than dwelling occupancies, that are likely to require examination, adjustment, servicing, or maintenance while energized, shall be field marked to warn qualified persons of potential electric arc flash hazards. This marking shall be located so as to be clearly visible to qualified persons before examination, adjustment, servicing, or maintenance of such equipment. **(See Figure 5-32)**

Design Tip: FPN 1 to **110.16** refers users to **NFPA 70E** for assistance in determining exposure, planning safe work practices, and selecting personal protective equipment. **FPN 2** to **110.16** refers users to **ANSI Z535.4** for guidelines pertaining to designing safety signs and labels to provide warnings, etc.

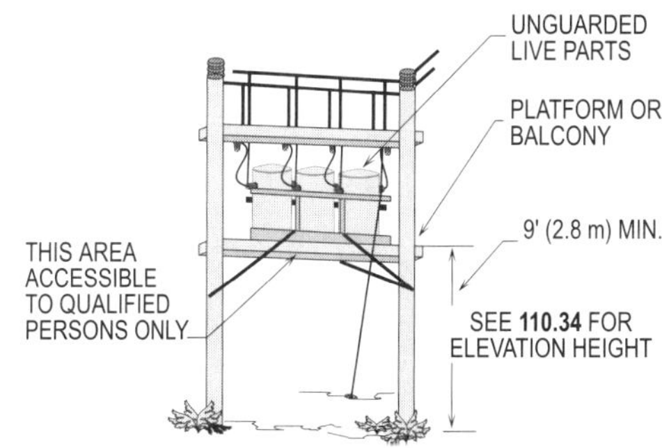

Figure 5-30. Electrical equipment elevated at least 9 ft (2.8 m) above grade shall be considered guarded by height.

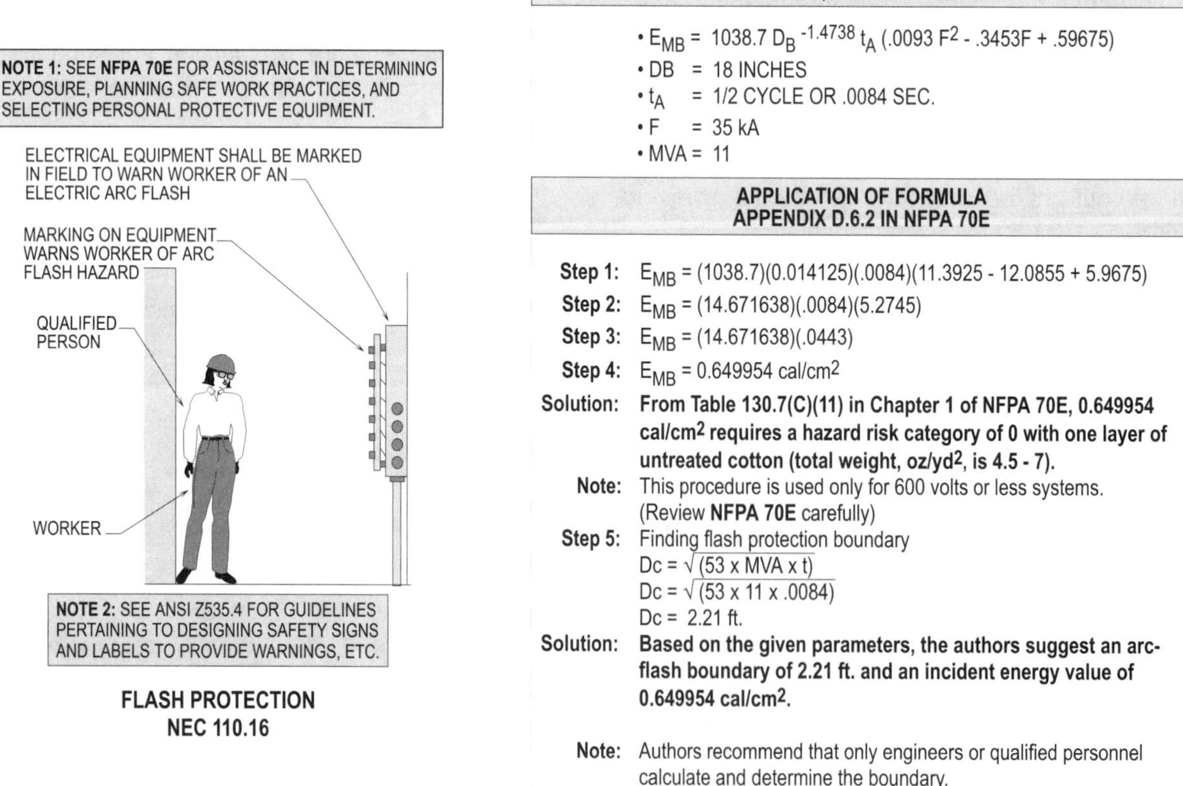

IDENTIFICATION OF DISCONNECTING MEANS
NEC 110.22, NEC 230.70(B), AND NEC 408.4

Figure 5-31. All circuits and disconnects shall be identified. OSHA regulations make **110.22** mandatory and retroactive for existing installations and for all new expanded or modernized systems - applying to switches as well as circuit breakers.

FLASH PROTECTION
NEC 110.16

- E_{MB} = 1038.7 D_B $^{-1.4738}$ t_A (.0093 F^2 - .3453F + .59675)
- DB = 18 INCHES
- t_A = 1/2 CYCLE OR .0084 SEC.
- F = 35 kA
- MVA = 11

APPLICATION OF FORMULA
APPENDIX D.6.2 IN NFPA 70E

Step 1: E_{MB} = (1038.7)(0.014125)(.0084)(11.3925 - 12.0855 + 5.9675)

Step 2: E_{MB} = (14.671638)(.0084)(5.2745)

Step 3: E_{MB} = (14.671638)(.0443)

Step 4: E_{MB} = 0.649954 cal/cm^2

Solution: From Table 130.7(C)(11) in Chapter 1 of NFPA 70E, 0.649954 cal/cm^2 requires a hazard risk category of 0 with one layer of untreated cotton (total weight, oz/yd^2, is 4.5 - 7).

Note: This procedure is used only for 600 volts or less systems. (Review **NFPA 70E** carefully)

Step 5: Finding flash protection boundary
Dc = $\sqrt{(53 \times MVA \times t)}$
Dc = $\sqrt{(53 \times 11 \times .0084)}$
Dc = 2.21 ft.

Solution: Based on the given parameters, the authors suggest an arc-flash boundary of 2.21 ft. and an incident energy value of 0.649954 cal/cm^2.

Note: Authors recommend that only engineers or qualified personnel calculate and determine the boundary.

Figure 5-32. This illustration shows that markings on electrical equipment shall be provided to warn employees of hazardous energy from an arcing fault.

Chapter 5. Working Clearances

Section   **Answer**

1. A minimum working space of ______ in. width shall be required in front of electrical equipment operating at 600 volts or less.
   (a) 24
   (b) 30
   (c) 36
   (d) 48

2. If rear access is necessary to work on deenergized parts on the back of enclosed equipment, a minimum working space of ______ in. horizontally shall be required.
   (a) 30
   (b) 36
   (c) 48
   (d) 60

3. Luminaires shall have a headroom clearance of at least ______ ft to give personnel sufficient room to stand in front of electrical equipment.
   (a) 4-1/2
   (b) 5-1/2
   (c) 6-1/2
   (d) 7-1/2

4. At least one entrance with a minimum of ______ in. x ______ ft shall be provided for access to the working space of high-voltage equipment.
   (a) 24 x 6-1/2
   (b) 24 x 7-1/2
   (c) 30 x 6-1/2
   (d) 30 x 7-1/2

5. Equipment doors and hinged panels shall have at least a ______ opening provided in the workspace. (600 volts or less)
   (a) 30°
   (b) 45°
   (c) 75°
   (d) 90°

6. A 600 volt or less switchgear with a bus rated ______ amps and over 6 ft wide shall have an exit at each end. (General Rule)
   (a) 800
   (b) 1000
   (c) 1200
   (d) 1500

7. Live electrical parts shall be considered protected where elevated at least ______ ft above grade. (600 volts or less)
   (a) 6
   (b) 8
   (c) 10
   (d) 12

8. The minimum clearance for a motor control center supplied with 12,470 volts and installed opposite a concrete wall is ______ ft.
   (a) 3
   (b) 4
   (c) 5
   (d) 6

9. The height required for unguarded live parts operating at 35,000 volts shall be ______ ft.
   (a) 9
   (b) 9 1/2
   (c) 10
   (d) 12

10. Personnel doors shall open in the direction of egress where intended for entrance to and egress from the working space less than ______ ft from the nearest edge of the working space. (Over 600 volts)
    (a) 6
    (b) 10
    (c) 20
    (d) 25

11. The minimum clearance in front of electrical equipment with an ungrounded wall opposite the equipment shall be _____ in.
    (a) 30
    (b) 36
    (c) 48
    (d) none of the above

12. The minimum clearance in front of electrical equipment opposite other electrical equipment with a voltage of 151 volts to 600 volts shall be _____ in.
    (a) 30
    (b) 36
    (c) 48
    (d) none of the above

13. Covers for an underground junction box that weigh over _____ lb shall be considered as accessible to only qualified personnel.
    (a) 20
    (b) 50
    (c) 100
    (d) 150

14. The minimum clearance required for a switchgear supplied with 4160 volts and installed opposite an insulated wall shall be _____ ft. (Corner grounded)
    (a) 4
    (b) 6
    (c) 9
    (d) 10

15. The minimum clearance required for equipment enclosures supplied with 13,800 volts and are mounted opposite one to another shall be _____ ft. (System is corner grounded.)
    (a) 4
    (b) 6
    (c) 9
    (d) 10

16. Other equipment that is associated with the electrical installation and is located above or below the electrical equipment shall be permitted to extend not more than _____ in. beyond the front of the electrical equipment.
    (a) 6
    (b) 12
    (c) 18
    (d) 24

17. The space equal to the width and depth of the equipment and extending from the floor to a height of _____ ft above the equipment or to the structural ceiling, whichever is lower, shall be dedicated to the electrical installation. (600 volts or less)
    (a) 4
    (b) 6
    (c) 10
    (d) 25

18. A wall, screen, or fence that is less than 6 ft high shall not be considered as preventing access to equipment unless other features are provided with a degree of isolation equivalent to a _____ ft fence. (Over 600 volts)
    (a) 7
    (b) 8
    (c) 9
    (d) 10

19. The floors of vaults in contact with the earth shall be of concrete that is not less than _____ in. thick. (Over 600 volts)
    (a) 1
    (b) 2
    (c) 4
    (d) 6

20. Where metal-enclosed equipment is accessible to the general public and the bottom of the enclosure is less than _____ ft above the floor or grade level, the enclosure door or hinged cover shall be kept locked. (Over 600 volts)
    (a) 8
    (b) 10
    (c) 12
    (d) 15

6

Services

The service equipment consists of two main parts: the service conduit and service conductors, the panelboard and the overcurrent protection devices. The wiring methods for the service consists of four main parts: the service, the feeders, the subfeeders, and the branch circuits. The size of the service-entrance conductors and equipment can be calculated by adding the total volt-amps of all the branch circuit loads and dividing by the configuration of voltage utilized.

The utility company supplies power to a building utilizing two methods: an overhead service drop or an underground lateral. The service conductors usually enter a building between the meter base, CT can or splice can, and service equipment. Conductors shall be permitted to be installed in conduit or cables designed for such use.

NUMBER OF SERVICES PERMITTED TO A BUILDING
ARTICLE 230, PART I

Only one service drop or lateral shall be permitted to be installed to a building. However, a second service drop or lateral shall be permitted to supply large loads and larger buildings. High-rise buildings permit additional service(s) to be installed to supply large loads that are present in multiple-occupancy buildings. Any additional service drops or laterals for the building shall have a reciprocal plaque denoting the locations of each service.

NUMBER OF SERVICES
230.2 AND FIGURES 6-1 THRU 6-7

To accommodate the load requirements for a facility, the following seven conditions shall permit more than one service to be installed.

FIRE PUMPS
230.2(A)(1)

To ensure against the interruption of power to fire pumps, a additional service shall be permitted to be installed. Fire pump overcurrent protection shall be set above the pump motor's locked rotor current rating, including other accessory loads per **230.90(A), Ex. 4,** with running overload protection being eliminated per **240.4(A)**. Disconnecting means rated 1000 amps or more supplied by a 277/480 volt service does not require ground-fault protection per **695.6(H)** and **240.13(3). (See Figure 6-1)**

EMERGENCY LIGHTING OR POWER SYSTEMS
230.2(A)(2)

Emergency lighting or power systems shall be permitted to be supplied by a additional service. Emergency lighting or power systems will provide the needed power for lighting and equipment for the safety of occupants to exit if the main service is interrupted for any reason, such as loss of the normal power. For further information, see **700.12(D)**. **(See Figure 6-2)**

SPECIAL OCCUPANCIES
230.2(B)(1)

By special permission, more than one service drop or underground lateral shall be permitted to be installed, provided there is no space available for service equipment to be mounted and overcurrent protection devices are readily accessible in each occupancy. For further information, see **230.40, Ex. 1** and **230.71(A)**.

When multiple tenants occupy individual units in an apartment complex or other types of buildings, the service drop or underground lateral shall be permitted to be classified as a multiple occupancy and served by more than one service. **(See Figure 6-3)**

BUILDINGS COVERING A LARGE AREA
230.2(B)(2)

Two or more services shall be permitted for buildings that cover a large area. Buildings that cover large areas usually have loads located a great distance from the service equipment. Service drops or underground laterals shall be permitted to supply buildings with panelboards and switchboards located in different areas because that is more practical than trying to use feeders, which will present voltage drop problems due to their long length. **(See Figure 6-4)**

> **Design Tip:** High-rise buildings are not included in this exception.

CAPACITY REQUIREMENTS
230.2(C)(1) AND (C)(2)

Load requirements in excess of 2000 amps shall be permitted to be supplied with more than one service to accommodate large loads that have been added. An additional service shall be permitted if the utility company requires more than one because the size of the loads being added and the way they are served. **(See Figure 6-5)**

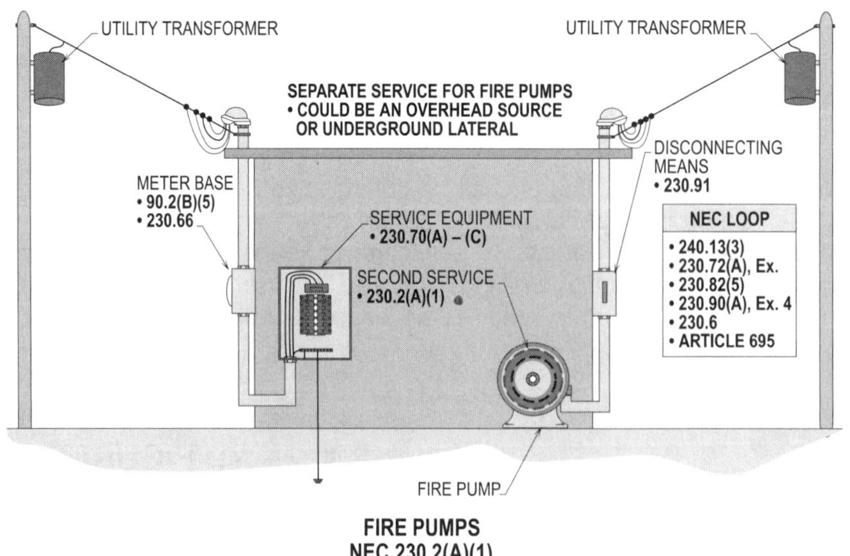

Figure 6-1. To ensure against the interruption of power to fire pumps, a additional service shall be permitted to be installed.

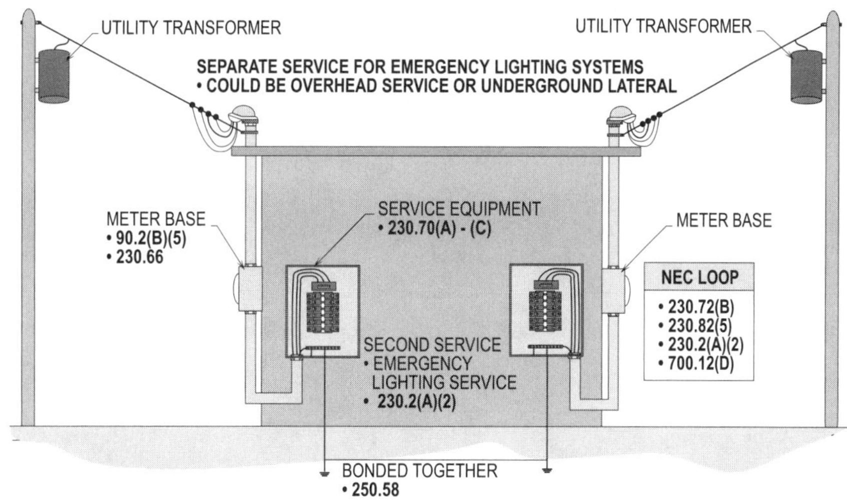

Figure 6-2. Emergency lighting or power systems shall be permitted to be supplied by a additional service.

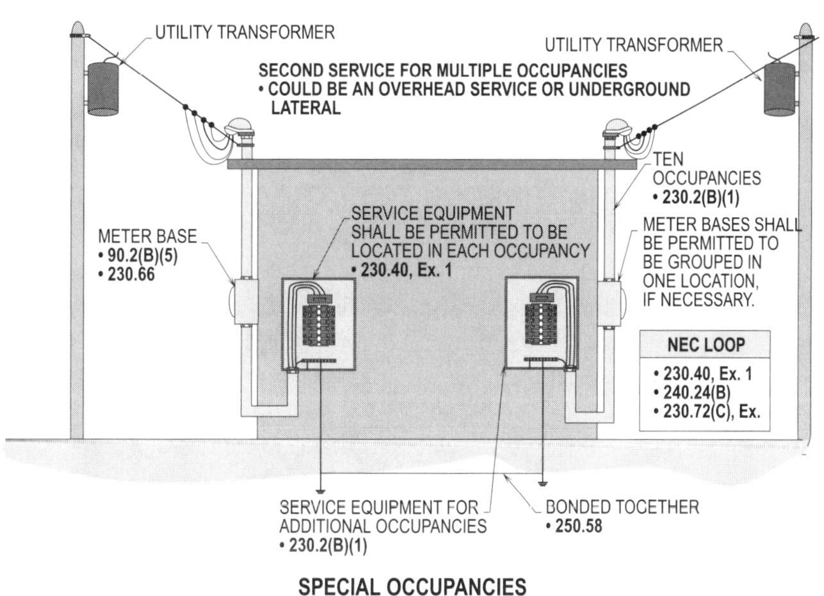

Figure 6-3. By special permission, more than one service drop or ungrounded lateral shall be permitted to be installed, provided there is space available for service equipment to be mounted and overcurrent protection devices are readily accessible to each occupancy.

DIFFERENT CHARACTERISTICS 230.2(D)

By special permission, a building of different voltages shall be permitted to have two or more service drops or underground laterals. A building operating on a special rate schedule permits an additional service to supply specific loads or equipment such as computers, etc. **(See Figure 6-6)**

For example, a service drop or lateral may consist of a 277/480 volt service and a 120/208 volt service supplying a building. The 120/208 volt service supplies general-purpose outlets and equipment, while the 277/480 volt service supplies special equipment.

UNDERGROUND SETS OF CONDUCTORS 230.2

A building shall be permitted to be supplied with more than one set of underground laterals. Sections **230.40, 230.71** and **230.72** require all laterals to be grouped and located adjacent to one another. When 1/0 AWG and larger conductors are used, this type of installation shall be permitted. At the supply end, the 1/0 AWG conductor shall be connected in parallel and shall be permitted to be run and terminated to six or less main circuit breakers or disconnecting switches used exclusively for supplying an individual service. **(See Figure 6-7)**

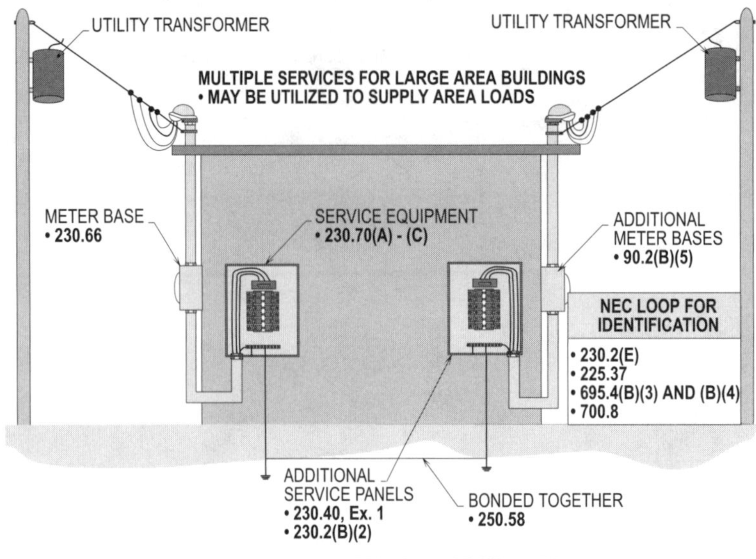

Figure 6-4. Two or more services shall be permitted for buildings that cover a large area.

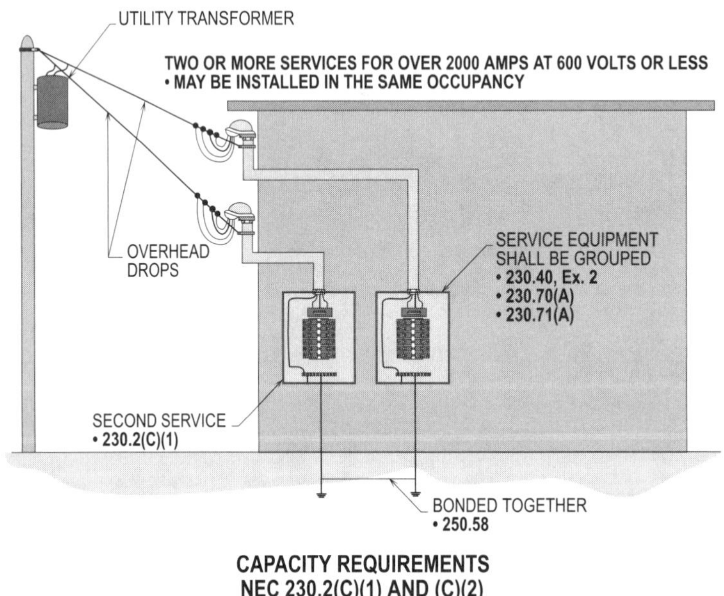

Figure 6-5. Load requirements in excess of 2000 amps shall be permitted to be supplied with more than one service.

ONE BUILDING OR OTHER STRUCTURE NOT TO BE SUPPLIED THROUGH ANOTHER
230.3

Service conductors supplying a building or structure shall not be permitted to pass through the interior of another building or structure.

Service conductors installed in a duct or conduit and under at least 2 in. (50 mm) of concrete or encased with at least 2 in. (50 mm) of concrete or brick covering shall be considered outside of the building and shall be permitted to pass under, in, or on a building or structure to supply another building or structure. See **230.6** for the rules and regulations pertaining to this requirement.

Design Tip: A feeder with an overcurrent protection device ahead of it does not fall under this rule. Therefore, feeders shall be permitted to be routed through the building to another building without the need to apply **230.6**.

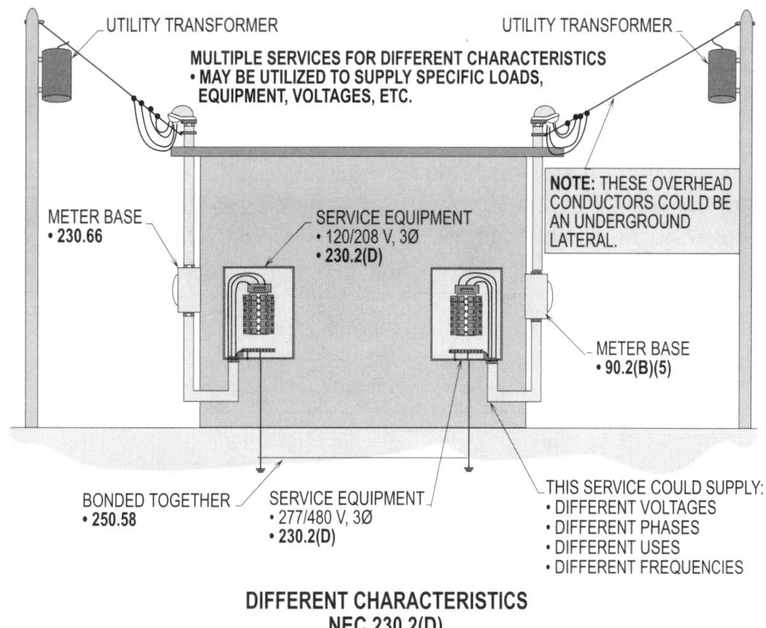

Figure 6-6. By special permission, a building of different voltages shall be permitted to have two or more service drops or underground laterals.

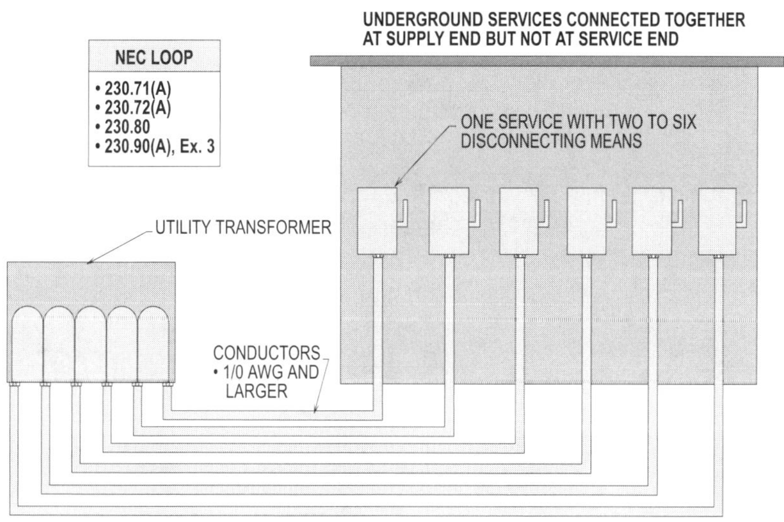

Figure 6-7. A building shall be permitted to be supplied with more than one set of underground laterals.

SELECTING INSULATION AND AMPACITY OF SERVICE CONDUCTORS
ARTICLE 230, PART I

Table 310.16 lists the different ampacities of conductors based on 60°C, 75°C, or 90°C terminals per **110.14(C)**. These allowable ampacity ratings are applied and used based on conductors consisting of not more than three current-carrying conductors in a raceway and ambient temperatures of 30°C and 86°F. Ampacity ratings vary with the type of insulation used to insulate the conductor. The type of insulation selected and the terminal rating of the overcurrent protection device used determine the ampacities of the conductors. The designer shall consider and allow for adjustment factors due to four or more current-carrying conductors and correction factors if the surrounding ambient exceeds 86°F. All these factors shall be evaluated when selecting conductors to calculate allowable ampacities to supply service loads.

CONDUCTORS CONSIDERED OUTSIDE THE BUILDING
230.6

Service conductor installed in a duct or conduit and under 2 in. (50 mm) of concrete or encased with at least 2 in. (50 mm) of concrete or brick covering shall be considered outside of the building. In other words, these conductors have never entered the building and an overcurrent protection device does not have to be installed ahead of these conductors. Service conductors installed in a conduit and buried at least 18 in. (450 mm) in the earth beneath a building or other structure shall be considered outside the building. **(See Figure 6-8)**

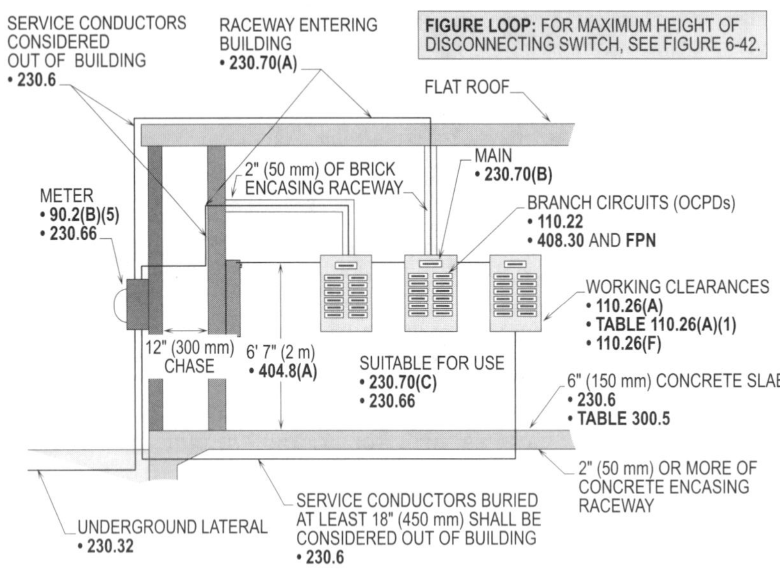

Figure 6-8. Service conductors installed in a duct or conduit and under 2 in. (50 mm) of concrete or encased with at least 2 in. (50 mm) of concrete or brick covering shall be considered outside of the building.

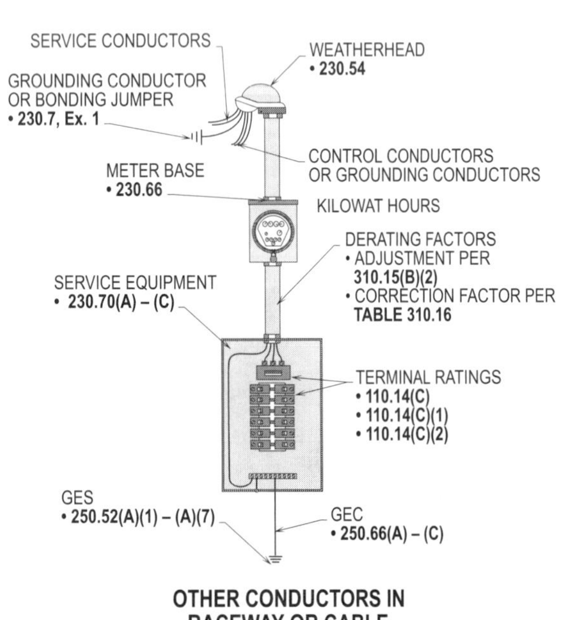

Figure 6-9. When installing service-entrance conductors in a service raceway or service-entrance cable, no other conductors shall be permitted unless an exception is applied.

OTHER CONDUCTORS IN RACEWAY OR CABLE
230.7

When installing service-entrance conductors in a service raceway or service-entrance cable, no other conductors shall be permitted. The following are two exceptions to the rule:

- When installing service-entrance conductors, the grounding conductors and bonding jumpers shall be permitted to occupy the same service raceway.
- Control conductors for load management with overcurrent protection (OCP) shall be permitted to occupy the same service raceway. **(See Figure 6-9)**

WIRING METHODS
230.43

The following wiring methods shall be permitted to be used when designing and installing service-entrance conductors:

- Open wiring on insulators
- Type IGS cable
- Rigid metal conduit (RMC)
- Electrical metallic tubing (EMT)
- Intermediate metal conduit (IMC)
- Electrical nonmetallic tubing (ENT)
- Service-entrance cables
- Wireways
- Busways
- Auxiliary gutters
- Rigid nonmetallic conduit (RNC)
- Cable bus
- Type MC cable
- Mineral-insulated and metal-sheathed cables
- Flexible metal conduit and LFMC
  - In lengths not exceeding 6 ft (1.8 m) per **250.102(A), (B), (C),** and **(E)**
- Liquidtight flexible nonmetallic conduit

CLEARANCE FROM BUILDING OPENINGS
230.9(C)

A clearance of 3 ft (900 mm) shall be required from windows designed to be opened, including porches, platforms, etc. A clearance of 3 ft (900 mm) shall not be required for service conductors attached above windows, since they shall be considered out of reach. Such clearance shall not be required for windows that do not open. **(See Figure 6-10)**

VEGETATION AS SUPPORT
230.10

Vegetation such as trees shall not permitted to be used as a means of support for overhead service conductors. **(See Figure 6-11)**

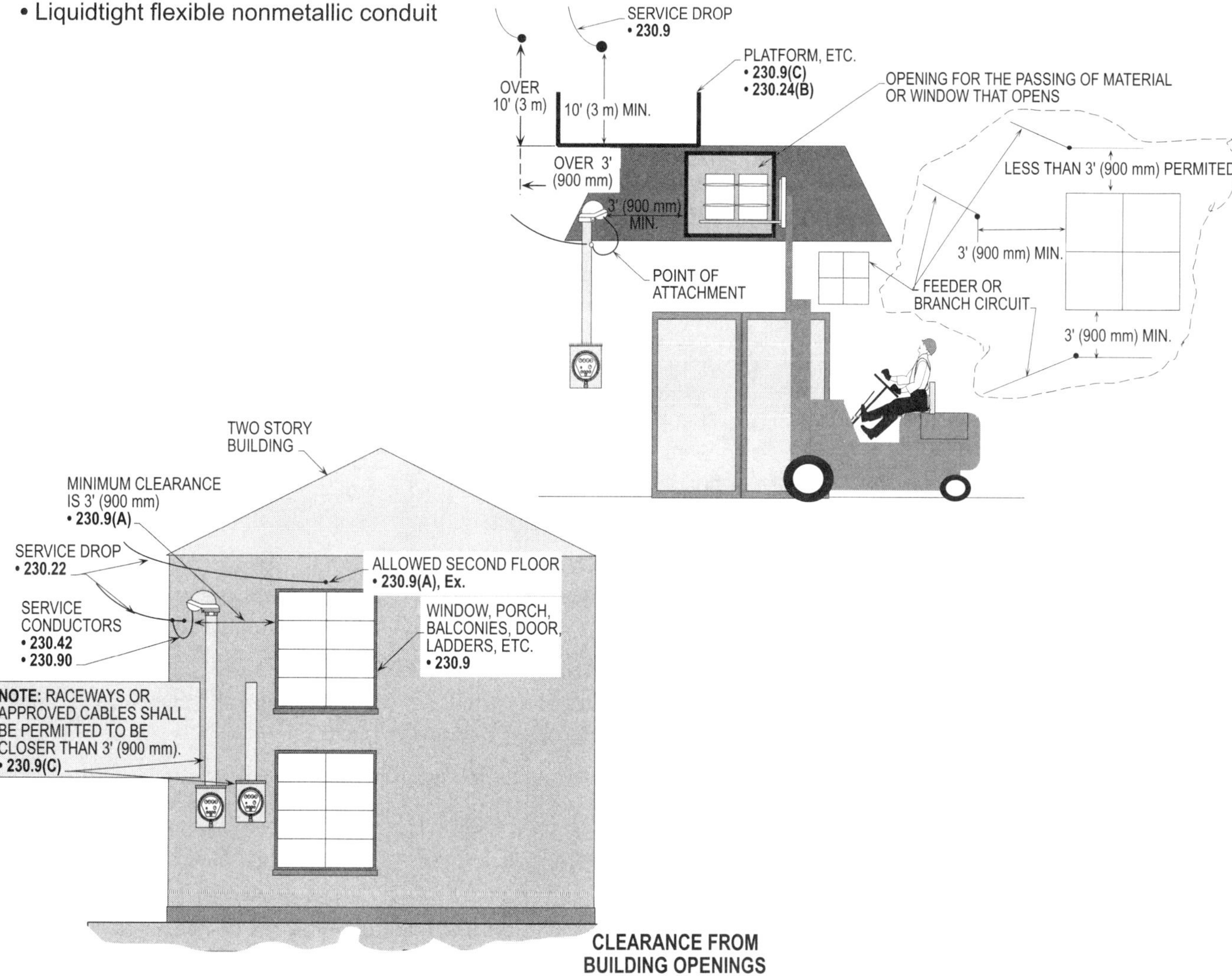

Figure 6-10. A clearance of 3 ft (900 mm) shall be required from windows designed to be opened, including porches, platforms, etc.

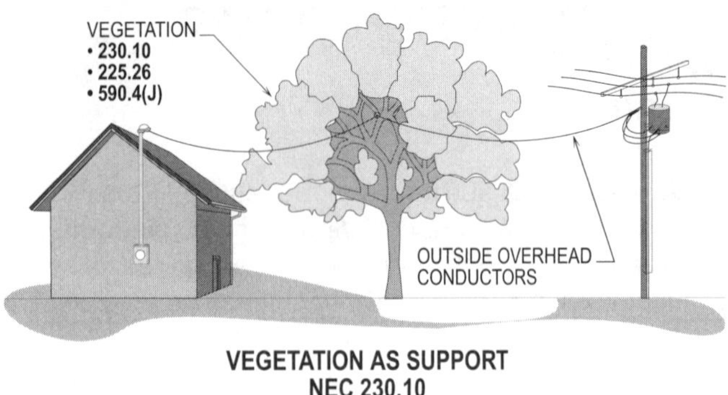

Figure 6-11. This illustration shows a tree being used as a support for overhead conductors – "NOT ALLOWED."

OVERHEAD SERVICES
ARTICLE 230, PART II

Service drops to a building shall be permitted to be supplied from an overhead supply from the utility pole to the attachment to the building. Such an installation shall be considered an overhead service drop. A minimum clearance from finished grade and the size and rating of the service drop conductors shall be observed when installing these conductors. The designer and installer shall consider who and what has access under these conductors. Such access determines the height of service drops from finished grade.

INSULATION OR COVERING
230.22

Service-entrance conductors shall be insulated to protect against short-circuit conditions (phase-to-phase) or ground-faults (phase-to-ground) when the service drop consists of individual conductors that supply the load of the building. The conductor insulation shall be approved for the type of service drop to be installed. Where multiconductors are installed for the service drop, a grounded (neutral) conductor shall be permitted to be uninsulated per **230.22, Ex.** Thermoplastic or thermosetting insulation shall be installed where applicable; review **Table 310.13** for the different types of insulation and their specific conditions of use. **[See Figures 6-12(a) and (b)]**

SIZE AND RATING
230.23

Service-drop conductors shall not be smaller than 8 AWG copper or 6 AWG aluminum. Service-drop conductors shall be permitted to be 12 AWG hard-drawn copper where installed to serve small loads of only one branch circuit. Such installations of small loads are phone booths, small polyphase motors, etc., that do not draw large currents.

Design Tip: The service-drop grounded (neutral) conductor shall not be smaller than the minimum size calculated per **220.61** for normal current flow and for ground-fault conditions per **250.24(C)(1)** and **(C)(2)**.

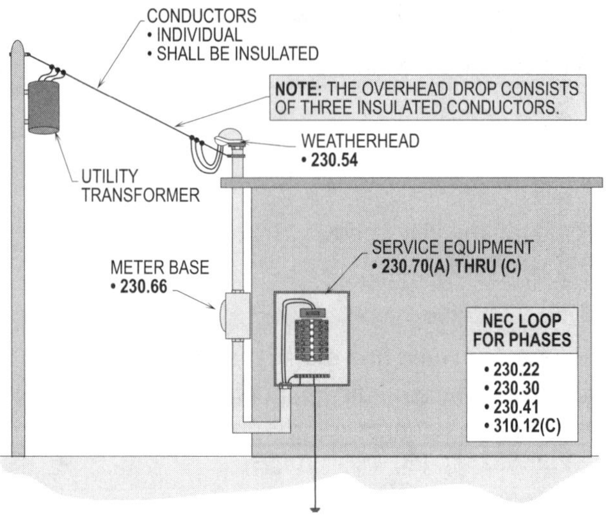

Figure 6-12(a). The conductor insulation shall be approved for the type of service drop to be installed.

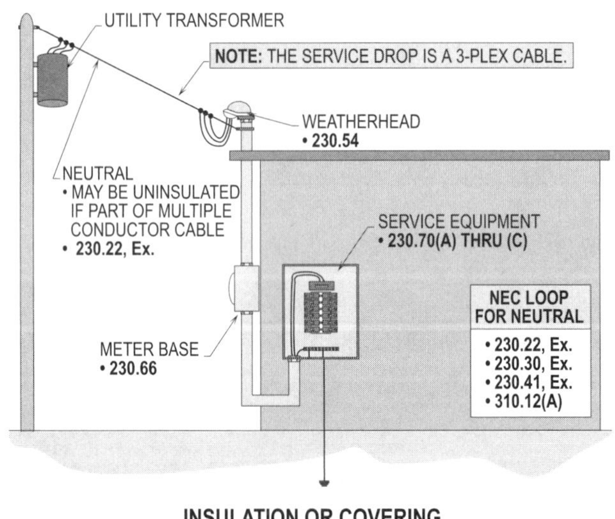

Figure 6-12(b). Where multiconductors are installed for the service drop, a grounded (neutral) conductor shall be permitted to be uninsulated.

CLEARANCES
230.24

Service-drop conductors shall not be readily accessible for voltages of 600 volts or less. To ensure the protection of conductors and the safety of the general public, proper clearances shall be provided underneath such conductors based on travel by people or vehicles.

ABOVE ROOFS
230.24(A)

A clearance of 8 ft (2.5 m) shall be provided for service-drop conductors passing over roofs. At least a 3 ft (900 mm) clearance shall be maintained in all directions for vertical clearances. These clearances shall be permitted to be reduced by applying any one of the following three exceptions: **[See Figures 6-13(a) thru (c)]**

OVERHANGING PORTION OF THE ROOF
230.24(A), Ex. 3

Service-drop conductors shall be permitted to be installed for overhanging roofs, provided no more than 6 ft (1.8 m) of such conductors does not pass over more than 4 ft (1.2 m) of the roof. The attachment of the service-drop conductors shall have a clearance of at least 18 in. (450 mm) from the conduit and roof line. **Note,** for size of conduit containing service-entrance conductors, check with local electrical ordinances and utility specifications. **[See Figure 6-13(c)]**

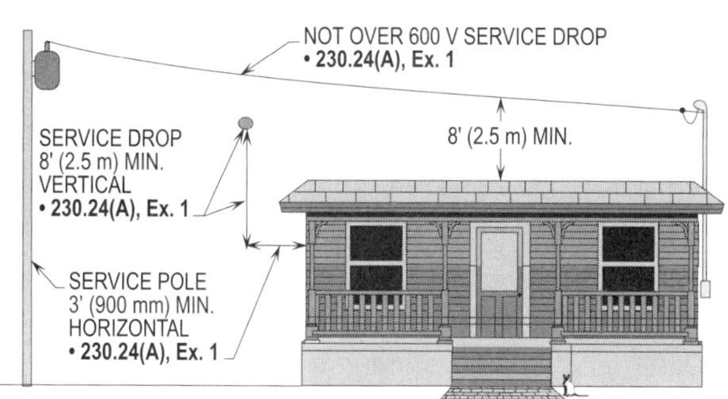

Figure 6-13(a). Service-drop conductors shall have minimum clearances, complying with **230.24(B),** where they are installed above roofs subject to accessibility.

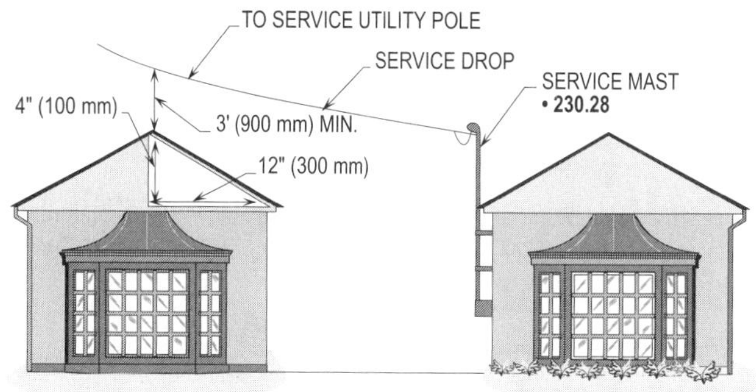

Figure 6-13(b). Service-drop conductors shall be permitted to pass above roofs where voltages between conductors do not exceed 300 volts and are installed at a height of 3 ft (900 mm), where the roof has a slope of at least 4 in. (100 mm) by 12 in. (300 mm).

ROOFS SUBJECT TO PEDESTRIAN OR VEHICULAR TRAFFIC
230.24(A), Ex. 1

Service-drop conductors shall have minimum clearances complying with **230.24(B)** where they are installed above roofs subject to pedestrian or vehicular traffic. Conductors shall have a vertical clearance of at least 10 ft (3 m) above flat roofs that are accessible to pedestrians. **[See Figure 6-13(a)]**

VOLTAGE NOT EXCEEDING 300 VOLTS
230.24(A), Ex. 2

Service-drop conductors shall be permitted to pass above roofs where voltages between conductors do not exceed 300 volts and are installed at a height of 3 ft (900 mm), where the roof has a slope of at least 4 in. (100 mm) by 12 in. (300 mm). Service-drop conductors shall be installed at a clearance of 8 ft (2.5 m) when the voltage between conductors exceed 300 volts, regardless of the slope of the roof. **[See Figure 6-13(b)]**

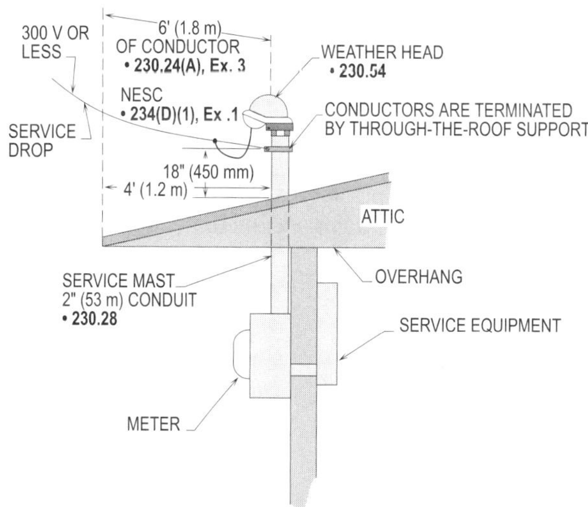

Figure 6-13(c). Service-drop conductors shall be permitted to be installed for roof overhangs provided no more than 6 ft (1.8 m) of such conductors does not pass over more than 4 ft (1.2 m) of the roof.

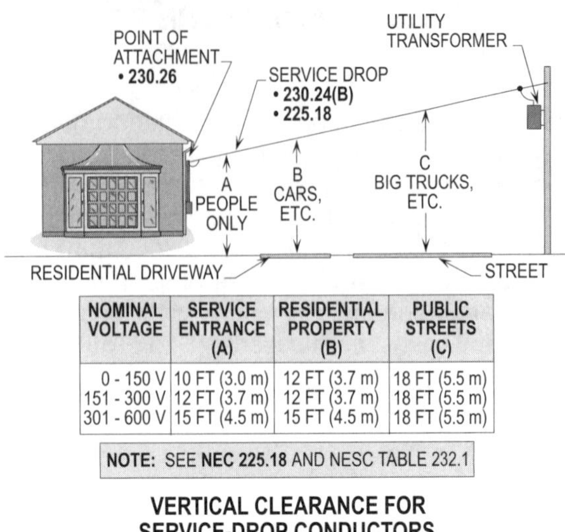

NOMINAL VOLTAGE	SERVICE ENTRANCE (A)	RESIDENTIAL PROPERTY (B)	PUBLIC STREETS (C)
0 - 150 V	10 FT (3.0 m)	12 FT (3.7 m)	18 FT (5.5 m)
151 - 300 V	12 FT (3.7 m)	12 FT (3.7 m)	18 FT (5.5 m)
301 - 600 V	15 FT (4.5 m)	15 FT (4.5 m)	18 FT (5.5 m)

NOTE: SEE **NEC 225.18** AND NESC TABLE 232.1

VERTICAL CLEARANCE FOR SERVICE-DROP CONDUCTORS NEC 230.24(B)

Figure 6-14. The vertical clearances from the finished grade to the service-drop conductors shall be determined by the voltage-to-ground.

VERTICAL CLEARANCE FOR SERVICE-DROP CONDUCTORS 230.24(B)

The vertical clearances from the finished grade to the service-drop conductors shall be determined by the voltage-to-ground. The following are the clearances required for each service voltage for service-drop conductors crossing or penetrating property or roof. Note that there are three levels of voltage (due to ground) on which these clearances shall be based. **(See Figure 6-14)**

150 VOLTS OR LESS TO GROUND

Service-drop conductors crossing property where people have access shall have a clearance of at least 10 ft (3 m) from finished grade. This vertical height shall be considered safe where only people have access under such conductors.

OVER 150 VOLTS TO 300 VOLTS-TO-GROUND

Service-drop conductors crossing residential property or driveways shall have a clearance of 12 ft (3.7 m) from finished grade. Only a car or pickup truck, etc. should have access under these conductors.

OVER 300 VOLTS-TO-GROUND

Service-drop conductors crossing residential property or driveways shall have a clearance of 15 ft (4.5 m) from finished grade where the voltage exceeds 300 volts-to-ground. The higher voltage-to-ground dictates a greater height to protect people and equipment from such voltage.

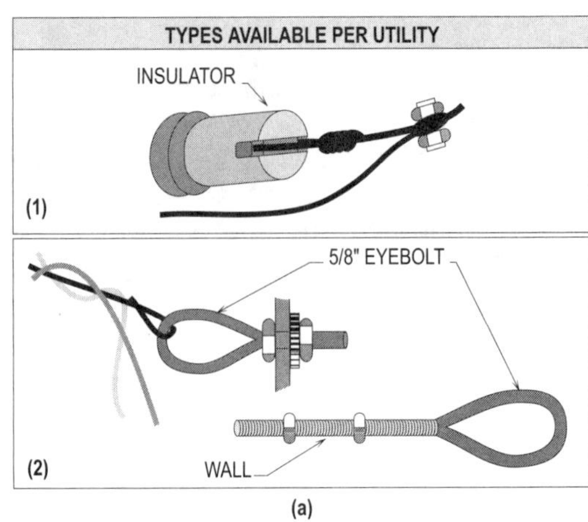

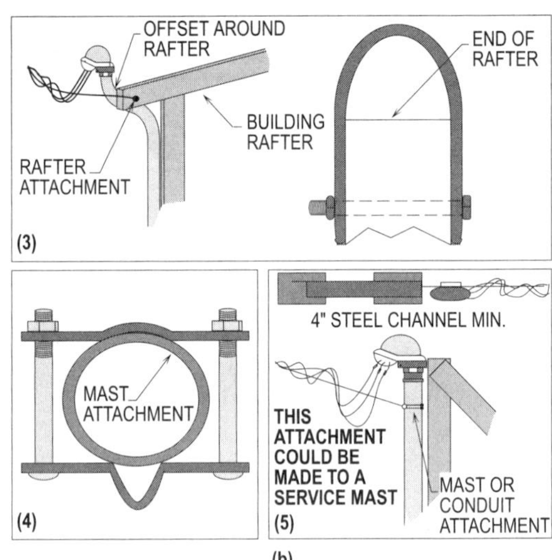

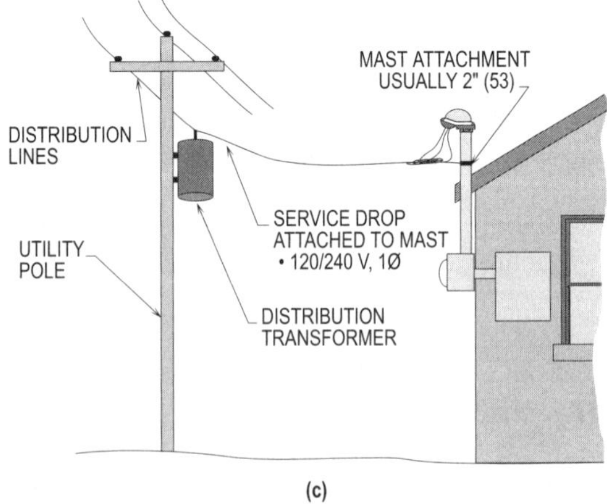

Figures 6-15(a), (b), and **(c).** Service-drop cables shall be permitted to be installed by attachment to the service mast pipe or to the building.

SUBJECT TO TRUCK TRAFFIC

Service-drop conductors crossing public streets, alleys, roads, parking areas subject to traffic, driveways on other than residential property, and other land such as cultivated, grazing, forest, and orchard shall have a clearance of 18 ft (5.5 m) from finished grade. This higher height shall be required to ensure safety for vehicles with greater heights and loads passing under such conductors.

POINT OF ATTACHMENT
230.26

The minimum point of attachment for service-drop conductors shall be 10 ft (3 m) from finished grade. The point of attachment shall be installed to provide the minimum clearances per **230.24**. The minimum point of attachment for service-drop conductors shall be installed based on the voltage level to ground.

> **Design Tip:** The point of attachment shall be made to the attachment fitting itself or the drip loop, whichever is greater.

MEANS OF ATTACHMENT
230.27

Service drop cables shall be permitted to be installed by attachment to the service mast pipe or to the building. The point of attachment for service drop conductors shall be permitted to be on a wall or rafter, with fittings approved for the attachment usually dictated by utilities. **[See Figures 6-15(a) through (c)]**

SERVICE MASTS AS SUPPORTS
230.28

Service drop cables shall be supported with service masts of adequate strength to prevent damage to building structures. Service masts of adequate strength will help prevent the strain on pipe where service-drop conductors are covered with ice or snow or subject to high winds. At least a 2 in. (53) metal rigid or intermediate metal conduit is required by most utility companies.

SUPPORTS OVER BUILDINGS
230.29

Poles shall be installed on each side of the building for service drop conductors passing over buildings. The clearance height of these poles shall be installed by the rules and regulations of **230.24**. Service drop conductors shall be permitted to be supported by an A-frame, cradle, or cross-arm that will provide a sufficient means to attach conductors to a building. **(See Figure 6-16)**

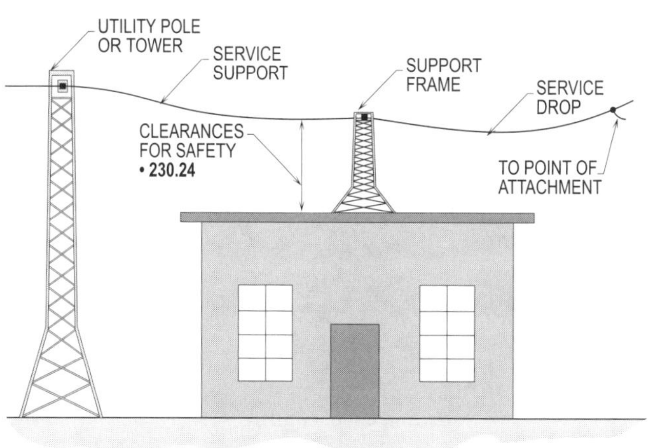

Figure 6-16. Poles shall be permitted to be installed on each side of the building for service drop conductors passing over buildings.

UNDERGROUND SERVICE-LATERAL CONDUCTORS
ARTICLE 230, PART III

A lateral is an underground service supplying electrical power to a building. Laterals shall be permitted to be terminated to a pad-mounted transformer or run underground and up a pole and terminated to an overhead transformer. An underground lateral shall be permitted to be installed inside or outside of a building, with terminations in a terminal box or meter can. The conductors run between the terminal box or meter can and service equipment shall be considered service-entrance conductors and shall be installed as such. **(See Figure 6-17)**

INSULATION
230.30

The conductor insulation shall be approved for the purpose if they are to be used as an underground service lateral. The rule calls for ungrounded (phase) conductors to be insulated so they will be protected from short-circuits and ground-faults.

UNINSULATED GROUNDED (NEUTRAL) CONDUCTOR
230.30, Ex.

A copper grounded (neutral) conductor shall be permitted to be bare when used for an underground service lateral utilizing a cable assembly or installed in conduit. If soil conditions will not deteriorate the copper, a bare copper conductor shall be permitted to be installed directly in the ground. Aluminum or copper clad bare conductors shall be permitted to be installed only when used in a cable assembly routed in conduit or approved for direct burial. **(See Figure 6-18)**

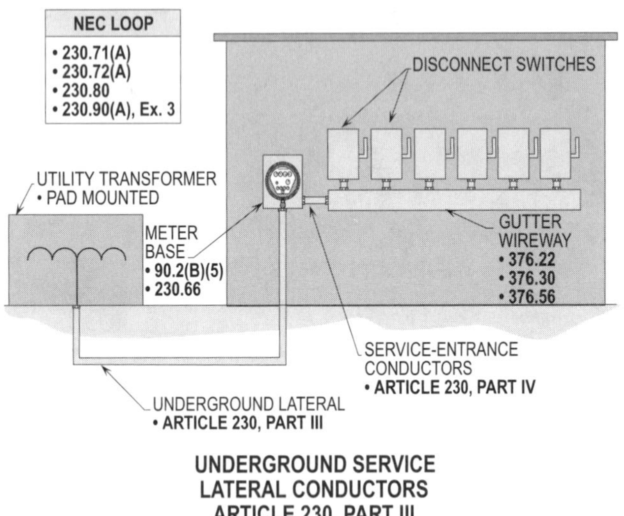

Figure 6-17. A lateral is an underground service supplying electrical power to a building.

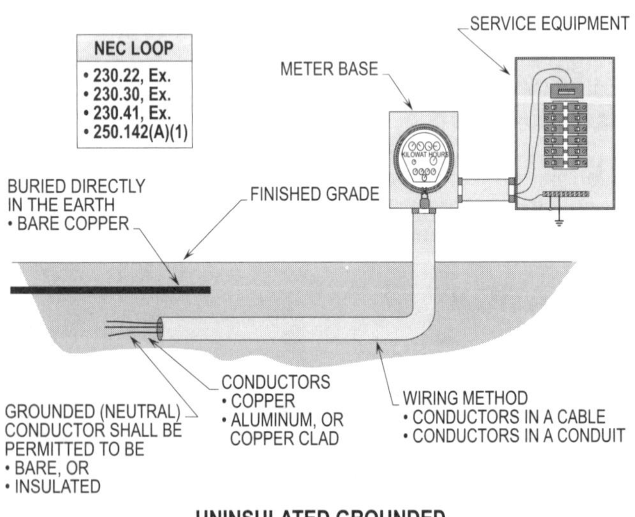

Figure 6-18. A copper grounded (neutral) conductor shall be permitted to be bare when used for an underground service lateral utilizing a cable assembly or installed in conduit.

SIZE AND RATING
230.31

Service lateral conductors shall not be permitted to be smaller than 8 AWG copper or 6 AWG aluminum. However, service-drop conductors shall be permitted to be 12 AWG hard-drawn copper where installed to serve small loads of only one branch circuit . Such installation of small loads are phone booths, small polyphase motors, etc., that require very small currents. **(See Figure 6-19)**

Design Tip: Service-drop grounded (neutral) conductor shall not be permitted to be smaller than the minimum size calculated by **220.61**, and sized for fault-currents per **250.24(C)(1)** and **250.24(C)(2)** as required by **230.31(C)**.

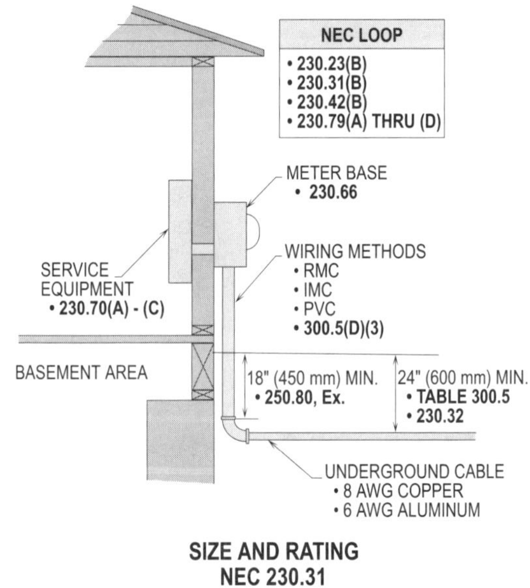

Figure 6-19. Service lateral conductors shall not be permitted to be smaller than 8 AWG copper or 6 AWG aluminum.

SERVICE-ENTRANCE CONDUCTORS
ARTICLE 230, PART IV

Service-entrance conductors shall be sized and installed large enough to supply the calculated loads of the premises. Service-entrance conductors installed in a raceway or cable shall be insulated to protect from short circuits and ground faults, except the grounded (neutral) conductor shall be permitted to be bare under certain conditions of use.

NUMBER OF SERVICE-ENTRANCE CONDUCTOR SETS
230.40

Only one set of service-entrance conductors shall be supplied by a service drop or service lateral. The following five exceptions allow service-entrance conductors to be installed by more than one set:

MORE THAN ONE OCCUPANCY
230.40, Ex. 1

Each occupancy or group of occupancies in a multiple occupancy building shall be permitted to have their service equipment supplied with one set of service-entrance conductors. Note that one to six service mains shall be permitted to be installed at such locations per **230.71(A)** and **230.72(A)**.

TWO TO SIX DISCONNECTS IN SEPARATE ENCLOSURES
230.40, Ex. 2

Two to six service disconnecting means in separate enclosures shall be permitted to be installed with one set of service-entrance conductors from a single service drop or underground lateral.

SINGLE-FAMILY DWELLING UNITS
230.40, Ex. 3

Single-family dwelling units and separate structures shall be permitted to have one set of service-entrance conductors routed to their service equipment from a single service drop or underground lateral.

TWO-FAMILY DWELLING OR MULTIFAMILY DWELLING
230.40, Ex. 4

A set of service-entrance conductors shall be permitted to be run to a two-family or multifamily dwelling to supply "common area" circuits such as lighting, central alarm, and signal circuits.

SUPPLY SIDE OF THE NORMAL SERVICE DISCONNECTING MEANS
230.40, Ex. 5

A set of service-entrance conductors shall be permitted to be connected to the supply side of the normal service disconnecting means to supply each or several systems covered per **230.82(5)** or **(6)**.

INSULATION OF SERVICE-ENTRANCE CONDUCTORS
240.41

The conductor insulation shall be approved for the purpose when utilized in an underground service lateral to supply power to a building or structure.

SIZE AND RATING
230.42(A) THRU (C)

Service-entrance conductors shall be installed to carry the total calculated load of the premises per **Article 220**. Service-entrance conductors shall be sized and selected from the ampacities per **Tables 310.16 through 310.19**, including all applicable Notes.

The ampacity for ungrounded (phase) conductors used for service-entrance shall be sized and selected according to the minimum requirements as follows:

- For limited loads of not less than 15 amps, a 15 amp service-entrance disconnecting means shall be permitted to be used per **230.79(A)**.

> **Design Tip:** Limited loads such as a motor on a single branch circuit shall be permitted to be supplied with a 15 amp service-entrance disconnecting means.

- Not more than two two-wire branch circuits shall be permitted to be supplied with a 30 amp disconnecting means and a 10 AWG or 8 AWG copper service-entrance conductors per **230.79(B)**.
- A minimum 100 amp, three-wire service disconnect means shall be installed for each one-family dwelling per **230.79(C)**. The minimum size service-entrance conductors that shall be permitted to be installed are 4 AWG THWN copper and 2 AWG THWN aluminum if **Table 310.15(B)(6)** is applied.

> **Design Tip:** For other types of insulation permitted, see **Table 310.15(B)(6)**.

- A minimum 60 amp service disconnect means shall be installed for all other installations per **230.79(D)**. The minimum size service-entrance conductors that are usually installed are 6 AWG THW or THWN copper or 4 AWG THW or THWN aluminum.

The ampacity for grounded (neutral) conductors for service-entrance shall be permitted to be sized and selected by the requirements listed in **220.61** and **250.24(C)(1)**. **(See Figure 6-20)**

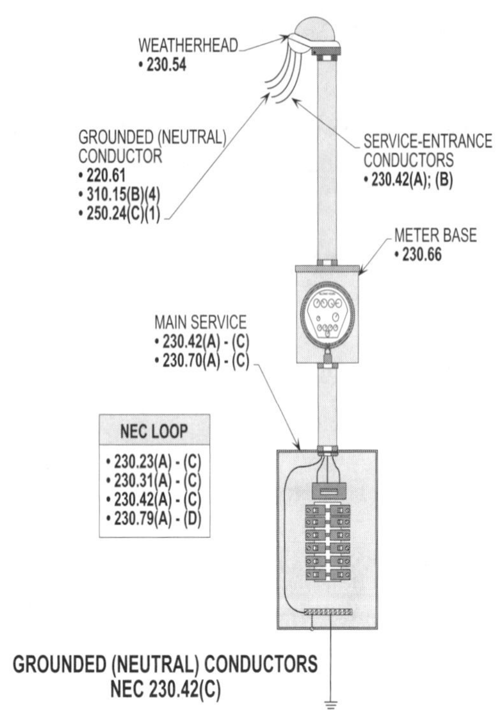

Figure 6-20. Service-entrance conductors shall be installed to carry the total calculated load of the premises per **Article 220**. Service-entrance conductors shall be sized and selected from the ampacities per **Tables 310.16 through 310.19** and **Table 310.13**.

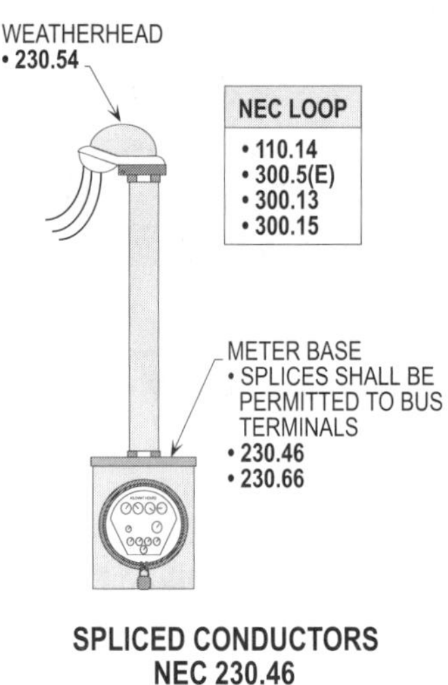

Figure 6-21. Metering equipment enclosures shall be permitted to be installed with spliced service-entrance conductors.

- Various sections and fittings shall be permitted to be installed for the connection and assembly of busways per **230.46**. **(See Figure 6-25)**
- Listed underground splice kits shall be permitted to be installed for existing service-entrance conductors per **230.46**. **(See Figure 6-26)**

SPLICED CONDUCTORS
230.46

Service-entrance conductors shall be permitted to be spliced or tapped in the following locations in accordance with **110.14, 300.5(E), 300.13** and **300.15**:

- Metering equipment enclosures shall be permitted to be installed with spliced service-entrance conductors per **230.46**. **(See Figure 6-21)**
- Service-entrance conductors shall be permitted to be tapped for two to six disconnecting means if grouped in a common location per **230.46**. **(See Figure 6-22)**
- Service-entrance conductors shall be permitted to be spliced where installing one wiring method to another per **230.46**. **(See Figure 6-23)**
- Service-entrance conductors shall be permitted to be spliced where an existing meter for a building has been relocated from inside to the outside of the building per **230.46**. **(See Figure 6-24)**

PROTECTION OF OPEN CONDUCTORS AND CABLES AGAINST DAMAGE – ABOVE GROUND
230.50

Service-entrance cable shall be installed and protected by rigid metal conduit, IMC, EMT, or PVC (Schedule 80), or by other approved methods if subject to physical damage. Service-entrance cable shall be provided with physical protection if installed near awnings, shutters, driveways, or coal chutes.

UNDERGROUND SERVICE-ENTRANCE CONDUCTORS
230.50(A)

Service-entrance cable shall be installed and protected by rigid metal conduit, IMC, EMT, or PVC (Schedule 80), or by other approved methods if subject to physical damage. Such wiring methods are found in **230.43(1) through (16)** in the NEC.

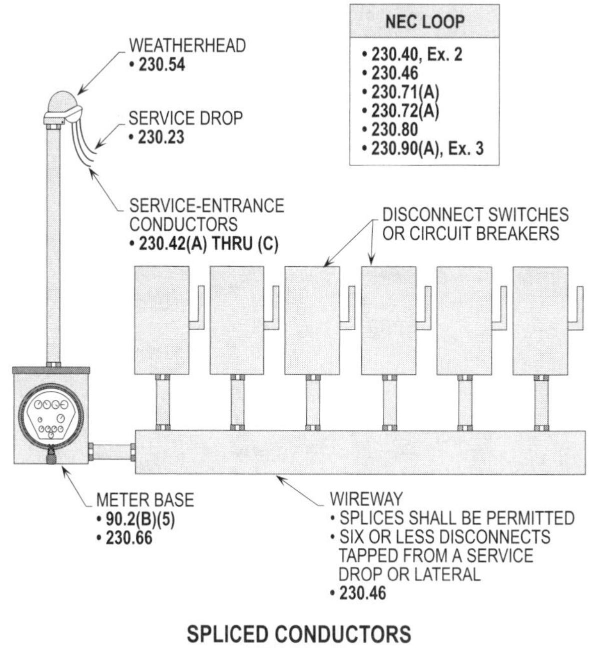

Figure 6-22. Service-entrance conductors shall be permitted to be tapped for two to six disconnecting means if grouped in a common location.

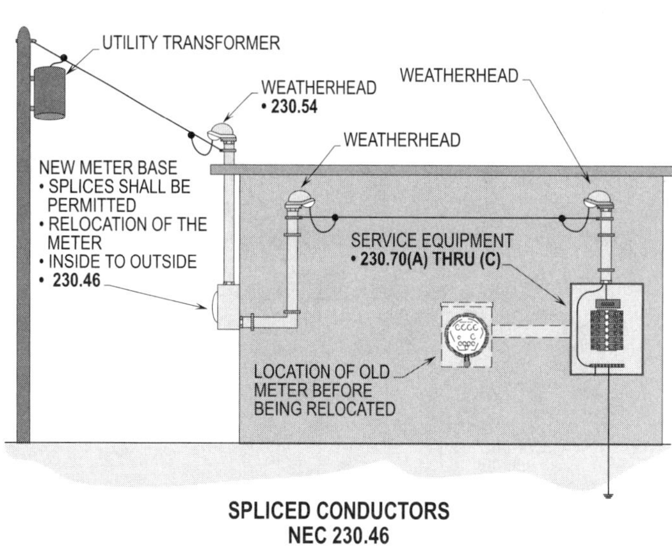

Figure 6-24. Service-entrance conductors shall be permitted to be spliced where an existing meter for a building is relocated from inside to the outside of the building.

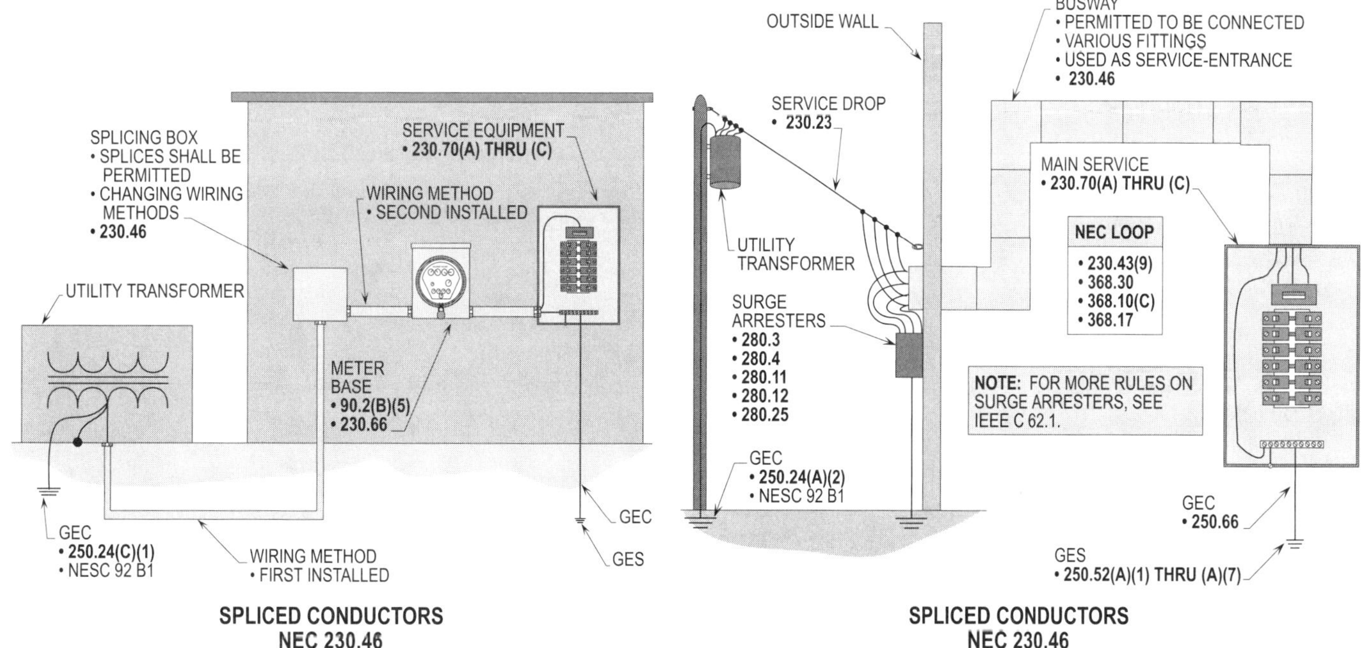

Figure 6-23. Service-entrance conductors shall be permitted to be spliced where splicing one wiring method to another.

Figure 6-25. Various sections and fittings shall be permitted to be installed for the connection and assembly of busways.

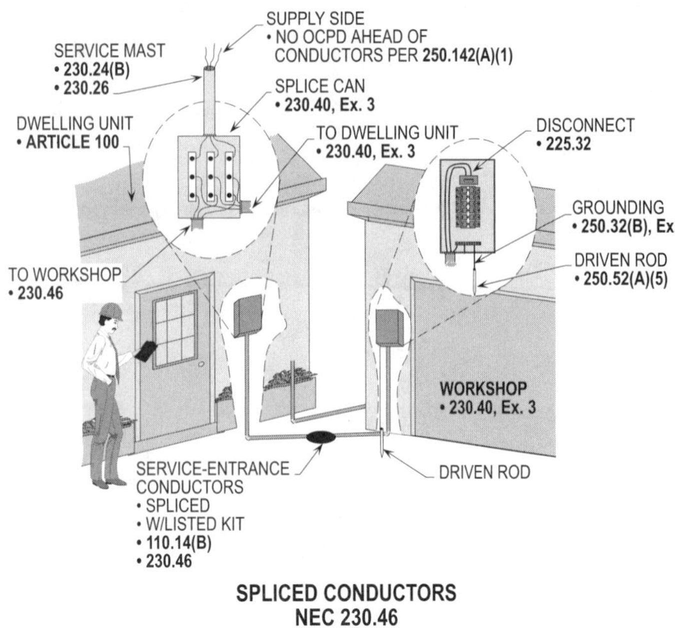

Figure 6-26. Listed underground splice kits shall be permitted to be installed for existing service-entrance conductors.

OTHER THAN SERVICE CONDUCTORS
230.50(B)

Open wiring on insulators shall be installed 10 ft (3 m) above finished grade or be protected where exposed to physical damage.

SERVICE CABLES
230.51(A)

Service-entrance cable shall be supported at 30 in. (750 mm) intervals and within 12 in. (300 mm) of the service weatherhead when supporting cable directly to the wall. **(See Figure 6-27)**

OTHER CABLES
230.51(B)

Other cables shall be supported at 15 ft (4.5 m) intervals on insulators if not approved to be supported directly to the wall. Other cables shall be installed with a 2 in. (50 mm) setoff clearance from the wall. **(See Figure 6-27)**

INDIVIDUAL OPEN CONDUCTORS
230.51(C)

Individual open wiring on insulators shall be supported at specific intervals when installed inside of a building. Open

wiring on insulators shall be spaced from the wall surface with a clearance to prevent the insulation from being damaged. Individually supported conductors shall have minimum clearance and support scheme complying with **Table 230.51(C)**. **(See Figure 6-28)**

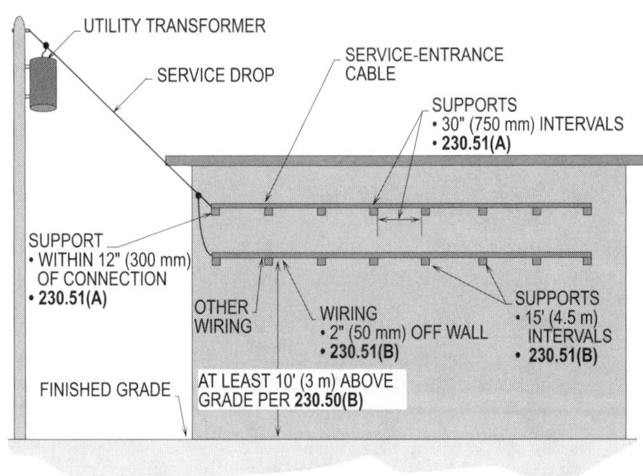

Figure 6-27. Service-entrance cables shall be supported within 30 in. (750 mm) intervals and within 12 in. (300 mm) of the service weatherhead when supporting cable directly to the wall. Other cables shall be supported at 15 ft (4.5 m) intervals on insulators if not approved to be supported directly to the wall. Other cables shall be installed with a 2 in. (50 mm) set off clearance from the wall.

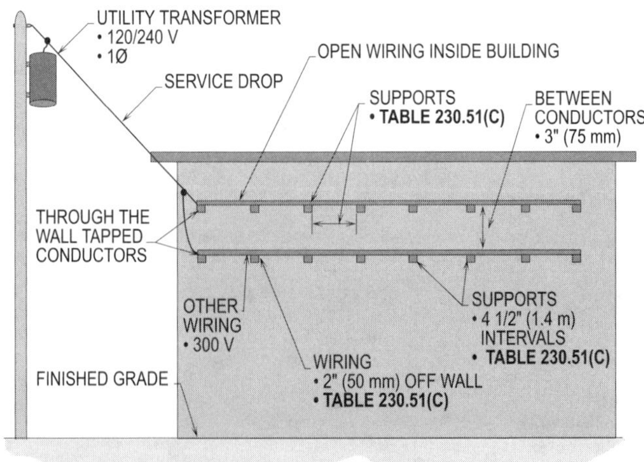

Figure 6-28. Individual open wiring on insulators shall be supported at specific intervals when installed inside a building.

INDIVIDUAL CONDUCTORS ENTERING BUILDINGS OR OTHER STRUCTURES
230.52

Roof bushings shall be installed where individual conductors enter a building through a roof. Nonabsorbent insulating tubes shall be installed where individual conductors enter a building through a wall. To prevent rain from entering the tubes, bushings shall be installed in a slanted direction and service-entrance conductors shall be provided with drip loops. **(See Figure 6-29)**

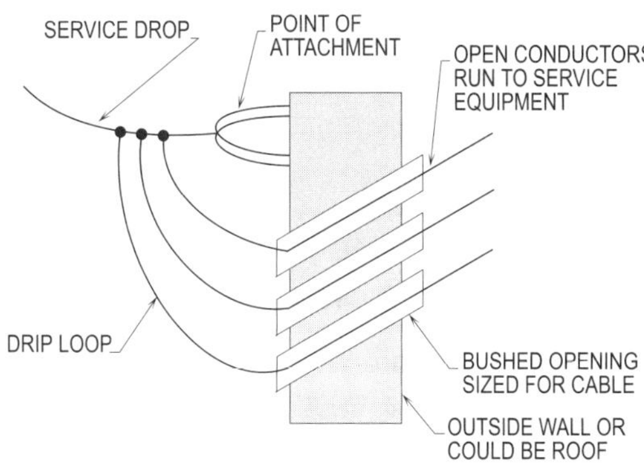

Figure 6-29. Roof bushings shall be installed where individual conductors enter a building through a wall.

RACEWAYS TO DRAIN
230.53

When installing rigid metal conduit, IMC, EMT, PVC, etc. all fittings shall be raintight, with a drain hole provided in the service ELL at the bottom of the run to protect service-entrance conductors. A means to drain condensation shall always be provided to prevent corrosion of metal. **(See Figure 6-30)**

OVERHEAD SERVICE LOCATIONS
230.54

When installing service cable or individual conductors in a raceway, the service entrance shall be installed above the service-drop conductors.

RAINTIGHT SERVICE HEAD
230.54(A)

If used, a raintight (weatherproof) service head shall be installed at the top of the service-entrance raceway.

Individual holes through the service head allow for the termination and protection of the service-drop and service-entrance conductors after they are spliced together. The fittings for the service head shall be placed or equipped so as to prevent moisture from entering or accumulating within.

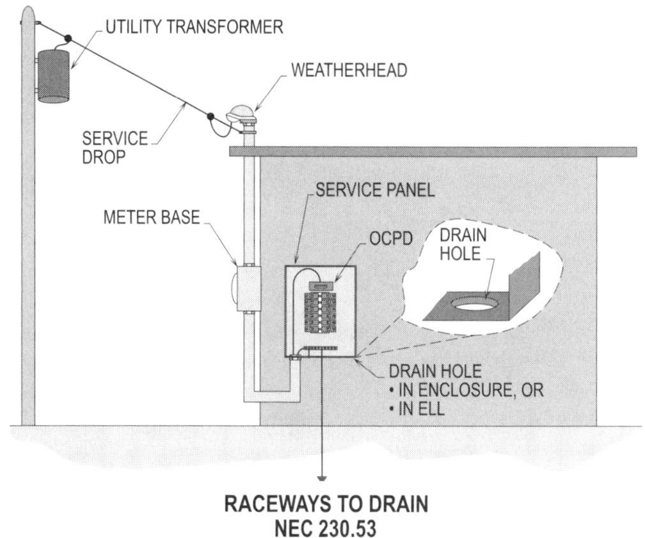

Figure 6-30. When installing rigid metal conduit, IMC, EMT, PVC, etc. all fittings shall be raintight, with a drain hole provided in the service ELL at the bottom of the run to protect service-entrance conductors.

SERVICE CABLE EQUIPPED WITH RAINTIGHT SERVICE HEAD OR GOOSENECK
230.54(B)

Service-entrance cable shall be installed in the form of a gooseneck and then taped and painted or plastic coated for protection from the weather. The entrance of rain through a service head shall be kept to a minimum. When installing weatherproof fittings, care shall be taken to make good tight connections and use sealing material properly. **(See Figure 6-31)**

SERVICE HEADS AND GOOSENECKS ABOVE SERVICE-DROP ATTACHMENT
230.54(C)

The service drop shall not be permitted to be installed above the service-entrance conductors where the overhang of the roof is too low and prevents proper installation of the service drop. The **Ex.** to **230.54(C)** requires the service head to be connected within 24 in. (600 mm) from the point of attachment when it is impracticable to locate it above the point of attachment.

Design Tip: When connecting the service drop to the point of attachment, a straight-line distance is used to determine the 24 in. (600 mm) distance. **[See Figures 6-32(a) and (b)]**

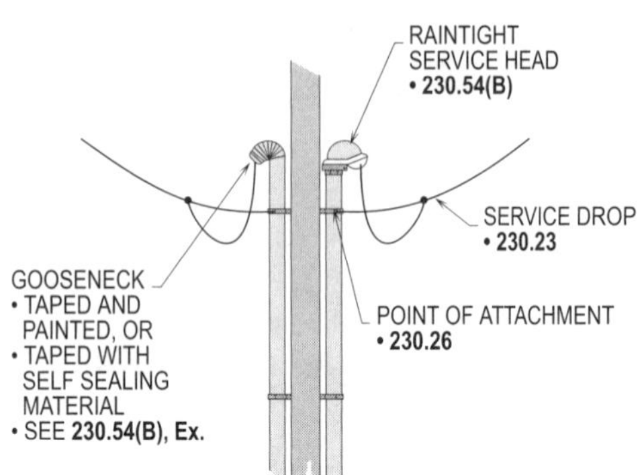

Figure 6-31. Service-entrance cable shall be installed in the form of a gooseneck and then taped and painted or plastic coated for protection from the weather.

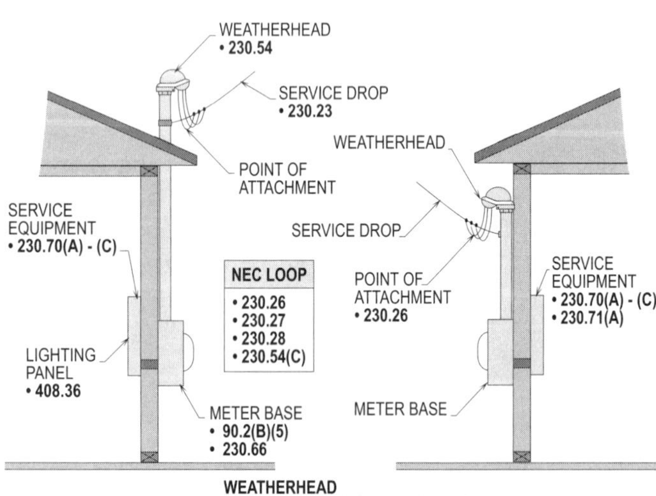

Figure 6-32(a). The service drop shall not be permitted to be installed above the service-entrance conductors where the overhang of the roof is too low and prevents proper installation of the service drop.

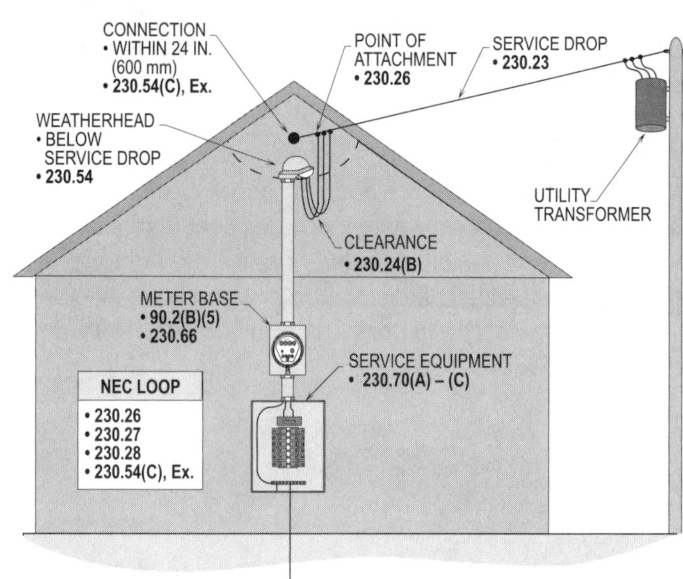

Figure 6-32(b). When connecting the service drop to the point of attachment, a straight-line distance is used to determine the 24 in. (600 mm) distance. **Note:** Get the approval of the AHJ for the above installation.

SERVICE EQUIPMENT TERMINATION (RECOMMENDED PROCEDURE)

Terminal boxes, cabinets, disconnecting switches, panelboards, switchgear, or similar enclosures shall be permitted to be installed for the termination of any service raceway or cable located in the building. Live parts of the service equipment shall be enclosed to protect personnel from injury. When using open-type switchboards, a terminating box shall not be required for the termination of service-entrance conductors. Conductors shall be protected with bushings where leaving the conduit.

SERVICE CONDUCTOR WITH THE HIGHER VOLTAGE-TO-GROUND 230.56

When installing a three-phase, four-wire, delta-connected service, where the midpoint of one phase winding is grounded, one phase leg will have a higher voltage-to-ground. An outer finish marked orange or identified by other effective means shall be used for the conductor at the higher voltage-to-ground. This identification will help prevent an electrician from connecting 120 volt loads to the 208 volts-to-ground (high-leg), since great damage may occur to 120-volt equipment if supplied by the high leg. **(See Figure 6-33)**

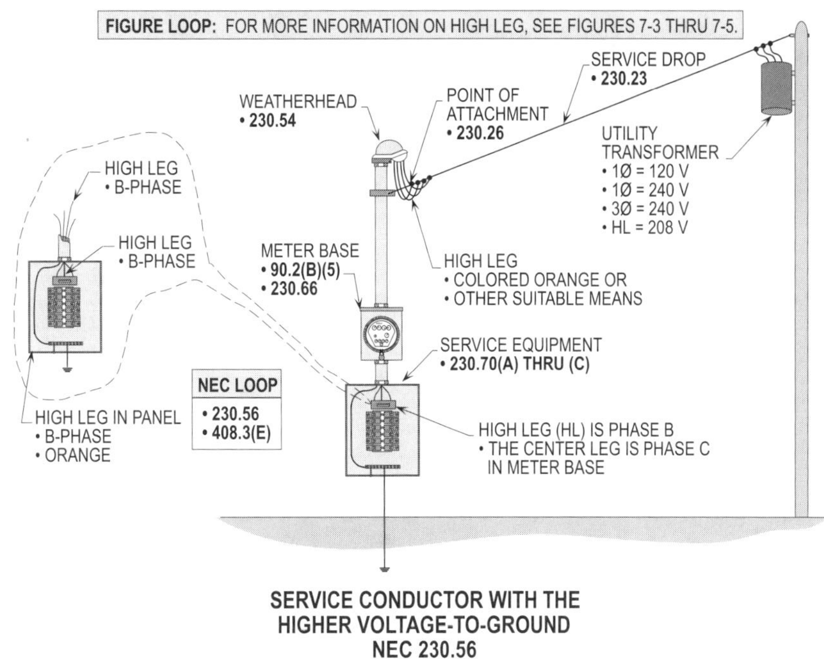

Figure 6-33. When installing a three-phase, four-wire, delta-connected service, where the midpoint of one phase winding is grounded, one phase leg will have a higher voltage-to-ground. An outer finish marked orange or identified by other effective means shall be used for the conductor with a higher voltage-to-ground.

SERVICE EQUIPMENT - GENERAL ARTICLE 230, PART V

Energized parts of service equipment shall be enclosed so that personnel are not exposed to accidental contact with live parts. Energized parts shall be guarded for the protection of personnel where not enclosed, since an electrical worker could make contact with live parts or create a short circuit or ground-fault condition. Service equipment shall be grounded and bonded to ensure the safe and effective operation of the electrical system and clear any circuit subjected to a short circuit.

SERVICE EQUIPMENT - ENCLOSED OR GUARDED
230.62

Energized parts of service equipment shall be enclosed so that personnel are not exposed to accidental contact with live parts while near or working on such equipment. Sections **110.27(A), 110.33,** and **110.34(A)** requires energized parts to be guarded to protect electrical personnel from dangerous arcs due to short-circuits and to the threat of electrical shock hazards. **(See pages 6 through 9 of Chapter 5)**

GROUNDING AND BONDING (RECOMMENDED ITEMS TO BE GROUNDED AND BONDED)

Service equipment and elements shall be grounded per **250.80**. The following elements of the service equipment shall be grounded in such a manner as to provide an effective grounding path:

- Service equipment
- Meter base housing
- Service disconnecting means and enclosure
- Metal sheath of cable
- Metal armor of cable
- Grounded (neutral) conductor
- Grounded (phase) conductor

(**See Chapter 11** for grounding rules)

AVAILABLE SHORT-CIRCUIT CURRENT
110.9 AND 110.10

Service equipment shall be designed and installed to withstand the available fault-current that could be delivered to service equipment terminals from the utilities or customer's transformer. Available fault currents are caused by phase-to-phase short circuits or phase-to-ground faults.

Circuit breakers, fuses, and other current-carrying elements shall be designed and installed to have an interrupting capacity equal to the short-circuit current that the equipment may be called upon to carry. Equipment shall be capable of handling and carrying high currents until the circuit is cleared. Current-limiting fuses or service-rated circuit breakers shall be permitted to be installed to help alleviate this problem. The let-through current of these fuses shall be limited to a value that will not damage the service equipment and its components.

Overcurrent protection devices shall be installed and selected to ensure proper interrupting rating and provide protection of components from the power company's transformer in case of a short-circuit condition. The overcurrent protection shall be designed and installed so that the interrupting rating has the ability to interrupt or open the circuit under short-circuit or ground-fault conditions.

The rating of the supply voltage and the resistance of the conductors, due to length and size, may limit the available fault current to a safe level if designed properly. A greater short circuit or ground fault will occur at the service equipment than a short circuit downstream on a feeder or branch circuit. Fault current is reduced by the accumulated resistance of the conductors and circuit elements located between the service and equipment. The point where the short circuit and ground fault occurs in the electrical system determines the amount of fault current that will flow.

Higher values of fault current will be produced for short runs of conductors to the service equipment. This would be the case if the transformer is located close to the building. Lower amounts of fault current will be produced, in most cases, to a safe value for long runs of conductors from the transformer to the service equipment. In this case, the transformer is located some distance away from the building and the service equipment. **(See Figure 6-34)**

MARKING
230.66

Service equipment rated at 600 volts or less shall be marked to identify it as suitable for use as service equipment. Individual meter socket enclosures shall not be considered as service equipment.

SERVICE EQUIPMENT –
DISCONNECTING MEANS
ARTICLE 230, PART VI

A means to disconnect the service shall be provided at each service. All ungrounded (phase) conductors shall be disconnected from the power supply. The service disconnecting means (one or more) shall be permanently marked, and each grouped in a common location when more than one service supplies a building.

LOCATION
230.70(A)

The service disconnecting means shall be located in a readily accessible location for all premises. The service disconnecting means shall be located as close as possible to where the service-entrance conductors enter the premises. Service disconnecting means shall not be permitted to be located in bathrooms. The service disconnecting means shall be located in accordance with **230.70(A)(1)** where a remote-control device(s) is used to activate the service disconnecting means. **(See Figure 6-35)**

> **Design Tip:** The service disconnecting means shall be permitted to be located on the outside or inside wall; such location complies with **230.70(A)**.

MAXIMUM NUMBER OF DISCONNECTS
230.71

No more than six fusible switches or six circuit breakers mounted in a single enclosure, in a group of separate enclosures, or in or on a panelboard or switchboard shall be installed for each set of service-entrance conductors. A power panel shall be permitted to have up to six circuit devices in a suitable enclosure. Six circuit breakers in a single enclosure shall be permitted to be used to supply air conditioners, water heaters, heating units, ranges, dryers, and general purpose circuits for lighting, receptacles, and appliances. **(See Figure 6-36)**

A main circuit breaker shall be permitted to be installed ahead of other circuit breakers in a panel, and one throw of the hand shuts the power OFF to all other circuit breakers and loads served. The disconnecting means for the service shall also be permitted to be six separate disconnecting switches with fuses or six separate circuit breakers, each in single enclosures.

> **Design Tip:** The entire service shall be designed so the service devices are shut OFF with six throws of the hand. **(See Figure 6-36)**

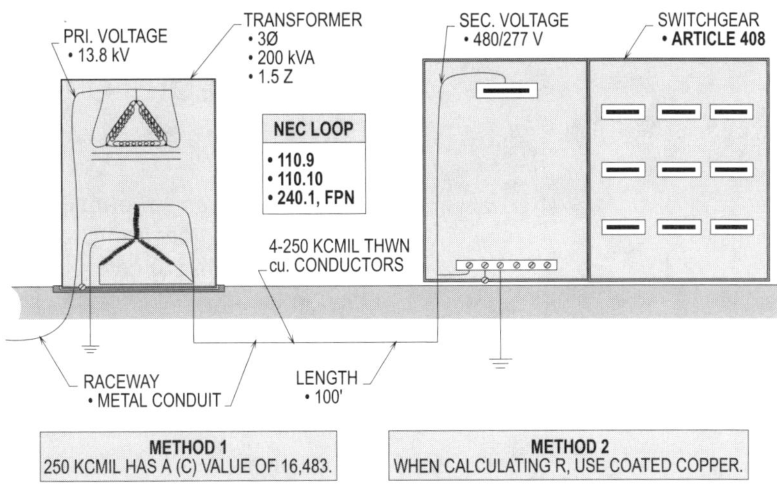

AVAILABLE SHORT-CIRCUIT CURRENT (ASCC) AT THE TERMINALS OF THE SERVICE EQUIPMENT

Calculating ASCC at service equipment using METHOD 1

Step 1: Calculating FLC of transformer
FLC = (kVA x 1000) ÷ (V x $\sqrt{3}$)
FLC = (200 kVA x 1000) ÷ (480 V x 1.732)
FLC = 241 A

Step 2: Calculating ASCC at transformer
ASCC = TLFC ÷ Z
ASCC = 241 A ÷ .015 (1.5%)
ASCC = 16,067 A

Step 3: Calculating F value using
manufacturer Table of C values
Table B of Appendix
F = 1.732 x L x I(ASCC) ÷ C x V(L TO L)
F = 1.732 x 100' x 16,067 ÷ 16,483 x 480
F = 2,782,804 ÷ 7,911,840
F − .352

Step 4: Calculating M value
M = 1 ÷ 1 + F
M = 1 ÷ 1 + .352
M = .740

Step 5: Calculating line to line
ASCC at service equipment
ASCC = ASCC x M
ASCC = 16,067 x .740
ASCC = 11,890 A

Solution: **The ASCC at the terminals of the
service equipment is 11,890 amps.**

AVAILABLE SHORT-CIRCUIT CURRENT (ASCC) AT THE TERMINALS OF THE SERVICE EQUIPMENT

Calculating ASCC at service equipment using METHOD 2

Step 1: Calculating ASCC of transformer
ACSS = TFLC ÷ Z
ASCC = 241 A ÷ .015 (1.5%)
ASCC = 16,067 A

Step 2: Calculating R of 100' of 250 KCMIL
Table A of Appendix
R = .054 x 100 ÷ 1,000
R = .0054

Step 3: Calculating R of transformer
R = V to GRD. ÷ ASCC
R = 277 V ÷ 16,067 A
R - .01724

Step 4: Calculating Total R
R of 250 KCMIL   = .0054
R of transformer   = .01724
Total R   = .02264

Step 5: Calculating ASCC of service equipment
ASCC = V to GRD. ÷ total R
ASCC = 277 V ÷ .02264
ASCC = 12,235 A

Solution: **The ASCC at the terminals of the
service equipment is 12,235 amps.**

**AVAILABLE SHORT-CIRCUIT CURRENT
NEC 110.9 AND 110.10**

Figure 6-34. Service equipment shall be designed and installed to withstand the available fault current that could be delivered to service equipment terminals from the utility's or customer's transformer.

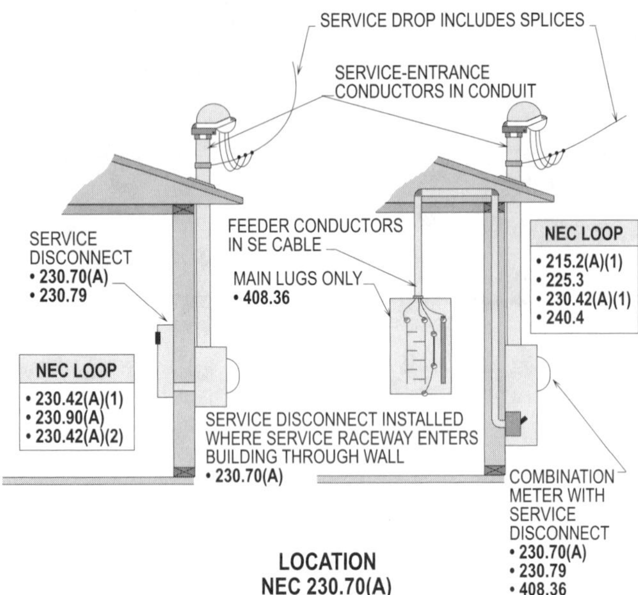

Figure 6-35. The service disconnecting means shall be located in a readily accessible location for all premises.

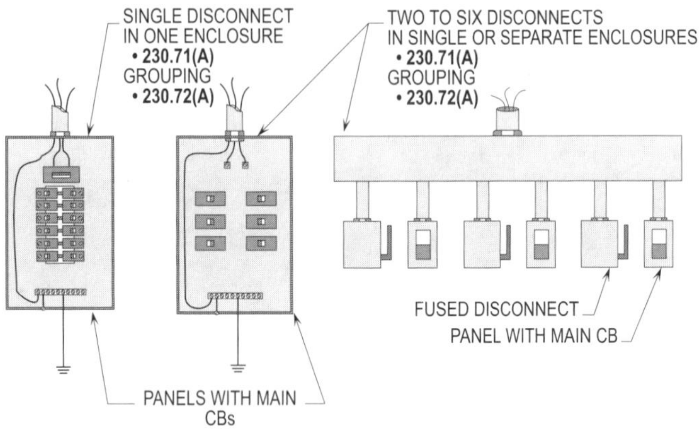

Figure 6-36. No more than six fusible switches or circuit breakers mounted in a single enclosure, in a group of separate enclosures, or in or on a panelboard or switchboard shall be installed for each set of service-entrance conductors.

GROUPING OF DISCONNECTS
230.72

The service disconnecting means shall be grouped in a common location and be readily accessible so the entire service may be disconnected with six throws of the hand. Dwelling units, apartment complexes, commercial facilities, or other types of premises shall be permitted to have their service designed with two to six disconnecting means. The disconnecting means shall be marked to indicate the load they serve. Disconnects shall be grouped in a common location where six or fewer are installed. A main disconnect shall be installed ahead of the disconnects if more than six disconnects are utilized. The following are two exceptions that allow switches to be ungrouped: **(See Figure 6-37)**

- Services installed per **230.2**
- Remote water pumps   .

SIMULTANEOUS OPENING OF POLES
230.74

The disconnecting means shall disconnect all service-entrance conductors from the power supply simultaneously. Ungrounded (phase) conductors shall be disconnected by a disconnecting switch or circuit breaker that is capable of disconnecting such conductors without rupturing. Grounded (neutral) conductors shall be permitted to be disconnected by the service-entrance disconnecting means. **(See Figure 6-38)**

DISCONNECTION OF GROUNDED CONDUCTOR
230.75

If the grounded (neutral) conductor is removed for any reason, a means shall be provided to disconnect the grounded (neutral) conductor safely. The disconnection of the grounded (neutral) conductor is usually removed from the terminal bar or bus manually by using proper tools designed for such purpose. Note, the grounded (neutral) conductor shall be permitted to be terminated in a different vertical section than the ungrounded (phase) conductors if the switchboard is of the multisection type. **(See Figure 6-39)**

RATING OF SERVICE DISCONNECTING MEANS
230.79

The main disconnecting means shall have a rating large enough to carry the calculated load per **Article 220**. In no case shall the rating be lower than as follows, per **Subdivisions (A), (B),** and **(C)** to **230.79**.

ONE CIRCUIT INSTALLATION
230.79(A)

The service disconnecting means shall be installed at a rating of 15 amps for single branch circuit loads supplying limited loads only.

TWO-CIRCUIT INSTALLATION
230.79(B)

The service disconnecting means shall be installed at a rating of 30 amps for one or two two-wire branch circuit loads. Note, these loads are small loads that require small amounts of current (amps) to operate.

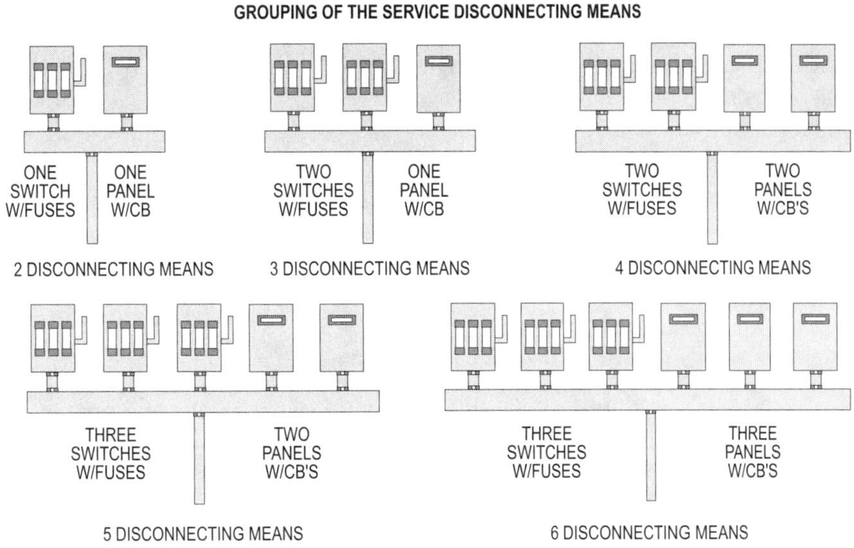

GROUPING OF THE SERVICE DISCONNECTING MEANS

ONE SWITCH W/FUSES | ONE PANEL W/CB

2 DISCONNECTING MEANS

TWO SWITCHES W/FUSES | ONE PANEL W/CB

3 DISCONNECTING MEANS

TWO SWITCHES W/FUSES | TWO PANELS W/CB'S

4 DISCONNECTING MEANS

THREE SWITCHES W/FUSES | TWO PANELS W/CB'S

5 DISCONNECTING MEANS

THREE SWITCHES W/FUSES | THREE PANELS W/CB'S

6 DISCONNECTING MEANS

GROUPING OF DISCONNECTS
NEC 230.72

Figure 6-37. The service disconnecting means shall be grouped in a common location and be readily accessible so the entire service may be disconnected with six throws of the hand.

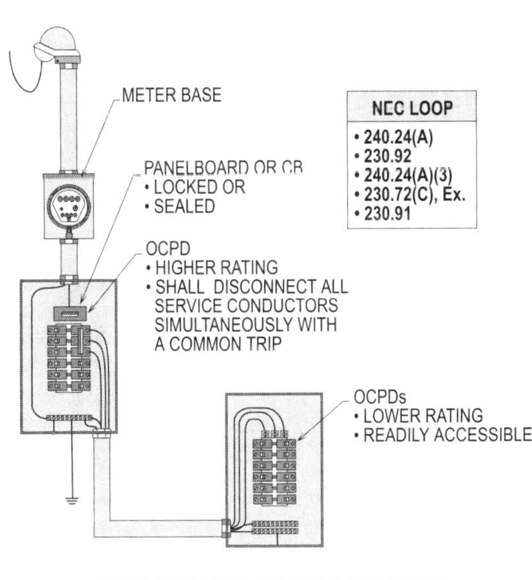

SIMULTANEOUS OPENING OF POLES
NEC 230.74

Figure 6-38. The disconnecting means shall disconnect all service-entrance conductors from the power supply simultaneously.

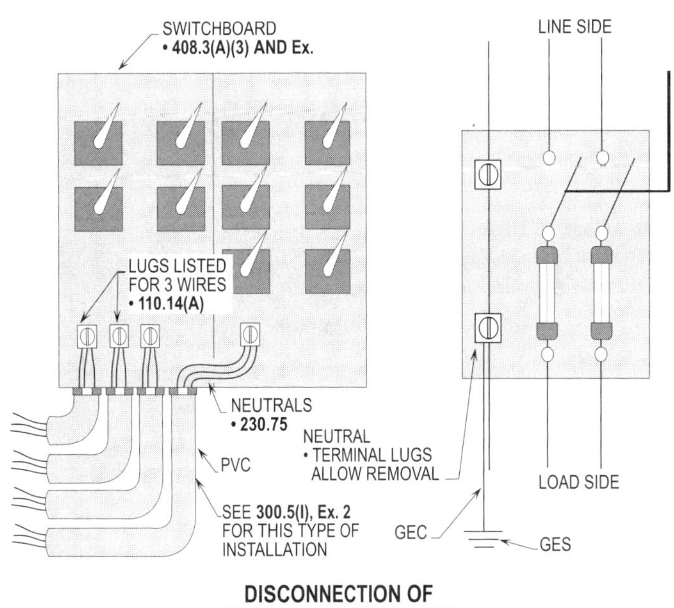

DISCONNECTION OF GROUNDED CONDUCTOR
NEC 230.75

Figure 6-39. If the grounded (neutral) conductor is removed for any reason, a means shall be provided to disconnect the grounded (neutral) conductor safely.

ONE-FAMILY DWELLING
230.79(C)

A three-wire service disconnecting means shall be installed at a rating of 100 amps for each dwelling unit. The service disconnecting switch shall be installed at a rating of at least 60 amps for all other installations below 100 amps. For applications of this rule, check with the AHJ.

COMBINED RATING OF DISCONNECTS
230.80

The combined rating for one to six switches or circuit breakers shall not be permitted to be less than the required rating for a single switch or circuit breaker. For further information, see **230.90(A), Ex. 3.**

For example, a number of conductors servicing six service disconnects has an ampacity of 1200 amps. This service shall be permitted to have six switches protected with 200 amp fuses each. Now, consider a number of conductors supplying six service disconnects that have a total ampacity of 765 amps. The total number of overcurrent protection devices (6) shall not be permitted to exceed 700 or 800 amps. (General Rule)

EQUIPMENT CONNECTED TO SUPPLY SIDE OF SERVICE DISCONNECT
230.82

The following are certain types of equipment that shall be permitted to be connected to the supply (line) side of the service disconnecting means:

- Cable limiters and other current-limiting devices
- Meters and meter sockets not rated over 600 volts with grounded and bonded enclosures
- Meter disconnect switches not rated over 600 volts with grounded and bonded enclosures
- Instrument transformers (current and voltage), high-impedance shunts, load management devices, surge arresters, and Type 1 surge-protective devices
- Taps for specific use
- Solar photovoltaic systems, fuel cell systems, or interconnected electric power production sources
- Control circuits
- Ground-fault protection systems or Type 2 surge protective devices

Design Tip: There should not be any overcurrent protection devices between the transformer, conductors, and loads served. Designers must not confuse this tap rule with **240.21** and **240.90,** which deal with conductors having an upstream overcurrent protection device.

TRANSFER EQUIPMENT RATED AS SERVICE EQUIPMENT

A transfer switch shall be used to disconnect the main service before transferring and connecting the alternate source of power. For further information, see **Article 100.** Transfer switches shall be permitted to be of the three-pole or four-pole type. If a three-pole transfer switch is used, the alternate source shall not be considered a separately derived system. When a four-pole transfer switch is used, because the grounded (neutral) conductor can be switched, the alternate source of power shall be considered a separately derived system. **(See Figure 6-40)**

NUMBER OF SUPPLIES
225.30

A separate disconnecting means shall be installed if more than one building is on the same property and under the same management. In other words, each separate building shall have a disconnecting means to completely disconnect all ungrounded (phase) conductors supplying or passing through such building or structure per **225.31. (See Figure 6-41)**

LOCATION
225.32

Each building shall have the disconnecting means installed in a location, either outside or inside, in or on each building in a readily accessible location and as close as possible where the conductors enter in the building. This disconnect shall be permitted to be provided with overcurrent protection or be of the nonfused or nonautomatic type.

SUITABLE FOR SERVICE EQUIPMENT
225.36

The disconnecting means shall be rated as a type suitable for use as service equipment. A set of three-way or four-way snap switches shall be permitted to be installed to disconnect garages or outbuildings on residential property. For further information regarding this rule, see **225.36, Ex.** and **250.32(A), Ex.**

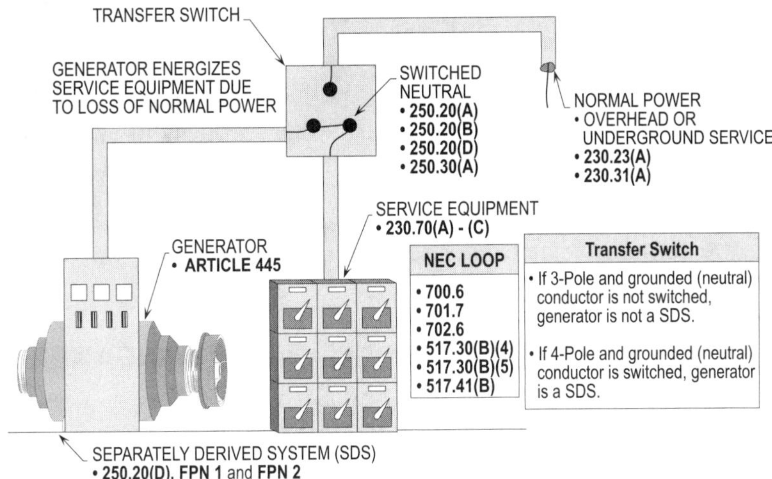

Figure 6-40. A transfer switch shall used to disconnect the main service before transferring and connecting the alternate source of power.

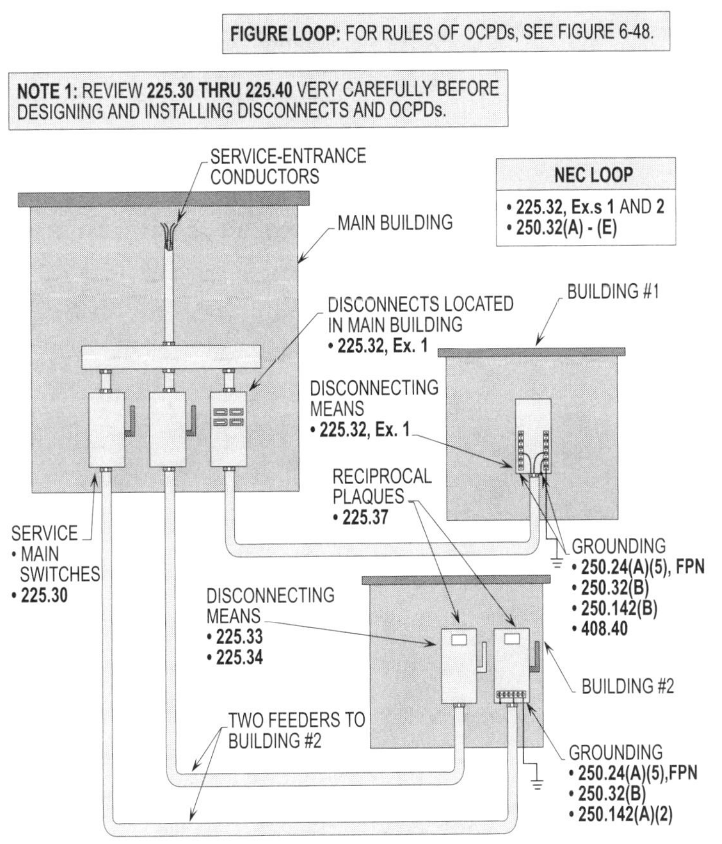

Figure 6-41. A separate disconnecting means shall be installed if more than one building is on the same property and under the same management. However, disconnects shall be permitted to be located in the main building per **Ex. 1** to **225.32**.

SERVICE EQUIPMENT – OVERCURRENT PROTECTION ARTICLE 230, PART VII

The general rule requires each ungrounded (phase) conductor to be provided with overcurrent protection devices to protect the service equipment. The rating of service-entrance conductors shall not be permitted to be exceeded when sizing overcurrent protection devices to protect the allowable ampacities of such conductors. However, there are exceptions that permit the overcurrent protection device to exceed the ampacity of the conductors.

UNGROUNDED CONDUCTOR
230.90(A)

The ampacity rating of service-entrance conductors shall not be permitted to be exceeded when installing overcurrent protection devices. A set of fuses shall be permitted where all the fuses are used to protect all ungrounded (phase) conductors of a circuit. Single-pole circuit breakers, grouped in accordance with **230.71(B)**, shall be considered one protective device. **(See Figure 6-42)**

MOTORS
230.90(A), Ex. 1

Motor conductors shall be permitted to be overfused for the start-up current per **430.52, 430.62**, and **430.63**. Overcurrent protection devices shall be sized to carry the high inrush current until the motor accelerates up to its running speed. **(See Figure 6-43)**

FUSES AND CIRCUIT BREAKERS
230.90(A), Ex. 2

When the setting or rating does not correspond to the allowable capacity of service conductors, the next higher standard fuse or circuit breaker shall be permitted to be installed. This type of installation shall be permitted to be installed per **240.4(B)**. The standard ratings of fuses or circuit breakers are listed in **240.6(A)**. **(See Figure 6-44)**

SIX CIRCUIT BREAKERS OR SIX SETS OF FUSES
230.90(A), Ex. 3

The number of overcurrent protection devices permitted for a service shall be one to six circuit breakers or six sets of fuses installed in a service-entrance panel. **Note,** it shall be permitted to have up to six separate enclosures with a single circuit breaker or single set of fuses in each. **(See Figure 6-45)**

For example, a 400 amp service entrance could be installed with six 70 amp circuit breakers. The total calculated rating would be 420 amps, which exceeds the 400 amp rating of the service entrance. The service shall be permitted to be protected at 420 amps per **240.4(B)**, but the ampacity of the service conductors is limited to the load, which shall not be permitted to exceed 400 amps per **230.42(A)** and **(B)**.

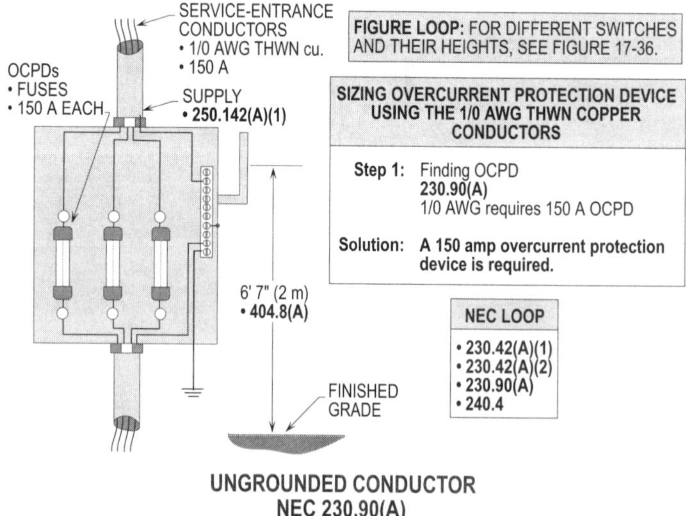

Figure 6-42. The ampacity rating of service-entrance conductors shall not be permitted to be exceeded when installing overcurrent protection devices.

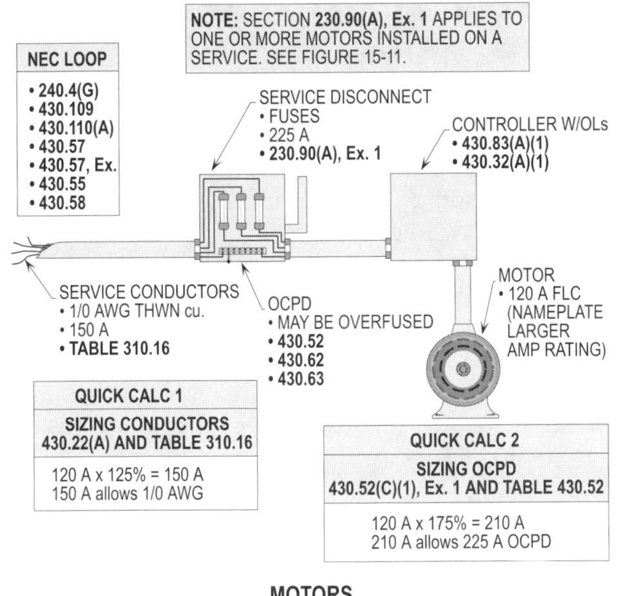

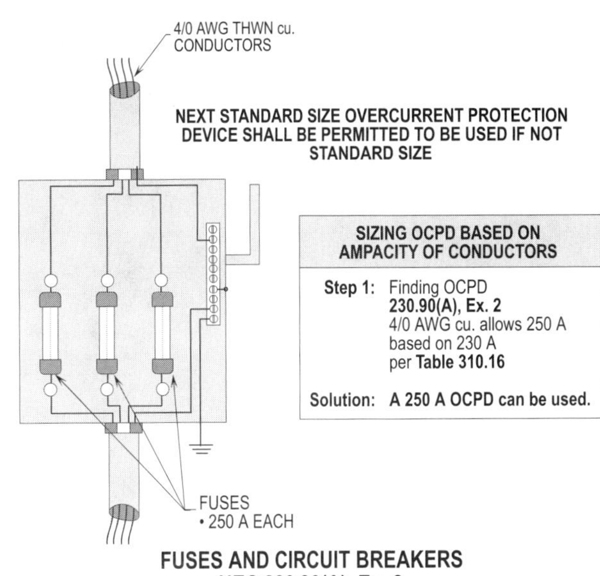

Figure 6-43. Overcurrent protection devices shall be sized to carry the high inrush current until the motor accelerates up to its running speed.

Figure 6-44. When the setting or rating does not correspond to the allowable capacity of service conductors, the next higher standard fuse or circuit breaker shall be permitted to be installed.

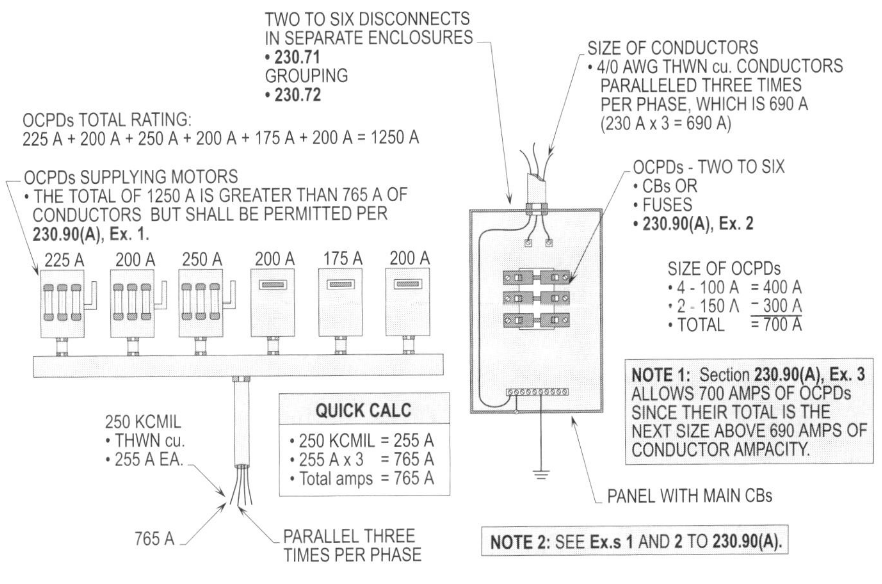

Figure 6-45. The number of overcurrent protection devices permitted for a service shall be one to six circuit breakers or six sets of fuses installed in a service-entrance panel or separate panelboards and switches.

FIRE PUMPS
230.90(A), Ex. 4

The overcurrent protection device shall be sized high enough to carry the locked-rotor current for a service supplying a fire pump. The overcurrent protection device shall not clear until a short circuit or ground fault develops. This designing technique allows the motor to pump water as long as possible to fight the fire. **(See Figure 6-46)**

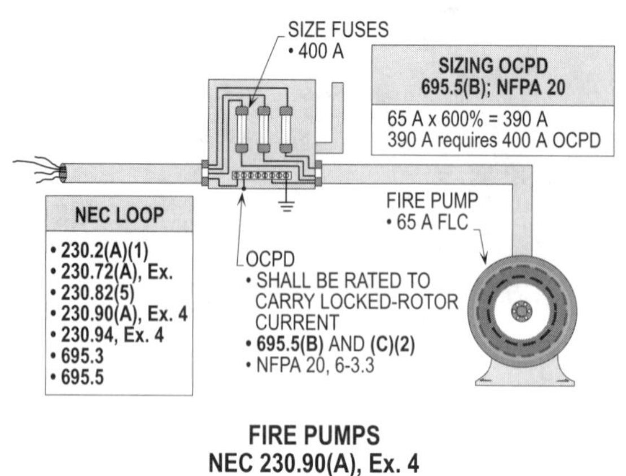

Figure 6-46. The overcurrent protection device shall be sized high enough to carry the locked-rotor current for a service supplying a fire pump.

TABLE 310.15(B)(6) TO AMPACITY TABLES OF 0 - 2,000 VOLTS
230.90(A), Ex. 5

This exception permits smaller-size service-entrance conductors to be installed than the calculated load of the dwelling unit. However, to obtain this smaller rating than calculated by the standard or optional method, 310.15(B)(6) to ampacity Tables 0-2000 volts shall be applied. **(See Figure 6-47)**

NOT IN GROUNDED CONDUCTOR
230.90(B)

Grounded (neutral) conductor or ungrounded (phase) conductors shall be permitted to be fused unless it is installed in a common trip circuit breaker. An individual fuse shall not be permitted to be installed for the grounded (neutral) conductor or ungrounded (phase) conductor, for all circuit conductors would not open at the same time if an individual fuse blew.

LOCATION
230.91

The service overcurrent protection device shall be an integral part of the service disconnecting means and located immediately adjacent to the service. A fused disconnecting switch or individual circuit breaker in a panel would comply with this requirement.

UNDER SINGLE MANAGEMENT
225.32, Ex. 1

Overcurrent protection devices shall be permitted to be grouped in one building where a property consists of more than one building or structure under single management per **240.24(B)(1)**. Feeders shall be permitted to be installed to supply power to each of the other buildings or structures, provided all occupants of that building have access to the disconnecting means to each building or structure. Overcurrent protection devices shall be permitted to be installed at the first building, with the disconnecting means installed at each of the other buildings or structures. **(See Figure 6-48)**

SERVICE AND FEEDER OVERCURRENT DEVICES
240.24(B)(1)

The main service-entrance conductors shall be permitted to be tapped without a main disconnecting means installed ahead of such disconnects. The disconnecting means and overcurrent protection devices for the service shall be provided with access for each occupant or user. However, the designer shall refer to **240.24(B)(1)**, which allows the overcurrent protection devices to be grouped in a single location under certain conditions of use. **(See Figure 6-48)**

LOCKED SERVICE OVERCURRENT DEVICES
230.92

Fuses and circuit breakers supplying an individual feeder or branch circuit and enclosures that are locked or sealed, additional fuses or circuit breakers shall be permitted to be installed, but they shall have a lower rating and shall be accessible to the user. The enclosure for fuses and circuit breakers shall be permitted to be locked in the closed position or individually locked, if the downstream overcurrent protection devices are readily accessible. The question is "readily accessible" to whom? For such interpretations, see **240.24(A)** and **(B)**, **230.72(C), Ex.**, and **Article 100**. **(See Figure 6-49)**

PROTECTION OF SPECIFIC CIRCUITS
230.93

A connection ahead of the meter shall be permitted to be installed ahead of the main to supply a flat-rate circuit for a water heater, heating equipment, or other such appliances. Overcurrent protection devices protecting the circuit shall be permitted to be locked to prevent access to unauthorized personnel. **(See Figure 6-50)**

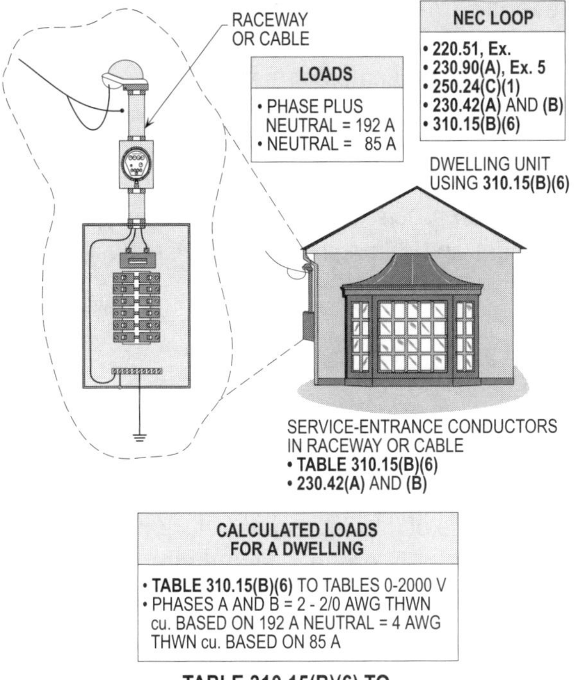

TABLE 310.15(B)(6) TO AMPACITY TABLES OF 0 - 2000 VOLTS NEC 230.90(A), Ex. 5

Figure 6-47. This exception permits smaller-size service-entrance conductors to be installed than the calculated load of the dwelling unit.

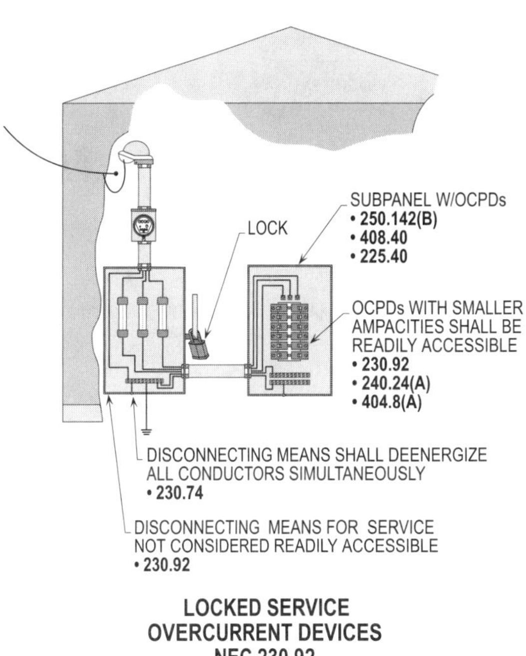

LOCKED SERVICE OVERCURRENT DEVICES NEC 230.92

Figure 6-49. Fuses and circuit breakers supplying an individual feeder or branch circuit located in enclosures that are locked or sealed, additional fuses or circuit breakers shall be permitted to be installed, but they shall have a lower rating and shall be accessible to the user.

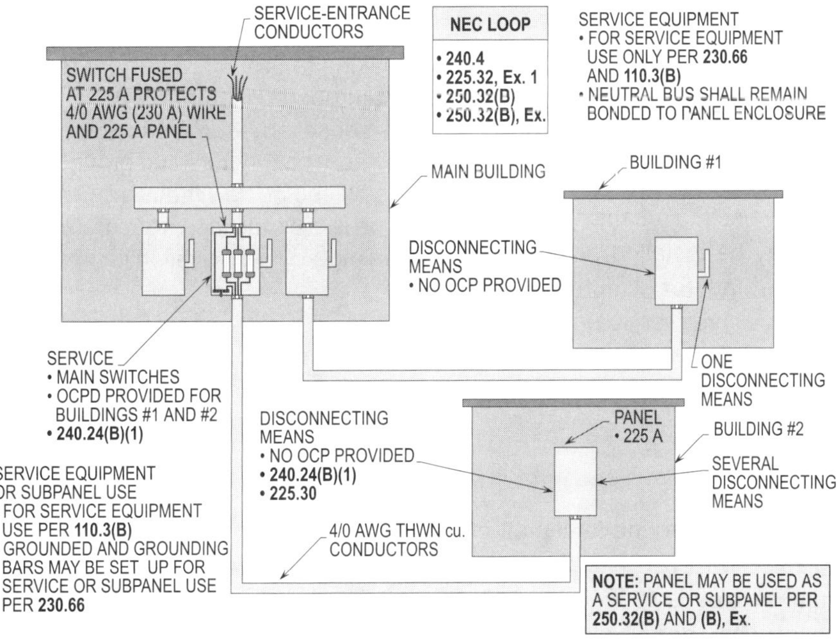

Figure 6-48. Overcurrent protection devices shall be permitted to be grouped in one building where a property consists of more than one building or structure under single management. (For disconnecting rules, see **Figure 6-41**)

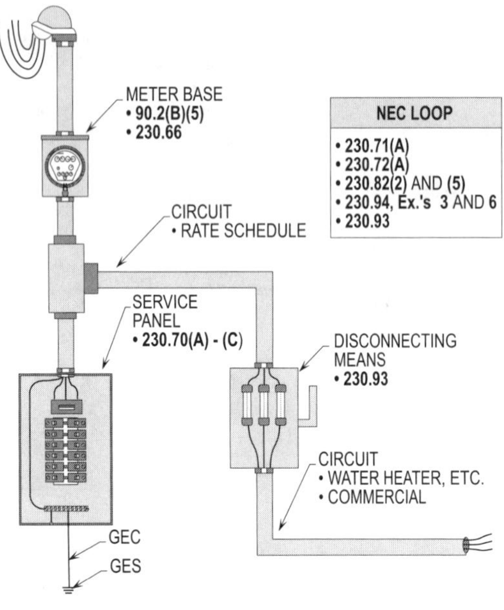

Figure 6-50. A connection ahead of the meter shall be permitted to be installed ahead of the main to supply a flat-rate circuit for a water heater, heating equipment, or other such appliances.

GROUND-FAULT PROTECTION OF EQUIPMENT
230.95

Solidly grounded wye electrical services of more than 150 volts-to-ground, having a service disconnecting means rated 1000 amps or more, must be protected by ground-fault protection. The service disconnecting means shall be rated for the largest fuse that may be installed or the highest continuous current trip setting for which the actual overcurrent device rating (circuit breaker) may be adjusted. **(See Figure 6-51)**

SETTING
230.95(A)

The maximum setting permitted for ground-fault protection shall not be permitted to exceed 1200 amps, when the service disconnecting means exceeds 1000 amps or more. The ground-fault protection shall permit all ungrounded (phase) conductors of the faulted circuit to open, until such circuit(s) is clear of the low-magnitude fault.

> **Design Tip:** Ground-fault protection shall not be required for six overcurrent protection devices that are not equal to or greater than 1000 amps.

To prevent damage to overcurrent protection devices, busbars, and service equipment, ground-fault systems shall clear a fault of 3000 amps or more within one second.

The system will be cleared once the low-magnitude fault currents are detected and the overcurrent protection device is tripped open. Such trip will prevent low-magnitude faults from burning and damaging the electrical system beyond use.

FUSES
230.95(B)

When installing a switch and fuse combination, the fuses shall be capable of interrupting any current higher than the interrupting capacity of the switch during a time until the ground-fault protective system causes the switch to open and clear the faulted circuit.

PERFORMANCE TESTING
230.95(C)

The following two types of ground-fault detection are used to sense low-magnitude ground-faults:

- Donut method
- Window method

When applying the donut method to sense any current leaking to ground, a current relay shall be placed over the main bonding jumper. When applying the window method to sense any current leaking to ground, the ungrounded (phase) conductors and grounded (neutral) conductors shall be placed in the window (ground-fault sensor). The window type is the most accurate method because it reads the most leakage current available.

> **Design Tip:** The system will be cleared once the low-magnitude fault currents are detected and the overcurrent protection device is tripped open. It is important that the sensor of the relay be set correctly.

SERVICE EQUIPMENT RATED OVER 600 VOLTS, NOMINAL
ARTICLE 230, PART VIII

High-voltage supply systems are installed to reduce the cost of conductors, raceways, elements, and equipment. Low-voltage supply systems require a greater cost of installation. The current values will decrease if the voltage applied is raised. The current values will increase, if the voltage applied is lowered. A high-voltage supply system installed for a large building will decrease the cost of installation by using small conductors, raceways, elements, and equipment. A low-voltage supply system installed for a small facility will increase the cost of installation for conductors, raceways, elements, and equipment. Load requirements determine the voltage level selected to supply power to the service equipment and downstream panelboards.

SERVICE POINT
ARTICLE 100

The service point is the point of connection between the facilities of the serving utility and the premises wiring. The service point is located ahead of the service equipment and is installed on the line side of the meter housing and service equipment. This part of the service equipment is located on the service side of the service point. **(See Figure 6-52)**

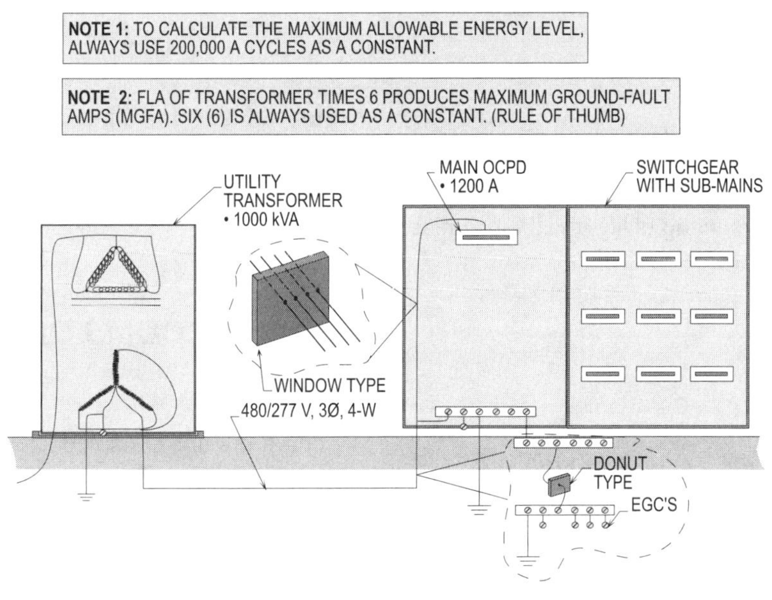

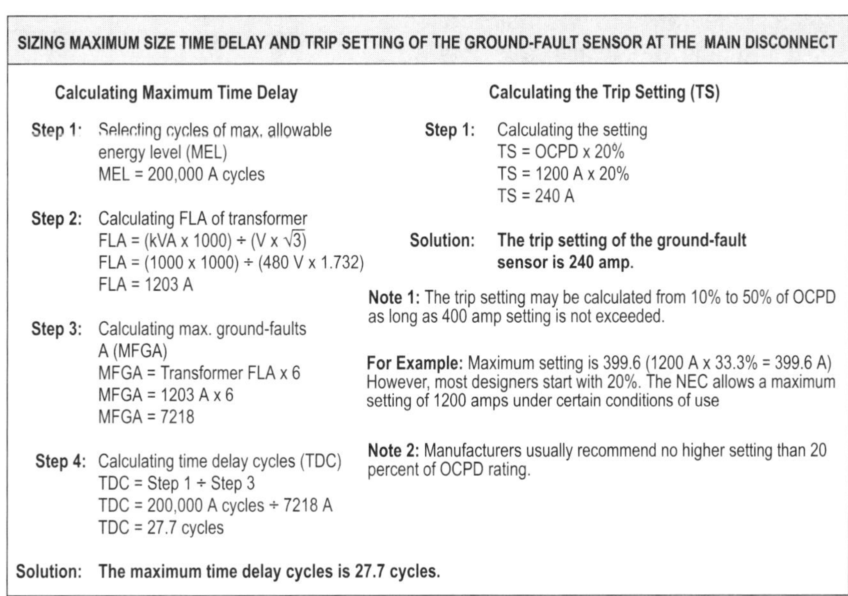

GROUND-FAULT PROTECTION OF EQUIPMENT
NEC 230.95

Figure 6-51. Solidly grounded wye electrical services of more than 150 volts-to-ground having a service disconnecting means rated 1000 amps or more shall be protected by ground-fault protection.

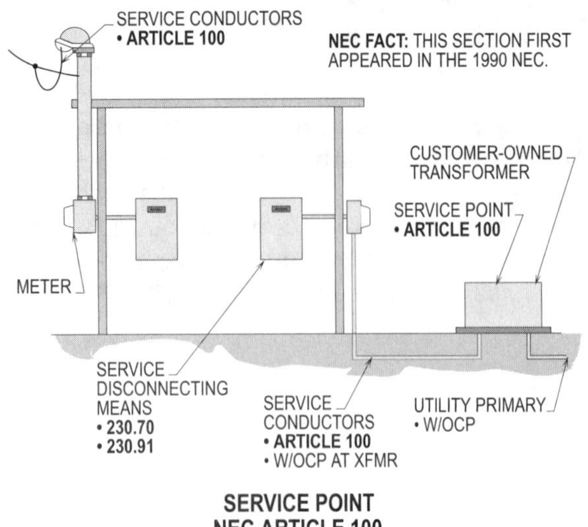

Figure 6-52. The service point is the point of connection between the facilities of the serving utility and the premises wiring.

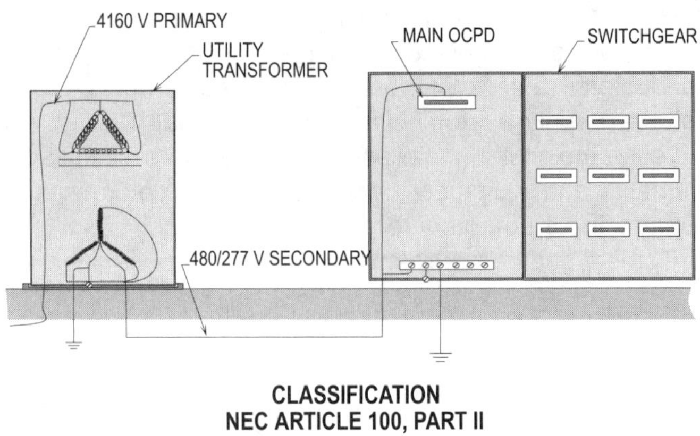

Figure 6-53. High-voltage primaries for buildings requiring large services are generally used with step-down transformers to supply loads located at various places in the building.

CLASSIFICATION
ARTICLE 100, PART II

High-voltage primaries for buildings requiring large services are generally used with step-down transformers to supply loads located at various places in the building. Lighting, receptacles, and other types of loads are supplied by step-down transformers. If the transformer belongs to the owner, the point of service is located at the transformer if installed outside or within a building per **Article 100**. In this case, the supply conductors to the primary side of the transformer belong to the utility company. The service conductors are the conductors installed between the service point and the first disconnecting means for the building or structure. **(See Figure 6-53)**

LOCKED ROOMS OR ENCLOSURES
110.34(C)

A permanent and conspicuous warning sign shall be posted where the voltage exceeds 600 volts. The signs shall be posted to prevent unauthorized personnel from coming in contact with energized parts. The sign shall read as follows:

DANGER - HIGH VOLTAGE - KEEP OUT

Design Tip: The sign shall have a warning of "High Voltage" and give a command "Keep Out" or "Authorized Personnel Only," etc. **(See Figure 6-54)**

WIRING METHODS
230.202(B)

The following wiring methods shall be permitted to be installed for high-voltage service-entrance conductors to protect them from physical damage per **300.37** and **300.50**:

- Rigid metal conduit (RMC)
- Intermediate metal conduit (IMC)
- Electrical metallic tubing (EMT)
- Rigid nonmetallic conduit (RNC)
- Cable trays
- Busways
- Cablebus
- Other identified raceways
- Type MV cables

ISOLATING SWITCHES
230.204

Isolating switches shall be installed for certain types of disconnecting means. Isolating switches shall be installed to open the circuit when there is no load. Equipment on the load side shall be permitted to be serviced by personnel without danger of the circuit being energized. **(See Figure 6-55)**

Design Tip: Isolating switches shall not be required to be installed for rollout or pullout equipment from switchgears or panel enclosures.

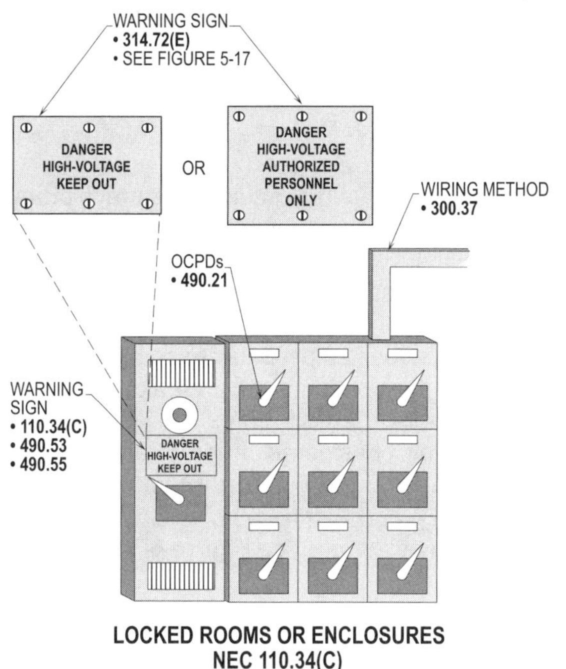

Figure 6-54. A permanent and conspicuous warning sign shall be posted where the voltage exceeds 600 volts.

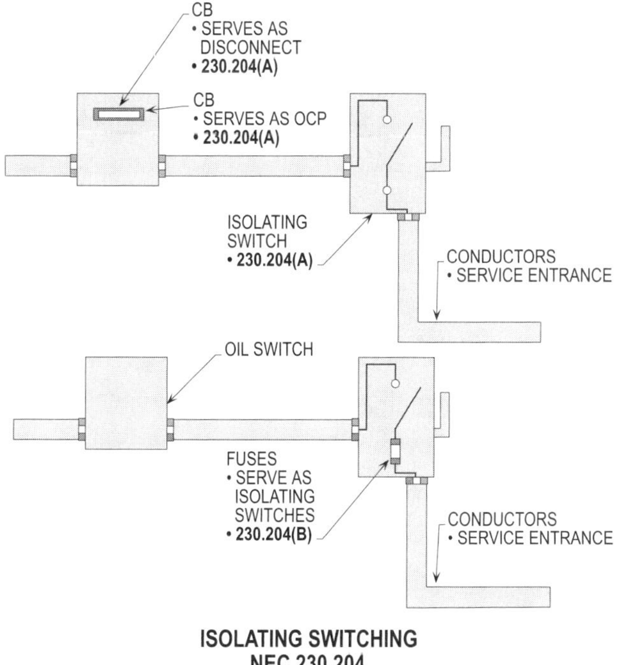

Figure 6-55. Isolating switches shall not be required to be installed for certain types of disconnecting means.

DISCONNECTING MEANS
230.205

All ungrounded (phase) conductors shall be disconnected simultaneously (common trip) by the service disconnecting means. The service disconnecting means shall be capable of being closed on a fault not less than the maximum available short-circuit current imposed on the terminals of the service equipment.

The characteristics of a fuse may contribute to the fault closing rating of the disconnecting means and serve as an aid for such closing. The disconnecting means shall be permitted to be installed ahead of or near the defined service point per **Article 100** in the NEC.

PROTECTION REQUIREMENTS
230.208

Service conductors operating at over 600 volts shall be provided with an overcurrent protection device in each ungrounded (phase) conductor. Short-circuit conditions shall be protected by properly sized overcurrent protection devices. Overload conditions shall be protected against by adding a second stage of overcurrent protection.

Overcurrent protection devices shall be permitted to be installed on the line side of the service disconnecting means or as an integral part of the service disconnecting means.

A fuse shall not exceed three (300 percent) times the ampacity of the continuous rating of the conductor to be installed. A circuit breaker shall have trip settings not to exceed six (600 percent) times the ampacity of the continuous rating of the conductor to be installed. [Also see **490.21(A) through (E)**]

High-voltage supply systems have different characteristics from low-voltage supply systems. A short circuit will clear much faster for a high-voltage system than for a low-voltage system. Percentages shall be permitted to be selected as listed above; higher percentages shall be permitted to be used to size overcurrent protection devices to protect service conductors from short-circuit conditions.

The higher voltage will clear a phase-to-phase or phase-to-ground short circuit very rapidly. This is not always the case when using lower voltage systems rated at 600 volts or less. **(See Figure 6-56)**

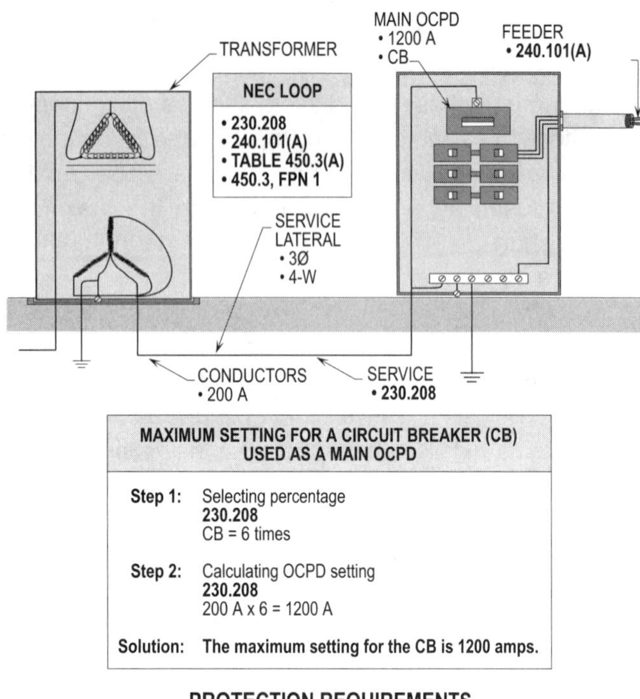

MAXIMUM SETTING FOR A CIRCUIT BREAKER (CB) USED AS A MAIN OCPD

Step 1:	Selecting percentage **230.208** CB = 6 times	
Step 2:	Calculating OCPD setting **230.208** 200 A x 6 = 1200 A	
Solution:	**The maximum setting for the CB is 1200 amps.**	

PROTECTION REQUIREMENTS
NEC 230.208

NOTE: SEE **490.21(A) thru (E)** FOR OVERLOAD PROTECTION PROCEDURES.

MAXIMUM RATING OF THE FUSES

Step 1:	Selecting percentage **230.208** Fuses = 3 times	
Step 2:	Calculating OCPD setting **230.208** 200 A x 3 = 600 A	
Solution:	**The rating of each fuse is 600 amps.**	

PROTECTION REQUIREMENTS
NEC 230.208

Figure 6-56. Service conductors operating at over 600 volts shall be provided with an overcurrent protection device in each ungrounded (phase) conductor.

OVERCURRENT PROTECTION AND REQUIREMENTS FOR FEEDER CONDUCTORS 240.100 AND 240.101

Feeders shall have a short-circuit overcurrent protection device in each ungrounded (phase) conductor. The protection device(s) shall be capable of detecting and interrupting all values of current that may occur at their location in excess of their trip setting or melting point. A fuse rated in continuous amperes not exceeding three times the ampacity of the conductor or a breaker having a trip setting of not more than six times the ampacity of the conductor shall be considered as providing the required short-circuit protection. For further information, see **490.21(A)** and **(B)** and **Article 490,** which deal with high-voltage electrical systems. **(See Figure 6-57)**

> **Design Tip:** The operating time of the protective device, the available short-circuit current, and the conductor used will need to be coordinated to prevent damaging or dangerous temperatures in conductor insulation under short-circuit or ground-fault conditions.

Branch circuits shall have a short-circuit protective device in each ungrounded (phase) conductor or comply with **490.21(A)** and **(B)**. The protective device(s) shall be capable of detecting and interrupting all values of current that may occur at their location in excess of their trip setting or melting point per **240.100(A)**. In other words, the overcurrent protection devices shall be sized to protect the conductors at their ampacities if not oversized to allow equipment with high-inrush current to start, run, and operate.

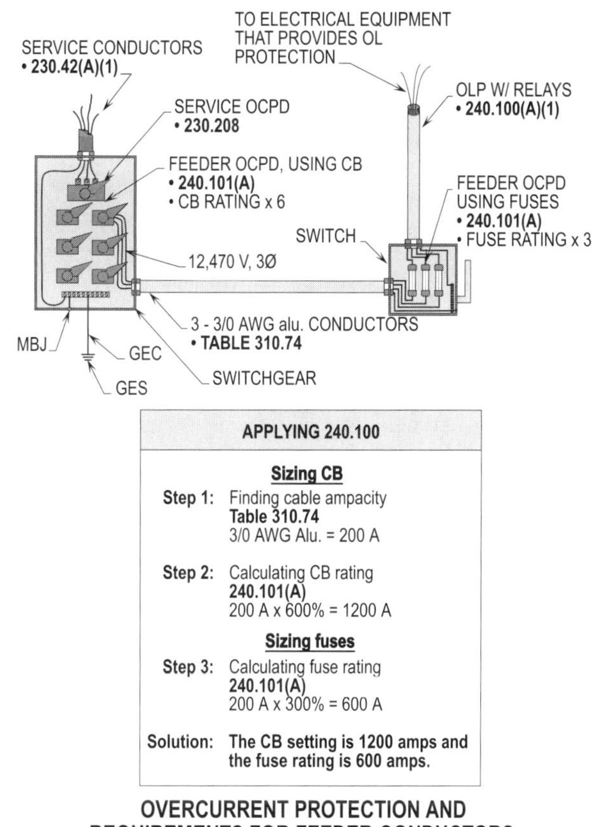

Figure 6-57. The above illustration shows the correct procedure for sizing circuit breakers and fuses to be used to protect high-voltage electrical conductors.

Chapter 6. Services

Section   Answer

1. Service-drop conductors shall be permitted to be ______ AWG hard-drawn copper where installed to serve small loads of only one branch circuit, such as a phone booth.
   (a) 14   (b) 12
   (c) 10   (d) 8

2. Service-drop conductors crossing property where people have access shall have a clearance of at least ______ ft from finished grade.
   (a) 10   (b) 12
   (c) 15   (d) 18

3. The minimum point of attachment for service-drop conductors shall be ______ ft from finished grade.
   (a) 10   (b) 12
   (c) 15   (d) 18

4. At least a ______ amp service disconnecting means shall be installed for a single-family dwelling.
   (a) 50   (b) 60
   (c) 100   (d) 125

5. Individual open wiring on insulators shall be installed ______ ft above finished grade or be protected where exposed to physical damage.
   (a) 6   (b) 8
   (c) 10   (d) 12

6. When installing a three-phase, four-wire, delta-connected service where the midpoint of one phase winding is grounded, an outer finish marked ______ or identified by other effective means shall be used for the conductor with a higher voltage-to-ground.
   (a) red   (b) yellow
   (c) blue   (d) orange

7. A high-voltage fuse shall not be permitted to exceed ______ times the ampacity of the continuous rating of the conductor to be installed.
   (a) 2   (b) 3
   (c) 4   (d) 6

8. Service conductors shall be considered outside the building when installed in a conduit and encased in at least ______ in. of concrete or brick.
   (a) 2   (b) 4
   (c) 6   (d) 8

9. A clearance of ______ ft shall be required from windows designed to be opened, including porches, platforms, etc.
   (a) 2   (b) 3
   (c) 5   (d) 6

10. Service-drop conductors shall not be permitted to be smaller than _____ copper or _____ aluminum.
    (a) 10; 8   (b) 8; 10
    (c) 8; 6   (d) 6; 8

11. Service-drop conductors shall be permitted to pass above roofs where voltages between conductors does not exceed 300 volts and are installed at a height of 3 ft, where the roof has a slope of at least _____ in. by _____ in.
    (a) 4; 10   (b) 4; 12
    (c) 6; 10   (d) 6; 12

12. Service-drop conductors shall be permitted to be installed for overhanging roofs provided no more than _____ ft of such conductors does not pass over more than _____ ft of the roof.
    (a) 6; 4   (b) 8; 4
    (c) 10; 6   (d) 8; 6

13. Service-entrance cable shall be supported within _____ in. intervals and within _____ in. of the service weatherhead when supporting cable directly to the wall.
    (a) 20; 6   (b) 30; 6
    (c) 20; 12   (d) 30; 12

14. Other cables shall be supported at _____ ft intervals on insulators if not approved to be supported directly to the wall.
    (a) 10   (b) 12
    (c) 15   (d) 20

15. The service disconnecting means shall be grouped in a common location and be readily accessible so the entire service can be disconnected with _____ throws of the hand.
    (a) 6   (b) 8
    (c) 10   (d) 12

16. The service disconnecting means shall be installed at a rating of _____ amps for single branch-circuit loads supplying limited loads.
    (a) 10   (b) 15
    (c) 20   (d) 30

17. The service disconnecting means shall be installed at a rating of _____ amps for one or two two-wire branch-circuit loads.
    (a) 10   (b) 15
    (c) 20   (d) 30

18. Solidly grounded wye electrical services of more than 150 volts-to-ground having a service disconnecting means rated _____ amps or more shall be protected by ground-fault protection.
    (a) 600   (b) 1000
    (c) 1200   (d) 1500

19. The maximum setting for ground-fault protection shall not be permitted to exceed _____ amps when the service disconnecting means exceeds 1000 amps or more.
    (a) 600   (b) 1000
    (c) 1200   (d) 1500

20. A circuit breaker shall have a trip setting not exceeding _____ times the ampacity
of the continuous rating of the conductor to be installed. (Over 600 volts)
    (a) 3   (b) 4
    (c) 6   (d) 8

21. Load requirements in a facility that are in excess of _____ amps shall be
permitted to be supplied with more than one service.
    (a) 500   (b) 1000
    (c) 1500   (d) 2000

22. A clearance of _____ ft shall be required for service-drop conductors passing
over roofs. At least a _____ ft clearance shall be maintained in all directions
for vertical clearances.
    (a) 8; 3   (b) 8; 6
    (c) 10; 3   (d) 10; 6

23. Service-drop conductors shall have a vertical clearance of at least _____ ft
above flat roofs that are accessible to pedestrians.
    (a) 6   (b) 7
    (c) 8   (d) 10

24. Service-drop conductors shall be permitted to be installed for overhanging
roofs provided the attachment of the service drop has a clearance of at least
_____ in. from the conduit and roof line.
    (a) 6   (b) 12
    (c) 18   (d) 24

25. Service-drop conductors crossing residential property or driveways shall have
a clearance of _____ ft from finished grade.
    (a) 10   (b) 12
    (c) 15   (d) 18

26. Service-drop conductors crossing residential property or driveways shall have
a clearance of _____ ft from finished grade where the voltage exceeds 300
volts-to-ground.
    (a) 10   (b) 12
    (c) 15   (d) 18

27. Service-drop conductors crossing public streets, alleys, roads, and parking
areas subject to traffic shall have a clearance of _____ ft from finished grade.
    (a) 10   (b) 12
    (c) 15   (d) 18

28. A minimum _____ amp, three-wire service disconnecting means shall be
installed for a single family dwelling.
    (a) 60   (b) 100
    (c) 150   (d) 200

29. Other cables used for servce conductors shall be installed with a _____ in.
setoff clearance from the wall.
    (a) 2   (b) 6
    (c) 10   (d) 12

_____________ _____________

30. No more than ______ fusible switches or circuit breakers mounted in a single enclosure, in a group of separate enclosures, or in or on a panelboard or switchboard shall be permitted to be installed for each set of service-entrance conductors.

    (a) 2   (b) 4

    (c) 6   (d) 10

Switchboards and Panelboards

Panelboards are intended to be installed and mounted in cabinets or cutout boxes placed "in or against a wall or partition." Panelboards are used for the control of small-capacity circuits.

Switchboards are not intended to be installed and mounted in cabinets or cutout boxes. Switchboards usually are installed to stand on the floor. Switchboards are used for the control of high-capacity circuits.

The space inside enclosures shall have sufficient space to terminate conductors to lugs of overcurrent protection devices or busbars. This space shall protect the conductors and equipment from physical damage when they are being terminated.

Table 312.6(A) shall be utilized to provide the proper space to terminate conductors formed into a L-bend configuration.

Table 312.6(B) shall be utilized for conductors that are formed into a S- or Z-bend configuration.

The minimum wire-bending space at terminals and correct width of wiring gutters in inches shall be determined by the number of conductors connected to each terminal per **408.55, Ex.s 1 through 4**.

SWITCHBOARD AND PANELBOARDS
ARTICLE 408, PARTS II AND III

Switchboards are designed with busbars for fuses or circuit breakers that are used for protecting and supplying feeders and branch circuits. Panelboards are designed with a permanently installed main circuit breaker or single set of fuses for disconnecting the number of circuit breakers or fuseholders installed to protect and supply feeders or branch circuits.

SUPPORT AND ARRANGEMENT OF BUSBARS AND CONDUCTORS
408.3

Busbars shall be held firmly in place in the enclosure and shall be arranged in such a way to allow adequate space for conductors to be terminated. Busbars shall be sized with enough capacity to handle the normal currents in amps of the electrical system. In addition, they shall carry safely any type of short-circuit or ground-fault condition.

CONDUCTORS AND BUSBARS ON A SWITCHBOARD OR PANELBOARD
408.3(A)

Conductors and busbars on a switchboard, panelboard, or control board shall be so located as to be free from physical damage. Service sections of such boards shall be separated from the other electrical components. These components shall be properly secured in an approved manner. **Note,** conductors entering a certain section shall terminate in such section, except where it is not practical for control, interconnecting conductors, or other conditions of use.

OVERHEATING AND INDUCTIVE HEATING
408.3(B)

The arrangement of busbars and conductors shall be such as to avoid overheating due to inductive effects. The inductive effect of each conductor is neutralized when grouping all conductors together. To accomplish the minimizing of heating effects, the ungrounded (phase) conductors, grounded (neutral) conductors, equipment grounding conductors, where used, shall be in the same enclosure, raceway, cable, cord, etc. **(See Figure 7-1)**

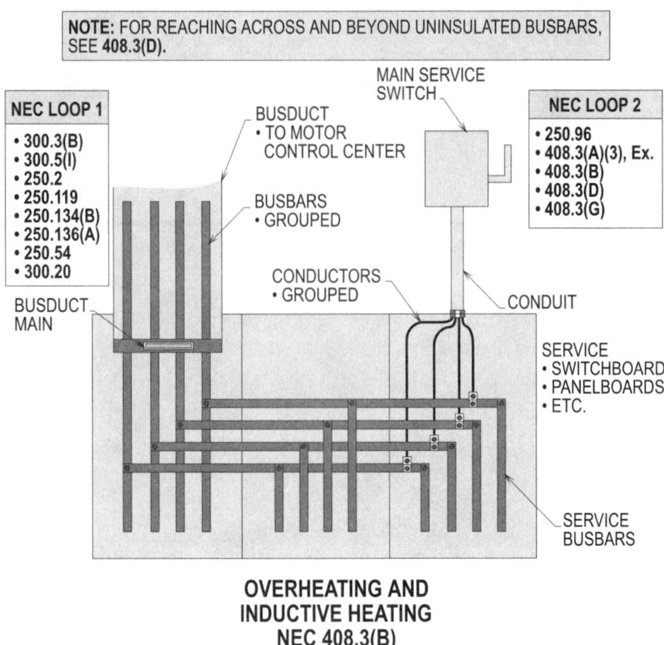

Figure 7-1. The arrangement of busbars and conductors shall be such as to avoid overheating due to inductive effects.

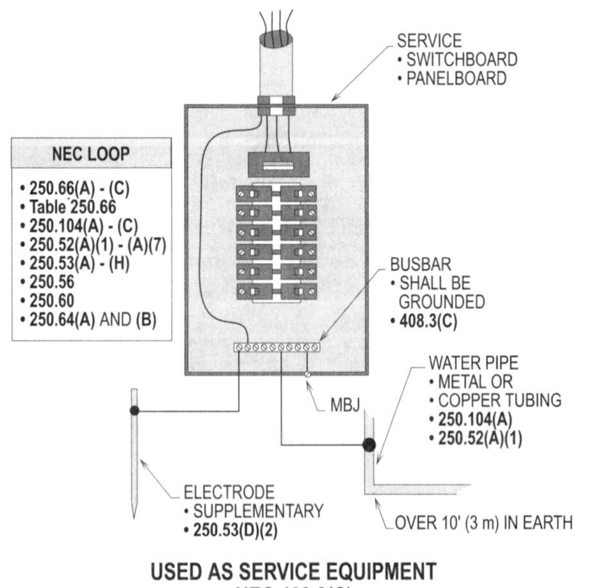

Figure 7-2. Each switchboard, panelboard, or control board used as service equipment shall be provided with a grounding bar bonded to the switchboard frame or the casing of such boards.

USED AS SERVICE EQUIPMENT
408.3(C)

Each switchboard, panelboard, or control board used as service equipment shall be provided with a grounding bar bonded to the switchboard frame or the casing of such boards. For further information, see **250.28(C)**.

The main bonding jumper shall be sized per **250.28(D)** and installed within the panelboard or any section of a switchboard to which the grounded service conductor on the supply side of the switchboard or panelboard frame is connected. All sections of a switchboard shall be bonded together with an equipment bonding conductor that is sized per **Table 250.122** or **Table 250.66**. **Note,** a screw or strap shall be permitted to be used instead of a bonding jumper per **250.28(A)**. **(See Figure 7-2)**

PHASE ARRANGEMENT
408.3(E)

Switchboards or panelboards supplied from a three-phase, four-wire, delta-connected system, where the midpoint of one phase winding (neutral tap) is grounded, the voltage to the grounded neutral will have a higher voltage-to-ground than the other two phases-to-ground.

For example, the higher voltage-to-ground shall be identified by orange color or tape or tagged according to **110.15** and **230.56** and shall be connected to Phase B on a 240 volt, three-phase, four-wire delta connected system. The higher voltage-to-ground identification (high-leg, wild-leg, stinger leg, or red-leg) shall be designed to prevent 120 volt loads from being connected to 208 volts obtained from the high-leg. The 240 volt loads shall be permitted to be derived between the phases, while the 120 volt loads shall be derived from two of the phases-to-ground as follows: **(See Figure 7-3)**

Phase-to-Phase V	Phase-to-Ground V
• A to B is 240 V	• A to N is 120 V
• A to C is 240 V	• B to N is 208 V
• B to C is 240 V	• C to N is 120 V

Design Tip: The center lug of utility meters shall not be connected to the B phase of a three-phase, 240 volt, four-wire, delta system. For metering purposes only, this three-phase high leg shall be connected to Phase C. A CT can with current transformers equipped with a remote meter shall be connected to B phase (high leg) in the CT can for meter conductors. However, it shall be terminated to Phase C in the remote meter base. Check with local utilities for verification of this rule. **(See Figure 7-4)**

The phase arrangement on three-phase buses shall be A, B, and C from the front to back, top to bottom or left to right as viewed from the front of the panelboard, switchboard, or control board. The arrangement shall be viewed as Phases C, B, and A from the back of such boards. With this arrangement of the phases, the high leg will always be placed as B and in the center. **(See Figure 7-5)**

Design Tip: The following methods of color coding shall be permitted to be used for phase arrangement but is no longer required in the NEC, except for the high leg:

• Phase A - black, Phase B - red, Phase C - blue
Used for 120/208 volt or less systems
• Phase A - brown, Phase B - orange, Phase C - yellow
Used for over 277/480 volt systems
• Phase A - black, Phase B - orange, Phase C - blue
Used for 120/240 volt, three-phase systems
• Phase A - black, Phase B - red or blue
Used for 120/240 volt, single-phase systems

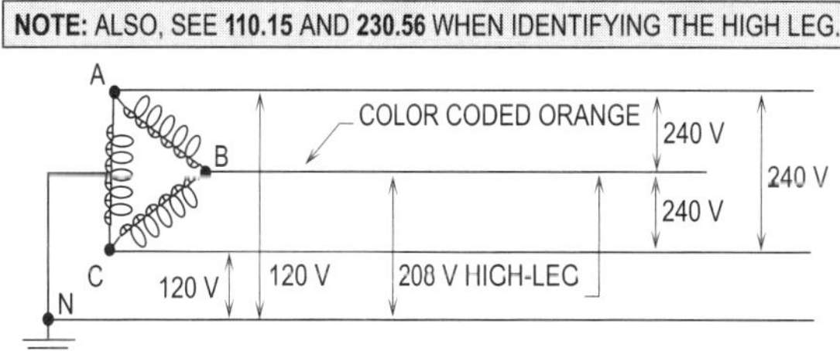

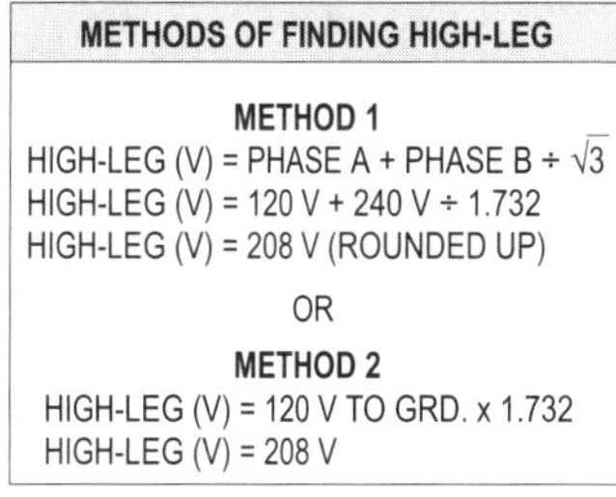

Figure 7-3. Switchboards or panelboards supplied from a three-phase, four-wire, delta-connected system, where the midpoint of one phase winding (neutral tap) is grounded, the voltage to the grounded neutral will have a higher voltage-to-ground than the other two phases-to-ground.

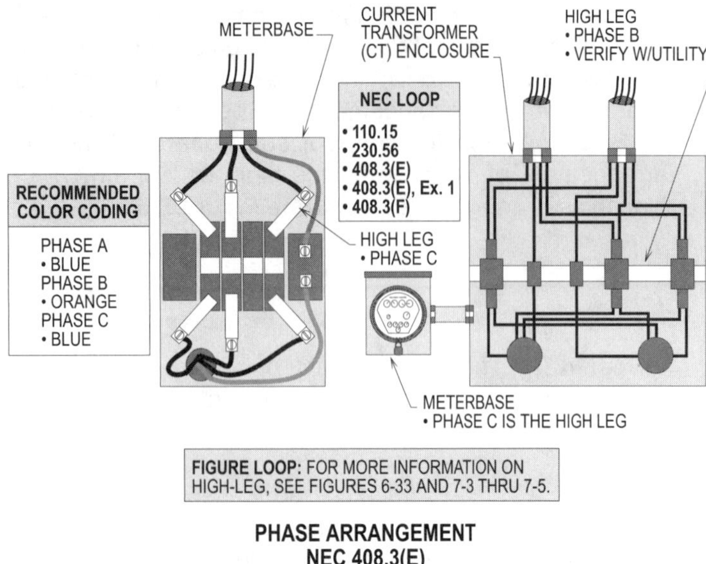

Figure 7-4. The center lug of utility meters shall not be connected to the B phase of a three-phase, 240 volt, four-wire, delta system. In a self-contained meter base, the C phase is the high leg.

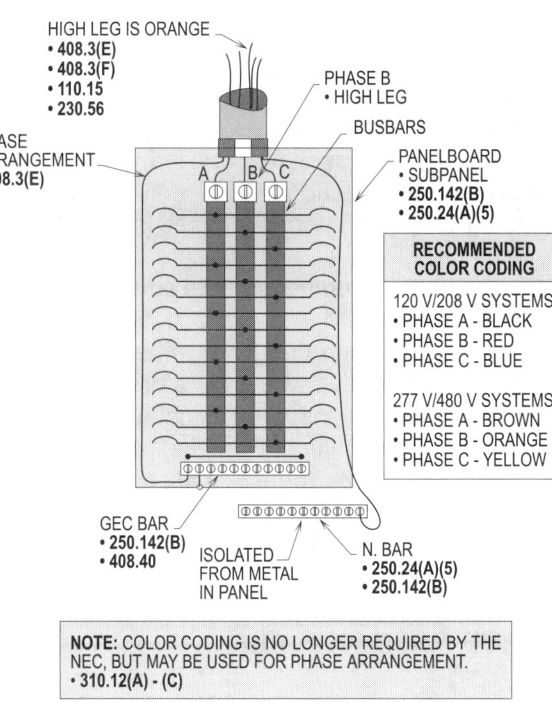

Figure 7-5. The phase arrangement on three-phase buses shall be A, B, and C from front to back, top to bottom or left to right as viewed from the front of the panelboard, switchboard, or control board.

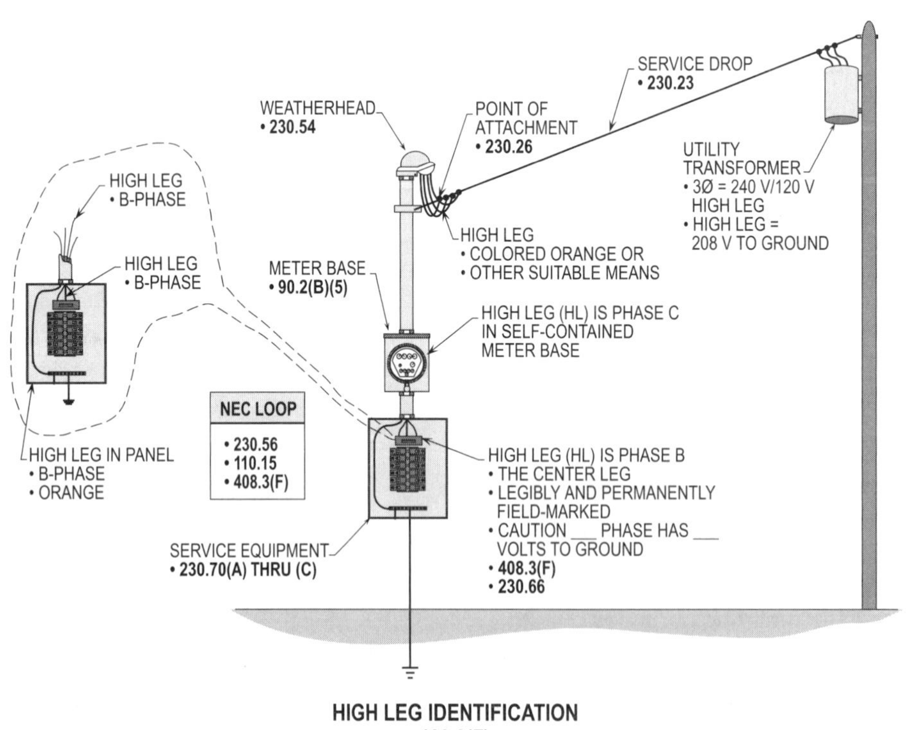

Figure 7-6. A switchboard or panelboard containing a four-wire, delta-connected system where the midpoint of one phase winding is grounded shall be legibly and permanently field marked.

HIGH LEG IDENTIFICATION
408.3(F)

A switchboard or panelboard containing a four-wire, delta-connected system where the midpoint of one phase winding is grounded shall be legibly and permanently field marked as follows:

CAUTION: _____ PHASE HAS _____ VOLTS TO GROUND

See **Figure 7-6** for a detailed illustration pertaining to high leg identification.

MINIMUM WIRE-BENDING SPACE
408.3(G)

Panelboards shall comply with the provision of **408.3(G)**, which requires the minimum gutter space in the board to meet the clearance rules listed in **Tables 312.6(A)** and **(B)** for L-, S-, or Z-bends.

If the lugs in the panelboard are removable, the clearance for L-bends shall be permitted to be reduced from the dimensions in **Table 312.6(A)** based on the number and size of conductors connected to each lug. **(See Figure 7-7)**

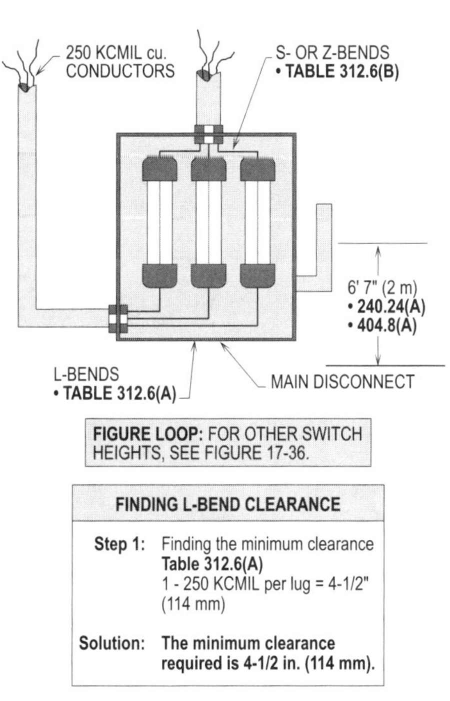

Figure 7-7. Panelboards shall comply with the provision of **408.3(G),** which requires the minimum gutter space in the board to meet the clearance rules listed in **Tables 312.6(A)** and **(B)** for L-, S-, or Z-bends.

If the lugs in the panelboard are removable, the clearance for S- or Z-bends shall be permitted to be reduced from the dimensions in **Table 312.6(B)** based on the number and size of conductors connected to each lug. **(See Figure 7-8)**

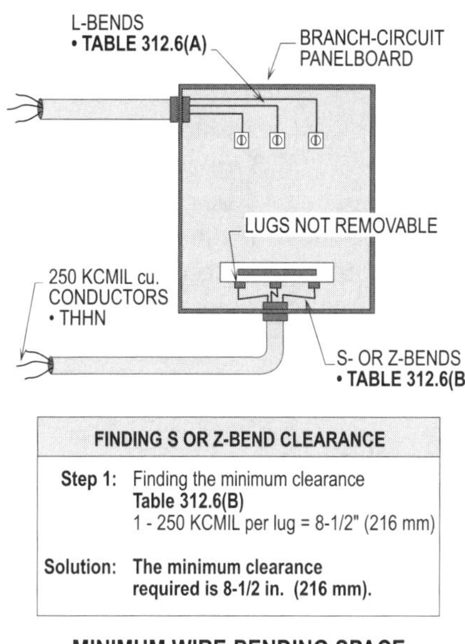

Figure 7-8. To prevent damage to conductors because of overbending, panelboards shall comply with the provision of **408.3(G),** which requires the minimum gutter space in the board to meet the clearance rules in **Tables 312.6(A)** or **(B)** for L-, S-, or Z-bends.

CLEARANCE FOR CONDUCTOR ENTERING BUS ENCLOSURES
408.5

Where conduits or other raceways enter the switchboard, sufficient space shall be provided to permit installation of conductors in the enclosure. The entering conduit or raceways, including their end fittings, shall not rise more than 3 in. (75 mm) above the bottom of the enclosure. The minimum clearance required between the bottom of the switchboard enclosure and busbars, with their supports, or other obstruction shall be 8 in. (200 mm) for insulated busbars, and 10 in. (250 mm) for noninsulated busbars. This clearance rule, if complied with, will prevent the overbending of conductors between entry and termination points. **(See Figure 7-9)**

DEDICATED EQUIPMENT SPACE
110.26(F)(1)(a)

Switchboards shall be installed in a dedicated space equal to the width and depth of the equipment and extending from the floor to a height of 6 ft (1.8 m) or to the structural ceiling (roof), whichever is lower. No piping, ducts, or equipment foreign to the electrical installation shall be installed in this dedicated space since leakage of water or condensation could damage the equipment enclosure. Any space extending 6 ft (1.8 m) above the dedicated space shall not be considered dedicated space. Sprinkler protection shall be permitted to be installed in this dedicated space to protect the equipment from fire hazards. For further information, see **NFPA 13, 4-4.14**.

Suspended ceilings with removable panels shall be permitted within the 6 ft (1.8 m) zone per **110.26(F)(1)(a), Ex.** This dedicated space shall be considered by electricians to be an area where they may install and connect cables and raceways without foreign items (other than electrical) blocking their access directly above equipment. **(See Figure 7-10)**

SWITCHBOARDS
ARTICLE 408, PART II

Switchboards that have exposed live parts shall be installed in dry locations accessible only to qualified personnel. Totally enclosed type switchboards shall be permitted to be installed anywhere acceptable to the AHJ. Switchboards may be used for service equipment, with feeders supplying power to subpanels.

Subpanels are supplied by feeders from the main service equipment. Subpanels are installed where the wiring of branch circuits extends lengths of great distances from the service equipment. Branch-circuit wiring is connected to the centralized subpanel, which reduces the amount of wiring and copper cost required to wire electrical apparatus located in the area.

A subpanel serving lighting and receptacle outlets plus special appliance loads shall be permitted to be supplied from one of the single-pole overcurrent protection devices in the power panel. Double-pole overcurrent protection devices are usually utilized to supply larger fixed electrical appliances such as a heating unit, an A/C unit, or a piece of processing equipment. **(See Figure 7-11)**

LOCATION OF SWITCHBOARDS
408.20 AND 408.17

Switchboards shall be installed to reduce the possibility of communicating fire to adjacent combustible material. Switchboards installed over a combustible floor shall have suitable protection provided. In addition, only qualified persons under the proper supervision of competent personnel shall have access to service, repair or pull maintenance on the parts of such equipment.

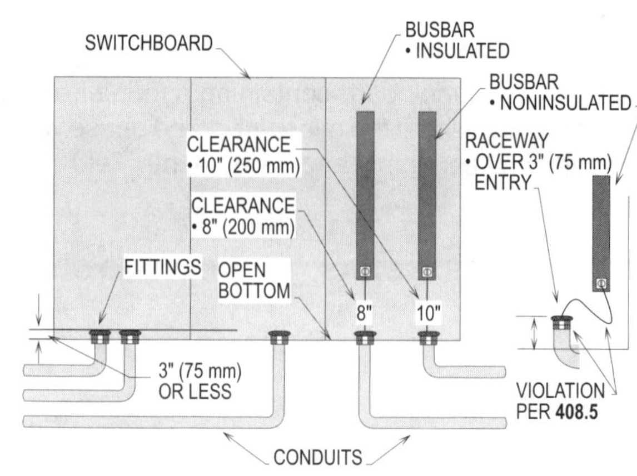

CLEARANCE FOR CONDUCTOR ENTERING BUS ENCLOSURES NEC 408.5

Figure 7-9. Where conduits or other raceways enter a switchboard, sufficient space shall be provided to permit installation of conductors in the enclosure without damaging their insulation.

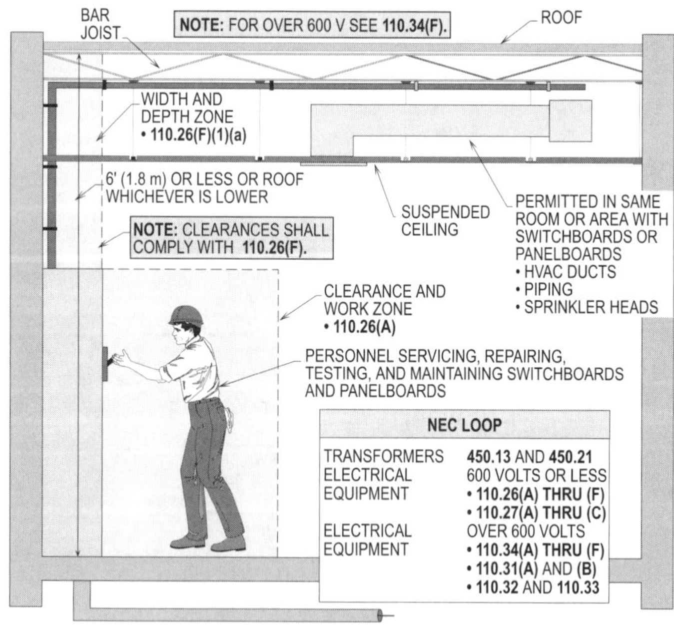

DEDICATED EQUIPMENT SPACE NEC 110.26(F)(1)(a)

Figure 7-10. Switchboards shall be installed in a dedicated space extending 6 ft (1.8 m) above the equipment from the floor or to the structural ceiling (roof), whichever is lower. No piping, ducts, or equipment foreign to the electrical installation shall be installed in this dedicated space.

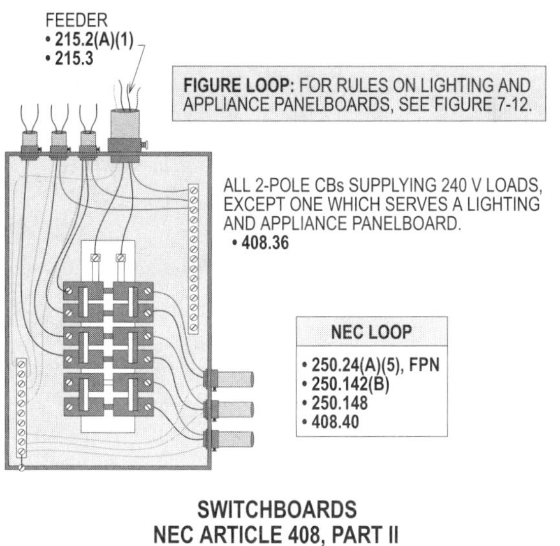

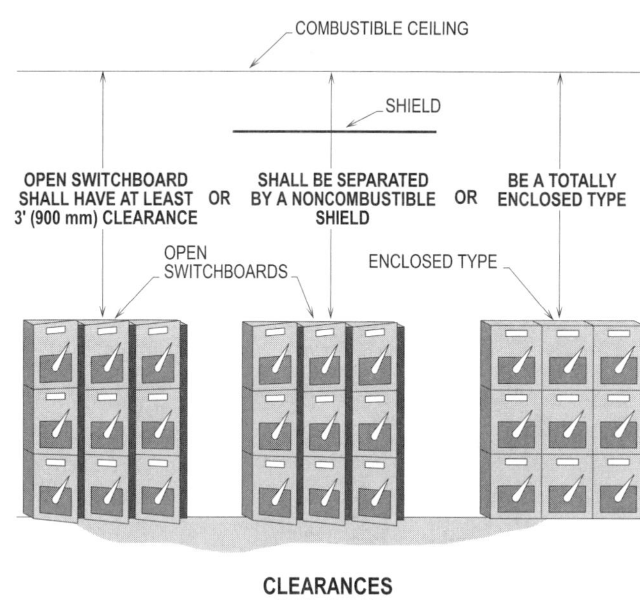

Figure 7-11. A subpanel serving lighting and receptacle outlets plus special appliance loads shall be permitted to be supplied from one of the single-pole overcurrent protection devices in the panelboard. Double-pole overcurrent protection devices are usually utilized to supply larger fixed electrical appliances such as a heating unit, an A/C unit, a large computer, a piece of processing equipment, or a water heater.

CLEARANCES
408.18

Switchboards that are not of the totally enclosed type shall have a clearance of 3 ft (900 mm) from the top of the switchboard to a ceiling constructed of wood, ceiling paper, and any other type of material that will burn. This clearance shall not be required for totally enclosed type switchboards or for a fireproof shield provided above the switchboard. Clearance around and above switchboards shall comply with the requirements of **110.26. (See Figure 7-12)**

GROUNDING OF INSTRUMENTS, RELAYS, METERS, AND INSTRUMENT TRANSFORMERS ON SWITCHBOARDS
408.22

Switchboard frames and structures supporting switching equipment shall be grounded. Frames of two-wire, DC single-polarity switchboards that are insulated shall not be permitted to be grounded. Proper grounding provides equipotential planes of electrical raceways, cables, and enclosures.

Figure 7-12. Switchboards that are not of the totally-enclosed type shall have a clearance of 3 ft (900 mm) from the top of the switchboard to a ceiling constructed of wood, ceiling paper, and any other type of material that will burn.

PANELBOARDS
ARTICLE 408, PART III

Panelboards shall have a rating equal to or higher than the minimum capacity required for the load. When designing a panelboard with a calculated load of 212 amps, the panelboard shall be sized with a rating of at least 212 amps. A panelboard of 225 amps is the next standard size above 212 amps and complies with such rules.

OVERCURRENT PROTECTION
408.36

A panelboard shall be protected by an overcurrent protective device having a rating not greater than that of the panelboard. The overcurrent protective device protecting the panelboard shall be located within or at any point on the supply side of the panelboard. **(See Figure 7-13)**

Individual protection shall be permitted to be omitted where a panelboard is used as service equipment with multiple disconnecting means per **230.71**. Where a panelboard is protected by three or more main circuit breakers or sets of fuses, the circuit breakers or sets of fuses shall not supply a second bus structure within the same panelboard assembly. **(See Figure 7-14)**

Individual protection shall be permitted to be omitted where a panelboard is protected on its supply side by two main circuit breakers or two sets of fuses having a combined rating not greater than that of the panelboard. This panelboard shall not contain more than 42 overcurrent devices. A single-pole circuit breaker counts as one circuit breaker. A double-pole circuit breaker counts as two circuit breakers. A three-pole circuit breaker counts as three circuit breakers. **(See Figure 7-15)**

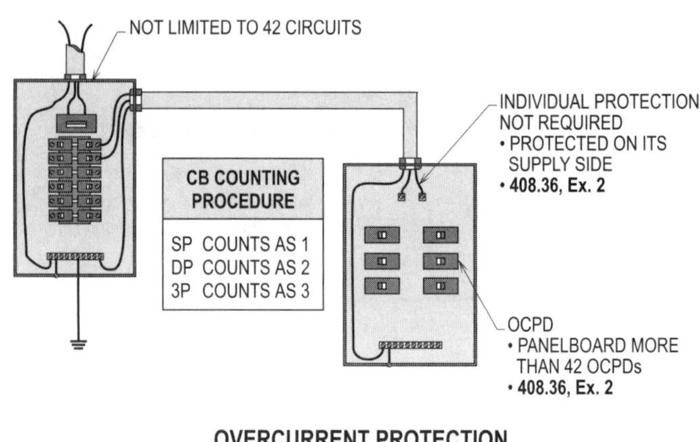

OVERCURRENT PROTECTION
NEC 408.36, Ex. 2

Figure 7-15. Individual protection shall be permitted to be omitted where a panelboard is protected on its supply side by two main circuit breakers or two sets of fuses having a combined rating not greater than that of the panelboard.

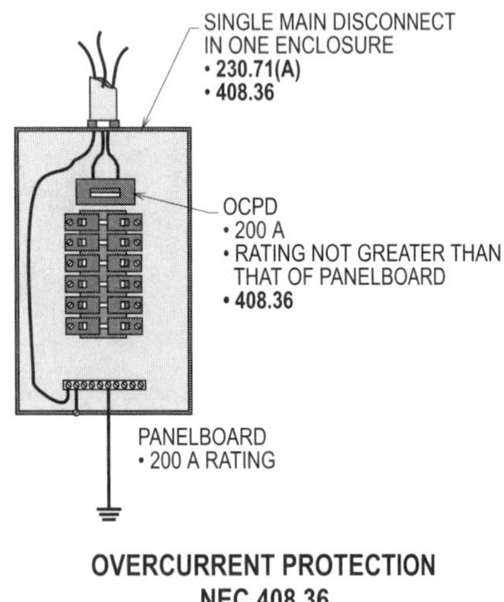

OVERCURRENT PROTECTION
NEC 408.36

Figure 7-13. A panelboard shall be protected by an overcurrent protective device having a rating not greater than that of the panelboard.

SPLIT-BUS PANELBOARD
408.36, Ex. 3

The lower section served by circuit breaker 6 is used to supply power to the lighting and receptacle loads plus the small appliance loads. The five remaining mains in the top section are used to supply loads such as a heating unit, an A/C unit, a dryer, a cooktop, or an oven, a range, or a water heater. **(See Figure 7-16)**

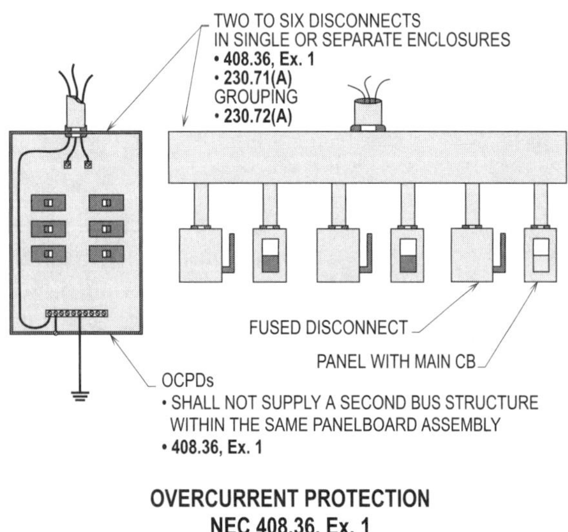

OVERCURRENT PROTECTION
NEC 408.36, Ex. 1

Figure 7-14. Individual protection shall be permitted to be omitted where a panelboard is used as service equipment with multiple disconnecting means per **230.71**.

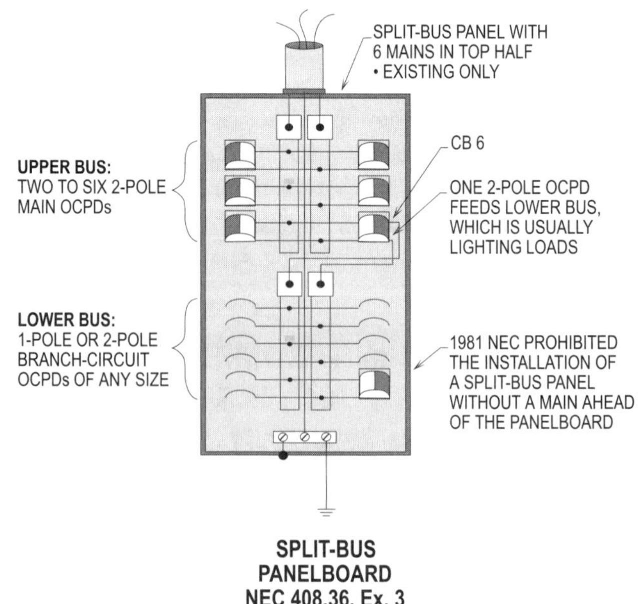

SPLIT-BUS
PANELBOARD
NEC 408.36, Ex. 3

Figure 7-16. A split-bus panelboard has six mains in the upper section, with one main feeding the lower section.

SNAP SWITCHES RATED AT 30 AMPS OR LESS
408.36(A)

Panelboards equipped with snap switches rated 30 amps or less shall be protected by overcurrent protection devices not exceeding the panelboards' rating. Note that a 225 amp panelboard with a 225 amp in-line main shall be permitted to be used for this type of installation. **(See Figure 7-17)**

SUPPLIED THROUGH A TRANSFORMER
408.36(B)

Where a lighting and appliance panelboard is supplied from the secondary of a transformer, the panelboard shall be provided with protection from the secondary side of the transformer. However, under certain design techniques and conditions of use, the overcurrent protection device shall be permitted to be installed on the primary side for a two-wire primary to a two-wire secondary per **408.36(B), Ex. (See Figure 7-18)**

For example, if the load is supplied by a single-phase, two-wire transformer with a single voltage, the secondary shall be permitted to be protected by the overcurrent devices in the primary or supply side of the transformer, provided the primary is protected as covered in **450.3(B)** and at a value that does not exceed the value determined by multiplying the transformer voltage ratio of the secondary-to-primary by the panelboard rating per **240.4(F)** and **240.21(C)(1)**.

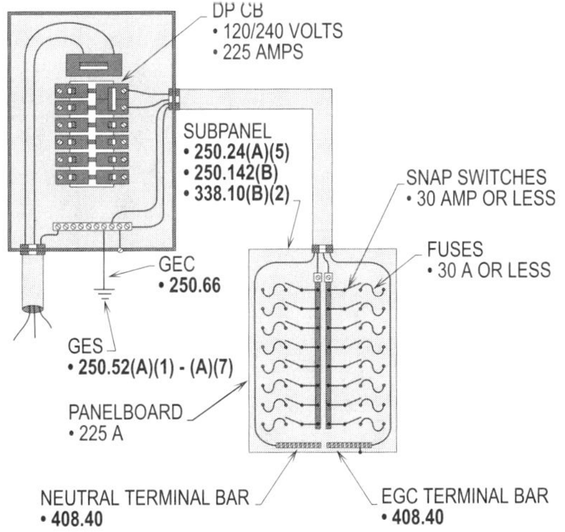

SNAP SWITCHES RATED AT
30 AMPS OR LESS
NEC 408.36(A)

Figure 7-17. Panelboards equipped with snap switches rated 30 amps or less shall be protected by overcurrent protection devices not exceeding the panelboards' rating.

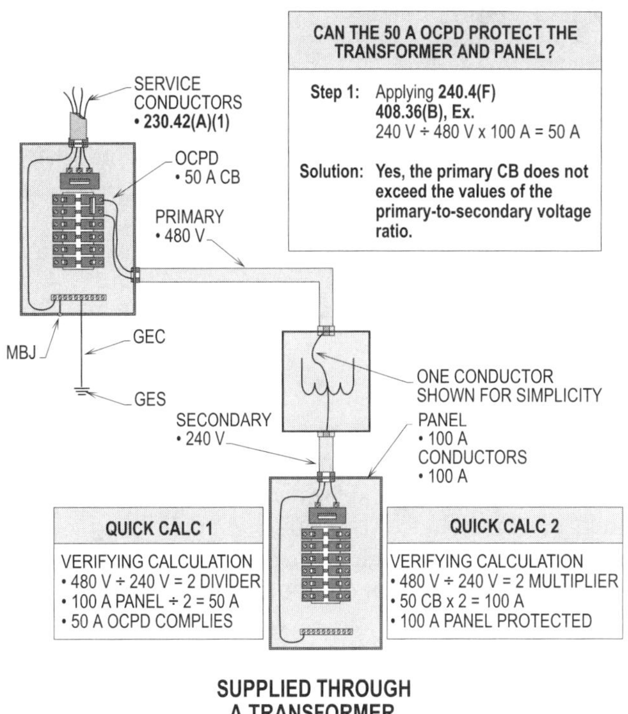

SUPPLIED THROUGH
A TRANSFORMER
NEC 408.36(B), Ex.

Figure 7-18. Where a lighting and appliance panelboard is supplied from the secondary of a transformer, the overcurrent protection device in the primary side shall be permitted to protect the panelboard and secondary side from overloads. (General rule)

DELTA BREAKERS
408.36(C)

Lighting and appliance panelboards shall not be permitted to be installed with delta breakers. In other words, a three-phase disconnect shall never be terminated to the buses of panelboards having less than three buses. The reasoning behind this requirement was that in the past, homeowners were overloading the rating of the panel, due to the absence of an in-line main.

BACK-FED DEVICES
408.36(D)

Plug-in type overcurrent protection devices or plug-in type main lug assemblies that are back-fed shall be secured in place by an additional fastener other than the pull-to-release type. The past history of such installed devices has proved that only plugging them in was not enough support to secure them properly.

ENCLOSURE
408.38

Panelboards shall be mounted in cabinets, cutout boxes, or enclosures designed for the purpose. Dead-front panelboards shall be installed unless accessible only to qualified persons.

GROUNDING OF PANELBOARDS
408.40

An equipment grounding terminal bar shall be secured inside the cabinet for the termination of all equipment grounding conductors used to ground the metal of equipment, junction boxes, pull boxes, etc. The equipment grounding terminal bar ensures that the equipment grounding conductors are bonded to the cabinet and panelboard frame and if done correctly, provide an effective path to ground.

Panelboard enclosures or cases shall be grounded. Where installed with a metal conduit such as rigid metal conduit, electrical metallic tubing, intermediate metal conduit (any metal raceway), metal-clad cables, etc., the metal of these wiring methods shall be permitted to be used for equipment grounding if they comply with **250.118**. An equipment grounding conductor shall be installed with a nonmetallic wiring method such as nonmetallic sheathed cable (romex or rope) or PVC. **(See Figure 7-19)**

The equipment grounding terminal bar and the neutral bar in service equipment shall be bonded to the panelboard case per **250.130** and **250.66** and connected to the grounding electrode system per **250.52(A)(1)** through **(A)(6)** by the grounding electrode conductor. The equipment grounding terminal bar and the neutral bar in other than service equipment shall not be permitted to be connected and bonded together in the panelboard enclosure per **250.142(B)**. Also see **250.24(A)(5)**. The neutral installed in the sub-panelboard enclosure shall be isolated back to the service equipment. Isolated equipment grounding conductors shall be permitted to pass through the panelboard enclosure per **250.146(D)**. All equipment grounding conductors and grounded (neutral) conductors in a subpanel shall be terminated to separate terminal bars per **408.40**. **(See Figure 7-20)**

> **Design Tip:** Equipment grounding conductors and grounded (neutral) conductors shall only terminate to a common bar at the service equipment, separately derived systems, and separately fed buildings from a main building supply.

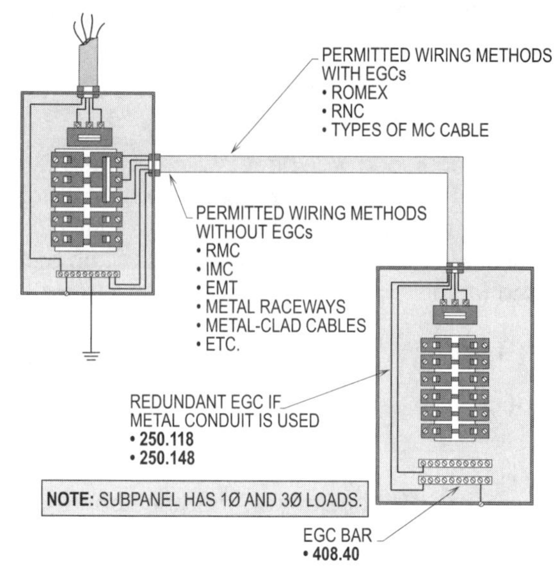

GROUNDING OF PANELBOARDS
NEC 408.40

Figure 7-19. An equipment grounding terminal bar shall be secured inside the cabinet for the termination of all equipment grounding conductors used to ground the metal of equipment, junction boxes, pull boxes, etc.

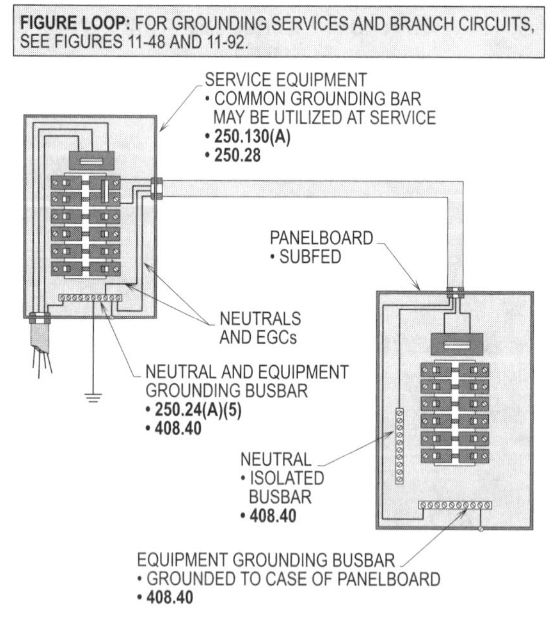

GROUNDING OF PANELBOARDS
NEC 408.40

Figure 7-20. The equipment grounding terminal bar and the neutral bar in other than service equipment shall not be permitted to be connected and bonded together in the panelboard enclosure.

ISOLATED EQUIPMENT GROUNDING CONDUCTOR
408.40, Ex.

Isolated equipment grounding conductors shall be permitted to pass through the panelboard with the power circuit conductors to reduce electromagnetic interference (electrical noise) if present. An isolated equipment grounding conductor shall be installed and connected directly to the equipment grounding terminal bar and neutral bar in the main service equipment or power source (separately derived system) to reduce the problem of electrical noise. Such noise, if not corrected, may disturb the operation of sensitive electronic equipment. **(See Figure 7-21)**

> **Design Tip:** This **Ex.** is necessary for isolating computer grounding systems that are cord-and-plug connected per **250.146(D)**. Also see **250.96(B)** for a similar rule when isolating the grounding system for hard-wired sensitive electronic equipment.

ISOLATED RECEPTACLES
250.146(D)

When electrical noise becomes a problem in the grounding circuit, a receptacle on the grounding circuit in which the grounding terminal or receptacle has been insulated from the mounting means or yoke shall be allowed. In this case, an insulated equipment grounding conductor shall be routed to the isolated terminal with the circuit conductors. This insulated equipment grounding (IEG) conductor shall be permitted to pass through one or more panelboards in the same premises without being connected thereto as covered in **408.40, Ex.** Such IEG shall be permitted to be terminated at the equipment grounding terminal at a separately derived system or the service equipment. Note that a grounding jumper (bonding jumper) shall be permitted to be used from grounded boxes to the grounding terminals of the receptacle, including boxes on conduit circuits, EMT circuits and some metal-clad circuits unless the box is surface mounted so that the mounting screws may be tightened to make a secure grounding connection between the device yoke and the box. See **250.146(A)** and **(B)** for this rule.

> **Design Tip:** The grounding scheme described above is mostly used for grounding computers or solid-state counters on pieces of sensitive electronic equipment. If metal conduits or cables are used to contain the circuit conductors, the metal parts should be insulated from the equipment and then the insulated equipment grounding conductor run back from the equipment to the source as permitted in **250.146(D)**.

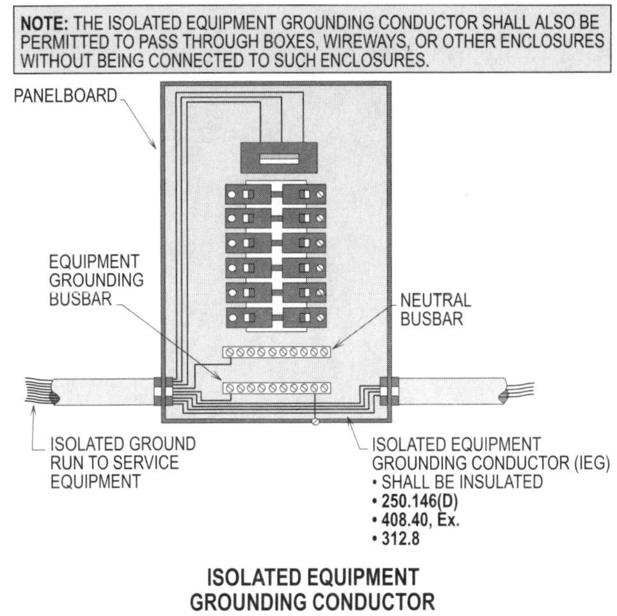

Figure 7-21. Isolated equipment grounding conductors shall be permitted to pass through the panelboard with the power circuit conductors to reduce electromagnetic interference (electrical noise) if present.

RUNNING CIRCUITS TO COMPENSATE FOR HARMONIC CURRENTS

Another concern that has recently become a problem is overloaded grounded (neutral) conductors on multiwire branch circuits. This overload is the result of harmonic currents drawn by today's typical business equipment and electronic discharge lighting systems.

To address the inherent overload produced by many of today's loads, a special AC or MC cable is available with either an oversized grounded (neutral) conductor for each three-phase, four-wire circuit or a full-sized grounded (neutral) conductor (same size as the ungrounded (phase) conductors) for each ungrounded (phase) conductor in the cable. It is important to note that use of such cables does not eliminate the harmonic currents, but simply accommodates the additional neutral heating caused by the harmonic currents drawn by such loads. **[See Figures 7-22(a) and (b)]**

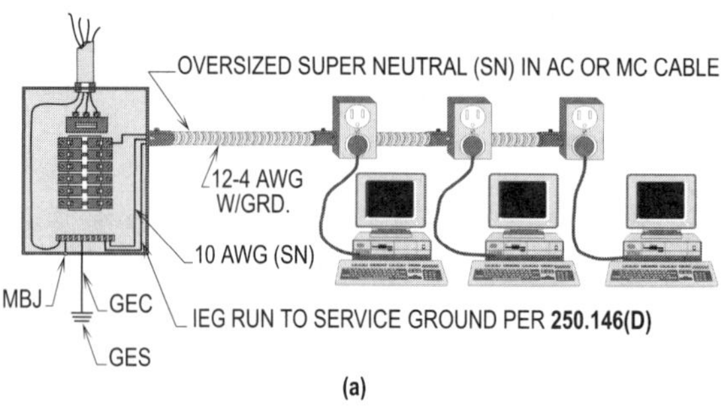

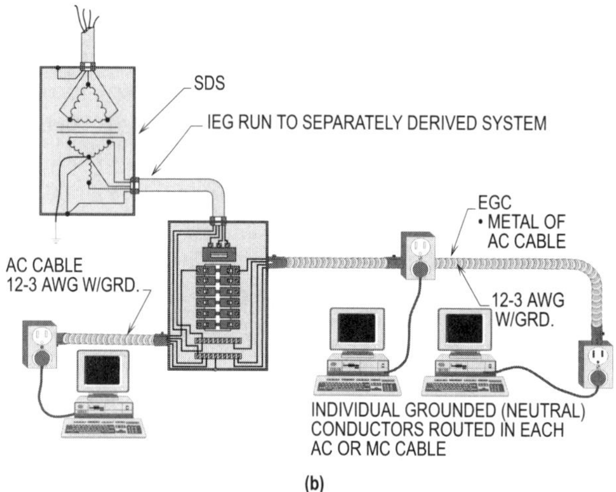

**RUNNING CIRCUITS TO COMPENSATE FOR
HARMONIC CURRENTS**

Figures 7-22(a) and **(b).** AC and MC cables are available with an extra or oversized grounded (neutral) conductor plus an equipment grounding conductor, to supply (as nearly as possible) clean and noise-free power to sensitive electronic equipment. **Note**, circuits can also be run in metal or nonmetallic raceways.

For hard-wire circuits, see **Figures 11-113** and **11-114** of Chapter 11.

Chapter 7. Switchboards and Panelboards

Section   Answer

__________   __________   1. Piping, ducts, or equipment foreign to the electrical installation shall be permitted to be installed in the dedicated space extending _____ ft from the floor above the equipment to the structural ceiling (roof), whichever is lower. (General Rule)
        (a) 3   (b) 4
        (c) 5   (d) 6

__________   __________   2. The minimum clearance required between the bottom of the switchboard enclosure and busbars, with their supports or other obstruction shall be _____ in. for insulated busbars.
        (a) 5   (b) 10
        (c) 12   (d) 15

__________   __________   3. Individual protection shall not be required for a panelboard protected on its supply side by _____ main circuit breakers having a combined rating not greater than that of the panelboard.
        (a) 1   (b) 2
        (c) 3   (d) 4

__________   __________   4. Panelboards equipped with snap switches rated at _____ amperes or less shall have overcurrent protection of 200 amperes or less.
        (a) 15   (b) 20
        (c) 30   (d) 50

__________   __________   5. A split-bus panelboard has _____ mains in the top section with one main feeding the bottom section.
        (a) 2   (b) 3
        (c) 4   (d) 6

__________   __________   6. A double-pole circuit breaker shall be counted as _____ overcurrent devices.
        (a) 2   (b) 3
        (c) 4   (d) 6

__________   __________   7. Switchboards shall be installed in a dedicated space extending _____ ft from the floor and above the equipment to the structural ceiling (roof), whichever is lower.
        (a) 6   (b) 20
        (c) 25   (d) 50

__________   __________   8. The minimum clearance required between the bottom of the switchboard enclosure and busbars, with their supports, or other obstruction shall be _____ in. for noninsulated busbars.
        (a) 8   (b) 10
        (c) 12   (d) 18

__________   __________   9. Where conduits enter a switchboard bottom, the conduits, including their end fittings, shall not rise more than _____ in. above the bottom of the enclosure.
        (a) 1   (b) 2
        (c) 3   (d) 6

_____________ _____________

10. Switchboards that are not of the totally enclosed type shall have a clearance of _______ ft from the top of the switchboard to a ceiling constructed of wood, ceiling paper, and any other type of material that will burn.

(a) 3 (b) 6

(c) 10 (d) 25

8

Conductors

In selecting the proper size conductors and overcurrent protection devices for supplying power in a circuit from the source to the load, it is important that the designer apply the appropriate rules of the *National Electrical Code*. The overcurrent protection device (OCPD) for conductors and equipment shall be sized in such a manner to open the circuit if currents reach a value that causes excessive or dangerous temperature in conductors or insulation. Such overcurrent protection device shall be selected, sized, and wired to protect both conductors and equipment, or a second stage of protection shall be provided. The allowable ampacities of conductors may vary depending upon their conditions of use, which are mainly based on the number of conductors in a raceway or cable and exposure to surrounding ambient temperatures.

TEMPERATURE LIMITATION OF CONDUCTORS
310.10, FPNs (1) THRU (4)

The conductor's allowable ampacity listed in **Tables 310.16 through 310.19** in the NEC are based on four determining factors [**310.10, FPNs (1) through (4)**] and they are as follows:

- Ambient temperature
- Heat generated internally
- Heat dissipated into ambient medium
- Adjacent load-carrying conductors

AMBIENT TEMPERATURE
310.10, FPN (1)

The ambient temperature may vary along the conductor length as well as from time to time, and correction factors to **Table 310.16** shall be used.

As stated above, the temperature surrounding the conductor may vary from 86°F (30°C) along the conductor length. However, if the ambient temperature is not 86°F, the allowable ampacity of the conductor as listed in **Table 310.16** shall be corrected according to the correction factors below the Table.

For example, three current-carrying copper conductors in a raceway exposed to a 40°C (104°F) ambient temperature shall have their allowable ampacities derated by 91 percent per correction factors to **Table 310.16**. **(See Figure 8-1)**

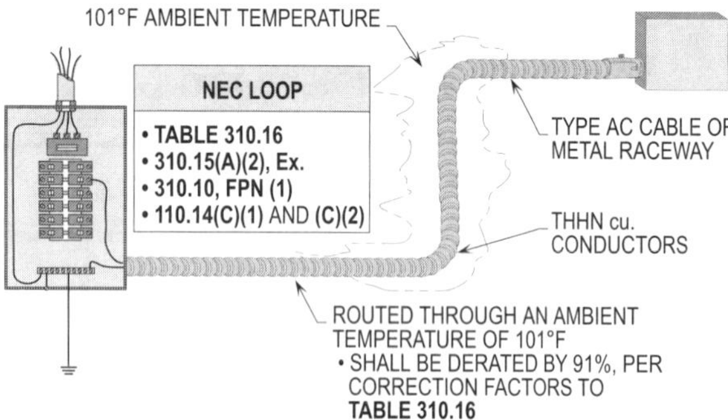

**TEMPERATURE LIMITATION
OF CONDUCTORS
NEC 310.10, FPN (1)**

Figure 8-1. Conductors routed through an ambient temperature greater than 86°F shall be derated per correction factors to **Table 310.16,** based upon surrounding temperatures. (This rule applies to conductors in cables and conduits.)

WHEN TO APPLY CORRECTION FACTORS
310.15(A)(2), Ex.

Ambient temperature will not affect the allowable ampacity of conductors per **Table 310.16** if that portion of the conductors routed through the higher ambient is 10 ft (3 m) or less and does not exceed 10 percent of the total length of such conductors.

Design Tip: The higher allowable ampacities shall be permitted to be utilized for the lower ampacity if the length is no greater than 10 ft (3 m) or no greater than 10 percent of the circuit conductors. The 10 ft (3 m) or 10 percent of conductor length represents exposure to ambient temperatures above 86°F. **(See Figure 8-2)**

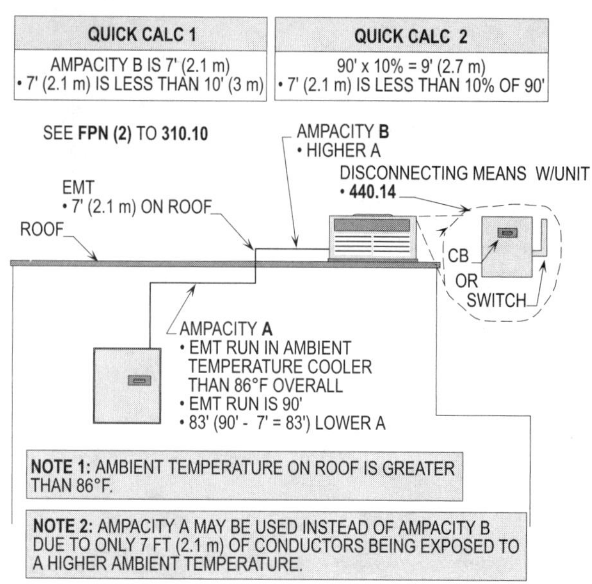

**WHEN TO APPLY CORRECTION FACTORS
NEC 310.15(A)(2), Ex.**

Figure 8-2. Where the total length of conductors in the EMT run (Ampacity A) is 83 ft or less, the derating factors do not have to be applied to the above circuit conductors.

HEAT GENERATED INTERNALLY
310.10, FPN (2)

With regard to heat generated internally in the conductor as the result of load current flow, including fundamental and harmonic currents, the allowable ampacity of a conductor shall be sized to carry the load without overheating its insulation beyond the listed temperature of such insulation per **Tables 310.13** and **310.16**. A properly sized conductor serves as a heat sink to remove heat away from the terminals of devices and equipment. Conductors not sized properly may cause terminals to develop greater heat than they are designed for, and improper operation may occur. To ensure that conductors are adequate to supply the load without overheating, the loads used to size such conductors shall be calculated per **210.19(A)(1)** and **215.2(A)(1)**.

HOW TO CALCULATE LOADS
210.19(A)(1) AND 215.2(A)(1)

Loads that are continuous shall have their full-load current, in amps, sized at 125 percent, and this value is added to noncontinuous loads at 100 percent, if any are present. **(See Figure 8-3)**

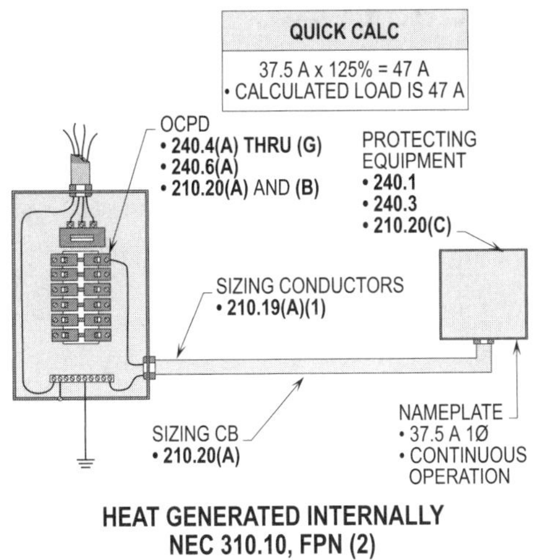

Figure 8-3. Conductors shall be sized large enough to carry the calculated load supplied (47 amps) without overloading its insulation and termination points.

DISSIPATION INTO AMBIENT MEDIUM
310.10, FPN (3)

The rate of heat dissipation into the ambient medium surrounding the conductor per **110.14(C)(1)** and **(C)(2)** depends upon the surrounding temperature.

As stated above, it is clear that, if the allowable ampacities of conductors are not selected according to the lowest temperature rating of devices and equipment, the higher temperature of insulation will hold heat and deteriorate internal working parts.

For example, the allowable ampacity of a 3 AWG THWN copper conductor terminated to a 75°C terminal is 100 amps based upon the 75°C column of **Table 310.16**. Note that it shall not be permitted to be terminated and used at 110 amps, which would be the allowable ampacity of the 90°C column per **Table 310.16**. (**Note**, 85 amps at 60°C terminals.)

MATCHING TEMPERATURE MARKINGS
110.14(C)(1) AND (C)(2)

The reason for the above rule is as follows:

If a conductor with 90°C insulation is connected to a terminal of 60°C, the 90°C insulation must be heated above 90°C before heat can be dissipated into the surrounding ambient medium. This heat held in the conductor by the higher insulation must be dissipated. As current heats up the material of the conductor, it is always seeking a cooler spot, and this cooler area is usually the inside elements of the

equipment and the circuit's overcurrent protection device. By matching the 90°C allowable ampacity of the conductor to the 60°C ampacities of the overcurrent protection device and equipment, the unwanted heating effects of internal components due to excessive current flow are limited. **(See Figure 8-4)**

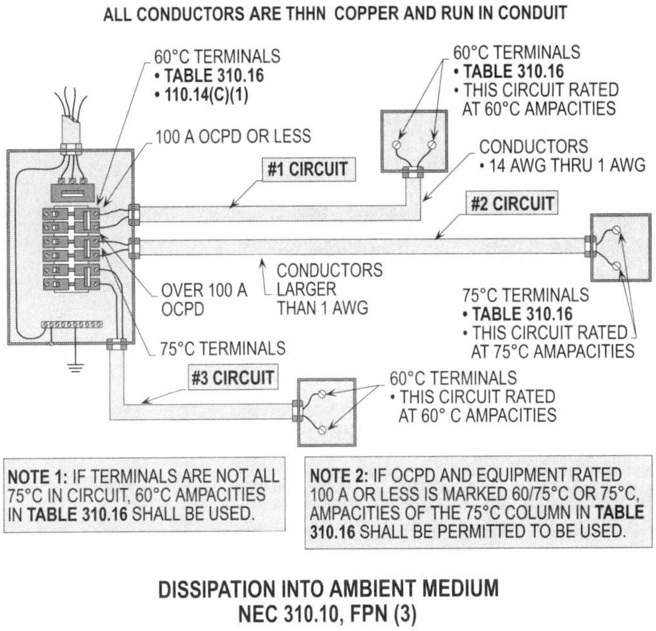

Figure 8-4. The above illustration shows the correct procedure for selecting allowable ampacities of conductors based upon terminal ratings. **(See Figures 8-11 and 14-16)**

ADJACENT LOAD-CARRYING CONDUCTORS
310.10, FPN (4)

The heating effect of adjacent load current-carrying conductors, known as the proximity effect per **310.15(B)(2)(a)** to ampacity Tables 0 - 2000 volts shall be considered.

As stated above, the proximity heating effect of conductors occurs where conductors are bundled or pulled together in raceways and cables. When conductors are installed in this manner, their ability to dissipate heat is greatly reduced and adjustment factors per **310.15(B)(2)(a)** shall be applied to adjust for such heating and prevent damaging insulation and components of devices and equipment.

WHEN TO APPLY ADJUSTMENT FACTORS 310.15(B)(2)(a) TO TABLES 0-2000 VOLTS

If four or more current-carrying conductors are pulled through a raceway or bundled in a cable for a distance greater than 24 in. (600 mm), the allowable ampacity of such conductors shall be reduced (derated) by the factors (percentages) listed in **Table 310.15(B)(2)(a)**.

For example, nine current-carrying conductors in a raceway shall have their ampacities listed in **Table 310.16** multiplied by 70 percent so they will not overheat beyond their insulation ratings if they should become fully loaded. **(See Figure 8-5 and Figures 8-13 and 8-16)**

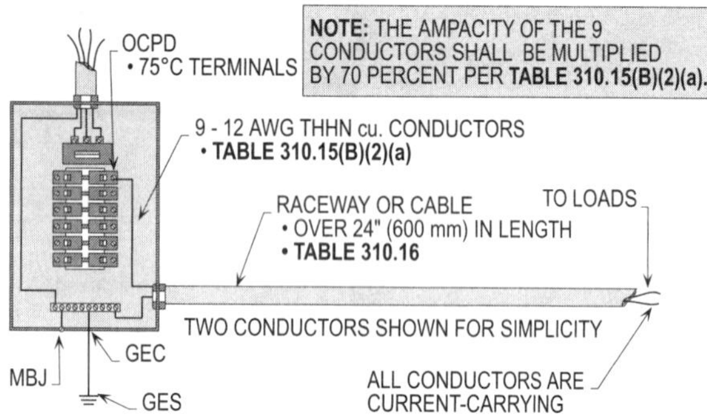

Figure 8-5. Adjustment factors per **310.15(B)(2)(a)** shall be applied where there are four or more current-carrying conductors in a raceway or cable.

DEFINITIONS
ARTICLE 100

There are certain definitions that shall be understood when calculating the allowable ampacities of conductors based upon their conditions of use.

AMPACITY
ARTICLE 100

The ampacity of a conductor is the current, in amps, that a conductor can carry continuously without exceeding its temperature rating. The condition of use for a conductor shall be considered when there are too many current-carrying conductors in a raceway or cable or surrounding ambient temperature that exceeds 86°F. Note that both of these conditions may exist in a circuit.

CONTINUOUS LOAD
ARTICLE 100

A continuous load is where the current in amps may operate for a period of three hours or more. Such continuous loads are as follows:

- Indoor lighting
- Outdoor lighting
- Advertising signs
- Special outlets
- Certain types of equipment, etc.

UNGROUNDED CONDUCTORS
ARTICLE 100 AND 310.12(C)

An ungrounded (phase) conductor is a circuit conductor that is not grounded but carries current to the load. Ungrounded (phase) conductors in a service, feeder, or branch circuit are known as the ungrounded (phase) conductors. *Ungrounded* is defined as not connected to ground or to a conductive body that extends the ground connection.

GROUNDED CONDUCTORS
ARTICLE 100 AND 310.12(A)

A grounded (neutral) conductor is a system or circuit conductor that is intentionally grounded. Such a grounded conductor shall be permitted to be a neutral per **220.61(A)** through **(C)** or a corner grounded delta conductor per **250.26** and **240.22**.

> **Design Tip:** A corner grounded delta-system conductor shall always be considered current-carrying because it is an ungrounded (phase) conductor. The neutral may or may not be current-carrying per **310.15(B)(4)(c)** to ampacity Tables 0-2000 volts.

EQUIPMENT GROUNDING CONDUCTORS
ARTICLE 100 AND 310.12(B)

The equipment grounding conductor (EGC) connects all the metal parts of enclosures and equipment together. It bonds such enclosures and equipment by terminating at the grounding bus at the service equipment or separately derived system. Equipment grounding conductors shall not be considered current-carrying conductors except during a ground fault per **310.15(B)(3)** and **(B)(5)**.

See Figure 8-6 for a detailed illustration of recommended color coding techniques.

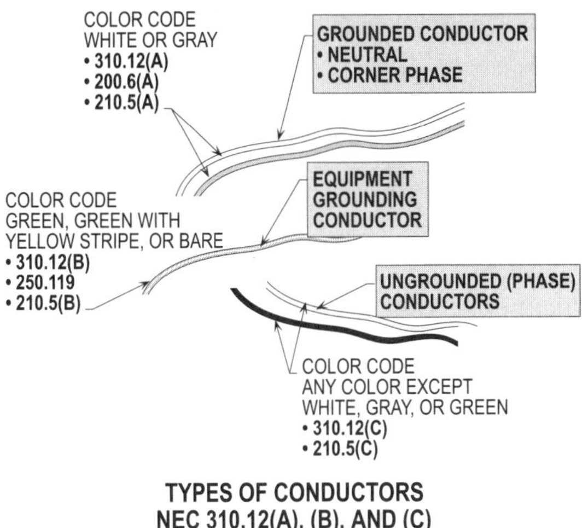

Figure 8-6. The recommendation to color code ungrounded (phase) conductors was deleted in the 1975 NEC.

OVERCURRENT
ARTICLE 100, 240.3, AND 240.4

Overcurrent is any current in excess of the rated current of equipment or the ampacity of a conductor. Such overcurrents are caused by any one of the following conditions:

- Short circuits
  (phase-to-phase)
- Ground faults
  (phase-to-ground)
- Overloads
  (slow heat buildup)

See Figure 8-7 for a detailed illustration pertaining to short-circuits, ground faults, and overloads in an electrical system.

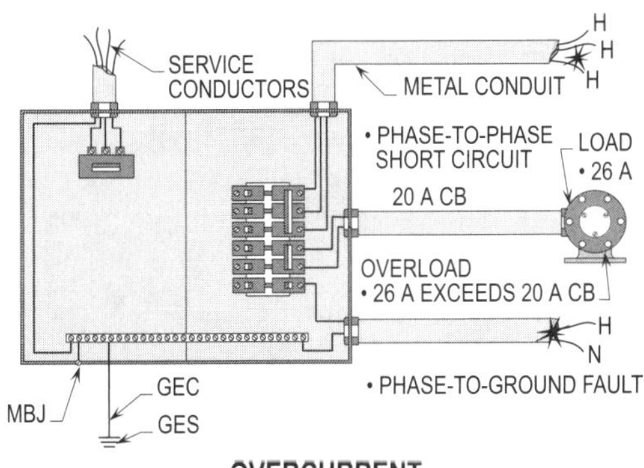

Figure 8-7. The above illustration shows the difference between short circuits, ground faults, and overloads in an electrical circuit.

CURRENT-CARRYING CONDUCTORS 310.15(B)(4)(c) TO TABLE 310.16 AND 725.51(A) AND (B)

All ungrounded (phase) conductors shall be considered current-carrying. Neutral conductors and control circuits are current-carrying under certain conditions of use. A neutral of a four-wire, three-phase wye system shall be considered current-carrying when it carries the major portion of the load involving nonlinear related loads.

WHEN IS A NEUTRAL CURRENT-CARRYING 310.15(B)(4)(c)

If the neutral on Phases A, B, and C has a load of 100 amps respectively and 51 amps (100 A x 51% = 51 A) is harmonic related, the neutral shall be considered a current-carrying conductor. (**See Figure 8-8**)

Design Tip: Any load over 50 percent is considered a major portion. However, 51 percent is normally used as the multiplier to determine the major portion in amps. (Rule of thumb method)

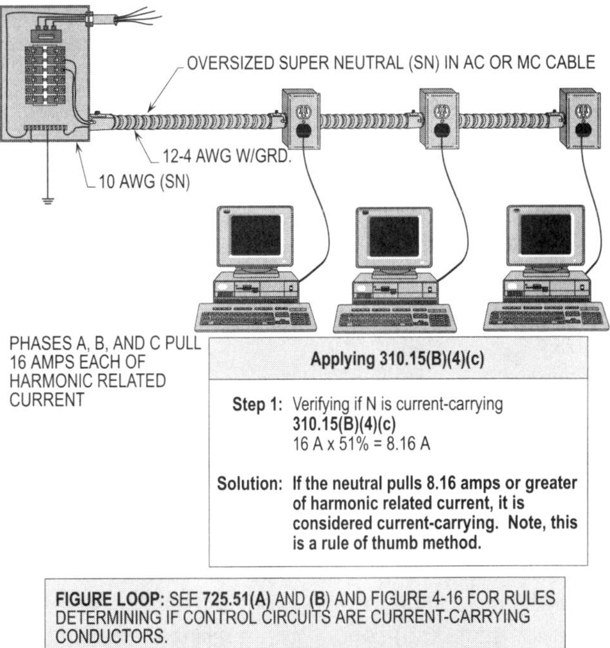

Figure 8-8. If the neutral in the above illustration is sharing harmonic loads that are considered the major portion of the circuit currents of Phases A, B, and C, the neutral shall be considered current-carrying. (Use larger amps of A, B, or C.)

WHEN ARE CONTROL CIRCUITS CURRENT-CARRYING
725.51(A) AND (B)

Control circuits shall be considered current-carrying when they carry continuously more than 10 percent of their allowable ampacities.

Twenty control circuit conductors carrying continuous current of 2 amps (20 A x 10% = 2 A) shall not be considered current-carrying conductors. Therefore, there could be 15 current-carrying conductors in a raceway; if three are classified as control circuits and three neutrals, all of which fall within the above rules, only 9 conductors shall be considered current-carrying. **(See Figure 4-16)**

CONDUCTOR IDENTIFICATION
310.12

A color code scheme was recommended for ungrounded (phase) conductors prior to the 1975 NEC. Editions after the 1975 NEC only required color coding for an ungrounded (phase) conductor or grounded (neutral) conductor and the equipment grounding conductor. The high leg of a delta system had to be colored orange or marked properly.

Conductors used in electrical wiring systems shall be identified properly to protect personnel working on such systems. It is essential to know which conductor by color represents the ungrounded (phase) conductor, the grounded (neutral) conductor, and equipment grounding conductor in an electrical circuit, because it is by color coding that conductors are connected to color-coded terminals of equipment.

GROUNDED CONDUCTORS
310.12(A)

Insulated conductors of 6 AWG and smaller wire, when used as grounded (neutral) conductors, shall have a white or gray colored insulation. Where the insulated conductors are larger than 6 AWG, they shall be identified by a white or gray colored insulation, or by a distinctive marking (white) at the terminals at the time they are being installed.

The grounded (neutral) conductor used in service-entrance raceways and cables shall be permitted to be uninsulated per **230.41, Ex.**

See **240.22** for further information concerning grounded (neutral) conductors and overcurrent protection devices.

EQUIPMENT GROUNDING CONDUCTORS
310.12(B)

An equipment grounding conductor shall be permitted to be in the same raceway, cord, or cable or may otherwise be run with circuit conductors. It shall be a part of the cable or cord, or be run in the same raceway with the circuit conductors to keep the impedance to the lowest value. When using NM, NMC, or UF cables, these cables shall contain the equipment grounding conductor as part of the cable to ground noncurrent-carrying metal parts.

The equipment grounding conductor shall be permitted to be bare, covered, or insulated. Where individual covered or insulated equipment grounding conductors are run, it is required that they shall have a continuous green outer finish or a continuous green outer finish with one or more yellow stripes. Bare conductors shall be permitted to always be used for such grounding.

When the equipment grounding conductor is larger than 6 AWG copper or aluminum, this conductor shall be permanently marked at the time of installation, at each end or any other accessible point. It shall be identified by any one of the following means:

- Stripping the insulation from the entire exposed length.

- Coloring the insulation or covering green at the termination.

- Marking the insulation or covering with green tape or green adhesive labels at the termination.

See **250.119** for a detailed description of identifying the equipment grounding conductor in a circuit.

UNGROUNDED CONDUCTORS
310.12(C)

The ungrounded (phase) conductors shall be permitted to be identified with any color of insulation or tagging except white, gray, or green.

Design Tip: Until the 1975 NEC, the color coding scheme for ungrounded (phase) conductors was usually as follows:

- 120/240 volt, single-phase system
  - Phase A - Black
  - Phase B - Blue or red

- 120/240 volt, three-phase delta system
  - Phase A - Black
  - Phase B - Orange
  - Phase C - Blue

- 120/208 volt, three-phase wye system
  - Phase A - Black
  - Phase B - Red
  - Phase C - Blue

- 277/480 volt, three-phase wye system
  - Phase A - Brown
  - Phase B - Orange
  - Phase C - Yellow
  - Known as BOY system

See Figure 8-9 for a detailed illustration of recommended color code procedures for Phases A, B, and C.

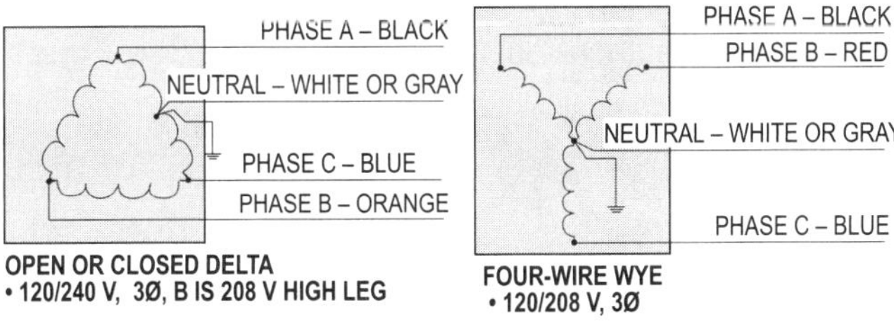

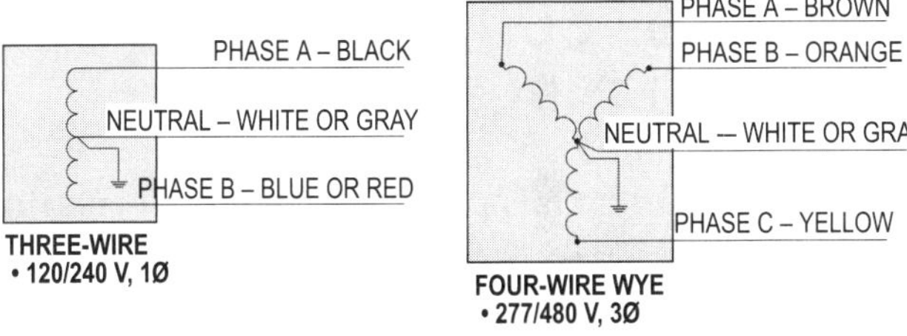

CONDUCTOR IDENTIFICATION
NEC 310.12(A), (B), AND (C)

Figure 8-9. Until the 1975 NEC was published, ungrounded (phase) conductors were recommended to be color coded.

Section **210.5(C)** requires the ungrounded (phase) conductors of multiwire circuits where more than one nominal voltage system exists in a building's wiring system to be identified by phase or line and system at all termination, connection, and splice points. The identification of each conductor, based on the voltage, shall be marked with tape, tagging, spray paint where permitted or other effective means accepted by the AHJ.

Design Tip: This is not a color code requirement; it is a means of identification so each system's voltage may easily be determined and identified.

The method utilized for conductors originating within each branch circuit panelboard or similar branch circuit distribution equipment shall be documented in a manner that is readily available, or shall be permanently posted at each panelboard housing such branch circuits. **(See Figure 8-10)**

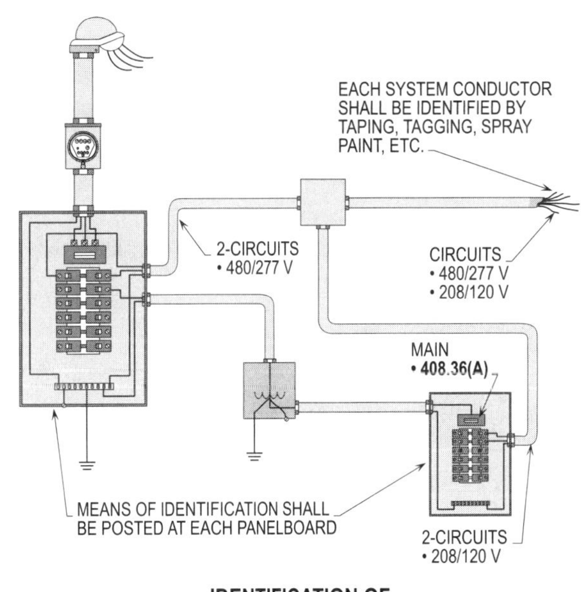

Figure 8-10. Where there is more than one system voltage supplying a premises, proper identification of circuits and panelboards shall be provided.

CONDUCTORS FOR GENERAL WIRING TABLE 310.13(A)

Conductors used for general wiring shall be insulated unless they are permitted to be otherwise due to the type of installation. The conductor insulation shall be of a type that is approved for the voltage, operating temperature, and location of use per **Table 310.13** in the NEC. Insulated conductors shall be rated for terminating to terminals or lugs rated at 60°C or 75°C or any combination of such per **110.14(C)** and **Table 310.16**. **(See Figure 8-20)**

TERMINAL RATINGS
110.14(C)(1)(a) AND (C)(1)(b)

The procedure used in verifying a terminal's overcurrent protection device rating is to check the rating listed on the overcurrent protection device to see if it is 60°C, 60°C/75°C, or 75°C. Where the rating of the overcurrent protection device is 100 amps or less, the terminal rating is 60°C if it is not marked as mentioned above. Overcurrent protection devices rated over 100 amps are rated at 75°C and may be loaded to the 75°C ampacities of conductors, which are found in **Tables 310.16 through 310.19** in the NEC. Note that motors shall be permitted to be cabled with 75°C terminals and ampacities.

For example: What is the allowable ampacity of a 4 AWG THHN copper conductor connected to an overcurrent protection device supplying power to equipment where all terminals are rated at 75°C? **(See Figure 8-11** and **110.14(C)** for further details.)

Step 1: Finding ampacity
Text in book 75°C terminals -
Table 310.16
4 AWG THHN cu. has an
allowable ampacity of 85 A

Solution: **A 4 AWG THHN cu. conductor may be loaded to 85 amp with 75°C terminals. Note:** See **110.14(C)(1)**.

Design Tip: Conductors shall be selected according to the lowest temperature rating of the equipment, and the allowable ampacities used accordingly.

For example, a piece of equipment with 60°C terminals shall be permitted to be supplied by 90°C THHN conductors. However, the allowable ampacity of the 90°C conductor shall be used at 60°C ampacities per **Table 310.16**.

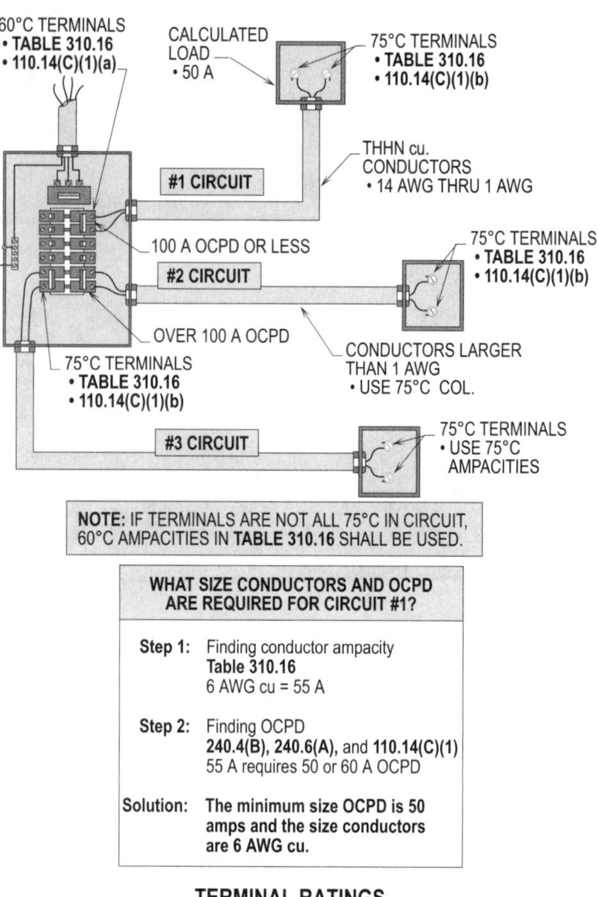

Figure 8-11. The allowable ampacity of conductors shall be determined by the markings on the overcurrent protection devices and electrical equipment supplied. **(See Figures 8-4 and 14-16)**

LONG- AND SHORT-TIME RATINGS OF CONDUCTORS
310.10 AND TABLE 310.16

There are two current ratings of conductors in an electrical system and they are as follows:

- Long-time current rating
- Short-time current rating

The long-time current rating of conductors shall be determined by selecting the allowable ampacity of the conductor based upon its size, material, and insulation.

For example: What is the allowable ampacity of one of three 8 AWG THHN copper conductors in a metal raceway system used at 60°C?

Step 1: Finding the rating
Table 310.16
8 AWG THHN cu. = 40 A

Solution: **The 8 AWG THHN copper conductors have an allowable ampacity of 40 amps at 60°C terminals.**

The short-time current of copper conductors shall be determined by dividing the CM rating of the conductor by 42.3 where the conductors are in the same raceway or cable assembly. If a single conductor is installed in a raceway or routed by itself, the CM rating shall be divided by 30. This method of calculating short-time current ratings is a rule of thumb method and should be used only as such.

For example: What is the short-time current rating of a 8 AWG THHN copper conductor in a conduit system with other conductors?

Step 1: Finding CM
Table 8, Ch. 9
8 AWG THHN cu. = 16,510 CM

Step 2: Finding the rating
I = CM ÷ 42.3
I = 16,510 CM ÷ 42.3
I = 390 A

Solution: **The short-time current rating is 390 amps for each 8 AWG in the conduit at 60°C terminals.**

For example: What is the short-time current rating of a 6 AWG THHN copper grounding electrode conductor that is run in a raceway to a water pipe?

Step 1: Finding the CM
Table 8, Ch. 9
6 AWG THHN cu. = 26,240 CM

Step 2: Finding the rating
I = CM ÷ 30
I = 26,240 CM ÷ 30
I = 875 A

Solution: **The short-time current rating of the 6 AWG is 875 amps.**

Design Tip: The short-time current rating in amps is based on a duration of 5 seconds or less.

See Figure 8-12 for a detailed illustration of calculating and determining long- and short-time current ratings of conductors using the rule of thumb method.

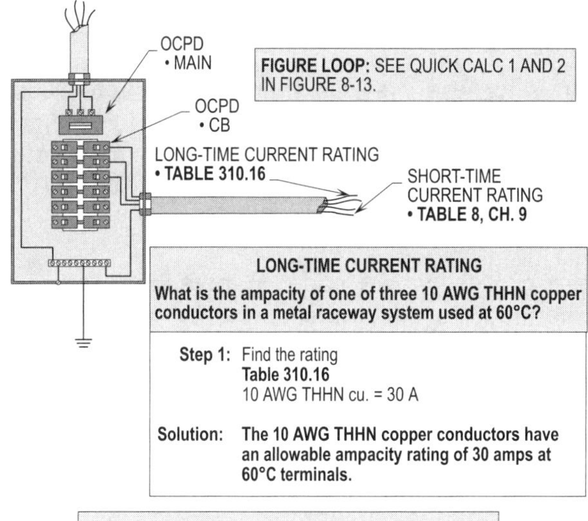

Figure 8-12. Calculating the long-time current ratings and short-time current ratings of conductors.

DERATING CIRCUIT CONDUCTORS TABLES 310.16 AND 310.19

Table 310.16 in the NEC shall not permit more than three current-carrying conductors to be enclosed in a raceway, cable, or buried side by side in the earth. The Table states that the allowable ampacities of conductors listed are limited to a surrounding temperature of 86°F (30°C) or less. Cables or raceways enclosing more than three current-carrying conductors and routed through an ambient temperature of over 86°F are subject to derating per **310.15(B)(2)(a)** to **Table 310.16** and the correction factors beneath the Table.

THREE OR MORE CURRENT-CARRYING CONDUCTORS 310.15(B)(2)(a) TO AMPACITY TABLES 0-2000 VOLTS

Cables or raceways enclosing four or more current-carrying conductors shall be derated per **310.15(B)(2)(a)** to ampacity Tables 0-2000 volts. Conductors considered current-carrying by the NEC are all ungrounded (phase) conductors.

Neutrals for wye systems shall be considered current-carrying where the major portion of the load (51 percent or greater) consists of harmonic currents such as electric discharge lighting, PC's, etc., per **310.15(B)(4)(c)** to ampacity Tables 0-2000 volts. **(See Figure 8-8)**

Control circuits that are loaded to more than 10 percent of the conductors allowable ampacity ratings shall be considered current-carrying per **725.51(A)**, **(B),** and **(C)**.

For example: What is the current-carrying ampacity of 9 - 12 AWG THHN copper conductors pulled through 3/4 in. (21) EMT? (All current-carrying)

Step 1:	Finding amperage of conductor **Table 310.16** 12 AWG THHN cu. = 30 A
Step 2:	Finding derating factors **Table 310.15(B)(2)(a)** 9 conductors requires 70%
Step 3:	Finding allowable ampacity **Table 310.15(B)(2)(a)** 30 A x 70% = 21 A
Solution:	**The allowable ampacity is limited to 21 amps for each conductor.**

See **Figure 8-13** for a detailed illustration of how to calculate and determine the allowable ampacities when applying the adjustment factors of **Table 310.15(B)(2)(a)**.

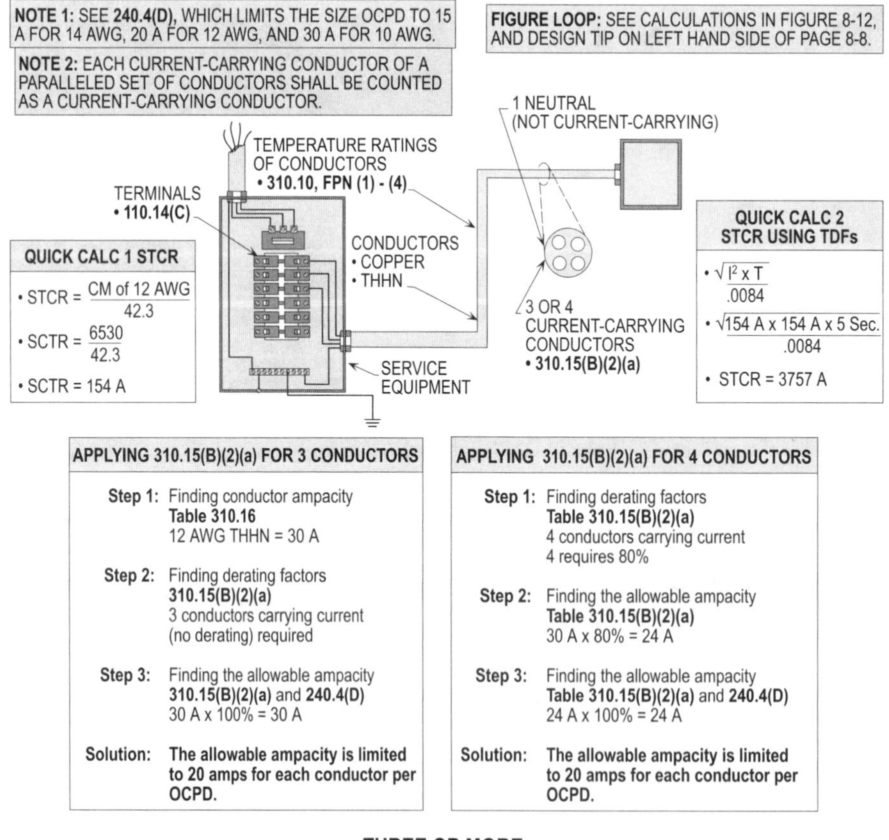

Figure 8-13. Determining the allowable ampacities of conductors based upon three or four or more current-carrying conductors pulled through a raceway or installed in a cable.

DERATING FOR AMBIENT TEMPERATURE CORRECTION FACTORS TABLE 310.16

Conductors routed through ambient temperatures exceeding 86°F shall be derated according to the correction factors of **Table 310.16**. The derating factors are listed in the ampacity correction factor chart below **Table 310.16**, and they shall be selected based on the ambient temperature that the conductors are exposed to. The ampacity correction factors shall be based on the material, insulation, and the size of the conductors utilized.

For example: What is the allowable ampacity of 4 - 10 AWG THHN copper conductors routed through an attic in an ambient temperature of 120°F?

Step 1: Finding amperage of conductors
Table 310.16
10 AWG THHN cu. = 40 amps

Step 2: Finding correction factors
Table 310.16
120°F requires 82%

Step 3: Finding allowable ampacity
Ampacity correction factors -
Table 310.16
40 A x 82% = 32.8 A

Solution: The allowable ampacity is limited to 32.8 amps for each conductor.

See **Figure 8-14** for a detailed illustration for calculating the allowable ampacities of conductors when correction factors due to ambient temperature are applied.

APPLYING BOTH DERATING FACTORS TABLE 310.15(B)(2)(a) AND AMPACITY CORRECTION FACTORS TO TABLE 310.16

Four or more current-carrying conductors enclosed in a cable or raceway and routed through an ambient temperature above 86°F shall have their allowable ampacities derated twice.

Too many current-carrying conductors in a cable or raceway have problems dissipating heat into the surrounding ambient medium. The load on the conductors must heat the cable or raceway above the surrounding ambient temperature before harmful heat can be dissipated. **[See 310.10, FPNs (1) and (4)]**

Derating the current-carrying ampacity of the conductors prevents overloading and deterioration of insulation.

For example: What is the allowable ampacity of 6 - current-carrying 14 AWG THHN copper conductors routed through an ambient temperature of 105°F?

Step 1: Finding amperage of conductors
Table 310.16
14 AWG THHN cu. = 25 A

Step 2: Finding derating factor
Table 310.15(B)(2)(a)
to Table 310.16
6 conductors requires 80%

Step 3: Finding derating factor
Correction factors to Table 310.16
105°F requires 87%

Step 4: Finding allowable ampacity
Table 310.15(B)(2)(a)
and correction factors
25 A x 80% x 87% = 17.4 A

Solution: The allowable ampacity is limited to 17.4 amps for each conductor.

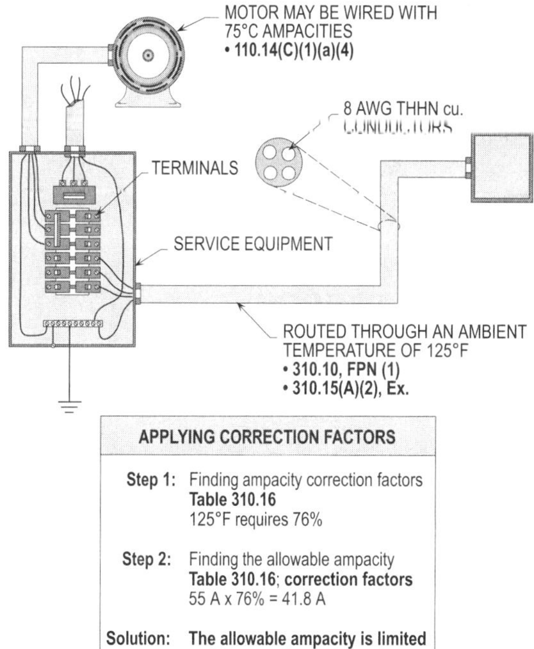

Figure 8-14. Calculating the allowable ampacity of conductors where they are routed through ambient temperatures above 86°F.

See Figure 8-15 for a detailed illustration of calculating and determining ampacities for four or more current-carrying conductors based upon an ambient temperature above 86°F.

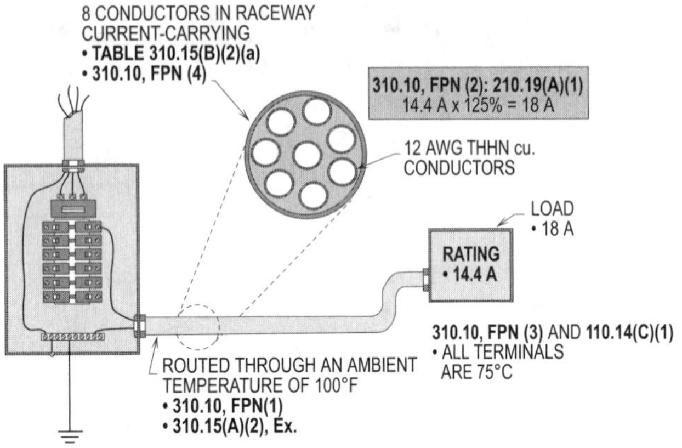

FINDING ALLOWABLE AMPACITY OF CONDUCTORS

Step 1: Finding amperage of conductors
Table 310.16
12 AWG THHN = 30 A

Step 2: Applying adjustment factor
Table 310.15(B)(2)(a)
8 conductors require 70%

Step 3: Applying correction factors
Table 310.16
100°F requires 91%

Step 4: Finding allowable ampacity
Table 310.15(B)(2)(a)
& Correction Factors
30 A x 70% x 91% = 19 A

Step 5: Finding OCPD
Table 310.16 and **240.4(D)**
12 AWG conductor with 19 A requires 20 A OCPD

Solution: **Section 240.4(D) and Table 310.16 requires a 20 amp overcurrent protection device on 12 AWG conductors.**

APPLYING BOTH DERATING FACTORS
TABLE 310.15(B)(2)(a)
AMPACITY CORRECTION FACTORS TO TABLE 310.16

Figure 8-15. Determining the allowable ampacity and size overcurrent protection device for conductors routed through an ambient temperature of more than 86°F with more than three current-carrying conductors (total of eight) in a cable or raceway.

APPLYING 50 PERCENT LOAD DIVERSITY FACTOR TABLE B.310.11

A 50 percent load diversity shall be permitted to be applied to the conductor ampacity when all the conductors are not loaded at the same time, if permitted by the AHJ.

See Figure 8-16 for a detailed illustration of applying derating factors with or without load diversity.

WHEN TO APPLY CORRECTION FACTORS 310.15(A)(2), Ex.

Ambient temperature will not affect the allowable ampacity of conductors per **Table 310.16** if that portion of the conductors routed through the higher ambient is 10 ft (3 m) or less and does not exceed 10 percent of the total length of such conductors.

Design Tip: The higher (allowable) ampacity shall be permitted to be utilized for the lower (allowable) ampacity if the length is no greater than 10 ft (3 m) or no greater than 10 percent of the circuit conductors or branch circuit elements, whichever is less.

For example: What size overcurrent protection device is required for 5 current-carrying 12 AWG THHN copper conductors routed through an ambient temperature of 102°F?

Step 1: Finding amperage of conductors
Table 310.16
12 AWG THHN cu. = 30 A

Step 2: Finding derating factors
Table 310.15(B)(2)(a)
5 conductors require 80%
102°F requires 91%

Step 3: Finding allowable amperage
Table 310.15(B)(2)(a)
30 A x 80% x 91% = 21.8 A

Step 4: Finding OCPD
Table 310.16; 240.4(D)
12 AWG conductor with
21.8 A requires 20 A OCPD

Note: This load of 20 amps is calculated per **210.19(A)(1)** and **210.20(A)**.

Solution: **Section 240.4(D) and Table 310.16 requires a 20 amp overcurrent protection device on 12 AWG conductors.**

As stated above, the temperature surrounding the conductor may vary from 86°F along the conductor length. However, if the ambient temperature is different than 86°F, the allowable ampacity of the conductor as listed in **Table 310.16** shall be corrected according to the correction factors below the Table. A derating factor (adjustment) shall be applied for four or more current-carrying conductors.

For example, three current-carrying copper conductors in a raceway exposed to a 40°C ambient temperature shall have their allowable ampacities derated by 91 percent per correction factors to **Table 310.16**.

APPLYING 50% LOAD DIVERSITY

What is the allowable ampacity of 16 - 12 AWG THHN cu. conductors, with diversity, located in the same raceway?

Step 1: Table 310.16
12 AWG THHN cu = 30 A

Step 2: Table B.310.11
16 conductors = 70%
30 A x 70% = 21 A

Solution: **The allowable ampacity of each conductor is 21 amps; however, only 8 can have this ampacity rating.**

APPLYING 50% LOAD DIVERSITY

What is the allowable ampacity of each conductor when applying 50% load diversity?

Step 1: FPN to Table B.310.11
$W = I^2 \times R \times \text{\# of conductors} \times 50\%$
$W = 30^2 \times 1 \times 16 \times 50\% = 7200\ W$

Step 2: $A = W \div \text{\# of conductors} = \sqrt{W} = A$
$A = 7200\ W \div 16 = \sqrt{450} = 21.2\ A$

Solution: **The allowable ampacity for 8 conductors is 21.2 amps with 50 percent diversity. See QUICK CALC for ampacity, using formula.**

APPLYING DERATING FACTORS

What is the allowable ampacity of 16 - 12 AWG THHN cu. conductors, without diversity, located in the same raceway?

Step 1: Table 310.16
12 AWG THHN cu = 30 A

Step 2: Table 310.15(B)(2)(a)
16 conductors = 50%
30 A x 50% = 15 A

Solution: **The allowable ampacity of 16 conductors is 15 amps.**

QUICK CALC 1
APPLYING TABLE B.310.11, FPN
(LOADING 8 WIRES)

- $A_2 = \sqrt{.5 \times N \div E \times A_1}$
- $A_2 = \sqrt{.5 \times 16 \div 16}$
- $A_2 = \sqrt{707 \times 30\ A}$
- $A_2 = 21.2\ A$

QUICK CALC 2
LOADING 12 WIRES

- $A_2 = \sqrt{.5 \times N \div E \times A_1}$
- $A_2 = \sqrt{.5 \times 16 \div 12 \times A_1}$
- $A_2 = 81\% \times 30\ A \times 70\%$
- $A_2 = 17\ A$

APPLYING 50 PERCENT
LOAD DIVERSITY FACTOR
TABLE B.310.11

Figure 8-16. Applying derating factors to determine allowable ampacities of conductors with load diversity.

BRANCH-CIRCUIT ELEMENTS 422.10(A) AND (B), AND 422.11(A) THRU (G)

The branch-circuit conductors and overcurrent protection devices are found after the derating factors have been applied. The branch-circuit elements shall be determined based upon calculated load, and this load shall be equal to or less than the allowable ampacity of the conductors after derating factors have been applied.

For example: What size overcurrent protection device is required for 4 - 14 AWG THWN copper current-carrying conductors in a raceway or cable?

Step 1: Finding amperage of conductors
Table 310.16
14 AWG THWN cu. = 25 A

Step 2: Finding derating factors
Table 310.15(B)(2)(a)
4 conductors requires 80%

Step 3: Finding allowable amperage
Table 310.15(B)(2)(a)
25 A x 80% = 20 A

Solution: **Section 240.4(D) and Table 310.16 require a 15 amp overcurrent protection device on 14 AWG conductors.**

For example: What size overcurrent protection device is required for 4 - 10 AWG THHN copper conductors that are run through an ambient temperature of 125°F? (Only three are current-carrying.)

Step 1: Finding amperage of conductors
Table 310.16
10 AWG THHN = 40 A

Step 2: Finding derating factors
Correction factors to Table 310.16
125°F requires 76%

Step 3: Finding allowable amperage
Correction factors to Table 310.16
40 A x 76% = 30.4 A

Solution: **Section 240.4(D) and Table 310.16 require a 30 amp overcurrent protection device on 10 AWG conductors.**

See Figure 8-17 for a detailed illustration of sizing and selecting elements of a branch circuit based upon conditions of use.

INDIVIDUAL CIRCUITS
210.19(A)(1) AND 210.20(A)

Individual branch circuits supply power to special appliance loads. Special appliance loads are loads such as heating units, water heaters, cooking equipment, etc. These loads require larger circuits to supply the greater amperage or VA ratings. Loads for special appliances served by individual circuits shall be determined by multiplying the amperage or VA by 100 percent for noncontinuous loads and 125 percent for continuous loads.

The following Sections in the NEC outline requirements that shall be used for sizing the branch-circuit elements of individual circuits.

- Disposals **430.22(A); 430.52; 422.10; 422.11**
- Compactor **430.22(A); 430.52; 422.10; 422.11**
- Dishwasher **430.22(A); 430.52; 422.10; 422.11**
- Cooking equip. **220.55; Table 220.55; 422.33(B)**
  - Ranges
  - Cooktops
  - Ovens
- Water heater **422.13**
- Heating units **424.3(B)**
- A/C units **440.22(A); 440.32; 440.62(B)**
  - Central
  - Window
- Motors **430.6(A)(1); 430.22(A); 430.52; Table 430.52**
- Welders **630.11; 630.12**
- X-ray **660.6(A); 517.73**

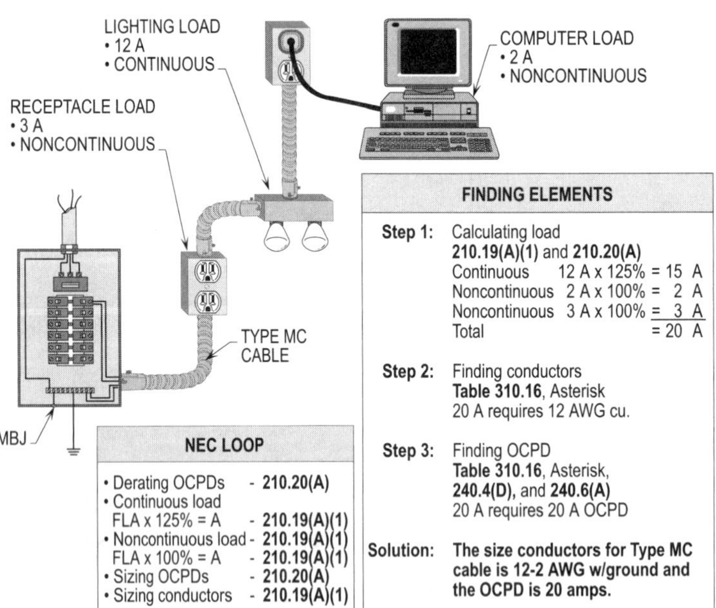

Figure 8-17. The size conductors and overcurrent protection devices shall be sized at 125 percent of the continuous loads plus 100 percent of the noncontinuous loads and derating factors, if necessary.

For example: What size THWN copper conductors and overcurrent protection device are required for an individual branch circuit to a small processing machine with a nameplate current rating of 42 amps (continuous load)?

Step 1:   Finding amperage
210.19(A)(1); 210.20(A)
42 A is the circuit current rating

Step 2:   Calculating load
210.19(A)(1); 210.20(A)
42 A x 125% = 52.5 A

Step 3:   Selecting conductors and OCPD
Table 310.16; 240.6(A)
52.5 A requires 60 A OCPD
65 A conductor allows 70 A OCPD
per **240.4(B)**

Solution: **A 60 amp overcurrent protection device and 6 AWG THWN copper conductors shall be permitted based on load, or 70 amp based on conductor ampacity.**

See Figure 8-18 for a detailed illustration of how to calculate and size the elements of an individual circuit.

CONDUCTORS IN PARALLEL
310.4

These requirements apply to copper and aluminum conductors connected in parallel so as to form a common conductor. Conductors sized 1/0 AWG or larger, comprising each phase or neutral, shall be permitted to be paralleled, that is, both ends of the paralleled conductors are connected together so as to form a single conductor. In paralleling conductors, the following conditions for the conductors shall be complied with:

- Be of the same length.
- Be of the same conductor material.
- Be the same circular-mil area.
- Be the same type of insulation.
- Be terminated in the same manner.
- Where run in separate raceways or cables, the raceways or cables shall have the same electrical characteristics.

See Figure 8-19 for a detailed illustration pertaining to the rules of paralleling conductors 1/0 AWG and larger.

Exceptions that permit conductors of any size to be run in parallel are as follows:

- Conductors supplying control power to indicating instruments, contactors, relays, solenoids and similar control devices, or for frequencies of 360 Hz and higher, may be run in parallel in sizes smaller than 1/0 AWG provided that the following conditions are complied with:
  - They are in the same raceway or cable.
  - The ampacity of each conductor is sufficient to carry the entire load shared by the parallel conductors.
  - If one parallel conductor should happen to become disconnected, the ampacity of each conductor is not exceeded.

Design Tip: When necessary to run equipment grounding conductors with paralleled conductors, these equipment grounding conductors shall be treated the same as the other conductors and sized by the provisions of **Table 250.122**.

LOCATIONS
310.8

There are locations, besides those exposed directly to the weather, in which conductors shall be considered to be installed in wet locations. In such locations, conductor insulation shall be either suitable for wet locations or lead covered. However, if lead covered, the insulation need not be suitable for wet locations. Such locations are as follows:

- Underground.
- In concrete slabs or masonry in direct contact with the earth, or masonry subject to moisture.
- In wet locations.
- Where condensation or accumulation of moisture within the raceway is likely to occur.

The following types of insulation are moisture-resistant:

- RHW; RHW-2; MTW
- THHW
- TW
- THW; THW-2
- THWN; THWN-2; ZW
- XHHW; XHHW-2
- Lead covered or other
- Types listed for use in wet locations

<table>
<tr><td>

APPLYING 210.19(A)(1)

What size OCPD and THWN copper conductors are required for the branch circuit?

Step 1: Calculating OCPD and conductors
210.19(A)(1) and **210.20(A)**
Conductors   OCPD
11 A x 125% = 13.75 A
6 A x 100% = 6.00 A
Total load = 19.75 A

Step 2: Selecting OCPD and conductors
Table 310.16, 240.4(B), and **240.4(D)**
19.75 A requires 20 A OCPD and
12 AWG THWN cu conductors.

Solution: A 20 amp OCPD and 12 AWG THWN cu. conductors are required.

Note: The above calculation does not include any adjustment or correction factors.

</td><td>

APPLYING 210.19(A)(1) AND CORRECTION FACTORS

What size OCPD and THHN copper conductors are required for the branch circuit, based upon an ambient temperature of 135°F?

Step 1: Finding load
210.19(A)(1) and **210.20(A)**
Conductors   OCPD
11 A x 125% = 13.75 A
6 A x 100% = 6.00 A
Total load = 19.75 A

Step 2: Applying corrections factors
Table 310.16, using 12 AWG THHN cu.
30 A x 71% = 21.3 A

Step 3: Finding conductor and OCPD
Table 310.16, 240.4(B), 240.4(D), and **240.6(A)**
21.3 A supplies 19.75 A load
19.75 A requires 20 A OCPD

Solution: A 20 amp OCPD and 12 AWG THHN cu. conductors are required.

</td></tr>
</table>

INDIVIDUAL CIRCUITS
NEC 210.19(A)(1)
NEC 210.20(A)

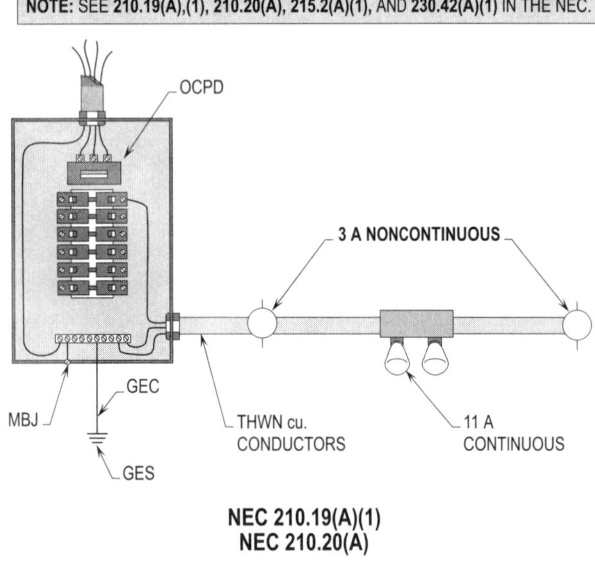

Figure 8-18. Calculating the load and sizing the elements of an individual circuit.

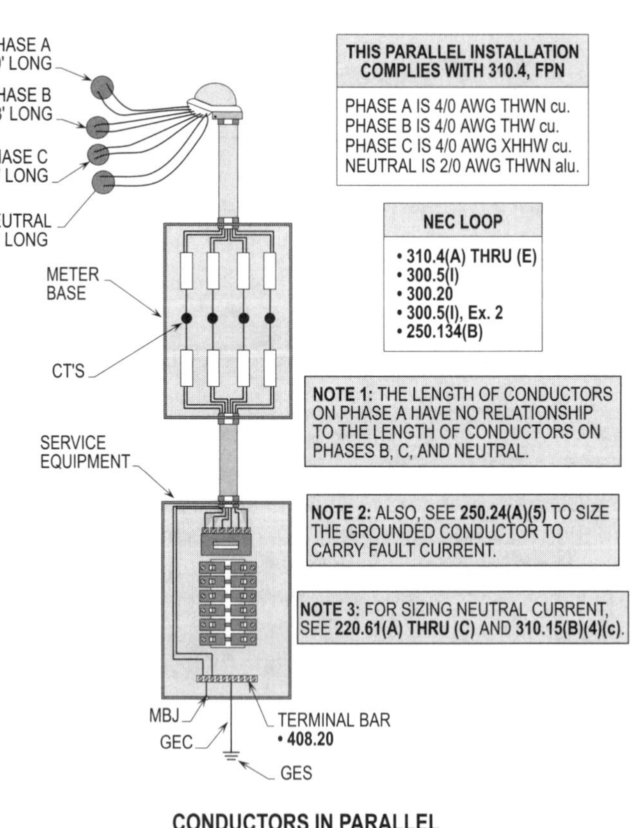

Figure 8-19. Conductors of one phase do not have to be the same as another phase, grounded (phase) conductor, or grounded (neutral) conductor.

CONDUCTOR CONSTRUCTIONS AND APPLICATIONS
310.13(A)

Table 310.13(A) in the NEC lists the different types of insulated conductors used for general wiring. The conductors listed shall be permitted to be used for any voltage up to and including 600 volts.

Conductor types shall be permitted to be suitable for both wet and dry locations. Some types are suitable only for dry locations. Conductors listed as suitable for dry locations shall not be permitted to be used in wet locations. Note that these are insulations for conductors that are listed for dry and damp locations. Conductors in **Table 310.13** shall be permitted to be used in any of the wiring methods recognized in **Chapter 3**, if utilized as specified in their respective Tables. **(See Figure 8-20)**

Insulation resistance may vary a great deal when exposed to different temperatures. The resistance is higher at lower temperatures and steadily decreases as the temperature rises. In other words, conductors exposed to high temperature may cause insulation to have a much lower insulation reading when tested.

Design Tip: Thermoplastic insulation may stiffen at temperatures colder than minus 10°C (plus 14°F), therefore, care must be exercised during installation. Thermoplastic insulation may also be deformed at normal temperatures where subjected to pressure, requiring care to be exercised during installation and at points of support.

TYPICAL AMBIENT DESIGN TEMPERATURE
1971 NEC

The 1971 NEC included Table 310-20(c), which contained typical ambient temperatures and which could be used for correcting ampacities of conductors according to the correction factors below Table 310-16. This Table was deleted in the 1975 NEC. However, the information concerning typical ambient temperatures in this Table may be used as a guideline when actual ambient temperatures are unknown. **(See Figure 8-21)**

TABLE 310.13(A) - CONDITIONS OF USE					
	COLUMN 2	COLUMN 3	COLUMN 4	COLUMN 5	COLUMN 6
TYPE LETTER	INSULATION	MAXIMUM OPERATING TEMPERATURE	APPLICATIONS PROVISIONS	SIZES AVAILABLE	OUTER COVERING
THHN	FLAME-RETARDANT, HEAT RESISTANT THERMOPLASTIC	90°C	DRY AND DAMP LOCATIONS	14-1000	NYLON JACKET OR EQUIVALENT
THWN	FLAME-RETARDANT, MOISTURE, AND HEAT RESISTANT THERMOPLASTIC	75°C	DRY AND WET LOCATIONS	14-1000	NYLON JACKET OR EQUIVALENT
XHHW	FLAME-RETARDANT, MOISTURE RESISTANT THERMOSET	90°C 75°C	DRY AND DAMP LOCATIONS WET LOCATIONS	14-2000	NONE
THW	FLAME-RETARDANT, MOISTURE, AND HEAT RESISTANT THERMOPLASTIC	75°C 90°C	DRY, DAMP, AND WET LOCATIONS ELECTRIC DISCHARGE LIGHTING EQUIPMENT • 410.31	14 - 2000	NONE
TW	FLAME-RETARDANT, MOISTURE RESISTANT THERMOPLASTIC	60°C	DRY AND WET LOCATIONS	14 - 2000	NONE

CONDUCTOR CONSTRUCTIONS AND APPLICATIONS NEC TABLE 310.13(A)

Figure 8-20. The above Table lists the most used insulations for conductors utilized in the general wiring of electrical systems. **Note:** THHN is used for hot temperatures, and XHHW is used for cold temperatures.

For Conductors Routed Outside:

- In conduit, use 50°C operating temperature
- In cable trays, use 45°C operating temperatures
- Note that the above operating temperatures are recommended practices.

TABLE 310.20(C) TO 1971 NEC		
LOCATION	TEMPERATURE	MINIMUM RATING OF REQUIRED CONDUCTOR INSULATION
WELL VENTILATED NORMALLY HEATED BUILDINGS	30°C (86°F)	* SEE NOTE BELOW
BUILDINGS WITH MAJOR HEAT SOURCES AS POWER STATIONS OR INDUSTRIAL PROCESSES	50°C (113°F)	75°C (167°F)
POORLY VENTILATED SPACES SUCH AS ATTICS	45°C (113°F)	75°C (167°F)
FURNACES AND BOILER ROOMS (MIN.) (MAX.)	40°C (104°F) 60°C (140°F)	75°C (167°F) 90°C (194°F)
OUTDOORS IN SHADE IN AIR	40°C (104°F)	75°C (167°F)
IN THERMAL INSULATION	45°C (113°F)	75°C (167°F)
DIRECT SOLAR EXPOSURE	45°C (113°F)	75°C (167°F)
PLACES ABOVE 60°C (140°F)		110°C (230°F)

NOTE: 60°C FOR UP TO AND INCLUDING 8 AWG COPPER AND UP TO AND INCLUDING 6 AWG ALUMINUM AND COPPER-CLAD ALUMINUM. 75°C FOR OVER 8 AWG COPPER AND 6 AWG ALUMINUM OR COPPER-CLAD ALUMINUM.

TYPICAL AMBIENT DESIGN TEMPERATURE
1971 NEC

Figure 8-21. The above table can be used with its ambient temperatures based upon specific locations when applying the correction factors to **Table 310.16**. **Note,** this table is a suggested table and not a mandatory table when designing and selecting ampacities of conductors based upon conditions of use.

DETERMINING AMPACITIES OF ELECTRICAL SYSTEMS OVER 600 VOLTS

The ampacities for solid dielectric insulated conductors rated over 2000 through 35,000 volts shall be permitted to be utilized in **Tables 310.67 through 310.86** and **310.60**. The ampacities of conductors in these tables are based on the Neher-McGrath method of calculating current-carrying capacity. However, the formula in **310.60(B)(1)** shall be permitted to be used to calculate the ampacities of conductors. This method can be very complex and time consuming, and such calculations shall be performed by proper engineering supervision. In some installations, the Neher-McGrath method of calculating ampacities of conductors will produce lower ampacity ratings. As a result, a lower installation cost for the consumer is possible.

DETERMINING CONDUCTOR AMPACITIES IN CABLES TABLES 310.67 THROUGH 310.86

Figure 310.60 to **Tables 310.67 through 310.86** contains cables and duct banks of different numbers with detail identification numbers that can be correlated to specific ampacities tables, and their correct ampacities selected based upon voltage and size.

For example, **Table 310.85** covers the ampacities of three triplexed single insulated copper conductors, and **Table 310.86** covers the ampacities of three triplexed single insulated aluminum conductors. For both Tables, the ampacity is given in two columns, one for 2001-5000 volts and a second for 5001-35,000 volts. **(See Figure 8-22)**

DETERMINING CONDUCTOR AMPACITIES IN DUCT BANKS TABLES 310.67 THRU 310.86

Table 310.77 shall be permitted to be used to determine the ampacities of conductors routed through duct banks that are buried in the ground. Such ampacities are based upon the voltage, number, configuration and installation arrangement of duct banks. **(See Figure 8-23)**

DERATING DUE TO OBSTRUCTIONS 310.60(C)(2)(a) TO TABLES 310.67 THRU 310.86

Section **310.60(C)(2)(a)** to **Tables 310.67 through 310.86** contains information concerning derating ampacities of conductors on the basis of burial depth. The rule does not require derating of conductor ampacities where the duct bank burial depth to miss the obstruction is less than 25 percent of the total run located below the 30 in. (750 mm) restriction. **(See Figure 8-24)**

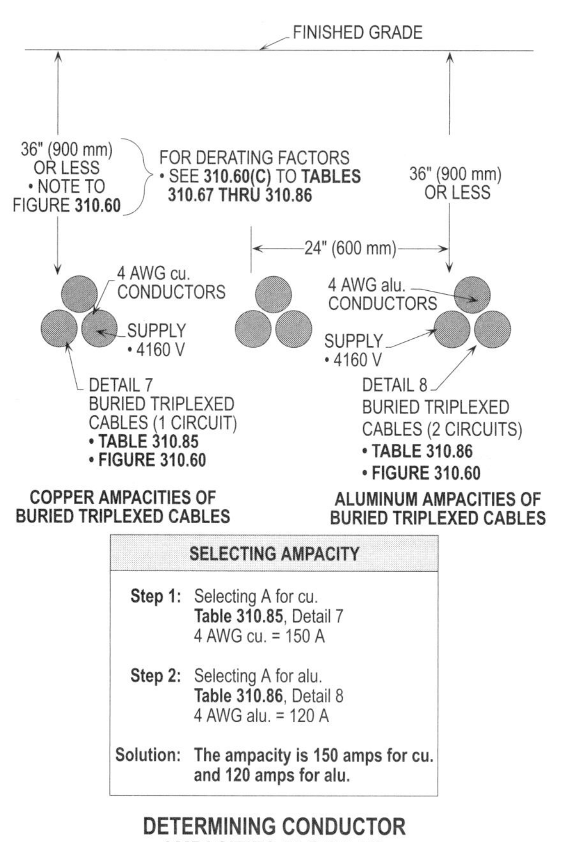

Figure 8-22. Determining ampacities of conductors utilizing the specific ampacities of **Tables 310.67 through 310.86**.

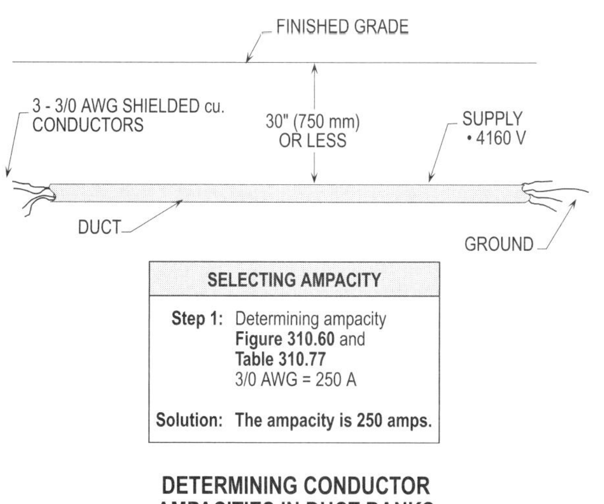

Figure 8-23. Determining ampacities of conductors utilizing the specific ampacities of **Tables 310.67 through 310.86**.

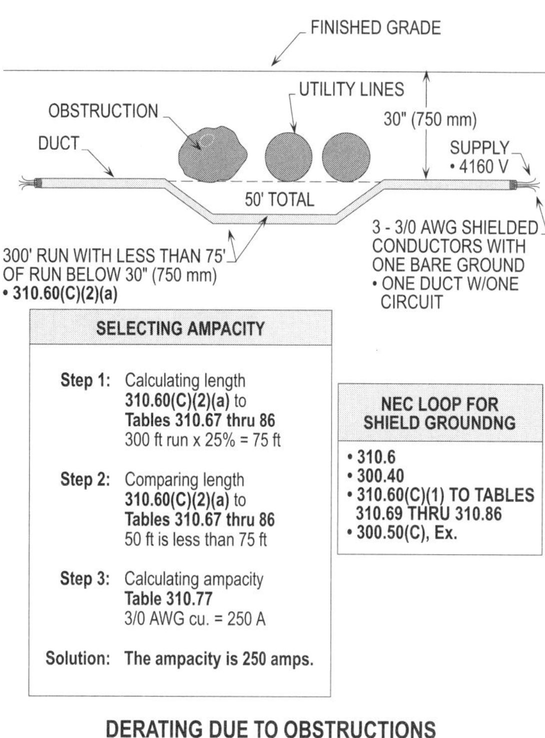

Figure 8-24. The above installation does not require the conductors' ampacities to be derated due to greater burial depths to miss the obstruction.

DERATING DUE TO OBSTRUCTIONS 310.60(C)(2)(b) TO TABLES 310.67 THRU 310.86

Section **310.60(C)(2)(b)** states that where the burial depth is deeper than 30 in. (750 mm) and 25 percent of the total run in a specific underground duct is exceeded, an ampacity derating factor of 6 percent per increased foot of depth for all values of RHO shall be utilized. See **Figure 310.60 to Tables 310.67 through 310.86** for rules pertaining to the installation of cables. **(See Figure 8-25)**

SPACING OF CONDUITS 310.60(C)(3) AND FIGURE 310.60

For the spacing of conduits entering enclosures, see the requirements of **310.60(C)(3)**. Note that the spacing of such conduits per **Figure 310.60** shall be permitted to be reduced without derating the ampacity of the conductors.

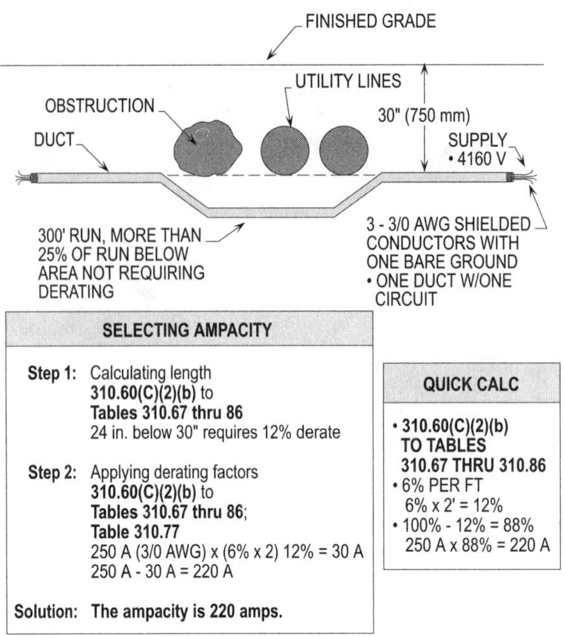

Figure 8-25. The above installation requires the conductors' ampacities to be derated as shown.

Chapter 8. Conductors

Section   Answer

_____________ _____________ 1. Conductors routed through an ambient temperature greater than _____ shall be derated per correction factors based upon surrounding temperatures.
    (a) 85° F   (b) 86° F
    (c) 87° F   (d) 90° F

_____________ _____________ 2. Ambient temperature will not affect the ampacity of conductors if that portion of the conductors routed through the higher ambient is _____ ft or less and does not exceed 10 percent of the total length.
    (a) 7   (b) 9
    (c) 10   (d) 12

_____________ _____________ 3. Loads that are continuous shall have their full-load currents sized at _____ percent, and this value added to noncontinuous loads at 100 percent. (Branch-circuit conductors)
    (a) 80   (b) 100
    (c) 125   (d) 150

_____________ _____________ 4. If 4 or more current-carrying conductors are pulled through a raceway or bundled in a cable for a distance greater than _________ in., the allowable ampacity of such conductors shall be reduced (derated).
    (a) 6   (b) 12
    (c) 18   (d) 24

_____________ _____________ 5. A continuous load is where the current in amps can operate for a period of time of _____ hours or more.
    (a) 3   (b) 4
    (c) 5   (d) 6

_____________ _____________ 6. An _____ conductor is a circuit conductor that is not grounded but carries current to the load.
    (a) equipment grounding   (b) grounded
    (c) ungrounded   (d) isolated

_____________ _____________ 7. The ungrounded (phase) conductors shall be permitted to be identified with any color of insulation or tagging except _____.
    (a) white   (b) gray
    (c) green   (d) all of the above

_____________ _____________ 8. Where installed in raceways, conductors of size _____ AWG and larger shall be stranded.
    (a) 8   (b) 6
    (c) 4   (d) 2

_____________ _____________ 9. Copper and aluminum conductors shall be permitted to be connected in parallel in sizes _____ AWG or larger.
    (a) 1   (b) 1/0
    (c) 2/0   (d) 4/0

_______________ _______________   **10.** Control circuits shall be current-carrying when they carry continuously more than _______ percent of their ampacity.

     (a) 10   (b) 20

     (c) 30   (d) 50

_______________ _______________   **11.** Insulated conductors of _______ AWG and smaller, when used as grounded conductors, shall have a continuous white or natural gray insulation.

     (a) 10   (b) 8

     (c) 6   (d) 4

_______________ _______________   **12.** Where the rating of the overcurrent protection device is _______ amps or less, the terminal rating shall be 60°C if it is not otherwise marked.

     (a) 60   (b) 100

     (c) 150   (d) 200

_______________ _______________   **13.** Conductor ampacities shall be derated where the duct bank burial depth to miss the obstruction is less than _______ percent of the total run located below the 30 in. restriction.

     (a) 10   (b) 15

     (c) 20   (d) 25

_______________ _______________   **14.** An ampacity derating factor of ___________ percent increased foot of depth for all values of RHO shall be utilized if burial depth is greater than, as in **310.60(C)(2)(a)** to **Tables 310.67 through 310.86**.

     (a) 3   (b) 6

     (c) 12   (d) 18

_______________ _______________   **15.** Where an insulated grounded conductor is larger than _______ AWG, it shall be identified by a white or gray colored insulation or by a distinctive marking at its terminal.

     (a) 6   (b) 4

     (c) 2   (d) 1

_______________ _______________   **16.** What size THHN copper conductors and overcurrent protection device are required for a circuit with a calculated load of 80 amps connected to 60°C terminals?

_______________ _______________   **17.** What is the long-time and short-time current rating of a 6 AWG THHN copper conductor in a conduit system with other conductors used at 60°C?

_______________ _______________   **18.** What is the ampacity rating for 4 - 10 AWG THHN copper conductors that are current-carrying?

_______________ _______________   **19.** What is the ampacity rating for 4 - 10 AWG THHN copper conductors (three current-carrying) routed through an ambient temperature of 125°F?

_______________ _______________   **20.** What is the ampacity rating for 8 - 10 AWG THHN copper conductors (all current-carrying) routed through an ambient temperature of 120°F?

_______________ _______________   **21.** What size overcurrent protection device is required for 4 current-carrying 12 AWG THHN copper conductors in a cable?

_______________ _______________   **22.** What size THWN copper conductors and overcurrent protection device are required for an individual branch circuit to a small processing machine with a nameplate current rating of 38 amps (continuous load)?

23. What size THHN copper conductors (2) and overcurrent protection device are required for an branch circuit with an 6 amp noncontinuous load and 10 amp continuous load, based upon an ambient temperature of 125°F?

24. What is the ampacity for a 6 AWG copper buried triplexed cables with a voltage supply of 4160? (Terminals are 90°C per **110.40.**)

25. What is the ampacity for 3 - 4/0 AWG shielded copper conductors with a voltage of 4160 with a 300 ft run of conduit, with 50 ft of run below 30 in. and 25 percent?

Overcurrent Protection Devices

The purpose of overcurrent protection devices is to monitor the current in a circuit and keep it at a level that will prevent overheating of conductors, elements, and equipment. Excessive current flowing in an electrical circuit generates heat, which raises the circuit's temperature. Such temperature depends entirely upon the amount of current flowing through the electrical circuit.

For example, if the temperature of the conductor is high due to excessive current flow, insulation may melt and cause short circuits or ground faults to occur. Current flowing through a conductor generates heat that is proportional to the square of the current. In other words, if the current is doubled, the amount of heat is increased to four times the original amount. Therefore, overcurrent protection devices shall be sized and selected to match the current-carrying capacity (ampacity) of the conductor and elements of the equipment or stages of circuit protection shall be provided.

PROTECTION OF EQUIPMENT
240.3

For overcurrent protection of appliances, motors, generators, etc., it is necessary to refer to the different Articles listed in this Section. In all installations, there are three main parts to be protected and they are as follows:

- Circuit conductors,
- Circuit elements, and
- Equipment.

The fuse or circuit breaker protecting the installation shall be of a rating small enough to protect both, according to the rules and regulations in the *National Electrical Code.* **(See Figure 9-1)**

Design Tip: If the overcurrent protection devices are sized greater than the ampacity of the conductors to allow a motor or A/C unit to start and run [or any load per **240.4(A) through (G)**], a second stage of protection shall be required to protect such equipment.

Note that the second stage of protection shall be permitted to be overloads in the motor controller or thermal protectors in the compressor used in an A/C unit. Time delay fuses that are properly sized shall also be permitted to be utilized to provide overload protection.

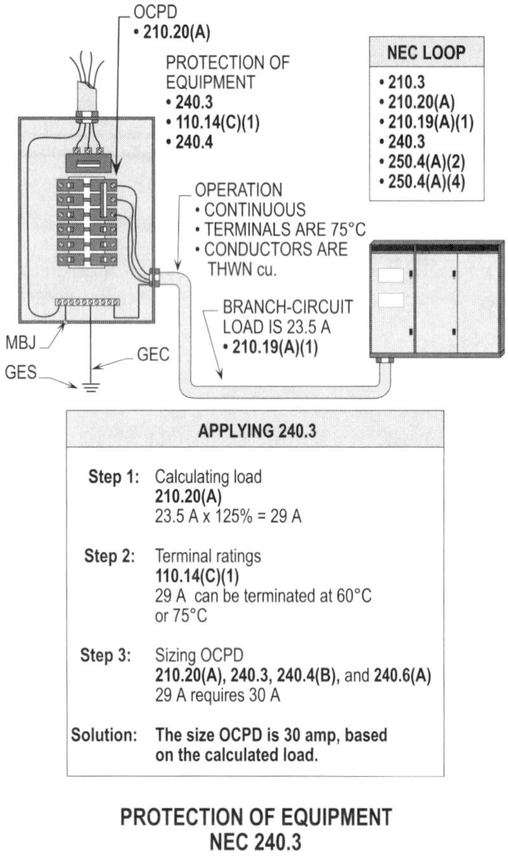

conductors, for different insulations, and for different ambient-temperature conditions. The overcurrent protection device shall be sized to protect the insulation of the conductors from damage caused by the current reaching an excessive value.

For example, a 2 AWG copper conductor with THWN insulation has an ampacity of 115 amps. A 100 or 110 amp fuse or circuit breaker setting would be the largest size permitted for protection of this conductor per **240.4(B)**. **(See Figure 9-2)**

Figure 9-1. The overcurrent protection device shall be sized and selected to properly protect the electrical equipment served.

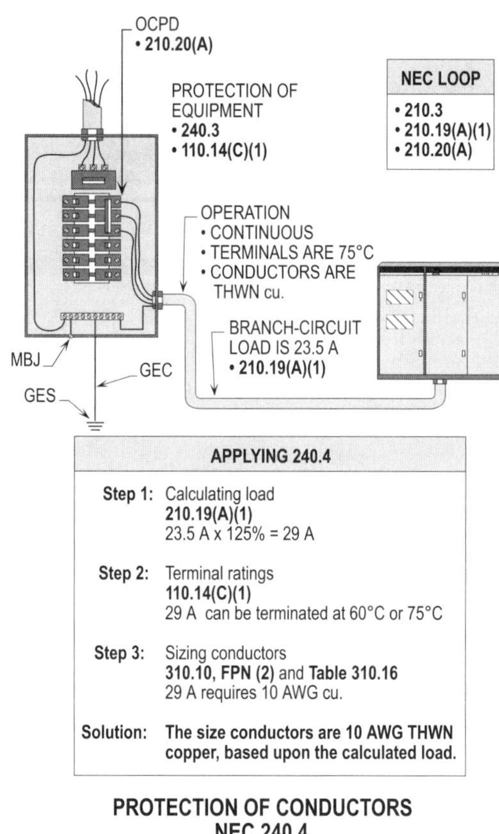

Figure 9-2. The conductors shall be sized based upon calculated load and selected to protect conductors and equipment.

PROTECTION OF CONDUCTORS 240.4

The general rule is that conductors shall be protected against overcurrent by a fuse or circuit breaker setting rated no higher than the ampacity of the conductor. Conductors have specific current-carrying ampacities for different sizes of

For further information, see **110.14(C)(1)** and **(C)(2)** to determine whether terminals and equipment ratings are mated or unmated.

However, there are exceptions [**Subdivisions (A) through (G)**] to the general rule, and in some cases the NEC allows the fuse size or circuit breaker setting to be above the ampacity of the conductor. Such exceptions listed in Parts are as follows:

- Power loss hazard
- Devices rated 800 amps or less
- Devices rated over 800 amps
- Tap conductors
- Motor-operated appliance circuit conductors
- Motor and motor control circuit conductors
- Phase converter supply conductors
- AC and refrigeration equipment circuit conductors
- Transformer secondary conductors
- Capacitor circuit conductors
- Electric welder circuit conductors
- Remote-control, signaling, and power-limited circuit conductors
- Fire alarm system circuit conductors

POWER LOSS HAZARD
240.4(A)

Conductor overload protection shall not be required where the interruption of the circuit would create a hazard, such as in a material-handling magnet circuit. However, short-circuit protection shall be provided to clear phase-to-phase shorts or ground faults should they occur.

DEVICES RATED 800 AMPS OR LESS
240.4(B)

If the standard current ratings of fuses or nonadjustable circuit breakers do not conform to the ampacity of the conductors being used, it shall be permitted to use the next larger standard rating when below 800 amps.

For example, the ampacity rating of a 2 AWG copper conductor with THWN insulation is 115 amps. There is no standard 115 amp fuse or circuit breaker setting. The nearest standard size above 115 amps is 125 amps. The 125 amp fuse or circuit breaker shall be permitted to be used in this case for protection of the conductor from short circuits. **(See Figure 9-3)**

DEVICES RATED OVER 800 AMPS
240.4(C)

If the overcurrent protection device is greater than 800 amps, the ampacity of the conductors shall be equal to or greater than the ampacities of the overcurrent protection device per **240.6(A)**. Note that **240.6(A)** permits an overcurrent protection device that is not listed in **240.6(A)** to be used if listed for such use. **(See Figure 9-4)**

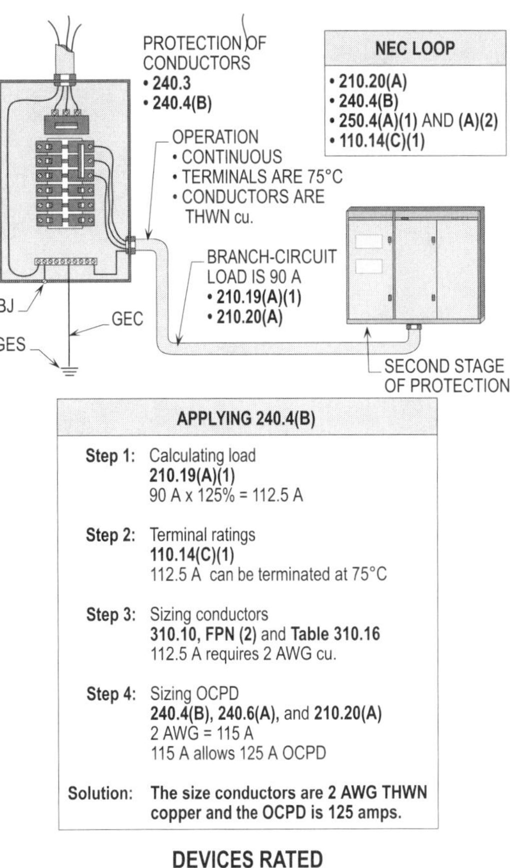

Figure 9-3. Where the ampacity of a conductor(s) does not correspond with a standard device per **240.6(A)**, the next higher rating shall be permitted to be used if 800 amps or less in rating.

TAP CONDUCTORS
240.4(E)

Smaller conductors shall be permitted to be tapped from larger conductors under certain conditions. Sections **240.5, 240.21(A),** and **210.19(A)(4)** in the NEC have tap rules, which permit a 18 in., 25 ft, and 100 ft tap respectively. Review these requirements carefully before attempting to make such a tap.

MOTOR-OPERATED APPLIANCE
CIRCUIT CONDUCTORS
240.4(G)

Motor-operated appliance circuit conductors are normally protected against overcurrent by the provisions listed in **Part II** of **Article 422**. **(See Figure 9-5)**

Design Tip: These conductors and equipment are not protected from overload if the overcurrent protection device exceeds 125 percent of the load served.

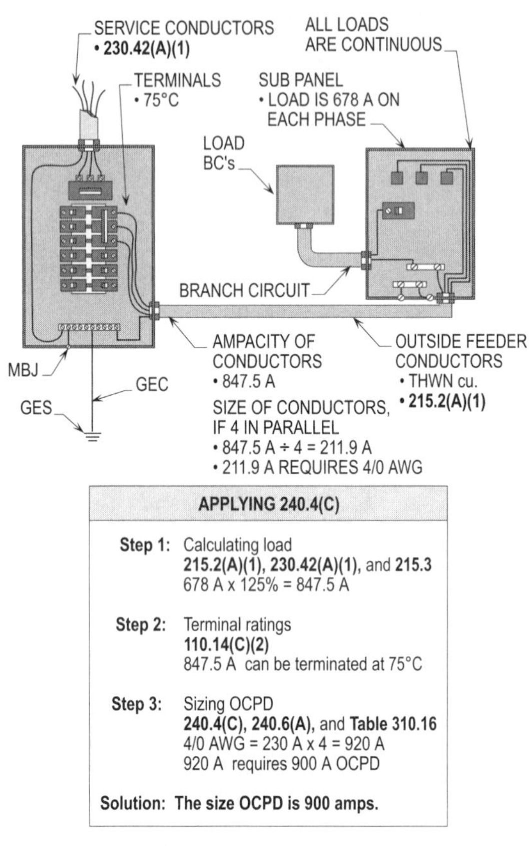

APPLYING 240.4(C)

Step 1: Calculating load
215.2(A)(1), 230.42(A)(1), and 215.3
678 A x 125% = 847.5 A

Step 2: Terminal ratings
110.14(C)(2)
847.5 A can be terminated at 75°C

Step 3: Sizing OCPD
240.4(C), 240.6(A), and Table 310.16
4/0 AWG = 230 A x 4 = 920 A
920 A requires 900 A OCPD

Solution: The size OCPD is 900 amps.

DEVICES RATED OVER 800 AMPS
NEC 240.4(C)

Figure 9-4. Where the ampacity of a conductor(s) does not correspond with a standard device per **240.6(A)**, the next lower size shall be used if rated over 800 amps.

MOTOR AND MOTOR-CONTROL CIRCUIT CONDUCTORS 240.4(G)

Motor circuits are another exception listed in **Subdivision (G)** to the general rule where the overcurrent protection device shall be permitted to be sized above the conductor's ampacity. Motor circuits shall be sized according to the requirements of **Article 430**. A study of these requirements will reveal that for motor circuits, a fuse size or circuit breaker setting in excess of the ampacity of the conductor shall be permitted by the NEC. This exception is intended to provide fuse or circuit breaker protection large enough to hold the high momentary inrush current required for starting and running the driven load. **(See Figure 9-6)**

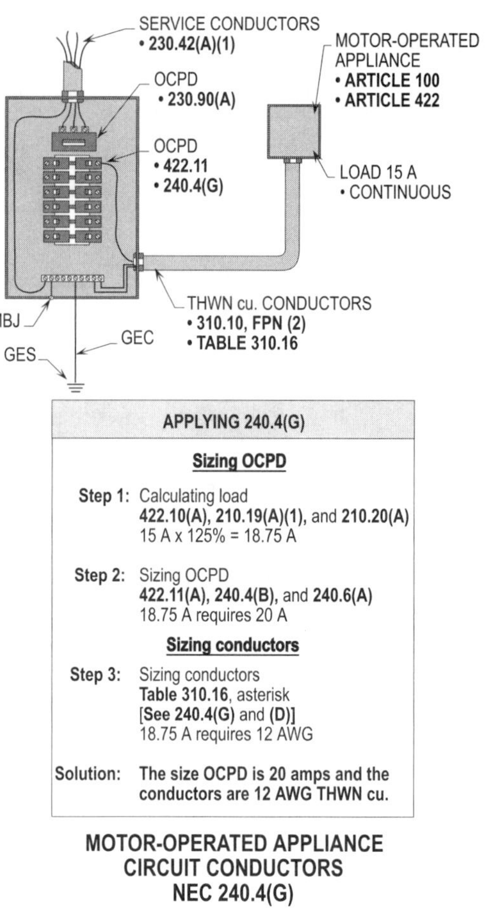

APPLYING 240.4(G)

Sizing OCPD

Step 1: Calculating load
422.10(A), 210.19(A)(1), and 210.20(A)
15 A x 125% = 18.75 A

Step 2: Sizing OCPD
422.11(A), 240.4(B), and 240.6(A)
18.75 A requires 20 A

Sizing conductors

Step 3: Sizing conductors
Table 310.16, asterisk
[See 240.4(G) and (D)]
18.75 A requires 12 AWG

Solution: The size OCPD is 20 amps and the conductors are 12 AWG THWN cu.

MOTOR-OPERATED APPLIANCE CIRCUIT CONDUCTORS NEC 240.4(G)

Figure 9-5. The above illustration shows the procedure for calculating the load to be used for sizing the overcurrent protection device and conductors supplying an appliance load.

PHASE CONVERTER SUPPLY CONDUCTORS 240.4(G)

Phase converter supply conductors for motor related loads and nonmotor loads shall be permitted to be protected against overcurrent by the rules and regulations of **455.7**. Before sizing the overcurrent protection device for such loads, review **Article 455,** which contains rules much different from the requirements of **Article 430** for sizing elements of motor circuits. **(See Figure 9-7)**

A/C AND REFRIGERATION EQUIPMENT CIRCUIT CONDUCTORS 240.4(G)

Circuit conductors supplying air conditioning and refrigeration equipment shall be protected against overcurrent by the provisions of **Parts III** and **VI of Article 440**. Note that these rules shall be used only for hermetically sealed motors and not for individual motors, per **Article 430**. **(See Figure 9-8)**

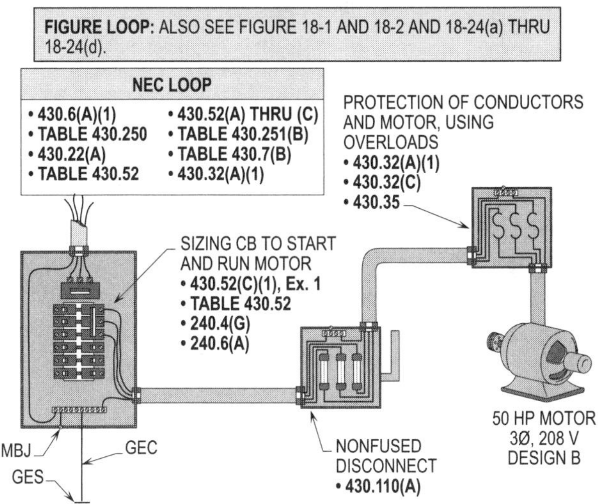

APPLYING 240.4(G)

Sizing minimum size CB

Step 1: Finding FLA
430.6(A)(1) and **Table 430.250**
50 HP = 143 A

Step 2: Finding percentages
430.52(C)(1) and **Table 430.52**
Min. = 250%

Step 3: Calculating size
430.52(C)(1) and **Table 430.52**
143 A x 250% = 357.5 A

Step 4: Selecting CB
240.4(G), 430.52(C)(1), and **240.6(A)**
357.5 A requires 350 A

Solution: **The minimum size circuit breaker is 350 amps.**

Sizing next size CB

Step 1: Finding percentages
430.52(C)(1), Ex. 1 and **240.6(A)**
Next size above 357.5 A is 400 A

Solution: **The next standard size circuit breaker above 357.5 amps is 400 amps.**

Sizing maximum size CB

Step 1: Finding percentages
430.52(C)(1), Ex. 2(c) and **240.6(A)**
Max. = 300%

Step 2: Calculating size
430.52(C)(1), Ex. 2(c)
143 A x 300% = 429 A

Step 3: Selecting CB
240.4(C) and **240.6(A)**
429 A requires 400 A

Solution: **The next standard size circuit breaker below 429 amps is 400 amps.**

Sizing conductors

Step 1: Calculating size
430.22(A)
143 A x 125% = 178.8 A

Step 2: Selecting conductors
310.10, FPN (2) and **Table 310.16**
178.8 A requires 3/0 AWG cu.

Solution: **The size conductors required are 3/0 AWG THWN copper.**

MOTOR AND MOTOR-CONTROL
CIRCUIT CONDUCTORS
NEC 240.4(G)

Figure 9-6. Section **430.52(C)(1), Ex. 1** allows the next size overcurrent protection device to be used if the percentages in **Table 430.52** times the motor's FLA does not correspond to a standard device per **240.6(A)**.

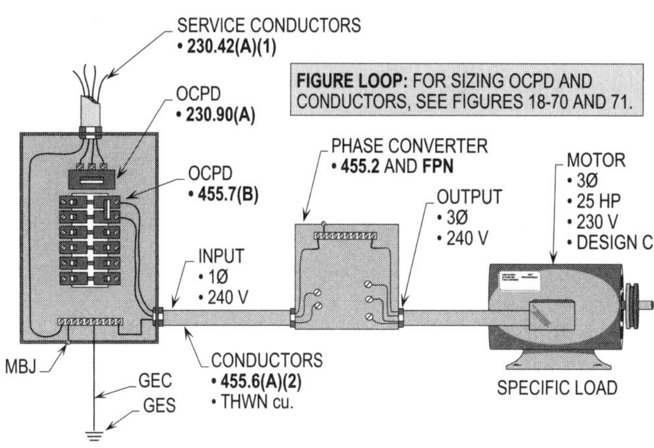

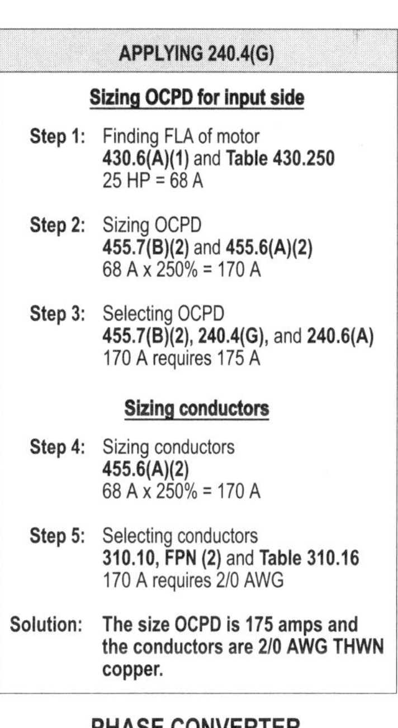

APPLYING 240.4(G)

Sizing OCPD for input side

Step 1: Finding FLA of motor
430.6(A)(1) and **Table 430.250**
25 HP = 68 A

Step 2: Sizing OCPD
455.7(B)(2) and **455.6(A)(2)**
68 A x 250% = 170 A

Step 3: Selecting OCPD
455.7(B)(2), 240.4(G), and **240.6(A)**
170 A requires 175 A

Sizing conductors

Step 4: Sizing conductors
455.6(A)(2)
68 A x 250% = 170 A

Step 5: Selecting conductors
310.10, FPN (2) and **Table 310.16**
170 A requires 2/0 AWG

Solution: **The size OCPD is 175 amps and the conductors are 2/0 AWG THWN copper.**

PHASE CONVERTER
SUPPLY CONDUCTORS
NEC 240.4(G)

Figure 9-7. The overcurrent protection device to start and run a motor, using a phase converter, shall be permitted to have overcurrent protection devices sized with a greater rating than the circuit conductors.

TRANSFORMER SECONDARY CONDUCTORS 240.4(F)

Conductors supplied by the secondary side of a single-phase transformer having a two-wire (single voltage) secondary or a three-wire delta secondary shall be considered as protected by overcurrent protection provided on the primary (supply) side of the transformer, providing this protection is in accordance with **450.3** and does not exceed the value determined by multiplying the secondary conductor ampacity by the secondary-to-primary transformer voltage ratio. [Also, see **240.21(C)(1)** and **408.36(B), Ex.**]

Note that the primary side shall be a two-wire for a two-wire single-phase secondary and a three-wire for a three-wire delta secondary. **[See Figures 9-9(a) and (b)]**

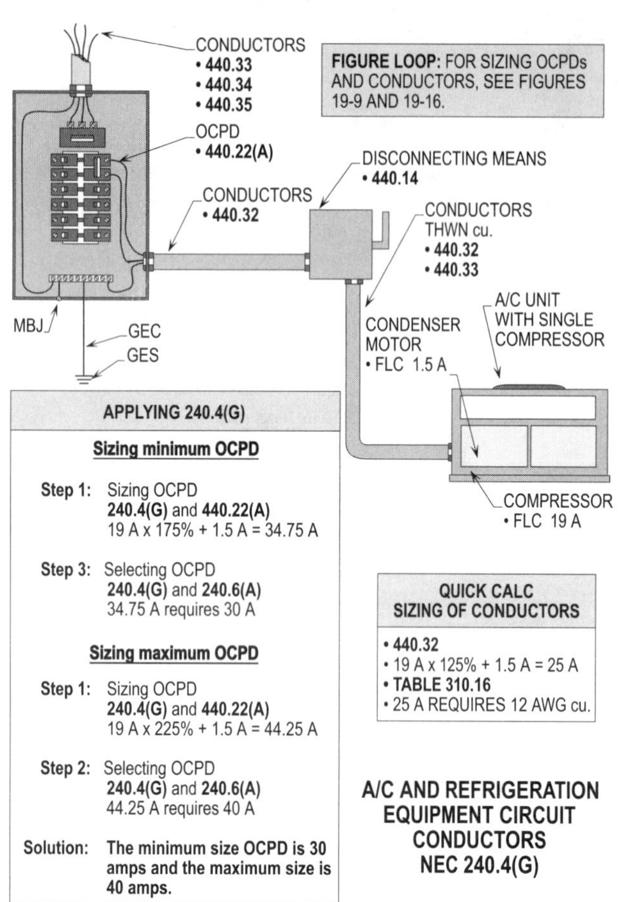

Figure 9-8. The overcurrent protection device for an A/C unit shall be permitted to be sized greater than the ampacity of the supply conductors.

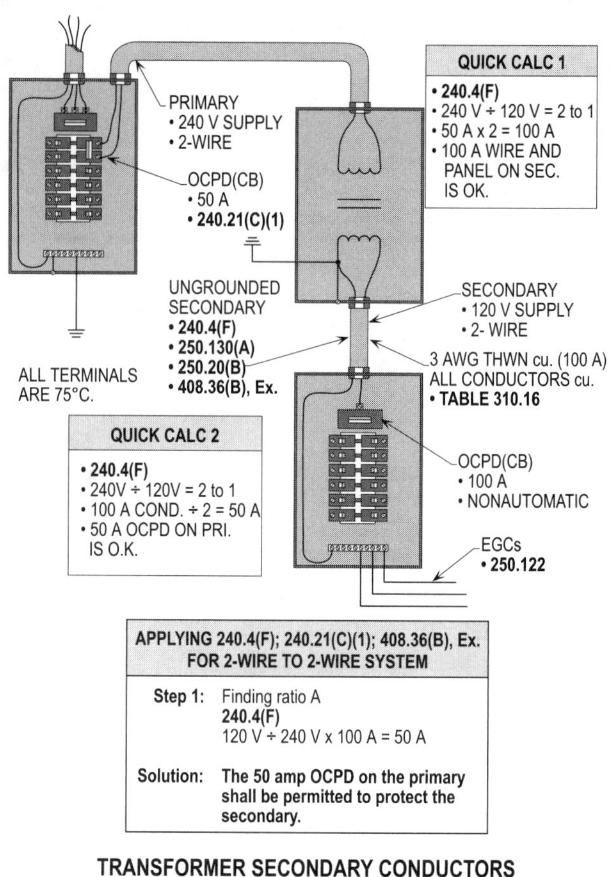

Figure 9-9(a). As shown above, under certain design conditions, the primary overcurrent protection device shall be permitted to be used to protect primary and secondary sides of two-wire to two-wire transformers.

CAPACITOR CIRCUIT CONDUCTORS 240.4(G)

Capacitors draw a high inrush current when connected to the line. Circuit protection shall be permitted to be sized high enough to hold the inrush current, which normally results in overfusing the conductors. Section **460.8(B)** permits overcurrent protection devices to be sized above the ampacities of these conductors when they supply capacitors. **(See Figure 9-10)**

ELECTRIC WELDER CIRCUIT CONDUCTORS 240.4(G)

Circuit conductors supplying welders shall be protected against overcurrent by the provisions of **Parts II** and **III** of **Article 630**. AC transformer and DC rectifier arc welders and conductors shall be protected per **630.12(A)** and **(B)**. Motor-generator arc welders and conductors shall also be protected per **630.12(A)** and **(B)**. Resistance welders and conductors shall be protected per **630.32(A)** and **(B)**. **(See Figure 9-11)**

REMOTE-CONTROL, SIGNALING, AND POWER-LIMITED CIRCUIT CONDUCTORS 240.4(G)

Remote-control circuits have overcurrent protection sized up to three times the ampacity of the conductors per **725.43** and **725.45(C)**. Motor-control circuits shall be permitted to fused above their ampacity. A good example of a motor-control circuit is a circuit to a push-button station (start-stop) derived from a magnetic motor controller.

For motor-control circuits, the NEC allows the protection for the control circuit to be sized up to 300 or 400 percent of the ampacity of the control circuit conductors. Such protection is based upon control circuits remaining in the control enclosure or leaving the enclosure to supply a remote feed stop-and-start station, etc.

For further information on sizing overcurrent protection devices for motor-control circuits, see **430.72(B)(1)** and **(B)(2), Table 430.72(B)** and **725.45(C). (See Figure 9-12)**

FIRE ALARM CIRCUIT CONDUCTORS 240.4(G)

Circuit conductors used in fire alarm systems shall be protected against overcurrent conditions by the provisions of **Parts II** and **III** of **Article 760**. For further information on sizing and selecting overcurrent protection devices for fire alarm circuits, see **760.43, 760.45** and **760.121** in the NEC. These circuits are special and such rules and regulations must be studied and well understood. **(See Figure 9-13)**

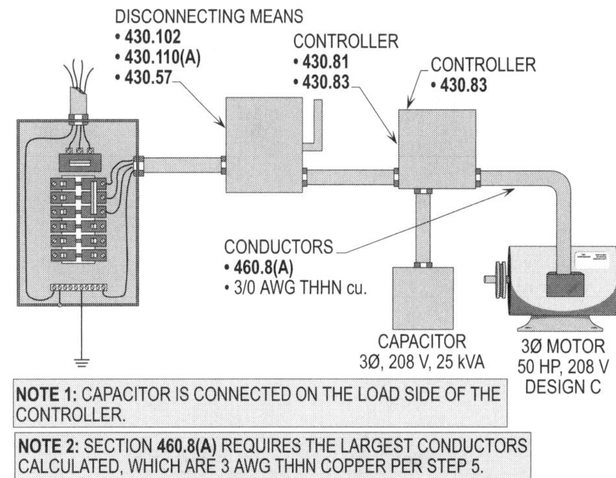

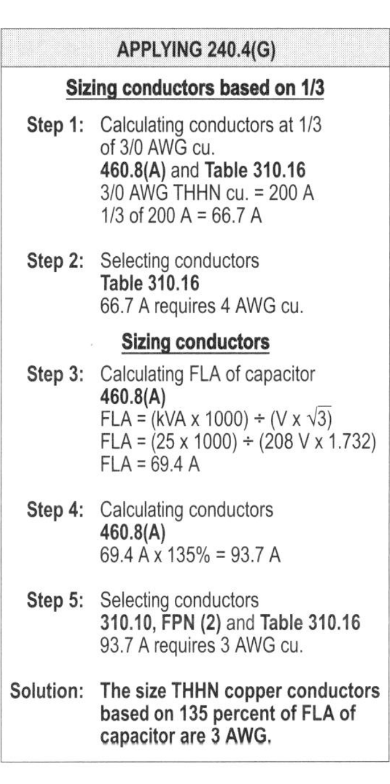

APPLYING 240.4(G)

Sizing conductors based on 1/3

Step 1: Calculating conductors at 1/3 of 3/0 AWG cu.
460.8(A) and **Table 310.16**
3/0 AWG THHN cu. = 200 A
1/3 of 200 A = 66.7 A

Step 2: Selecting conductors
Table 310.16
66.7 A requires 4 AWG cu.

Sizing conductors

Step 3: Calculating FLA of capacitor
460.8(A)
FLA = (kVA x 1000) ÷ (V x $\sqrt{3}$)
FLA = (25 x 1000) ÷ (208 V x 1.732)
FLA = 69.4 A

Step 4: Calculating conductors
460.8(A)
69.4 A x 135% = 93.7 A

Step 5: Selecting conductors
310.10, FPN (2) and **Table 310.16**
93.7 A requires 3 AWG cu.

Solution: The size THHN copper conductors based on 135 percent of FLA of capacitor are 3 AWG.

CAPACITOR CIRCUIT CONDUCTORS NEC 240.4(G)

Figure 9-10. As shown above, the conductors supplying power to capacitors are sized and selected, based either on the FLA of the capacitor times 135 percent or on 1/3 of the ampacity of the conductor serving the motor, whichever is greater.

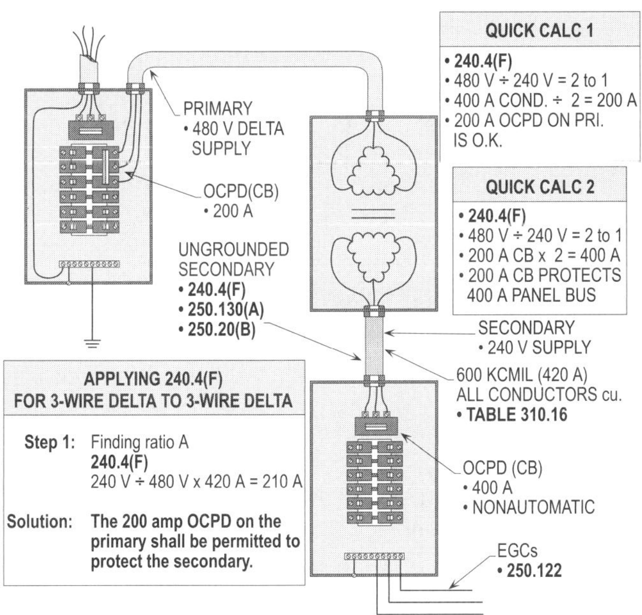

TRANSFORMER SECONDARY CONDUCTORS NEC 240.4(F)

Figure 9-9(b). As shown above, under certain design conditions, the primary overcurrent protection device shall be permitted to be used to protect the primary and secondary sides of a three-wire to three-wire delta-connected transformer. **(See Figure 20-8)**

PROTECTION OF FIXTURE WIRES AND CORDS 240.5

A 20 amp overcurrent protection device shall be permitted as adequate protection for fixture wires or flexible cords or for tinsel cord, in sizes 16 AWG or 18 AWG, respectively. Fixture wire taps that comply with **210.19(A)(3)** and **(A)(4)** shall be permitted to be protected when tapped from 30, 40, and 50 amp branch circuits.

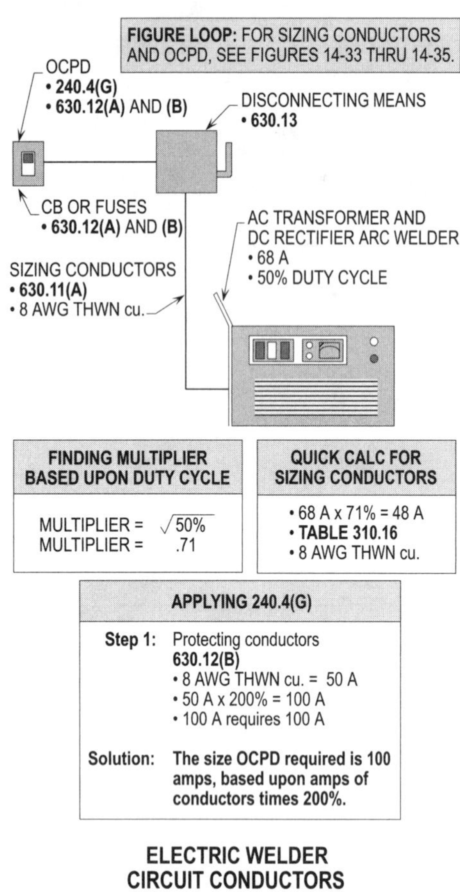

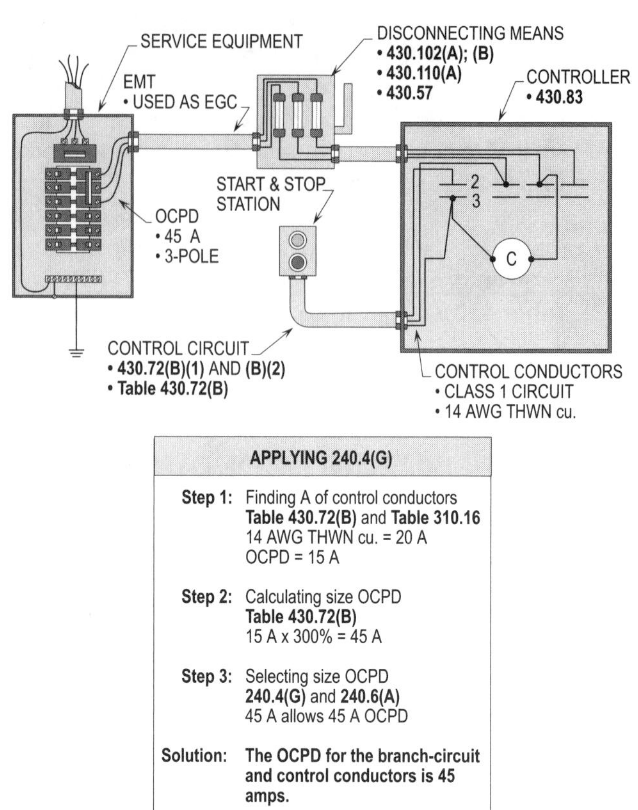

Figure 9-11. As shown above, to allow welders to operate, the overcurrent protection device shall be permitted to be sized greater than the ampacity of the conductors supplying the welder.

Figure 9-12. As shown above, control circuit conductors shall be permitted to be protected by the branch circuit overcurrent protection device which is sized greater than the conductor ampacity rating.

FLEXIBLE CORD
240.5(B)(1)

Where flexible cord is approved and used with a specific listed appliance or luminaire, it shall be considered to be protected.

Design Tip: For 20 amp circuits, 18 AWG, up to a tapped length of 50 ft, and 16 AWG, up to a tapped length of 100 ft, shall be permitted to be used as fixture whips per **410.117(C)** and outside lighting standards per **240.5(B)(2)** and **410.30(B)**.

FIXTURE WIRE
240.5(B)(2)

Fixture wire shall be permitted to be protected by the overcurrent device of the branch circuit if sized and selected per **Article 210** and complies as follows:

- 20 amp circuits, 18 AWG up to 50 ft (15 m)
- 20 amp circuits, 16 AWG up to 100 ft (30 m)
- 20 amp circuits, 14 AWG and greater
- 30 amp circuits, 14 AWG and greater
- 40 amp circuits, 12 AWG and greater
- 50 amp circuits, 12 AWG and greater

See Figure 9-14 for a detailed illustration of protecting fixture wire from overcurrent conditions.

EXTENSION CORD SETS
240.5(B)(3)

16 AWG and larger flexible cord conductors used in listed extension cord sets that are connected to a 20 amp branch circuit per **Article 210** shall be permitted to be protected from overcurrent conditions. In other words, a second stage of protection shall not be required to be provided in addition to the branch-circuit overcurrent protection device. **(See Figure 9-15)**

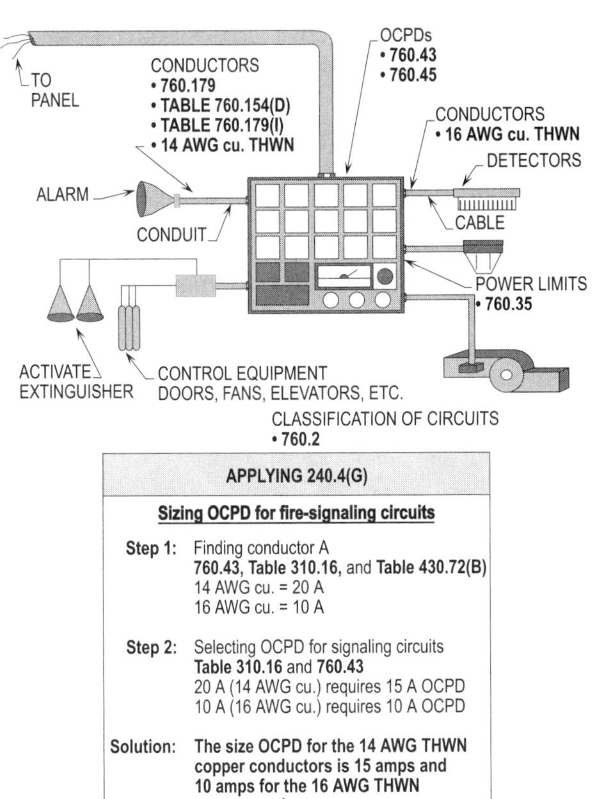

APPLYING 240.4(G)
Sizing OCPD for fire-signaling circuits
Step 1: Finding conductor A 760.43, Table 310.16, and Table 430.72(B) 14 AWG cu. = 20 A 16 AWG cu. = 10 A
Step 2: Selecting OCPD for signaling circuits Table 310.16 and 760.43 20 A (14 AWG cu.) requires 15 A OCPD 10 A (16 AWG cu.) requires 10 A OCPD
Solution: The size OCPD for the 14 AWG THWN copper conductors is 15 amps and 10 amps for the 16 AWG THWN copper conductors.

**FIRE ALARM
CIRCUIT CONDUCTORS
NEC 240.4(G)**

Figure 9-13. Fire-signaling circuits shall be protected from short circuits and ground faults.

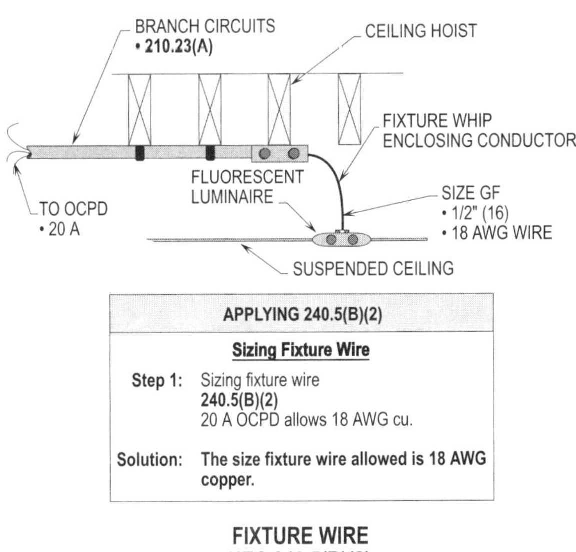

APPLYING 240.5(B)(2)
Sizing Fixture Wire
Step 1: Sizing fixture wire 240.5(B)(2) 20 A OCPD allows 18 AWG cu.
Solution: The size fixture wire allowed is 18 AWG copper.

**FIXTURE WIRE
NEC 240.5(B)(2)**

Figure 9-14. As shown above, fixture wire in whips shall be permitted to be smaller in rating than the branch-circuit overcurrent protection device and conductors.

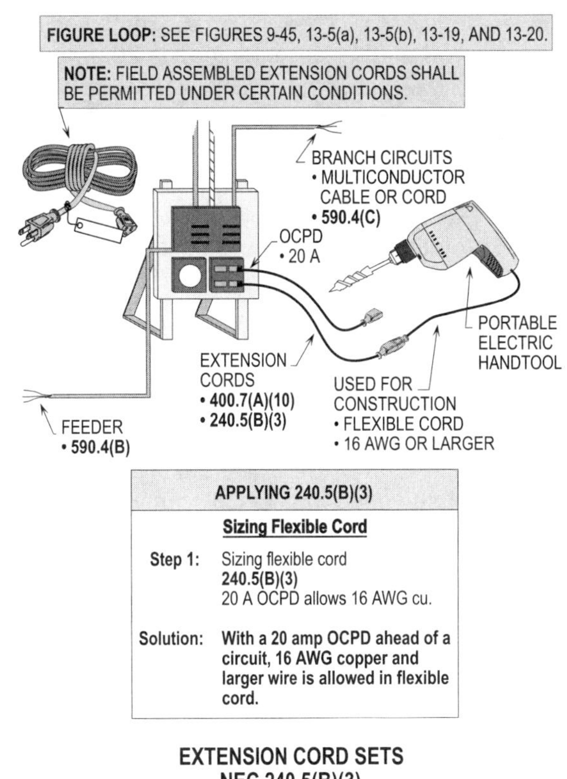

APPLYING 240.5(B)(3)
Sizing Flexible Cord
Step 1: Sizing flexible cord 240.5(B)(3) 20 A OCPD allows 16 AWG cu.
Solution: With a 20 amp OCPD ahead of a circuit, 16 AWG copper and larger wire is allowed in flexible cord.

**EXTENSION CORD SETS
NEC 240.5(B)(3)**

Figure 9-15. As shown above, 16 AWG and larger wire shall be permitted to be used in extension cord sets. **(Also, see Figure 13-19)**

STANDARD AMPERE RATINGS 240.6(A)

Standard ratings for fuses and circuit breakers are available as follows: 15, 20, 25, 30, 35, 40, 45, 50, 60, 70, 80, 90, 100, 110, 125, 150, 175, 200, 225, 250, 300, 350, 400, 450, 500, 600, 700, 800, 1000, 1200, 1600, 2000, 2500, 3000, 4000, 5000, and 6000 amps. For fuses only, 1, 3, 6, and 10 amps are also available as standard sizes to be used for the protection of smaller circuits such as motors rated at 8 amps or less. As for circuit breakers, the ratings above are not circuit breaker sizes but settings.

For example, a circuit breaker (CB) manufactured with a 100 amp frame shall be permitted to have settings of 15 through 100 amps. A frame size of 225 amps shall be permitted have settings of 110 through 225 amps, when installed in a 225 amp panelboard.

Design Tip: The setting of a circuit breaker is actually the tripping point.

For example, a circuit breaker with a 20 amp setting, will trip open when the current exceeds 20 amps. A 30 amp fuse will open the circuit when the circuit current in amps exceeds 30 amps.

See Figure 9-16 for a detailed illustration of selecting an overcurrent protection device based on load.

The NEC does not prohibit the use of a nonstandard rated overcurrent protection device in an electrical system.

For example, a 1500 amp circuit breaker or fuse shall be permitted to be used for a circuit with a calculated load of 1575 amps when listed for such use per **240.6(A)**. This prevents the designer from having to round down to 1200 amps per **240.4(C)** and losing 375 amps.

ADJUSTABLE TRIP CIRCUIT BREAKERS 240.6(B) AND (C)

The rating of an adjustable trip circuit breaker having external means for adjusting the long time pickup setting for amp ratings is the maximum setting allowed by the device. Circuit breakers that have removable and sealable covers over the adjusting means, or are located behind bolted equipment enclosure doors, or are located behind locked doors accessible only to qualified personnel, shall be permitted to have ampere ratings equal to the adjusted (set) longtime pickup settings.

For example, if one of these rules is not applied and a feeder supplies a panelboard, the conductors for the circuit shall be capable of handling the manufacturer setting of such a circuit breaker.

Design Tip: It is not the intent of this section to prohibit the use of nonstandard ampere ratings for fuse and inverse time circuit breakers that have been approved by the manufacturer for such use per **240.6(A)**. **(See Figure 9-17)**

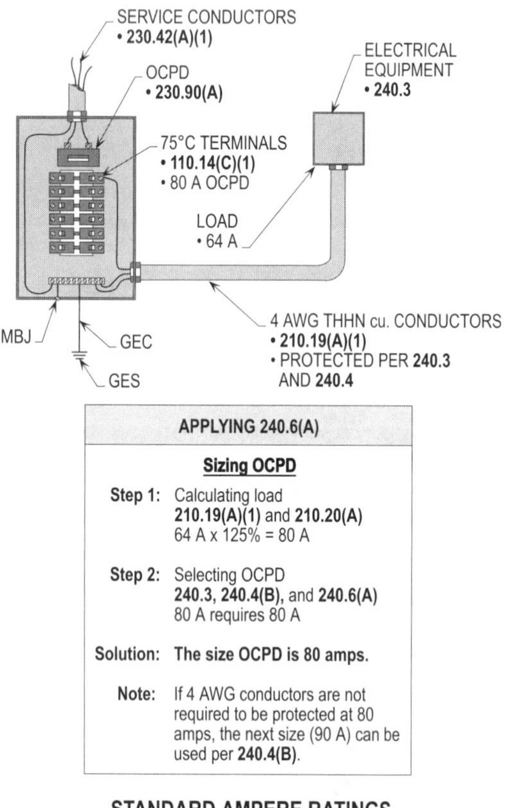

STANDARD AMPERE RATINGS
NEC 240.6(A)

Figure 9-16. As shown above, a 80 amp overcurrent protection device will protect the 85 amp conductors (4 AWG THWN cu.) and 80 amp calculated load from short circuits, ground faults, and overloads.

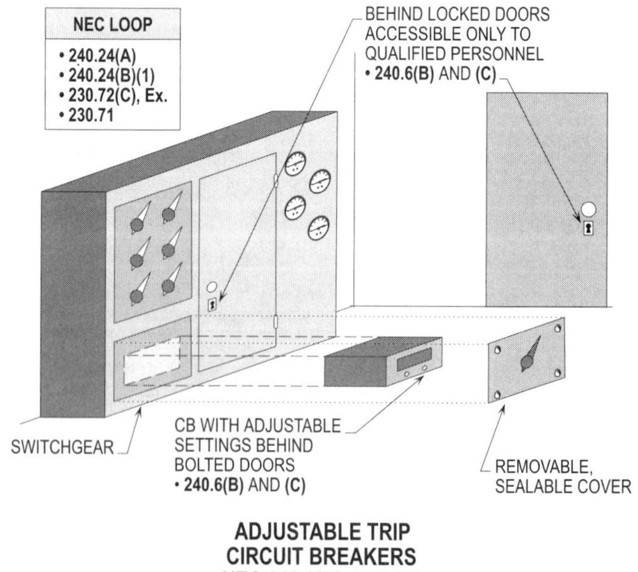

ADJUSTABLE TRIP
CIRCUIT BREAKERS
NEC 240.6(B) AND (C)

Figure 9-17. As illustrated above, the use of adjustable trip circuit breakers is allowed only if access for adjustment is limited to qualified personnel.

SUPPLEMENTARY OVERCURRENT PROTECTION 240.10

Electrical equipment that comes with overcurrent protection within the equipment that is provided by the manufacturer is permitted. Just because there is overcurrent protection in the equipment does not mean these devices have to be readily accessible per **240.24(A)(2)**. Such supplementary devices are designed for the purpose of detecting overcurrents and protection of individual parts or circuits of the appliance or utilization equipment, and they do not serve as a substitute for the branch-circuit protection. **(See Figure 9-18)**

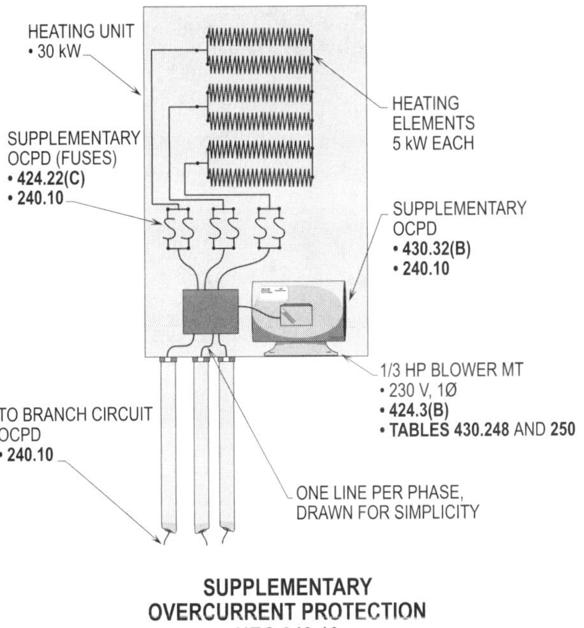

Figure 9-18. Supplementary overcurrent protection devices shall not be required to be readily accessible. However, the upstream branch-circuit overcurrent protection device shall be readily accessible.

ELECTRICAL SYSTEM COORDINATION 240.12

In industrial applications where an orderly shutdown is necessary, the coordination of overcurrent devices has always been a problem for electrical personnel. Often an orderly shutdown is needed to protect personnel from hazard(s) as well as preventing damage to equipment. To provide such coordination, the following shall be followed:

• Coordinated short-circuit protection, and

• Overload indication shall be based on monitoring systems or devices.

For example, a wye system supplying power to a number of loads can have a resistance of 12 amps installed in the ground circuit, which will limit the ground fault current to a predetermined value. A relay shall be permitted to be connected across the ground resistor, or in series with it, which actuates an alarm if a low amperage ground fault should occur in one phase. Such fault would not cause a shut down, but would alert personnel that a ground fault has developed and must be located. If necessary, arrangements shall be made to orderly shut down the loads served and find and clear the faulted circuit. **(See Figure 9-19)**

Design Tip: Coordination is defined as properly localizing a fault condition to restrict outages to the equipment affected, accomplished by choice of selective fault-protection devices. The monitoring system may cause the condition to go to alarm, allowing corrective action or an orderly shutdown, thereby minimizing personnel hazard and equipment damage.

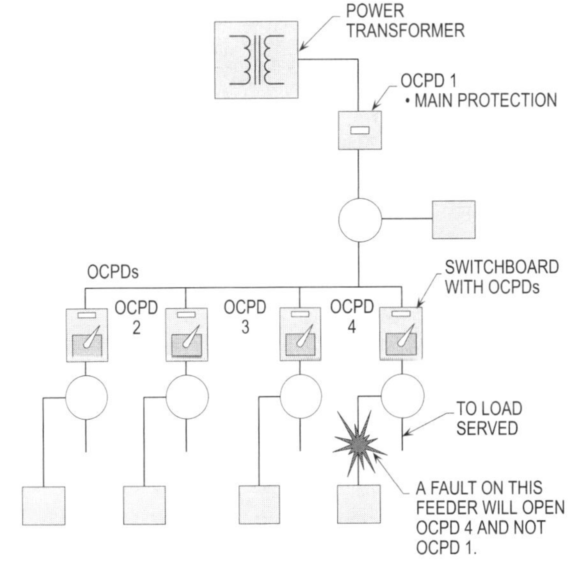

Figure 9-19. Electrical system coordination shall be provided in electrical circuits so a downstream overcurrent protection device during a fault condition will not open an upstream overcurrent protection device.

USING HANDLE TIES 240.15(B)

Circuit breakers shall open all ungrounded (phase) conductors of the circuit.

For example, when one of the trip elements operates, it trips the circuit breaker, which opens all poles of the circuit. A circuit breaker used for short-circuit protection shall have a trip unit for each ungrounded (phase) conductor, and a pole for each ungrounded (phase) conductor that opens the circuit under short-circuit and ground fault conditions.

> **Design Tip:** For ungrounded two-wire circuits, two single-pole circuit breakers with identified handle ties shall be permitted to be used in place of a two-pole circuit breaker.

Circuit breakers for three-phase circuits shall be three-pole circuit breakers with a simultaneously common trip. Circuit breakers with identified handle ties shall be permitted to be used as follows:

- Except where limited by the provisions of **210.4(B)**, individual single-pole circuit breakers, with or without identified handle ties, shall be permitted to be utilized as the protection for each ungrounded (phase) conductor of multiwire branch circuits that serve only single-phase line-to-neutral loads,

- In grounded systems, individual single-pole circuit breakers with identified handle ties shall be permitted to be utilized as the protection for each ungrounded (phase) conductor for line-to-line connected loads for single-phase circuits or three-wire DC circuits, and

- For line-to-line loads in four-wire, three-phase systems or five-wire, two-phase systems having a grounded (neutral) conductor and no conductor operating at a voltage greater than allowed in **210.6**, individual single-pole circuit breakers with identified handle ties shall be permitted to be utilized as the protection for each ungrounded (phase) conductor.

See Figure 9-20 for a detailed illustration of using handle ties with single-pole circuit breakers.

LOCATION IN CIRCUIT
240.21

The general rule for locating fuses or circuit breakers in a circuit is that they shall be installed at the source of the circuit.

For example, where a branch circuit taps to a feeder, the fuses or circuit breaker protecting the branch circuit shall be provided at the point where the tap is made to the feeder.

The following requirements shall be used when making taps and locating overcurrent protection devices in circuits. The following are exceptions to the rule.

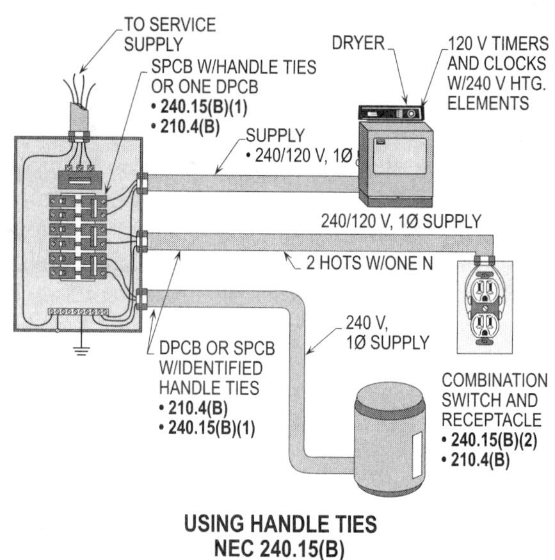

Figure 9-20. In certain installations and types of equipment, single-pole circuit breakers shall be permitted to be used with identified handle ties.

BRANCH-CIRCUIT CONDUCTORS
240.21(A)

Taps to individual outlets and circuit conductors shall be permitted to be protected by the branch-circuit overcurrent protection devices when complying with the requirements of **210.19** and **210.20**.

Section **210.19** permits taps of smaller conductors to be made to larger branch-circuit conductors for certain purposes, including taps for a small range.

For example, a 14 AWG conductor shall be permitted to tap to a 20, 25, or 30 amp branch circuit. A 12 AWG shall be permitted to tap to a 40 or 50 amp branch circuit, and fusing shall not be required at the point of supply. **(See Figure 9-21)**

TAPS NOT OVER 10 FT (3 m) LONG
240.21(B)(1)

An example of applying this rule is a tap feeding a panelboard. The tap shall not be permitted to be over 10 ft (3 m) long and shall terminate at such panelboard. Mechanical protection by conduit, tubing, or metal gutter shall be required, and conductors shall be sized to carry the total load. If these requirements are followed, no overcurrent protection device shall be required at the point of such tap. **(See Figure 9-22)**

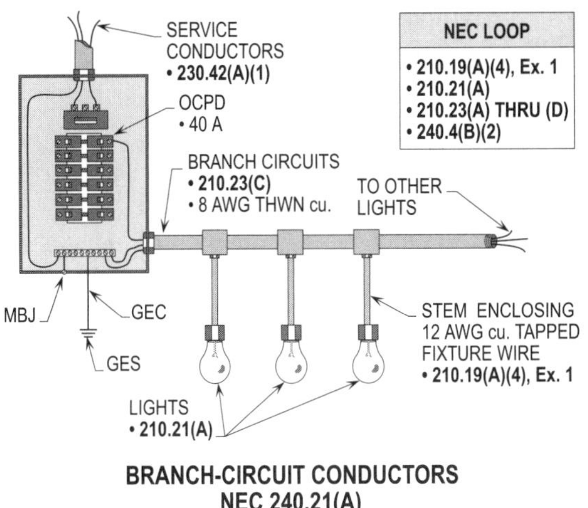

BRANCH-CIRCUIT CONDUCTORS
NEC 240.21(A)

Figure 9-21. The above illustration shows 12 AWG THWN copper fixture wire being tapped from a branch circuit having a 8 AWG THWN copper supply conductor.

TAPS NOT OVER 25 FT (7.5 m) LONG 240.21(B)(2)

A tap from a larger conductor to a smaller conductor shall be permitted to extend over 20 ft (6 m), up to a distance of 25 ft (7.5 m), provided the current-carrying capacity of the tap is at least 1/3 that of the larger conductor. Overcurrent protection shall not be required at the point of the tap, but the tap shall have overcurrent protection at the end of the run. Such tap shall be properly sized and protected from physical damage and enclosed in a raceway. **(See Figure 9-23)**

TAPS SUPPLYING A TRANSFORMER [PRIMARY PLUS SECONDARY NOT OVER 25 FT (7.5 m) LONG] 240.21(B)(3)

Transformer taps with primary plus secondary conductors that are not over 25 ft (7.5 m) in length shall be permitted to be made without overcurrent protection at the point of such taps. The following requirements shall be applied when applying this tap rule:

- Tap conductor ampacity is at least 1/3 that of the feeder,
- Secondary conductor ampacity is at least 1/3 that of the feeder, based on the primary-to-secondary transformer ratio,
- Total length of tap is not over 25 ft (7.5 m), primary plus secondary,
- All conductors are protected from physical damage, and

- Secondary conductors terminate at a fuse or circuit breaker sized to protect the secondaries.

See Figure 9-24 for a detailed illustration of applying the requirements of this tap rule.

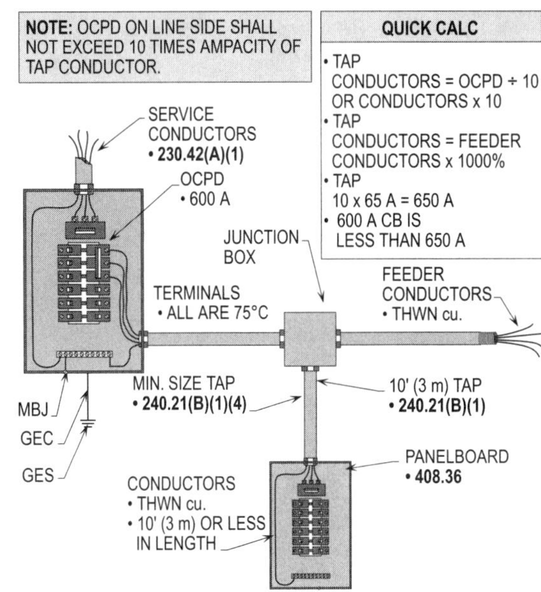

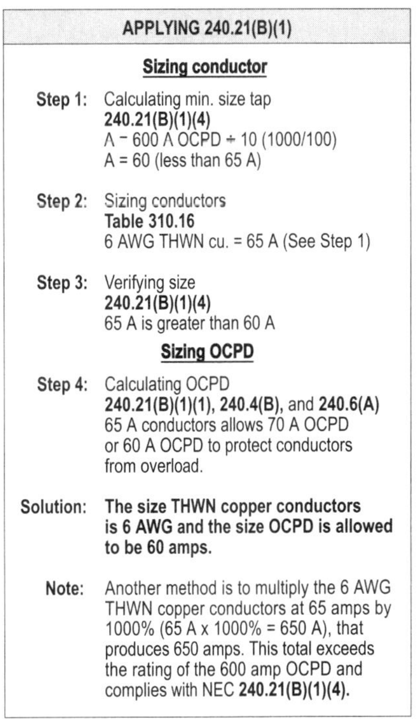

TAPS NOT OVER 10 FT (3 m) LONG
NEC 240.21(B)(1)

Figure 9-22. The above illustration shows the proper procedure for making a tap using the 10 ft (3 m) tap rule.

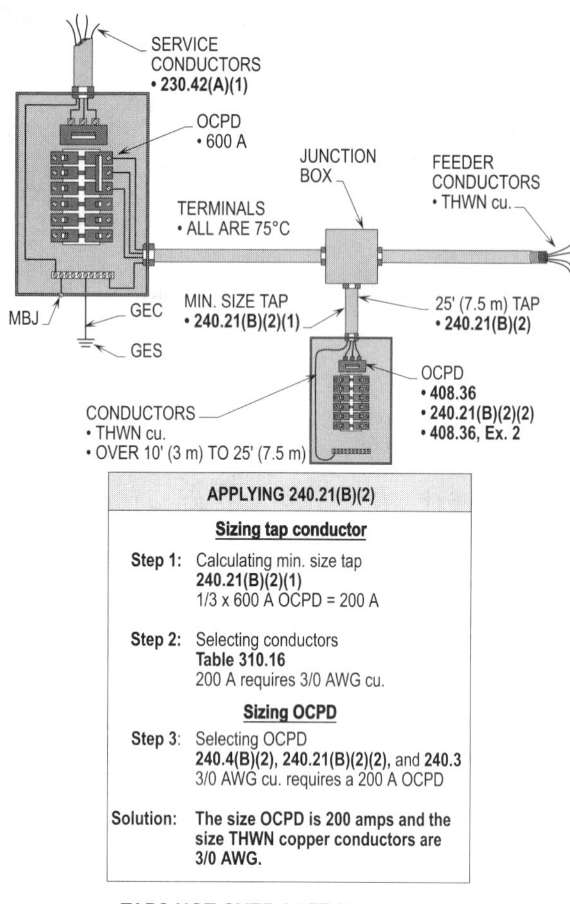

Figure 9-23. The above illustration shows the proper procedure for making a tap using the 25 ft (7.5 m) tap rule.

TAPS OVER 25 FT (7.5 m) LONG
240.21(B)(4)

For taps in high bay manufacturing buildings, the protection for the tap shall be permitted to be at the end of the tapped conductors, if the tap is not over 25 ft (7.5 m) long horizontally and not over 75 ft (22.5 m) long vertically. Note that the total run, both horizontally and vertically, is limited to 100 ft (30 m) or less in length. The following requirements shall be complied with when applying this tap rule:

- The ampacity of the tap conductors is at least 1/3 of the rating of the overcurrent device protecting the feeder conductors,

- The tap conductors terminate at a single circuit breaker or a single set of fuses that will limit the load to the ampacity of the tap conductors. This single overcurrent device shall be permitted to supply any number of additional overcurrent devices on its load side,

- The tap conductors are protected from physical damage by being enclosed in a raceway or other approved means,

- The tap conductors are continuous from end-to-end without splices,

- The tap conductors are sized 6 AWG copper or 4 AWG aluminum or larger,

- The tap conductors do not penetrate walls, floors, or ceilings, and

- The tap is at least 30 ft (9 m) from the floor.

See Figure 9-25 for a detailed illustration of applying the requirements of this tap rule.

TRANSFORMER SECONDARY CONDUCTORS
240.21(C)

Overcurrent protection devices for circuits shall be located at the point where the service to those circuits originates. However, it shall be permitted to connect conductors to the secondary side of transformers. Such conductors shall be designed and installed by the rules and regulations of **240.21(B)** and **(C)** in the NEC. Overcurrent protection shall be provided by **Tables 450.3(A)** or **(B)**.

PROTECTION BY PRIMARY OVERCURRENT DEVICE
240.21(C)(1)

For a detailed illustration of procedures for applying this rule, see **240.4(F)** and **Figures 9-9(a)** and **9-9(b)**.

TRANSFORMER SECONDARY CONDUCTORS NOT OVER 10 FT (3 m) LONG
240.21(C)(2)

Conductors shall be permitted without overcurrent protection at the transformer secondary where all the following conditions are complied with:

- Conductors do not exceed 10 ft (3 m) in length.

- Conductors shall have a current rating not less than the combined calculated loads of the circuits supplied by the conductors. Their ampacity shall not be permitted to be less than the rating of the overcurrent protection device at the termination of the secondary conductors.

- The conductors shall not be permitted to extend beyond the switchboard, panelboard, disconnecting means, or control devices they supply.

- Conductors shall be enclosed in a raceway that will extend from the connection to the enclosure of an enclosed switchboard, panelboard, or control devices, or to the back of an open switchboard.

See Figure 9-26 for the proper procedure for connecting conductors to the secondary of transformers.

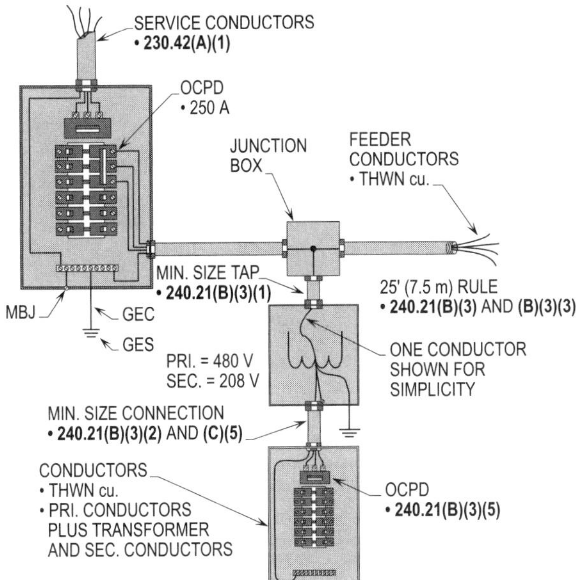

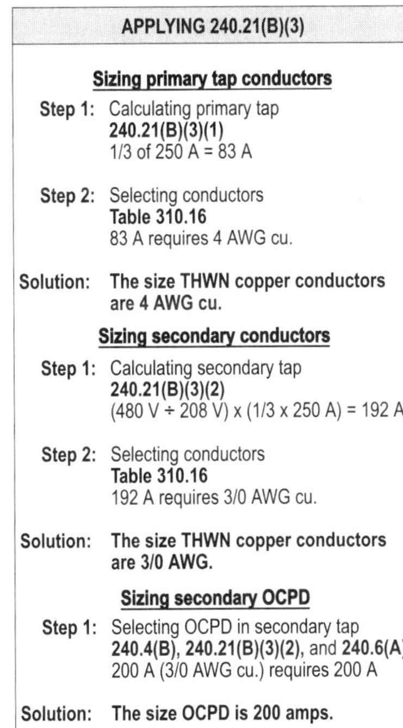

APPLYING 240.21(B)(3)

Sizing primary tap conductors

Step 1: Calculating primary tap
240.21(B)(3)(1)
1/3 of 250 A = 83 A

Step 2: Selecting conductors
Table 310.16
83 A requires 4 AWG cu.

Solution: The size THWN copper conductors are 4 AWG cu.

Sizing secondary conductors

Step 1: Calculating secondary tap
240.21(B)(3)(2)
(480 V ÷ 208 V) x (1/3 x 250 A) = 192 A

Step 2: Selecting conductors
Table 310.16
192 A requires 3/0 AWG cu.

Solution: The size THWN copper conductors are 3/0 AWG.

Sizing secondary OCPD

Step 1: Selecting OCPD in secondary tap
240.4(B), 240.21(B)(3)(2), and **240.6(A)**
200 A (3/0 AWG cu.) requires 200 A

Solution: The size OCPD is 200 amps.

**TAPS SUPPLYING A TRANSFORMER
[PRIMARY PLUS SECONDARY
NOT OVER 25 FT (7.5 m) LONG]
NEC 240.21(B)(3)**

Figure 9-24. The above illustration shows the correct sizing of a primary tap plus transformer and the sizing of a secondary connection supplying a panelboard.

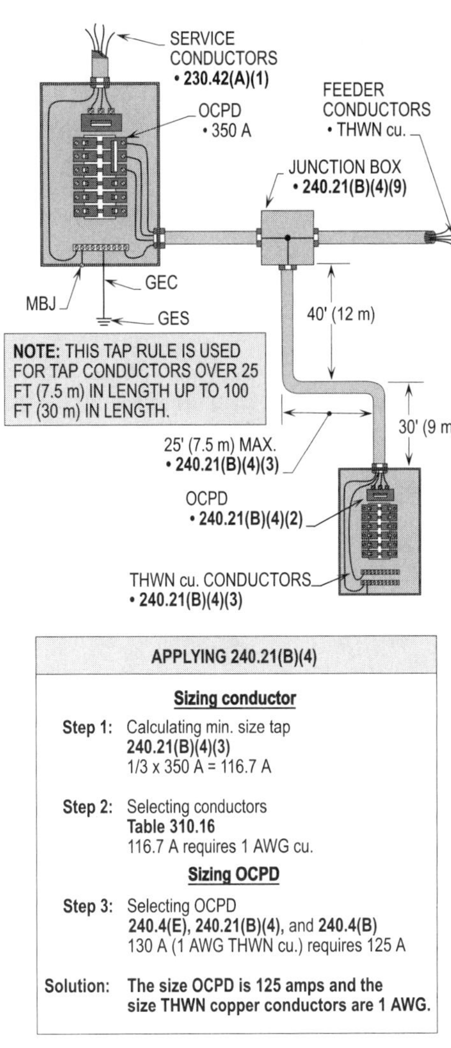

APPLYING 240.21(B)(4)

Sizing conductor

Step 1: Calculating min. size tap
240.21(B)(4)(3)
1/3 x 350 A = 116.7 A

Step 2: Selecting conductors
Table 310.16
116.7 A requires 1 AWG cu.

Sizing OCPD

Step 3: Selecting OCPD
240.4(E), 240.21(B)(4), and **240.4(B)**
130 A (1 AWG THWN cu.) requires 125 A

Solution: The size OCPD is 125 amps and the size THWN copper conductors are 1 AWG.

**TAPS OVER 25 FT (7.5 m) LONG
NEC 240.21(B)(4)**

Figure 9-25. The above illustration shows the correct procedure for making a tap using the requirements of the 100 ft (30 m) tap rule.

INDUSTRIAL INSTALLATION SECONDARY CONDUCTORS NOT OVER 25 FT (7.5 m) LONG 240.21(C)(3)

Conductors shall be permitted to be connected to the transformer secondary of a separately derived system for industrial installations, without overcurrent protection at the connection, where all the following conditions are complied with:

- Conditions of maintenance and supervision ensure that only qualified persons service the systems.

- The length of the secondary conductors does not exceed 25 ft (7.5 m) in length.

- The ampacity of the secondary conductors is not less than the secondary current rating of the transformer, and the sum of the ratings of the overcurrent devices does not exceed the ampacity of the secondary conductors.

- All overcurrent protection devices are grouped.

- The secondary conductors are protected from physical damage by being enclosed in an approved raceway or by other approved means.

See Figure 9-27 for applying the requirements of this tap rule.

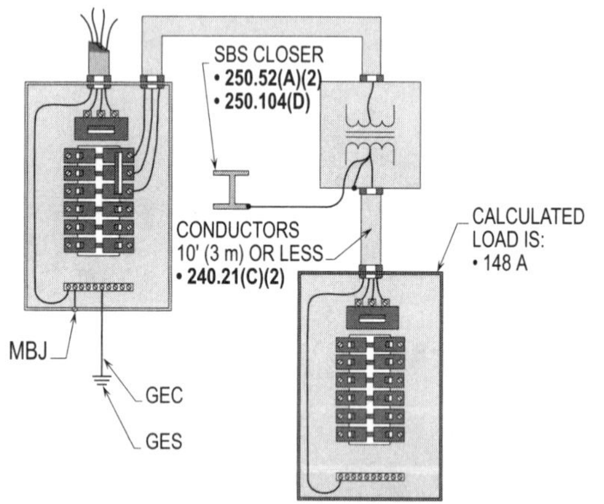

SIZING THWN cu. CONDUCTORS	SIZING OCPD
Step 1: Calculating min. size **240.21(C)(2)** Calculated load is 148 A	**Step 1:** Calculating OCPD **240.4(E), 240.21(C)(2), and 240.4(B)** 1/0 AWG cu. = 150 A OCPD rated at 150 A protects conductors from overload
Step 2: Sizing conductors **Table 310.16** 1/0 AWG THWN cu. = 150 A	**Solution:** The size OCPD is permitted to be 150 amps.
Step 3: Verifying size **240.21(C)(2)** 150 A is greater than 148 A	
Solution: The size THWN copper conductors are 1/0 AWG rated at 150 amps.	

**TRANSFORMER SECONDARY CONDUCTORS
NOT OVER 10 FT (3 m) LONG
NEC 240.21(C)(2)**

Figure 9-26. The above illustration shows the procedure for sizing a 10 ft (3 m) connection from the secondary of a transformer.

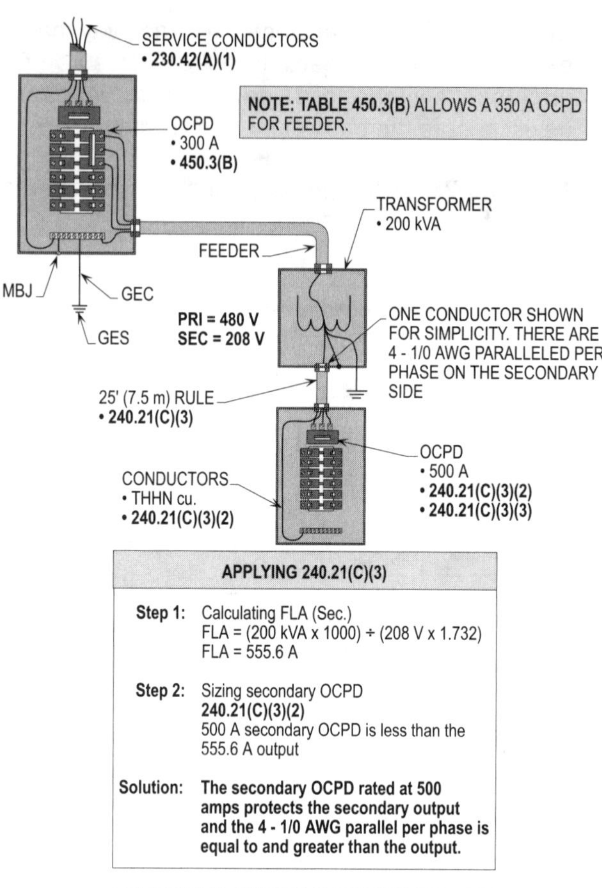

APPLYING 240.21(C)(3)	
Step 1:	Calculating FLA (Sec.) FLA = (200 kVA x 1000) ÷ (208 V x 1.732) FLA = 555.6 A
Step 2:	Sizing secondary OCPD **240.21(C)(3)(2)** 500 A secondary OCPD is less than the 555.6 A output
Solution:	**The secondary OCPD rated at 500 amps protects the secondary output and the 4 - 1/0 AWG parallel per phase is equal to and greater than the output.**

**INDUSTRIAL INSTALLATION SECONDARY
CONDUCTORS NOT OVER 25 FT (7.5 m) LONG
NEC 240.21(C)(3)**

Figure 9-27. The above illustration shows the procedures to be applied when a 25 ft (7.5 m) installation is made from the secondary side of a transformer.

OUTSIDE SECONDARY CONDUCTORS 240.21(C)(4)

Outside conductors shall be permitted to be connected at the transformer secondary, without overcurrent protection at the connection. However, the following conditions shall be complied with:

- The conductors are protected from physical damage in an approved manner.

- The conductors terminate at a single circuit breaker or a single set of fuses that limits the load to the ampacity of the conductors. This single overcurrent protection device shall be permitted to supply any number of overcurrent protection devices on its load side.

- The overcurrent device for the conductors is an integral part of a disconnecting means or is located immediately adjacent thereto.

- The disconnecting means for such conductors are installed in a readily accessible location either outside a building or structure or inside nearest the point of entrance of the conductors.

See Figure 9-28 for applying the requirements for this type of installation.

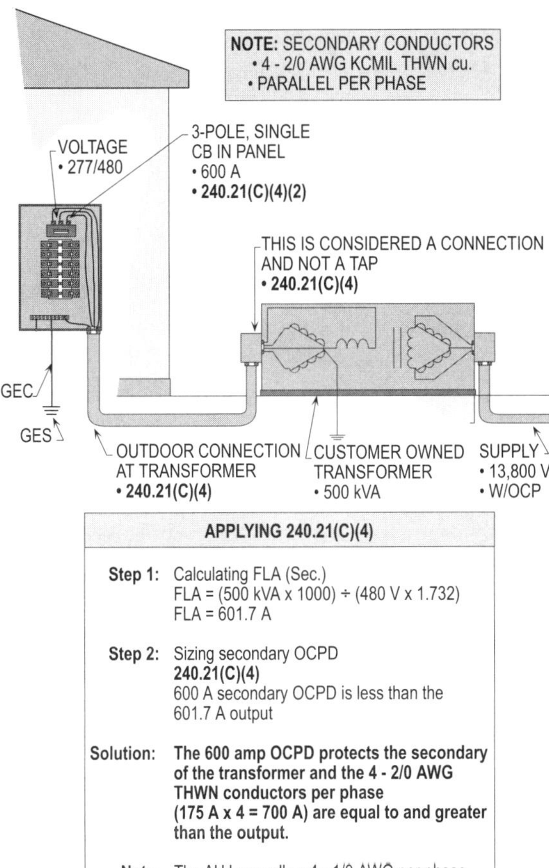

APPLYING 240.21(C)(4)	
Step 1:	Calculating FLA (Sec.) FLA = (500 kVA x 1000) ÷ (480 V x 1.732) FLA = 601.7 A
Step 2:	Sizing secondary OCPD **240.21(C)(4)** 600 A secondary OCPD is less than the 601.7 A output
Solution:	**The 600 amp OCPD protects the secondary of the transformer and the 4 - 2/0 AWG THWN conductors per phase (175 A x 4 = 700 A) are equal to and greater than the output.**
Note:	The AHJ may allow 4 - 1/0 AWG per phase (150 A x 4 = 600 A).

**OUTSIDE SECONDARY CONDUCTORS
NEC 240.21(C)(4)**

Figure 9-28. The above illustration shows the procedure to be applied when making an outside connection from the secondary side of a customer-owned transformer.

SERVICE CONDUCTORS
240.21(D)

Service-entrance conductors shall be permitted to be protected by overcurrent protection devices by the provisions of **230.91**. Generally, the service overcurrent device is an integral part of the disconnecting means or it shall be installed adjacent.

For example, switches and fuses shall be either "integral" or "adjacent." A set of fuses used as service overcurrent protection shall be permitted to be separate from the disconnect, but if this is the case, the fuses shall be located "immediately adjacent to" such disconnect.

Note that circuit breakers are combined with such disconnecting means and shall provide overcurrent protection for the service-entrance conductors. **(See Figure 9-29)**

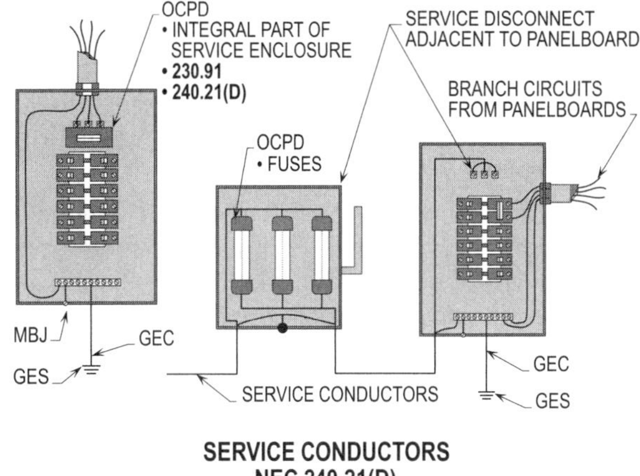

**SERVICE CONDUCTORS
NEC 240.21(D)**

Figure 9-29. The service-entrance conductors shall be protected by overcurrent protection devices which are either integral to the equipment or adjacent to disconnect switch.

BUSWAY TAPS
240.21(E)

Section **368.17** permits a reduction in size of a busway. In such cases, an additional overcurrent protection device shall not be required for smaller busways at the point of the tap. When making a tap from a larger busway, the smaller busway shall comply with the following:

- Have a current-carrying capacity of at least 1/3 of the overcurrent setting,
- Not be over 50 ft (15 m) long, and
- Not be in contact with combustible material.

See Figure 9-30 for a detailed illustration of applying the requirements of this tap rule.

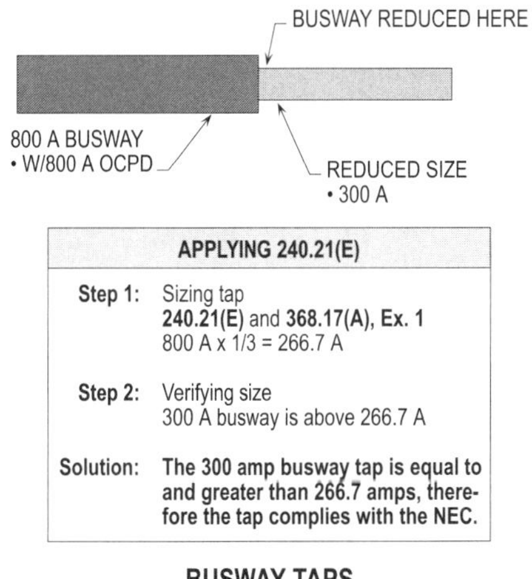

APPLYING 240.21(E)	
Step 1:	Sizing tap **240.21(E) and 368.17(A), Ex. 1** 800 A x 1/3 = 266.7 A
Step 2:	Verifying size 300 A busway is above 266.7 A
Solution:	**The 300 amp busway tap is equal to and greater than 266.7 amps, therefore the tap complies with the NEC.**

**BUSWAY TAPS
NEC 240.21(E)**

Figure 9-30. The above is a detailed illustration of applying the requirements for a busway tap.

MOTOR CIRCUIT TAPS
240.21(F)

Where more than one motor is on a feeder, each motor circuit is tapped from the feeder conductors. The tap shall be protected by the motor branch-circuit protection. The motor branch-circuit protection shall be located where the tap conductors terminate, and the following requirements shall be complied with:

- If the length of the tap is 10 ft (3 m) or less, conductors with ampacity less than 1/3 that of the feeder shall be permitted to be used, provided that the conductors are in a raceway or enclosed within a controller. If field installed, the overcurrent protection device on the line side of the tap conductors shall not be permitted to exceed 10 times the tapped conductors' ampacity.

- If the length of the tap is 25 ft (7.5 m) or less, conductor ampacity shall be at least 1/3 that of the feeder, provided that the conductors are protected from physical damage. Conduit, EMT, flexible metal conduit, or AC or MC cable, etc., provide suitable protection for the conductors.

- In high-bay manufacturing facilities, a tap to a feeder shall be permitted to be up to 100 ft (30 m) in length, for a horizontal and vertical run combined, but the horizontal parts of a run shall be limited to 25 ft (7.5 m) or less in length. The tap conductors shall comply with the following:
    - Have an ampacity 1/3 that of the feeder,
    - Be unspliced,
    - Be suitably protected from physical damage or installed in raceways,
    - Be at least 6 AWG copper or 4 AWG aluminum and
    - Not penetrate walls, floors, or ceilings.
    - Tap shall not be made less than 30 ft (9 m) from the floor.

See Figure 9-31 for a detailed illustration of applying the requirements of this tap rule.

CONDUCTORS FROM
GENERATOR TERMINALS
240.21(G)

Generator conductors shall have an ampacity equal to at least 115 percent of the nameplate current rating of the generator. The following are exceptions to the general rule:

- If the design or operation of the generator is such as to prevent overloading, an ampacity of 100 percent shall be permitted to be utilized.

- Where an integral overcurrent protection device is provided by the manufacturer, with conductors terminated to the device.

See Figure 9-32 for a detailed illustration of applying the requirements of this tap rule. Note that if the conductors are not sized based on 115 percent of the generator's output in amps, an overcurrent protection device shall be provided at the generator.

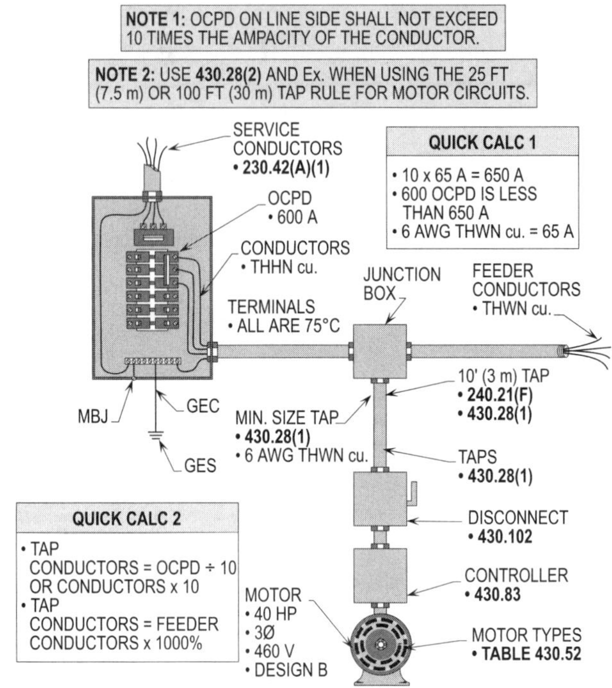

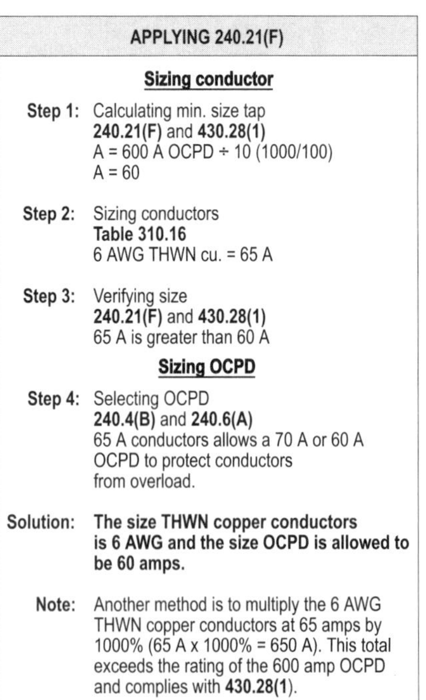

APPLYING 240.21(F)

Sizing conductor

Step 1: Calculating min. size tap
240.21(F) and 430.28(1)
A = 600 A OCPD ÷ 10 (1000/100)
A = 60

Step 2: Sizing conductors
Table 310.16
6 AWG THWN cu. = 65 A

Step 3: Verifying size
240.21(F) and 430.28(1)
65 A is greater than 60 A

Sizing OCPD

Step 4: Selecting OCPD
240.4(B) and 240.6(A)
65 A conductors allows a 70 A or 60 A OCPD to protect conductors from overload.

Solution: **The size THWN copper conductors is 6 AWG and the size OCPD is allowed to be 60 amps.**

Note: Another method is to multiply the 6 AWG THWN copper conductors at 65 amps by 1000% (65 A x 1000% = 650 A). This total exceeds the rating of the 600 amp OCPD and complies with **430.28(1)**.

MOTOR CIRCUIT TAPS
NEC 240.21(F)

Figure 9-31. The above illustration shows the proper procedure for making a 10 ft (3 m), 25 ft (7.5 m), or 100 ft (30 m) tap for motor circuits. Note that the tap illustrated is the 10 ft (3 m) tap rule.

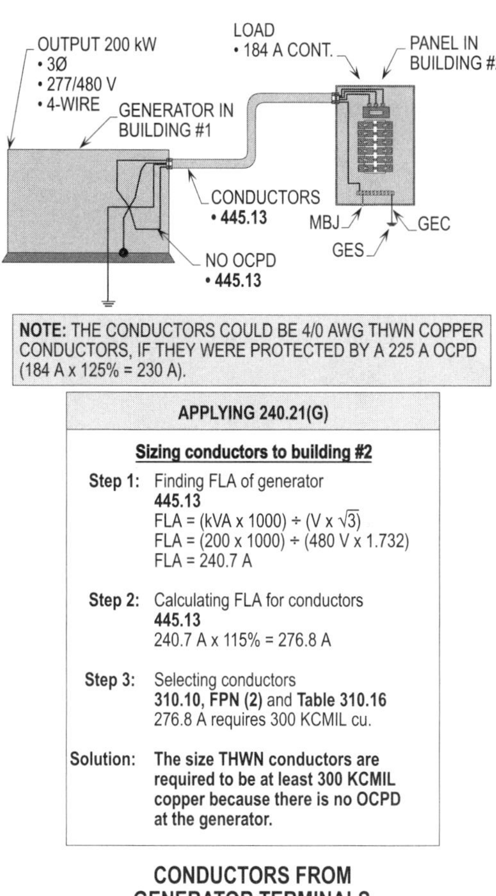

**CONDUCTORS FROM
GENERATOR TERMINALS
NEC 240.21(G)**

Figure 9-32. The above illustration shows the proper procedure to apply when tapping from a generator.

SUPERVISED INDUSTRIAL INSTALLATIONS 240.90

Many of these rules apply directly to large transformers, secondary conductors. The existing NEC treated conductors connected to a transformer secondary in the same way as tap conductors with no upstream protection. This is technically incorrect and creates difficulties for large industrial installations. These difficulties do not come up in most commercial and smaller industrial installations, because transformers of the size where a problem begins to be seen either do not exist or are installed indoors near their associated switchgear or are associated with services that are not subject to the same rules.

In large installations, with outdoor transformers in the 1000 kVA class and larger and supplying indoor secondary switchgear, it is extremely difficult and undesirable, if not impossible, for the secondary conductors to be 25 ft (7.5 m) or less in length.

CONNECTIONS UP TO 100 FT (30 m) 240.92(C)(1)(1)

Unprotected lengths of secondary conductors shall be permitted to be up to 100 ft (30 m) if the transformer primary overcurrent device is sized at a value (reflected to the secondary by the transformer phase voltage ratio) of not more than 150 percent of the secondary conductor ampacity. **(See Figure 9-33)**

PROTECTED BY A DIFFERENTIAL RELAY 240.92(C)(1)(2)

Secondary conductors shall be permitted to be protected by a differential relay with a trip setting equal to or less than the conductor ampacity. Note that a differential relay [per **240.100(A)(1)**] provides superior short-circuit protection at a trip-open value that is usually well below the conductor ampacity. **(See Figure 9-33)**

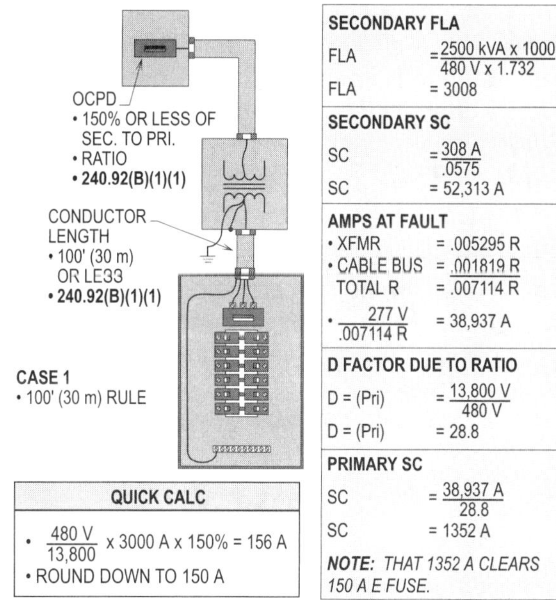

DESIGN TIP: Consider a 2500 kVA, 5.75% Z transformer with a 13.8 kV to 480/277 V ratio and 500 mVA available short-circuit current on the primary for 100 circuit feet of 3000 amp cable bus, and a three-phase bolted fault at the end of the bus (worst case). About 38,937 amps will flow from the system or 1352 amps on the primary, which will clear a typical 150 E fuse and meet the maximum 150 percent requirement within .42 seconds. This time vs. current value is well within the rating of the secondary conductors. (Cable bus has a resistance of .001819 and the XFMR has a resistance of .005295.)

**SHORT-CIRCUIT AND
GROUND-FAULT PROTECTION
240.92(C)(1) THRU (C)(3)**

Figure 9-33. The above illustration shows conductors being protected from short-circuit and ground fault conditions.

ENGINEERING SUPERVISION
240.92(C)(1)(3)

Conductors shall be permitted to be protected with greater lengths if calculations are made under engineering supervision and it is determined that the secondary conductors will be protected within recognized time versus current limits for all short-circuit and ground fault conditions that could occur. **(See Figure 9-33)**

GROUNDED (NEUTRAL) CONDUCTOR
240.22

No overcurrent protection device shall be permitted to be connected in series with any conductor that is intentionally grounded, unless the overcurrent device operates to open the ungrounded (phase) conductors simultaneously (common trip) with such grounded (neutral) conductor. This rule prohibits a fuse from being installed in the grounded (neutral) conductor, because the fuse could blow and open the grounded (neutral) conductor and the ungrounded (phase) conductors would still be energized. This can cause hazardous voltage levels to appear on the electrical elements of the circuit. Note that grounded (neutral) conductors shall be permitted to be grounded (phase) conductors or grounded (neutral) conductors.

COMMON TRIP
240.22(1)

Where the overcurrent device opens all conductors of the circuit, including the grounded conductor, and is so designed that no pole can operate independently, the grounded conductor shall be permitted to be protected by an overcurrent protection device.

OVERLOAD PROTECTION
240.22(2)

Where required by **430.36** or **430.37** for motor running (overload) protection.

For example, running overload protection of a three-phase motor fed by a three-wire, three-phase circuit with one leg grounded may have such protection provided by installing a properly sized fuse. These fuses are usually of the time delay type, so they shall be permitted to be sized at 125 percent or less of the motors FLC in amps per **430.32(A)(1)**.

See Figure 9-34 for a detailed illustration of providing overcurrent protection in the grounded conductor.

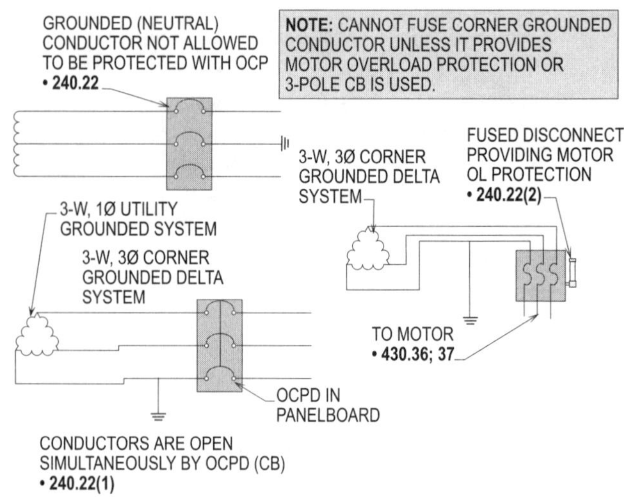

GROUNDED (NEUTRAL) CONDUCTOR
NEC 240.22

Figure 9-34. The grounded (neutral) conductor shall not be permitted to be protected by a circuit breaker unless it opens simultaneously (common trip) with all circuit conductors. To be fused, it shall provide overload protection to a motor.

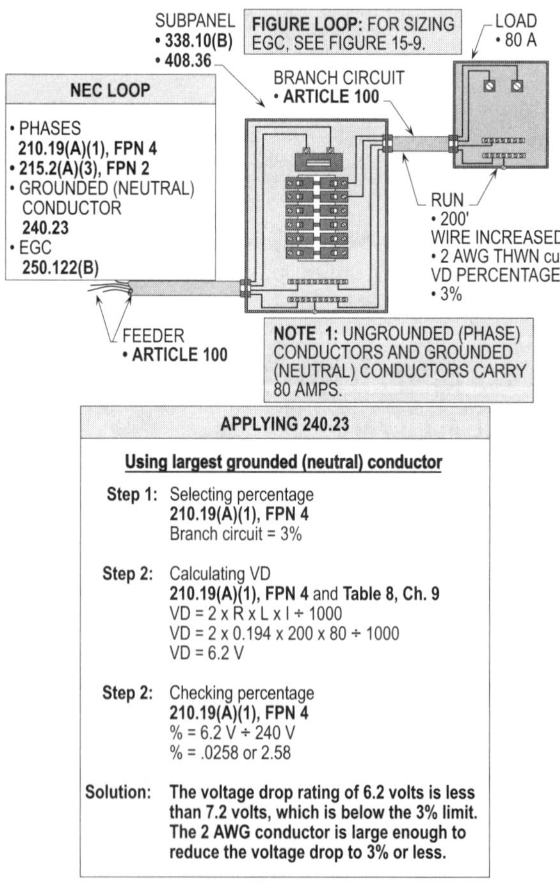

CHANGE IN SIZE OF
GROUNDED (NEUTRAL) CONDUCTOR
NEC 240.23

Figure 9-35. It is permissible to increase the size of the grounded (neutral) conductor if there is a similar change in the ungrounded (phase) conductors.

CHANGE IN SIZE OF GROUNDED (NEUTRAL) CONDUCTOR
240.23

If a change should occur in the size of the ungrounded (phase) conductor, a similar change shall be permitted to be made in the size of the grounded (neutral) conductor if necessary.

For changing the size of the ungrounded (phase) conductor due to voltage drop (VD), see **210.19(A)(1), FPN 4** and **215.2(A)(3), FPN 2**. For changing the size of the equipment grounding conductor, see **250.122(B)**. **(See Figures 9-35 and 15-9)**

ACCESSIBILITY OF OVERCURRENT PROTECTION DEVICES
240.24(A)

For safety, the requirement that overcurrent protection devices be readily accessible is mandatory. They shall be located so that they may be readily reached in case of emergencies or for servicing, without reaching over objects, climbing on chairs, ladders, etc.

See Figure 9-36 for a detailed illustration of overcurrent protection devices being readily accessible. However, there are four exceptions to the accessibility rule, and they are as follows:

- For busways, **368.17(C)** permits overcurrent protection for a plug-in to be out of reach from the finished grade.
- For luminaires and appliances that are manufactured with built-in fuses, the fuses shall not be required to be readily accessible per **240.10**.
- For services, the overcurrent protection device shall be permitted to be located at the beginning of the run, at the point where the service-entrance tap is made to the service drop or lateral. In such cases, the main overcurrent protection device shall not be required to be readily accessible inside the facility. See **225.40** and **230.92** for further details concerning this type of installation.
- For overcurrent protection devices installed adjacent to motors, appliances, or other types of equipment that are out of reach, such overcurrent protection devices shall not be required to be readily accessible. However, these devices shall be capable of being reached by a portable means. For further information, refer to **404.8(A), Ex. 2**.

See Figure 9-37 for a detailed illustration for applying the requirements of these rules.

Figure 9-36. Overcurrent protection devices shall be readily accessible without having to remove or climb over objects such as chairs, desks, etc.

OCCUPANCY
240.24(B)

The general rule does not allow overcurrent protection devices to be located where they are exposed to physical damage. Furthermore, they shall not be installed where ignitible materials are near enough to catch fire in case of accidental sparking or overheating of such devices.

SERVICE AND FEEDER OVERCURRENT DEVICES
240.24(B)(1)

In a multiple-occupancy building where electric service and electrical maintenance are provided by the building management and are under continuous building management supervision, the service overcurrent device and feeder overcurrent devices supplying more than one occupancy shall be permitted to be accessible to authorized management personnel only.

SERVICE, FEEDER, AND BRANCH-CIRCUIT OVERCURRENT DEVICES
240.24(B)(2)

Section **240.24(B)(2)** permits guest rooms or guest suites (without permanent provisions for cooking) of hotels and motels intended for transient occupancy to have overcurrent protection devices for such occupancies shall be permitted to be accessible to authorized management personnel only. **(See Figure 9-38)**

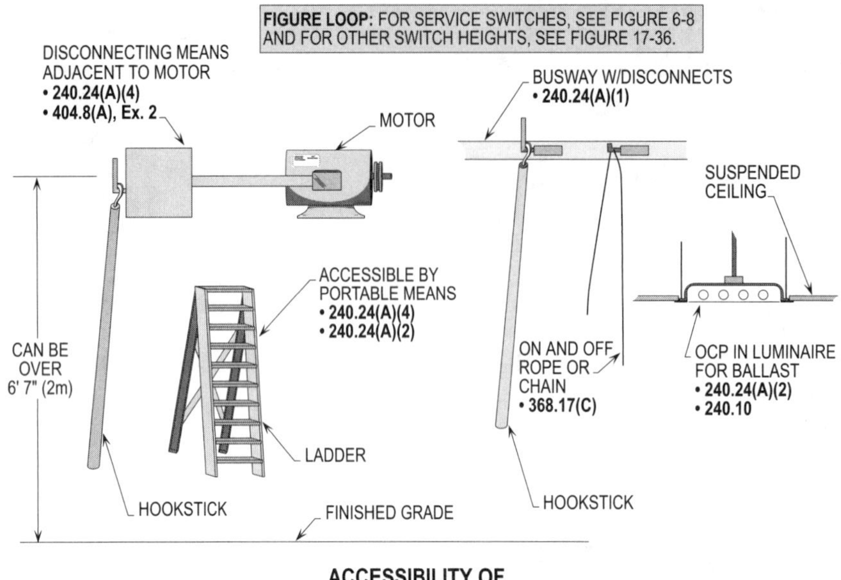

Figure 9-37. The above illustration shows installations where the overcurrent protection devices shall not be required to be readily accessible.

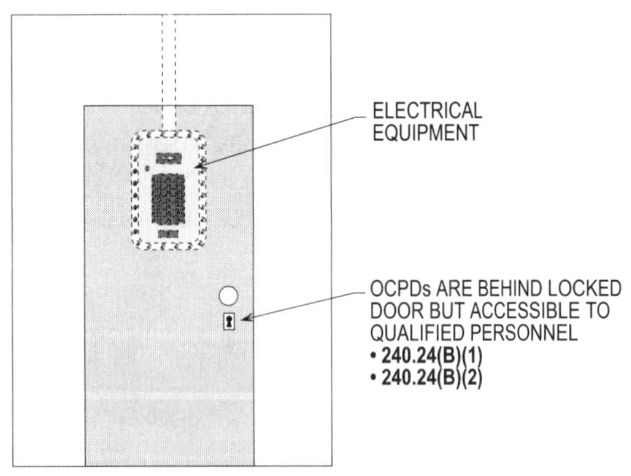

Figure 9-38. Overcurrent protection devices shall be readily accessible to each occupant in a premises. However, they shall be permitted to be behind locked doors if qualified personnel are present.

FUSES
ARTICLE 240, PARTS V AND VI

Fuses for conductors with specific types of insulations are available to supply power to equipment. Conductors are sized to carry the load of the equipment in amps without deteriorating the insulation per **310.10**. The overcurrent protection devices shall be sized to protect the conductors and equipment from short circuit, ground faults, and overloads.

The calculated load for fuses shall be determined by calculating the loads at noncontinuous duty (100 percent) or continuous duty (125 percent) and applying demand factors, where applicable. Depending on the type of fuse used, a fuse will hold five times its rating for different periods of time.

SIZING
210.20(A), 215.3, AND 230.42(A)(1)

Fuses shall be sized based upon the operating characteristics of the load served. The NEC defines a continuous load as a load that operates continuously for three hours or more.

The load current in amps shall be calculated by multiplying the 160 amp continuous load current by 125 percent (160 A x 125% = 200 A) to obtain the size, which is 200 amps. **(See Figure 9-39)**

Design Tip: An exception to the general rule allows the fuse to carry 100 percent of the continuous load current if the fuse and enclosure has been rated by specific design for 100 percent rated current operation. See the exceptions to **210.20(A)**, **215.3**, and **230.42(A)(2)**.

Fuses are equipped with a time-delay feature that is designed to allow loads requiring high inrush currents to start and run. Such loads are motors, compressors, welders, etc. For further information pertaining to these loads, see **240.4(A) through (G)**.

For example, a 200 amp time-delay fuse will hold about 5 times its rating (200 A x 5 = 1000 A), which is 1000 amps. Such fuse will hold this value of 1000 amps for about 10 seconds based upon its time characteristics. A 200 amp nontime-delay fuse will hold 1000 amps for 2 seconds or less without blowing and opening the circuit because of inrush current. Therefore, a piece of equipment with an inrush current that is less than 1000 amps will start and run without blowing the fuse under normal starting and running conditions. **[See Figures 9-40(a) and (b)]**

TIME AND TEMPERATURE RELATIONSHIPS

A fuse temperature has the same heating effect as a conductor. The ampacity is rated at a standard ambient temperature for fuses and conductors. A fuse is usually 25°C, with the ampacity derated for temperatures above the standard ambient temperature.

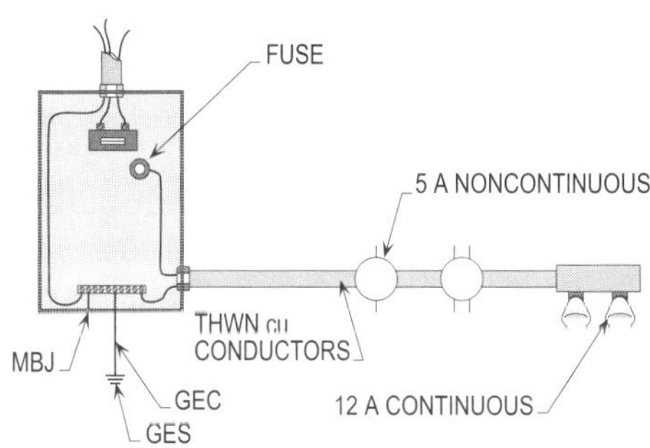

APPLYING 210.19(A)(1)

Sizing fuse

Step 1: Calculating fuse
210.20(A)
12 A x 125% = 15 A
5 A x 100% = 5 A
Total load = 20 A

Step 2: Selecting fuse
Table 310.16 & asterisk, 240.4(B), and 240.6(A)
20 A requires 20 A fuse

Sizing conductors

Step 3: Selecting conductors
Table 310.16 & asterisk, 240.4(B), 240.6(A), and 240.4(D)
20 A requires 12 AWG THWN

Solution: **A 20 amp fuse is required and 12 AWG THWN copper conductors are required.**

SIZING
NEC 210.20(A), 215.3, AND 230.42(A)(1)

Figure 9-39. The above illustration shows the procedure for calculating the size fuse and conductors based upon continuous operation.

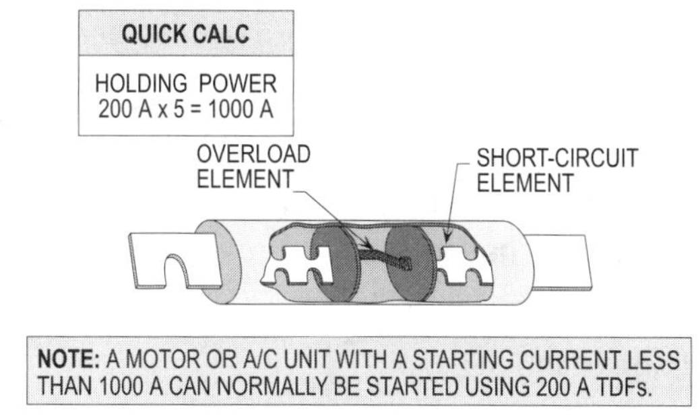

OPERATION CHARACTERISTICS
NEC 240.60(C)

Figure 9-40(a). The above illustration shows that a fuse will hold 5 times its rating for 2 to 10 seconds based upon the type used.

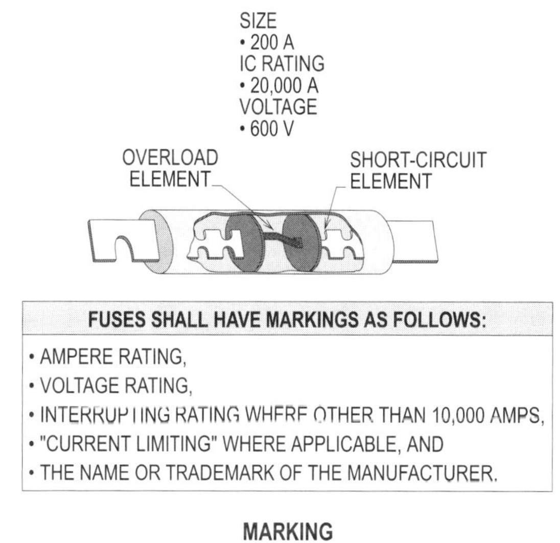

MARKING
NEC 240.60(C)

Figure 9-40(b). The above is a detailed illustration of how fuses shall be marked.

Depending on the type and class being installed, fuses consist of various current ratings that are usually based on specific time periods without opening.

For example, time-delay fuses shall withstand current higher than the rated current (overload currents) for minimum periods of time.

In a very short increment of time, a circuit shall be opened when short circuits or ground faults occur within a circuit without damaging the electrical elements of the circuit.

MARKING
240.60(C)

Fuses are required to be plainly marked, either by printing on the fuse barrel or by a label attached to the barrel. So that fuses can be sized, selected, and installed correctly, the following information shall be provided:

- Ampere rating
- Voltage rating
- Interrupting rating where other than 10,000 amps
- "Current limiting" where applicable
- The name or trademark of the manufacturer

DISCONNECTING MEANS
FOR FUSES
240.40

Except as noted below, each set of cartridge fuses shall have an individual disconnect on the supply side, so that any one set of fuses can be individually deenergized.

For other types of fuses and for thermal cutouts, the rule applies where voltages are over 150 volts-to-ground.

If the fuses or cutouts are accessible only to qualified persons, such as maintenance electricians, these requirements shall not be permitted to be applied.

For example, one disconnect shall be permitted to serve a number of fused circuits, and it shall not be necessary to provide a disconnect for each circuit.

Where a sequence of meters, fuses, and switches are utilized for services, the fuses shall be permitted to be installed ahead of the switch per **240.40** and **230.82**.

For special cases, one disconnect shall be permitted for several motors on one machine, as with metal- or wood-working machines per **240.40** and **430.112, Ex.**

Note that each motor shall be provided with an overcurrent protection device that allows it to start and run per **430.52**. In addition, overload protection shall be required for each motor per **430.32(A)(1)**.

Section 240.40 also permits a single disconnect where several elements of a fixed space heater are protected by individual fuses within the heating unit enclosure per **424.22(C)**. **(See Figure 9-41)**

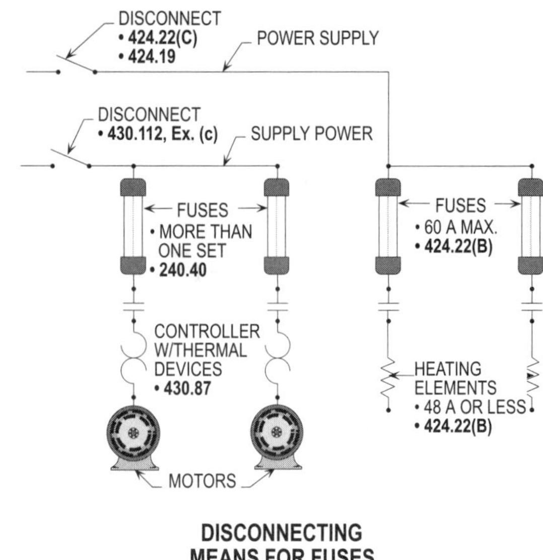

Figure 9-41. The above illustration shows different cases where one disconnect shall be permitted to be used to disconnect more than one set of fuses.

CIRCUIT BREAKERS
ARTICLE 240, PART VII

Circuit breakers and conductors with insulations are available to supply power to equipment. Conductors shall be sized to carry the load of the equipment in amps without deteriorating the insulation per **310.10**. The overcurrent protection devices shall be sized to protect the conductors and equipment from short circuits, ground faults, and overloads.

All circuit breakers and conductors shall be designed and installed according to the latest provisions of the NEC to ensure protection and proper installation.

The calculated load for circuit breakers shall be determined by calculating the loads at noncontinuous operation (100 percent) and continuous operation (125 percent) and applying demand factors, where applicable. Depending on the frame size of the unit, a circuit breaker can hold approximately three times its rating for different periods of time.

OPERATION CHARACTERISTICS

A circuit breaker operates either by thermal or magnetic principles or any combination of such. It is a device that protects conductors from excessive current. Basically, a circuit breaker is an automatic switch that trips open and clears the circuit when the current passing through it exceeds its rating.

For example, if a current of 32 amps flows through the elements of a 30 amp circuit breaker, the element in the circuit breaker, due to excessive heat, will bend and trip open the circuit.

A large current (amps) will generate more heat than a small amount of current, so naturally a circuit breaker will trip faster during a large overload condition than during a smaller one.

> **Design Tip:** A circuit breaker equipped with a magnetic plate operates both thermally and magnetically. The thermal protection clears overloads, which are currents that are developed slowly on the circuit conductor. The magnetic operation clears short circuits and ground faults, which are large currents that occur suddenly.

What makes circuit breakers so convenient to use is their ability to be used as switches and because spare ones are not necessary should a short circuit, ground fault, or overload occur on a circuit protected by such.

SIZING
210.20(A), 215.3, AND 230.42(A)(1)

Circuit breakers shall be sized based upon the operation characteristics of the load served. The NEC defines a continuous load as a load that operates continuously for three hours or more.

The load current in amps shall be calculated by multiplying the 160 amp continuous load current by 125 percent (160 A x 125% = 200 A) to obtain the size circuit breaker, which is 200 amps.

> **Design Tip:** An exception to the general rule allows the circuit breaker to carry 100 percent of the continuous load current if the circuit breaker and enclosure has been rated by specific design for 100 percent rated current operation. See the exceptions to **210.20(A), 215.3,** and **230.42(A)(2)**.

Circuit breakers are equipped with a time-delay feature that is designed to allow loads requiring high inrush currents to start and run. Such loads are motors, compressors, welders, etc. For further detailed information pertaining to these loads, see **240.4(A) through (G)**.

For example, a 200 amp circuit breaker will hold about 3 times its rating (200 A x 3 = 600 A), which is 600 amps. Such circuit breaker will hold this value (600 amps) for a number of seconds based upon its frame size. A 200 amp circuit breaker will hold 600 amps for about 35 seconds without tripping open the circuit. Therefore, a piece of equipment with an inrush current that is less than 600 amps will start and run under normal starting conditions without tripping the circuit breaker. **(See Figure 9-42)**

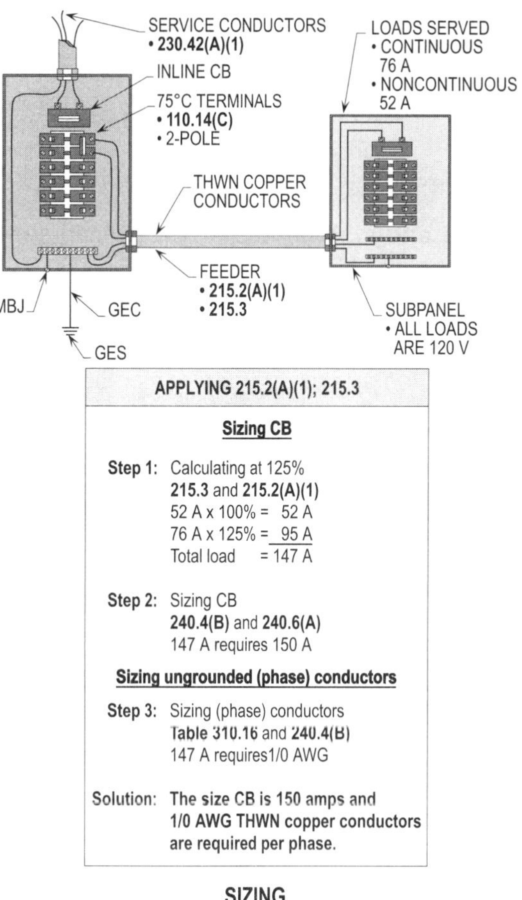

Figure 9-42. The above illustration shows the procedure for sizing the circuit breaker and conductors based upon continuous operation.

METHOD OF OPERATION
240.80

Operation by hand simply means that the circuit breaker has a handle that the operator can grasp to open or close the circuit. Circuit breakers that are pneumatic or electrically controlled may become inoperative, or the contact might stick in the make position and, without a handle to deenergize the circuit breaker, there would be no way to open the circuit. Should the electrical or pneumatic control fail to operate the circuit, it can be opened manually using the handle on the circuit breaker.

INDICATING
240.81

When the circuit breaker is turned ON, the handle shall be installed so that it is in the UP position. Circuit breakers shall be clearly marked in the open OFF or closed ON position. Circuit breaker handles that operate vertically instead of rotationally or horizontally shall be installed with the ON position of the handle in the UP position.

MARKING
240.83

The following markings shall be applied before installing circuit breakers:

- **Durable and visible.** The ampere rating shall be durable and visible after installation.

- **Location.** The ampere rating shall be molded, stamped, etched, or similarly marked into the handles of circuit breakers rated 100 amps or less.

- **Interrupting rating.** Interrupting current (IC) rating other than 5000 amps shall be marked with IC rating.

- **Used as switches.** Circuit breakers used to switch 120 or 277 volt fluorescent lighting circuits shall be listed and shall be marked SWD or HID. Circuit breakers used to switch high-intensity discharge lighting circuits shall be listed and shall be marked HID.

- **Voltage marking.** Voltage rating not less than the nominal system voltage shall be marked on circuit breakers for the capability to interrupt fault currents between phases or phase-to-ground.

> **Design Tip:** Circuit breakers used for supplementary protection shall not require an interrupting rating marked on the device.

See Figure 9-43 for a detailed illustration pertaining to the rules for marking circuit breakers.

APPLICATIONS
240.85

Circuit breakers shall be permitted to be used with grounded, ungrounded, or grounded neutral systems when identified with a straight voltage marking.

For example, a circuit breaker shall be permitted to be used with grounded, ungrounded, or grounded neutral systems when identified with a straight voltage marking of 480 volts, etc.

Circuit breakers shall be permitted to be used with grounded neutral systems when identified with a slash voltage marking.

For example, a circuit breaker shall be permitted to be used with grounded neutral systems when identified with a slash voltage marking of 480/277 volts.

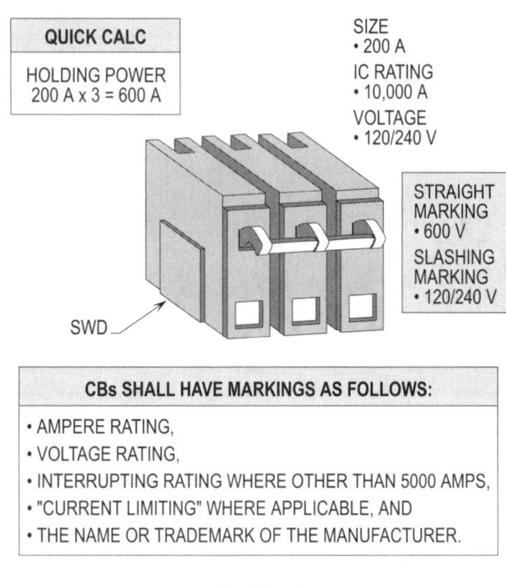

Figure 9-43. The above illustration shows the markings that shall be placed on circuit breakers. Note that a circuit breaker will hold 3 times its rating for a period based upon its frame size and rating. A 200 amp circuit breaker will start a load with a starting current of 600 amps or less. (200 A x 3 = 600 A)

GROUND FAULT PROTECTION OF EQUIPMENT
230.95, 240.13, AND 215.10

Where fault currents are available and due to the damaging effects of ground faults, it has become necessary to prevent injury to personnel and damage to equipment by providing ground fault protection for equipment. The following rules shall be followed.

SIZE
230.95, 240.13, AND 215.10

Ground fault protection of equipment shall be required for grounded wye connected services of more than 150 volts-to-ground and overcurrent protection devices rated 1000 amps or greater.

> **Design Tip:** This rule applies mainly for 480/277 volt wye services. However, such protection is not prohibited for 208/120 volt wye systems.

SETTING
230.95(A), 240.13, AND 215.10

The maximum setting permitted for ground fault protection shall not be permitted to exceed 1200 amps, when the service disconnecting means exceeds 1000 amps or more. Ground fault protection shall allow all ungrounded (phase) conductors of the faulted circuit to open simultaneously if a low magnitude fault develops in an ungrounded (phase) conductor.

> **Design Tip:** Ground fault protection shall not be required for 6 - 700 amp disconnects, but shall be required for 5 - 700 amp disconnects and one 1000 amp disconnect.

For example, ground fault protection shall be required per **230.95** if a 900 amp fuse is installed in a 1200 amp disconnect and an adjustable circuit breaker is set below 1000 amps but could later be set at 1000 amps or greater.

CLEARING TIME
230.95(A), 240.13, AND 215.10

To prevent damage to overcurrent protection devices, busbars, and service equipment, the ground fault system shall be required to clear a fault of 3000 amps or more within one second. The following two types of ground fault detection are used to sense low-magnitude ground faults:

- Donut method
- Window method

When applying the donut method to sense any current leaking to ground, a current relay shall be placed over the main bonding jumper. When applying the window method to sense any current leaking to ground, the ungrounded (phase) conductors and grounded (neutral) conductors shall be placed in the window (ground fault sensor). The objective is to sense current leaking to ground from a ungrounded (phase) conductor.

> **Design Tip:** The system will be cleared once the low-magnitude fault currents are detected and the overcurrent protection device is tripped open. This clearing procedure prevents damage to equipment due to low magnitude arcs produced by such leakage current.

See Figure 9-44 for a detailed illustration for applying such rules and regulations.

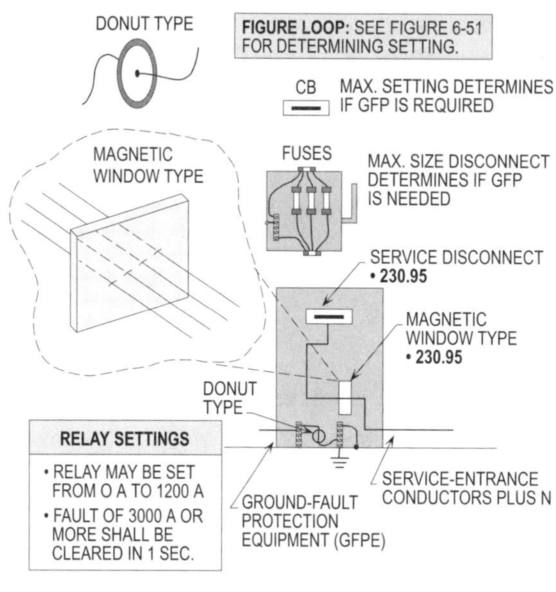

Figure 9-44. The above illustration shows the relay settings and relay requirements when installing ground fault protection for equipment.

PERFORMANCE TESTING
230.95(C), 240.13, AND 215.10

Performance testing shall be performed at time of installation for the ground fault protection system. The test shall be run according to approved test instructions that are furnished with the equipment.

A written record of the test shall be required to be available to the AHJ before final approval of the installation.

GROUND FAULT PROTECTION FOR PERSONNEL
590.6

Ground fault circuit interrupters shall be required to protect personnel where a leakage current might develop due to a faulty electrical hand tool. Leakage currents can be caused by frayed insulation, damaged equipment, moisture, etc.

FUNCTION

A leakage current is the result of a current-carrying path developing between a current-carrying wire, equipment, and ground. The function of a GFCI is to detect small currents where overcurrent protection devices such as fuses and circuit breakers cannot open or deenergize the circuit due to their possessing a higher rating. These currents of a few thousandths of an ampere can be detected by a GFCI-protective device that allows the circuit to open or deenergize in a fraction of a second. Class A GFCIs will allow a circuit to open or deenergize when a ground fault current of about 5 mA (milliamperes) is detected.

TYPES

The following are four basic categories of GFCI-protected devices:

- GFCI receptacles
  - Flush-mounted receptacles
  - Surface-mounted receptacles
  - Plug-in receptacles
- GFCI circuit breakers
- Portable GFCIs
- GFCI-protected extension cord sets

Flush-mounted and surface-mounted type receptacles are designed and available in terminated or feed-through models. The design of terminated models will protect only their own receptacles. The design of feed-through models will protect their own receptacles and all other receptacles installed on the load side of the GFCI. Plug-in receptacles are only available to be installed in a terminated model.

Circuit breakers and GFCIs are combined devices for providing overcurrent protection and fault current protection. The circuit breaker may be installed as an ON and OFF switch for the branch circuit when circuits or components need servicing.

Portable equipment contains both receptacles and circuit breakers that contain GFCIs. The circuit breaker assembly supplying such equipment may or may not be a GFCI protective device.

Extension cord sets are usually protected by a GFCI in the attachment cord body.

See Figure 9-45 for the requirements of applying the rules for GFCI protection.

OVERCURRENT PROTECTION DEVICES FOR SYSTEMS OVER 600 VOLTS

Overcurrent protection devices shall be sized properly to protect the elements of electrical systems rated over 600 volts. High-voltage electrical systems behave differently from systems of 600 volts or less. When short circuits or ground faults develop on high-voltage circuits, great damage to conductors, elements, and equipment can occur. Therefore, it is imperative to protect such systems from these hazards.

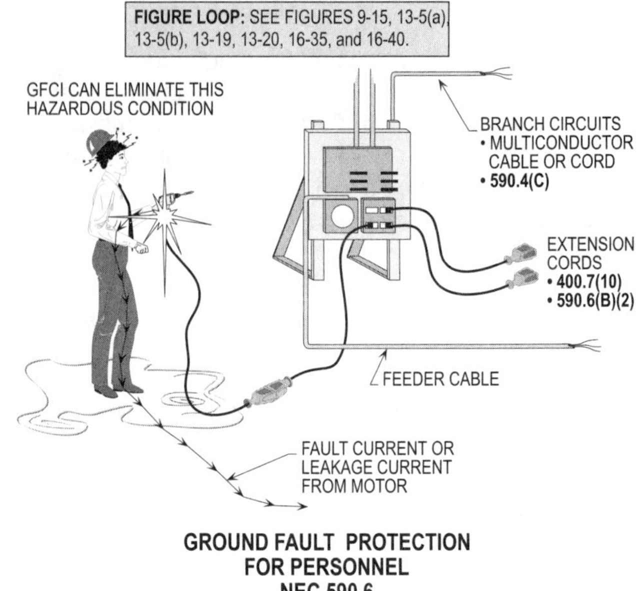

Figure 9-45. GFCI devices can detect ground faults of 5 mA, clear the circuit, and prevent electrocution of workers using electric hand tools.

FEEDERS AND BRANCH CIRCUITS 240.100(A), (B), AND (C)

Branch circuits shall have a short-circuit protective device in each ungrounded (phase) conductor or comply with **490.21(A)** and **(B)**. The protective device(s) shall be capable of detecting and interrupting all values of current that can occur at their location in excess of their trip setting or melting point per **240.100(A)(1)**. In other words, the overcurrent protection devices shall be sized to protect the conductors at their ampacities, if not oversized to allow equipment with high inrush current to start, run, and operate.

Note that the protective device(s) shall be capable of detecting and interrupting all values of current that can occur at their location in excess of their trip setting or melting point per **240.100(B)**.

Feeder and branch circuit conductors shall have overcurrent protection in each ungrounded (phase) conductor located at the point where the conductor receives its supply or at a location in the circuit determined under engineering supervision that includes, but is not limited to, considering the appropriate fault studies and time-current coordination analysis of the protective devices and the conductor damage curves.

OVERCURRENT RELAYS AND CURRENT TRANSFORMERS
240.100(A)(1)

Circuit breakers used for overcurrent protection of three-phase circuits shall have a minimum of three overcurrent relays operated from three current transformers. The separate overcurrent relay elements (or protective functions) shall be permitted to be part of a single electronic protective relay unit. On three-phase, three-wire circuits, an overcurrent relay in the residual circuit of the current transformers shall be permitted to replace one of the phase relays.

An overcurrent relay operated from a current transformer that links all phases of a three-phase, three-wire circuit shall be permitted to replace the residual relay and one of the phase-conductor current transformers. Where the neutral conductor is not regrounded on the load side of the circuit as permitted in **250.184(B)**, the current transformer shall be permitted to link all three-phase conductors and the grounded circuit conductor (neutral).

FUSES
240.100(A)(2)

When fuses are used, they shall be connected in series with each ungrounded (phase) conductor.

CONDUCTOR PROTECTION
240.100(C)

The operating time of the protective device, the available short-circuit current, and the conductor used shall be coordinated to prevent damaging or dangerous temperatures in conductors or conductor insulation under short-circuit conditions.

ADDITIONAL REQUIREMENTS FOR FEEDERS
240.101(A) AND (B)

Feeders shall have a short-circuit overcurrent protection device in each ungrounded (phase) conductor. The protective device(s) shall be capable of detecting and interrupting all values of current that can occur at their location in excess of their trip setting or melting point. A fuse rated in continuous amperes not exceeding three times the ampacity of the conductor or a breaker having a trip setting of not more than six times the ampacity of the conductor shall be considered as providing the required short-circuit protection. For further information, see **490.21(A)** and **(B)** and **Article 490,** which deals with high-voltage electrical systems. **[See Figures 9-46(a) through (d)]**

Design Tip: The operating time of the protective device, the available short-circuit current, and the conductor used will need to be coordinated to prevent damaging or dangerous temperatures in conductors or conductor insulation under short-circuit or ground-fault conditions.

OTHER ACCESSORIES

When selecting the arrangement of the overcurrent protection devices for systems over 600 volts, see **490.21(A)** and **(B)**. Switchboard requirements are found in **Part III** of **Article 490**. Wiring methods and techniques are outlined in **300.37** and **300.50,** based on aboveground or underground installations. Cable preparation, shielding, and grounding requirements are listed in **300.40, 310.6,** and **310.60(C)(1)** in the NEC.

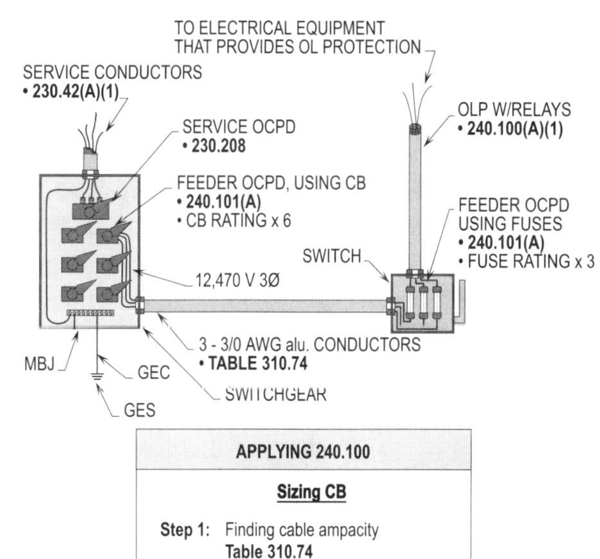

Figure 9-46(a). The above illustration shows methods of providing overload protection.

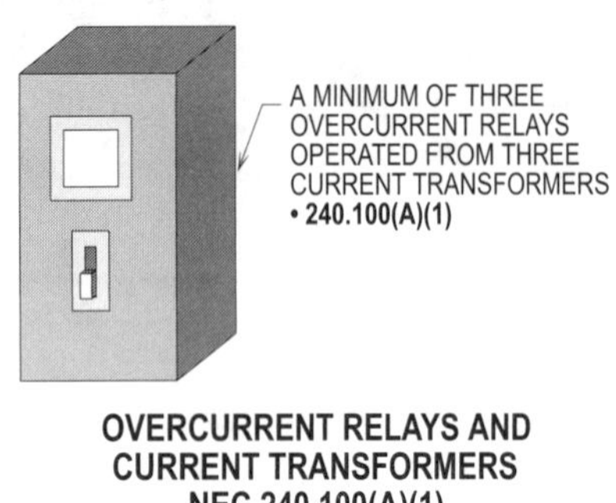

**OVERCURRENT RELAYS AND
CURRENT TRANSFORMERS
NEC 240.100(A)(1)**

Figure 9-46(b). The above illustration shows the correct procedure for sizing circuit breakers and fuses to be used to protect high-voltage electrical systems for feeders.

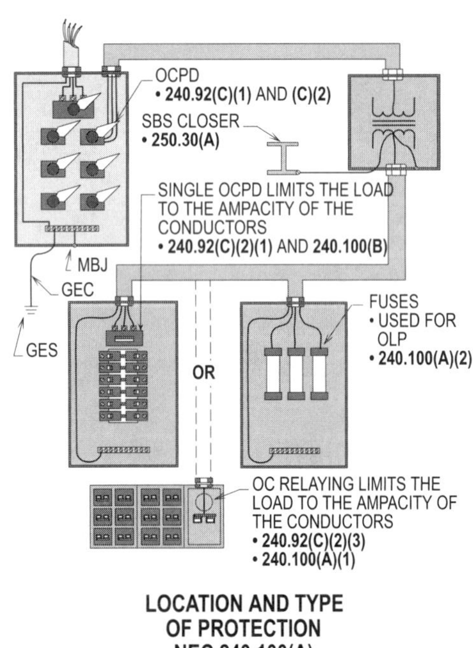

**LOCATION AND TYPE
OF PROTECTION
NEC 240.100(A)**

Figure 9-46(c). The overload protection (OLP) for the protection of the feeder conductors shall be permitted to be provided by installing overcurrent relays, which will provide the ultimate protection scheme.

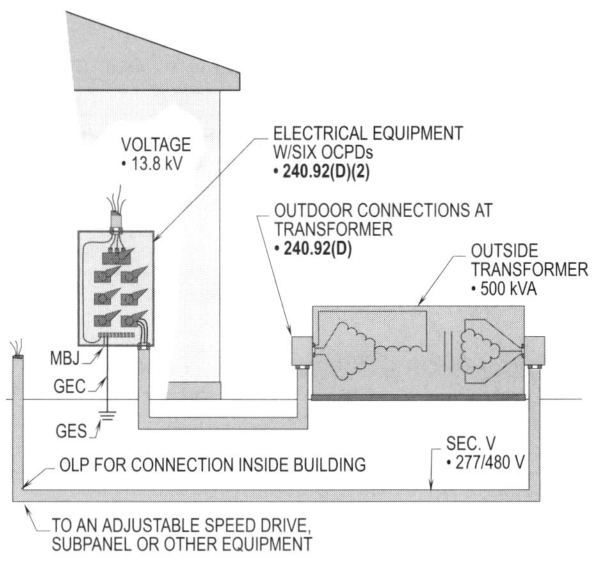

**OUTSIDE FEEDER TAPS
NEC 240.92(D)**

Figure 9-46(d). The above illustration shows alternate means allowed for protecting conductors connected to a transformer located outside.

Chapter 9. Overcurrent Protection Devices

Section   Answer

1. If the overcurrent protection device is greater than _____ amps, the ampacity of the conductors shall be equal to or greater than the ampacities of the overcurrent protection device.
   (a) 200   (b) 400
   (c) 600   (d) 800

2. 16 AWG and larger flexible cord conductors used in field-assembled extension cord sets that are connected to a _____ amp branch circuit shall be considered protected from overcurrent conditions.
   (a) 15   (b) 20
   (c) 25   (d) 30

3. In grounded systems, individual single-pole circuit breakers with _____ handle ties shall be permitted to be utilized as the protection for each ungrounded (phase) conductor for line-to-line connected loads for single-phase circuits or three-wire DC circuits.
   (a) approved   (b) identified
   (c) listed   (d) labeled

4. Generator conductors shall have an ampacity equal to at least _____ percent of the nameplate current rating of the generator. (General Rule)
   (a) 100   (b) 110
   (c) 115   (d) 125

5. For a branch circuit, a continuous load is a load that operates for _____ hours or more.
   (a) 1   (b) 2
   (c) 3   (d) 4

6. Circuit breakers shall be permitted to be used with grounded, ungrounded, or grounded neutral systems when identified with a _____ voltage marking.
   (a) straight   (b) approved
   (c) listed   (d) slash

7. A tap feeding a panelboard shall not be permitted to be over _____ ft long. (Maximum tap)
   (a) 10   (b) 15
   (c) 20   (d) 25

8. Transformer (primary) tap conductor ampacity shall be at least __________ of the feeder's overcurrent protection device. (Primary plus transformer and secondary)
   (a) 1/4   (b) 1/3
   (c) 1/2   (d) 3/4

_____________ _____________ **9.** Ground fault protection of equipment shall be required for grounded wye-connected services of more than 150 volts-to-ground and overcurrent protection devices rated _____ amps or greater.
    (a) 600   (b) 800
    (c) 900   (d) 1000

_____________ _____________ **10.** To prevent damage to overcurrent protection devices, busbars, and service equipment, the ground fault system shall be required to clear a fault of _____ amps or more within one second.
    (a) 1000   (b) 2000
    (c) 3000   (d) 5000

_____________ _____________ **11.** If the standard current ratings of fuses or nonadjustable circuit breakers do not conform to the ampacity of the conductors being used, it shall be permitted to use the next larger standard rating when below _____ amps.
    (a) 100   (b) 400
    (c) 800   (d) 1000

_____________ _____________ **12.** A _____ amp overcurrent protection device shall be considered as being adequate protection for fixture wire, in size 16 AWG or 18 AWG.
    (a) 15   (b) 20
    (c) 30   (d) 40

_____________ _____________ **13.** Standard ampere ratings for fuses are available in sizes of _____ amps.
    (a) 3   (b) 6
    (c) 10   (d) all of the above

_____________ _____________ **14.** A tap from a larger conductor to a smaller conductor shall be permitted to extend up to a distance of _____ ft, provided the current-carrying capacity of the tap is at least 1/3 that of the larger conductor.
    (a) 10   (b) 20
    (c) 25   (d) 50

_____________ _____________ **15.** For taps in high-bay manufacturing buildings, the protection for the tap shall be permitted to be at the end of the tapped conductors, if the tap is not over 25 ft long horizontally and not over _____ ft total length.
    (a) 50   (b) 100
    (c) 150   (d) 200

_____________ _____________ **16.** Fuses shall be plainly marked with _____.
    (a) ampere rating   (b) voltage rating
    (c) the name of the manufacturer   (d) all of the above

_____________ _____________ **17.** Before circuit breakers are used as switches, they shall be plainly marked _____.
    (a) SW   (b) SWD
    (c) HIT   (d) for the location

_____________ _____________ **18.** Ground fault detection such as the _____ method is used to sense low-magnitude ground faults.
    (a) donut   (b) window
    (c) all of the above   (d) none of the above

19. Where a feeder is installed in a high-bay manufacturing building, the tap shall be made no less than _____ ft from the floor.
    (a) 30   (b) 40
    (c) 50   (d) 60

20. Circuit breakers rated at _____ amperes or less and 600 volts or less shall have the ampere rating stamped, etched, or similarly marked into their handles.
    (a) 50   (b) 100
    (c) 200   (d) 800

21. What size overcurrent protection device is required to protect a continuous branch-circuit load of 22.5 amps?

22. What size THWN copper conductors are required to protect a continuous branch-circuit load of 22.5 amps?

23. What size overcurrent protection device is required to protect a subpanel with a load of 649 amps on each phase paralleled four times? (All load are continuous)

24. What size overcurrent protection device (CB) and THHN copper conductors are required to protect a motor-operated appliance with a 13 amp continuous load?

25. What is the minimum, next, and maximum size circuit breaker for a 208 volt, 40 HP, Design B, three-phase motor?

26. What size overcurrent protection device (CB) and THWN copper conductors are required for a phase converter with a 230 volt, 20 HP, Design C, three-phase motor?

27. What is the minimum and maximum size circuit breaker allowed to start and run an A/C unit with a compressor having a full-load current rating of 18 amps?

28. What size overcurrent protection device is required to protect a two-wire to two-wire transformer with a 240 volt primary and a 120 volt secondary with 4 AWG THWN copper conductors?

29. What size overcurrent protection device is required to protect a three-wire to three-wire transformer with a 480 volt primary and a 240 volt secondary with 500 KCMIL copper conductors?

30. What size THWN copper conductors are required for a 208 volt, 20 kVA, three-phase capacitor based on the FLA of the capacitor times 135 percent or 1/3 of the conductors ampacity (3/0 AWG THWN)?

31. What size overcurrent protection device is required for an AC transformer and DC rectifier arc welder rated at 56 amps with an 80 percent duty cycle supplied with 8 AWG THWN copper conductors?

32. What size overcurrent protection device for the branch-circuit and control conductors is required for a Class 1 control circuit supplied by 14 AWG THWN copper conductors?

33. What size overcurrent protection device for fire-signaling circuits is required for an alarm supplied by 16 AWG THWN copper conductors and a detector supplied by 14 AWG copper conductors?

34. What size fixture wire is required when supplied by a 30 amp branch circuit?

35. What size THWN copper conductors and overcurrent protection device are required for a tap using the 10 ft tap rule, with the overcurrent protection device for the feeder being tapped rated at 400 amps?

36. What size THWN copper conductors and overcurrent protection device are required for a tap using the 25 ft tap rule, with the overcurrent protection device for the feeder being tapped rated at 400 amps?

37. What size overcurrent protection device is required for the secondary, and what size THWN copper conductors is needed for the primary (480 V) and secondary (208 V) taps and connection that are over 25 ft long? The primary overcurrent protection device is rated at 200 amps.

38. What size overcurrent protection device and THWN copper conductors are required for a tap using the 100 ft tap rule, with an overcurrent protection device for the feeder being tapped rated at 300 amps?

39. What size overcurrent protection device and THWN copper conductors are required for a tap using the 10 ft tap rule for a motor, with an overcurrent protection device for the feeder being tapped rated at 400 amps? (Note that terminals are 60°C)

40. What size THWN copper conductors are required when tapping a 277/480 volt, three-phase, four-wire generator with an 250 kVA rating?

10

Equipment Over 600 Volts

Through the years, high-voltage electrical systems have increased at an accelerating rate. The task of knowing all there is to know about high-voltage systems is particularly difficult for designers, installers, maintainers, and inspectors. The term *high voltage*, as used in this chapter, refers to systems operating above 600 volts nominal, which is measured phase-to-phase. However, the electrical industry usually refers to systems of 601 volts to 15 kV as medium-voltage circuits, and voltage systems above 69 kV are known as high-voltage circuits.

Because of the potential hazards involved with high-voltage systems, extreme care and detailed planning should go into the design, installation, and maintenance of such systems, specifically the application of Article 490 of the *National Electrical Code* as well as other pertinent and applicable articles. The information found in this chapter will cover most of these requirements.

SERVICE POINT
ARTICLE 100

A service point is the "point of connection between the facilities of the serving utility and the premises wiring." All equipment on the load side of that point is subject to the rules of the NEC. Equipment on the supply side is the responsibility of the utility company and is not regulated by the NEC. This definition of service point shall be used to establish where the originating point of the service conductors is located. (Also, see **230.200** of the NEC.)

The definition of service point does identify where the NEC becomes applicable, and does pinpoint the origin of service conductors.

Conductors between the service point of a particular installation and the service disconnect are identified as service conductors and are subject to NEC rules regarding service conductors per **Article 100**.

For example, this rule implies that conductors ahead of the service-entrance equipment or between the metering enclosureare considered service conductors. **(See Figure 10-1)**

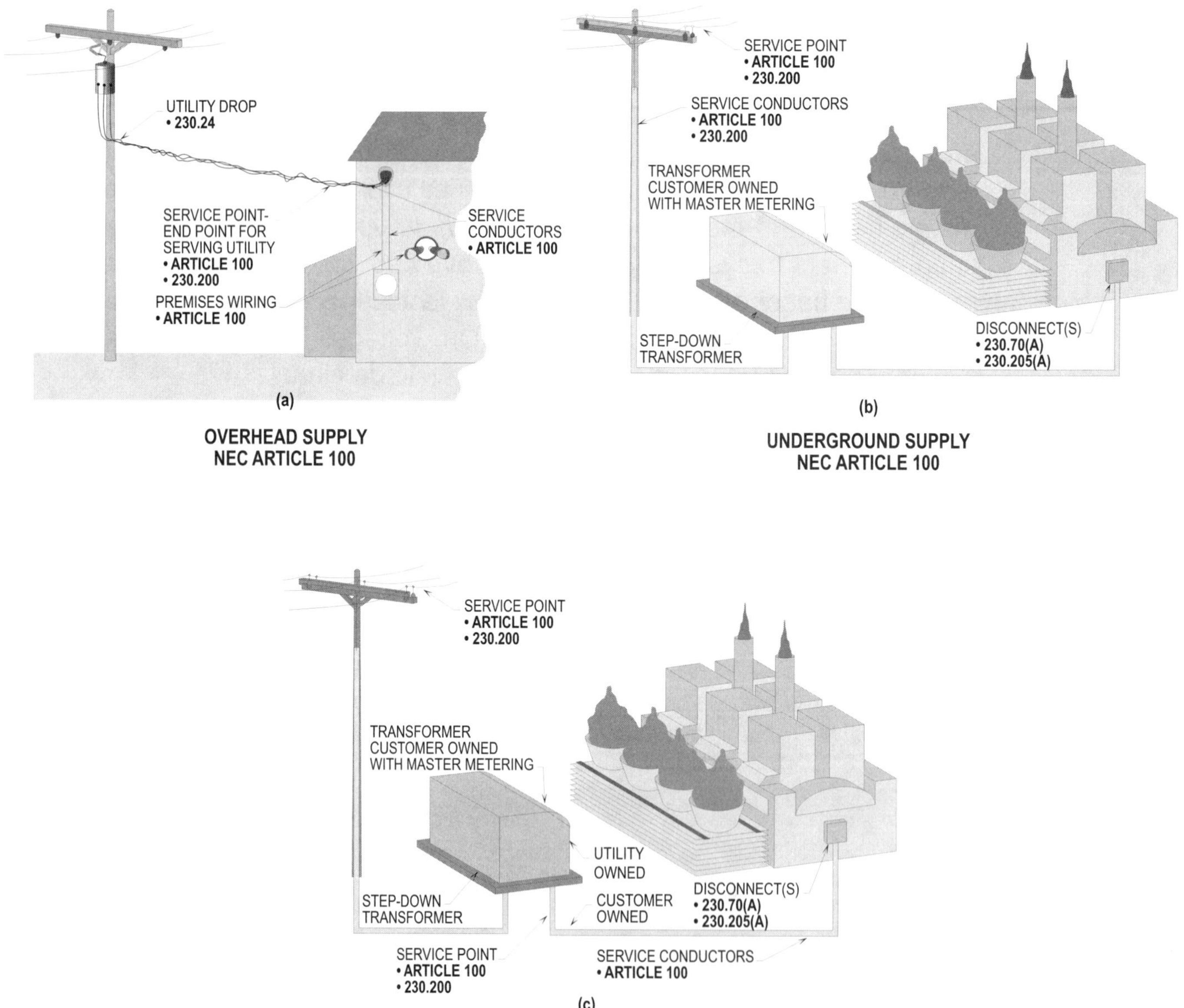

Figure 10-1. Service point is the point of connection between the facilities of the servicing utility and the premises wiring.

SERVICE CONDUCTORS

For very large buildings, such as large hotels, office buildings, and industrial sites, high-voltage primaries may be installed to step-down transformers located within the facility. Here, the primaries are the service conductors, except in the case where they enter a vault, or where they enter metal-enclosed switchgear, or where they enter a locked room accessible only to qualified personnel, in which case the secondaries within the building are the service conductors, not the primaries entering the building. If the transformers are located outside the building, the secondaries entering the building are considered the service conductors.

These rules must be applied when the secondary voltage is 600 volts or less. For secondary voltages over 600 volts, either the primaries or the secondaries can be classified as the service conductors, if serving an industrial complex or large commercial site. **Article 100** of the NEC shall be applied to classify such conductors.

SERVICE-ENTRANCE CONDUCTORS 230.202

This rule specifies the wiring methods that are permitted for use as service-entrance conductors where it has been established that primary conductors are the service conductors or where the secondary conductors are the service conductors and operate at more than 600 volts.

The basic types of wiring methods permitted for high-voltage service entrances are:

- **(1)** Rigid metal conduit (RMC)
- **(2)** Intermediate metal conduit (IMC)
- **(3)** Electrical metallic tubing (EMT)
- **(4)** PVC
- **(5)** Cable trays
- **(6)** Busways
- **(7)** Cablebus
- **(8)** Other identified raceways
- **(9)** Metal clad cable (Type MC)
- **(10)** Medium voltage cables (Type MV)
- **(11)** Bare conductors
- **(12)** Bare busbars
- **(13)** Direct burial cables

See Figure 10-2 for a detailed illustration.

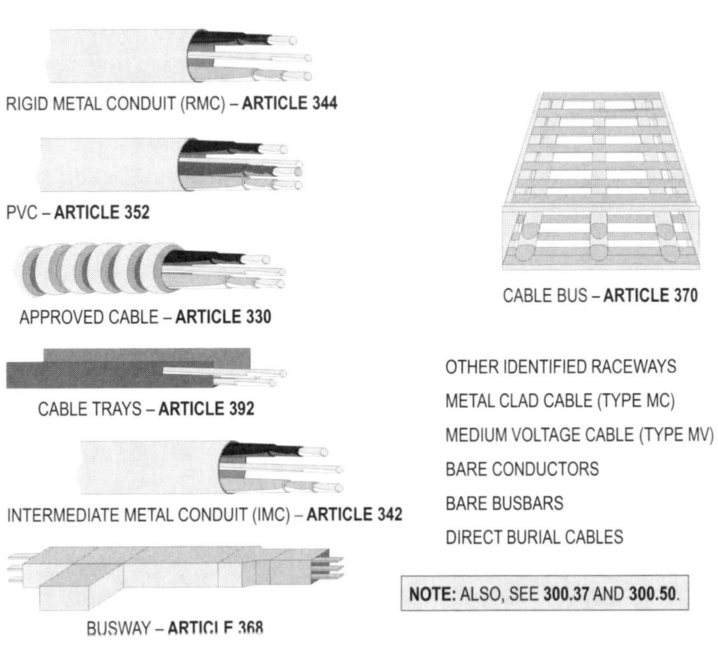

Figure 10-2. The above wiring methods shall be permitted to be used to enclose high-voltage conductors that supply high-voltage electrical systems. **Note:** Always check with the AHJ for permitted wiring methods.

WARNING SIGNS
110.34(C) AND 314.72(E)

A warning sign used to warn unauthorized persons who might make contact with live parts shall always include the phrase "KEEP OUT" immediately after the phrase **"DANGER – HIGH VOLTAGE."**

This is an mandatory factor in utilizing such a sign to act as an effective deterrent. Such wording shall always be used on signs to warn personnel that high-voltage (over 600 V) electrical systems and equipment are present and be extremely careful. (**See Figure 10-3**)

WARNING SIGNS

The main reason that live parts of electrical systems rated over 600 volts shall be posted with warning signs is to prevent unauthorized persons from coming in contact with live parts (phases). High-voltage systems are very dangerous due to the threat of arcing or flashing across and making contact with grounded objects such as a person's body. When an individual is shocked by high voltage, there is usually an explosion at the point where the current leaves the body.

This is due to the high pressure created by the higher voltage, which forces the current to flow through the lower resistance of the body. Electrical current flowing through the resistance of one's body behaves like the current flowing through a resistive heater element. Heat is produced in the body just as in the heater element, and this production of heat can severely damage the elements inside the body. Therefore, only qualified persons are allowed to be near or work on high-voltage systems.

This warning sign rule shall be applied to all pull and junction boxes housing conductors over 600 volts, which include the following wiring methods:

(1) All boxes shall provide a complete enclosure for contained conductors or cables.

(2) All covers on boxes shall be securely fastened in place to prevent easy access of unauthorized personnel. Covers on underground boxes that weigh over 100 lb are considered to be complying with this requirement. Boxes shall have covers mounted properly, and all openings shall have seals to prevent the entering of dirt, lint, etc. per **314.72(E)**. Covers for boxes shall be permanently marked, **"DANGER – HIGH VOLTAGE – KEEP OUT."** The marking shall be placed on the outside of the box cover and shall be readily visible and legible.

Note, a warning that high voltage is present is required along with a command. Examples of the command are:

DANGER – HIGH VOLTAGE – KEEP OUT

DANGER – HIGH VOLTAGE – AUTHORIZED PERSONNEL ONLY

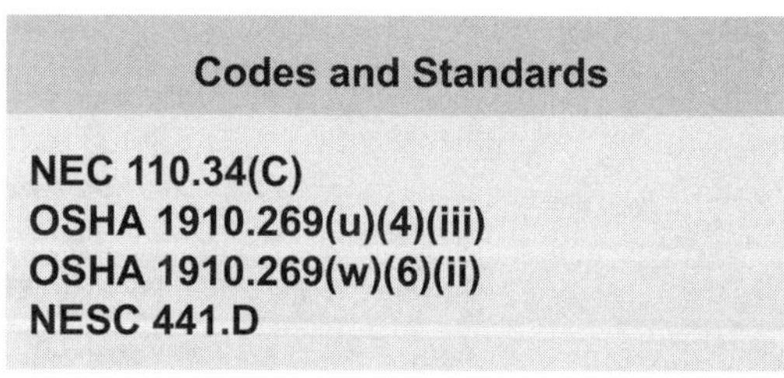

Codes and Standards
NEC 110.34(C)
OSHA 1910.269(u)(4)(iii)
OSHA 1910.269(w)(6)(ii)
NESC 441.D

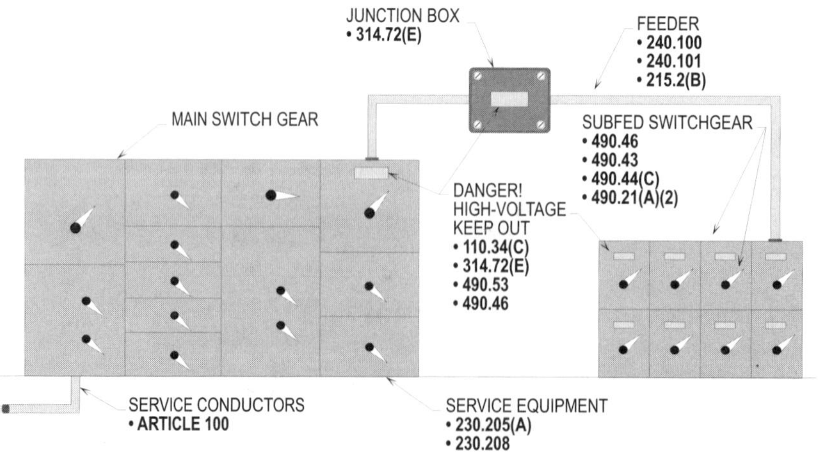

Figure 10-3. The covers on enclosures and junction boxes enclosing high-voltage elements shall be permanently marked "DANGER — HIGH VOLTAGE — KEEP OUT" and such markings shall be readily visible and legible.

ISOLATING SWITCHES
230.204, 490.21(E), AND 490.22

For certain types of disconnects, an isolating switch is required. Sections **230.204** and **490.21** specify the different types of disconnects approved for high-voltage services.

An air-break isolating switch shall be used between an oil switch or an air, oil, vacuum, or sulfur hexachloride circuit breaker and the supply conductors. However, removable truck panels or metal-enclosed units can be used that provide the disconnection of all live parts when in the removed position.

The purpose of this supply side disconnect is to ensure safety for personnel while repairing or pulling maintenance on such equipment.

Section **230.204(A)** through **(D)** shall be permitted to be utilized to determine if an isolation switch is required ahead of the overcurrent protection device protecting the service conductors and equipment. **(See Figure 10-4)**

Note, review **490.22** for more information on these requirements.

WHERE REQUIRED
230.204(A), 490.21(E), AND 490.22

When oil switches or air or oil circuit breakers are used as the disconnecting means, an additional isolating switch shall be installed ahead of the oil switch or the circuit breaker. *Isolating switch* is the term used for a switch that is to be opened only when there is no load on the circuit. An isolating switch does not have to have a rating high enough to break the load current. Such a switch shall open after the oil switch or the circuit breaker is opened. Opening the isolating switch deenergizes the oil switch or circuit breaker so it can be safely handled or worked on when servicing or repairs are required.

High-voltage service equipment of a type that can be rolled or pulled out from its switchgear or panel enclosure is available. An interlock permits the equipment to be removed only if the switch is in the open position. When this type of equipment is used, an isolating switch is not required. **[See Figure 10-5(a)]**

For the racking in or out of a circuit breaker, see **Figure 10-5(b)**.

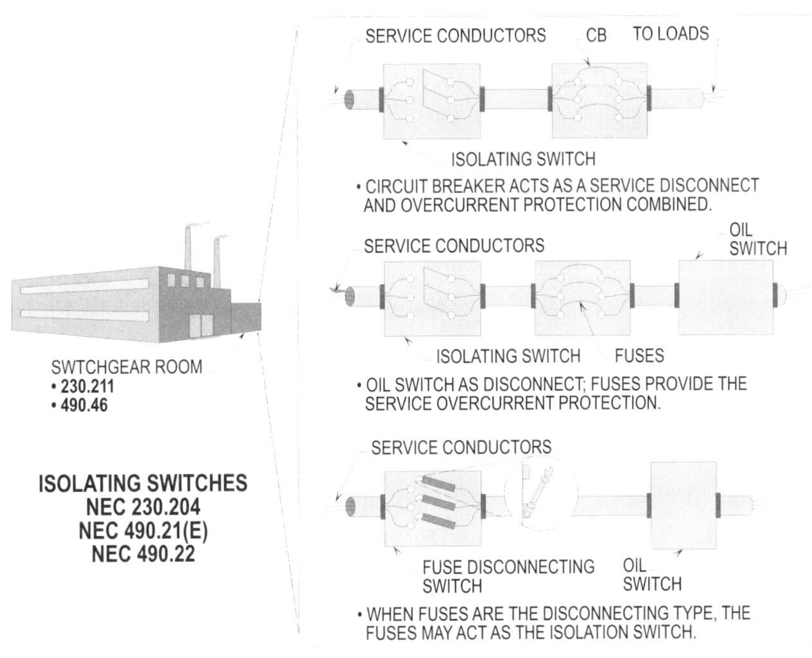

Figure 10-4. Except where roll-out equipment is used, an isolating switch is required when an oil switch or a circuit breaker is used as a service disconnect.

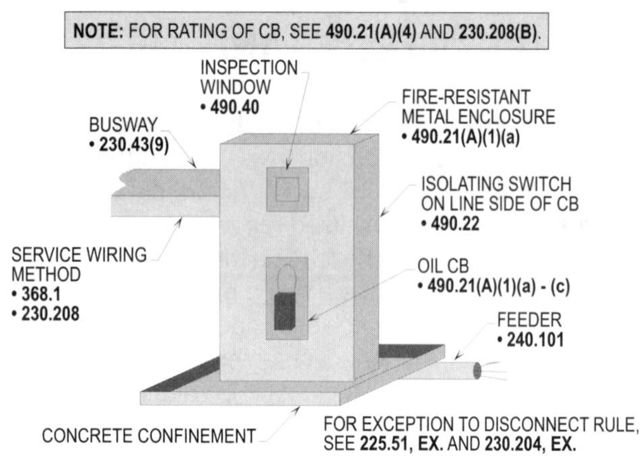

Figure 10-5(a). Isolating switch on line side of a circuit breaker used to disconnect the service conductors.

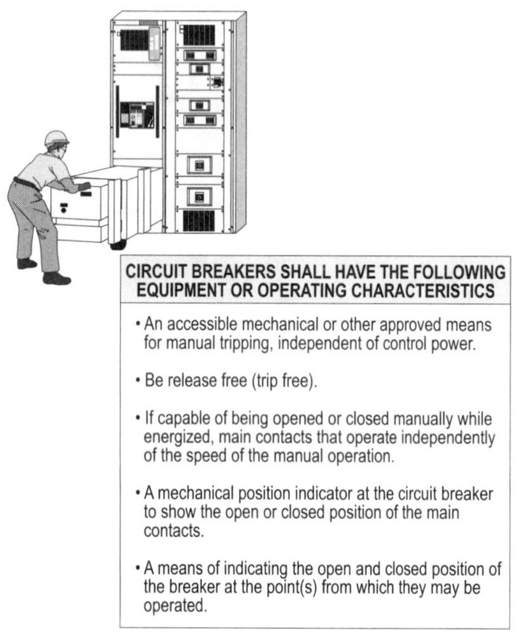

Figure 10-5(b). The above illustration shows an electrician racking out a circuit breaker.

FUSE AS ISOLATING SWITCH
230.204(B) AND 490.22

If a nonautomatic oil switch and fuses are used as the disconnecting means, and the fuses are of a type that open and close like a switch, no isolating switch is required. However, the disconnecting fuses shall be mounted in the line ahead of the oil switch. **(See Figure 10-6)**

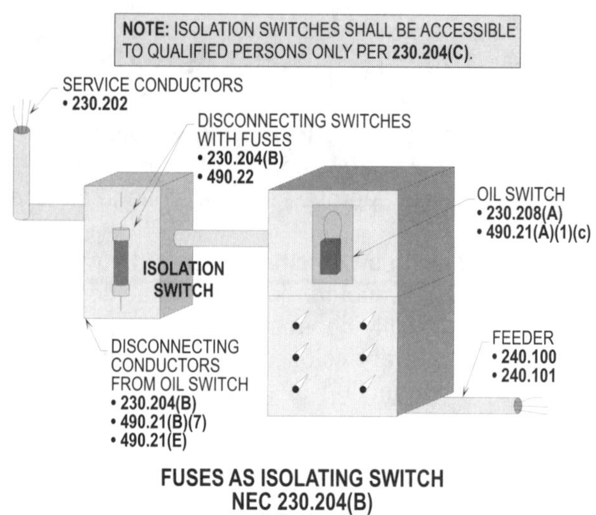

Figure 10-6. Disconnect with fuses used to disconnect (by isolating switch) and protect the service conductors supplying the service equipment.

ACCESSIBLE TO QUALIFIED PERSONNEL ONLY
230.204(C) AND 490.32

Isolating switches shall be accessible to qualified personnel only. The main reason for this requirement is isolating switches are dangerous when opened under load. Therefore, they shall never be accessible to unqualified personnel or the general public.

CONNECTION TO GROUND
NEC 230.204(D) AND 490.37

Isolating switches must provide means for readily connecting the load side controls to a grounding electrode system, equipment ground busbar, or grounded structural steel. One way of accomplishing this is to use a double throw switch for the isolating switch, so arranged that in one position the load connects the line, and in the other position the load connects the grounding connection. **(See Figure 10-7)**

GROUNDING

The purpose of the grounding mentioned above serves as a safety factor in case a switch accidentally becomes energized.

When shielded-type high-voltage cables have great lengths, the cable and the shield become a fairly good sized capacitor, and such grounding diverts the charge away, thus preventing what might have been a painful shock to personnel serving such cables and equipment. See **310.60**, **310.60(C)(1)**, and **300.40**.

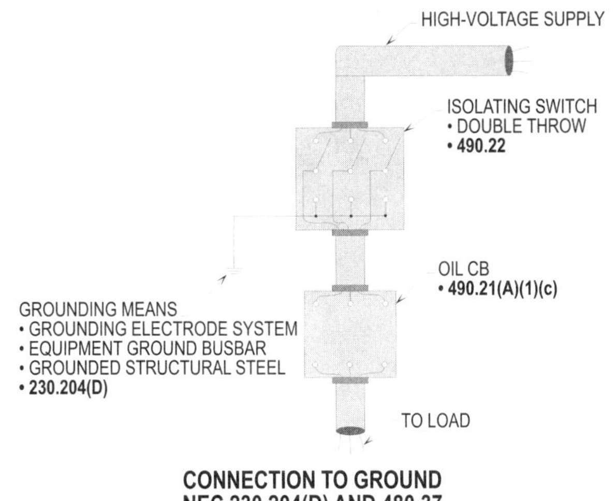

Figure 10-7. The above is one acceptable method for grounding the load side of an open isolating switch used to disconnect an oil circuit breaker.

DISCONNECTING MEANS
230.205

A high-voltage disconnecting means and overcurrent protection device shall comply with certain rules and regulations to qualify for service equipment use.

LOCATION
230.205(A)

The basic rule requires a high-voltage service disconnect means to be located "nearest the point of entrance of the service conductors" into the building or structure being supplied. This rule is described for 600 volts or less in **230.70(A)** and **230.208;** for over 600 volts the reference is **230.205(A).**

Section **230.205(C)** permits a feeder to a separate building on a multibuilding property under single management to have its disconnect in another building on the property. However, such a remote disconnect shall be capable of being electrically opened by a control device in the building feeder circuit. In addition, the control device shall be marked to show its function and provide visual indication of the ON or OFF status of the remote service disconnect. (**See Figure 10-8**)

TYPE
230.205(B)

The service disconnecting means shall simultaneously disconnect all ungrounded conductors and shall be capable of being closed on a fault equal to at least the maximum available short-circuit current in the circuit. See **230.208** and **230.205** for types permitted to be used for this purpose.

OVERCURRENT DEVICES AS
DISCONNECTING MEANS
230.206

The different types of disconnects approved for high-voltage services are specified in **230.205**, and protection techniques are found in **230.208**, *Protection Requirements*. The service disconnecting means providing the overcurrent protection shall also comply with these requirements.

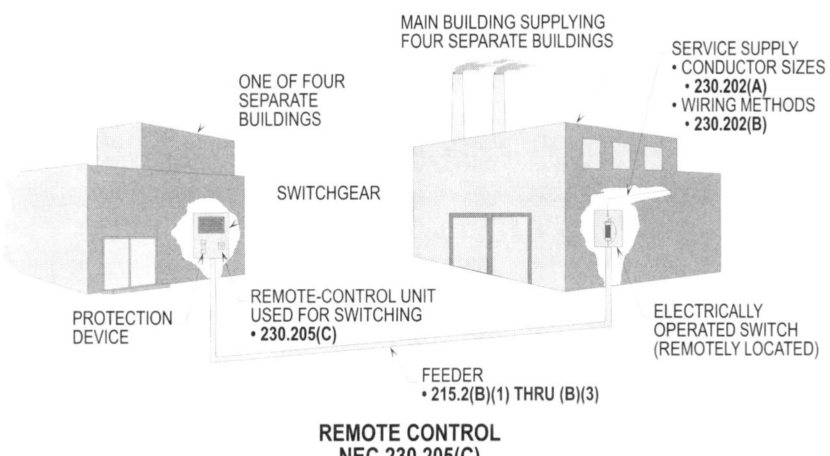

Figure 10-8. The above shows a remote-control unit that is used to deenergize the electrically operated switch in the main building.

OVERCURRENT PROTECTION DEVICES

Overcurrent protection devices can be used as a disconnecting means only where the circuit breaker or alternative for it specified in **230.208** for service overcurrent devices meets the requirements specified in **230.205**. For feeder requirements, see **240.100** and **240.101**.

Codes and Standards
NEC 230.206
NESC 171.   (where applicable)
NESC 161.A (where applicable)

EQUIPMENT IN SECONDARIES
230-207 PER 1993 NEC

In existing installations, this rule allows the disconnecting means with overcurrent protection in the secondaries to be omitted if there is a circuit breaker or load interrupter in the primary side of the line and the secondaries are connected to a common bus. If there is a vault, this primary circuit breaker or load interrupter shall be operable from outside the vault, and in addition it shall be of proper rating to protect the secondaries.

Only one set of secondary conductors shall be installed. If two or more sets of secondary conductors are utilized, the rule shall not be applied. **(See Figure 10-9)**

OVERCURRENT PROTECTION DEVICES

Section 230-207 in the 1993 NEC and **240.4(F)** are the only NEC requirements that permit the primary overcurrent protection device to provide protection for the secondary conductors. If one of these requirements cannot be applied, the secondary conductors shall have overcurrent protection provided in the secondary side.

See **450.3(A)** for protecting primary and secondary of high voltage transformers rated at over 600 volts and **450.3(B)** for 600 volts or less.

Note, because this type of protection does exist in older installations, it must have been installed before the 1996 NEC in order to comply. For this reason, it is mentioned so that the user is aware that this rule did once exist.

PROTECTION REQUIREMENTS
230.208, 240.100, AND 240.101(A)

High-voltage systems differ from low-voltage systems: high-voltage systems clear a short circuit faster. If a short circuit occurs on a high-voltage system, as soon as the insulation of a conductor fails, the circuit device trips open. Therefore, the elements and wiring methods of high-voltage systems (over 600 V) are treated differently from those systems of 600 volts or less.

An overcurrent protection device shall be provided in each ungrounded (phase) conductor for service conductors

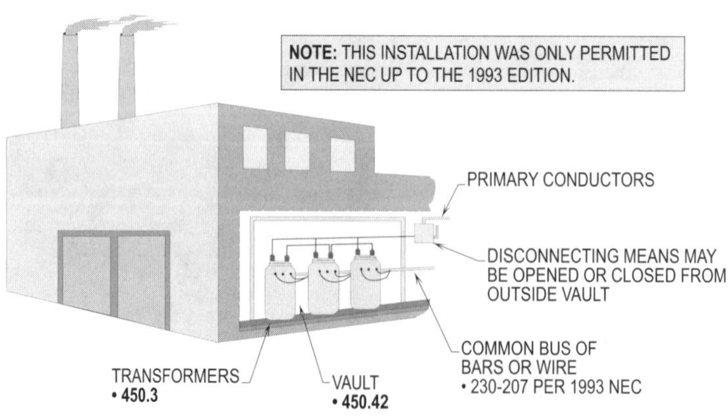

Figure 10-9. Disconnecting means with overcurrent protection can be omitted in the secondary side of the transformer, if a disconnecting means is located outside of the vault with only one set of secondary conductors connected to a common bus.

operating at over 600 volts. This overcurrent protection device protects against short-circuit currents; however, it does not protect against overload currents. The overcurrent protection device must be installed on the load side of the service disconnecting means or as an integral part of the service disconnecting means. A fuse rated in continuous amps, not to exceed three times the ampacity of the conductor, shall be permitted to be installed. A circuit breaker trip setting of not more than six times the ampacity of the conductors shall be permitted to be installed to protect against short-circuit currents. (**See Figure 10-10**)

Note, the rules for overcurrent protection and for disconnects are contained in this Section. There are two cases covered, and they are as follows:

(1) When the service equipment is in a vault or consists of metal-enclosed switchgear.

(2) When the service equipment is not in a vault or does not consist of metal-enclosed switchgear.

A vault is a specially constructed room, with roof, walls, floor of concrete or brick, and fire-resistant doors. It shall be constructed according to the specifications of **450.41 through 450.48**. Otherwise, it is not classified as a vault. (**See Figure 10-11**)

Metal-enclosed switchgear shall be in an enclosure constructed of metal of "substantial thickness."

Codes and Standards
NEC 230.208
NESC 171.   (where applicable)
NESC 161.A (where applicable)

IN VAULT OR CONSISTING OF METAL-ENCLOSED SWITCHGEAR 230.208(A) AND 490.21(E)

When the service equipment is installed in a vault or consists of metal-enclosed switchgear, the overcurrent protection and disconnecting means shall be permitted to be any one of the following:

(1) Oil switch and fuses,

(2) Load interrupter and fuses,

(3) Oil fuse cutout,

(4) Automatic-trip circuit breaker, or

(5) Isolating switch and fuses.

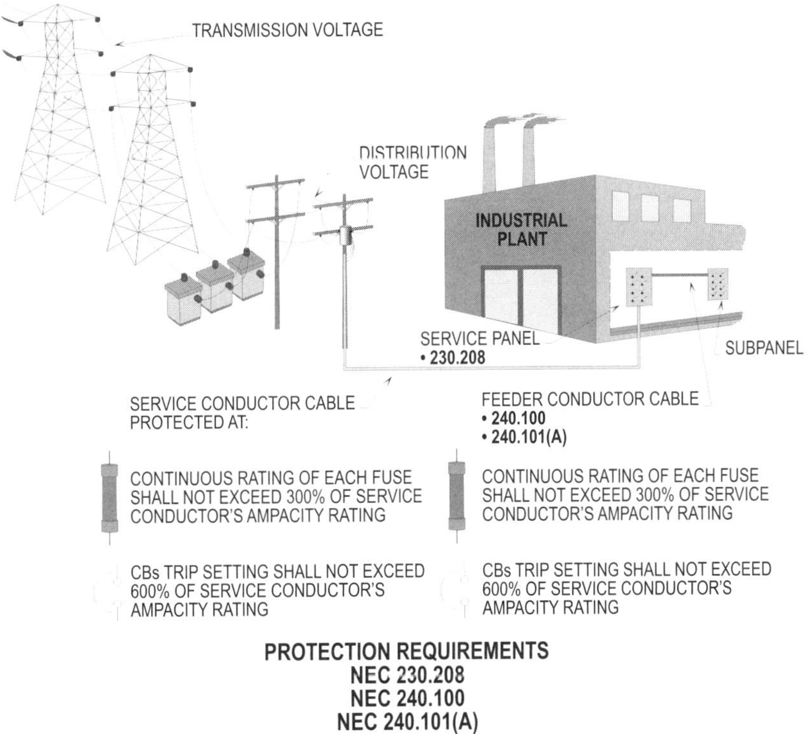

Figure 10-10. Maximum ratings or settings of the overcurrent protection device permitted for high-voltage service and feeder conductors.

If an oil fuse cutout is used, it could be a disconnect type, a high-voltage fuse mounted with a hinge on one end so that it can be opened and closed, thus serving as a disconnect as well as providing overcurrent protection.

A load interrupter is a switch or circuit breaker that is not intended to provide overload protection; therefore, fuses are required when these are used. In this case, the interrupter is the disconnect, and the fuses are sized to provide the overload protection.

An isolating switch is not rated to interrupt load current and shall not be opened under load. When an isolating switch is used as primary disconnect, it shall be interlocked with a disconnect in the secondary side in such a way that it would not be possible to open the primary isolating switch, without the secondary switch is in the open position. (**See Figure 10-12**)

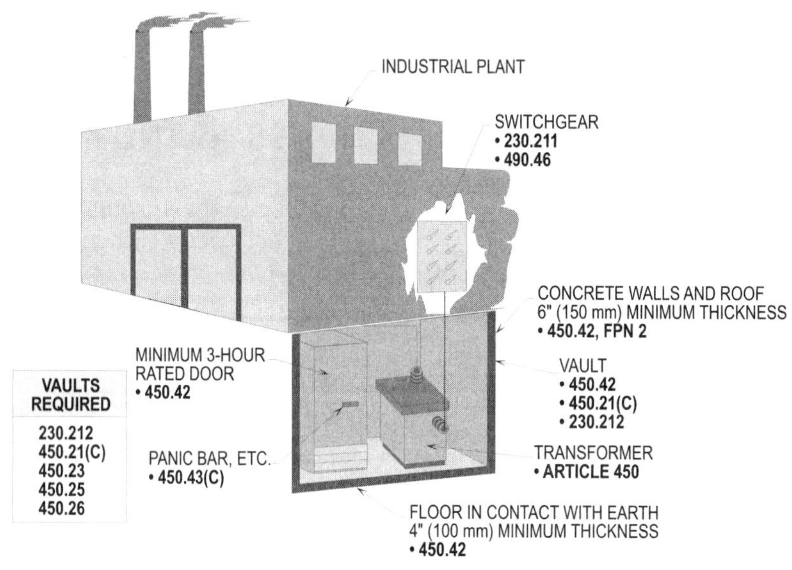

Figure 10-11. Some transformers are required to be installed in vaults because of their characteristics of use.

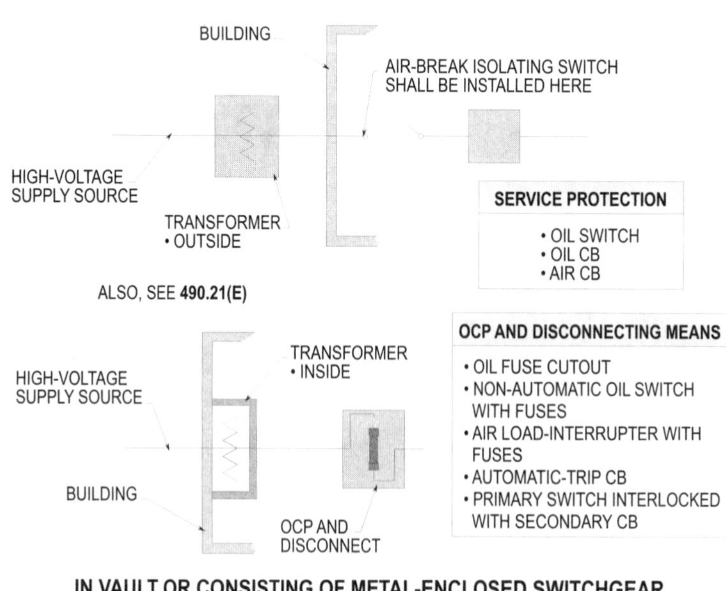

Figure 10-12. Overcurrent protection devices and disconnecting means that are permitted to be used for protecting and disconnecting high-voltage elements of electrical systems.

NOT IN VAULT OR NOT CONSISTING OF METAL-ENCLOSED SWITCHGEAR ARTICLE 490, PART II

When the service equipment is not in a vault or does not consist of metal-enclosed switchgear, the overcurrent protection and disconnecting means may be:

(1) Load interrupter and fuses located outside the building, or

(2) Automatic-trip circuit breaker located outside the building.

FUSES 490.21(B) AND 490.22

Interrupting rating of fuses shall be at least equal to the maximum possible short-circuit current available to the circuit.

CIRCUIT BREAKERS 490.21(A) AND 490.22

Circuit breakers shall be the type that will trip independently of the circuit breaker handle. With the handle closed, the breaker can trip internally without moving the handle. Thus, the breaker is "free to open in case it is closed on an overload." If a circuit breaker is used, it shall be this type. Circuit breakers must have an interrupting capacity at least equal to the maximum short circuit current that can be delivered to its terminals. (**See Figure 10-13**)

ENCLOSED OVERCURRENT DEVICES 230.208(B)

Sections **215.2(B)(1)** through **(B)(3)** and **215.3** require derating of 600 volt or less feeder conductors and overcurrent devices in cases of "long continuous" loads. For services rated over 600 volts, the derating requirement for sizing the overcurrent protection device does not apply to continuous operated loads per **230.208(B)**.

SURGE ARRESTERS (LIGHTNING ARRESTERS) 230.209 AND ARTICLE 280

Surge arresters installed in accordance with the requirements of **Article 280** shall be installed on each ungrounded overhead service conductor on the supply side of the service equipment.

A surge arrester is defined as a protective device for limiting surge voltage by discharging or bypassing surge current. It also prevents continued flow of follow current while remaining capable of repeating these functions as needed.

Such surge arresters break down at voltages higher than the supply voltage, permitting the higher voltage and currents to flow to ground, thus protecting the equipment on the electrical system.

Once the surge passes to ground, the arrester "heals itself," shutting off follow current from the supply system. (**See Figure 10-14**)

SERVICE EQUIPMENT – GENERAL PROVISIONS 230.210

Service equipment, including instrument transformers, shall conform to **Article 490, Part I** of the NEC.

METAL-ENCLOSED SWITCHGEAR 230.211 AND 490.47

Metal-enclosed switchgear must consist of a substantial metal structure with a sheet-metal enclosure. If properly built, the above switchgear will protect personnel from energized parts of the service elements. Where installed over a wood floor, suitable protection shall be provided, as required per **408.17**. This rule is intended to prevent the heat from the service elements from igniting combustible material and starting a firc. (**See Figure 10-15**)

FEEDERS SUPPLYING TRANSFORMERS 215.2(B)(1)

Section **215.2(B)(1)** governs the rules and regulations for determining the size of feeder conductors supplying transformers rated over 600 volts. This rule requires the ampacity of such conductors shall not be less than the sum of the nameplate ratings of the transformers supplied. (**See Figure 10-16**)

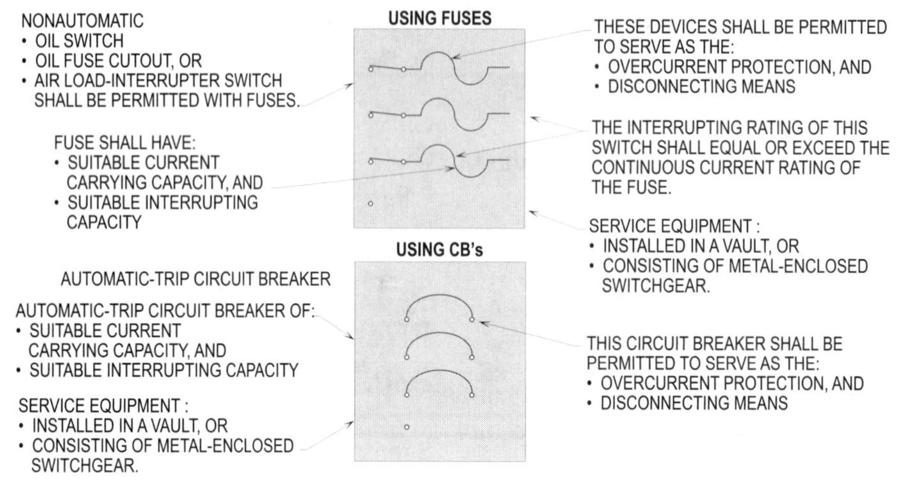

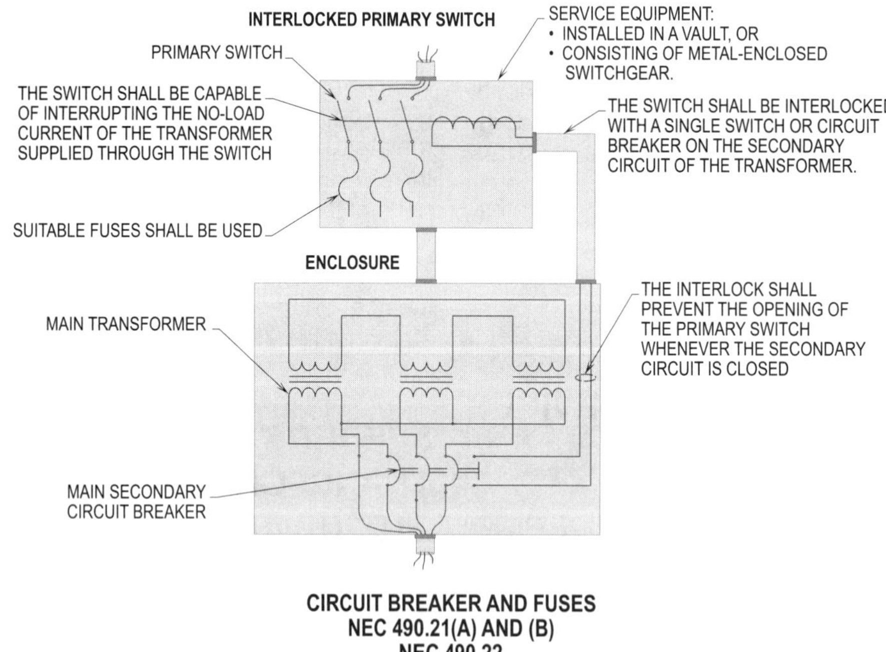

Figure 10-13. Using fuses and circuit breakers in a special adaption of switch-disconnect and fused overcurrent protection for high voltage electrical systems.

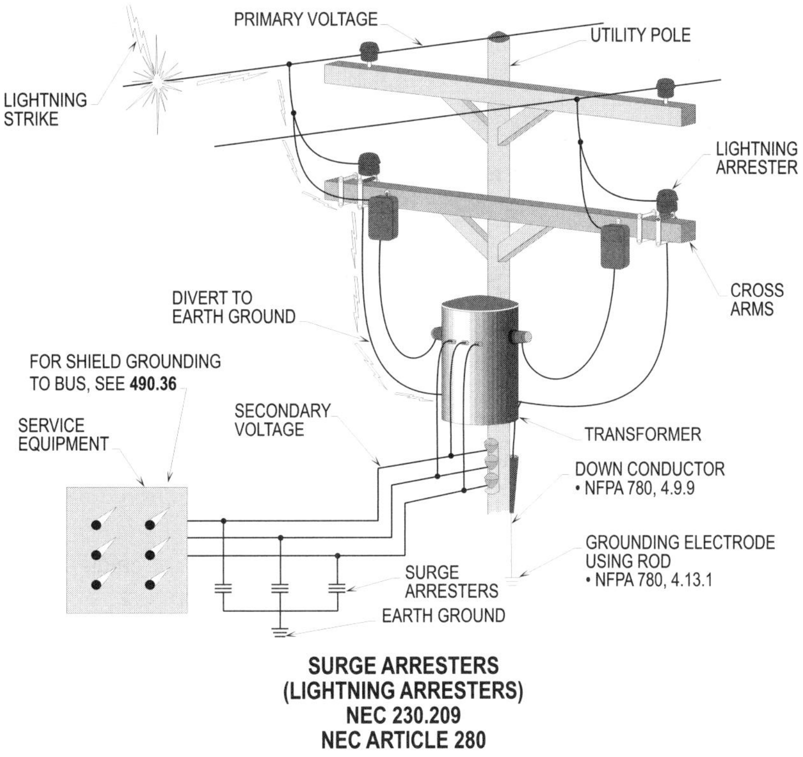

Figure 10-14. Dangerous high voltage surges are bypassed by using an arrester.

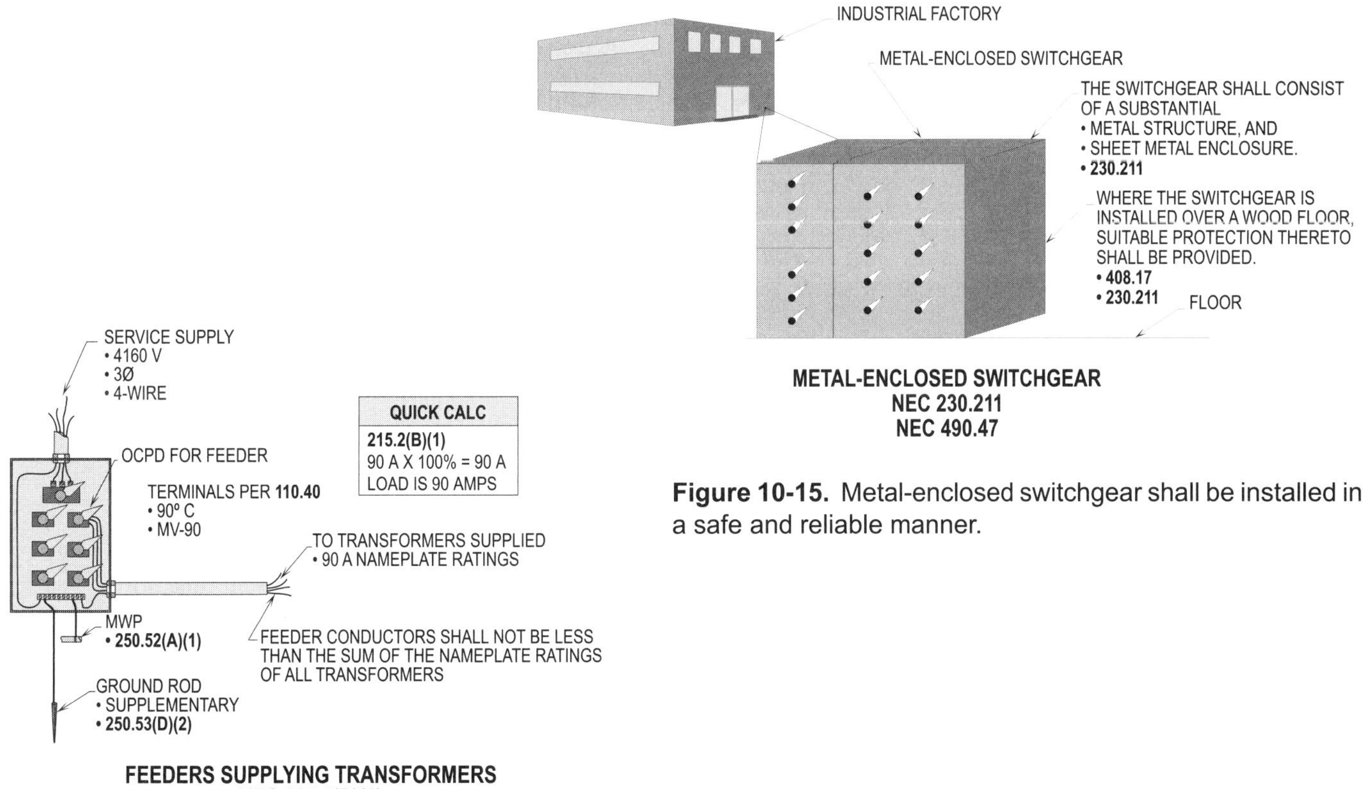

Figure 10-15. Metal-enclosed switchgear shall be installed in a safe and reliable manner.

Figure 10-16. The above illustrates the procedure for calculating the load for a feeder supplying transformers.

FEEDERS SUPPLYING TRANSFORMERS AND UTILIZATION EQUIPMENT NEC 215.2(B)(2)

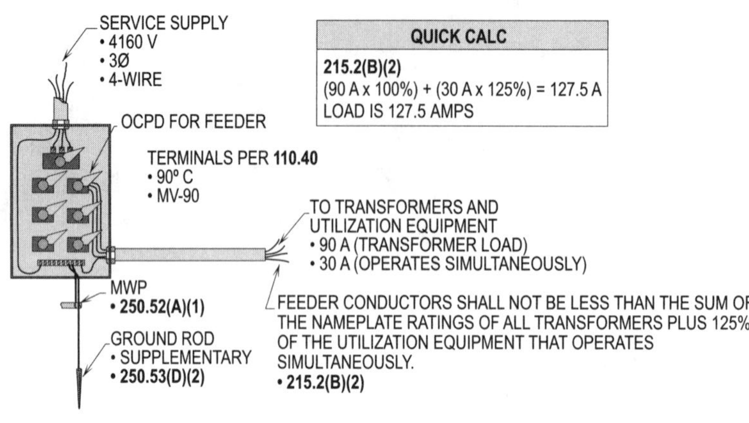

Figure 10-17. The above illustrates the procedure for calculating the load to supply utilization equipment and transformers.

FEEDERS SUPPLYING TRANSFORMERS AND UTILIZATION EQUIPMENT 215.2(B)(2)

Section **215.2(B)(2)** governs the rules and regulations for determining the size of feeder conductors supplying transformers and utilization equipment rated over 600 volts. This rule requires such conductors to be not less than the sum of the nameplate ratings of the transformers, plus 125 percent of the designed potential load of all utilization equipment that is capable of operating simultaneously. **(See Figure 10-17)**

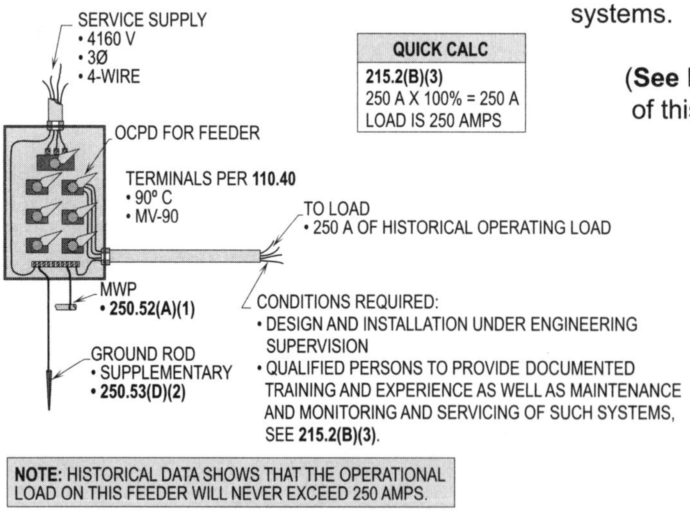

SUPERVISED INSTALLATIONS NEC 215.2(B)(3)

Figure 10-18. The above illustrates the procedure for calculating the load for a feeder supplying a supervised installation.

SUPERVISED INSTALLATIONS 215.2(B)(3)

Section **215.2(B)(3)** governs the rules and regulations for determining the size of feeder conductors for supervised installations where such sizes are determined by qualified person(s) under engineering supervision.
Supervised installations are considered those portions of a facility where the following conditions are complied with:

- Conditions of design and installation are provided under engineering supervision.
- Qualified persons with documented training and experience in systems over 600 volts, provide maintenance, monitoring, and servicing of such systems.

(**See Figure 10-18** for a detailed description of this section.)

SERVICES
230.208

Service-entrance conductors shall have a short-circuit protective device in each ungrounded conductor on the load side or be installed as an integral part of the service-entrance disconnect. The protective device shall be capable of detecting and interrupting all values of current in excess of its trip setting or melting point that can develop at its terminals. A fuse rated in continuous amperes shall not exceed three times the ampacity of the conductor, or a circuit breaker with a trip setting of not more than six times the ampacity of the conductors. Protected devices sized with these percentages shall be considered capable of providing short-circuit protection.

Such protection shall be installed as follows:

(1) Either on load side of service disconnect,

(2) Or as an integral part of the service equipment.

Section **230.208(B)** indicates that there are overcurrent protection devices that are not required to be derated 80 percent of their rating where they are supplying continuous duty loads (three hours or more per **Article 100**).

Service-entrance conductors shall be protected from dangerous overloads, which can damage insulation from short circuits and ground faults.

To accomplish the above, there are three sections to review when calculating ampacities to size the overcurrent protection devies and conductors. These sections are **230.42, 230.90, 230.208,** and **230.202(A).**

See Figure 10-19 for the maximum percentages permitted to be applied for selecting fuses and circuit breakers for the protection of service-entrance conductors.

FEEDERS
240.100 AND 240.101(A)

For short-circuit protection of high-voltage feeders, fuse rating shall be permitted to be up to three times the conductor ampacity. Circuit breaker setting shall be permitted to be up to six times conductor ampacity. These values apply for short-circuit protection only, and are not meant to provide overload protection. An overload feature of a lower rating could be included in the same fuse or circuit breaker.

The melting time-current characteristics of fuse units, refill units, and links for power fuses must be determined as follows:

(1) The current-responsive element ratings 100

amperes or less must melt in 300 seconds at a rms current within the range of 200 to 250 percent of the continuous current rating of the fuse unit, refill unit, or fuse link.

(2) The current-responsive element with ratings above 100 amperes must melt in 600 seconds at a rms current within the range of 220 to 264 percent of the continuous current rating of the fuse unit, refill unit, or fuse link.

(3) The melting time-current characteristics of a power fuse at any current greater than the 200 to 240 or 264 percent listed in (1) or (2) above must be listed by the manufacturer's published time current curves, since the current-responsive element is a distinctive feature of each manufacturer.

(4) The maximum steady-state rms current must not exceed the minimum melting time by more than 20 percent.

Note, E-rated fuses are assigned melting times at 200 percent or more of their continuous-current rating. The 300 percent for fuses times 200 percent is essentially 600 percent that in effect is the same as circuit breakers. It is easily seen why **230.208** permits the rating of 300 percent for fuses and 600 percent for circuit breakers.

See Figure 10-20 for the maximum percentages allowed to be applied for selecting fuses or circuit breakers for the protection of feeder conductors.

Codes and Standards

NEC 240.100
NESC 161.A   (where applicable)

BRANCH CIRCUITS
240.100

High-voltage branch circuits shall have a short-circuit protective device in each ungrounded (phase) conductor. Or, as an alternative, if circuit breakers are used for protection, two overcurrent relays operated from two current transformers in each of two phases shall be permitted.

This rule is intended to provide protection of personnel from high-voltage electrical systems. The protective device shall be capable of detecting and interrupting all values of current that might occur at their location at levels exceeding their trip setting or melting point.

The ampacities of high-voltage conductors are listed in **Tables 310.77** through **310.86** of the NEC. **Table 310.77** is utilized for copper conductors in isolated conduit. *Isolated* is defined as "not readily accessible to personnel unless special means for access are used."

See Figure 10-21 for the maximum percentages permitted to be applied for selecting fuses and circuit breakers for the protection of branch-circuit conductors.

Standards to review
NEC 240.100
NESC 161.A   (where applicable)

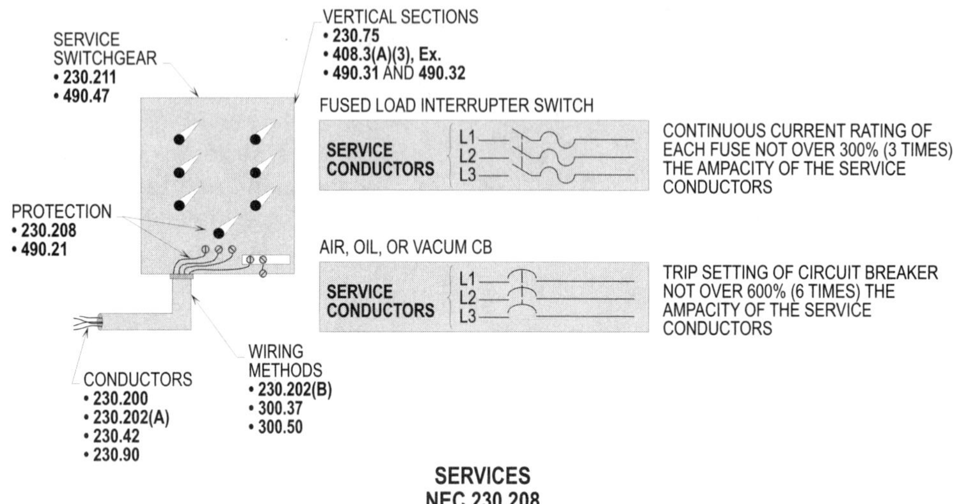

Figure 10-19. The maximum setting of fuses and circuit breakers protecting service-entrance conductors shall be determined by multiplying the ampacity of the service conductors by percentages as permitted per **230.208**.

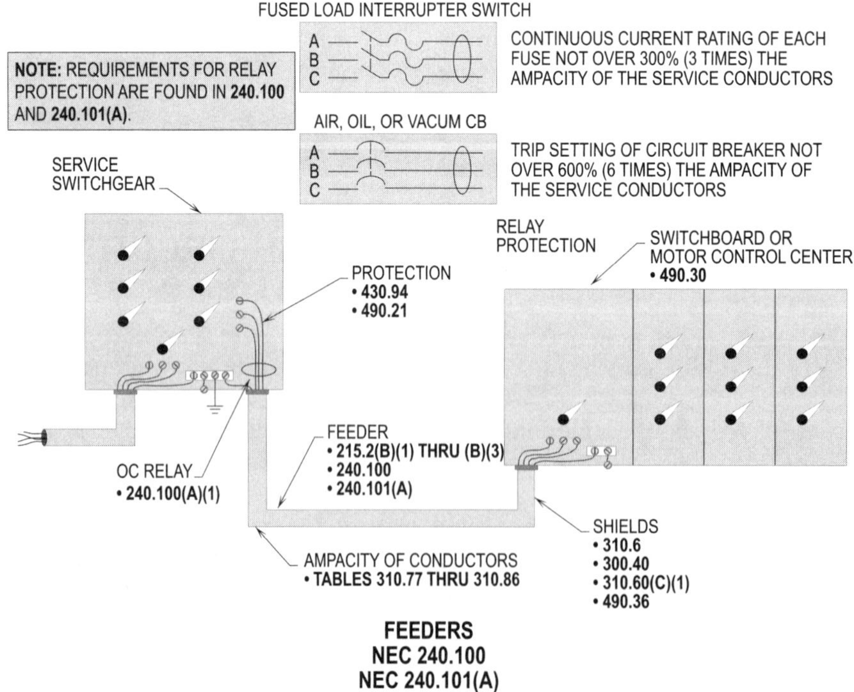

Figure 10-20. The maximum setting of fuses and circuit breakers protecting feeder conductors shall be determined by multiplying the ampacity of the feeder conductors by percentages as permitted per **240.100** and **240.101**.

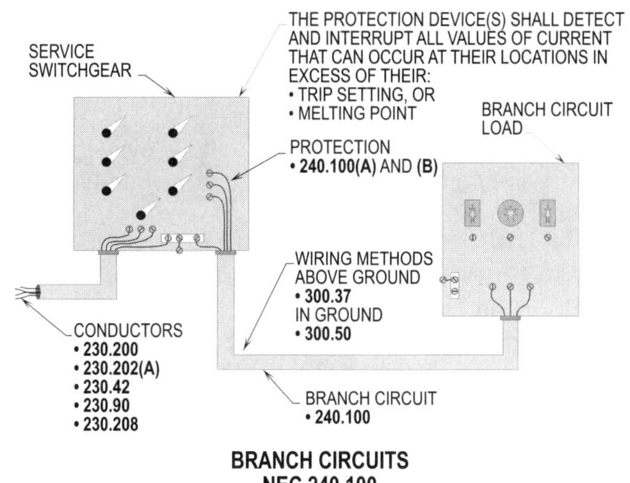

Figure 10-21. The general rule requires a high-voltage branch circuit to be protected by a short-circuit protective device in each ungrounded (phase) conductor.

Note, when grounding high-voltage electrical systems, circuits, and equipment, review **250.180** through **250.190** very carefully.

GROUNDING HIGH-VOLTAGE SYSTEMS
250.180

Special consideration must be given to grounding alternating-current systems and circuits of 1 kV (1000 volts) and over. The reason for grounding high-voltage systems (over 600 V) is the same reason as for grounding low-voltage systems (600 V or less). System and circuits are solidly grounded for several reasons, and they are as follows:

(1) To limit the voltage due to lightning,

(2) To limit the voltage due to line surges,

(3) To limit the voltage due to unintentional contact of the supply with higher voltage lines,

(4) To stabilize the voltage to ground during normal operation, and

(5) To facilitate the operation of the overcurrent device in case of ground fault.

Basically, it is up to the designer whether a high-voltage electrical system is grounded or not. Generally, it is the method in which the electrical system is utilized that really determines if it operates grounded or ungrounded. (**See Figure 10-22) Note,** for substation grounding, see Figure 10-37 in this book.

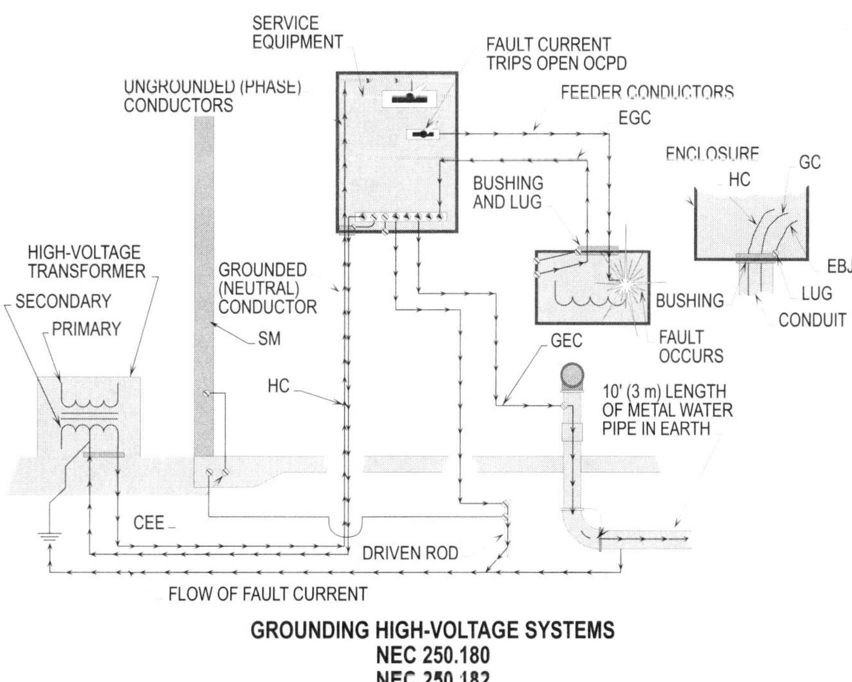

Figure 10-22. Solidly grounded electrical systems are grounded to protect from lighting surges, line surges, to regulate the voltage to ground, and to facilitate the operation of the overcurrent protection devices.

WIRING METHODS – COVERS REQUIRED
300.31

Any boxes, fittings, and similar enclosures shall have covers suitable for preventing accidental contact with energized parts or physical damage to parts or insulation.

Sections **110.34(C)** and **314.72(E)** require boxes to be closed with suitable covers and securely fastened in place. Underground box covers that weigh over 100 lb (45 kg) shall be considered as complying with this requirement.

Note, this rule can be applied to manhole covers. Covers for boxes shall be permanently marked **"DANGER – HIGH VOLTAGE – KEEP OUT."** The marking shall be on the outside of the box cover and is required to be readily visible. Letters shall be block type at least 1/2 in. (13 mm) in height. See **110.34(C)** for rules pertaining to enclosures housing high-voltage conductors and elements. **(See Figure 10-23)**

Codes and Standards
NEC 300.31
OSHA 1910.269(u)(4)(iii)
OSHA 1910.269(w)(6)(ii)
NESC 441.D (where applicable)

WIRING METHODS – CONDUCTORS OF DIFFERENT SYSTEMS
300.32

Conductors of high-voltage and low-voltage systems shall not occupy the same wiring enclosure or pull and junction box. **(See Figure 10-24)**

Section **300.3(C)(2)(d)** permits conductors of high-voltage (over 600 V) and low-voltage (600 V or less) systems to occupy the same wiring enclosure for motors, switchgear and control assemblies, and other similar enclosures. **(See Figure 10-25)**

Section **300.3(C)(2)(e)** permits the mixing of conductors of such systems in manholes if low-voltage conductors are separated from high-voltage conductors in a manner acceptable to the AHJ. **(See Figure 10-26)**

MIXING CONDUCTORS

The rules for mixing conductors of high voltage (over 600 V) and low voltage (600 V or less) together can be summed up as follows:

Conductors over 600 volts shall not be permitted to be installed in the same raceway or enclosure with conductors of 600 volts or less, except:

(1) For motor, switchgear and control assemblies, and other similar enclosures.

(2) In manholes.

For more detailed information on mixing conductors of low- and high-voltage systems, see **300.3(C)(2)(a) through (e).**

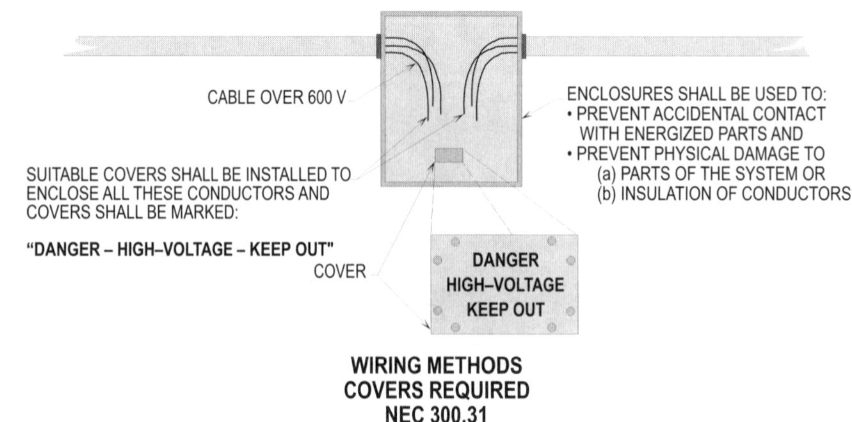

Figure 10-23. Properly marked covers are required to enclose high-voltage cables.

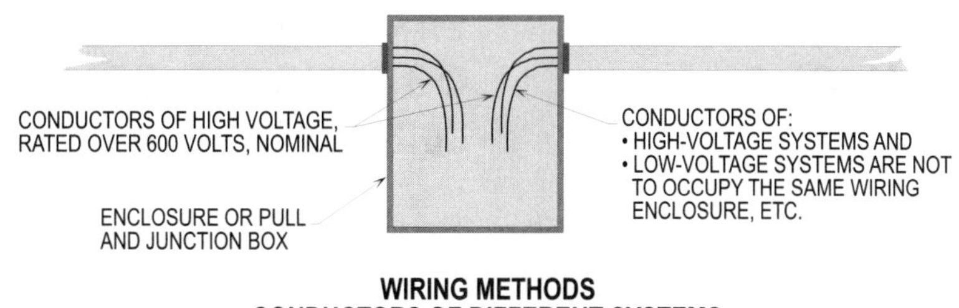

Figure 10-24. Conductors of different systems are not to occupy the same enclosures that are used to enclose the high-voltage conductors.

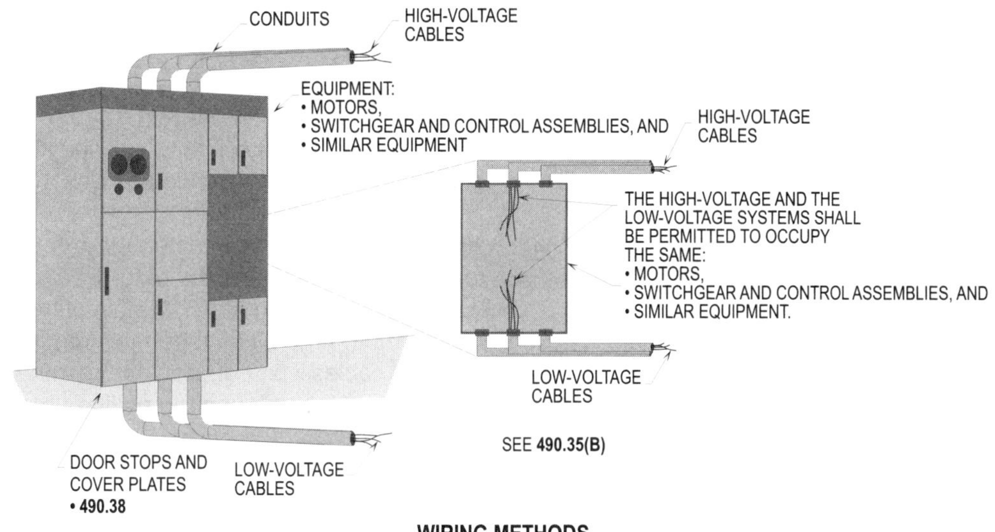

Figure 10-25. High-voltage and low-voltage cables shall be permitted to occupy the same enclosures, such as motors, switchgear and control assemblies, and other similar equipment.

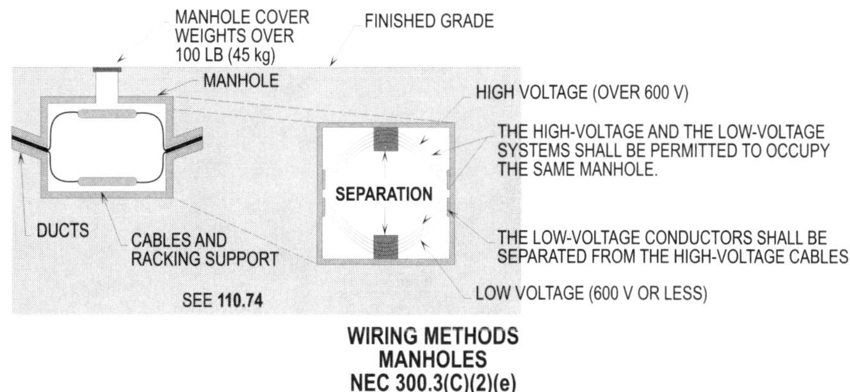

Figure 10-26. High-voltage and low-voltage cables shall be permitted to occupy the same manhole if they are properly separated.

WIRING METHODS – RACEWAY INSTALLATIONS 300.18(A)

Raceways, except those used for exposed work and having removable covers, shall first be installed as a complete raceway system without the conductors. Pull wires, if used, must not be installed until the raceway system is in place. Approved pulling compound may be used as a lubricant in inserting conductors in raceways. Cleaning agents or lubricants having a deleterious effect on conductor insulation and coverings must not be used.

Although this requirement is found in **300.18(A)**, it is still a good rule to apply even for high voltage conductors. Carefully review the **Exception** to 300.18(A). **(See Figure 10-27)**

EXTRA PROTECTION

The following procedure will add extra protection to the deteriorating effect of pulling compounds and cleanness on the inside portion of raceways:

(1) It is recommended for ducts encased in concrete to have a wire brush, mandrel, and cleaning rags pulled through them before pulling conductors.

(2) After conductors are installed in underground duct banks, unused ducts should be plugged and those with conductors should be sealed with an approved electrical sealing compound.

WIRING METHODS – CONDUCTOR BENDING RADIUS 300.34 AND 314.71

The bending radius of high-voltage conductors shall comply with the following:

(1) Shielded or lead-covered conductors shall not be bent to a radius of less than 12 times overall diameter.

(2) Nonshielded conductors shall not be bent to a radius of less than 8 times overall diameter.

Section **314.71(A)** and **(B)** requires that straight pulls, the length of a box to be 48 times the outside diameter of the largest cable or conductor entering the box. For angle or U pulls, the minimum width and length are at least 36 times outside diameter of the largest cable or conductor plus the sum of the outside diameters, over sheath, of all other cables entering the same wall. **[See Figures 10-28(a) and 28(b)]**.

Note, when pulling conductors off a reel, be sure that the reels are placed so that the natural bends of the cable as it leaves the reels can enter the raceways without reversing their natural bends. Damage may occur to the shielding of shielded cables if precautions are not taken to prevent over-bending the cable during the pulling process.

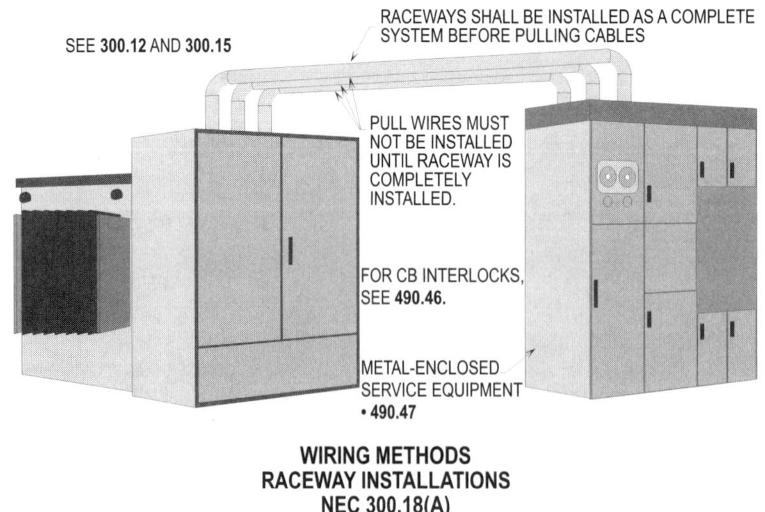

Figure 10-27. Raceway systems enclosing high-voltage cables shall be installed completely between enclosures before pulling pull wires and cables.

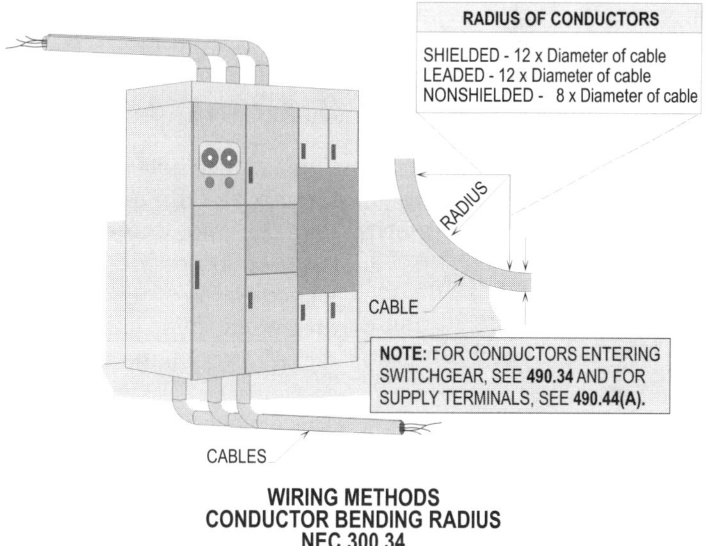

Figure 10-28(a). Conductors for systems over 600 volts shall not exceed a certain bending radius to prevent the damaging of insulation, shields, etc.

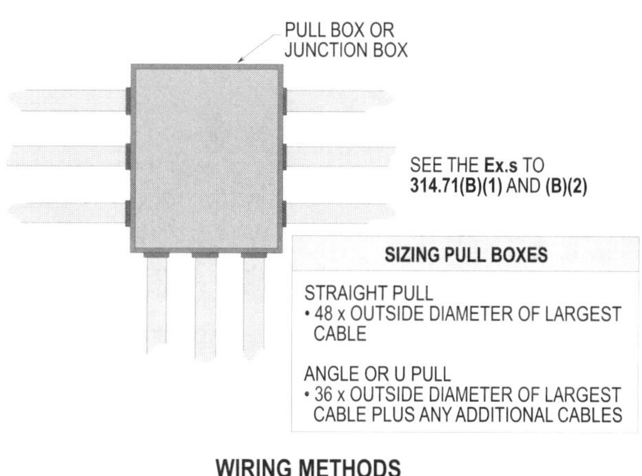

Figure 10-28(b). The above rules shall be applied when sizing boxes to enclose high-voltage cables and conductors.

WIRING METHODS – PROTECTION AGAINST INDUCTION HEATING 300.35 AND 300.20

AC circuits pulled in metal ducts can cause circulating currents in the wall of the duct. Induced heating of the ducts can be avoided by grouping all conductors of the same circuit together in the same conduit or cable, or by binding single conductor cables together in circuit groups.

Where an alternating current flows in a conductor, it sets up a varying magnetic field around the conductor. The magnetic field sets up induced currents in the surrounding metal, such as conduit. Where conductors of a circuit are grouped together in the same conduit, the field set up by one conductor cancels that of the other(s), for the currents are at all times in opposite directions. There will be no magnetic field, and therefore no induced currents in the metal wall of the conduit.

If one phase of an AC circuit is enclosed in a conduit, there is no opposite field to cancel its field, and such current set up in the metal conduit might be of sufficient magnitude to raise the temperature of the conduit to a dangerous value. The NEC requires where AC conductors are pulled in a metal conduit or raceway, all conductors of a circuit shall be grouped together in the same raceway.

Note, for nonmetallic and aluminum conduit this requirement does not apply, since induced currents will not set up in nonconducting material. Therefore, conductors can be grouped or run individual (same place) in raceways such as PVC as permitted in **300.5(I), Ex. 2**.

Where a single AC conductor passes through a separate hole in the steel wall of a box or enclosure, heating of the steel will result due to hysteresis. To avoid such heating, all conductors of the same circuit shall be grouped together and enter through one bushing. However, if it is necessary for each conductor to enter through an individual hole, heating can be avoided by cutting slots in the metal between the individual holes. These slots will break the magnetic path in the steel. Enclosures that are made of nonferrous metals such as PVC, aluminum, brass, and bronze do not require slots when individual conductors pass through individual holes per **300.35** and **300.20**. (**See Figure 10-29**)

GROUNDING
250.20(C), 250.180, AND 250.184(B)

High-voltage wiring and equipment installations shall be grounded in accordance with the provisions listed in **Article 250** of the NEC. (**See Figure 10-30**)

Note, the supply transformer does not have to be grounded per **250.20(C)** and **250.130(B)**. However, metal enclosures enclosing electrical elements and conductors shall be grounded with an equipment grounding means and should be sized per **250.122**.

WIRING METHODS – SIZE OF PULL AND JUNCTION BOXES
314.71

Pull and junction boxes shall provide adequate space and dimensions for the installation of conductors and accessories.

Terminal housing supplied with motors shall comply with the provisions listed in **430.12**. Motors that are provided with terminal housings per **314.71, Ex.**, shall be of metal and of substantial construction.

When these terminal housings enclose wire-to-wire connections, they shall have minimum dimensions and usable volumes in compliance with **430.12(C)** and Table **430.12(B)** of the NEC. (**See Figure 10-31**)

WIRING METHODS – STRAIGHT PULLS
314.71(A)

The length of the box shall be not less than 48 times the outside diameter, over sheath, of the largest conductor or cable entering the box.

Note: 48 times is a minimum, as in many cases, one may increase this length for ease in handling of the cables and for splicing. (**See Figure 10-32**)

WIRING METHODS – ANGLE OR U PULLS
314.71(B)

The distance between each cable or conductor entry inside the box and the opposite wall of the box shall not be less than 36 times the outside diameter, over sheath, of the largest cable or conductor. This distance shall be increased for additional entries by the amount of the sum of the outside diameters, over sheath of all other cables or conductor entries through the same wall of the box.

The distance between a cable or conductor entry and its exit from the box shall be not less than 36 times the outside diameter, over sheath, of that cable or conductor.

EXCEPTION 1 TO 314.71(B)

Where a conductor or cable entry is in the wall of a box opposite to a removable cover, the distance from that wall to the cover shall be in conformance with provisions of **300.34**. (**See Figure 10-33**)

Section **300.34** requires the conductor to be bent to a radius of not less than 8 times the overall diameter for nonshielded conductors or 12 times the diameter for shielded or lead-covered conductors during or after the installation has been made. (**See Figure 10-33**)

EXCEPTION 2 TO 314.71(B)

Where cables are nonshielded and not lead covered, the distance of 36 times the outside diameter shall be permitted to be reduced to 24 times the outside diameter. [Also see **314.71(B)(2)** and **Ex.**]

WIRING METHODS – REMOVABLE SIDES
314.71(C)

One or more sides of any pull box shall be removable so personnel will have reasonable access to repair or service splices or conductors.

WIRING METHODS – CONSTRUCTION AND INSTALLATION REQUIREMENTS
314.72

Boxes enclosing high-voltage conductors shall be constructed of suitable material to substantially support and hold such conductors. Boxes shall be located in areas where the conductors inside can be maintained in a safe and reliable manner.

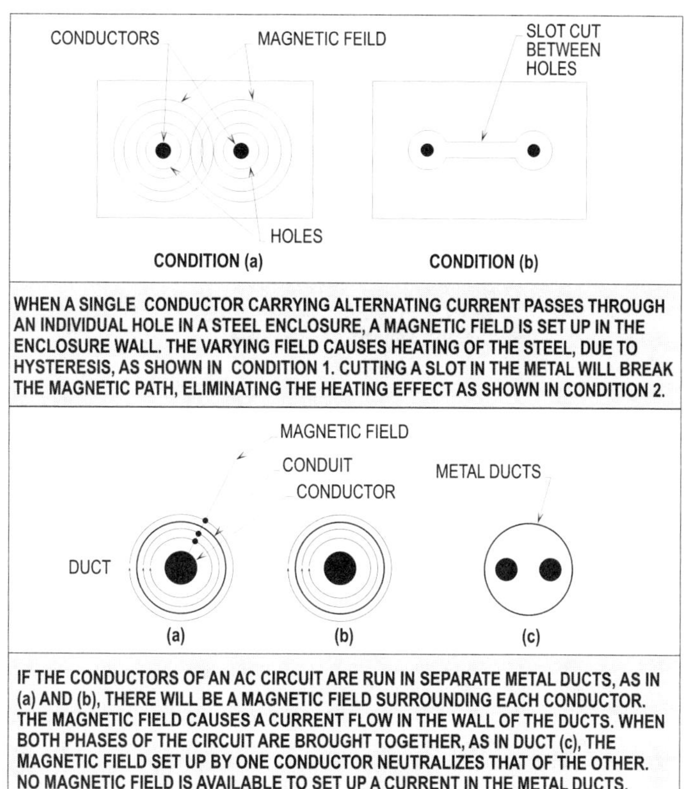

Figure 10-29. Conductors and cables shall be routed in raceways and ducts to prevent induction heating capable of damaging insulation.

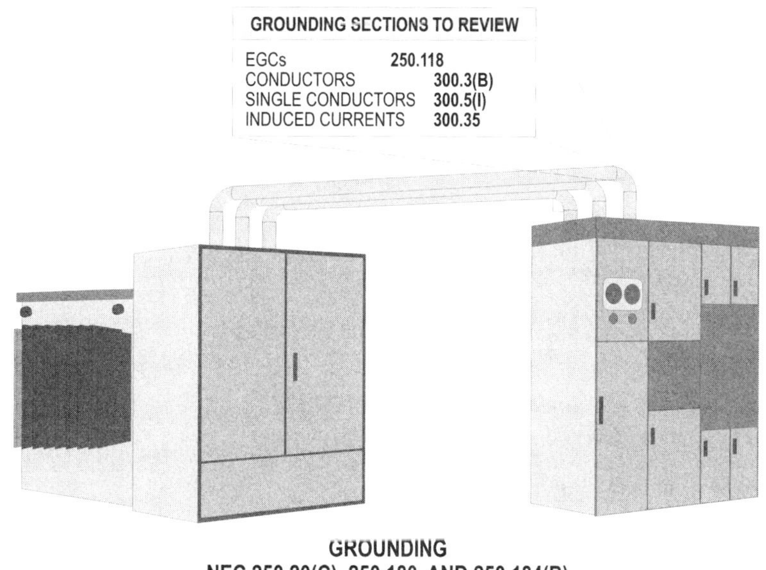

Figure 10-30. High-voltage electrical systems and wiring methods shall comply with **Article 250** of the NEC, as well as other pertinent sections.

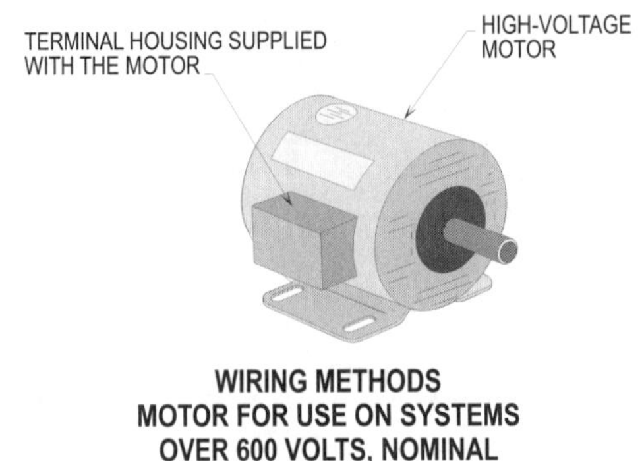

Figure 10-31. The terminal housing shall comply with the provisions of **430.12**.

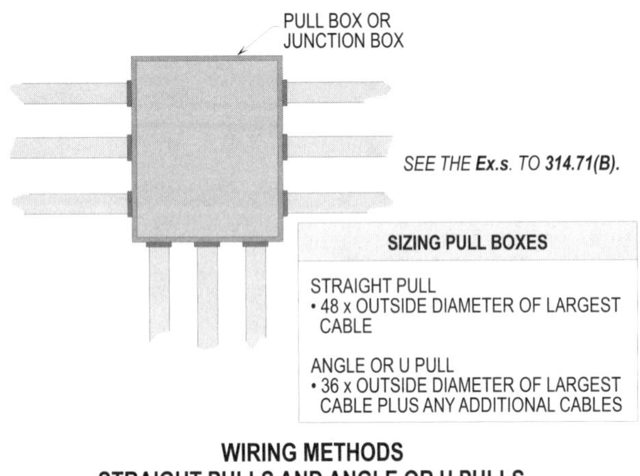

Figure 10-32. The above rules shall be applied when sizing boxes to enclosed high-voltage cables and conductors.

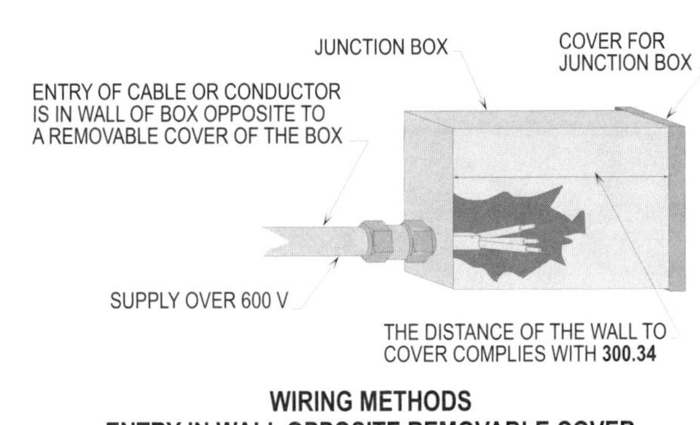

Figure 10-33. A certain distance shall be maintained where a cable enters the wall of a juncture box opposite to a removable cover.

WIRING METHODS – CORROSION PROTECTION 314.72(A)

Boxes shall be made of material inherently resistant to corrosion or shall be suitably protected, both internally and externally, by enameling, galvanizing, plating, or other effective means.

For example, the following wiring methods and equipment shall be considered protected against corrosion:

(1) Certain materials, like brass and aluminum, are corrosion-resistant in themselves.

(2) Equipment made of such materials does not require a corrosion-resistant coating.

(3) Steel or iron conduit and boxes shall be galvanized or cadmium-coated.

(4) Boxes or cabinets marked "Raintight" or "Outdoor type" shall be permitted to be used outdoors.

(5) Enameled conduit and fittings shall not be permitted to be used outdoors, nor indoors in wet or corrosive locations.

(6) In wet indoor locations, conduit, cable, and fittings shall be stooled off the surface at least 1/4 in. (6 mm).

WIRING METHODS – PASSING THROUGH PARTITIONS 314.72(B)

Suitable bushings, shields, or fittings having smooth rounded edges shall be provided where conductors or cables pass through partitions and at other locations where necessary. The insulation of conductors must not be subjected to physical damage. **(See Figure 10-34)**

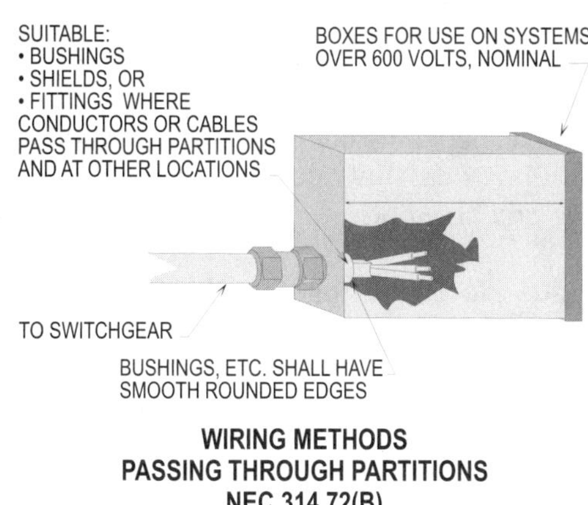

Figure 10-34. Protection of cables or conductors passing through partitions is required to prevent damaging insulation.

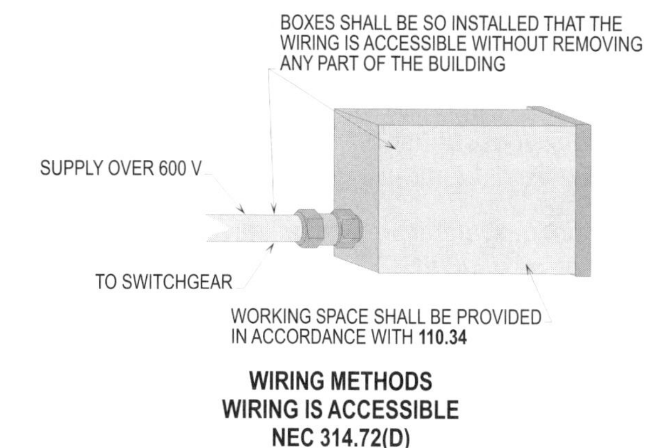

Figure 10-35. Wiring in boxes enclosing high-voltage cables and conductors shall be installed so that wiring is accessible.

PARTITIONS

The following methods will aid in helping prevent the spread of fire through partitions:

(1) When holes are made in fire-resistant walls, partitions, or ceilings, the fire resistance is destroyed.

(2) When a hole must be made for a cable to pass through, space around the cable shall be plugged with a fire-resistant material.

(3) The material shall be equal to the fire-resistant rating of the wall.

WIRING METHODS – COMPLETE ENCLOSURE 314.72(C)

Boxes shall provide a complete enclosure for the contained conductors or cables. Section **300.31** requires suitable covers to be installed on all enclosures, and **300.11** requires enclosures to be properly secured.

WIRING METHODS – WIRING IS ACCESSIBLE 314.72(D)

Boxes shall be so installed that the wiring is accessible without removing any part of the building. Working space shall be provided in accordance with the rules listed in **110.34**. **(See Figure 10-35)**

WIRING METHODS – SUITABLE COVERS 314.72(E)

Boxes shall be closed by a suitable cover securely fastened in place. Underground box covers that weigh over 100 lb (45 kg) shall be considered as meeting this requirement.

For example: This "100 lb (45 kg)" rule will satisfy the requirement for a manhole cover being securely fastened in place. Covers for boxes shall be permanently marked **"DANGER – HIGH VOLTAGE – KEEP OUT."** The marking shall be on the outside of the box cover and shall be readily visible. Letters shall be the block type at least 1/2 in. (13 mm) in height to be considered legible to read adequately. **(See Figure 10-36)**

WIRING METHODS – SUITABLE FOR EXPECTED HANDLING 314.72(F)

Boxes and their covers shall be capable of withstanding the handling to which they may be subjected.

The recommended construction specifications for boxes and covers can be summed up as follows:

(1) Shall be corrosion-resistant.

(2) Galvanized, cadmium-plated, or painted equipment will comply with this rule.

(3) Smaller sheet-metal boxes shall be at least 16 gauge.

(4) Larger boxes (over 100 cu. in.) shall comply with requirements for cabinets and cutout boxes. However, this rule does not apply for hinging.

(5) Malleable iron boxes, cast aluminum, brass, and bronze boxes shall have a wall thickness of at least 3/32 in. (2.38 mm).

(6) Other cast metal boxes shall have a wall thickness of at least 1/8 in. (3.17 mm).

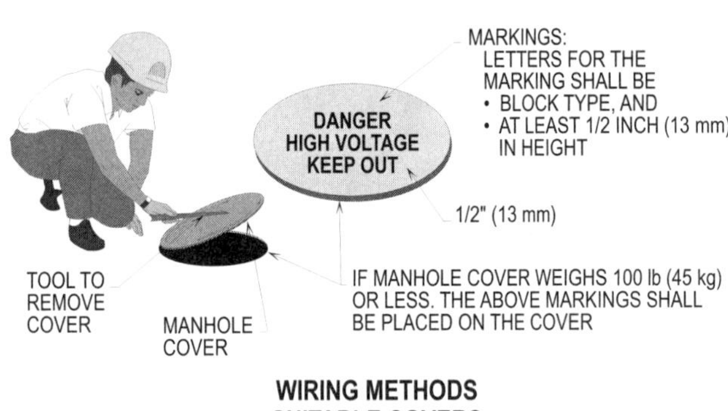

Figure 10-36. Manhole covers or underground box covers that weigh over 100 lb (45 kg) shall be considered accessible only to qualified personnel.

GROUND GRID
NFPA 70B - 9.1.5 THRU 9.1.7

A grid grounding scheme consists of buried conductors and driven ground rods that are interconnected to form a continuous grid network.

Ground grids are usually used for the grounding of lines and equipment located in the substation. The equipment and apparatus shall be bonded and grounded to the ground grid using large conductors to minimize the grounding resistance that limits the potential between equipment and ground surface.

Fences shall be constructed inside the ground grid and connected as needed to provide a safe bonding scheme. Sometimes, large conductors are run underneath substations and spaced properly to create a very low resistance to ground. **(See Figure 10-37)**

FEEDERS ROUTED TO FACILITIES
215.2(B)(1) THRU (B)(3)

Feeders can be run from the switchgear installed in the building that is normally located in the substation. **(See Figure 10-38)**

When calculating the load for sizing feeders, see Figures 10-16 and 10-18 in this chapter.

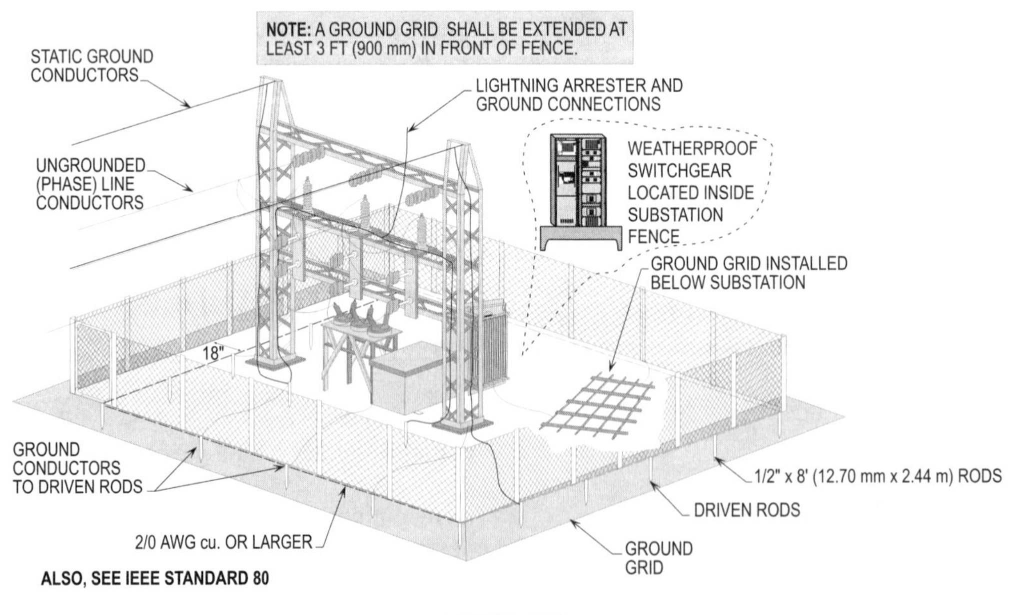

Figure 10-37. To create equipotential planes between ungrounded (phase) line conductors and equipment, elements are grounded to a counterpoise ground and driven rods or ground grid.

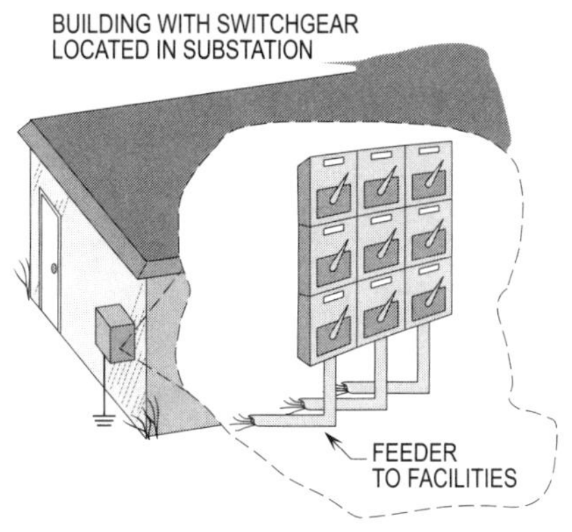

Figure 10-38. Feeders can be run from the switchgear installed in the building that is located in the substation.

Chapter 10. Equipment Over 600 Volts

Section   **Answer**

1. Underground box covers that weigh over _____ lb shall be considered as being closed by a suitable cover fastened in place.
   (a) 50   (b) 100
   (c) 150   (d) 200

2. Service-entrance conductors (over 600 volts) shall not be smaller than _____ AWG.
   (a) 6   (b) 4
   (c) 2   (d) 1

3. Isolating switches shall be permitted to be accessible to _____.
   (a) general public   (b) unqualified persons
   (c) qualified persons   (d) all of the above

4. For short-circuit protection of high-voltage feeders, the fuse rating shall be permitted to be up to _____ times the conductor ampacity.
   (a) 2   (b) 3
   (c) 5   (d) 6

5. For short-circuit protection of high-voltage feeders, the circuit breaker rating shall be permitted to be up to _____ times the conductor ampacity.
   (a) 2   (b) 3
   (c) 5   (d) 6

6. Shielded or lead-covered high-voltage conductors shall not be bent to a radius of less than _____ times overall diameter.
   (a) 6   (b) 8
   (c) 12   (d) 15

7. Other types of high-voltage conductors that are not shielded or lead-covered shall not be bent to a radius of less than _____ times overall diameter.
   (a) 6   (b) 8
   (c) 12   (d) 15

8. The length of the box for a straight pull shall be not less than _____ times the outside diameter, over sheath, of the largest conductor or cable entering the box.
   (a) 18   (b) 24
   (c) 36   (d) 48

9. The distance between each conductor entry inside the box and the opposite wall of the box for an angle pull shall be not less than _____ times the outside diameter, over sheath, of the largest conductor.
   (a) 18   (b) 24
   (c) 36   (d) 48

_______________   _____________

10. Feeders supplying transformers and utilization equipment shall not be less than the sum of the transformer plate ratings of the transformer plus ______ percent of the designed potential load of all utilization equipment that is capable of operating simultaneously.

    (a) 80   (b) 100

    (c) 110   (d) 125

Grounding

Equipment and circuits that are grounded and bonded properly are protected from lightning storms and excessive surges of voltages. Property is also protected and safety for people is provided. There are three grounding schemes in a well designed and installed grounding system, and they are as follows:

- Circuit and system grounding
- Equipment grounding
- Bonding supply and load side

There are types of conductors per the NEC that, if sized and selected correctly, may be used as a network to ground and bond electrical systems for safety.

Before attempting to design and install the elements of an effective grounding scheme, the following definitions of *ground, grounded (grounding), grounded conductor, grounding conductor, bonding (bonded),* and *equipment grounding conductor* per **Article 100** and **250.2** must be reviewed very carefully. These definitions are related to the grounding and bonding of the electrical system and equipment.

Ground is defined as the earth.

Grounded (grounding) is defined as connected (connecting) to earth or to a conducting body that extends the ground connection.

Grounded conductor is defined as a system or circuit conductor that is intentionally grounded.

Grounding conductor is a conductor used to connect equipment or the grounded circuit of a wiring system to a grounding electrode or electrodes.

Bonding (bonded) is defined as connected to establish electrical continuity and conductivity.

Equipment grounding conductor (EGC) is the conductive path installed to connect normally noncurrent-carrying metal parts of equipment together and to the system grounded conductor or to the grounding electrode conductor or both.

When grounding and bonding electrical systems for safety, such grounding and bonding procedures shall be done at the following locations:

- Service equipment
- Separately derived system
- Separate buildings or structures
- Load side of overcurrent protection device

This chapter deals with grounding and bonding of these locations as well as other methods of grounding.

GENERAL REQUIREMENTS FOR GROUNDING AND BONDING
250.4

The following types of systems have general requirements that are to be applied for grounding and bonding:

- Grounded systems
- Ungrounded systems

GROUNDED SYSTEMS
250.4(A)

The following general requirements for grounding and bonding shall be accomplished for grounded systems to protect property as well as provide safety for people:

- Electrical system grounding
- Grounding of electrical equipment
- Bonding of electrical equipment
- Bonding of electrically conductive materials and other equipment
- Effective ground fault current path

ELECTRICAL SYSTEM GROUNDING
250.4(A)(1)

Systems and circuits shall be grounded to limit voltage due to lightning, line surges, or unintentional contact with higher voltage lines. Systems and circuits shall be grounded to stabilize the voltage-to-ground during normal operation. Systems and circuits shall be solidly grounded to facilitate overcurrent device operation in case of ground faults. **(See Figure 11-1)**

GROUNDING OF ELECTRICAL EQUIPMENT
250.4(A)(2)

Normally noncurrent-carrying conductive materials that enclose electrical conductors or equipment, or forming part of such equipment, shall be connected to earth so as to limit the voltage-to-ground on these materials.

Circuits and equipment shall be grounded to facilitate overcurrent device operation in case of insulation failure or ground faults. To ensure this, the grounding path from circuit equipment and metal enclosures shall be continuous and not subject to damage. This path shall be capable of safely handling fault currents that may be imposed on it. Impedance shall be sufficiently low enough to keep the voltage-to-ground at a minimum and facilitate the opening of overcurrent protection devices ahead of the circuit. **(See Figure 11-2)**

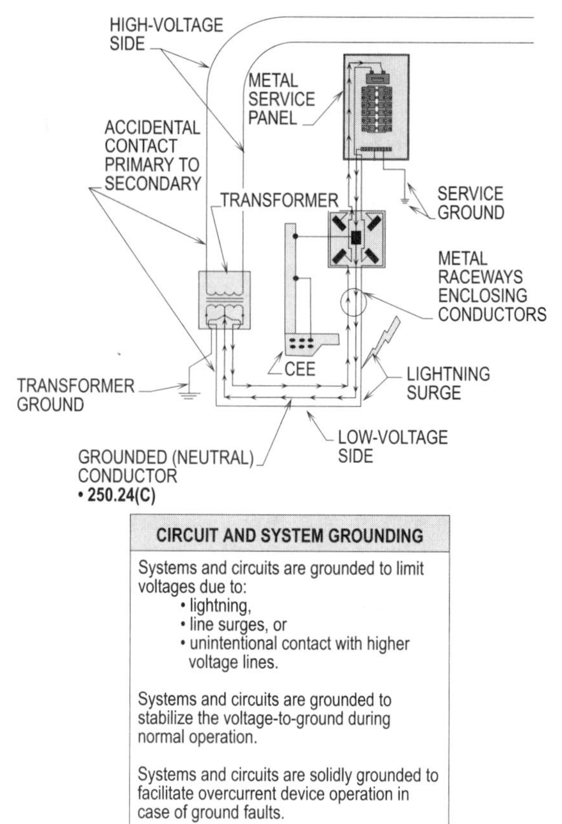

Figure 11-1. This illustration shows the benefits of grounding an electrical system for safety on the supply side.

BONDING OF ELECTRICAL EQUIPMENT
250.4(A)(3)

Normally noncurrent-carrying conductive materials that enclose electrical conductors or equipment, or forming part of such equipment, shall be connected together and to the electrical supply source in such a manner so as to establish an effective ground fault current path. **(See Figure 11-3)**

BONDING OF ELECTRICALLY CONDUCTIVE MATERIALS AND OTHER EQUIPMENT
250.4(A)(4)

Additionally, noncurrent-carrying electrically conductive materials, such as metal water piping, gas piping, and structural steel members that are likely to become energized shall be connected together and to the electrical supply source in such a manner that an effective ground fault current path is established. **(See Figure 11-3)**

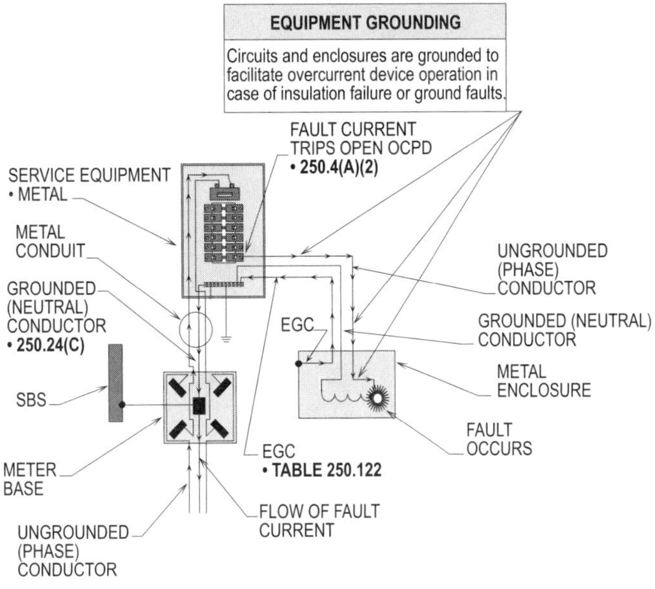

Figure 11-2. This illustration shows the benefits of equipment grounding on the load side.

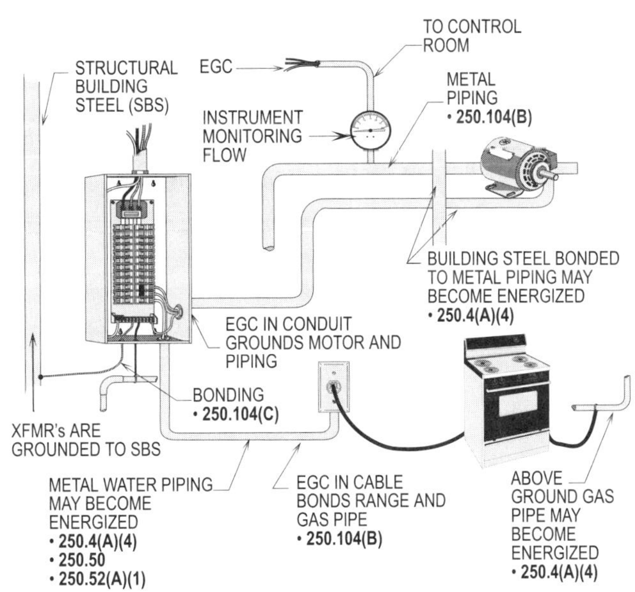

Figure 11-3. This illustration shows methods of how to bond electrically conductive materials (other than metal water piping) and other equipment.

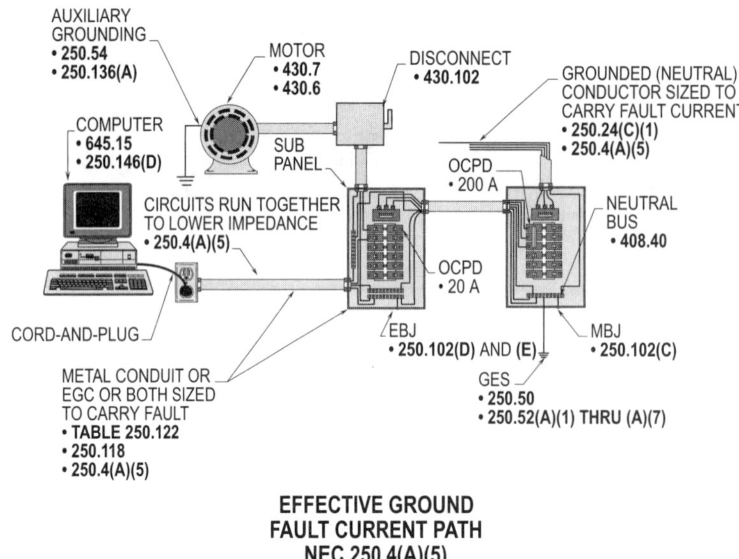

Figure 11-4. This illustration shows that an effective ground fault current path from circuits, equipment, and conductor enclosures shall be installed in a manner that creates a permanent, low-impedance circuit capable of safely carrying the maximum ground fault current likely to be imposed on it from any point on the wiring system.

EFFECTIVE GROUND FAULT CURRENT PATH
250.4(A)(5)

An effective ground fault current path from circuits, equipment, and conductor enclosures shall be installed in a manner that creates a permanent, low-impedance circuit facilitating the operation of the overcurrent device or ground detector for high-impedance grounded systems. It shall be capable of safely carrying the maximum ground fault current likely to be imposed on it from any point on the wiring system. **(See Figure 11-4)**

Note, the earth shall not be used as the sole equipment grounding conductor or effective ground fault current path.

UNGROUNDED SYSTEMS
250.4(B)

The following general requirements for grounding and bonding shall be accomplished for ungrounded systems to protect property as well as provide safety for people:

- Grounding electrical equipment
- Bonding of electrical equipment
- Bonding of electrically conductive materials and other equipment
- Path for fault current

GROUNDING ELECTRICAL EQUIPMENT
250.4(B)(1)

Noncurrent-carrying conductive materials that enclose

electrical conductors or equipment, or forming part of such equipment, shall be connected to earth in such a manner so as to limit the voltage imposed by lightning or unintentional contact with higher voltage lines and limit the voltage to ground on these materials. **(See Figure 11-5)**

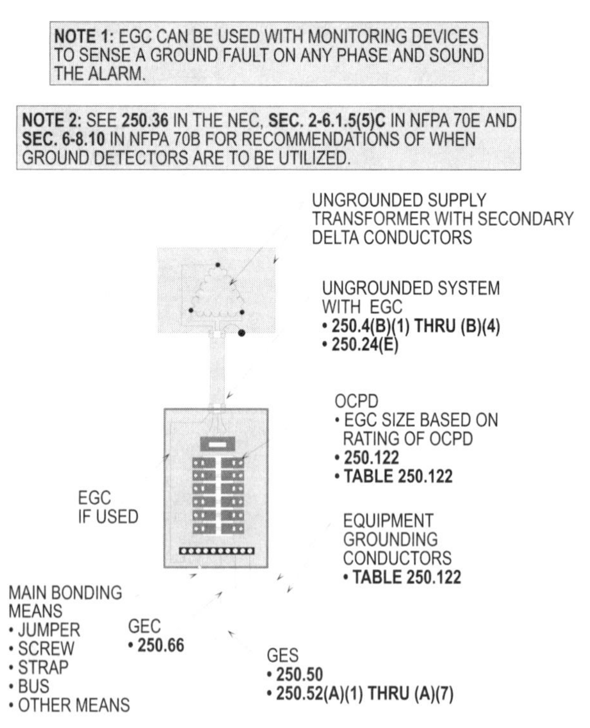

Figure 11-5. This illustration shows the grounding and bonding requirements for an ungrounded system.

BONDING OF ELECTRICAL EQUIPMENT
250.4(B)(2)

Noncurrent-carrying conductive materials that enclose electrical conductors or equipment, or forming part of such equipment, shall be connected together and to the supply system grounded equipment in such a manner so as to create a permanent, low-impedance path for ground fault current that is capable of carrying the maximum fault current likely to be imposed on it.

BONDING OF ELECTRICALLY CONDUCTIVE MATERIALS AND OTHER EQUIPMENT
250.4(B)(3)

Electrically conductive materials, such as metal water piping, gas piping, and structural steel members that are likely to become energized shall be bonded together and to the supply system grounded equipment in such a manner to create a permanent, low-impedance path for grounding current that is capable of carrying the maximum fault current likely to be imposed on it.

PATH FOR FAULT CURRENT
250.4(B)(4)

Electrical equipment, wiring, and other electrically conductive material likely to become energized shall be installed in a manner that creates a permanent, low-impedance circuit from any point on the wiring system to the electrical supply source so as to facilitate the operation of overcurrent devices should a second ground fault from a different phase develop on the wiring system.

Note, the earth shall not be used as the sole equipment grounding conductor or effective ground fault current path.

OBJECTIONABLE CURRENT OVER GROUNDING CONDUCTORS
250.6(A) THRU (E)

Unbalanced loads along with multiple grounding points may have objectionable current flowing in grounding conductors within the circuits of the wiring system. When the grounded (neutral) conductor is intentionally connected to earth ground at the supply transformer and the service equipment of the building, there will be two parallel paths present for unbalanced current to travel over. The two parallel paths of unbalanced current will return back to the utility supply transformer through the grounded (neutral) conductor and through the ground path of the earth soil. More current returns from the transformer to the service equipment over the faulted ungrounded (phase) conductor and trips open the overcurrent protection device of the faulted circuit conductor. The following conditions of use shall be considered for objectionable current over grounding conductors: **(See Figure 11-6)**

- Arrangement to prevent objectionable current
- Alterations to stop objectionable current
- Temporary currents not classified as objectionable currents
- Limitations to permissible alterations
- Isolation of objectionable direct current ground currents

ARRANGEMENT TO PREVENT OBJECTIONABLE CURRENT
250.6(A)

Depending on the resistance of the ground path of the grounded (neutral) conductor and earth soil, about 10 percent of the current will flow through the ground and approximately 90 percent will flow over the grounded (neutral) conductor. Balancing the load of the system as well as possible reduces the unbalanced current (objectionable) so that excessive amounts will not flow over the grounded (neutral) conductor. **(See Figure 11-7)**

> **Design Tip:** The grounded (neutral) conductor only carries the unbalanced current between the ungrounded (phase) conductors per **220.61(A)**. The grounded (neutral) conductor shall be permitted to be increased in size due to nonlinear loading on the ungrounded (phase) conductors per **220.61(C), FPN 2**. For transformer requirements pertaining to nonlinear loading, see **450.3, FPN 2**.

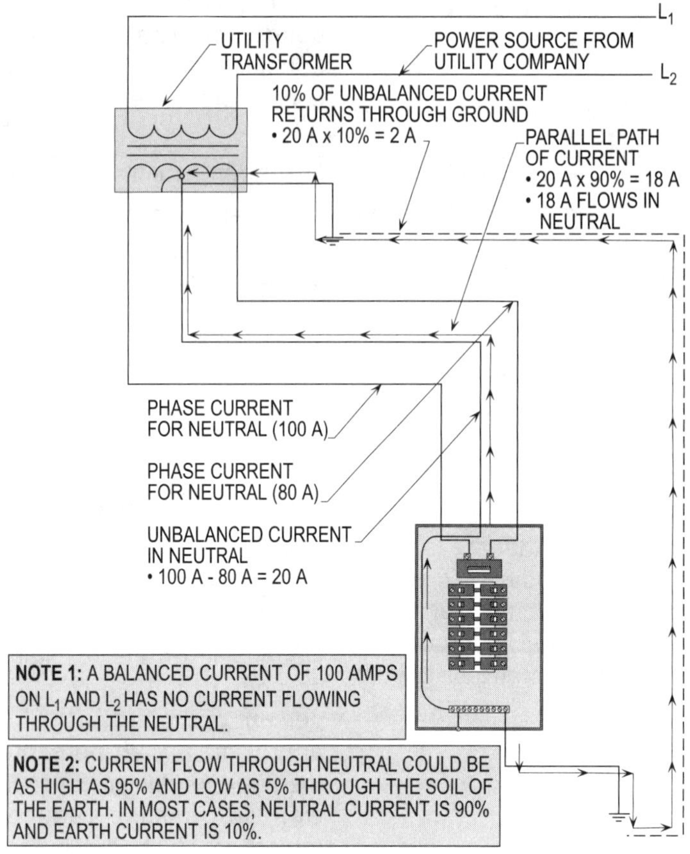

Figure 11-6. This illustration shows a condition where objectionable current flow in the earth will not be a problem.

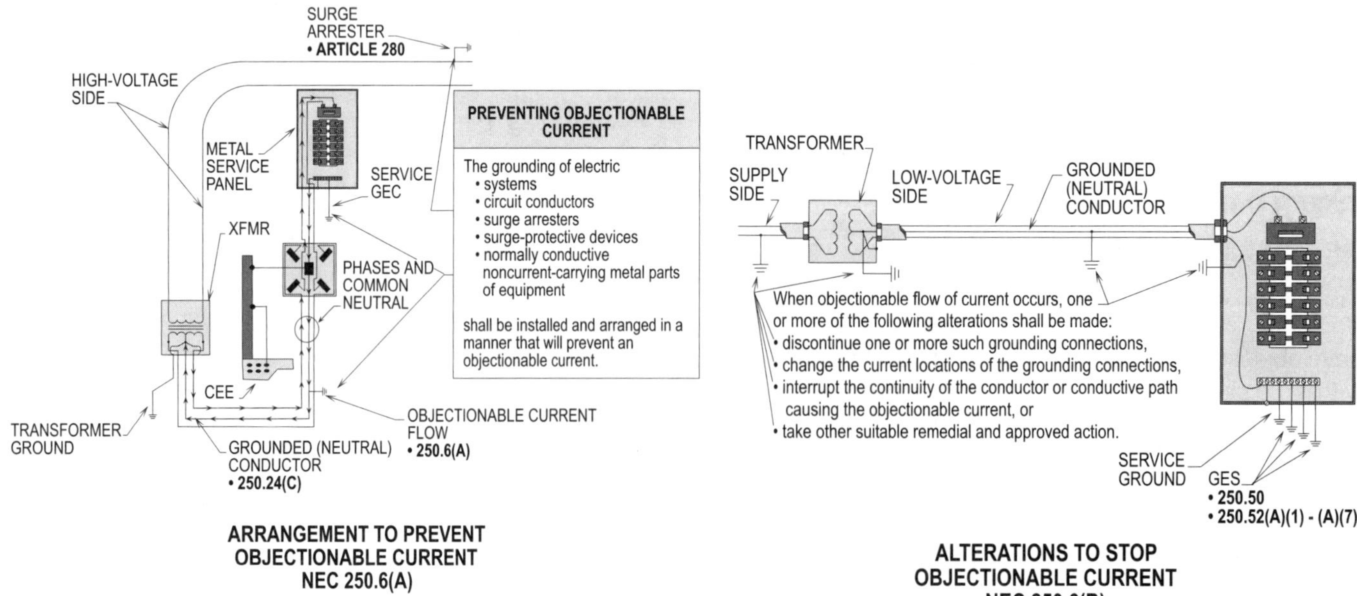

Figure 11-7. This illustration shows a connection to ground of the grounded (neutral) conductor that is not considered an objectionable current flow.

Figure 11-8. This illustration shows methods used to help eliminate objectionable current flow.

ALTERATIONS TO STOP OBJECTIONABLE CURRENT
250.6(B)

Objectionable current could occur from multiple grounds being installed on the same system. Objectionable current could flow from one ground connection to another, after such current has entered from one or more of the grounding points in the grounding system. If objectionable current is present, one or more of the following steps will help solve this problem:

- One or more grounding connections should be disconnected, but do not disconnect all of them.

- Placement of grounds should be altered.

- Break the conductive path of interconnecting grounding connections, so as to eliminate the continuity between grounding connections.

- Consult the authority having jurisdiction for suitable methods that are acceptable to help solve the problem. **(See Figure 11-8)**

> **Design Tip:** There are conditions where a grounding conductor shall be permitted to be removed without disrupting the safety of the grounding system.

For example, a grounding conductor connecting the metal of an electrical enclosure to the structural steel for the diversion of static electricity buildup on such equipment.

TEMPORARY CURRENTS NOT CLASSIFIED AS OBJECTIONABLE CURRENTS
250.6(C)

Temporary currents develop from accidental shorts, such as ground fault currents, that flow over grounding conductors and shall not be considered objectionable currents. **(See Figure 11-9)**

> **Design Tip:** Currents that flow over equipment grounding conductors, bonding jumpers, and grounded (neutral) conductors shall be considered temporary currents and shall not be defined as objectionable current flow per **250.6(C)**, for they will only flow until the OCPD of the circuit trips open.

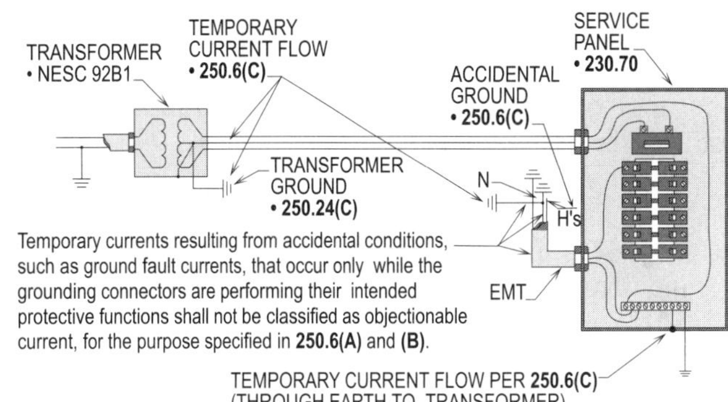

Figure 11-9. This illustration shows examples of temporary currents that are not considered objectionable currents.

LIMITATIONS TO PERMISSIBLE ALTERATIONS
250.6(D)

Currents generated from electronic equipment shall not be considered objectionable currents where such currents introduce noise or data errors. Currents that have such electrical noise or voltage fluctuations that produce dirty power conditions over the circuits shall also not be considered objectionable current flow. **(See Figure 11-10)**

> **Design Tip:** To correct noise problems for sensitive electronic equipment, it shall be permitted to isolate the supply conduit with a nonmetallic spacer. This spacer shall be located at the sensitive electronic equipment and shall not be permitted to take the place of the safety ground (equipment grounding conductor).

ISOLATION OF OBJECTIONABLE DIRECT CURRENT GROUND CURRENTS
250.6(E)

Where isolation from undesirable DC ground current is required, such as in the area of cathodic protected systems, a listed solid state AC coupling/DC isolating device shall be permitted to be placed in the equipment grounding conductor path to provide an effective return path for AC ground fault current and also blocking DC current. **(See Figure 11-11)**

Note, it is necessary to restrict the flow of DC cathodic protection current to only the metallic parts to be protected. Such listed AC coupling and DC blocking devices permits flow of the AC current while blocking DC. These devices are evaluated for their ability to withstand the rated fault currents. They are also evaluated for connections in the electrical grounding system in accordance with the instructions for the equipment.

Design Tip: Equipment installed in series with the equipment grounding conductor shall be approved by the AHJ or be third-party approved by a UL (type) 508 shop.

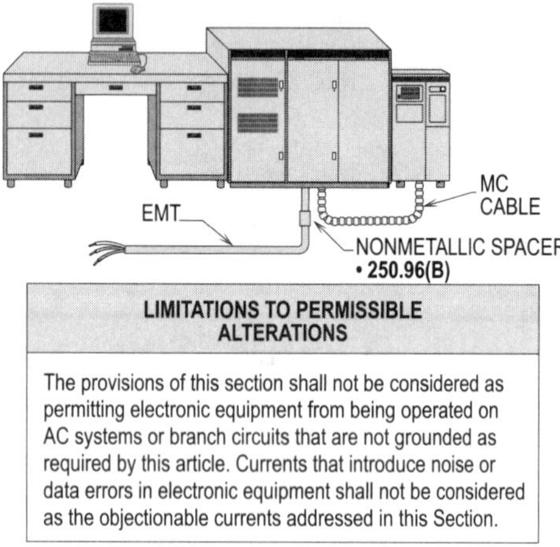

Figure 11-10. This illustration shows a conduit system that is altered to help reduce electrial noise generated from the operation of the electronic equipment.

CONNECTION OF GROUNDING AND BONDING EQUIPMENT
250.8

The following means shall be permitted for the connection of grounding conductors and bonding jumpers:

- Listed pressure connections
- Terminal bars
- Pressure connections listed as grounding and bonding equipment
- Exothermic welding process
- Machine screw-type fasteners that engage not less than two threads or are secured with a nut
- Thread-forming machine screws that engage not less than two threads in the enclosure
- Connections that are part of a listed assembly
- Other listed means

See **Figure 11-12** for a detailed illustration for the connection of grounding conductors and bonding jumpers.

Note, connection devices or fittings that depend solely on solder shall not be permitted to be used.

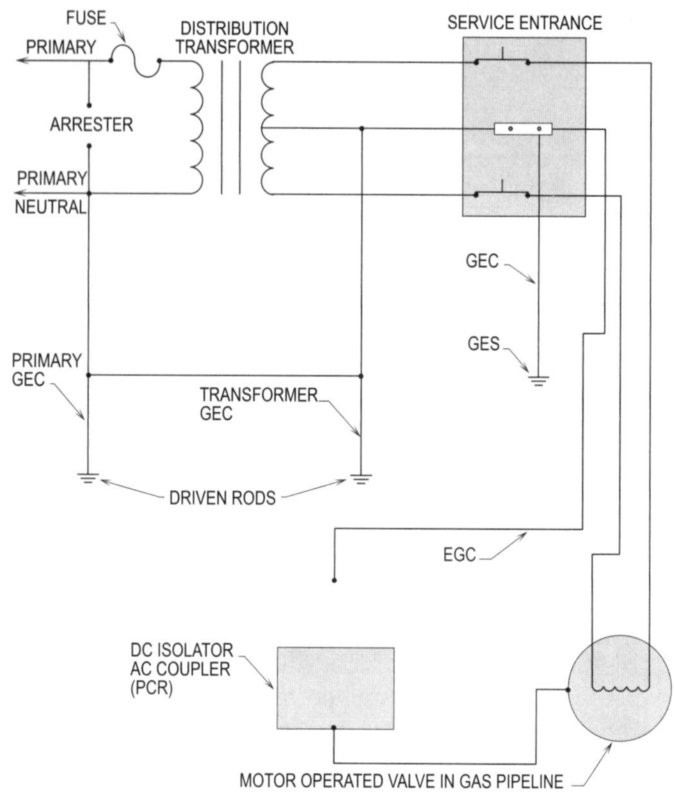

Figure 11-11. This illustration shows an AC coupling and DC blocking device used for cathodic protection.

PROTECTION OF GROUND CLAMPS AND FITTINGS
250.10

Unless approved for general use without protection, ground clamps and fittings shall be installed as follows:

- Be located so that they will not be subject to damage or
- Be enclosed in metal, wood, or equivalent protective covering.

When exothermic welding (CAD-welding) is used, the resistance of the connection should be checked with a ductor. If a low resistance is measured, the problems of a high-resistance connection should be eliminated. **(See Figure 11-13)**

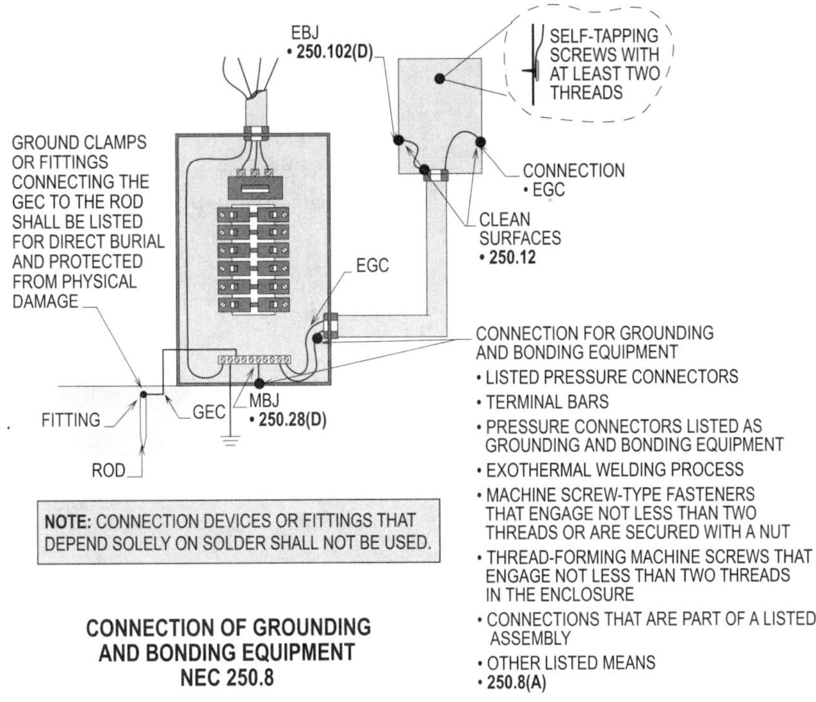

Figure 11-12. This illustration shows the permitted methods for connection of grounding conductors and bonding jumpers.

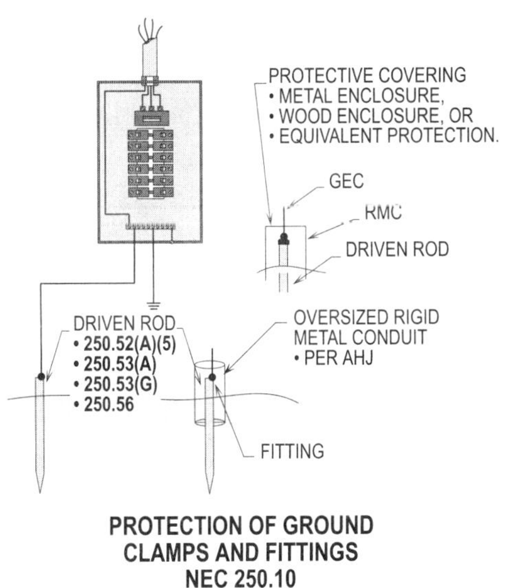

Figure 11-13. This illustration shows the methods that shall be permitted to be used to protect the grounding clamp from physical damage.

AC SYSTEMS TO BE GROUNDED 250.20

The following AC systems shall be grounded:

- AC systems of less than 50 volts.
- AC systems of 50 to 1000 volts.
- AC systems of 1 kV and over.
- Separately derived systems.

See Figure 11-14 for a detailed discription when applying these requirements.

AC SYSTEMS OF LESS THAN 50 VOLTS 250.20(A)

Circuits and equipment operating at less than 50 volts are found in **Article 720**. Class 1, 2, and 3 circuits including rules pertaining to their installation procedures are listed in **725.41(A)** and **725.121**.

AC systems of less than 50 volts shall be grounded under the following conditions:

- A transformer installed to supply low voltage receives its supply from a transformer exceeding 150 volts-to-ground.

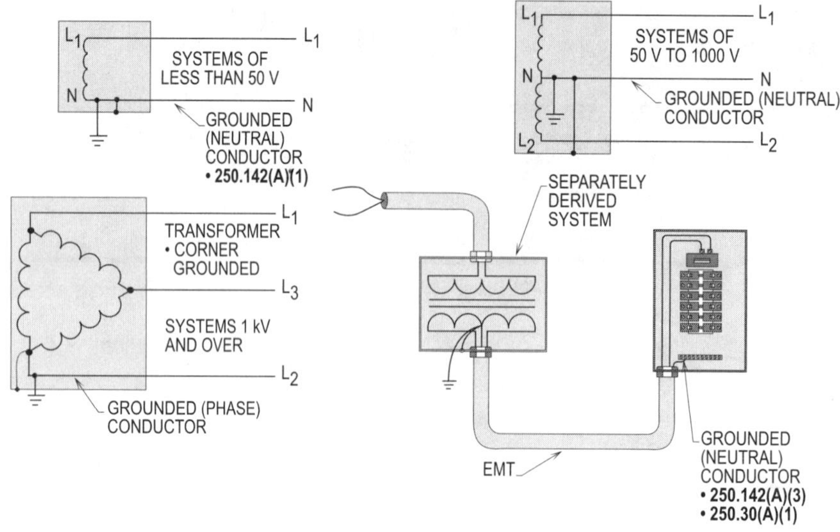

Figure 11-14. This illustration shows electrical systems that shall be grounded under certain conditions of use.

For example, a circuit of 277 volts-to-ground supplying the primary side of such a transformer is considered a circuit of over 150 volts-to-ground, and its secondary shall be grounded.

- A transformer installed to receive its supply from a transformer with an ungrounded system. This condition includes a supply voltage obtained from a transformer that has an ungrounded secondary.

- Where low-voltage overhead conductors are installed outside and not inside.

For example, AC conductors of 50 volts that are run outside overhead shall have one conductor grounded.

See Figure 11-15 for a detailed illustration of AC systems of less than 50 volts.

AC SYSTEMS OF 50 TO 1000 VOLTS 250.20(B)

AC systems of 50 to 1000 volts shall be grounded under any one of the following conditions:

- The maximum voltage-to-ground on the ungrounded (phase) conductors does not exceed 150 volts. This voltage-to-ground circuit is usually a 120 volt circuit derived from a 120/240 or 120/208 volt system.

- The system is nominally rated as 120/208 and

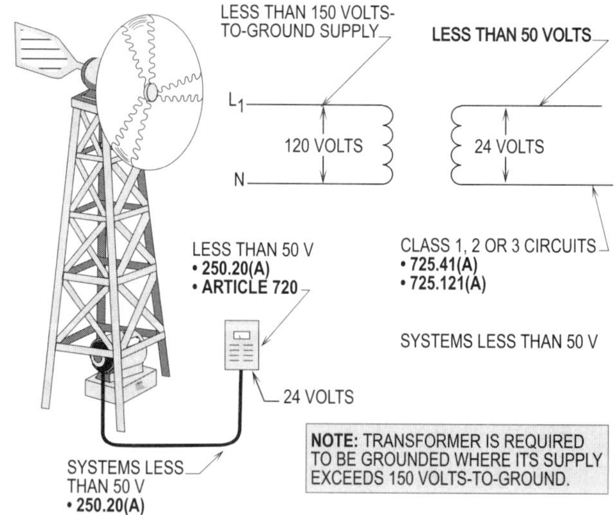

Figure 11-15. This illustration shows examples of systems less than 50 volts.

277/480 volt, three-phase, four-wire wye and is connected so that the grounded (neutral) conductor is used as a circuit conductor. The voltage-to-ground and between phases for 50 to 1000 volts may be any level. Wye systems of 2400/4160 volts are also used to supply circuits requiring a higher voltage served from a wye hookup.

- The system is nominally rated as 240/120 volt, three-phase, four-wire delta-connected, in which the midpoint of one phase is used as a circuit conductor, with one phase conductor having a higher voltage-to-ground than the other two. **(See Figure 11-16)**

The following types of AC systems that are rated 50 volts to 1000 volts shall be grounded:

- 120 volt, two-wire, single-phase
- 120/240 volt, three-wire, single-phase
- 120/208 volt, four-wire, three-phase wye
- 277/480 volt, four-wire, three-phase wye
- 480 volt corner grounded, three-phase delta
- 240 volt, three-wire, three-phase delta
- 480 volt, three-wire, three-phase delta
- 600 volt, three-wire, three-phase delta

AC SYSTEMS OF 1 kV AND OVER 250.20(C)

AC systems of 1 kV and over shall be grounded if supplying mobile or portable equipment per **250.188**. Other types of AC systems of 1 kV and over that are installed do not have to be grounded. Where such systems are grounded, they shall comply with the applicable provisions of **Article 250**. **(See Figure 11-17)**

SEPARATELY DERIVED SYSTEMS 250.20(D)

A separately derived system is derived from an generator, converter windings, or transformer to reduce the voltage from high-voltage to low-voltage levels or vice versa in a building. Transformers of 4160 or 13,800 volts have voltages that may be reduced to utilization voltage of 480 volts or 120/208 volts to supply loads that are located at different floor levels within a building. **(See Figure 11-18)**

Where a generator is used as an alternative source of power and is provided with transfer equipment that includes a grounded (neutral) conductor, the generator shall be classified as a separately derived system and grounded per **250.30(A)**. **(See Figure 11-19)**

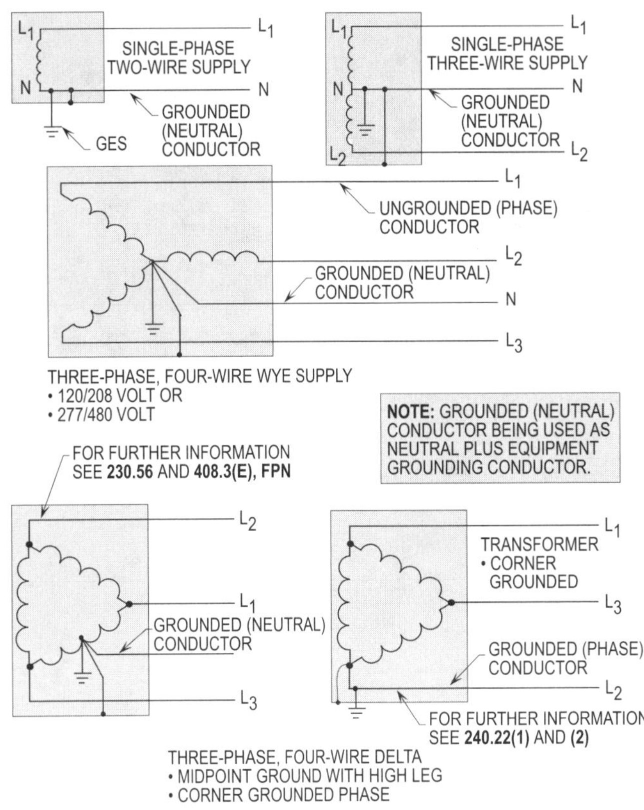

Figure 11-16. This illustration shows AC systems of 50 volts to 1000 volts that shall be grounded when the grounded (neutral) conductor or ungrounded (phase) conductor is used as an equipment grounding conductor during a ground-fault condition.

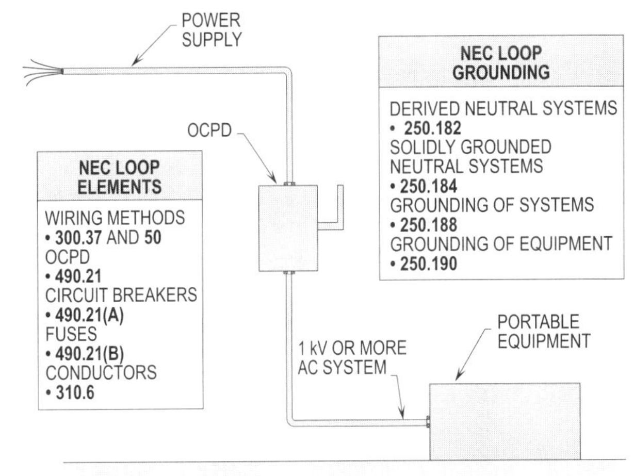

Figure 11-17. This illustration shows that AC systems of 1 kV and over shall be grounded if supplying mobile or portable equipment.

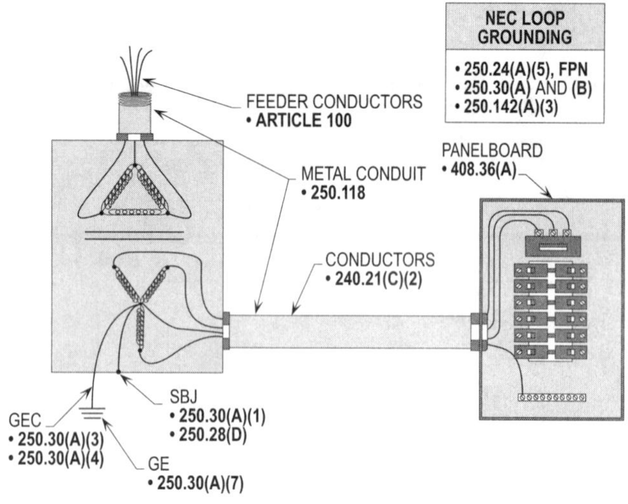

SEPARATELY DERIVED SYSTEMS
NEC 250.20(D)

Figure 11-18. This illustration shows a separately derived system; it is derived from a generator, converter windings, or transformers to reduce the voltage from high voltage to low voltage or vice versa in a building.

AC SYSTEMS OF 50 VOLTS TO 1000 VOLTS NOT REQUIRED TO BE GROUNDED 250.21(A)

The following AC systems of 50 volts to 1000 volts shall not be required to be grounded:

- Circuits installed for industrial electric furnaces or any other means of heating metals for refining, melting, or tempering.

- Separately derived systems used exclusively for rectifiers supplying only adjustable speed industrial drives. Such systems are used for speed control in industrial facilities and shall not be permitted to be utilized for anything else.

- Separately derived systems installed with primaries not exceeding 1000 volts used exclusively for secondary control circuits where conditions of maintenance and supervision ensure that only qualified persons will service the installation. However, the continuity of control power is required with ground detectors installed on the control system to sound an alarm if one phase should become grounded and a ground fault condition occurs. **(See Figure 11-20)**

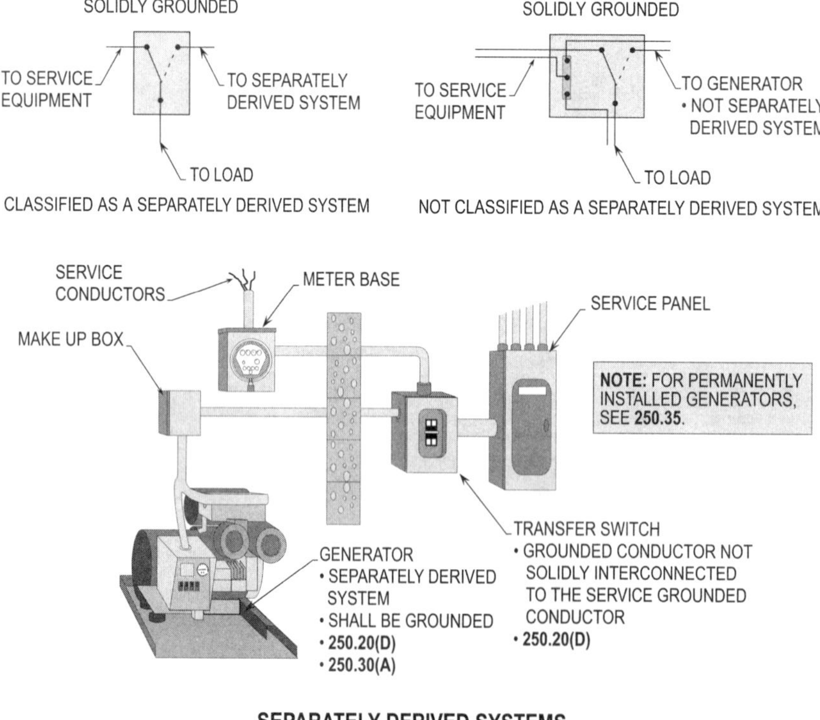

SEPARATELY DERIVED SYSTEMS
250.20(D)

Figure 11-19. This illustration shows if the generator is classified as a separately derived system or not.

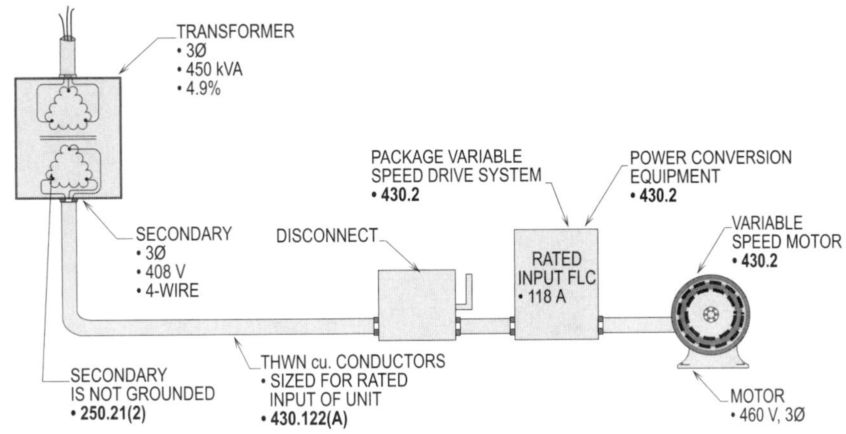

Figure 11-20. This illustration shows a system shall not be required to be grounded because of its supplying rectifiers powering and adjustable speed industrial drive.

CIRCUITS NOT TO BE GROUNDED 250.22

The following circuits shall not be grounded due to their conditions of use:

- Circuits for electric cranes operating over combustible fibers in Class III locations per **503.155**. Combustible fibers are easily ignited with sparking or arcing devices. This rule was designed to eliminate this problem and hazardarous condition.

- Circuits operating in health care facilities such as anesthetizing locations per **517.160(A)(2)**. In operating rooms, circuits shall not be permitted to be grounded to ensure against ground faults.

- Circuits for electrolytic cells shall not be permitted to be grounded per **Article 668.**

- Secondary circuits for lighting systems operating at less than 30 volts shall not be permitted to be grounded per **411.5(A)**.

- Secondary circuits for lighting systems that are under water shall not be permitted to be grounded per **680.23(A)(2)**.

See Figures 11-21(a), (b), and **(c)** for a detailed illustration when applying these requirements.

SYSTEM GROUNDING CONNECTIONS 250.24(A)

The grounding electrode conductor shall be installed to bond and ground the service equipment terminal neutral bus to the grounding electrode system per **250.52(A)(1) through**

(A)(7). The grounding electrode conductor is installed to connect equipment grounding conductors, grounded (neutral) conductors, and the service equipment enclosure to a reliable earth ground. Bonding jumpers shall be installed by using a wire, bus, screw, or other suitable conductor per **250.28(A)**. Grounding electrodes bonded together for two or more services shall be considered a single grounding electrode system that will bond the metal enclosures of equipment of all sources together to create equipotential planes. Note that the head of a screw used to bond in the above items shall be green in color and be visible after the installation per **250.28(B)**. **(See Figure 11-22)**

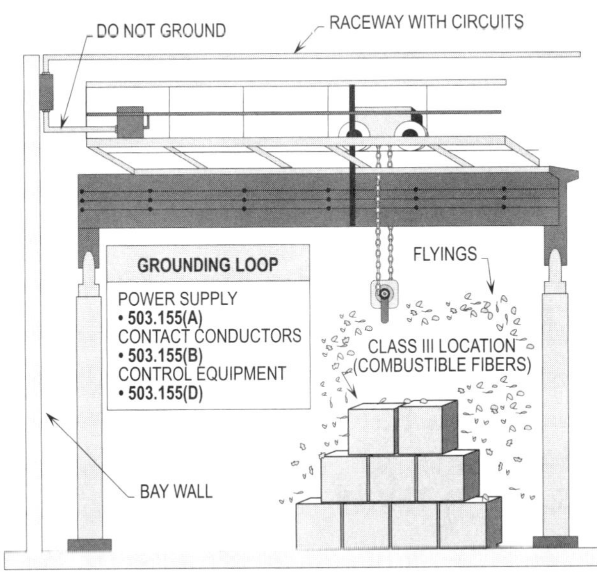

Figure 11-21(a). This illustration shows a circuit that shall not be permitted to be grounded.

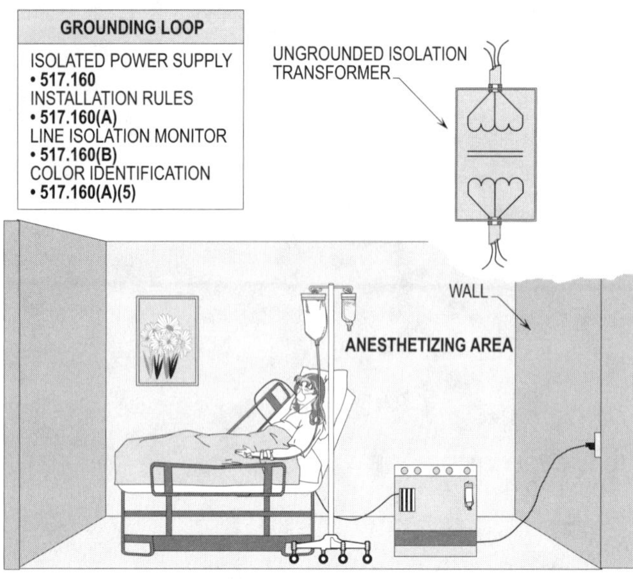

Figure 11-21(b). This illustration shows a circuit that shall not be permitted to be grounded.

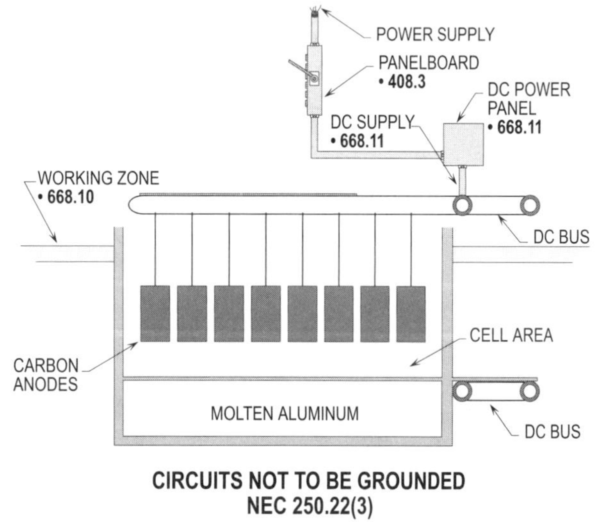

Figure 11-21(c). This illustration shows a circuit that shall not be permitted to be grounded.

GROUNDING SERVICE-SUPPLIED AC SYSTEMS
250.24(A)(1) THRU (A)(5)

AC grounded systems shall be grounded at each service by a grounding electrode conductor that is usually from the neutral bus terminal to a grounding electrode system that could be a cold water pipe, driven rod, or other electrode, or any combination per **250.52(A)(1) through (A)(7)**. The

grounding electrode conductor shall be connected to the grounded service conductor when applying one or more of the following grounding requirements:

- General
- Outdoor transformer
- Dual fed services
- Main bonding jumper as wire or busbar
- Load-side grounding connections

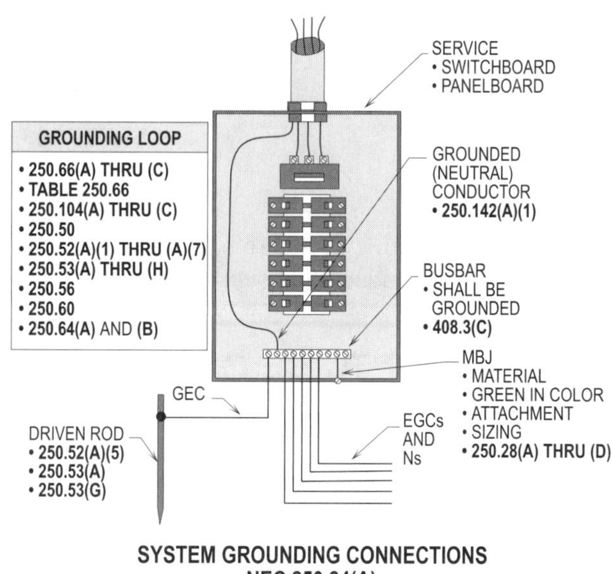

Figure 11-22. The grounding electrode conductor shall be installed in such a manner so as to ground the grounded busbar, the equipment grounding conductors, and grounded (neutral) conductor to the grounding electrode system to form a single-point ground connected to earth ground.

GENERAL
250.24(A)(1)

The grounding electrode conductor shall be installed to bond and ground the service equipment terminal neutral bus to the grounding electrode system per **250.52(A)(1) through (A)(7)**. The grounding electrode conductor shall be installed to connect equipment grounding conductors, grounded (neutral) conductors, and the service equipment enclosure to a reliable earth ground. **(See Figure 11-23)**

OUTDOOR TRANSFORMER
250.24(A)(2)

For transformers that supply the service and are located outside the building, at least one additional grounding connection shall be installed from the grounded service conductor to a grounding electrode, either at the transformer or elsewhere outside the building. **(See Figure 11-23)**

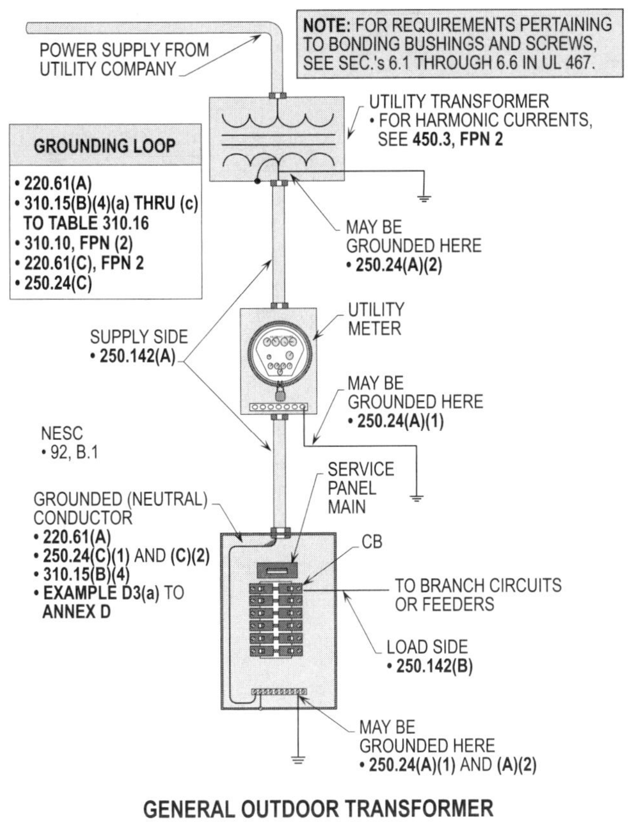

Figure 11-23. This illustration shows that AC grounded systems shall be grounded at each service and the outdoor supply transformer.

DUAL FED SERVICES
250.24(A)(3)

A single grounding electrode conductor connection to the tie point of the grounded circuit conductor(s) from each power source shall be permitted to be installed where services are dual fed (double ended) in a common enclosure or grouped together in service enclosures and employing a secondary tie. **(See Figure 11-24)**

MAIN BONDING JUMPER AS WIRE OR BUSBAR
250.24(A)(4)

Where the main bonding jumper (wire or busbar) is installed from the grounded (neutral) conductor terminal bar, the bus to the equipment grounding terminal bar or bus in the service equipment, the grounding electrode conductor shall be permitted to be connected to the equipment grounding terminal bar or bus to which the main bonding jumper is connected.

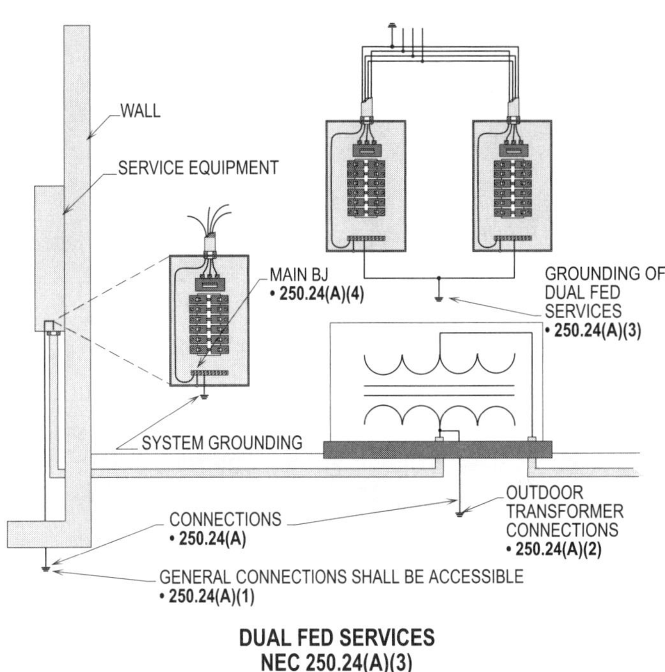

Figure 11-24. This illustration shows the requirements for grounding and bonding dual fed services from a service-entrance supply.

LOAD SIDE GROUNDING CONNECTIONS
250.24(A)(5)

Unless otherwise permitted, a grounded (neutral) conductor shall not be connected to normally noncurrent-carrying metal parts of equipment, or to equipment grounding conductors, or be reconnected to ground on the load side of the service disconnecting means. **(See Figure 11-25)**

Design Tip: All equipment grounding conductors and equipment bonding jumpers on the load side shall be sized from **Table 250.122** based on the rating of the overcurrent protection device ahead of the circuit conductors.

CALCULATING FAULT CURRENT
110.9 AND 110.10

By dividing the length of wire between supply and load by 1000 and multiplying by the resistance, the amount of fault current that will flow may be calculated. The resistance of each length is added together and divided into the voltage-to-ground to derive the fault current at the location of such short. **(See Figure 11-26)**

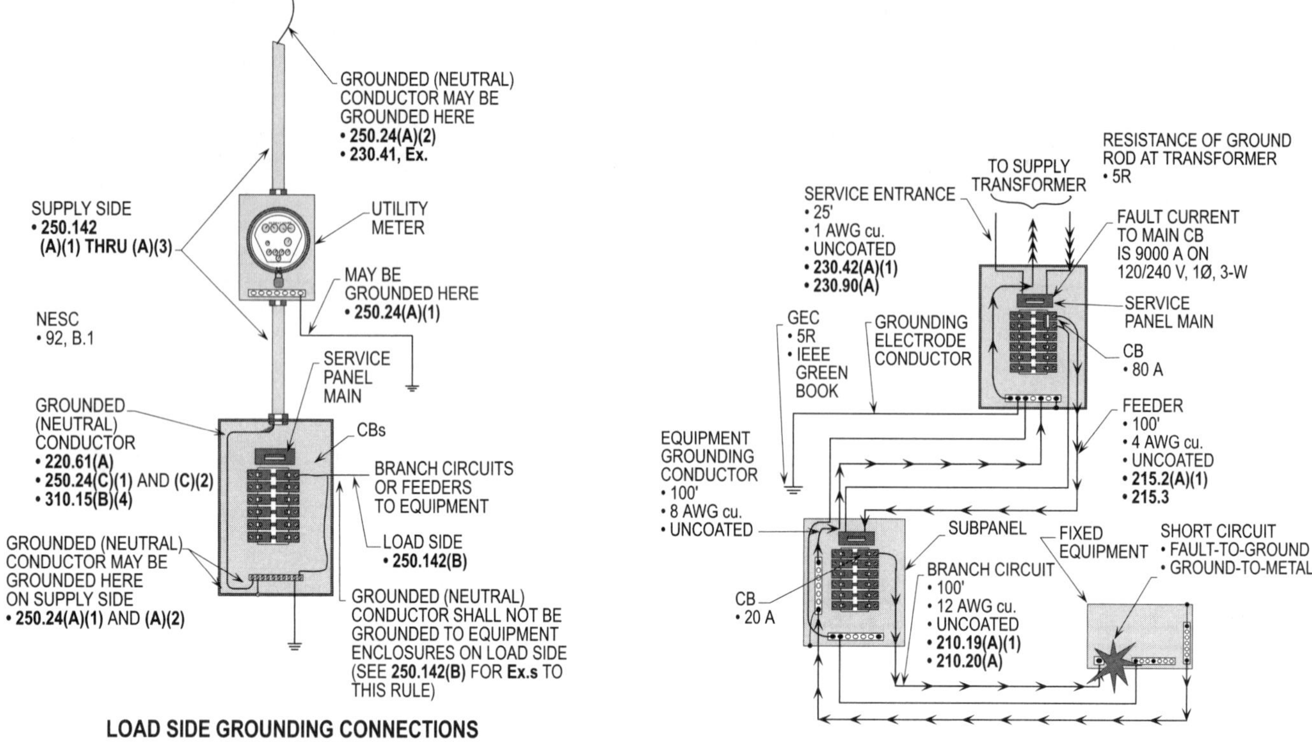

Figure 11-25. This illustration shows that the grounded conductor shall not be connected to normally noncurrent-carrying metal parts of equipment, or to equipment grounding conductor(s), or be reconnected to ground.

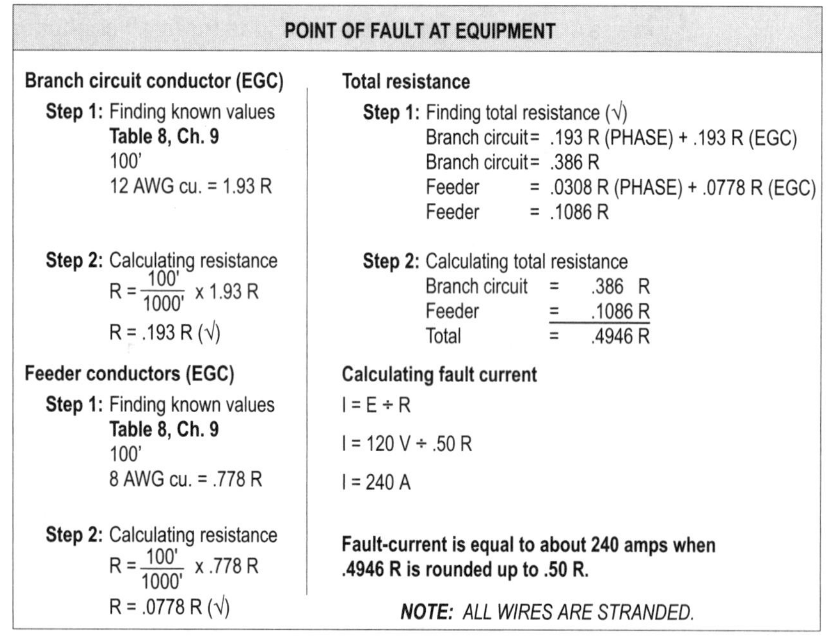

POINT OF FAULT AT EQUIPMENT

Branch circuit conductor (EGC)

Step 1: Finding known values
Table 8, Ch. 9
100'
12 AWG cu. = 1.93 R

Step 2: Calculating resistance
$R = \dfrac{100'}{1000'} \times 1.93\ R$
R = .193 R (√)

Feeder conductors (EGC)

Step 1: Finding known values
Table 8, Ch. 9
100'
8 AWG cu. = .778 R

Step 2: Calculating resistance
$R = \dfrac{100'}{1000'} \times .778\ R$
R = .0778 R (√)

Total resistance

Step 1: Finding total resistance (√)
Branch circuit = .193 R (PHASE) + .193 R (EGC)
Branch circuit = .386 R
Feeder = .0308 R (PHASE) + .0778 R (EGC)
Feeder = .1086 R

Step 2: Calculating total resistance
Branch circuit = .386 R
Feeder = .1086 R
Total = .4946 R

Calculating fault current

I = E ÷ R

I = 120 V ÷ .50 R

I = 240 A

Fault-current is equal to about 240 amps when .4946 R is rounded up to .50 R.

NOTE: ALL WIRES ARE STRANDED.

CALCULATING FAULT CURRENT
NEC 110.9
NEC 110.10

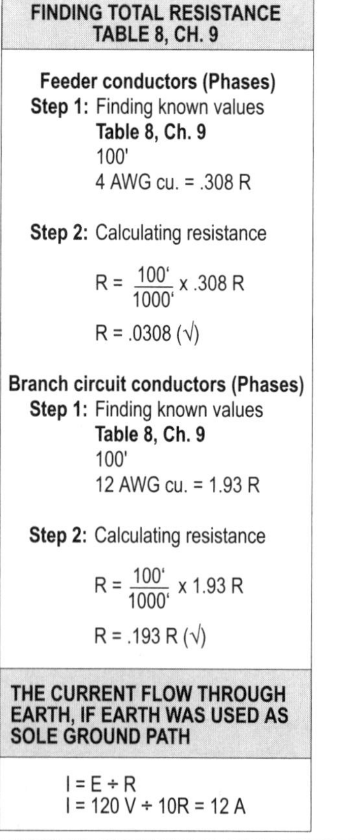

FINDING TOTAL RESISTANCE TABLE 8, CH. 9

Feeder conductors (Phases)
Step 1: Finding known values
Table 8, Ch. 9
100'
4 AWG cu. = .308 R

Step 2: Calculating resistance
$R = \dfrac{100'}{1000'} \times .308\ R$
R = .0308 (√)

Branch circuit conductors (Phases)
Step 1: Finding known values
Table 8, Ch. 9
100'
12 AWG cu. = 1.93 R

Step 2: Calculating resistance
$R = \dfrac{100'}{1000'} \times 1.93\ R$
R = .193 R (√)

THE CURRENT FLOW THROUGH EARTH, IF EARTH WAS USED AS SOLE GROUND PATH

I = E ÷ R
I = 120 V ÷ 10R = 12 A

NOTE: THIS AMOUNT OF CURRENT WILL NOT NORMALLY FLOW, IT WILL USUALLY BE MUCH LESS.

Figure 11-26. This illustration shows the procedures for calculating the fault current at the point of the fault.

MAIN BONDING JUMPER
250.24(B)

For a grounded system, an unspliced main bonding jumper shall be used to connect the equipment grounding conductor(s) and the service disconnect enclosure to the grounded (neutral) conductor within the enclosure for each service disconnect in accordance with **250.28**.

GROUNDED CONDUCTOR BROUGHT TO SERVICE EQUIPMENT
250.24(C)

The grounded (neutral) conductors shall be run to each service disconnecting means and shall be bonded to each disconnecting means enclosure for AC systems that operate at less than 1000 volts and are grounded at any point. **(See Figure 11-27)**

> **Design Tip:** A good example of this rule is where six disconnects are installed at the service equipment and three are used with neutral connections and three are not. The three without neutral connections are still required to have the grounded (neutral) conductor run to each disconnect to provide a low-impedance path for the fault current to return over and trip the cutout fuses on the primary side of the power transformer.

The grounded (neutral) conductor(s) shall be installed in accordance with the following conditions of use:

- Sizing
- Routing
- Parallel conductors
- High impedance

GROUNDED CONDUCTORS BROUGHT TO ASSEMBLY LISTED FOR USE AS SERVICE EQUIPMENT
250.24(C), Ex.

Where more than one service disconnecting means is located in a single assembly that is listed for use as service equipment, the grounded (neutral) conductor shall be permitted to be run to the assembly common grounded conductor(s) terminal or bus. The assembly shall include a main bonding jumper for connectng the grounded (neutral) conductor(s) to the assembly enclosure.

ROUTING AND SIZING
250.24(C)(1)

The grounded (neutral) conductor shall be installed with the ungrounded (phase) conductors and shall not be permitted to be smaller than the required grounding electrode conductor per **Table 250.66**, but shall not be required to be

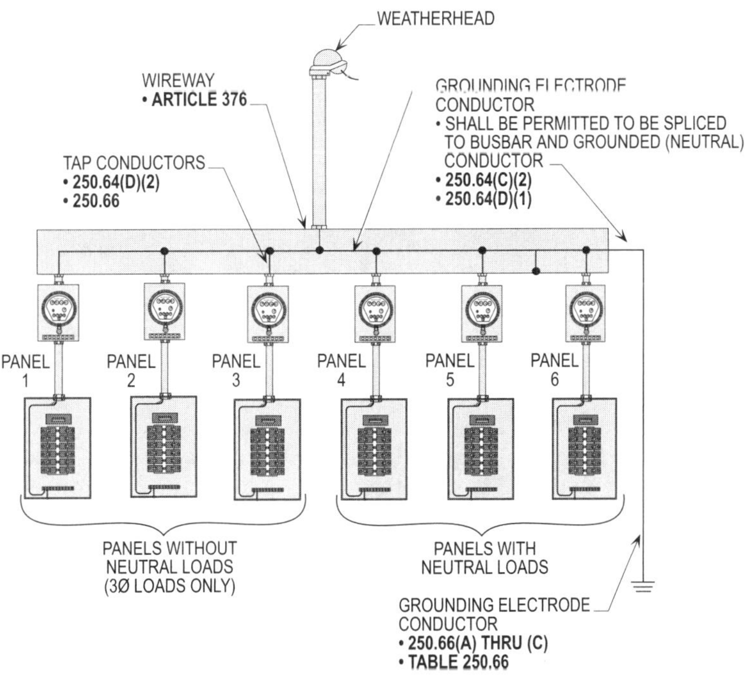

Figure 11-27. The grounded (neutral) conductor shall be run to each service disconnecting means, even if it is not used to supply single-phase loads with neutral connections.

larger than the largest ungrounded service (phase) conductor. The grounded (neutral) conductor shall not be permitted to be smaller than 12-1/2 percent of the area of the largest service-entrance (phase) conductor larger than 1100 KCMIL copper or 1750 KCMIL aluminum. **(See Figure 11-28)**

It is estimated that the amount of ground fault current that will flow in the system is approximately 5 to 10 percent in the ground and 90 to 95 percent on the grounded (neutral) conductor between the supply transformer and service equipment. The rating of the grounded (neutral) conductor shall be calculated at least 12-1/2 percent of the largest ungrounded (phase) conductor. However, the equipment grounding conductor should be calculated at not less than 25 percent of the largest ungrounded (phase) conductor to ensure equipment grounding conductors provide safe and dependable fault-ground paths. For an overcurrent protection device to clear a circuit safely, a fault current of at least 6 to 10 times its rating shall be available.

Note that fault current will travel from the point of fault through the grounded (phase or neutral) conductor to the supply transformer and then will return through the phase that was faulted to ground and open the overcurrent protection device. The overcurrent protection will rapidly trip a faulted phase that has a fault current of 6 to 10 times its rating.

For example: What is the (appropriate) amount of fault current that is required to clear a 150 amp OCPD?

Step 1:	Calculating percentage
	250.24(C)
	150 A x 6 = 900 A
	150 A x 10 = 1500 A

Solution: The minimum fault current is 900 amps, and the maximum fault current is 1500 amps.

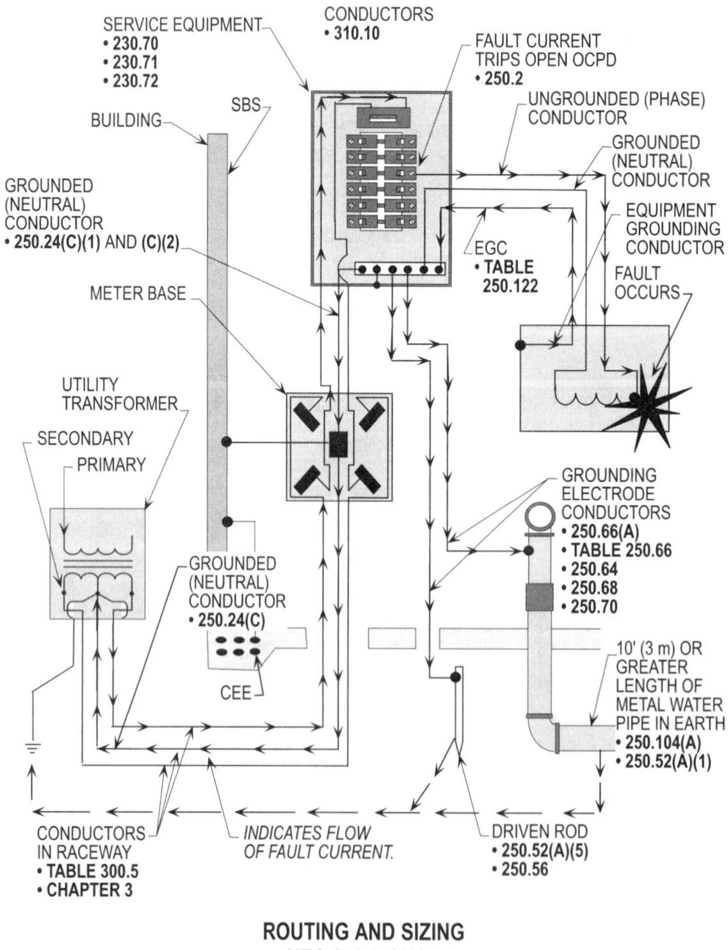

Figure 11-28. The grounded (neutral) conductor (fault current return path) shall be installed with the ungrounded (phase) conductors and run to each service if the secondary of the utility's power transformer is grounded.

Grounded (neutral) conductors shall be installed to provide an effective path for fault currents to travel over where phase-to-ground faults occur in the electrical system. The grounded (neutral) conductors shall be sized as large as the grounding electrode conductor per **250.66** and **Table 250.66**. The grounded (neutral) conductor shall be sized at least 12-1/2 percent of the area of the largest ungrounded (phase) conductor where the service conductors are installed larger than 1100 KCMIL copper or 1750 KCMIL aluminum.

Note: Two-phase or three-phase conductors shall not be permitted to be run to the service without installing a grounded (neutral) conductor when the utility company's secondary is grounded. An example of this rule is where all the service loads are three-phase, 480 volt and a step-down separately derived system is installed to supply single-phase loads of 120/208 volt.

For example: What size copper grounded (neutral) conductor is required based on the service-entrance conductors being rated 250 KCMIL THWN-THHN copper?

Step 1: Finding the grounded (neutral) conductor
250.142(A)(1); 250.24(C)(1); Table 250.66
250 KCMIL requires 2 AWG cu.

Solution: The size grounded (neutral) conductor is 2 AWG copper.

PARALLEL CONDUCTORS
250.24(C)(2)

The grounded (neutral) conductor shall be based on the total circular mil area of the service-entrance conductors when installed in parallel. Grounded (neutral) conductors that are installed in two or more raceways shall be based on the size of the ungrounded service-entrance (phase) conductor in the raceway, but not smaller than 1/0 AWG.

Note, see **310.4** for grounded (neutral) conductors that are connected in parallel.

For example: In a paralleled three-phase, four-wire service with 4 - 300 KCMIL THWN copper conductors per phase, what size grounded (neutral) conductor is required when installed in a single conduit run, using RMC?

Step 1: Finding total KCMIL
250.24(C)(1)
300 KCMIL x 4 per phase = 1200 KCMIL

Step 2: Finding KCMIL to size grounded (neutral) conductor
250.24(C)(2); 250.66
1200 KCMIL x .125 = 150 KCMIL

Step 3: Finding CM rating
Table 8, Ch. 9
150 KCMIL requires 3/0 AWG cu.

Step 4: Sizing grounded (neutral) conductor for the conduit run
250.24(C)(2); Table 8, Ch. 9; 310.4

Solution: At least 3/0 THWN copper grounded (neutral) conductors are required per 250.24(C)(2) and 310.4.

For example: In a paralleled three-phase, four-wire service with 3 - 700 KCMIL THWN copper conductors per phase, what size grounded (neutral) conductor is required in each run of conduit?

Step 1: Finding total KCMIL
250.24(C)(1)
700 KCMIL x 1 = 700 KCMIL

Step 2: Finding KCMIL to size grounded (neutral) conductor
250.24(C)(2); 250.66
700 KCMIL requires 2/0 AWG cu.

Step 3: Dividing KCMIL in each conduit run
250.24(C)(2)
12 1/2% rule does not apply

Step 4: Sizing grounded (neutral) conductor for each conduit run
250.24(C)(2); Table 8, Ch. 9; 310.4
700 KCMIL requires 2/0 AWG cu.

Solution: At least 1 - 2/0 AWG THWN copper grounded (neutral) conductor is required in each conduit per 250.24(C)(2) and 310.4.

HIGH IMPEDANCE
250.24(C)(3)

The grounded (neutral) conductor on a high-impedance grounded neutral system shall be grounded per **250.36**.

GROUNDING ELECTRODE CONDUCTOR
250.24(D)

A grounding electrode conductor shall be used to connect the equipment grounding conductors, the service equipment enclosures, and where the system is grounded, the grounded service conductor to the grounding electrode or electrodes. The grounding electrode conductor shall be sized per **250.66**. **(See Figure 11-29)**

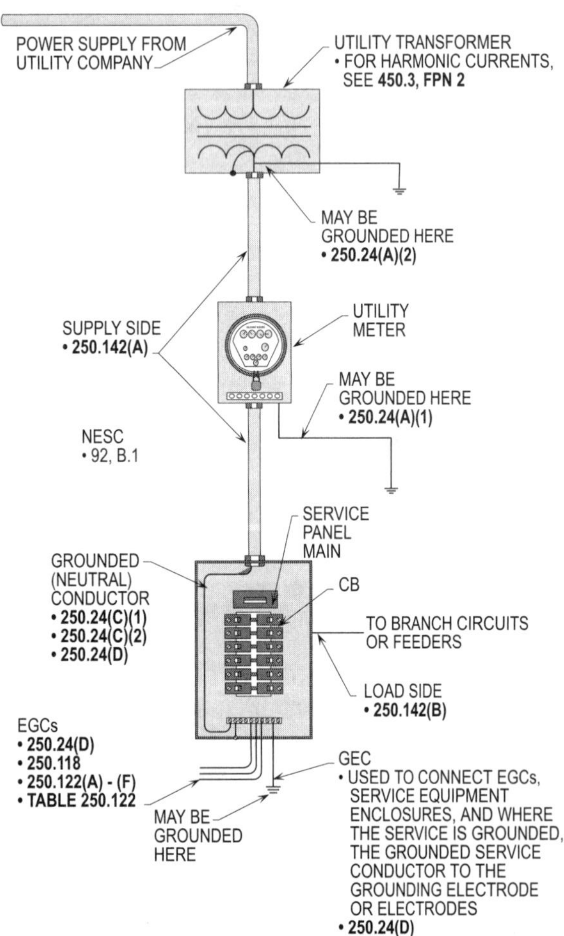

Figure 11-29. This illustration shows the procedure for using a grounding electrode conductor to connect equipment grounding conductors and the service equipment enclosures, and where the system is grounded, the grounded service conductor to the grounding electrode or electrodes.

Where the three-phase, three-wire critical load is relatively large compared with loads that require a grounded (neutral) circuit conductor, a high-impedance grounded service power supply is sometimes used. This arrangement requires an on-site transformer for loads that require a grounded (neutral) circuit conductor.

Note that high-impedance grounded neutral system connections shall be made per **250.36**. High-impedance grounded systems should not be used unless they are equipped with ground fault indicators or alarms, or both, and qualified persons are available to quickly locate and remove ground faults. If ground faults are not promptly removed, the service reliability will be reduced.

UNGROUNDED SYSTEM GROUNDING CONNECTIONS
250.24(E)

A premises wiring system that is supplied by an AC service supply that is ungrounded shall have at the service a grounding electrode conductor connected to the grounding electrode system. The grounding electrode conductor shall be connected to a metal enclosure of the service conductors at any accessible point from the load end of the service drop or service lateral to the service disconnecting means.

On an ungrounded system, none of the circuit conductors of the system are intentionally grounded. Note that the bonding together of all conductive enclosures and equipment in each circuit with equipment grounding conductors shall be done. These equipment grounding conductors shall be run with or enclose the circuit conductors, and they shall provide a permanent, low-impedance path for ground fault currents to be detected. **(See Figure 11-30)**

CONDUCTOR TO BE GROUNDED – AC SYSTEMS
250.26

The grounded (neutral) conductor, whether a grounded phase or neutral, is usually installed with a white or gray insulation or otherwise identified. The following systems shall be grounded and have a grounded (phase) conductor or grounded (neutral) conductor: **(See Figure 11-31)**

- 120 volt, two-wire, single-phase – one neutral conductor
- 120/240 volt, three-wire, single-phase – the neutral conductor
- 120/208 volt, four-wire, three-phase wye – the common conductor
- 277/480 volt, four-wire, three-phase wye – the common conductor

- 240 volt, three-wire, three-phase delta – the common conductor
- 480 volt, three-wire, three-phase delta – the common conductor
- 600 volt, three-wire, three-phase delta – the common conductor
- 480 volt corner grounded delta – one-phase conductor

MAIN BONDING JUMPER AND SYSTEM BONDING JUMPER 250.28(A) THRU (D)

An unspliced main bonding jumper (for grounded systems) shall be installed to connect the equipment grounding conductors and the service disconnecting enclosure to the grounded (neutral) conductor of the system within the enclosure for each service disconnect. The following conditions of use shall be complied with when installing main bonding jumpers and system bonding jumpers:

- Material
- Construction
- Attachment
- Size

MATERIAL
250.28(A)

Unspliced main bonding jumpers and system bonding jumpers shall be of copper or other corrosion-resistant material. Main bonding jumpers and system bonding jumpers shall be installed as a wire, bus, screw, or similar suitable conductor. **(See Figure 11-32)**

CONSTRUCTION
250.28(B)

When a screw is used as the main bonding jumper or system bonding jumper, it shall be identified with a green-colored finish that is visible with the screw installed. **(See Figure 11-32)**

ATTACHMENT
250.28(C)

Section **250.8** specifies the manner in which the main bonding jumper and system bonding jumper shall be connected for electrical circuits and equipment. Section **250.70** specifies the manner by which grounding electrodes shall be attached together after sizing and selecting such grounding conductors. **(See Figure 11-32)**

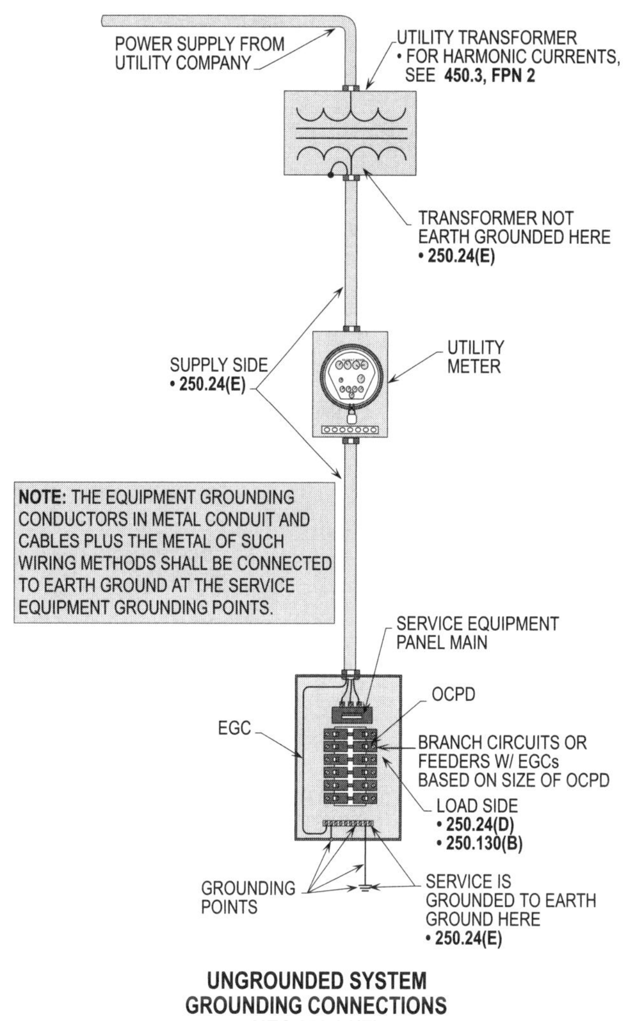

Figure 11-30. This illustration shows the procedures for installing an ungrounded service supply.

SIZE
250.28(D)(1) THRU (D)(3)

The size of the main bonding jumper and the system bonding jumper shall be based on the size of the service-entrance conductors and shall be sized per **Table 250.66**. These bonding jumpers shall be sized at least 12-1/2 percent of the area of the largest ungrounded (phase) conductor where the service conductors are installed larger than 1100 KCMIL copper or 1750 KCMIL aluminum.

The main bonding jumper for each enclosure (service consists of more than a single enclosure) shall be sized per **250.28(D)(1)** based on the largest ungrounded service (phase) conductor serving that enclosure.

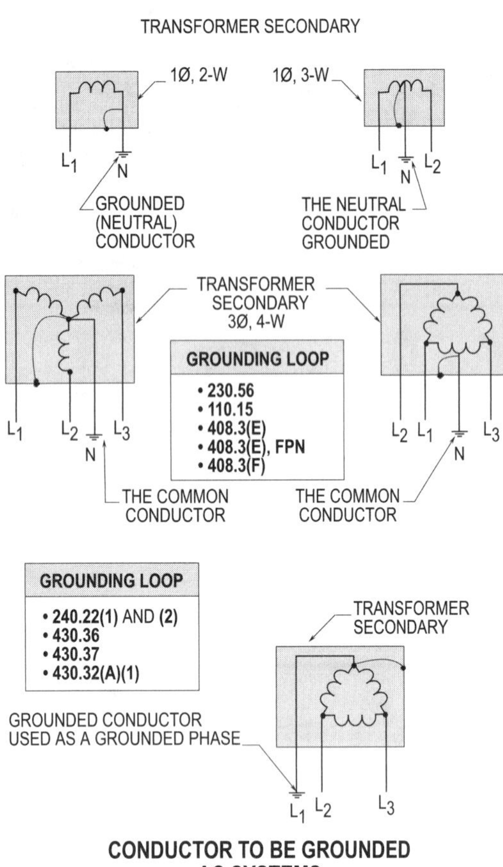

Figure 11-31. The grounded (neutral) conductor, whether a grounded phase or neutral, is usually installed with a white or gray insulation or otherwise identified.

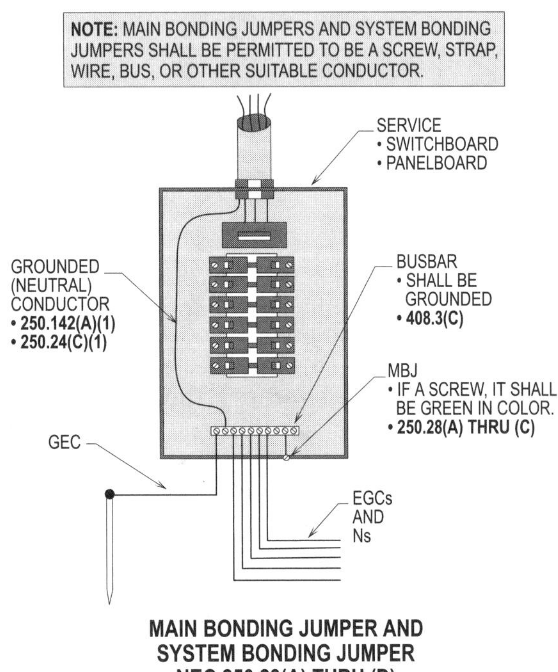

Figure 11-32. This illustration shows the items permitted to be used as a main bonding jumper.

For example: What size copper main bonding jumper is required based on the service-entrance conductors being rated at 250 KCMIL THWN-THHN copper?

Step 1: Finding the main bonding jumper
250.28(D); Table 250.66
250 KCMIL requires 2 AWG cu.

Solution: The size of the main bonding jumper is required to be 2 AWG copper.

The system bonding jumper for each enclosure (separately derived system supplies more than a single enclosure) shall be sized per **250.28(D)(1)** based on the largest ungrounded feeder (phase) conductor serving that enclosure, or a single system bonding jumper shall be installed at the source and sized per **250.28(D)(1)** based on the equivalent size of the largest supply conductor determined by the largest sum of the areas of the corresponding conductors of each set.

For example: In a paralleled three-phase, four-wire service with 3 - 700 KCMIL THWN copper conductors per phase, what size main bonding jumper is required to ground the busbar to the service equipment enclosure?

Step 1: Finding total KCMIL
250.28(D)
700 KCMIL x 3 = 2100 KCMIL

Step 2: Finding KCMIL to size MBJ
250.28(D)
2100 KCMIL x .125 = 262.5 KCMIL

Step 3: Sizing MBJ
250.28(D); Table 8, Ch. 9
262.5 KCMIL requires 300 KCMIL

Solution: The size of the main bonding jumper is required to be 300 KCMIL copper.

GROUNDING SEPARATELY DERIVED AC SYSTEMS
250.30

Low-voltage and high-voltage feeder-circuits are sometimes installed from floor-to-floor in a high-rise building, with transformers installed on each floor to reduce the voltage to 120/240 or 120/208 volts for general-use lighting and receptacle loads in large building applications. Such grounding, since the 1978 NEC, may be installed either at the transformer or at the load served that is connected and supplied from the secondary side per **240.21(C)(2), (C)(3), and 240.92(B)(1) through (B)(3). (See Figure 11-33)**

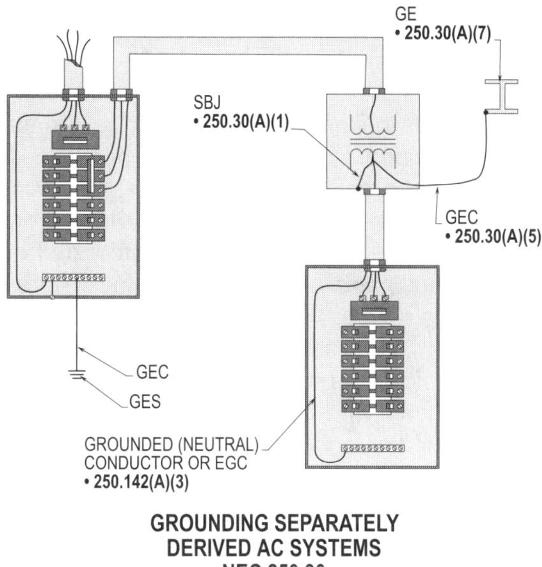

**GROUNDING SEPARATELY
DERIVED AC SYSTEMS
NEC 250.30**

Figure 11-33. The system bonding jumper and grounding electrode conductor is designed and installed based on the derived ungrounded (phase) conductors supplying the panelboard, switch, or other equipment connected from the secondary side of the transformer.

GROUNDED SYSTEMS
250.30(A)

A separately derived system (SDS) that is grounded shall comply with the following:

- System bonding jumper
- Equipment bonding jumper size
- Grounding electrode conductor, single separately derived system
- Grounding electrode conductor, multiple separately derived system
- Installation
- Bonding
- Grounding electrode
- Grounded conductor

SYSTEM BONDING JUMPER
250.30(A), Ex.

The grounding connection requirements for high-impedance grounded neutral systems shall be made per **250.36** and **250.186**.

SYSTEM BONDING JUMPER
250.30(A)(1)

The system bonding jumper (unspliced) shall be designed and installed based on the derived ungrounded (phase) conductors supplying the panelboard, switch, or other equipment connected from the secondary side of the transformer and sized per **250.28(D)** and **250.102(C)**. The system bonding jumper shall be sized per **250.66** and **Table 250.66** from the ungrounded (phase) conductors up to 1100 KCMIL for copper and 1750 KCMIL for aluminum. The system bonding jumper shall be sized at least 12-1/2 percent (.125) of the area of the largest ungrounded (phase) conductor where the service conductors are installed larger than 1100 KCMIL copper or 1750 KCMIL aluminum. The system bonding jumper shall be installed and connected at any single point on the separately derived system from the source to the first system disconnecting means or overcurrent protection device. If the ungrounded (phase) conductors are larger than 1100 KCMIL for copper and 1750 KCMIL for aluminum, the system bonding jumper will normally be larger than the grounding electrode conductor.

For example: What size copper system bonding jumper is required to bond and ground the secondary of a separately derived system having 4 - 3/0 AWG THWN copper conductors connected to its secondary?

Step 1: Finding SBJ
250.30(A)(1); Table 250.66
3/0 AWG THWN cu. requires
4 AWG cu.

Solution: **The size of the system bonding jumper is required to be 4 AWG copper.**

For example: In a paralleled three-phase, four-wire service with 4 - 600 KCMIL THWN copper conductors per phase, what size system bonding jumper is required to bond and ground the separately derived system?

Step 1:	Finding total KCMIL for SBJ	
	250.30(A)(1); 250.28(D)	
	600 KCMIL x 4 = 2400 KCMIL	

Step 1: Finding total KCMIL for SBJ
250.30(A)(1); 250.28(D)
600 KCMIL x 4 = 2400 KCMIL

Step 2: Finding KCMIL for SBJ
250.30(A)(1); 250.28(D)
2400 KCMIL x .125 = 300 KCMIL

Step 3: Sizing SBJ
250.30(A)(1); 250.28(D);
Table 8, Ch. 9
300 KCMIL requires 300 KCMIL

Solution: The size of the system bonding jumper is required to be 300 KCMIL copper.

DUAL FED (DOUBLE ENDED)
250.30(A)(1), Ex.1

A single system bonding jumper connection to the point of the grounded circuit conductors from each power source shall be permitted for separately derived systems that are dual fed (double ended) in a common enclosure and employing a secondary tie. **(See Figure 11-34)**

ADDITIONAL SYSTEM BONDING JUMPER
250.30(A)(1), Ex. 2

An additional system bonding jumper connection (more than one) shall be permitted to be made only where doing so will not create a parallel path for the grounded circuit conductor. The additional system bonding jumper shall not be permitted to be smaller than the other system bonding jumper but shall not be required to be larger than the ungrounded (phase) conductor(s). **(See Figure 11-35)**

This section states that no parallel path may be formed when installing an additional system bonding jumper at equipment supplied by a separately derived system. The basic rule requires the system bonding jumper and the grounding electrode conductor connection to the grounded system conductor to be made at the same point. There is an exception to this rule, but only for cases where there won't be a parallel path, such as a nonmetallic conduit (PVC) run back to the transformer's source and so located to not share conductive contact with common structural elements or equipment grounding conductors. The purpose of this rule is to keep the grounded (neutral) conductor current confined and flowing over insulated electrical circuit conductors.

BONDING JUMPER – CLASS 1, CLASS 2, OR CLASS 3 CIRCUITS
250.30(A)(1), Ex. 3

The system bonding jumper shall be sized not smaller than the derived ungrounded (phase) conductors and shall not be smaller than 14 AWG copper or 12 AWG aluminum for a system that supplies a Class 1, Class 2, or Class 3 circuit, and is derived from a transformer that is rated not more than 1000 volt-amperes. **(See Figure 11-36)**

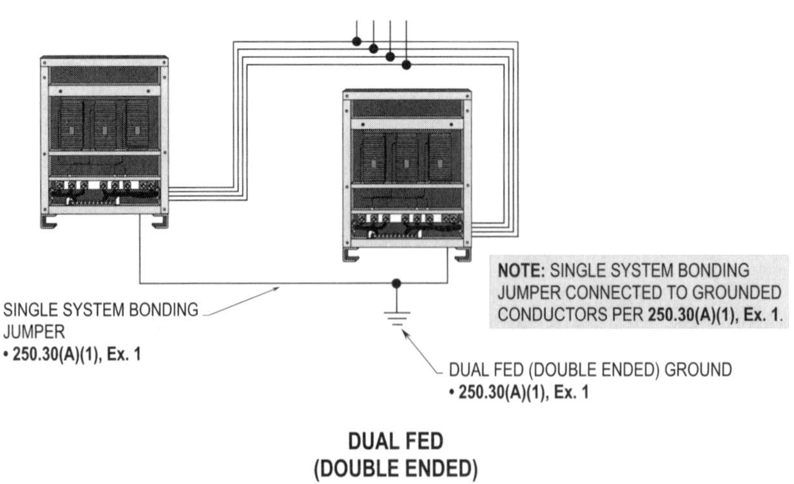

Figure 11-34. This illustration shows a single system bonding jumper connection for separately derived systems that are dual fed (double ended).

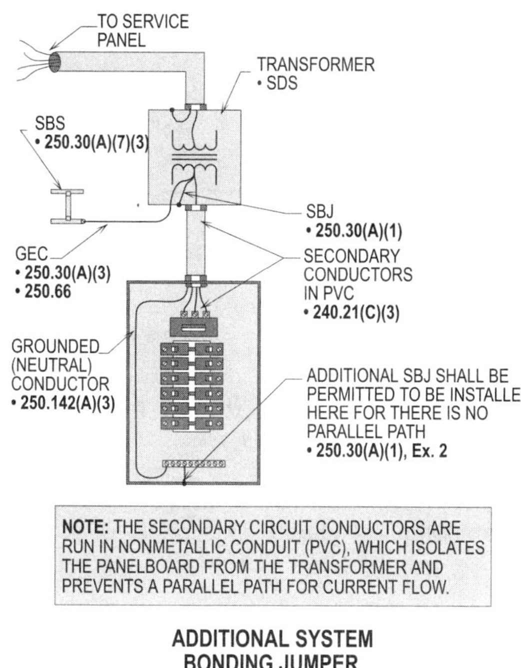

Figure 11-35. This illustration shows the grounding and bonding necessary when an additional system bonding jumper is used.

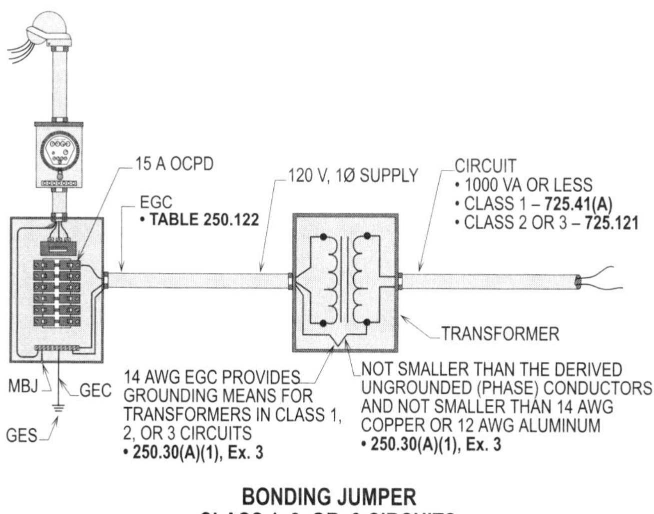

Figure 11-36. This illustration shows the bonding requirements for a small transformer using the equipment grounding conductor, instead of installing a grounding electrode conductor and connecting it to a grounding electrode.

EQUIPMENT BONDING JUMPER SIZE 250.30(A)(2)

Where an equipment bonding jumper (wire) is run with the derived ungrounded (phase) conductors from the source of a separately derived system to the first disconnecting means, it shall be sized in accordance with **250.102(C)**, based on the size of the derived ungrounded (phase) conductors.

GROUNDING ELECTRODE CONDUCTOR, SINGLE SEPARATELY DERIVED SYSTEM 250.30(A)(3)

The grounding electrode conductor shall be designed and installed based on the derived ungrounded (phase) conductors supplying the panelboard, switch, or other equipment connected from the secondary of the transformer and shall be sized per **Table 250.66**. The grounding electrode conductor shall be installed and connected at any point on the separately derived system from the source to the first system disconnecting means or overcurrent protection device. When the KCMIL rating is greater than 1100 for copper and 1750 for aluminum, the grounding electrode conductor will usually be smaller than the bonding jumper.

A common continuous grounding electrode conductor shall be permitted to be extended from the grounding electrode system and run through the building and the connection made at an accessible location near the separately derived system required to be grounded per **250.30(A)(4)**.

For example: What size copper grounding electrode conductor is required to bond and ground the secondary of a separately derived system having 4 - 3/0 AWG THWN copper conductors connected to its secondary?

Step 1:   Finding size GEC
250.30(A)(3); Table 250.66
3/0 AWG THWN cu. requires
4 AWG cu.

Solution: The size of the grounding electrode conductor is required to be 4 AWG copper.

For example: In a paralleled three-phase, four-wire service with 4 - 600 KCMIL THWN copper conductors per phase, what size grounding electrode conductor is required to bond and ground the separately derived system?

Step 1: Finding total KCMIL for GEC
250.30(A)(3); 250.28(D)
600 KCMIL x 4 = 2400 KCMIL

Step 2: Sizing GEC
250.30(A)(3); Table 250.66
2400 KCMIL requires 3/0 AWG cu.

Solution: The size of the grounding electrode conductor is required to be 3/0 AWG copper.

SYSTEM BONDING JUMPER
250.30(A)(3), Ex. 1

The system bonding jumper (wire or busbar) specified in **250.30(A)(1)** shall be permitted to connect the grounding electrode conductor to the equipment grounding terminal, bar or bus, provided the equipment grounding terminal, bar or bus is of sufficient size for the separately derived system. **(See Figure 11-37)**

GROUNDING ELECTRODE CONDUCTOR, LISTED EQUIPMENT
250.30(A)(3), Ex. 2

Where a separately derived system originates in listed equipment suitable as service equipment, the grounding electrode conductor from the service or feeder equipment to the grounding electrode conductor shall be permitted for the separately derived system, provided the grounding electrode conductor is of sufficient size for the separately derived system. The grounding electrode connection for the separately derived system shall be permitted to be made to the bus where the equipment ground bus internal to the equipment is not smaller than the required grounding electrode conductor for the separately derived system. **(See Figure 11-38)**

GROUNDING ELECTRODE CONDUCTOR, CLASS 1, CLASS 2, OR CLASS 3 CIRCUITS
250.30(A)(3), Ex. 3

For a Class 1, Class 2, or Class 3 remote-control or signaling transformer that is rated 1000 VA or less and has a grounded secondary conductor bonded to the metal case of the

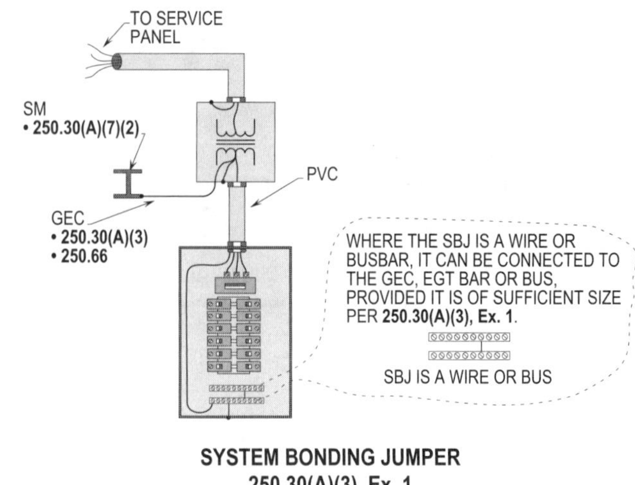

Figure 11-37. This illustration shows the requirements for the system bonding jumper.

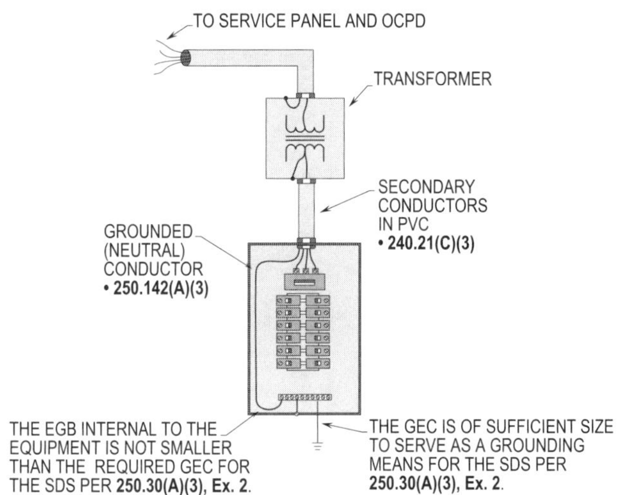

Figure 11-38. This illustration shows the requirements for grounding to the equipment grounding bus in the equipment listed as service equipment.

transformer, no grounding electrode conductor shall be required to be installed. However, the metal transformer case shall be properly grounded by a grounded metal raceway or equipment grounding conductor that supplies its primary, or by an equipment grounding conductor that connects the case back to the grounding electrode for the primary system.

At least a 14 AWG copper conductor shall be used to bond and ground the transformer secondary to the transformer frame. This leaves the supply raceway, equipment grounding conductor, or both to the transformer to provide the grounding return path back to the common service ground.

GROUNDED ELECTRODE CONDUCTOR, MULTIPLE SEPARATELY DERIVED SYSTEMS
250.30(A)(4)

A properly sized continuous common grounding electrode conductor shall be permitted to ground all separately derived systems on each floor of a high-rise building. The continuous common grounding electrode conductor shall be sized in accordance with **250.66** based on the total area of the derived ungrounded (phase) conductors smaller than 3/0 AWG copper or 250 KCMIL aluminum. This connection shall be made at the same point on the separately derived system where the system bonding jumper is installed.
(See Figure 11-39)

GROUNDING ELECTRODE CONDUCTOR, SYSTEM BONDING JUMPER
250.30(A)(4), Ex. 1

The system bonding jumper (wire or busbar) specified in **250.30(A)(1)** shall be permitted to connect the grounding electrode conductor to the equipment grounding terminal, bar or bus, provided the equipment grounding terminal, bar or bus is of sufficient size for the separately derived system.
(See Figure 11-40)

GROUNDING ELECTRODE CONDUCTOR, CLASS 1, CLASS 2, OR CLASS 3 CIRCUITS
250.30(A)(4), Ex. 2

For a Class 1, Class 2, or Class 3 remote-control or signaling transformer that is rated 1000 VA or less and has a grounded secondary conductor bonded to the metal case of the transformer, no grounding electrode conductor shall be required to be installed. However, the metal transformer case shall be properly grounded by a grounded metal raceway or equipment grounding conductor that supplies its primary, or by an equipment grounding conductor that connects the case back to the grounding electrode for the primary system.

At least a 14 AWG copper conductor shall be used to bond and ground the transformer secondary to the transformer frame. This leaves the supply raceway, equipment grounding conductor, or both to the transformer to provide the grounding return path back to the common service ground.
(See Figure 11-41)

COMMON GROUNDING ELECTRODE CONDUCTOR – SIZE
250.30(A)(4)(a)

The common grounding electrode conductor tap shall not be smaller than 3/0 AWG copper or 250 KCMIL minimum.

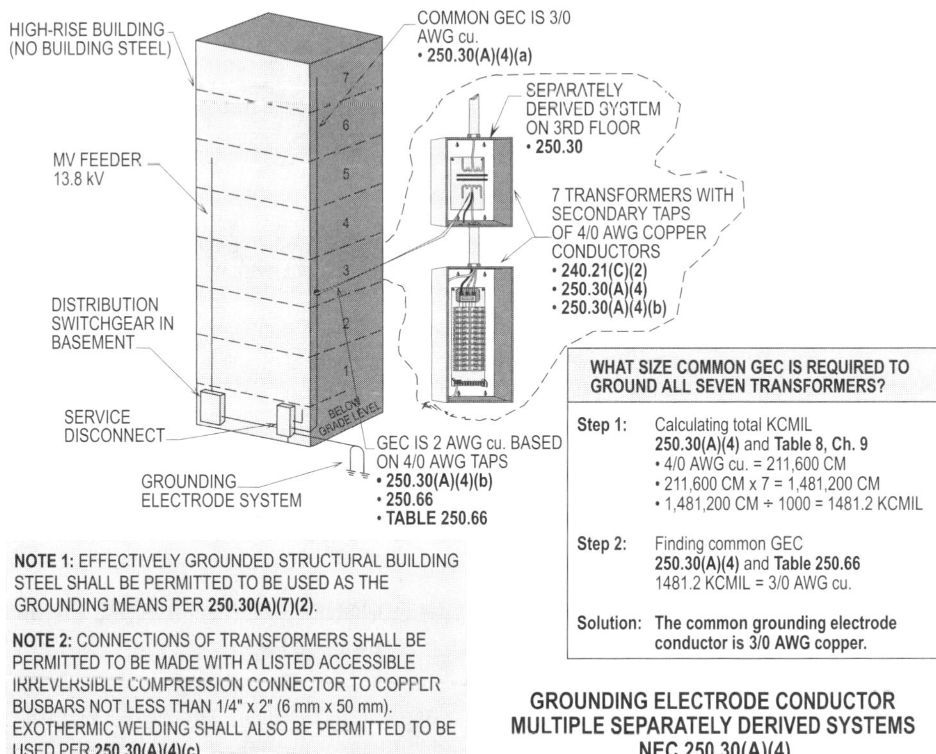

Figure 11-39. This illustration shows a properly sized continuous common grounding electrode conductor serving as grounding means on each floor of a high-rise building.

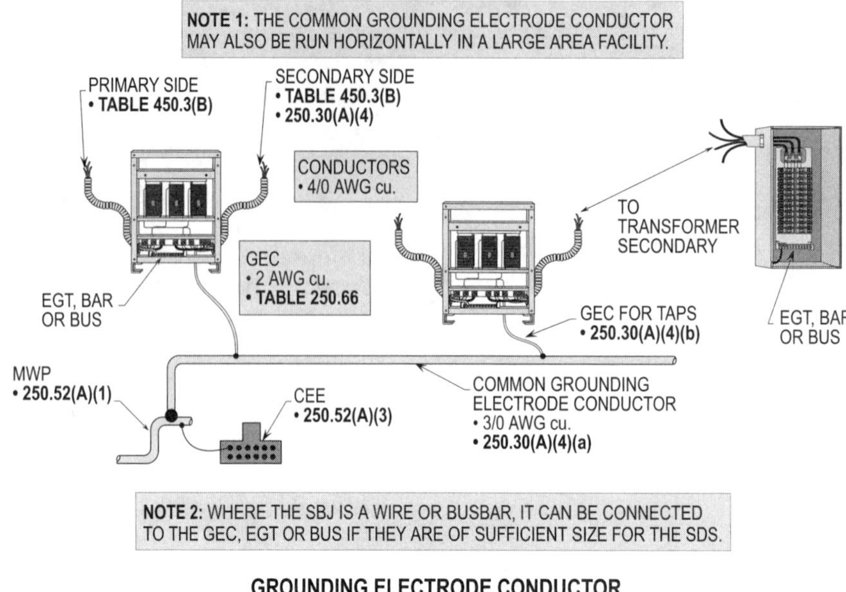

Figure 11-40. This illustration shows the requirements for connecting the system bonding jumper and grounding the separately derived system to the common grounding electrode.

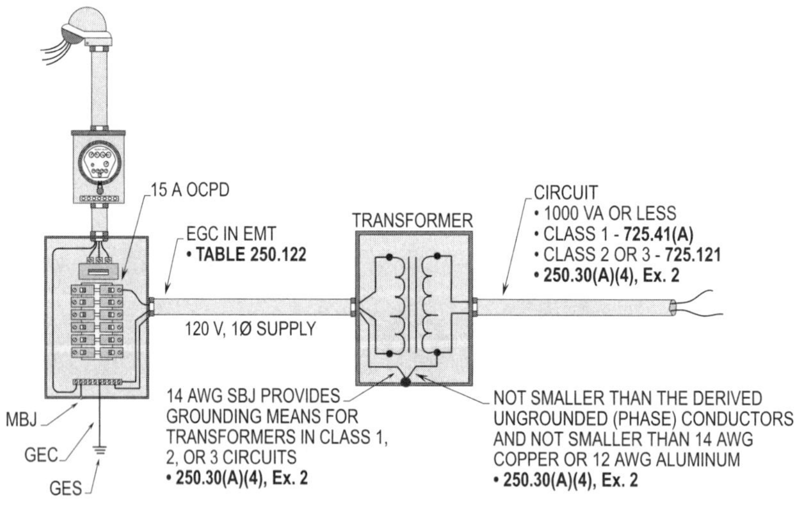

Figure 11-41. This illustration shows that no grounding electrode conductor is required to be installed.

TAP CONDUCTOR SIZE
250.30(A)(4)(b)

Each tap conductor from the separately derived system to the continuous common grounding electrode conductor shall be sized per **250.66** based on the derived ungrounded (phase) conductors of the separately derived system that it supplies. **(See Figure 11-42)**

TAP CONDUCTOR SIZE – LISTED EQUIPMENT
250.30(A)(4)(b), Ex.

Where a separately derived system originates in listed equipment suitable as service equipment, the grounding electrode conductor from the service or feeder eqiupment to the grounding electrode shall be permitted for the separately derived system, provided the grounding electrode

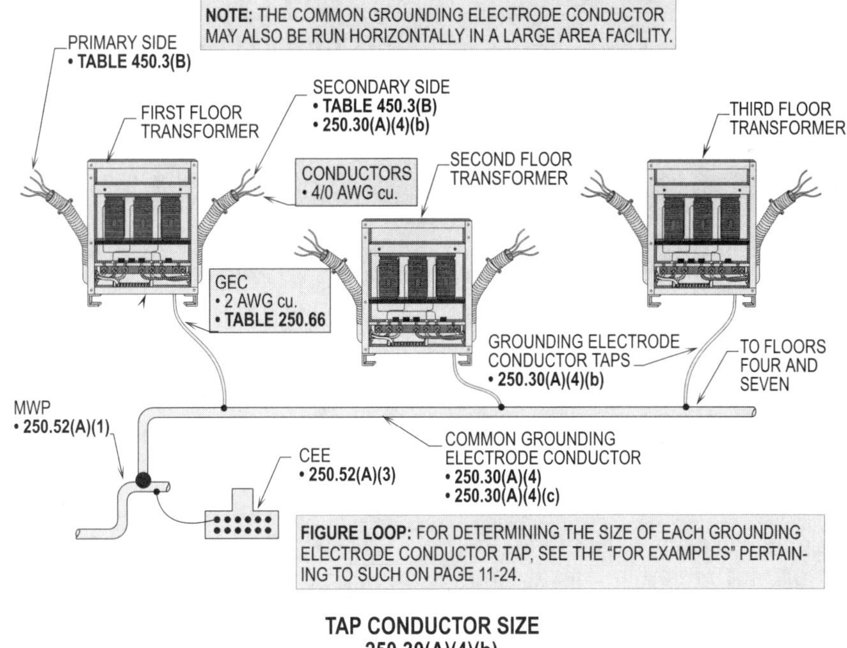

Figure 11-42. This illustration shows that each tap conductor from the separately derived system to the continuous common grounding electrode conductor shall be sized per **250.66** based on the derived ungrounded (phase) conductors of the separately derived system that it supplies.

conductor is of sufficient size for the separately derived system. The grounding electrode connection for the separately derived system shall be permitted to be made to the bus where the equipment ground bus internal to the equipment is not smaller than the required grounding electrode conductor for the separately derived system. **(See Figure 11-43)**

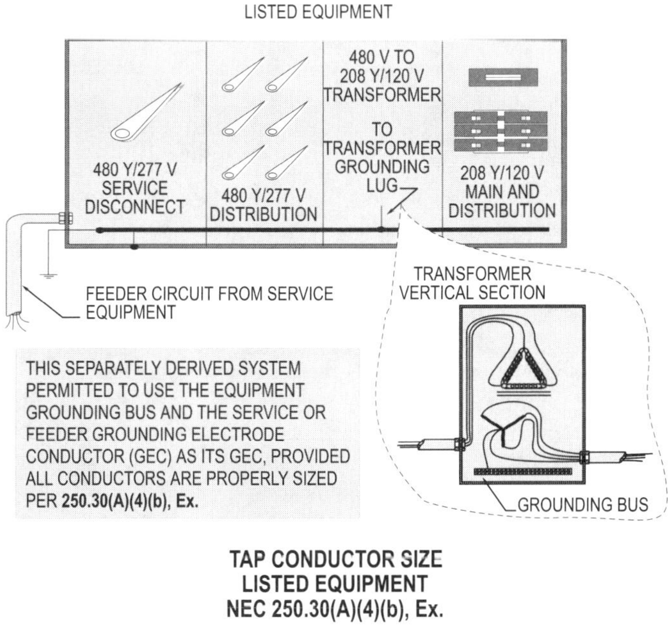

Figure 11-43. The grounding electrode conductor shall be as near as possible and preferably in the same area as the grounding electrode conductor connection to the system.

CONNECTIONS
250.30(A)(4)(c)

Connections of the grounding electrode conductor from the separately derived system to the continuous common grounding electrode conductor shall be made at an accessible location by a connector listed for the purpose, by listed connections to aluminum or copper busbars not less than 1/4 in. x 2 in. (6 mm x 50 mm), or by the exothermic welding process.

Tap conductors that are connected to the common grounding electrode conductor shall be connected in such a manner that the common grounding electrode conductor remains without a splice or joint.

INSTALLATION
250.30(A)(5)

The taps to each separately derived system and the continuous common grounding electrode conductor shall comply with **250.64(A), (B), (C),** and **(E)**, which covers the installation requirements for grounding electrode conductors.

BONDING
250.30(A)(6)

Structural steel and metal piping shall be bonded, if available, and connected to the grounded (neutral) conductor of a separately derived system per **250.104(D)**.

GROUNDING ELECTRODE
250.30(A)(7)

The grounding electrode conductor shall be as near as possible and preferably in the same area as the grounding electrode conductor connection to the system. From the following choices, one shall be selected and installed in the order that they are listed:

- Nearest grounded metal water pipe within 5 ft (1.5 m) from the point of entrance into the building per **250.52(A)(1)**

- Nearest grounded structural building steel per **250.52(A)(2)**

Note: Metal water pipe located in the area shall be bonded to the grounded (neutral) conductor per **250.104(D)(1)**.

See Figure 11-44 for a detailed illustration pertaining to the installation of the grounding electrode conductor.

OTHER ELECTRODES
250.30(A)(7), Ex. 1

Any of the electrodes identified in **250.52(A)** shall be permitted to be used where the electrodes specified by **250.30(A)(7)** are not available.

The grounding electrode conductor shall not be required to be installed larger than 3/0 AWG for copper or 250 KCMIL for aluminum when connecting to the nearest building steel or nearest metal water pipe system. The grounding electrode conductor shall not be required to be installed larger than 6 AWG copper or 4 AWG aluminum when connecting to a driven rod or other made electrodes.

BUILDINGS OR STRUCTURES SUPPLIED BY A FEEDER(S) OR BRANCH CIRCUIT(S)
250.32

These requirements apply to two or more buildings or structures supplied from one service. Basically, the grounded system of each building or structure shall have a grounding electrode that is also connected to the metal enclosure of the disconnecting means that is actually defined as a feeder. However, a grounding electrode shall be provided at each building or structure and connected to the disconnecting means as well as the grounded circuit conductor of the AC supply on the supply side of the building or structure disconnecting means. **(See Figure 11-45)**

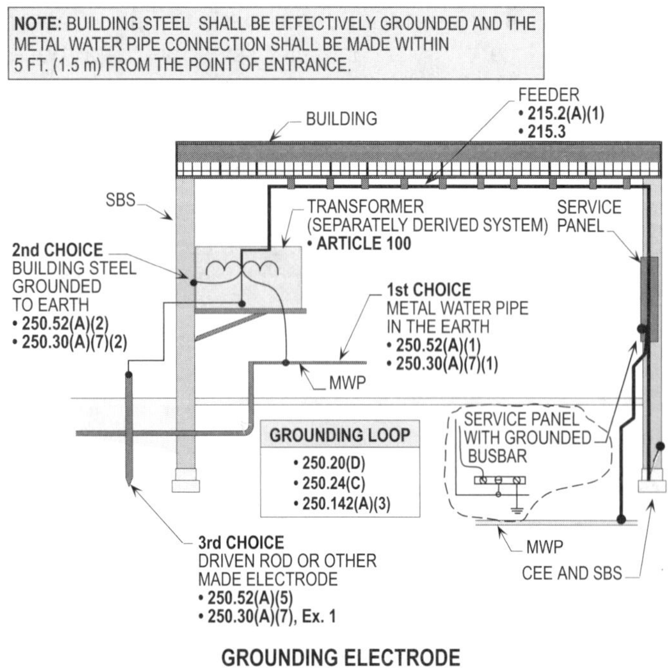

Figure 11-44. The grounding electrode conductor shall be as near as possible and preferably in the same area as the grounding electrode conductor connection to the system.

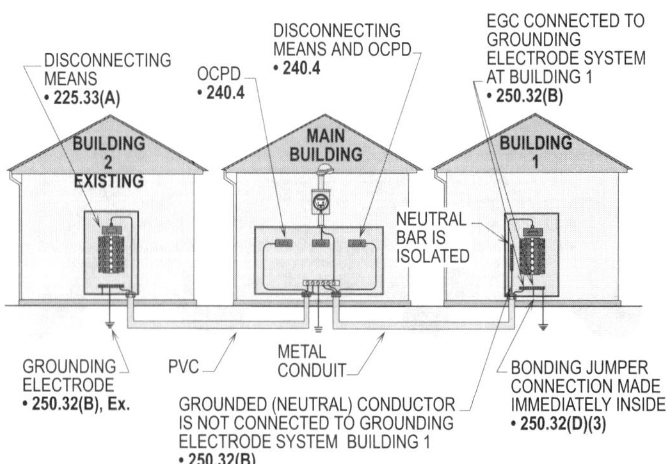

Figure 11-45. This illustration shows the use of the grounded (neutral) conductor and the equipment grounding conductor in a feeder supplying another building or structure.

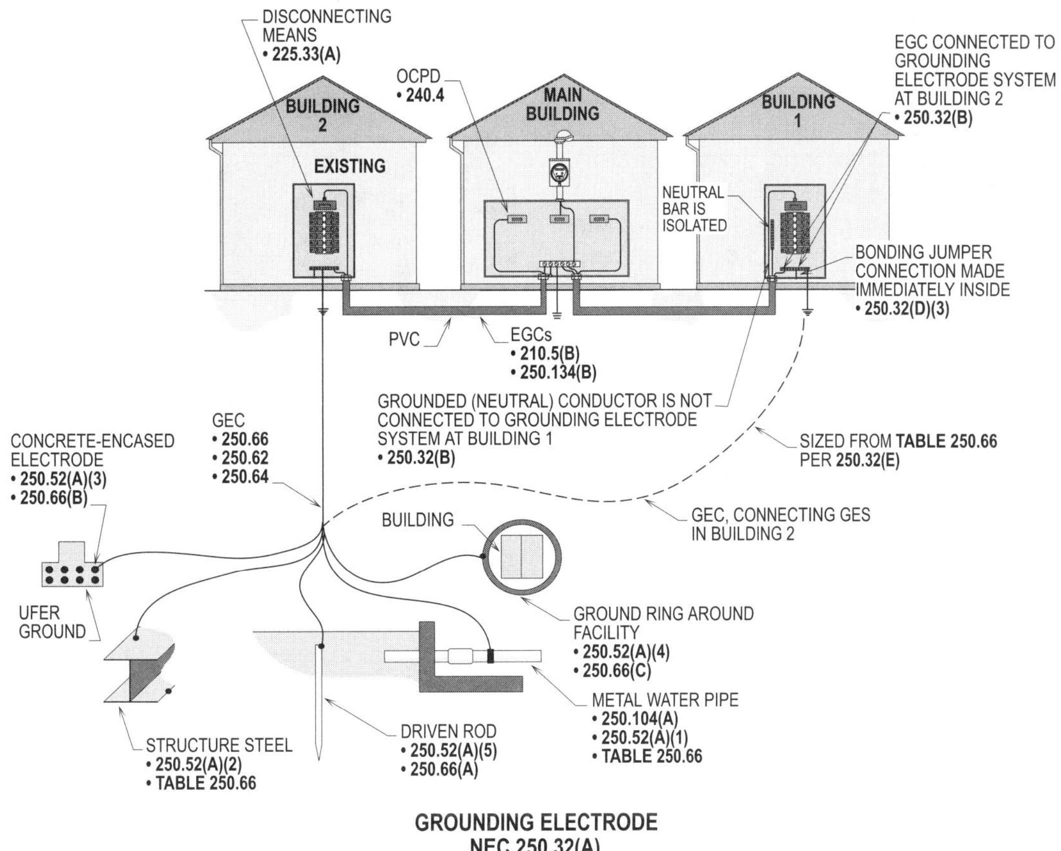

Figure 11-46. If any of the electrodes that are listed in **250.52(A)(1) through (A)(7)** are available, they shall be bonded together to form the grounding electrode system.

GROUNDING ELECTRODE
250.32(A)

Where two or more buildings or structures are supplied from one service by a feeder or branch circuit, any of the grounding electrodes that are listed in **250.52(A)(1) through (A)(7)** shall be bonded together to form the grounding electrode system, if one or all are available. **(See Figure 11-46)**

GROUNDING ELECTRODE – ONE BRANCH CIRCUIT
250.32(A), Ex.

A grounding electrode at a separate building or structure shall not be required to be installed where only one branch circuit, including a multiwire branch circuit, serves the building or structure and the branch circuit includes an equipment grounding conductor for grounding noncurrent-carrying parts of all equipment.

For further information, see **250.32(D)** for applying the requirements when installing the disconnecting means per **225.32, Ex.s 1** and **2**. **(See Figure 11-47)**

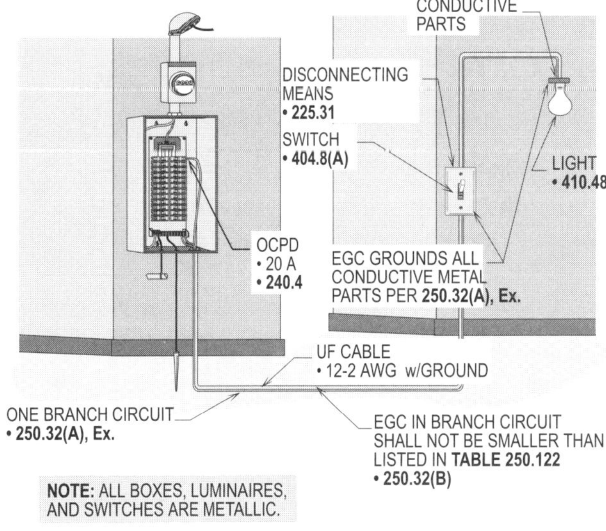

Figure 11-47. A grounding electrode shall not be required to be installed at a separate building or structure where only one branch circuit, including a multiwire branch circuit, is run and the branch circuit includes an equipment grounding conductor for grounding noncurrent-carrying parts of all equipment.

GROUNDED SYSTEMS
250.32(B)

Where a building or structure is supplied from a service in another building by more than one branch circuit, a grounding electrode shall be installed at the additional building(s) or structure(s) being served. The equipment grounding conductor run to the other building or structure shall be sized per **Table 250.122**.

Where livestock is housed, the equipment grounding conductor shall be insulated or covered where routed with the feeder or branch circuit and bonded to the metal case of the enclosure. Note that the neutral bus and grounded (neutral) conductor is isolated from the case. See **250.142(B)** and **408.40** for rules pertaining to this type of installation. **(See Figure 11-48)**

Design Tip: An individual equipment grounding conductor shall be installed to provide an effective return path for the fault current to travel over instead of using a grounded (neutral) conductor for such use. This particular installation is used to prevent the grounded (neutral) conductor and equipment grounding conductor from joining together at both ends, which if done, provides a parallel path for stray currents to travel over; this is not desirable for such an installation and causes problems for certain types of equipment.

For example: A 150 amp (75°C terminals) panelboard in building 1 is supplied by 3 - 1/0 AWG THWN copper conductors and the neutral calculated load is 48 amps. The feeder is protected by a 150 amp overcurrent protection device (75°C terminals). What size grounded (neutral) conductor and equipment grounding conductor is required using THWN copper conductors in the feeder?

Step 1: Finding neutral in feeder
215.2(A)(1); Table 310.16
48 A load requires 8 AWG cu.

Step 2: Finding EGC in feeder
250.122(A); Table 250.122
150 A OCPD requires 6 AWG cu.

Solution: **The size of the neutral is 8 AWG copper and the equipment grounding conductor is 6 AWG copper.**

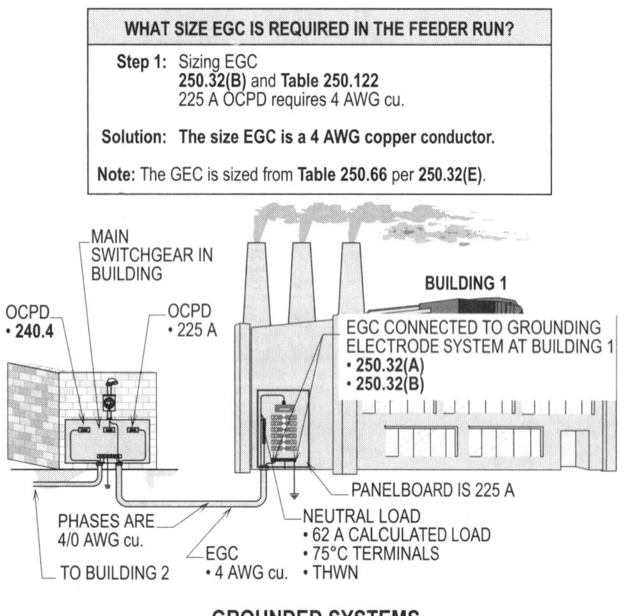

Figure 11-48. This type of installation is used to prevent the grounded (neutral) conductor and equipment grounding conductor from joining together at both ends, which if done, provides a parallel path for stray currents to travel over.

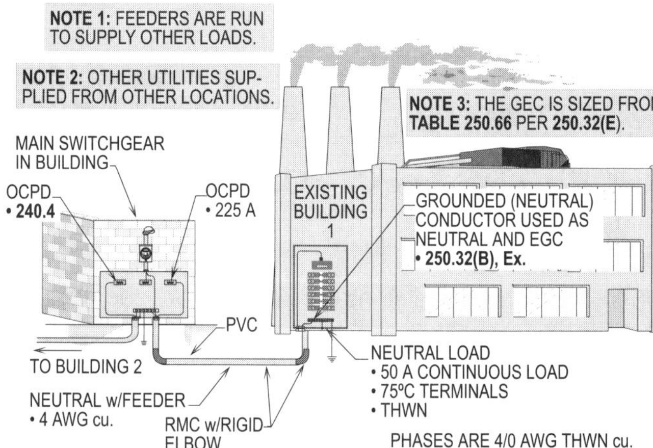

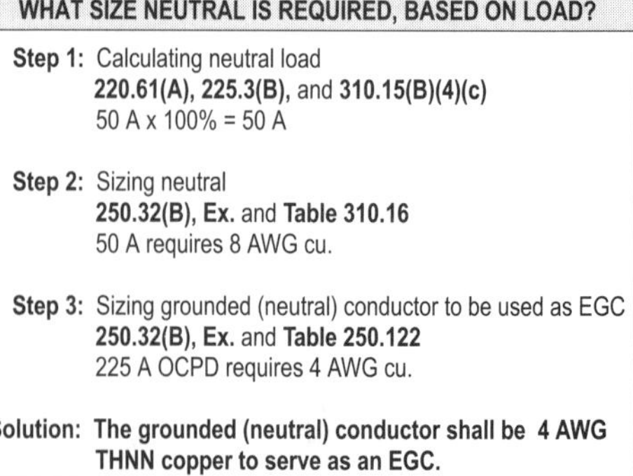

Figure 11-49. This illustration shows the proper method for sizing the grounded conductor to be used as an equipment grounding conductor.

GROUNDED CONDUCTOR
250.32(B), Ex.

For existing premises wiring systems only, the grounded (neutral) conductor shall be permitted to be connected to the building or structure disconnecting means and to the grounding electrodes where all of the following conditions are met:

- An equipment grounding conductor is not installed.
- There are no continuous metallic paths bonded to the grounding system.
- Ground fault protection of equipment has not been installed.

Where the grounded (neutral) conductor is used in this manner, the size shall not be smaller than the larger of either of the following:

- The sizing required per **220.61**.
- The sizing required per **250.122**.

See **Figure 11-49** for the proper method of sizing the grounded (neutral) conductor to be used as an equipment grounding conductor.

For example: A 225 amp (75°C terminals) panelboard in building 1 is supplied by 3 - 4/0 AWG THWN copper conductors and the neutral load is 50 amps. The feeder is protected by a 225 amp overcurrent protection device (75°C terminals). What size grounded (neutral) conductor is required to service as an equipment grounding conductor? (Loads are continuous)

Step 1:	Calculating neutral load
	220.61(A); 225.3(B); 310.15(B)(4)(c)
	50 A x 100% = 50 A

Step 2:	Sizing neutral
	250.32(B)(2); Table 310.16
	50 requires 8 AWG cu.

Step 3:	Sizing neutral to be used as EGC
	250.32(B)(2); Table 250.122
	225 A OCPD requires 4 AWG cu.

Solution: The grounded (neutral) conductor is required to be 4 AWG copper to serve as an EGC.

UNGROUNDED SYSTEMS
250.32(C)

A grounding electrode for an ungrounded system shall be connected only to the service equipment enclosure where installed at one or more buildings. The feeder or branch circuit shall be grounded to an electrode at the other building being served from the service of the main building. All equipment grounding conductors at the separate building shall be connected to a grounding bus that is bonded to the enclosure. Such enclosure is bonded and grounded to the grounding electrode conductor and grounding electrode system.

PORTABLE AND VEHICLE-MOUNTED GENERATORS
250.34

Portable is defined as equipment that is easily carried from one location to another. Mobile equipment is capable of being moved as on wheels or rollers, as when mounted on a vehicle or placed on a trailer. Under certain conditions of use, the frame of a portable generator shall not be required to be connected to ground, such as to a ground rod, water pipe, structural steel, etc.

PORTABLE GENERATORS
250.34(A)

The frame of a portable generator shall not be required to be connected to a grounding electrode as derived in **250.52** if it supplies only the equipment on the generator or cord-and-plug connected equipment connected to receptacles mounted on the generator, provided all the following conditions are complied with:

- An equipment grounding conductor is installed to bond the receptacles to the frame of the generator.
- The equipment grounding conductor in the cord is installed to bond the exposed normally non-current-carrying metal parts of the equipment connected to the frame of the generator.

See Figure 11-50 for a detailed illustration of when to apply these requirements.

VEHICLE-MOUNTED GENERATORS
250.34(B)

The frame of the vehicle shall not be required to be connected a grounding electrode as outlined in **250.52** for a system supplied by a generator on the vehicle, provided:

- The generator frame is bonded to the vehicle frame.
- The generator supplies only equipment mounted to the vehicle, or the generator supplies cord-and-plug connected equipment through receptacles mounted on the vehicle, or both equipment located on the vehicle and cord-and-plug connected equipment through receptacles mounted on the vehicle or on the generator.

- Exposed metal parts of the equipment served are bonded to the generator frame either directly or through the receptacles.

See Figure 11-51 for a detailed illustration when applying these requirements

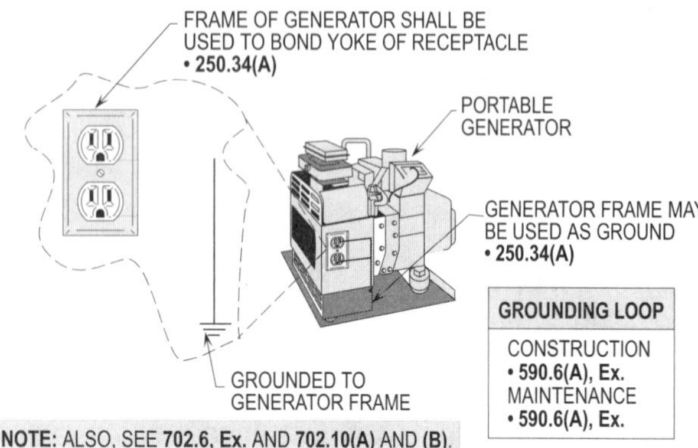

Figure 11-50. The frame of a portable generator shall not be required to be connected to a grounding electrode as specified in **250.52** if it supplies only the equipment on the generator or cord-and-plug connected equipment connected to receptacles mounted on the generator.

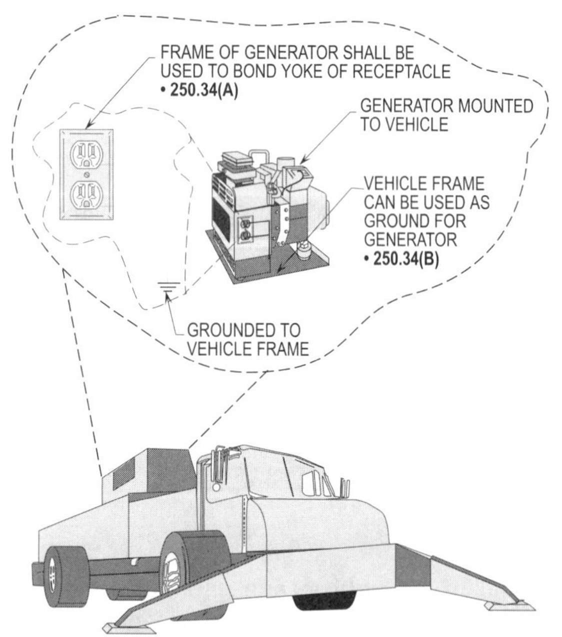

Figure 11-51. The frame of a generator shall not be required to be connected to a grounding electrode as specified in **250.52** if it supplies only the equipment supplied from the generator or cord-and-plug connected equipment connected to receptacles mounted and bonded to the frame of the vehicle.

HIGH-IMPEDANCE GROUNDING 250.36

High-impedance grounding is accomplished by inserting a resistance or reactance between the grounding electrode conductor and the grounded (neutral) conductor. Ground fault current is limited to a safe value that will not cause excessive damage to components and equipment should such a condition occur. Under ground fault conditions, ground-fault current may be monitored using ground fault relays and trip the system overcurrent protection devices open only if necessary. Ungrounded systems generally have high-impedance grounding installed so the system has a reference to ground without deenergizing the circuits, if a ground fault occurs on one of the ungrounded (phase) conductors.

GROUNDING ELECTRODE SYSTEM 250.50

All grounding electrodes that are present at each building or structure served, the following (first six) electrodes shall be bonded together to form the grounding electrode system:

- Metal underground water pipe
- Metal frame of the building or structure
- Concrete-encased electrode
- Ground ring
- Rod and pipe electrodes
- Plate electrodes

If one of the above electrodes is not available, one or more the following electrodes shall be installed and used:

- Ground ring
- Rod and pipe electrodes
- Plate electrodes
- Other local metal underground systems or structures

See Figures 11-52(a) and (b) for a detailed illustration when applying these requirements.

ELECTRODES PERMITTED FOR GROUNDING 250.52

The following types of electrodes shall be permitted for grounding:

- Metal underground water pipe
- Metal frame of the building or structure
- Concrete-encased electrode
- Ground ring

- Rod and pipe electrodes
- Plate electrodes
- Other local metal underground systems or structures

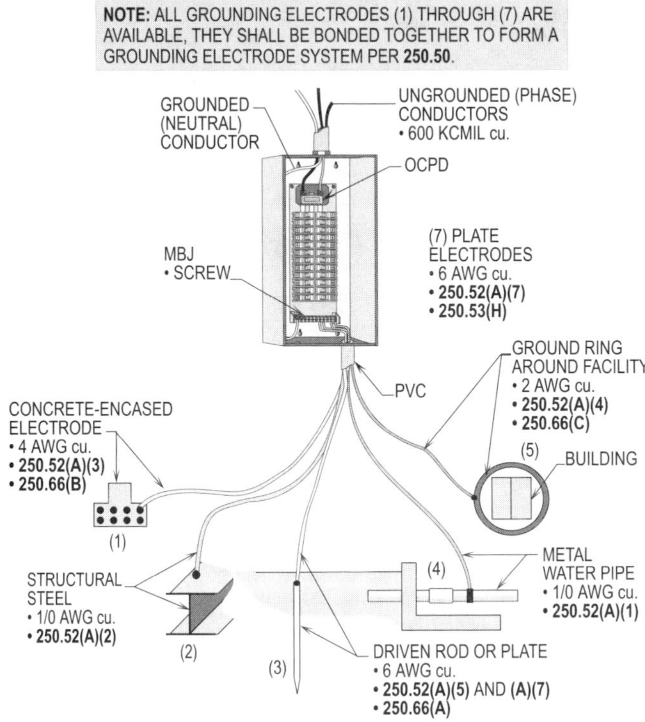

Figure 11-52(a). This illustration shows the types of grounding electrodes that shall be bonded together to form the grounding electrode system if available.

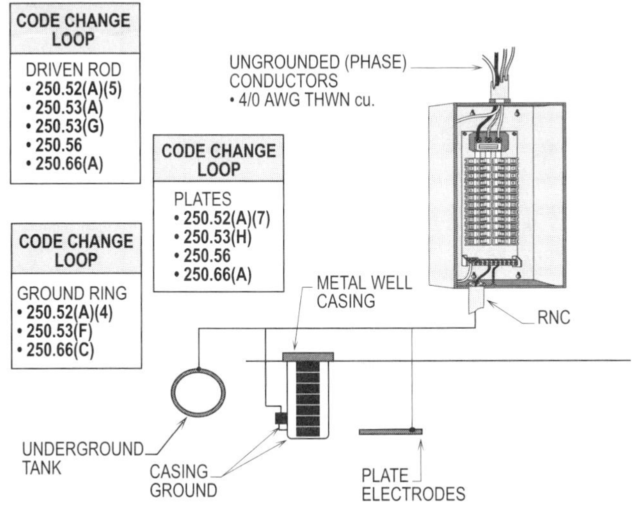

Figure 11-52(b). This illustration shows the types of grounding electrodes that shall be bonded together to form the grounding electrode system if available.

METAL UNDERGROUND WATER PIPE 250.52(A)(1)

Metal water pipe with lengths of at least 10 ft (3 m) long in the earth shall be connected to the grounded (neutral) bar in the service equipment enclosure. The grounded (neutral) bar shall be bonded to the service equipment enclosure, the grounded (neutral) conductor, the grounded (neutral) conductors, and the equipment grounding conductors by a bonding jumper sized per **Table 250.66**. Metal water piping shall be electrically continuous or made electrically continuous by bonding around insulated joints or sections.

The grounding connection to the metal water pipe shall be made at a point no more than 5 ft (1.52 m) where the piping enters a building. It is recommended that the grounding conductor from the supplementary driven ground rod be terminated to the common grounded bar in the service equipment panel. **(See Figure 11-53)**

INDUSTRIAL, COMMERCIAL, AND INSTITUTIONAL BUILDINGS 250.52(A)(1), Ex.

The grounding electrode conductor shall be permitted to be installed further than 5 ft (1.52 m) in industrial, commercial, and institutional buildings where conditions of maintenance and supervision ensure that only qualified personnel will service the installation and the entire length of the interior metal water pipe that is being used for the conductor that is exposed. Under these conditions, such metal water pipes shall be permitted to be used as grounding electrode conductors to ensure continuity. **(See Figure 11-54)**

METAL FRAME OF THE BUILDING OR STRUCTURE 250.52(A)(2)

The metal frame of the building or structure shall be permitted to be used a grounding electrode where any of the following methods are used to a earth connection:

- A single structural metal member has 10 ft (3 m) or more in direct contact with the earth or encased in concrete that is in direct contact with the earth.
- The structural metal frame is bonded to one or more of the grounding electrodes found in **250.52(A)(3)** or **(A)(4)**.
- The structural metal frame is bonded to one or more of the grounding electrodes in **250.52(A)(5)** or **(A)(7)** that comply with **250.56**.
- Other approved means of establishing a connection to earth.

See Figure 11-55 for a detailed illustration of structural steel used as a grounding electrode.

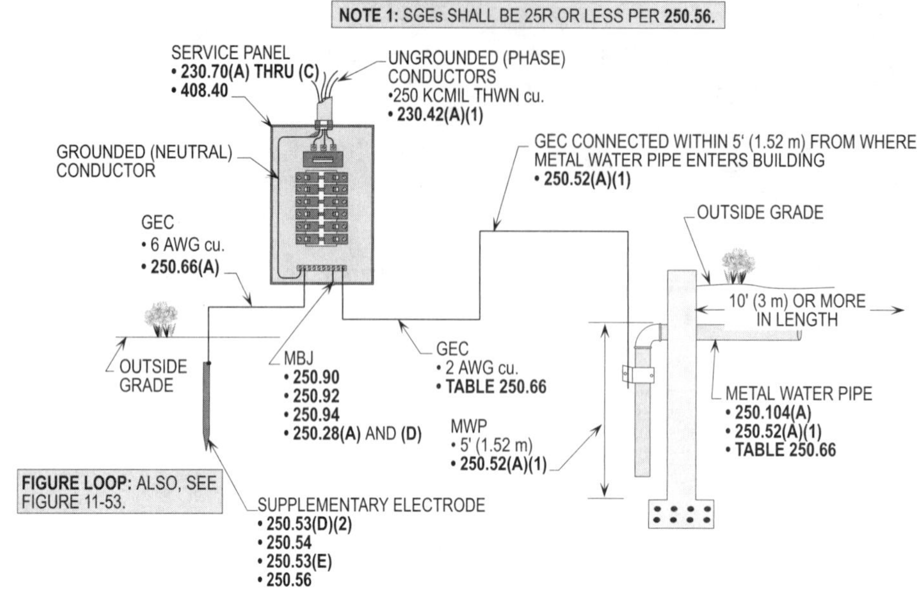

Figure 11-53. In most all installation (existing), the grounding electrode conductor connects the service equipment grounded (neutral) conductor to a copper tubing or metal water piping system or a driven rod or all other electrodes, if available.

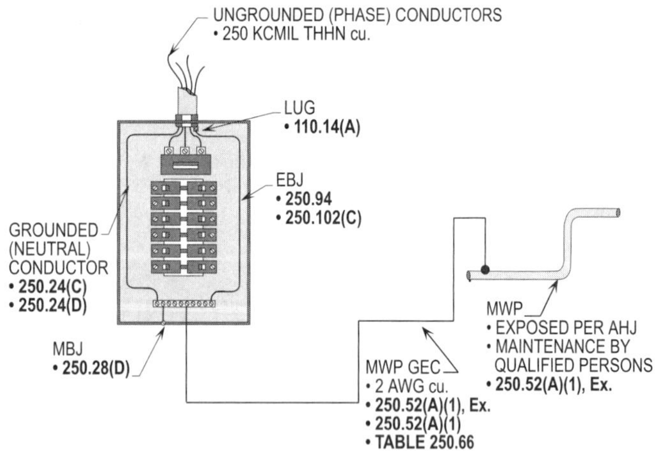

Figure 11-54. In industrial, commercial, and institutional buildings, where the metal water pipe is exposed and maintained by qualified personnel, the grounding electrode conductor shall be permitted to be connected at any accessible location.

CONCRETE-ENCASED ELECTRODE
250.52(A)(3)

Concrete-encased electrodes consist of 1/2 in. x 20 ft (13 mm x 6 m) lengths of reinforcing rebar located in the foundation. The 20 ft (6 m) length of rebar shall be permitted to be in one continuous piece or in many pieces spliced together to form a 20 ft (6 m) or more continuous length. The metal reinforcing rebars shall be of the conductive type. A concrete-encased electrode shall also be permitted to be a 4 AWG bare copper conductor at least 20 ft (6 m) long that is installed in the footing of the foundation. The 20 ft (6 m) long 4 AWG conductor shall be located in at least 2 in. (50 mm) of concrete located horizontally near the bottom or vertically, and with that portion of the foundation or footing. The reinforcing rebars or 4 AWG bare copper conductor installed properly usually provides a resistance of about 3 to 5 ohms. **(See Figure 11-56)**

GROUND RING
250.52(A)(4)

A bare copper conductor not smaller than 2 AWG and at least 20 ft (6 m) shall be permitted to be installed in the earth to form a ground ring grounding system that encircles the building or structure. **(See Figure 11-57)**

Design Tip: A more reliable and dependable ground may be obtained by burying the conductor about 3 ft (900 mm) from the building and periodically driving ground rods and attaching them to the bare conductor by listed connecting methods.

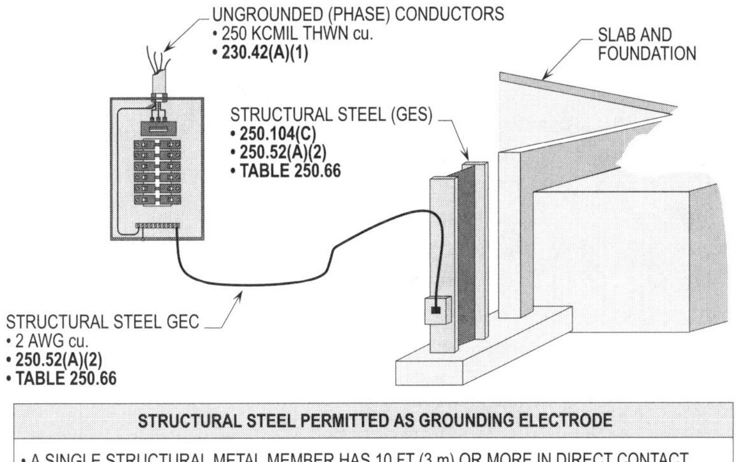

Figure 11-55. This illustration shows structural building steel used as a grounding electrode.

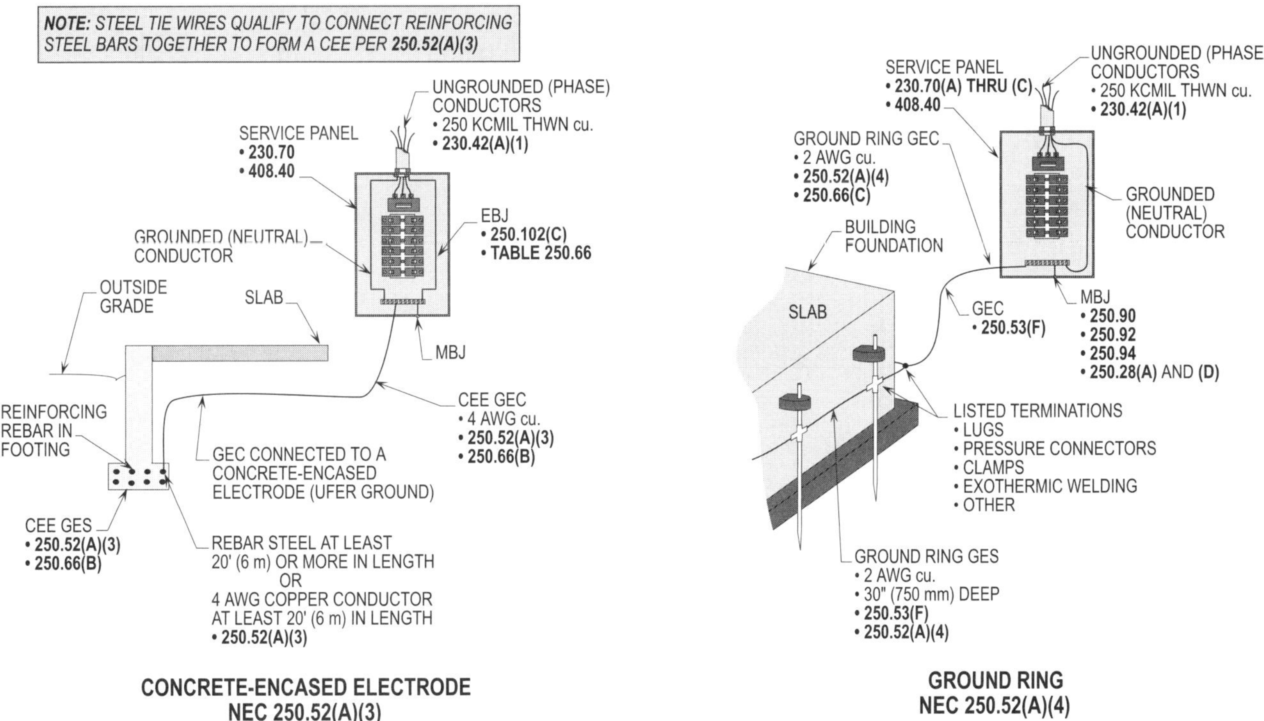

Figure 11-56. Concrete-encased electrodes consist of 1/2 in. x 20 ft (13 mm x 6 m) lengths of reinforcing rebar located in the foundation. A concrete-encased electrode shall be permitted to be a 4 AWG bare copper conductor at least 20 ft (6 m) long, which shall be installed in the footing of the foundation.

Figure 11-57. A bare copper conductor not smaller than 2 AWG and at least 20 ft (6 m) shall be permitted to be installed in the earth to form a ground ring grounding system that encircles the building or structure.

ROD AND PIPE ELECTRODES
250.52(A)(5)

A driven electrode used to ground the service shall be permitted to be a 1/2 in. x 8 ft (13 mm x 2.5 m) copper (brass) rod or a 3/4 in. x 10 ft (21 mm x 3 m) galvanized pipe. A copper rod is usually the type of driven electrode that is installed. Driven rods are utilized to connect the grounded (neutral) bar terminal to earth ground. Where the water pipe system in the ground is nonmetallic (PVC) and converts to metal pipe or copper tubing above ground, a driven rod shall be permitted to be used to bond and ground the metal water pipe system and connect the electrical system to earth ground. **(See Figure 11-58)**

The metal water pipe system shall be bonded into the grounding electrode system, whether or not the piping in the ground is nonmetallic (PVC) or metal. The following are acceptable made electrodes that shall be permitted to be used:

- 1/2 in. x 8 ft (13 mm x 2.5 m) copper rod
- 3/4 in. x 10 ft (21 mm x 3 m) metal pipe or conduit
- 5/8 in. x 8 ft (15.87 mm x 2.5 m) rebar
- 5/8 in. (16 mm) or larger stainless steel rod or 1/2 in. x 8 ft (13 mm x 2.5 m) if listed for such use

PLATE ELECTRODES
250.52(A)(7)

Plate electrodes shall be a minimum of 2 sq. ft (0.186 sq. m), and if made of iron or steel they shall be at least 1/4 in. (6.4 mm) thick. If made of a nonferrous metal such as copper, they shall be 0.06 in. (1.5 mm) in thickness. **(See Figure 11-58)**

OTHER LOCAL METAL UNDERGROUND SYSTEMS OR STRUCTURES
250.52(A)(8) AND FIGURE 11-52(b)

Other local metal underground systems and structures such as piping systems, underground tanks, and underground metal well casings that are efficiently bonded to a metal water pipe.

GROUNDING ELECTRODE SYSTEM INSTALLATION
250.53

To complete the grounding electrode system, the following grounding electrodes have specific installation requirements applied:

- Rod, pipe, and plate electrodes
- Electrode spacing
- Bonding jumper
- Metal underground water pipe
- Supplemental electrode bonding connection size
- Ground ring
- Rod and pipe electrodes
- Plate electrode

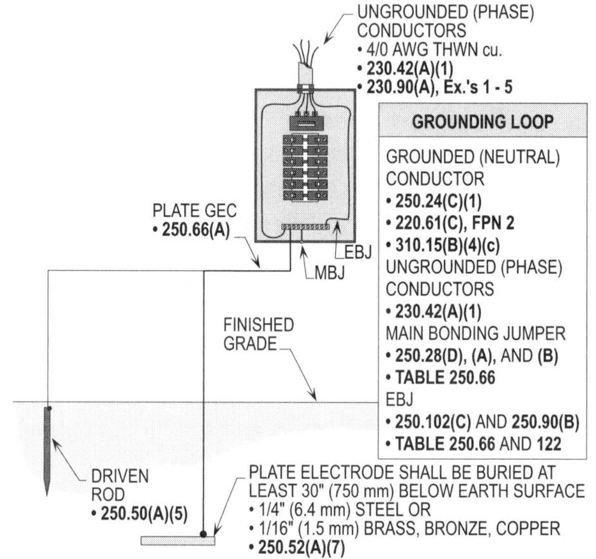

Figure 11-58. Rod and plate electrodes shall be installed when none of the electrodes found in **250.52(A)(1) through (A)(4)** are available.

ROD, PIPE, AND PLATE ELECTRODES
250.53(A)

Rod, pipe, and plate electrodes shall be driven 8 ft (2.5 m) into the earth, and plate electrodes shall be embedded below the permanent moisture level where practicable. Rod, pipe, and plate electrodes shall be free from nonconductive coatings such as paint and enamel.

ELECTRODE SPACING
250.53(B)

Where more than one rod, pipe, or plate electrode is used, each electrode of one grounding system (including that used for air terminals) shall not be less than 6 ft (1.83 m) from any other electrode of another grounding system. Two or more grounding electrodes that are effectively bonded together shall be considered a single grounding electrode system.

BONDING JUMPER
250.53(C)

Where bonding jumper(s) are used to connect the grounding electrodes together to form the grounding electrode system, the bonding jumper(s) shall be installed in accordance with **250.64(A), (B),** and **(E)**; they shall be sized in accordance with **250.66** and shall be connected in accordance with **250.70**.

METAL UNDERGROUND WATER PIPE
250.53(D)

The continuity of the grounding path for the bonding connection to the interior metal water piping shall not be permitted to rely on water meters, filtering devices, or similar equipment. **(See Figure 11-59)**

> **Design Tip:** Many rural areas do not have water meters but may have water filtering or softening equipment that could be removed. Such removal would break or impair the grounding path or bonding connection of the metal interior water piping. Proper bonding and joining together of the metal piping can be accomplished by using correctly sized bonding jumpers.

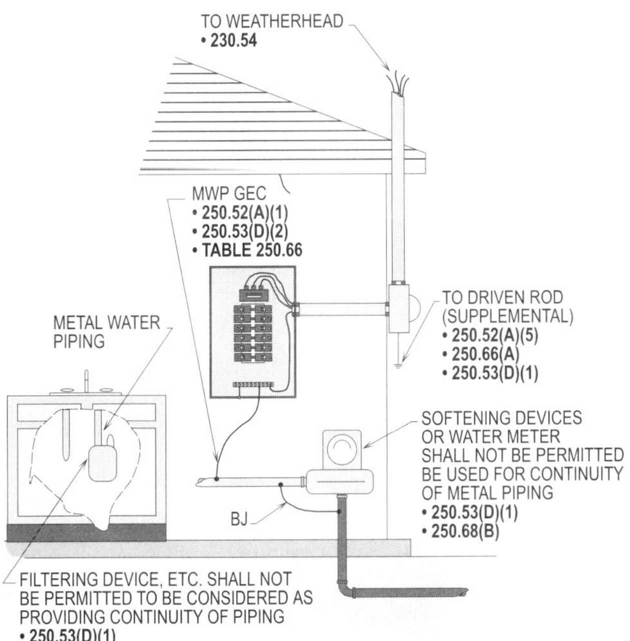

Figure 11-59. This illustration makes it clear that continuity of the grounding path for the bonding connection to the interior metal water piping shall not be permitted to rely on water meters, filtering devices, or similar equipment.

The copper tubing or metal water pipe in a new installation may be installed in the earth or above grade. Either installation requires the copper or metal water pipe to be supplemented by an additional electrode. The additional grounding electrode to supplement the metal water pipe shall be permitted to be any of the electrodes listed in **250.52(A)(2) through (A)(8)**. **(See Figure 11-60)**

SUPPLEMENTAL ELECTRODE BONDING CONNECTION SIZE
250.53(E)

Where a rod, pipe, or plate electrode is used as the supplemental electrode, that portion of the bonding jumper that is the sole connection to the supplemental grounding electrode shall not be required to be larger than 6 AWG copper wire or 4 AWG aluminum wire.

GROUND RING
250.53(F)

Where a ground ring is used as the grounding electrode, the ground ring shall be buried not less than 30 in. (750 mm) below the earth's surface.

ROD AND PIPE ELECTRODES
250.53(G)

Rod and pipe electrodes shall be installed where there is at least 8 ft (2.44 m) of length in contact with the soil. Where a rock bottom is encountered, the electrode shall be permitted to be driven at an angle not to exceed 45 degrees from the vertical or shall be permitted to be buried at least 30 in. (750 mm) deep. **(See Figure 11-61)**

PLATE ELECTRODE
250.53(H)

Where a plate electrode is used as the grounding electrode, the plate electrode shall be buried not less than 30 in. (750 mm) below the earth's surface.

AUXILIARY GROUNDING ELECTRODES
250.54

Auxiliary grounding electrodes (one or more) shall be permitted to be installed to augment the equipment grounding conductors specified in **250.118** and shall not be required to comply with the electrode bonding requirements of **250.50** or **250.53(C)** or the resistance requirements of **250.56**, but the earth shall not be permitted to be used as an effective ground fault current path as specified in **250.4(A)(5)** and **250.4(B)(4)**. **(See Figure 11-62)**

METAL UNDERGROUND WATER PIPE
NEC 250.53(D)

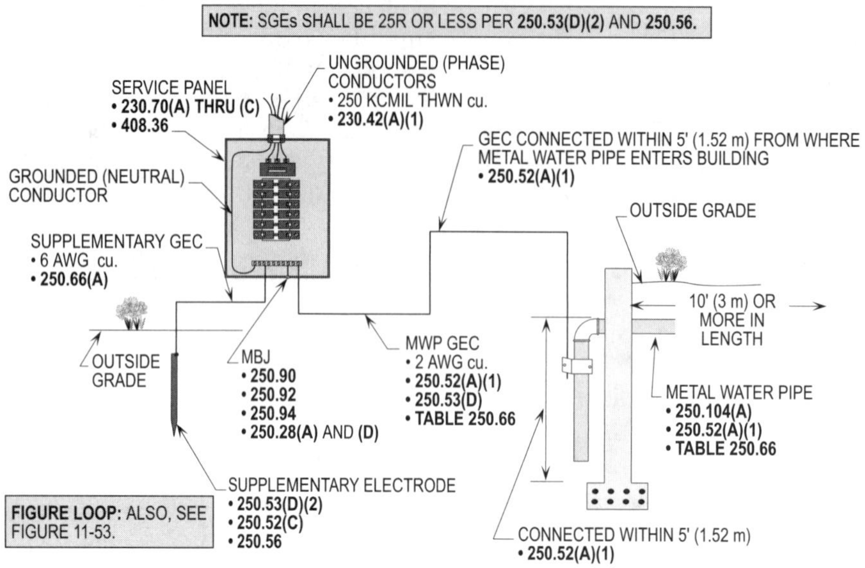

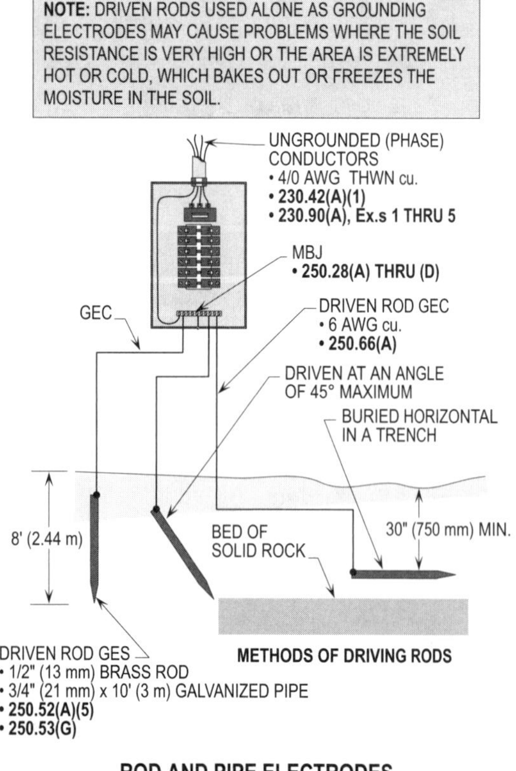

Figure 11-60. In most all installations (existing), the grounding electrode conductor connects the service equipment grounded (neutral) conductor to a copper tubing or a metal water pipe system, which is supplemented by a driven rod or other electrodes (shall be permitted for this purpose) if available.

ROD AND PIPE ELECTRODES
NEC 250.53(G)

Figure 11-61. This illustration shows the methods by which rod and pipe electrodes are to be installed.

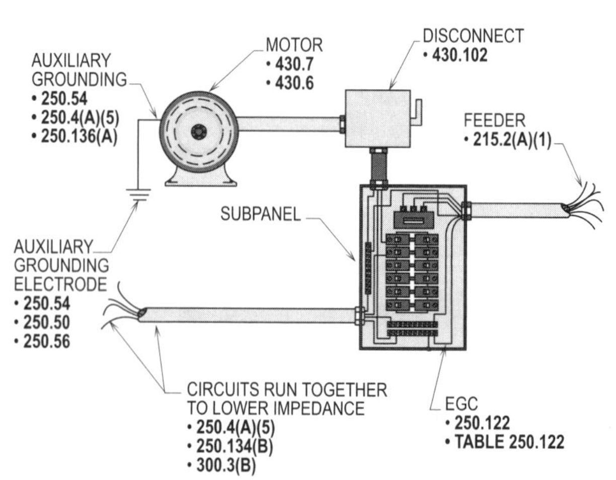

Design Tip: This, for instance, requires an equipment grounding conductor to be used because the earth resistance in almost every case is too high to properly allow overcurrent protection devices to operate when a ground fault occurs.

AUXILIARY GROUNDING ELECTRODES
250.54

Figure 11-62. Auxiliary grounding electrodes (one or more) shall be permitted to be used to supplement the equipment grounding conductor but are never to be used as the sole grounding means.

RESISTANCE OF ROD, PIPE, AND PLATE ELECTRODES
250.56

Rod, pipe, and plate electrodes shall have a resistance to ground of 25 ohms or less, wherever practicable. When the resistance is greater than 25 ohms, two or more electrodes shall be permitted to be connected in parallel or extended to a greater length. Note that such an electrode or supplementary grounding electrode that measures more than 25 ohms shall be augmented by one additional electrode of a type permitted by **250.52(A)(4) through (A)(8)**.

Continuous metal water piping systems usually have a ground resistance of less than 3 ohms. Metal frames of buildings normally have a good ground and usually have a resistance of less than 25 ohms. As pointed out in **250.52(A)(2)**, the metal frame of a building, if effectively grounded, shall be permitted to be used as an electrode. Local metallic water systems and well casings make good grounding electrodes in most all types of installations. **[See Figure 11-63(a), (b), and (c)]**

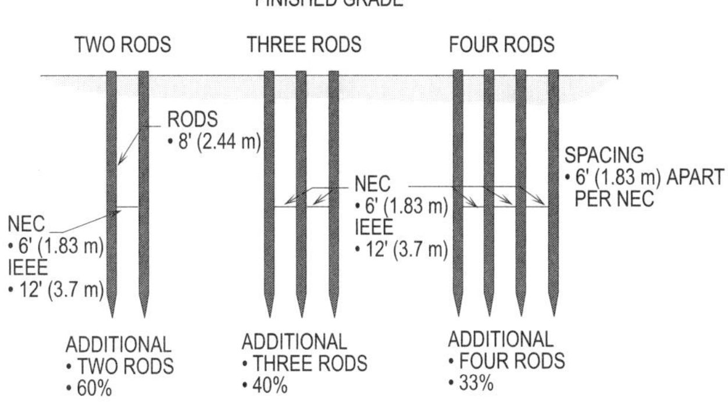

Figure 11-63(b). Rod, pipe, and plate electrodes shall have a resistance to ground of 25 ohms or less, wherever practicable. When the resistance is greater than 25 ohms, two or more electrodes shall be permitted to be connected in parallel or extended to a greater length.

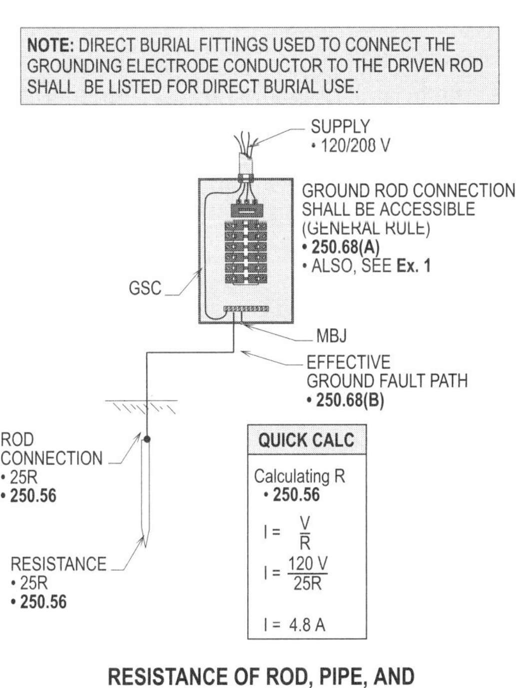

Figure 11-63(a). This illustration shows the procedure for calculating the amps to a driven rod, based on 25 ohms.

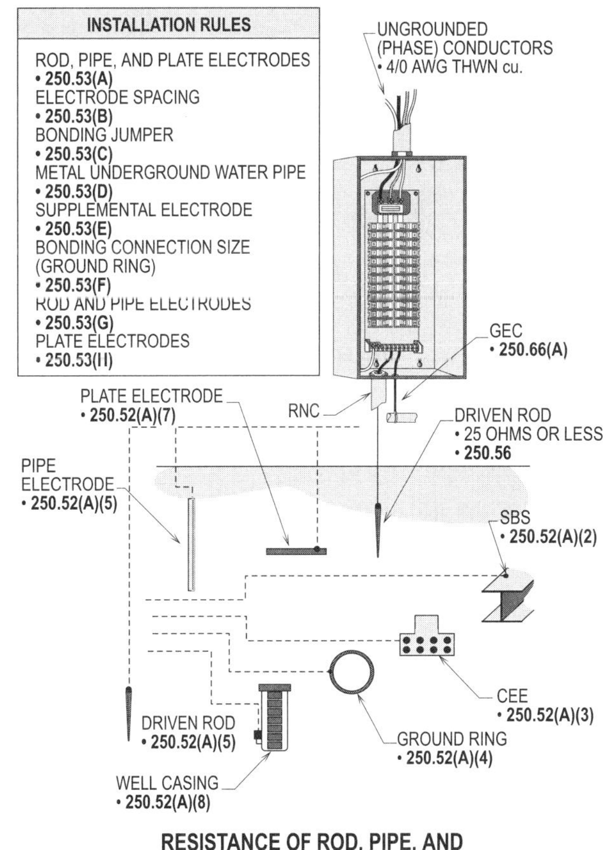

Figure 11-63(c). If a resistance of 25 ohms is not obtained, a rod, pipe, or plate electrode or supplementary grounding electrode shall be augmented by one additional electrode per **250.52(A)(4) through (A)(8)**.

SOIL TREATMENT

When rod, pipe, or plate electrodes are used, grounding may be greatly improved by the use of chemicals, such as magnesium sulfate, copper sulphate, or rock salt. A trench, a basin, or a doughnut-type hole system may be dug around the ground rod, into which the chemicals are placed. Rain and snow will dissolve the chemicals and allow them to penetrate the soil and saturate the rod, providing lower resistance. **(See Figure 11-64)**

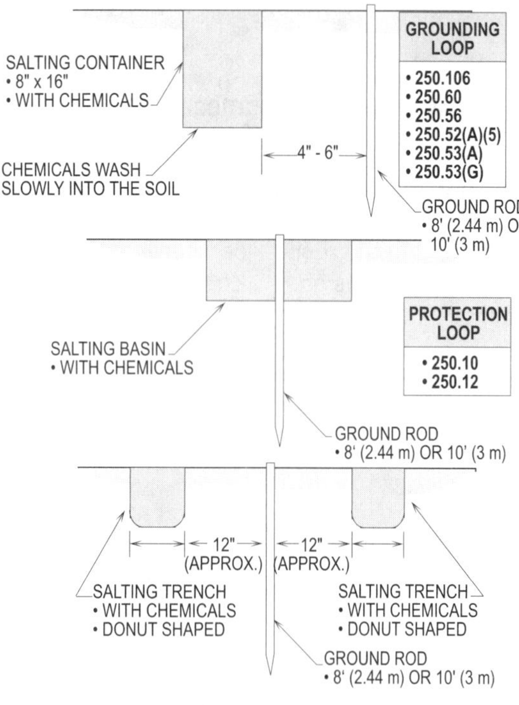

Figure 11-64. When made electrodes are used, grounding may be greatly improved by the use of chemicals, such as magnesium sulfate, copper sulphate, or rock salt.

USE OF AIR TERMINALS
250.60

Lightning down conductors, and rod, pipe, and plate electrodes that are connected to lightning rods for grounding, shall not be permitted to be used in place of electrodes for grounding electrical wiring systems and equipment. However, they shall be permitted to be bonded together to limit the difference of potential that might appear between them. See **250.60** and **250.106**. **(See Figure 11-65)**

See **250.106, 800.100(D), 810.21(J),** and **820.100(D)** for required bonding and grounding of other interconnected systems. **(See Figure 11-66)**

Design Tip: Difference of potential between two different grounding systems on the same building, if tied together, will reduce the potential difference between them.

For example, lightning rods and the electrical system grounding electrodes are bonded together for this reason.

GROUNDING ELECTRODE CONDUCTOR MATERIAL
250.62

The grounding electrode conductor shall be of copper, aluminum, or copper-clad aluminum and shall be installed as solid or stranded, insulated, covered, or bare. The grounding electrode conductor selected shall be resistant to any corrosive condition existing at the installation or be suitably protected against corrosion. **(See Figure 11-67)**

GROUNDING ELECTRODE CONDUCTOR INSTALLATION
250.64

The following wiring methods shall be permitted to be utilized for bonding, grounding, and enclosing grounding electrode conductors:

- Aluminum or copper-clad aluminum conductors
- Securing and protection against physical damage
- Continuous
- Service with multiple disconnecting means enclosures
- Enclosures for grounding electrode conductors
- Installation to electrode(s)

See Figure 11-68 for a detailed illustration when applying these requirements.

ALUMINUM OR COPPER-CLAD ALUMINUM CONDUCTORS
250.64(A)

Bare or aluminum or copper-clad aluminum conductors shall not be permitted to be used in the following locations:

- Where in direct contact with masonry or the earth
- Where subject to corrosive conditions

Aluminum or copper-clad aluminum grounding conductors shall not be permitted to be installed within 18 in. (450 mm) of the earth where such conductors are installed outside.

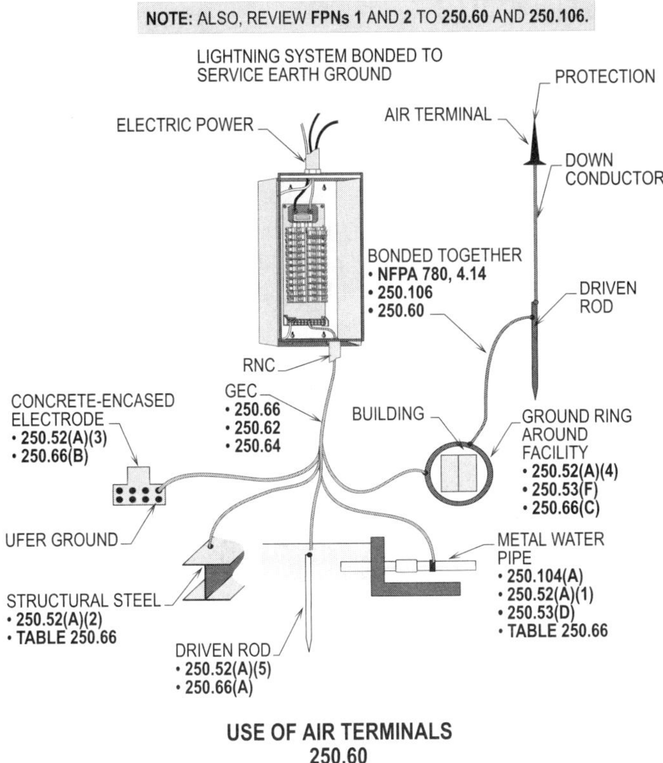

Figure 11-65. This illustration shows the requirements for installing and grounding lightning protection systems.

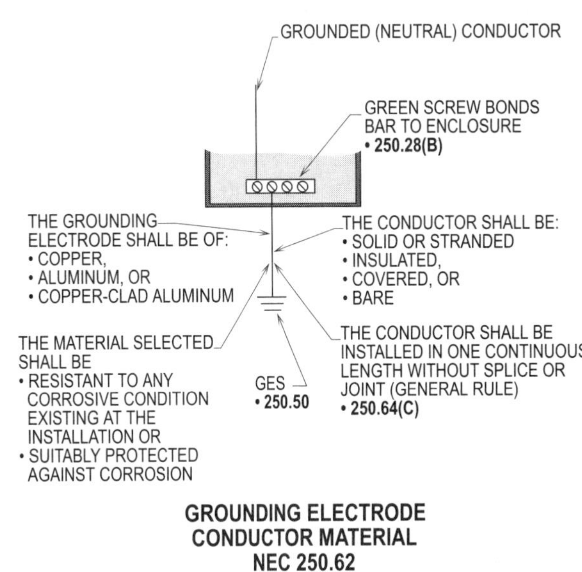

Figure 11-67. This illustration shows the materials that the grounding electrode may be made of and that may be used to ground the service equipment to a grounding electrode.

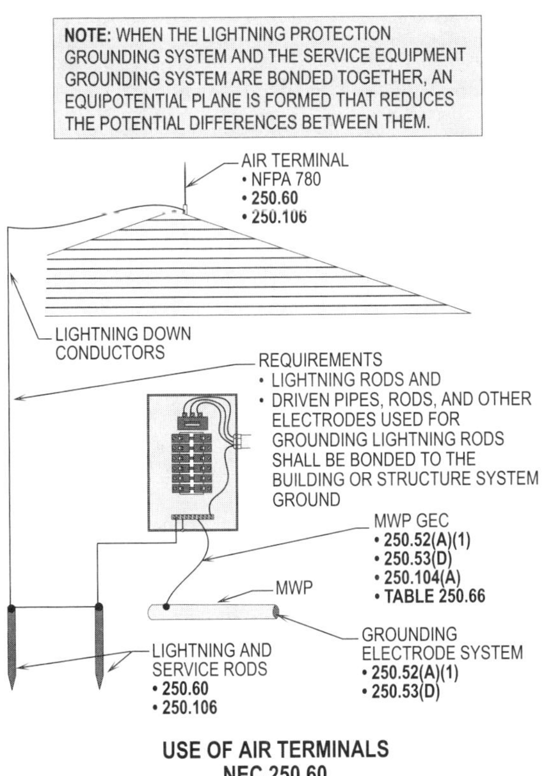

Figure 11-66. This illustration shows the requirements for installing and grounding lightning protection systems.

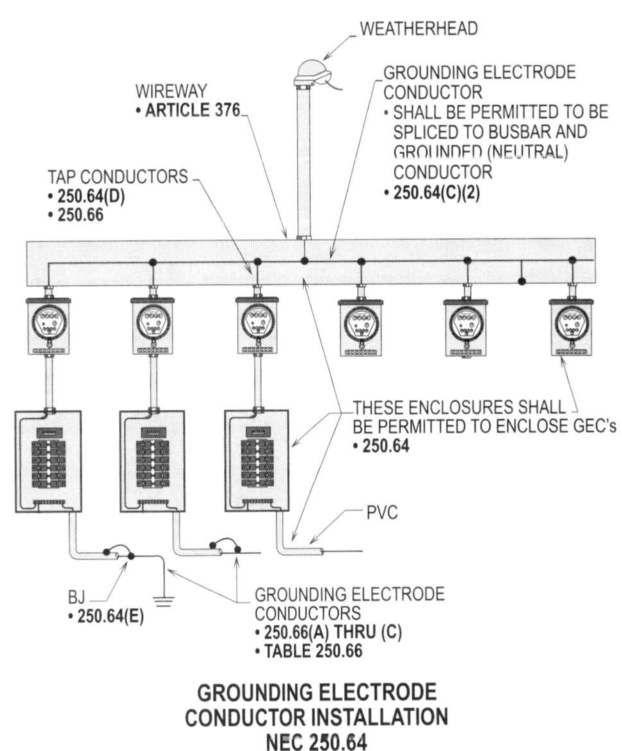

Figure 11-68. These enclosures shall be permitted to enclose grounding electrode conductors and also serve as a grounding and bonding means.

SECURING AND PROTECTION AGAINST PHYSICAL DAMAGE
250.64(B)

The following requirements shall be used to select and install the grounding electrode conductor that is used to connect the service enclosure to the grounding electrode system:

- Grounding electrode conductors 4 AWG or larger shall be securely fastened to the surface. Grounding electrode conductors not exposed to physical damage shall not require additional protection.

- A 6 AWG grounding electrode conductor shall be permitted to be run along the building surface construction without metal covering if not exposed to physical damage. Rigid metal conduit, intermediate metal conduit, PVC (schedule 80), electrical metallic tubing, or cable armor shall be permitted to be used where the grounding electrode conductor is exposed to physical damage.

- Rigid metal conduit, intermediate metal conduit, PVC (schedule 80), electrical metallic tubing, or cable armor shall be permitted to be used for installing a grounding electrode conductor smaller than 6 AWG. In most cases, a 8 AWG grounding electrode conductor is required to be installed per **Table 250.66** since there is no overcurrent protection device ahead of service conductors.

See Figure 11-69 for a detailed illustration pertaining to these requirements.

CONTINUOUS
250.64(C)

The grounding electrode conductor shall be installed in one continuous length without a splice or joint, except as permitted in the following:

- Splicing shall be permitted only by irreversible compression-type connectors listed for the purpose or by the exothermic welding process. **(See Figure 11-70)**

- A grounding electrode conductor shall be permitted to be spliced in busbars if they are located as meter enclosures, etc. **(See Figure 11-71)**

SERVICE WITH MULTIPLE DISCONNECTING MEANS ENCLOSURES
250.64(D)

Where a service consists of more than a single enclosure as permitted in **230.71(A)**, the following grounding electrode connections shall be made:

- Grounding electrode conductor taps
- Individual grounding electrode conductors
- Common location

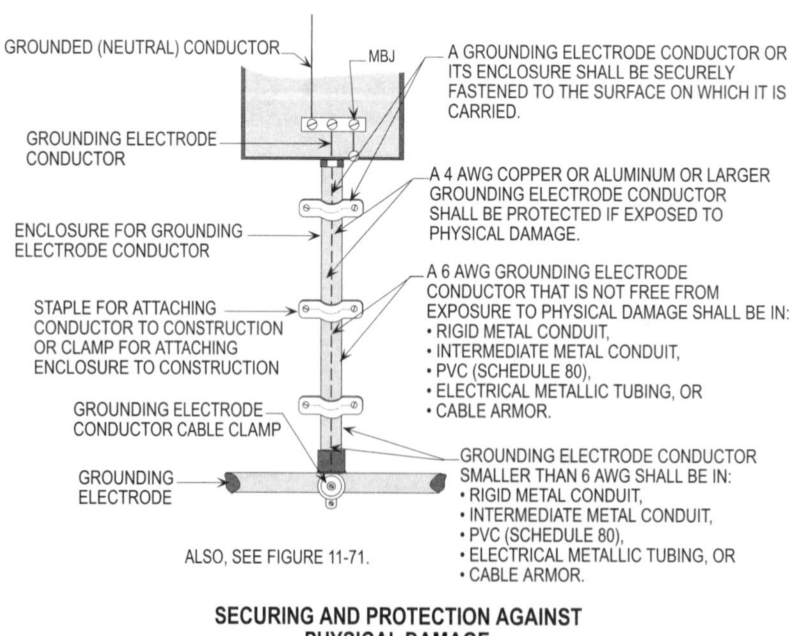

Figure 11-69. This illustration shows the requirements for installing and protecting the grounding electrode conductor with specific wiring methods.

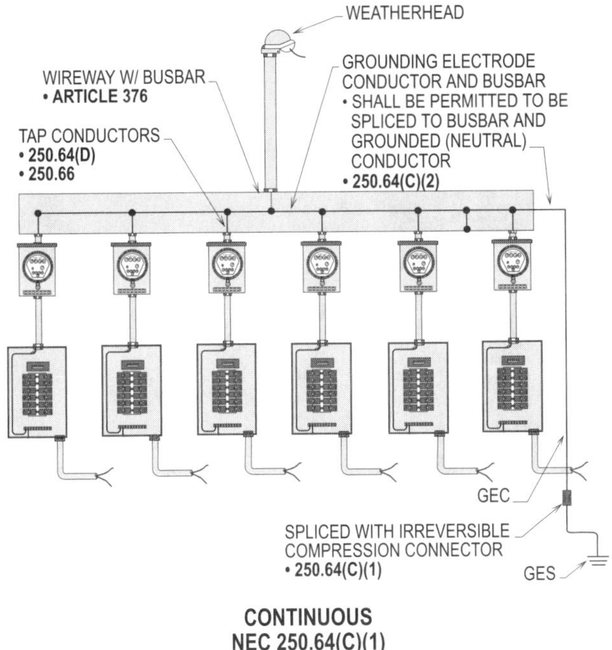

Figure 11-70. A grounding electrode conductor shall be permitted to be spliced in busbars located in meters, gutter, enclosures, or wireways, or with irreversible compression connectors.

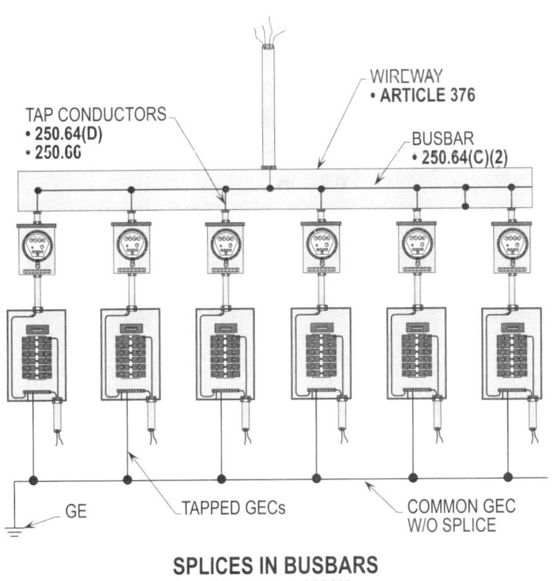

Figure 11-71. A grounding electrode conductor shall be permitted to be connected to taps in a service consisting of more than a single enclosure. The tap conductors shall be connected to the grounding electrode conductor in such a manner that the grounding electrode conductor remains without a splice or joint.

GROUNDING ELECTRODE CONDUCTOR TAPS
250.64(D)(1)

A common grounding electrode conductor and grounding electrode conductor taps shall be installed where the service is installed per **230.40, Ex. 2**. The common grounding electrode conductor shall be sized per **Table 250.66**, based on the largest ungrounded (phase) service-entrance conductor(s). The common grounding electrode conductor shall be sized per **Table 250.66, Note 1**, if the service-entrance conductors connect directly to a service drop or service lateral. The grounding electrode conductor taps shall be sized per **250.66**, based on the largest ungrounded (phase) conductor serving that enclosure. The tap conductor shall be connected to the common grounding electrode conductor by exothermic welding or with connectors listed as grounding and bonding equipment. There shall be no splice or joint to the common grounding electrode conductor. **(See Figure 11-72)**

INDIVIDUAL GROUNDING ELECTRODE CONDUCTORS
250.64(D)(2)

An individual grounding electrode conductor shall be connected between the grounded (neutral) conductor in each service equipment disconnecting means enclosure and the grounding electrode system. Each individual grounding electrode conductor shall be sized per **250.66**, based on the ungrounded (phase) service-entrance conductors supplying each individual disconnecting means. **(See Figure 11-72)**

COMMON LOCATION
250.64(D)(3)

A wireway or other acceptable enclosure on the supply side of the service disconnecting means shall be used for a grounding electrode conductor connected to the grounded service conductors. The grounding electrode conductor shall be sized per **250.66**, based on the ungrounded (phase) service-entrance conductors at the common location where the connection is made. **(See Figure 11-72)**

ENCLOSURES FOR GROUNDING ELECTRODE CONDUCTORS
250.64(E)

If a ferrous metal enclosure is used over the grounding electrode, it shall be electrically continuous from its origin to the grounding electrode. It shall be securely fastened to the grounding clamp so that there is electrical continuity from where it originates to the grounding electrode. If a

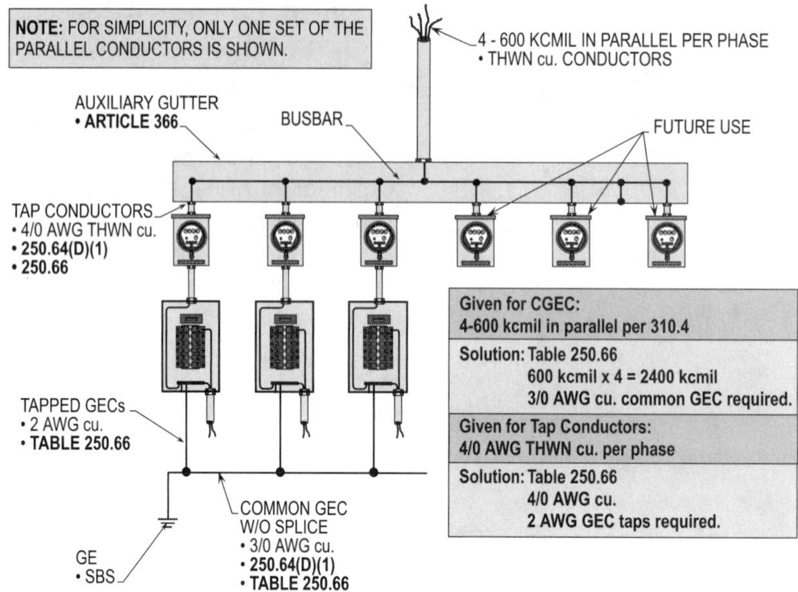

Figure 11-72. This illustration shows the requirements for a common grounding electrode conductor and grounding electrode conductor taps.

ferrous metal enclosure is used for physical protection for the grounding electrode conductor and it is not continuous from the enclosure to the grounding electrode, it shall be bonded at both ends. The bonding jumper shall be the same size as or larger than the required enclosed grounding electrode conductor. **(See Figure 11-73)**

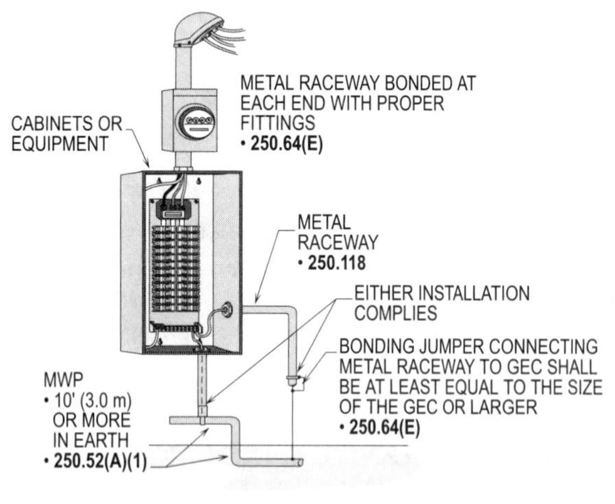

Figure 11-73. This illustration shows requirements for bonding a raceway containing the grounding electrode conductor.

INSTALLATION TO ELECTRODE(S) 250.64(F)

Grounding electrode conductor(s) and bonding jumpers interconnecting grounding electrodes shall be installed per **250.64(F)(1) through (F)(3)**. The grounding electrode conductor shall be sized for the largest grounding electrode conductor required among all the electrodes connected to it.

A grounding electrode conductor shall be permitted to be run to any convenient grounding electrode that is available in the grounding electrode system where the other electrode(s), if any, are connected by bonding jumpers per **250.53(C)**.

The grounding electrode conductor shall be permitted to be run to one or more grounding electrode(s) individually.

Bonding jumper(s) from grounding electrode(s) shall be permitted to be connected to an aluminum or copper busbar not less than 1/4 in. x 2 in. (6 mm x 50 mm). The busbar for such installation shall be securely fastened and shall be installed in an accessible location. These connections shall be made by a listed connector or by the exothermic welding process.

SIZE OF ALTERNATING CURRENT GROUNDING ELECTRODE CONDUCTOR
TABLE 250.66

The following grounding electrodes shall be grounded by conductors sized and selected as listed in **Table 250.66** and **250.66(A) through (C)**.

- Connections to metal water pipe
- Connections to structural steel
- Connections to rod, pipe, or plate electrodes
- Connections to concrete-encased electrodes
- Connections to ground rings

CONNECTIONS TO METAL WATER PIPE
250.66, TABLE 250.66, 250.104(A), 250.52(A)(1), AND 250.53(D)

The procedure for selecting the grounding electrode conductor to ground the service to a metal water pipe shall be determined by the size of the service-entrance conductors. **(See Figure 11-53)**

For example: What size copper GEC is required to ground a service to a metal water pipe supplied by 250 KCMIL copper conductors?

> **Step 1:** Finding size GEC
> **Table 250.66**
> 250 KCMIL cu. = 2 AWG cu.
>
> **Solution: The size grounding electrode conductor is required to be 2 AWG copper.**

CONNECTIONS TO STRUCTURAL BUILDING METAL
250.66, TABLE 250.66, 250.104(C), AND 250.52(A)(2)

The procedure for selecting the grounding electrode conductor to ground the service to structural metal shall be determined by the size of the service-entrance conductors. **(See Figure 11-55)**

For example: What size copper grounding electrode conductor is required to ground a service to structural metal supplied by 250 KCMIL copper conductors?

> **Step 1:** Finding the GEC
> **Table 250.66**
> 250 KCMIL cu. requires 2 AWG cu.
>
> **Solution: The size grounding electrode conductor is required to be 2 AWG copper.**

CONNECTIONS TO ROD, PIPE, OR PLATE ELECTRODES
250.66(A)

The service equipment shall be permitted to be grounded with a rod, pipe, or plate electrode where there are no other electrodes available, per **250.52(A)(1)** through **(A)(4)** and **(A)(8)**. A driven rod or supplementary grounding electrode with a resistance of 25 ohms or less is considered low enough to allow the grounded system to operate safely and function properly. If the driven rod is used as a supplementary grounding electrode to the metal water pipe system, it should be connected to the grounded terminal bar in the service equipment panelboard. For further information, see **250.52(A)(5)** through **(A)(8)** and **250.53(A), (G),** and **(H)** for the application of this rule when using one of the electrodes listed there. The grounding electrode conductor shall not be required to be larger than 6 AWG copper or 4 AWG aluminum where connected to electrodes such as driven rods. **(See Figure 11-58)**

For example: What is the current flow in a 6 AWG copper grounding electrode conductor connecting the common grounded terminal bar in the service equipment to a driven rod? (The supply voltage is 120/208 volt, three-phase system)

> **Step 1:** Finding amperage
> **250.56**
> $I = 120\ V \div 25\ R$
> $I = 4.8\ A$
>
> **Solution: The normal current flow is about 4.8 amps.**

CONNECTIONS TO CONCRETE-ENCASED ELECTRODES
250.66(B)

Lengths of rebar 1/2 in. (13 mm) in diameter and at least 20 ft (6 m) long in length shall be permitted to be used as a grounding electrode to ground the service equipment. The size of the grounding electrode conductor shall be a 4 AWG or larger copper conductor per **250.66(B)** and **250.52(A)(3)**. The rebar system shall be permitted to be one length that is 1/2 in. x 20 ft (13 mm x 6 m) long, or a number of such that are spliced together to form a 20 ft (6 m) length. This grounding method is known in the electrical industry as the UFER ground.

A concrete-encased electrode shall also be permitted to be a 4 AWG copper conductor instead of 1/2 in. x 20 ft (13 mm x 6 m) rebar. The 4 AWG copper conductor shall be located within 2 in. (50 mm) of concrete located horizontally near the bottom or vertically, and with that portion of the foundation.

The size of the grounding electrode shall be at least a 4 AWG copper per **250.66(B)**. The 4 AWG copper grounding electrode conductor is selected based on the 4 AWG copper grounding electrode installed in the foundation. **(See Figure 11-56)**

CONNECTIONS TO GROUND RINGS
250.66(C)

A bare copper conductor not smaller than 2 AWG and at least 20 ft. (6 m) long shall be installed to create a ground ring that will encircle a building or structure. This ring shall be permitted to be utilized as the main grounding electrode if necessary or just bonded in as one per **250.50, 250.52(A)(4),** and **250.53(C). (See Figure 11-57)**

GROUNDING ELECTRODE CONDUCTOR AND BONDING JUMPER CONNECTION TO GROUNDING ELECTRODES
250.68

The following methods shall be adhered to when connecting the grounding electrode conductor to grounding electrodes:

- Accessibility
- Effective grounding path

ACCESSIBILITY
250.68(A)

The 1987 *National Electrical Code* made it very clear that installers and inspectors shall work together to see that grounding clamps and fittings used to connect the grounding electrode conductor or bonding jumpers to water pipes or building steel are accessible. All installations built before the adoption of the 1978 NEC did not necessarily require the grounding clamps or fittings to be accessible. This revision in the 1978 NEC came about because grounding clamps or fittings were removed or damaged during construction and never fixed or replaced. A missing ground clamp could cause the grounded (neutral) conductor to float, and proper voltage regulation would not be accomplished. **(See Figure 11-74)**

ACCESSIBILITY NOT REQUIRED
250.68(A), Ex.s 1 AND 2

Grounding clamps, fittings, or other approved methods that connect the grounding electrode conductor to concrete-encased electrodes, ground rings, plates, or driven rods are not required to be accessible. However, they shall be approved (listed) for direct burial.

Exothermic or irreversible compression connections used at terminations, together with the mechanical means used to attach such terminations to fire proofed structural metal whether or not the mechanical means is reversible, shall not be required to be accessible.

EFFECTIVE GROUNDING PATH
250.68(B)

The connection of the grounding electrodes or bonding jumpers to the grounding electrode shall be in a manner that ensures a permanent and effective grounding path. When metal piping systems are used as a grounding electrode, effective bonding conductors of sufficient length shall be installed around insulated joints and sections and around any equipment that is likely to be disconnected for repairs or replacement. **(See Figure 11-75)**

METHODS OF GROUNDING AND BONDING CONDUCTOR CONNECTION TO ELECTRODES
250.70

When connecting grounding or bonding conductors to grounding fittings by suitable lugs, pressure connectors, clamps, or other listed means, including exothermic welding should be used. Soldering shall never be used. The ground clamps shall be of a material that is compatible for both the grounding electrode and the grounding electrode conductor. No more than one conductor shall be connected to an electrode unless the connector is listed for the purpose per **110.14(A)**. Grounded fittings used to connect the grounding conductor shall be permitted to be any of the following:

- Exothermic welding
- Listed lugs
- Listed pressure connectors
- Listed clamps

Direct burial grounding clamps shall be utilized on grounding electrodes in the earth, such as pipes, rods, rebar, encased electrodes, etc. **(See Figure 11-76)**

BONDING
250.90

Bonding is defined as connected to establish electrical continuity and conductivity. Such bonding jumpers shall be sized to handle these larger available fault currents when necessary to prevent damaging components and equipment.

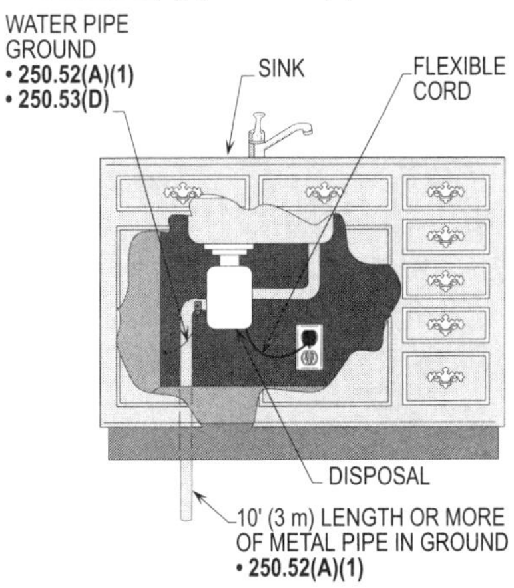

NOTE: WHEN SIZING GECs TO BE CONNECTED TO METAL WATER PIPES, USE THE RECOMMENDED GUIDE BELOW:

MINIMUM SIZE WATER PIPE ELECTRODES AND GROUNDING ELECTRODE CONDUCTORS	
SIZE OF GEC's	SIZE OF WATER PIPE, IN INCHES
8 AWG	1/2" (16)
6 AWG	3/4" (21)
4 AWG	1" (27)
2 AWG	1 1/4" (35)
1/0 AWG	1 1/2" (41)
2/0 AWG	2" (53)
3/0 AWG OR 4/0 AWG	2 1/2" (63)

THE SIZE OF THE METAL WATER PIPES USED TO CONNECT THE GEC's ARE RECOMMENDED SIZES.

Figure 11-74. The 1987 NEC made it very clear that installers and inspectors shall work together to see that grounding clamps and fittings used to connect the grounding electrode conductor to metal water pipes or building steel are accessible. All installations built before the adoption of the 1978 NEC did not necessarily require the grounding clamps or fittings to be accessible.

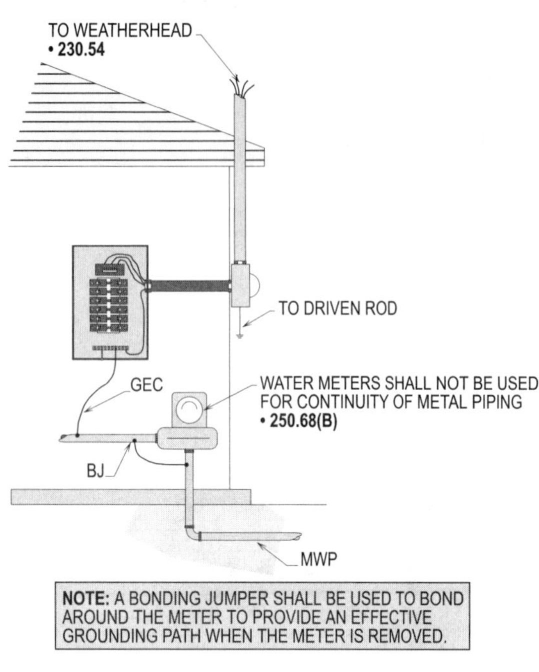

Figure 11-75. This illustration shows that water meters and other devices shall not be permitted to be used for continuity of the metal water piping.

SERVICES
250.92

The following bonding requirements shall be considered for services:

- Bonding of services
- Method of bonding at the service

BONDING OF SERVICES
250.92(A)

The following noncurrent-carrying metal parts of service equipment shall be bonded together:

- Service raceways, cable trays, cablebus framework, auxiliary gutters, or service cable armor or sheath except as permitted per **250.84**.
- All service enclosures containing service conductors, including meter fittings, meter base enclosure or CT cans, interposed in the service raceway or armor.

Service-entrance raceways, meter base enclosures, service equipment enclosures, and all other metal enclosures of the service equipment shall be bonded together to ensure a continuous metal-to-metal grounding system. **(See Figure 11-77)**

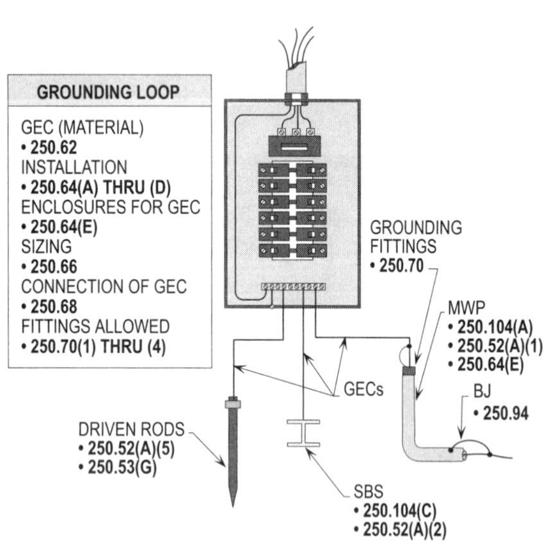

Figure 11-76. When connecting grounding conductors to grounding fittings by suitable lugs, use pressure connectors, clamps, or other listed means, including exothermic welding.

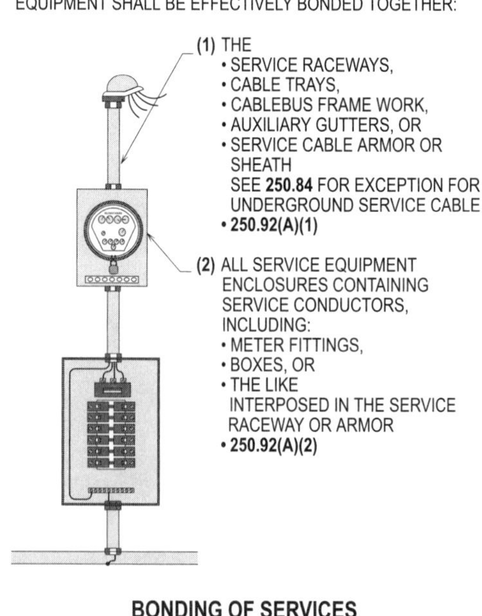

Figure 11-77. All metal raceways, metal-clad of cables, meter bases, and metal equipment enclosures shall be bonded together for safety.

METHOD OF BONDING AT THE SERVICE 250.92(B)

Any one of the following five methods can be utilized for the bonding of service equipment:

- Grounded service conductor
- Threaded connections
- Threadless couplings and connectors
- Other listed devices such as bonding type locknuts, bushings, and bushings with bonding jumpers
- Bonding jumpers

See Figure 11-78 for a detailed illustration for applying these requirements.

METHOD OF BONDING AT THE SERVICE 250.92(B)(1) THRU (B)(4)

A bonding conductor shall be used to connect and bond the raceway to the enclosure where smaller concentric or eccentric knockouts are removed, leaving larger ones. A metal bushing with a lug on the threaded raceway is used to terminate the bonding jumper to a lug connected to the enclosure or grounded busbar terminal. Any one of the following methods shall be permitted to be utilized for the bonding of service equipment:

- Grounded service conductor
- Threaded connections
- Threadless couplings and connectors
- Other listed devices

GROUNDED SERVICE CONDUCTOR 250.92(B)(1)

The grounded service conductor shall be permitted for bonding equipment by one of the following methods per **250.8**:

- Listed pressure connectors
- Terminal bars
- Pressure connectors listed as grounding and bonding equipment
- Exothermic welding process
- Machine screw-type fasteners that engage not less than two threads or are secured with a nut
- Thread-forming machine screws that engage not less than two threads in the enclosure
- Connections that are part of a listed assembly
- Other listed means

See Figure 11-79 for a detailed illustration when applying these requirements.

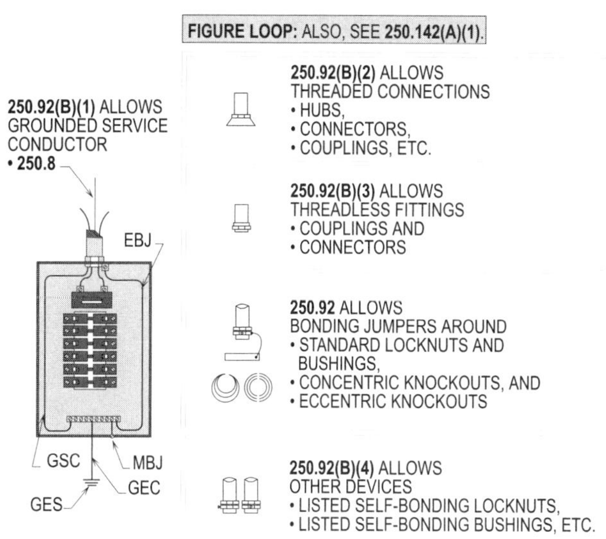

Figure 11-78. This illustration show the proper methods for bonding raceways at the service equipment

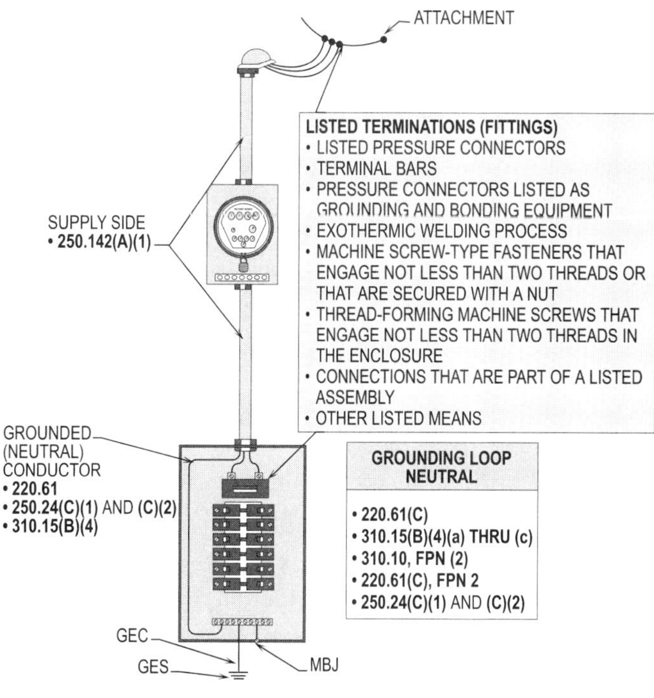

Figure 11-79. This illustration shows the fittings that shall be permitted to be used to bond and connect the grounded (neutral) conductor to the metal enclosure housing the components of the service.

THREADED CONNECTIONS
250.92(B)(2)

Threaded connections, such as threaded couplings or threaded bosses, shall be permitted to be installed where the service equipment is threaded into the meter base or the service equipment enclosure. Such threaded connections shall be made up wrench-tight. **(See Figure 11-80)**

THREADLESS COUPLINGS AND CONNECTORS
250.92(B)(3)

Threadless couplings and connectors installed for rigid metal conduit, intermediate metal conduit, and electrical metallic tubing shall be made up tight. Standard locknuts and bushings shall not be permitted to be installed for bonding the raceway to the enclosure. Only locknuts that are approved for bonding such as extra thick locknuts and bushings equipped with a set screw shall be permitted for this purpose. **(See Figure 11-81)**

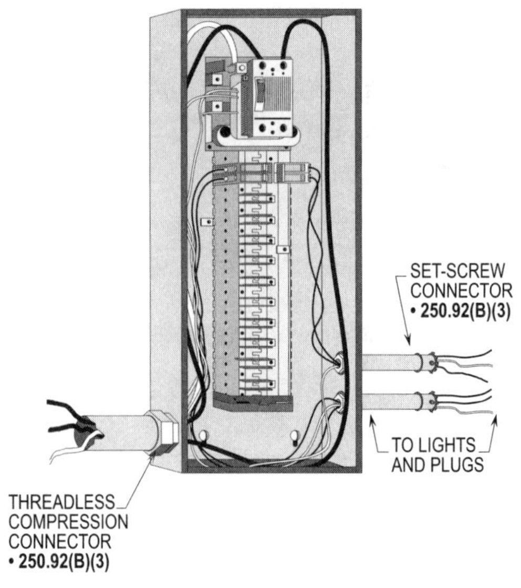

**THREADLESS COUPLINGS
AND CONNECTORS
NEC 250.92(B)(3)**

Figure 11-81. This illustration shows threadless connectors used as a bonding means for grounding the electrical metallic tubing to the metal panelboard.

OTHER LISTED DEVICES
250.92(B)(4)

A wedge fitting (bushing) or locknut fitting is an application of other approved devices that shall be permitted to be used. A grounding wedge is equipped with a set screw that tightens the metal conduit and ensures proper bonding of the service equipment enclosure.

Design Tip: Two locknuts (one is self-bonding) with a bushing are an acceptable bonding means. However, two standard locknuts shall not be permitted to be used as a bonding means per **250.92(B)(3)**. Note that bonding jumpers shall be sized by the provisions listed in **250.102(C)** and **Table 250.66**. **(See Figure 11-82)**

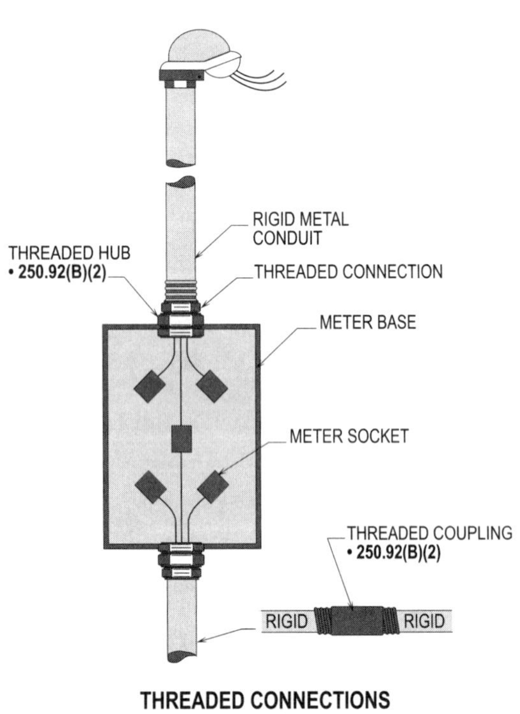

**THREADED CONNECTIONS
NEC 250.92(B)(2)**

Figure 11-80. This illustration shows a threaded hub used as the bonding means for grounding the rigid metal conduit to the metal of the meter base enclosure.

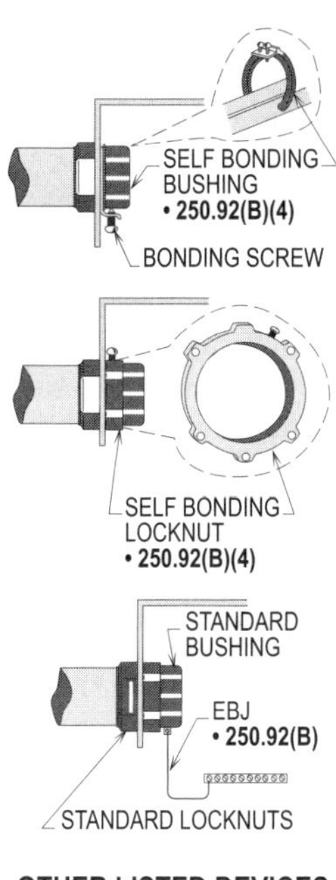

Figure 11-82. This illustration shows self-bonding devices such as locknuts and bushings to bond and ground rigid metal conduit to the metal enclosure enclosing electrical components of equipment.

BONDING FOR OTHER SYSTEMS
250.94

Intersystem bonding and grounding conductors shall be connected to an intersystem bonding termination that is provided external to enclosures at the service equipment and at the disconnecting means for any additional buildings or structures. The intersystem bonding termination for the intersystem bonding and grounding conductors shall be accessible for the connection and inspection and have the capacity for connection of not less than three intersystem bonding conductors. The intersystem bonding termination shall consist of one of the following:

- A set of terminals (listed as grounding and bonding equipment) securely mounted to the meter enclosure and electrically connected to the metal enclosure.

- A bonding bar located near the service equipment enclosure, meter enclosure, or raceway for service conductors. The bonding bar shall be connected with a 6 AWG or larger copper conductor to an equipment grounding conductor(s) in the service equipment enclosure, meter enclosure, or exposed nonflexible metallic raceway.

- A bonding bar located near the grounding electrode conductor. A 6 AWG or larger copper conductor shall be used for connecting the bonding bar to the grounding electrode conductor.

See Figure 11-83 for a detailed illustration for applying these requirements.

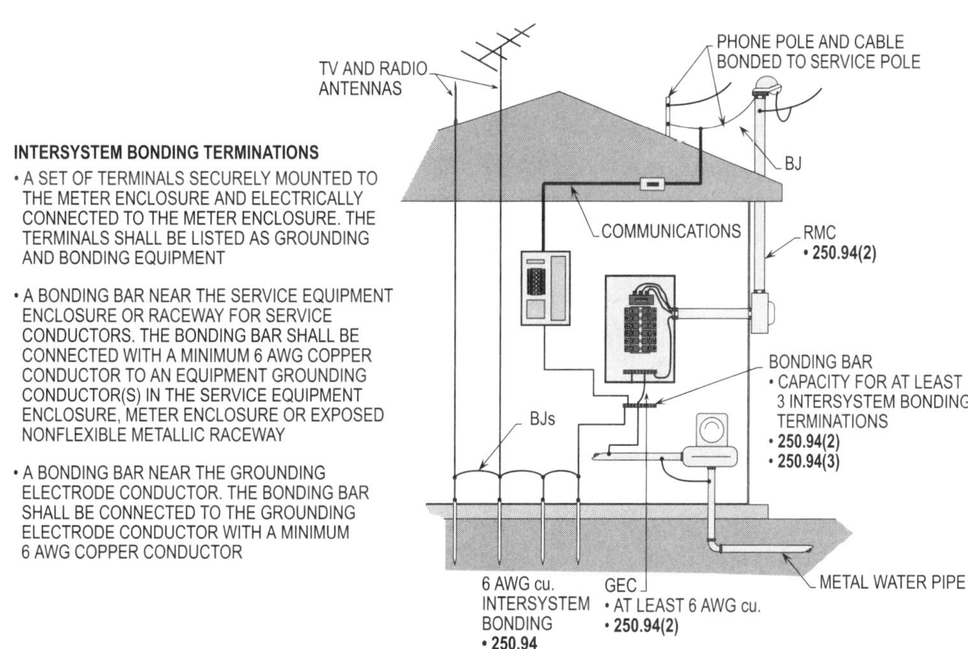

Figure 11-83. This illustration shows the methods for connecting the electrical power system grounding to telecommunications system grounding and bonding system to ensure there is no difference of potential between the two systems.

BONDING OTHER ENCLOSURES
GENERAL
250.96(A)

Metal raceways, cable trays, cable armor, etc. shall be bonded to metal enclosures, boxes, cabinets, etc. Metal raceways shall be installed as a complete run from a metal box to metal box per **300.10**. Such wiring methods shall be terminated with listed connectors that comply with **300.12, 300.13,** and **300.15** in the NEC. Note that when metal raceways, armor cables, cable sheaths, enclosures, frames, and other metal noncurrent-carrying parts that serve as the equipment grounding conductors, whether using a supplementary grounding conductor or not, proper bonding shall be required to ensure the safe handling of such fault currents that may be imposed and to ensure electrical continuity. Nonconductive paint or similar coatings shall be removed at the threads or other contact points to ensure clean surfaces. Fittings or bonding jumpers are used for such a purpose. **(See Figure 11-84)**

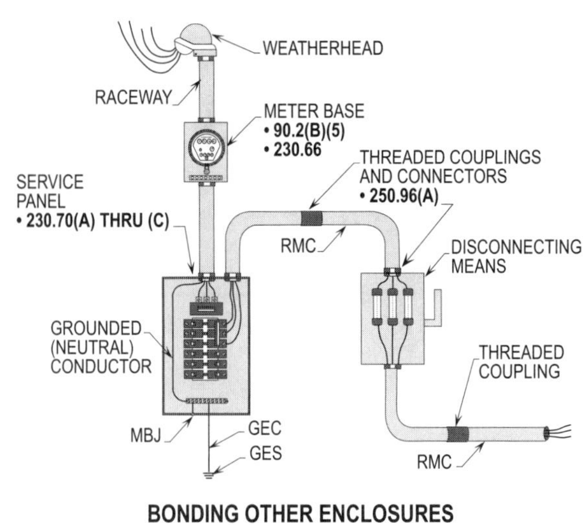

BONDING OTHER ENCLOSURES
GENERAL
NEC 250.96(A)

Figure 11-84. This illustration shows the metal weatherhead and raceway to the meter base, raceway between the panelboard, raceway to the disconnecting means, and the raceway from the disconnect supplying the load is properly connected together to form a continuous unbroken grounding path.

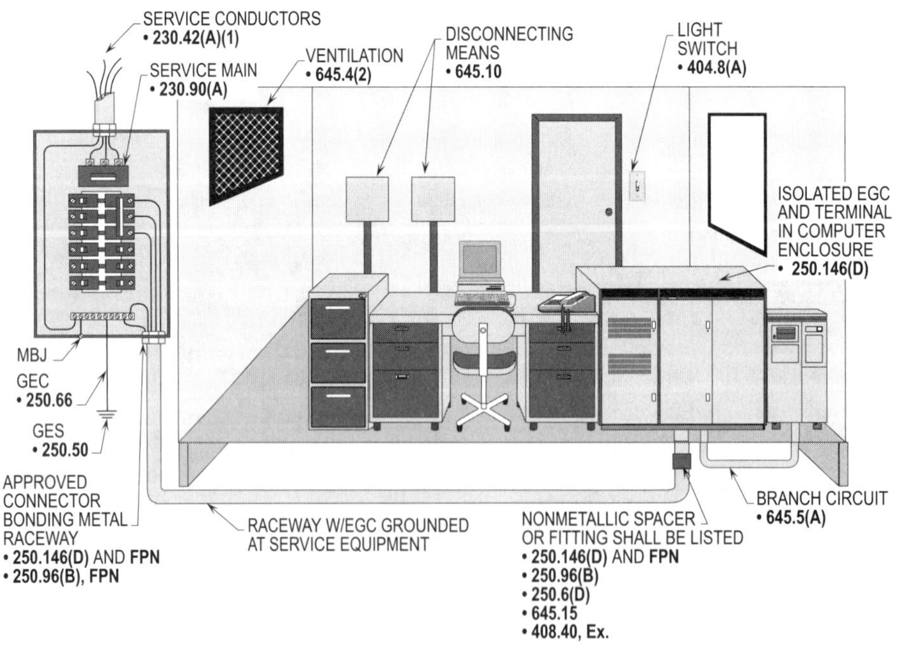

ISOLATED GROUNDING CIRCUITS
NEC 250.96(B)

See Figure 11-85. Reduction of electromagnetic interference that creates unwanted noise in sensitive electronic equipment may be reduced by installing a nonmetallic spacer or fitting in the metal raceway at the point of connection to such equipment.

ISOLATED GROUNDING CIRCUITS 250.96(B)

A nonmetallic spacer or fitting shall be permitted to be installed between the metal raceway supplying sensitive electronic equipment at the point of connection. The equipment grounding conductor shall be isolated, and the metal raceway grounded. Electromagnetic interference that creates unwanted noise may be reduced in sensitive electronic equipment when this type of installation is applied. **(See Figure 11-85)**

BONDING FOR OVER 250 VOLTS 250.97

When on the load side, one of the following wiring methods shall be used for bonding circuits over 250 volts-to-ground, not including services:

- Install the following wiring methods between sections of metal conduit, intermediate metal conduit, electrical metallic tubing, or other type of metal raceways.

    (a) Threaded couplings

    (b) Threadless couplings

- Install the following wiring methods between metallic raceways, metallic boxes, cabinet enclosures, etc.

    (a) Jumpers

    (b) Two locknuts and bushing

    (c) Threaded (bosses) hubs

    (d) Other approved devices

Metal raceways and conduits enclosing conductors at more than 250 volts-to-ground (other than service conductors) shall have electrical continuity assured by one of the methods outlined in **250.92(B)(2) through (B)(4)**. Note that the grounded (neutral) conductor shall not be permitted to be used (shall be permitted for the supply side of services) except as permitted for separate buildings and separately derived systems per **250.24(A)(5), FPN, 250.92(B)(1),** and **250.142(A)(1), (A)(2),** and **(A)(3). (See Figure 11-86)**

ECCENTRIC OR CONCENTRIC KNOCKOUTS 250.97, Ex.

If oversized eccentric or concentric knockouts are not encountered, the following items shall be permitted to be utilized:

- Threadless fittings
- Two locknuts
- One locknut
- Listed fittings

See Figure 11-87 for a detailed illustration when applying these requirements.

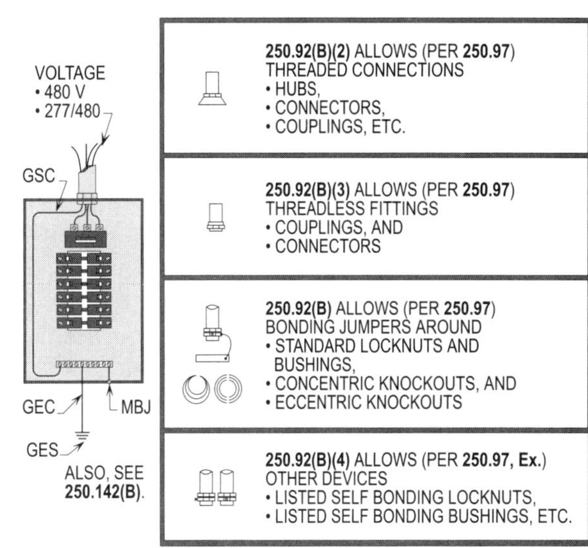

Figure 11-86. This illustration shows the methods that shall be permitted to be used to bond and ground metal conduits to metal enclosures.

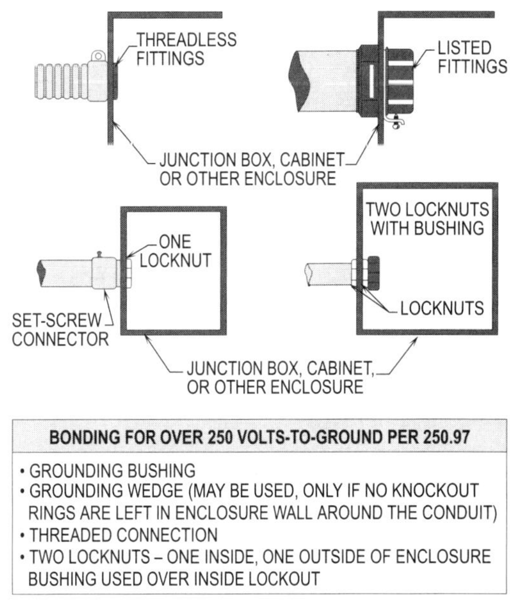

Figure 11-87. This illustration shows fittings that shall be permitted to be used to bond and ground the raceways and enclosures together to form a continuous grounding path.

THREADLESS FITTINGS
250.97, Ex. (1)

Threadless couplings and connectors shall be permitted to be used with metal sheath cables. However, they shall be listed as required per **300.15** and **300.15(F)**.

TWO LOCKNUTS
250.97, Ex. (2)

Two locknuts shall be permitted to be used with rigid metal conduit or intermediate metal conduit, one locknut on the outside and one on the inside of the enclosure when entering boxes or cabinets. Appropriate bushings inside the enclosure, etc. shall be used to protect the insulation of the conductors. Grounding connections and circuit conductors that are not made up tight are the cause of many arcing faults and electrical fires.

ONE LOCKNUT
250.97, Ex. (3)

One locknut on the inside of the boxes or cabinets provides proper bonding where the fitting shoulders seat firmly on the box or cabinet. These include electric metallic tubing connectors, cable connectors, and flexible metal conduit connectors, etc.

LISTED FITTINGS
250.97, Ex. (4)

A wedge fitting (bushing) or locknut fitting is an application of other listed devices that shall be permitted to be used. A grounding wedge is equipped with a set-screw that tightens the metal conduit and ensures proper bonding of the equipment enclosure.

SIZE – EQUIPMENT BONDING JUMPER ON SUPPLY SIDE OF SERVICE
250.102(C)

There has always been considerable confusion as to when to use **Table 250.66** and **Table 250.122**. By reviewing the headings of each Table there should be no confusion. **Table 250.66** is used for grounding electrode conductors for AC systems and sized based on the fact that no overcurrent protection device is ahead of the conductors.

Table 250.122 is used for equipment grounding conductors for grounding raceways and equipment and sized based on an overcurrent protection device ahead of the circuit conductors.

For example: What size copper equipment bonding jumper is required to bond the service raceway (having 3 - 250 KCMIL THWN copper conductors) to the grounded busbar in a panelboard?

Step 1: Sizing EBJ on supply side
250.102(C); Table 250.66
250 KCMIL THWN cu. requires 2 AWG cu.

Solution: The size equipment bonding jumper required on the supply side is 2 AWG copper.

For example: What size individual equipment bonding jumper is required to bond each service raceway (RMC) containing 3 - 700 KCMIL THWN copper conductors paralleled three times per phase?

Step 1: Sizing EBJ on supply side
250.102(C); Table 250.66
700 KCMIL THWN cu. requires 2/0 AWG cu.

Solution: The size equipment bonding jumper required for each conduit on the supply side is 2/0 AWG copper.

For example: What size equipment bonding jumper (single conductor) is required to bond all service raceways (RMC) containing 3 - 700 KCMIL THWN copper conductors paralleled three times per phase?

Step 1: Sizing EBJ on supply side
250.102(C)
700 KCMIL x 3 = 2100 KCMIL

Step 2: Finding KCMIL to size EBJ
2100 KCMIL x .125 = 262.5 KCMIL

Step 3: Sizing EBJ
Table 8, Ch. 9
262.5 KCMIL requires 300 KCMIL

Solution: The size equipment bonding jumper required for all service raceways on the supply side is 300 KCMIL copper.

SIZE – EQUIPMENT BONDING JUMPER ON LOAD SIDE OF SERVICE
250.102(D)

Table 250.122 lists the size of the equipment bonding jumpers on the load side of the service overcurrent protection devices. The bonding jumper shall be a single conductor sized per **Table 250.122** based on the largest overcurrent protection device that supplies the electrical circuits. If the bonding jumper connects two or more raceways or cables, the same rules apply and proper size conductors shall be selected. Note that it is not necessary to size the equipment bonding jumper larger than the ungrounded (phase) conductors, but in no case shall it be smaller than 14 AWG.

For example: What size equipment bonding jumper is required to bond the raceway of a feeder-circuit having a 225 amp OCPD ahead of its conductors?

Step 1:   Sizing EBJ on load side
            250.102(D); Table 250.122
            225 A OCPD requires 4 AWG cu.

Solution: **The size equipment bonding jumper required on the load side is 4 AWG copper.**

INSTALLATION
250.102(E)

It shall be permitted to install the equipment bonding jumper either inside or outside of the raceway or enclosure. When it is installed outside of the raceway or enclosure, it shall not be over 6 ft (1.8 m) in length and shall be routed with raceway or enclosure. If the equipment bonding jumper is inside the raceway, refer to **310.12(B), 250.119,** and **250.148** for the conductor identification requirements. Note that if the equipment bonding jumper is routed on the outside of the raceway, listed grounding fittings shall be used to terminate raceway and equipment bonding jumper.

BONDING OF PIPING SYSTEMS AND EXPOSED STRUCTURAL STEEL
250.104

The rules of this section require that certain piping systems and structural steel be bonded and grounded properly to ensure the safety of the electrical system.

METAL WATER PIPING – GENERAL
250.104(A)(1)

Regardless of whether the water piping is supplied by nonmetallic pipe to the building or structure, proper bonding of the interior metal water piping to the service equipment enclosure shall be required. Metal water piping that is installed in or attached to a building or structure shall be bonded to the service equipment enclosure, the grounded (neutral) conductor at the service, the grounding electrode conductor where of sufficient size, or the one or more grounding electrodes used. The bonding jumper to metal water piping shall be sized per **Table 250.66**, except as permitted in **250.104(A)(2)** and **(A)(3)**. **(See Figure 11-88)**

METAL WATER PIPING – BUILDINGS OF MULTIPLE OCCUPANCY
250.104(A)(2)

In multiple occupancies, if the interior piping is metal and the piping systems of all occupancies are not tied together, the metal piping shall be isolated from all other occupancies due to the use of nonmetallic pipe as a supply line. Each individual metal water pipe system shall be bonded separately to the panelboard or switchboard enclosure in each apartment. Bonding jumper points of attachment shall be accessible and sized per **Table 250.122** based on the rating of the overcurrent protection device for the circuit supplying the occupancy. This ensures that the metal piping system in each occupancy is at ground potential and that the threat of electrical shock and fires is eliminated. **(See Figure 11-89)**

METAL WATER PIPING – MULTIPLE BUILDINGS OR STRUCTURES SUPPLIED BY A FEEDER(S) OR BRANCH CIRCUIT(S)
250.104(A)(3)

The interior metal water piping system shall be bonded to the building or structure disconnecting means enclosure where located at the building or structure, or to the equipment grounding conductor run with the supply conductors, or to the one or more grounding electrodes used. The bonding jumper shall be sized per **Table 250.66** based on the size of the feeder or branch circuit conductors that supply the building. The bonding jumper shall not be required to be larger than the largest ungrounded (phase) feeder or branch circuit conductors supplying the building. **(See Figure 11-90)**

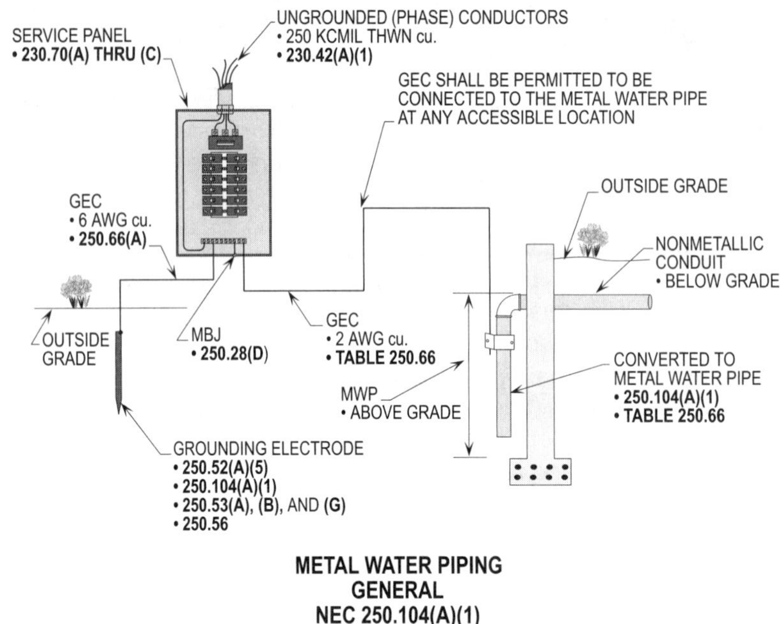

Figure 11-88. Regardless of whether the water piping is supplied by nonmetallic pipe to the building or structure, proper bonding of the interior metal water piping to the service equipment enclosure shall be required.

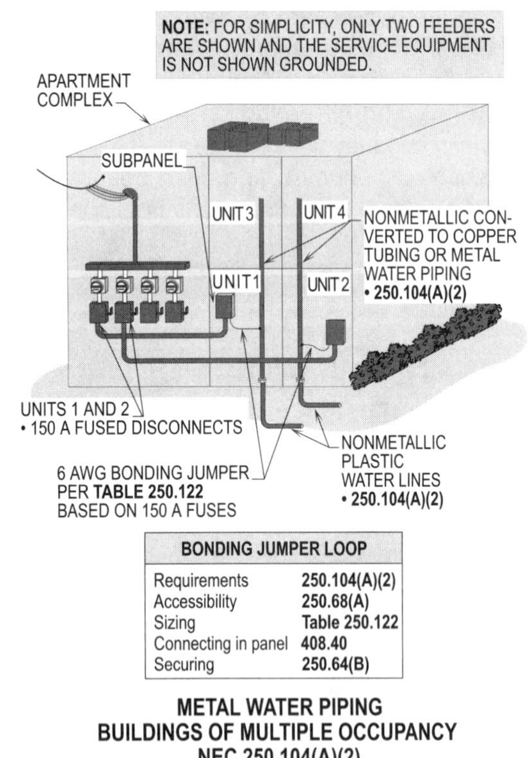

BONDING JUMPER LOOP	
Requirements	250.104(A)(2)
Accessibility	250.68(A)
Sizing	Table 250.122
Connecting in panel	408.40
Securing	250.64(B)

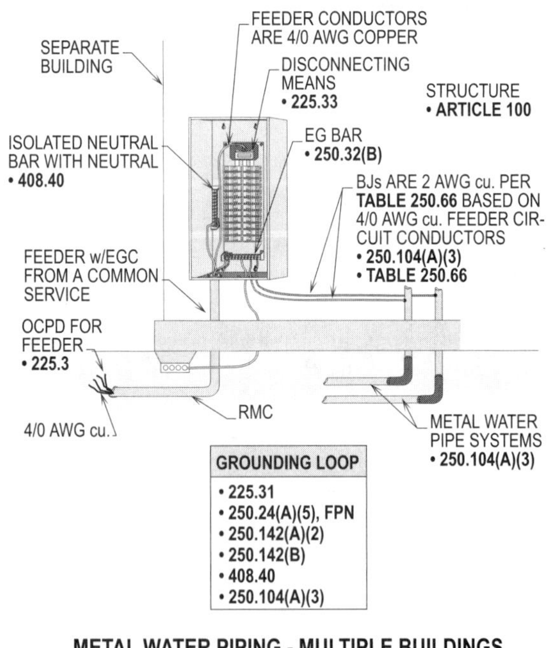

GROUNDING LOOP
• 225.31
• 250.24(A)(5), FPN
• 250.142(A)(2)
• 250.142(B)
• 408.40
• 250.104(A)(3)

Figure 11-89. This illustration shows the procedure for sizing the bonding jumper to bond and ground the copper tubing or metal water piping to the grounded busbar in the subpanel in apartment units 1 and 2.

Figure 11-90. This illustration shows the procedure for sizing the bonding jumper to connect the equipment grounding conductors (run with the feeder-circuit conductors) and the grounded busbar to the concrete-encased electrode of the separate building supplied from a common service.

OTHER METAL PIPING
250.104(B)

All metal piping systems and gas piping that is likely to be subjected to being energized shall be bonded to the service equipment ground at the service enclosure to the common grounding conductor, if it is of sufficient size, or to one or more of the other grounding electrodes. This bonding jumper shall be sized per **Table 250.122. (See Figure 11-91)**

Design Tip: With circuit conductors that are likely to energize metal piping, it is permitted to use the equipment grounding conductor run with the circuit conductors to bond and ground such piping system.

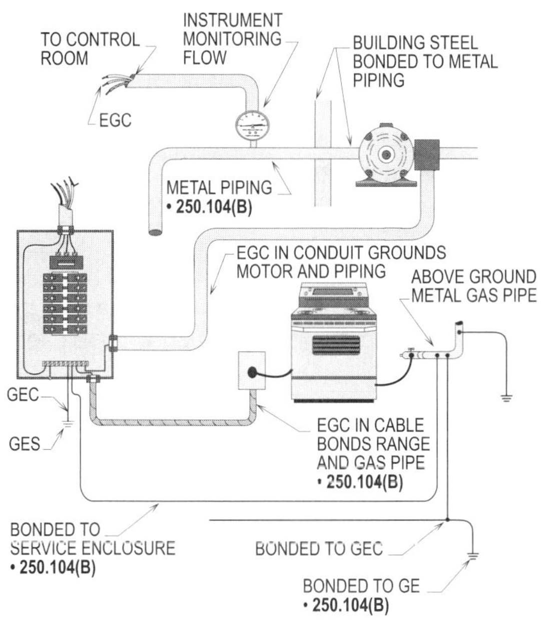

OTHER METAL PIPING
NEC 250.104(B)

Figure 11-91. Metal piping and gas piping that may be energized shall be bonded and grounded with the equipment grounding conductor in the supplying branch circuit or feeder that could energize such piping.

STRUCTURAL METAL
250.104(C)

Exposed building steel frames that are not intentionally grounded shall be bonded to either the service enclosure, the neutral bus, the grounding electrode conductor, or a grounding electrode. The bonding conductor shall be sized per **Table 250.66**, and installed per **250.64(A), (B), and (E)**. **(See Figure 11-92)**

Design Tip: For additional safety, it is recommended to bond all metal air ducts and metal piping systems on the premises to reduce voltage differences.

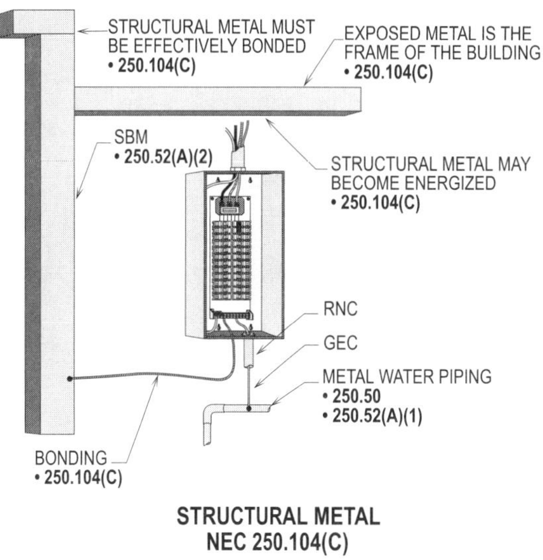

STRUCTURAL METAL
NEC 250.104(C)

Figure 11-92. Exposed interior building steel frames that are not intentionally grounded and are likely to become energized shall be bonded to either the service enclosure, the neutral bus, the grounding electrode conductor, or a grounding electrode.

SEPARATELY DERIVED SYSTEM
250.104(D)

The following shall be bonded to separately derived systems:

- Metal water piping system(s)
- Structural metal

METAL WATER PIPING(S)
250.104(D)(1)

When there is a separately derived system utilizing a grounding electrode near an available point of metal water pipe that is in the area supply by such derived system, the metal water pipe system shall be bonded to the derived system's grounded (neutral) conductor. Each bonding jumper shall be sized per **Table 250.66** based on the largest ungrounded (phase) conductor of the separately derived system. **(See Figure 11-93)**

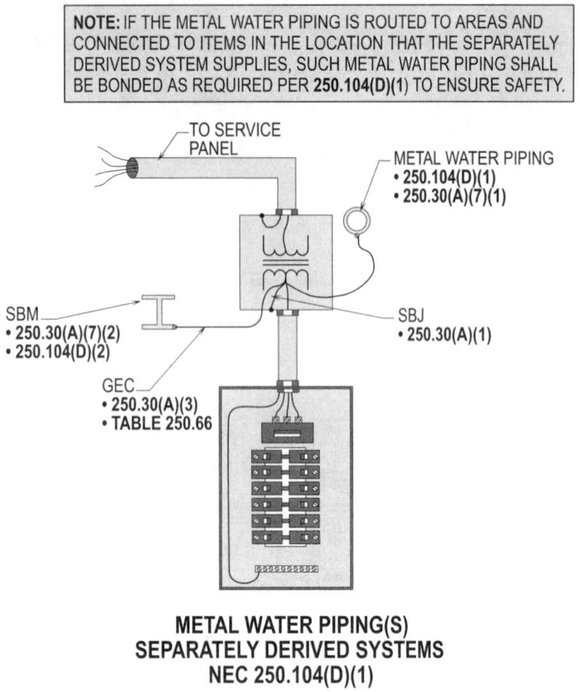

NOTE: IF THE METAL WATER PIPING IS ROUTED TO AREAS AND CONNECTED TO ITEMS IN THE LOCATION THAT THE SEPARATELY DERIVED SYSTEM SUPPLIES, SUCH METAL WATER PIPING SHALL BE BONDED AS REQUIRED PER **250.104(D)(1)** TO ENSURE SAFETY.

**METAL WATER PIPING(S)
SEPARATELY DERIVED SYSTEMS
NEC 250.104(D)(1)**

Figure 11-93. This illustration shows that metal water pipe located in the area of the separately derived system shall be bonded and grounded to the grounded (neutral) conductor located in the transformer's grounding point.

STRUCTURAL METAL
250.104(D)(2)

Exposed structural metal shall be bonded to the grounded conductor of each separately derived system where it is interconnected to form the building frame (i.e., exists in the area). The connection to the separately derived system shall be made at the same point where the grounding electrode conductor is connected. Each bonding jumper shall be sized per **Table 250.66**, based on the largest ungrounded (phase) conductor of the separately derived system.

TYPES OF EQUIPMENT GROUNDING CONDUCTORS
250.118

The equipment grounding conductor run with or enclosing the circuit conductors shall be one or more, or a combination, of the following:

- A copper, aluminum, or copper-clad aluminum conductor
- Rigid metal conduit (RMC)
- Intermediate metal conduit (IMC)
- Electrical metallic tubing (EMT)
- Listed flexible metal conduit (FMC)
- Listed liquidtight flexible metal conduit (LTFC)
- Flexible metal tubing (FMT)
- Type AC cable (BX)
- Mineral-insulated sheathed cable (MI)
- Type MC cable (MC)
- Cable trays
- Cablebus framework
- Other listed electrically continuous metal raceways and listed auxiliary gutters
- Surface metal raceways listed for grounding

See Figure 11-94 for a detailed illustration pertaining to the various types of wiring methods that are permitted to be used as equipment grounding conductors. **Sections 250.118(5)a through d** and **(6)a through e** permit FMC and LTFC to be used as follows:

The maximum length that shall be permitted to be used in any combination of flexible metal conduit, flexible metallic tubing, and liquidtight flexible metal conduit in any grounding run is 6 ft (1.8 m) or less.

The circuit conductors in a 6 ft (1.8 m) run shall be protected by overcurrent devices not to exceed 20 amperes, but shall also be permitted to be less than 20 amps. The flexible metal conduit or tubing shall be terminated only in fittings listed for grounding. **(See Figure 11-95)**

Liquidtight flexible metal conduit shall be permitted to be used in sizes up to 1-1/4 in. (35) that have a total length not to exceed 6 ft (1.8 m) in the ground return path. Fittings for terminating the flexible conduit shall be listed for grounding. Liquidtight flexible metal conduit 3/8 (12) and 1/2 in. (16) trade size shall be protected by overcurrent devices rated at 20 amps or less, and sizes of 3/4 in. (21) through 1-1/4 in. (35) trade size shall be protected by overcurrent devices rated at 60 amps or less. **(See Figure 11-96)**

IDENTIFICATION OF EQUIPMENT GROUNDING CONDUCTORS
250.119

Equipment grounding conductors shall be permitted to be installed as bare, covered, or insulated, unless otherwise permitted elsewhere in the NEC. Equipment grounding conductors smaller than 6 AWG copper or aluminum shall have a continuous outer finish that is either green or green with one or more yellow stripes, unless otherwise permitted elsewhere in the NEC.

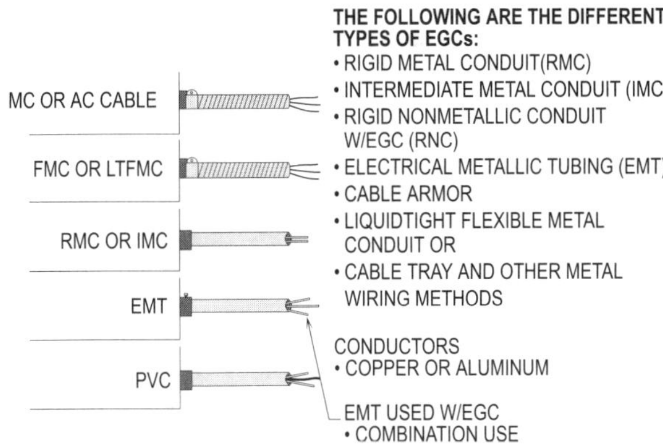

Figure 11-94. The various types of wiring methods permitted to be used as equipment grounding conductors. [See **250.118(1) through (14)** for a complete list of wiring methods.]

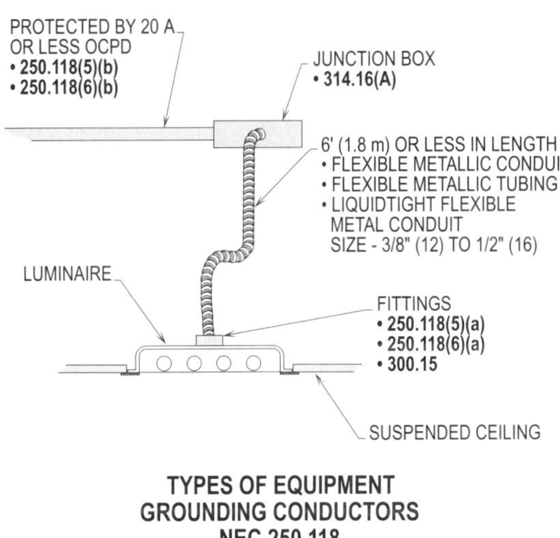

Figure 11-95. This illustration shows the maximum length permitted for listed flexible metal conduit and listed liquidtight flexible metal conduit to be used as an equipment grounding means.

CONDUCTORS LARGER THAN 6 AWG 250.119(A)

Equipment grounding conductors that are insulated or covered shall be permitted to be installed larger than 6 AWG copper or aluminum if color coded at each end and at every point where the conductor is accessible. The following methods shall be permitted to be used to identify an equipment grounding conductor larger than 6 AWG:

• Stripping the insulation or covering from the entire length exposed.

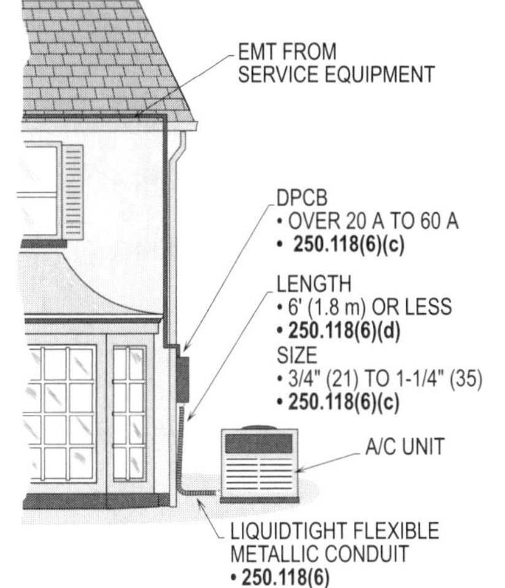

Figure 11-96. This illustration shows the maximum length that listed liquidtight flexible metal conduit shall be permitted to be used as a grounding means without an equipment grounding conductor routed with the circuit conductors inside the flex.

• Coloring the exposed insulation or covering green at the termination.

• Marking the exposed insulation or covering with green colored tape or green adhesive labels at the termination.

Note, equipment grounding conductors that are larger than 6 AWG shall not be required to be marked in conduit bodies that contain no splices or unused hubs.

See Figure 11-97 for a detailed illustration when applying these requirements.

EQUIPMENT GROUNDING CONDUCTOR INSTALLATION 250.120(A) THRU (C)

When an equipment grounding conductor is used in a raceway, cable tray, cable armor, cablebus framework, or cable sheath, and where it is routed in a raceway or cable, the installation shall conform to the applicable provisions in the NEC based upon this type wiring system.

All terminations and joints shall be approved for such use. All joints and fittings shall be made up tight by using tools suitable for the purpose.

Sections **250.130(A)** and **(B)** provide for using the equipment grounding conductor as a separate conductor.

Equipment grounding conductors shall be permitted to be bare, insulated aluminum, or copper-clad aluminum. Bare conductors shall not be permitted to come in direct contact with masonry or the earth or where subject to corrosive conditions. Aluminum or copper-clad aluminum conductors shall not be permitted to be terminated within 18 in. (450 mm) of the earth.

If the grounding conductors are routed in the hollow spaces of a building that is protected, sizes smaller than 6 AWG shall not be required to be in raceways.

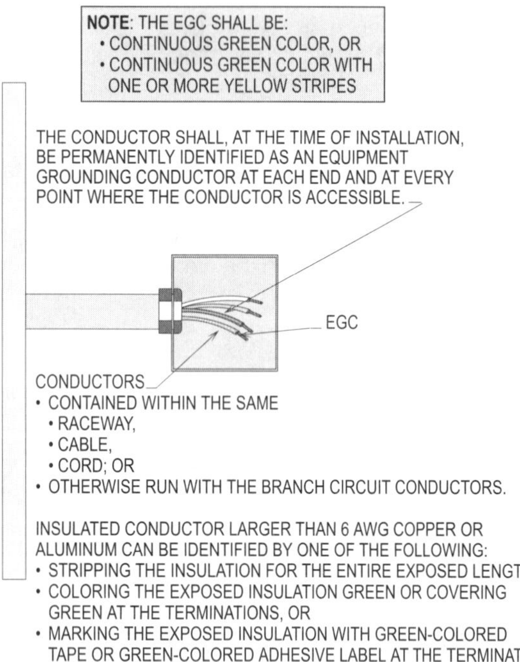

CONDUCTORS LARGER THAN 6 AWG
NEC 250.119(A)

See Figure 11-97. This illustration shows the methods of identifying and using the equipment grounding conductor when its insulation is other than green in color.

SIZING OF EQUIPMENT GROUNDING CONDUCTORS
250.122 AND TABLE 250.122

When an equipment grounding conductor is used in a raceway, cable tray, cable armor, or cable sheath, and where it is routed in a raceway or cable, the installation and size of the conductor shall conform to the applicable provisions in the NEC. The size of the equipment grounding conductor shall be based on the size of the overcurrent protection device (OCPD) protecting the circuit conductors per **Table 250.122**. **(See Figure 11-98)**

The equipment grounding conductor is the conductor used to ground the noncurrent-carrying metal parts of equipment. The function of the equipment grounding conductor is to keep the equipment elevated above ground, at zero potential, and provides a path for the ground fault current. Equipment grounding conductors protect elements of circuits and equipment and also ensure the safety of personnel from electrical shock. The size of the equipment grounding conductor shall be based on the size of the overcurrent protection device protecting the circuit conductors.

For example: What size copper equipment grounding conductor is required to supply a subpanel, if the OCPD protecting the feeder-circuit is rated at 400 amps?

Step 1: Sizing EGC
Table 250.122
400 A OCPD requires 3 AWG cu.

Solution: The size equipment grounding conductor is required to be 3 AWG copper.

GENERAL
250.122(A)

Equipment grounding conductors 6 AWG and smaller shall have their insulation for its entire length colored green or green with one or more yellow stripes. Note that the equipment grounding conductor shall be permitted to be bare under certain conditions of use. The insulation or covering shall be stripped from its entire length or length exposed. [For this requirement, see **250.119(A)**.]

Equipment grounding conductors that are insulated or covered shall be permitted to be installed larger than 6 AWG copper or aluminum if color coded at each end and at every point where the conductor is accessible. The following methods shall be permitted to be used to identify an equipment grounding conductor larger than 6 AWG.

- Stripping the insulation or covering from the entire length exposed.
- Coloring the exposed insulation or covering green at the terminations.
- Marking the exposed insulation or covering with green-colored tape or green adhesive labels at the terminations.

See Figure 11-99 for a detailed illustration when applying these requirements.

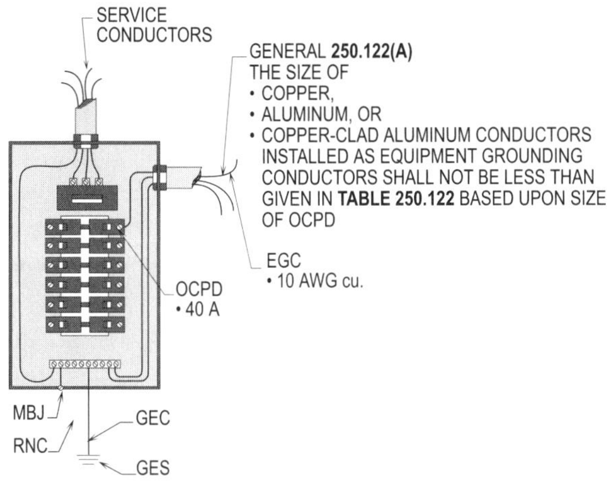

Figure 11-98. This illustration shows the equipment grounding conductor shall be sized based on the overcurrent protection device ahead of the circuit conductors supplying the electrical equipment.

INCREASED IN SIZE
250.122(B)

Where ungrounded (phase) conductors are increased in size, equipment grounding conductors (if installed) shall be increased in size proportionately according to the circular mil area of the ungrounded (phase) conductors. **(See Figure 11-100)**

MULTIPLE CIRCUITS
250.122(C)

Single equipment grounding conductors that are run with multiple circuits in the same raceway, cable, or cable tray shall be permitted to be sized based on the largest overcurrent protection device protecting the circuit conductors in the raceway, cable, or cable tray.

For example: What size equipment grounding conductor is required for multiple circuits in the same raceway with the following known values:

- 3 - 10 AWG copper conductors (30 A OCPD)
- 2 - 12 AWG copper conductors (20 A OCPD)
- 3 - 14 AWG copper conductors (15 A OCPD)

Step 1: Sizing EGC
250.122(C)
30 A OCPD requires 10 AWG cu.

Solution: A 10 AWG copper equipment grounding conductor is required for multiple circuits in the same raceway.

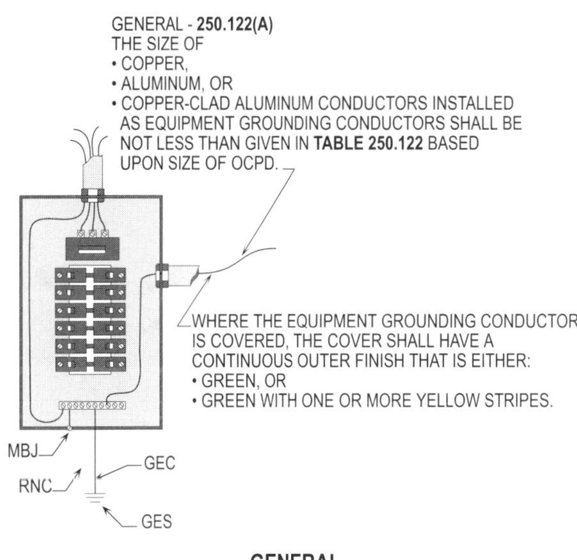

Figure 11-99. This illustration shows marking techniques that shall be permitted to be used when identifying the equipment grounding conductor routed through a raceway or cable.

MOTOR CIRCUITS
250.122(D)

The equipment grounding conductor shall be sized based on the rating of the overcurrent protection device per **250.122(A)**. Where the overcurrent protection device is a motor short-circuit protector or an instantaneous trip breaker, the equipment grounding conductor shall be sized not smaller than that given by **250.122(A)** using the maximum permitted rating of a dual element time-delay fuse per **430.52(C)(1), Ex. 1**. **(See Figure 11-101)**

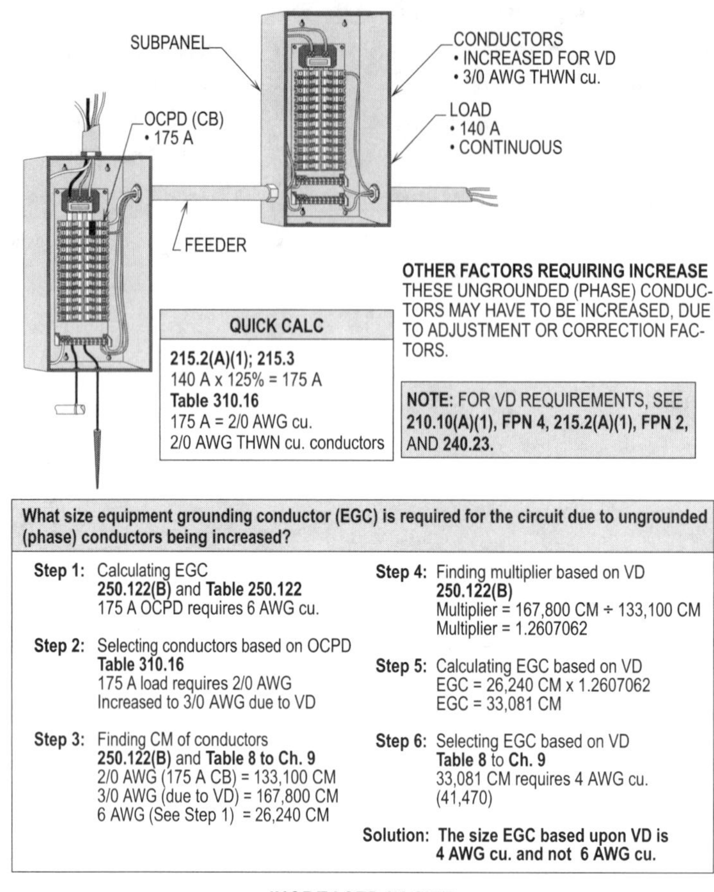

Figure 11-100. This illustration shows the equipment grounding conductor being increased in size proportionately according to the circular mil area of the ungrounded (phase) conductors.

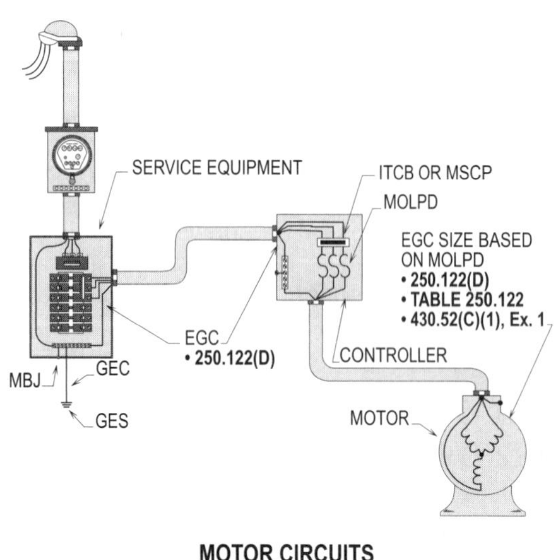

Figure 11-101. This illustration shows that under certain design conditions, the size of the equipment grounding conductor shall be permitted to be determined by the rating of the motor overload protective device (MOLPD) in the controller.

FLEXIBLE CORD AND FIXTURE WIRE 250.122(E)

The equipment grounding conductor in a flexible cord, with the largest circuit conductor 10 AWG or smaller, and equipment grounding conductors that are part of flexible cords or used with fixture wires per **240.5** shall not be permitted to be smaller than 18 AWG copper and not smaller than the circuit conductors. The equipment grounding conductor shall be sized per **Table 250.122** where larger than 10 AWG.

CONDUCTORS IN PARALLEL 250.122(F)

Circuit conductors that are connected in parallel and routed through separate conduits shall have a separate equipment grounding conductor run in each conduit. The equipment grounding conductor shall be sized per **Table 250.122**, based on the rating of the overcurrent protection device protecting the circuit.

For example: What size copper equipment grounding conductors are required to ground the metal parts of a piece of equipment supplied by 3 - 3/0 AWG THWN copper conductors per phase connected to a 600 amp OCPD in the panelboard?

Step 1: Sizing EGC's in each conduit
250.122(F); Table 250.122
600 A OCPD requires 1 AWG cu.

Solution: The size equipment grounding conductors are required to be 1 AWG copper for conductors installed in parallel.

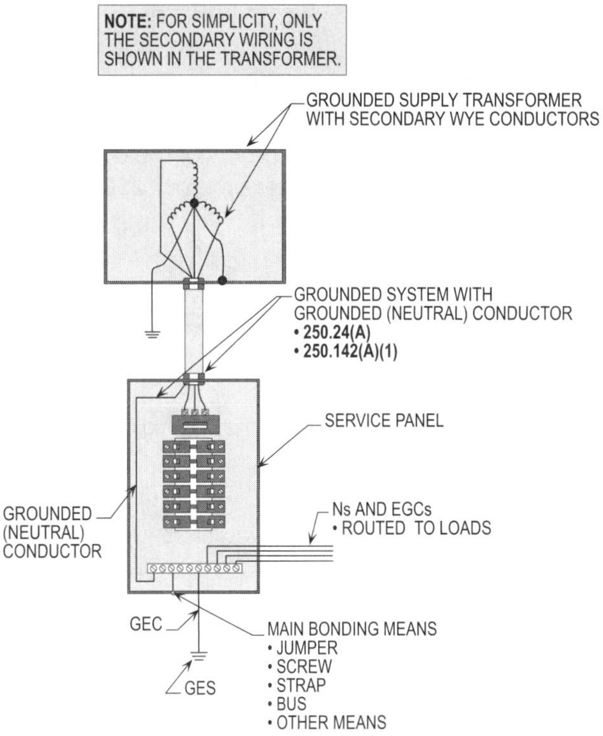

EQUIPMENT GROUNDING CONDUCTOR CONNECTIONS
250.130

Equipment grounding conductor connections for a separately derived system and for service equipment shall be made in accordance with **250.30(A)(1)**. The equipment grounding conductor connections shall be made at service equipment and separately derived systems as follows:

- For grounded systems
- For ungrounded systems

FOR GROUNDED SYSTEMS
250.130(A)

The equipment grounding conductor shall be bonded to the grounded (neutral) conductor and the grounding electrode conductor at the service equipment neutral busbar terminal. The grounded conductor (may be a neutral) shall be installed and connected to the grounding electrode system per **250.50** and **250.52(A)(1) through (A)(8)**. Note that the grounded (neutral) conductor shall be connected to the busbar where the grounding electrode conductor is terminated per **250.24(A)**. **(See Figure 11-102)**

FOR UNGROUNDED SYSTEMS
250.130(B)

The equipment grounding conductor shall be bonded to the grounding electrode conductor at the service equipment neutral busbar terminal. A grounded (neutral) conductor is not present to be connected to ground. Equipment grounding conductors installed per **250.118** shall be permitted to be used to bond and connect all metal noncurrent-carrying parts of the wiring system to the grounding electrode conductor at the service equipment. **(See Figure 11-103)**

Figure 11-102. This illustration shows the utility transformer and the panelboard both connected to earth ground for safety.

NONGROUNDING RECEPTACLE REPLACEMENT OR BRANCH CIRCUIT EXTENSIONS
250.130(C)(1) THRU (C)(5)

The basic rule for replacement of receptacles is that grounding-type receptacles shall be used to replace existing receptacles where there is an equipment grounding conductor routed with the branch circuit. A nongrounding-type receptacle shall be used with a branch circuit that has no equipment grounding conductor or grounding means. The equipment grounding conductor shall connect the green grounding terminal of the receptacle to the grounding electrode conductor or to the closest metal water pipe per **250.104(A)**, **250.50**, **250.52(A)(1)** and **250.53(D)**. **(See Figure 11-104)**

The NEC allows five methods by which a nongrounding receptacle shall be permitted to be replaced with a circuit without an equipment grounding conductor, and they are as follows:

- Install a nongrounding type receptacle at each receptacle location.

- A GFCI receptacle protecting each outlet at receptacle location.
- A GFCI receptacle protecting a single outlet and additional outlets downstream.
- An equipment grounding conductor routed from the receptacle to a metal water pipe per **250.53(D)** in the NEC.
- A GFCI CB protecting all outlets.

See Figure 11-105 for a detailed illustration when applying these requirements.

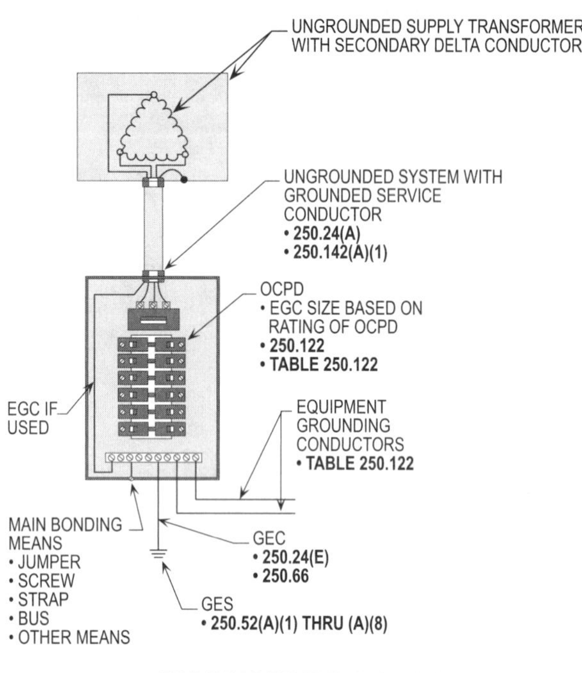

Figure 11-103. This illustration shows an ungrounded transformer being used to supply uninterrupted power to an industrial plant.

EQUIPMENT FASTENED IN PLACE OR CONNECTED BY PERMANENT WIRING METHODS (FIXED) – GROUNDING
250.134

Noncurrent-carrying metal parts of equipment, raceways, cables, and other enclosures, if grounded, shall be connected to an equipment grounding conductor by one of the following methods where required to be grounded for the safety of circuits, equipment, and personnel:

- Equipment grounding conductor types
- With circuit conductors

See Figure 11-106 for a detailed illustration when applying these requirements.

EQUIPMENT GROUNDING CONDUCTOR TYPES
250.134(A)

The following wiring methods shall be permitted to be installed as the equipment grounding conductor to ground electrical enclosures and equipment cases:

- Rigid metal conduit (RMC)
- Electrical metallic tubing (EMT)
- Type AC cable
- Approved metal conduit or cables

See Figure 11-107 for a detailed illustration when applying these requirements.

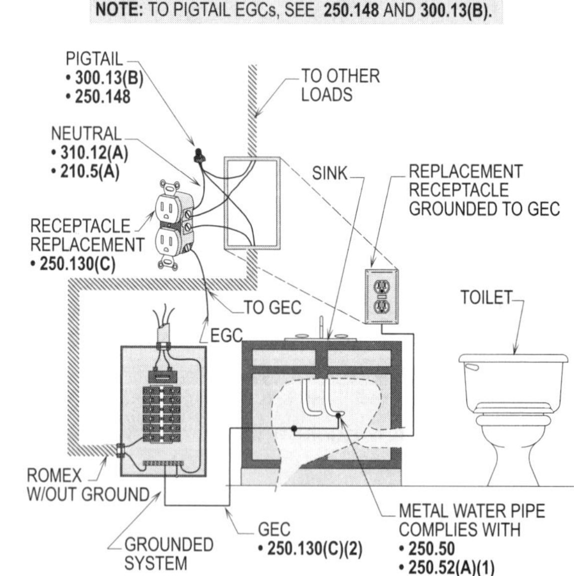

Figure 11-104. This illustration shows one method in which a grounding-type receptacle shall be permitted to be grounded to the grounding electrode conductor. Note that the branch circuit is not equipped with an equipment grounding conductor.

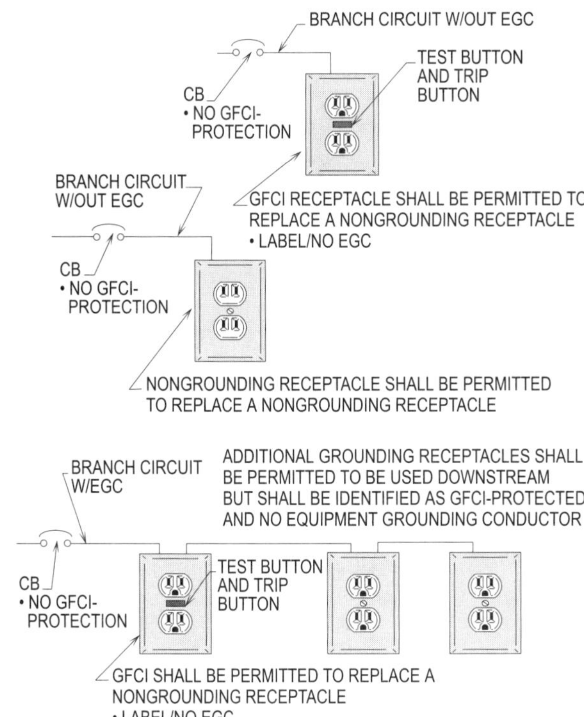

NONGROUNDING RECEPTACLE REPLACEMENTS OR BRANCH CIRCUIT EXTENSIONS NEC 250.130(C)(1) THRU (C)(5)

Figure 11-105. This illustration shows methods that shall be permitted to be used to replace and protect receptacles when there is no equipment grounding conductor available in the branch circuit wiring method.

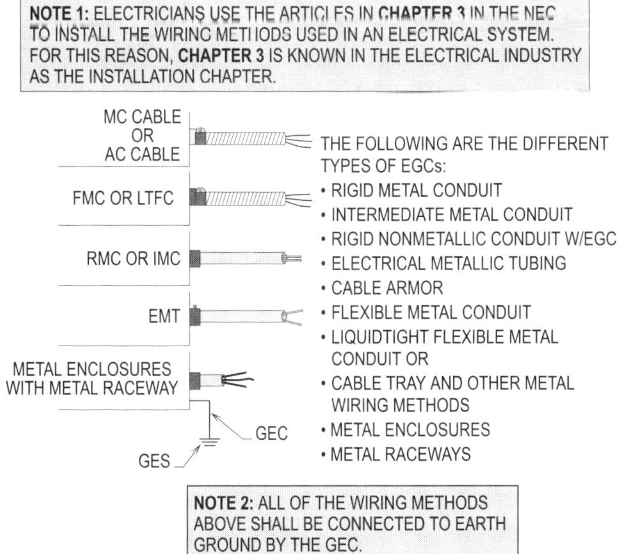

EQUIPMENT FASTENED IN PLACE OR CONNECTED BY PERMANENT WIRING METHODS (FIXED) – GROUNDING NEC 250.134

Figure 11-106. This illustration shows wiring methods that shall be permitted to be used to earth ground equipment that is fastened in place.

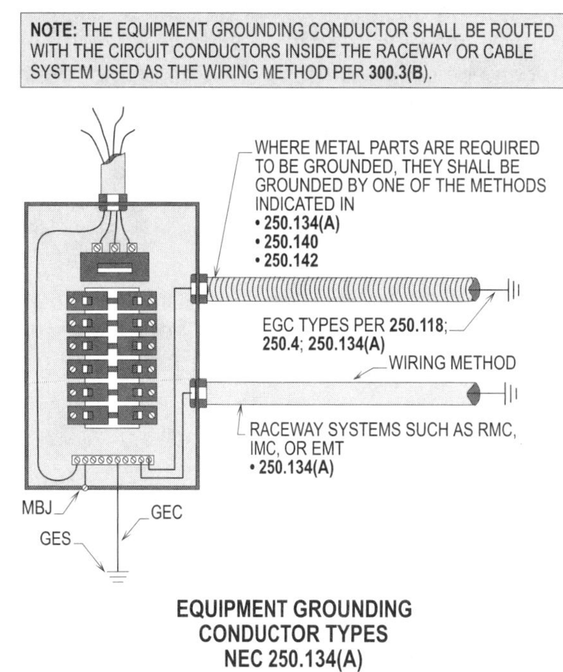

EQUIPMENT GROUNDING CONDUCTOR TYPES NEC 250.134(A)

Figure 11-107. This illustration shows wiring methods that shall be permitted to be used to earth ground the noncurrent-carrying metal parts of electrical related equipment and enclosures.

WITH CIRCUIT CONDUCTORS 250.134(B)

An equipment grounding conductor contained within the same raceway, cable, or cord shall be permitted to be carried along with circuit conductors and installed to ground electrical enclosures and equipment. Equipment grounding conductors shall be permitted to be bare, covered, or insulated. Insulated equipment grounding conductors shall be identified green or green with one or more yellow stripes. For more detailed information on this subject (with circuit conductors), see **300.3(B)**, **300.5(I)**, and **300.20**. **(See Figure 11-108)**

CORD-AND-PLUG CONNECTED EQUIPMENT 250.138

The metal cases of portable cord-and-plug connected stationary equipment, if grounded, shall be connected to an equipment grounding conductor by any one of the following wiring methods:

- By means of an equipment grounding conductor

- By means of a separate flexible wire or strap

Metal enclosures housing conductors shall be permitted to be used as a grounding means if an approved grounding type attachment plug is utilized with one fixed member making contact for the purpose of bonding and grounding the metal enclosure. However, the attachment plug shall be of a type that is approved for grounding from a grounding-type receptacle, and such cord is required to carry an equipment grounding conductor. One end shall be attached to the grounding terminal of the attachment plug, and the other end shall terminate to the frame of the portable equipment. For proper grounding, it is important that the metal enclosure of such conductors be attached to the attachment plug and to the equipment by approved connectors.

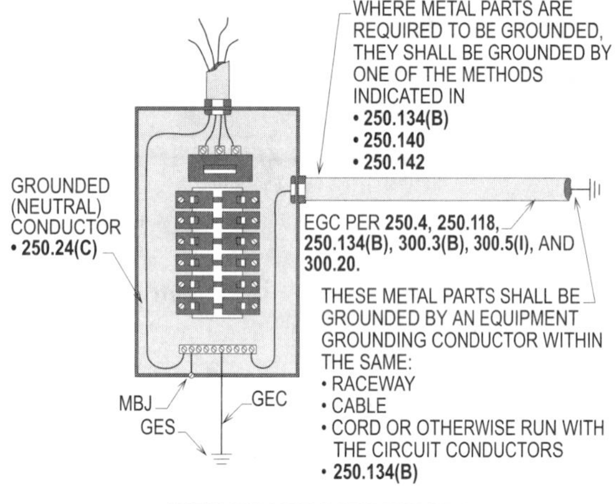

Figure 11-108. This illustration shows the methods by which the equipment grounding conductors shall be routed to ensure a low-impedance path for clearing ground faults.

BY MEANS OF AN EQUIPMENT GROUNDING CONDUCTOR
250.138(A)

The equipment grounding conductor in a cord is used to connect the grounding prong of a plug to ground the equipment. Equipment grounding conductors shall be permitted to be bare or insulated with a green color or green with one or more yellow stripes. Such grounding conductors shall be permitted to be routed in the cable or flexible cord assembly, provided that it terminates in an approved grounding type attachment plug having a fixed grounding-type contact member. **(See Figure 11-109)**

The **Ex.** to **250.138(A)** allows a movable self-restoring type such as an approved attachment plug that is equipped with a hinged grounding prong with a spring. Note that this spring may be folded out of the way and still restore itself back to the normal position on the grounding-type receptacle.

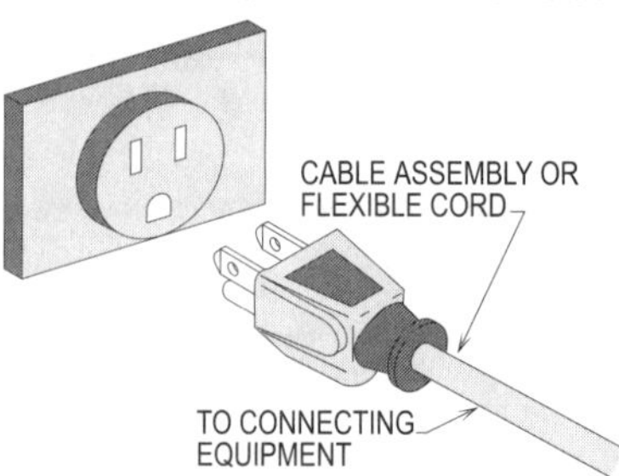

Figure 11-109. This illustration shows a flexible cord with an attachment cap and equipment grounding conductor used to ground the connected equipment to the receptacle and its supply.

BY MEANS OF A SEPARATE FLEXIBLE WIRE OR STRAP
250.138(B)

A separate flexible wire or strap shall be permitted to be used to connect to an equipment grounding conductor. This strap should be protected from physical damage as well as practical to ensure proper bonding and grounding.

FRAMES OF RANGES AND CLOTHES DRYERS
250.140

A new branch-circuit installation for ranges, cooktops, ovens, clothes dryers, including junction boxes or outlet boxes that are part of the circuit shall be permitted to be bonded and grounded with an equipment grounding conductor. However, an isolated grounded (neutral) conductor shall also be installed. For further information concerning this type of installation, see **250.114, 250.134,** and **250.138**. **(See Figure 11-110)**

Existing branch-circuit installations for the frames of ranges, cooktops, ovens, clothes dryers, and junction boxes or outlet boxes that are part of the circuit shall be permitted to be bonded and grounded with the grounded (neutral) conductor under the following conditions:

- The supply circuit is 120/240 volt, single-phase, three-wire or 208/120 volt derived from a three-phase, four-wire, wye-connected system.

- The grounded (neutral) conductor installed is no smaller than 10 AWG copper or 8 AWG aluminum.

- The grounded (neutral) conductor is insulated and part of NM cable (Romex) or PVC conduit.

- Grounding contacts of receptacles furnished as part of the equipment are bonded to the equipment.

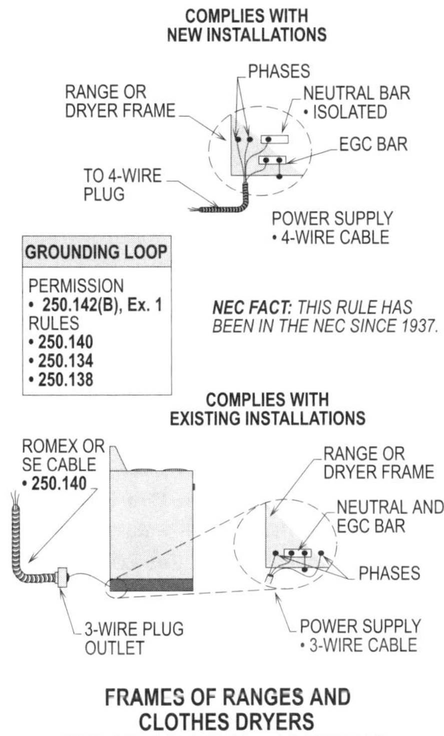

Figure 11-110. This illustration shows the procedure for installing a range, cooktop, oven, or dryer on existing and new branch-circuit installations.

USE OF GROUNDED CIRCUIT CONDUCTOR FOR GROUNDING EQUIPMENT
250.142

Under certain conditions, all metal parts of enclosures used to install the service equipment shall be permitted to be grounded by the grounded (neutral) conductor on the supply side of the system. The service weatherhead, service raceway, service meter base, and the service equipment enclosure are included when using this type of grounded system.

The grounded (neutral) conductor shall be installed and isolated from the other system circuit conductors and from metal enclosures or metal conduits when installed on the load side of the system.

SUPPLY SIDE EQUIPMENT
250.142(A)

The grounded (phase or neutral) conductor shall be permitted to be used as a current-carrying conductor and grounding means on the supply side of the service disconnecting means and secondary side of a separately derived system as follows:

- On the supply side of service equipment per **250.142(A)(1)** and **250.24(C)(1)**.

- On the supply side of the main service disconnecting means for separate buildings and structures per **250.142(A)(2)** and **250.32(B)** and **Ex.**

- On the supply side of the disconnect or overcurrent protection device of a separately derived system per **250.142(A)(3)** and **250.30(A)(1)**.

See Figure 11-111 for a detailed illustration when applying these requirements.

LOAD-SIDE EQUIPMENT
250.142(B)

The grounded (neutral) conductor shall not be permitted to be used as an equipment grounding conductor on the load side except as follows:

- Frame of ranges, wall-mounted ovens, counter-mounted cooking units, and clothes dryers per **250.140**. (Only for existing branch-circuits)

- Where one or more buildings or structures are supplied from a common AC grounded service, each grounded service at each individual building or structure shall be separately grounded per **250.32**.

- If no service ground fault protection is provided, all meter socket enclosures located near the service disconnecting means shall be permitted to be grounded by the grounded (neutral) conductor on the load side of the service disconnect. The grounded (neutral) conductor shall not be permitted to be smaller than the size specified in **Table 250.122**.

- DC systems shall be permitted to be grounded at the first disconnecting means or overcurrent protection device per **250.164(B)(2)**.

See Figure 11-112 for a detailed illustration when applying these requirements.

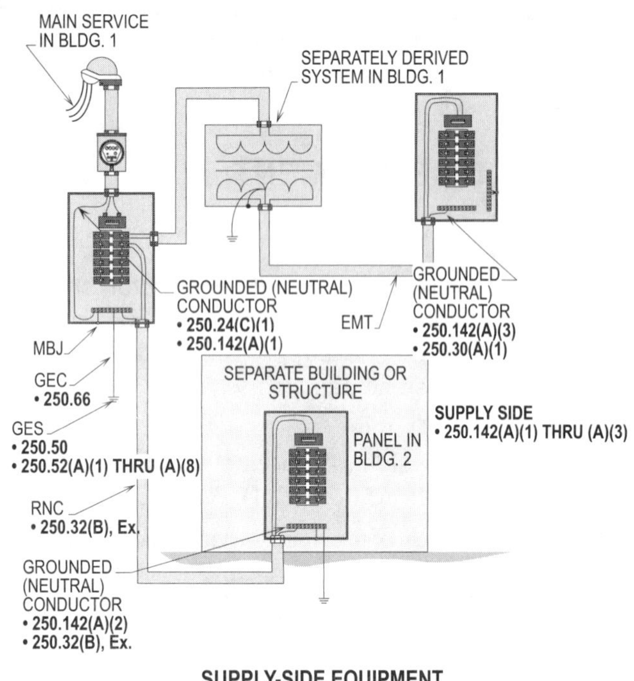

Figure 11-111. This illustration shows three installations where the grounded (neutral) conductor shall be permitted to be used as a current-carrying neutral plus equipment grounding conductor.

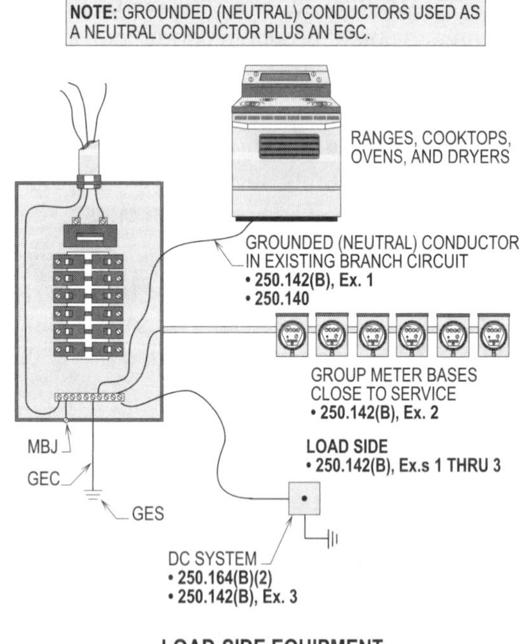

Figure 11-112. This illustration shows an installation where the grounded (neutral) conductor is used as a neutral conductor plus an equipment grounding conductor.

CONNECTING RECEPTACLE GROUNDING TERMINAL TO BOX 250.146

A good connection to ground shall be made by connecting the receptacle grounding terminals to the box. A bonding jumper shall be permitted to be used to connect the grounding terminal of a grounding-type receptacle to a metal grounded box. The equipment bonding jumper shall be sized based on the rating of the overcurrent protection device per **Table 250.122**. The following are four exceptions when a connection shall be permitted to be made without using a bonding jumper:

- Surface-mounted box
- Contact devices or yokes
- Floor boxes
- Isolated receptacles

SURFACE-MOUNTED BOX 250.146(A)

If a surface-mounted box such as a handy box complies with **250.146(B)**, then a bonding jumper providing metal-to-metal contact is not required. In other words, the #6-32 screws on the yoke of the device shall be considered an acceptable means of bonding the box and yoke together. Note that a bonding jumper shall be required if metal-to-metal contact is not provided by the connection above. A listed exposed work cover shall be permitted to be the grounding and bonding means if the device is attached to the cover with at least two fasteners that are permanent or have a thread locking or screw locking means and the cover mounting holes are located on a flat non-raised portion of the cover. **(See Figure 11-113)**

CONTACT DEVICES OR YOKES 250.146(B)

A flush-mounted box shall be permitted to be installed with self-grounding screws to ground receptacles with a bonding jumper. Note that these contact devices and yokes are specifically listed for proper grounding when used in conjunction with supporting screws to ensure adequate grounding between the yoke and flush-type mounted boxes, receptacles, or switches. **(See Figure 11-114)**

FLOOR BOXES 250.146(C)

The metal yoke of the receptacle requires no bonding jumper to be installed for floor boxes. Floor boxes are designed to provide a good contact between the metal box. However, such floor boxes shall be specifically designed and listed for such purpose. **(See Figure 11-115)**

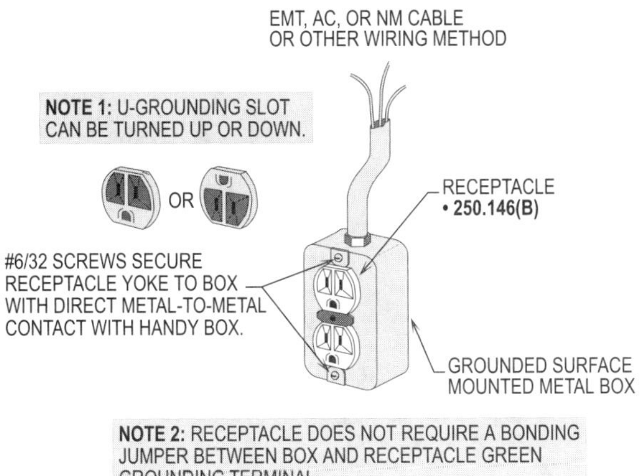

Figure 11-113. This illustration shows the surface box and the receptacle yoke making metal-to-metal contact.

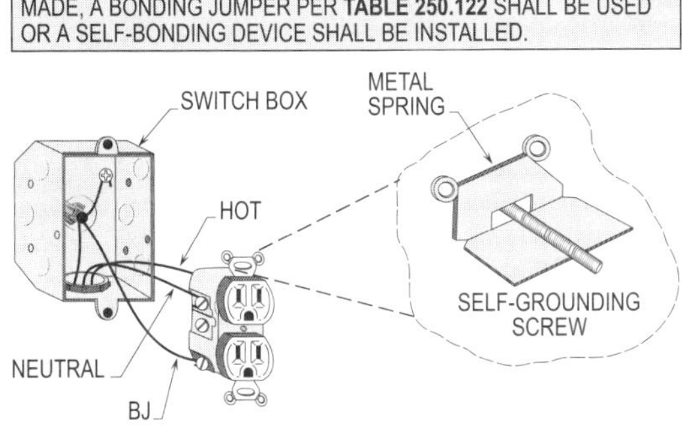

Figure 11-114. This illustration shows a self-bonding device used to make metal-to-metal contact between switch box and receptacle yoke.

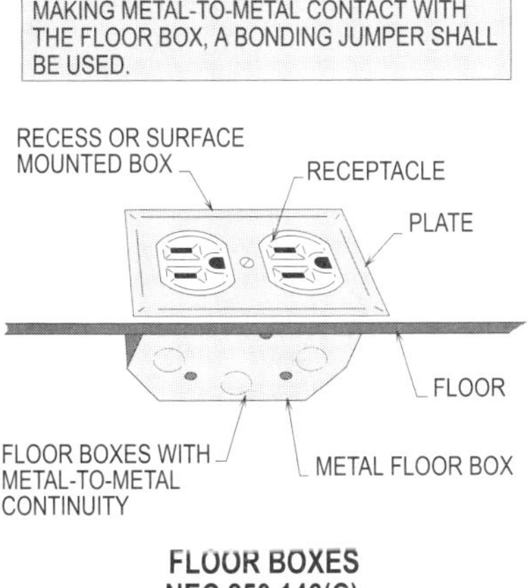

Figure 11-115. This illustration shows the technique for bonding and grounding a receptacle yoke to a metal floor plate.

ISOLATED RECEPTACLES
250.146(D)

An isolated equipment grounding conductor shall be used to connect the receptacle isolated ground terminal back to the grounded connection at the service equipment or separately derived system to reduce electromagnetic interference.

This rule applies when electrical noise known as electromagnetic interference occurs in the grounding circuit; an insulated equipment grounding conductor shall be permitted to be run to the isolated terminal with the circuit conductors. This insulated grounding conductor shall be permitted to pass through one or more panelboards, boxes, wireways, or other enclosures in the same building without being connected thereto as allowed in **408.40, Ex.** Note that it still has to terminate at the equipment grounding terminal at a separately derived system or service panel. Also, see **250.96(B)**. **(See Figure 11-116)**

CONTINUITY AND ATTACHMENT OF EQUIPMENT GROUNDING CONDUCTORS TO BOXES
250.148

Where more than one equipment grounding conductor enters a box, all such conductors shall be spliced within the box with devices suitable for the purpose. Equipment grounding conductors shall be connected and bonded to metal boxes to ensure grounding continuity. In other words, all grounding conductors entering the box shall be made electrically and mechanically secure, and the pigtail from them serves as the grounding connection to the device being installed. If the device is removed, the continuity of the equipment grounding conductors will not be disrupted. **(See Figure 11-117)**

COMPUTER GROUNDING USING A SPACER
645.15 AND 250.96(B)

Problems often occur with computers due to electrical noise within the electrical circuits. Special grounding methods shall be installed to minimize the electrical noise. A nonmetallic spacer shall be permitted to be installed between sensitive electronic equipment and the service panel or source per **250.6(D)** and **250.96(B)**. An insulated equipment grounding conductor shall be installed with circuit conductors. **(See Figure 11-118)**

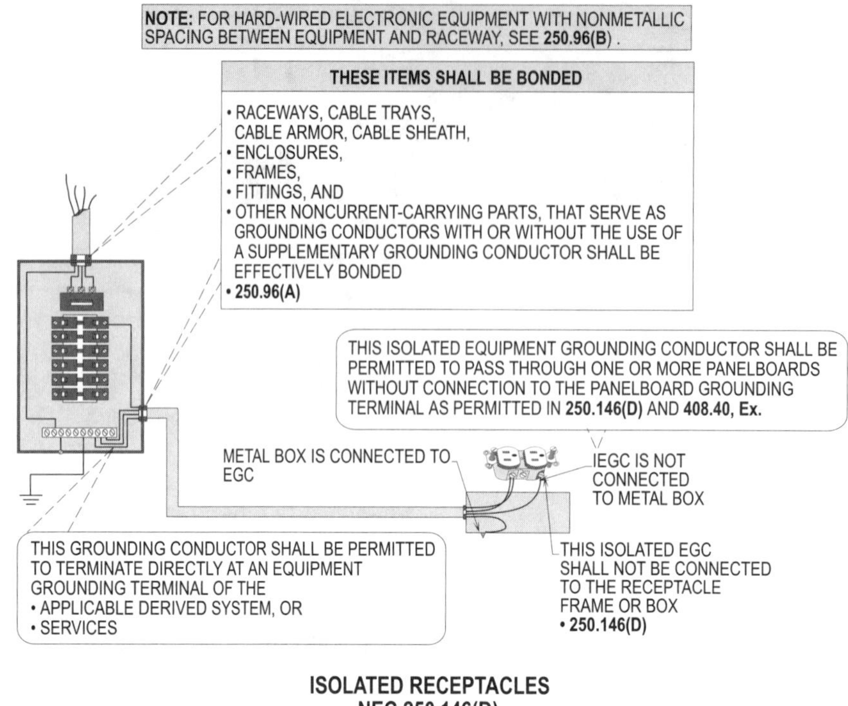

Figure 11-116. This illustration shows the techniques for bonding and grounding the yoke of an isolation receptacle to a metal box.

Note that when installing a grounding electrode conductor to a driven rod or other type of electrode that is not connected to the grounding electrode system of the building, the computer is not completely isolated; this creates a grounding hazard. The grounding electrode conductor has no low-impedance return path for the fault current to return to the power source and trip open the overcurrent protection device. This is due to the high resistance of the earth.

RADIAL GROUNDING
645.15

More grounding noise is sensed when there is not an equipotential plane between the computer and other grounded metal.

To eliminate such noise, the computer shall be grounded with a radial system that clears the ground system of ground loops. Ground loops produce paths for circulation of the ground currents, and these currents create unwanted electrical noise that causes computers to malfunction.

Grounding for a computer shall be installed in accordance with **Article 250**. The grounding electrode conductor at the service equipment or subfed transformer shall be connected to all exposed noncurrent-carrying parts of the computer enclosure. The equipment grounding conductor is used to provide this function of grounding to the enclosure of the computer. **(See Figure 11-119)**

HIGH-FREQUENCY REFERENCE GRID
645.15

A high-frequency reference grid system shall be permitted to be connected to a computer grounding point. A single ground point shall be grounded radially and tied to a high-frequency reference grid for computer equipment. The computer shall be grounded by the grid at the same ground point. The grounding electrode is then grounded to the computer ground point. This type of grounding scheme is installed to help minimize electrical noise and complies with the rules and regulations of the NEC.

The power conductors, grounds, and grounded (neutral) conductor shall be installed in the same conduit or cable except for the reference grid ground; if need be, it shall be permitted to be routed away from these circuit conductors.

SWIMMING POOLS
ARTICLE 680, PART II

Swimming pools and decorative fountains shall be grounded and bonded to prevent electrical shock to people wading and swimming in the pool water. Proper sized bonding and grounding conductors are essential to ensure the safety of persons having a good time in or around the wet areas of the swimming pool.

WHERE MORE THAN ONE EGC BRANCH CIRCUITS ENTERS A BOX, ALL SUCH CONDUCTORS SHALL BE:
• SPLICED, OR
• JOINED WITHIN THE BOX OR TO THE BOX WITH DEVICES SUITABLE FOR THE USE.

NOTE THAT THE ARRANGEMENT OF THESE GROUNDING CONDUCTORS SHALL BE SUCH THAT THE DISCONNECTION OR REMOVAL OF A:
• RECEPTACLE,
• FIXTURE, OR
• OTHER DEVICE FED FROM THE BOX, WILL NOT INTERFERE WITH THE GROUNDING CONTINUITY.

THE ARRANGEMENT OF THESE GROUNDING CONDUCTORS SHALL BE SUCH THAT THE DISCONNECTION OR REMOVAL OF THE DEVICES WILL NOT INTERRUPT THE GROUNDING CONTINUITY.
• 250.148(B)

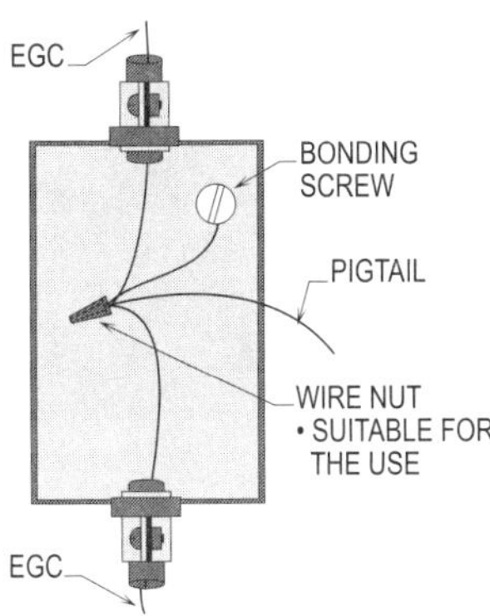

CONTINUITY AND ATTACHMENT OF EQUIPMENT GROUNDING CONDUCTORS TO BOXES NEC 250.148

Figure 11-117. This illustration shows that equipment grounding conductors shall be bonded and grounded to boxes, enclosures, etc. with a method that will ensure uninterrupted continuity.

EQUIPOTENTIAL BONDING
680.26

The equipotential bonding shall be installed to reduce voltage gradients in the pool area.

BONDED PARTS
680.26(B)

The conductive pool shell, perimeter surfaces, metallic components, underwater lighting, metal fittings, electrical equipment, and metal wiring methods and equipment shall be bonded together using solid copper conductors that are insulated, covered, or bare and not smaller than 8 AWG or with rigid metal conduit of brass or other identified corrosion-resistant metal. The 8 AWG or larger solid copper bonding conductor is installed to reduce voltage gradients in the pool area. This conductor shall not be extended or attached to remote panelboards, service equipment, or electrodes.

CONDUCTIVE POOL SHELLS
680.26(B)(1)

The conductive pool shell shall be bonded. The following shall be considered conductive materials due to water permeability and porosity:

• Poured concrete,

• Pneumatically applied or sprayed concrete, and

• Concrete block with painted or plastered coatings.

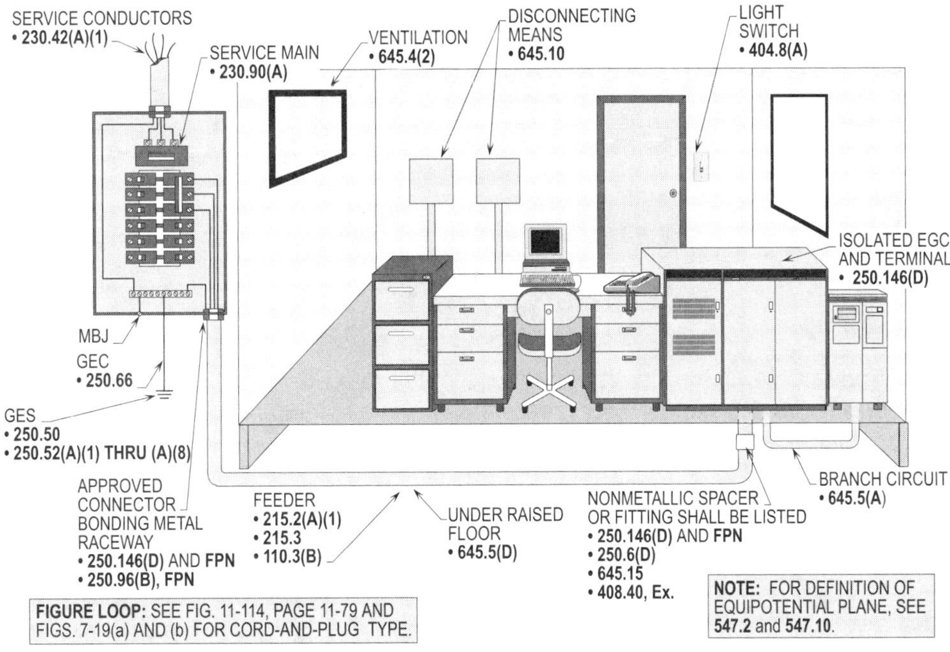

COMPUTER GROUNDING USING A SPACER NEC 645.15 NEC 250.96(B)

Figure 11-118. A nonmetallic spacer shall be permitted to be installed between sensitive electronic equipment and the service panel or source. However, an equipment grounding conductor shall be run to such equipment.

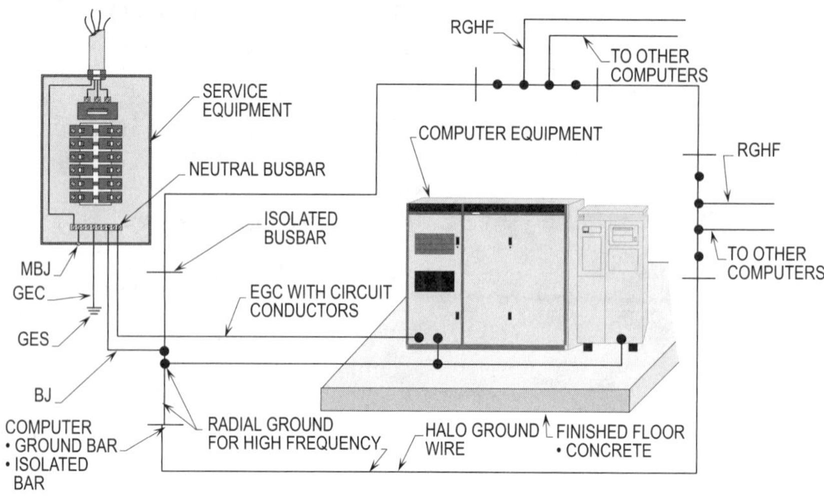

**SERVICE EQUIPMENT SUPPLYING COMPUTER
NEC 645.15**

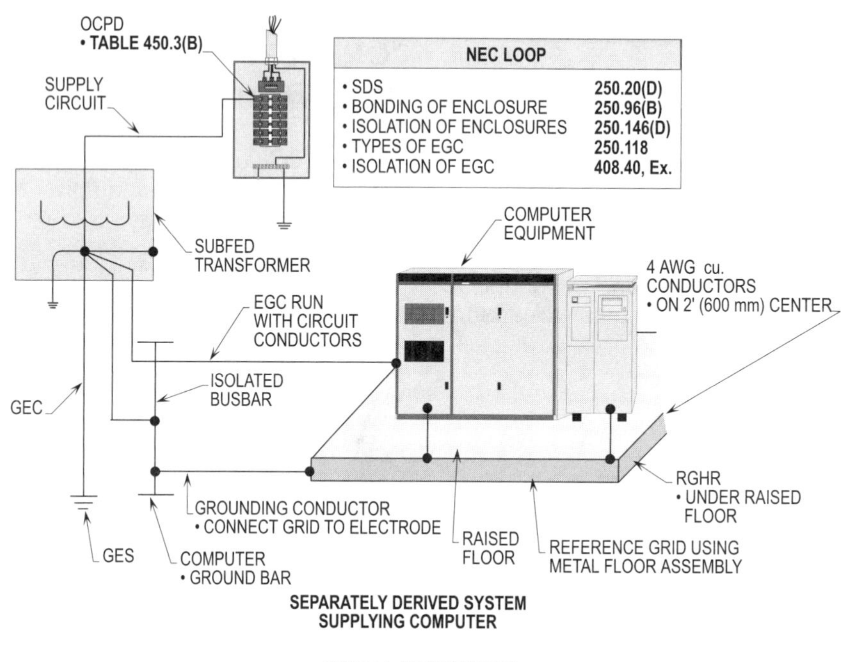

**SEPARATELY DERIVED SYSTEM
SUPPLYING COMPUTER**

**RADIAL GROUNDING
NEC 645.15**

Figure 11-119. Ground loops produce paths for circulation of the ground currents, and these currents create unwanted electrical noise that causes computers to malfunction.

Note, vinyl liners and fiberglass composite shells shall not be considered to be conductive materials.

Unencapsulated structural reinforcing steel shall be bonded together by steel tie wires or the equivalent where structural reinforcing steel is encapsulated in a nonconductive compound. A copper conductor grid shall be installed as follows:

- The grid shall be constructed of 8 AWG or larger bare solid copper conductors bonded to each other at all points of crossing.

- The grid shall conform to the contour of the pool, also the pool deck.

- The grid shall be arranged in a 12 in. x 12 in. (300 mm x 300 mm) network of conductors in a uniformly spaced perpendicular grid patterns, with a tolerance of 4 in. (100 mm).

- The grid shall be secured within or under the pool no more than 6 in. (150 mm) from the outer contour of the pool shell.

PERIMETER SURFACES
680.26(B)(2)

The perimeter shall extend for 3 ft (1 m) horizontally beyond the inside walls of the pool and shall include the following:

- Unpaved surfaces,
- Poured concrete, and
- Other types of paving.

The perimeter surface shall be bonded and attached to the pool reinforcing steel or copper conductor grid at a minimum of four (4) points uniformly spaced around the perimeter of the pool. The structural steel reinforcing steel shall be bonded per **680.26(B)(1)(a)**. Where structural reinforcing steel is not available or is encapsulated in a nonconductive compound, a copper conductor(s) shall be installed where the following conditions are complied with:

- At least one 8 AWG or larger bare solid copper conductor shall be provided.
- The conductors shall follow the contour of the perimeter surface.
- Only listed splices shall be permitted.
- The copper conductor shall be 18 in. to 24 in. (450 mm to 600 mm) from the inside walls of the pool.
- The copper conductor shall be secured within or under the perimeter surface 4 in. to 6 in. (100 mm to 150 mm) below the subgrade.

See **Figure 11-120** for a detailed illustration pertaining to the equipotential bonding of a swimming pool.

GROUNDING
680.6

All metal parts around the pool shall be bonded and grounded to ground potential and thus reduce the threat of shock hazard. The equipment and parts that shall be grounded include the following items:

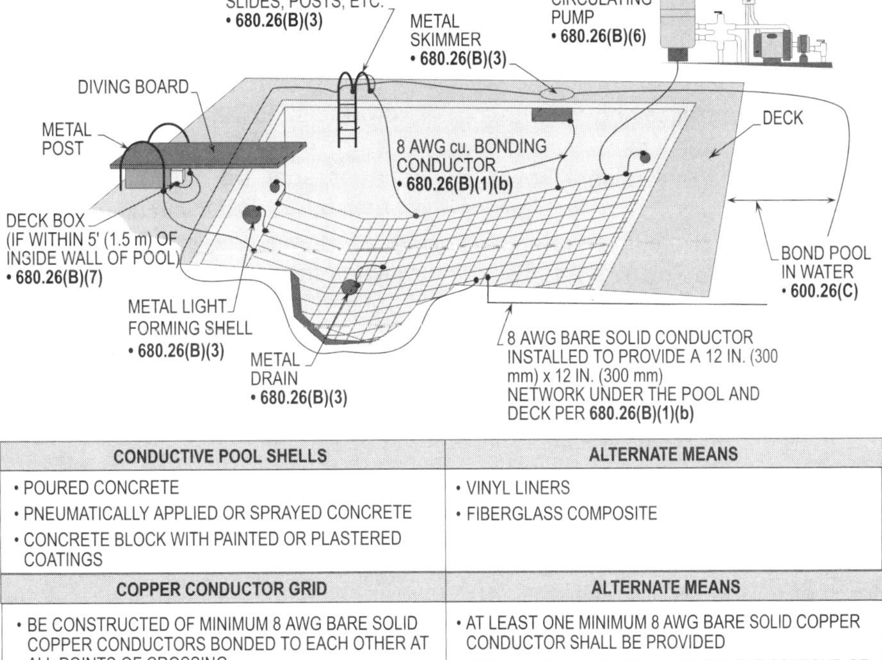

CONDUCTIVE POOL SHELLS	ALTERNATE MEANS
• POURED CONCRETE • PNEUMATICALLY APPLIED OR SPRAYED CONCRETE • CONCRETE BLOCK WITH PAINTED OR PLASTERED COATINGS	• VINYL LINERS • FIBERGLASS COMPOSITE
COPPER CONDUCTOR GRID	**ALTERNATE MEANS**
• BE CONSTRUCTED OF MINIMUM 8 AWG BARE SOLID COPPER CONDUCTORS BONDED TO EACH OTHER AT ALL POINTS OF CROSSING • CONFORM TO THE CONTOUR OF THE POOL AND THE POOL DECK • BE ARRANGED IN A 12 IN. (300 mm) BY 12 IN. (300 mm) NETWORK OF CONDUCTORS IN A UNIFORMLY SPACED PERPENDICULAR GRID PATTERN WITH TOLERANCE OF 4 IN. (100 mm)	• AT LEAST ONE MINIMUM 8 AWG BARE SOLID COPPER CONDUCTOR SHALL BE PROVIDED • THE CONDUCTORS SHALL FOLLOW THE CONTOUR OF THE PERIMETER SURFACE • THE REQUIRED CONDUCTOR SHALL BE 18 TO 24 IN. (450 TO 600 mm) FROM THE INSIDE WALL OF THE POOL • THE REQUIRED CONDUCTOR SHALL BE SECURED WITHIN OR UNDER THE PERIMETER SURFACE 4 TO 6 IN. (100 TO 150 mm) BELOW THE SUBGRADE

EQUIPOTENTIAL BONDING
NEC 680.26(A) THRU (C)

Figure 11-120. The bonding of all metal parts and the grounding of the elements keep the interconnected system and steel at the same potential, which creates an equipotential plane.

- Through-wall lighting assemblies and underwater luminaires, other than those low-voltage lighting products listed for the application without a grounding conductor

- All electrical equipment located within 5 ft (1.5 m) of the inside walls of the pool

- All electrical equipment associated with the recirculating pump

- Junction boxes

- Transformer enclosures

- Ground fault circuit interrupters

- Panelboards that are not part of the service equipment and they supply electrical equipment associated with the pool

The equipment grounding conductor used to ground the metal parts of equipment, junction boxes, etc. shall be sized from **Table 250.122**, based on the rating of the overcurrent protection device protecting the branch-circuit conductors.

Grounding of the junction box and wet-niche luminaires shall be accomplished by a 12 AWG copper insulated equipment grounding conductor. The grounding conductor shall be sized according to the rating of the overcurrent protection device per **Table 250.122**. It shall not be smaller than 12 AWG and shall be unspliced unless the exception is applied.

The grounding conductor shall be installed and run with the circuit conductors in rigid metal conduit, intermediate metal conduit, or rigid nonmetallic conduit from the panelboard to the deck box. The main function of the 12 AWG equipment grounding conductor is to clear the overcurrent protection device in case of a ground fault. **(See Figure 11-121)**

Flexible cords used to ground wet-niche luminaires shall be provided with a 12 AWG or larger insulated copper equipment grounding conductors connected to a terminal in the supply deck box. The type of connection provides continuity when the luminaire is removed for servicing.

> **Design Tip:** The 12 AWG equipment grounding conductor is used to clear a ground fault, while the 8 AWG bonding jumper is utilized to bond the metal parts of the pool to the grid system.

HOT TUBS, SPAS, AND HYDROMASSAGE TUBS
ARTICLE 680, PARTS IV AND VII

Hot tubs, spas, and hydromassage tubs shall be bonded and grounded for the same reasons as swimming pools for the safety of those using them. Such tubs shall be permitted to be located inside or outside of the dwelling unit or building.

BONDING
680.42(B)

Bonding by metal-to-metal mounting on a common frame or base shall be permitted. The metal band or hoops that are used to secure wooden staves shall not be required to bonded per **680.26**. **(See Figure 11-122)**

GROUNDING
680.43(F)

The following electrical equipment shall be grounded:

- All electrical equipment located within 5 ft (1.5 m) of the inside wall of the spa or hot tub

- All electrical equipment associated with the circulating system of the spa or hot tub

HYDROMASSAGE BATHTUBS
680.70, 680.71, AND 680.72

Hydromassage bathtubs are treated in the same manner as conventional bathtubs. The wiring methods for hydromassage bathtubs shall comply with **Chapters 1 through 4** in the NEC. All elements for hydromassage bathtubs shall be on an individual branch circuit(s) and protected by a readily accessible GFCI. See **410.10(D)** for hanging luminaires over and around the tub. Location of switches for luminaires and receptacle outlets are treated in the same manner as a regular bathtub. **(See Figure 11-123)**

> **Design Tip:** See **404.4** and **406.8(C)** for locating and installing switches and receptacles that are not associated with the hydromassage tub. Access to the circulating motor for servicing is required per **Article 100, 430.14(A)**, and **680.73**.

GROUNDING ANTENNA SYSTEMS PER NEC AND UL

Satellite dishes, CATV, and antenna systems shall be grounded for safety. Inspectors sometimes find that these systems are not bonded and grounded properly when making an inspection of such systems. Many times they are not grounded into the grounding electrode system, but are grounded independently with different potentials to ground. **(See Figure 11-124)**

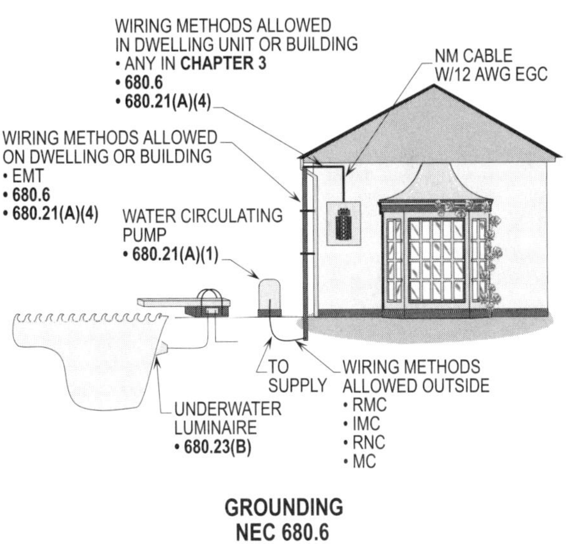

Figure 11-121. This illustration shows the types of wiring methods permitted to be installed for swimming pools.

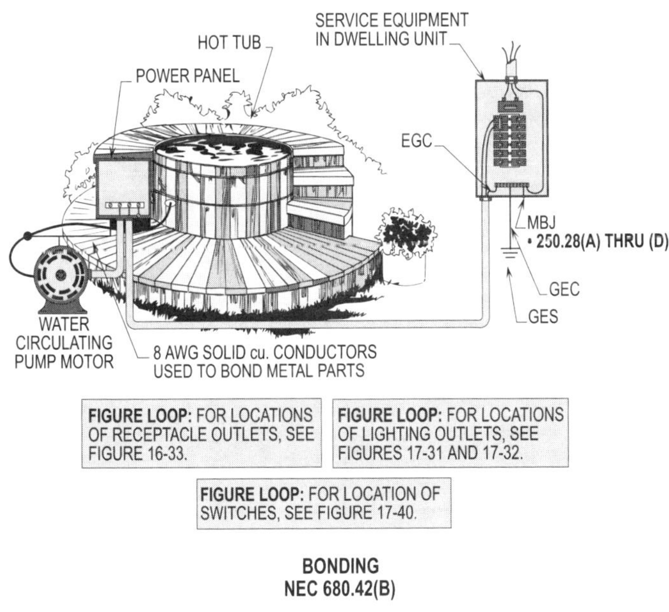

Figure 11-122. A 8 AWG solid copper conductor shall be used to ground all noncurrent-carrying parts associated with the hot tub including metal piping, etc. located within 5 ft (1.5 m) of the tub.

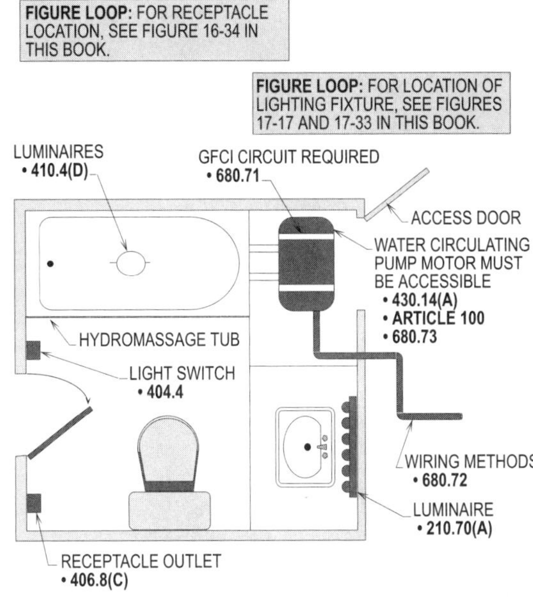

Figure 11-123. Hydromassage bathtubs are treated in the same manner as regular bathtubs except they shall be on an individual branch circuit(s) and protected by a readily accessible GFCI to protect the users from electrical shock. Luminaires shall comply with **410.10(D)**. Switches and receptacles shall meet the provisions of **404.4** and **406.8(C)**.

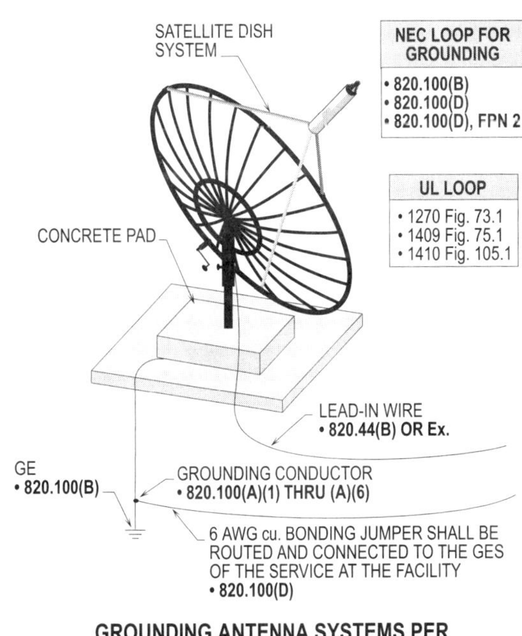

Figure 11-124. The illustration shows the proper grounding of a satellite dish system.

Chapter 11. Grounding

Section   Answer

________ ________   **1.** AC systems of _____ kV and over shall be grounded if supplying mobile or portable equipment.
        (a) 1   (b) 2
        (c) 3   (d) 6

________ ________   **2.** It is permitted to install the equipment bonding jumper either inside or outside of the raceway or enclosure in lengths of _____ ft or less. (Using flex)
        (a) 3   (b) 5
        (c) 6   (d) 10

________ ________   **3.** The grounding electrode conductor shall be installed within _____ ft where a metal water pipe enters a commercial building.
        (a) 3   (b) 5
        (c) 6   (d) 10

________ ________   **4.** All electrical equipment located within _____ ft of the inside walls of the pool shall be bonded and grounded.
        (a) 2   (b) 3
        (c) 5   (d) 6

________ ________   **5.** A grounded conductor, whether a grounded phase or neutral, is usually installed with a __________ or natural gray insulation or otherwise identified.
        (a) black   (b) purple
        (c) green   (d) white

________ ________   **6.** The bonding jumper shall be sized at least __________ percent of the area of the largest conductor where the service conductors are installed larger than 1100 KCMIL copper or 1750 KCMIL aluminum.
        (a) 6-1/2   (b) 12-1/2
        (c) 15-1/2   (d) 20-1/2

________ ________   **7.** Existing branch-circuit installations for frames or ranges, cooktops, ovens, clothes dryers, and junction boxes or outlet boxes that are part of the circuit shall be permitted to be bonded and grounded with grounded (neutral) conductor provided it is installed no smaller than __________ copper.
        (a) 10   (b) 8
        (c) 6   (d) 4

________ ________   **8.** A(n) __________ spacer fitting shall be permitted to be installed between the metal raceway and sensitive electronic equipment at the point of connection to reduce electrical noise.
        (a) listed   (b) metallic
        (c) identified   (d) nonmetallic

________ ________   **9.** When an equipment bonding jumper is installed outside of the raceway or enclosure, it shall not be over __________ ft in length. (Using flex)
        (a) 3   (b) 5
        (c) 6   (d) 10

10. Liquidtight flexible metal conduit 3/8 in. and 1/2 in. trade sizes shall be protected by overcurrent devices rated at __________ amps or less.
    (a) 15   (b) 20
    (c) 30   (d) 60

11. The metal frame of a building or structure shall have __________ ft or more of a single structural metal member in direct contact with the earth.
    (a) 5   (b) 6
    (c) 10   (d) 15

12. Grounding of the junction box and wet-niche luminaire for a pool is accomplished by installing a __________ AWG copper insulated equipment grounding conductor.
    (a) 12   (b) 10
    (c) 8   (d) 6

13. The rating of a grounded conductor shall be at least __________ percent of the largest ungrounded phase conductor.
    (a) 10   (b) 12 1/2
    (c) 15 1/2   (d) 25

14. Which of the following methods can be utilized for the bonding of service equipment?
    (a) grounded service conductor   (b) threaded connections
    (c) bonding jumpers   (d) all of the above

15. The interior metal water piping, located more than __________ ft from the point of entrance to the building, shall not be permitted to be used as part of the grounding electrode system. (Splicing other GECs to it)
    (a) 5   (b) 6
    (c) 10   (d) 20

16. Metal water pipe lengths of at least __________ ft long in the earth shall be connected to the grounded neutral bar in the service-equipment enclosure.
    (a) 5   (b) 10
    (c) 15   (d) 20

17. A concrete-encased electrode no smaller than __________ AWG bare copper conductor at least 20 ft long shall be permitted to be installed in the footing of the foundation.
    (a) 8   (b) 4
    (c) 2   (d) 1

18. A bare copper conductor not smaller than 2 AWG and at least 20 ft in length shall be installed at least __________ ft below the earth to create a ground ring grounding system for encircling the building or structure.
    (a) 2-1/2   (b) 4-1/2
    (c) 5-1/2   (d) 6-1/2

19. A driven electrode used to ground the service can be a __________ in. x __________ ft copper (brass) rod.
    (a) 1/2, 8   (b) 1/2, 10
    (c) 3/4, 8   (d) 3/4, 10

20. Made electrodes shall have a resistance to ground of _________ ohms or less.
    (a) 10   (b) 20
    (c) 25   (d) 50

21. Liquidtight flexible metal conduit can be used in sizes up to 1-1/4 in. that has a total length not to exceed _________ ft from the ground return path and with fittings for terminating the flexible conduit.
    (a) 3   (b) 5
    (c) 6   (d) 10

22. Where a main bonding jumper is a screw only, the screw shall be identified with a _____ finish.
    (a) white   (b) green
    (c) gray   (d) black

23. The common grounding electrode conductor for multiple separately derived systems shall not be smaller than _____ AWG copper.
    (a) 6   (b) 2
    (c) 1/0   (d) 3/0

24. Rod and pipe electrodes shall not be less than _____ ft in length.
    (a) 4   (b) 6
    (c) 8   (d) 10

25. Steel plate electrodes shall be at least _____ in. in thickness.
    (a) 1/16   (b) 1/4
    (c) 5/8   (d) 3/4

26. A ground ring shall be buried at a depth below the earth's surface of not less than _____ in.
    (a) 24   (b) 30
    (c) 36   (d) 48

27. A ground rod shall be permitted to be driven at an angle up to _____ degrees if a rock bottom is encountered.
    (a) 15   (b) 30
    (c) 45   (d) 60

28. Where used outside, aluminum grounding conductors shall not be terminated within _____ in. of the earth.
    (a) 18   (b) 24
    (c) 30   (d) 36

29. A _____ AWG or larger copper grounding electrode conductor shall be protected where exposed to physical damage.
    (a) 6   (b) 4
    (c) 2   (d) 1

30. The grounding electrode conductor shall not be permitted to be larger than _____ AWG copper where connected to a ground rod.
    (a) 8   (b) 6
    (c) 4   (d) 2

31. What size copper bonding jumper and grounding electrode conductor is required to bond the secondary side of a separately derived system having 2/0 AWG THWN copper conductors?

32. What size copper main bonding jumper is required for service equipment supplied with phase conductors of 500 KCMIL copper?

33. What size copper equipment bonding jumper is required where three raceways enter a switchgear, with each raceway individually bonded to the enclosure supplied with phase conductors of 250 KCMIL copper?

34. What size copper EBJ is required where three raceways enter a switchgear with only one bonding jumper used to bond the raceway? Service conductors are 250 KCMIL copper, paralleled 3 times. (Supply side)

35. What size copper grounding electrode conductor is required to be installed to a metal water pipe supplied with phase conductors of 2/0 AWG THWN copper?

36. What size copper grounding electrode conductor is required to be installed for a concrete-encased electrode supplied with phase conductors of 3/0 AWG THWN copper?

37. What size copper equipment grounding conductor is required to be installed for an A/C unit supplied by an 80 amp circuit breaker?

12

Designing Wiring Methods

Boxes and conduits shall be sized and selected to accommodate the number of conductors required to supply power to various pieces of electrical equipment in residential, commercial, and industrial locations. Boxes, in addition to housing conductors, shall have enough fill area to provide sufficient space for receptacles, switches, dimmers, and combination devices without damaging the conductors insulation. When boxes and conduits are not sized properly, conductors are sometimes packed in undersized boxes and are pinched by large-size devices that intrude into the box's fill area. Therefore, conductors, devices, fittings, and clamps shall be counted and the appropriate box selected. Boxes and conduits shall be sized based on the number of conductors of either the same size and having the same characteristics or of different sizes and different characteristics.

BOX FILL CALCULATIONS
314.16(B)

The box size selected shall be based on the number and size of conductors, devices, and fittings that are contained within the box. The designer, installer, and inspector shall count the number of items that are in the box and select the size box from **Table 314.16(A)**, based upon the maximum number of conductors or the minimum cubic inch capacity rating. The column for the maximum number of conductors shall be used when the conductors are the same size. The appropriate cubic inch capacity column shall be applied for combination conductors in the same box.

CONDUCTOR FILL
314.16(B)(1)

Conductors passing through the box unbroken and not pulled into a loop, coil, or spliced together with scotchlocks shall be counted as 1 conductor. Conductors passing through boxes are utilized to supply power to boxes supporting luminaires, receptacles, etc. Spliced conductors shall be counted as 1 conductor for each.

For example: What is the count used for 10 conductors passing straight through an octagon box and the count for 10 that are spliced?

Step 1: Counting conductors
314.16(B)(1)
10 passing through counted as 10

Solution: Ten conductors are counted for the ten conductors passing through the box.

Step 2: Counting conductors
314.16(B)(1)
10 spliced x 2 count as 20

Solution: Twenty conductors are counted for the ten spliced conductors in the box.

See Figures 12-1(a) and (b) for counting the number of conductors passing through or spliced in a box.

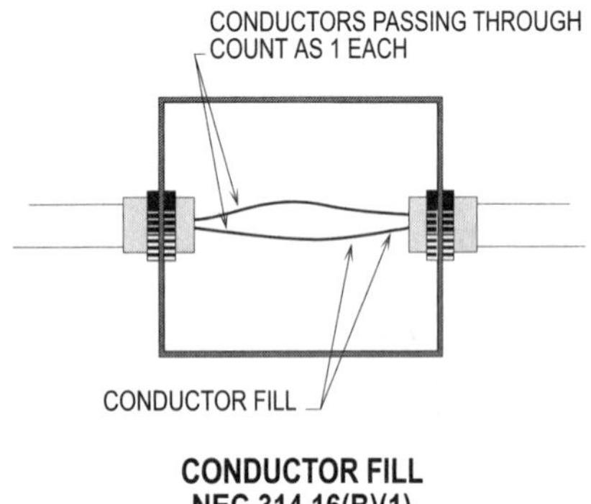

Figure 12-1(a). Conductors passing straight through the box shall be counted as 1 each.

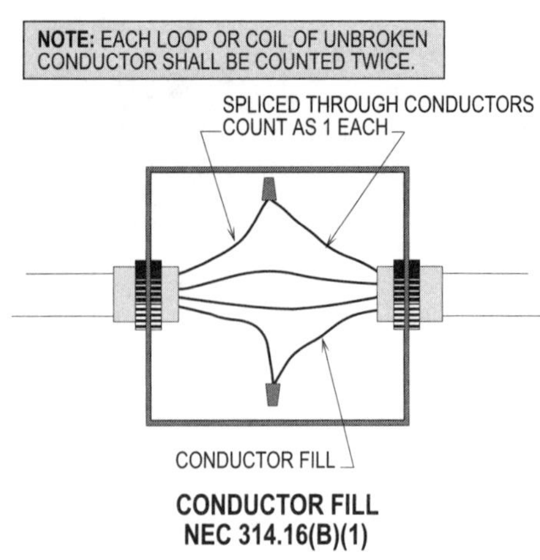

Figure 12-1(b). Spliced conductors shall be counted as 1 for each conductor spliced in the box.

CONDUCTOR FILL
314.16(B)(1)

A conductor that is not entirely within the box shall not be required to be counted as a conductor used toward the box fill. A pigtail from a splice within the box to a receptacle or a switch mounted to the box, or a bonding jumper from the receptacle yoke bonded to the box, is an example of such a conductor. **(See Figure 12-2)**

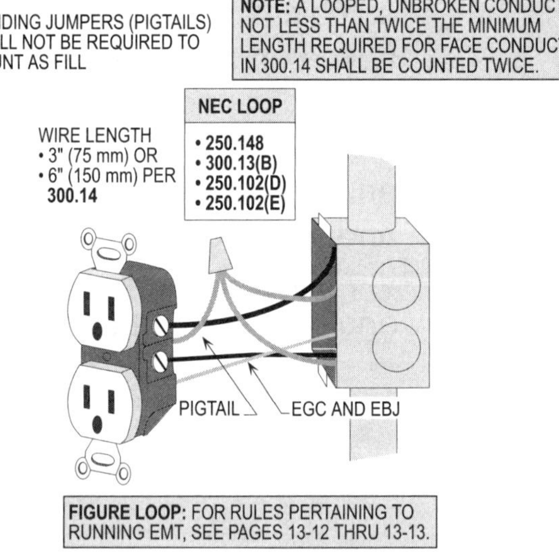

Figure 12-2. A pigtail or bonding jumper shall not be required to be counted toward the fill space of the box.

CONDUCTOR FILL
314.16(B)(1), Ex.

Wires from luminaires shall be counted in determining the size of the box. Fixture wire sizes from 18 AWG to 12 AWG or larger shall be counted to determine the box size used to support the luminaire. The **Ex.** to **314.16(B)(1)** permits the canopy of a luminaire to be used to house conductors or four or less fixture wires in sizes smaller than 14 AWG plus an equipment grounding conductor. It is not necessary to count these fixture wires toward the fill of the ceiling box supporting the luminaire or ceiling fan where there is a canopy.

Review and verify that both the luminaire box and luminaire canopy can be used for fill area. If so, a smaller ceiling box shall be permitted to be installed to support the luminaire.

> **Design Tip:** Until the 1984 NEC was published, fixture wires were not counted in determining the fill space of a box.

Note, the 1984 *National Electrical Code* required fixture wires 14 AWG and larger to be counted, and the 1987 NEC required fixture wire in sizes 18 AWG and 16 AWG to be counted. However, these requirements shall not be permitted to be retroactive to existing installations that would require larger boxes.

> **For example:** What is counted toward the fill space of two boxes where there are 2 - 14 AWG fixture wires without the use of a canopy and 3 - 16 AWG fixture wires plus an equipment grounding conductor utilizing a canopy?
>
> **Step 1:** Counted conductors
> (fixture wire only)
> **314.16(B)(1), Ex.**
> 2 - 14 AWG fixture wires
> count as 2
>
> **Solution: Two fixture wires are counted as 2 conductors in the box.**
>
> **Step 1:** Counted conductors
> (fixture wire only)
> **314.16(B)(1), Ex.**
> 3 - 16 AWG fixture wires count is 0
>
> **Solution: The fixture wires plus the equipment grounding conductor do not count toward the fill of the box.**

See Figure 12-3 for the procedure used to count fixture wires.

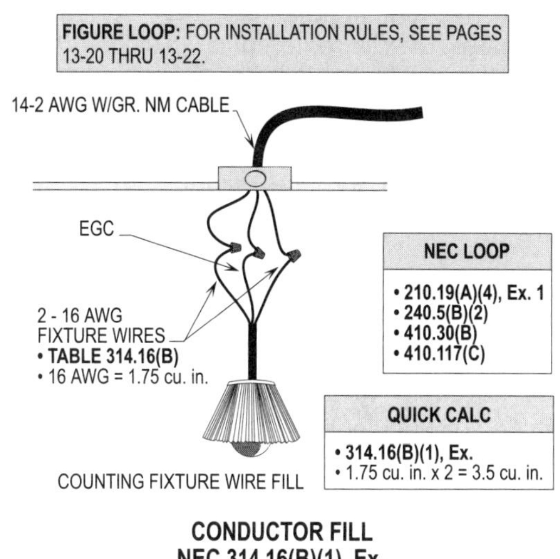

Figure 12-3. Fixture wires shall be counted in determining the size box to support a luminaire.

Note, fixture wires smaller than 14 AWG that are inside a luminaire canopy shall not be required to be counted toward the fill space for sizing a box.

CLAMP FILL
314.16(B)(2)

Boxes with one or more cable clamps that are installed to support cables in the box shall have 1 conductor added toward the fill calculation to determine the size box. The conductor used to represent the one or more cable clamps shall be based on the largest conductor in any one cable entering the box. The outer covering of the cable that is used to protect the insulation of the conductors shall be extended 1/4 in. (6 mm) past the clamp in the box to prevent the clamp from damaging the insulation of the conductor. When the clamp is tightened to support the cable, this extended covering beyond the clamp will prevent the insulation of the conductor from being pinched or nicked. The arcing or sparking of a bare conductor to metal that is grounded may cause a fire hazard. **(See Figure 12-4)**

Note, see the next design tip below "support fitting fill" as to the time when the NEC required cable clamps, luminaire studs, and hickeys to be counted as 1 conductor.

SUPPORT FITTINGS FILL
314.16(B)(3)

Boxes that contain luminaire studs or hickeys shall have 1 conductor added for each fitting. The conductor for each fitting is selected and based on the largest conductor entering the box. By adding a conductor for each fitting in the box, a larger box is required to be selected and helps avoid an undersized box. Conductors in boxes shall have sufficient room in the box per **110.7** and **314.16** to breathe and dissipate heat and prevent the overheating of its insulation per **310.10, FPN (1) through (4)**.

For example, a box contains cables of 14-2 AWG and 12-2 AWG w/ground and is equipped with a luminaire stud and hickey. The size conductors used for the count of the stud and hickey shall be determined by each one being counted as a 12 AWG. In other words, the two fittings count as 12 AWG conductors and shall be based on the largest conductor entering the box that is the 12-2 AWG w/ground cable. **(See Figure 12-5)**

Design Tip: Luminaire studs, hickeys, and cable clamps were counted as 1 conductor by the 1978 NEC and previous editions. It did not matter how many were present, the total number counted as 1 conductor. The 1981 NEC required luminaire studs and hickeys to be counted as 1 each. However, these rules are not retroactive to installations before the publication of the 1981 NEC. The designer must be careful when routing larger conductors with smaller conductors through or in the same box, because a larger box may be needed due to such protection.

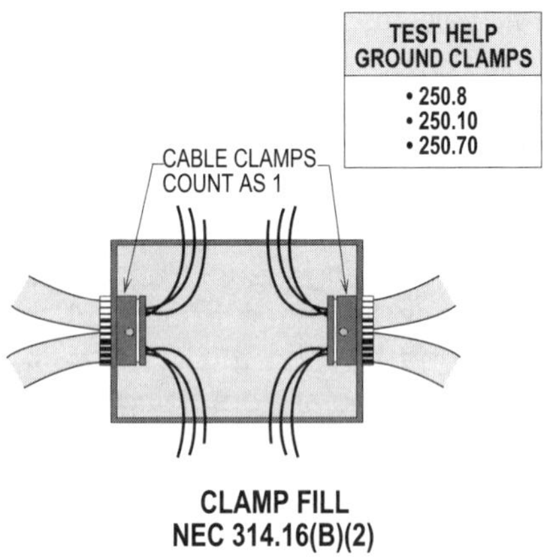

CLAMP FILL
NEC 314.16(B)(2)

Figure 12-4. Boxes with one or more cable clamps shall have 1 conductor counted toward the fill space of the box.

DEVICE OR EQUIPMENT FILL
314.16(B)(4)

A receptacle or switch that is mounted on a strap or yoke shall be counted as 2 conductors in determining the fill of the box. The count of 2 shall be based on the conductor size that is connected to the receptacle or switch. One strap or yoke could have one, two, or three receptacles mounted on it and still be counted as 2 conductors. **Article 100** defines a receptacle as a single receptacle having a contact device, with no other contact device on the same strap. A multiple receptacle is a single device with two or more receptacles mounted to it. A duplex device has two receptacles, and a despar device may have three or four receptacles mounted on it. The same rules apply for switches. A device or utilization equipment wider than a single 2 in. (50 mm) device box shall have double volume allowances provided for each gang required for mounting. **(See Figure 12-6)**

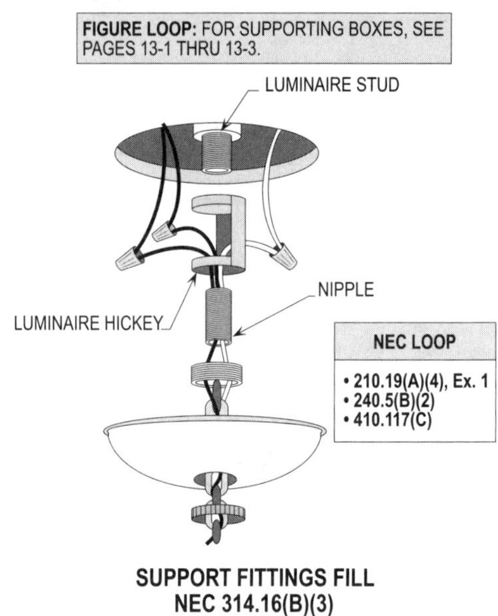

SUPPORT FITTINGS FILL
NEC 314.16(B)(3)

Figure 12-5. Boxes containing luminaire studs or hickeys shall have 1 conductor added for each fitting.

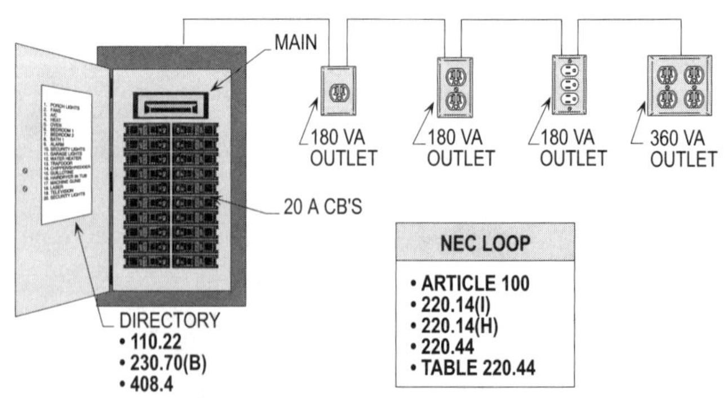

DEVICE OR EQUIPMENT FILL
NEC 314.16(B)(4)

Figure 12-6. The yoke or strap with mounted devices shall be counted as 2 conductors based on the size conductors connecting to its terminals.

EQUIPMENT GROUNDING CONDUCTOR FILL
314.16(B)(5)

Equipment grounding conductors passing through or spliced together in the box count shall be counted as 1. It doesn't matter how many equipment grounding conductors are installed in a box. All the equipment grounding conductors added together shall be counted as 1. It is a good recommendation to place the equipment grounding conductors into the first box. The equipment grounding conductors may be insulated or bare but shall be bonded to the box if the box is the metal type. Should the bare copper of one of the equipment grounding conductors come in contact with the metal of the box, it would not create a problem because it is bonded to the box anyway per **250.148(A)**.

For example: What is the count toward the fill of the box where there are 6 equipment grounding conductors in the box that are used to ground the metal enclosures of the equipment?

Step 1: Counting conductors
314.16(B)(5)
6 EGCs count as 1

Solution: **One conductor is counted for the 6 equipment grounding conductors in the box.**

The only time that equipment grounding conductors in a box count more than one is when an isolation equipment grounding conductor (IEGC) is run to an isolation receptacle and used to ground the metal enclosure of a PC. The isolation equipment grounding conductor is routed with the equipment grounding conductor in the same conduit and bypasses the metal yoke (strap) of the receptacle and connects directly to the metal enclosure of the PC. All the regular equipment grounding conductors shall be counted as 1 plus 1 for the isolation equipment grounding conductor, for a total of 2. See **250.146(D)** and **408.40, Ex.** for the requirements of installing the isolation equipment grounding conductor.

Design Tip: An isolation equipment grounding conductor shall be permitted to be used to ground the metal of any type of cord-and-plug connected sensitive electronic equipment to reduce electromagnetic noise.

For example: How many conductors are counted for 1 equipment grounding conductor to a metal box and 1 isolation equipment grounding conductor for a PC?

Step 1: Counting conductors
314.16(B)(5)
1 EGC counts as 1
1 IEGC counts as 1

Solution: **There are 2 conductors counted for the equipment grounding conductor and isolation equipment grounding conductor in the box.**

See Figures 12-7(a) and (b) for a detailed illustration of installing equipment grounding conductors and isolation grounding conductors in boxes.

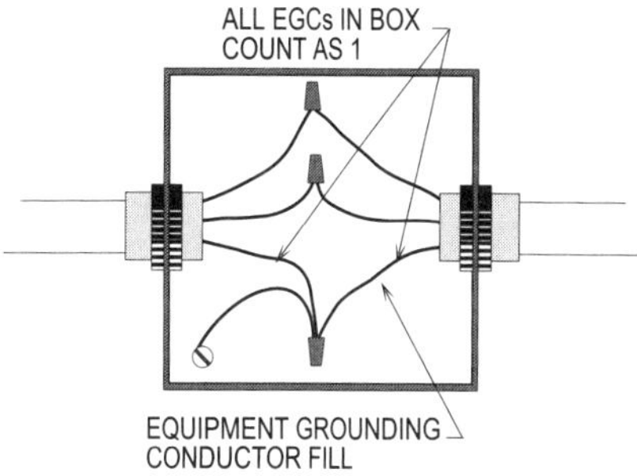

**EQUIPMENT GROUNDING
CONDUCTOR FILL
NEC 314.16(B)(5)**

Figure 12-7(a). Equipment grounding conductors passing through or spliced together in the box shall be counted as 1.

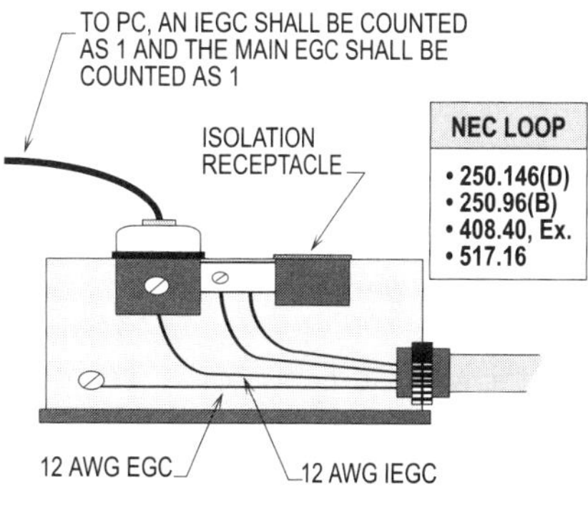

**EQUIPMENT GROUNDING
CONDUCTOR FILL
NEC 314.16(B)(5)**

Figure 12-7(b). One additional conductor shall be added to the count where an isolation equipment grounding conductor is contained in the box.

BOXES
314.16(A) AND Table 314.16(A)

Boxes are usually required to be mounted to support luminaires, devices, fans, etc. Boxes are available in four main types from **Table 314.16(A)**. The four types most used are octagonal, round, square, or device. The octagonal or round box is used for mounting luminaires or ceiling fans or for splicing conductors. Square boxes are used with plaster rings to mount and support luminaires, ceiling fans, or devices. They are also used for the splicing of conductors.

Device boxes are used to support the yoke (strap) of receptacles, switches, or combinations of receptacles, switches, pilot lights, etc. They are also used to splice conductors. **Table 314.16(A)** lists the sizes of these boxes in cubic inches (cu. in.) and the size conductors of the same size in rating from 18 AWG through 6 AWG. See **314.28(A)(1)** and **(A)(2)** to size boxes for conductors sized 4 AWG and larger.

Based upon the number of the same size conductors or cubic inch ratings of combination conductors (18 AWG through 6 AWG), the box size shall be determined from **Table 314.16(A)**.

The requirements of **314.16(A)(1)** and **(B)(1) through (B)(5)** shall be applied to calculate the number of conductors or cubic inch value of each conductor used to select the size box from **Table 314.16(A)**.

OCTAGON BOXES
Table 314.16(A)

Octagon boxes are used mainly to mount and properly support luminaires. They are also utilized for splicing the conductors of branch circuits or feeders supplying power to various pieces of electrical apparatus such as ranges, cooktops, ovens, heating units, A/C units, etc. Special octagon boxes that are listed shall be permitted to support ceiling fans. Octagon boxes that are installed to support luminaires shall not be permitted to support luminaires that weigh more than 50 lb (23 kg) per **314.27(B)**. Ceiling fans that are supported by listed octagon boxes shall weigh 35 lb (16 kg) or less per **422.18** and **314.27(D)**. [See Figure 12-8(a) and (b)]

Design Tip: Reference **314.23** and **314.27** for the rules pertaining to the proper methods of supporting boxes.

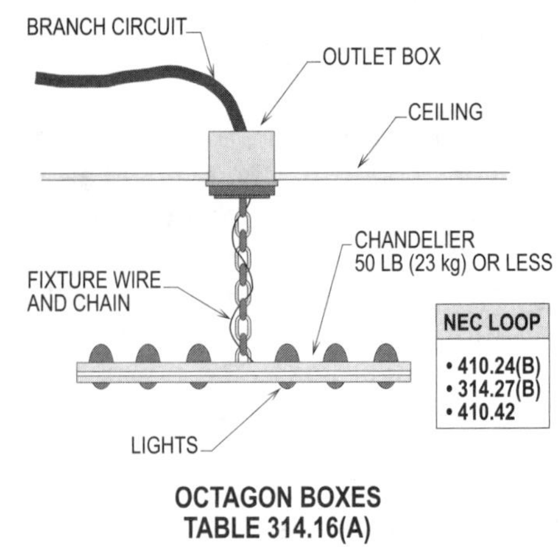

Figure 12-8(a). Outlet boxes shall not be permitted to support luminaires weighing more than 50 lb (23 kg).

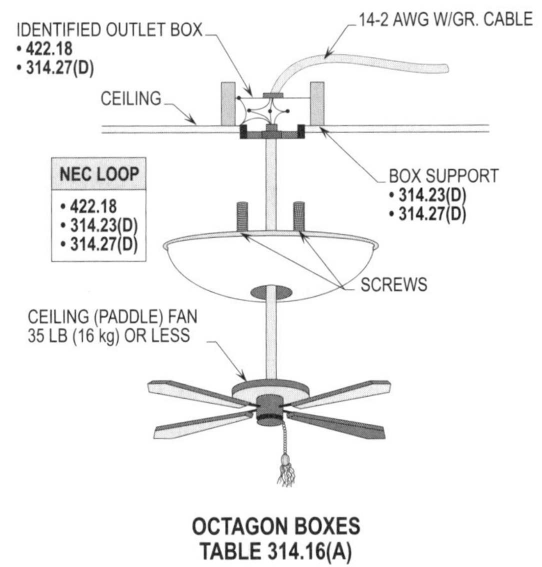

Figure 12-8(b). Outlet boxes shall not support ceiling (paddle) fans weighing more than 35 lb (16 kg).

SAME SIZE CONDUCTORS
314.16(B)(1) THROUGH (B)(5) AND Table 314.16(A)

Boxes containing conductors that are the same size shall have their total fill space determined by adding the number of conductors plus additional conductors for each fitting or device. The cubic inch value listed in **Table 314.16(B)** for the largest conductor in the box shall be used to determine the fill capacity for the fittings. For devices, use the conductor size that is connected to its terminals.

For example: What size octagon box is required to support a luminaire supplied with 2 - 14-2 AWG w/ground nonmetallic sheathed cables (romex) with a luminaire stud, hickey, and 2 - 14 AWG fixture wires? Neutrals are spliced with a pigtail, and romex connectors are used instead of romex clamps.

Step 1:	Same size conductors	
	314.16(B)(1) through (B)(5)	
	(Conductor fill)	
	314.16(B)(1)	
	2 - 14 AWG hots	= 2
	2 - 14 AWG neutrals	= 2
	(EGC fill)	
	314.16(B)(5)	
	2 - 14 AWG EGCs	= 1
	(Support fitting fill)	
	314.16(B)(3)	
	1 luminaire stud	= 1
	1 hickey	= 1
	(Conductor fill)	
	314.16(B)(1)	
	1 pigtail	= 0
	(Clamp fill)	
	314.16(B)(2)	
	2 romex connectors	= 0
	(Fixture wire fill)	
	314.16(B)(1), Ex.	
	2 - 14 AWG fixture wires	= 2
	Total	= 9

Step 2:	Selecting box
	Table 314.16(A)
	9 - 14 AWG conductors requires
	4" x 2-1/8" (100 mm x 54 mm) box

Solution: **A 4 in. x 2-1/8 in. (100 mm x 54 mm) octagon box is required.**

See Figure 12-9 for a step-by-step procedure on sizing and selecting octagon boxes where the conductors are the same size.

Problem: What size octagon box is required to support a luminaire with a canopy supplied with 2 - 12-2 AWG w/ground nonmetallic sheathed cables (romex) with a luminaire stud, hickey, and 4 - 18 AWG fixture wires, plus EGCs?

Step 1:	Counting conductors	
	314.16(B)(1) thru (B)(5)	
	(Conductor fill)	
	2 - 12 AWG hots	= 2
	2 - 12 AWG neutrals	= 2
	(EGC fill)	
	2 - 12 AWG EGCs	= 1
	(Support and fitting fill)	
	1 luminaire stud	= 1
	1 hickey	= 1
	(Fixture wire fill)	
	3 - 18 AWG fixture wires	= 0
	plus EGC	= 0
	Total	= 7

Step 2:	Selecting box
	Table 314.16(A)
	7 - 12 conductor requires a
	4" x 2-1/8" (100 mm x 54 mm) box

Solution: **A 4 in. x 2-1/8 in. (100 mm x 54 mm) octagon box is required.**

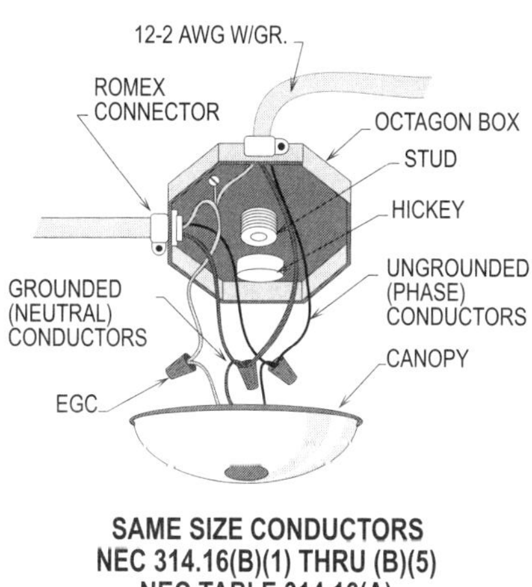

Figure 12-9. Sections **314.16(B)(1) through (B)(5)** and **Table 314.16(A)** are utilized to size and select an octagon box based on the box containing the same size conductors. **Note,** the canopy is used for the fill of the fixture wires.

DIFFERENT SIZE CONDUCTORS 314.16(B)(1) through (B)(5) AND Table 314.16(B)

Boxes containing conductors that are not the same size shall be calculated based on the cubic inch rating of each conductor listed in **Table 314.16(B)**. The cubic inch rating of each conductor shall be added together, plus the number of fittings or devices. The cubic inch rating used for each fitting shall be based on the largest conductor entering the box. The cubic inch rating for yokes or straps shall be based on the size conductor that is connected to the switch or receptacle.

Fixture wires in sizes 18 AWG or 16 AWG shall be determined by applying the combination calculation per **Table 314.16(B)**.

There are installations where different size conductors are spliced in boxes and fed out of the boxes to supply power to loads. The boxes are used as junction points for the branch circuits between the panelboard and loads.

For example: What size octagon box is required to support a luminaire with 2 - 12-2 AWG w/ground nonmetallic sheathed cables (rope) with a luminaire stud, two cable clamps, two pigtails, and 2 - 18 AWG fixture wires? (No luminaire canopy)

Step 1: Combination conductors
314.16(B)(1) through (B)(5);
Table 314.16(B)
(Conductor fill)
314.16(B)(1)
2 - 12 AWG hots
2.25 cu. in. x 2   = 4.5   cu. in.
2 - 12 AWG neutrals
2.25 cu. in. x 2   = 4.5   cu. in.
(EGC fill)
314.16(B)(5)
2 - 12 AWG EGCs
2.25 cu. in. x 1   = 2.25   cu. in.
(Supporting fitting fill)
314.16(B)(3)
1 luminaire stud
2.25 cu. in. x 1   = 2.25   cu. in.
(Cable clamp fill)
314.16(B)(2)
2 cable clamps
2.25 cu. in. x 1   = 2.25   cu. in.
(Conductor fill)
314.16(B)(1)
2 pigtails   = 0   cu. in.
(Fixture wire fill)
314.16(B)(1), Ex.
2 - 18 AWG fixture wires
1.5 cu. in. x 2   = 3.0   cu. in.
Total   = 18.75 cu. in.

Step 2: Selecting Box
Table 314.16(A)
18.75 cu. in. requires 4" x 2-1/8"
(100 mm x 54 mm) box

Solution: A 4 in. x 2-1/8 in. (100 mm x 54 mm) octagon box is required.

For example: What size octagon box with cable clamps is required for 2 - 14 AWG and 2 - 12 AWG w/ground romex cables that are spliced in the box between the panelboard and the loads?

Step 1: Combination conductors
314.16(B)(1) through (B)(5);
Table 314.16(B)
(Conductor fill)
314.16(B)(1)
2 - 12 AWG hots
2.25 cu. in. x 2   = 4.5 cu. in.
2 - 12 AWG neutrals
2.25 cu. in. x 2   = 4.5 cu. in.
2 - 14 AWG hots
2   cu. in. x 2   = 4   cu. in.
2 - 14 AWG neutrals
2   cu. in. x 2   = 4   cu. in.
(EGC fill)
314.16(B)(5)
2 - 12 AWG EGCs
2.25 cu. in. x 1   = 2.25 cu. in.
2 - 14 AWG EGCs
2   cu. in. x 0   = 0
(Cable clamp fill)
314.16(B)(2)
2 cable clamps
2.25 cu. in. x 1   = 2.25 cu. in.
Total   = 21.5   cu. in.

Step 2: Selecting box
Table 314.16(A)
21.5 cu. in. requires 4" x 2-1/8"
(100 mm x 54 mm)

Solution: A 4 in. x 2-1/8 in. (100 mm x 54 mm) octagon box is required.

See Figure 12-10 for a step-by-step procedure on sizing and selecting octagon boxes where the conductors are different sizes.

The rise on the plaster ring may be 1/4, 1/2, or 3/4 in. to provide more space inside to mount receptacles and switches. This extra space in the box due to the rise of the plaster ring permits a receptacle or switch with a greater depth to be installed. Plaster rings are also used to extend the surface of the box inside the wall or ceiling to the face of the wall or ceiling where a receptacle (device) or luminaire may be mounted. Square boxes are utilized to splice circuit conductors to supply power to various electrical loads in dwelling units or buildings.

SAME SIZE CONDUCTORS
314.16(B)(1) THROUGH (B)(5) AND
Table 314.16(A)

Section **314.16(A)** lists the items in or out of the box and the number of conductors that shall be added for each item based on the condition of use. Where all conductors are the same size, each item in or out of the box shall be counted based on these same size conductors.

Problem: What size octagon box is required for a 12-2 AWG w/ground cable and 3 - 16 AWG fixture wires?

Step 1:	Counting conductors **314.16(B)(1) thru (B)(5); Table 314.16(B)** **(Conductor fill)**

1 -12 AWG hot	2.25 cu. in. x 1	=	2.25
1 -12 AWG neutral	2.25 cu. in. x 1	=	2.25
(EGC fill)			
1 -12 AWG EGC	2.25 cu. in. x 1	=	2.25
1 -16 AWG EGC		=	0
(Fixture wire fill)			
2 -16 AWG fix. wires	1.75 cu. in. x 2	=	3.5
Total			= 10.25

Step 2:	Selecting box **Table 314.16(A)** 10.25 cu. in. requires 4" x 1-1/4" box (100 mm x 32 mm)
Solution:	**A 4 in. x 1-1/4 in. (100 mm x 32 mm) octagon box is required.**

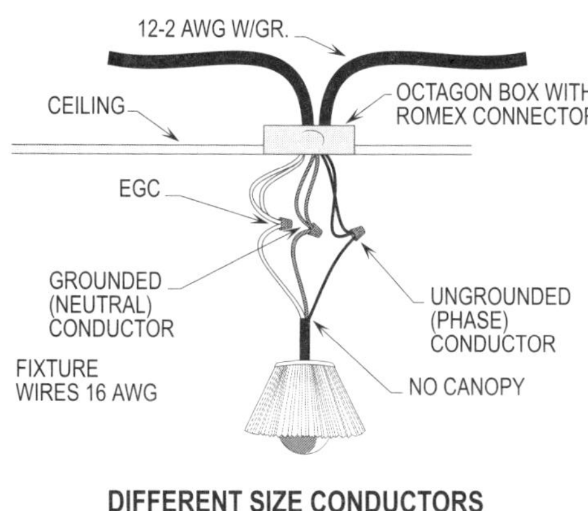

Figure 12-10. Sections **314.16(B)(1) through (B)(5)** and **Table 314.16(B)** are utilized to size and select the box based on an octagon box containing different conductor size. Multiplying the cubic inch of each conductor times the number derives the box size.

SQUARE BOXES
Table 314.16(A)

Square boxes have greater cubic inch ratings than octagon boxes. Square boxes are usually installed where more than four romex cables enter and leave the box. Square boxes may be used with plaster rings that are designed to accommodate the mounting of a receptacle, switch, light, etc.

For example: What size square box is required for 2 - 12-2 AWG w/ground romex cables connecting to a receptacle and 4 - 12-2 AWG w/ground romex cables that pass through the box? (Use an extension ring.)

Step 1:	Same size conductors **314.16(B)(1) through (B)(5)** **(Conductor fill)** **314.16(B)(1)**

2 - 12 AWG hots	= 2
2 - 12 AWG neutrals	= 2
4 - 12 AWG hots (through)	
4 x 2	= 8
4 - 12 neutrals (through)	
4 x 2	= 8
(EGC fill)	
314.16(B)(5)	
10 - 12 AWG EGCs	= 1
Total	= 21

Step 2:	Selecting box **Table 314.16(A)** 21 - 12 AWG requires 4-11/16" x 1-1/4" (120 mm x 32 mm) with ring
Solution:	**A 4-11/16 in. x 1-1/4 in. (120 mm x 32 mm) square box with extension ring is required. Note: Extension ring allows 22 - 12 AWG conductors to be installed per 314.16(A).**

See Figure 12-11 for a step-by-step procedure on sizing and selecting square boxes where the conductors in the box are the same size.

Problem: What size square box is required to contain 4 -12-2 AWG w/ground cables supported in the box with cable clamps?

Step 1:	Counting conductors **314.16(B)(1) thru (B)(5)** **(Conductor fill)**	
	4 -12 AWG hots	= 4
	4 -12 AWG neutrals	= 4
	(EGC fill)	
	4 12 AWG EGCs	= 1
	(Clamp fill)	
	2 cable clamps	= 1
	Total	= 10

Step 2:	Selecting box **Table 314.16(A)** 10 -12 AWG conductors requires 4-11/16" x 1-1/4" (120 mm x 32 mm)

Solution: A 4-11/16 in. x 1-1/4 in. (120 mm x 32 mm) square box is required.

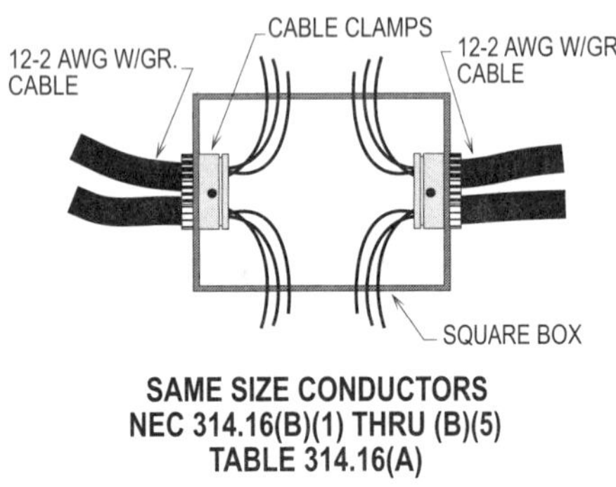

Figure 12-11. Sections **314.16(B)(1) through (B)(5)** and **Table 314.16(B)** are utilized to size and select a square box based on the box containing the same size conductors.

DIFFERENT SIZE CONDUCTORS 314.16(B)(1) THRU (B)(5) AND TABLE 314.16(B)

Section **314.16(B)** refers to **Table 314.16(B)**, which requires the cubic inch rating of each conductor in the box to be used for determining the fill space in cubic inches. This total rating shall be used for selecting the box size. Section **314.16(B)** refers to **314.16(A)** for selecting the cubic inch rating of each fitting, cable clamp, etc. in the box, based on the larger conductor entering the box. The cubic inch rating for each yoke shall be based on the conductor size that is connected to the device terminal.

For example: What size square box is required where 4 - 14 AWG and 4 - 12 AWG romex cables w/grounds are spliced and routed to loads in the dwelling unit or building?

Step 1:	Combination conductors **314.16(B)(1) through (B)(5);** **Table 314.16(B)** **(Conductor fill)** **314.16(B)(1)**		
	4 - 14 AWG hots		
	2 cu. in. x 4	= 8	cu. in.
	4 - 14 AWG neutrals		
	2 cu. in. x 4	= 8	cu. in.
	4 - 12 AWG hots		
	2.25 cu. in. x 4	= 9	cu. in.
	4 - 12 AWG neutrals		
	2.25 cu. in. x 4	= 9	cu. in.
	(EGC fill) **314.16(B)(5)**		
	4 - 12 AWG EGCs		
	2.25 cu. in. x 1	= 2.25	cu. in.
	Total	= 36.25	cu. in.

Step 2:	Selecting box **Table 314.16(A)** 36.25 cu. in. requires 4-11/16" x 2-1/8" (120 mm x 54 mm)

Solution: A 4-11/16 in. x 2-1/8 in. (120 mm x 54 mm) square box is required.

See Figure 12-12 for a step-by-step procedure for sizing and selecting square boxes where the conductors are different sizes.

DEVICE BOXES TABLE 314.16(A)

Device boxes are used to mount switches to control luminaires and other electrical loads. They also are used to mount receptacles for the cord-and-plug connection of electrical appliances. The depth space (fill) in the box shall be sized to accommodate the size device installed. A dimmer switch (device) will take up more depth (space) in the box than a regular-sized toggle switch. A GFCI receptacle in most cases takes up greater depth (space) for installation than other size devices. The designer and installer shall select the proper size device box to accommodate not only conductors and fittings but also the depth of the device; how much fill it requires shall also be considered.

Problem: What size square box with cable clamps is required for 2 - 14 AWG and 2 - 12 AWG w/grounds romex cables that are spliced in the box between the panelboard and the loads?

Step 1:	Counting conductors				
	314.16(B)(1) thru (B)(5); Table 314.16(B)				
	(Conductor fill)				
	2 - 12 AWG hots	2.25 cu. in. x 2	=	4.5	cu. in.
	2 - 12 AWG neutrals	2.25 cu. in. x 2	=	4.5	cu. in.
	2 - 14 AWG hots	2 cu. in. x 2	=	4.	cu. in.
	2 - 14 AWG neutrals	2 cu. in. x 2	=	4.	cu. in.
	(EGC fill)				
	2 - 12 AWG EGCs	2.25 cu. in. x 1	=	2.25	cu. in.
	2 - 14 AWG EGCs	2. cu. in. x 1	=	0	
	(Clamp fill)				
	2 cable clamps	2.25 cu. in. x 1	=	2.25	cu. in.
	Total			= 21.5	cu. in.
Step 2:	Selecting box				
	Table 314.16(A)				
	21.5 cu. in. requires 4-11/16" x 1-1/4" (120 mm x 32 mm)				
Solution:	**A 4-11/16 in. x 1-1/4 in. (120 mm x 32 mm) square box is required.**				

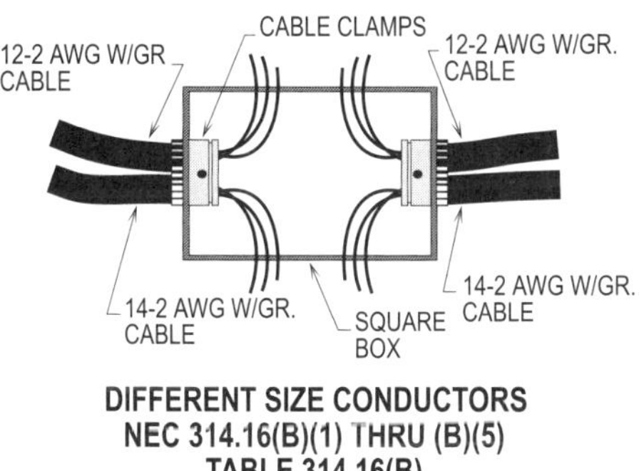

DIFFERENT SIZE CONDUCTORS
NEC 314.16(B)(1) THRU (B)(5)
TABLE 314.16(B)

Figure 12-12. Section **314.16(B)** and **Table 314.16(B)** is utilized to size and select a square box based on the box containing different size conductors.

SAME SIZE CONDUCTORS 314.16(B)(1) THRU (B)(5) AND TABLE 314.16(A)

Section **314.16(A)** contains specific rules that shall be applied where the device box has the same size conductors with the various types of fittings, devices, etc. all present in the box. Devices shall be counted as 2 conductors to allow more fill (space) to accommodate the different size devices that are available in the electrical industry today. The count of 2 conductors shall be based upon the size conductors that are connected to the terminals of the device mounted on the yoke (strap).

For example: What size device box with cable clamps is required for 2 - 14-2 AWG w/ground romex cables and 1 duplex receptacle?

Step 1:	Same size conductors	
	314.16(B)(1) through (B)(5)	
	(Conductor fill)	
	314.16(B)(1)	
	2 - 14 AWG hots	= 2
	2 - 14 AWG neutrals	= 2
	(EGC fill)	
	314.16(B)(5)	
	2 - 14 AWG EGCs	= 1
	(Device fill)	
	314.16(B)(4)	
	1 receptacle	= 2
	(Clamp fill)	
	314.16(B)(2)	
	2 cable clamps	= 1
	Total	= 8
Step 2:	Selecting box	
	Table 314.16(A)	
	8 - 14 requires 3" x 2" x 3-1/2"	
	(75 mm x 50 mm x 90 mm)	
Solution:	**A 3 in. x 2 in. x 3-1/2 in. (75 mm x 50 mm x 90 mm) device box is required.**	

See Figure 12-13 for a step-by-step procedure for sizing and selecting device boxes where the conductors are the same size in the box.

DIFFERENT SIZE CONDUCTORS 314.16(B)(1) THRU (B)(5) AND TABLE 314.16(B)

Section **314.16(B)** requires the cubic inch rating of each conductor from **Table 314.16(B)** to be utilized where conductors are not the same size in the box. The cubic inch rating of each conductor shall be multiplied by the number in the box, and the total cubic inch value shall be used to determine the size device box from **Table 314.16(A)**.

For example: What size device or square box with cable clamps is required for 1 - 14-2 AWG w/ground passing through, 1 - 10-2 AWG w/ground passing through, and 4 - 12-2 AWG w/ground that are spliced with pigtails and connected to a dimmer switch?

Step 1: Sizing box with different size conductors
314.16(B)(1) through (B)(5);
Table 314.16(B)
(Conductor fill)
314.16(B)(1)
2 - 14 AWG hots (spliced through)
2   cu. in. x 2   = 4   cu. in.
2 - 14 AWG neutrals (spliced through)
2   cu. in. x 2   = 4   cu. in.
(EGC fill)
314.16(B)(5)
2 - 14 AWG EGCs
(spliced through)   = 0
(Conductor fill)
314.16(B)(1)
4 - 12 AWG hots (spliced)
2.25 cu. in. x 4   = 9   cu. in.
4 - 12 AWG neutrals (spliced)
2.25 cu. in. x 4   = 9   cu. in.
(EGC fill)
314.16(B)(5)
4 - 12 AWG EGCs (spliced)
2.25 cu. in. x 0   = 0
(Conductor fill)
314.16(B)(1)
2 - 10 AWG hots (spliced)
2.5 cu. in. x 2   = 5   cu. in.
2 - 10 AWG neutrals (spliced)
2.5 cu. in. x 2   = 5   cu. in.
(EGC fill)
314.16(B)(5)
2 - 10 AWG EGCs (spliced)
2.5 cu. in. x 1   = 2.5 cu. in.
(Device fill)
314.16(B)(4)
1 - dimmer switch (device)
2.25 cu. in. x 2   = 4.5 cu. in.
Total   = 43 cu. in.

Step 2: Selecting box with different size conductors
Table 314.16(A)
43 cu. in. requires a 4-11/16" x 1-1/4" (120 mm x 32 mm) square box with extension ring and plaster ring

Solution: **A 4-11/16 in. x 1-1/4 in. (120 mm x 32 mm) square box w/extension ring and plaster ring is required.**

See Figure 12-14 for a step-by-step procedure for sizing and selecting device boxes where the conductors in the box are different sizes.

Problem: What size device box is required for 2 - 12 AWG hots, 2 - 12 AWG neutrals, a receptacle, 1 bonding jumper, and 1 pigtail?

Step 1: Counting conductors
314.16(B)(1) thru (B)(5)
(Conductor fill)

Device	= 2
2 - 12 AWG hots	= 2
2 - 12 AWG neutrals	= 2
2 - 12 AWG pigtails	= 0
1 - 12 AWG EGC (BJ)	= 0
Total	= 6

Step 2: Selecting box
Table 314.16(A)
6 - 12 AWG conductors requires 3" x 2" x 2-3/4" (75 mm x 50 mm x 70 mm)

Solution: **A 3 in. x 2 in. x 2-3/4 in.** (75 mm x 50 mm x 70 mm) **device box is required.**

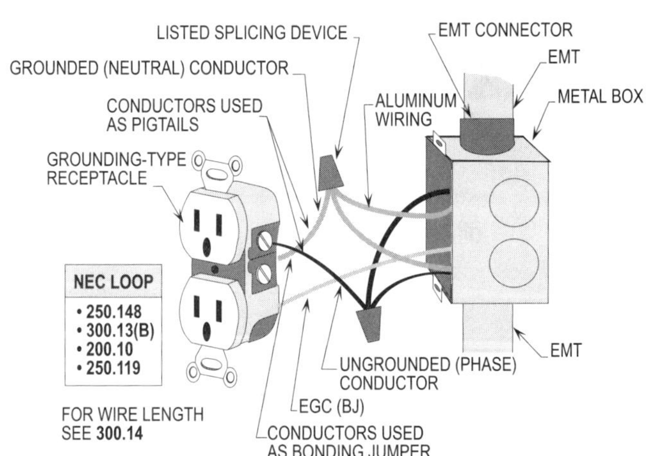

Figure 12-13. Sections **314.16(B)(1) through (B)(5)** and **Table 314.16(A)** are utilized to size and select a device box based on the box containing the same size conductors.

<table>
<tr><td>

Problem: What size square box is required to contain 2 - 10 AWG conductors passing through, 4 - 12 AWG conductors, and 1 receptacle? There are 2 pigtails and 1 bonding jumper in the box.

Step 1:   Counting conductors
314.16(B)(1) thru (B)(5); Table 314.16(B)
(Conductor fill)
2 - 10 AWG hots   2.5   cu. in. x 2 =   5   cu. in.
2 - 12 AWG hots   2.25 cu. in. x 2 =   4.5 cu. in.
2 - 12 AWG neutrals 2.25 cu. in. x 2 =   4.5 cu. in.
(Device fill)
1 receptacle   2.25 cu. in. x 2 =   4.5 cu. in.
(Conductor fill)
2 pigtails   = 0
1 bonding jumper   = 0
Total   = 18.5 cu. in.

Step 2:   Selecting box
Table 314.16(A)
18.5 cu. in. requires 4" x 1-1/2" (100 mm x 38 mm) square box

Solution: **A 4 in. x 1-1/2 in. (100 mm x 38 mm) square box with plaster ring is required.**

</td></tr>
</table>

OTHER BOXES
314.16(A)(2)

Other boxes include boxes that are not listed in **Table 314.16(A)**. They are junction boxes with greater fill (space) area that allows more conductors to be routed straight through or spliced in the box. Junction boxes come in many sizes and are built to house the different size conductors in **Table 314.16(A)**. If a box size selected from **Table 314.16(A)** with an extension ring will not accommodate the number of conductors ranging from 18 AWG to 6 AWG installed in the box, then a junction box shall be used. See the chart in the Appendix for the standard size junction boxes available to contain branch circuit and feeder conductors.

SAME SIZE CONDUCTORS
314.16(B)(1) THRU (B)(5) AND 314.16(A)

Because there are not any tables available with the number of conductors permitted in junction boxes for conductors that are the same size, a calculation applying the cubic inch value of each conductor shall be done. By multiplying the cubic inch rating of each conductor by the number in the box and using this total cubic inch rating, the proper size junction box may be selected. The cubic inch rating of each conductor shall be found in **Table 314.16(B)** for conductors 18 AWG through 6 AWG. The cubic inch fill space of a junction box shall be found by multiplying the dimensions of the box.

For example, a 4 in. x 4 in. x 4 in. (100 mm x 100 mm x 100 mm) junction box has a cubic inch (cu. in.) rating of 64 cu. in. (4 in. x 4 in. x 4 in. = 64 cu. in.). This box will contain any arrangement of conductors not exceeding 64 cu. in. in total rating.

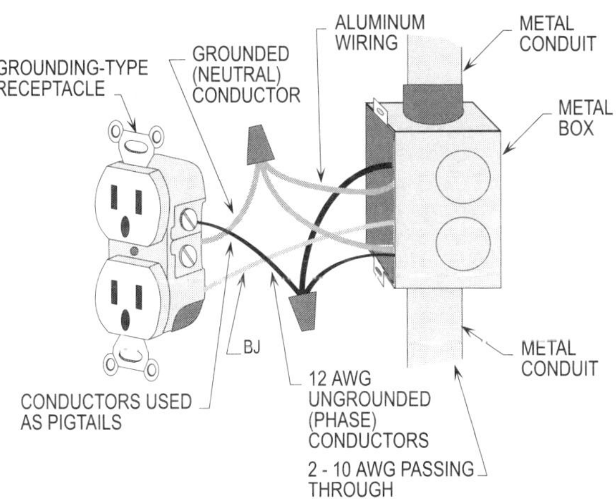

**DIFFERENT SIZE CONDUCTORS
NEC 314.16(B)(1) THRU (B)(5)
TABLE 314.16(B)**

Figure 12-14. Sections **314.16(B)(1) through (B)(5)** and **Table 314.16(B)** are utilized to size and select a device box based on the box containing different size conductors. The box shall be sized based upon the number of cubic inches of each conductor.

When an existing panelboard in a remodeling job in a dwelling unit is used for a junction box, it shall comply with **240.24(D)** and **(E)**, which prohibits a panelboard with overcurrent protection devices in a clothes closet around combustible material. Section **230.70(A)** prohibits the service disconnecting means to be located in a bathroom of dwelling units, commercial, or industrial locations.

For example: What size junction box is required to contain 6 - 12-2 AWG w/ground, 4 - 12-2 AWG w/ground and 2 - 12 AWG without ground cables where all the conductors in the cables are spliced?

Step 1: Sizing box with the same size conductors
314.16(A)(2)
(Conductor fill)
314.16(B)(1)
12 - 12 AWG hots (spliced)
2.25 cu. in. x 12 = 27 cu. in.
12 - 12 AWG neutrals (spliced)
2.25 cu. in. x 12 = 27 cu. in.
(EGC fill)
314.16(B)(5)
12 - 12 AWG EGCs (spliced)
2.25 cu. in. x 1 = 2.25 cu. in.
(Conductor fill)
314.16(B)(1)
8 - 12 AWG hots (spliced)
2.25 cu. in. x 8 = 18 cu. in.
8 - 12 AWG neutrals (spliced)
2.25 cu. in. x 8 = 18 cu. in.
(EGC fill)
314.16(B)(5)
8 - 12 AWG EGCs (spliced)
2.25 cu. in. x 0 = 0
(Conductor fill)
314.16(B)(1)
4 - 12 AWG hots (spliced)
2.25 cu. in. x 4 = 9 cu. in.
4 - 12 AWG neutrals (spliced)
2.25 cu. in. x 4 = 9 cu. in.
(EGC fill)
314.16(B)(5)
4 - 12 AWG EGCs (spliced)
2.25 cu. in. x 0 = 0
Total = 110.25 cu. in.

Step 2: Selecting box for the same size conductors
314.16(B), Chart
6" x 6" x 4" = (150 mm x 150 mm x 100 mm) box = 144 cu. in.
144 cu. in. will contain 110.25 cu. in.

Solution: A 6 in. x 6 in. x 4 in. (150 mm x 150 mm x 100 mm) junction box is required.

See **Figure 12-15** for a step-by-step procedure for sizing and selecting junction boxes with the same size conductors.

Problem: What size junction box is required to contain 4 raceways with 8 - 12 AWG in each and 2 raceways with 4 - 12 AWG in each?

Step 1: Counting sq. in.
314.16(B)(1) thru (B)(5); Table 314.16(B)
(Conductor fill using other boxes)
8 - 12 AWG conductors 2.25 cu. in. x 8 = 18
8 - 12 AWG conductors 2.25 cu. in. x 8 = 18
8 - 12 AWG conductors 2.25 cu. in. x 8 = 18
8 - 12 AWG conductors 2.25 cu. in. x 8 = 18
4 - 12 AWG conductors 2.25 cu. in. x 4 = 9
4 - 12 AWG conductors 2.25 cu. in. x 4 = 9
Total = 90

Step 2: Selecting box
Box chart
6" x 4" x 4" (150 mm x 100 mm x 100 mm) box = 96 cu. in.
96 cu. in. will contain 90 cu. in.

Solution: **A 6 in. x 4 in. x 4 in. (150 mm x 100 mm x 100 mm) junction box is required.**

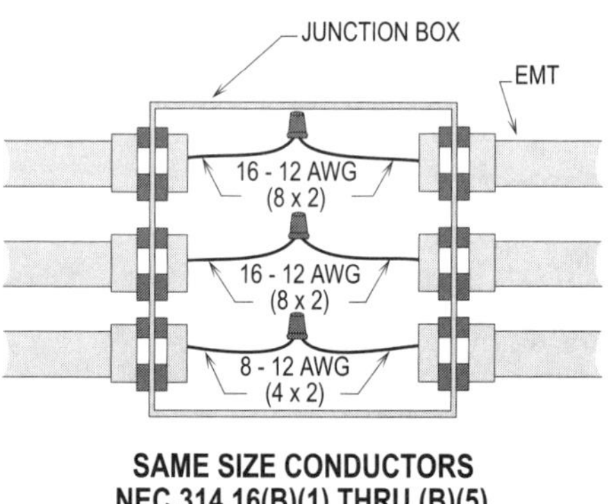

Figure 12-15. Sizing and selecting a junction box to contain the same size conductors.

DIFFERENT SIZE CONDUCTORS 314.16(A)(2)

To select the proper size junction box housing different size conductors, the cubic inch rating of each conductor from **Table 314.16(B)** shall be selected and multiplied by the number of conductors for each cubic inch rating. The total calculation of the cubic inch ratings shall be used to select the proper size junction box.

For example: What size junction box is required to house the following different size conductors?

Step 1: Sizing box with different size conductors
314.16(B)(1) through (B)(5);
Table 314.16(B)
(Conductor fill)
314.16(B)(1)
8 - 10 AWG hots (spliced)
2.5 cu. in. x 8   = 20 cu. in.
8 - 10 AWG neutrals (spliced)
2.5 cu. in. x 8   = 20 cu. in.
(EGC fill)
314.16(B)(5)
 8 - 10 EGCs (spliced)
2.5 cu. in. x 0   = 0
(Conductor fill)
314.16(B)(1)
20 - 14 AWG hots (spliced)
2 cu. in. x 20   = 40 cu. in.
20 - 14 AWG neutrals (spliced)
2 cu. in. x 20   = 40 cu. in.
(EGC fill)
314.16(B)(5)
20 - 14 AWG EGCs (spliced)
2 cu. in. x 0   = 0
(Conductor fill)
314.16(B)(1)
12 - 12 AWG hots (spliced)
2.25 cu. in. x 12   = 27 cu. in.
12 - 12 AWG neutrals (spliced)
2.25 cu. in. x 12   = 27 cu. in.
(EGC fill)
314.16(B)(5)
12 - 12 AWG EGCs (spliced)
2.25 cu. in. x 0   = 0
(Conductor fill)
314.16(B)(1)
4 - 6 AWG hots (spliced)
5   cu. in. x 4   = 20 cu. in.
4 - 6 AWG neutrals (spliced)
5   cu. in. x 4   = 20 cu. in.
(EGC fill)
314.16(B)(5)
4 - 6 AWG EGCs (spliced)
5   cu. in. x 1   = 5 cu. in.
Total   = 219 cu. in.

Step 2: Selecting box with different size conductors
Chart for junction boxes
8" x 8" x 4" (200 mm x 200 mm x 100 mm) box = 256 cu. in.
256 cu. in. will contain 219 cu. in.

Solution: An 8 in. x 8 in. x 4 in.
(200 mm x 200 mm x 100 mm)
junction box is required.

See Figure 12-16 for a step by step procedure for sizing and selecting junction boxes with different size conductors.

Problem: What size junction box is required to contain the following different size conductors?

Step 1: Counting conductors
314.16(B)(1) thru (B)(5); Table 314.16(B)
(Conductor fill using other boxes)
16 - 10 AWG conductors   2.5 cu. in. x 16 = 40
20 - 14 AWG conductors   2   cu. in. x 20 = 40
24 - 12 AWG conductors   2.25 cu. in. x 24 = 54
10 - 6 AWG conductors   5   cu. in. x 10 = 50
Total   = 184

Step 2: Selecting box
Box chart
8" x 6" x 4" (200 mm x 150 mm x 100 mm) = 192 cu. in.
192 cu. in. will contain 184 cu. in.

Solution: A 8 in. x 6 in. x 4 in. (200 mm x 150 mm x 100 mm) junction box is required.

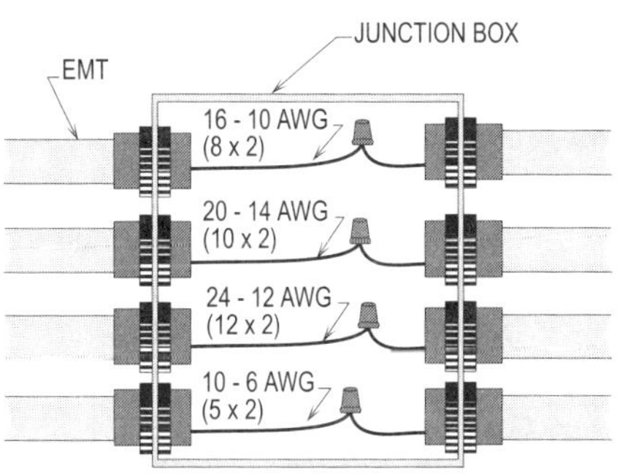

Figure 12-16. Sizing and selecting a junction box to contain different size conductors.

PLASTER RINGS AND EXTENSION RINGS 314.16(A)(1)

Plaster rings and extension rings are accessories that are used in conjunction with octagon, round, square, or device boxes to extend the face of the box to mount luminaires, ceiling fans, or devices, or to provide more fill space in the box.

For example, the face of an octagon box needs to be extended 1/4 in., 1/2 in., or 3/4 in. to the surface of a sheetrock ceiling. A 1/2 in. plaster ring with provisions for mounting a luminaire may be used to extend the face of the octagon box to the surface edge of the sheetrock.

Extension rings are used to extend the face of boxes a greater distance than plaster rings because plaster rings come with only limited extension height to extend the face of a box to a given height.

For example, one or two extension rings may be used to extend the face of a box to the surface edge of a wall or ceiling where a plaster ring is mounted and a device, luminaire, etc. is mounted to the plaster ring. Most designers, installers, and inspectors interpret **314.22** to permit one extension ring to be used to provide additional fill space in the box. Check with the local inspector for his or her interpretation of this rule.

The reason for this requirement of limiting the number of extension rings for fill space is to allow access to the inside box area for servicing the spliced conductors. **(See Figure 12-17)**

CONDUIT BODIES
314.16(C)

Conduit bodies containing conductors smaller than 6 AWG shall not be permitted to be less than twice the cross-sectional area of the largest conduit to which they can be connected. The number of conductors that are permitted in conduit bodies shall be determined by the square inch area per **Table 5** and **Table 4** of **Chapter 9**.

For example: What is the cross-sectional area of an LB (conduit body) that is connected to a 3/4 in. conduit?

Step 1: Finding cross-sectional area
Table 4, Ch. 9
3/4" (21) EMT conduit = .533 sq. in.

Step 2: Calculating size LB
314.16(C)
.533 sq. in. x 2 = 1.066 sq. in.

Solution: **The LB (conduit body) has to have a cross-sectional area of 1.066 sq. in.**

Design Tip: A larger LB may be used with reducing bushings to accomplish this requirement where needed.

Conduit bodies with less conduit entries shall not be permitted to contain splices, taps, or devices unless they comply with **314.16(B)** and are supported properly. **(See Figure 12-18)**

Problem: What size square box with extension ring is required to contain 21 - 12 AWG conductors? (use min. size)

Step 1: Counting conductors
314.16(B)(1) thru (B)(5); Table 314.16(A)
(Using extension rings for greater fill area)
21 - 12 AWG conductors requires
4-11/16" x 1-1/4" (120 mm x 32 mm) with extension ring

Solution: **Use a 4-11/16 in. x 1-1/4 in. (120 mm x 32 mm) square box with an extension ring. Note: With a ring, 22 - 12 AWG conductors may be housed.**

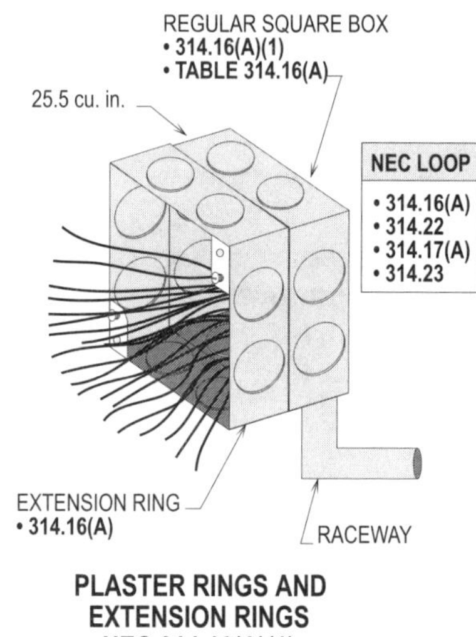

Figure 12-17. Increasing the fill space using an extension ring.

JUNCTION BOXES
314.28

Junction boxes housing conductors 4 AWG and larger that are pulled through raceways shall be sized based upon a straight pull or an angle pull.

STRAIGHT PULLS
314.28(A)(1)

In a straight pull (straight through), the raceway is connected to one side of the box and another raceway is connected to the opposite side, with the junction box used as a junction between the raceways. Junction boxes utilized in a straight pull shall be sized by multiplying the largest raceway in the run by not less than the multiplier 8 per **314.28(A)(1)**.

Sometimes junction boxes will have more than two raceways connected to its sides (walls). Junction boxes with more than two raceways connected to their sides shall be sized by multiplying the largest raceway by 8.

For example: What is the minimum length for a straight pull consisting of 1 run of 2 in. (53) raceways?

Step 1: Finding the multiplier
314.28(A)(1)
Multiplier = 8

Step 2: Calculating length
314.28(A)(1)
2" (53) raceway x 8 = 16"

Solution: **The minimum length of the junction box is 16 in.**

For example: What is the minimum length for a junction box with a straight pull that has 4 in. (103), 2 in. (53), and 1 in. (27) raceways connected to its sides?

Step 1: Finding the multiplier
314.28(A)(1)
Multiplier = 8

Step 2: Calculating length
314.28(A)(1)
4" (103) x 8" = 32"

Solution: **The minimum length is 32 in.**

Design Tip: The minimum length of the junction box is 16 in. long, while the width of the wall that the raceways are connected to shall be sized to accommodate the number of raceways plus the locknuts and bushings. The proper length and width is usually provided when a standard size junction box is selected.

See Figure 12-19 for a step-by-step procedure for sizing junction boxes for straight pulls.

ANGLE OR U PULLS
314.28(A)(2)

An angle pull is where the raceways enter and leave the junction box from the top wall to one of the side walls or from one of the side walls to the bottom wall. Junction boxes in angle pulls shall be sized by multiplying the largest raceway in the run by the multiplier 6 per **314.28(A)(2)**. In a U pull, the raceway leaves the same wall and forms a U pull configuration.

For example: What is the minimum length of a junction box for an angle pull that has one run of 4 in. (103) raceways?

Step 1: Finding the multiplier
314.28(A)(2)
Multiplier = 6

Step 2: Calculating length
314.28(A)(2)
4" (103) x 6 = 24"

Solution: **The minimum length is 24 in. x 24 in.**

WHAT WOULD BE THE A AND B DIMENSION OF THE LB IF IT CONTAINS 6 - 6 AWG		QUICK CALC 2
QUICK CALC 1		TABLE 314.16(B) 6 AWG = 5.00 cu. in. FILL AREA 30 cu. in. ÷ 5.00 cu. in. = 6 6 - 6 AWG MAY BE SPLICED IN 1" (27) LB
LB DIMENSION A **314.28(A)(2)** 1" (27) EMT x 6 = 6" • LBA = 6"	LB DIMENSION B **TABLE 312.6(A)** 6 - 6 AWG REQUIRES 1-1/2" • LBB = 1-1/2"	
LB = 6" x 1-1/2"		

CONDUIT BODIES
NEC 314.16(C)

Figure 12-18. Conduit bodies containing 6 AWG or smaller conductors shall be at least twice the cross-sectional area of the largest conduit to which they can be connected. This rule is designed to provide proper room in the LB.

Junction boxes having more than one run of raceways connecting to their walls to form an angle pull shall be sized by multiplying the largest raceway by not less than 6 and adding the remaining raceways to this value.

For example: What is the minimum length of a junction box for an angle pull that has a 4 in. (103), 3 in. (78), 2 in. (53), and 1 in. (27) raceway connected to the right wall and bottom wall?

> **Step 1:** Finding the multiplier
> **314.28(A)(2)**
> Multiplier = 6
>
> **Step 2:** Calculating length
> **314.28(A)(2)**
> 4" (103) x 6 =24"
> 3" (78)
> 2" (53)
> 1" (27)
> 30"
>
> **Solution: The minimum length is 30 in. x 30 in.**

See **Figure 12-20** for a step-by-step procedure for sizing junction boxes for angle pulls.

SIZING CONDUITS
ANNEX C

Conduits are used to enclose conductors to supply power to various types of electrical equipment. The same size conductors with the same type of insulation may be pulled through conduits. They may also be routed through conduits with different sizes and types of insulation. **Tables** in **Chapter 9** shall be used to size conduits based on the types of insulation and size of conductors that are pulled through the conduit system.

ENCLOSING SAME SIZE
CONDUCTORS
TABLES C.1 THRU C.12(A), ANNEX C

Conduits with conductors that have the same type of insulation and are the same size shall be permitted to be selected by using **Tables C.1 through C.12(A), Annex C**. The procedure for selecting the size of the conduit is to take the size of the conductors and align them up in one column of the correct Tables and select the proper size in the appropriate column based on the type of insulation and number of conductors. This is the easiest and fastest method to use in determining the size conduit housing the same size conductors.

For example: What size EMT conduit is required for 10 - 12 AWG XHHW copper conductors routed through the conduit to a pull box?

> **Step 1:** Finding size conduit
> **Table C.1, Annex C**
> 10 - 12 AWG requires 3/4" (21)
>
> **Solution: A 3/4 in. (21) conduit is required.**

See **Figure 12-21** for a step-by-step procedure for sizing conduits enclosing the same size conductors.

Problem: What is the minimum length for a junction box with a straight pull that has 4 in. (103) raceway on opposite sides and a second box that has 3 1/2 in. (91), 2 in. (53), and 1 in. (27) raceways connected to its sides?

Step 1:	Finding the multiplier **314.28(A)(1)** Multiplier = 8
Step 2:	Calculating length (3 1/2" Raceway) **314.28(A)(1)** 3 1/2" (91) x 8 = 28"
Step 3:	Calculating length (4" Raceway) **314.28(A)(1)** 4" (103) x 8 = 32"
Solution:	**The minimum length is 28 in. and 32 in., respectively, for each box.**

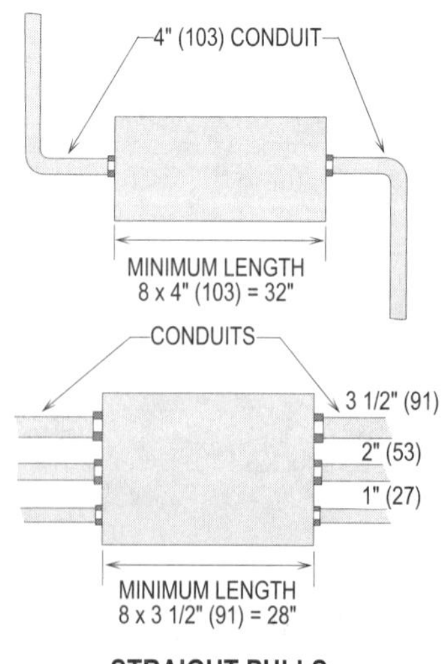

Figure 12-19. The size junction box for a straight pull shall be found by multiplying the largest conduit by not less than 8.

Problem: What is the minimum length of a junction box for an angle pull that has a 3 in. (78), 3 in. (78), 2 in. (53), and 1 in. (27) EMT that is connected to the right wall and bottom wall?

Step 1: Finding the multiplier
314.28(A)(2)
Multiplier = 6

Step 2: Calculating length
314.28(A)(2)
3" (78) x 6 = 18"

	3" (78)
	2" (53)
	1" (27)
	24"

Solution: **The minimum length is 24 in. x 24 in.**

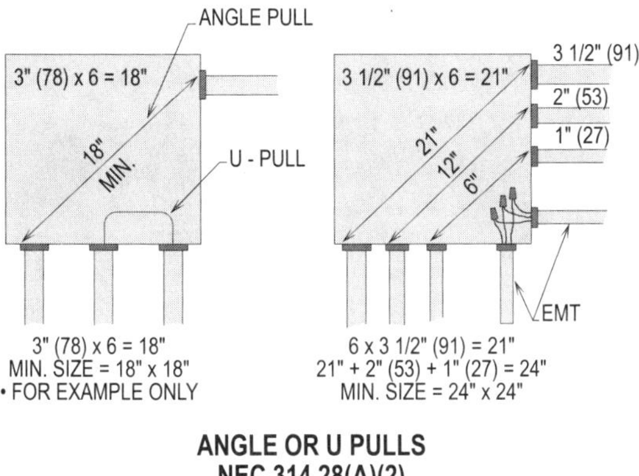

Figure 12-20. The size junction box for a angle pull shall be found by multiplying the largest conduit by not less than 6 and adding this number to the remaining conduits. Calculation will determine the size junction box.

ENCLOSING DIFFERENT SIZE CONDUCTORS
TABLES 4 AND 5 TO CHAPTER 9

Conduits enclosing different size conductors shall be sized by selecting the square inch area of each conductor per **Table 5** in **Chapter 9** and multiplying by the number. This total is used to select the size conduit per **Table 4** in **Chapter 9**.

For example: What size EMT conduit is required to enclose 2 - 4 AWG, 1 - 6 AWG, and 1 - 8 AWG THWN copper conductors?

Step 1: Finding sq. in. area
Table 5, Ch. 9
4 AWG = .0824 sq. in.
6 AWG = .0507 sq. in.
8 AWG = .0366 sq. in.

Step 2: Calculating sq. in. area
Table 5, Ch. 9
.0824 sq. in. x 2 = .1648 sq. in.
.0507 sq. in. x 1 = .0507 sq. in.
.0366 sq. in. x 1 = .0366 sq. in.
Total   = .2521 sq. in.

Step 3: Selecting size conduit
Table 4, Ch. 9
.2521 sq. in. requires .346 sq. in.

Solution: **A 1 in. (27) conduit is required.**

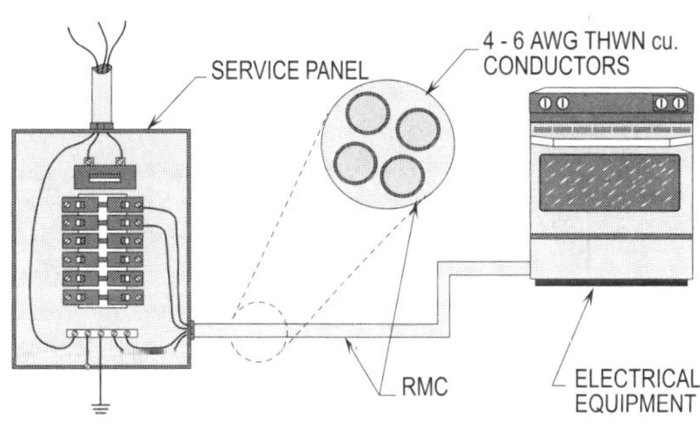

Problem: What size rigid metal conduit is required to enclose 4 - 6 AWG THWN copper conductors?

Step 1: Finding size conduit
Table C.8, Annex C
4 - 6 AWG requires 3/4" (21)

Solution: A 3/4 in. (21) RMC is required.

QUICK CALC	
TABLE 4, CH. 9	= 40%
3/4" (21)	= .220
TABLE 5, CH. 9	
6 AWG	= .0507 sq. in.
.220 (40%) ÷ .0507 sq. in.	= 4.339
4.339	= 4 conductors
3/4" (21)	= 4 conductors

ENCLOSING SAME
SIZE CONDUCTORS
NEC TABLE C.1 THRU C.12(A), ANNEX C

Figure 12-21. Tables C.1 through C.12(A) in **Annex C** shall be permitted to be used to size the conduit enclosing the same size conductors.

See **Figure 12-22** for an illustrated procedure for sizing conduits with different size conductors.

> **Problem:** What size EMT is required to enclose 12 - 14 AWG, 4 - 12 AWG, and 4 - 10 AWG THWN copper conductors?
>
> **Step 1:** Finding sq. in. area
> **Table 5, Ch. 9**
> 14 AWG = .0097 sq. in. area
> 12 AWG = .0133 sq. in. area
> 10 AWG = .0211 sq. in. area
>
> **Step 2:** Calculating sq. in. area
> **Table 5, Ch. 9**
> .0097 sq. in. x 12 = .1164 sq. in.
> .0133 sq. in. x 4 = .0532 sq. in.
> .0211 sq. in. x 4 = .0844 sq. in.
> Total = .254 sq. in.
>
> **Step 3:** Selecting size EMT
> **Table 4, Ch. 9**
> .254 sq. in. requires .346 sq. in.
>
> **Solution:** A 1 in. (27) EMT is required.

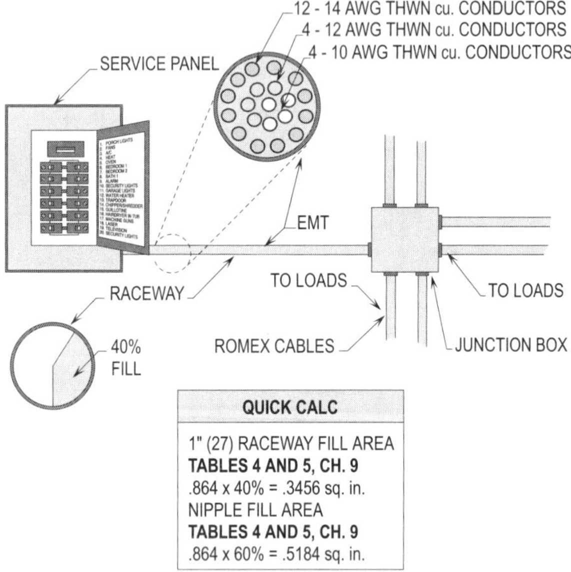

Figure 12-22. Table 5 and **Table 4** in **Chapter 9** shall be permitted to be used to size the conduit enclosing different size conductors.

SIZING NIPPLES
NOTE (4) TO TABLES, CHAPTER 9

The difference in a nipple and a conduit is that a nipple is 24 in. (600 mm) or less in length. A conduit system is any run over 24 in. (600 mm) in length.

A conduit system with more than two conductors shall be permitted to have a 40 percent fill per **Table 4, Chapter 9**, while a nipple shall be permitted to have a 60 percent fill per **Note (4)** to **Chapter 9**.

Nipples shall be sized by finding the square inch area of each conductor and multiplying by the number. The total 100 percent fill in square inches in **Table 5, Chapter 9** shall be selected based upon the conduit size, and this value shall be multiplied by 60 percent. If the total square inch area of all the conductors is less or equal to the total of the square inch area produced by applying the 60 percent factor, the size of the nipple shall be selected based upon that conduit size used in the calculation.

> **For example:** What size EMT nipple is required to enclose 36 - 12 AWG THWN copper conductors that are installed between a panelboard and junction box?
>
> **Step 1:** Finding sq. in. area
> **Table 5, Ch. 9**
> 12 AWG = .0133 sq. in.
>
> **Step 2:** Calculating sq. in. area
> **Table 5, Ch. 9**
> .0133 sq. in. x 36 = .4788 sq. in.
>
> **Step 3:** Finding conduit at 100% total fill
> **Table 4, Ch. 9**
> 1" (27) conduit that has a 100% total of .864 sq. in.
> (diameter squared x .7854 = sq. in. area)
> (1.049 x 1.049 x .7854 = .864 sq. in.)
>
> **Step 4:** Applying 60% fill
> **Note 4 to Ch. 9**
> sq. in. area x 60% = fill area
> .864 x 60% = .518 sq. in.
>
> **Step 5:** Selecting the nipple
> **Table 4, Ch. 9**
> .518 sq. in. is greater than .4788 sq. in.
>
> **Solution:** The size nipple is 1 in. (27).
> Note: The derating rules of Table 310.15(B)(2)(a) do not apply to nipples per Ex. 3 to 310.15(B)(2)(a).

See **Figure 12-23** for an illustrated procedure for sizing nipples applying the 60 percent fill rule.

Problem: What size EMT nipple is required to enclose 9 - 12 AWG, 9 - 14 AWG, 12 - 10 AWG, and 14 - 8 AWG THHN copper conductors?

Step 1: Finding sq. in. area
Table 5, Ch. 9
12 AWG = .0133 sq. in.
14 AWG = .0097 sq. in.
10 AWG = .0211 sq. in.
8 AWG = .0366 sq. in.

Step 2: Calculating sq. in. area
Table 5, Ch. 9
.0133 sq. in. x 9   = .1197 sq. in.
.0097 sq. in. x 9   = .0873 sq. in.
.0211 sq. in. x 12   = .2532 sq. in.
.0366 sq. in. x 14   = .5124 sq. in.
Total   = .9726 sq. in.

Step 3: Finding conduit at 100% total fill
Table 4, Ch. 9
1-1/2" (41) conduit has a 100% total of 2.036 sq. in.

Step 4: Applying 60% fill
Note (4) to Ch. 9
sq. in. area x 60% = fill area
2.036 x 60% = 1.2216 sq. in.

Step 5: Selecting the nipple
Table 4, Ch. 9
1.2216 sq. in. is greater than .9726 sq. in.

Solution: A 1-1/2 in. (41) nipple is required.

Figure 12-23. Table 5 and **Table 4** in **Chapter 9** and **Note (4)** to **Tables** in **Chapter 9,** shall be permitted to be used to calculate the size of nipples. Nipples are 24 in. (600 mm) or less in length, and a raceway is over 24 in. (600 mm) in length.

GUTTER SPACE
312.7 AND 366.56

Gutter space is used to enter or leave enclosures that enclose conductors that are terminated to serve the electrical equipment or are tapped or spliced to serve other loads. Certain clearances and space requirements to protect the conductors and prevent overcrowding are essential per **312.7** and **366.56**.

AUXILIARY GUTTERS
366.22

The number of current-carrying conductors permitted in an auxiliary gutter without derating per **Table 310.15(B)(2)(a)** to **Table 310.16** is 30 or less per **366.6(A)**. The size of an auxiliary gutter shall determined by dividing the total square inch area of the conductors by 20 percent fill area for installing conductors. The total square inch area of the conductors shall be found by multiplying the square inch area per **Table 5, Ch. 9** of each conductor, which is based on the size and insulation of each conductor placed in the auxiliary gutter.

For example: What size auxiliary gutter is required to house 3 - 350 KCMIL THWN copper conductors (feeder-circuit) that have 3 - 1/0 AWG, 3 - 1 AWG, and 3 - 2 AWG THWN copper conductors spliced to them?

Step 1: Finding sq. in. area
Table 5, Ch. 9
350 KCMIL   = .5242
1/0 AWG   = .1855
1   AWG   = .1562
2   AWG   = .1158

Step 2: Calculating sq. in. area
Table 5, Ch. 9
.5242 sq. in. x 3   = 1.5726 sq. in.
.1855 sq. in. x 3   = .5565 sq. in.
.1562 sq. in. x 3   = .4686 sq. in.
.1158 sq. in. x 3   = .3474 sq. in.
Total   = 2.9451 sq. in.

Step 3: Sizing gutter
366.22(A)
sq. in. area divided 20% = total fill
2.9451 divided 20% = 14.7255 sq. in.

Step 4: Selecting gutter
Chart
4" x 4" = 16"

Step 5: Applying 75% fill for splices
sq. in. of gutter x 75% = fill area
16 sq. in. x 75% = 12 sq. in.

Solution: A 4 in. x 4 in. auxiliary is required with only 12 in. of the gutter space used for splicing conductors.

See Figure 12-24 for a detailed procedure for designing, sizing, and selecting auxiliary gutters.

PANELBOARDS
408.3(G)

Panelboards shall comply with the provisions of **408.3(G)**, which requires the minimum gutter space in the panel to meet the clearances in **Table 312.6(A)** for L-bends and **Table 312.6(B)** for S- or Z-bends.

Electricians shall enter panelboards using an L-, S-, or Z-bends so that there is adequate room to terminate the conductors without damaging the insulation, which could later create a ground fault condition and cause a power failure.

For example: What are the minimum clearances between the terminating lugs and the side of a panelboard employing an L-bend using 4/0 AWG THWN copper conductors?

Step 1: Finding the minimum clearances
Table 312.6(A)
1 - 4/0 AWG per lug = 4" (102 mm)

Solution: The minimum clearance required is 4 in. (102 mm).

Panelboards utilizing S- or Z-bends require greater clearance between lugs and the sides (walls) where conductors are terminated. S- or Z-bends are harder to handle, bend, shape, and make ready for termination. L-bends are a lot easier to handle and connect than S- or Z-bends.

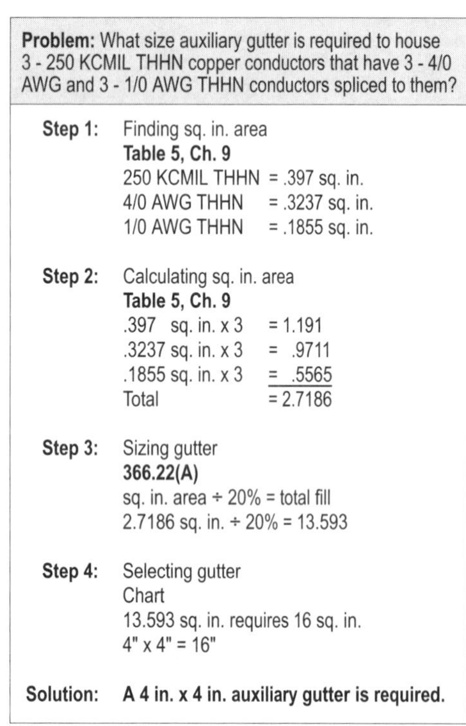

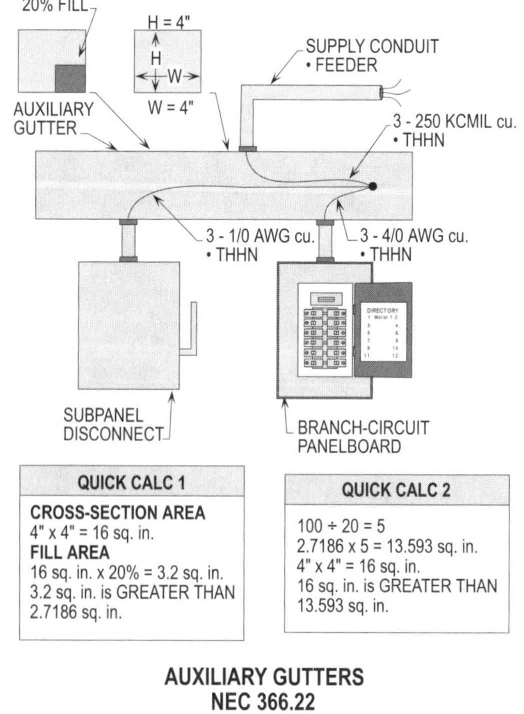

Figure 12-24. Table 5 and **Table 4** in **Chapter 9** and **366.22** shall be used to calculate and size auxiliary gutters.

For example: What is the minimum clearance between the terminating lugs and the sides of the panelboard employing 4/0 AWG THHN copper conductors?

Step 1: Finding the minimum clearance
Table 312.6(B)
1 - 4/0 AWG per lug = 7" (178 mm)

Solution: **The minimum clearance required is 7 in. (178 mm).**

If the lugs in the panelboard are removable, the clearance shall be permitted to be reduced from the dimension in **Table 312.6(B)** based on the number and size of conductors connected to each lug.

For example: What is the minimum clearance for 4/0 AWG THHN copper conductors terminating to the lugs in a panelboard entering from the bottom wall? (Lugs removable)

Step 1: Finding the minimum clearance
Table 312.6(B)
1 - 4/0 AWG per lug
7"(178 mm) - 1"(25.4 mm) = 6"(152 mm)

Solution: **The minimum clearance required is 6 in. (152 mm).**

See Figure 12-25 for an illustrated procedure for determining the minimum clearance of conductors between lugs and sides (walls) in a panelboard.

CABLE TRAYS
ARTICLE 392

Cable trays shall be permitted to be used for the installation of a number of cables that are installed in a single run from one location to another. Steel or aluminum cable tray sections are used with the ends bolted together to form a single run of the required length. Where installing and routing cables in cable trays, the cables are laid on racks, troughs, or hangers.

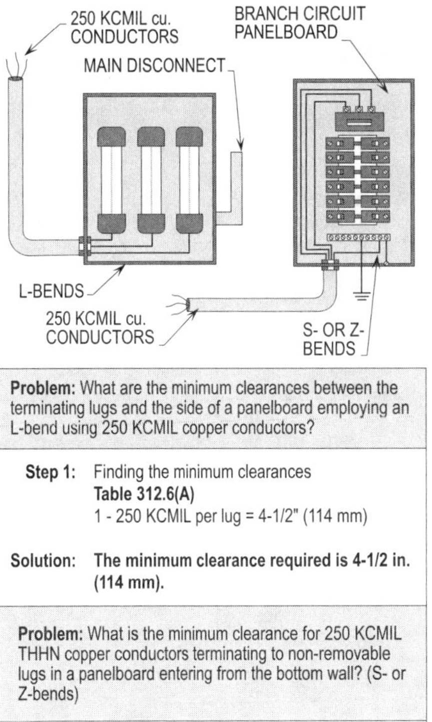

Problem: What are the minimum clearances between the terminating lugs and the side of a panelboard employing an L-bend using 250 KCMIL copper conductors?

Step 1: Finding the minimum clearances
Table 312.6(A)
1 - 250 KCMIL per lug = 4-1/2" (114 mm)

Solution: The minimum clearance required is 4-1/2 in. (114 mm).

Problem: What is the minimum clearance for 250 KCMIL THHN copper conductors terminating to non-removable lugs in a panelboard entering from the bottom wall? (S- or Z-bends)

Step 1: Finding the minimum clearance
Table 312.6(B)
1 - 250 KCMIL per lug = 8-1/2" (216 mm)

Solution: The minimum clearance required is 8-1/2 in. (216 mm).

PANELBOARDS
NEC 408.3(G)

Figure 12-25. The distance between lugs and enclosures walls for L-, S-, or Z-bends shall be permitted to be determined in **Tables 312.6(A)** and **312.6(B)**. These Tables are based upon bends provide safe terminating space for conductors.

MULTICONDUCTOR CABLES
392.9

Where sizing and installing multiconductor cables in ladder, ventilated-trough, solid-bottom, or ventilated channel-type trays, the requirements of **392.9** shall be complied with.

ANY MIXTURE OF CABLES
392.9(A)

The diameter of all cables 4/0 AWG and larger shall be added together, and the total of each cable shall not be permitted to exceed the width of ladder or ventilated-trough trays. All cables 4/0 AWG and larger shall be installed side-by-side. Cables smaller than 4/0 AWG shall not be required to be installed side-by-side per **Table 392.9, Column 1.** When installing cables rated larger than 4/0 AWG and smaller than 4/0 AWG, the requirements per **Table 392.9, Column 2** shall be used. **(See Figure 12-26)**

Note that cables 1/0 AWG through 4/0 AWG shall be permitted to be stacked in a angular and square configuration.

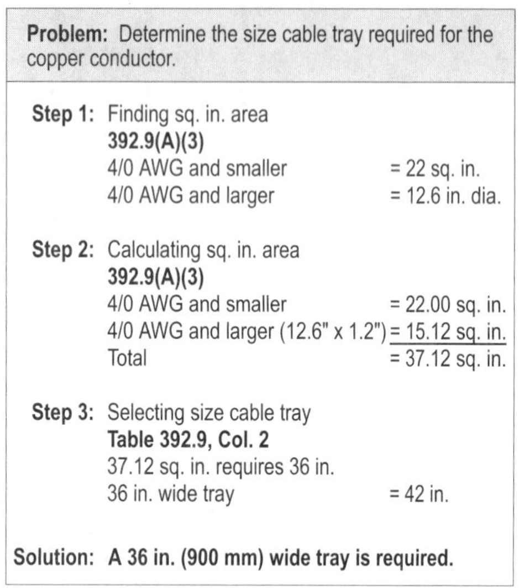

Problem: Determine the size cable tray required for the copper conductor.

Step 1: Finding sq. in. area
       392.9(A)(3)
       4/0 AWG and smaller   = 22 sq. in.
       4/0 AWG and larger   = 12.6 in. dia.

Step 2: Calculating sq. in. area
       392.9(A)(3)
       4/0 AWG and smaller   = 22.00 sq. in.
       4/0 AWG and larger (12.6" x 1.2") = 15.12 sq. in.
       Total   = 37.12 sq. in.

Step 3: Selecting size cable tray
       Table 392.9, Col. 2
       37.12 sq. in. requires 36 in.
       36 in. wide tray   = 42 in.

Solution: A 36 in. (900 mm) wide tray is required.

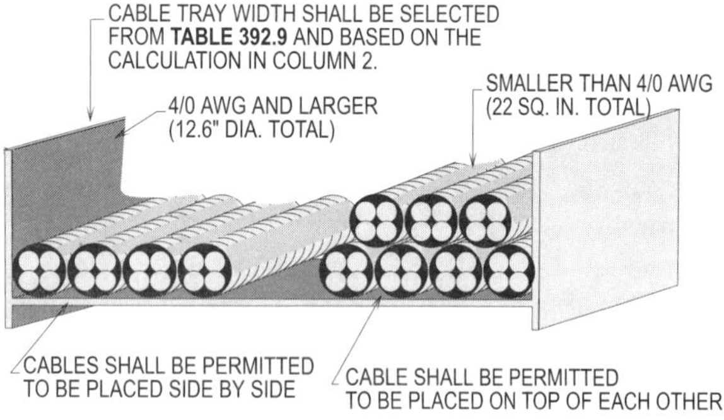

Figure 12-26. When installing cables rated larger than 4/0 AWG and smaller than 4/0 AWG, the requirements per **Table 392.9, Column 2** shall be used.

MULTICONDUCTOR CONTROL AND/OR SIGNAL CABLES ONLY 392.8(B)

The total cross-sectional areas of all cables in cable trays with an inside depth of 6 in. (150 mm) or less containing control and/or signal cables shall not be permitted to exceed 50 percent of the cross-sectional area of the cable tray.

SOLID-BOTTOM CABLE TRAYS CONTAINING ANY MIXTURE 392.9(C)

The diameter of all cables that are rated 4/0 AWG and larger shall not be permitted to exceed 90 percent of the solid-bottom tray width. Cables smaller than 4/0 AWG shall meet the requirements of **Table 392.9, Column 3**. A combination of cables that are rated 4/0 AWG and larger or less than 4/0 AWG shall meet the requirements of **Table 392.9, Column 4**.

SOLID-BOTTOM CABLE TRAY – MULTICONDUCTOR CONTROL AND/OR SIGNAL CABLES ONLY 392.9(D)

For control and/or signal cables are installed in cable trays with an inside depth of 6 in. (150 mm) or less, the total cross-sectional areas of the cables shall not be permitted to exceed 40 percent of the cross-sectional area of the tray.

VENTILATED CHANNEL CABLE TRAYS 392.9(E)

The total cross-sectional areas of all cables shall not be permitted to exceed 1.3 square inches (850 sq. mm) for 3 in. (75 mm) wide trays, 2.5 square inches (1600 sq. mm) for 4 in. (100 mm) wide trays, or 3.8 square inches (2450 sq. mm) for 6 in. (150 mm) wide trays when installing ventilated channel-type trays.

SINGLE CONDUCTOR CABLES 392.10

When designing and installing single-conductor cables in ladder, ventilated-trough, or ventilated channel-type trays the requirements in **392.10** shall be applied.

When installing cables in ladder or ventilated-trough trays, the total diameter of all cables 1000 KCMIL and larger shall not be permitted to exceed the width of the cable tray. Cables smaller than 1000 KCMIL shall be installed per **Table 392.10(A), Column 1**.

A combination of cables that are rated 1000 KCMIL and larger or less than 1000 KCMIL shall meet the requirements of **Table 392.10(A), Column 2. (See Figure 12-27)**

Problem: Determine the size cable tray required for the copper conductors.

Step 1: Finding sq. in. area
Table 5, Ch. 9
300 KCMIL THHN = .4608 sq. in.
1000 KCMIL THHN = 1.3478 sq. in.

Step 2: Calculating sq. in. area
392.10(A)(2); Table 392.10(A), Col. 1
.4608 sq. in. x 14 = 6.4512 sq. in.
1.3478 sq. in. x 10 = 13.478 sq. in.
Total = 19.9292 sq. in.

Step 3: Selecting size cable tray
Table 392.10(A), Col. 1
19.9292 sq. in. requires 24 in. width
24 in. wide tray requires 26 sq. in.

Solution: A 24 in. (600 mm) wide cable tray has 26 sq. inches per
Table 392.10(A).

SIZING CASE 1	SIZING CASE 2
Step 1: Calculating cable tray for case 1 **392.9(B)** 18 x .35 sq. in. = 6.3 sq. in. 35 x .15 sq. in. = 5.25 sq. in. Total = 11.55 sq. in.	**Step 1:** Calculating solid bottom tray **392.9(B)** 16 x .170 = 2.72 sq. in.
Step 2: Applying multiplier **392.9(B)** 11.55 sq. in. ÷ 50% = 23.1 sq. in.	**Step 2:** Selecting cable tray **Table 392.9** 2.72 sq. in. requires 6 in.
Step 3: Applying depth of tray **392.9(B); Table 392.9** 23.1 sq. in. ÷ 4" = 5.775 in.	**Solution:** The size cable tray required is 6 in. (150 mm) per the NEC.
Solution: A cable tray with a 6 in. (150 mm) width is required.	
Note: In the case of a solid bottom tray, the multiplier is 40% and the results would be 11.55 ÷ 40% = 28.88 ÷ 4" = 7.2 in., in which a 12 in. (300 mm) cable tray is required per **Table 392.9.**	

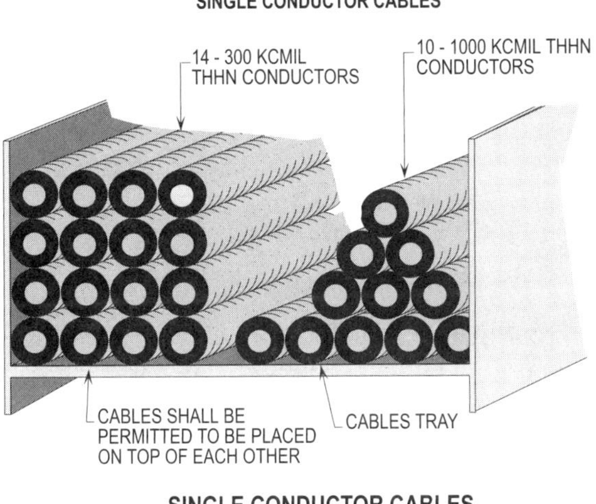

SINGLE CONDUCTOR CABLES
NEC 392.10(A)(2)
TABLE 392.10(A), COLUMNS 1 AND 2

Figure 12-27. When installing cables in ladder or ventilated-trough trays, the total diameter of all cables 1000 KCMIL and larger shall not be permitted to exceed the width of the cable tray.

The total diameter of all cables installed in ventilated channel-type trays shall not be permitted to exceed the inside width of the channel.

CONTROL AND SIGNALING CABLES
NEC 392.9(B) AND (D)

Figure 12-28. The above is the procedure for calculating and sizing cable trays for supporting control and signaling cables using a solid or open cable tray system.

CONTROL AND SIGNALING CABLES 392.9(B) AND (D)

Ladder or ventilated-trough trays shall be permitted to be filled with control or signaling cables or both up to 50 percent of the cross-sectional area of their trays up to 6 in. (150 mm) deep. Solid-bottom trays shall be calculated using the same procedures, except 40 percent is utilized as the allowable multiplier. **(See Figure 12-28)**

DETERMINING AMPACITIES OF CONDUCTORS

The derating factors of **Table 310.15(B)(2)(a)** to **Tables 0 through 2000 volts** covers and applies only to multiconductor cables that have more than three current-carrying conductors. Derating is based upon the number of conductors in the cable, and not the number of conductors in the cable tray.

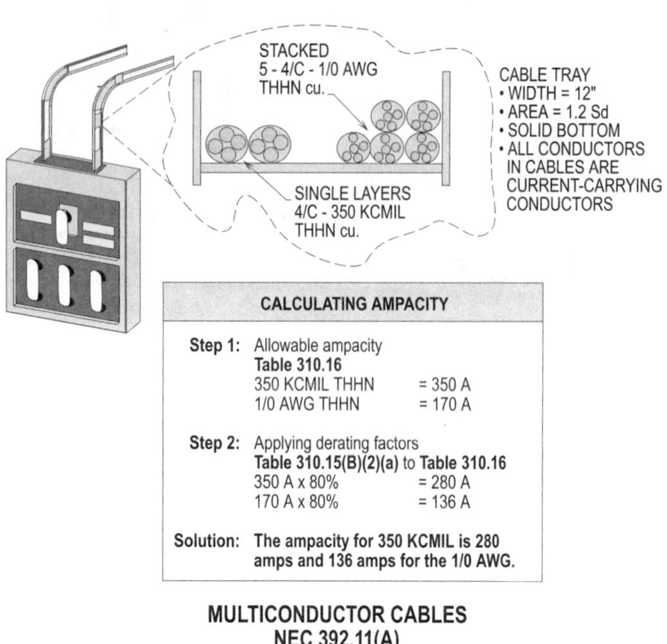

Figure 12-29(a). Calculating the ampacity of multiconductor cables supported by a cable tray.

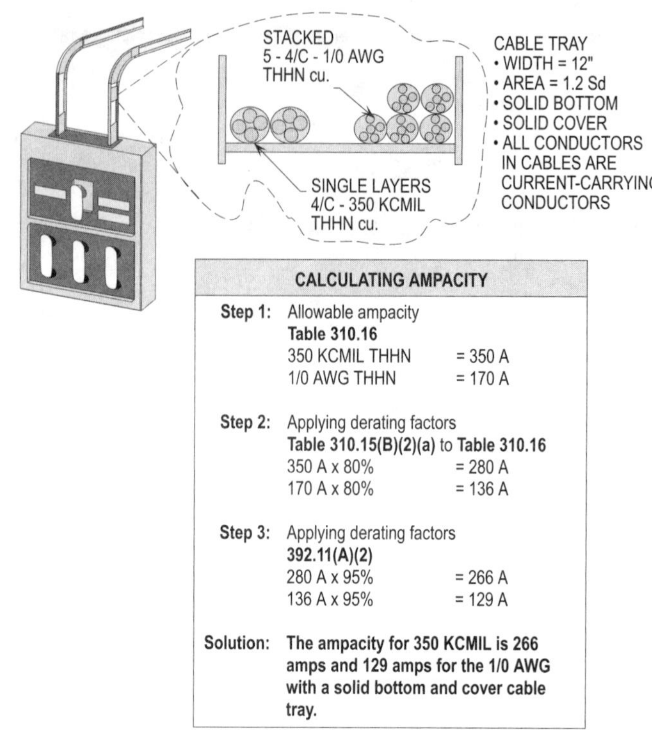

Figure 12-29(b). Calculating the ampacity of multiconductor cables supported by a solid-bottom cable tray with a solid cover.

MULTICONDUCTOR CABLES
392.11(A)

Tables 310.16 and **310.18** cover the allowable ampacities that apply to multiconductor cables nominally rated 2000 volts or less, provided they are installed according to the requirements of **392.9**.

Ampacities shall be cut to 95 percent of the ampacity in **Tables 310.16** and **310.18** where cable trays are continuously covered for more than 6 ft (1.8 m) with solid cover that is not ventilated per **392.11(A)(2)**.

When multiconductor cables in uncovered trays are installed in a single layer and maintain a spacing of one conductor diameter between cables, their ampacity shall be determined according to **310.15(B)(2)(a)** as required per **392.11(A)(3)** (ambient corrected ampacities of multiconductor cables rated 0-2000 volts). Refer also to **Table B.310.3** in Annex B for another ampacity Table to be used under certain conditions. **[See Figures 12-29(a) and (b)]**

SINGLE CONDUCTOR CABLES
392.11(B)(1) AND (B)(2)

The following will apply to the ampacity of single conductor cables or single conductor cables that are twisted or bound together, such as triplex or quadraplex, etc. Note that this applies to nominally rated cables of 2000 volts or less.

If single conductor cables, 600 KCMIL or larger, are installed per **392.10** and in uncovered cable trays, the ampacities of the single conductor cables shall not exceed 75 percent of the ampacities as shown in **Tables 310.17** and **310.19**. Where cable trays are continuously covered for more than 6 ft (1.8 m) and unventilated covers are used, ampacities for 600 KCMIL and larger cables shall not exceed the ampacity of 70 percent of the level ampacities shown in **Tables 310.17** and **310.19**. **[See Figures 12-30(a) and (b)]**

When cables are installed according to **392.10**, single conductor cables from 1/0 AWG through 500 KCMIL installed in uncovered cable trays shall have their ampacities calculated at 65 percent of the allowable ampacities in **Tables 310.17** and **310.19**. If the cable trays are continuously covered for more than 6 ft (1.8 m) with solid

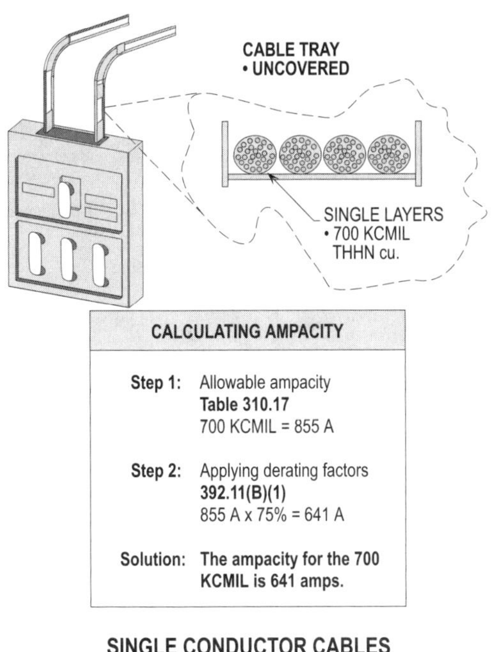

Figure 12-30(a). Single conductors that are 600 KCMIL or larger shall have their ampacities in **Table 310.17** derated by 75 percent to obtain allowable ampacities.

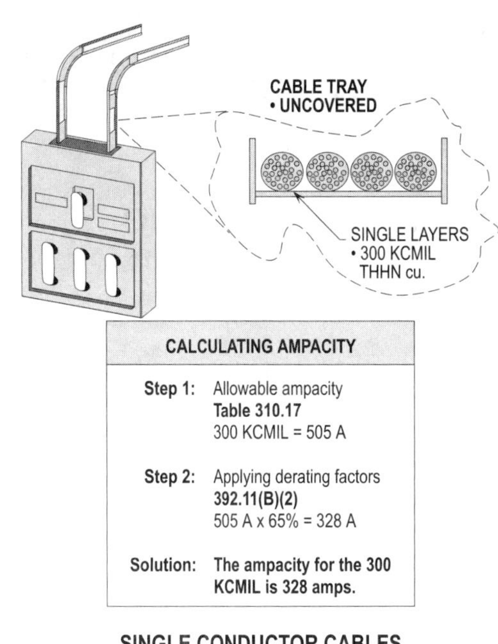

Figure 12-31(a). Single conductors that are 1/0 AWG through 500 KCMIL shall have their ampacities in **Table 310.17** derated by 65 percent to obtain allowable ampacities.

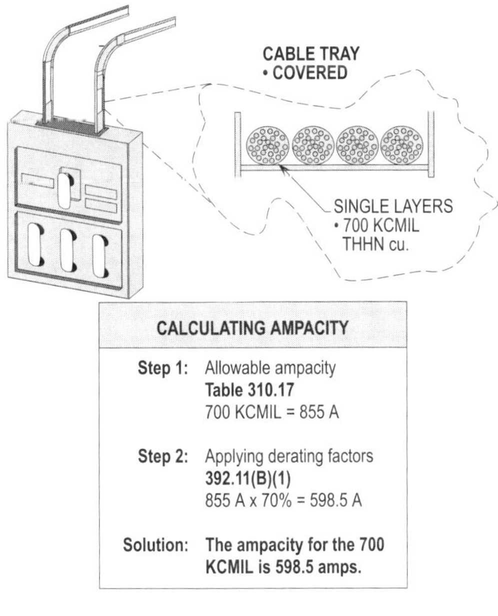

Figure 12-30(b). Single conductors that are 600 KCMIL or larger shall have their ampacities in **Table 310.17** derated by 70 percent to obtain allowable ampacities.

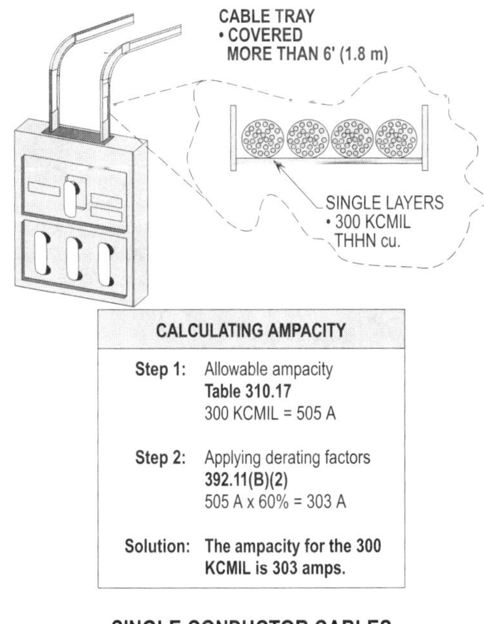

Figure 12-31(b). Single conductors that are 1/0 AWG through 500 KCMIL shall have their ampacities in **Table 310.17** derated by 60 percent to obtain allowable ampacities.

unventilated covers, for 1/0 AWG through 500 KCMIL single conductor cables, the ampacities shall not be permitted to exceed 60 percent of the allowable ampacities as covered in **Tables 310.17** and **310.19**. **[See Figures 12-31(a) and (b)]**

GROUNDING AND BONDING CABLE TRAYS
392.5(E) AND 392.6(A)

Cable trays shall be permitted to have mechanically discontinuous segments between cable tray runs and/or between cable tray runs and equipment. Such discontinuous segments shall be bonded and grounded for safety to protect wiring methods, equipment, and personnel from electrical shock and fire hazards.

In industrial facilities where conditions of maintenance and supervision ensure that only qualified people will service the installation, cable trays that are designed to support cables and other loads shall be permitted to be used to support such systems.

For conduits, cables, etc. that bond to cable trays, a listed clamp or adapter shall be used. If a listed cable tray clamp or adapter is used, a nearby support, such as a support of 3 ft (900 mm) shall not be required. **[See Figures 12-32(a), (b),** and **(c)]**

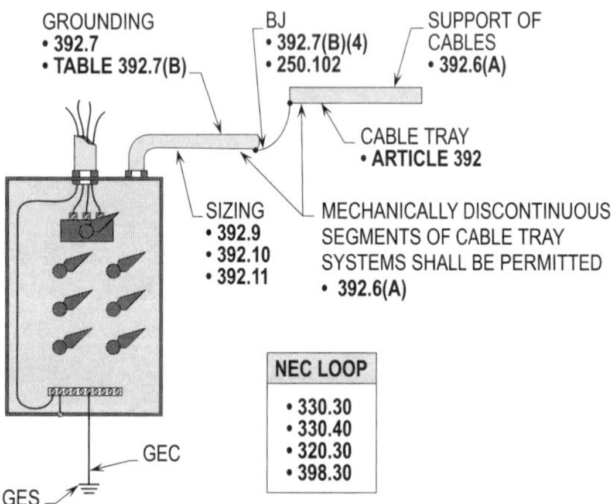

GROUNDING AND BONDING CABLE TRAYS
NEC 392.5(E) AND NEC 392.6(A)

Figure 12-32(a). Discontinuous segments of cable trays, such as for changes of direction or elevation, shall be permitted.

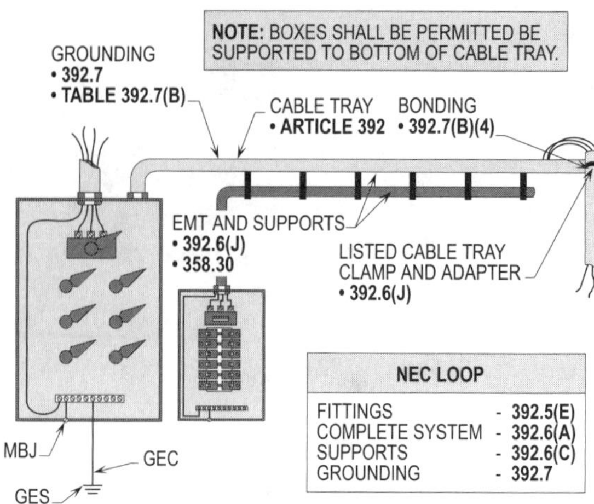

GROUNDING AND BONDING CABLE TRAYS
NEC 392.6(J)

Figure 12-32(b). The procedure by which conduits and cables may be run, connected, and supported to or by cable tray systems.

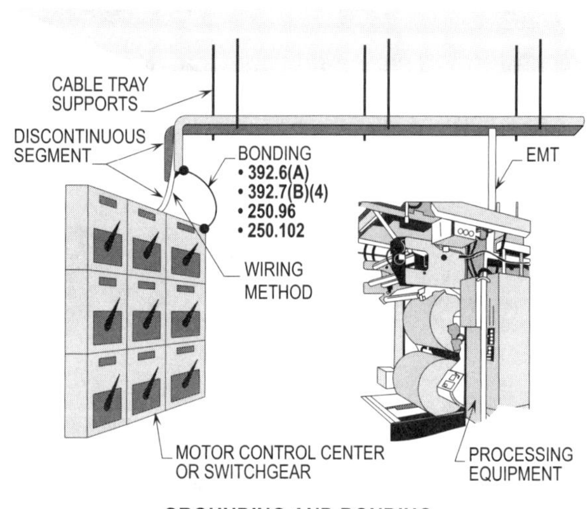

GROUNDING AND BONDING CABLE TRAYS
NEC 392.5(E) AND NEC 392.6(A)

Figure 12-32(c). This illustration shows discontinuous segments of the cable tray and wiring methods shall be permitted.

EXPANSION FITTINGS
352.44

Where expansion and contraction due to temperature difference might be encountered, approved rigid nonmetallic conduit expansion joints shall be used. The substances of PVC expand and contract with temperature changes, which will put undue strain on the conduit. Properly designed expansion joints absorb this expansion and contraction, avoiding damage to the conduit and fittings. **(See Figure 12-33)**

For Example: RNC is run between two panels. It runs 40 ft up the wall, then 90's into the attic and runs 16 ft. How many expansion joints with 6 in. expansion capacity are needed for the run if the temperature for the run over the year is an average of 100°F?

Step 1: Finding number of joints
352.44; Table 352.44
100°F = 4.06

Step 2: Finding total charge
Table 352.44 and text
- 200 ft run /100 ft   = 2
- 4.06 x 2   = 8.12
- 8.12 ÷ 6   = 1.35

Solution: **Two expansion joints should be used and installed at about 100 ft intervals.**

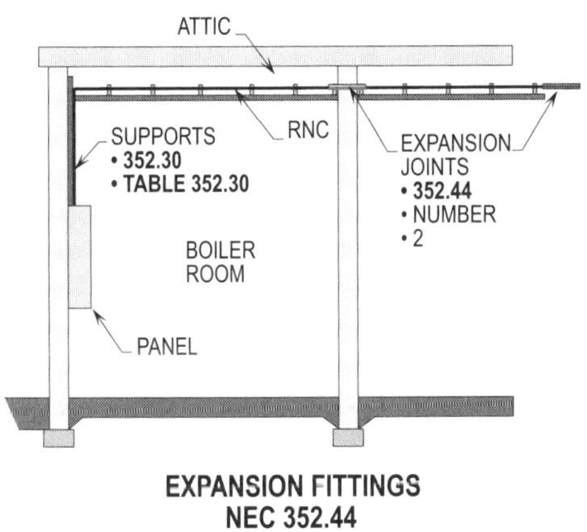

Figure 12-33. Installing expansion joints in a run of RNC.

Chapter 12. Designing Wiring Methods

Section   Answer

1. Conductors passing through the box unbroken and not pulled into a loop or spliced together shall be counted as _____ each in determining the fill of the box.
   (a) 0   (b) 1
   (c) 2   (d) 3

2. Equipment grounding conductors passing through or spliced together in the box shall be counted as _____ in determining the fill of the box. (general rule)
   (a) 1   (b) 2
   (c) 3   (d) 4

3. A conductor that is not entirely within the box shall be required to be counted as _____ conductor used toward the box fill.
   (a) 0   (b) 1
   (c) 2   (d) 3

4. A receptacle or switch that is mounted on a strap or yoke is counted as _____ conductor in determining the fill of the box.
   (a) 0   (b) 1
   (c) 2   (d) 3

5. Boxes with one or more cable clamps that are installed to support cables in the box shall have _____ conductor added in determining the fill of the box.
   (a) 0   (b) 1
   (c) 2   (d) 3

6. Boxes that contain fixture studs shall have _____ conductors added in determining the fill of the box.
   (a) 1   (b) 2
   (c) 3   (d) 4

7. Octagon boxes, unless listed for such, shall not support luminaires that weigh more than _____ lb.
   (a) 20   (b) 25
   (c) 35   (d) 50

8. Octagon boxes shall not support ceiling fans that weigh more than _____ lbs. or less.
   (a) 20   (b) 25
   (c) 35   (d) 50

9. Junction boxes housing conductors _____ AWG and larger that are pulled through raceways shall be sized based upon a straight pull or an angle pull.
   (a) 1   (b) 2
   (c) 3   (d) 4

_______ _______ **10.** A conduit system is any run over ______ in. long in length.
    (a) 6   (b) 12
    (c) 18   (d) 24

_______ _______ **11.** The size of an auxiliary gutter shall be determined by dividing the total square inch area of the conductor by ______ percent fill area for installing conductors.
    (a) 10   (b) 20
    (c) 25   (d) 40

_______ _______ **12.** Ladder or ventilated trough trays may be filled with control or signaling cables or both up to ______ percent of the cross-sectional area of their trays up to 6 in. deep.
    (a) 20   (b) 40
    (c) 50   (d) 60

_______ _______ **13.** When installing cables in ladder or ventilated trough trays, the total diameter of all cables ______ KCMIL and larger shall not exceed the width of the cable tray.
    (a) 600   (b) 1000
    (c) 1500   (d) 2000

_______ _______ **14.** Junction boxes in angle pulls shall be sized by multiplying the largest raceway in the run by the multiplier ______.
    (a) 6   (b) 8
    (c) 10   (d) 12

_______ _______ **15.** Junction boxes in straight pulls shall be sized by multiplying the largest raceway in the run by the multiplier ______.
    (a) 6   (b) 8
    (c) 10   (d) 12

_______ _______ **16.** A nipple shall be permitted to have a ______ percent fill area.
    (a) 40   (b) 50
    (c) 60   (d) 75

_______ _______ **17.** A nipple is ______ in. or less in length.
    (a) 6   (b) 12
    (c) 18   (d) 24

_______ _______ **18.** Ceiling (paddle) fans that are supported to octagon boxes (identified) shall not weigh more than ______ lb.
    (a) 25   (b) 35
    (c) 40   (d) 50

_______ _______ **19.** Luminaires that are supported by a octagon box (general rule) shall not weigh more than ______ lb.
    (a) 25   (b) 35
    (c) 40   (d) 50

_______ _______ **20.** The fill area for installing conductors in an auxiliary gutter shall be determined by dividing the total square inch area by ______ percent.
    (a) 20   (b) 40
    (c) 60   (d) 80

Section   Answer

21. What is the count used for 8 conductors passing through an octagon box and the count for 8 that are spliced?

22. What is the count toward the fill of the box where there are 8 equipment grounding conductors in the box that are used to ground the metal enclosures of the equipment?

23. How many conductors are counted toward the fill for 1 equipment grounding conductor to a metal box and 1 isolation equipment grounding conductor for a PC?

24. What size octagon box (w/canopy) is required to support a luminaire supplied with 2 - 14-2 AWG w/ground nonmetallic sheathed cables with a fixture stud, hickey, and 4 - 18 AWG fixture wires, plus EGCs?

25. What size octagon box, with cable clamps, is required for 2 - 12 AWG and 2 - 14 AWG w/ground romex cables that are spliced in the box between the panelboard and the loads?

26. What size square box is required for 2 - 14-2 AWG w/ground romex cables that are supported within the box with two cable clamps?

27. What size square box, with cable clamps, is required for 2 - 12 AWG and 2 - 10 AWG w/grounds romex cables that are spliced in the box between the panelboard and the loads?

28. What size device box is required for 1 - 10 AWG straight through conductor, 2 - 10 AWG hots, 2 - 10 AWG neutrals, 2 EGCs, 1 bonding jumper, and 1 pigtail? (Romex is the wiring method)

29. What size device box is required for 2 - 12 AWG conductors (phases and neutrals), 4 - 10 AWG conductors (phases and neutrals) and 1 - 30 amp receptacle? There is 1 pigtail and 1 bonding jumper in the box. (EMT is the wiring method)

30. What size junction box is required for 4 raceways with 10 - 12 AWG in each and 2 raceways with 8 - 12 AWG in each that are spliced together?

31. What size junction box is required for 2 raceways with 24 - 14 AWG, 2 raceways with 18 - 12 AWG, 2 raceways with 14 - 10 AWG, and 2 raceways with 8 - 6 AWG conductors that are spliced together?

32. What size square box, with extension ring, is required for 26 - 12 AWG conductors? (Use minimum size)

33. What is the cross-sectional area of an LB (conduit body) that is connected to a 2 in. EMT?

34. What is the minimum length for a junction box, with a straight pull, that has 4 in., 2 in., and 1-1/2 in. raceways connected to its sides?

35. What is the minimum length of a junction box for an angle pull that has a 3-1/2 in., 2 in., and 1 in. raceway that is connected to the right wall and bottom walls?

36. What size EMT conduit is required for 4- 8 AWG THWN copper conductors?

___________ ___________ **37.** What size EMT conduit is required for 10-14, 6-12, and 6-10 AWG THWN copper conductors?

___________ ___________ **38.** What size EMT nipple is required for 10-14, 10-12, 8-10, and 12-8 AWG THWN copper conductors?

___________ ___________ **39.** What size auxiliary gutter is required to house 3 - 500 KCMIL THWN copper conductors that have 3 - 4/0 and 3 - 2/0 AWG THWN conductors spliced to them?

___________ ___________ **40.** What are the minimum clearances between the terminating lugs and the side of a panelboard employing an L-bend, using 500 KCMIL copper conductors?

___________ ___________ **41.** What is the minimum clearance for 4/0 AWG THHN copper conductors terminating to the lugs in a panelboard, entering from the bottom wall? (S- or Z-bends with non-removable lugs)

___________ ___________ **42.** What is the ampacity for 4 - 2/0 AWG THHN multiconductor copper conductors and 4 - 300 KCMIL THHN multiconductor copper conductors supported by a solid-bottom cable tray?

___________ ___________ **43.** What is the ampacity for 4 - 2/0 AWG THHN multiconductor copper conductors and 4 - 300 KCMIL THHN multiconductor copper conductors supported by a solid-bottom cable tray with a solid cover?

___________ ___________ **44.** What is the ampacity for single layer 600 KCMIL THHN copper conductors installed in a cable tray that is uncovered?

___________ ___________ **45.** What is the ampacity for single layer 600 KCMIL THHN copper conductors installed in a cable tray that is covered?

___________ ___________ **46.** What is the ampacity for single layer 500 KCMIL THHN copper conductors installed in a cable tray that is uncovered?

___________ ___________ **47.** What is the ampacity for single layer 500 KCMIL THHN copper conductors installed in a cable tray that is covered?

13

Installing Wiring Methods

Wiring methods used to wire in residential, commercial, and industrial locations are cable systems, conduit systems, and tubing systems that enclose and protect the conductors. Wiring methods installed according to the provisions of the *National Electrical Code*, whether in walls, ceilings, floors, or attics assure safety. Good workmanship and proper inspection procedures work together in ensuring safe installations. For rules and regulations pertaining to designing wiring methods, see Chapter 12 of this book.

SUPPORTING BOXES
314.23

Boxes and junction boxes used to support luminaires, ceiling fans, and devices shall be mounted in such a manner to adequately support the box. Section **314.23(A)** requires the box to be fastened to the surface to which it is mounted, unless the surface does not provide an adequate means of support. Section **314.23(B)** permits enclosures to be structurally mounted to a framing member of the dwelling unit or building. The box shall be permitted to be directly mounted to the framing member or by using a metal or wooden brace. Any of the methods used shall provide rigid support where luminaires and devices are mounted to the box.

Metal braces shall be made of at least 0.020 in. (0.51 mm) of metal, and wooden braces shall not be less than 1 in. x 2 in. (25 mm x 50 mm) in cross-section, per **314.23(B)(2)**.

Nails shall be permitted to be used to mount boxes when the nails pass through the inside of the box within 1/4 in. (6 mm) of the back or ends of the box. This provision of **314.23(B)(1)** indicates that screws otherwise shall be used to support a box to framing members.

See Figures 13-1(a) and **(b)** for methods of mounting boxes to framing members.

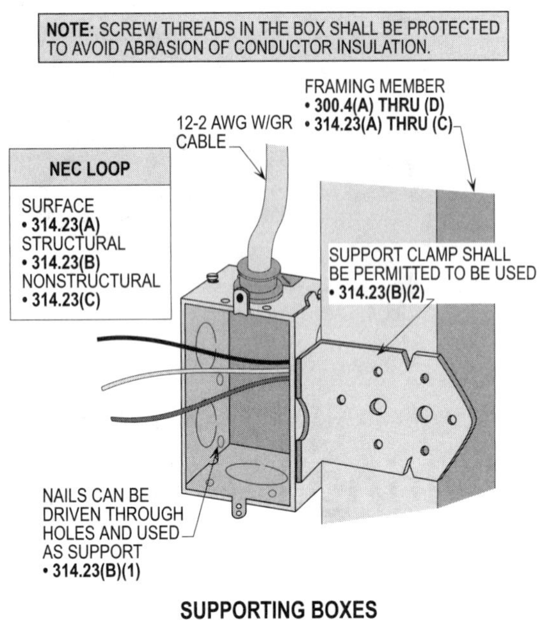

Figure 13-1(a). Mounting methods that shall be permitted to be used to support boxes to the framing members.

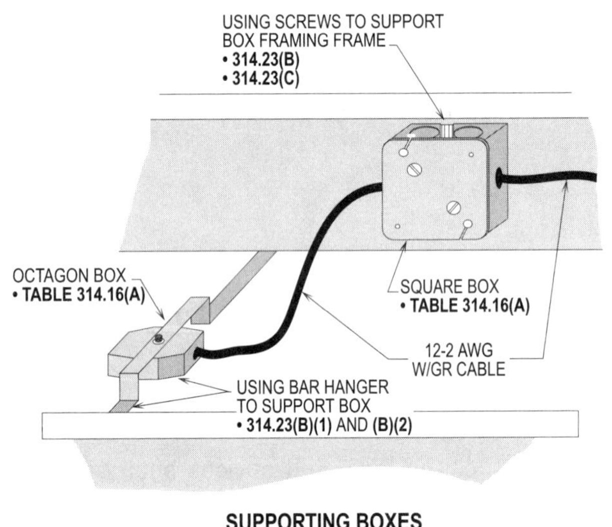

Figure 13-1(b). Using a bar hanger and screw to support boxes to framing members.

DEVICE BOXES
Table 314.16(A)

Device boxes are used to support receptacles, switches, or a combination of receptacles and switches, etc. mounted on a single yoke (strap). Device boxes are available in different sizes to accommodate the many sizes of switches and dimmers or receptacles and GFCI receptacles. **Table**

314.16(A) has a list of device boxes with different depths that are designed to contain conductors, receptacles, switches, GFCI receptacles, dimmers, etc. with oversized dimensions.

The size of device boxes, along with the proper depth, shall be chosen to house the width and depth of the receptacle or switch. The device box not only shall have adequate space for the device mounted to it, but it shall provide space for the number of conductors entering and leaving the box.

OTHER BOXES
314.16(A)(2)

Boxes 100 cubic inches or less in size, conduit bodies with more than two conduit entries, and boxes that are nonmetallic shall be durably and legibly marked with their cubic inch fill area. The fill capacity of each box shall be determined by multiplying each conductor by the cubic inch rating of each conductor from **Table 314.16(B)**. The cubic inch fill area of a junction box shall be found by multiplying the dimensions of the box.

For example, a box with dimensions of 4 in. x 4 in. x 4 in. has a cubic inch fill area of 64 sq. in. (4" x 4" x 4" = 64").

WITHOUT SUPPORTING DEVICES, LUMINAIRES, OR LAMPHOLDERS
314.23(E)

Boxes with threaded enclosures or hubs shall be permitted to be supported with two or more conduits that are threaded wrench tight into the enclosure or hubs. Each conduit shall be secured within 3 ft (900 mm) of the enclosure, or within 18 in. (450 mm) of the enclosure if all conduit entries are on the same side. The box is limited to 100 cu. in. (1650 cu. mm) or less in size. **(See Figure 13-2)**

WITH SUPPORTING DEVICES, LUMINAIRES, OR LAMPHOLDERS
314.23(F)

Boxes with threaded sides or hubs shall be permitted to be used to support devices or luminaires without the boxes being supported from within to the frame, wall, pole, structure, etc. However, for this method of supporting to be applied, the conduit shall be threaded wrench tight into the box and be supported within 18 in. (450 mm) of the box. The box is limited to 100 cu. in. (1650 cu. mm) or less in size.

For example, a box with threaded sides and with a threaded conduit entering and leaving is placed underneath the diving board of a swimming pool. The threaded conduit shall be supported within 18 in. (450 mm) of the box and has a toggle switch mounted to it to control the pool light. The box does not have to be independently supported, because the conduit with the 18 in. (450 mm) supports shall be permitted to serve as the required supporting means. **(See Figure 13-3)**

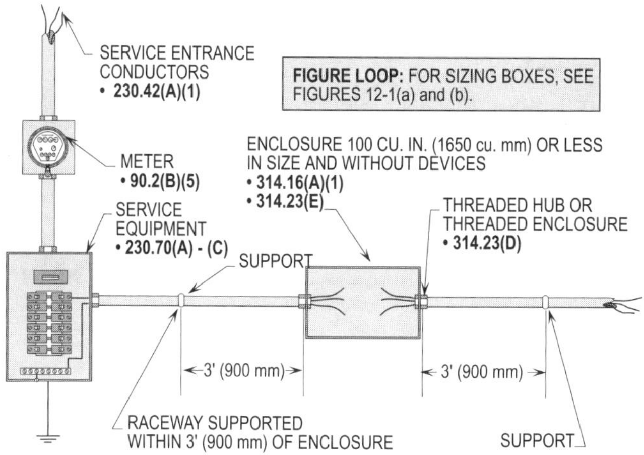

Figure 13-2. If supported within 3 ft (900 mm) on two or more sides, enclosures without devices do not require additional supports.

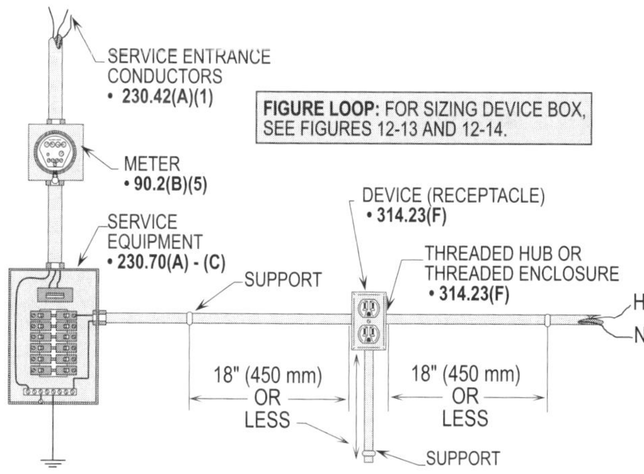

Figure 13-3. If supported within 18 in. (450 mm) on two or more sides, enclosures with devices do not require additional supports. (**Note,** threaded hubs shall be identified.)

CABLE SYSTEMS
ARTICLES 320 THROUGH 340

Multiconductor cables are flexible assemblies with protective coverings for protection of the conductors. Armored cable (BX), metal-clad cable (MC), nonmetallic sheathed cable (romex), and service-entrance cable are four types of multiconductor cables used as wiring methods for the wiring of modern-day electrical systems.

ARMORED CABLE: TYPE AC
ARTICLE 320

Type AC cable (BX) is a fabricated assembly of insulated conductors in a flexible metallic enclosure. To ensure a reliable path for fault current to return from the point of fault to the grounded (neutral) conductor connected to the grounded busbar in the main service panelboard, a bonding conductor or spinal is in contact with the armor. When installing cables in wet locations, a special type of armored cable (ACL) shall be used, but the cable shall not be permitted to be buried in the ground.

USES PERMITTED
320.10

Type AC cable shall be permitted to be installed in the following locations:

- For feeders and branch circuits in both exposed and concealed work
- In cable trays
- In dry locations
- Embedded in plaster finish on brick or other masonry, except in damp or wet locations
- Where run or fished in air voids of masonry block or tile walls where such walls are not exposed or subject to excessive moisture or dampness.

Note, this is not an all-inclusive list.

USES NOT PERMITTED
320.12

Type AC cable shall not be permitted to be installed in the following locations:

- Where subject to physical damage
- In damp or wet locations
- In air voids of masonry block or tile walls where such walls are exposed or subject to excessive moisture or dampness
- In areas with corrosive fumes or vapors
- Embedded in plaster finish or brick or other masonry

> in damp or wet locations
>
> **Design Tip:** Check area to see if only flexible connections are permitted for the wiring connection to flexible equipment.

IN ACCESSIBLE ATTICS
320.23

When attics are accessible without stairways or pull-down ladders, armored cable shall be protected by guard strips only when installed within 6 ft (1.8 m) of the scuttle-hole opening. Guard strips shall not be required for AC cable that is routed along the sides of the joists or rafters or studded through drilled holes.

Guard strips shall be used to protect cables that are installed below 7 ft (2.1 m) in an attic. These strips are normally 1 in. x 2 in. (25 mm x 50 mm) wooden strips nailed directly to the structure. Guard strips shall be used to protect cables run on top of the floor joists. Guard strips shall not be required for cables installed above 7 ft (2.1 m) or cables run on the side of floor joists, rafters, and studs. **[See Figures 13-13(a) and (b)]**

SECURING AND SUPPORTING
320.30

Type AC cable shall be supported and secured by staples, cable ties, straps, hangers, or small fittings. Type AC cable shall be secured with 12 in. (300 mm) and secured and supported at intervals not exceeding 4-1/4 ft (1.4 m).

Horizontal runs of Type AC cable installed in wooden or metal framing members shall be considered supported where such support does not exceed 4-1/2 ft (1.4 m) intervals.

Type AC cable shall be permitted to be unsupported in the following locations:

- When fished between access points through concealed spaces in finished buildings or structures
- Where flexibility is needed, in lengths not more than 2 ft (600 mm)
- In lengths not more than 6 ft (1.8 m) within an accessible ceiling

BOXES AND FITTINGS
320.40

When AC cable is terminated in a box, cabinet, or piece of equipment, the conductors shall be protected from abrasion by a connector. An insulating bushing shall be used with the connector. This bushing slips over the conductor and under the armor of the cable for additional protection.

METAL-CLAD CABLE: TYPE MC
ARTICLE 330

Type MC cable is a factory assembly of one or more insulated circuit conductors with or without optical-fiber members enclosed in an armor of interlocking metal tape or a smooth or corrugated metallic sheath. The following are types of MC cable:

- Interlocked metal
- Corrugated metallic sheath
- Smooth sheath

> **Design Tip:** All three types of MC cable are intended for above ground installation except when marked for direct burial.

USES PERMITTED
330.10

When installing Type MC cable and where not subject to physical damage, the following installations shall be permitted:

- For services, feeders, and branch circuits
- For power, lighting, control, and signal circuits
- Indoors or outdoors
- Where exposed or concealed
- Direct buried where identified for such use
- In cable trays where identified for such use
- In any raceway
- As aerial cable on a messenger
- In hazardous (classified) locations as permitted
- In dry locations and embedded in plaster finish on brick or other masonry except in damp or wet locations
- Wet locations where any of the following conditions are complied with:
  - **(a)** Metallic covering is impervious to moisture
  - **(b)** Lead-sheath or moisture-impervious jacket is provided under the metal covering
  - **(c)** Insulated conductors under the metallic covering are listed for use in wet locations and a corrosion-resistant jacket is provided over the metallic sheath
- Where single conductor cables are used, all phase conductors and, where used, the grounded (neutral) conductor shall be grouped together to minimize

induced voltage on the sheath

USES NOT PERMITTED
330.12

Type MC cable shall not be permitted to installed in the following conditions:

- Where subject to physical damage

- Where exposed to any of the following destructive corrosive conditions, unless the metallic sheath or armor is resistant to the conditions or is protected by material resistant to the following conditions:

   (a) Direct burial in the earth or embedded in concrete unless identified for direct burial

   (b) Exposed to cinder fill, strong chlorides, caustic alkalis, or vapors of chlorine or of hydrochloric acids

SECURING AND SUPPORTING
330.30

Type MC cable shall be supported and secured by staples, cable ties, straps, hangers, or similar fittings or other approved means.

Type MC cable shall be secured at intervals not exceeding 6 ft (1.8 m). MC cable containing four or fewer conductors sized no larger than 10 AWG shall be secured within 12 in. (300 mm). MC cable shall be supported at intervals not exceeding 6 ft (1.8 m).

Type MC cable shall be permitted to be unsupported in the following locations:

- Cable that is fished between access points

- Lengths not more than 6 ft (1.8 m) within an accessible ceiling

NONMETALLIC-SHEATHED CABLE: TYPES NM, NMC, AND NMS
ARTICLE 334

Romex or rope is the term used in the field when installing nonmetallic-sheathed cable. This cable has two, three, or four conductors with a green insulated or bare conductor used for grounding equipment. The conductors are protected by an outer nonmetallic jacket. There are three types of nonmetallic-sheathed cable:

- NM cable
- NMC cable
- NMS cable

Type NM cable has a flame-retardant and moisture-resistant outer jacket and is restricted to being installed for inside locations. Type NMC cable has a flame-retardant and moisture-resistant outer jacket and also is fungus-resistant and corrosion-resistant and shall be permitted to be installed for outside locations. Type NMS cable is a factory assembly of insulated power, communications, and signaling conductors enclosed within a common sheath of moisture-resistant, flame-retardant, nonmetallic material.

> **Design Tip:** Nonmetallic-sheathed cable shall be permitted to be used for economical wiring of single-family dwellings, multifamily dwellings, and small commercial buildings.

USES PERMITTED
334.10

Types NM, NMC, and NMS shall be permitted to be installed in the following locations:

- In one- and two-family dwellings

- In multifamily dwellings of Types III, IV, and V construction except as prohibited per **334.12**

- In other structures of Types III, IV, and V construction except as prohibited per **334.12**. Cables shall be concealed within walls, floors, or ceilings that provide a thermal barrier of at least 15-minute finish rating

- Cable trays in structures permitted to be Types III, IV, or V where the cables are identified for such use

Type NM cable shall be permitted for the following locations:

- For both exposed and concealed work in normally dry locations

- Where installed or fished in air voids in masonry block or tile walls

Type NMC cable shall be permitted for the following locations:

- For both exposed and concealed work in dry, moist, damp, or corrosive locations

- In outside and inside walls of masonry block or tile

- In a shallow chase in masonry, concrete, or adobe protected against nails or screws by a steel plate at least 1/16 in. (1.59 mm) thick and covered with plaster, adobe, or similar finish

Type NMS cable shall be permitted for the following locations:

- For both exposed and concealed work in normally dry locations

- Where installed or fished in air voids in masonry block

or tile walls

USES NOT PERMITTED
334.12

Type NM, NMC, and NMS shall not be permitted to be installed in the following locations:

- In any dwelling unit not specifically in **334.10(1) through (3)**
- Where exposed in dropped or suspended ceilings in other than one- and two-family and multifamily dwellings
- As service-entrance cable
- In commercial garages having hazardous (classified) locations per **511.3**
- In theaters and similar locations
- In motion picture studios
- In storage battery rooms
- In hoistways or on elevators or escalators
- Where embedded in poured cement, concrete, or aggregate
- In hazardous (classified) locations

Type NM and NMS shall not be permitted to be installed in the following locations:

- Where exposed to corrosive fumes or vapors
- Where embedded in masonry, concrete, adobe, fill, or plaster
- In a shallow chase in masonry, concrete, or adobe and covered with plaster, adobe, or similar finish
- In wet or damp locations

EXPOSED WORK
334.15

Exposed cable systems shall be run along the surface of finished areas, or running boards shall be provided and cables shall closely follow such surfaces and be supported per **334.30**.

Nonmetallic-sheathed cable shall be protected from physical damage by the following:

- Conduit
- Electrical metallic tubing
- Schedule 80 PVC rigid nonmetallic conduit
- Other means

Where nonmetallic-sheathed cable passes through a floor, the cable shall be enclosed at least 6 in. (150 mm) above the floor by one of the following means:

- Rigid metal conduit

- Intermediate metal conduit
- Electrical metallic tubing
- Schedule 80 PVC rigid nonmetallic conduit
- Approved means

Cable systems mentioned above that are installed in unfinished basements and crawl spaces shall be permitted to be routed through the center of framing members (studs) without additional supports. Cable systems that are not run through the center of the framing members shall be permitted to be placed in a cut notch in the framing member and protected by a 1/16 in. (1.59 mm) thick steel plate. Cable systems smaller than 8-3 AWG and 6-2 AWG shall be secured to the face or placed and supported on running boards.

The NM cable sheath shall extend through the conduit or tubing into the box not less than 1/4 in. (6 mm). The NM cable shall be secured within 12 in. (300 mm) of the point where the cable enters the conduit or tubing.

THROUGH OR PARALLEL TO FRAMING MEMBERS
334.17

Framing members that are notched and have the above raceways installed in the notched area shall not be required to have a 1/16 in. (1.52 mm) thick plate placed over them for protection.

SECURING AND SUPPORTING
334.30

Nonmetallic sheathed cable shall be supported and secured every 4-1/2 ft (1.4 m) and within 12 in. (300 mm) of every outlet or fitting. Cables installed within 6 ft (1.8 m) of a scuttle hole or trap door with a ladder shall not be permitted to be installed unless the cable is run along the sides of floor joists, rafters, or studs. A guard strip shall be required when the cable is installed on the top of joists or rafters. **[See Figures 13-7 and 13-13(a) and (b)]**

SERVICE-ENTRANCE CABLE: TYPES SE AND USE
ARTICLE 338

The following are three types of service-entrance cable that can be used as a wiring method for the wiring of modern-day electrical systems:

- SE
- USE

• ASE

Type SE cable has a flame-retardant, moisture-resistant covering that is unarmored and will not withstand severe mechanical abuse. Type USE cable has a moisture-resistant covering that is unarmored and aids in preventing deterioration of the cable when it is buried in the earth. Type ASE cable has an armored covering to provide additional protection of the cable.

Type SE cable is available in sizes ranging from 12 AWG to 4/0 AWG. The cable may be installed with two insulated conductors and one bare conductor or with three insulated conductors and one bare conductor.

USES PERMITTED – SERVICE-ENTRANCE CONDUCTORS
338.10(A)

Service-entrance cable shall be permitted to be used for service-entrance wiring but shall be installed as required per **Article 230**.

USES PERMITTED – BRANCH CIRCUITS OR FEEDERS
338.10(B)

Service-entrance cable shall be permitted to be used for general interior wiring, but the grounded (neutral) conductor shall be insulated when installed with the type of insulation approved for the purpose, such as thermoset or thermoplastic.

When installing the grounded (neutral) conductor in service-entrance cable, the grounded (neutral) conductor shall be permitted to be insulated or bare where the uninsulated conductor is installed as an equipment grounding conductor and not as a circuit conductor or grounded (neutral) conductor.

The temperature limitations for the insulation on the conductors is rated either:

- 60°C
- 75°C
- 90°C

  For example, TW insulation has a 60°C rating and THWN insulation has a 75°C rating. These ratings are applied per **Table 310.16** and are matched to the terminal rating per **110.14(C)(1)** and **110.14(C)(2)**.

Conductors with 60°C rated insulation are used on most cooking appliances, such as ranges, cooktops, and ovens. When installing conductors with 90°C rated insulation, the load shall be limited to the current rating capacity of conductors rated with 60°C rated ampacities. The 90°C ampacities are only used for derating purposes due to equipment adjustment or correction factors or both.

> **Design Tip: Table 310.15(B)(2)(a) to Ampacity Tables 0 to 2000 volts** and **correction factors** at the bottom of the **Tables** shall be observed when designing and installing service-entrance conductors.

METAL CONDUITS
ARTICLES 342 THRU 358

Metal conduit provides excellent protection for conductors. When designing wiring systems utilizing metal conduit, the manufacturer's suggestions and applicable sections of the NEC should be followed. Depending on the type used, metal conduit shall be permitted to be installed outdoors or indoors in damp or dry areas.

INTERMEDIATE METAL CONDUIT: TYPE IMC
ARTICLE 342

Intermediate metal conduit (IMC) is a steel threadable raceway of circular cross-section designed for the physical protection and routing of conductors and cables and for use as an equipment grounding conductor when installed with its integral or associated coupling and appropriate fittings.

USES PERMITTED
342.10

Intermediate metal conduit (IMC) shall be listed and shall be permitted to be used for all atmospheric conditions and occupancies.

Intermediate metal conduit, elbows, couplings, and fittings shall be permitted to be installed in the following:

- Where installed in concrete
- Where installed in direct contact with the earth
- Where installed in areas subject to severe corrosive influences

Intermediate metal conduit shall be protected by an encased concrete layer at least 2 in. (50 mm) thick to be buried in a cinder fill, or it shall be buried at least 18 in. (450 mm) under the cinder fill to ensure protection.

All supports, bolts, straps, screws, etc. shall be of corrosion-resistant materials or protected against corrosion by corrosion-resistant materials.

SIZE
342.20

Intermediate metal conduit shall be permitted to be installed in a minimum size of 1/2 in. (16) and a maximum size of 4 in. (103).

REAMING AND THREADING
342.28

When cutting and threading intermediate metal conduit in the field, a standard cutting die with a 3/4 in. (1 in 16) taper per foot shall be used. To protect the wire insulation from abrasion, all cut conduit ends shall be reamed to remove rough edges. For more details on this requirement, see the standard for pipe threads, general purpose, which is ANSI B.1.20.1.

SECURING AND SUPPORTING
342.30

Intermediate metal conduit shall be supported every 10 ft (3 m) and within 3 ft (900 mm) of each outlet box, junction box, device box, etc. Where structural members do not permit fastening within 3 ft (900 mm), the distance shall be supported at 5 ft (1.5 m). Intermediate metal conduit shall not be required to be supported in the following locations:

- Where approved, a distance of 3 ft (900 mm) shall not be required at the service head for above-the-roof termination of a service mast.

- A distance of 20 ft (6 m) between supports shall be permitted be used for vertical rises from machines, motors, etc. if the conduit is made up with threaded couplings, the conduit is firmly supported and securely fastened at the top and bottom of the riser, and no other means of intermediate support is readily available.

- Horizontal runs supported by openings through framing members at intervals not greater than 10 ft (3 m) and securely fastened within 3 ft (900 mm) of termination points shall be permitted.

- Where oversized concentric or eccentric knockouts are not encountered, Type IMC shall not be permitted to be unsupported where the raceway is not more than 18 in. (450 mm) and remains in unbroken lengths (without coupling).

COUPLINGS AND CONNECTORS
342.42

Threadless couplings and connectors shall be made tight where installed with intermediate metal conduit. Threadless couplings and connectors shall not be permitted to be used on threaded ends unless listed for the purpose. Where buried in masonry or concrete and installed in wet locations the following types shall be used:

- Concretetight
- Raintight

Running threads shall not be permitted to be used for connection at couplings. The galvanized coating would be removed and the conduit would rust if running threads were installed at the connection of couplings. The conduit at the joint would also be weaken if installed with running threads and be subject to breakage if stepped on or run over with a wheel barrel full of concrete during the pouring of the slab.

BUSHINGS
342.46

A bushing shall be provided to protect the wire from abrasion where a conduit enters a box, fitting, or other enclosure unless the box, fitting, or enclosure design provides equivalent protection.

RIGID METAL CONDUIT: TYPE RMC
ARTICLE 344

Rigid metal conduit is a threadable raceway of circular cross-section designed for the physical protection and routing of conductors and cables and for use as an equipment grounding conductor when installed with its integral or associated coupling and appropriate fittings. RMC is generally made of steel (ferrous) with protective coatings or of aluminum (nonferrous). Special use types are red brass and stainless steel. (See **Page 13-12** and **Figure 13-14**)

USES PERMITTED
344.10

Galvanized steel and stainless steel rigid metal conduit (RMC) shall be listed and shall be permitted to be used for all atmospheric conditions and occupancies.

Red brass rigid metal conduit shall be permitted to be installed for direct burial swimming pool applications.

Aluminum rigid metal conduit shall be permitted to be

installed where judged suitable for the environment. Ferrous raceways and fittings shall be permitted to be installed only indoors and in occupancies not subject to severe corrosive influences where protected from corrosion solely by enamel.

Rigid metal conduit, elbows, couplings, and fittings shall be permitted to be installed in the following:

- Where installed in concrete
- Where installed in direct contact with the earth
- Where installed in areas subject to severe corrosive influences

Aluminum rigid metal conduit shall be provided with approved supplementary corrosion protection where encased in concrete or in direct contact with the earth.

Rigid metal conduit shall be protected by an encased concrete layer at least 2 in. (50 mm) thick to be buried in a cinder fill, or it shall be buried at least 18 in. (450 mm) under the cinder fill to ensure protection.

All supports, bolts, straps, screws, etc. shall be of corrosion-resistant materials or protected against corrosion by corrosion-resistant materials.

SIZE
344.20

Rigid metal conduit shall be permitted to be installed in a minimum size of 1/2 in. (16) and a maximum size of 6 in. (155). The following exception permits the conduit size to be installed smaller than 1/2 in. (16):

- A rigid metal conduit shall be permitted to enclose leads to connect a motor that is separated from the motor box per **430.245(B)**.

REAMING AND THREADING
344.28

When cutting and threading rigid metal conduit in the field, a standard cutting die with a 3/4 in. (1 in 16) taper per foot shall be used. To protect the wire insulation from abrasion, all cut conduit ends shall be reamed to remove rough edges.

SECURING AND SUPPORTING
344.30

Rigid metal conduit shall be supported every 10 ft (3 m) and within 3 ft (900 mm) of each outlet box, junction box, device box, etc. Where structural members do not permit fastening within 3 ft (900 mm), the distance shall be supported at 5 ft (1.5 m). Rigid metal conduit shall not be required to be supported in the following locations:

- Where approved, a distance of 3 ft (900 mm) shall not be required at the service head for above-the-roof termination of a service mast.

- A distance of 20 ft (6 m) between supports shall be permitted be used for vertical rises from machines, motors, etc. if the conduit is made up with threaded couplings, is firmly supported, and is securely fastened at the top and bottom of the riser, and no other means of intermediate support is readily available.

- Horizontal runs supported by openings through framing members at intervals not greater than 10 ft (3 m) and securely fastened within 3 ft (900 mm) of termination points shall be permitted.

- Where oversized, concentric or eccentric knockouts are encountered, Type RMC shall be permitted to be unsupported where the raceway is not more than 18 in. (450 mm) and remains in unbroken lengths (without coupling).

COUPLINGS AND CONNECTORS
344.42

Threadless couplings and connectors shall be made tight where installed with rigid metal conduit. Where buried in masonry or concrete and installed in wet locations, the following types shall be used:

- Concretetight
- Raintight

Running threads shall not be permitted to be used for connection at couplings. The galvanized coating would be removed and the conduit would rust if running threads were installed at the connection of couplings. The conduit at the joint would also be weaken if installed with running threads and be subject to breakage if stepped on or run over with a wheel barrel full of concrete during the pouring of the slab.

BUSHINGS
344.46

A bushing shall be provided to protect the wire from abrasion where a conduit enters a box, fitting, or other enclosure unless the box, fitting, or enclosure design provides equivalent protection.

FLEXIBLE METAL CONDUIT: TYPE FMC
ARTICLE 348

Flexible metal conduit is a raceway of circular cross-section made of helically wound, formed, interlocked metal strip.

USES PERMITTED
348.10

Flexible metal conduit shall be listed and shall be permitted to be installed in both exposed and concealed locations.

USES NOT PERMITTED
348.12

Flexible metal conduit shall not be permitted to be installed in the following locations:

- In wet locations
- In hoistways, except for the installation of control circuits and short extensions on elevator cars
- In storage-battery rooms
- In hazardous (classified) locations, except as permitted in other Articles in the NEC
- In oil or gasoline areas without the installation of wire insulation is approved for such locations
- Where embedded in poured concrete or aggregate or installed underground
- Where subject to physical damage

SIZE
348.20

Flexible metal conduit shall be permitted to be installed in a minimum size of 1/2 in. (16) and a maximum size of 4 in. (103). Flexible metal conduit in 3/8 in. (12) sizes shall be permitted to be used as a wiring method for the following installations:

- To enclose the motor leads to connect a motor that is separated from the motor box per **430.245(B)**
- Part of a listed assembly does not exceed 6 ft (1.8 m) in length for tap connections to luminaires per **410.117(C)** or for utilization equipment
- For installing manufactured wiring systems per **604.6(A)**
- In hoistways per **620.21(A)(1)**
- As part of a listed assembly to connect wired luminaire sections per **410.137(C)**

SECURING AND SUPPORTING
348.30

Flexible metal conduit shall be supported within 12 in. (300 mm) of each box, cabinet, etc. and supported at intervals not exceeding 4-1/2 ft (1.4 m). Flexible metal conduit shall not be required to be supported in the following locations:

- Where flexible metal conduit is fished between access points
- Where flexibility is necessary after installation, lengths shall not exceed the following:
    - 3 ft (900 mm) for trade sizes 1/2 through 1-1/4 (16 through 35)
    - 4 ft (1200 mm) 1-1/2 through 2 (41 through 53)
    - 5 ft (1500 mm) 2-1/2 (63) and larger
- Part of a listed assembly does not exceed 6 ft (1.8 m) in length for tap connections to luminaires per **410.117(C)**

GROUNDING AND BONDING
348.60

Flexible metal conduit shall be permitted to be installed as a equipment grounding conductor per **250.118(5)**. Lengths of 6 ft (1.8 m) or less and protected at 20 amps or less may be installed for grounding of equipment per **348.60**. An equipment grounding conductor shall be installed to connect equipment where flexibility is needed after installation. Where required, an equipment bonding jumper shall be installed around flexible metal conduit per **250.102(D)** and **(E)**, and such bonding jumper shall be sized per **Table 250.122** based upon the size of the branch-circuit overcurrent protection device.

LIQUIDTIGHT FLEXIBLE METAL CONDUIT: TYPE LFMC
ARTICLE 350

Liquidtight flexible metal conduit is a raceway of circular cross section having an outer liquidtight, nonmetallic, sunlight-resistant jacket over an inner flexible metal core with associated couplings, connectors, and fittings for the installation of electric conductors.

USES PERMITTED
350.10

Liquidtight flexible metal conduit shall be permitted to be installed in both exposed or concealed locations as follows:

- Where conditions of installation, operation, or maintenance require flexibility or protection from liquids, vapors, or solids
- For installations permitted by **501.10(B)**, **502.10(A)(2)**, **502.10(B)(2)**, **503.10(A)** and **504.20** and in other hazardous (classified) locations where specifically approved, and by **553.7(B)**
- Where listed and marked for direct burial

USES NOT PERMITTED
350.12

Liquidtight flexible metal conduit shall not be permitted to be installed where:

- Subject to physical damage
- An operating temperature will be produced in excess of that for which the material is approved for any combination of ambient and conductor temperatures

SIZE
350.20

Liquidtight flexible metal conduit shall be permitted to be installed in a minimum size of 1/2 in. (16) and a maximum size of 4 in. (103). The exception allows 3/8 in. (12) conduit to be installed per **348.20(A)**.

SECURING AND SUPPORTING
350.30

Liquidtight flexible metal conduit shall be supported within 12 in. (300 mm) of each box, cabinet, etc. and supported at intervals not exceeding 4-1/2 ft (1.4 m). Horizontal runs shall be supported through framing members at intervals not exceeding 4-1/2 ft (1.4 m) and within 12 in. (300 mm) of each termination point. Liquidtight flexible metal conduit shall not be required to be supported in the following locations:

- Where liquidtight flexible metal conduit is fished between access points
- Where flexibility is needed after installation, lengths shall not exceed the following:
    - 3 ft (900 mm) for trade sizes 1/2 through 1/4 (16 through 35)
    - 4 ft (1200 mm) for trade sizes 1-1/2 through 2 (41 through 53)
    - 5 ft (1500 mm) for trade sizes 2-1/2 (63) and larger
- Part of a listed assembly does not exceed 6 ft (1.8 m) in length for tap connections to luminaires per **410.117(C)**

RIGID POLYVINYL CHLORIDE CONDUIT TYPE PVC
ARTICLE 352

Schedule 40 and Schedule 80 are the two types of rigid polyvinyl chloride conduit normally used in the industry. A rigid polyvinyl chloride conduit is a rigid nonmetallic conduit of circular cross section, with integral or associated couplings, connectors, and fittings for the installation of electrical conductors and cables. **[See Figures 13-10(a) and (b)]**

SCHEDULE 40

Nonmetallic materials such as fiber, polyvinyl chloride, or polyethylene are used so as to make Schedule 40 PVC waterproof, rustproof, and rotproof.

SCHEDULE 80

Schedule 80 PVC is durable and constructed of a thicker and heavier plastic material than Schedule 40 PVC and shall be permitted to be used above and below grade.

USES PERMITTED
352.10

Listed PVC conduit shall be permitted to be installed in any one of the following locations:

- Concealed in walls, floors, and ceilings
- Where subject to severe corrosive influences per **300.6** and where subject to chemicals
- In cinder fill
- In wet locations. Corrosion-resistant materials such as supports, bolts, straps, etc. shall be installed or be protected by approved corrosion-resistant materials
- In dry and damp locations not prohibited per **352.12**
- In exposed work not subjected to physical damage if identified for such use

> **Design Tip:** Schedule 80 PVC shall be permitted be installed and utilized in exposed locations subjected to physical damage per **300.5(D)**. However, Schedule 40 PVC shall not be permitted be used for this particular installation.

- Underground installations. See **300.5** and **300.50(B)** in the NEC. Conduits listed for the purpose shall be permitted to be installed underground in continuous lengths from a reel.
- PVC conduit shall be permitted to support nonmetallic conduit bodies not larger than the largest trade size of an entering raceway.

USES NOT PERMITTED
352.12

PVC conduit (general rule) shall not be permitted to be installed in the following locations:

- Hazardous (classified) locations. See **503.10(A), 504.20, 514.8, 515.8,** and **501.10(B)**
- Support of luminaires or other equipment
- Subject to physical damage unless identified for such use
- For ambient temperatures in excess of 50°C (122°F) unless listed otherwise
- For insulation temperature limitations not exceeding the listing for the conduit
- Theaters and similar locations, except as provided per **518.4** and **520.5**

SIZE
352.20

PVC conduit shall be permitted to be installed in a minimum size of 1/2 in. (16) and a maximum size of 6 in. (155).

SECURING AND SUPPORTING
352.30

PVC conduit shall be supported within 3 ft (900 mm) of each outlet box, junction box, device box, etc. PVC conduit shall be supported so that movement from thermal expansion or contraction will be permitted. When installing rigid nonmetallic conduit that is larger than 1 in. (27), it shall be supported per **Table 352.30**.

Where oversized concentric or eccentric knockouts are not encountered, PVC conduit shall be permitted to be unsupported where the raceway is not more than 18 in. (450 mm) and remains in unbroken lengths (without coupling).

Design Tip: Listed PVC shall be permitted to be supported in accordance to the listing other than the spacing requirements per **Table 352.30**. For expansion fittings, see **352.44** and page 12-29.

TUBING
ARTICLES 358 THRU 362

Tubing provides protection for conductors. When designing wiring systems utilizing tubing, the manufacturer's suggestions and applicable sections of the NEC should be followed.

ELECTRICAL METALLIC TUBING
ARTICLE 358

Electrical metallic tubing is an unthreaded thin-wall raceway of circular cross section designed for the physical protection and routing of conductors and cables and for use as an equipment grounding conductor when installed utilizing appropriate fittings. Electrical metallic tubing is generally made of steel (ferrous) with protective coatings or of aluminum (nonferrous).

USES PERMITTED
358.10

Electrical metallic tubing shall be permitted to be installed in both exposed and concealed locations.

Electrical metallic tubing, elbows, couplings, and fittings shall be permitted to be installed in the following:

- Where installed in concrete
- Where installed in direct contact with the earth
- Where installed in areas subject to severe corrosive influences

All supports, bolts, straps, screws, etc. shall be of corrosion-resistant materials or protected against corrosion by corrosion-resistant materials.

USES NOT PERMITTED
358.12

Electrical metallic tubing shall not be permitted to be installed in the following locations:

- Where subject to severe physical damage
- Protected from corrosion solely by enamel
- Protected by a encased noncinder concrete layer at least 2 in. (50 mm) thick to be buried in a cinder concrete or cinder fill, or buried at least 18 in. (450 mm) under the cinder fill
- In any hazardous (classified) locations
- For the support of luminaire or other equipment except conduit bodies no larger than the largest trade size of the tubing
- Where practicable, dissimilar metals in contact anywhere in the system shall be avoided to eliminate the possibility of galvanic action

SIZE
358.20

Electrical metallic tubing shall be permitted to be installed in a minimum size of 1/2 in. (16) and a maximum size of 4 in. (103). The following exception permits the conduit size to be installed smaller than 1/2 in. (16):

- Electrical metallic tubing shall be permitted to enclose leads to connect a motor that is separated from the motor box per **430.245(B)**.

SECURING AND SUPPORTING
358.30

Electrical metallic tubing shall be supported every 10 ft (3 m) and within 3 ft (900 mm) of each outlet box, junction box, device box, etc. Horizontal runs supported by openings through framing members at a distance not greater than 10 ft (3 m) and within 3 ft (900 mm) of termination points shall be permitted. Electrical metallic tubing shall not be required to be supported in the following locations:

- Where structural members do not permit fastening within 3 ft (900 mm), the distance shall be supported at 5 ft (1.5 m).

- Electrical metallic tubing shall be permitted to be fished for unbroken lengths (without coupling).

Where oversized concentric or eccentric knockouts are not encountered, electrical metallic tubing shall be permitted to be unsupported where the raceway is not more than 18 in. (450 mm) and remains in unbroken lengths (without coupling).

COUPLINGS AND CONNECTORS
358.42

Threadless couplings and connectors shall be made tight where installed with electrical metallic conduit. Where buried in masonry or concrete and installed in wet locations the following types shall be used:

- Concretetight
- Raintight

ROUGH-IN

Rough-in wiring methods are used in slabs, walls, ceilings, attics, and floors between the service panelboard and boxes that are used to support devices and luminaires or contain spliced conductors.

Rough-in is the stage of construction where all the wiring methods are left open in ditches, walls, ceilings, attics, floors, etc. for inspection before being covered up. Wiring methods in the rough-in stage should never be insulated over or sheet-rocked until the inspector leaves a card of approval, usually green or blue; a card of disapproval or rejection is normally red. **(See Figure 13-4)**

Boxes and other enclosures are mounted to the framing members. Cable or conduit systems are routed from the service panel and connected properly per **300.15**. Cable and conduit systems are to be continuous and comply with **300.10** and **300.12**.

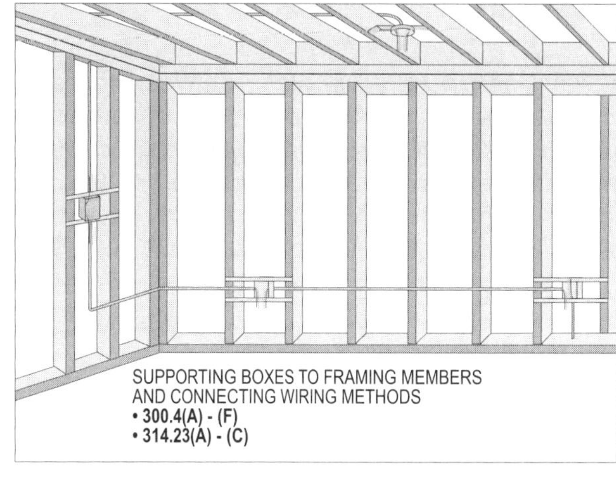

Figure 13-4. Wiring methods installed in walls, ceilings, attics, and floors are not allowed to be covered up with insulation or sheetrock until the rough-in inspection has been made by the authority having jurisdiction.

TEMPORARY POLE

A temporary pole (construction pole) consists of the meter base and panelboard, with receptacle outlets mounted beneath the panelboard. The temporary pole must be installed in the earth at a depth [usually 3 ft (900 mm)] that will hold the pole with mounted equipment substantially. Braces shall be installed to support the pole and the service drop between the pole and the utility pole. The braces shall be arranged so that they are opposite in direction from the point of attachment of the service drop conductors. Areas that have service conductors underground, rather than overhead, will utilize a temporary pole or pedestal. A pedestal has the meter and the overcurrent protection devices installed within; in other words, it is self contained. **[See Figures 13-5(a) and (b)]**

Procedures for design, installation, and inspections of temporary power poles are as follows:

- Check support of pole
- Check bracing of pole

- Check grounding of pole
- Check for main(s) in panelboard
- Check for GFCI protection of receptacles
- Check point of attachment for service drop conductors
- Check size of service-entrance conductors
- Check for weatherproof equipment

SLAB
TABLE 300.5

Raceways are installed in slabs between the service equipment panelboard and the outlet boxes used to support devices such as receptacles and switches. Raceways are installed in the slab to connect subpanels to the service panelboard and allow capacity for more circuits to be installed. Raceways are sometimes routed in the slab to supply power to special pieces of equipment such as heating units, A/C units, etc. For future additions, such as a detached garage or a small tool house out back, a raceway shall be installed in the slab between the panelboard and the foundation and extended outside for later use. Raceways installed in the slab shall be considered supported where they are tied to rebar reinforcing steel. Boxes and enclosures shall be supported to the framing members per **314.23(B)** and **(C)**. **(See Figure 13-6)**

Procedures for design, installation, and inspection of wiring methods are as follows:

- Check that raceways are tied to the steel and supported.
- Check floor boxes for support by raceways and tie wire.
- Check ends of raceways for concrete caps.
- Check raceways to verify they are complete and fittings are tight.
- Check that raceways are supported where they stub-up in the walls.

WALLS
300.4

Wiring methods installed horizontally or run parallel to framing members (studs) shall be supported. Wiring methods run horizontally through drilled holes shall be considered properly supported by the framing members. Wiring methods run parallel to framing members shall be supported by the provisions of the NEC pertaining to each type of wiring method.

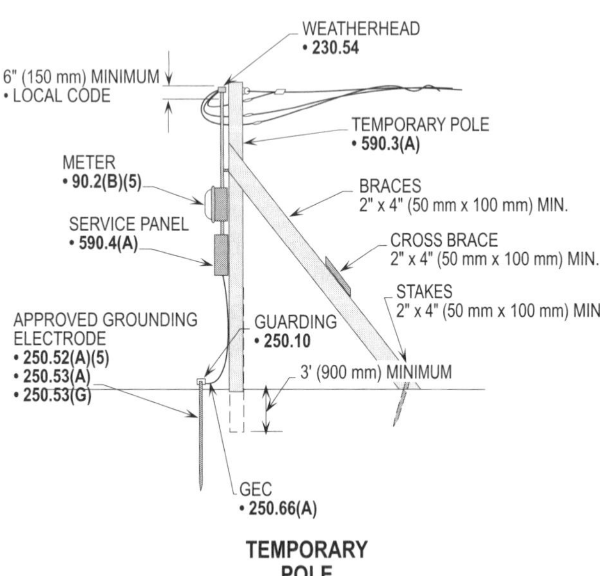

Figure 13-5(a). The above is a recommended procedure for installing an overhead temporary power pole to be used on a construction site.

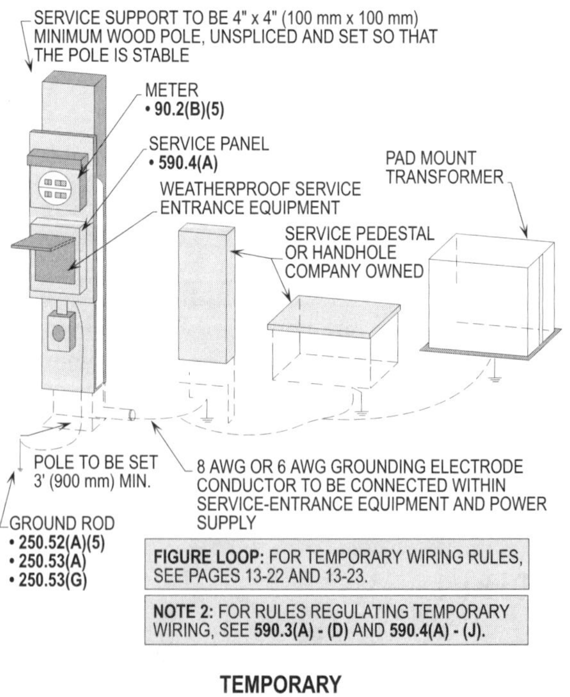

Figure 13-5(b). The above is a recommended procedure for installing an underground temporary pole (pedestal) to be used on construction sites. **Note,** for temporary wiring rules, see pages 13-22 and 13-23.

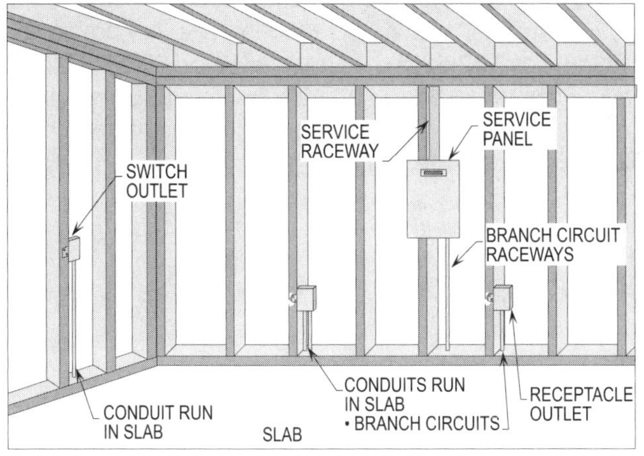

Figure 13-6. Conduits shall be permitted to be run in the slab to connect outlets and other enclosures to the service equipment panel.

CABLE SYSTEMS
ARTICLES 320, 330, 334, AND 338

Cable systems shall be supported and securely fastened where they are routed through studs, joists, or similar wood members per **320.17, 330.17, 334.17,** and **338.10(B)(4)**. Exposed cable systems running along the surface of finished areas or running boards shall closely follow such surfaces and shall be supported per **320.17, 330.17, 334.17,** and **338.10(B)(4)**. AC cable, nonmetallic sheathed cable, and service-entrance cable shall be supported at intervals not exceeding 4-1/2 ft (1.4 m) and within 12 in. (300 mm) of each cabinet, box, or fitting. **(See Figure 13-7)**

Cable systems mentioned above that are installed in unfinished basements or crawl spaces shall be permitted to be routed through the center of framing members (studs) without additional supports. Cable systems that are not run through the center of the framing members shall be permitted to be placed in a cut notch in the framing member and protected by a 1/16 in. (1.6 mm) thick steel plate. Cable systems smaller than 8-3 AWG and 6-2 AWG shall be secured to the face or placed and supported on running boards. **(See Figure 13-8)**

Procedures for design, installation, and inspection of wiring methods are as follows:

- Check supports that are supporting cables.
- Check spacing of supports.
- Check to verify that proper fittings for terminating cables are utilized.
- Check loose fitting supports of cables at boxes, equipment, etc.
- Check cables for exposure to damage.
- Check type of installation:
    (a) Cables run through drilled holes in wall studs
    (b) Cables run through notched cuts on the edge of wall studs
    (c) Cables run on the sides of wall studs
    (d) Cables run on the bottom of wall studs
    (e) Cables run on the top of wall studs

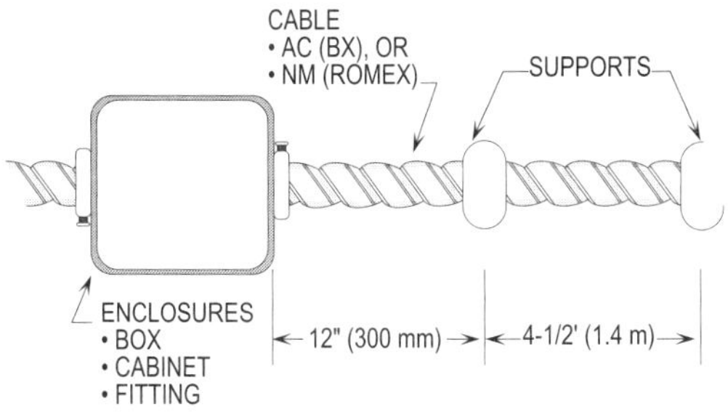

Figure 13-7. Nonmetallic sheathed cable and AC cable shall be supported within 12 in. (300 mm) of each box cabinet or fitting and at intervals not exceeding 4-1/2 ft (1.4 m).

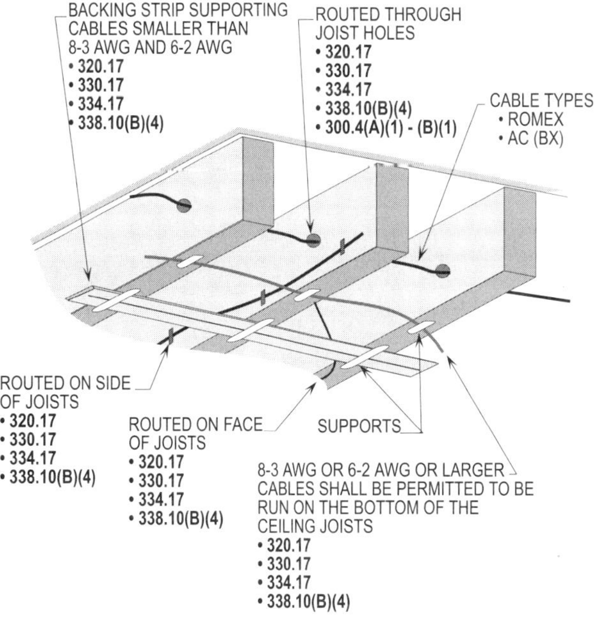

Figure 13-8. The method in which cables are routed through framing members in an unfinished basement shall be determined by the size and number of conductors in the cable.

RACEWAY SYSTEMS
ARTICLES 342, 344, 348, 350, AND 352

Raceway systems run horizontally through drilled holes in the framing members shall be considered supported. Those that are run along the surface, on top, on bottom, or parallel to the framing members shall be supported per **342.30, 344.30, 348.30, 350.30,** and **352.30**.

Raceway systems such as rigid metal conduit and IMC are required to be supported at intervals every 10 ft (3 m) and securely fastened within 3 ft (900 mm) of each box, enclosure, and fitting for sizes 1/2 in. (16) and 3/4 in. (21) per **Table 344.30(B)(2)**. See **Table 344.30(B)(2)** for spacing of supports for larger raceways using threaded connections.

For example: What is the maximum distance between supports for 1 in. (27) IMC using threaded connections?

Step 1: Finding distance for supports
342.30(B)(2) and **Table 344.30(B)(2)**
1" (27) IMC = 12' (3.7 m)

Solution: Supports are required at 12 ft (3.7 m) intervals.

Framing members that are notched and have the above raceways installed in the notched areas shall not be required to have a 1/16 in. (1.6 mm) thick plate placed over them for protection. **(See Figure 13-9)**

PVC conduit shall be supported and securely fastened every 3 ft (900 mm) and within 3 ft (900 mm) of every box, cabinet or fitting per **352.30(A)**. Schedule 40 or 80 PVC conduit shall be supported and securely fastened within 3 ft (900 mm) of each box, enclosure, and fitting and at intervals per **352.30(B)** and **Table 352.30**.

For example: What is the maximum distance between supports for 1-1/2 in. (41) PVC conduit?

Step 1: Finding distance for supports
352.30(B) and **Table 352.30**
1-1/2" (41) PVC = 5' (1.5 m)

Solution: Supports are required at 5 ft (1.5 m) intervals.

See Figures 13-10(a) and **(b)** for the installation requirements for PVC conduit.

Procedures for design, installation, and making inspections are as follows:

- Check supports supporting raceways.

- Check spacing of supports.

- Check and verify that proper fittings for terminating raceways are utilized.

- Check loose fittings and support of raceways at boxes, equipment, etc.

- Check type of installation such as:

  (a) Raceways run through drilled holes in wall studs

  (b) Raceways run through notched cuts on the edge of the wall studs

  (c) Raceways run on the sides of the wall studs

  (d) Raceways run on the bottom of wall studs

  (e) Raceways run on the top of wall studs

CEILINGS
300.4

Wiring methods run through the center of the ceiling rafters shall be considered supported. When they are routed on the top, bottom, or face they shall be supported by the provisions in the NEC. Wiring methods supported by the NEC shall be considered to be safe and reliable.

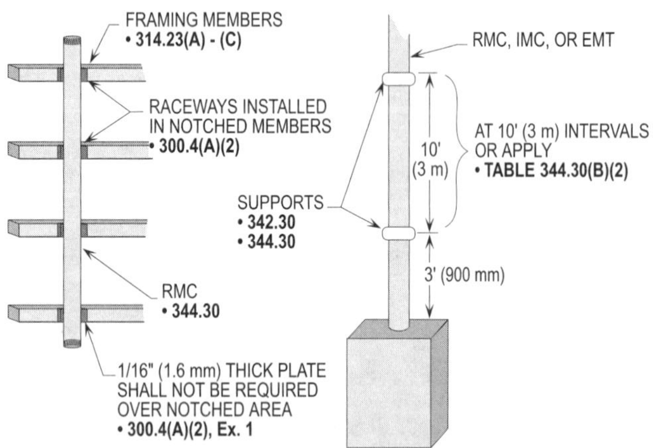

Figure 13-9. Installation requirements and supporting rules pertaining to metal raceways.

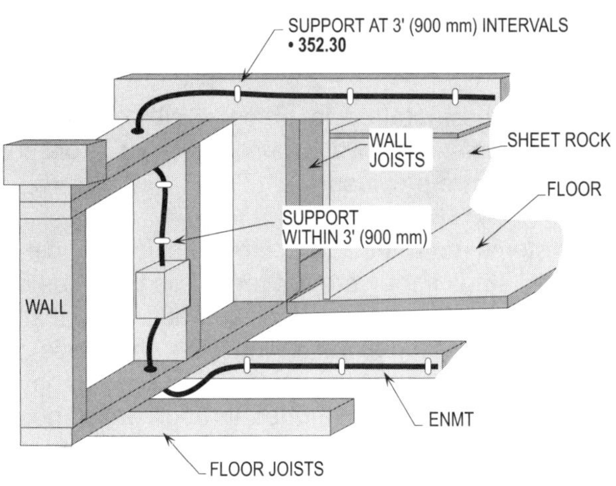

Figure 13-10(a). PVC conduit shall be supported and securely fastened at intervals not exceeding 3 ft (900 mm) and within 3 ft (900 mm) of each box, cabinet, fitting, etc.

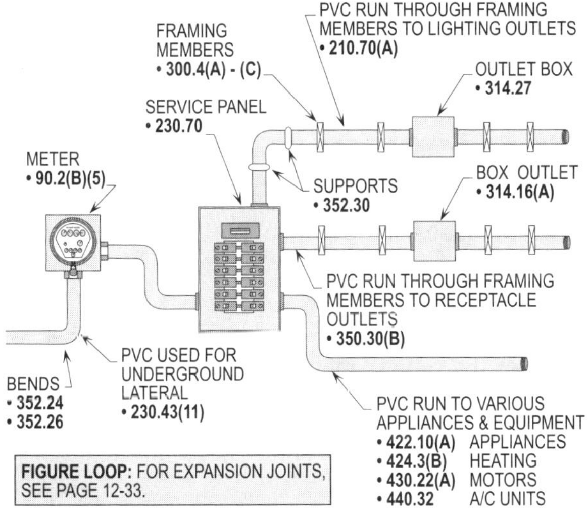

Figure 13-10(b). PVC conduit shall be permitted to be used to wire residential, commercial, and industrial locations.

CABLE SYSTEMS
ARTICLES 320, 330, 334, AND 338

Cable systems shall be permitted to be run through the center of rafters and joists, or they may be run on the surface (face). Cables shall be considered supported where the center of the rafter or ceiling joist is drilled and the cable is pulled through each drilled joist. Cables routed on the face shall be supported per **320.23(A), 330.23, 334.23,** and **338.10(B)(4)**.

AC cable, nonmetallic sheathed cable, and service-entrance cable shall be supported and securely fastened within 12 in. (300 mm) of every box, cabinet, or fitting and at intervals not exceeding 4-1/2 ft (1.4 m).

Some inspectors, by special permission, allow the cables to be supported at both ends if the cables are pulled tight and supported in the middle of the cable run.

> **Design Tip:** This method of supporting is only allowed where the cable is installed on the top of the ceiling studs. The inspector shall always be consulted if supports are installed other than by the provisions in the NEC.

Ceiling joists shall be permitted to be notched and the cables installed in the notch if protected by a 1/16 in. (1.6 mm) steel plate. Sometimes notching the ceiling joist is easier than drilling a hole in the center of the joist. **(See Figure 13-11)**

Procedures for design, installation, and inspection of cables are as follows:

- Check supports supporting cables.
- Check spacing of supports.
- Check and verify that proper fittings for terminating cables are utilized.
- Check loose fittings and supports of cables at boxes, equipment, etc.
- Check cables for exposure to damage.
- Check type of installation such as:
  - **(a)** Cables run through drilled holes in ceiling joists
  - **(b)** Cables run through notched cuts on the edge of the ceiling joist
  - **(c)** Cables run on the sides of the ceiling joists
  - **(d)** Cables run on the bottom of the ceiling joists
  - **(e)** Cables run on the top of the ceiling joists

RACEWAY SYSTEMS
ARTICLES 342, 344, 348, 350, AND 352

Raceway systems are usually installed in notches that are cut into the face of the framing members. It is much easier to run the raceway system in the notched area than to drill and run the raceway through the center of the joist. Where the construction will permit, it is easier to route the raceway system on the top or bottom of the ceiling joist. The raceway systems shall be supported as previously mentioned. **(See Figure 13-12)**

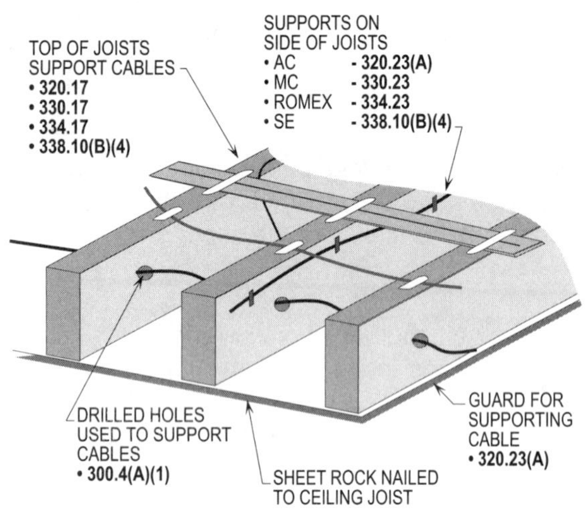

Figure 13-11. The above includes the rules for installing cables in attics.

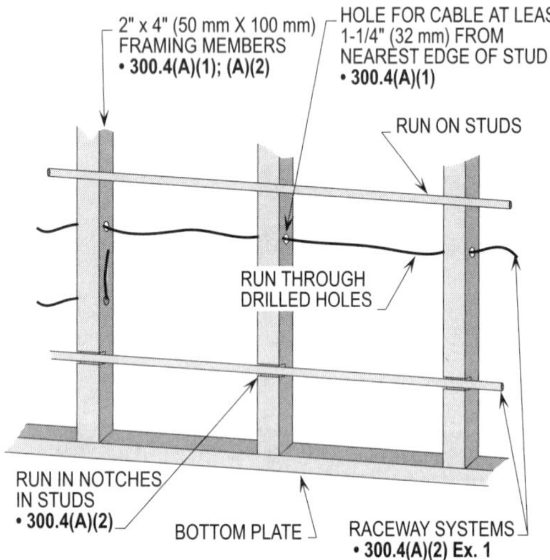

Figure 13-12. Raceway systems may be run on the studs, in notches in the studs, or through drilled holes in the studs.

ATTICS
320.23

There are two height levels where cable systems shall be protected when they are installed in attics.

For example, cables installed above 7 ft (2.1 m) shall not be required to be protected. However, cables installed at or below 7 ft (2.1 m) shall be protected under certain conditions of use. Raceway systems shall be considered protected from physical damage in most all types of installations.

CABLE SYSTEMS
ARTICLES 320, 330, 334, AND 338

Cable systems installed in attics shall be considered inaccessible unless stairways, pulldown ladders, or scuttle-hole openings are provided.

Cable systems run within 6 ft (1.8 m) of a scuttle hole shall have guard strips if they are not routed on the sides of the ceiling joists. Cable systems installed at or less than 7 ft (2.1 m) and run on top of the ceiling joists shall have guard strips installed to protect the cables if exposed to physical damage. Cables that are studded through drilled holes or run on the sides of the ceiling joists shall not be required to be protected with guard strips even when they are at or below the 7 ft (2.1 m) limitation.

Cables installed above 7 ft (2.1 m) shall not be required to be protected with guard strips. The proper supports are required as mentioned above, to protect the cables from being damaged.

Procedures for design, installation, and inspection of cables are as follows:

- Check supports supporting cables.
- Check spacing of supports.
- Check to verify that proper fittings to terminate cables are utilized.
- Check loose fittings and supports of cables at boxes, equipment, etc.
- Check cables for exposure to damage.
- Check the type of installation such as:
  (a) Cables run through drilled holes in ceiling joists
  (b) Cables run in cuts (notches) in the edge of ceiling joists
  (c) Cables run on the sides of ceiling joists
  (d) Cables run on the bottom of ceiling joists
  (e) Cables run on the top of ceiling joists

See Figures 13-13(a) and **(b)** for cable systems installed in attics with installation requirements.

RACEWAY SYSTEMS
ARTICLES 342, 344, 348, 350, AND 352

Raceway systems installed in attics and on the top of a joist or on the side shall not be required to be protected. They shall be supported according to the NEC as mentioned above for walls and ceilings.

Inspectors usually require electrical nonmetallic conduit to be protected in the attic in the same manner as nonmetallic sheathed cable. See **320.23** and **334.23**.

See Figure 13-14 for raceways installed in attics with installation requirements listed.

Some cities require raceways to enclose the conductors instead of cables so that arcs and sparks will be contained. If a short circuit or ground fault should occur in the electrical system, such arcing and sparking is contained in the raceway system which will help prevent fires.

Procedures for design, installation, and inspection of raceway systems are as follows:

- Check supports supporting raceways.

- Check spacing of supports.

- Check to verify that proper fittings for connecting raceways are utilized.

- Check loose fittings and supports of raceways at boxes, equipment, etc.

- Check raceways for exposure to damage.

- Check type of installation such as:

  (a) Raceways run through drilled holes in ceiling joists

  (b) Raceways run through notched cuts on the edge of the ceiling joists

  (c) Raceways run on the sides of ceiling joists

  (d) Raceways run on the bottom of the ceiling joists

  (e) Raceways run on the top of the ceiling joists

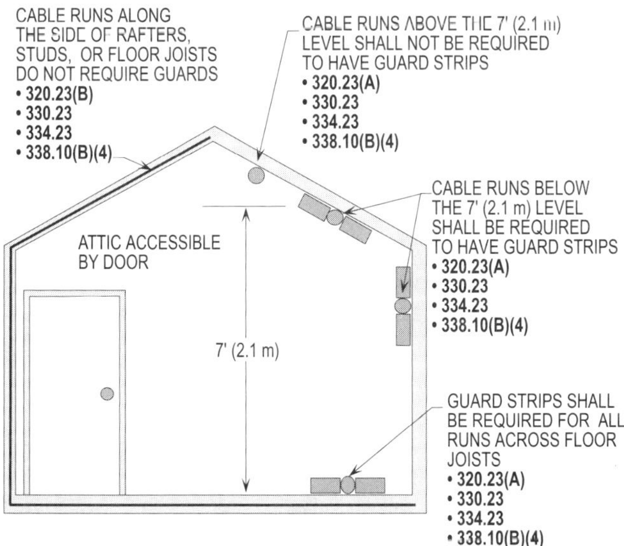

Figure 13-13(a). Requirements for installing cables in accessible attics where using a door or pulldown stairs.

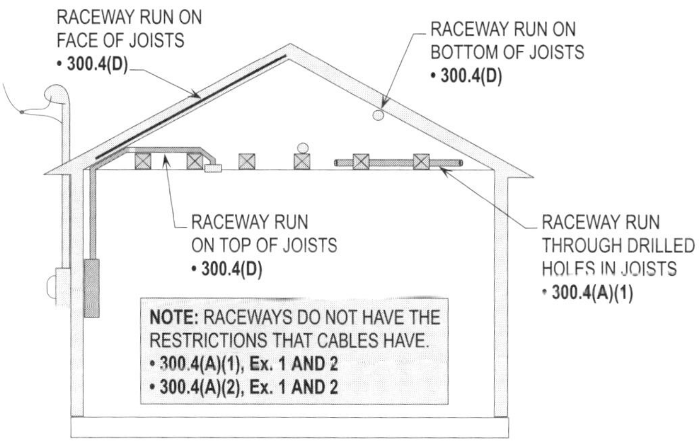

Figure 13-13(b). Requirements for installing cables in accessible attics where using a scuttle hole.

Figure 13-14. Requirements for installing raceways in attics where using raceway systems.

FLOORS
300.4

Cable systems are only installed in floor areas where the foundation is the pier and beam type. The cables shall be permitted to be pulled through drilled holes in the center of the floor joists, stapled to the sides, or routed on the bottom. Guard strips shall be required only if the cable is exposed to physical damage.

The supporting requirements for cables installed under the floor on the sides, bottoms, or drilled holes of floor joists are, the same as mentioned for walls or ceilings. **(See Figure 13-15)**

Procedures for making inspections are as follows:

- Check supports supporting cables.
- Check spacing of supports.
- Check loose fittings and supports of cables at boxes, equipment, etc.
- Check cables that are exposed to physical damage.
- Check type of installation such as:
  (a) Cables run through drilled holes in floor joists
  (b) Check cables routed in cut notches in floor joists
  (c) Check cables routed on sides of floor joists
  (d) Check cables routed on bottom of floor joists

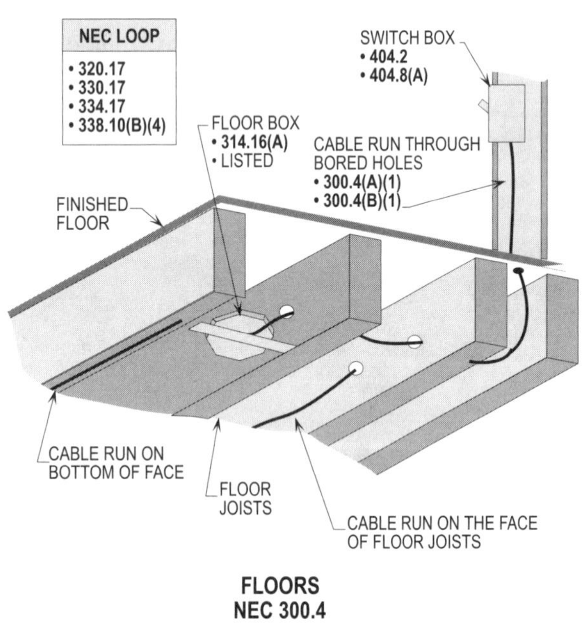

Figure 13-15. Requirements for installing cables in floors, where cables are used as the wiring method.

FLEXIBLE CORDS AND EXTENSION CORDS
400.7, 400.8, AND ARTICLE 590

Flexible cords shall be permitted to be used for wiring certain types of utilization equipment. Flexible cords shall not be permitted as a permanent wiring method to supply power to receptacles, luminaires, etc. Extension cords shall be permitted to be used only on a temporary basis and shall not be permitted to be substituted as a permanent wiring method. **(See Figure 13-16)**

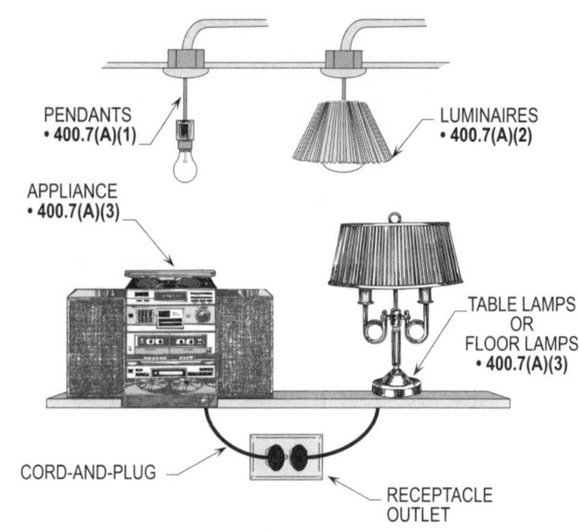

Figure 13-16. The above are approved installations using flexible cord.

FLEXIBLE CORDS
400.7 AND 400.8

Flexible cords are approved to be attached to luminaires, appliances and similar equipment. Luminaires and appliances shall not be permitted to be permanently wired into the circuit with flexible cords. Flexible cords for portable equipment shall be equipped with caps and plugs and plugged into a receptacle. There are three types of flexible cords used, as follows:

- Not permitted for hard usage
- Permitted for hard usage
- Permitted for extra-hard usage

Types of flexible cords and their permitted use are listed in **Table 400.4**. The ampacities of the different types of flexible cords are listed in **Table 400.5(A)**.

> **Design Tip:** Flexible cords are to be used very carefully where they are used for temporary power reasons. After their use , the AHJ will require such cords to be removed due to the threat of fire. They are only to be used for temporary use and not for hard-wired systems.

For example: What type of three-conductor flexible cord and ampacity rating are required where the cord is subject to abrasion and diggings? The load on the cord is 16.5 amps and the cord is the hard-service type (noncontinuous load).

Step 1: Finding type of cord
Table 400.4
Type S, SE, or SEO may be used

Solution: Flexible cord is Type S, SE, or SEO.

Step 1: Finding the ampacity of cord
Table 400.5(A)
16.5 A requires a 12 AWG cord

Solution: A 12-3 AWG w/ground flexible cord is required.

Flexible cord shall be permitted to be used to cord-and-plug connect electrical apparatus, and they are as follows:

- Pendants
- Luminaires
- Appliance or portable luminaire connections
- Elevator cables
- Cranes and hoists
- Stationary equipment requiring frequent interchange
- Noise or vibration utilization equipment
- Appliances requiring maintenance
- Connection of moving parts
- Where specifically permitted elsewhere in the NEC

Flexible cords shall not be permitted to be used as a permanent wiring method under certain conditions of use. They are not to be permanently attached at both ends or be concealed, or to pass through walls, ceilings, floors, or openings in doors. They are never to be used as a substitute for permanent wiring methods.

Flexible cord shall not be permitted as a wiring method and they are as follows:

- Not permitted as a substitute for building wiring
- Not permitted to pass through walls, ceilings, or floors
- Not permitted to pass through doorways, windows, or similar openings
- Not permitted to be permanently attached to building surfaces
- Not permitted to be concealed behind walls, ceilings, or floors
- Not permitted where subject to physical damage

See Figure 13-17 for a detailed description on how flexible cords shall not be permitted to be used.

Design Tip: Temporary power taps per **UL White Book, 90.7** and **110.3(B)** shall be permitted for the cord-and-plug connection to sensitive electronic equipment. Temporary power taps allow several pieces of equipment, such as a PC hard disk, printer, etc., to be cord-and-plug connected into one self-protected multioutlet strip that is plugged into a receptacle outlet.

See Figure 13-18 for a detailed illustration of a temporary power tap being used for cord-and-plug connections when connecting a PC and its accessories.

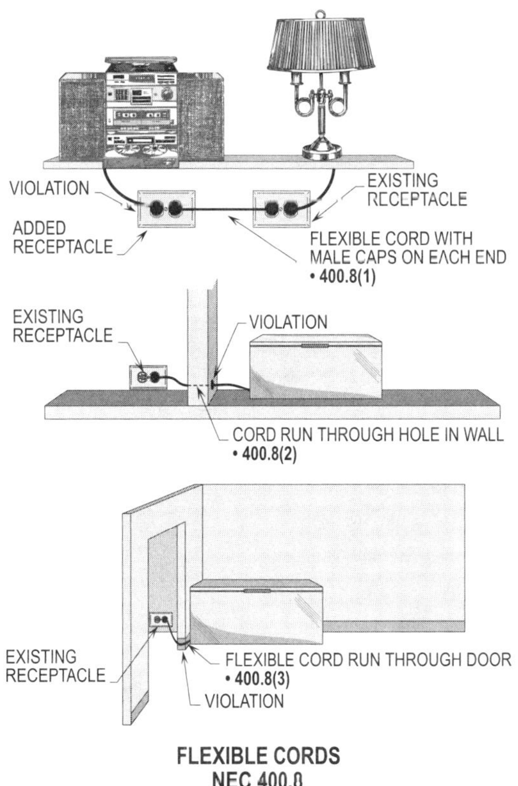

Figure 13-17. The above are installations where flexible cords shall not be permitted to be used.

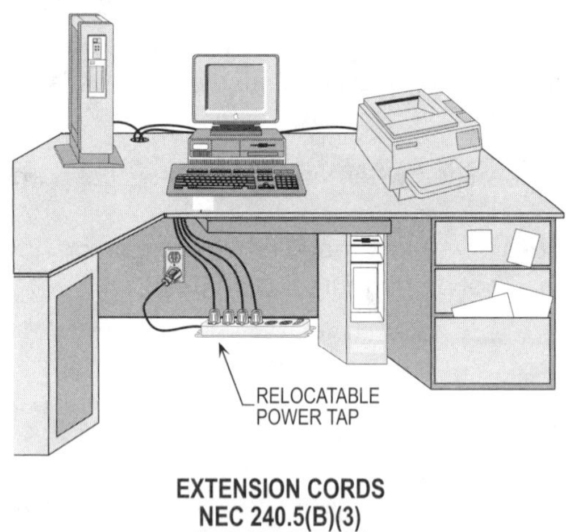

Figure 13-18. A number of sensitive electronic pieces of equipment shall be permitted to be plugged into a temporary power tap per **UL White Book, 90.7** and **110.3(B)**.

EXTENSION CORDS
240.5(B)(3)

This section allows electricians to make up extension cords in the field using flexible cords, if separately listed cord bodies and components are utilized.

Note that listed manufactured cords shall be permitted to be used as before. The NEC now agrees with OSHA that extension cords shall be permitted to be made up in the field. **(See Figure 13-19)**

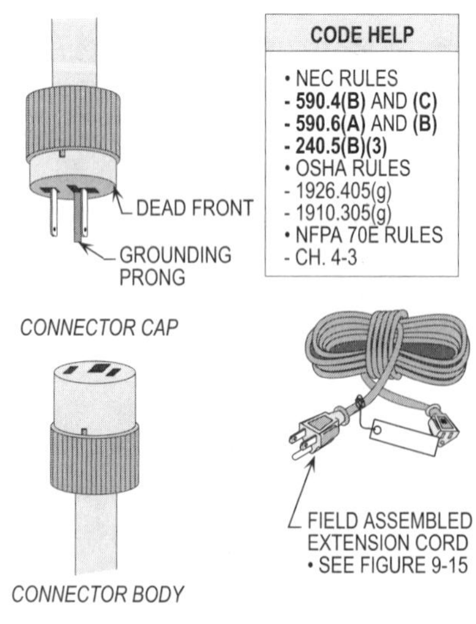

Figure 13-19. Extension cords shall be permitted to be field wired using flexible cord, if listed bodies and components are utilized.

TEMPORARY WIRING
ARTICLE 590

Temporary wiring shall be permitted to be used on construction sites and for remodeling, maintenance, repair, or demolition activities. The following requirements give the authorities having jurisdiction something to use for inspecting such wiring.

FEEDERS
590.4(B)

Temporary feeders shall comply with **Article 240**. They shall also originate in an approved method for the distribution of power. Cord and cable assemblies shall be permitted to be used if they are of a type identified in **Table 400.4** for hard usage or extra-hard usage.

For the purpose of this section, Type NM and Type NMC cables (romex) shall be permitted to be used in any dwelling, building, or structure without any height limitations.

BRANCH CIRCUITS
590.4(C)

All branch circuits shall originate in an approved power outlet or panelboard. Conductors shall be permitted within cable assemblies, or within multiconductor cords or cables, of a type identified in **Table 400.4** for hard usage or extra-hard usage. All conductors shall be protected as provided in **240.4, 240.5,** and **240.100**. For the purposes of this section, Type NM and Type NMC cables shall be permitted to be used in any dwelling, building, or structure without any height limitations.

Branch circuits installed for the purposes specified in **590.3(B)** or **(C)** shall be permitted to be run as single insulated conductors. Where the wiring is installed in accordance with **590.3(B)**, the voltage-to-ground shall not be permitted to exceed 150 volts, the wiring shall not be permitted to be subject to physical damage, and the conductors shall be supported on insulators at intervals of not more than 10 ft (3 m) or for festoon lighting; the conductors shall be arranged so that excessive strain is not transmitted to the lampholders. **(See Figure 13-20)**

Note, for GFCI protection of receptacles and assured equipment grounding conductor program, see pages 16-20 of Chapter 16 of this book.

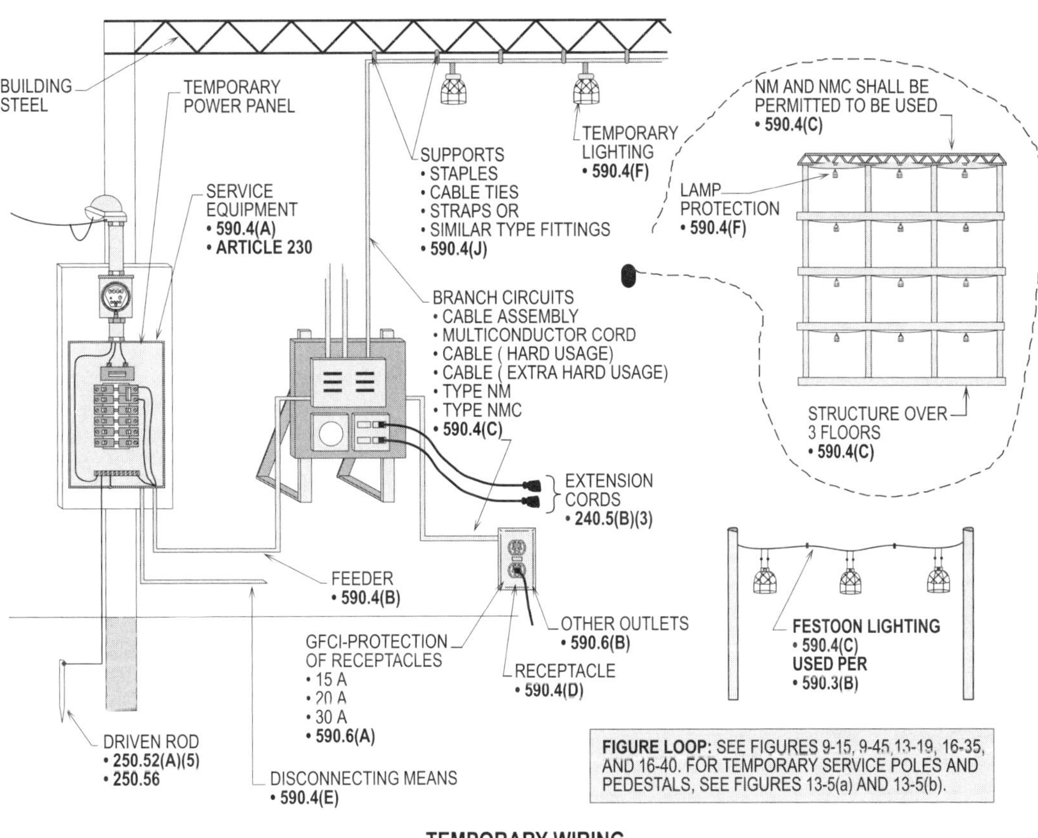

Figure 13-20. The illustration shows the rules pertaining to temporary wiring.

Chapter 13. Installing Wiring Methods

Section   **Answer**

1. Nails shall be permitted to be used to mount boxes when the nails pass through the inside of the box within _____ in. of the back or ends of the box.
   - (a) 1/16
   - (b) 1/8
   - (c) 1/4
   - (d) 1/2

2. If supported within _____ ft on two or more sides, enclosures without devices do not require additional supports.
   - (a) 2
   - (b) 3
   - (c) 4
   - (d) 5

3. Nonmetallic-sheathed cable (romex) and AC cable (BX) shall be supported within _____ in. of each box, cabinet, or fitting and at intervals not exceeding 4-1/2 ft.
   - (a) 12
   - (b) 18
   - (c) 24
   - (d) 36

4. Type AC cable installed in attics above _____ ft shall not be required to be protected.
   - (a) 4
   - (b) 5
   - (c) 6
   - (d) 7

5. Nonmetallic-sheathed cable run within _____ ft of a scuttle hole shall have guard strips if they are not routed on the sides of the ceiling joists.
   - (a) 5
   - (b) 6
   - (c) 7
   - (d) 10

6. Rigid metal conduit shall be permitted to be installed in a minimum size of _____ in.
   - (a) 1/2
   - (b) 3/4
   - (c) 1
   - (d) 2

7. When cutting and threading rigid metal conduit in the field, a standard cutting die with a _____ in. taper per foot shall be used.
   - (a) 3/8
   - (b) 1/2
   - (c) 3/4
   - (d) 1

8. Intermediate metal conduit shall be permitted to be installed in a maximum size of _____ in.
   - (a) 3
   - (b) 4
   - (c) 5
   - (d) 6

9. EMT shall be supported every _____ ft and within 10 ft of each outlet box, junction box, device box, etc.
   - (a) 1
   - (b) 2
   - (c) 3
   - (d) 6

10. Electrical metallic tubing shall be permitted to be installed in a minimum
    size of _______ in.
    (a) 3/8   (b) 1/2
    (c) 3/4   (d) 1

11. Flexible metal conduit shall be supported within 12 in. of each box, cabinet,
    etc. and supported at intervals not exceeding _______ ft.
    (a) 2   (b) 3
    (c) 4   (d) 4-1/2

12. Rigid nonmetallic conduit shall be permitted to be installed in a maximum size
    of _______ in.
    (a) 2   (b) 3
    (c) 4   (d) 6

13. Temporary poles must be installed in the earth at a depth usually _______ ft that
    will hold the pole with mounted equipment substantially.
    (a) 3   (b) 4
    (c) 5   (d) 6

14. Schedule _______ is durable and shall be permitted to be exposed to physical
    damage.
    (a) 20   (b) 40
    (c) 80   (d) 90

15. AC cable shall be supported within _______ in. of every box, cabinet, or fitting
    and at intervals not exceeding 4-1/2 ft.
    (a) 6   (b) 12
    (c) 18   (d) 24

16. Cable systems that are not run through the center of the framing members can
    be placed in a cut notch in the framing member and protected by a steel plate
    of _______ in.
    (a) 1/16   (b) 1/8
    (c) 3/16   (d) 1/4

17. Where a cable wiring method is installed through bored holes in joists, holes
    shall be bored so that the edge of the hole is not less than _______ in. from the
    nearest edge of the wood member.
    (a) 3/4   (b) 1
    (c) 1-1/4   (d) 1-1/2

18. Type AC cable shall be permitted to be unsupported for lengths not more than
    _______ ft within an accessible ceiling.
    (a) 3   (b) 4
    (c) 5   (d) 6

19. Type MC cable shall be secured at intervals not exceeding _______ ft.
    (a) 3   (b) 4
    (c) 6   (d) 10

20. Intermediate metal conduit shall be protected by an encased concrete layer at
    least _______ in. thick to be buried in a cinder fill.
    (a) 1   (b) 2
    (c) 3   (d) 4

Branch Circuits

Branch-circuit conductors extend between the final overcurrent protection device protecting the circuit conductors supplying power to equipment and outlets. The equipment supplied is either permanently installed and hard-wired or cord-and-plug connected to properly design and select receptacle outlets.

Branch circuits in this chapter are designed to supply the following type of occupancies:

- Residential
- Commercial
- Industrial

RESIDENTIAL

Branch circuits utilized in residential occupancies shall be calculated differently from those in commercial and industrial locations. The general purpose circuits are calculated at 3 VA per square foot of the dwelling unit.

For example, a dwelling unit of 2000 sq. ft has a volt-amp rating of 6000 VA (2000 sq. ft x 3 VA = 6000 VA). This VA rating shall be used to supply all the general purpose lighting and receptacle outlets in the dwelling unit.

Another example would be the 2-20 amp small appliance circuits that are used to supply countertop receptacles and other wall outlets in the kitchen, pantry, dining room, and breakfast room. Individual circuits such as a 12 kW range circuit shall be calculated as 8 kW per **Table 220.55, Col. C**. Other loads in the dwelling unit shall be permitted to be reduced in VA due to their operation and use.

GENERAL PURPOSE CIRCUITS
220.12 AND TABLE 220.12

The calculations for loads in various occupancies shall be based on VA (volt-amperes) per square foot. In calculating the VA per square foot, the outside dimensions of the building shall be used. They do not include the area of open porches and attached garages with dwelling unit occupancies. However, if there is an unused basement, it should be assumed that it will be finished later, so it shall be included in the calculation so that the capacity of the wiring system will be adequate to serve such loads at a later date.

The load values used in **Table 220.12** shall be considered at 100 percent power factor. If less than 100 percent power factor, equipment is installed with sufficient capacity. Such equipment shall be figured in to take care of the additional higher current values.

> **Design Tip:** The number of 15 or 20 amp branch circuits for a dwelling unit shall be determined by multiplying the square footage by 3 VA per sq. ft per **220.12** and **Table 220.12** and dividing by the size overcurrent protection device times the voltage of the circuit. **(See Figure 14-1)**

SMALL APPLIANCE CIRCUITS
220.52, 210.11(C)(1), AND 210.52(B)(1)

A minimum of 2 - 20 amp, 1500 VA small appliance circuits shall be required to supply receptacle outlets that are located in the kitchen, pantry, breakfast room, and dining room. The two small appliance circuits shall be routed to the kitchen countertop(s), and the outlets proportioned among the two circuits as evenly as possible to prevent unbalanced loading of the circuits. Unbalanced loading may trip open the overcurrent protection device because too many portable appliances are plugged into the same small appliance circuit. **(See Figure 14-2)**

LAUNDRY CIRCUIT
220.52(B), 210.11(C)(2), AND 210.52(F)

At least 1 - 20 amp, 1500 VA laundry circuit shall be required to supply receptacle outlets in the laundry room. All laundry equipment shall be located within 6 ft (1.8 m) of the receptacle outlet per **210.50(C)**. Sometimes a 20 amp duplex receptacle is used to cord-and-plug connect a washing machine and gas dryer. No other outlets shall be permitted to be supplied by this 20 amp, 2-wire small appliance circuit. **(See Figure 14-3)**

Finding Number of 15 Amp Circuits

Step 1: Finding VA per sq. ft
220.12 and **Table 220.12**
The VA per sq. ft is 3

Step 2: Finding total VA
Table 220.12
2700 sq. ft x 3 VA = 8100 sq. ft

Step 3: Finding the number of circuits
210.11(A)
8100 VA ÷ 1800 VA (15 A OCPD x 120 V) = 4.5
(Rounded up to 5)

Solution: The number of 15 amp, 2-wire circuits is 5.

GENERAL PURPOSE CIRCUITS
NEC 220.12 AND TABLE 220.12

Figure 14-1. Determining the number of 15 amp, 2-wire circuits to supply power to the general purpose lighting and receptacle outlets throughout the dwelling unit.

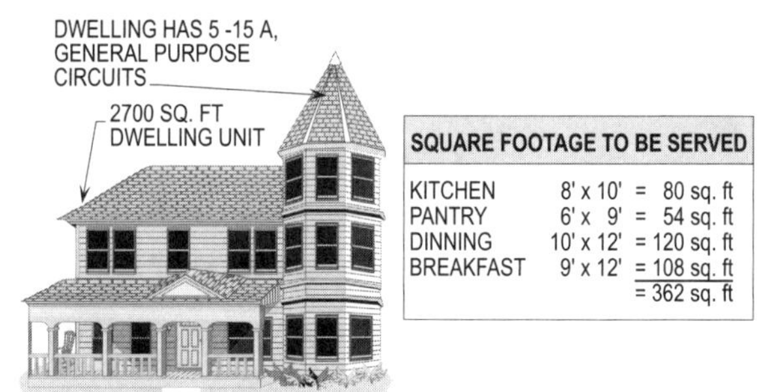

SQUARE FOOTAGE TO BE SERVED

KITCHEN	8' x 10'	= 80 sq. ft
PANTRY	6' x 9'	= 54 sq. ft
DINNING	10' x 12'	= 120 sq. ft
BREAKFAST	9' x 12'	= 108 sq. ft
		= 362 sq. ft

FINDING NUMBER OF OUTLETS

Step 1: Finding number of 20 amp small appliance circuits
210.11(A) and **Table 220.12**
20 A OCPD x 2400 VA (120 VA ÷ 3) = 800 sq. ft

Step 2: Checking sq. ft area
210.11(A); **Table 220.12**
• Two 20 A small appliance circuits required
• One 20 A small appliance circuit (800 sq. ft)
• 800 sq. ft x 2 circuits = 1600 sq. ft
• 1600 sq. ft is greater than 362 sq. ft

Solution: **Because the 362 sq. ft area does not exceed 1600 sq. ft, only two small appliance circuits are required.**

SMALL APPLIANCE CIRCUITS
NEC 220.52(A)
NEC 210.11(C)(1)
NEC 210.52(B)(1)

Figure 14-2. Determining the number of outlets that are permitted on a 20 amp small appliance circuit.

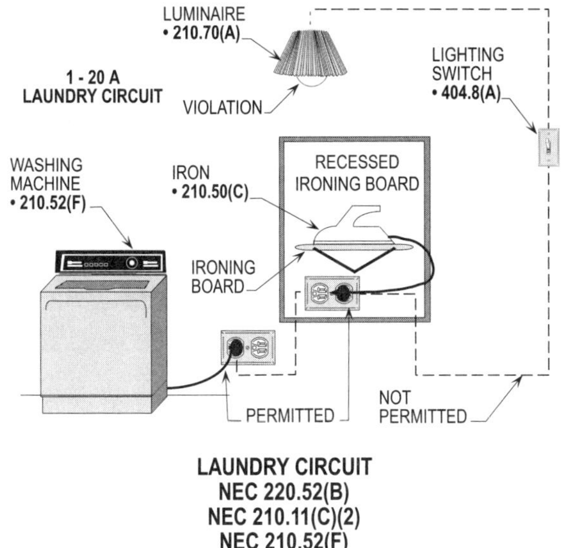

Figure 14-3. At least 1 - 20 amp, 1500 VA laundry circuit shall be required to supply receptacle outlets in the laundry room.

INDIVIDUAL CIRCUITS
210.19(A)(1)

Individual branch circuits in dwelling units shall be calculated at 100 percent or 125 percent of full-load amps of the appliance served or by applying demand factors based upon the type of appliance.

RESIDENTIAL COOKING EQUIPMENT
220.55 AND TABLE 220.55

The procedure for calculating the branch circuit loads for ranges, cooktops, and ovens shall be determined by applying the demand factors in **Table 220.55**. The demand factors listed in Columns A, B, and C are based on the size of the range, cooktop, or oven. When the kW rating exceeds 12 or there is more than one unit, one of the **Footnotes** shall be used in conjunction with **Table 220.55**.

COLUMN A
TABLE 220.55

Column A in **Table 220.55** shall be used for cooking equipment rated less 3.5 kW. These units vary in rating and have limited use.

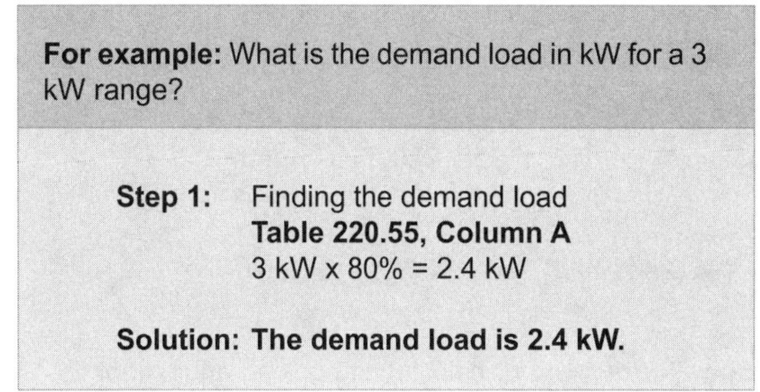

See **Figure 14-4** for calculating cooking equipment loads per **Column A** in **Table 220.55**.

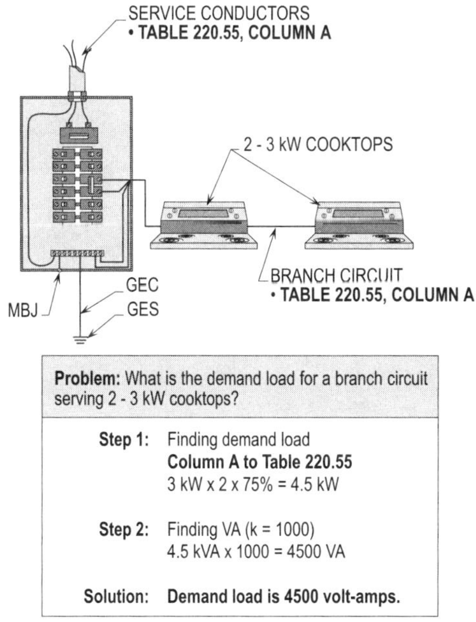

Figure 14-4. The maximum demand load shall be determined by the number of cooking units times the percentage factor applied in **Table 220.55, Column A.**

COLUMN B
TABLE 220.55

Column B in **Table 220.55** shall be used for cooking equipment rated from 3.5 kW to 8.75 kW. These size units are usually the types in use today.

For example: What is the demand load in kW for an 8.75 kW range?

    Step 1:   Finding the demand load
        Table 220.55, Column B
        8.75 kW x 80% = 7 kW

    Solution: The demand load is 7 kW.

See **Figure 14-5** for calculating cooking equipment loads per **Column B** in **Table 220.55**.

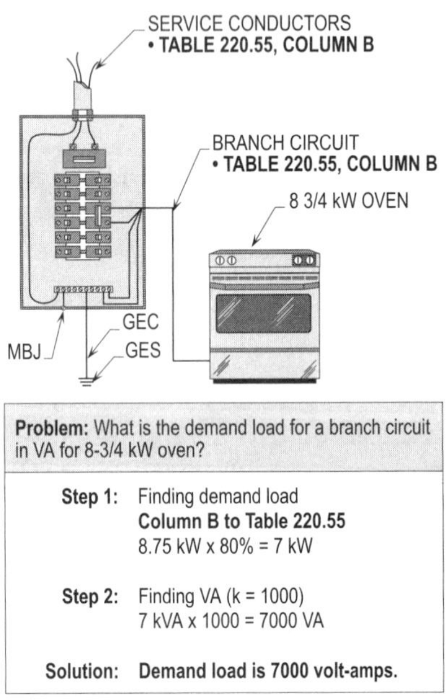

Problem: What is the demand load for a branch circuit in VA for 8-3/4 kW oven?

Step 1:   Finding demand load
        Column B to Table 220.55
        8.75 kW x 80% = 7 kW

Step 2:   Finding VA (k = 1000)
        7 kVA x 1000 = 7000 VA

Solution: Demand load is 7000 volt-amps.

TABLE 220.55, COLUMN B

Figure 14-5. The maximum demand load for a branch circuit shall be determined by the number of cooking units times the percentage factors applied in **Table 220.55, Column B**.

COLUMN C
TABLE 220.55

Column C in **Table 220.55** shall be used for cooking equipment rated from over 8.75 kW to 12 kW. These units are usually rated 9 to 12 kW, respectively.

For example: What is the demand load in kW for a 12 kW range?

    Step 1:   Finding the demand load
        Table 220.55, Column C
        12 kW range = 8 kW

    Solution: The demand load is 8 kW.

See **Figure 14-6** for calculating cooking equipment loads per **Column C** in **Table 220.55**.

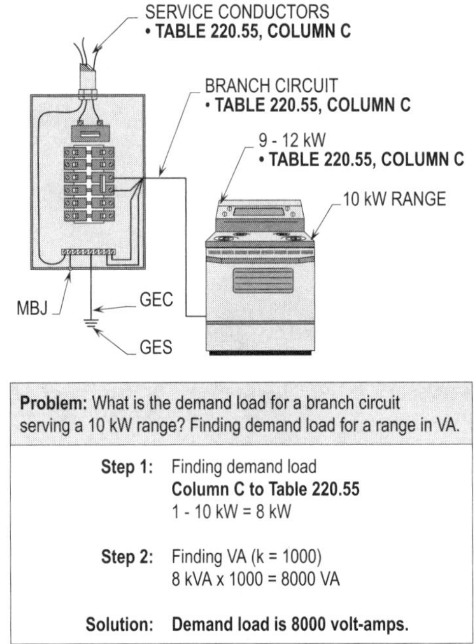

Problem: What is the demand load for a branch circuit serving a 10 kW range? Finding demand load for a range in VA.

Step 1:   Finding demand load
        Column C to Table 220.55
        1 - 10 kW = 8 kW

Step 2:   Finding VA (k = 1000)
        8 kVA x 1000 = 8000 VA

Solution: Demand load is 8000 volt-amps.

TABLE 220.55, COLUMN C

Figure 14-6. The demand load in kW for a branch circuit serving cooking equipment in **Column C** is already determined per **Table 220.55**.

NOTE 1
TABLE 220.55

Note 1 to **Table 220.55** shall be applied where the kW rating of the cooking equipment is over 12 kW but not over 27 kW. All kW ratings that exceed 12 kW shall be multiplied by 5 percent. The kW rating of 1 range is listed in **Column C** in **Table 220.55** and shall be multiplied by this total demand. Units of sizes larger than 12 kW are usually combination units found in kitchens with limited space.

For example: What is the demand load in kW for a 18 kW range?

Step 1:   Finding percentage
Note 1 to **Table 220.55**
18 kW - 12 kW = 6 kW
6 kW x 5% = 30%

Step 2:   Finding the demand load
Table 220.55, Col. C
8 kW x 130% = 10.4 kW

Solution: The demand load is 10.4 kW.

See Figure 14-7 for calculating cooking equipment loads per **Note 1** to **Table 220.55**.

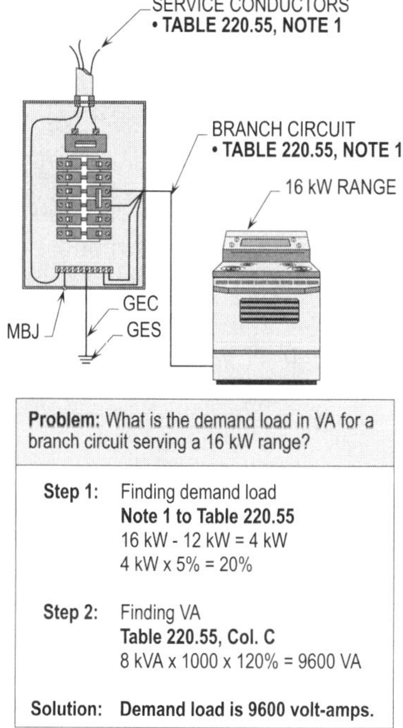

Figure 14-7. For cooking equipment rated over 12 kW to 27 kW in **Column C**, **Note 1** requires an increase of 5 percent for each kW over 12 kW. This percentage times 8 kW will determine the demand load to be used to size the elements.

NOTE 2
TABLE 220.55

For cooking equipment of unequal values rated over 12 kW to 27 kW in **Column C**, **Note 2** shall be applied and the value calculated by adding the kW ratings of all units and dividing by the number of units; all ranges below 12 kW shall be calculated at 12 kW. When an average rating is found, the number of units shall be increased by 5 percent for each kW exceeding 12 kW to derive the allowable kW.

See Figure 14-8 for demand factors to be applied per Table **220.55, Column C, Note 2**. **Note,** the demand factor selected from **Table 220.55, Column C** shall be based upon the number of units.

Design Tip: It is permissible per **Note 3** to **Table 220.55** to add all pieces of cooking equipment with ratings over 1-3/4 kW through 8-3/4 kW together and multiply by the percentage of **Columns A** or **B**; whichever produces the smaller kW rating shall be permitted to be used. This calculation shall be permitted to be applied by permission of the AHJ if it provides the smaller kW rating of all the methods available in **Table 220.55** and **Notes**.

NOTE 4
TABLE 220.55

Note 4 to **Table 220.55** shall be permitted to be applied to a counter-installed cooktop with one or two wall-mounted ovens. The total kW of each cooking unit shall be totaled, and all kW exceeding 12 kW shall be multiplied by 5 percent. The kW rating of 1 range, not 3 as listed in **Column C** in **Table 220.55**, shall be multiplied by this total. The advantage of this rule is allowing one branch circuit to be run, instead of three individual circuits, one to each unit.

For example: What is the demand load in kW for a 10 kW cooktop, an 8 kW oven, and a 6 kW oven connected to a 240 volt, single-phase, branch circuit?

Step 1:   Finding percentage
Note 4 to **Table 220.55**
10 kW + 8 kW + 6 kW = 24 kW
24 kW - 12 kW = 12 kW
12 kW x 5% = 60%

Step 2:   Finding the demand load
Table 220.55, Col. C
8 kW x 160% = 12.8 kW

Solution: The demand load is 12.8 kW.

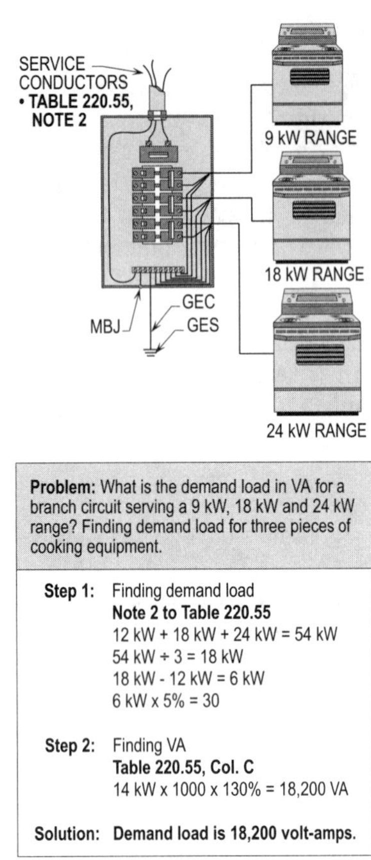

Problem: What is the demand load in VA for a branch circuit serving a 9 kW, 18 kW and 24 kW range? Finding demand load for three pieces of cooking equipment.

Step 1: Finding demand load
Note 2 to Table 220.55
12 kW + 18 kW + 24 kW = 54 kW
54 kW ÷ 3 = 18 kW
18 kW - 12 kW = 6 kW
6 kW x 5% = 30

Step 2: Finding VA
Table 220.55, Col. C
14 kW x 1000 x 130% = 18,200 VA

Solution: Demand load is 18,200 volt-amps.

TABLE 220.55, NOTE 2

Figure 14-8. Cooking equipment of unequal values rated over 12 kW to 27 kW in **Column C**, **Note 2** shall be calculated by adding the kW ratings of all units and dividing by the number of units; all ranges below 12 kW shall be calculated at 12 kW. When an average rating is determined, the number of units shall be increased by 5 percent for each kW exceeding 12 kW.

See Figure 14-9 for calculating cooking equipment loads for taps per **Ex. 1** to **210.19(A)(3)**.

SIZING TAPS
210.19(A)(3), Ex. 1

A tap to connect a cooktop and ovens shall be permitted to be made from a 50 amp branch circuit when the tap conductors are sized from the kW rating of each piece of cooking equipment per **Note 4** to **Table 220.55**. Whenever taps are made from larger conductors with smaller conductors, the tap shall comply with **240.21(A)**.

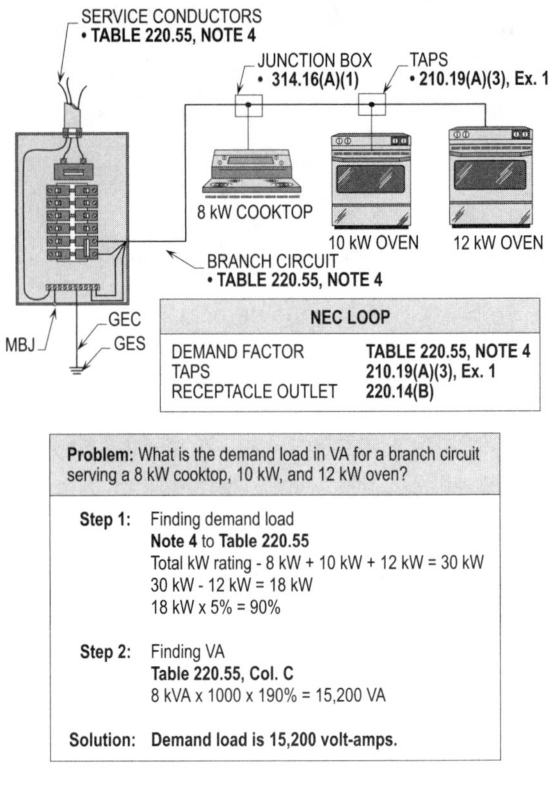

NEC LOOP	
DEMAND FACTOR	TABLE 220.55, NOTE 4
TAPS	210.19(A)(3), Ex. 1
RECEPTACLE OUTLET	220.14(B)

Problem: What is the demand load in VA for a branch circuit serving a 8 kW cooktop, 10 kW, and 12 kW oven?

Step 1: Finding demand load
Note 4 to Table 220.55
Total kW rating - 8 kW + 10 kW + 12 kW = 30 kW
30 kW - 12 kW = 18 kW
18 kW x 5% = 90%

Step 2: Finding VA
Table 220.55, Col. C
8 kVA x 1000 x 190% = 15,200 VA

Solution: Demand load is 15,200 volt-amps.

TABLE 220.55, NOTE 4

Figure 14-9. The demand load for a cooktop and 2 or less wall-mounted ovens shall be determined by finding the amperage rating of each unit in kW. Each kilowatt that exceeds 12 kW shall be increased by 5 percent to obtain the multiplier. The multiplier times the demand for 1 unit in **Column C** in **Table 220.55** derives the demand load for the elements of the circuit.

For example, can a 14 AWG tap be made from a 12 AWG branch circuit and be down sized and used as a switch leg? Naturally, the answer is no, because the tap cannot meet the provisions of **240.21(A) through (G)** for making an approved tap. However, **240.21(A)** and **210.19(A)(3), Ex. 1** applied together permit a smaller tap to be made for cooking equipment circuits tapped from conductors of a larger sized branch circuit.

See Figure 14-10 for calculating cooking equipment loads for taps per **Ex. 1** to **210.19(A)(3)**.

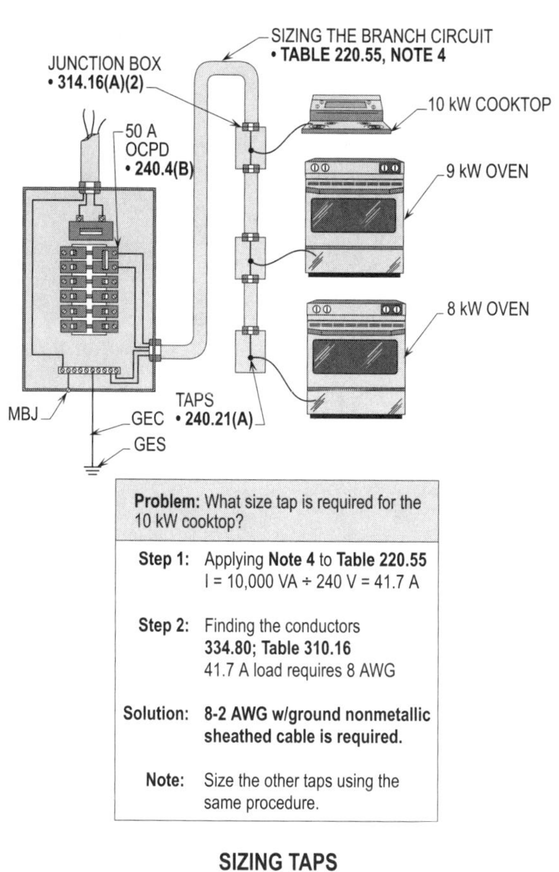

Problem: What size tap is required for the 10 kW cooktop?

Step 1: Applying **Note 4** to **Table 220.55**
I = 10,000 VA ÷ 240 V = 41.7 A

Step 2: Finding the conductors
334.80; Table 310.16
41.7 A load requires 8 AWG

Solution: **8-2 AWG w/ground nonmetallic sheathed cable is required.**

Note: Size the other taps using the same procedure.

SIZING TAPS
NEC 210.19(A)(3), Ex. 1

Figure 14-10. Determining the load in kW and amps to size the elements for a branch circuit that taps to cooking equipment.

ELECTRIC CLOTHES DRYERS
220.54

Dryer equipment loads shall be at least 5000 VA or the nameplate rating, whichever is larger. This value is used for sizing the branch-circuit load. When installing four or fewer dryers the load shall be calculated at 100 percent. When installing five or more dryers the load shall be calculated by the percentages listed in **Table 220.54** based on the number of dryers being installed. When calculating the load for a 4500 VA dryer, the load shall be calculated at 5000 VA for the branch circuit that is used to cord-and-plug connect the unit. Dryers installed in dwelling units shall comply with provisions of **220.14(B)** and **220.54**. Dryers shall be calculated at 100 percent for noncontinuous operation and 125 percent for continuous operation if they are installed as commercial dryers.

For example: What is the load for a 6000 W dryer when sizing the branch circuit in a dwelling unit?

Step 1:   Finding VA
220.14(B); 220.54
6000 VA x 100% = 6000 VA

Solution: The branch circuit load is 6000 VA.

See Figure 14-11 for calculating dryer equipment loads per **Table 220.54**.

Note that an existing branch circuit in a dwelling unit can be used as a wiring method to supply an electric dryer. If NM cable is used, the neutral shall be insulated. However, the neutral shall be permitted to be uninsulated if SE cable is the existing wiring method.

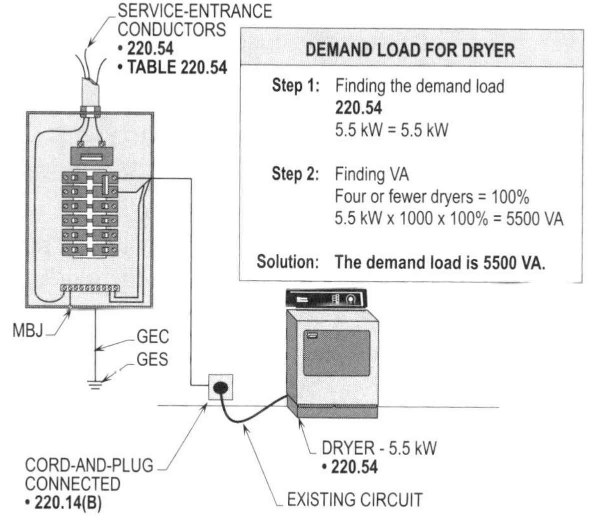

DEMAND LOAD FOR DRYER

Step 1: Finding the demand load
220.54
5.5 kW = 5.5 kW

Step 2: Finding VA
Four or fewer dryers = 100%
5.5 kW x 1000 x 100% = 5500 VA

Solution: The demand load is 5500 VA.

ELECTRIC CLOTHES DRYERS
NEC 220.54
TABLE 220.54

Figure 14-11. The demand load for household dryers shall be figured at 5 kVA or the nameplate rating, whichever provides the greater rating. Four or fewer dryers shall be calculated at 100 percent of the nameplate rating. Five or more dryers shall be permitted to have a percentage applied based on the number of units per **Table 220.54**. The Table is based on five dryers being used at different times and the load being limited to about 80 percent of total. **Table 220.54** is used for dwelling units in apartment complexes.

RATINGS
210.3

Branch circuits shall be classified by the rating or setting of the overcurrent protection device protecting the circuit. Two or more outlets shall be permitted to be protected by the following overcurrent protection devices based upon the calculated load per **210.19(A)**:

- 15 amp
- 20 amp
- 30 amp
- 40 amp
- 50 amp

PERMISSIBLE LOADS
210.23

The rating or setting of the overcurrent protection device shall not be permitted to be exceeded by the load on an individual branch circuit. If the load is continuous, the load shall be multiplied by 125 percent. A fastened-in-place appliance shall be permitted to be connected to a general purpose circuit if its amp rating does not exceed 50 percent of the branch circuit.

15 AND 20 AMPERE BRANCH CIRCUITS
210.23(A)

A 15 or 20 amp branch circuit shall be permitted to supply luminaires and/or utilization equipment in residential, commercial, or industrial locations.

For example: What is the total VA rating for a 15 or 20 amp, 120 volt, 2-wire branch circuit supplying a noncontinuous load?

Finding VA of 20 amp branch circuit

Step 1: Finding VA
210.23(A)
VA = 15 A x 120 V
VA = 1800 VA

Solution: The total VA is 1800.

Step 1: Finding VA
210.23(A)
VA = 20 A x 120 V
VA = 2400 VA

Solution: The total VA is 2400.

For example: What is the total VA rating for a 15 or 20 amp, 240 volt, 3-wire branch circuit supplying noncontinuous loads?

Step 1: Finding VA
210.23(A)
VA = 15 A x 240 V
VA = 3600

Solution: The total VA is 3600.

Step 1: Finding VA
210.23(A)
VA = 20 A x 240 V
VA = 4800

Solution: The total VA is 4800.

The rating of any one cord-and-plug connected utilization equipment shall not be permitted to exceed 80 percent of the branch circuit rating if connected to a general purpose circuit supplying 2 or more outlets.

For example: What size overcurrent protection device is required to be installed for a 7 amp, 120 volt, single-phase compactor that is cord-and-plug connected per **422.16(B)(2)**?

Step 1: Finding usable amperage of 15 or 20 amp branch circuit
210.23(A)
(If continuous)
A = 15 A x 80%
A = 12 A
A = 20 A x 80%
A = 16 A

210.23(A)
(If noncontinuous)
A = 15 A x 100%
A = 15 A
A = 20 A x 100%
A = 20 A

Solution: The OCPD of 15 amp shall be permitted to be used.

The 7 amp full-load current rating of the compactor does not exceed the (15 A x 80% = 12 A) loading range of the 15 amp overcurrent protection device. The 7 amp full-load current rating of the compactor does not exceed the (20 A x 80% = 16 A) loading range of the 20 amp overcurrent protection device. Therefore, the 20 amp overcurrent protection device shall be permitted to be used to supply the 7 amp compactor. **Note,** the AHJ may require the 15 amp OCPD to be used.

Fixed appliances (fastened-in-place) shall be permitted to draw up to 50 percent of the rating of a branch circuit supplying 2 or more general purpose outlets that serve lighting and receptacle loads.

For example: Can an 8 amp, 120 volt, single-phase A/C window unit be connected to an existing branch circuit?

Step 1:   Finding A of branch circuit
210.23(A)
A = 50% of 20 A OCPD
A = 10 A

Step 2:   Calculating A for A/C unit
210.23(A); 440.62(B); (C); 440.32
A = 8 A x 125%
A = 10 A

Step 3:   Verifying permissive A
210.23(A); 440.62(B); (C); 210.3
A = 20 A OCPD x 80%
A = 16 A

Solution: **Yes, the A/C window unit rated at 8 amps shall be permitted to be connected to the 20 amp branch circuit.**

A 20 amp overcurrent protection device shall be permitted to protect a fastened-in-place appliance with a rating of 10 amps or less after applying the 50 percent rule. The remaining 50 percent of the overcurrent protection device is used to supply lighting and/or cord-and-plug connected appliances. **(See Figure 14-12)**

30 AMPERE BRANCH CIRCUITS
210.23(B)

A 30 amp branch circuit shall be permitted to be installed to supply fixed lighting units with heavy-duty lampholders in other than a dwelling unit(s) or utilization equipment in any occupancy. The rating of any individual cord-and-plug connected appliance shall not be permitted to exceed 80 percent of the branch circuit rating. The rating of any individual cord-and-plug connected appliance shall not be permitted to draw more than 24 amps (30 A x 80% = 24 A) when connected to an 30 amp overcurrent protection device. **(See Figure 14-13)**

For example: What is the load for a 23 amp dishwasher used at continuous duty?

Step 1:   Finding amperage
220.18(A)
A = 23 A x 125%
A = 28.75

Step 2:   Finding branch circuit
210.23(B)
28.75 A requires 30 A

Solution: **The branch circuit load is 28.75 amps.**

A 30 amp branch circuit shall be permitted to supply a single appliance that is used at continuous operation in any type occupancy.

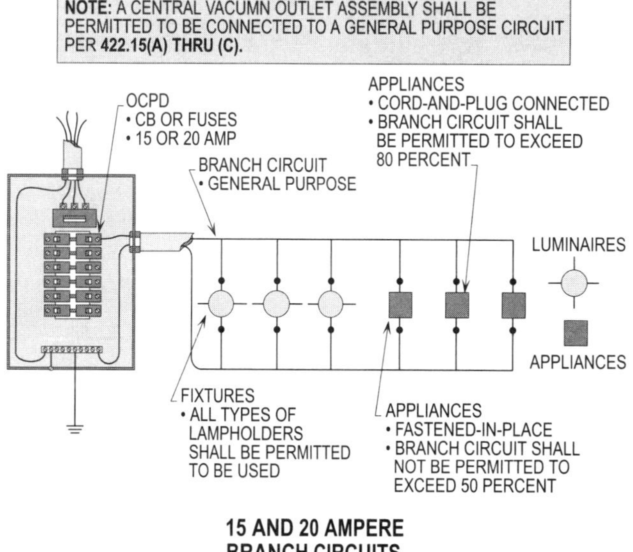

Figure 14-12. A 20 amp overcurrent protection device shall be permitted to protect a fastened-in-place appliance with a rating of 10 amps or less after applying the 50 percent rule. The remaining 50 percent of the overcurrent protection device is used to protect lighting and/or cord-and-plug connected appliances.

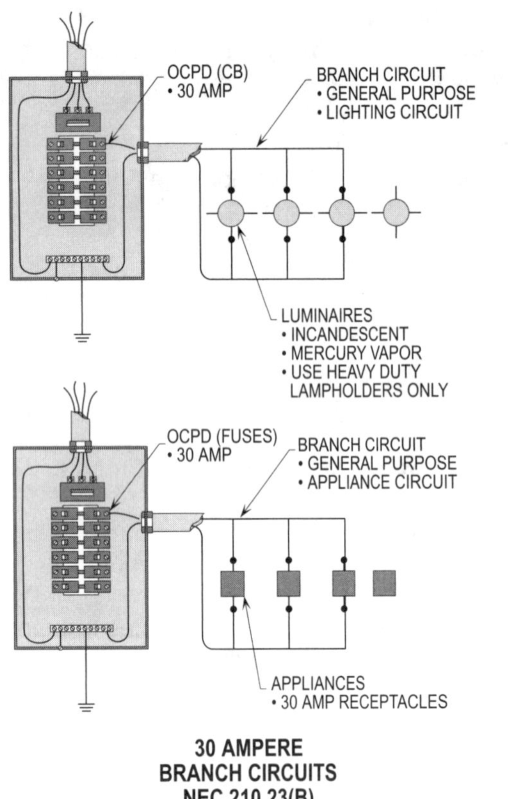

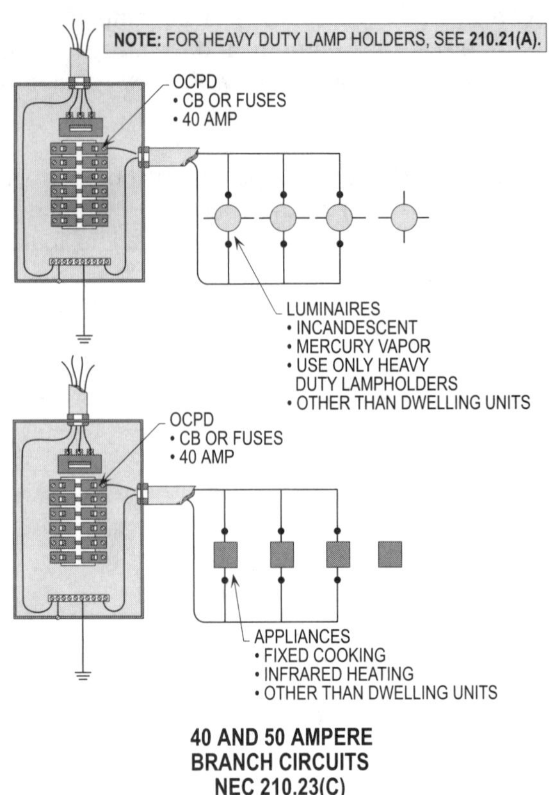

Figure 14-13. A 30 amp branch circuit shall be permitted to be installed to supply fixed lighting units with heavy-duty lampholders in other than a dwelling unit(s) or utilization equipment in any occupancy.

40 AND 50 AMPERE BRANCH CIRCUITS 210.23(C)

A 40 or 50 amp branch circuit shall be permitted to supply cooking appliances that are fastened in place in any occupancy. Fixed lighting units with heavy-duty lampholders or infrared heating units shall be permitted for such circuits except for other than dwelling units. Equipment such as a water heater, dryer, or heating unit shall be permitted to be supplied by a 40 or 50 amp branch circuit. **(See Figure 14-14)**

For example: What is the load for a 37 amp water heater used at continuous operation in a commercial building?

Step 1: Finding amperage
422.13
A = 37 A x 125%
A = 46.25

Solution: The branch circuit load is 46.25 amps.

Figure 14-14. A 40 or 50 amp branch circuit shall be permitted to supply cooking appliances that are fastened in place in any occupancy.

BRANCH CIRCUITS LARGER THAN 50 AMPERES 210.23(D)

Branch circuits larger than 50 amps shall supply only nonlighting outlet loads. A maximum load of 50 amps shall be used for multioutlet branch circuits installed for lighting. A combination of loads exceeding 50 amps shall be permitted be used for multioutlet branch circuits that are not connected to lighting units, such as plugs for welders.

CONDUCTORS ARTICLE 310

Branch circuit conductors used for general wiring shall be rated for the following insulations per **Table 310.16**:

- 60°C
- 75°C
- 90°C

The temperature rating of conductors and their conditions of use are listed in **Table 310.13(A)**. Not more than 3 current-carrying conductors in a raceway or cable run in an ambient temperature of not more than 30°C or 86°F shall be used for branch circuit conductors based on **Table 310.16**. The types of insulation available are listed in **Table 310.16** for copper and aluminum conductors. Conductor ampacities shall be determined by condition of use and by the terminal ratings of overcurrent protection devices and equipment per **110.14(C)(1)** and **(C)(2)**. **(See Figure 8-4** of **Chapter 8)**

60°C CONDUCTORS
TABLE 310.16, COLUMN 2 AND
110.14(C)(1)(a)

The following type of conductors shall be permitted to be installed when using **Table 310.16, Column 2** for 60°C ampacities:

- TW
- UF

The temperature rating of the conductors are rated at 60°C per **Table 310.13(A)** and **Table 310.16.** The conductors shall be terminated to 60°C terminals and 60°C ampacities shall be used.

> **Design Tip:** All conductors with the "W" (60°C or 75°C) rating insulation shall be permitted to be installed in dry, damp, or wet locations per **310.8(B)** and **(C)**.

Overcurrent protection devices rated at 100 amps or less shall be permitted to be terminatcd with conductors 14 AWG through 1 AWG with 60°C ampacities per **110.14(C)(1)(a)**. The maximum size of 1 AWG Is listed so that a 60°C conductor ampacity will match to terminals of overcurrent protection devices and equipment rated at 100 amps.

> **For example:** What size THHN copper conductor is required to supply a piece of equipment operating at 90 amps? (Load already calculated at continuous or noncontinuous operation.)
>
> **Step 1:** Finding load
> **Table 310.13(A)** and **Table 310.16**
> Load = 90 A
>
> **Step 2:** Finding conductor A at 60°C
> **Table 310.16**
> 95 A = 2 AWG THHN
>
> **Solution: The conductor is required to be 2 AWG THHN.**

The allowable ampacity for each conductor shall be rated at 60°C when installing nonmetallic-sheathed cable (romex or rope) per **334.80**. However, these conductors have insulation rated at 90°C that permits such conductors to be used for derating purposes. **(See Figure 14-15)**

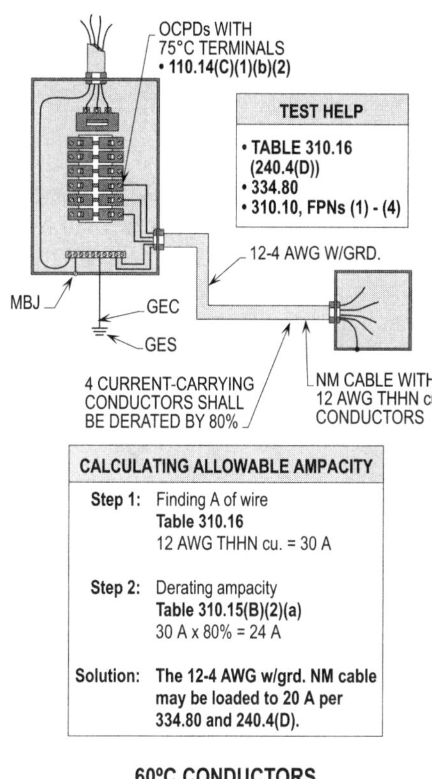

Figure 14-15. The allowable ampacity for each conductor is rated at 60°C, when installing nonmetallic-sheathed cable (romex or rope) per **334.80** and **334.112**. However, these conductors have insulation rated at 90°C, which permits such conductors to be used for derating purposes.

75°C CONDUCTORS
TABLE 310.16, COLUMN 3 AND
110.14(C)(2)(b)

The following types of conductors shall be permitted to be installed when using **Table 310.16, Column 3** for 75°C ampacities:

- RHW
- THHW
- THW
- THWN
- XHHW
- USE
- ZW

Overcurrent protection devices rated over 100 amps shall be permitted to be terminated with conductors larger than 1 AWG with 75°C ampacities per **110.14(C)(2)(b)**. Conductors shall be permitted to be installed with higher temperature ratings if the ampacities are matched to the terminals of the overcurrent protection device and equipment. The terminals shall be marked by one of the following temperature ratings:

- 60°C
- 60°C/75°C

Note that 60°C ampacities shall only be permitted to be applied to terminals marked 60°C. 75°C ampacities shall only be permitted to be applied to terminals marked 75°C. Note that 60°C ampacities shall only be permitted be applied using 75°C ampacities where conductors 14 AWG through 1 AWG are installed. 75°C terminals shall be permitted be used and connected with any size conductor listed in the 75°C column.

For example: What size amperage rating is allowed for a 4 AWG copper conductor using the ampacities of the 60°C column and 75°C column, respectively?

Step 1:   Finding amperage
Table 310.16, Columns 2 and 3
60°C = 70 A
75°C = 85 A

Solution: **The allowable amperage rating is 70 amp for 60°C and 85 amps for 75°C. Note: The 60°C terminal rating reduces the 4 AWG THHN copper conductor to only 70 amps, and not 85 or 95 amps.**

90°C CONDUCTORS
TABLE 310.16, COLUMN 4

The following types of conductors shall be permitted to be used to wire modern-day electrical systems. See **Table 310.16, Column 4** and **Table 310.13(A)** for ampacities and conditions of use for 90°C rated conductors:

- TBS
- SA
- SIS
- FEP
- FEPB
- MI
- RHH
- RHW-2
- THHN
- THHW
- THW-2
- THWN-2
- USE-2
- XHH
- XHHW
- XHHW-2
- ZW-2

These conductors with the 90°C rated insulation shall be permitted to be connected to terminals rated at 60°C, 75°C, and 90°C. Devices and equipment with 90°C terminals shall be used for connecting higher ampacity conductors that are rated 90°C. However, such overcurrent protection devices and equipment that are mated (matched) are not available as of today.

For example: What is the allowable ampacity required for a 6 AWG THWN copper conductor connected to a 60°C OCPD installed in a panelboard?

Step 1:   Finding amperage and condition of use
Table 310.16, Column 4 and **Table 310.13(A)**
A = 75

Step 2:   Finding allowable ampacity
Table 310.16, Column 2
A = 55

Solution: **The allowable ampacity is limited to 55 amps because of the 60°C terminals.**

Design Tip: Overcurrent protection devices rated at 100 amps or less shall be terminated with conductors 14 AWG through 1 AWG with 60°C ampacities per **110.14(C)(1)(a)**, if not otherwise marked. Overcurrent protection devices rated over 100 amps shall be permitted to be terminated with conductors larger than 1 AWG with 75°C ampacities per **110.14(C)(2)(b)**.

The following ampacity values shall be used for the temperature ratings of terminals using a 1/0 AWG THHN copper conductor:

- 60°C = 125 amps
- 75°C = 150 amps
- 90°C = 170 amps

A load of 125 amps or less on 60°C terminals shall be permitted to be served by a 1/0 AWG THHN copper conductor. A load of 150 amps or less on 75°C terminals shall be permitted to be served by a 1/0 AWG THHN copper conductor. A load of 170 amps or less on 90°C terminals shall be permitted to be served by a 1/0 AWG THHN copper conductor. The ampacities shall be matched to the terminals of the overcurrent protection device and equipment. (**See Figure 14-16**)

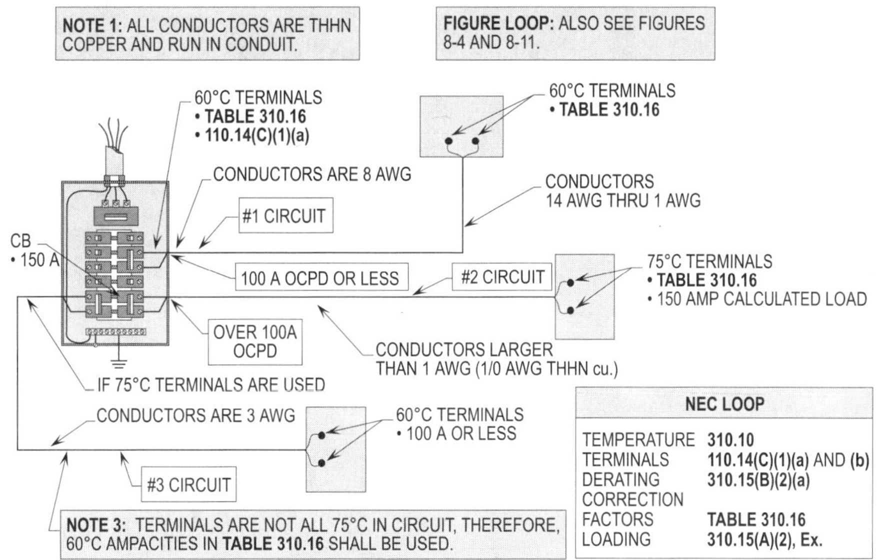

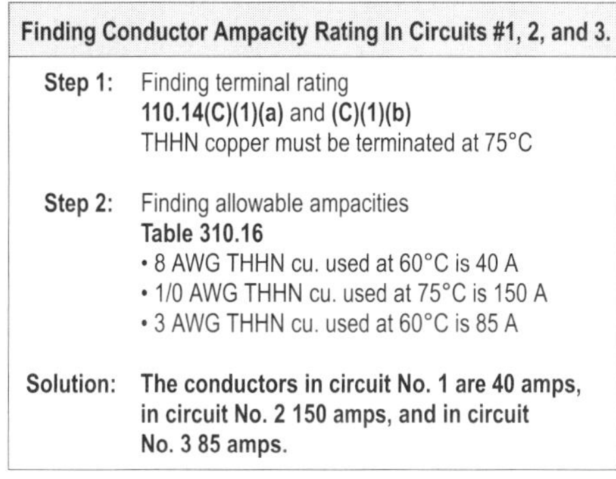

TEMPERATURE LIMITATIONS
NEC 110.14(C)(1)(a) AND (C)(2)(b)

Figure 14-16. Determining the allowable ampacity rating of terminals and sizing conductors based on 60°C/75°C or 75°C ampacities from **Columns 2** and **3 in Table 310.16**. This rule prevents terminals from being overheated.

AMBIENT TEMPERATURES
310.15(B)(2)(a)

Where there are not more than 3 current-carrying conductors in a raceway or cable, the allowable ampacities listed in **Table 310.16** shall be used. If four or more current-carrying conductors or surrounding temperature exceeding 86°F are present, derating factors shall be applied, based upon their conditions of use. See the top of **Table 310.16** for the conditions that shall be applied before selecting the allowable ampacities of such conductors.

For example: What is the allowable ampacity for 6 - 12 AWG THHN cu. conductors that are all current-carrying? (**Note:** Adjustment factors shall be applied.)

Step 1: Calculating ampacity
Table 310.16
12 AWG THHN cu. = 30 A

Step 2: Applying derating factors
310.15(B)(2)(a); 310.10, FPN (4)
30 A x 80% = 24 A

Solution: The allowable ampacity is 24 amps.

For example: What is the allowable ampacity for 4 - 12 AWG THHN copper conductors (three current-carrying) that are in an ambient temperature of 102°F? (**Note:** Correction factors shall be applied.)

Step 1:	Calculating ampacity **Table 310.16** 12 AWG THHN cu. = 30 A
Step 2:	Applying derating factors **Asterisk to Table 310.16** 30 A x 91% = 27.3 A
Solution:	**The allowable ampacity is 27.3 amps.**

The ampacity of conductors shall be derated at least three times when applying the following rules for branch circuits and determining the allowable ampacities:

- More than 3 current-carrying conductors
  - Adjustment factors

- Ambient temperature exceeds 86°F
  - Correction factors

- Continuous duty loads
  - FLA x 125 percent

More than 3 current-carrying conductors in a raceway shall be derated by the percentages listed in **310.15(B)(2)(a)**. Conductors routed through ambient temperatures exceeding 86°F shall be derated by the percentages according to the Ampacity Correction Factors of **Table 310.16**. If the load is continuous (three hours or more), the value shall be multiplied by 125 percent per **210.19(A)(1)**.

For example: What is the allowable ampacity for 4 - 12 AWG THHN cu. conductors that are all current-carrying and routed through an ambient temperature of 102°F?

Step 1:	Calculating ampacity **Table 310.16** 12 AWG THHN cu. = 30 A
Step 2:	Applying adjustment factors **310.15(B)(2)(a)** 30 A x 80% = 24 A
Step 3:	Applying correction factors **240.4(D); Table 310.16** 24 A x 91% = 21.8 A
Solution:	**The allowable ampacity is 21.8 amps.**

For example: What size overcurrent protection device is required to serve a calculated continuous load of 70.5 amps using 4 AWG THHN copper conductors with an ampacity of 85 amps?

Step 1:	Finding ampacity **210.19(A)(1); Table 310.16** 70.5 A requires 80 A OCPD 4 AWG = 85 A = 80 A OCPD
Solution:	**A 80 amp overcurrent protection device is required.**

Design Tip: An 80 amp overcurrent protection device shall be permitted to be installed, because 56.4 amps x 125 percent is 70.5 amps and **240.4(B)** permits the next size overcurrent protection device above 70.5 amps to be used. **(See Figure 14-17)**

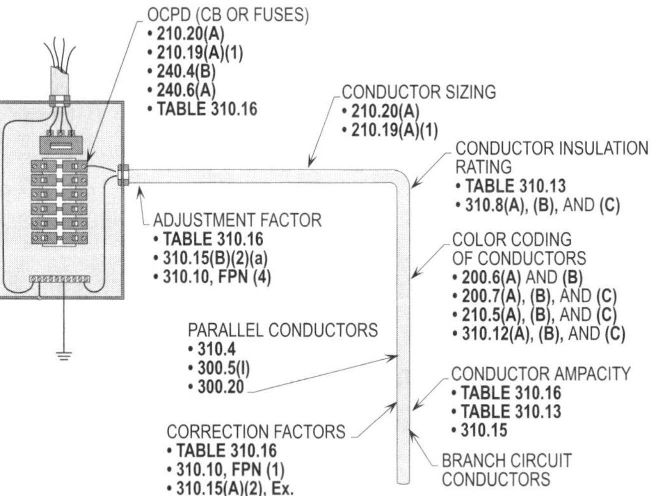

Figure 14-17. More than 3 current-carrying conductors in a raceway shall be derated by the percentages listed in **Table 310.16** and **Table 310.15(B)(2)(a)**. Conductors routed through ambient temperatures exceeding 86°F shall be derated by the percentages according to the **Ampacity Correction Factors** of **Table 310.16**. If the load is continuous (three hours or more), the value shall be multiplied by 125 percent per **210.19(A)**.

BRANCH CIRCUIT VOLTAGE LIMITATIONS
210.6

The following voltage limitations are divided into three categories, and each category of voltage ratings is designed to supply certain loads:

- 120 volts between conductors
- 277 volts-to-ground
- 600 volts between conductors

120 VOLTS BETWEEN CONDUCTORS
210.6(B)

120 volts between conductors is not a restricted voltage and shall be permitted to be used to supply the following loads in any type of occupancy:

- Terminals of lampholders applied within their voltage ratings.

- Ballasts for fluorescent or high intensity discharge (HID) luminaires.

- Cord-and-plug connected or permanently connected appliances.

Cord-and-plug connected or permanently (hard-wired) connected appliances rated over 1440 VA are usually supplied by individual circuits. These appliances are generally connected to general purpose circuits that supply more than one outlet, etc. Such appliances shall be permitted to be any one of the following types of equipment:

- Heating units
- A/C units
- Welders
- Water heaters
- Processing machine
- Etc.

(See Figure 14-18)

277 VOLTS-TO-GROUND
210.6(C) AND 225.7(C)

Circuits exceeding 120 volts between conductors and not exceeding 277 volts-to-ground shall be permitted to supply any one of the following types of electrical apparatus:

- Listed electric-discharge luminaires

- Listed incandescent luminaires

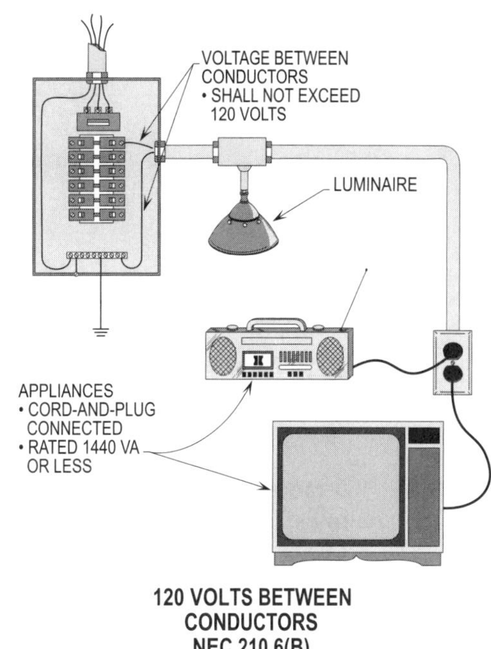

Figure 14-18. Cord-and-plug connected or permanently (hard-wired) connected appliances rated over 1440 VA are usually supplied by individual circuits.

- Mogul-base screw-shell lampholders

- Other than screw-shell type lampholders applied within their voltage ratings

- Ballasts for fluorescent or high intensity discharge (HID) luminaires

- Cord-and-plug connected or permanently connected appliances or other utilization equipment

[See Figures 14-19(a) through (d)]

> **Design Tip:** Listed electric-discharge or incandescent luminaires are no longer required to be installed at a minimum height of 8 ft (2.5 m) above finished grade when supplied by 480/277 volt, three-phase, four-wire system.

Luminaires for illumination shall be permitted to be installed for outdoor areas of industrial establishments, office buildings, schools, stores, and other commercial or public buildings where the luminaires are supplied by 480/277 volt circuits per **225.7(C)**. However, luminaires shall not be permitted to be located within 3 ft (900 mm) from windows, platforms, fire escapes, etc.

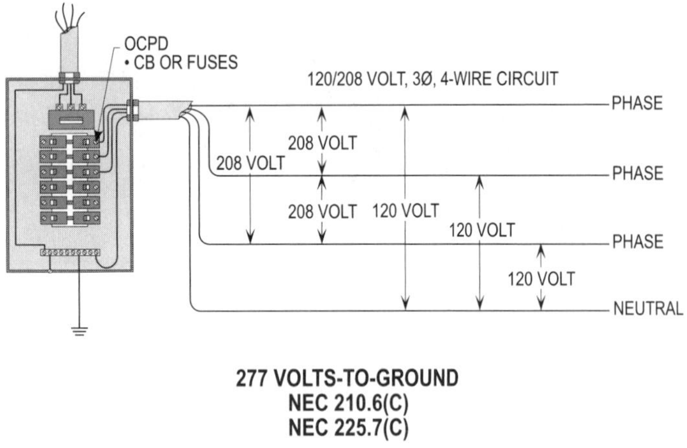

Figure 14-19(a). The above is a diagram of a 120/208 volt, three-phase, four-wire circuit.

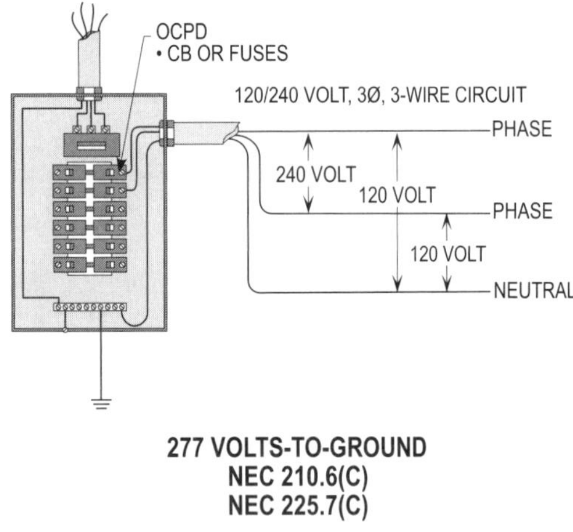

Figure 14-19(b). The above is a diagram of a 120/240 volt, single-phase, three-wire circuit.

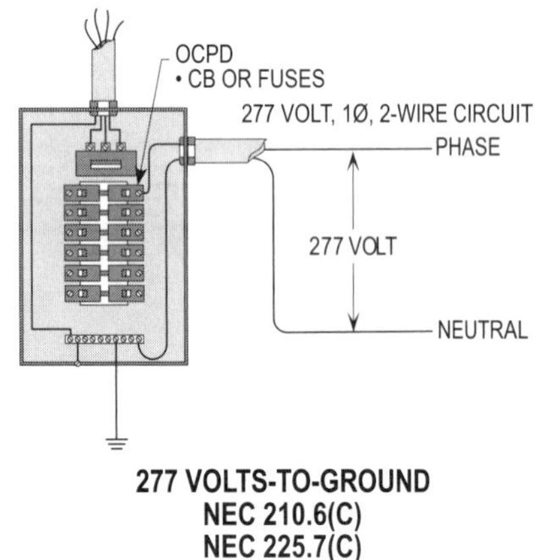

Figure 14-19(c). The above is a diagram of a 277 volt, single-phase, two-wire circuit.

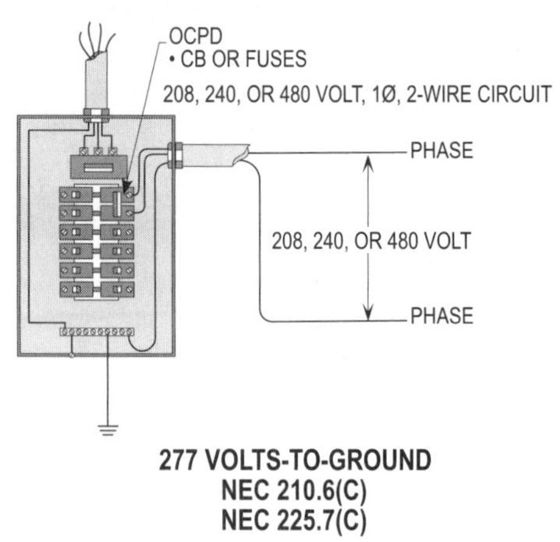

Figure 14-19(d). The above is a diagram of a 208, 240, or 480 volt, single-phase, two-wire circuit.

600 VOLTS BETWEEN CONDUCTORS 210.6(D) AND 225.7(D)

Circuits exceeding 277 volts-to-ground and not exceeding 600 volts between conductors shall be permitted to supply the following types of electrical apparatus:

- Ballasts for electric-discharge luminaires where mounted by one of the following methods:

  (a) At a height not less than 22 ft (6.7 m) on poles or similar structures for the illumination of outdoor areas such as highways, roads, bridges, athletic fields, or parking lots

  (b) At a height not less than 18 ft (5.5 m) on other structures, such as tunnels

- Utilization equipment that is cord-and-plug connected or permanently connected

- Luminaires powered from direct-current systems

These circuits are usually derived by the following types of electrical systems:

- Ungrounded 480 volt, three-wire systems
- Corner grounded 480 volt, three-wire systems

Ungrounded 480 volt, three-wire systems will have 480 volts-to-ground; if one of the legs is accidentally grounded, the power is not lost. Corner grounded 480 volt, three-wire systems will have 480 volts-to-ground because one leg is intentionally grounded. Circuits exceeding 277 volts and not exceeding 600 volts between conductors shall be permitted to supply the auxiliary equipment of electric-discharge lamps as permitted in **225.7(D)**.

DETERMINING AMPERAGE

The amperage for single-phase branch circuits shall be determined by dividing the VA by the supply voltage ($I = VA \div V$). The amperage for three-phase branch circuits shall be determined by dividing the VA by the supply voltage times the square root of three ($I = VA \div V \times 1.732$). The 1.732 is determined by taking the $\sqrt{3}$. These amperage ratings shall be used to select the conductors and overcurrent protection devices and other elements of branch circuits.

FINDING AMPERAGE
SINGLE-PHASE CIRCUITS

The amperage for single-phase branch circuits shall be determined by dividing the VA rating of electrical equipment by the supply voltage of the circuit. (See **page 4-2** of **Chapter 4**)

For example: What is the amperage rating of an 2400 VA calculated load connected to a 120 volt, single-phase branch circuit?

Step 1:   Finding amperage
$I = VA \div V$
$I = 2400 \text{ VA} \div 120 \text{ V}$
$I = 20 \text{ A}$

Solution: **The branch-circuit amperage is 20 amps.**

FINDING AMPERAGE
THREE-PHASE CIRCUITS

The amperage for three-phase branch circuits shall be determined by dividing the VA rating for the electrical equipment by the supply voltage times 1.732. The amperage is evenly (as possible) distributed on legs 1, 2, and 3 when dividing the VA by the voltage times 1.732. (See **page 4-3** of **Chapter 4**)

In some cases, the values shown on the three-phase chart are used when determining the three-phase amperage instead of multiplying the supply voltage by 1.732. However, other calculations with the full square root of 3 times the voltage will be used. (See **page 4-4** of **Chapter 4**)

For example: What is the amperage rating of an 8960 VA calculated load connected to a 208 volt, three-phase branch circuit?

Step 1:   Finding amperage
$I = VA \div V \times 1.732$
$I = 8960 \text{ VA} \div 360 \text{ V } (208 \times 1.732)$
$I = 25 \text{ A}$

Solution: **The branch-circuit amperage is 25 amps.**

COMMERCIAL AND INDUSTRIAL

Branch circuits used in commercial and industrial locations shall be calculated differently from those in residential dwelling units. Most loads in commercial and industrial locations are used continuous for three hours or more without being interrupted. Therefore, such branch circuits supplying these loads shall be calculated at 125 percent of their rating. However, there are loads that operate at noncontinuous operation; some are thermostatically controlled, and these do not fall under such rules.

LIGHTING LOADS
ARTICLE 220, PART III

Incandescent or electric-discharge luminaires shall be permitted to bc installed in or on a premise. Such loads shall be calculated at noncontinuous operation (100 percent) or continuous operation (125 percent). Noncontinuous operated loads shall be calculated at 100 percent when used for less than three hours at any given time. Continuous operated loads shall be calculated at 125 percent when used for more than three hours without employing an OFF or cycle period.

NONCONTINUOUS OPERATION
210.19(A)(1) AND 210.20(A)

Noncontinuous lighting shall be calculated at 100 percent of the total VA or amperage rating of the branch circuit. Luminaires and cord-and-plug connected table lamps or floor lamps used for various periods of time shall be permitted to be installed or connected to these circuits. The branch circuit will never be overloaded during their time of use if proper calculations are made and the correct size elements selected.

For example: How many outlets (1.5 A per outlet) are permitted to be connected to a 20 amp branch circuit used at noncontinuous operation?

Step 1: Finding amperage of outlets
210.11(A), 210.19(A)(1), and 210.20(A)
A = 180 VA x 100% ÷ 120 V
A = 1.5

Step 2: Finding number of outlets
210.11(A)
No. = 20 A OCPD ÷ 1.5 A
No. = 13

Solution: **The number of noncontinuous outlets permitted on a 20 amp branch circuit is 13.**

Design Tip: Applying the same procedure (15 A OCPD ÷ 1.5 A = 10), 10 outlets shall be permitted on a 15 amp branch circuit. The limitation of outlets on branch circuit alleviates the nuisance tripping of the overcurrent protection device.

CONTINUOUS OPERATION
210.19(A)(1) AND 210.20(A)

Continuous lighting loads shall be calculated at 125 percent of the total VA or amperage rating of the branch circuit. Incandescent or electric-discharge luminaires shall be permitted to be installed in or on commercial or industrial buildings for continuously operated lighting loads.

For example: How many lighting outlets are permitted to be installed on a 20 amp branch circuit used at continuous operation?

Step 1: Finding amperage of outlets
210.11(A), 210.19(A)(1), and 210.20(A)
A = 225 VA (180 VA x 125%) ÷ 120 V
A = 1.875

Step 2: Finding number of outlets
210.11(A)
No. = 20 A OCPD ÷ 1.875 A
No. = 10

Solution: **The number of continuous operated lighting outlets permitted on a 20 amp branch circuit is 10.**

Design Tip: Applying the same procedure (15 A OCPD ÷ 1.875 A = 8), 8 outlets shall be permitted on a 15 amp branch circuit. The limitation of outlets on branch circuit alleviates the nuisance tripping of the overcurrent protection device.

OTHER LOADS
210.20(A)

The rating of the branch-circuit overcurrent device serving continuous loads, such as store lighting and similar loads, shall not be less than the noncontinuous load plus 125 percent of the continuous load.

Design Tip: The minimum branch-circuit conductor size, without the application of any adjustment or correction factors, shall have an allowable ampacity equal to or greater than the noncontinuous load plus 125 percent of the continuous load. **(See Figure 14-20)**

For example: What is the maximum continuous operated load that can be connected to a 20 amp branch circuit?

Step 1: Finding amperage
210.19(A)(1) and 210.20(A)
A = 20 A x 80%
A = 16 A
A = 16 A x 125%
A = 20

Solution: **The branch-circuit load is limited to 16 amps.**

For example: What size overcurrent protection device is permitted to serve a continuous operated load of 14 amps and a noncontinuous load of 2.5 amps?

Step 1: Finding amperage
210.20(A) and 210.19(A)(1)
14 A x 125% = 17.5 A
2.5 A x 100% = 2.5 A
Total   = 20   A

Solution: **A 20 amp overcurrent protection device is required.**

Design Tip: The grounded (neutral) conductor shall not be required to be considered a current-carrying conductor if it is designed to carry only the unbalanced load from resistive loads connected to the ungrounded (phase) conductors.

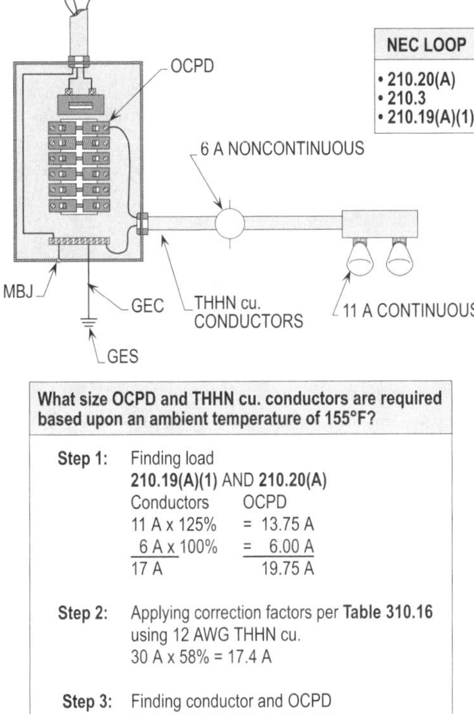

Figure 14-20. Both the overcurrent protection device and conductors are calculated at 100 percent for noncontinuous loads and 125 percent for continuous loads. These values are added together to obtain the total load.

ELECTRIC DISCHARGE LOADS
220.14, 210.20(A), AND 310.10, FPN (2)

Branch-circuit overcurrent protection devices shall be sized at 125 percent of the VA or amperage rating of each ballast installed to supply the ballast of electric discharge lighting units. Such electric discharge lighting units and loads are as follows:

- Fluorescent
- Mercury vapor
- High pressure sodium
- Low pressure sodium
- Metal halide

The total wattage of each bulb or lamp shall not be permitted to be calculated at 125 percent to determine the total lighting load. This type of calculation does not comply with the NEC and usually requires a larger lighting load than the VA rating of each ballast times the number times 125 percent.

Design Tip: The grounded (neutral) conductor shall be considered a current-carrying conductor if it is installed for electric-discharge lighting loads unless the loading does not comply with **310.15(B)(4)(c)**.

For example: What is the lighting load for a branch circuit supplying 12 - 1.5 amp ballasts that serve 24 - F25 CW lamps used at continuous operation?

Step 1:   Finding amperage
          220.18(B)
          I = ballast A x No. of outlets
          I = 1.5 A x 12
          I = 18 A

Step 2:   Applying percentage
          210.19(A)(1) and **210.20(A)**
          I = 18 A x 125%
          I = 22.5 A

Solution: **The branch circuit load is 22.5 amps.**

OUTLETS
220.14(D), (E), AND (L)

Branch circuits of 120 volts shall be calculated at 180 VA (1.5 A) per outlet where the VA rating of ballasts or the wattage of incandescent bulbs is unknown. Heavy-duty lampholders of 120 volts shall be calculated at 600 VA (5 A). Lampholders connected to a branch circuit having a rating in excess of 20 amps shall have a rating of not less than 660 watts (5.5 A) if of the admedium type and not less than 750 watts (6.25 A) if of any other type per **210.21(A)**. Outlets shall be calculated at noncontinuous operation (100 percent) and continuous operation (125 percent) times their VA rating to determine the branch circuit load. Overcurrent protection devices shall be calculated at 125 percent. **(See Figure 14-21)**

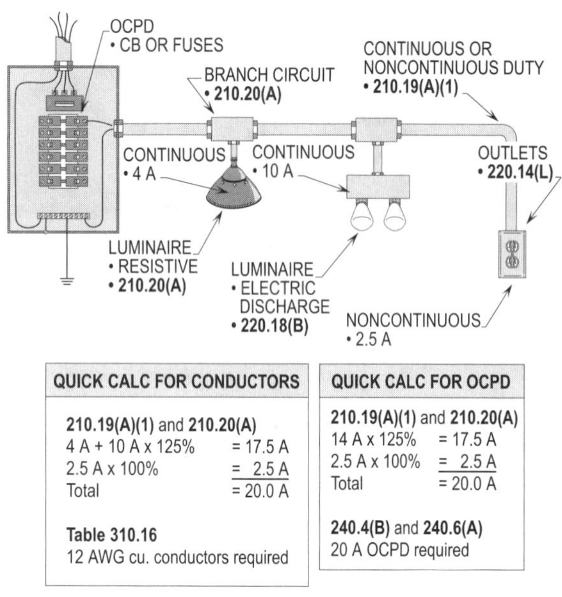

Figure 14-21. Outlets shall be calculated at noncontinuous operation (100 percent) and continuous operation (125 percent) times their VA rating to determine the branch-circuit load. Overcurrent protection devices shall be calculated at 125 percent.

For example: What size overcurrent protection device and THHN copper conductors (branch circuit) are required for 9 outlets used at continuous operation to supply luminaires?

Step 1:   Finding VA
           210.19(A)(1) and **210.20(A)**
           VA = No. of outlets x VA x 125%
           VA = 1620 VA (9 x 180 VA) x 125%
           VA = 2025

Step 2:   Finding amperage
           210.11(A)
           I = VA ÷ V
           I = 2025 VA ÷ 120 V
           I = 16.9

Step 3:   Finding conductor and OCPD
           Table 310.16 and Asterisk,
           240.4(B), 240.6(A), and **240.4(D)**
           12 AWG cu. = 30 A

Solution: **The branch-circuit conductors are**
           sized 12 AWG, and the overcurrent
           protection device is 20 amps per
           Asterisk to Table 310.16.

For example: What size overcurrent protection device and THHN copper conductors (branch circuits) are required for 11 outlets used at noncontinuous operation to supply luminaires?

Step 1:   Finding VA
           210.19(A)(1) and **210.20(A)**
           VA = No. of outlets x VA x 100%
           VA = 1980 VA (11 x 180 VA) x 100%
           VA = 1980

Step 2:   Finding amperage
           210.11(A)
           I = VA ÷ V
           I = 1980 VA ÷ 120 V
           I = 16.5

Step 3:   Finding conductor and OCPD
           Table 310.16 and Asterisk,
           240.4(B), 240.6(A), and **240.4(D)**
           12 AWG THHN cu. = 30 A
           Protected by 20 A OCPD

Solution: **The branch-circuit conductors are**
           sized 12 AWG, and the overcurrent
           protection is selected at 20 amps.

SHOW WINDOWS
220.14(G) AND 220.43(A)

Show window lighting loads shall be calculated at a minimum of 180 VA per outlet for branch circuits if the VA is unknown. The total VA rating for show window lighting loads shall be multiplied by 125 percent for each branch circuit. Show window lighting loads shall be multiplied by 125 percent because they can operate for a period of three hours or more to provide illumination for advertisement of goods.

For example: What is the number of outlets permitted to be installed on a 20 amp branch circuit supplying lighting loads in a show window?

Step 1: Finding No. of outlets
220.14(G), 210.19(A)(1), and **210.20(A)**
No. of outlets =
device x 120 V ÷ VA x 125%
No. of outlets =
(20 A x 120 V) ÷ 180 VA x 125%
(2400 VA ÷ 180 VA) x 125%
13.30 x 125%
No. of outlets = 10.7

Solution: The number of outlets permitted to be installed on a 20 amp branch circuit is 10.7. Note: Inspectors usually permit 180 VA to be used if the load for such units is unknown.

See **Figure 14-22** for calculating the load for show windows based upon either linear foot or individual outlets.

TRACK LIGHTING
220.43(B)

Lighting track loads shall be calculated at 150 VA for each 2 ft (600 mm) of track to determine the load of branch circuits. To properly balance the load, the VA rating is divided between the number of circuits supplying the length of lighting track.

For example: What is the VA rating for 30 ft of lighting track?

Step 1: Finding VA
220.43(B)
VA = Track length ÷ 2' x 150 VA
VA = (30' ÷ 2') x 150 VA
VA = 2250

Solution: The total VA rating to be used for calculating the service or feeder load is 2250 VA.

The load on multicircuit lighting tracks shall be balanced as adequately as possible using one of the following methods:

Calculating 2 circuits
2 branch circuits
150 VA for each 2 ft
VA = 150 VA ÷ 2
VA = 75

Calculating 3 circuits
3 branch circuits
150 VA for each 2 ft
VA = 150 VA ÷ 3
VA = 50

QUICK CALC
Number Of Receptacle Outlets 210.62
= 80' ÷ 12' = 6.7
= 7 RECEPTACLE OUTLETS

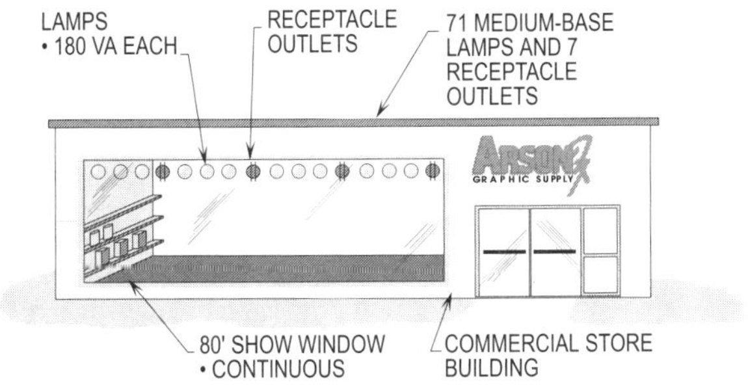

What is the lighting load for the show window based upon the number of lamp outlets and linear feet?

Finding load

Step 1: Calculating the load
220.43(A)
80 ft x 200 VA = 16,000 VA

Step 2: Calculating the load for the OCPD
230.42(A)(1)
16,000 VA x 125% = 20,000 VA

Solution: **The lighting load is 20,000 volt-amps.**

Finding number of outlets

Step 1: Number of outlets
220.14(G)
71 x 180 VA = 12,780 VA

Step 2: Calculating the load for the OCPD
230.42(A)(1)
12,780 VA x 125% = 15,975 VA

Solution: **The lighting load is 15,975 VA**

Note: The 20,000 VA calculation is used because it is greater than 15,975 VA.

SHOW WINDOWS
NEC 220.14(G)
NEC 220.43(A)

Figure 14-22. Calculating the load for a show window using largest load – either the number of outlets or the linear feet shall be used, whichever is greater.

Design Tip: Use the actual load on the track if possible.

The 150 VA multicircuit track lighting shall not apply to dwelling units per **220.43(B)**. The load for track lighting would be included in the 3 VA per sq. ft listed in **Table 220.12**. **(See Figure 14-23)**

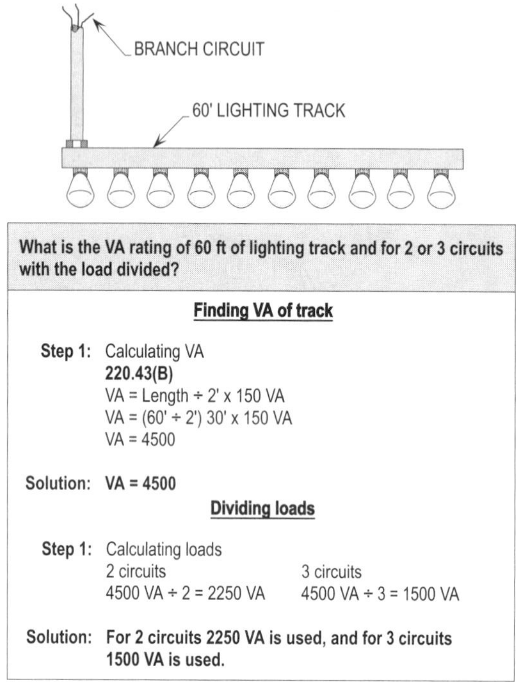

Figure 14-23. The 150 VA for each 2 ft (600 mm) of lighting track shall be used to calculate the load for service or feeder loads. The load of the track shall be divided as evenly as possible on each circuit supplying the track.

SIGN LOADS
ARTICLE 600, PART I

Each commercial building and each commercial occupancy accessible to pedestrians shall be provided with a sign circuit that is accessible at each entrance for pedestrians from a sidewalk, street, etc. Signs shall be considered to be in continuous operation (three hours or more) if installed and used in commercial occupancies.

REQUIRED BRANCH CIRCUIT
600.5(A)

Each commercial building and each commercial occupancy shall be required to be installed with at least one 20 amp branch circuit to supply an outlet for a sign or outline lighting that is located in an accessible location. This 20 amp branch circuit shall not be permitted to have any other loads connected to the overcurrent protection device protecting such circuits.

RATING
600.5(B)(1) AND (B)(2)

Branch circuits that supply signs and outline lighting systems containing incandescent and fluorescent forms of illumination shall be limited to 20 amps or less. Branch circuits that supply transformers for neon tubing installations shall be limited to 30 amps or less.

For example, transformers installed for channel letters and ballasts for electric discharge lamps shall be rated 20 amps or less for branch circuits. Transformers installed only to connect branch circuits for neon or channel letters shall be rated 30 amps or less.

CALCULATED LOAD
220.14(F)

When sizing the service or feeder calculation for a sign calculated at a minimum of 1200 VA, the load shall be multiplied by 125 percent for continuous operation if the sign burns for three hours or more. When sizing the load for the overcurrent protection device, the load shall be multiplied by 125 percent to obtain the load that is based upon continuous operation.

For example: What is the total load allowed for a 16 amp wall sign supplied by a 20 amp, 120 volt branch circuit?

Step 1:	Finding amperage **600.5(A)** OCPD = 20 A
Step 2:	Calculating amperage **220.14(F)** and **210.20(A)** 20 A x 80% = 16 A 16 A x 125% = 20 A
Step 3:	Selecting OCPD **220.14(F)** and **210.20(A)** 16 A load is permitted
Solution:	**The size overcurrent protection device required is 20 amps for a sign that burns three hours or more at 16 amps or less.**

See **Figure 14-24** for calculating the load for a sign.

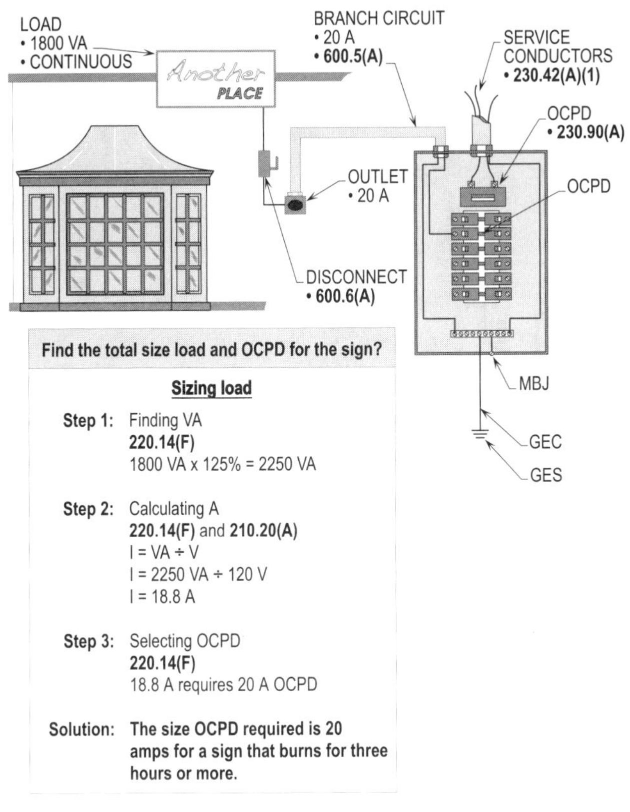

Figure 14-24. Sign loads that operate for three hours or more shall be calculated at 125 percent, and elements selected from this value.

RECEPTACLE LOADS
ARTICLE 220, PART III

Receptacle outlets shall be permitted to be installed for cord-and-plug connected appliances, utilization equipment, and table or floor lamps in dwelling units, apartments, condominiums, townhouses, and commercial or industrial locations. Cord-and-plug connected loads shall be connected in an arrangement not to load up the branch circuit. Cord-and-plug connected loads are general purpose, serving more than one outlet or individually serving one outlet from a branch circuit. Cord-and-plug connected items shall be installed and located in a manner to prevent the use of extension cords except for temporary use.

RECEPTACLE OUTLETS
220.14(I)

A general purpose branch circuit shall be calculated at 180 VA for each outlet where supplying more than one outlet utilizing cord-and-plug connected items. The number of outlets shall be calculated at 180 VA times noncontinuous operation at 100 percent and continuous operation at 125 percent. Note that overcurrent protection devices and conductors shall be sized at 125 percent. **(See Figure 14-25)**

For example: What is the load for 8 receptacle outlets that are supplying cord-and-plug connected loads used at noncontinuous operation?

Step 1:   Finding VA
          Load = No. of outlets x 180 VA x 100%
          Load = 8 x 180 VA x 100%
          Load = 1440 VA

Solution: The load is 1440 VA.

For example: How many noncontinuous duty receptacle outlets can be connected to a 20 amp general purpose branch circuit?

Step 1:   Finding amperage of outlets
          220.14(I)
          180 VA ÷ 120 V = 1.5 A

Step 2:   Finding number of outlets
          220.14(I) and **210.11(A)**
          20 A OCPD ÷ 1.5 A x 100% = 13

Solution: The number of outlets permitted on a 20 amp OCPD is 13.

For example: How many continuous duty receptacle outlets can be connected to a 20 amp general purpose branch circuit?

Step 1:   Finding amperage of outlets
          220.14(I)
          180 VA ÷ 120 V = 1.5 A

Step 2:   Finding number of outlets
          220.14(I) and **210.11(A)**
          20 A OCPD ÷ 1.5 A x 125% = 10

Solution: The number of outlets permitted on a 20 amp OCPD is 10.

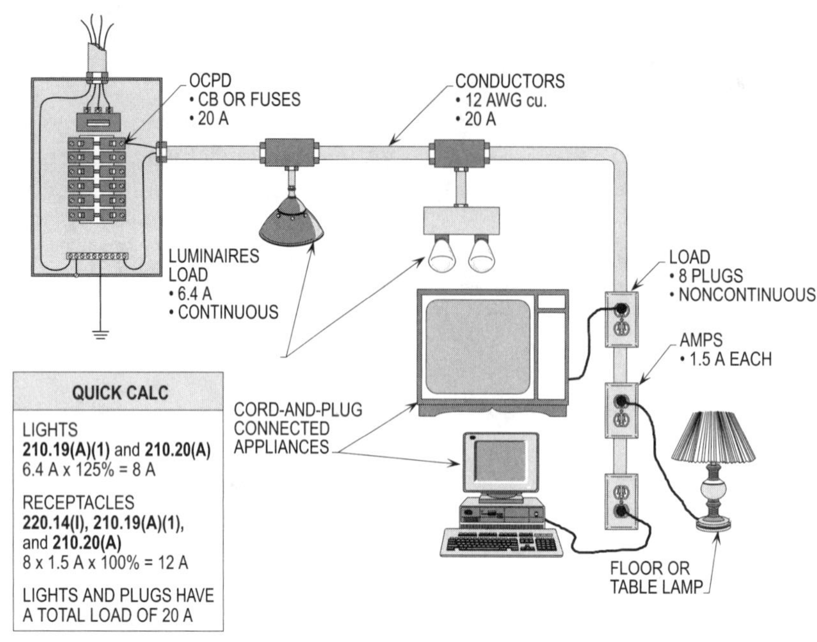

Figure 14-25. Receptacle outlets shall be calculated at 180 VA times noncontinuous operation at 100 percent and continuous operation at 125 percent.

INDIVIDUAL LOADS
210.19(A)(1)

Individual cord-and-plug connected loads for branch circuits shall be determined by multiplying the noncontinuous operated load by 100 percent and the continuous operated load by 125 percent. The amount of amperage for each cord-and-plug connected load shall be found by applying the power formula. The following shall apply when using the power formula:

I = amps	I = VA ÷ V
P = volt-amps	VA = I x V
E = voltage	V = VA ÷ I

Design Tip: When applying calculations in this book, the variation of the power formula is used, since the NEC recognizes volt-amps (VA) for load calculations per the **Examples** in **Annex D**.

For example: What size THWN copper conductors and overcurrent protection device is required for an individual branch circuit to a hot tub having a nameplate current rating of 42 amps?

Step 1: Finding amperage
680.9
42 A is the circuit current rating

Step 2: Calculating load
680.9
42 A x 125% = 52.5 A

Step 3: Selecting conductors
Table 310.16
52.5 A requires 6 AWG THWN cu.

Step 4: Selecting OCPD
240.4(B) and **680.9**
52.5 A load requires 60 A OCPD
65 A conductor allows 70 A OCPD

Solution: A 60 amp OCPD and a 6 AWG THWN copper conductor are permitted.

See Figure 14-26 for calculating the load to an individual cord-and-plug connected load.

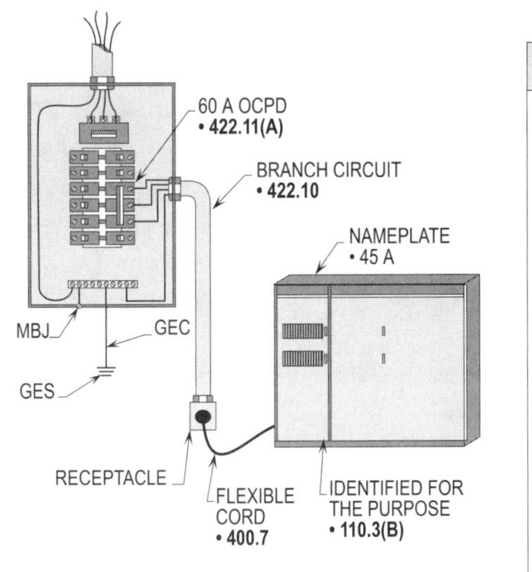
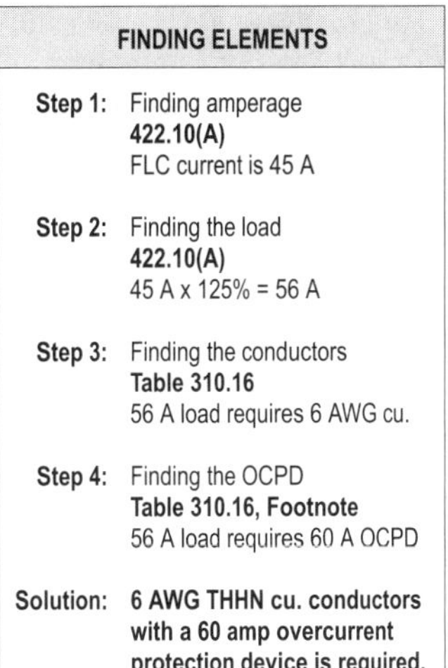

Figure 14-26. Individual loads shall be calculated at 100 percent for noncontinuous and 125 percent of continuous, times the nameplate FLA of the appliance.

MULTIWIRE BRANCH CIRCUITS 210.4

Multiwire branch circuits are used to supply power to line-to-neutral loads only when more than one ungrounded (phase) conductor is sharing the grounded (neutral) conductor. All conductors of a multiwire branch circuit shall originate from the same panelboard or similar distribution equipment and shall be switched by individual circuit breakers. Multiwire branch circuits shall be supplied by double-pole circuit breakers or individual circuit breakers with handle ties to provide safety from electrical shock while serving the circuits or equipment.

Section **300.13(B)** makes it mandatory that grounded (neutrals) conductors be pigtailed to prevent different voltage levels between the ungrounded (phase) conductors should the grounded (neutral) conductor become loose due to bad connections, etc.

Multiwire branch circuits shall be permitted to serve power to one outlet for an individual piece of equipment with the other circuit(s) supplying power to a number of outlets that are used for receptacles, lights, and other cord-and-plugged equipment. **(See Figure 14-27)**

All ungrounded (phase) conductors shall be simultaneously disconnected at the point where each multiwire branch circuit originates.

The ungrounded (phase) and grounded (neutral) conductors of each multiwire branch circuit shall be grouped by wire or similar means in at least one location within the panelboard or other point of origination.

COMMERCIAL COOKING EQUIPMENT 210.19(A)(1) AND 210.20(A)

Commercial cooking equipment shall be calculated at noncontinuous operation (100 percent) or continuous operation (125 percent) to determine the branch-circuit load. When installing more than one piece of cooking equipment, the demand factors shall be selected from **Table 220.56**.

For example: What is the load for a 10 kW cooking unit that is supplied by a 240 volt, single-phase branch circuit used at continuous operation?

Step 1: Finding kW for OCPD
210.20(A)
10 kW x 125% = 12.5 kW

Step 2: Finding kW for conductor
210.19(A)(1)
10 kW x 125% = 12.5 kW

Solution: The continuous operated load is 12.5 kW.

Design Tip: The overcurrent protection device shall be calculated at 125 percent, and the conductors shall be calculated at 125 percent. The overcurrent protection device shall be permitted to be increased and decreased in size per **240.4** or **240.4(E)**.

See **Figure 14-28** for calculating the load for a commercial cooking unit.

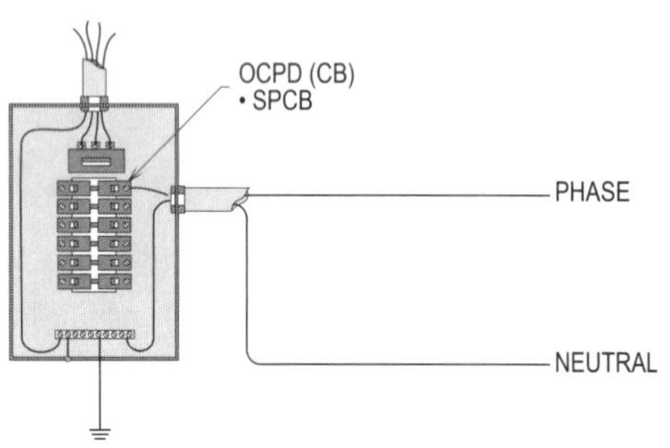

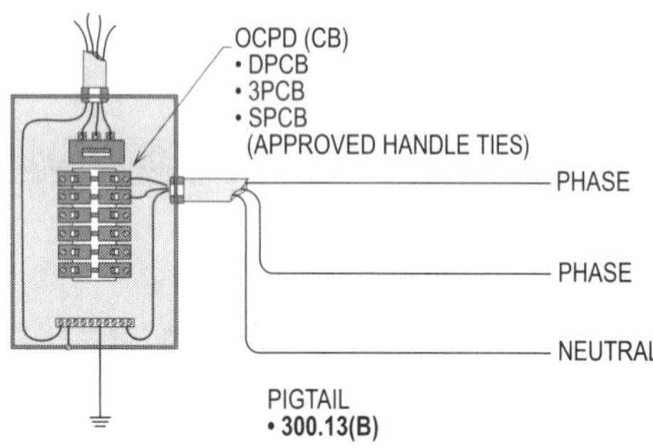

Figure 14-27. Multiwire branch circuits are used to supply power to line-to-neutral loads only when more than one ungrounded (phase) conductor is sharing the grounded (neutral) conductor.

WATER HEATER LOADS
ARTICLE 422, PART II

Water heaters are designed with elements to heat water at different stages of use. For larger amounts of hot water, more of the elements shall be connected into the circuit to heat the water to replace that which was used. When smaller amounts of hot water is needed, fewer elements are used in the circuit to heat the water.

CONDUCTORS
422.13 AND 422.10(A)

Storage-type water heaters having a capacity of 120 gallons (450 L) or less shall have a rating not less than 125 percent of the nameplate rating to size the branch circuit conductors.

For example: What size conductors are required to supply power to a 240 volt, single-phase water heater pulling 5000 VA?

Step 1: Finding amperage
5000 VA ÷ 240 V = 21 A

Step 2: Finding the loads
422.13 and **422.10(A)**
21 A x 125% = 26 A

Step 3: Finding the conductors
334.80 and **Table 310.16**
26 A load requires 10 AWG cu.

Solution: **The size conductors required are 10 AWG.**

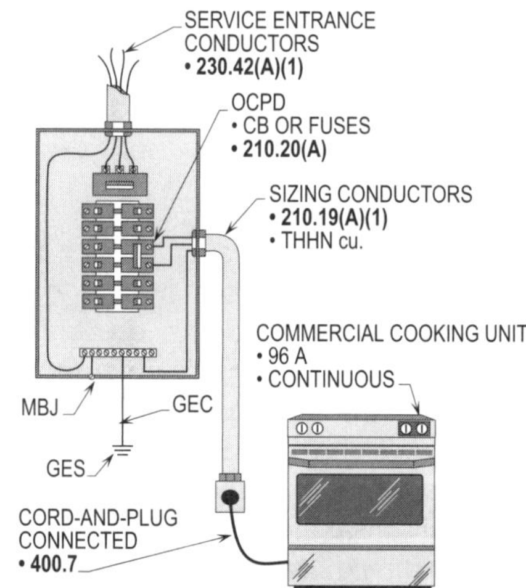

What size conductor and OCPD is required, based on load calculations?

Step 1: Finding the load
210.19(A)(1), **422.10(A)**, and **422.11(A)**
96 A x 125% = 120 A

Step 2: Finding conductor and OCPD
Table 310.16 and **240.6(A)**
Conductors = 1 AWG THHN cu.
OCPD = 125 A

Solution: **The conductors are 1 AWG THHN copper, and the size OCPD is 125 A based upon calculated load.**

COMMERCIAL COOKING EQUIPMENT
NEC 210.19(A)(1)
NEC 210.20(A)

Figure 14-28. Commercial cooking equipment shall be calculated at noncontinuous operation (100 percent) or continuous operation (125 percent) to determine the branch-circuit load.

OVERCURRENT PROTECTION
422.13 AND 422.11(A)

Overcurrent protection devices shall be sized not less than 125 percent of the heating load to prevent tripping open the overcurrent protection device and disconnecting all the elements connected into the circuit. Approximately 100 percent of the water heater's connected load is pulled from resistance heating elements.

For example: What size overcurrent protection device is required to supply power to a 240 volt, single-phase water heater pulling 5000 VA (21 A) at continuous operation?

Step 1:   Finding load for OCPD
           422.13; 422.11(A)
           21 A x 125% = 26 A

Step 2:   Finding the OCPD
           422.13; 240.4(B)
           26 A load requires a 30 A OCPD

Solution: **The overcurrent protection device is 30 amps.**

DISCONNECTING MEANS
422.31(B)

A disconnecting means shall not be required to be installed at the water heater when the overcurrent protection device is readily accessible. The overcurrent protection device shall be permitted to be installed as the disconnect where installed in a service panel that is readily accessible and is located outside or inside the building. An accessible cord-and-plug shall be permitted to serve as the disconnecting means for cord-and-plug connected water heaters where readily accessible per **422.33(A)**. However, such units shall be listed for cord-and-plug connection. **(See Figure 14-29)**

HEATING LOADS
ARTICLE 424, PART I

Heating elements in heating units are rated at 5 kW each. The elements are stacked in the heating unit to provide the rated kW for a particular size occupancy. Two stacked elements provide a 10 kW heating unit. Three stacked elements provide a 15 kW heating unit and so forth. Branch-circuit conductors and overcurrent protection devices shall be sized and selected based on the kW rating of each heating unit plus the blower motor per **424.3(B)**.

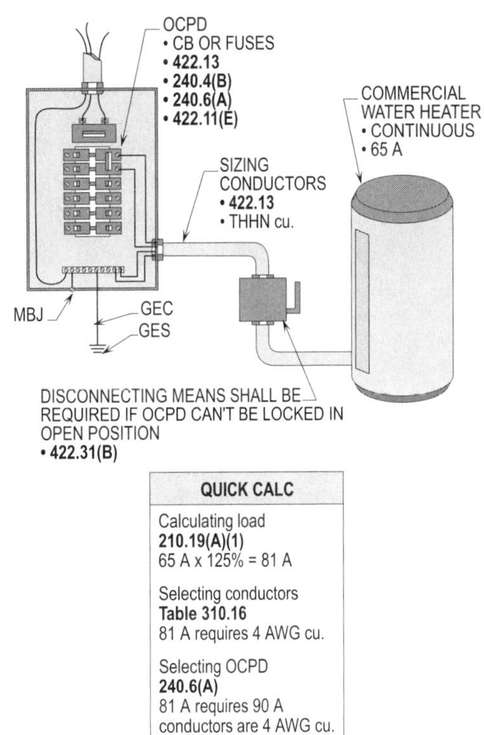

Figure 14-29. A disconnecting means shall not be required to be installed at the water heater when the overcurrent protection device is readily accessible.

Heating units with a number of elements shall be subdivided so that they may be more easily supplied from a local panelboard or from the service equipment. Individual branch circuits or a feeder shall utilized in routing power to the heating unit.

CONDUCTORS
424.3(B)

Heating units shall be calculated at 125 percent of the heating element load plus 125 percent of the blower motor load when present for sizing the branch-circuit conductors. The heating elements and blower motor load shall be at least 125 percent when sizing the ampacity of conductors.

For example: What size THWN copper conductors are required for a 25 kW, 240 volt, single-phase heating unit with a 4 amp blower motor?

Step 1:   Finding amperage
25 kVA x 1000 ÷ 240 V = 104 A

Step 2:   Finding the load
424.3(B)
104 A + 4 A x 125% = 135 A

Step 3:   Finding the conductors
Table 310.16
150 A load requires 1/0 AWG THWN cu.

Solution: **1/0 AWG THWN copper conductors are required.**

OVERCURRENT PROTECTION
424.3(B)

Heating units shall be calculated at 125 percent of the heating element load plus 125 percent of the blower motor load when present for sizing the overcurrent protection device. The next higher size rating overcurrent protection device shall be permitted to be installed if it does not correspond to this rating per **240.4(B)**. The next higher size overcurrent protection device prevents tripping open when the heating unit requires all the 5 kW rated elements to satisfy the heating load called for by the thermostat.

For example: What size overcurrent protection device is required for a 25 kW, 240 volt, single-phase heating unit with a 4 amp blower motor?

Step 1:   Finding load for OCPD
424.3(B)
104 A + 4 A x 125% = 135 A

Step 2:   Finding OCPD
240.6(A); 240.4(B)
150 A is the next higher standard size

Solution: **The size OCPD is 150 amps.**

DISCONNECTING MEANS
424.19

A disconnecting means shall be installed for a self-contained heating unit with a controller that energizes the circuits to the heating elements and blower motor. A fused or nonfused disconnect, an automatic breaker, or a nonautomatic circuit breaker used as the disconnecting means for a heating unit shall be located within sight and within 50 ft (15 m) of the heating unit per **Article 100, 424.19(A), and 430.102. (See Figure 14-30)**

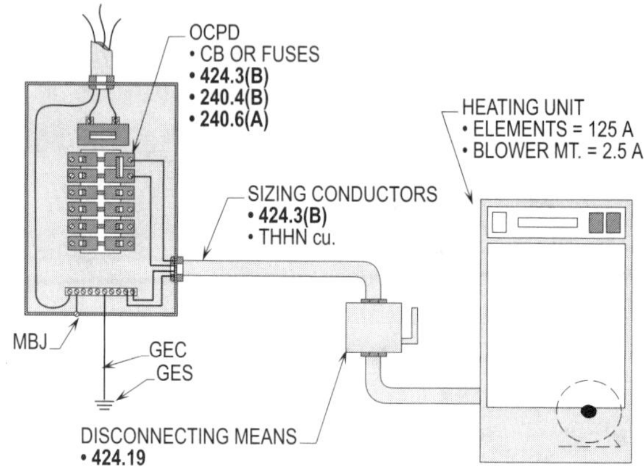

Figure 14-30. A disconnecting means shall be installed for a self-contained heating unit with a controller that energizes the circuits to the heating elements and blower motor.

AIR CONDITIONING LOADS
ARTICLE 440, PART III

Air conditioning units are rated in Btu or tons. One ton contains 12,000 Btu. A five ton compressor in a central air conditioning unit has 60,000 Btu (5 x 12,000 Btu = 60,000 Btu). The air conditioner load shall be calculated by the square footage and construction of the premises. The size air conditioner rated in Btu shall be selected from this calculation. The conductors shall be sized large enough to carry the load of the compressor, and the overcurrent protection device shall be selected to allow the compressor to start without tripping open, due to inrush current from high head pressures.

CONDUCTORS
440.32

Branch-circuit conductors supplying a single motor-compressor shall have an ampacity not less than 125 percent of either the motor-compressor rated load current or the branch circuit selection current, whichever is greater.

The branch-circuit conductors shall be sized large enough to prevent damage to the insulation of the conductor caused by an overload. Overload relays are installed and are usually adjusted to trip open on an overload exceeding 140 percent of full-load current rating of the compressor. This type of condition shall be compensated for by sizing and selecting the conductors at 125 percent of the compressor's full-load current rating. The condenser motor load shall be added at 100 percent when present.

For example: What size conductors are required to supply power to a 240 volt, single-phase air conditioning unit with a compressor rating of 6000 VA plus a 3 amp condenser motor? (Use THHN copper conductors terminated to 60°C terminals.)

Step 1:   Finding amperage of compressor
6000 VA ÷ 240 V = 25 A

Step 2:   Finding the total load
440.32
25 A x 125% + 3 A = 34 A

Step 3:   Finding the conductors
Table 310.16
34 A load requires 8 AWG cu.

Solution: 8 AWG THHN copper conductors are required.

OVERCURRENT PROTECTION
440.22(A)

To allow the compressor to start and run, the overcurrent protection device shall be sized properly. The overcurrent protection device shall have a rating or setting not exceeding 175 percent of the motor-compressor rated full-load current or the branch circuit selection current, whichever is greater to determine the minimum rating. The overcurrent protection device shall be permitted to be increased up to 225 percent of the full-load current if an air conditioner unit will not start using 175 percent. Condenser motors shall be calculated at 100 percent and added to this total, and the overcurrent protection device selected accordingly.

When air conditioning units are installed on the roof or on the outside of the premises where the sun goes down, the air conditioning unit may have trouble starting and operating during hot summer months. Overcurrent protection devices shall be sized by the listing on the nameplate of the unit. Fuses and circuit breakers (HACR) shall be installed where listed on the nameplate of the unit. For further information, see Underwriters Laboratory book, *Electrical Appliance and Utilization Equipment Directory*.

Note that overcurrent protection devices for A/C units installed on the roofs are already calculated and sized at the maximum value of 225 percent per **110.3(B)** and **440.22(A)**.

Conductors supplying A/C units on roofs are calculated already per **440.4(C)** and **440.32**, and the nameplate will list the size conductors or circuit needed.

For example: What minimum and maximum size overcurrent protection device is required to supply power to a 240 volt, single-phase air conditioning unit with a compressor rating of 6000 VA plus a 3 amp condenser motor?

Step 1:   Finding load for OCPD
440.22(A)
25 A x 175% + 3 A = 46.7 A

Step 2:   Finding OCPD
240.4(G) and **240.6(A)**
45 A is the next lower size permitted

Solution: The lower size OCPD is 45 amps.

Step 1:   Finding load for OCPD
440.22(A)
25 A x 225% + 3 = 59.25 A

Step 2:   Finding OCPD
240.4(G) and **240.6(A)**
50 A is the next higher size permitted

Solution: The higher size OCPD is 50 amps.

DISCONNECTING MEANS
440.14

The disconnecting means shall be located within 50 ft (15 m) and within sight from the air conditioning unit per **Article 100**. The disconnecting means shall be permitted to be installed on or within the air conditioning unit. The disconnecting means shall be permitted to be installed where capable of being locked in the open position where conditions of maintenance and supervision ensure that only qualified personnel will service the equipment. Note that **Ex. 1** to **440.14** applies only to industrial compressors and not to A/C units. A disconnecting means shall not be required where within sight of such equipment. **(See Figure 14-31)**

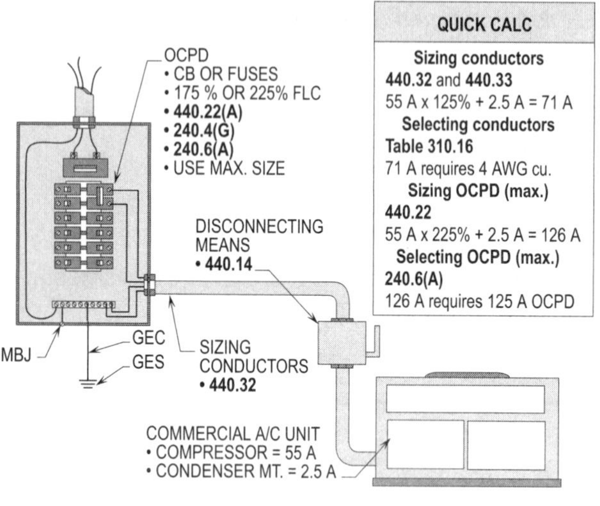

DISCONNECTING MEANS
NEC 440.14

Figure 14-31. The disconnecting means shall be located within 50 ft (15 m) and within sight from the air conditioning unit per **Article 100**. Also, see **Figures 19-9** and **19-16**.

MOTOR LOADS
ARTICLE 430, PARTS II AND IV

Motors are rated in horsepower (HP). The amount of work that a motor can perform depends upon its HP rating class, code letter, or design letter. A motor with a high horsepower rating may do more work than a motor with a low horsepower rating. The overcurrent protection devices and conductors used to supply power to motors shall be sized and based on the horsepower, voltage, and number of phases. The starting current of a motor shall be based on its code letter per **Table 430.7(B)** or design letter per **Tables 430.251(A)** and **(B)**.

CONDUCTORS
430.22(A)

Branch-circuit conductors supplying a single motor shall have an ampacity not less than 125 percent of the motor full-load current rating. Single-phase motors shall have their full-load current ratings selected from **Table 430.248** per **430.6(A)(1)**. Three-phase motors shall have their full-load current rating selected from **Table 430.250** per **430.6(A)(1)**.

For example: What size conductors are required to supply power to a 10 HP, 208 volt, single-phase motor? (Using THWN copper conductors)

Step 1:	Finding amperage **430.6(A)(1); Table 430.248** 10 HP = 55 A
Step 2:	Finding the load **430.22(A)** 55 A x 125% = 68.75 A
Step 3:	Finding the conductors **Table 310.16** 68.75 A load requires 4 AWG cu.
Solution:	**The branch-circuit conductors required are 4 AWG copper.**

For example: What size conductors are required to supply power to a 10 HP, 208 volt, three-phase motor? (Using THWN copper conductors)

Step 1:	Finding amperage **430.6(A)(1); Table 430.250** 10 HP = 30.8 A
Step 2:	Finding the load **430.22(A)** 30.8 A x 125% = 38.5 A
Step 3:	Finding the conductors **Table 310.16** 38.5 A load requires 8 AWG cu.
Solution:	**The branch-circuit conductors required are 8 AWG copper.**

OVERCURRENT PROTECTION
430.52 AND TABLE 430.52

When power is applied to the windings of a motor and it starts, the motor has high inrush currents. The amount of current required to drive a load is called running current. Inrush currents of the motor shall be held by an overcurrent protection device sized large to allow the motor to accelerate the driven load. The inrush current of most motors is four to six times the running current of the motor, based on its code letter per **Table 430.7(B)** or design letter per **Table 430.52**. Note that Design E motors can have 8-1/2 to 15 times FLA. Overcurrent protection devices shall not be permitted to exceed the percentages listed in **Table 430.52**. Section **430.52(C)(1)** requires the overcurrent protection device to based on the percentages for the type of device used in the columns of **Table 430.52**. If the percentage does not correspond to a standard device listed in **240.6(A)**, the higher standard size rating shall be permitted to be used per **430.52(C)(1), Ex. 1**. If the motor will not start and run, the percentage shall be permitted to be increased per **430.52(C)(1), Ex. 2(a) through (c)**. These percentages for the overcurrent protection device are as follows:

- Nontime-delay fuses shall not be permitted to exceed 400 percent of the full-load current for fuses rated 600 amps or less per **430.52(C)(1), Ex. 2(a)**.

- Time-delay fuses shall not be permitted to exceed 225 percent of the full-load current per **430.52(C)(1), Ex. 2(b)**.

- Circuit breakers shall not be permitted to exceed 400 percent of the full-load current for ratings of 100 amps or less, or shall not be permitted to exceed 300 percent of the full-load current for ratings over 100 amps per **430.52(C)(1), Ex. 2(c)**.

- Instantaneous trip circuit breakers shall not be permitted to exceed 1300 or 1700 percent of the full-load current per **430.52(C)(3), Ex. 1**.

For example: What size nontime-delay fuse, time-delay fuse, circuit breaker, and instantaneous circuit breaker are required for a 20 HP, 230 volt, three-phase motor with a full-load current rating of 54 amps per **Table 430.250**? (Nameplate amps is 49 amps)

Nontime-delay fuses

Step 1: Finding percentage
430.52(C)(1) and **Table 430.52**
300%

Step 2: Finding amperage
430.52(C)(1)
54 A x 300% = 162 A

Step 3: Finding NTDF (round down size)
430.52(C)(1), 240.4(G), and **240.6(A)**
162 A requires 150 A OCPD

Step 4: Finding NTDF (round up size)
430.52(C)(1), Ex. 1, 240.4(G), and **240.6(A)**
The next higher standard size above 162 A is 175 A OCPD

Step 5: Finding NTDF (max. size)
430.52(C)(1), Ex. 2(a)
52 A x 400% = 208 A
The largest size below 208 is 200 A OCPD

Solution: The next higher size nontime-delay fuses required are 175 amps.

<table>
<tr><td valign="top" width="50%">

Time-delay fuses

Step 1: Finding percentage
430.52(C)(1) and **Table 430.52**
175%

Step 2: Finding amperage
430.52(C)(1)
54 A x 175% = 94.5 A

Step 3: Finding TDF (round down size)
430.52(C)(1), 240.4(G), and **240.6(A)**
94.5 A requires 90 A OCPD

Step 4: Finding TDF (round up size)
430.52(C)(1), Ex. 1, 240.4(G), and **240.6(A)**
The next standard size above 94.5 A is 100 A OCPD

Step 5: Finding TDF (max. size)
430.52(C)(1), Ex. 2(b)
54 A x 225% = 121.5 A
The largest size below 121.5 A is 110 A OCPD

Solution: **The next higher size time-delay fuse required is 100 amps.**

</td><td valign="top" width="50%">

Circuit breaker

Step 1: Finding percentage
430.52(C)(1) and **Table 430.52**
250%

Step 2: Finding amperage
430.52(C)(1)
54 A x 250% = 135 A

Step 3: Finding CB (round down size)
430.52(C)(1), 240.4(G), and **240.6(A)**
135 A requires 125 A OCPD
The next size below 135 A is 125 A OCPD

Step 4: Finding CB (round up size)
430.52(C)(1), Ex. 1, 240.4(G), and **240.6(A)**
The next standard size above 135 A is 150 A OCPD

Step 5: Finding CB (max. size)
430.52(C)(1), Ex. 2(c)
54 A x 400% = 216 A
The largest size below 216 A is 200 A OCPD

Solution: **The next higher size circuit breaker required is 150 amps.**

</td></tr>
</table>

Instantaneous circuit breaker

Step 1: Finding percentage
430.52(C)(3) and **Table 430.52**
800%

Step 2: Finding amperage
430.52(C)(3)
54 A x 800% = 432 A

Step 3: Finding CB
430.52(C)(1), 240.4(G), and **240.6(A)**
The minimum setting is 432 A

Step 4: Finding CB
430.52(C)(3), Ex. 1
54 A x 1300% = 702 A
The maximum setting is 702 A

Solution: **The maximum size instantaneous circuit breaker setting is 702 amps.**

DISCONNECTING MEANS
430.102 AND 430.107

A disconnecting means shall be located within sight from the controller and shall disconnect the controller to allow personnel to service the motor without the danger of the branch circuit accidentally being energized. The disconnecting means shall be permitted to be installed by one of the following methods to disconnect a motor circuit:

- The disconnecting means shall be located adjacent to the controller within sight and located within 50 ft (15 m). The same rule shall be applied for the motor.

- The disconnecting means shall be capable of being locked in the open position if located adjacent to the controller. Under this rule, the motor shall not be required to be located within sight.

- An additional disconnecting means shall be provided within 50 ft (15 m) of the motor and within sight if the disconnecting means by the controller cannot be locked in the open position and the motor is located out of sight. **(See Figure 14-32)**

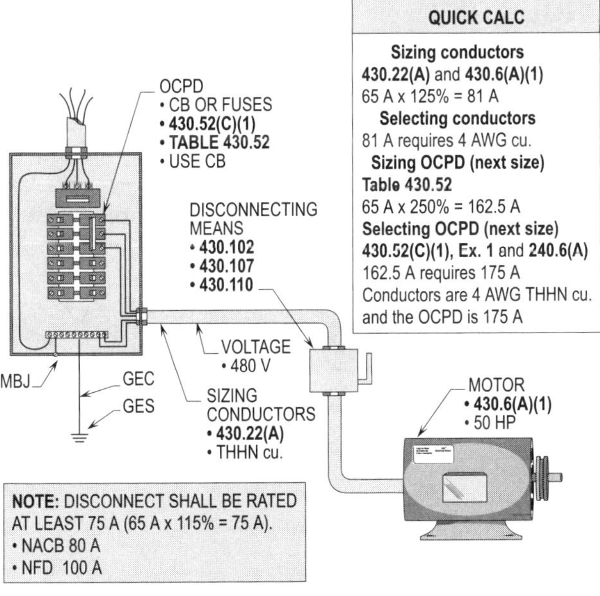

Figure 14-32. A disconnecting means shall be located within sight from the controller and shall disconnect the controller to allow personnel to service the motor without the danger of the branch circuit accidentally being energized. Also, see **Figures 18-1, 18-2,** and **18-24(a) through (d)**.

WELDER LOADS
ARTICLE 630, PARTS II AND III

There are three types of welders used in modern-day welding shops. The type used determines how the circuit elements shall be designed.

The procedure for calculating the full-load amps for welders is to obtain the duty-cycle factor and select the multiplier. The primary amps of the welder shall be multiplied by the multiplier to derive full-load amps to size the elements of the branch circuit. The following are the three types of welders that shall be permitted to be used:

- AC/DC arc welders
- Motor-generator arc welders
- Resistance welders

AC/DC ARC WELDERS
ARTICLE 630, PART III

Arc welders are used to perform welding operations through the medium of an arc drawn between the weld and a metal rod. The metal from the rod (electrode) is added to the weld, making the weld substantially strong.

CONDUCTORS
630.11(A)

When sizing the branch circuit conductors for AC/DC arc welders, the current-carrying capacity shall not be permitted to be less than the primary current of the welder times a duty cycle factor listed in **Table 630.11(A)**.

For example: What size THWN copper conductors are required to supply an AC transformer and DC rectifier arc welder rated at 68 amps with a 50 percent duty cycle?

Step 1:	Finding FLC **630.11(A)** Welder = 68 A	
Step 2:	Finding multiplier **630.11(A)** 50% = .71	
Step 3:	Calculating amps **630.11(A)** 68 A x 71% = 48.28 A	
Step 4:	Selecting conductors **Table 310.16** 48.28 A requires 8 AWG cu.	

Solution: The size THWN copper conductors are 8 AWG.

OVERCURRENT PROTECTION
630.12(A)

The welder's primary full-load current rating listed on the nameplate shall be selected at not more than 200 percent for sizing the overcurrent protection device. Branch-circuit conductors shall be protected at a rating not exceeding 200 percent of their allowable ampacities per **630.12(B)**. **(See Figure 14-33)**

For example: What size overcurrent protection device is required for the conductors supplying an AC transformer and DC rectifier arc welder rated at 68 amps with a 50 percent duty cycle?

Step 1:	Finding FLC **630.12(A), 630.12(B),** and **Table 310.16** Welder = 68 A x 71% = 48 A Conductors = 50 A	

Step 1: Finding FLC
630.12(A), 630.12(B), and **Table 310.16**
Welder = 68 A x 71% = 48 A
Conductors = 50 A

Step 2: Finding multiplier
630.12(A) and **630.12(B)**
Multiplier = 200%

Step 3: Calculating amps
630.12(A) and **630.12(B)**
68 A x 200% = 136 A

Step 4: Selecting OCPD
240.4(G), 240.6(A), and **630.12**
136 A requires 125 A

Step 5: Protecting conductors
630.22(B)
8 AWG THWN cu. = 50 A
50 A x 200% = 100 A
100 A requires 100 A

Solution: The size OCPD required is 100 amps based upon amps of conductors times 200 percent. The OCPD at the welder is 125 amps.

MOTOR-GENERATOR ARC WELDERS
ARTICLE 630, PART II

The arc welding principles are used to perform work for motor-generator arc welders. Two metal parts are welded together by a main electrode used to strike an arc that melts the electrode and supplies the metal necessary to join the metal parts together.

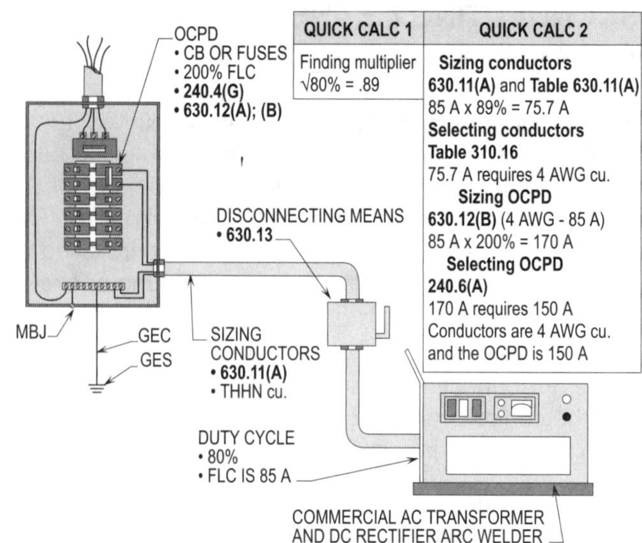

Figure 14-33. The welder's primary full-load current rating listed on the nameplate shall be selected at not more than 200 percent for the overcurrent protection device to be installed. Branch-circuit conductors shall be protected at a rating not exceeding 200 percent of their allowable ampacities.

CONDUCTORS
630.11(A)

When sizing the branch-circuit conductors for motor-generator arc welders, the current-carrying capacity shall not be permitted to be less than the rated primary current of the welder times a duty cycle factor listed in **Table 630.11(A)**.

For example: What size THWN copper conductors are required to supply a motor-generator arc welder rated at 76 amps having a 90 percent duty cycle?

Step 1: Finding FLC
630.11(A)
Welder = 76 A

Step 2: Finding multiplier
630.11(A)
90% = .96

Step 3: Calculating amps
630.11(A)
76 A x 96% = 72.96 A

Step 4: Selecting conductors
Table 310.16
72.96 A requires 4 AWG cu.

Solution: The size THWN copper conductors are 4 AWG.

OVERCURRENT PROTECTION
630.12(A)

The welder's primary full-load current rating listed on the nameplate shall be selected at not more than 200 percent for sizing the overcurrent protection device. Branch-circuit conductors shall be protected at a rating not exceeding 200 percent of their allowable ampacities per **630.12(B)**. **(See Figure 14-34)**

For example: What size overcurrent protection device is required for the conductors to supply a motor-generator arc welder rated at 76 amps having a 90 percent duty cycle?

Step 1: Finding FLC
630.12(A); 630.12(B); Table 310.16
Welder = 76 A x 96% = 73 A
Conductors = 85 A (4 AWG cu.)

Step 2: Finding multiplier
630.12(A); 630.12(B)
Multiplier = 200%

Step 3: Calculating amps
630.12(A); 630.12(B)
76 A x 200% = 152 A

Step 4: Selecting OCPD for welder
240.4(G); 240.6(A); 630.12(A)
152 A requires 150 A

Step 5: Selecting OCPD for conductors
630.12(B)
85 A x 200% = 170 A

Solution: The size OCPD required for the conductors is 150 amps.

RESISTANCE WELDERS
ARTICLE 630, PART III

Resistance welders use a heavy current that flows through the small area of material in contact with such material at a particular time. This type of welding is accomplished by pressing two metal parts together as they reach the molten state. Resistance welders do not add metal to the weld. Resistance welders may have high inrush current while welding together certain materials.

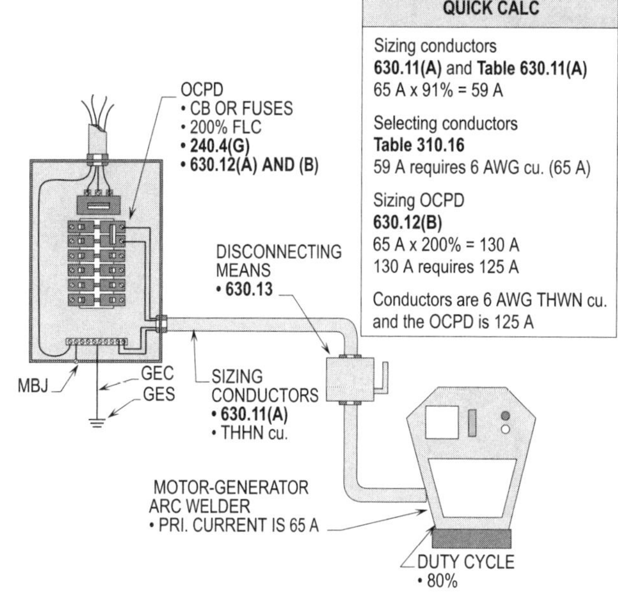

Figure 14-34. The welder's primary full-load current rating listed on the nameplate shall be selected at not more than 200 percent for the overcurrent protection device. Branch-circuit conductors shall be protected at a rating not exceeding 200 percent of their allowable ampacities.

CONDUCTORS
630.31(A)

When sizing the branch-circuit conductors for resistance welders, the current-carrying capacity shall not be permitted to be less than the primary current of the welder times a duty cycle factor listed in **Table 630.31(A)(2)**.

For example: What size THWN copper conductors are required to supply a resistance welder rated at 91 amps having a 40 percent duty cycle?

Step 1: Finding FLC
630.31(A)
Welder = 91 A

Step 2: Finding multiplier
630.31(A)
40% = .63

Step 3: Calculating amps
630.31(A)
91 A x 63% = 57.33 A

Step 4: Selecting conductors
Table 310.16
57.33 A requires 6 AWG cu.

Solution: The size THWN copper conductors are 6 AWG.

OVERCURRENT PROTECTION
630.32(A)

The welder's primary full-load current rating listed on the nameplate shall be selected at not more than 300 percent for sizing the overcurrent protection device. Branch-circuit conductors protected at a rating not exceeding this value shall be considered protected from overloading per **630.32(B)**. **(See Figure 14-35)**

For example: What size overcurrent protection device is required for the conductors supplying power to a resistance welder rated at 91 amps having a 40 percent duty cycle?

Step 1:   Finding FLC
630.32(A)
Welder = 91 A

Step 2:   Finding multiplier
630.32(A)
Multiplier = 300%

Step 3:   Calculating amps
630.32(A)
91 A x 300% = 273 A

Step 4:   Selecting OCPD
240.4(G), 240.6(A), and **630.12**
273 A requires 250 A

Step 5:   Protecting conductors
630.32(B)
6 AWG THWN cu. = 65 A
65 A x 300% = 195 A
195 A requires 175 A

Solution: The size OCPD required to protect the conductors is 175 amps. The size OCPD for the welder is 250 amps.

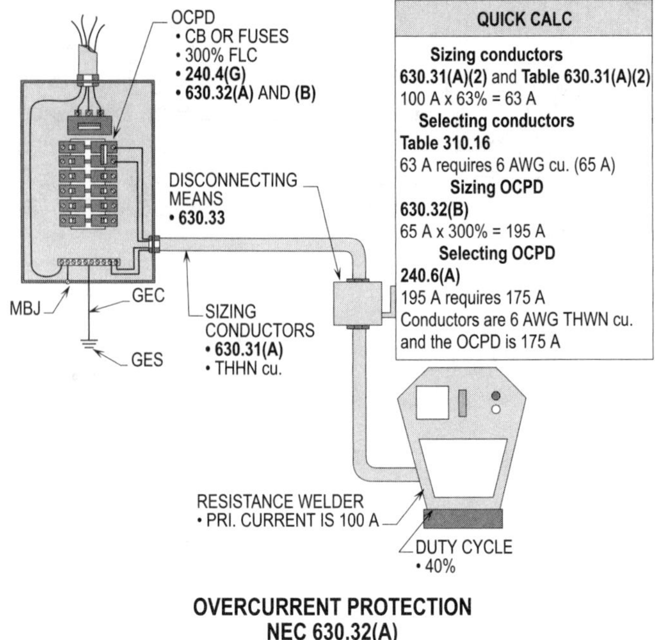

OVERCURRENT PROTECTION
NEC 630.32(A)

Figure 14-35. The welder's primary full-load current rating listed on the nameplate shall be selected at not more than 300 percent for sizing the overcurrent protection device. Branch-circuit conductors protected at a rating not exceeding this value shall be considered protected from overloading.

Chapter 14. Branch Circuits

Section   **Answer**

1. A minimum of 2-20 amp, _______ VA small appliance circuits are required to supply receptacle outlets that are located in the kitchen, pantry, breakfast room, and dining room.
   - (a) 1000
   - (b) 1200
   - (c) 1500
   - (d) 1800

2. At least 1-20 amp, _______ VA laundry circuit is required to supply receptacle outlets in the laundry room.
   - (a) 1000
   - (b) 1500
   - (c) 1800
   - (d) 2400

3. A tap to connect a cooktop and ovens can be made from a _______ amp branch circuit when the tap conductors are sized from the kW rating of each piece of cooking equipment per **Note 4** to **Table 220.55**.
   - (a) 30
   - (b) 40
   - (c) 50
   - (d) 60

4. Dryer equipment loads shall be at least _______ VA or the nameplate rating, whichever is larger, when selecting a VA for the branch circuit load.
   - (a) 1500
   - (b) 3000
   - (c) 4500
   - (d) 5000

5. A _______ amp branch circuit shall be permitted to be installed to supply fixed lighting units with heavy-duty lampholders in other than a dwelling unit(s) or utilization equipment in any occupancy.
   - (a) 30
   - (b) 40
   - (c) 50
   - (d) 60

6. Continuous lighting loads in commercial buildings shall be calculated at _______ percent of the total VA or amperage rating of the branch circuit.
   - (a) 80
   - (b) 100
   - (c) 110
   - (d) 125

7. Show window lighting loads shall be calculated at a minimum of _______ VA per outlet for branch circuits if the VA is unknown.
   - (a) 180
   - (b) 200
   - (c) 250
   - (d) 300

8. When installing four or fewer dryers, the load shall be calculated at _______ percent.
   - (a) 110
   - (b) 110
   - (c) 115
   - (d) 125

9. A fastened-in-place appliance shall be permitted to be connected to a general purpose circuit if its amp rating does not exceed _______ percent of the branch circuit.
   - (a) 50
   - (b) 60
   - (c) 75
   - (d) 80

__________ __________

10. Overcurrent protection devices rated over _____ amps are allowed to be terminated with conductors larger than 1 AWG with 75°C ampacities.

    (a) 50   (b) 60
    (c) 75   (d) 100

__________ __________

11. General purpose circuits shall be calculated at _____ VA per sq. ft based on the square footage of the dwelling unit.

    (a) 2   (b) 3
    (c) 3.5   (d) 4

__________ __________

12. The rating of any one cord-and-plug connected utilization equipment shall not exceed _____ percent of the branch circuit rating if connected to a general purpose circuit supplying two or more outlets.

    (a) 50   (b) 75
    (c) 80   (d) 100

__________ __________

13. Branch circuits larger than _____ amps shall supply only nonlighting outlet loads.

    (a) 20   (b) 30
    (c) 40   (d) 50

__________ __________

14. Cord-and-plug connected or permanently (hard-wired) connected receptacles rated over _____ VA are usually supplied by individual circuits.

    (a) 1440   (b) 1660
    (c) 2250   (d) 2440

__________ __________

15. Lighting track loads shall be calculated at _____ VA for each 2 ft of track to determine the load for branch circuits.

    (a) 150   (b) 180
    (c) 200   (d) 220

__________ __________

16. How many 15 amp circuits are allowed to supply power to the general purpose lighting and receptacle outlets in a 3000 sq. ft dwelling unit?

__________ __________

17. What is the demand load in VA for a branch circuit serving a 9 kW range?

__________ __________

18. What is the demand load in VA for a branch circuit serving 2-3 kW cooktops?

__________ __________

19. What is the demand load in VA for a branch circuit serving a 8.5 kW oven?

__________ __________

20. What is the demand load in VA for a branch circuit serving a 18 kW range?

__________ __________

21. What is the demand load in VA for a service or feeder serving a 10 kW, 18 kW, and 20 kW range?

__________ __________

22. What is the demand load in VA for a branch circuit serving a 9 kW cooktop, 8 kW, and 14 kW oven?

__________ __________

23. What size nonmetallic-sheathed cable tap is required for a 12 kW cooktop?

__________ __________

24. What is the demand load in VA for a branch circuit serving a 4500 VA dryer?

__________ __________

25. What is the demand load in VA for a branch circuit serving a 6500 VA dryer?

26. What is the total VA rating for a 15 amp, 120 volt, two-wire branch circuit supplying a continuous load?

27. What is the total VA rating for a 15 amp, 240 volt, two-wire branch circuit supplying a continuous load?

28. What is the load in amps for a 24 amp dishwasher used at continuous duty?

29. What is the load in amps for a 32 amp water heater used at continuous operation in a commercial building?

30. What size amperage rating is allowed for a 2 AWG copper conductor using the ampacities of the 60°C column and 75°C column?

31. What is the allowable ampacity for 8-10 AWG THHN copper conductors that are all current-carrying?

32. What is the allowable ampacity for 3-10 AWG THHN copper conductors that are in an ambient temperature of 105°F?

33. What is the allowable ampacity for 6-10 AWG THHN copper conductors that are all current-carrying in an ambient temperature of 105°F?

34. What is the amperage rating of an 2800 VA calculated load connected to a 120 volt, single-phase branch circuit?

35. What is the amperage rating of an 9250 VA calculated load connected to a 208 volt, three-phase branch circuit?

36. How many outlets are permitted to be connected to a 20 amp branch circuit used at noncontinuous operation?

37. How many outlets are permitted to be connected to a 20 amp branch circuit used at continuous operation?

38. What size overcurrent protection device is allowed to serve a continuous operated load of 13.5 amps and noncontinuous load of 3 amps in an ambient temperature of 155°F?

39. What is the lighting load in amps for a branch circuit supplying 10-1.5 amp ballasts that serve 24-F25, CW lamps used at continuous operation?

40. What size overcurrent protection device and THHN copper conductors are required for 12 outlets used at noncontinuous operation to supply luminaires used at 120 volts?

41. What size overcurrent protection device and THHN copper conductors are required for 10 outlets used at continuous operation to supply luminaires?

42. How many outlets are permitted to be installed on a 20 amp branch circuit supplying lighting loads in a show window?

43. What is the lighting load in VA for 70 ft of show window? (Lighting is used at continuous duty.)

_____________ _____________ **44.** What is the show window lighting load in VA for 70 medium-base lamps?
(Lamps are used at continuous duty.)

_____________ _____________ **45.** What is the VA rating for 25 ft of lighting track?
(Lighting is used at noncontinuous duty.)

15

Feeders

Feeders are all circuit conductors between the service equipment, the source of a separately derived system or other power supply source and the final overcurrent protection devices protecting the branch-circuit conductors and equipment. Feeders may be installed to supply subpanels in dwelling units, apartments, and commercial and industrial occupancies. Feeders may also be installed in commercial and industrial occupancies and routed through the plant, where taps are sometimes made to these conductors and connected to supply panelboards, disconnect switches, control centers, and other types of electrical equipment.

LOADS
ARTICLE 220, PART III

The rules and regulations of **Article 220, Part III** shall be used to size and select the feeder conductors, overcurrent protection devices, and other pertinent elements. Demand factors shall be permitted to be applied to loads that are grouped, and percentages shall be permitted to be applied that will reduce the load. These percentages shall be based on how the equipment is used in the electrical system.

VOLTAGE
220.5(A)

The ampacity of electrical components for feeders shall be determined by the voltage ratings listed in **220.5(A)**. Unless another voltage is specified, the following voltage ratings shall be used to calculate feeder loads:

- 120 volt
- 120/240 volt
- 208 Y/120 volt
- 240 volt
- 347 volt
- 480 Y/277 volt
- 480 volt
- 600 Y/347 volt
- 600 volt

UNGROUNDED (PHASE) CONDUCTORS 215.2(A)(1) AND 215.3

Ungrounded (phase) conductors that are used in feeders shall be sized and selected based on the total amps or the total VA divided by the voltage of the circuit. The total amps or VA of the feeder loads shall be calculated at continuous or noncontinuous operation or a combination of both. Also, loads with demand factors shall be calculated at percentages less than those used for continuous or noncontinuous loads, and such percentages shall be derived for how the loads are used on the feeder. The smallest size feeder shall not be less than 30 amps per **215.2(A)(2)**. To comply with this minimum rating, 10 AWG copper or 8 AWG aluminum conductors shall be used, per asterisk below **Table 310.16** and **240.4(D)**.

Ungrounded (phase) conductors shall be considered current-carrying conductors for derating purposes when applying adjustment or correction factors per **310.15(B)(2)(a)** and **Table 310.16**.

CALCULATING AMPS

If the VA is known, amps may be found by dividing the VA by the configuration of voltage for the feeder.

For example: What is the ampacity for a feeder load of 18,400 VA that is supplied by a 120/240 volt, single-phase system?

Step 1:   Finding amperage
$I = VA \div A$
$I = 18{,}400 \text{ VA} \div 240 \text{ V}$
$I = 77 \text{ A}$

Solution: The feeder ampacity is 77 amps.

For example: What is the ampacity for a feeder load of 18,400 VA supplied by a 120/208 volt, three-phase, four-wire system?

Step 1:   Finding amperage
$I = VA \div (V \times 1.732)$
$I = 18{,}400 \text{ VA} \div 360 \text{ V}$
$I = 51.1 \text{ A}$

**Solution: The feeder ampacity is 51 amps.
Note: The .1 amp is dropped
per 220.5(B).**

The feeder components shall be selected from the 51 amps per phase. The 51 amps per phase shall be connected to Phase A, Phase B, and Phase C. This calculation shall be verified by dividing the VA by the voltage per phase and dividing the total by 3.

For example: What is the ampacity for a feeder load of 18,400 VA supplied by a 120/208 volt, three-phase system?

Step 1:   Finding amperage
$I = VA \div V$
$I = 18{,}400 \text{ VA} \div 120 \text{ V}$
$I = 153.3$
$I = 153.3 \text{ A} \div 3$
$I = 51 \text{ A}$

Solution: The feeder ampacity is 51 amps.

See Figure 15-1 for calculating amps based on volt-amps and voltage.

POWER FACTOR

Power factor (PF) is the ratio of actual power used in a circuit to the apparent power drawn from the line. Actual power is the true power used to produce heat or work. Actual power is also known in the industry as real, true, or useful power.

Terms:

- Power factor   = PF
- Watts   = W
- Volts   = V
- Amps   = I

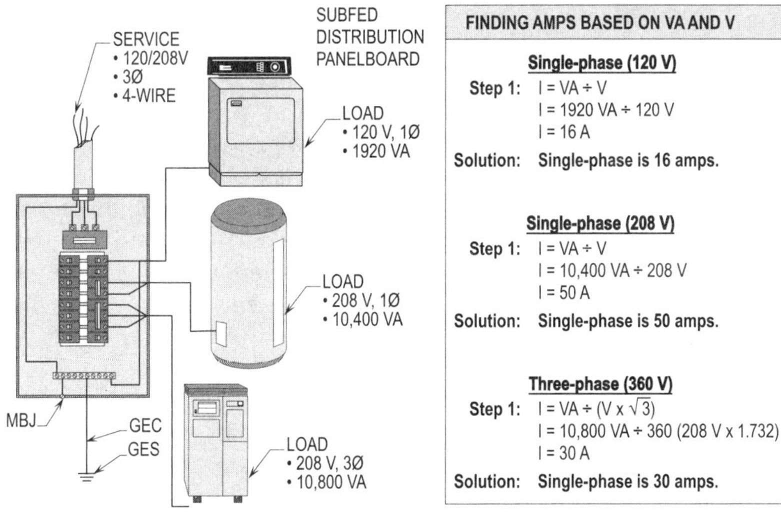

Figure 15-1. Finding the amps of a feeder based upon volt-amps and voltage.

ACTUAL POWER – SINGLE-PHASE

The actual power in watts in a pure resistance circuit shall be found by multiplying volts times amps.

For example: A 240 volt, single-phase feeder with a 51 amp load has a load of 12,240 watts.

Step 1: $W = V \times I$
$W = 240\ V \times 51\ A$
$W = 12{,}240$

Solution: Actual power equals 12,240 watts.

If the circuit above has a poor power factor of 75 percent because of inductive loads, the load in amps shall be calculated as follows:

For example: A 240 volt, single-phase feeder of 51 amps having a power factor of 75 percent has a load of 9180 watts.

Step 1: $W = V \times I \times PF$
$W = 240\ V \times 51\ A \times 75\%$
$W = 9180$

Solution: Actual power equals 9180 watts.

ACTUAL POWER – THREE-PHASE

The actual power in watts in a three-phase circuit with inductive loads shall be found by multiplying $V \times \sqrt{3} \times I \times PF$.

For example: The actual power in watts for a three-phase circuit with a load of 51 amps that is supplied by a 208 volt circuit with a power factor of 75 percent is 13,770 watts. (Using 360 V)

Step 1: $W = (V \times \sqrt{3}) \times I \times PF$
$W = (208\ V \times 1.732) \times 51\ A \times 75\%$
$W = 13{,}770$

Solution: Actual power equals 13,770 watts.

Note, the actual power is 18,360 watts (208 V x 1.732 x 51 A = 18,360 W) if the power factor is 100 percent instead of 75 percent.

See Figure 15-2 for calculating amps for a feeder having poor power factor.

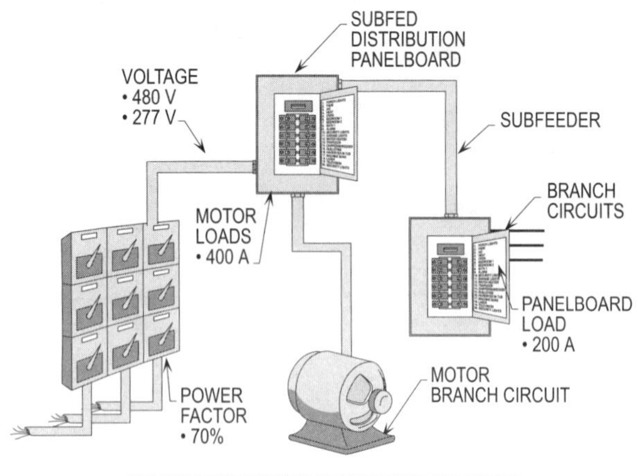

Figure 15-2. Calculating amps for a feeder based upon poor power factor.

NONCONTINUOUS OPERATED LOADS
215.2(A)(1) AND 215.3

Noncontinuous operated loads shall be calculated at 100 percent or where appropriate demand factors shall be permitted to be applied to specific loads. Based upon their condition of use, the total VA rating for such loads may be derived. Noncontinuous loads are classified as loads that operate for a period of time less than three hours per **Article 100**. Portable or fixed cord-and-plug connected or permanently connected loads may be classified as noncontinuous loads.

For example: What is the VA rating for a connected load of 34,500 VA to size the elements of the feeder at noncontinuous operation?

Step 1:   Finding VA
215.2(A)(1); 215.3
VA = VA x 100%
VA = 34,500 VA x 100%
VA = 34,500

Solution: **The total VA for the feeder is 34,500 VA.**

CONTINUOUS OPERATED LOADS
215.2(A)(1) AND 215.3

Continuous operated loads are classified as loads that operate for a period of three hours or more. Continuous loads do not operate at varying or intermittent operation. A continuous operated load shall be supplied continuously for a period of three hours or more without the circuit being interrupted. An industrial processing machine used in a facility to perform a work task for a work day of eight hours falls under such use and classification.

The total VA rating for sizing conductors to such a machine shall be obtained by multiplying the continuous load by 125 percent. Feeder conductors shall be increased in size due to voltage drop, ambient temperature, or too many current-carrying conductors in a raceway or cable.

For example: What is the VA rating for a continuous operated processing machine with a connected load of 34,500 VA ?

Step 1:   Finding VA
215.2(A)(1); 215.3
VA = VA x 125%
VA = 34,500 VA x 125%
VA = 43,125 VA

Solution: **The total VA for the feeder is 43,125 VA.**

See Figure 15-3 for calculating the load in amps based upon continuous or noncontinuous operation.

DEMAND FACTORS
ARTICLE 220, PART II

Demand factors shall be permitted to be applied to the VA rating for specific loads, depending on their conditions of use. The percentages for which demand factors shall be permitted to be applied are listed in the NEC. Feeders supplying loads under specific condition of use shall be permitted to be reduced by applying demand factors to obtain the total rating. Loads shall be permitted to be reduced by applying a demand factor are as follows:

Types of loads per NEC:

• Lighting	- **Table 220.42**
• Receptacles	- **Table 220.44** and **220.14(I)**

- Dryers   - **Table 220.54**
- Ranges   - **Table 220.55**
- Kitchen equipment   - **Table 220.56**
- Dwelling units   - **220.82(B) and (C)**
- Existing dwelling units   - **220.83**
- Multifamily dwelling units   - **Table 220.84**
- Schools   - **Table 220.86**
- Existing loads   - **220.87**
- Restaurants   - **Table 220.88**
- Farms   - **Table 220.102** and **Table 220.103**

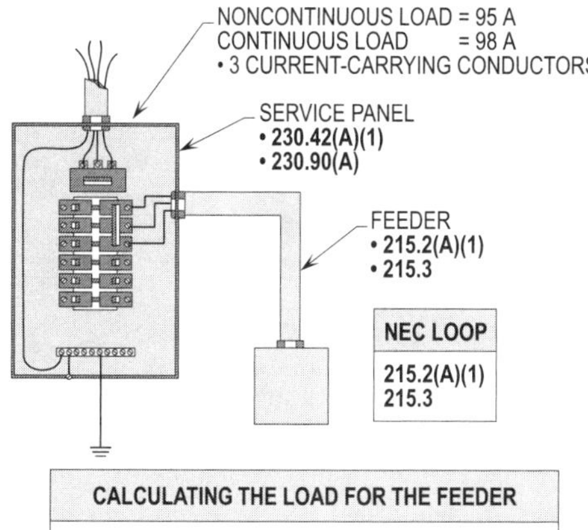

Figure 15-3. Calculating the load in amps for a feeder based upon continuous and noncontinuous operation.

APPLYING DEMAND FACTORS FOR RECEPTACLE LOADS 220.44, TABLE 220.44, AND TABLE 220.42

The VA rating for receptacle loads shall be calculated at 100 percent for the first 10,000 VA, and the remaining VA shall be calculated at 50 percent per **Table 220.44**. **Table 220.44** shall only be permitted to be applied to receptacles that are used at noncontinuous operation. The demand factors of **Table 220.42** shall be permitted to be used to determine the load for receptacle outlets that are figured at 180 VA and then added to the lighting load.

For example: What is the load for a number of general purpose receptacles used at noncontinuous operation that are utilized to supply a connected load of 24,800 VA?

Step 1:   Finding VA
       Table 220.44
       First 10,000 VA at 100%
       10,000 VA x 100%   = 10,000 VA
       Remaining VA at 50%
       14,800 x 50%   = 7,400 VA
       Total load   = 17,400 VA

Solution: The total VA load is 17,400 VA.

See **Figure 15-4** for calculating the load in amps for a feeder with loads having demand factors.

GROUNDED (NEUTRAL) CONDUCTOR ARTICLE 220, PART III AND ARTICLE 310

The grounded (neutral) conductor shall be intentionally grounded at the service equipment. The grounded (neutral) conductor shall be sized to carry the maximum unbalanced current. The largest load between the grounded (neutral) conductor and any one ungrounded (phase) conductor is the maximum unbalanced current that shall be carried. A demand factor shall be permitted be applied to the grounded (neutral) conductor, under specific conditions of use.

SIZING 220.61(A) THRU (C)

The feeder grounded (neutral) conductor load shall be the maximum unbalanced load connected between the grounded (neutral) conductor and any one ungrounded (phase) conductor. The first 200 amps of neutral current shall be calculated at 100 percent. All resistive loads on the grounded (neutral) conductor exceeding 200 amps shall be permitted to have a demand factor of 70 percent applied, and this value added to the first 200 amps taken at 100 percent. All inductive neutral current shall be calculated at 100 percent with no demand factor applied. The feeder grounded (neutral) conductor load shall be 70 percent of the demand load for cooking equipment or dryer loads that, for example, are installed in dwelling units. **(See Figure 15-5)**

The neutral current in amps for a three-wire, two-phase or five-wire, two-phase system shall be multiplied by 140 percent which is derived by taking the square root of 2. By multiplying the grounded (neutral) conductor amps by 140 percent ($\sqrt{2}$ = 141%), the unbalanced current in the grounded (neutral) conductor will be collected from all ungrounded (phase) conductors. The grounded (neutral) conductor shall be multiplied by 140 percent instead of 141 percent (rounded down), based on the ungrounded (phase) conductor with the highest ampacity per **220.61(A), Ex**. Basically, if calculated using this procedure, the grounded (neutral) conductors are not overloaded because 120 volt loads are switched in and out on the circuits at different intervals of time.

For example: What is the grounded (neutral) conductor load for a five-wire, two-phase feeder with each phase rated at 150 amps?

Step 1: Finding amperage
220.61(A), Ex.
150 A x 140% = 210 A

Solution: The grounded (neutral) conductor load is 210 amps.

What is the demand load for the 82 kW cooking equipment load?

Finding demand load

Step 1: Calculating percentage
Table 220.56
17 pieces allows 65%

Step 2: Applying demand factors
Table 220.56 and 220.56
82 kW x 65% = 53.3 kW

Solution: The demand load is 53.3 kV.

Note: 53.3 kW is greater than the two largest loads of 18 kVA.

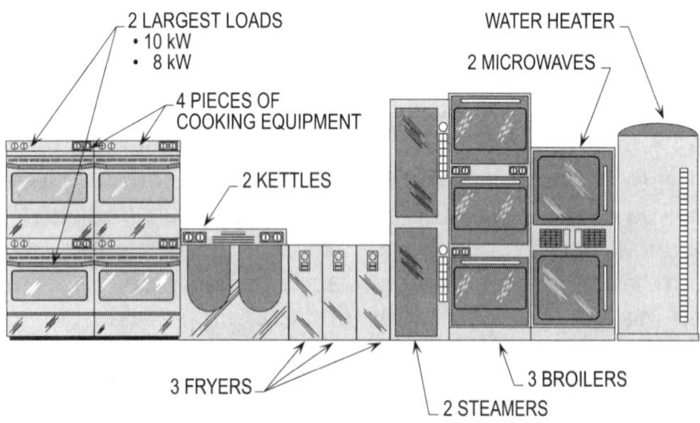

Figure 15-4. Calculating the load in amps for a feeder with 17 pieces of commercial cooking equipment.

For example: What is the neutral load for a feeder with an inductive load of 400 amps?

Step 1: Calculating neutral load
220.61(C) and **310.15(B)(4)(c)**
First 200 A x 100%   = 200 A
Remaining 200 A x 100%   = 200 A
Total load   = 400 A

Solution: The total inductive neutral load is 400 amps.

For example: What is the neutral load for a feeder with a resistive load of 400 amps?

Step 1: Calculating neutral load
220.61(B) and **310.15(B)(4)(c)**
First 200 A x 100%   = 200 A
Remaining 200 A x 70%   = 140 A
Total load   = 340 A

Solution: The total resistive neutral load is 340 amps.

For example: What is the neutral load for a 9 kW range?

Step 1: Finding demand
Table 220.54, Col. C
9 kW = 8 kW

Step 2: Calculating neutral load
220.61(B)
8 kW x 70% = 5.6 kW

Solution: The total neutral range load is 5.6 kW

Figure 15-5. Calculating the feeder grounded (neutral) conductor load for an inductive, resistive, and range load.

UTILIZING THE GROUNDED (NEUTRAL) CONDUCTOR 310.15(B)(4)

Section **310.15(B)(4)** is divided into three main subdivisions to explain the loading conditions and use of the grounded (neutral) conductors. The following are three subdivisions to be used to derive loading conditions based on use as follows:

• Subdivision (a) – unbalanced current
• Subdivision (b) – common conductor carries
• Subdivision (c) – to determine if current-carrying

SUBDIVISION (a) 310.15(B)(4)(a)

The grounded (neutral) conductor shall not be considered a current-carrying conductor when carrying only the unbalanced current from other ungrounded (phase) conductors. When circuits are properly balanced, the grounded (neutral) conductor carries very little current.

When sizing the load for a two-wire circuit, the grounded (neutral) conductor carries the same amount of current as the ungrounded (phase) conductor. This type of installation has no unbalanced load, therefore the grounded (neutral) conductor carries full current.

For example: What is the grounded (neutral) conductor load for a single-phase, 120 volt, two-wire circuit supplying a load of 14 amps?

Step 1: Finding amperage
220.61 and **310.15(B)(4)(a)**
Ungrounded (phase) conductor = 14 A
Grounded (neutral) conductor = 14 A

Solution: The grounded (neutral) conductor carries a load of 14 amps.

When sizing the load for a three-wire circuit, the grounded (neutral) conductor carries the unbalanced load of the two ungrounded (phase) conductors. This type of installation has an unbalanced load unless both ungrounded (phase) conductors pull the same amount of current on each ungrounded (phase) conductor.

For example: What is the unbalanced grounded (neutral) conductor load for a three-wire circuit carrying 64 amps and 52 amps on the ungrounded (phase) conductors?

Unbalanced condition

Step 1: Finding amperage
220.61 and **310.15(B)(4)(a)**
Ungrounded (phase) conductor
Phase A = 64 A
Ungrounded (phase) conductor
Phase B = - 52 A
Total = 12 A

Balanced condition

Step 1: Finding amperage
220.61 and **310.15(B)(4)(a)**
Ungrounded (phase) conductor
Phase A = 64 A
Ungrounded (phase) conductor
Phase B = - 64 A
Total = 0 A

Solution: The grounded (neutral) conductor load is 12 amps for the unbalanced condition and 0 amps for the balanced condition.

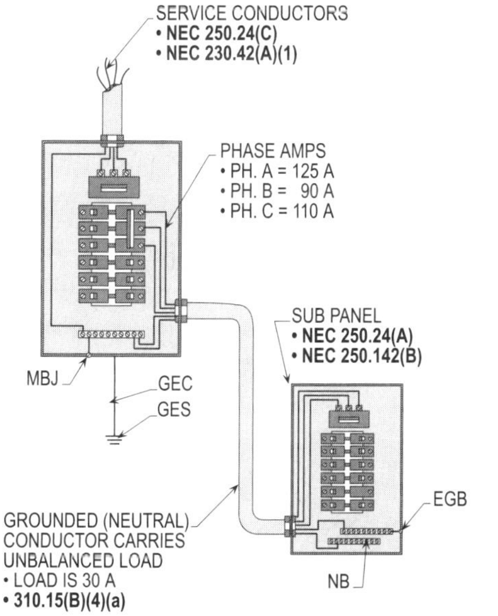

FINDING GROUNDED (NEUTRAL) CONDUCTOR LOAD

Step 1: Calculating grounded (neutral) conductor load
220.61; 310.15(B)(4)(a)

$$A = \sqrt{PH\ A^2 + PH\ B^2 + PH\ C^2 - (PH\ A \times PH\ B) - (PH\ B \times PH\ C) - (PH\ C \times PH\ A)}$$

$$A = \sqrt{125^2 + 90^2 + 110^2 - (125\ A \times 90\ A) + (90\ A \times 110\ A) + (110\ A \times 125\ A)}$$

$$A = \sqrt{15{,}625\ A + 8{,}100\ A + 12{,}100\ A - 11{,}250\ A + 9{,}900\ A + 13{,}750\ A}$$

$$A = \sqrt{35{,}825\ A - 34{,}900\ A}$$

$$A = \sqrt{925\ A}$$

$$A = 30\ A$$

Solution: The grounded (neutral) conductor load is 30 amps.

Note: If phases A and B were lost, the grounded (neutral) conductor current would be 110 amps and if Phase C were lost, the grounded (neutral) conductor current would be 35 amps (PH A = 125 A - PH B = 90 A = 35 A).

CALCULATING NEUTRAL CURRENT FOR THREE-PHASE CIRCUITS

Figure 15-6. Calculating grounded (neutral) conductor current for three-phase circuits.

CALCULATING NEUTRAL CURRENT FOR THREE-PHASE CIRCUITS

A specific formula shall be used to calculate the grounded (neutral) conductor current for three-phase feeders. Where currents on Phases A, B, and C are of different values, the grounded (neutral) conductor current shall be calculated as shown in **Figure 15-6.**

SUBDIVISION (b)
310.15(B)(4)(b)

The grounded (neutral) conductor of a three-wire, 120/208 volt feeder shall be the same size as the ungrounded (phase) conductors for a feeder derived from a four-wire, 120/208 volt system. The reason for this is the grounded (neutral) conductor of a three-wire circuit consisting of two-phase conductors. A three-wire system carries approximately the same amount of current as the ungrounded (phase) conductor. Therefore, a reduction in ampacity is not allowed.

For example: What is the grounded (neutral) conductor load for a 120/208 volt, single-phase circuit taken from a four-wire wye, three-phase system with 190 amps on phase A, 170 amps on phase B, and 90 amps for the neutral?

Step 1: Finding ampacity of neutral
220.61; 310.15(B)(4)(b)
Ungrounded (phase) conductor
A = 190 A
Ungrounded (phase) conductor
B = 170 A
Grounded (neutral) conductor
N = 190 A

Solution: **The grounded (neutral) conductor load is 190 amps based on the largest ungrounded (phase) conductor.**

SUBDIVISION (c)
310.15(B)(4)(c)

The grounded (neutral) conductor for a four-wire, three-phase system supplying nonlinear loads shall be the same size as the ungrounded (phase) conductors. The grounded (neutral) conductor shall be considered a current-carrying conductor due to the harmonic currents generated by these loads.

For example, harmonic loads generated by ballasts are approximately 25 percent per phase in a four-wire wye, three-phase system. Such harmonic currents are collected by the grounded (neutral) conductor from the ungrounded (phase) conductors in a four-wire system because they are all in phase. Because these loads produce third harmonics, harmonic currents present in the grounded (neutral) conductor add up to 75 percent that is derived from 25 percent of such current on each ungrounded (phase) conductor. Under this condition of use, the grounded (neutral) conductor shall not be permitted to be reduced by 70 percent for current from 120 volt loads exceeding 200 amps per **220.61(C).**

HARMONIC CURRENT
310.10, FPN (2)

Harmonic currents usually appear in power distribution equipment that supplies nonlinear loads. There are two basic types of nonlinear loads – single-phase and three-phase. Single-phase nonlinear loads are found in offices, while three-phase nonlinear loads are dominant in industrial plants. However, there are other nonlinear loads that present problems to the electrical system because of their harmonic content.

EFFECTS ON NEUTRAL
220.61(C)(2), FPN 2

In electrical systems with many single-phase nonlinear loads, the grounded (neutral) conductor current may actually exceed the ungrounded (phase) current. Excessive overheating may occur because there is no overcurrent protection device in the grounded (neutral) conductor to limit the current to a safe value.

In a four-wire system with single-phase nonlinear loads, there are certain odd-numbered harmonics called triplens. These odd multiples of harmonic current are known as the 3rd, 9th, 15th, etc., and they do not cancel but add together in the grounded (neutral) conductor of a wye connected circuit.

Design Tip: Under normal conditions of use for a balanced linear load, the fundamental 60 Hz portion of the ungrounded (phase) conductor currents will cancel in the grounded (neutral) conductor. Therefore, there is not an excessive heating problem in the grounded (neutral) conductor.

SIZING GROUNDED (NEUTRAL) CONDUCTOR ELEMENTS

Grounded (neutral) conductors, busbars, and connecting lugs shall be sized to carry the full rating of ungrounded (phase) conductor currents. Such elements easily become overloaded due to grounded (neutral) conductors being overloaded with the triplen harmonics.

Design Tip: Triplen harmonics are in addition to the normal grounded (neutral) conductor current. [See **FPN 2** to **220.61(C)(2)**]

DEMAND FACTORS
220.61(B)

A demand factor of 70 percent shall be applied to all grounded (neutral) conductor loads exceeding 200 amps for nonlinear loads. Nonlinear related loads shall be calculated at 100 percent.

For example: What is the load for the grounded (neutral) conductor if it exceeds 200 amps and has more than 50 percent of its load harmonically related? The ungrounded (phase) conductors are carrying a total grounded (neutral) conductor load of 275 amps.

Step 1:   Finding amperage
310.15(B)(4)(c)
Ungrounded (phase) conductors
• 275 amps

Step 2:   Calculating amperage
220.61(C)
First 200 A x 100%   = 200 A
Next 75 A x 100%   = 75 A
Total   = 275 A

Solution: The grounded (neutral) conductor load is 275 amps.

The grounded (neutral) conductor shall be considered a current-carrying conductor because of the harmonic currents generated by these loads, and **310.15(B)(2)(a)** shall be applied for four or more current-carrying conductors in a conduit, cable, etc.

For example: What is the grounded (neutral) conductor load for 120 volt loads having harmonic currents of 400 amps per phase?

Step 1:   Finding amperage
310.15(B)(4)(c)
Ungrounded (phase) conductors
• 400 A

Step 2:   Calculating amperage
220.61(C)
400 A x 100% = 400 A

Solution: The grounded (neutral) conductor load is 400 amps. Note: No reduction of ampacity shall be permitted due to harmonic currents.

VOLTAGE DROP
ARTICLE 215

Conductors are sometimes increased in size to prevent excessive voltage drop (VD) due to long runs between the overcurrent protection devices and the load served. Due to the long runs of feeder conductors, they shall be increased in size to compensate for poor voltage drop. The voltage drop on the feeder conductors should not exceed 2 to 3 percent at the farthest outlet supplying power to the loads. The voltage drop on the feeder conductors should not exceed 5 percent overall.

SINGLE-PHASE CIRCUITS
215.2(A)(3), FPN 2

The following formula shall be applied when calculating the voltage drop in a two-wire or three-wire, single-phase or three-phase feeder:

Resistive method

• $VD = 2 \times R \times L \times I \div 1000$

Circular-mil method

• $VD = 2 \times R \times L \times I \div CM$

The following values shall be used when calculating the voltage drop:

VD = voltage drop

R = resistivity for conductor material
Use **Chapter 9, Table 8, Columns 6** or **8** (uncoated ohm/MFT), or use 12.9 for copper and 21.2 for aluminum

L = one-way length of circuit conductor in feet

For example: Using the resistivity method to determine the voltage drop and size elements of a single-phase feeder having the following characteristics:

- VD is held to 3%
- Voltage is 240
- Length is 300 ft.
- Load is 180 amps
- OCPD is 200 amps
- Conductors are 3/0 AWG THWN copper

Finding the VD using the resistivity method

Step 1: Selecting percentage
215.2(A)(3), FPN 2
Feeder = 3%

Step 2: Calculating VD
215.2(A)(3), FPN 2 and **Table 8, Ch. 9**
VD = 2 x R x L x I ÷ 1000
VD = 2 x .0766 x 300 x 180 ÷ 1000
VD = 8.2728

Step 3: Calculating allowable VD
VD = supply V x 3%
VD = 240 V x 3%
VD = 7.2 V

Step 4: Checking percentage
215.2(A)(3), FPN 2
% = VD ÷ V
% = 8.2728 ÷ 240
% = .03447 or 3.447

Solution: The voltage drop rating of 8.2728 is greater than 7.2 V and a larger conductor shall be used to reduce the 3.447% to 3% or less.

Lowering VD using larger 4/0 AWG conductor

Step 1: Selecting percentage
215.2(A)(3), FPN 2
Feeder = 3%

Step 2: Calculating VD
215.2(A)(3), FPN 2 and **Table 8, Ch. 9**
VD = 2 x R x L x I ÷ 1000
VD = 2 x .0608 x 300 x 180 ÷ 1000
VD = 6.5664

Step 3: Checking percentage
215.2(A)(3), FPN 2
% = 6.5664 V ÷ 240 V
% = .02736 or 2.736

Solution: The voltage drop rating of 6.5664 volts is less than 7.2 volts, which is well below the 3% limit. The 4/0 AWG conductors are large enough to reduce the voltage drop to 3% or less.

FIGURE LOOP: FOR SIZING GROUNDED (NEUTRAL) CONDUCTOR AND EGC TO CORRECT VD, SEE FIGURES 9-35 AND 15-9.

SINGLE-PHASE CIRCUITS
NEC 215.2(A)(3), FPN 2

Figure 15-7(a). Calculating the VD using the resistivity method and using the larger conductor.

For example: Using the CM method to determine the voltage drop and size elements of a single-phase feeder having the following characteristics:

- VD is held to 3%
- Voltage is 240
- Length is 300 ft.
- Load is 180 amps
- OCPD is 200 amps
- Conductors are 3/0 AWG THWN copper

Finding the VD using the CM method

Step 1: Selecting percentage
215.2(A)(3), FPN 2
Feeder = 3%

Step 2: Calculating VD
215.2(A)(3), FPN 2 and **Table 8, Ch. 9**
VD = 2 x R x L x I ÷ CM
VD = 2 x 12.9 x 300 x 180 ÷ 167,800
VD = 8.3027

Step 3: Calculating allowable VD
VD = supply V x 3%
VD = 240 V x 3%
VD = 7.2 V

Step 4: Checking percentage
215.2(A)(3), FPN 2
% = VD ÷ V
% = 8.3027 ÷ 240
% = .0345 or 3.45

Solution: The voltage drop rating of 8.3027 is greater than 7.2 V and a larger conductor shall be used to reduce the 3.45% to 3% or less.

Lowering VD using larger 4/0 AWG conductor

Step 1: Selecting percentage
215.2(A)(3), FPN 2
Feeder = 3%

Step 2: Calculating VD
215.2(A)(3), FPN 2 and **Table 8, Ch. 9**
VD = 2 x R x L x I ÷ 211,600
VD = 2 x 12.9 x 300 x 180 ÷ 211,600
VD = 6.584

Step 3: Checking percentage
215.2(A)(3), FPN 2
% = 6.584 V ÷ 240 V
% = .0274 or 2.74

Solution: The voltage drop rating of 6.584 volts is less than 7.2 volts, which is well below the 3% limit. The 4/0 AWG conductors are large enough to reduce the voltage drop to 3% or less.

FIGURE LOOP: FOR SIZING GROUNDED (NEUTRAL) CONDUCTOR AND EGC TO CORRECT VD, SEE FIGURES 9-35 AND 15-9. FOR CALCULATING VD IN 3Ø CIRCUITS, SEE FIGURE 15-8.

SINGLE-PHASE CIRCUITS
NEC 215.2(A)(3), FPN 2

Figure 15-7(b). Calculating the VD using the CM method and using the larger conductor.

I = current in conductor in amperage

CM = conductor area in circular mils
See **Chapter 9, Table 8**

1000 = length of conductors based on **Table 8, Chapter 9**

Conductors routed at greater lengths will have the resistance of each conductor increased that will oppose the flow of current. By running larger conductors, the diameter of each conductor is increased in size, creating a greater path for the flow of current and less opposition to the movement of electrons that keeps the voltage high at the conductor end.

As listed above, there are two methods that are used by designers and installers to calculate voltage drop in a feeder. One is the resistivity concept, considered the most accurate. The second is the circular mil (CM) method, which has been used for many years. The resistivity method is utilized by selecting the resistivity of the conductors. The CM method consists of selecting the CM rating of the conductors from **Table 8, Ch. 9** in the NEC. These values, whichever chosen, are inserted into the formula with other data and the voltage drop calculated. **[See Figures 15-7(a) and (b)]**

THREE-PHASE CIRCUITS
215.2(A)(3), FPN 2

The voltage drop for three-phase circuits shall be calculated by multiplying the voltage drop by .866. The same formula for finding voltage drop in single-phase circuits shall be used to determine the voltage drop in three-phase circuits. Voltage drop multiplied by .866 is found by dividing $\sqrt{3}$ by 2 (1.732 ÷ 2 = .866). The .866 is basically produced from the additional conductor (third conductor), which is derived from a three-phase system instead of a two-phase system. **(See Figure 15-8)**

FINDING THE SIZE EQUIPMENT GROUNDING CONDUCTOR
250.122(B)

Where current-carrying conductors have to be larger to compensate for voltage drop, the equipment grounding conductors shall be adjusted proportionally to the circular-mil area.

Design Tip: If a single equipment grounding conductor is run with multiple circuits in the same raceway, it shall be sized for the largest overcurrent device protecting the conductors in the raceway or cable.

See Figure 15-9 for a detailed procedure for calculating and selecting the size equipment grounding conductor to be installed.

For example: Using the resistivity method to determine the voltage drop and size elements of a three-phase feeder having the following characteristics:

- VD is held to 3%
- Voltage is 480/277
- Length is 700 ft.
- Load is 180 amps
- OCPD is 200 amps
- Conductors are 4/0 AWG THWN copper

Finding the VD using the resistivity method

Step 1: Selecting percentage
215.2(A)(3), FPN 2
Feeder = 3%

Step 2: Calculating VD
215.2(A)(3), FPN 2 and Table 8, Ch. 9
VD = 2 x R x L x I ÷ 1000
VD = 2 x .0608 x 700 x 180 ÷ 1000
VD = 15.3216

Step 3: Applying .866 for three-phase
VD = VD x .866
VD = 15.3216 x .866
VD = 13.269

Step 4: Calculating allowable VD
VD = supply V x 3%
VD = 480 V x 3%
VD = 14.4 V

Step 5: Checking percentage
215.2(A)(3), FPN 2
% = VD ÷ V
% = 13.269 ÷ 480
% = .0276 or 2.76

Solution: The voltage drop rating of 13.269 is less than 14.4 V, therefore the 4/0 AWG THWN copper shall be used.

THREE-PHASE CIRCUITS
NEC 215.2(A)(3), FPN 2

Figure 15-8. Calculating the VD using the resistivity method for a three-phase feeder.

FINDING THE SIZE GROUNDED (NEUTRAL) CONDUCTOR
240.23

Where a change in the size of the ungrounded (phase) conductors occurs due to voltage drop, a similar change may have to be made in the size of the grounded (neutral) conductor.

For example, if the ungrounded (phase) conductor supplying a 120 volt load of 175 amps were increased from 2/0 AWG THWN copper to 3/0 AWG THWN copper due to voltage drop problems, the grounded (neutral) conductor would have to be increased from 2/0 AWG copper to 3/0 AWG copper also.

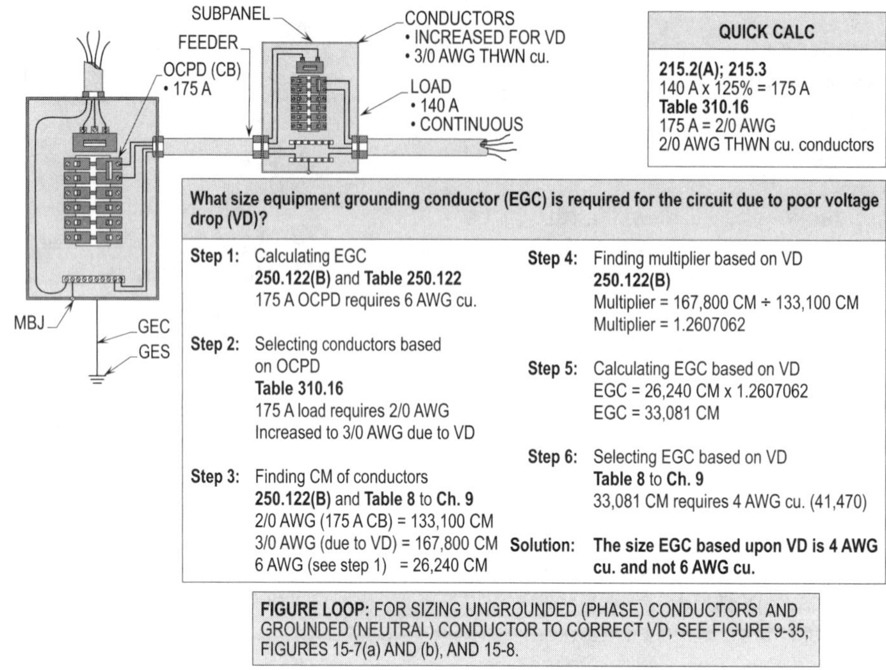

Figure 15-9. Calculating the size equipment grounding conductor based upon poor VD in the feeder due to excessive length of conductors.

SIZING CONDUCTORS BASED ON VD
215.2(A)(3), FPN 2

Sometimes, it is desirable to size and select the conductors based on the percentage of VD needed for the circuit to operate properly. Such conductors may be found by dividing R (12.9 for copper and 21.2 for aluminum) x L (length) x I (amps) by VD percent.

These rules and examples are also used for branch circuits to determine the amount of voltage drop where there are excessive long runs of conductors between the protection devices and load served. (**See Figure 15-10**)

CALCULATION OF A FEEDER
215.2(A)(1) AND 215.3

A feeder may consist of loads such as lighting, receptacles, appliances, heating, air conditioning, and motors. Each load shall be evaluated and calculated at continuous or noncontinuous operation.

If loads with demand factors are present, they shall be separated into individual loads and demand factors shall be applied based on their condition of use in the electrical system.

See Figure 15-11 for calculation procedures pertaining to a feeder with combination loads.

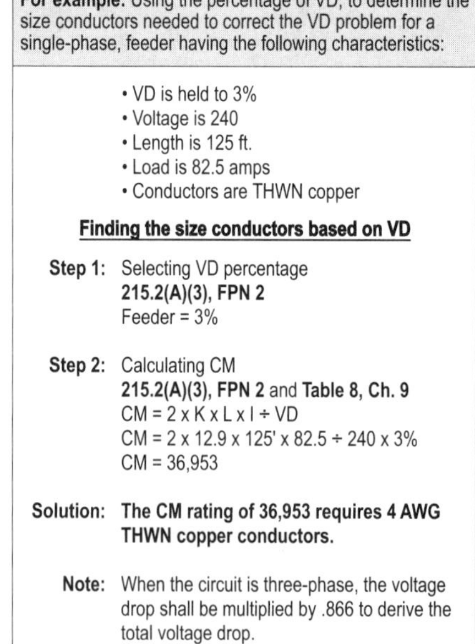

Figure 15-10. Calculating the size conductors based on VD to size conductors needed to correct the VD problem for a feeder.

<table>
<tr><td>

SIZING FEEDER OCPD BASED ON AMPACITY OF CONDUCTORS

Step 1: Calculating loads
215.2(A)(1), Table 430.248, 220.50, and **430.63**
- Lighting load
  40 A x 125% = 50 A
- Receptacle load
  37.5 A x 100% = 37.5A
- Appliance load
  30 A x 125% = 37.5A
- Heat or A/C load
  10 kW x 1000 x
  100% ÷ 240 V = 42 A
- Motor load; **Table 430.248**
  5 HP = 28 A x 100% = 28 A
  5 HP = 28 A x 100% = 28 A
  1 HP = 16 A x 100% = 16 A
  2 HP = 24 A x 100% = 24 A
- Largest motor load
  28 A x 25% = 7 A
  Total load =270 A

Step 2: Selecting conductors
310.10, FPN (2) and **Table 310.16**
270 A requires 300 AWG KCMIL THWN cu.

Step 3: Selecting OCPD based on conductors
215.3; 240.4(B), 430.63, and **240.6(A)**
300 KCMIL THWN cu. = 285 A
285 A = 300 A OCPD

Solution: The size OCPD is allowed to be a 300 amp CB, based upon amps of conductors.

Note: If the largest OCPD for any motor of the group is calculated and added to the remaining loads, the same OCPD is produced as in Step 3 above.

</td><td>

SIZING FEEDER OCPD BASED ON LARGEST OCPD FOR ANY MOTOR OF THE GROUP

Step 1: Calculating loads
215.3; Table 430.248
- Lighting load
  40 A x 125% = 50 A
- Receptacle load
  37.5 A x 100% = 37.5A
- Appliance load
  30 A x 125% = 37.5A
- Heat or A/C load
  10 kW x 1000 x
  100% ÷ 240 V = 42 A
- Motor load
  430.62(A), 430.52(C)(1), and **Table 430.52**
  5 HP = 28 A x 250% = 70 A
  5 HP = 28 A x 100% = 28 A
  1 HP = 16 A x 100% = 16 A
  2 HP = 24 A x 100% = 24 A
  Total load =305 A

Step 2: Selecting OCPD
430.62(A) and **240.6(A)**
300 A is the next standard size below 305 A

Solution: The size OCPD is a 300 amp.

Note: The OCPD for the feeder is the same size (300 A) whether sized from the ampacity of the conductors or largest OCPD for any one motor of the group.

</td></tr>
</table>

Note: For simplicity, a one line diagram is shown.

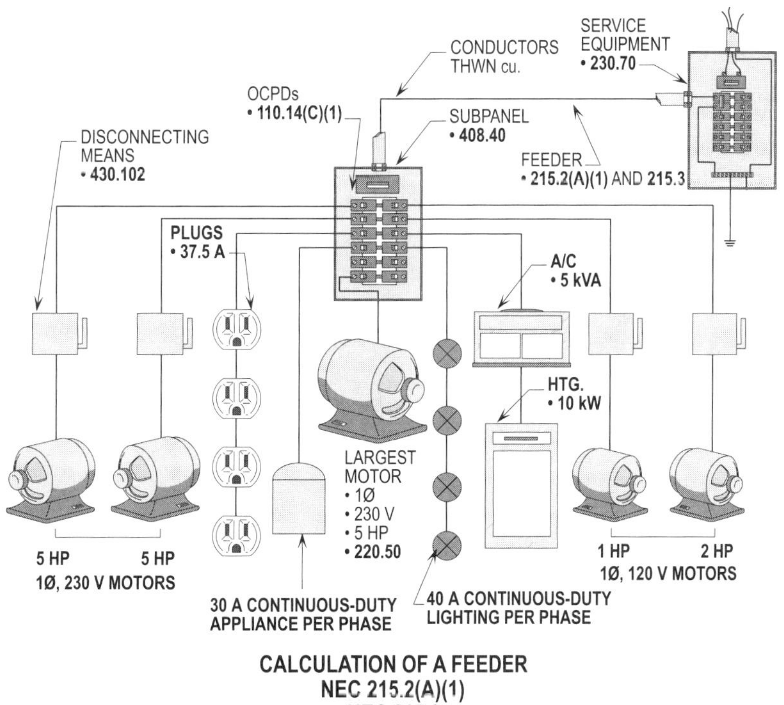

Figure 15-11. Calculating a feeder with loads such as lighting, receptacle, appliances, heating, air conditioning, and motors with continuous and noncontinuous loads.

Design Tip: For more detailed rules on sizing taps, sizing conductors, and overcurrent protection schemes, see the appropriate chapter in this book pertaining to such subjects. Also, see the calculation chapters for a variety of calculations pertaining to various types of buildings and loads.

OPTIONAL CALCULATION FOR A FEEDER
220.87, Ex.

If the maximum demand data for a one year period is not available, the calculated load shall be permitted to be based on the maximum demand (measure of average power demand over a 15-minute period) continuously recorded over a minimum 30-day period using a recording ammeter or power meter connected to the highest loaded ungrounded (phase) conductor of the feeder or service, based on the initial loading at the start of the recording. The recording shall reflect the maximum demand of the feeder or service being taken when the building or space is occupied and shall include by measurement or calculation, the larger of the heating or cooling equipment load and other loads that may be periodic in nature due to seasonal or similar conditions.

Design Tip: The existing demand at 125 percent plus the new load calculated at continuous and noncontinuous operation shall not be permitted to exceed the ampacity of the feeder rating. The feeder shall have overcurrent protection provided as required by **240.4** and **230.90**. (See Figure 15-12)

OPTIONAL CALCULATION FOR A SERVICE
220.87

The optional calculation shall be permitted to be used to determine if additional loads may be added to existing feeder or service conductors. This rule allows the actual maximum demand figures to be utilized when the following conditions are complied with:

- The maximum demand data is available for one year.

- The existing demand at 125 percent plus the new load does not exceed the ampacity of such service.

- Overcurrent protection devices comply with **240.4** and **230.90**.

See Figure 15-13 for a detailed illustration for applying such a procedure.

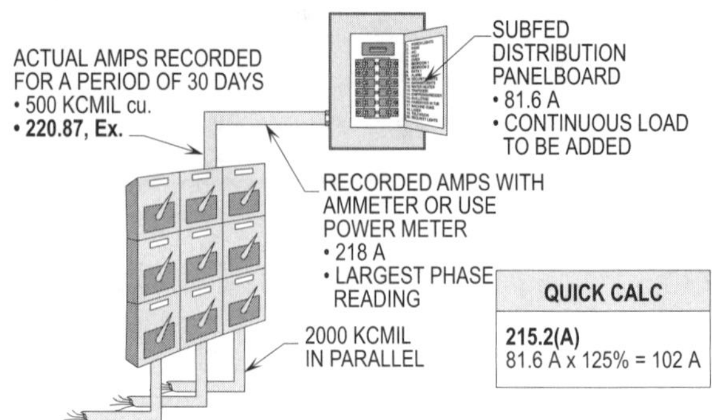

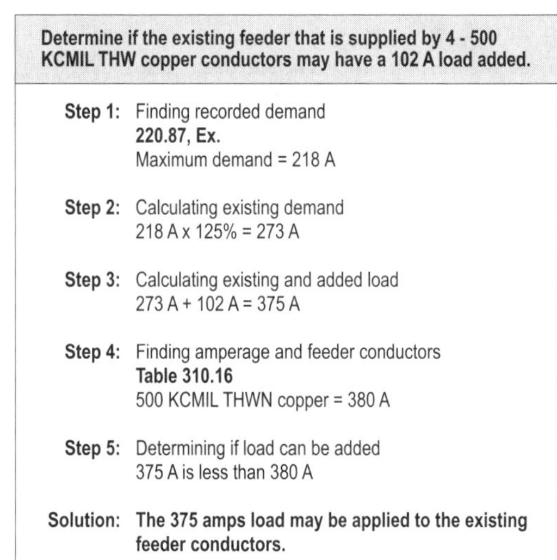

Determine if the existing feeder that is supplied by 4 - 500 KCMIL THW copper conductors may have a 102 A load added.

Step 1: Finding recorded demand
220.87, Ex.
Maximum demand = 218 A

Step 2: Calculating existing demand
218 A x 125% = 273 A

Step 3: Calculating existing and added load
273 A + 102 A = 375 A

Step 4: Finding amperage and feeder conductors
Table 310.16
500 KCMIL THWN copper = 380 A

Step 5: Determining if load can be added
375 A is less than 380 A

Solution: The 375 amps load may be applied to the existing feeder conductors.

OPTIONAL CALCULATION FOR A FEEDER
NEC 220.87, Ex.

Figure 15-12. Calculating the load in amps for a feeder using the optional calculation.

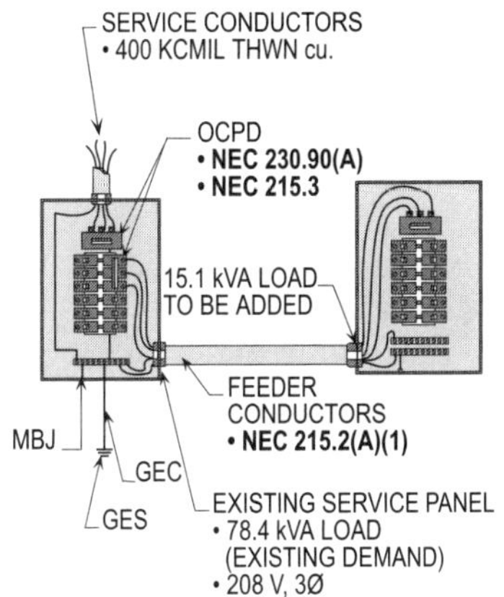

CALCULATING LOAD IN AMPS

Step 1: Finding demand for service
220.87
Maximum demand = 78.4 kVA

Step 2: Calculating existing demand
78.4 kVA x 125% = 98 kVA

Step 3: Calculating total kVA
215.2(A)(1)
98 kVA + 15.1 kVA = 113.1 kVA

Step 4: Calculating amperage
Table 310.16
400 KCMIL THWN copper = 335 A
113.1 kVA x 1000 = 113,100 VA
113,100 VA ÷ (208 x 1.732) = 314 A
314 A is less than 335 A

Solution: **The 15.1 kVA demand load may be applied to the existing service without upgrading the elements.**

**OPTIONAL CALCULATION
FOR A SERVICE
NEC 220.87**

Figure 15-13. The above illustration is a calculation for a service to verify if a load may be added.

Chapter 15. Feeders

Section   **Answer**

1. Continuous loads are classified as loads that operate for a period of _____ hours or more.

   (a) 1   (b) 2
   (c) 3   (d) 5

2. All inductive neutral current, even if noncontinuous, shall be calculated at _____ percent with no demand factor applied.

   (a) 80   (b) 100
   (c) 125   (d) 150

3. The voltage drop on the feeder conductors should not exceed _____ percent overall, if the VD on the branch circuit does not exceed 2 percent.

   (a) 2   (b) 3
   (c) 4   (d) 5

4. Noncontinuous operated loads shall be calculated at _____ percent, or demand factors can be applied to specific loads under certain conditions.

   (a) 75   (b) 80
   (c) 100   (d) 125

5. The VA rating for receptacle loads shall be calculated at 100 percent for the first 10,000 VA, and the remaining VA shall be calculated at _____ percent per **Table 220.44**.

   (a) 25   (b) 50
   (c) 60   (d) 75

6. The feeder neutral load shall be calculated at 100 percent for the first 200 amps, and the remaining load shall have a demand factor of _____ percent applied.

   (a) 50   (b) 60
   (c) 70   (d) 75

7. The voltage drop on feeder conductors should not exceed 2 percent to _____ percent at the farthest outlet supplying power to the loads.

   (a) 2   (b) 3
   (c) 4   (d) 5

8. If the maximum demand data for a one year period is not available, the calculated load shall be permitted to be based on the maximum demand (measure of average power demand over a 15 minute period) continuously recorded over a minimum _____ day period using a recording ammeter or power meter connected to the highest loaded phase of the feeder or service, based on the initial loading at the start of the recording .

   (a) 7   (b) 10
   (c) 24   (d) 30

9. The neutral current in amps for a three-wire, two-phase or five-wire, two-phase system shall be multiplied by _______ percent.
   (a) 140 (b) 150
   (c) 175 (d) 200

10. The feeder grounded (neutral) conductor load shall be _______ percent of the demand load for cooking equipment.
    (a) 50 (b) 65
    (c) 70 (d) 75

11. What is the ampacity for a feeder load of 20,800 VA supplied by a 120/240 volt, single-phase system?

12. What is the ampacity for a feeder load of 20,800 VA supplied by a 120/208 volt, three-phase system?

13. What is the ampacity for a 240 volt, single-phase feeder with a 58 amp load?

14. What is the ampacity for a 240 volt, single-phase feeder with a 58 amp load with a power of 75 percent?

15. What is the ampacity for a 208 volt, three-phase feeder with a 58 amp load with a power factor of 75 percent?

16. What is the VA rating for a connected load of 32,800 VA to size the elements of the feeder at noncontinuous operation?

17. What is the VA rating for a for a continuously operated processing machine with a connected load of 32,800 VA?

18. What is the load in VA for a number of general purpose receptacles used at noncontinuous operation that are utilized to supply a connected load of 28,400 VA?

19. What is the neutral load for a feeder with a inductive load of 300 amps?

20. What is the neutral load for a feeder with a resistive load of 300 amps?

21. What is the neutral load in kW for a 10 kW range?

22. What is the voltage drop using the resistivity method with the following characteristics? (Use larger conductor if necessary)

    • VD is held to 3%
    • Voltage is 240
    • Length is 300 ft
    • Load is 180 amps
    • OCPD is 200 amps
    • Conductors are 3/0 AWG THWN copper

Section   Answer

23. What is the voltage drop using the CM method with the following characteristics? (Use larger conductor if necessary)

    • VD is held to 3%
    • Voltage is 240
    • Length is 300 ft
    • Load is 180 amps
    • OCPD is 200 amps
    • Conductors are 3/0 AWG THWN copper

24. What size feeder overcurrent protection device is required for the following loads based on the ampacity of conductors?

    • Lighting load = 30 A
      (continuous operation)
    • Receptacle load = 35 A
    • Appliance load = 30 A
      (continuous operation)
    • Heating load = 10 kW
    • A/C load = 6 kVA

Motor loads
    • 5 HP, 230 V, single-phase
    • 5 HP, 230 V, single-phase
    • 2 HP, 120 V, single-phase
    • 1 HP, 120 V, single-phase

25. Determine whether an existing service supplied by 4 - 400 KCMIL THWN copper conductors can have an 86 amp load added to a service with a 198 amp recorded amperage rating for a period of 30 days.

16

Receptacle Outlets

The rules for installing receptacle outlets in one- and two-family dwelling units are more stringent than for commercial or industrial occupancies. Where the required number of receptacles are not installed in a dwelling unit, the power needed to serve the number of appliances in use today is not available. Therefore, in dwelling units only, a specific number shall be installed. However, this rule does not apply to commercial and industrial locations. Receptacles are installed as needed for cord-and-plug electrical items. Extension cords used improperly create a fire hazard. For this reason, the *National Electrical Code* requires outlets in all types of occupancies to be installed in such a manner as to cord-and-plug connect appliance loads without the use of extension cords.

GFCI protection of 15 or 20 amp, 125 volt receptacle outlets shall be required in specified areas both indoors and outdoors to protect personnel from electric shock while using cord-and-plug connected electric hand tools, radios, TVs, stereos, etc. This rule applies to certain areas in dwelling units, commercial locations, and industrial locations.

GROUNDING RECEPTACLES
406.3(A)

Receptacles installed on 15 or 20 amp general purpose branch circuits shall be the grounding type. They are required to be the three-wire type with a brass terminal for the ungrounded (phase) conductor, a silver terminal for the grounded (neutral) conductor, and a green terminal for the equipment grounding conductor. The brass terminal is for the short slot and represents the connection for the ungrounded (phase) conductor. The U slot is for the equipment grounding conductor terminal on the receptacle, while the long slot represents the grounded (neutral) conductor terminal. Any color conductor but white, gray, green, green with a yellow stripe, or bare shall be permitted to be connected to the ungrounded (phase) brass terminal per **210.5(C)** and **310.12(C)**. Only white or gray shall be connected to the grounded (neutral) silver terminal per **200.6(A)**, **210.5(A)**, and **310.12(A)**. A conductor with a green or green with yellow stripe shall be the only color permitted to be connected to the green equipment grounding conductor terminal per **250.134(B)**, **210.5(B)**, **250.119(A)**, and **310.12(B)**. The insulation of the equipment grounding conductor shall be permitted to be stripped per **250.119(A)**. **(See Figure 16-1)**

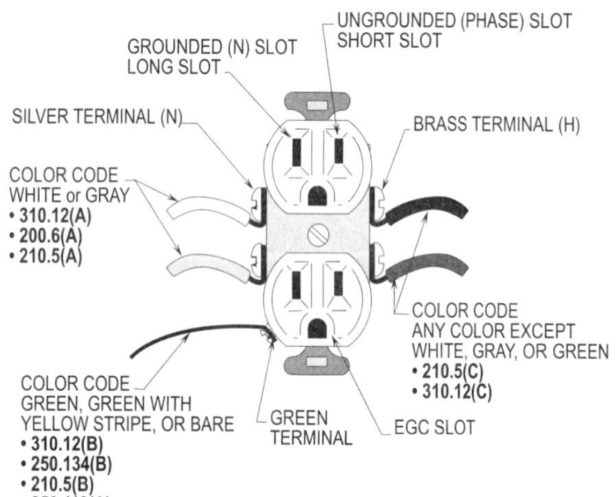

Figure 16-1. Identifying the components and correct color coding for receptacles.

Receptacles shall be rated for the voltage and current based upon the size of the general purpose branch circuit to which they are connected.

Receptacles shall be permitted to be connected to an individual branch circuit, a general purpose branch circuit, a general purpose branch circuit with two or more receptacles, or a combination of lighting and receptacle outlets.

A receptacle shall be permitted to be connected to an individual branch circuit, but to do so, it shall be rated not less than the ampacity of the branch circuit. The size of the overcurrent protection device shall be utilized in determining the rating of the branch circuit.

For example, a 20 amp overcurrent protection device (circuit breaker or fuse) classifies the branch circuit as 20 amp even if the conductors of the circuit are 10 AWG per **210.19(A)(1)**, **210.20(A)**, and **210.23**. Sometimes conductors are increased for voltage drop correction, adjustment, or correction factors that create a larger conductor than normally required. **(See Figure 16-2)**

An individual branch circuit supplies power to a single load that is cord-and-plug connected to a single receptacle. To determine the size of the load permitted to be connected to the receptacle, refer to **Table 210.21(B)(2)**, which lists the maximum cord-and-plug connected load that shall be permitted to be plugged into a receptacle.

For example: What is the maximum size load that may be cord-and-plug connected to a receptacle supplied by a general purpose branch circuit?

Step 1: Max. load allowed
Table 210.21(B)(2)
Col. 1 Branch circuit = 15 or 20 A
Col. 2 Receptacle = 15 A
Col. 3 Max. load = 12 A

Solution: Max. load is 12 amps.

Table 210.21(B)(3) lists the amperage rating of a receptacle that shall be permitted be connected to a branch circuit based on the rating of the overcurrent protection device and conductors.

For example: What is the minimum and maximum size receptacle in amps that may be connected to a 20 amp general purpose branch circuit?

Step 1: Max. size receptacle allowed
Table 210.21(B)(3)
Col. 1 Branch circuit = 20 A
Col. 2 Receptacle = 15 or 20 A

Solution: Max. size receptacle is 15 or 20 amps.

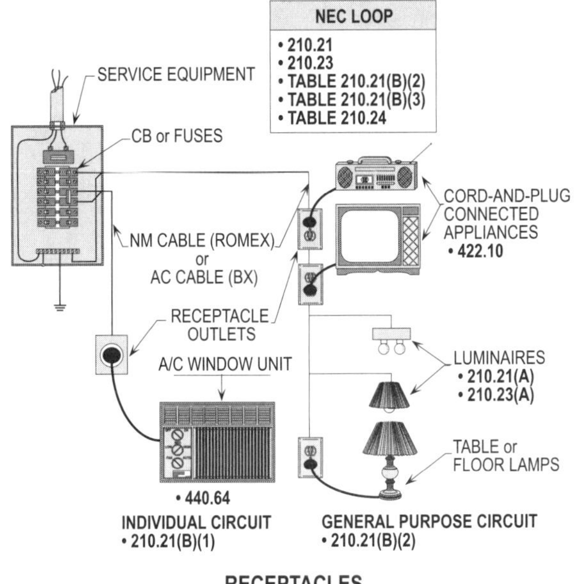

RECEPTACLES
NEC 210.21(B)

Figure 16-2. An individual branch circuit has only one outlet and a single load whereas a general purpose branch circuit has more than one outlet and load that may be cord-and-plug connected or hard-wired.

NONGROUNDING RECEPTACLES
406.3(A), Ex.

The general rule requires receptacles of the grounding type to be installed in existing dwelling units. Conductors of copper should be used to connect the terminals of the receptacle to the branch circuit conductors. Otherwise, copper and aluminum type receptacles shall be installed listed for such use.

Aluminum conductors of a branch circuit should be pigtailed to a 6 in. (150 mm) copper jumper and connected to the terminals of the receptacle. This is required because the existing terminals of receptacles are not usually rated for aluminum connections. Unless certain rules are applied, nongrounding receptacles shall be connected to branch circuits not having an equipment grounding conductor.

If an existing branch circuit is wired with an AC cable and the existing receptacle outlet is the nongrounding type, a grounding receptacle shall be used to replace the existing nongrounding receptacle, since an AC listed cable may be used as an equipment grounding means per **406.3(D)** and **250.118(8)**. Only one conductor shall be placed under the

bending screw on the terminal of the receptacle unless it is approved for more than one conductor per **110.3(B)** and **110.14(A)**. **(See Figure 16-3)**

REPLACEMENT OF RECEPTACLES
250.130(C)

The basic rule for replacement of receptacles is that grounding type receptacles shall be used to replace existing receptacles. There is no problem in applying this requirement as long as the branch circuit is equipped with an equipment grounding conductor or grounding means that complies with **250.118**. A nongrounding type receptacle shall be used with a branch circuit that has no equipment grounding conductor or grounding means. There are two conditions in the NEC that allow a GFCI receptacle to be used. Either a GFCI receptacle shall be permitted to be used to replace and protect a single receptacle outlet, or a GFCI feed-through receptacle shall be permitted to be used to protect additional outlets downstream. The receptacle outlets downstream shall be permitted to be of the grounding type but shall be identified as being protected by a GFCI circuit with no ground. **Section 250.130(C)** permits an equipment grounding conductor sized from the overcurrent protection device of the branch circuit per **Table 250.122** to be utilized. The equipment grounding conductor shall connect the green grounding terminal of the receptacle to the closest metal water pipe as required by **250.104(A)**, **250.52(A)(1)**, and **250.53(D)**. The equipment grounding conductor shall not be permitted to be connected to a driven rod.

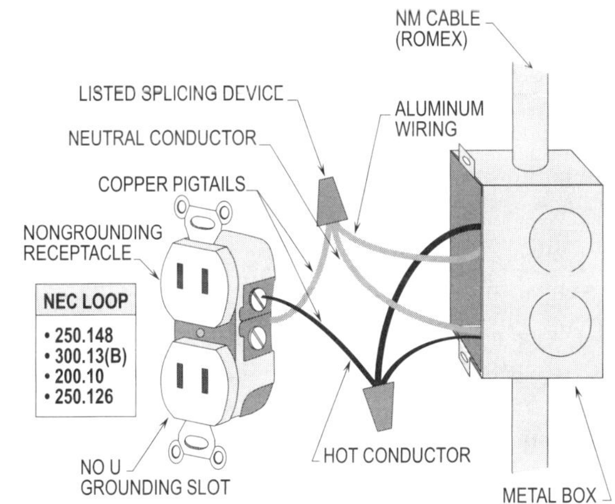

NONGROUNDING RECEPTACLES
NEC 406.3(A), Ex.

Figure 16-3. Nongrounding receptacles are used where the branch circuit is not provided with an equipment grounding conductor. If the branch circuit wiring method is aluminum, a copper pigtail shall be used to terminate the aluminum conductor to the terminals of the receptacle. **Note:** The splicing device shall be listed.

Design Tip: Check with the inspector for his or her interpretation of the point of connection to the metal water pipe. The authority having jurisdiction may require the connection of the equipment grounding conductor to be made within 5 ft (1.52 m) of where the metal water pipe enters the dwelling unit per **250.52(A)(1)**. However, with the metal water pipe in the earth and entering the premises from T's from more than one point, the AHJ may allow connection to any one of such entries.

The NEC allows five methods by which a nongrounding receptacle shall be permitted to be replaced, and they are as follows:

- A nongrounding receptacle.
- A GFCI receptacle protecting a single outlet.
- A GFCI receptacle protecting a single outlet and additional outlets downstream.
- An equipment grounding conductor routed from the receptacle to a metal water pipe per **250.52(A)(1)**.
- A GFCI circuit breaker protecting all outlets.

See Figure 16-4 for a detailed illustration on methods used for replacing nongrounding receptacles on branch circuits without an equipment grounding conductor.

NUMBER ON A CIRCUIT
210.11(A)

Section **210.11(A)** is used to determine the number of outlets permitted on a branch circuit. Section **210.11** refers to **220.12** and **Table 220.12**, which require 3 VA per sq. ft for a dwelling unit to determine the total VA to select the number of branch circuits.

For example, a 3000 sq. ft dwelling unit has a total of 9000 VA (3000 sq. ft x 3 VA = 9000 VA) to supply general purpose receptacle outlets. Using this concept, there are 4 - 120 volt, 20 amp branch circuits provided to supply general purpose receptacle outlets. (See below)

To determine the number of outlets permitted on a branch circuit, the overcurrent protection device shall be multiplied by 120 volts per **210.11(B)** and divided by 3 VA per **Table 220.12**. This number may be verified as follows:

No. = (3000 sq. ft x 3 VA) ÷ (20 A OCPD x 120 V)
No. = 3.75
3.75 requires 4 - 20 amp branch circuits

For example: How many receptacle outlets in a dwelling unit may be connected to a 20 amp general purpose branch circuit?

Step 1:	Finding VA **210.11(B)** 20 A OCPD x 120 V = 2400 VA
Step 2:	Finding sq. ft **Table 220.12** 2400 VA ÷ 3 VA = 800 sq. ft
Step 3:	Finding number of outlets allowed **220.14(J); 220.12; 210.11(B)** Allows any number to be installed
Solution:	**There is no limit to the number of outlets permitted in the 800 sq. ft area. Note that local codes may limit the number of outlets that can be connected to a branch circuit.**

In commercial and industrial locations, the number of receptacle outlets permitted on a general purpose branch circuit are calculated differently from the number in dwelling units.

For example: In a commercial or industrial area, the limited number of 13 outlets on a general purpose branch circuit is based on the following procedure.

Step 1:	Finding amperage of outlets **220.14(J)** 180 VA ÷ 120 V = 1.5 A
Step 2:	Finding number of outlets **220.14(J)** 20 A OCPD ÷ 1.5 A = 13 A 20 A circuit is limited to 13
Solution:	**The number of outlets permitted on a 20 amp branch circuit is 13.**

Design Tip: Using the same procedure, 10 outlets shall be permitted on a 15 amp overcurrent protection device. The limitation of outlets presents nuisance tripping of the overcurrent protection device.

Municipal electrical boards may amend the NEC and allow different numbers of outlets on a 15 or 20 amp branch circuit. The two methods listed in determining the number of receptacle outlets on a branch circuit in dwellings, commercial locations, and industrial locations are the most common methods used. **(See Figure 16-5)**

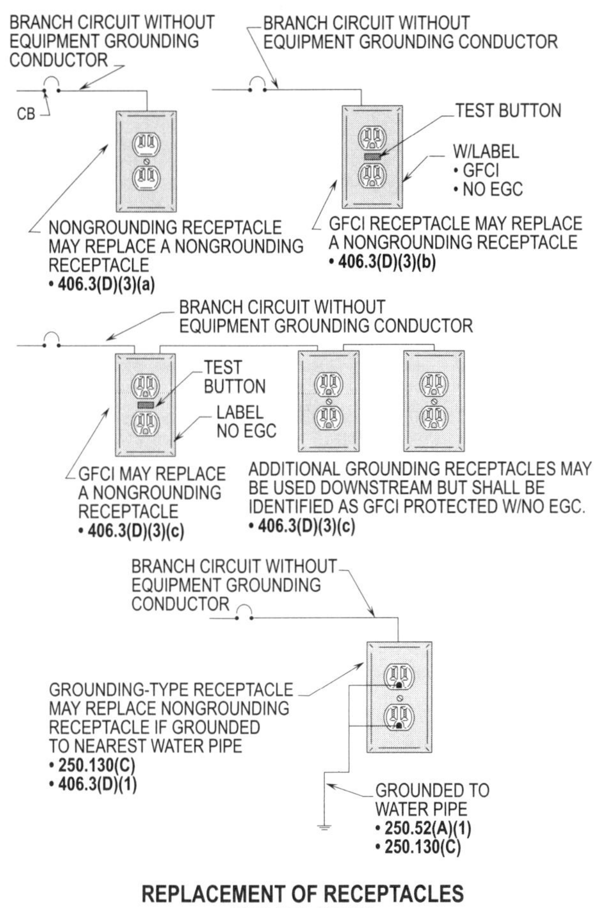

Figure 16-4. Methods for replacing nongrounding receptacles on branch circuits without an equipment grounding conductor.

DWELLING UNIT LOCATIONS 210.52(A) THRU (H)

The provisions for installing the required number of receptacle outlets in dwelling units are listed in **210.52(A) through (H)**. The basic rule requires receptacle outlets to be installed where flexible cords are used to connect electrical appliances and equipment. This Section, with **Subdivisions (A) through (H)**, lists the exact location and number to be installed. The number of receptacle outlets required differs according to whether installation is inside or outside the dwelling unit.

How many general-use receptacle outlets in a dwelling unit are permitted to be connected on a 15 amp, 120 volt branch circuit in a dwelling unit?

Step 1: Finding sq. ft
210.11(B); Table 220.12
Circuit sq. ft = V x device ÷ VA per sq. ft
Circuit sq. ft = (120 V x 15 A CB) ÷ 3 VA
Circuit sq. ft = 600

Step 2: Finding outlets
220.14(J); 210.11(B)
Any number of outlets permitted in the 600 sq. ft area.

Solution: Outlets can be any number.

How many outlets in a commercial building used at noncontinuous operation are permitted to be connected to a 15 amp CB or fuse on a 120 volt branch circuit?

Step 1: Finding outlets
220.14(J); 210.11(B)
Circuit sq. ft = device ÷ 1.5 A
Circuit sq. ft = 15 A CB ÷ 1.5 = 10

Solution: Outlets = 10.

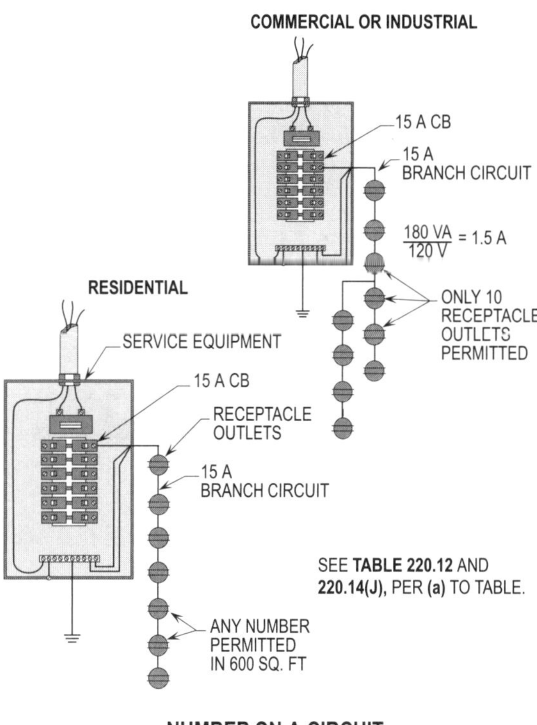

Figure 16-5. For determining the number of receptacle outlets on a 15 amp branch circuit using the "any number permitted" method or the limited method is at the discretion of the AHJ. **Note:** The number per circuit may vary from city to city or from state to state based upon the electrical ordinances of each area.

WALL RECEPTACLE OUTLETS
210.52(A)(1) THRU (A)(3)

Section **210.52(A)** lists the requirements for installing receptacle outlets in dwelling units, including single-family dwellings, duplexes, single-family dwellings in apartment complexes, townhouses, and condominiums. Per **Article 100** in the NEC a dwelling unit is defined as a single unit, providing complete and independent living facilities for one or more persons, including permanent provisions for:

- Living
- Sleeping
- Sanitation
- Cooking

See Figure 16-6 for a detailed illustration of a dwelling unit.

Receptacles that are an integral part of a luminaire, appliance, or cabinet shall not be counted as one of the required receptacle outlets. Those outlets installed more than 5 ft 6 in. (1.7 m) above the finished floor are not permitted to be counted per **210.52(4)**. In other words, they shall not be counted as one of the required outlets for normal use. In these cases, additional outlets shall be installed to comply with the number of outlets required per **210.52(A)(1)**. **(See Figure 16-7)**

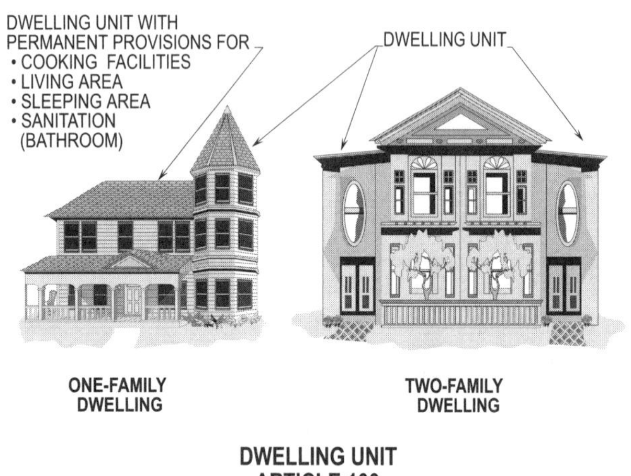

Figure 16-6. A dwelling unit in a one-family house or duplex contains provisions for permanent cooking and sanitation as well as facilities for living and sleeping. These provisions shall be permitted to be located in one or more rooms to be classified as dwelling unit per **Article 100**.

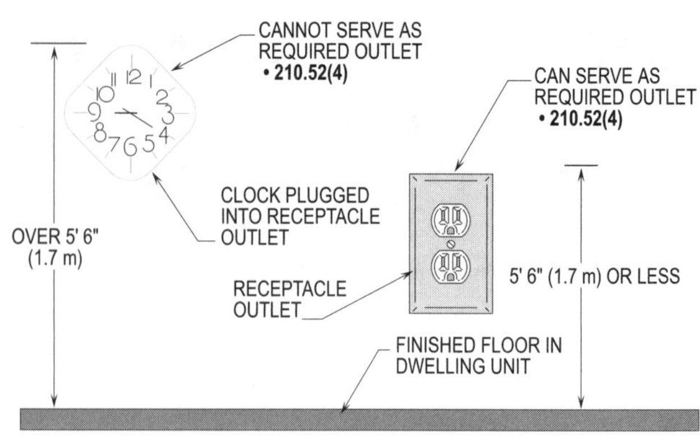

Figure 16-7. Receptacle outlets located more than 5 ft 6 in. (1.7 m) from the floor in a dwelling unit shall not serve as one of the required outlets listed in **210.52**.

Section **210.52** gives permission for a receptacle to be installed in a listed factory-assembled baseboard heater, so as to prevent extension cords from being plugged into the receptacle outlet above the baseboard heater. Heat from the electric heater elements may overheat the insulation of the cord and cause a fire hazard from the arcing and sparking that can occur. The **FPN** recommends that baseboard heaters be installed according to the instructions accompanying the heater unit per **90.7** and **110.3(B)**.

> **Design Tip:** This rule applies strictly to electric baseboard heater with heating elements. **(See Figure 16-8)**

Receptacle outlets shall be installed on walls in every inhabitable room, including the hallway, of every dwelling unit, except the bathroom. The procedure for properly spacing the receptacles along the floor line of an unbroken wall is to provide an outlet within 6 ft (1.8 m) of the cord-and-plug connected equipment.

The first outlet shall be installed 6 ft (1.8 m) from the door entering the room. The measurement for installing this outlet shall be permitted to be made on either side of the door. From the outlet measured at 6 ft (1.8 m), additional outlets measured at 12 ft (3.7 m) intervals along the unbroken wall shall be installed. Outlets installed in this manner allow electrical appliances with 6 ft (1.8 m) cords to be cord-and-plug connected without the use of an extension cord. Installing the first outlet at 6 ft (1.8 m) and each additional outlet every 12 ft (3.7 m) thereafter complies with this requirement and makes it possible to cord-and-plug connect appliances without creating a fire hazard. Any wall space of 2 ft (600 mm) or more requires an outlet for a receptacle to be installed.

Design Tip: The wall space of 2 ft (600 mm) is based on trimmed-out walls, etc. and not from a 2 ft (600 mm) rough-in measurement of framing members.

A wall space is a wall unbroken along the floor line by a door, bookcase, fireplace, or window that extends all the way to the floor line. A floor receptacle outlet located close to the wall shall be permitted to be counted as one of the outlets required by **210.52(A)(3)**. **(See Figure 16-9)**

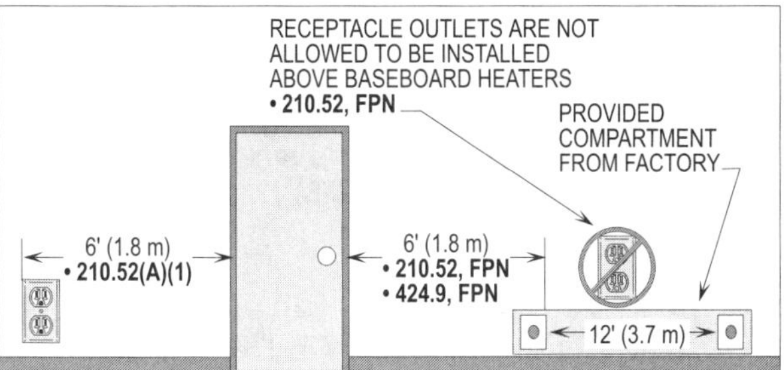

Figure 16-8. Electric baseboard heaters shall not be located below a receptacle outlet. Partitions from the factory meant for installing receptacles shall be permitted to be put in the baseboard heater to comply with the 12 ft (3.7 m) requirement of **210.52(A)(1)**.

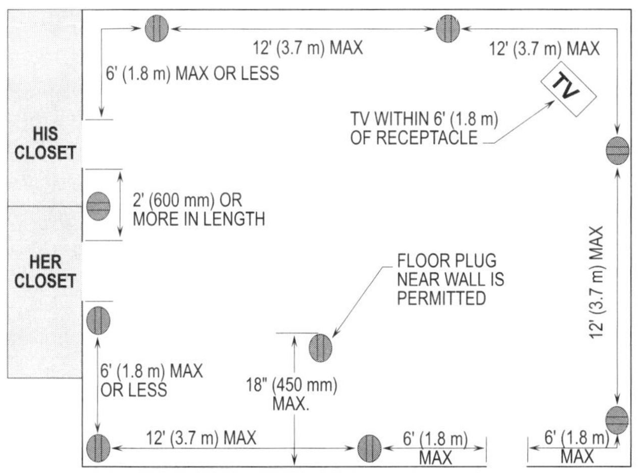

Figure 16-9. Receptacle outlets shall be installed so that there is no point on the wall greater than 6 ft (1.8 m) from a receptacle outlet.

Railings used for room dividers, where furniture with floor or table lamps is backed up to the railing, shall have receptacle outlets installed for cord-and-plug connections such as table or floor lamps. This rule is to prevent the use of extension cords, which are usually run through walls, doors, under carpets, etc. per **400.8(1) through (7)**. **[See Figure 16-10(a)]**

The sliding portion of an exterior door shall not be considered wall space. Fixed panels, including the fixed portion of a sliding glass door unit in an exterior wall, shall be considered wall space for the purpose of spacing the required number of receptacle outlets per **210.52(A)(2)(2)**. The NEC allows a floor receptacle to be used to provide the required number of receptacle outlets per **210.52(A)(3)**. **[See Figure 16-10(b)]**

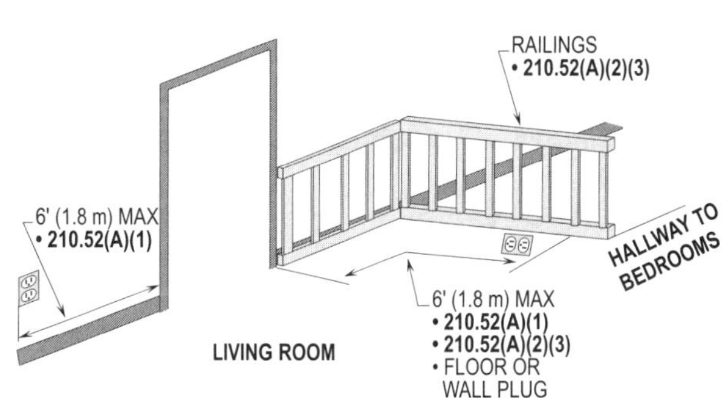

Figure 16-10(a). Railings used as room dividers shall have receptacle outlets installed per **210.52(A)(2)(3)** to cord-and-plug connect floor lamps, table lamps, etc.

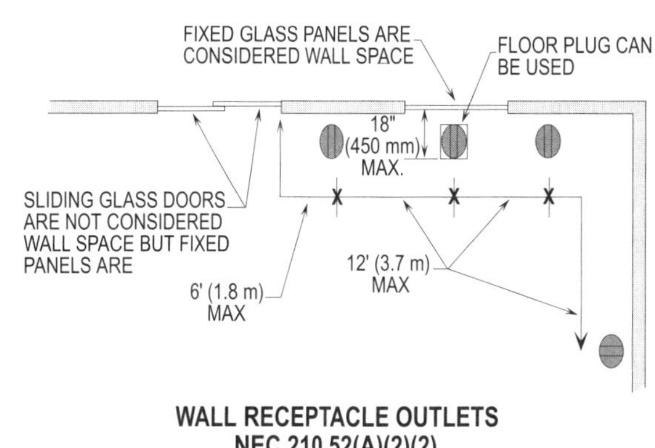

Figure 16-10(b). Sliding glass doors or panels shall not be considered wall space when spacing receptacles to comply with **210.52(A)**. However, fixed panels shall be considered wall space.

RECEPTACLE OUTLETS FOR SMALL-APPLIANCE CIRCUITS 210.52(B), 210.52(C), 210.11(C)(1), AND 220.52(A)

A minimum of 2 - 20 amp, 120 volt, 1500 VA small-appliance circuits shall be required to supply receptacle outlets that are located in the kitchen, pantry, breakfast room, and dining room. The 2 small-appliance circuits shall be routed to the kitchen countertop(s), and the outlets proportioned among the 2 circuits as evenly as possible to prevent unbalanced loading of the circuits. Unbalanced loading may trip open the overcurrent protection device if too many portable appliances are plugged into the same small-appliance circuit. **(See Figure 16-11)**

The 2 - 20 amp small-appliance circuits wired with 12-2 AWG w/ground, nonmetallic-sheathed cable (romex or rope) shall only supply receptacle outlets located in the kitchen, pantry, breakfast room, and dining room. All other receptacle outlets shall be served by the general purpose branch circuits or individual branch circuits per **Table 220.12** and **210.23(A),** or **210.19(A)(1)**. Switched receptacle outlets connected to the general purpose branch circuits per **210.52(B)(1), Ex. 1** and **210.70(A)(1), Ex. 1** to serve swag lights, table lamps, floor lamps, etc. shall also be allowed. These are in addition to those required in **210.52(B)**. All receptacle outlets shall be installed so that no point along the floor line of an unbroken wall is further than 6 ft (1.8 m) from an outlet per **210.52(A)(1)**.

Exception 2 to **210.52(B)(1)** allows a receptacle outlet to be installed from an individual branch circuit to supply a compressor motor for refrigeration equipment.

Exception 2 to **210.52(B)(2)** permits receptacle outlets to be installed on the small appliance circuits to supply loads such as timers, clocks, burner ignition systems on gas ovens, cooktops, and ranges.

> **Design Tip:** This rule makes it clear that such mentioned motor loads shall be permitted to be supplied by an individual 15 amp branch circuit per **210.19(A)(1)**, general purpose motor branch circuit per **430.53(A)**.

There is an **Ex.** to **210.52(B)(2)**, which permits another outlet to be supplied by the small-appliance circuits. The **Ex. 1** permits a clock outlet to be served by any one of the small appliance circuits. To apply this exception, the clock outlet shall be located in the kitchen, pantry, dining room, or breakfast room. **(See Figure 16-12)**

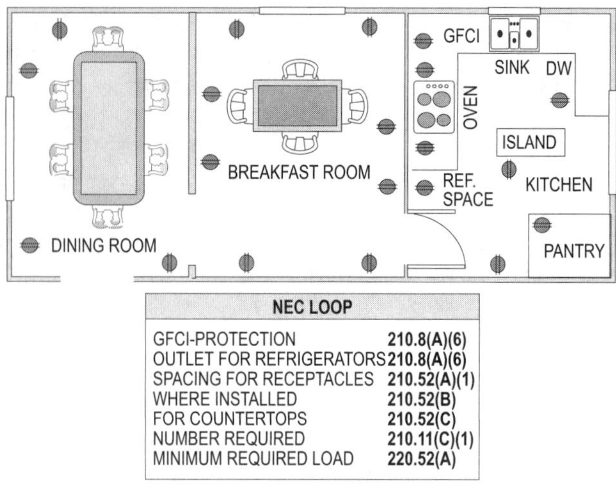

NEC LOOP	
GFCI-PROTECTION	210.8(A)(6)
OUTLET FOR REFRIGERATORS	210.8(A)(6)
SPACING FOR RECEPTACLES	210.52(A)(1)
WHERE INSTALLED	210.52(B)
FOR COUNTERTOPS	210.52(C)
NUMBER REQUIRED	210.11(C)(1)
MINIMUM REQUIRED LOAD	220.52(A)

RECEPTACLE OUTLETS FOR SMALL-APPLIANCE CIRCUITS NEC 210.52(B)

Figure 16-11. A minimum of 2 - 20 amp small-appliance circuits shall be provided to serve the receptacle outlets located in the kitchen, pantry, breakfast room, and dining room.

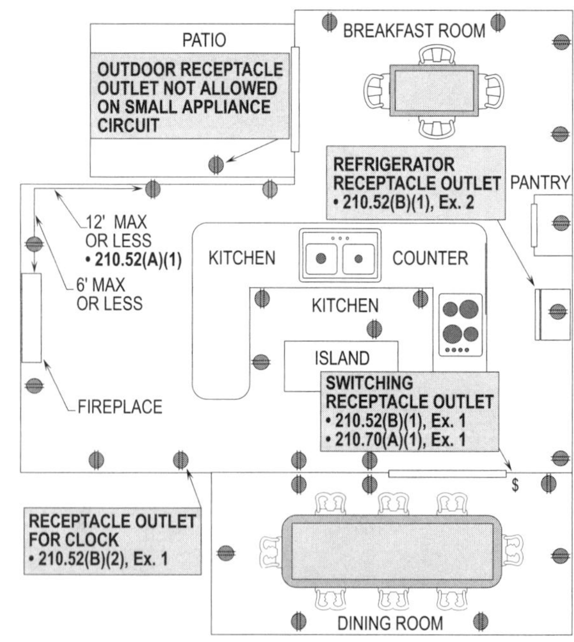

RECEPTACLE OUTLETS FOR SMALL-APPLIANCE CIRCUITS NEC 210.52(B)(1), Ex. 1

Figure 16-12. A clock outlet shall be permitted to be installed on the small-appliance circuits. In addition, a switched receptacle outlet shall be permitted to be installed on the general purpose branch circuit.

RECEPTACLE OUTLETS OVER COUNTERTOPS
210.52(C)

There are three types of countertops that may be present in dwelling units, and they are as follows:

- countertop with a wall behind it per **210.52(C)(1)**
- island countertops per **210.52(C)(2)**
- countertop with no wall behind it (peninsular) per **210.52(C)(3)**

WALL COUNTERTOP SPACES
210.52(C)(1)

A receptacle outlet shall be provided at each countertop space 12 in. (300 mm) or wider in the kitchen and dining room areas. No point along the wall shall be permitted to be more than 24 in. (600 mm) from a receptacle outlet. Each countertop shall be treated separately in providing the number and spacing of outlets. Countertops divided by a sink, a range, a cooktop, an oven, or a refrigerator shall have receptacle outlets installed on each side where there is 12 in. (300 mm) or more of countertop space along the wall.

The procedure for laying out the location of these outlets is to measure 24 in. (600 mm) and then at 4 ft (1.2 m) intervals until the last receptacle outlet is no further than 24 in. (600 mm) in any direction from an outlet. Receptacle outlets will be accurately provided for each countertop space by applying these measurements. Any receptacle outlets rendered inaccessible by the position of installed appliances such as a refrigerator shall not be considered as one of the required outlets. **(See Figure 16-13)**

ISLAND COUNTERTOP SPACES
210.52(C)(2)

An island countertop stands alone with no wall behind or beside it. Island countertops with a short dimension of 12 in. (300 mm) or greater or a long dimension of 24 in. (600 mm) or greater shall have at least one receptacle installed. **(See Figure 16-14)**

> **Design Tip:** See **210.52(B)(2)**, which lists the requirements for installing the small-appliance circuits for countertops in the kitchen and dining room areas. Refer to **210.8(A)(6)** for countertop receptacle outlets requiring GFCI protection.

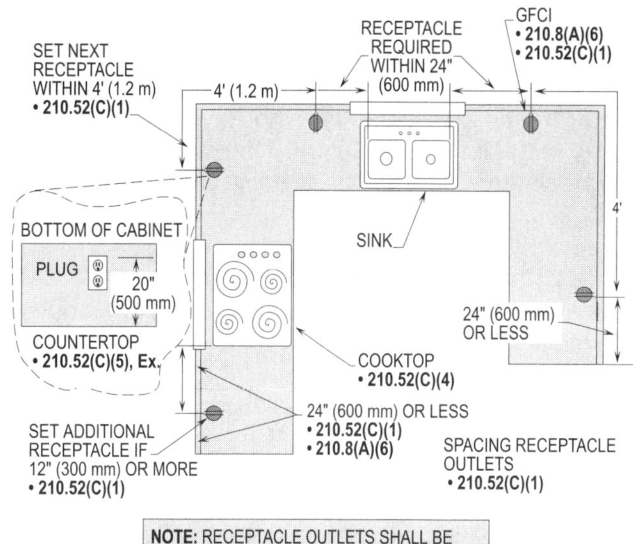

Figure 16-13. Receptacle outlets installed at kitchen countertops shall be located so no point along the countertop space is more than 24 in. (600 mm), measured horizontally, from a receptacle outlet.

PENINSULAR COUNTERTOP SPACES
210.52(C)(3)

Receptacle outlets shall be installed at each peninsular countertop. A peninsular is a countertop without a wall on either side. Receptacle outlets shall be installed for each peninsular with a long dimension of 24 in. (600 mm) or greater and a short dimension of 12 in. (300 mm) or greater. A peninsular countertop shall be measured from the connecting edge, and receptacle outlets installed accordingly. The receptacle outlets shall be installed above or within 12 in. (300 mm) below the countertop or a tombstone type receptacle shall be permitted to be installed on the countertop.

> **Design Tip:** If the countertop has an overhang of more than 6 in. (150 mm), a receptacle outlet shall not be permitted to be installed below the countertop per **210.52(C)(5), Ex. (See Figure 16-14)**

RECEPTACLE OUTLETS IN BATHROOMS
210.52(D)

A bathroom is not just a room but is an area including a basin(s) with a toilet, a tub, or shower per **Article 100**.

A utility sink or wet bar sink shall not be considered a bathroom, because no other fixtures mentioned are present.

> **Design Tip:** The key to defining a bathroom is that a basin and at least one or more of the other fixtures shall be present.

Section **210.52(D)** requires at least one receptacle outlet to be installed in bathrooms within 3 ft (900 mm) of the outside edge of each basin in a dwelling unit. The receptacle outlet shall be located on a wall or partition that is adjacent to the basin or basin countertop or installed on the side or face of the basin cabinet not more than 12 in. (300 mm) below the countertop. Section **210.8(A)(1)** requires all 15 or 20 amp, 125 volt receptacles to be GFCI protected for the safety of personnel using electric grooming tools. **(See Figure 16-15)**

> **Design Tip:** Each basin is required to have a receptacle outlet installed within 3 ft (900 mm) and located in such a manner so as to serve the sink area safely, without the use of extension cords.

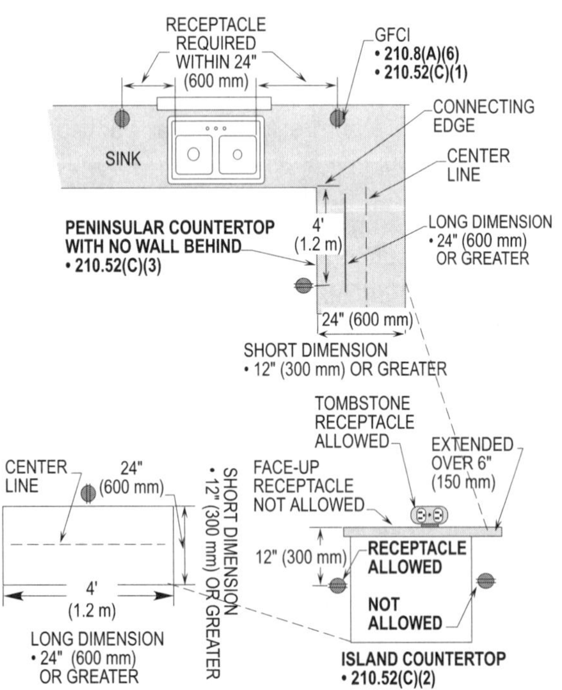

Figure 16-14. Installation requirements for receptacle outlets on island or peninsular countertops.

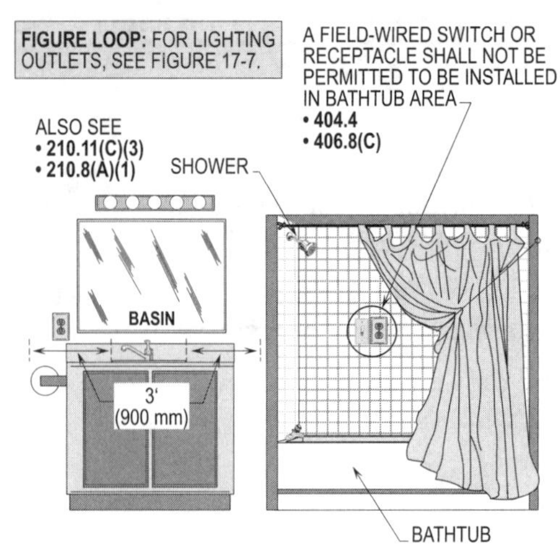

Figure 16-15. At least one receptacle outlet to serve the bathroom area shall be installed within 3 ft (900 mm) of the outside edge of each basin in a dwelling unit.

RECEPTACLE OUTLETS OUTDOORS 210.52(E)

At least two receptacle outlets shall be installed outdoors for one- and two-family dwellings. There shall be at least one receptacle outlet installed in the front and the back of such units. Section **210.8(A)(3)** requires such outdoor receptacle outlets to be GFCI protected to protect personnel from electrical shock when using such tools as drill motors, lawn mowers, hedge clippers, etc.

All outdoor receptacles with direct grade-level access shall be GFCI protected. *Direct grade-level* access is defined as a location 6 ft 6 in. (2 m) or less while standing from grade level and readily accessible to the user. Outdoor receptacle outlets shall be located on outside walls or open porches. Receptacle outlets located in the eaves of dwellings to connect Christmas lights or on balconies to cord-and-plug connect radios, VCRs, stereos, or TVs shall be GFCI protected even if they are located over 6 ft 6 in. (2 m) from direct grade level.

At least one receptacle outlet shall be installed within the perimeter for balconies, decks, and porches that are accessible from inside the dwelling unit. The receptacle shall be installed not more than 6-1/2 ft (2 m) above the balcony, deck, or porch surface.

Section **406.8(B)(1)** requires unattended receptacle outlets to be listed weather-resistant type with raintight covers to protect the components. Unattended outlets are those serving items such as outdoor Christmas lights, water pumps, etc. For receptacle outlets located outdoors for townhouses and condominiums with zero lot lines, see **210.8(A)(3)** and **210.52(E)**. **(See Figure 16-16)**

For example, a 20 amp small-appliance laundry circuit (20 A OCPD x 80% = 16 A) should be loaded with no more than 16 amps if used for long periods of time, such as three hours or longer. **(See Figure 16-17)**

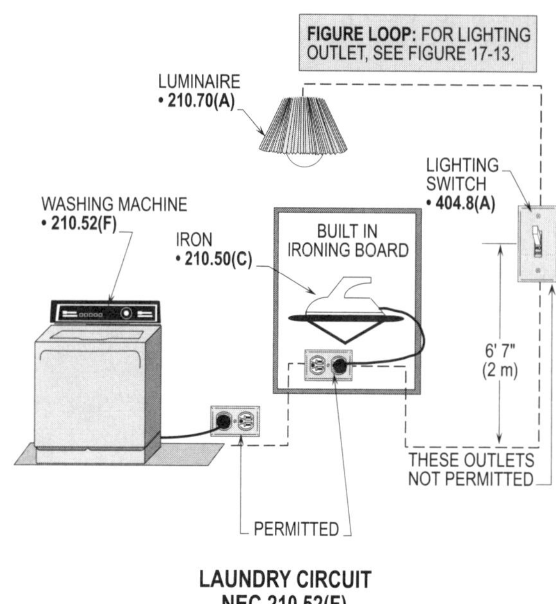

Figure 16-16. Receptacle outlets mounted on a wall or located on an open front porch or stoop of a dwelling unit shall be GFCI protected. **Note:** All the receptacles above shall be GFCI protected.

Figure 16-17. The 20 amp laundry circuit shall be permitted to be used only to serve laundry equipment. The laundry circuit shall be permitted to be routed to the garage, basement, utility room, or wherever the laundry room is located.

LAUNDRY CIRCUIT
210.52(F)

All laundry receptacle outlets shall be supplied by a 20 amp circuit utilizing a 12-2 AWG nonmetallic-sheathed cable with ground. This cable is usually nonmetallic-sheathed cable or armored cable.

> **Design Tip:** A conduit system shall be permitted to be used with individual conductors pulled in after the conduit is installed. For the installer to use this wiring method, a local code usually requires it.

At least one receptacle outlet shall be installed for the laundry area. All laundry equipment shall be located within 6 ft (1.8 m) of the receptacle outlet per **210.50(C)**. Sometimes a 20 amp duplex receptacle is used to cord-and-plug connect a washing machine and a gas dryer. No other outlets shall be supplied by this 20 amp, 2-wire small-appliance circuit.

RECEPTACLE OUTLETS IN BASEMENTS AND GARAGES
210.52(G)

At least one receptacle outlet shall be installed in the basement and garage, attached or detached, with electric power. GFCI protected receptacles shall be installed in basements and garages to protect people from electric shock when using electric hand tools.

RECEPTACLES IN BASEMENTS
210.52(G)

At least one GFCI protected receptacle outlet per **210.8(A)(5)** shall be installed in the basement, when it is left unfinished. An unfinished basement means that the basement is used for storage or a workshop. If more than one receptacle outlet is installed, then all of the receptacle outlets shall be GFCI protected. These outlets are to protect people from electric shock when using electrical hand tools such as electric drills, saws, sanders, etc. If the basement is finished into one or more habitable rooms, each separate unfinished portion of the basement shall have a receptacle outlet installed.

Exception to **210.8(A)(5)** does not require GFCI protection where a receptacle supplies only a permanently installed fire alarm or burglar alarm system.

At least one GFCI protected receptacle per **210.8(A)(4)** shall be installed in a crawl space at or below grade level. This receptacle shall be permitted to be used for serving HVAC equipment or be used for a drop light, hand tool, etc.

See Figure 16-18 and **210.70(A)(3)** for the requirements of lighting outlets.

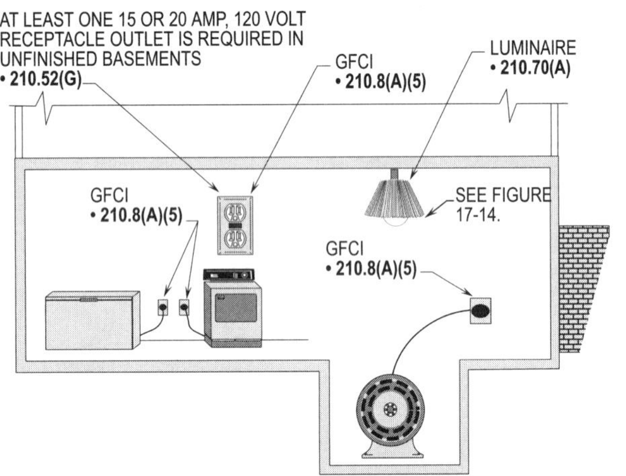

Figure 16-18. At least one 15 or 20 amp, 125 volt receptacle shall be installed in an unfinished basement.

RECEPTACLES IN GARAGES
210.52(G)

At least one GFCI receptacle per **210.52(G)** and **210.8(A)(2)** shall be installed in an attached or unattached garage.

A GFCI receptacle outlet shall not be required in a detached garage unless electric power is routed to the garage. Refer to **210.70(A)(2)(a)** for luminaire requirements. **(See Figure 16-19)**

RECEPTACLES IN HALLWAYS
210.52(H)

Hallways in dwelling units that are 10 ft (3 m) or more in length without passing through a doorway shall have a receptacle outlet installed. This receptacle outlet may be

used for the connection of plugged-in appliances such as table lamps, floor lamps, vacuum cleaners, etc. The length of the hallway shall be measured along its center line to determine whether a receptacle outlet is required. The receptacle outlet should be located so it serves the hallway in the most convenient manner. **(See Figure 16-20)**

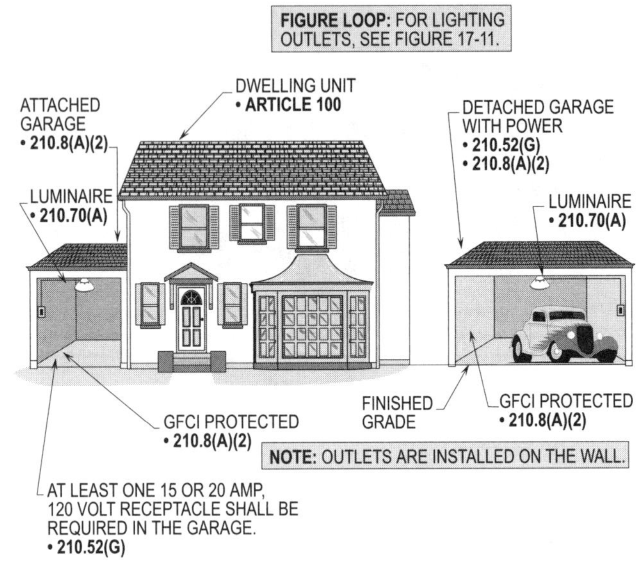

Figure 16-19. At least one 15 or 20 amp receptacle outlet shall be installed in the garage at 5 ft 6 in. (1.7 m) or less from the finished floor to be considered a required outlet.

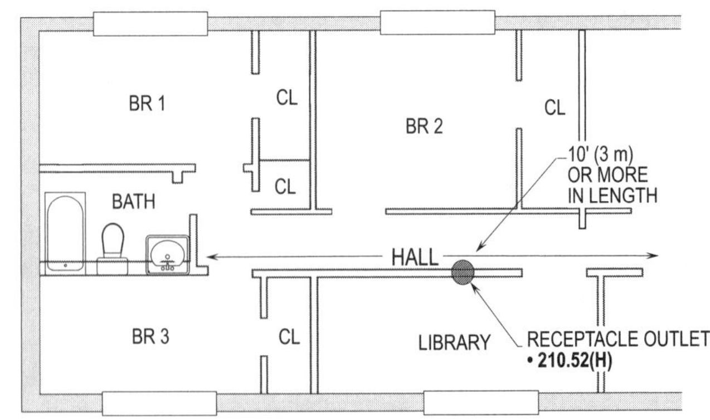

Figure 16-20. At least one receptacle outlet shall be installed in dwelling unit hallways that are 10 ft (3 m) or more in length without passing through a doorway.

GFCI PROTECTION OF RECEPTACLES
210.8(A)(1) THRU (A)(8)

A GFCI circuit breaker or receptacle detects any imbalance current in the circuit such as current leaking to ground through the body of a person using an electric hand tool. This imbalance of current trips open the GFCI unit at about 4 to 6 milliamps plus or minus 1 milliamp. The action of the GFCI prevents any harm to the person using the receptacle under adverse conditions of use.

It only takes a relatively small amount of current flowing through the body to be fatal. Electrocution may occur at about 380 milliamp. GFCI protection protects people from such hazards while working with electrical hand tools.

> **Design Tip:** GFCI protection of receptacles shall be required in bathrooms, garages, outdoors, basements, crawl spaces, kitchens, wet bars, and boathouses. **(See Figure 16-21)**

PROTECTION IN BATHROOMS
210.8(A)(1)

All 15 or 20 amp, 125 volt receptacles shall be provided with GFCI protection if located in the bathroom. (For the definition of bathroom, see **Article 100**.) The receptacles shall be supplied by a circuit protected by a GFCI circuit breaker or by a GFCI receptacle, or be served by a fed-through GFCI protected receptacle.

The NEC requires at least one receptacle outlet to be installed within 3 ft (900 mm) of the outside edge of each sink per **210.52(D)** and **210.8(A)(1)** and also requires this receptacle and all others, if present, to be GFCI protected. Sections **404.4** and **406.8(C)** do not permit a switch outlet to be installed in the bathtub area, where the elements of such devices may be sprayed with water.

> **Design Tip:** If a 20 amp, 120 volt laundry receptacle is installed in the bathroom, it is required to be GFCI protected per **210.8(A)(1)**. **(See Figure 16-22)**

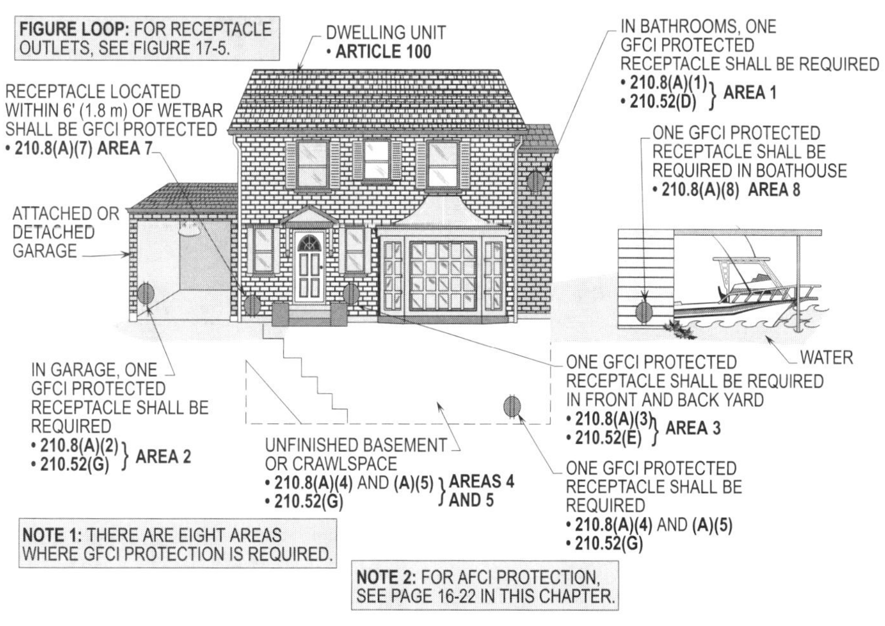

Figure 16-21. One or more GFCI protected receptacles shall be required for the bathroom, kitchen, garage, crawl space, unfinished basement, and the outdoors (front and back) of a dwelling unit. GFCI protected receptacles are also required within 6 ft (1.8 m) of the wet bar sink. At least one GFCI protected receptacle outlet shall be required for a boathouse.

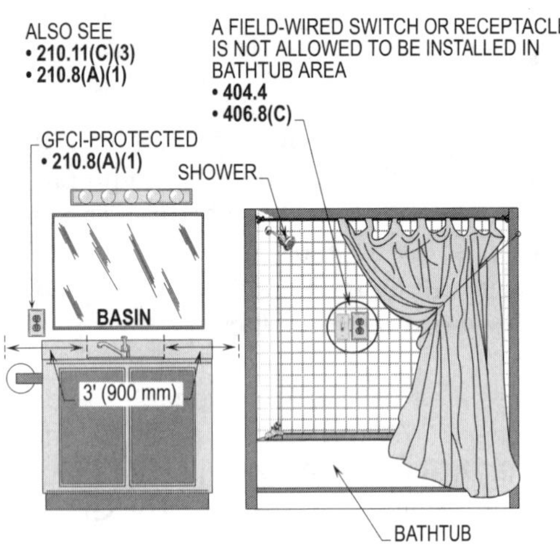

Figure 16-22. At least one GFCI receptacle outlet shall be installed within 3 ft (900 mm) of each basin in a bathroom.

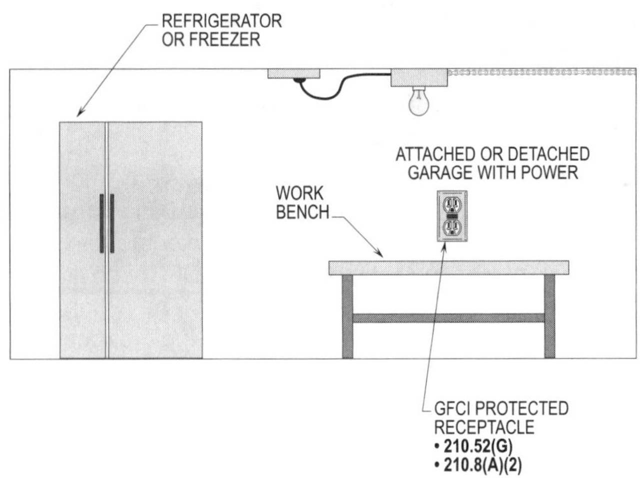

Figure 16-23. All 15 or 20 amp, 125 volt receptacles installed in garages shall be GFCI protected. **Note:** At least one shall be installed.

PROTECTION IN GARAGES
210.8(A)(2)

All 15 or 20 amp, 125 volt receptacles installed in garages shall have GFCI protection for the safety of personnel. Since the concrete slab is in direct contact with the earth, the garage shall be considered a hazardous area for people using electric hand tools. Leakage current from a faulty electric hand tool may flow through the human body, through the concrete slab, and through the earth to complete the circuit.

The NEC requires only one 15 or 20 amp, 125 volt receptacle outlet to be installed. If there are more installed, such as over a workbench, they shall be GFCI protected.

See Figure 16-23 for a detailed illustration pertaining to receptacles installed in garages.

PROTECTION IN ACCESSORY BUILDINGS
210.8(A)(2)

Accessory buildings (unfinished or finished) for a dwelling unit not intended to be used as a habitable room and limited to storage areas, work areas, or similar purposes, shall have GFCI protection for all 125 volt, single-phase, 15 and 20 amp receptacles. **(See Figure 16-24)**

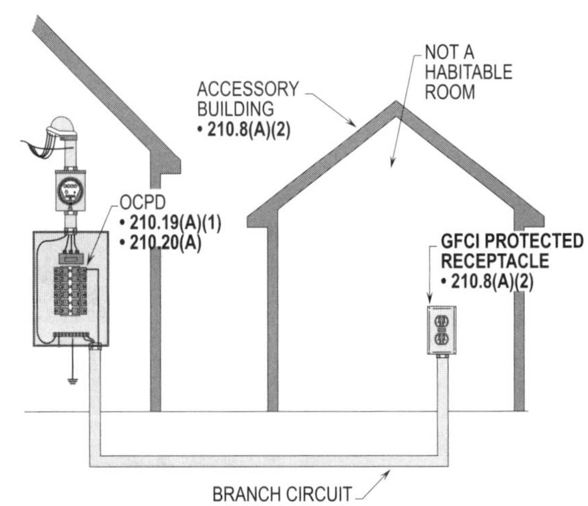

Figure 16-24. GFCI protection shall be provided for all 125 volt, single-phase, 15 and 20 amp receptacles for accessory buildings (unfinished or finished) for a dwelling unit.

PROTECTION OUTDOORS
210.8(A)(3)

All 15 or 20 amp, 125 volt receptacle outlets located outdoors that are installed 6 ft 6 in. (2 m) or less from direct grade level shall be GFCI-protected. Direct grade level is defined as being readily accessible and located 6 ft 6 in. (2 m) or less from grade level, while standing.

Receptacles at 6 ft 6 in. (2 m) or less from grade and located in the front and back yard of houses shall be GFCI protected.

Receptacles located in post lights at 6 ft 6 in. (2 m) or less shall be GFCI protected. Receptacle outlets located above 6 ft 6 in. (2 m), such as in eaves for the connection of Christmas lights or on balconies, shall also be GFCI protected. **(See Figure 16-25)**

At least one receptacle outlet shall be installed within the perimeter of the balcony, deck, or porch area if accessible from inside the dwelling unit. The receptacle shall not be located more than 6 ft 6 in. (2 m) above the balcony, deck, or porch surface.

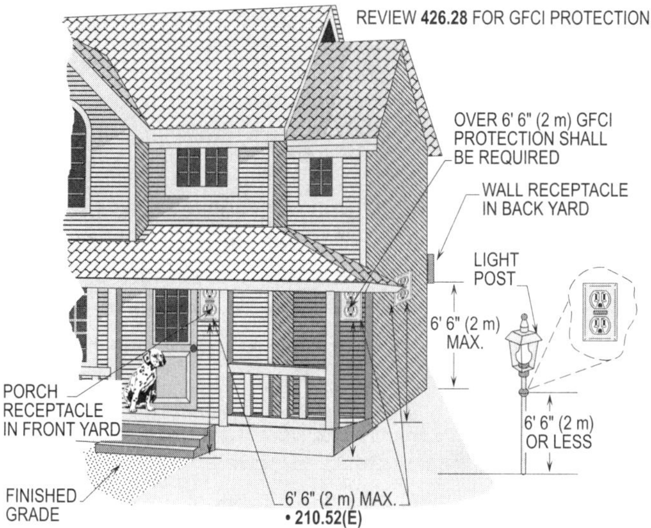

Figure 16-25. At least one receptacle outlet located outdoors in the front and back shall be installed 6 ft 6 in. (2 m) or less from finished grade and be readily accessible to the user. These outlets shall be GFCI protected.

PROTECTION IN BASEMENTS AND CRAWL SPACES
210.8(A)(4) AND 210.8(A)(5)

All 15 or 20 amp, 125 volt receptacle outlets installed in crawl spaces at or below grade level to serve electrical equipment shall be GFCI protected per **210.8(A)(4)**. Most mechanical codes require a receptacle outlet to be installed for the servicing of mechanical and related equipment.

See Figure 16-26 for rules pertaining to the installation and protection of receptacles in crawl spaces and basements.

At least one 15 or 20 amp, 125 volt receptacle outlet shall be installed in unfinished basements of dwelling units. These basements are unfinished and are not habitable. If they were habitable, they would be finished as a bedroom, den, game room, etc. and occupied by people. An unfinished basement is not trimmed out and may be utilized as a workshop area or storage area.

The **Exception** to **210.8(A)(5)** does not require a permanently installed fire alarm or burglar alarm system receptacle to be GFCI protected.

PROTECTION OVER COUNTERTOPS
210.8(A)(6) AND (A)(7)

All 15 or 20 amp, 125 volt receptacle outlets installed over kitchen countertops shall be GFCI protected. Receptacle outlets included are those located over or below the countertop and on islands or peninsulas that are supplied by the small appliance circuits per **210.11(C)(1)**, **210.52(B)**, and **210.52(C)**. **(See Figure 16-27)**

> **Design Tip:** If a pull chain luminaire with a receptacle outlet is installed over the kitchen sink, it shall be GFCI protected per **210.8(A)(6)**.

Receptacle outlets located over utility sinks shall not be required to be GFCI protected, because the sink does not have an additional fixture such as a toilet, a tub, or shower per **Article 100**. Receptacle outlets installed within 6 ft (1.8 m) of laundry, utility, and wetbar sinks shall be GFCI protected to protect personnel using blenders and other drink-related appliances per **210.8(A)(7)**.

Utility sinks are usually used to clean mops, etc., while wetbar sinks are utilized to entertain guests at parties and other social related functions in the home.

> **Design Tip:** A receptacle used to cord-and-plug connect a refrigerator shall not be required to be GFCI protected per **210.8(A)(6)**. **(See Figure 16-28)**

PROTECTION FOR BOATHOUSES
210.8(A)(8)

All receptacle outlets rated at 15 or 20 amps, 125 volts that are installed in boathouses to provide power for electric hand tools to work on boats shall be GFCI protected. Personnel using electric hand tools are exposed to wet locations due to the presence of water. GFCI protected circuits protect personnel from electrical shock while using electricity around such hazardous conditions. See **555.19(B)(1)** for such protection pertaining to marinas and boatyards. **(See Figure 16-29)**

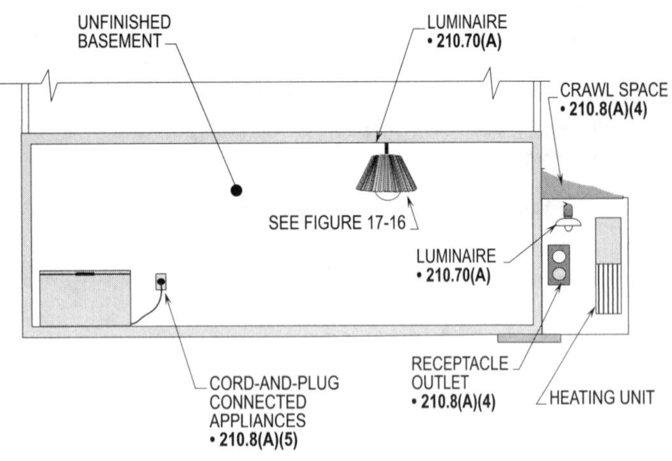

PROTECTION IN BASEMENTS AND CRAWL SPACES
NEC 210.8(A)(4)
NEC 210.8(A)(5)

Figure 16-26. All 15 or 20 amp, 125 volt receptacle outlets installed in crawl spaces at or below grade level to serve electrical equipment shall be GFCI protected.

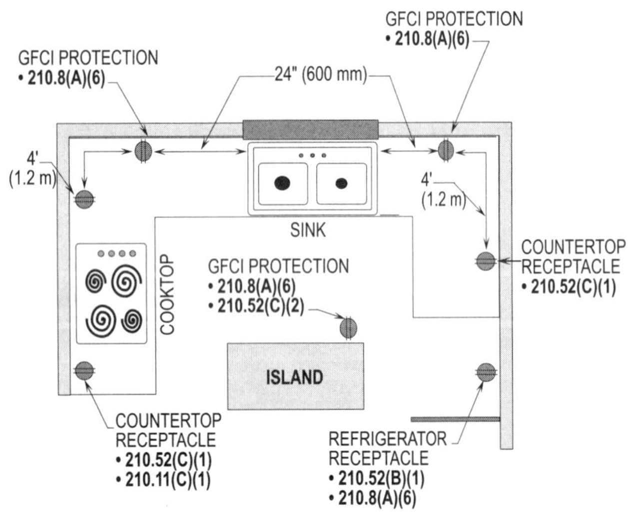

PROTECTION OVER COUNTERTOPS
NEC 210.8(A)(6)

Figure 16-27. All 15 or 20 amp, 125 volt receptacles serving kitchen countertop surfaces located above or below shall be GFCI protected. This rule is to protect the user.

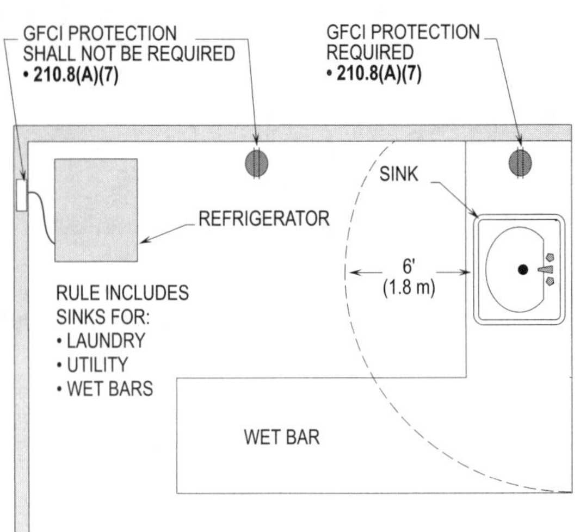

PROTECTION OVER COUNTERTOPS
NEC 210.8(A)(7)

Figure 16-28. All 15 or 20 amp, 125 volt receptacle outlets located within 6 ft (1.8 m) of a wet bar sink shall be GFCI protected except for a properly located refrigerator plug.

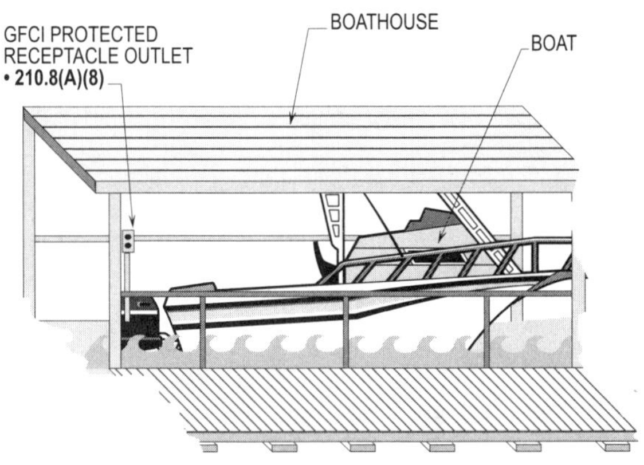

PROTECTION FOR BOATHOUSES
NEC 210.8(A)(8)

Figure 16-29. All 15 or 20 amp, 125 volt receptacle outlets shall be GFCI protected when installed in boathouses.

PROTECTION AROUND SWIMMING POOLS 680.22(A)(1) THRU (A)(5)

The following requirements for protection around swimming pools apply to residential, commercial, and industrial locations.

Receptacle outlets installed near swimming pools and fountains shall be properly located to minimize the threat of shock hazards to personnel in or around the pool. Receptacles shall not be permitted to be located within 6 ft (1.83 m) of the inside walls of the pool per **680.22(A)(2)** and **(A)(3)**.

At least one 15 or 20 amp, 125 volt receptacle outlet shall be installed at a minimum of 6 ft (1.83 m) from and not more than 20 ft (6 m) from the inside walls of the pool or fountain per **680.22(A)(2), (A)(3)** and **(A)(5)**. Receptacle outlets that are installed behind enclosed rooms with hinged or sliding doors, windows, or other barriers shall be permitted within the 6 ft (1.83 m) limitation of a swimming pool per **680.22(A)(5)**. The receptacle outlet located within the 20 ft (6 m) boundary of the swimming pool shall be GFCI-protected per **680.22(A)(4)**. [See Figures 16-30(a) and (b)]

Design Tip: See **680.22(A)(1)** that permits a single locking receptacle outlet to be located not less than 6 ft (1.83 m) from the inside walls of the pool. This receptacle outlet shall be GFCI protected per **680.22(A)(1)**. A cord-and-plug connected recirculating pool pump can easily be removed in winter. **(See Figure 16-31)**

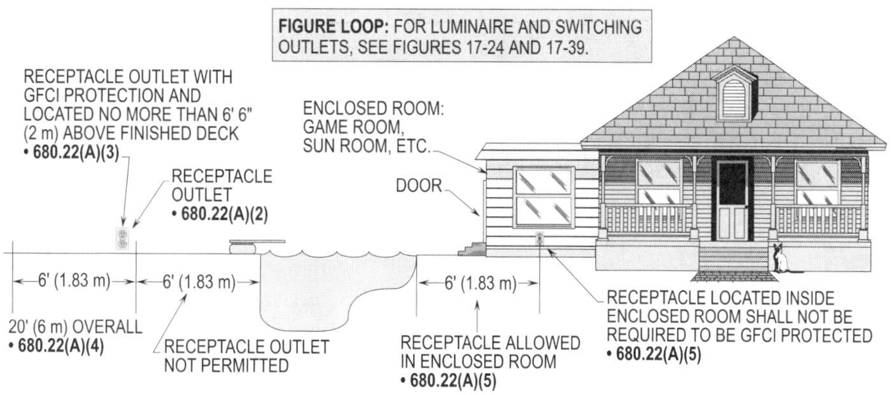

Figure 16-30(a). At least one GFCI protected receptacle shall be located in an area within 6 ft (1.83 m) to 20 ft (6 m) from the inside wall of the pool. Receptacles located inside enclosed rooms and within 6 ft (1.83 m) of the inside walls of the pool shall not be required to be GFCI protected. Personnel inside the enclosed room are not exposed to the pool area.

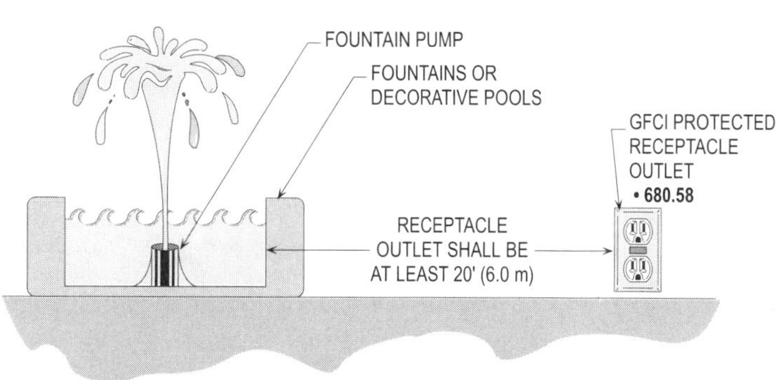

Figure 16-30(b). Receptacle outlets for fountains or decorative pools shall be located at least 20 ft (6.0 m) from the receptacle to the inside walls. **Note:** This receptacle is not required, but if installed, shall be GFCI protected.

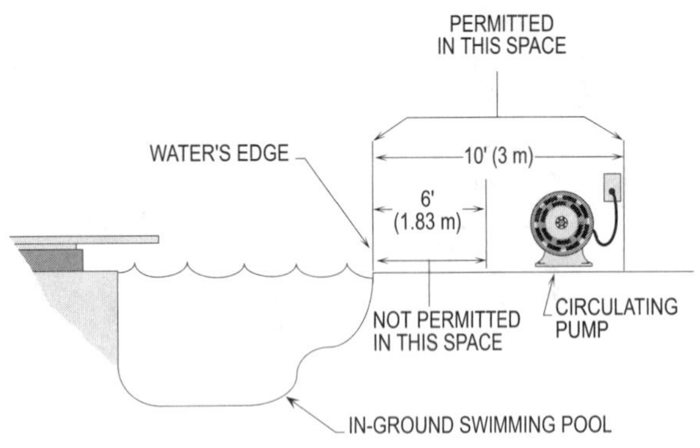

PROTECTION AROUND SWIMMING POOLS
NEC 680.22(A)(1)

Figure 16-31. A single locking type GFCI protected receptacle outlet shall be permitted to be installed between 6 ft (1.83 m) and 10 ft (3 m) from the inside walls of a swimming pool or fountain to supply power to a water pump motor.

PROTECTION FOR STORABLE POOLS
680.32

Storable pools shall be installed at least 6 ft (1.83 m) from any receptacle outlet per **680.22(A)(1)**. Electrical equipment associated with storage pools shall be provided with GFCI protection for the safety of personnel using the pool. See **90.7** and **110.3(B)**.

Design Tip: The cord supplying power from the receptacle outlet to the pool equipment shall also have GFCI protection. The supply cord may be longer than 3 ft (900 mm) per **680.7(A)**. There are UL listed supply cords of 25 ft (7.5 m) in length and approved to be used to cord-and-plug connect filter pumps to storable pools. The 3 ft (900 mm) limitation does not apply to cords used to connect filter pumps to storable pools. The GFCI protection shall be permitted to be provided by a GFCI protected receptacle or a remote GFCI protected receptacle or circuit breaker per **680.32**. **(See Figure 16-32)**

PROTECTION FOR SPAS OR HOT TUBS
680.43(A)(1) AND (A)(2)

Receptacle outlets shall be located at least 6 ft (1.83 m) from the inside walls of a spa or hot tub to ensure the safety of the user. Receptacle outlets installed 6 ft (1.83 m) to 10 ft (3 m) from the inside walls of the spa or hot tub shall be GFCI protected by a GFCI protected receptacle or circuit breaker.

At least one receptacle outlet shall be provided in this area for the protection of personnel using cord-and-plug connected radios, TVs, stereos, etc. Receptacle outlets supplying power to a spa or hot tub shall be GFCI protected per **680.43(A)(3)**. **(See Figure 16-33)**

Design Tip: See **680.42** for the installation requirements of spas or hot tubs that are located outdoors and apply **Parts I** and **II** of **Article 680** that pertain to swimming pools.

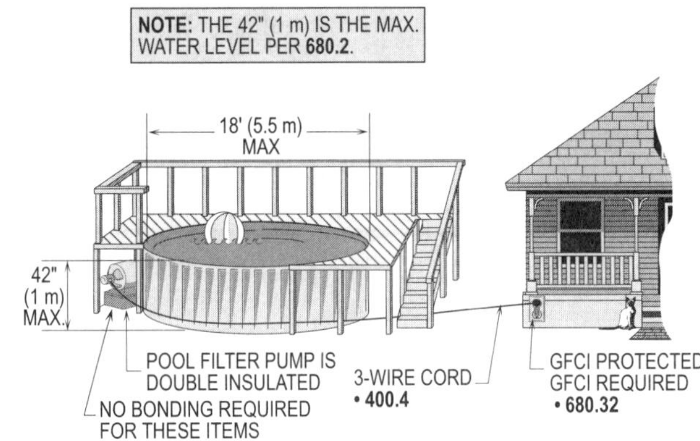

PROTECTION FOR STORABLE POOLS
NEC 680.32

Figure 16-32. Electrical equipment that is cord-and-plug connected shall be served by a GFCI protected receptacle outlet or circuit.

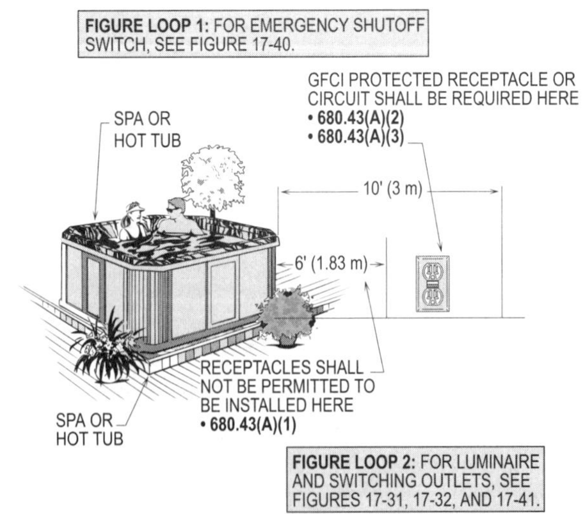

PROTECTION FOR SPAS OR HOT TUBS
NEC 680.43(A)(1) AND (A)(2)

Figure 16-33. Receptacle outlets installed around a spa or hot tub that are located inside the dwelling unit. For spa or hot tubs located outside, see **680.40**.

PROTECTION FOR HYDROMASSAGE TUBS
680.71

Receptacle outlets installed near hydromassage tubs shall be provided with GFCI protection per **210.8(A)(1)**. Hydromassage tubs are treated just like regular bathtubs per **680.72**. All electrical elements for hydromassage tubs shall be supplied by GFCI circuits per **680.71**. The wiring methods shall comply with the provisions of **Chapters 1 through 4** in the NEC, which are to be applied generally. Circulating motors shall comply with the requirements of **430.14(A)** and **680.73** for accessibility, so that proper maintenance and adequate ventilation may be provided. **(See Figure 16-34)**

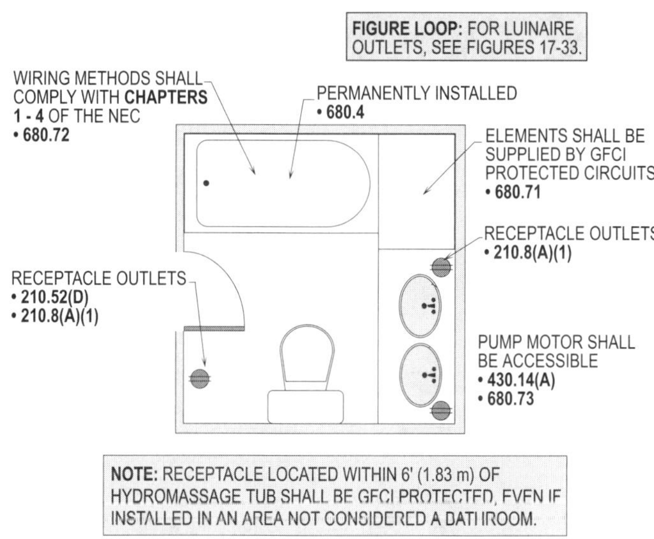

Figure 16-34. Hydromassage bathtubs are treated as regular bathtubs and shall comply with **680.70, 680.71, 680.72,** and **680.73** and other pertinent Sections of the NEC.

RECEPTACLES ON CONSTRUCTION SITES
590.6

All 15, 20, and 30 amp, 125 volt receptacles used by personnel during construction, remodeling, maintenance, repair, or demolition purposes shall be GFCI protected. GFCI protection shall be provided so that workers are protected from electrical shock. There are three methods by which this may be accomplished.

- A GFCI circuit breaker shall be permitted to be used to supply power to all receptacle outlets. The GFCI circuit breaker protects all receptacle outlets downstream.

- A GFCI receptacle shall be permitted to be installed at each outlet. This provides protection at each location of use and may be reset immediately when tripped open due to a faulty hand tool, etc.

- A portable GFCI receptacle shall be permitted to be used by workers that is protected by a regular circuit breaker. Each worker carries a portable GFCI and plugs all electric hand tools into a plain receptacle outlet. If the GFCI receptacle trips for any reason, it can be reset at the work location without workers losing time. **(See Figure 16-35)**

The authority having jurisdiction (AHJ) has the authority to approve a written procedure that requires three-wire extension cords with equipment grounding conductors. Personnel shall be designated to enforce the program utilizing the following procedures:

- All equipment grounding conductors shall be tested for continuity and shall be electrically continuous.

- Each receptacle and attachment plug shall be tested for correct attachment of the equipment grounding conductor.

- All required tests shall be performed as follows:
  (a) Before first use on site
  (b) When there is evidence of damage
  (c) Before equipment is returned to service following any repairs
  (d) At intervals not exceeding 3 months

Note, the assured equipment grounding conductor program only applies to industrial sites where conditions of maintenance and supervision ensure that only qualified personnel are involved per **590.6(B)(2)** and **590.6(A), Ex.**

Receptacle outlets shall not be connected to any branch circuits supplying power to lighting outlets per **590.4(D)**. The reason for this rule is to prevent a ground fault on an electric hand tool from opening the overcurrent protection device on the circuit and putting the worker in the dark, which could be dangerous.

Design Tip: Section **590.6(B)(2)(a)(1)** and **OSHA 1926, Subpart K** must be studied carefully and applied properly to protect the users of such cords and provide a safe workplace for employees building or remodeling dwelling units.

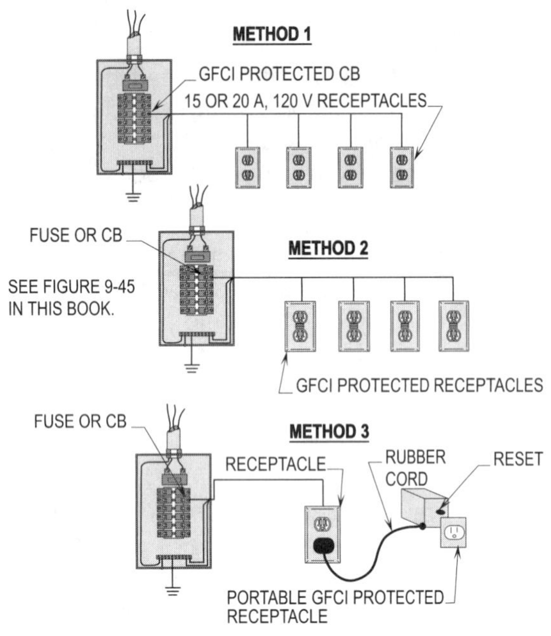

RECEPTACLES ON CONSTRUCTION SITES
NEC 590.6(A)

Figure 16-35. Three methods of providing GFCI protection for construction sites to protect personnel.

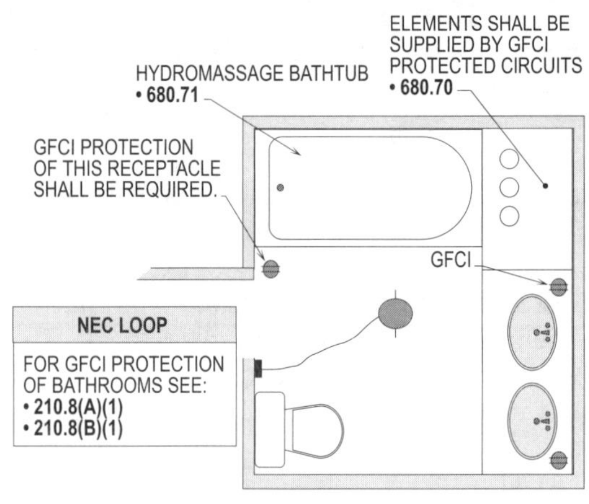

RECEPTACLES INSTALLED IN BATHROOMS OF COMMERCIAL AND INDUSTRIAL LOCATIONS
NEC 210.8(B)(1)

Figure 16-36. In the bathrooms of commercial and industrial locations, GFCIs shall be installed for the protection of personnel.

RECEPTACLES INSTALLED IN BATHROOMS OF COMMERCIAL AND INDUSTRIAL LOCATIONS
210.8(B)(1)

In the bathrooms of commercial and industrial locations, GFCIs shall be installed for the protection of personnel. This will include all 125 volt, 15 and 20 ampere circuits in the bathroom. Note that such receptacle outlets shall not be required to be installed, but if they are, they shall be GFCI protected. By definition, a bathroom is an area including a basin and at least one of the following: a tub, a toilet, a shower, or a combination. **(See Figure 16-36)**

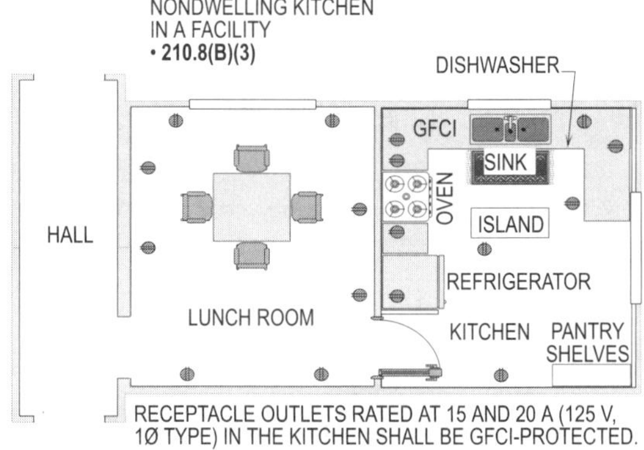

RECEPTACLES INSTALLED IN KITCHENS OF COMMERCIAL AND INDUSTRIAL BUILDINGS
NEC 210.8(B)(2)

Figure 16-37. All 125 volt, 15 or 20 amp receptacles installed on roofs shall have ground fault protection for personnel.

RECEPTACLES INSTALLED IN KITCHENS OF COMMERCIAL AND INDUSTRIAL BUILDINGS
210.8(B)(2)

All 125, 15 or 20 amp receptacles installed in kitchens of commercial and industrial buildings shall have ground fault protection for personnel. **(See Figure 16-37)**

RECEPTACLES INSTALLED ON ROOFTOPS OF COMMERCIAL AND INDUSTRIAL BUILDINGS
210.8(B)(3)

All 125 volt, 15 or 20 amp receptacles installed on roofs shall have ground-fault protection for personnel. Note that this does not apply to receptacles installed on the roofs of dwelling units. **(See Figure 16-38)**

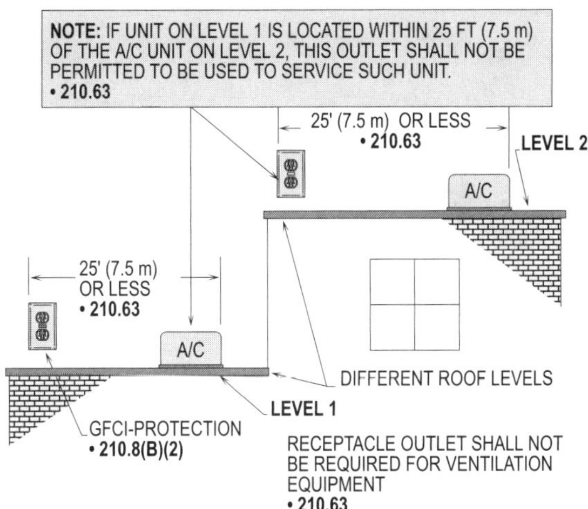

Figure 16-38. All 125, 15 or 20 amp receptacles installed in kitchens of commercial and industrial buildings shall have ground fault protection for personnel.

RECEPTACLES INSTALLED OUTDOORS
210.8(B)(4)

All 125 volt, 15 and 20 amp receptacles installed outdoors shall have GFCI protection. **(See Figure 16-39)**

RECEPTACLES USED FOR MAINTENANCE ACTIVITIES
590.6(A)

All 125 volt, 15, 20 and 30 amp receptacles used for electric hand tools, etc. to pull maintenance on electrical apparatus and equipment shall be GFCI protected. **(See Figure 16-40)**

AFCI PROTECTED OUTLETS
210.12(A) AND (B)

All 120 volt, single-phase, 15- and 20-ampere branch circuits supplying outlets in the following rooms of a dwelling unit shall be protected by a listed arc-fault circuit interrupter, combination type, installed to provide protection of the branch circuit:

- Family rooms,
- Dining rooms,
- Living rooms,
- Parlors,
- Libraries,
- Dens,
- Bedrooms,
- Sun rooms,
- Recreation rooms,
- Closets,
- Hallways, or
- Similar rooms or areas

See **Figure 16-41** for a detailed illustration pertaining to AFCI protection in dwelling units.

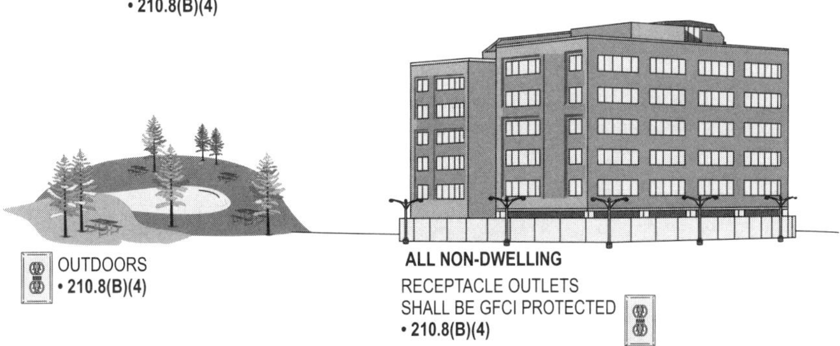

Figure 16-39. GFCI protection shall be required for all 125 volt, 15 and 20 amp receptacles installed outdoors.

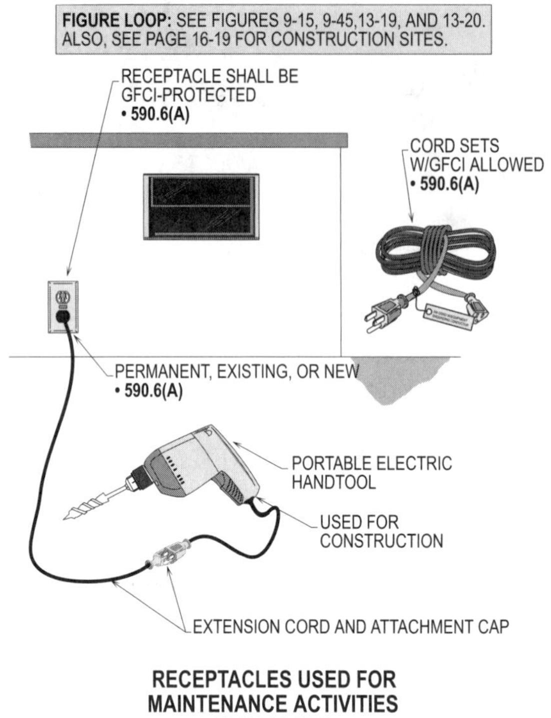

Figure 16-40. Employees using electric hand tools to pull maintenance on electrical equipment shall be fully protected by GFCI protection of power circuit and elements.

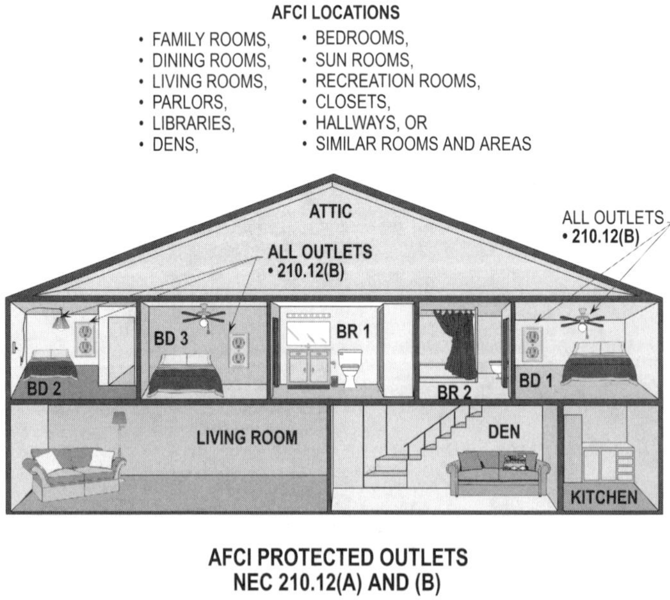

Figure 16-41. Certain outlets in bedrooms of dwelling units shall be AFCI protected.

Chapter 16. Receptacle Outlets

Section   **Answer**

1. A minimum of 2 - 20 amp, _______ VA small appliance circuits are required to supply receptacle outlets in dwelling units that are located in the kitchen, pantry, breakfast room, and dining room.
   - (a) 1000
   - (b) 1200
   - (c) 1500
   - (d) 1800

2. Receptacle outlets shall be installed so that there is no point on the wall greater than _______ ft from a receptacle outlet.
   - (a) 2
   - (b) 4
   - (c) 5
   - (d) 6

3. An island countertop with a short dimension of _______ in. or greater shall have a receptacle outlet for each 4 ft in a dwelling unit.
   - (a) 10
   - (b) 12
   - (c) 18
   - (d) 24

4. A bathroom is an area including a basin(s), with a _______.
   - (a) toilet
   - (b) tub
   - (c) shower
   - (d) all of the above

5. A receptacle outlet shall be installed, at a minimum of _______ ft from and not more than 20 ft from the inside walls of a pool. (General Rule)
   - (a) 5
   - (b) 6
   - (c) 10
   - (d) 12

6. Receptacle outlets installed by a hydromassage tub shall be provided with GFCI protection if they are located within _______ ft.
   - (a) 6
   - (b) 8
   - (c) 10
   - (d) 12

7. Receptacle outlets shall be installed for each peninsular countertop with a long dimension of _______ in. or greater and a short dimension of 12 in. or greater.
   - (a) 12
   - (b) 18
   - (c) 24
   - (d) 30

8. Grade level access, while standing, shall be considered _______ or less from finished grade.
   - (a) 5 ft 6 in.
   - (b) 6 ft 6 in.
   - (c) 7 ft 6 in.
   - (d) 8 ft

9. At least _______ GFCI protected receptacle shall be installed in a basement.
   - (a) 1
   - (b) 2
   - (c) 3
   - (d) 4

10. Receptacles located inside enclosed rooms and within _______ ft of the inside walls of the pool shall not be required to be GFCI protected.
    - (a) 4
    - (b) 5
    - (c) 6
    - (d) 10

___________ ___________ **11.** Receptacle outlets shall be installed so that there is no point on the wall is greater than _____ from a receptacle.
    (a) 4 ft   (b) 5 ft
    (c) 6 ft   (d) 10 ft

___________ ___________ **12.** Receptacle outlets in a dwelling unit shall not serve as one of the required outlets if located more than _____ from the floor.
    (a) 4 ft 6 in.   (b) 5 ft 6 in.
    (c) 6 ft 6 in.   (d) 7 ft 6 in.

___________ ___________ **13.** What is considered wall space when spacing receptacles?
    (a) sliding glass doors   (b) sliding panels
    (c) fixed panels   (d) none of the above

___________ ___________ **14.** Which of the following countertops are present in a dwelling unit?
    (a) wall type   (b) peninsular
    (c) island   (d) all of the above

___________ ___________ **15.** All receptacle outlets in a dwelling unit installed in kitchens shall be GFCI protected if located within:
    (a) 4 ft 0 in. of the sink   (b) 5 ft 6 in. of the sink
    (c) 6 ft 6 in. of the sink   (d) all of the above

___________ ___________ **16.** What is the minimum wall space for receptacle outlets to be installed in a dwelling unit?
    (a) 1 ft   (b) 2 ft
    (c) 6 ft   (d) 12 ft

___________ ___________ **17.** When installing receptacles in a dwelling unit for hallways, at least one receptacle outlet shall be installed at a minimum of _____ ft or more in length.
    (a) 2   (b) 6
    (c) 10   (d) 12

___________ ___________ **18.** Receptacle outlets shall be GFCI protected if they are located from the inside walls of an indoor spa or hot tub at _____.
    (a) 2 ft to 10 ft   (b) 6 ft to 10 ft
    (c) 10 ft to 20 ft   (d) 10 ft to 25 ft

___________ ___________ **19.** All laundry equipment shall be located with _____ ft of the receptacle outlet.
    (a) 6   (b) 8
    (c) 10   (d) 12

___________ ___________ **20.** At least one receptacle outlet shall be installed in bathrooms within _____ ft of the outside edge of each basin in a dwelling unit.
    (a) 2   (b) 3
    (c) 5   (d) 6

Lighting and Switching Outlets

Lighting outlets can be installed in specified locations to ensure the proper illumination for residential, commercial, and industrial locations. Lighting shall be provided by luminaires, or be controlled by a wall switch or by table lamps, floor lamps, swag lamps, etc., that are cord-and-plug connected into wall switch controlled receptacles. Pull-chain luminaires may be installed without wall switches in some locations and under certain conditions of use.

Luminaires supported by metallic or nonmetallic boxes are either ceiling mounted or wall mounted. For the convenience of the user, wall-switched receptacle outlets can be mounted to boxes installed in the wall, baseboard, or floor and may be used to cord-and-plug connect table or floor lamps.

Switches may be installed to operate luminaires and receptacles from more than one location in a residential, commercial, or industrial location. Switching outlets are mounted on the wall at convenient locations and heights to switch the lighting outlets or receptacle outlets on and off. Only in residential occupancies are lighting outlets and switches mandated by the *National Electrical Code* to be installed in specific locations to switch on and illuminate certain areas.

GROUNDING LUMINAIRES
410.42(A) AND 410.46

In residential, commercial, and industrial locations, the exposed metal parts of luminaires shall be connected to an equipment grounding conductor if the branch circuit is provided with an equipment grounding conductor. The equipment grounding conductor shall be selected from any of the wiring methods listed in **250.118** and sized per **Table 250.122,** based upon the size of the overcurrent protection device. Section **410-92** of the 1971 NEC and earlier editions required luminaires with metal parts to be used with metallic wiring systems such as metal conduit or metal-clad cables.

The metal-clad AC cables (BX) and metal-clad cables (MC) were used to ground the exposed metal parts. The metal of metal conduits such as rigid metal conduit and EMT was mostly used as a grounding means. Copper or aluminum conductors were also pulled in conduits and utilized as an additional grounding means. Because nonmetallic wiring systems were not always equipped with an equipment grounding conductor, the 1975 NEC in **410-18(a)** required the exposed metal parts of luminaires to be grounded with an approved grounding means per **250-91(b)**. Section **250-114(a)** required the metal of boxes supporting luminaires to be grounded with the metal of the conduit or cable or with an equipment grounding conductor. Since the 1975 NEC, luminaires with exposed metal parts have been required to be grounded in new or existing installations. **(See Figure 17-1)**

UNGROUNDED LUMINAIRES
410.42(B)

The branch-circuit wiring in older sites of residential, commercial, and industrial locations does not have an equipment grounding conductor in the nonmetallic sheathed cable (Romex) or in knob-and-tube wiring systems to ground the exposed metal parts of luminaires. Luminaires in older facilities were not required by the NEC to be grounded with an equipment grounding conductor until the 1975 edition of the code.

Part R and **410-91** through **410-93** in the 1971 NEC (grounding lighting fixtures) was deleted and relocated to **Part E** and **410-18(a)** and **(b)** of the 1975 NEC.

The **Ex.** to **410-93** in the 1971 NEC permitted luminaires with metal parts to be wired with nonmetallic raceways and nonmetallic sheathed cables. If a metal cable was used as a wiring method, it had to have a listed grounding means. To accomplish this rule, the cable had to be AC (BX) or metal clad (MC) of the grounding type.

In those days, nonmetallic wiring systems were systems that did not have an equipment grounding means. Such wiring systems were knob-and-tube, nonmetallic raceways, or nonmetallic sheathed cable (Romex).

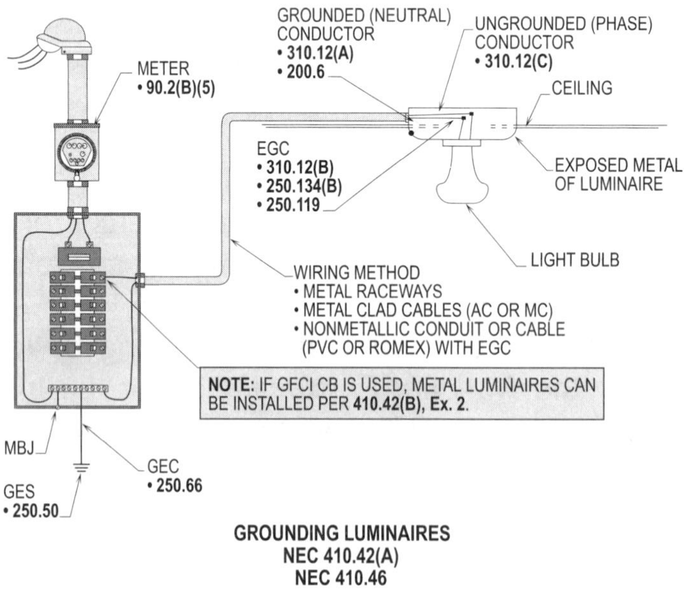

Figure 17-1. The metal parts of luminaires have been required to be grounded in new or existing work since the 1975 NEC. For exceptions to the rule, see **Ex.** to **410.18** in the 1975 NEC.

The **Ex.** to Sec. **410-93** in the 1971 NEC allowed metal luminaires and boxes that were mounted on nonconducting ceilings or walls and located not less than 8 ft (2.5 m) vertically or 5 ft (1.5 m) horizontally from grounded surfaces to be installed without grounding the metal of the luminaires and boxes. **(See Figure 17-2)**

> **Design Tip:** The 1971 NEC and earlier editions required metal boxes that were not grounded to be located 5 ft (1.5 m) from a bathtub or shower. It was the metal box and not the snap switch that required either grounding or being located 5 ft (1.5 m) from the bathtub or shower. (See **404.4** of the 2002 NEC.)

The 1975 NEC required the metal boxes to be grounded with an equipment grounding conductor that had to be installed in nonmetallic cables from the factory, or the use of metal conduits, or the metal cladding of cables had to provide such grounding. Therefore, the 5 ft (1.5 m) requirement was deleted and a grounding means was no longer required to be provided in the wiring method. See **404.4** and **406.8(C)** for rules pertaining to field wiring switches and receptacle outlets in the bathtub area.

> **Design Tip:** Boxes were not required for the mounting of luminaires until the 1928 edition of the NEC. Because of this requirement, older existing occupancies may not have boxes installed for the mounting and supporting of luminaires and switches.

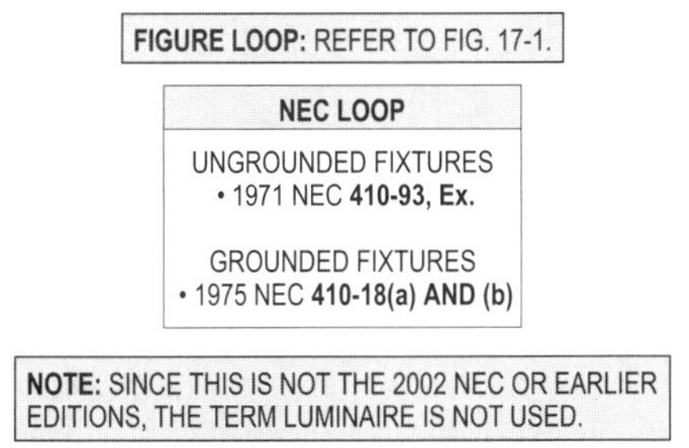

UNGROUNDED LUMINAIRES
NEC 410.42(B)

Figure 17-2. The metal exposed parts of luminaires were not required to be grounded by the 1971 NEC and earlier editions. The 1975 NEC and later editions do require exposed metal parts of luminaires to be grounded.

REPLACEMENT LUMINAIRES 410.42(B)

Luminaires with exposed metal parts shall not be used to replace luminaires on existing wiring systems that are not equipped with an equipment grounding means. Section **410-18(b)** in the 1975 NEC as well as recent editions require a luminaire with an insulated material to be used to replace an existing metal luminaire in an existing wiring system. If luminaires with exposed metal parts are used for replacements, a branch circuit with an equipment grounding conductor shall be provided per **410.42(B)**. **(See Figure 17-3)**

Care shall be taken when replacing an existing luminaire with a new one. If a luminaire with exposed metal parts is installed as a replacement, use an insulating nonmetallic type or any insulating type that isolates the metal of the luminaire should be used. The branch circuit shall have a wiring method that has an equipment grounding conductor or shall be installed as permitted by the one of the exceptions to **410.42(B)**. See **250.118, 250.122,** and **250.134(B)** for the selection, routing, and termination of the equipment grounding conductor. Note that the above rules and regulations apply to residential, commercial, and industrial locations.

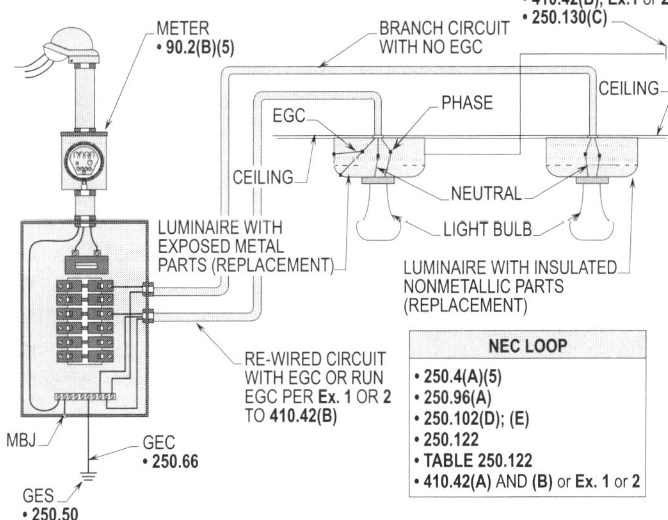

REPLACEMENT LUMINAIRES
NEC 410.42(B)

Figure 17-3. Since the publication of the 1975 NEC, replacement luminaires installed on branch circuits without an equipment grounding conductor must be of the insulated or nonmetallic type. The parts of luminaires that are not insulated to isolate the exposed metal parts of the luminaires shall be wired with a branch circuit having an equipment grounding conductor to ground all metal parts or an equipment grounding conductor can be installed, as permitted by **410.42(B), Ex. 1** or **2.**

NUMBER ON A CIRCUIT
210.11(A)

In residential dwelling units, the number of lighting outlets that may be connected to a 15 or 20 amp general purpose branch circuit can be determined by multiplying the rating of the branch circuit by 120 volts and dividing by 3 VA per square foot.

> **For example:** The number of lighting outlets allowed on a 15 amp branch circuit in a dwelling unit is determined by the following procedure.
>
> **Step 1:** Finding VA
> **210.11(A)**
> VA = V x OCPD
> VA = 120 V x 15 A
> VA = 1800
>
> **Step 2:** Finding number of outlets
> No. = VA ÷ 3 VA sq. ft
> No. = 1800 ÷ 3
> No. = 600 sq. ft
>
> **Solution:** **In a residential dwelling unit, there is no limit to the number of lighting outlets permitted in the 600 sq. ft area.**

The concept of the number of outlets allowed on a 15 or 20 amp general purpose branch circuit is more easily understood by referencing **220.12**, **Table 220.12,** and **210.11(A)** and **(B)**. **Table 220.12** requires 3 VA per sq. ft times the sq. ft area of a dwelling unit to determine the VA rating to supply the number of outlets for general purpose lighting and receptacle outlets. Notice that the reference [a], by "dwelling units," **Table 220.12,** refers to the footnote and verifies this procedure for determining the number of outlets on a branch circuit.

Inspection authorities who disagree with this concept and want to limit the number usually apply the 1.5 amp (180 VA ÷ 120 V = 1.5) method per **220.14(I)**. The rating of the overcurrent protection device of 15 amps is divided by 1.5 amp to determine the number of lighting outlets permitted on a 15 or 20 amp branch circuit. A 15 amp overcurrent protection device divided by 1.5 amps (15 A OCPD ÷ 1.5 A = 10) permits ten lighting outlets to be connected to a 15 amp general purpose branch circuit.

The authority having jurisdiction (AHJ) may permit any number of lighting outlets (high or low) by the local electrical ordinance to be connected to a 15 or 20 amp general purpose branch circuit.

For example, by applying such local codes, a 15 amp general purpose branch circuit is permitted to have ten outlets installed for lighting and a 20 amp circuit is permitted to have only 13 outlets for lighting. **[See Figure 17-4(a)]**

> **For example:** Finding the number of lighting outlets on a 20 amp general purpose branch circuit.
>
> **Step 1:** Finding VA
> **210.11(B)**
> VA = V x OCPD
> VA = 120 x 20
> VA = 2400
>
> **Step 2:** Finding number of outlets
> No. = VA ÷ 3 VA per sq. ft
> No. = 2400 ÷ 3 VA
> No. = 800 sq. ft
>
> **Solution:** **In a residential dwelling unit, there is not a limit to the number of lighting outlets permitted in the 800 sq. ft area.**

> **Design Tip:** There are cases where the AHJ wishes to limit the number of outlets on a circuit; see text for methods used to determine the number of lighting outlets allowed on a general purpose branch circuit.

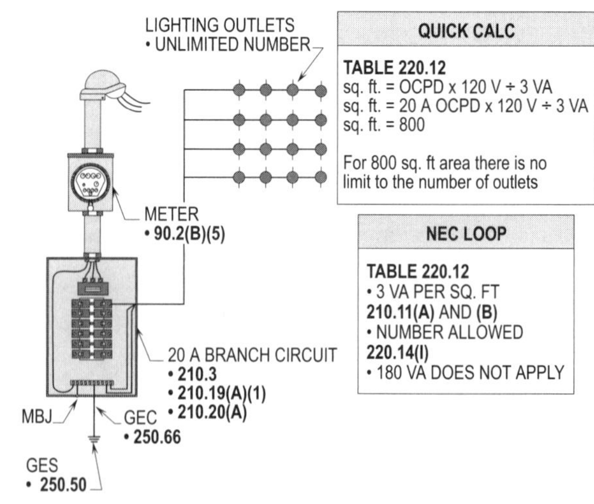

Figure 17-4(a). Section **210.11(B)** can be used to determine the number of lighting outlets permitted on a general purpose branch circuit for a dwelling unit.

NUMBER ON A CIRCUIT
210.11(A)

In commercial and industrial locations, the outlets on a general purpose branch circuit shall be calculated at 180 VA each or the load rating, whichever is greater. The number of outlets times 180 VA, times 100 percent, is used to calculate the load of a branch circuit supplying outlets of noncontinuous operation. The number of outlets times 180 VA times 125 percent is used to calculate the load of a branch circuit supplying outlets of continuous operation. Overcurrent protection devices and conductors shall be calculated at 125 percent and selected per **240.4** and **240.6(A)** for overcurrent protection devices and per **Table 310.16** for conductors.

For example: What is the VA rating for 13 receptacle outlets supplying cord-and-plug connected loads used at noncontinuous operation?

Noncontinuous operation

Step 1: Finding VA
220.14(I); 210.19(A)(1)
VA = No. of outlets x 180 VA x 100%
VA = 13 x 180 VA x 100%
VA = 2340

Solution: The VA is 2340 and the number of outlets at noncontinuous operation is limited to 13 as calculated.

For example: What is the VA rating for 10 receptacle outlets supplying cord-and-plug connected loads used at continuous operation?

Continuous operation

Step 1: Finding VA
220.14(I); 210.19(A)(1)
VA = No. of outlets x 180 VA x 125%
VA = 10 x 180 VA x 125%
VA = 2250

Solution: The VA is 2250 for 10 outlets used at continuous operation.

See Figure 17-4(b) for calculating the number of outlets allowed on a branch circuit for commercial and industrial locations.

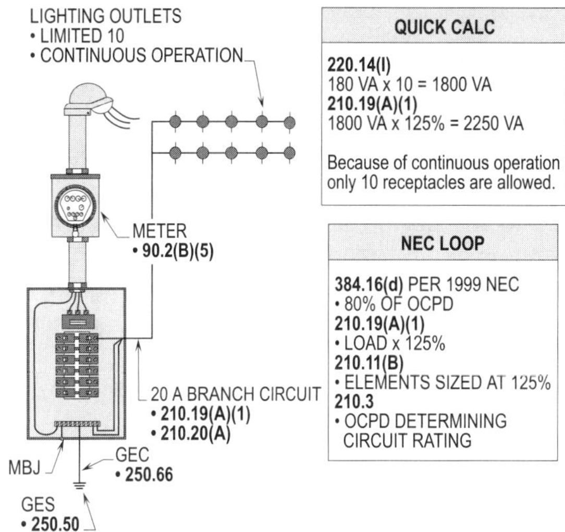

COMMERCIAL AND INDUSTRIAL RECEPTACLES
NEC 210.11(B)

Figure 17-4(b). Receptacles in commercial and industrial locations shall be calculated at continuous or noncontinuous operation based on each outlet at a minimum of 180 VA.

LIGHTING OUTLETS IN DWELLING UNITS
210.70(A)

At least one wall switch-controlled lighting outlet is required in all habitable rooms, halls, stairways, attached garages or detached garages with electric power, bathrooms, and outdoor exits and entrances to the dwelling unit. Kitchens and bathrooms shall have a lighting outlet on the ceiling or wall that is controlled by a wall switch. In addition, a luminaire shall be installed in an unfinished or finished basement, attic, or crawl space used for storage or for air handling equipment, etc. Note that in some cases, these rules pertaining to lighting outlets may be applied to commercial and industrial locations as well as residential. Such will be noted in the text when appropriate. **(See Figure 17-5)**

LIGHTING OUTLETS IN HABITABLE ROOMS
210.70(A)(1)

At least one lighting outlet shall be installed to provide lighting for the illumination of habitable rooms. Habitable rooms are rooms in the dwelling unit such as the bedroom, living room, den, dining room, breakfast room, etc. The lighting outlets may be installed in the ceiling or on the wall if the location in which they are mounted provides proper lighting. Section **210.70(A)(1), Ex. 1** allows wall-switched receptacle outlets to be mounted on the wall at a height such that floor lamps or table lamps may be cord-and-plug connected to provide proper lighting. The only rooms in a dwelling unit not allowed to have a wall-switched receptacle outlet to provide lighting are the kitchen and bathroom(s). The kitchen is required to have at least one lighting outlet mounted to the ceiling or wall that is switched by a wall switch. A pull-chain lighting outlet installed over the sink or a lighting outlet in the vent-a-hood is in addition to, and shall not be counted as, the lighting outlet required per **210.70(A)(1)**. **(See Figure 17-6)**

> **Design Tip: Ex. 2** to **210.70(A)(1)** allows lighting outlets to be controlled by occupancy sensors, listed for such use. This rule allows an occupancy sensor to control lighting outlets in habitable rooms of dwelling units, which includes bathrooms, hallways, stairways, and garages and at each outdoor entrance and exit. However, a manual override that will allow the sensor to function as a wall switch shall be provided.

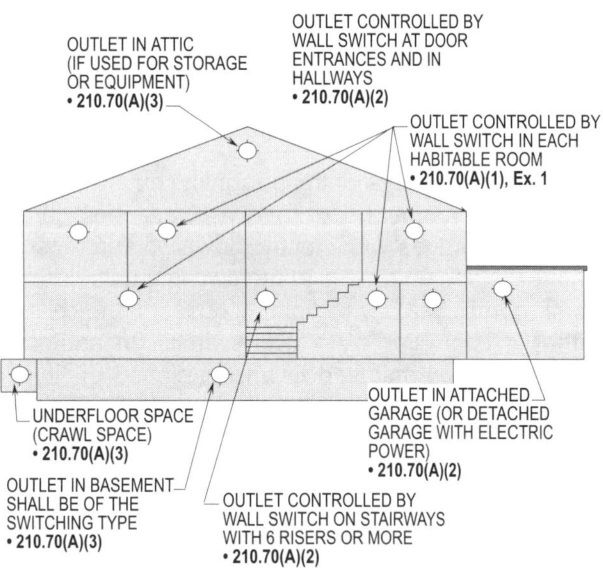

LIGHTING OUTLETS IN DWELLING UNITS
NEC 210.70(A)

Figure 17-5. At least one lighting outlet shall be installed in these locations to provide proper lighting for safety, pleasure, etc. For receptacle outlets, see **Figure 16-9**.

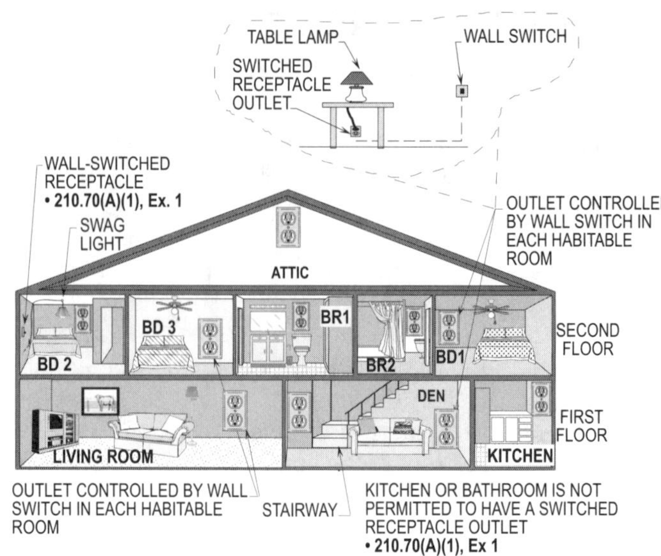

LIGHTING OUTLETS IN HABITABLE ROOMS
NEC 210.70(A)(1)

Figure 17-6. Habitable rooms in a dwelling unit shall be provided with a wall or ceiling lighting outlet switched by a wall switch. Rooms, except the kitchen and bathroom(s), shall be permitted to have a switched receptacle outlet with cord-and-plug connected table lamps or floor lamps to provide the necessary lighting.

LIGHTING OUTLETS IN BATHROOMS
210.70(A)(1)

Lighting outlets are required in bathrooms to provide lighting for bathing and personal care. The lighting outlets may be installed in the ceiling or on the wall above the mirror. Lighting outlets are sometimes installed over bathtubs or in showers to prevent shadows due to the location of the required lighting outlets per **210.70(A)(1)**. Luminaires installed over bathtubs or in showers are usually surface gasket or recessed type. At least one lighting outlet shall be provided in the bathroom, and it shall be wall-switched. Check with the inspector for the type that is allowed. **(See Figure 17-7)**

> **Design Tip:** The lighting outlet could be a combination vent/fan/heater/luminaire that complies with **210.70(A)(1)**.

WALL-SWITCHED RECEPTACLE LIGHTING OUTLETS
210.70(A)(1), Ex. 1

A wall-switched receptacle outlet may be used in lieu of a wall-switched lighting outlet in habitable rooms other than kitchens and bathrooms. A wall-switched receptacle with cord-and-plug connected table lamps or floor lamps that are used to provide lighting per **210.70(A)(1), Ex. 1** may be used. **(See Figure 17-8)**

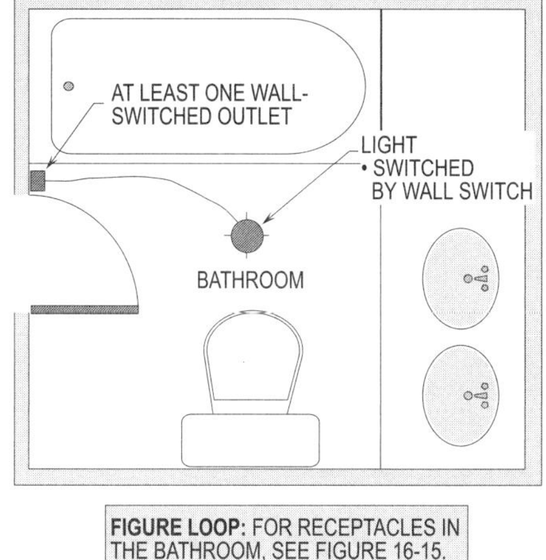

Figure 17-7. Bathrooms are required to have a wall-switched lighting outlet. Other outlets are in addition to this switch.

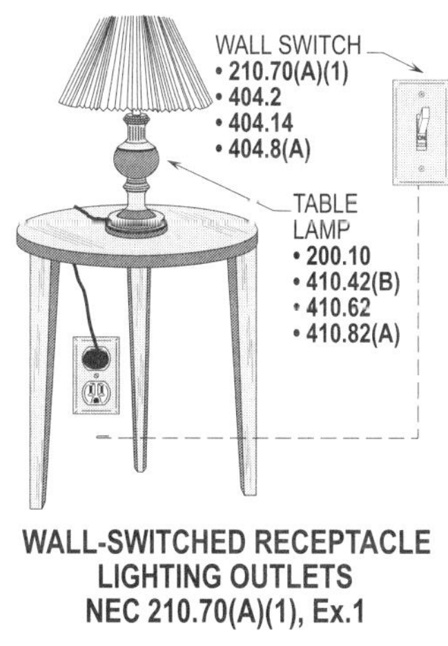

Figure 17-8. Wall-switched receptacles with table lamps or floor lamps plugged into an outlet may be used instead of a wall switch outlet in certain rooms.

LIGHTING OUTLETS IN HALLWAYS
210.70(A)(2)(a)

The hallways in dwelling units shall have a wall-switched ceiling or wall-mounted lighting outlet to provide proper lighting. Section **210.70(A)(2), Ex.** allows remote, central, or automatic control of lighting outlets installed in hallways.

The control method used to switch lighting outlets in hallways shall turn the lighting outlets on and off as needed to provide the necessary lighting. **(See Figure 17-9)**

LIGHTING OUTLETS IN STAIRWAYS
210.70(A)(2)(a)

Lighting outlets shall be installed in interior stairways for illumination, and a wall switch shall be provided at each level to control the lighting outlets. Where there is a difference between floor levels of six risers or more, a wall switch to control the lighting outlet or outlets shall be provided at each level. A lighting outlet at a door on a landing in a stairway that provides the proper lighting and switching complying with **210.70(A)(2)** is also required. **(See Figure 17-10)**

For remote, central, or automatic control of lighting for hallways and stairways in multifamily dwellings, it may be desirable to locate switches or use time clocks where they may not be intentionally or inadvertently turned to the OFF position.

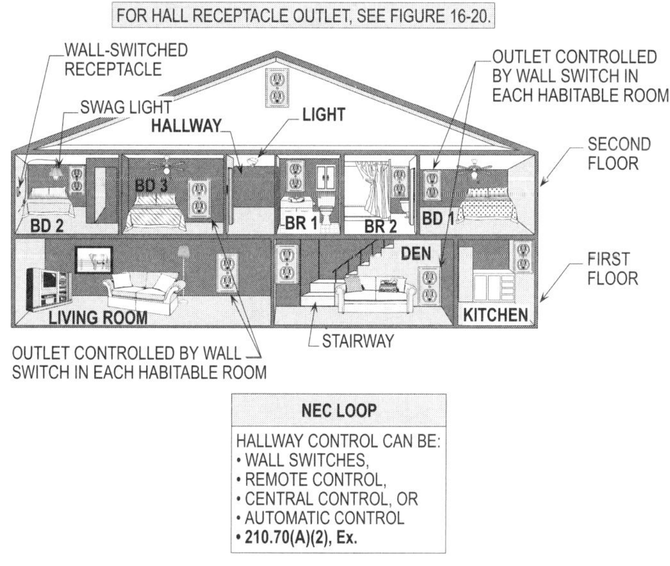

Figure 17-9. Hallways in dwelling units shall have a lighting outlet installed that is switched by a wall switch.

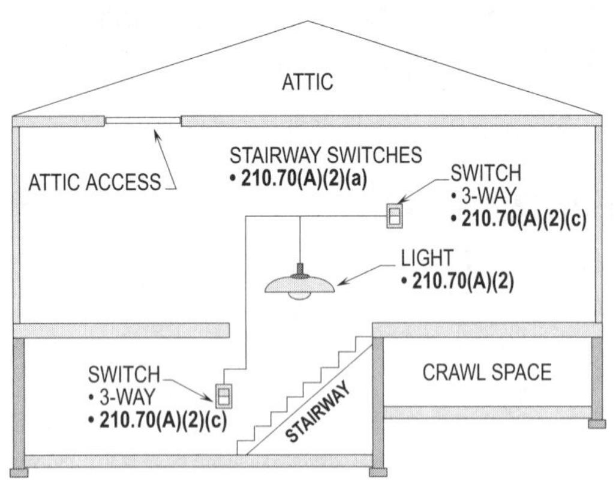

LIGHTING OUTLETS IN STAIRWAYS
NEC 210.70(A)(2)(a)

Figure 17-10. Stairways in dwelling units shall have wall-switched lighting outlets installed to provide proper illumination. Floor levels with six or more risers between them are considered different levels.

LIGHTING OUTLETS IN GARAGES
210.70(A)(2)(a) AND (b)

One lighting outlet shall be installed in the garage to provide lighting for parking vehicles. If a utility room is in the garage, there shall be lighting for washing and drying clothes. One or more of the lighting outlets shall be controlled by a wall-mounted switch.

A lighting outlet is not required at a vehicle door in an attached or unattached garage because it is not considered an outdoor entrance per **210.70(A)(2)**.

A detached garage with power routed to it requires a lighting outlet. If the detached garage has a walkway between the dwelling unit and garage, a lighting outlet is usually installed with a set of three-way switches to control the lighting outlet at either the dwelling or garage. **(See Figure 17-11)**

LIGHTING OUTLETS AT
OUTSIDE DOORS
210.70(A)(2)(b)

A lighting outlet is required at each outside door that is classified as an entrance or exit. This lighting outlet must be installed in a location that provides lighting at the door and steps, to prevent people from accidentally falling due to darkness. The lighting outlet may be mounted on the ceiling or wall and shall be wall-switched. The wall switch

should be located by the door in a location so that the control of the lighting outlet and luminaire can easily be found. **(See Figure 17-12)**

> **Design Tip:** The switch should not be located behind a closing door because of safety as well as easy access for switching purposes by the user.

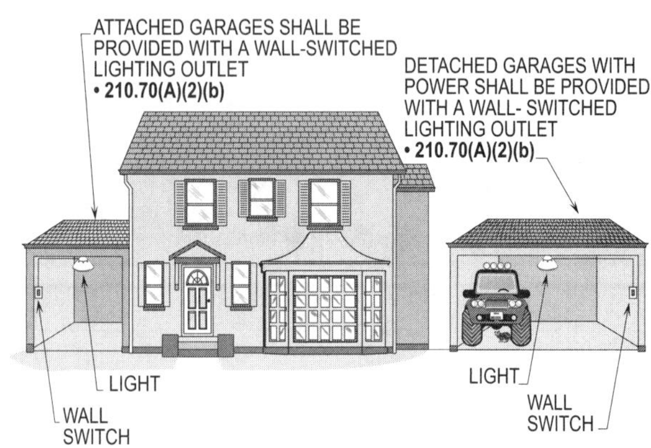

LIGHTING OUTLETS IN GARAGES
NEC 210.70(A)(2)(a) AND (b)

Figure 17-11. Attached garages shall be provided with a wall-switched lighting outlet. Detached garages with power are also required to have lighting outlets controlled by a wall switch. These switches are usually the three-way type. (For receptacle outlets see Figure 16-19)

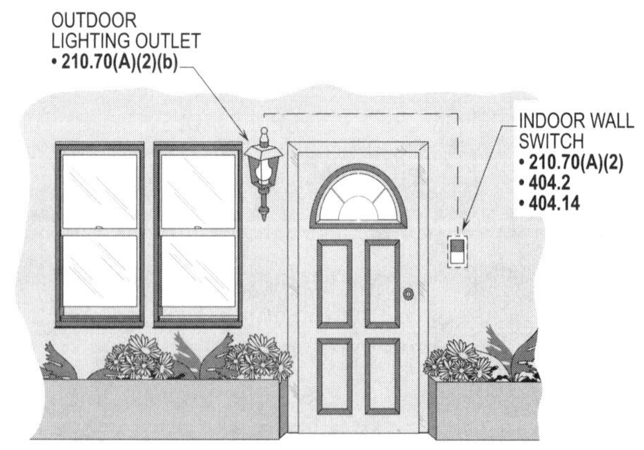

LIGHTING OUTLETS AT OUTSIDE DOORS
NEC 210.70(A)(2)(b)

Figure 17-12. Wall switches installed in the dwelling unit shall control a lighting outlet that is located at each door used as an entrance or exit.

LIGHTING OUTLETS IN UTILITY ROOMS
210.70(A)(3)

One wall-switched lighting outlet shall be required in utility rooms. The wall switch shall be located at the entry of the utility room. At least one lighting outlet is required if the utility room is used for storage or equipment that needs to be serviced. If the washing machine and clothes dryer are located in the utility room, lighting outlets are required for washing and drying clothes and servicing the machines. **(See Figure 17-13)**

> **Design Tip:** The lighting outlet located in an utility room does not necessarily require a wall switch to switch the lighting unit on or off. In other words, a properly located pull chain could serve as such a lighting outlet.

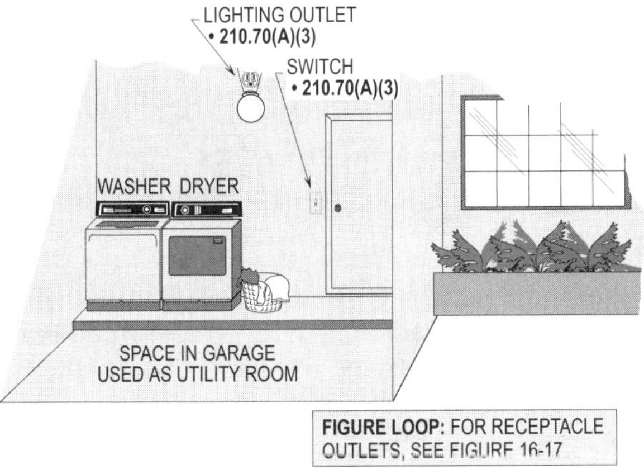

LIGHTING OUTLETS IN UTILITY ROOMS
NEC 210.70(A)(3)

Figure 17-13. A lighting outlet controlled by a pull chain or wall switch shall be installed in a utility room. The utility room may be located in the dwelling unit or garage.

The lighting outlet installed in the utility room may be surface or recess mounted. It could be cove lighting or any other wall-switched type lighting. The type of luminaire may be incandescent or fluorescent, whichever type the designer wants to install. The type really depends on the number of footcandles needed to illuminate the area.

LIGHTING OUTLETS IN BASEMENTS
210.70(A)(3)

At least one lighting outlet shall be required for the illumination of basements where there is storage or equipment installed that requires servicing. Such equipment can be air-handling equipment, refrigeration equipment, air conditioning equipment, etc. Table saws, routers, sanders,

etc. may be located in the basement for the purpose of a workshop. More than one lighting outlet may be required, in this case, to provide the proper lighting. Lighting for servicing the sump pump should be provided. The sump pump, with its equipment, is usually considered equipment requiring servicing, due to being located in a pit. **(See Figure 17-14)**

Section **210.70(A)(3)** requires the lighting outlet to be switched. However, it does not specifically state that it must be a wall-switched lighting outlet. If **210.70(A)(3)** required this lighting outlet to be wall-switched, it wouldn't comply with the various mechanical codes that permit it to be a pull-chain type under certain conditions of use.

> **Design Tip:** The *Uniform Mechanical Code* requires a wall switch, while the *Standard Mechanical Code* requires only a switched lighting outlet that could be a pull-chain type luminaire.

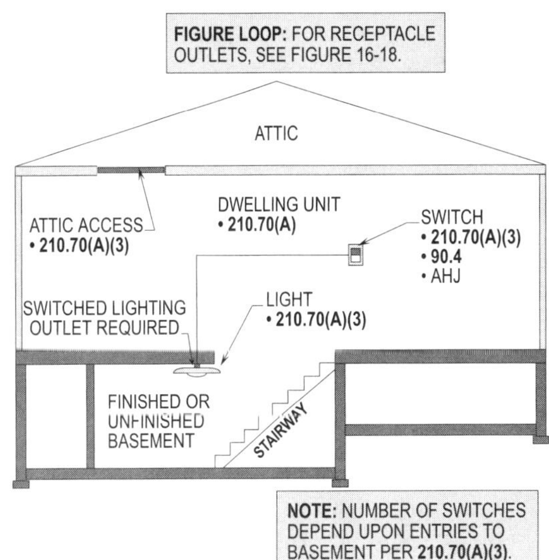

LIGHTING OUTLETS IN BASEMENTS
NEC 210.70(A)(3)

Figure 17-14. A switched lighting outlet shall be required in the basement of a dwelling unit to provide lighting for safe entrance, exiting, and servicing of equipment. This lighting outlet may be controlled by a pull chain or wall switch.

LIGHTING OUTLETS IN ATTICS
210.70(A)(3)

At least one lighting outlet shall be required in attics that are floored and used for storage or for electric equipment that requires servicing. A switch is required to turn the lighting outlet on and off. The switch may be incorporated into the luminaire or be controlled by a wall switch at the point of entry into the attic. Some inspectors permit a wall switch only, while others allow a pull-chain at the point of entry. Either complies, depending on which mechanical code is used. The pull-chain is usually controlled by an extended string from the luminaire at the point of entry. **(See Figure 17-15)**

> **Design Tip:** See *Uniform Mechanical Code*, *Southern Mechanical Code* or other appropriate codes for requirements concerning the switching rules for lighting outlets installed in attics to service the HVAC.

Note that **210.70(C)** requires a wall-switched lighting outlet for such use in commercial and industrial locations.

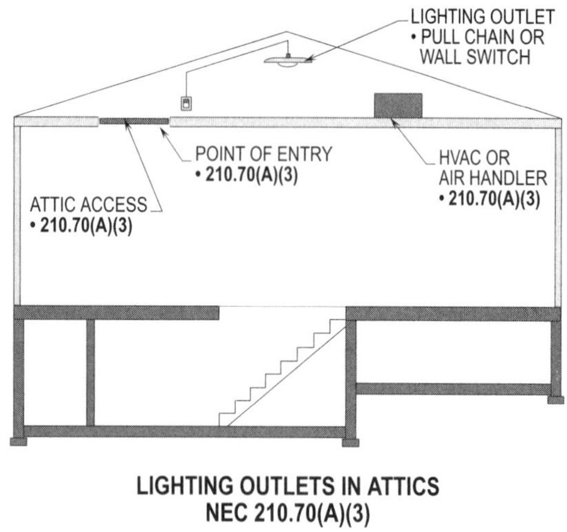

LIGHTING OUTLETS IN ATTICS
NEC 210.70(A)(3)

Figure 17-15. Attic space used for storage or equipment shall have a pull chain or wall-switched lighting outlet. **Note:** See local mechanical code for the type required and receptacle outlet requirements.

LIGHTING OUTLETS IN UNDERFLOOR SPACES
210.70(A) AND (C)

A switched lighting outlet shall be required at underfloor spaces or crawl spaces where the space is used for storage or for equipment that requires servicing. Crawl spaces are located at or below grade level. Underfloor spaces are usually located under pier-and-beam type dwelling units or those constructed on the side of a hill. Lighting outlets shall not be required where there is no storage or equipment needing service. **(See Figure 17-16)**

Note that **210.70(C)** requires a wall-switched lighting outlet for such use in commercial and industrial locations.

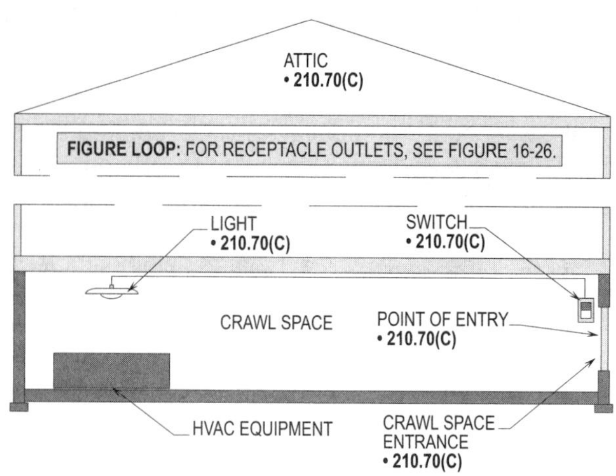

LIGHTING OUTLETS IN UNDERFLOOR SPACES
NEC 210.70(A) AND (C)

Figure 17-16. Underfloor spaces such as crawl spaces shall be provided with a lighting outlet that is controlled by a pull chain or wall switch. The lighting outlet shall not be required if the space is not used for storage or HVAC.

LIGHTING OUTLETS OVER BATHTUBS
410.10(D)

Hanging luminaires, track lighting, and ceiling (paddle) fans are not permitted to be hung over bathtubs. There is the hazard of electrical shock when changing lightbulbs and the danger of electrocution due to grabbing the hanging luminaire for support if the bather should slip when stepping from the tub. Due to these hazardous conditions, hanging units shall be installed at least 8 ft (2.5 m) vertically and shall be located at least 3 ft (900 mm) horizontally from the tub or shower threshold in all directions.

Luminaires that are marked for damp locations or for wet locations where subject to shower spray, shall be permitted to be located within the actual outside dimensions of the bathtub or shower to a height of 8 ft (2.5 m) vertically from the top of the bathtub or shower threshold.

The authority having jurisdiction usually requires luminaires to be surface mounted with a gasket or to be recessed if the bathtub is used for bathing and showering. Bathtubs used just for bathing are usually permitted to have regular surface-mounted luminaires installed. (Check with the AHJ.) **(See Figure 17-17)**

> **Design Tip:** The bathroom is not classified as a wet location, and these rules apply to hydromassage tubs and bathtubs where luminaires are concerned.

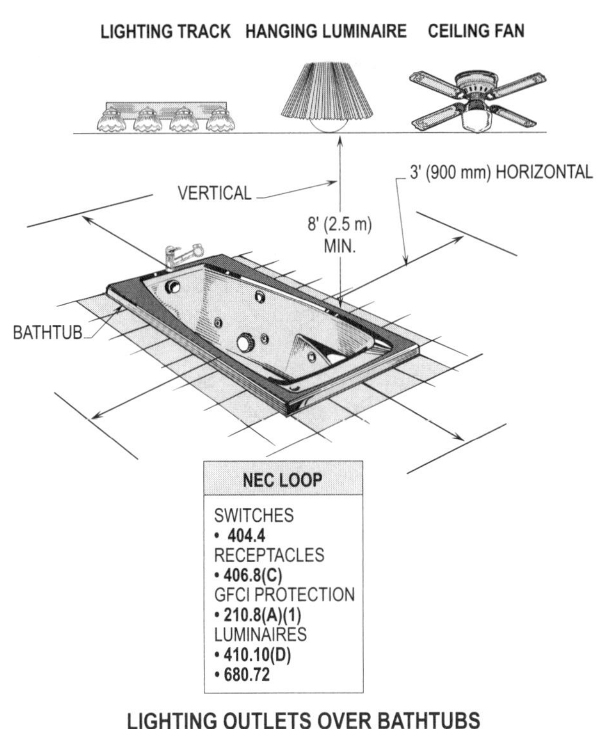

Figure 17-17. Hanging luminaires, ceiling fans, and track lighting are permitted to be hung over bathtubs if installed at least 8 ft (2.5 m) over the tub or shower threshold and located at least 3 ft (900 mm) from the tub or shower threshold in all directions.

LUMINAIRES IN CLOTHES CLOSETS
410.16(A) THRU (C)

In residential, commercial, and industrial locations, **210.70(A)** does not require a lighting outlet to be installed in clothes closets. If a lighting outlet is installed in a clothes closet, **410.16** lists the requirements for locating and installing the lighting outlets. Section **410.16** is divided into three subdivisions. Subdivision **(A)** lists the types of luminaires that are permitted to be used. Subdivision **(B)** lists the types of luminaires that are not permitted. Subdivision **(C)** deals with the location in the closet where the luminaire can be mounted.

TYPES PERMITTED
410.16(A)(1), (A)(2), AND (A)(3)

Section **410.16(A)** lists the types of luminaires that are permitted in clothes closets. Surface-mounted or recessed fluorescent luminaires are permited. Incandescent lamps, depending on wattage rating, get very hot, and the

elements, if accidentally broken, can fall into stored combustible materials and cause fires. The NEC requires incandescent luminaires to be provided with a lens that covers the luminaire completely to prevent such accidents.

Fluorescent lamps operate much cooler than incandescent bulbs and therefore can be mounted closer to the storage area than the incandescent type.

Surface-mounted fluorescent or LED luminaires identified as suitable for installation are permitted.

LUMINAIRE TYPES NOT PERMITTED
410.16(B)

Section **410.16(B)** does not permit incandescent luminaires with open or partially enclosed light bulbs (pull-chain or keyless) to be installed. Pendant luminaires or lampholders such as rosettes or the hanging type are not permitted.

LOCATION
410.16(C)(1) THRU (C)(5)

Section **410.16(C)** lists the dimensions at which luminaires shall be positioned from the storage area. The luminaires permitted are grouped into two types: the surface-mounted and recessed type. Surface-mounted luminaires are mounted to a ceiling box and recessed luminaires are recessed in the ceiling with an approved can (hat) listed for the purpose. Surface-mounted or LED luminaires of the incandescent type shall be mounted at least 12 in. (300 mm) from the storage space (area). The incandescent luminaire may be mounted on the wall above the door or ceiling per **410.16(C)(1)**.

Surface-mounted fluorescent luminaires shall have a clearance of at least 6 in. (150 mm) from the storage area. The luminaire may be mounted on the wall above the door or on the ceiling in the same way as the incandescent luminaire per **410.16(C)(1)**.

Recessed luminaires are available as an incandescent, LED, or fluorescent type. The minimum clearance of 6 in. (150 mm) or less shall be maintained from the storage area whether incandescent, LED, or fluorescent recessed luminaires are installed per **410.16(C)(3)** and **(C)(4)**. **(See Figure 17-18)**

Surface-mounted fluorescent or LED luminaires shall be permitted to be installed within the storage space where identified for this use.

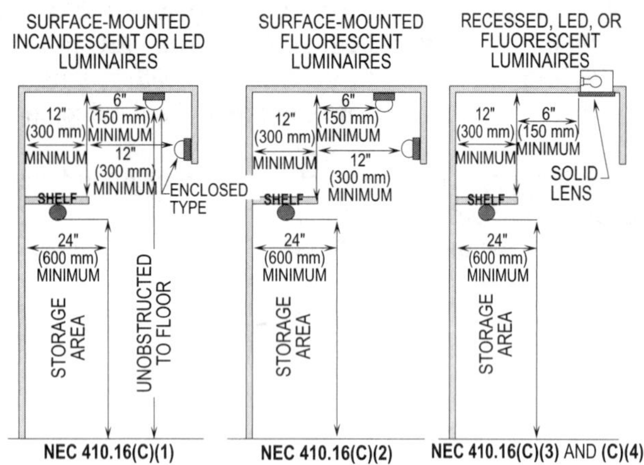

Figure 17-18. Installation requirements for luminaires located in clothes closets.

GUEST ROOMS OR GUEST SUITES IN HOTELS AND MOTELS
210.70(B)

At least one wall-switched controlled lighting outlet or receptacle for a table, floor, or hanging lamp shall be installed in the guest rooms or guest suites of hotels, motels, etc. **(See Figure 17-19)**

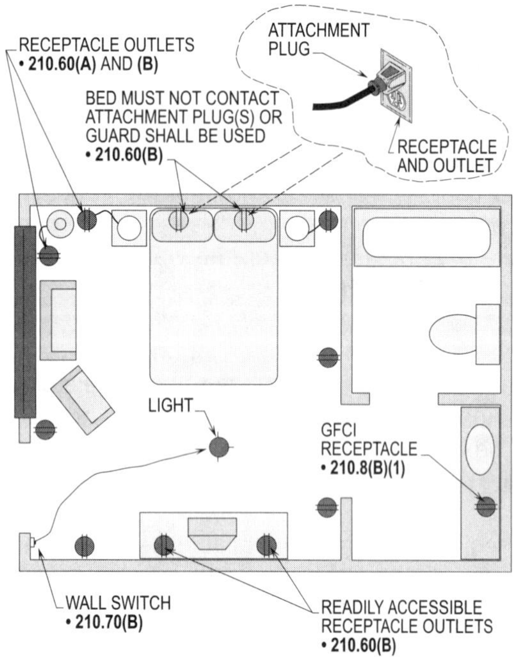

Figure 17-19. At least one wall-switched outlet shall be provided to switch on a lamp in the guest rooms of hotels and motels.

COMMERCIAL AND INDUSTRIAL ATTICS AND UNDERFLOOR SPACES
210.70(C)

One or more switch-controlled lighting outlets shall be installed near equipment requiring service, such as heating or air conditioning equipment installed in attics or underfloor spaces. The switch shall be installed at the entry point to the attic or underfloor space. This rule provides a wall-switched outlet at the point of entry so the user or maintenance person does not have to search for the switch, which could cause a safety problem. **(See Figure 17-20)**

ILLUMINATION FOR ELECTRICAL EQUIPMENT IN COMMERCIAL AND INDUSTRIAL LOCATIONS
110.26(D)

Illumination shall be provided for all working spaces about service equipment, switchboards, panelboards, or motor control centers installed indoors. Additional luminaires shall not be required where the workspace is illuminated by an adjacent light source. In electrical equipment rooms, the illumination shall not be controlled by automatic means only. **(See Figure 17-21)**

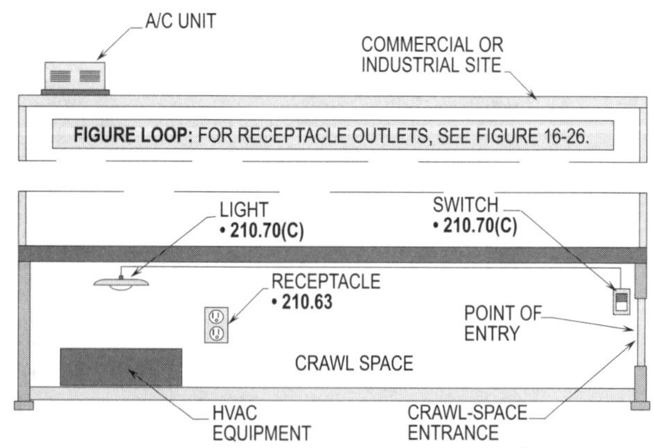

Figure 17-20. At least one wall-switched outlet shall be provided at crawl-space entry to switch light for storage and equipment service area.

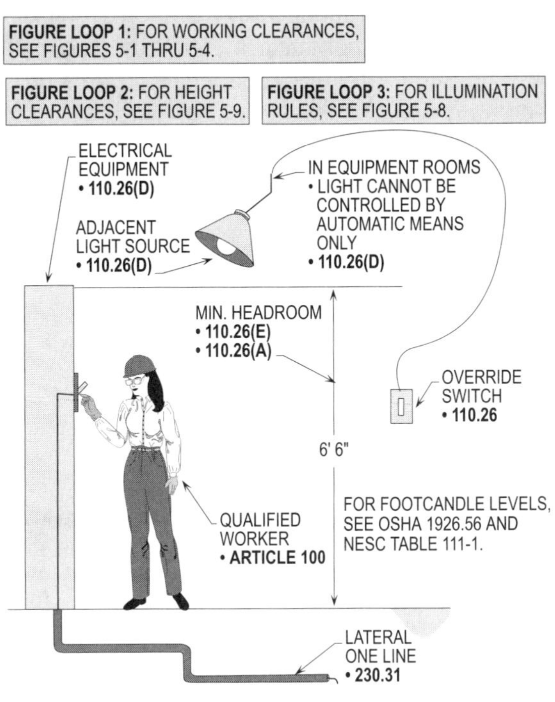

Figure 17-21. Lighting outlets providing illumination over electrical equipment may be controlled automatically. However, a regular switch for overriding purposes shall be provided.

RECESSED LUMINAIRES
410.115(C)

In residential, commercial, and industrial locations, recessed luminaires are found in new or existing installations. In such premises, per the 1987 NEC, all recessed luminaires were required to be equipped with a thermal protector to prevent overheating. There have been two types allowed since the 1987 NEC, and they are thermal protected (TP) and insulation covered (IC) with thermal protection. The thermal-protected type shall be clear of all insulation by 3 in. (75 mm) on the top and sides per **410.116(B)**.

The luminaire can (hat) shall have a clearance of at least 1/2 in. (13 mm) from combustible materials such as wooden rafters per **410.116(A)(1)**. The IC recessed listed luminaire may be covered with insulation and set against the wooden rafter for support per **410.116(A)(2)**.

In existing facilities with hung (suspended) ceilings, a suspended recessed luminaire may be used per the 1984 NEC. All other types of recessed incandescent luminaires shall have a clearance of at least 3 in. (75 mm) from the top to the sides and be located at 1/2 in. (13 mm) from combustible material. Fluorescent recessed luminaires are also permitted to be installed. **(See Figure 17-22)**

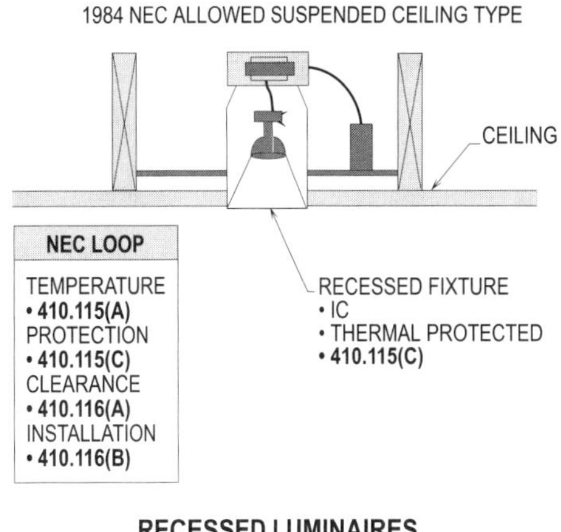

Figure 17-22. Recessed luminaires may be utilized with certain installation requirements being applied.

DISCHARGE LIGHTING
410.140(B)

Discharge lighting systems requiring open secondary voltage of more than 1000 volts are not permitted to be installed inside or outside dwelling units. Open secondary voltage is the secondary output side of a ballast or transformer that supplies power to the electric discharge lighting unit. The primary is the input side and is supplied by a 120 volt branch circuit that is protected by an overcurrent protection device per **210.20(A)**. **(See Figure 17-23)**

Design Tip: This rule has been interpreted to apply only to discharge lighting units installed inside. It clearly allows them to be installed both inside or outside of such premises.

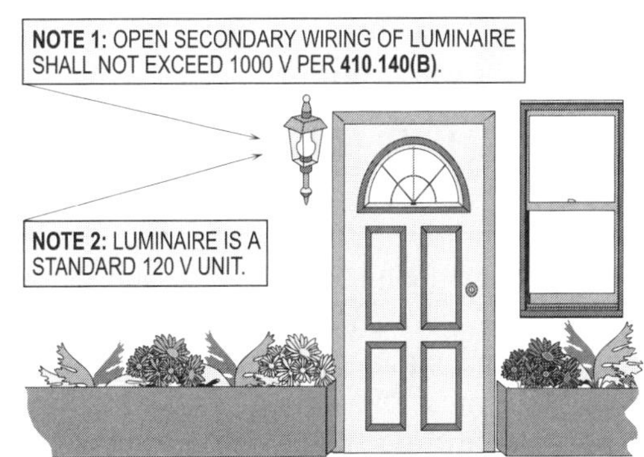

Figure 17-23. The open-circuit voltage of ballast and transformers supplying electric discharge lighting units shall not exceed 1000 volts when installed outside or inside dwelling units.

LIGHTING OUTLETS AT SWIMMING POOLS
680.22(C)

In residential, commercial, and industrial locations, lighting outlets used to support luminaires and ceiling fans shall be located at specific locations over or around swimming pools. The positioning and location of the lighting outlets are measured from the inside walls of the pool. This basic rule prohibits lighting outlets from being installed 5 ft (1.5 m) horizontally or 5 ft (1.5 m) vertically from the inside walls of the pool per **680.22(C)(3)**.

Lighting outlets with luminaires or ceiling (paddle) fans shall be located at least 12 ft (3.7 m) above the maximum water level of the pool per **680.22(C)(1)**. A swimming pool built at an existing premise may have luminaires located in the area above the 5 ft (1.5 m) horizontal boundary per **680.22(C)(3)**. The luminaire shall be existing (not new) and rigidly attached to the structure. Luminaires may be installed between 5 ft (1.5 m) and 10 ft (3 m) horizontally with GFCI protected circuits per **680.22(C)(4)**.

Luminaires located at safe heights over and around the pool protect personnel while they are servicing or changing lamps. Persons swimming in or playing near the pool will be protected from contacting live parts in a luminaire if it is located at these required heights. **(See Figure 17-24)**

Lighting outlets and ceiling (paddle) fans shall be installed over swimming pools located inside such structures. The lighting outlets supporting the luminaires or ceiling fans shall be located at least 7 ft 6 in. (2.3 m) above the maximum water level of the swimming pool. The branch circuit supplying power to these luminaires shall be GFCI protected per **680.22(C)(2)**. **(See Figure 17-25)**

UNDERWATER LUMINAIRES
680.23(A)(1) THRU (A)(8)

In residential, commercial, and industrial locations, underwater luminaires shall be installed in such a manner so as to prevent electrical shock to persons swimming in the pool. Luminaires installed underwater shall be located at least 18 in. (450 mm) below the normal water level per **680.23(A)(5)**. Section **680.23(A)(5)** permits specially designed underwater luminaires to be installed not less than 4 in. (100 mm) below the normal water level per **90.7** and **110.3(B)**.

Underwater luminaires shall be inherently equipped with thermal protection to protect against overheating when the luminaire relies on submersion in water for safe operation.

The luminaire shall be supplied by a branch circuit with 150 volts or less between conductors. The branch circuit shall include a grounded (neutral) conductor. There are three types and they are as follows:

- Wet-niche
- Dry-niche
- No-niche

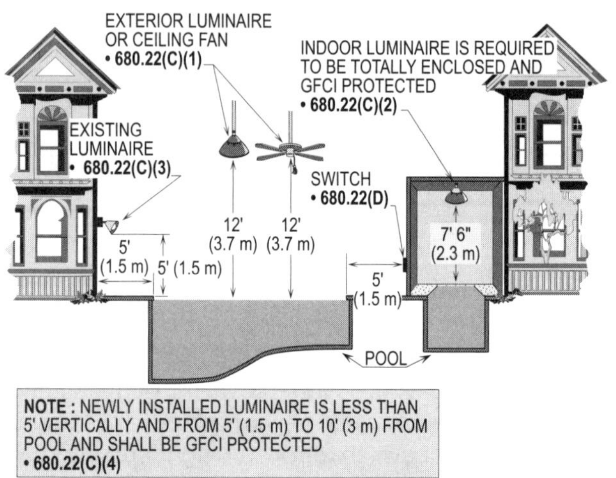

Figure 17-24. Location of luminaires around a swimming pool shall be a certain height and distance from the inside walls of the pool. This rule protects personnel from electrical hazards and shock.

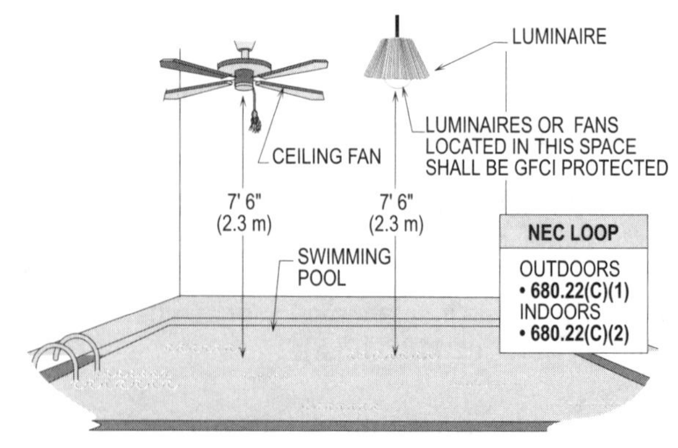

Figure 17-25. Installation requirements for installing luminaires and fans over swimming pools located inside.

WET-NICHE LUMINAIRES
680.23(B)(1) THRU (B)(5)

Where used in residential, commercial, and industrial locations, wet-niche luminaires shall be equipped with a metal or plastic forming shell that is approved for installation in swimming pool walls. The forming shell should be equipped with threaded entries to connect rigid metal conduit, IMC, LFNC, or RNC, and be made with metallic or nonmetallic material. The conduit shall extend from the shell to a junction box on the deck or yard area. The deck box shall be located under the diving board or in another protected area used for this purpose. The deck box shall be located at least 4 ft (1.2 m) from the inside walls of the pool and be at least 4 in. (100 mm) from the deck to the inside bottom of the deck box per **680.24(A)(2)(a)** and not less than 8 in. (200 mm) from the maximum water level. The measurement that produces the greater elevation is naturally chosen.

For example, if 4 in. (100 mm), measured from the deck to the inside bottom of the box, is greater than 8 in. (200 mm) from the maximum water level, the deck box shall be installed using the 4 in. (100 mm) measurement. **(See Figure 17-26)**

Design Tip: The deck box may be located in a flower bed at the side of the facility where an approved conduit is routed all the way between the forming shell and deck box. If rigid metal brass conduit is used, it shall be routed as a complete system. In the same manner, if PVC is used, the 8 AWG solid or stranded copper conductor shall be pulled unbroken from the shell to the deck box, wherever it is located.

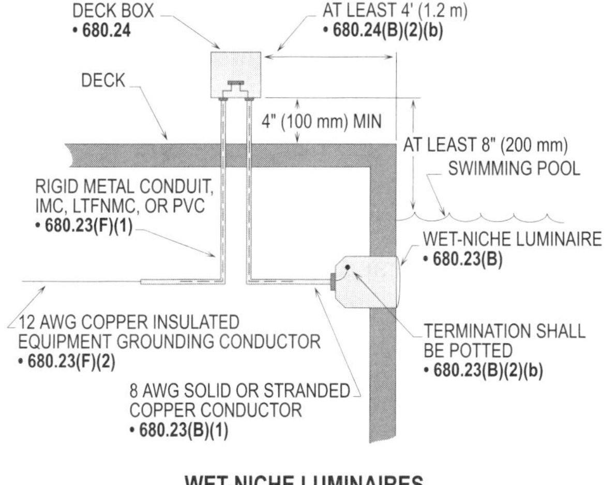

WET-NICHE LUMINAIRES
NEC 680.23(B)

Figure 17-26. The deck box shall be located at least 4 ft (1.2 m) from the inside wall of the pool and have a height of 4 in. (100 mm) measured from the deck or 8 in. (200 mm) measured from the water, whichever is greater.

The 4 ft (1.2 m) minimum distance of the deck box shall be permitted to be reduced with an effective barrier between the deck box and inside walls of the pool.

For example, a solid fence may be located 2 ft from the pool with the deck box located on the opposite side of such barrier and pool. **(See Figure 17-27)**

Where PVC is used to connect the shell to the junction box (deck box), an 8 AWG solid or stranded insulated copper conductor shall be run through the PVC and connected to the forming shell. The termination of the 8 AWG conductor shall be potted to prevent corrosion. The cord end and terminals within the wet-niche luminaire shall be sealed to prevent the entry of water per **680.23(B)(4)**. The equipment grounding conductor in the cord is used to ground the wet-niche luminaire when it is removed from the shell to be serviced. **(See Figure 17-28)**

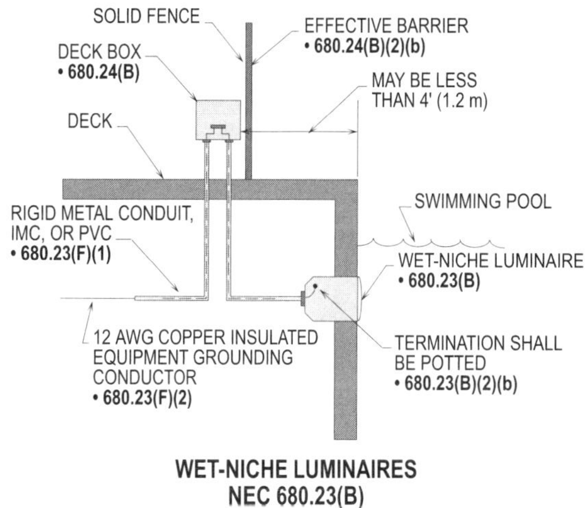

WET-NICHE LUMINAIRES
NEC 680.23(B)

Figure 17-27. The deck box located behind a solid permanent barrier may be located less than 4 ft (1.2 m) from the inside walls of the pool. The barrier prevents easy access if properly designed and isolates personnel between box and pool.

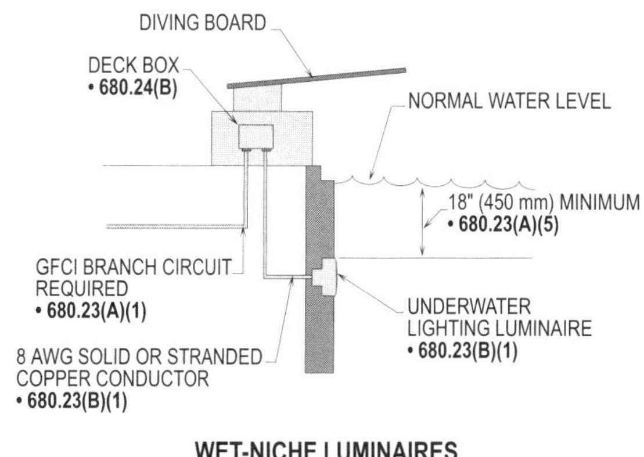

WET-NICHE LUMINAIRES
NEC 680.23(B)(1)

Figure 17-28. An 8 AWG insulated copper conductor shall be routed in PVC conduit to ground to the metal forming shell.

DRY-NICHE LUMINAIRES
680.23(C)

In the walls of pools installed in residential, commercial, and industrial locations, dry-niche luminaires may be installed instead of wet-niche luminaires. This type of luminaire is installed outside the walls of the pool in closed recesses that provide for drainage of water that might accumulate. Dry-niche luminaires shall be wired with approved rigid metal conduit, IMC, or PVC with an equipment grounding conductor for each conduit entry. The luminaire shall have adequate provisions for drainage of water per **680.23(C)(1)**. **(See Figure 17-29)**

NO-NICHE LUMINAIRES
680.23(D)

No-niche luminaires may be installed instead of wet- or dry-niche luminaires. No-niche luminaires have no exposed metal parts, and they contain impact-resistant polymeric lenses. If installed in residential, commercial, and industrial pools, they shall be required to be of the listed type per **680.23(D)**, **90.7**, and **110.3(B)**. **(See Figure 17-30)**

LUMINAIRES OVER
SPAS OR HOT TUBS
680.43(B)

Lighting outlets with luminaires and ceiling (paddle) fans located over spas or hot tubs shall have a clearance of 7 ft 6 in. (2.3 m) above the maximum water level. This clearance is required within 5 ft (1.5 m) in all directions. The supplying branch circuit shall be GFCI protected per **680.43(B)(1)(b)**. If luminaires and ceiling fans are located 12 ft (3.7 m) above spas or hot tubs, the supplying branch circuit is not required to be GFCI protected per **680.43(B)(1)(a)**. **(See Figure 17-31)**

Recessed and surface-mounted luminaires may be hung over a spa or hot tub, under certain conditions. Recessed luminaires with glass or plastic lens and nonmetallic trim suitable for wet locations are permitted to be hung less than 7 ft 6 in. (12.3 m) per **680.43(B)(1)(b)** and **(c)**. Luminaires installed in this manner will protect users of spas and hot tubs from serious electrical shock. Surface-mounted luminaires with glass or plastic globes and nonmetallic bodies suitable for wet locations per **680.43(B)(c)(1)** and **(2)** may also be used. Note that the above rules apply to residential, commercial, and industrial installed spas and hot tubs. **(See Figure 17-32)**

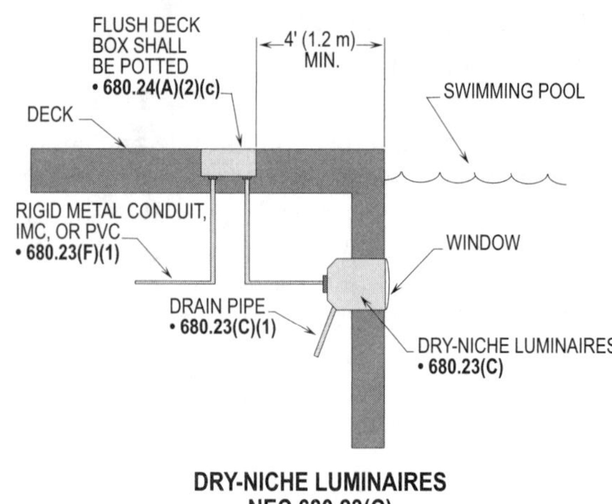

Figure 17-29. Installation requirements for dry-niche luminaires in swimming pool walls shall be carefully designed.

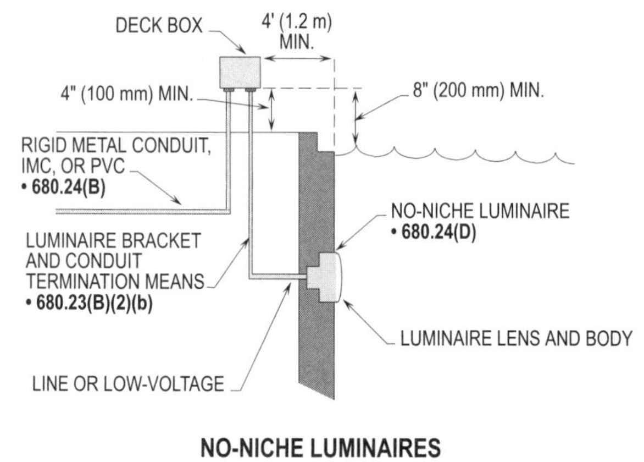

Figure 17-30. No-niche luminaires are now available in both line and low-voltage type. **Note:** All grounding connections are to be made to the mounting bracket to ensure grounding.

LUMINAIRES OVER
HYDROMASSAGE BATHTUBS
680.72 THRU 680.74

In residential, commercial, and industrial locations, hydromassage bathtubs are treated as conventional bathtubs.

Lighting outlets and luminaires shall be installed per **410.10(D)**. All the elements for hydromassage bathtubs shall be wired with GFCI protected circuits whether they are cord-and-plug connected or permanently hard wired per **680.72**. Receptacle outlets shall be GFCI protected per **210.8(A)(1)**. Motors used to circulate the water shall be accessible for maintenance and service by a removable cover or trap door per **430.14(A)** and **680.73**. **(See Figure 17-33)**

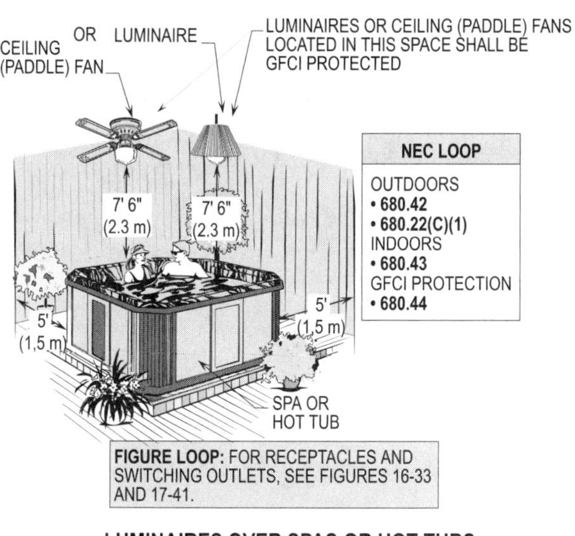

LUMINAIRES OVER SPAS OR HOT TUBS
NEC 680.43(B)

Figure 17-31. GFCI protected lighting outlets with luminaires or ceiling (paddle) fans shall be located at least 7 ft 6 in. (2.3 m) from the maximum water level of the spa or hot tub. **Note:** Luminaires or ceiling (paddle) fans located over 12 ft (3.7 m) above the maximum water level are not required to be GFCI protected.

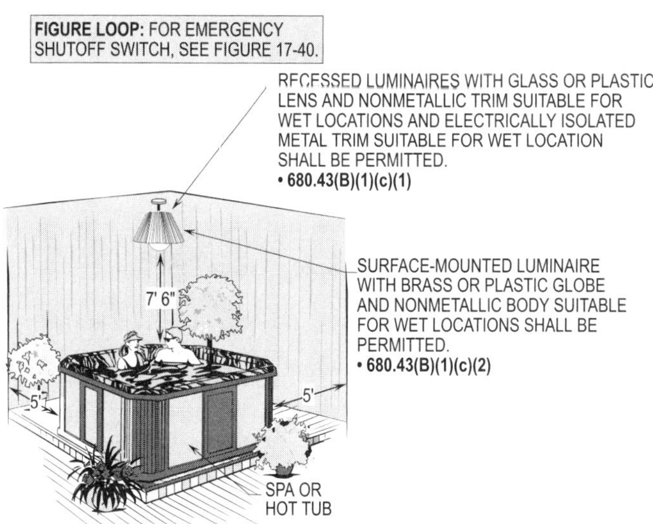

LUMINAIRES OVER SPAS OR HOT TUBS
NEC 680.43(B)(1)(c)(1)
NEC 680.43(B)(1)(c)(2)

Figure 17-32. Location and installation requirements for recessed surface-mounted luminaires over hot tubs or spas.

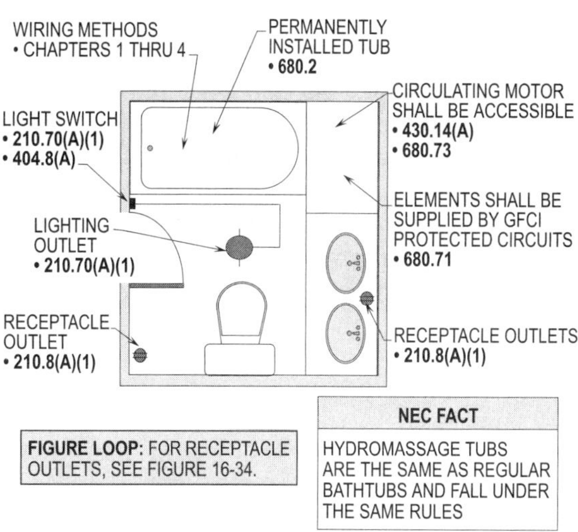

LUMINAIRES OVER HYDROMASSAGE BATHTUBS
NEC 680.70 THRU 680.74

Figure 17-33. Installation requirements for luminaires over a hydromassage bathtub.

SWITCHING OUTLETS
404.2

In residential, commercial, and industrial locations, switches are used to control wiring systems, providing easy access to complete circuits for energizing and deenergizing electrical equipment. A variety of types and styles is available to meet the various requirements and locations. Switches are designed so that they are ON when they are in the up position and OFF in the down position. This position indicates, in a precise manner, whether they are ON or OFF.

Switches shall disconnect only the ungrounded (phase) conductors to an electrical load. They shall not disconnect the grounded (neutral) conductor unless the phase and neutral conductors are disconnected simultaneously. Switches shall disconnect all ungrounded (phase) conductors from the terminals of a screw shell type lampholder. Switches shall be located and mounted at specific heights that are acceptable to the user per **404.8(A)**.

TYPES OF SWITCHES
404.2(A)

Single-pole switches have two terminals and are designed to switch only one ungrounded (phase) and return conductor. A single conductor must not be routed between the switch and the luminaire. Where nonmetallic-sheathed cable (Romex) or armored cable (BX) can be used as the wiring method, the white or natural gray conductor in the cable may be used as a switch leg, while the black conductor in the cable may be used as the return leg (switch leg) per **200.7(C)(2)**. The switch leg supplies power to the line side of the switch and the return line from the load side of the switch connects power to the luminaire or other type of load. The conductor of the other side (return) of the circuit shall be run with the switched (leg) conductor to prevent induced currents. Induced currents are only a problem where metal conduits or metal-clad cables are used as the wiring method. Since current in one conductor is equal to and opposite in direction to current in the other conductor, the inductive effect is canceled. **(See Figure 17-34)**

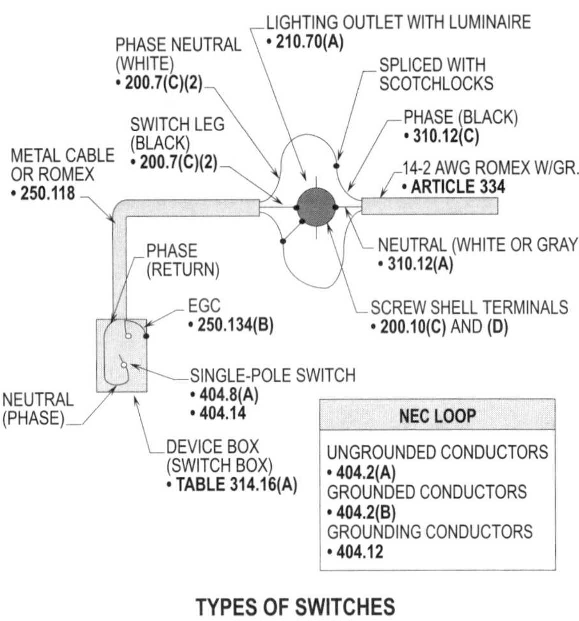

Figure 17-34. The white or gray conductor in a Romex cable (nonmetallic-sheathed cable) can be used to run the phase conductor to a switch and return as a black conductor (switch leg) to the luminaire.

Design Tip: Always run the switch leg and return leg in the same conduit or cable. See **300.3(B)** and **300.20** for more information on methods used to prevent inductive heating.

Three-way switches have three terminals to connect conductors. The odd-color terminal can be used for the connection of the phase supply or switch leg conductor. The other two of the same color can be used for the traveler conductors. Four-way switches have the same color terminals identified for the travelers. Travelers for three-way and four-way switches may be routed alone in metal conduits or metal-clad cables, where used as wiring methods in residential, commercial, and industrial locations.

The grounded (neutral) conductor does not have to be run with the travelers. A set of three-ways (two switches) may be used to switch one or more luminaires from two locations. A set of three-ways and a four-way may control one or more luminaires from three locations. For example, a set of three-ways with three four-ways will switch ON and OFF one or more luminaires in five different locations.

Design Tip: With each four-way switch added, the luminaire(s) may be controlled at additional locations. **(See Figure 17-35)**

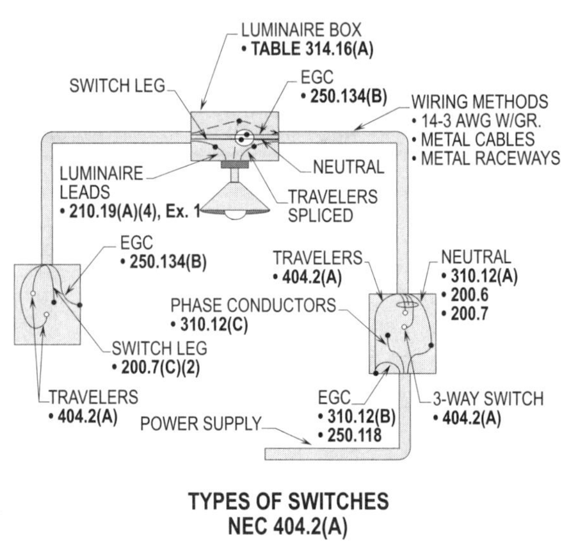

Figure 17-35. Three-way switches may be used to switch a luminaire from two different locations by using a set of travelers.

SWITCHING OUTLET HEIGHTS 404.8(A)

In residential, commercial, and industrial locations, switches shall be located at a height accessible to the user. To comply with this requirement, switches should be located at 6 ft 7 in. (2 m) or less from the toggle of the switch in the ON position. (Most inspectors apply this rule.) Switching outlet boxes are usually installed at 56 in. (1.4 m) to the center of the box to provide a height accessible to users. Lesser heights may be used where needed. **(See Figure 17-36)**

Design Tip: Switches may be located at any height below the 6 ft 7 in. (2 m) to the toggle in the ON position as long as they comply with **404.8(A)**. This rule is usually applied by electrical inspectors to ensure a maximum height.

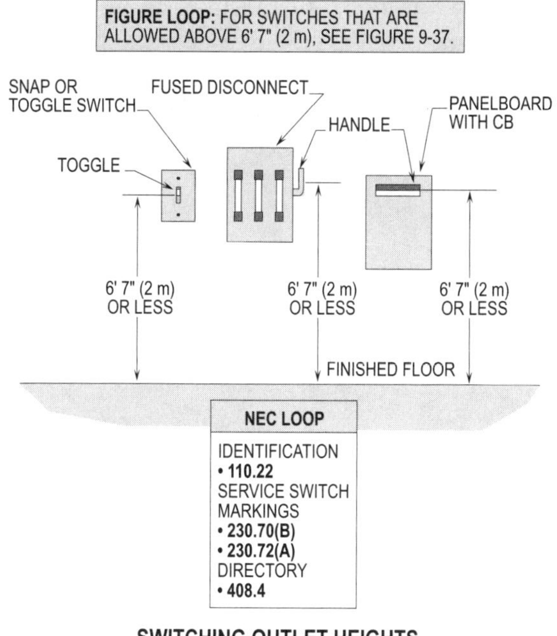

Figure 17-36. Enclosure with switching devices such as (toggle) snap switches, disconnects with fuses, or panelboards with circuit breakers may be located to a maximum height of 6 ft 7 in. (2 m) from the finished grade to the center of the toggle, disconnect handle, or circuit breaker handle in the ON position.

LOADING SWITCHES
404.14

General-use switches are basically divided into two types for energizing and deenergizing loads. AC general-use switches may be used to control resistive loads (incandescent lights) or electric discharge lighting (fluorescent lights). The switch rating shall be at least equal to the load it controls.

For example, a load for luminaires of 15 amps may be connected to a 15 amp AC rated switch. A load of 20 amps may be connected to a 20 amp AC rated switch.

AC general-use switches controlling motor loads shall be calculated at 125 percent of the full load current of the motor or be limited to not more than 80 percent of the switch rating.

For example, a motor load of 12 amps requires a 15 amp AC toggle switch (12 A x 125% = 15 A). Note that 80 percent of a 15 amp AC toggle switch (15 A x 80% = 12 A) allows a 12 amp load to be switched. **(See Figure 17-37)**

AC-DC general-use switches may be used to turn ON and OFF incandescent or inductive lighting loads. The rating of an AC-DC switch shall be twice the amperage rating of the inductive load at 50 percent of the amperage rating of the switch at the applied voltage.

For example, a 15 amp AC-DC toggle switch is required to supply a lighting load of 7.5 amps (7.5 A x 2 = 15 A). Note that 50 percent of a 15 amp AC-DC toggle switch (15 A x 50% = 7.5 A) allows a load of 7.5 amp to be switched. The above rules for switches apply to residential, commercial, and industrial locations where switches are used for such use. **(See Figure 17-38)**

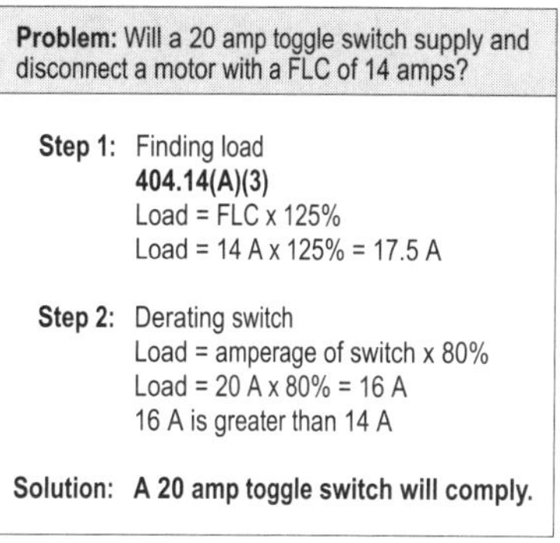

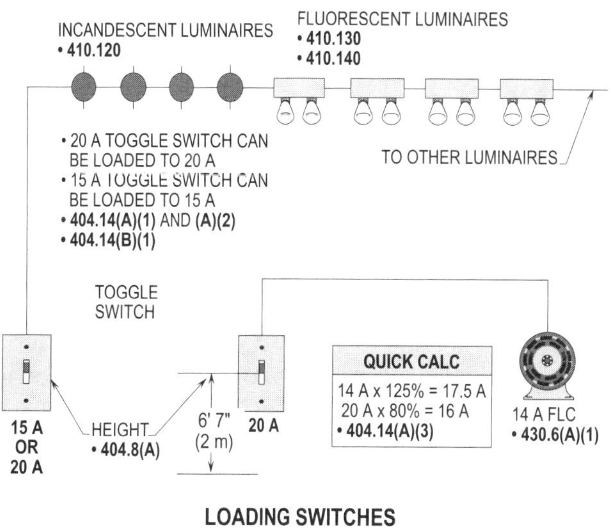

Figure 17-37. The size load that a toggle switch will supply and disconnect is determined by the load served. Switches supplying lighting loads can be loaded to 100 percent of their rating. Switches supplying motors are sized at 125 percent of the FLC.

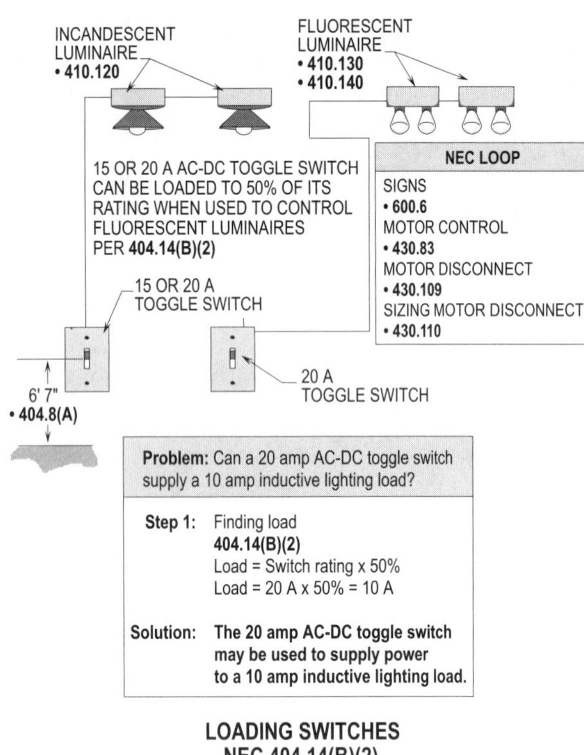

Figure 17-38. A 20 amp AC-DC toggle switch may be used at 100 percent of its rating to control incandescent lighting loads. It is limited to only 50 percent of its amp rating to control inductive lighting loads.

SWITCHING OUTLETS BY SWIMMING POOLS
680.22(D)

Switching devices installed in residential, commercial, and industrial locations shall be located at least 5 ft (1.5 m) from the inside walls of swimming pools. If the swimming pool and switching devices are separated by a permanent barrier, this rule does not apply. Switching devices such as toggle switches, safety switches, meters, overcurrent protection devices in panelboards, etc. shall not be permitted within 5 ft (1.5 m) of the inside walls of the pool. A distance of 5 ft (1.5 m) or less shall be permitted with proper barriers between switches and pool. **(See Figure 17-39)**

EMERGENCY SWITCH FOR SPAS AND HOT TUBS
680.41

A clearly labeled emergency shutoff switch for control of the recirculation system and jet system shall be installed at least 5 ft (1.5 m) away, adjacent to and within sight of the spa or hot tub. Note that this requirement does not apply to

single-family dwellings. For a maintenance disconnect, see **680.12**. **(See Figure 17-40)**

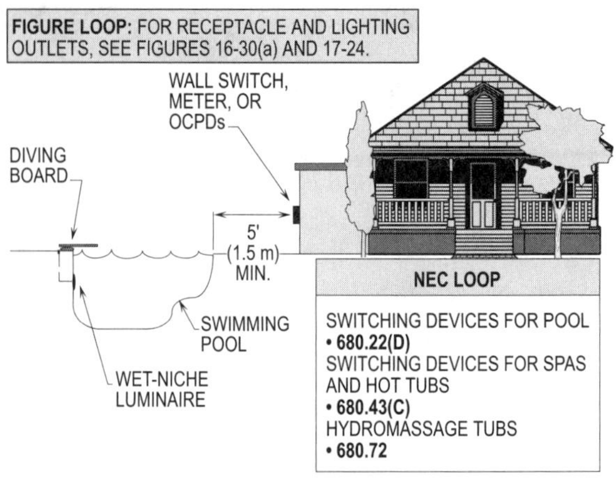

Figure 17-39. Switching devices are not allowed to be installed within 5 ft (1.5 m) of the inside walls of swimming pools. Switching devices include electrical apparatus such as wall switches, meter bases with meters, and panelboards with overcurrent protection devices, etc.

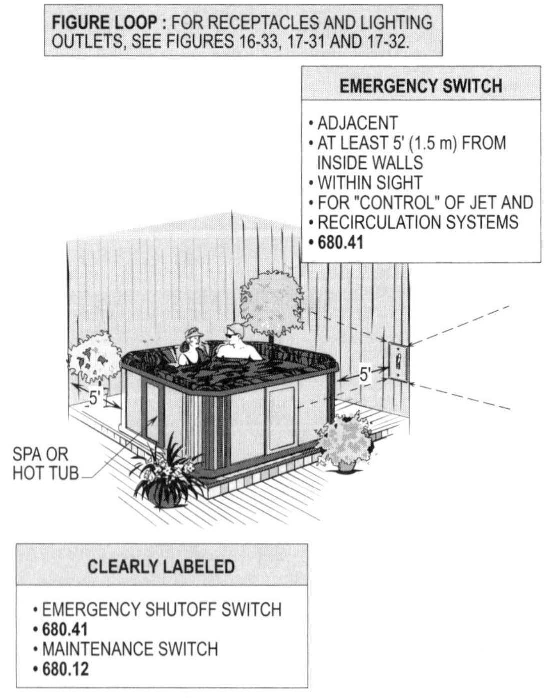

Figure 17-40. A clearly labeled emergency shutoff switch for control of the recirculation system and jet system shall be installed at least 5 ft (1.5 m) away, adjacent to and within sight of the spa or hot tub.

SWITCHING OUTLETS BY SPAS OR HOT TUBS
680.43(C)

Switching outlets with toggle switches installed in residential, commercial, and industrial locations shall be located at least 5 ft (1.5 m) from the inside walls of spas or hot tubs. This is to prevent someone from receiving an electrical shock from a faulty switch while turning a lighting load ON or OFF. Switches and controls associated with a spa or hot tub may be closer than 5 ft (1.5 m), where used as an approved (listed) assembly. **(See Figure 17-41)**

GROUNDED OR UNGROUNDED
404.12

Switching outlets and boxes that are used to support switches shall be connected to an equipment grounding conductor if they are metal per **250.148(A)**. Outlets that are nonmetallic are not required to be grounded per **250.148(B)**. A means shall be provided to ensure grounding continuity of the branch circuit. The equipment grounding conductor may be used for this purpose. Toggle switches, without metal plates and used to control one or more luminaires, are not required to be grounded per **110.3(B)** and **404.12**. **(See Figure 17-42)**

> **Design Tip:** See **410.42(A)** for the grounding of metal boxes if they are used. Nonmetallic boxes are not required to be grounded per **410.42(B)** for lighting outlets. See **250.96** and **404.12** for the grounding of device yokes and metal switch plates.

Note that the rules above shall be required for switching outlets and boxes used to support switches that are installed in residential, commercial, and industrial locations.

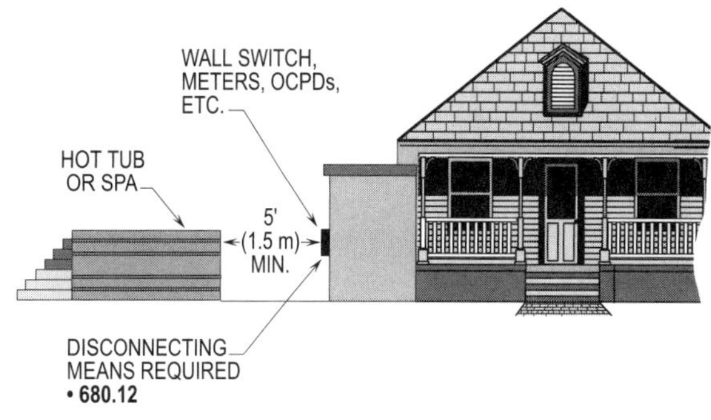

SWITCHING OUTLETS BY SPAS OR HOT TUBS
NEC 680.43(C)

Figure 17-41. Switching devices are not permitted to be installed within 5 ft (1.5 m) of hot tubs or spas. Switching devices include electrical apparatus such as wall switches, meter bases with meters per utility, and panelboards with overcurrent protection devices, etc. Check with local utility for requirements.

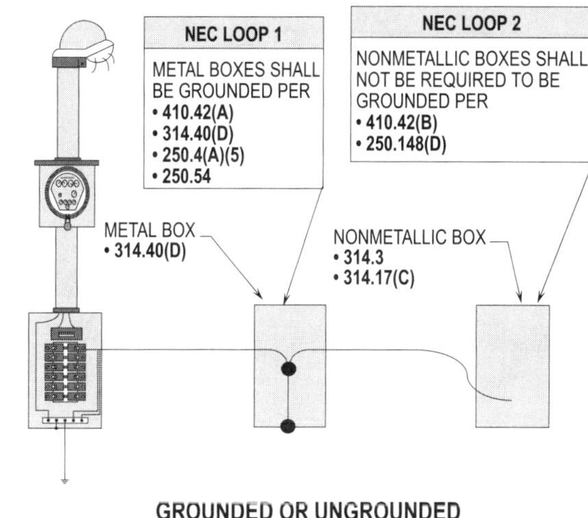

GROUNDED OR UNGROUNDED
NEC 404.12

Figure 17-42. Metal boxes are required to be connected to an equipment grounding conductor. Nonmetallic boxes are not required to be grounded with a grounded means. See **250.2(D)** and **250.54** for further explanation.

Chapter 17. Lighting and Switching Outlets

Section   **Answer**

1 Lighting outlets with luminaires or ceiling (paddle) fans located _______ ft above spas or hot tubs shall not be required to be GFCI protected.
    (a) 8   (b) 10
    (c) 12   (d) 16

2. AC general-use switches controlling motor loads shall be calculated at _______ percent of the full-load current (in amps) of the motor or not more than _______ percent of the switch rating.
    (a) 100; 75   (b) 125; 75
    (c) 100; 80   (d) 125; 80

3. Toggle switches supplying motors shall be sized at _______ percent of the FLC (in amps).
    (a) 100   (b) 125
    (c) 150   (d) 200

4. A distance of less than _______ ft is permitted, even if a proper barrier between switches and pool is installed.
    (a) 5   (b) 6
    (c) 10   (d) 12

5. Hanging luminaires shall be installed at least _______ ft over the tub and shall be located at least _______ ft from the tub in all directions in dwelling units.
    (a) 6; 3   (b) 6; 5
    (c) 8; 3   (d) 8; 5

6. The lighting outlets supporting luminaires or ceiling fans shall be located at least _______ ft _______ in. above the maximum water level of the swimming pool, if installed indoors.
    (a) 5; 6   (b) 7; 6
    (c) 5; 8   (d) 7; 8

7. Specially designed underwater lighting luminaires shall be installed not less than _______ in. below the normal water level.
    (a) 1   (b) 2
    (c) 3   (d) 4

8. An _______ AWG insulated copper conductor shall be routed in PVC conduit to ground to the metal forming shell.
    (a) 8   (b) 6
    (c) 4   (d) 2

9. Habitable rooms are such rooms in the dwelling units as:
    (a) bedroom   (b) living room
    (c) dining room   (d) all of the above

10. Surface-mounted luminaires of the incandescent type shall be mounted in dwelling units from the storage space (area) at least ______ in.
    (a) 6
    (b) 10
    (c) 12
    (d) 24

11. Surface-mounted luminaires of the fluorescent type shall be mounted in dwelling units from the storage space (area) at least ______ in.
    (a) 6
    (b) 12
    (c) 18
    (d) 24

12. Underwater lighting luminaires shall be located below the normal water level at least ______.
    (a) 6
    (b) 12
    (c) 18
    (d) 24

13. A wall switch to control the lighting outlets in dwelling units shall be provided at each level, where there is a difference between floor levels of ______ risers or more.
    (a) 4
    (b) 5
    (c) 6
    (d) 10

14. In a commercial location, the outlets on a general purpose branch circuit shall be calculated at ______ VA or the load rating, whichever is greater.
    (a) 100
    (b) 120
    (c) 180
    (d) 200

15. Surface-mounted LED luminaires of the incandescent type shall be mounted at least ______ in. from the storage space (area).
    (a) 6
    (b) 12
    (c) 18
    (d) 24

16. Recessed luminaires that are thermal protected (tp) shall be clear of all insulation by ______ in. on the top and sides.
    (a) 3
    (b) 4
    (c) 6
    (d) 12

17. A recessed luminaire can (hat) shall have a clearance of at least ______ in. from combustible materials such as wooden rafters.
    (a) 1/16
    (b) 1/8
    (c) 1/4
    (d) 1/2

18. Discharge lighting systems requiring open secondary voltage of more than ______ volts are not permitted to be installed inside or outside dwelling units.
    (a) 1000
    (b) 1200
    (c) 1500
    (d) 1800

19. The deck box for a swimming pool shall be located at least ______ ft from the inside walls of the pool.
    (a) 2
    (b) 3
    (c) 4
    (d) 5

20. The deck box for a swimming pool shall be 4 in. from the deck to the inside bottom of the deck box and not less than ______ in. from the maximum water level.
    (a) 6
    (b) 8
    (c) 10
    (d) 12

18
Motors

There are three currents that must be determined before designing and selecting the elements to make up circuits supplying power to motors. The first current that must be found is the full-load amps (FLA) from **Table 430.248** for single-phase motors and **Table 430.250** for three-phase motors. This current rating in amps, per **430.6(A)(1),** is used to size all the elements of the circuit except the overload protection. The second current to be determined is the nameplate amps on the motor per **430.6(A)(2)**. This current rating in amps is used to size the overloads (OLs) to protect the motor windings and conductors. The third current, per **430.7(A)(9)**, is the locked-rotor current (LRC), in amps, from **Tables 430.251(A)** and **(B)**. The overcurrent protection device shall be sized large enough to hold this current rating (LRC) in amps and permit the motor to start and run. When using the code letter to determine the locked-rotor current (starting current), see **Table 430.7(B)**.

Electrical systems containing AC and DC motors must have their circuits and elements designed per **Article 430** of the *National Electrical Code*. AC and DC motors are available in various types and sizes. To protect such motors and circuits and still allow them to operate, the conductors, controllers, starters, protection devices, and disconnecting means must be designed and installed properly.

BRANCH-CIRCUIT AND FEEDER CONDUCTORS
430.1

The following branch-circuit and feeder elements of a motor system are designed and installed based upon the characteristics of the motor involved:

- Motor branch-circuit and feeder conductors
- Motor branch-circuit and feeder overcurrent protection devices
- Motor overload protection
- Motor control circuits
- Motor controllers and motor control centers
- Motor disconnecting means

SIZING CONDUCTORS FOR SINGLE MOTORS
430.6(A)(1) AND 430.22(A)

Branch-circuit conductors supplying a single motor shall have an ampacity not less than 125 percent of the motor full-load current rating, in amps, per **Tables 430.247, 430.248, 430.249,** and **430.250,** respectively. For example, a 20 HP, 208 volt, three-phase motor per **Table 430.250** has a full-load current of 59.4 amps. The full-load amps (FLA) for sizing the conductors is determined by multiplying 59.4 amps x 125 percent which equals 74.25 amps.

A motor will normally have a starting current of 4 to 6 times the full-load current of the motor's FLA for motors marked with code letters A through G, and 8-1/2 to 15 times for NEMA B, high-efficiency motors. Design B, C, and D motors have a starting current of about 4 to 6 times their full-load amps when starting and driving a motor load.

There are heating effects on the conductors that develop when motors are starting and accelerating the driven load. To eliminate such effects, the conductor's current-carrying capacity is increased by taking 125 percent of the motor's full-load current rating in amps. For example, a motor with a FLC rating of 42 amps shall have conductors with a current-carrying capacity of at least 52.5 amps (42 A x 125% = 52.5 A) to safely carry the load when starting and also protect insulation due to overload conditions.

SIZING CONDUCTORS FOR SINGLE-PHASE MOTORS
430.22(A)

Section **430.6(A)(1)** of the NEC requires the full-load current in amps for single-phase motors to be obtained from **Table 430.248**. This FLC rating in amps is then multiplied by 125 percent per **Table 220.3** and **430.22(A)** to derive the total amps to select the conductors from **Table 310.16** to supply power to the motor windings. **(See Figure 18-1)**

SIZING CONDUCTORS FOR THREE-PHASE MOTORS
430.22(A)

Section **430.6(A)(1)** of the NEC requires the full-load current in amps for three-phase motors to be obtained from **Table 430.250**. This FLC rating in amps is multiplied by 125 percent per **Table 220.3** and **430.22(A)** to derive the total amps to select the conductors for the motor windings. **(See Figure 18-2)**

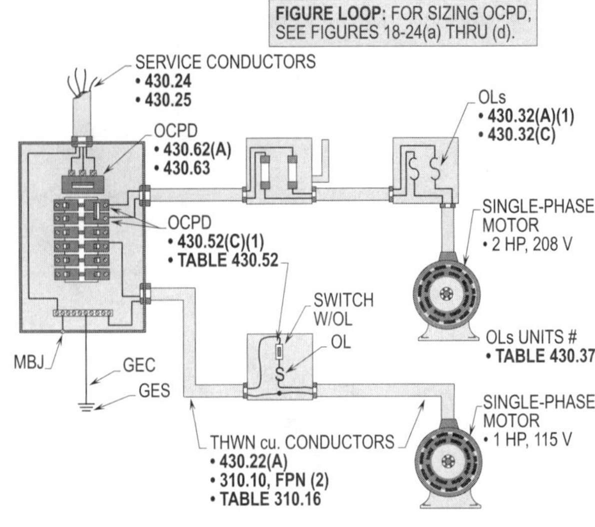

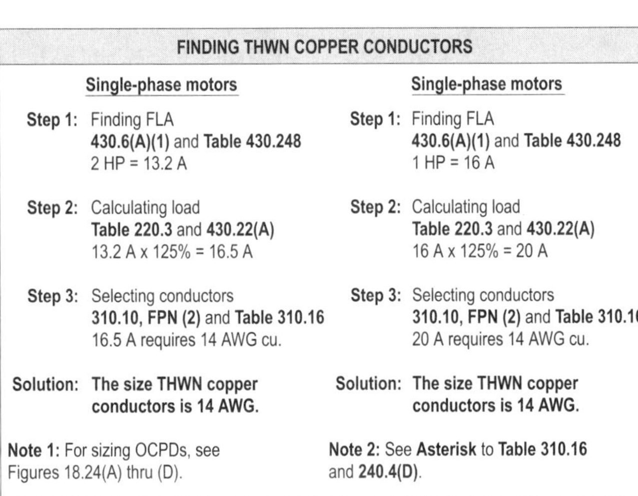

FINDING THWN COPPER CONDUCTORS

Single-phase motors	Single-phase motors
Step 1: Finding FLA 430.6(A)(1) and **Table 430.248** 2 HP = 13.2 A	**Step 1:** Finding FLA 430.6(A)(1) and **Table 430.248** 1 HP = 16 A
Step 2: Calculating load **Table 220.3** and **430.22(A)** 13.2 A x 125% = 16.5 A	**Step 2:** Calculating load **Table 220.3** and **430.22(A)** 16 A x 125% = 20 A
Step 3: Selecting conductors **310.10, FPN (2)** and **Table 310.16** 16.5 A requires 14 AWG cu.	**Step 3:** Selecting conductors **310.10, FPN (2)** and **Table 310.16** 20 A requires 14 AWG cu.
Solution: The size THWN copper conductors is 14 AWG.	**Solution:** The size THWN copper conductors is 14 AWG.
Note 1: For sizing OCPDs, see Figures 18.24(A) thru (D).	**Note 2:** See **Asterisk** to Table 310.16 and **240.4(D)**.

SIZING CONDUCTORS FOR SINGLE-PHASE MOTORS
NEC 430.22(A)

Figure 18-1. Determining the size branch-circuit conductors to supply single-phase motors.

SIZING CONDUCTORS FOR MULTISPEED MOTORS
430.22(B)

The circuit conductors for multispeed motors shall be sized large enough, to the controller, to supply the highest nameplate full-load current rating of the multispeed motor winding involved. A single overcurrent protection device is permitted to serve each speed for a multispeed motor per **430.22(B)**. The speed with the greater amps is used to size the overcurrent protection device and conductors. Overload protection shall be provided for each speed to protect each winding from excessive current during an overload condition. **(See Figure 18-3)**

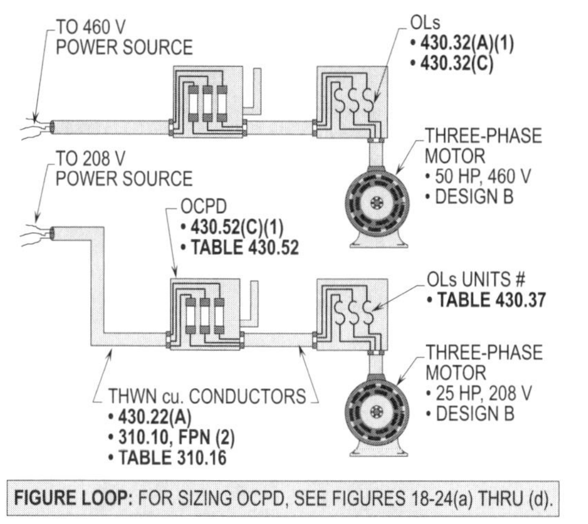

FINDING THWN COPPER CONDUCTORS

Three-phase motors	Three-phase motors
Step 1: Finding FLA **430.6(A)(1)** and **Table 430.250** 50 HP = 65 A	**Step 1:** Finding FLA **430.6(A)(1)** and **Table 430.250** 25 HP = 74.8 A
Step 2: Calculating load **Table 220.3** and **430.22(A)** 65 A x 125% = 81.25 A	**Step 2:** Calculating load **Table 220.3** and **430.22(A)** 74.8 A x 125% = 93.5 A
Step 3: Selecting conductors **310.10, FPN (2)** and **Table 310.16** 81.25 A requires 4 AWG cu.	**Step 3:** Selecting conductors **310.10, FPN (2)** and **Table 310.16** 93.5 A requires 3 AWG cu.
Solution: **The size THWN copper conductors is 4 AWG.**	**Solution:** **The size THWN copper conductors is 3 AWG.**

SIZING CONDUCTORS FOR THREE-PHASE MOTORS
NEC 430.22(A)

Figure 18-2. Determining the size branch-circuit conductors to supply three-phase motors.

SIZING CONDUCTORS FOR WYE START AND DELTA RUN MOTORS
430.22(C)

The branch-circuit conductors for wye-start and delta run connected motors shall be selected based on the full-load current on the line side of the controller. The selection of conductors between the controller and the motor shall be based on 58 percent (1 ÷ 1.732 = .58) of the motor's full-load current, in amps, times 125 percent for continuous use. **(See Figure 18-4)**

SIZING CONDUCTORS FOR DUTY CYCLE MOTORS
430.22(E)

Conductors for a motor used for short-time, intermittent, periodic, or varying duty do not require conductors to be

sized with a current-carrying capacity of 125 percent of the motor's full-load current, in amps. **Table 430.22(E)** permits the conductors to be sized with a percentage times the nameplate current rating, in amps, based upon the duty cycle classification of the motor.

When sizing conductors to supply individual motors that are used for short time, intermittent, periodic, or varying duty, the requirements of **Table 430.22(E)** shall apply. Varying heat loads are produced on the conductors by the starting and stopping duration of operation cycles, which permits conductor sizing changes. In other words, such conductors are never subjected to continuous operation due to ON and OFF periods, and therefore conductors are never fully loaded for long intervals of time. For this reason, conductors can be downsized. **(See Figure 18-5)**

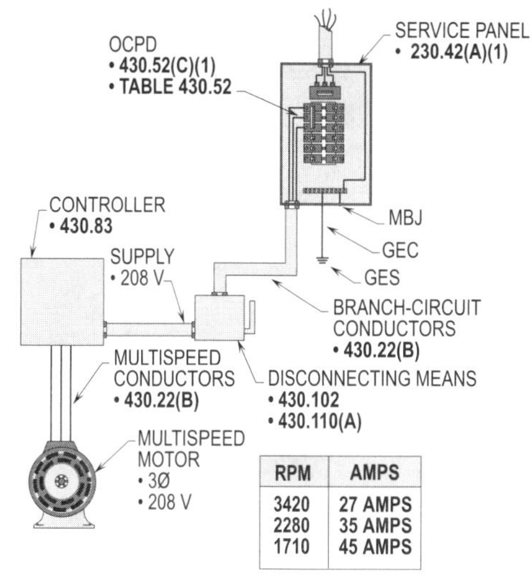

RPM	AMPS
3420	27 AMPS
2280	35 AMPS
1710	45 AMPS

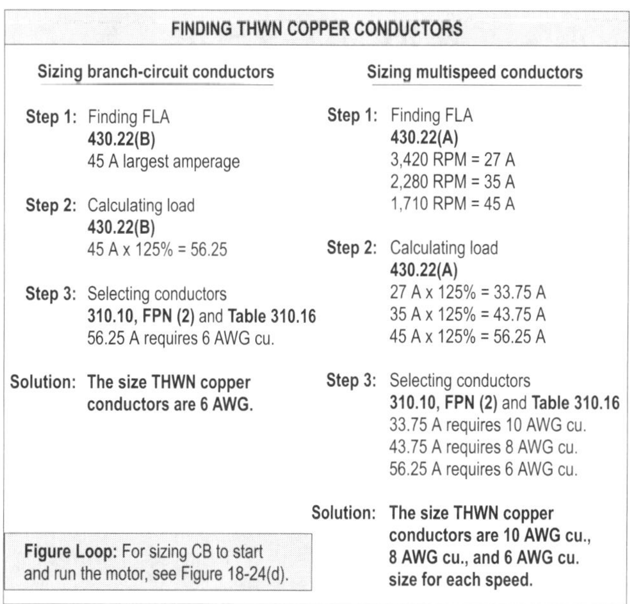

FINDING THWN COPPER CONDUCTORS

Sizing branch-circuit conductors	Sizing multispeed conductors
Step 1: Finding FLA **430.22(B)** 45 A largest amperage	**Step 1:** Finding FLA **430.22(A)** 3,420 RPM = 27 A 2,280 RPM = 35 A 1,710 RPM = 45 A
Step 2: Calculating load **430.22(B)** 45 A x 125% = 56.25	**Step 2:** Calculating load **430.22(A)** 27 A x 125% = 33.75 A 35 A x 125% = 43.75 A 45 A x 125% = 56.25 A
Step 3: Selecting conductors **310.10, FPN (2)** and **Table 310.16** 56.25 A requires 6 AWG cu.	**Step 3:** Selecting conductors **310.10, FPN (2)** and **Table 310.16** 33.75 A requires 10 AWG cu. 43.75 A requires 8 AWG cu. 56.25 A requires 6 AWG cu.
Solution: **The size THWN copper conductors are 6 AWG.**	**Solution:** **The size THWN copper conductors are 10 AWG cu., 8 AWG cu., and 6 AWG cu. size for each speed.**

SIZING CONDUCTORS FOR MULTISPEED MOTORS
NEC 430.22(B)

Figure 18-3. Determining the size branch-circuit conductors to supply multispeed motors.

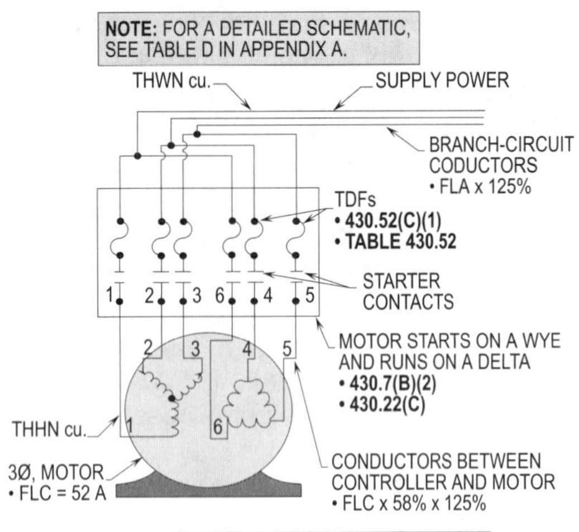

SIZING CONDUCTORS

Step 1: Calculating load
430.22(C)
BC = 52 A x 125% = 65 A
CM = 52 x 58% x 125% = 37.7 A

Step 2: Selecting conductors
Table 310.16
65 A = 6 AWG cu.
37.7 A = 8 AWG cu.

Solution: Branch circuit requires 6 AWG and conductors between the controller and motor require 8 AWG THHN copper conductors.

**SIZING CONDUCTORS FOR
WYE START AND DELTA RUN MOTORS
NEC 430.22(C)**

Figure 18-4. Determining the size conductors to supply motors starting on a wye and running on delta.

SIZING CONDUCTORS FOR ADJUSTABLE SPEED DRIVE SYSTEMS
430.122(A) AND 430.2

Power conversion equipment, when supplied from a branch circuit, includes all elements of the adjustable speed drive system. The rating in amps is used to size the conductors, which are based upon the power required by the conversion equipment. When the power conversion equipment provides overcurrent protection for the motor, no additional overload protection is required.

The disconnecting means can be installed in the line supplying the conversion equipment, and the rating of the disconnect shall not be less than 115 percent of the input current rating of the conversion unit.

Power conversion equipment requires the conductors to be sized at 125 percent of the rated input of such equipment.

Note, for more information on adjustable speed drives, see **430.120 through 430.128** at the end of this chapter.

Design Tip: Power conversion equipment contains solid state units that change the cycles or chop part of the waveforms to vary the speed of squirrel cage motors as needed for the application. **(See Figure 18-6)**

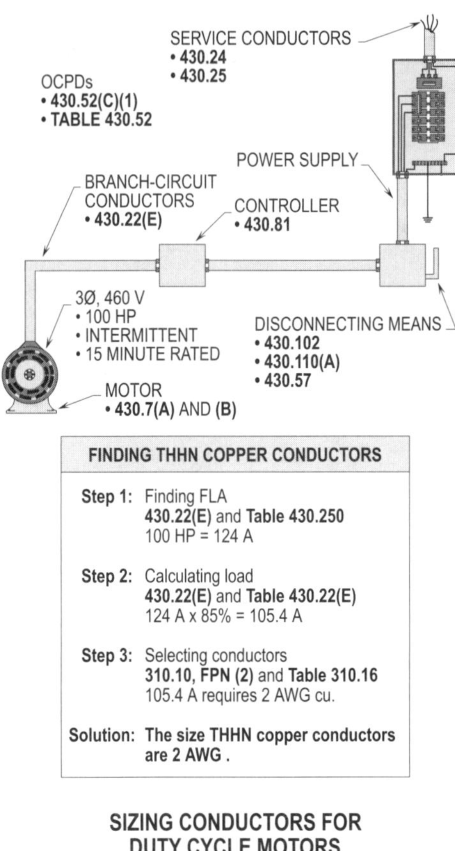

FINDING THHN COPPER CONDUCTORS

Step 1: Finding FLA
430.22(E) and **Table 430.250**
100 HP = 124 A

Step 2: Calculating load
430.22(E) and **Table 430.22(E)**
124 A x 85% = 105.4 A

Step 3: Selecting conductors
310.10, FPN (2) and **Table 310.16**
105.4 A requires 2 AWG cu.

Solution: The size THHN copper conductors are 2 AWG .

**SIZING CONDUCTORS FOR
DUTY CYCLE MOTORS
NEC 430.22(E)**

Figure 18-5. Determining the size conductors to supply duty cycle related motors.

SIZING CONDUCTORS FOR PART-WINDING MOTORS
430.22(D) AND 430.4

Induction or synchronous motors that have a part-winding start are designed so that at starting they energize the primary armature winding first. After starting, the remainder of the winding is energized in one or more steps. The purpose of this arrangement is to reduce the initial inrush current until the motor accelerates to its running speed.

The inrush current at start is locked-rotor current and at times can be quite high. A standard part-winding-start induction motor is designed so that only half of its winding is energized at start; then, as it comes up to speed, the other half is energized, so both halves are energized and carry equal current to drive the load.

Separate overload devices shall be used on a standard part-winding-start induction motor to protect the windings from excessive, damaging currents. This means that each half of the motor winding has to be individually provided with overload protection. These requirements are covered in **430.32** and **430.37**. Each half of the windings has a trip current value that is one half of the specified running current. As required by **430.52(C)(1)**, each of the two motor windings shall have branch circuit, short-circuit, and ground fault protection that is to be selected at not more than one half the percentages listed in **430.52(C)(1)** and **Table 430.52**. **(See Figure 18-7)**

> **Design Tip:** Section **430.4, Ex.** allows a single device with this one half rating, for both windings, provided that it will permit the motor to start and run. If a time-delay (dual element) fuse is used as a single device for both windings, its rating shall be permitted if it does not exceed 150 percent of the motor's full-load current.

WOUND-ROTOR SECONDARY
430.23

Wound-rotor motors are three-phase motors that are installed with two sets of leads. The main leads to the motor windings (field poles) is one set, and the secondary leads

to the rotor is the other set. The secondary leads on one end connect to the rotor through slip rings, and the other end of the leads connects through a controller and a bank of resistors. The speed of the motor varies when the amount of resistance in the motor circuit is varied. The rotor will turn slower when the resistance is greater in the rotor, and faster when such resistance is lowered.

SIZING CONDUCTORS FOR CONTINUOUS DUTY
430.23(A)

The motor shall have an ampacity not less than 125 percent of the full-load secondary current of the motor where secondary leads are installed between the controller and the motor. The secondary full-load current rating, in amps, is obtained from the manufacturer or found on the nameplate of the motor.

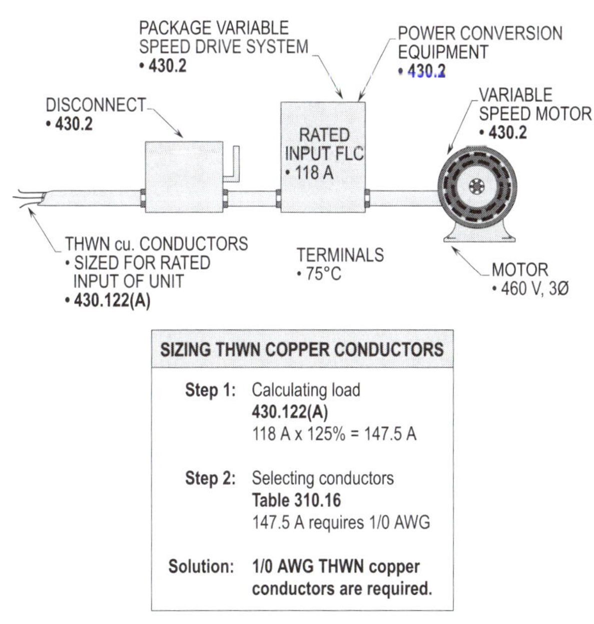

Figure 18-6. Determining the size conductors to supply power conversion equipment. **(See Figure 18-62)**

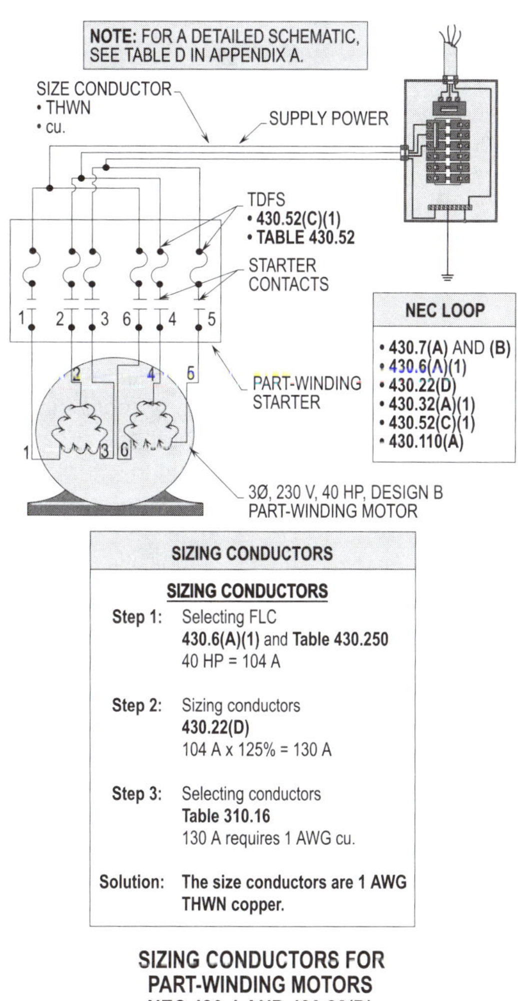

Figure 18-7. Determining the size conductors to supply part-winding motors.

SIZING CONDUCTORS FOR OTHER THAN CONTINUOUS DUTY
430.23(B)

When installing a motor to be used for short-time, intermittent, periodic, or varying duty, the secondary conductors shall be sized not less than 125 percent of the secondary current per **Table 430.22(E)**. The classification of service determines the correct percentages to select and apply, when sizing the conductors, based upon the cycles of the motor.

SIZING CONDUCTORS FOR RESISTORS, SEPARATED FROM CONTROLLER
430.23(C)

Where the secondary resistor is separate from the controller, the ampacity of the conductors between the controller and resistor shall not be less than the resistor duty classification percentages listed in **Table 430.23(C)**. **(See Figure 18-8)**

SIZING CONDUCTORS FOR SEVERAL MOTORS
430.24

The full-load current rating of the largest motor shall be multiplied by 125 percent to select the size of conductors for a feeder supplying a group of two or more motors. The remaining motors of the group shall have their full-load current ratings added to this value, and this total amperage is then used to size the conductors. **(See Figure 18-9)**

SIZING CONDUCTORS FOR DUTY CYCLE MOTORS
430.24, Ex. 1

Table 430.22(E) shall be used for sizing the amperage rating for a motor that is classified as either short-time, intermittent, periodic, or varying duty. The amperage rating shall be based on 100 percent of the full-load current rating on the motor's nameplate if rated for continuous operation.

The full-load current rating, in amps, of the largest motor shall be multiplied by 125 percent to select the size conductors for a feeder supplying a group of two or more motors. The remaining motors of the group shall have their total full-load current ratings added to the calculated amps of the duty cycle motor or to the amps of the 125 percent

motor, whichever is greater. The feeder conductors shall be sized by this total full-load current, in amperes. **(See Figure 18-10)**

SIZING CONDUCTORS SUPPLYING MOTORS AND OTHER LOADS
430.25

The motor load shall be calculated per **430.22(A)** or **430.24** when designing combination loads that consist of one or more motor loads on the same circuit with lights, receptacles, appliances, or any combination of such loads. For other than motor loads, **Article 220** and other applicable articles shall be used to calculate such loads. The ampacity required for the feeder conductors shall be equal to all the total loads involved. The overcurrent protection devices used to protect conductors and elements from short circuits and ground faults shall be sized per **430.62(A)** and **430.63**. **(See Figure 18-11)**

SIZING THE BRANCH-CIRCUIT PROTECTIVE DEVICE
TABLE 430.52, 430.52(C)(1), AND (C)(3)

The motor branch-circuit overcurrent device shall be capable of carrying the starting current of the motor. Short-circuit and ground fault current is considered to be properly taken care of when the overcurrent protection device does not exceed the values in **Table 430.52,** as permitted by the provisions of **430.52(C)(1)** with exceptions.

Different percentages are selected for particular devices based upon one of the four columns listed in **Table 430.52**. The percentages are used to size and select the proper size overcurrent protection device to allow a certain type of motor to start and run. The motor has a momentary starting current that is necessary for the motor to have power to start and drive the connected load at the driven equipment. Note that the overcurrent protection device sized per **430.52(C)(1)** provides protection from short circuits and ground faults. Overload protection shall be provided for conductors and motor windings per **430.32(A)(1)** and **430.32(C)**. **(See Figure 18-12)**

> **Design Tip:** In cases where the values for branch-circuit protective devices determined by **Table 430.52** do not correspond to the standard sizes or ratings of fuses, nonadjustable circuit breakers, or thermal devices, or possible settings of adjustable circuit breakers adequate to carry the starting currents of the motor, the next higher size rating or setting shall be permitted to be used.

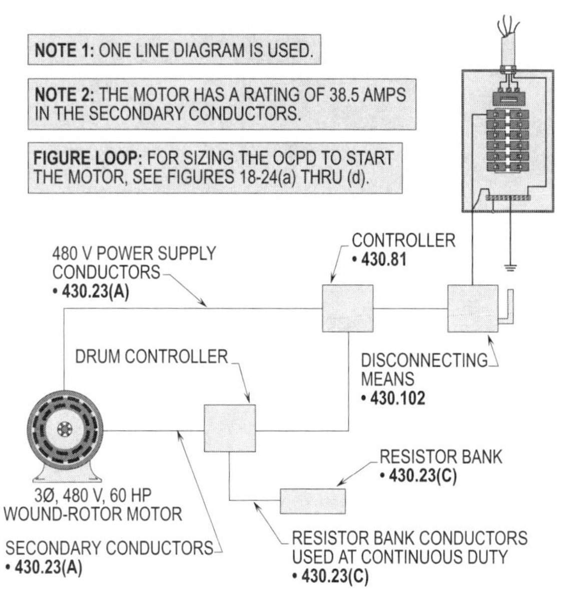

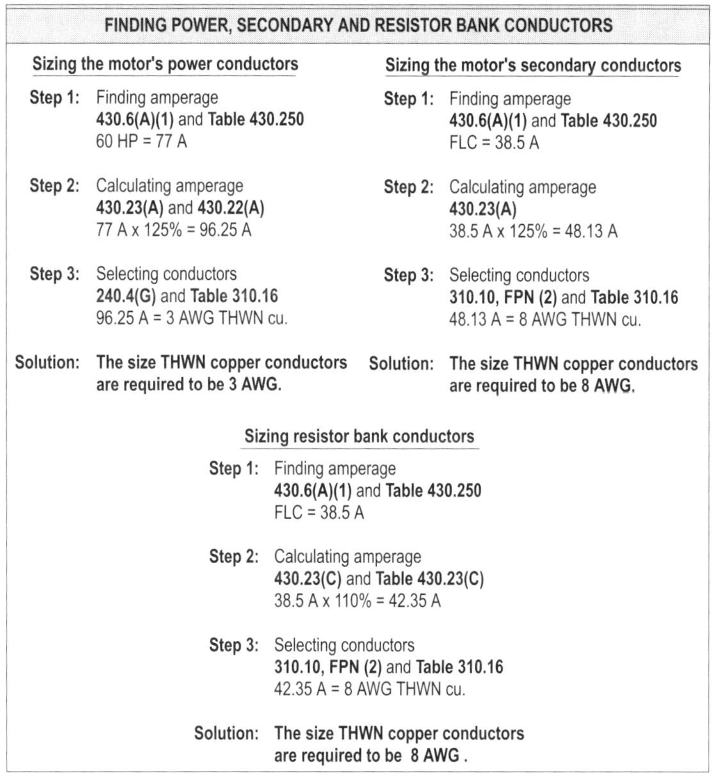

SIZING CONDUCTORS FOR RESISTORS, SEPARATED FROM CONTROLLER NEC 430.23(A), (B), AND (C)

Figure 18-8. Determining the size conductors to supply wound-rotor motors.

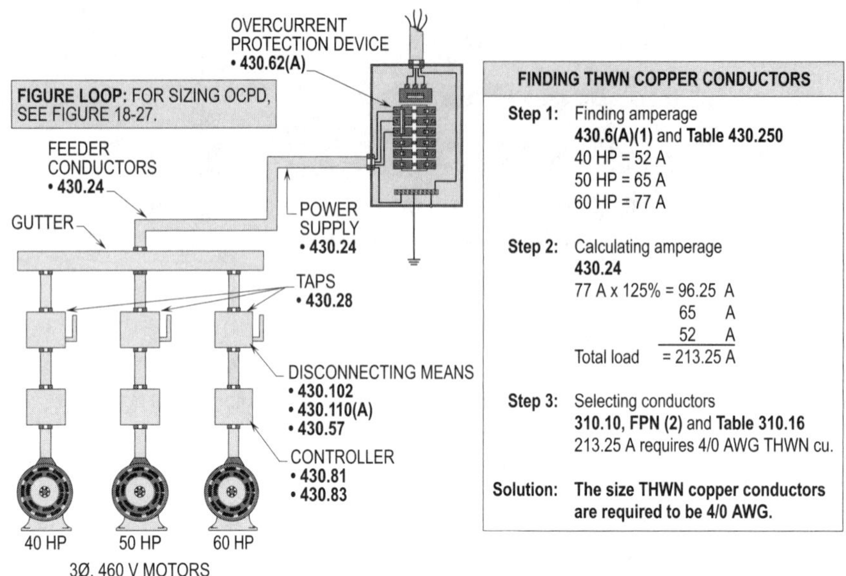

Figure 18-9. Determining the size conductors for a feeder to supply several motors.

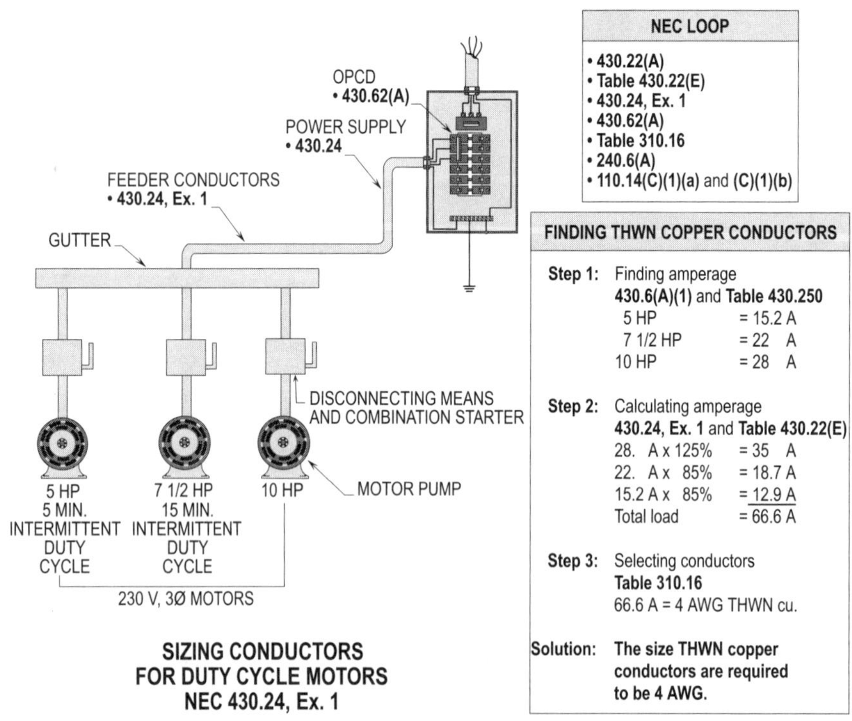

Figure 18-10. Determining the size conductors for a feeder to supply duty cycle related motors.

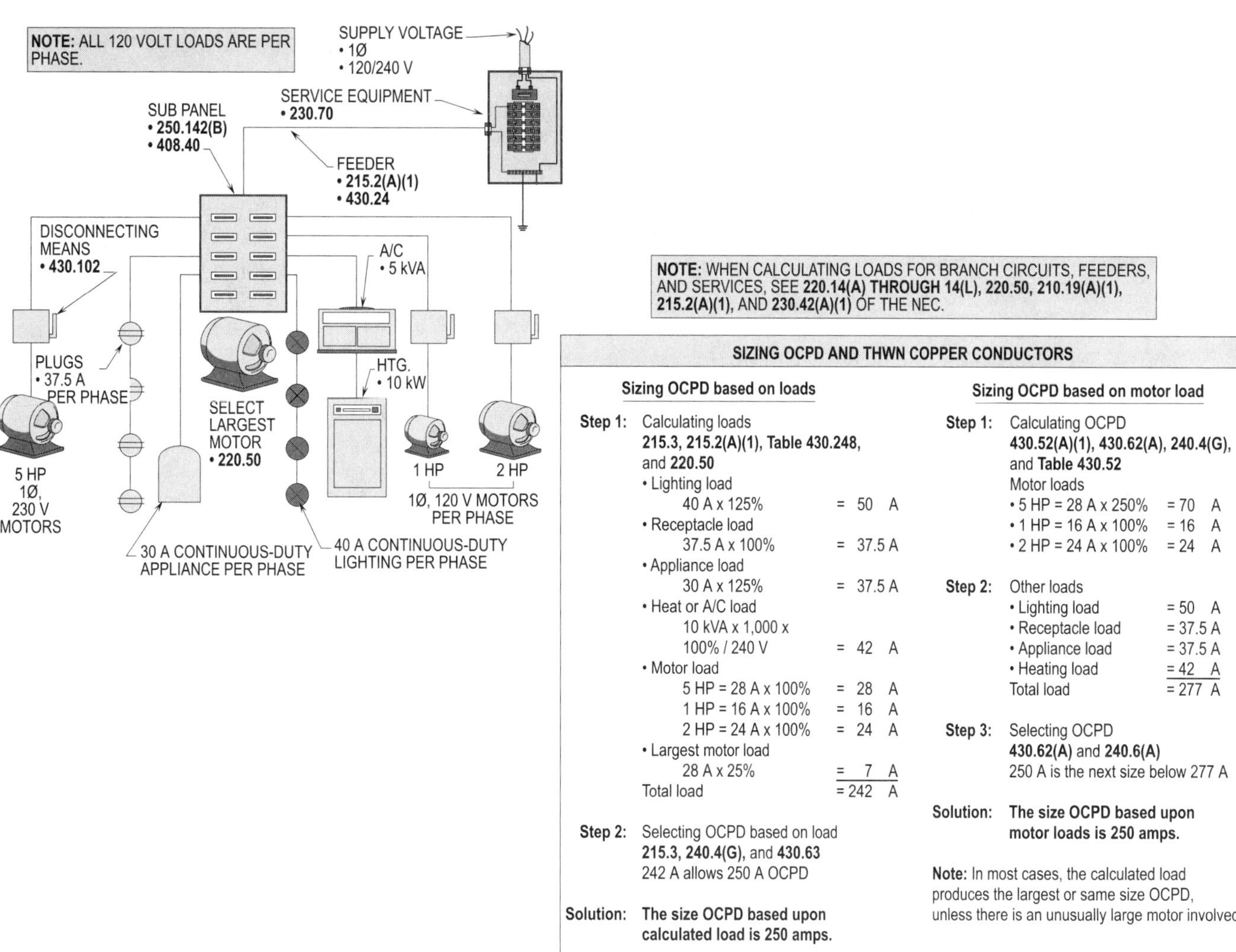

SIZING OCPD AND THWN COPPER CONDUCTORS

Sizing OCPD based on loads

Step 1: Calculating loads
215.3, 215.2(A)(1), Table 430.248, and **220.50**
- Lighting load
  40 A x 125% = 50 A
- Receptacle load
  37.5 A x 100% = 37.5 A
- Appliance load
  30 A x 125% = 37.5 A
- Heat or A/C load
  10 kVA x 1,000 x 100% / 240 V = 42 A
- Motor load
  5 HP = 28 A x 100% = 28 A
  1 HP = 16 A x 100% = 16 A
  2 HP = 24 A x 100% = 24 A
- Largest motor load
  28 A x 25% = 7 A
  Total load = 242 A

Step 2: Selecting OCPD based on load
215.3, 240.4(G), and 430.63
242 A allows 250 A OCPD

Solution: The size OCPD based upon calculated load is 250 amps.

Sizing OCPD based on motor load

Step 1: Calculating OCPD
430.52(A)(1), 430.62(A), 240.4(G), and **Table 430.52**
Motor loads
- 5 HP = 28 A x 250% = 70 A
- 1 HP = 16 A x 100% = 16 A
- 2 HP = 24 A x 100% = 24 A

Step 2: Other loads
- Lighting load = 50 A
- Receptacle load = 37.5 A
- Appliance load = 37.5 A
- Heating load = 42 A
  Total load = 277 A

Step 3: Selecting OCPD
430.62(A) and 240.6(A)
250 A is the next size below 277 A

Solution: The size OCPD based upon motor loads is 250 amps.

Note: In most cases, the calculated load produces the largest or same size OCPD, unless there is an unusually large motor involved.

Sizing conductors

Step 1: Calculating loads
215.2(A)(1)
- Lighting load
  40 A x 125% = 50 A
- Receptacle load
  37.5 A x 100% = 37.5 A
- Appliance load
  30 A x 125% = 37.5 A
- Heat or A/C load
  10 kVA x 1,000 x 100% ÷ 240 V = 42 A
- Motor load
  5 HP = 28 A x 100% = 28 A
  1 HP = 16 A x 100% = 16 A
  2 HP = 24 A x 100% = 24 A
- Largest motor load
  28 A x 25% = 7 A
  Total load = 242 A

Step 2: Selecting conductors
310.10, FPN (2) and Table 310.16
242 A requires 250 KCMIL

Solution: The size conductors are 250 KCMIL THWN cu.

Sizing neutral

Step 1: Calculating load
430.24 and 220.61
- Lighting load
  40 A x 100% = 40 A
- Receptacle load
  37.5 A x 100% = 37.5 A
- Appliance load
  30 A x 100% = 30 A
- Motor load
  1 HP = 16 A x 100% = 16 A
  2 HP = 24 A x 100% = 24 A
- Largest motor load
  24 A x 25% = 6 A
  Total load = 153.5 A

Step 2: Selecting conductors
310.10, FPN (2) and Table 310.16
153.5 A requires 2/0 AWG cu.

Solution: The size of the conductors based upon 100 percent are 2/0 AWG THWN copper.

Note: The AHJ may require the continuousl oads to be multiplied by 125 percent when calculating the neutral load.

**SIZING CONDUCTORS SUPPLYING
MOTORS AND OTHER LOADS
NEC 430.25**

Figure 18-11. Determining the size conductors for motors and other loads supplied by a feeder.

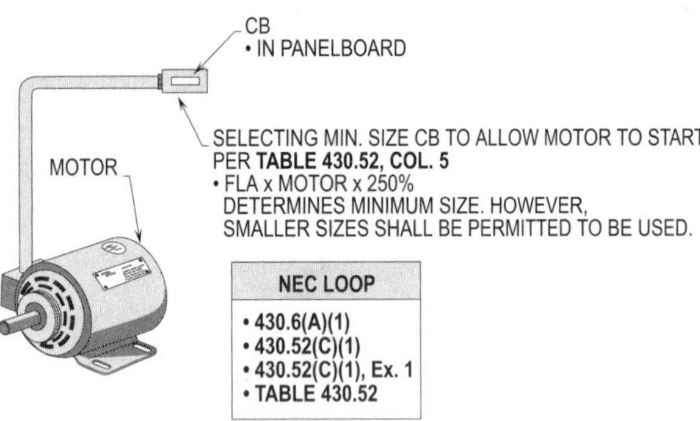

**SIZING THE BRANCH-CIRCUIT PROTECTIVE DEVICE
NEC TABLE 430.52
NEC 430.52(C)(1) AND (C)(3)**

Figure 18-12. Selecting the percentages to determine the minimum size (rounding down), next size (rounding up), and maximum size circuit breaker to allow a motor to start and run. **[See Figure 18-13 through 18-14(b)]**

APPLYING THE EXCEPTIONS
430.52(C)(1), Ex. 1 AND Ex. 2

There are Exceptions that permit larger overcurrent protection devices to be used where the overcurrent protection device, as specified in **Table 430.52,** will not permit the starting current of the motor to start and run. Where the motor fails to start and run because of excessive inrush starting currents, one of the following exceptions can be applied.

APPLYING Ex. 1

If the values of the branch-circuit, short circuit, and ground fault protection devices determined from **Table 430.52** do not conform to standard sizes or ratings of fuses, nonadjustable circuit breakers, or possible settings on adjustable circuit breakers, it does not matter if they are capable or not capable of adequately carrying the load involved; the next higher setting or rating shall be permitted. In other words, you can round up or round down the size of the overcurrent protection device automatically by choice. **(See Figure 18-13)**

APPLYING Ex. 2

If the ratings listed in **Table 430.52** and **Ex. 1** to **430.52(C)(1)** are not sufficient for the starting current of the motor, the following overcurrent protection devices with percentages shown can be used to start and run motors having high inrush starting currents. **[See Figure 18-14(a)]**

When nontime-delay fuses are used and they do not exceed 600 amperes in rating, it shall be permitted to increase the fuse size up to 400 percent of the full-load current, but never over 400 percent.

Time-element fuses (dual-element) shall not exceed 225 percent of the full-load current in amps, but they may be increased to up this percentage.

Inverse time-element breakers shall be permitted to be increased in rating. However:

- They shall not exceed 400 percent of full-load current of the motor for 100 amperes or less, or

- They may be increased to 300 percent where a full-load current is greater than 100 amperes.

See Figure 18-14(b) for a detailed illustration on selecting percentage for sizing overcurrent protection devices.

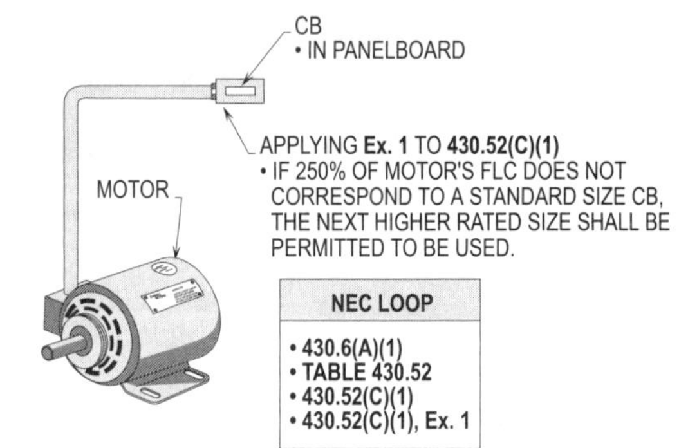

**SIZING THE BRANCH-CIRCUIT PROTECTIVE DEVICE
NEC TABLE 430.52
NEC 430.52(C)(1), Ex. 1**

Figure 18-13. Where the percentages of **Table 430.52** times the full-load current of motor in amps does not correspond to a standard size overcurrent protection device, the next higher size rating above this percentage shall be permitted to be used.

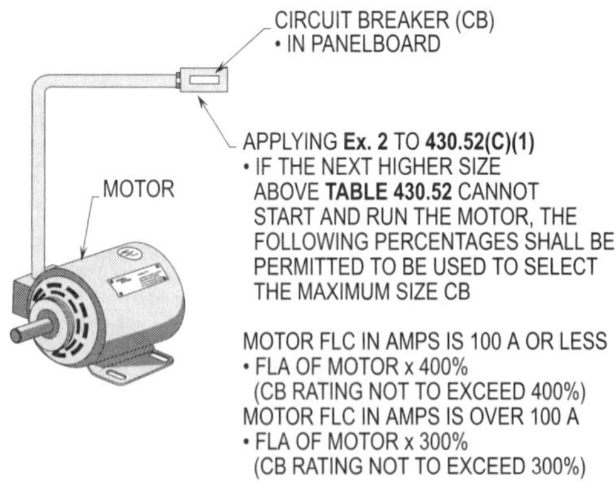

**SIZING THE BRANCH-CIRCUIT PROTECTIVE DEVICE
NEC TABLE 430.52
NEC 430.52(C)(1), Ex. 2**

Figure 18-14(a). When the percentages of **Table 430.52** and **430.52(C)(1), Ex. 1** won't allow the motor to start and run the driven load, the maximum size circuit breaker of **430.52(C)(1), Ex. 2(c)** shall be permitted to be used.

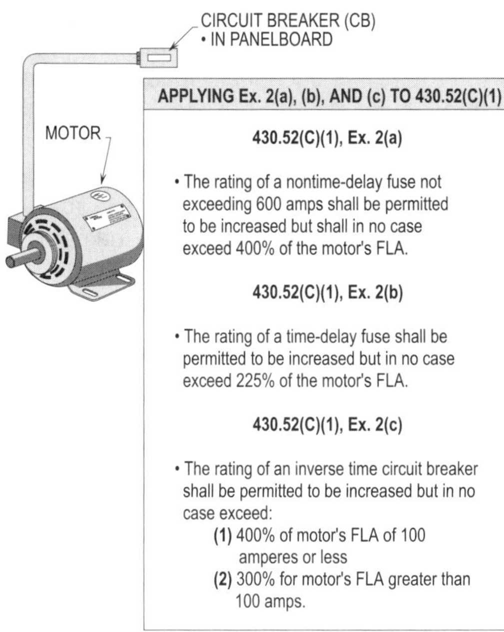

SIZING THE BRANCH-CIRCUIT PROTECTIVE DEVICE
NEC 430.52(C)(1), Ex. 2(a), (b) AND (c)

Figure 18-14(b). When the percentages of **Table 430.52** and **430.52(C)(1), Ex. 1** won't allow the motor to start and run, the maximum percentages of **430.52(C)(1), Ex. 2(a), (b),** and **(c)** shall be permitted to be applied.

USING INSTANTANEOUS TRIP CIRCUIT BREAKERS 430.52(C)(3), Ex. 1

An instantaneous trip circuit breaker shall be used only if it is adjustable, and it is a part of a combination controller that has overcurrent protection in each controller. Such combination, when used, has to be approved. An instantaneous trip circuit breaker is allowed to have a damping device, to limit the inrush current when the motor is started.

If the specified setting in **Table 430.52** is not sufficient for the starting current of the motor, the setting on an instantaneous trip circuit current may be increased, provided that in no instance it exceeds 1300 percent of the motor's full-load current rating, in amps, for motors marked as Class B, C, or D.

Design Tip: For Design E and Class B NEMA motors, the setting on the instantaneous trip circuit breakers shall be permitted to be adjusted up to 1700 percent to allow the motor to start and run.

See Figure 18-15 for adjusting the maximum trip settings on instantaneous trip circuit breakers to allow motors to start and accelerate their driven load.

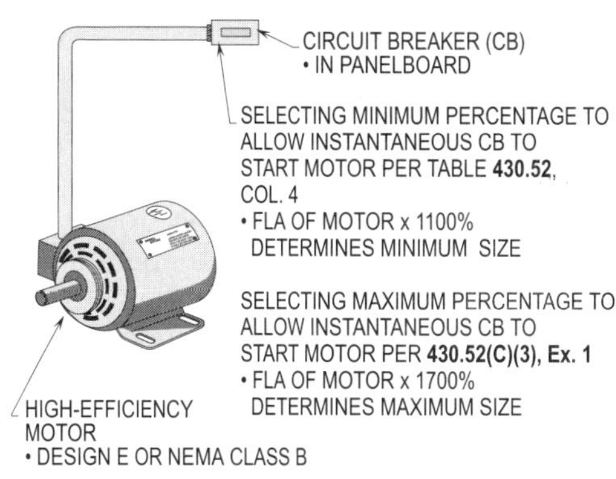

USING INSTANTANEOUS TRIP
CIRCUIT BREAKERS
NEC TABLE 430.52
NEC 430.52(C)(3), Ex. 1

Figure 18-15. Determining the minimum and maximum size setting on an instantaneous trip circuit breaker to allow a motor to start and run a driven load. Note that after the minimum size overcurrent protection device has been determined, a smaller size shall be permitted to be selected.

TYPES OF MOTORS
TABLE 430.52

The following are five types of motors to be considered when sizing overcurrent protection devices to allow motors to start and run:

- Single-phase AC squirrel-cage
- Three-phase AC squirrel-cage
- Wound-rotor
- Synchronous
- DC

SINGLE-PHASE AC SQUIRREL-CAGE MOTORS

Squirrel-cage motors are known in the electrical industry as induction motors. An induction motor operates on the same principles as the primary and secondary windings of a transformer. When power energizes the field windings, they serve as the primary by inducing voltage into the rotor that serves as the secondary windings. Squirrel-cage motors have two windings on the stator: one winding is the run winding, and the other is the starting winding. This additional starting winding on the stator is required for split-phase, single-phase, induction motors to have the capacity to start and run. The starting winding has a higher resistance to ground than the running winding, which creates a phase displacement between the two windings. It is this phase displacement between the two windings that gives split-phase motors the power to start.

The phase displacement is about 18 degrees to 30 degrees in angular phase displacement, which provides enough starting torque (twist or force) to start the motor. The motor operates on the running winding when the rotor starts turning and has established a running speed at about 75 percent to 80 percent of the motor's synchronous speed. The starting winding is disconnected by a centrifugal switch that is installed in the circuit of the starting winding. **(See Figure 18-16)**

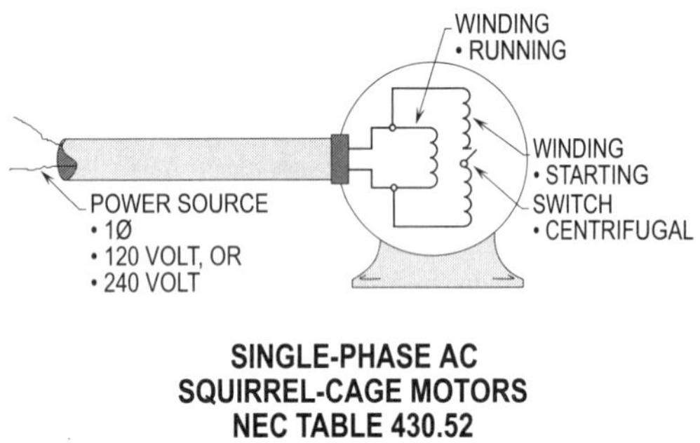

Figure 18-16. The above is an example of a single-phase squirrel-cage induction motor that is listed in **Table 430.52** and **Table 430.248** of the NEC.

THREE-PHASE SQUIRREL-CAGE MOTORS

Three-phase induction motors have three separate windings per pole on the stator that generate magnetic fields that are 120 degrees out of phase with each other. An additional starting winding is not required for three-phase motors to start and run. An induction motor will always have a peak phase of current. This is due to alternating current reversing its direction of flow. In other words, when alternating current of one phase reverses its direction of flow, a peak current will be developed on one phase and, as current reverses direction again, a second phase will peak, etc. Three-phase motors provide a smooth and continuous source of power once they are started and driving their load. **(See Figure 18-17)**

WOUND-ROTOR MOTORS

Wound-rotor motors are classified as three-phase induction motors. They are similar in design to squirrel-cage induction

motors. Wound-rotor motors are three-phase motors having two sets of leads. One set is the main leads to the motor windings (field poles) and the other set is the secondary leads to the rotor. The secondary leads are connected to the rotor through the slip rings, while the other ends of the leads are connected through a controller and a bank of resistors. The speed of the motor varies with the amount of resistance added in the motor circuit. The rotor will turn slower when the resistance is greater in the rotor, and vice versa. The resistance may be incorporated in the controller, or the resistor banks may be separate from the motor. **(See Figure 18-18)**

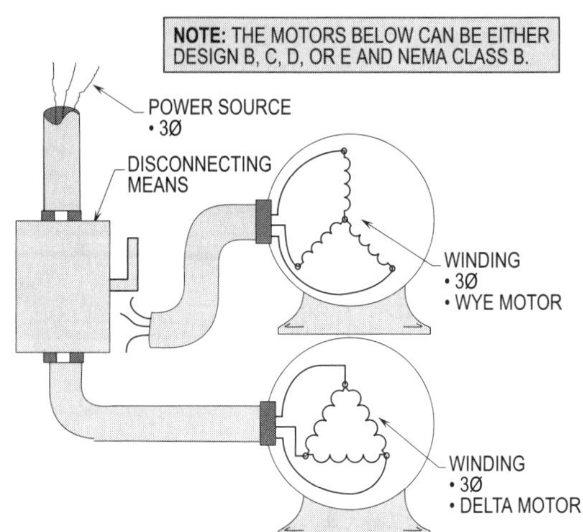

Figure 18-17. The above is an example of a three-phase squirrel-cage induction motor that is listed in **Table 430.52** and **Table 430.250** of the NEC.

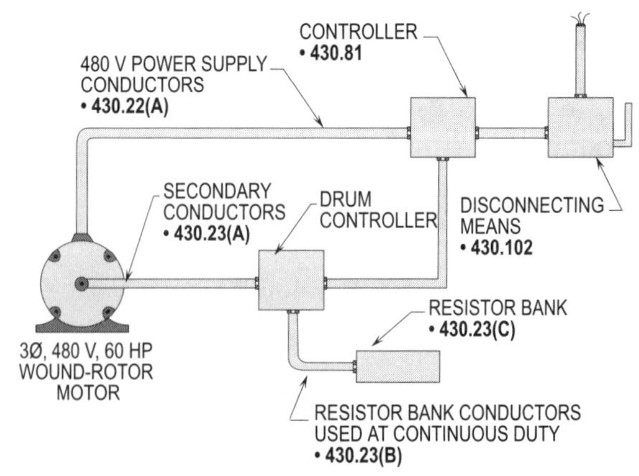

Figure 18-18. The above is an example of a three-phase wound-rotor motor listed in **Table 430.52** and **Table 430.250** of the NEC.

SYNCHRONOUS MOTORS

The following are two types of synchronous motors that are available:

- Nonexcited
- Direct-current excited

Synchronous motors are available in a wide range of sizes and types that are designed to run at designed speeds. A DC source is required to excite a direct-current excited synchronous motor. The torque required to turn the rotor of a synchronous motor is produced when the DC current of the rotor field locks in with the magnetic field of the stator AC current. **(See Figure 18-19)**

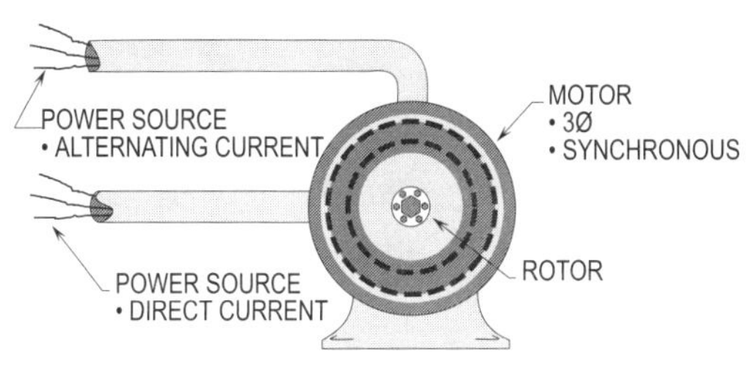

SYNCHRONOUS MOTORS
NEC TABLE 430.52

Figure 18-19. The above is an example of a three-phase synchronous motor that is listed in **Table 430.52** and **Table 430.250** of the NEC.

DC MOTORS

Direct current only is used to operate DC related motors. A DC motor is designed with the following two main parts:

- The stator
- The rotor

The stationary frame of the motor is called the stator. The armature mounted on the drive shaft is known as the rotor. By applying direct current to the rotor, the speed may be adjusted for a DC motor that drives the driven load at a specific speed. **(See Figure 18-20)**

CODE LETTERS
430.7(B) AND TABLE 430.7(B)

Code letters are installed on motors by manufacturers for calculating the locked-rotor current (LRC) in amps based upon the kVA per horsepower that is selected from the motor's code letter. Overcurrent protection devices shall be set above the locked-rotor current of the motor to prevent the overcurrent protection device from opening when the rotor of the motor is starting. The following two methods can be used to calculate and select the locked-rotor current of motors:

- Utilizing code letters to determine LRC
- Utilizing horsepower to determine LRC

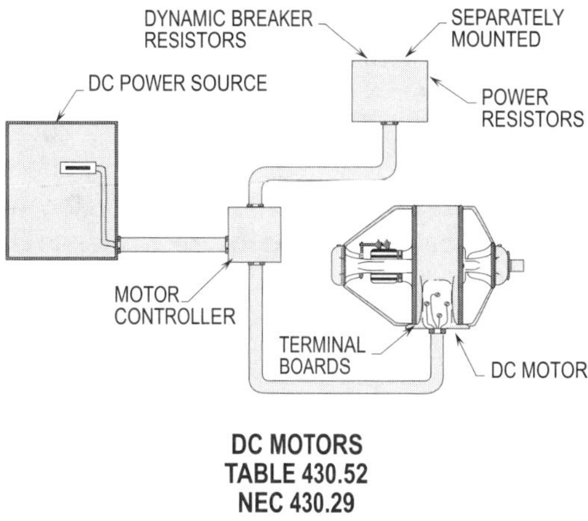

DC MOTORS
TABLE 430.52
NEC 430.29

Figure 18-20. The above is an example of a DC motor that is listed in **Table 430.52** and **Table 430.247** of the NEC.

UTILIZING CODE LETTERS TO FIND LRC
430.7(B) AND TABLE 430.7(B)

Code letters shall be marked on the nameplate, and such letters are used for designing locked-rotor current. Locked-rotor current for code letters is listed in **Table 430.7(B)** in kVA (kilovolt-amps) per horsepower, based upon a particular code letter.

For example: What is the locked-current rating for a three-phase, 208 volt, 20 horsepower motor with a code letter B marked on the nameplate of the motor?

Step 1: Finding LRC amps
Table 430.7(B)
A = (kVA per HP) x (1000 ÷ V x 1.732)
A = (3.54 x 20 x 1000) ÷ (208 V x 1.732)
A = 70,800 ÷ 360
A = 197

Solution: The locked-rotor current is 197 amps. Note that Table 430.7(B) must be used to find LRC's of the motor, based on their code letters per the 1996 NEC and earlier editions.

LOCKED-ROTOR CURRENT UTILIZING HORSEPOWER TABLES 430.251(A) AND (B)

The locked-rotor current of a motor may be found in **Tables 430.251(A)** and **(B)**. The locked-rotor current for single-phase and three-phase motors are selected from one of these Tables, based upon the phases, voltage, and horsepower rating of the motor. For motors with code letters A through G, round the nameplate current in amps up to an even number (unit of ten) and multiply by 6 to obtain the LRC of the motor. Note that code letters are not found in **Tables 430.251(A)** and **(B)**; they are not listed on the motor's nameplate anymore. Motors will be marked either as Design B, C, D, or E letter to indicate which locked-rotor currents are to be selected from **Tables 430.251(A)** and **(B)** based on horsepower, phases, and voltages.

For example: What is the locked-rotor current rating for a three-phase, 460 volts, 50 horsepower, Design B motor?

Table method using Design letter

Step 1: Finding LRC amps
Table 430.251(B)
50 HP requires 363 A

Solution: The locked-rotor current is 363 amps.

For example: What is the locked-rotor current of a motor with a nameplate current of 63 amps, based upon code letter A through G?

Rule of thumb method using code letter

Step 1: Finding even number (unit of ten)
Table 430.7(B)
Round up 63 A to 70 A

Step 2: Calculating LRC
Table 430.7(B)
70 A x 6 = 420 A

Solution: The locked-rotor current is 420 amps. Note: This method can be used only for code letters A thru G.

See Figures 8-21(a) and (b) for calculating and selecting the locked-rotor current of a motor.

Design Tip: Engineers and electricians shall select the locked-rotor current rating from **Tables 430.251(A)** and **(B)** when using Design B, C, D, or E motors. The overcurrent protection device shall be set above the locked-rotor current of the motor so the motor can start and run.

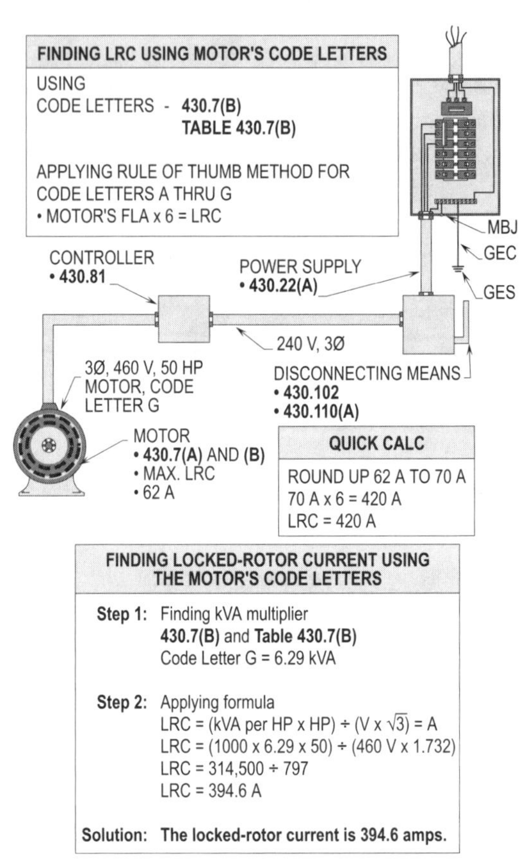

LOCKED-ROTOR CURRENT UTILIZING HORSEPOWER NEC TABLES 430.251(A) AND (B)

Figure 18-21(a). For motors having code letters instead of Design letters, the LRC shall be calculated per **Table 430.7(B)** using the code letter of the motor.

SIZING AND SELECTING OVERCURRENT PROTECTION DEVICES
TABLE 430.52, COLUMNS 2, 3, 4, AND 5

The overcurrent protection device shall be sized for the starting current of the motor and selected to allow the motor to start and run. The overcurrent protection device per **Table 430.52** shall protect the branch-circuit conductors from short circuits and ground faults. The following four overcurrent protection devices selected from **Table 430.52** will start most motors under normal starting conditions:

- Nontime-delay fuses per Column 2
- Time-delay fuses per Column 3
- Instantaneous trip circuit breakers per Column 4
- Inverse-time circuit breakers per Column 5

NONTIME-DELAY FUSES
TABLE 430.52, COLUMN 2

Nontime-delay fuses are installed with instantaneous trip features to detect short circuits and thermal characteristics to sense slow heat buildup in the circuit. A nontime-delay fuse will hold 5 times (500 percent) its rating for approximately 1/4 to 2 seconds based upon the type used.

For example: What is the holding time in amps for a nontime-delay fuse of 150 amps?

Step 1: Finding holding amps
A = fuse rating x 500%
A = 150 A x 500%
A = 750

Solution: The holding time in amps of a nontime-delay fuse is 750 amps. Note: This fuse will blow in 1/4 to 2 seconds so the motor will have to start and accelerate the load quickly.

See Figure 18-22(a) for a detailed illustration of sizing nontime-delay fuses to allow motors to start and run.

TIME-DELAY FUSES
USING MAXIMUM SIZE
TABLE 430.52, COLUMN 3

Time-delay fuses are also equipped with instantaneous trip features to detect short circuits and thermal characteristics to sense slow heat buildup in the circuit. Time-delay fuses are used because of their time-delay action to allow a motor to start. Time-delay fuses will hold 5 times (500 percent) of their rating, which will permit most motors to start and accelerate the driven load. Note that time-delay fuses that are sized at 125 percent or less of the motor's FLC rating can provide overload protection for the motor.

A time-delay fuse will hold 5 times its rating for 10 seconds, and this delayed action provides more acceleration time to allow the motor to start without tripping the overcurrent protection device.

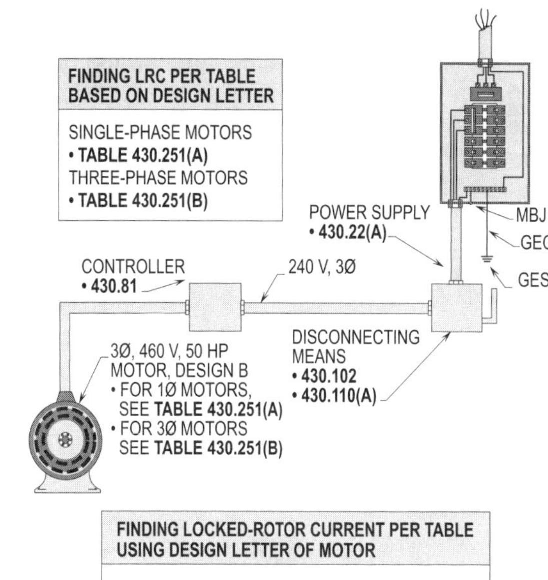

**LOCKED-ROTOR CURRENT UTILIZING HP
NEC TABLES 430.251(A) AND (B)**

Figure 18-21(b). Tables 430.251(A) and **(B)** shall be permitted to be used to determine the LRC in amps for motors with Design letters.

For example: What is the holding power in amps for a time-delay fuse of 150 amps?

Step 1: Finding holding amps
A = fuse x 500%
A = 150 A x 500%
A = 750
Solution: The rating of a time-delay fuse is 750 amps. Note: This fuse holds five times its rating for ten seconds without blowing and opening the circuit.

See Figure 18-22(b) for sizing time-delay fuses to hold the motors locked-rotor current, in amps.

INSTANTANEOUS TRIP
CIRCUIT BREAKERS
TABLE 430.52, COLUMN 4

Instantaneous trip circuit breakers are installed with instantaneous values of current to respond from short circuits only. Thermal protection is not provided for instantaneous trip circuit breakers. Instantaneous trip circuit breakers will hold about three times their rating on the low setting and five times their next setting, seven times their next setting, and approximately ten times their rating on the high setting. Certain types allow such settings to be adjusted from 0 to 1700 percent. **[See Figure 18-23(a)]**

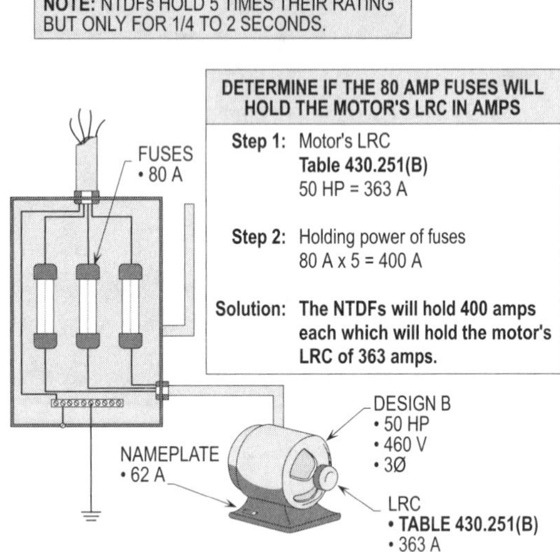

NONTIME-DELAY FUSES
NEC TABLE 430.52, COLUMN 2

Figure 18-22(a). Nontime-delay fuses will hold five times their rating, and when this rating is above the locked-rotor current it should allow the motor to start and run based upon LRC.

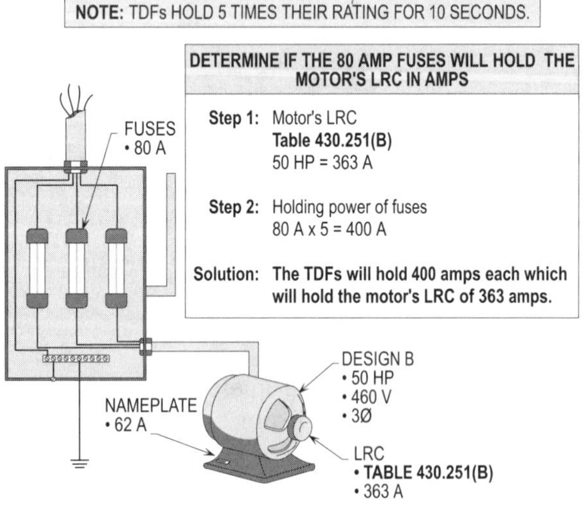

TIME-DELAY FUSES
USING MAXIMUM SIZE
NEC TABLE 430.52, COLUMN 3

Figure 18-22(b). Time-delay fuses will hold five times their rating, and when this rating is above the locked-rotor current, in amps, it should allow the motor to start and run based upon LRC.

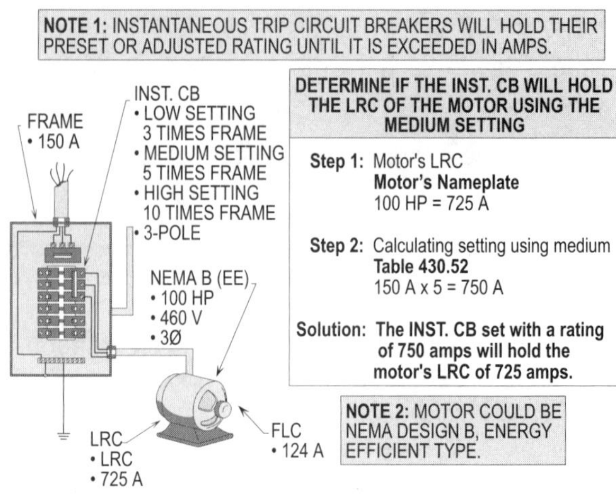

INSTANTANEOUS TRIP CIRCUIT BREAKERS
NEC TABLE 430.52, COLUMN 4

Figure 18-23(a). An instantaneous circuit breaker with its rating set above the locked-rotor current, in amps, of a motor will allow the motor to start and run.

INVERSE-TIME CIRCUIT BREAKERS
TABLE 430.52, COLUMN 5

Inverse-time circuit breakers are designed with instantaneous trip features to detect short circuits and thermal characteristics to sense slow heat buildup in the circuit. If heat should occur in the windings of the motor, the instantaneous values of current will be detected by the thermal action of the circuit breaker and will trip open the circuit if it is sized properly. The magnetic action of the circuit breaker will clear the circuit if short circuits or ground faults should occur on the circuit elements or equipment served.

Design Tip: Inverse-time circuit breakers will hold about three times their rating for different periods of time based upon their frame size. For example, a motor with a full-load current of 585 amps can be started with a 200 amp circuit breaker.

This can be verified by multiplying the 200 amp circuit breaker by 3, which is equal to 600 amps; 585 amps divided by 3 is equal to 195 amps. By rounding up to the next size circuit breaker per **430.52(C)(1), Ex. 1**, the size circuit breaker is 200 amps, per **240.6(A)**. Note that this size circuit breaker allows the motor to start and run. **[See Figure 18-23(b)]**

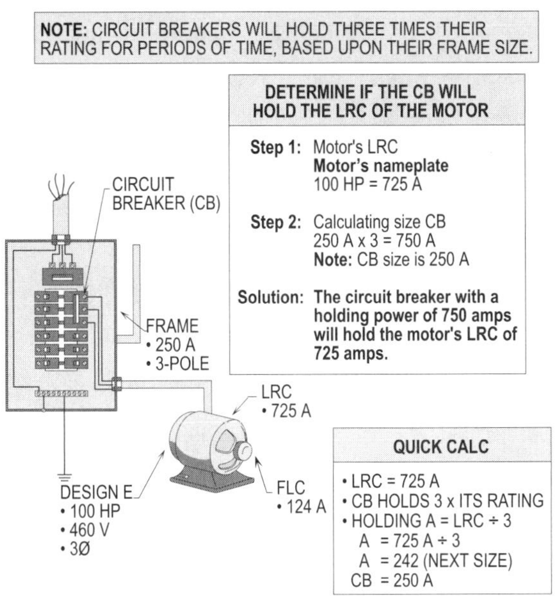

Figure 18-23(b). Circuit breakers sized at least three times their rating provide an amp rating above the locked-rotor current of the motor and will hold such LRC.

SIZING MAXIMUM OVERCURRENT PROTECTION DEVICE 430.52(C)(1), Ex. 2(a) THRU (c)

Where the rating specified in **Table 430.52** is not sufficient for the starting current of the motor, the following ratings (percentages) shall be applied:

- Nontime-delay fuses (400 percent)

- Time-delay fuses (225 percent)

- Inverse-time circuit breakers (400 and 300 percent)

- Instantaneous-trip circuit breakers (0-1700 percent based on Design letter or code letter)

SIZING OVERCURRENT PROTECTION DEVICES TO ALLOW MOTORS TO START AND RUN 430.52(C)(1) AND TABLE 430.52

The branch-circuit protection for a motor may be a fuse or circuit breaker located in the line at the point where the branch circuit originates. The fuse or circuit breaker is located either at a service cabinet or distribution panel or in the motor control center. When there is only one motor on a branch circuit, the fuse or circuit breaker is sized according to **Table 430.52** and **430.52(C)(1)**.

To use the table properly will require explanation. There is the matter of "Design letters." A Design letter provides certain electrical characteristics of a particular motor that are needed to size the overcurrent protection device to permit the motor to start and accelerate its load. To apply **Table 430.52**, it is necessary to take the following steps:

- Select the phase of the motor
  - (a) Single-phase
  - (b) Three-phase (poly-phase)

- Select type of motor
  - (a) Squirrel-cage induction
  - (b) Wound-rotor
  - (c) DC
  - (d) Synchronous

Note that motors can be single-phase or three-phase types.

- Select the Design letter of the motor
  - (a) Design B
  - (b) Design C
  - (c) Design D
  - (d) Design E (from Europe, IEC)

- Select the type overcurrent protection device
  - (a) Column 2 is for NTDFs
  - (b) Column 3 is for TDFs
  - (c) Column 4 is for circuit breakers with instantaneous trip settings or adjustments
  - (d) Column 5 is for circuit breakers with both instantaneous trip settings and thermal trip characteristics

See Figures 18-24(a) through (d) for sizing and selecting the size overcurrent protection devices per **Table 430.52** to allow motors to start and run their driven loads. Note that the minimum (rounded down) and next size overcurrent protection device will be sized for a particular size (rounded up) motor.

NONTIME-DELAY FUSES USING THE MAXIMUM SIZE 430.52(C)(1), Ex. 2(a)

If the minimum or next-size overcurrent protection device does not allow the motor to start and run, the maximum size rating of a nontime-delay fuse not exceeding 600 amps shall be permitted to be increased but shall in no case exceed 400 percent of the FLA of the motor. **[See Figure 18-25(a)]**

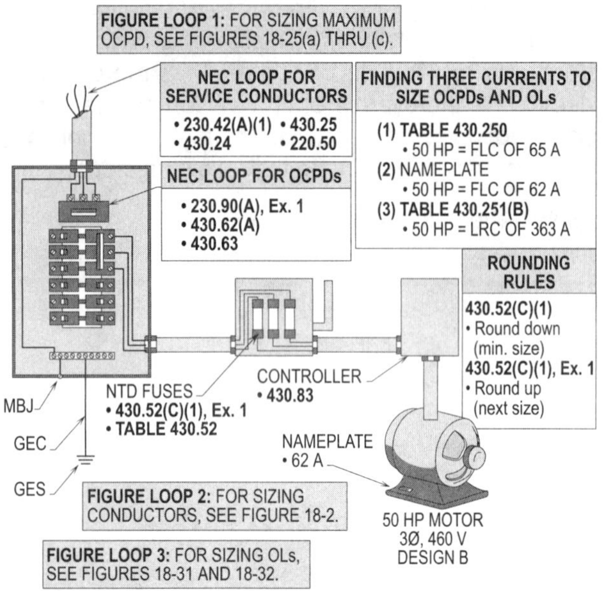

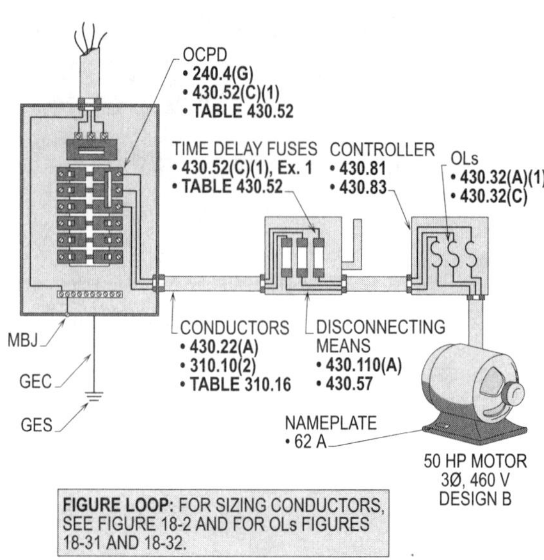

FINDING NONTIME-DELAY FUSES
Step 1: Finding FLA **430.6(A)(1)** and **Table 430.250** 50 HP = 65 A
Step 2: Finding percentage **430.52(C)(1)** and **Table 430.52** Minimum size = 300%
Step 3: Calculating minimum size **430.52(C)(1)** Minimum size = 65 A x 300% = 195 A
Step 4: Calculating next size **430.52(C)(1), Ex. 1** 195 A allows 200 A
Step 5: Selecting NTDFs **240.6(A)** Minimum size = 175 A Next size = 200 A
Solution: The minimum size nontime-delay fuse is 175 amps and the next size is 200 amps.

**SIZING OVERCURRENT PROTECTION DEVICES
TO ALLOW MOTORS TO START AND RUN
NEC 430.52(C)(1) AND Ex. 1**

FINDING TIME-DELAY FUSES
Step 1: Finding FLA **430.6(A)(1)** and **Table 430.250** 50 HP = 65 A
Step 2: Finding percentage **430.52(C)(1)** and **Table 430.52** Minimum size = 175%
Step 3: Calculating minimum size **430.52(C)(1)** Minimum size = 65 A x 175% = 113.75 A
Step 4: Calculating next size **430.52(C)(1), Ex. 1** 113.75 allows 125 A
Step 5: Selecting TDFs **240.6(A)** Minimum size = 110 A Next size = 125 A
Solution: The minimum size time-delay fuse is 110 amps and the next size is 125 amps.

**SIZING OVERCURRENT PROTECTION DEVICES
TO ALLOW MOTORS TO START AND RUN
NEC 430.52(C)(1) AND Ex. 1**

Figure 18-24(a). Determining the minimum and next-size nontime-delay fuses per **Table 430.52** and **430.52(C)(1)** and **Ex. 1** to start and run a motor. Note that a smaller overcurrent protection device than the minimum size (rounded down) shall be permitted to be used, if it will start the motor.

Figure 18-24(b). Determining the minimum and next-size time-delay fuses to start and run a motor. Note that a smaller time-delay fuse than the minimum size (rounded down) shall be permitted to be used, if it will start the motor.

TIME-DELAY FUSES
USING MAXIMUM SIZE
430.52(C)(1), Ex. 2(b)

To allow a motor to start and run, the rating of a time-delay fuse shall be permitted to be increased but shall in no case exceed 225 percent of the full-load current, in amps, of the motor. **[See Figure 18-25(b)]**

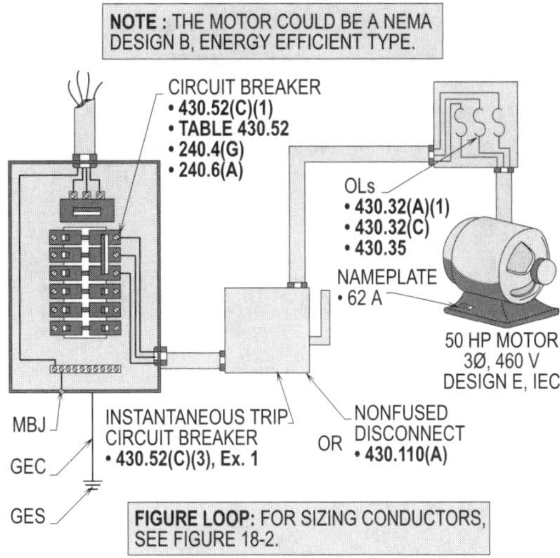

INSTANTANEOUS TRIP CIRCUIT BREAKER

Step 1: Finding FLA
430.6(A)(1) and **Table 430.250**
50 HP = 65 A

Step 2: Selecting the minimum setting
430.52(C)(3) and **Table 430.52**
Minimum size = 1100%

Step 3: Calculating minimum setting
430.52(C)(3)
Minimum size = 65 A x 1100% = 715 A

Step 4: Calculating maximum setting
430.52(C)(3), **Ex. 1**
65 A x 1700% = 1105 A

Step 5: Selecting instaneous trip circuit breaker
Minimum setting = 715 A
Maximum setting = 1105 A

Solution: **The minimum setting is 715 amps and the maximum setting is 1105 amps. However, a smaller setting shall be permitted to be used.**

SIZING OVERCURRENT PROTECTION DEVICES
TO ALLOW MOTORS TO START AND RUN
NEC 430.52(C)(3) AND Ex. 1

Figure 18-24(c). Determining the minimum and maximum setting for an instantaneous trip circuit breaker to start and run a motor. Note that a smaller minimum setting shall be permitted to be used, if it will start the motor.

INVERSE-TIME CIRCUIT BREAKERS 430.52(C)(1), Ex. 2(c)

The rating for inverse-time circuit breakers shall be permitted to be increased but shall in no case exceed 400 percent for a full-load current of 100 amps or less. Full-load current greater than 100 amps shall be permitted to be increased 300 percent. **[See Figure 18-25(c)]**

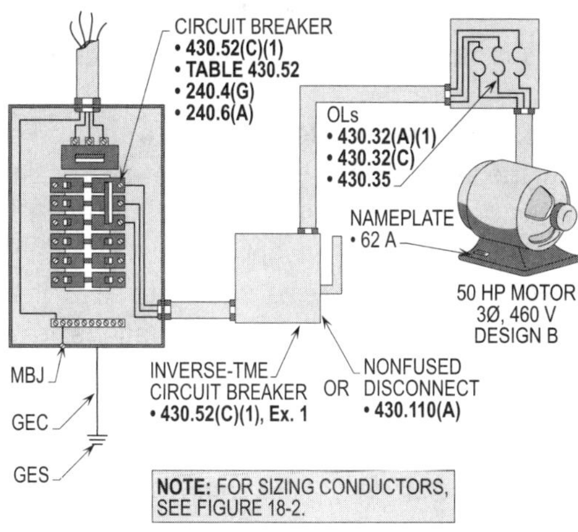

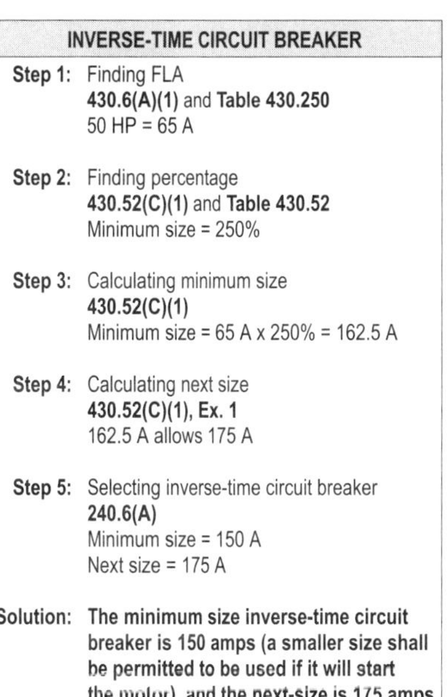

INVERSE-TIME CIRCUIT BREAKER

Step 1: Finding FLA
430.6(A)(1) and **Table 430.250**
50 HP = 65 A

Step 2: Finding percentage
430.52(C)(1) and **Table 430.52**
Minimum size = 250%

Step 3: Calculating minimum size
430.52(C)(1)
Minimum size = 65 A x 250% = 162.5 A

Step 4: Calculating next size
430.52(C)(1), **Ex. 1**
162.5 A allows 175 A

Step 5: Selecting inverse-time circuit breaker
240.6(A)
Minimum size = 150 A
Next size = 175 A

Solution: **The minimum size inverse-time circuit breaker is 150 amps (a smaller size shall be permitted to be used if it will start the motor), and the next-size is 175 amps.**

SIZING OVERCURRENT PROTECTION DEVICES
TO ALLOW MOTORS TO START AND RUN
NEC 430.52(C)(1) and Ex. 1

Figure 18-24(d). Determining the minimum and next-size inverse-time circuit breaker to start and run a motor.

SIZING OVERCURRENT PROTECTION DEVICE FOR TWO OR MORE MOTORS 430.62(A)

To determine the size overcurrent protection device to be installed for a feeder supplying two or more motors, the following procedures shall be applied:

- Apply **Table 430.52** to select largest motor
- Size largest overcurrent protection device for any one motor of group
- Add FLA of remaining motors
- Do not exceed this value with overcurrent protection device rating

See Figure 18-26 for a detailed procedure of sizing overcurrent protection device for a feeder motor circuit.

The largest overcurrent protection device shall be selected based on the motor's full-load current rating times the percentages selected from **Table 430.52**. The next higher standard size rating can be applied per **430.52(C)(1), Ex. 1**. The next higher standard size shall be used for the largest motor in the group. The full-load current ratings, in amps, of the remaining motors are added to the rating of the largest overcurrent protection device. The next lower standard size overcurrent protection device (round down) shall be selected from this total per **240.6(A)**, if this value does not correspond to a standard overcurrent protection device.

A larger overcurrent protection device shall not be permitted to be installed because there is no exception to **430.62(A)** to allow the next size above this rating to be selected. Note that **430.62(A)** only allows a smaller overcurrent protection device to be installed ahead of such feeder conductors, which is the next size below the calculation. **(See Figure 18-27)**

SIZING BRANCH CIRCUIT TO SUPPLY TWO OR MORE MOTORS
430.52 AND 430.53

Section **430.53** outlines the rules where two or more motors, or one or more motors, and other loads are installed and connected to one branch circuit protected by an individual overcurrent protection device.

MOTOR NOT OVER 1 HP
430.53(A)

Two or more motors may be installed without individual overcurrent protection devices if rated less than 1 horsepower each and the full-load current rating of each motor does not exceed 6 amps. Motors not rated over 1 horsepower shall be within sight of the motor, manually started, and portable. Sections **430.32** and **430.42** shall be applied for running overload protection for each motor if these conditions are not met.

The overcurrent protection device rated at 20 amps or less can protect a 120 volt or less branch circuit supplying these motors. Branch circuits of 600 volts or less can be protected by a 15 amp or less overcurrent protection device. **(See Figure 18-28)**

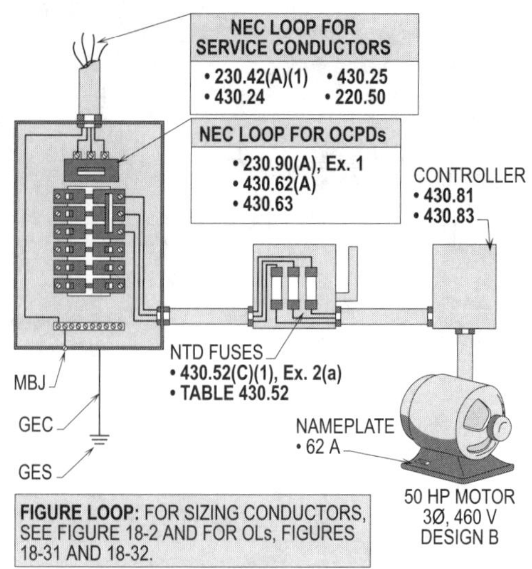

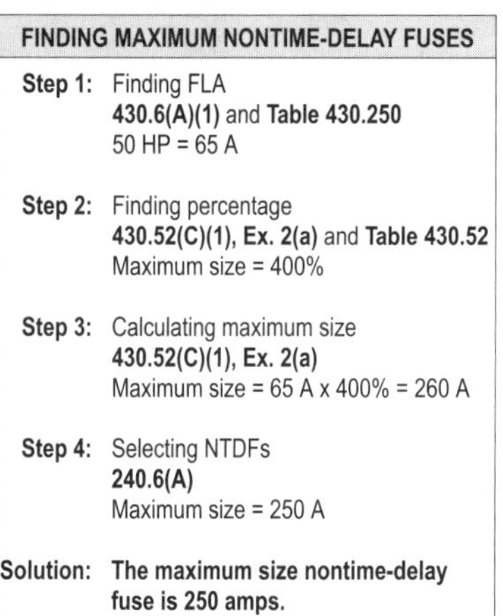

FINDING MAXIMUM NONTIME-DELAY FUSES

Step 1: Finding FLA
430.6(A)(1) and **Table 430.250**
50 HP = 65 A

Step 2: Finding percentage
430.52(C)(1), **Ex. 2(a)** and **Table 430.52**
Maximum size = 400%

Step 3: Calculating maximum size
430.52(C)(1), **Ex. 2(a)**
Maximum size = 65 A x 400% = 260 A

Step 4: Selecting NTDFs
240.6(A)
Maximum size = 250 A

Solution: The maximum size nontime-delay fuse is 250 amps.

NONTIME-DELAY FUSES
USING THE MAXIMUM SIZE
NEC 430.52(C)(1), Ex. 2(a)

Figure 18-25(a). Nontime-delay fuses shall be permitted to be increased to a maximum size of 400 percent of the motor's full-load current rating in amps. Note that for maximum setting of instantaneous trip circuit breaker, see **Figure 18-24(c)**.

SMALLEST RATED MOTOR PROTECTED
430.53(B)

The branch-circuit overcurrent protection device shall be permitted to protect the smallest rated motor of the group for two or more motors of different ratings if the largest motor is permitted to start. The smallest rated motor of the group shall have its overcurrent protection device set at no higher value than allowed per **Table 430.52**. The smallest rated motor and other motors of the group shall be provided with overload protection if necessary per **430.32**. **(See Figure 18-29)**

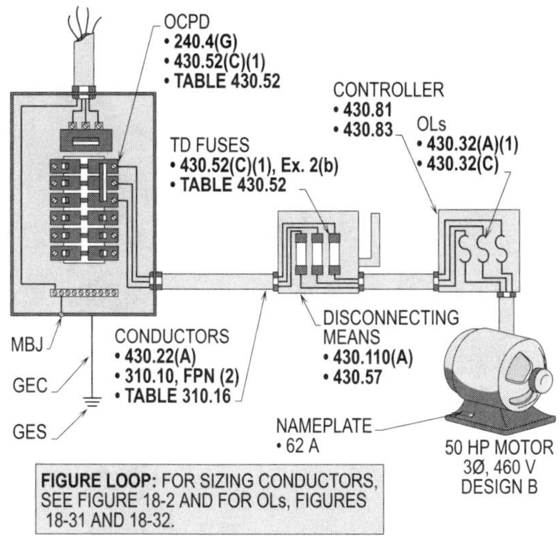

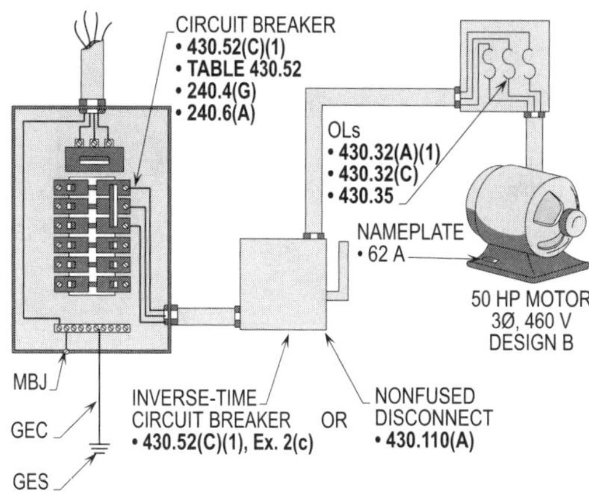

FINDING TIME-DELAY FUSES

Step 1: Finding FLA
430.6(A)(1) and Table 430.250
50 HP = 65 A

Step 2: Finding percentage
430.52(C)(1), Ex. 2(b) and Table 430.52
Maximum size = 225%

Step 3: Calculating maximum size
430.52(C)(1), Ex. 2(b)
Maximum size = 65 A x 225% = 146 A

Step 4: Selecting TDFs
240.6(A)
Maximum size = 125 A

Solution: **The maximum size time-delay fuse is 125 amps.**

**TIME-DELAY FUSES
USING MAXIMUM SIZE
NEC 430.52(C)(1), Ex. 2(b)**

Figure 18-25(b). Determining the maximum size time-delay fuse to start and run a motor (smaller size permitted).

INVERSE-TIME CIRCUIT BREAKER

Step 1: Finding FLA
430.6(A)(1) and Table 430.250
50 HP = 65 A

Step 2: Finding percentage
430.52(C)(1), Ex. 2(c) and Table 430.52
Maximum size = 400%

Step 3: Calculating maximum size
430.52(C)(1), Ex. 2(c)
Maximum size = 65 A x 400% = 260 A

Step 5: Selecting inverse-time circuit breaker
240.6(A)
Maximum size = 250 A

Solution: **The maximum size inverse-time circuit breaker is 250 amps.**

**INVERSE-TIME CIRCUIT BREAKERS
NEC 430.52(C)(1), Ex. 2(c)**

Figure 18-25(c). An inverse-time circuit breaker shall be permitted to be a maximum size of 400 percent of the motor's full-load current rating in amps because the motor's FLA rating is 100 amps or less. Note that a small circuit breaker shall be permitted to be used.

OTHER GROUP INSTALLATIONS
430.53(C)

Two or more motors of any size shall be permitted to be installed and connected to an individual branch circuit. However, the largest motor of the group shall be protected by the percentages listed in **Table 430.52** for sizing and selecting fuses and circuit breakers. Each motor controller and component installed in the group shall be approved for such use. The following are elements that shall be sized and selected properly:

- Overcurrent protection devices
- Controllers
- Running overload protection devices

The elements shall be permitted to be installed as a listed factory assembly or field installed as separate assemblies listed for such conditions of use.

Design Tip: Any number of motor taps shall be permitted to be installed where a fuse or circuit breaker is installed at the point where each motor is tapped to the line. This type of installation made with a branch circuit from a feeder per **430.28** and **430.53(D)** is often utilized. **(See Figure 18-30)**

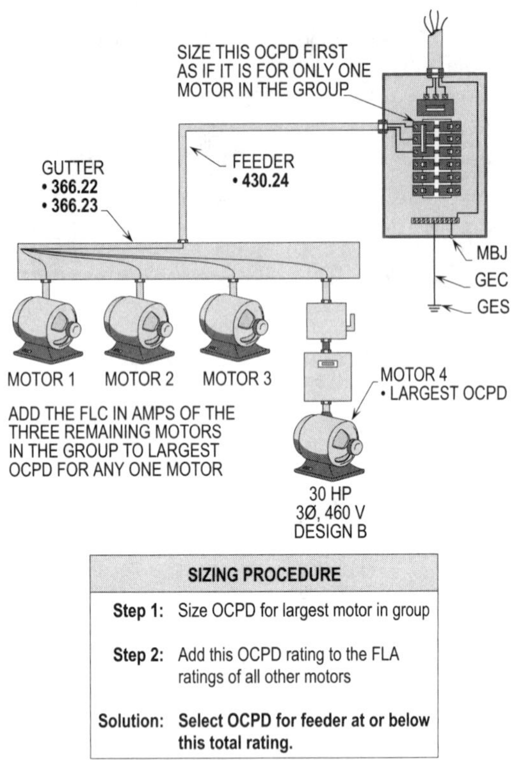

Figure 18-26. Determining overcurrent protection device for a feeder with several motors being protected from short-circuit and ground-fault conditions.

SIZING THE RUNNING OVERLOAD PROTECTION FOR MOTORS 430.32(A)(1)

Devices such as thermal protectors, thermal relays, or fusetrons may be installed to provide running overload protection for motors rated more than 1 horsepower. The service factor or temperature rise of the motor shall be used when sizing and installing the running overload protection for motors. The running overload protection is set to open at 115 percent or 125 percent of the motor's full-load current. Under certain conditions of use, the running overload protection shall be set at 115 percent when the motor is not marked with a service factor or temperature rise. Time-delay fuses selected and sized at these percentages provide overload or backup overload protection.

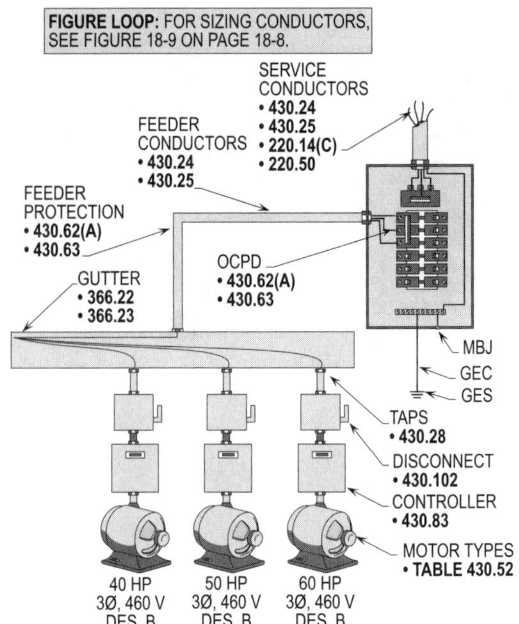

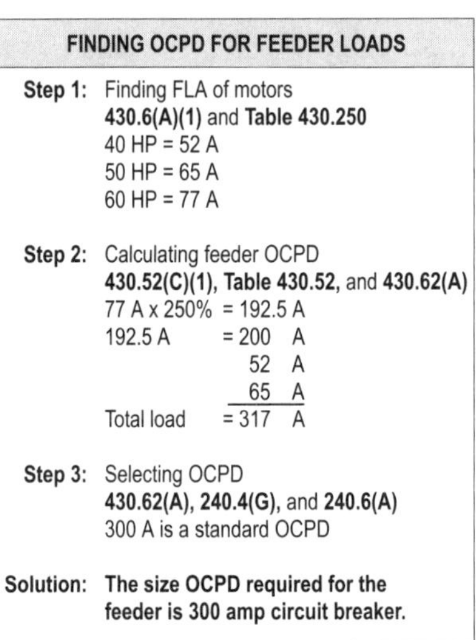

Figure 18-27. Sizing an overcurrent protection device for a feeder.

MINIMUM SIZE OVERLOAD PROTECTION DEVICE 430.32(A)(1)

The amperage for full-load current ratings listed in **Tables 430.247 through 430.250** shall not be used when sizing the running overload protection. The full-load current listed with the motor's nameplate shall be used to size the setting of the separate running overload protection.

The running overload protection shall be selected and rated no larger than the following minimum percentages based upon the full-load current rating, in amps, listed on the motor's nameplate:

- Motors, with a marked service factor not less than 1.15, use 125 percent x FLA

- Motors, with a marked temperature rise not over 40°C, use 125 percent x FLA

- All other motors, 115% x FLA

(See Figure 18-31)

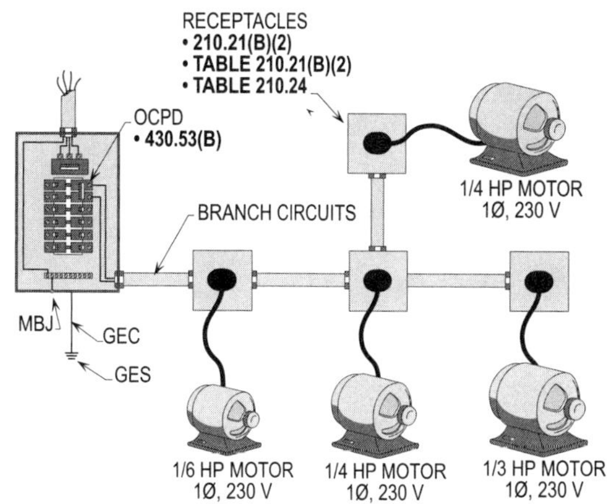

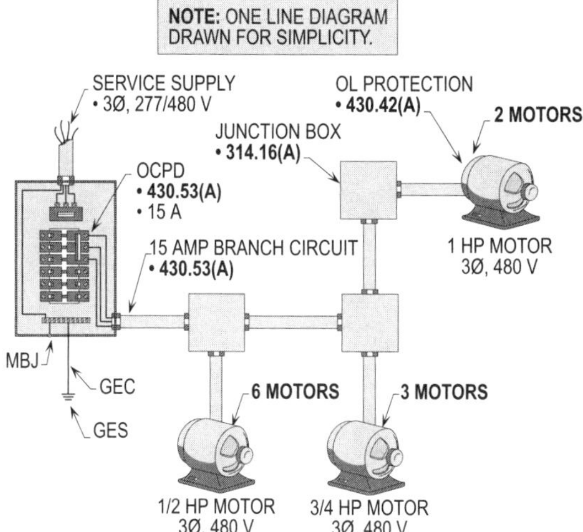

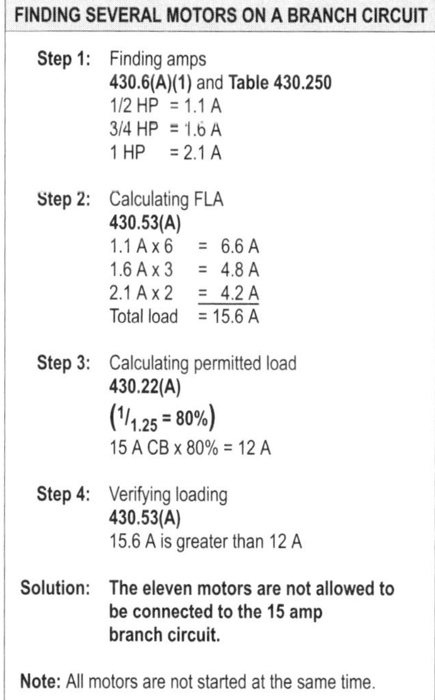

FINDING SEVERAL MOTORS ON A BRANCH CIRCUIT

Step 1: Finding amps
430.6(A)(1) and **Table 430.250**
1/2 HP = 1.1 A
3/4 HP = 1.6 A
1 HP   = 2.1 A

Step 2: Calculating FLA
430.53(A)
1.1 A x 6 = 6.6 A
1.6 A x 3 = 4.8 A
2.1 A x 2 = 4.2 A
Total load = 15.6 A

Step 3: Calculating permitted load
430.22(A)
$\left(^1/_{1.25} = 80\%\right)$
15 A CB x 80% = 12 A

Step 4: Verifying loading
430.53(A)
15.6 A is greater than 12 A

Solution: **The eleven motors are not allowed to be connected to the 15 amp branch circuit.**

Note: All motors are not started at the same time.

**MOTOR NOT OVER 1 HP
NEC 430.53(A)**

FINDING SEVERAL MOTORS ON A BRANCH CIRCUIT

Step 1: Finding amps
430.6(A)(1) and **Table 430.248**
1/6 HP = 2.2 A
1/4 HP = 2.9 A
1/4 HP = 2.9 A
1/3 HP = 3.6 A

Step 2: Calculating FLA
430.53(B)
2.2 A x 1 = 2.2 A
2.9 A x 1 = 2.9 A
2.9 A x 1 = 2.9 A
3.6 A x 1 = 3.6 A
Total load = 11.6 A

Step 3: Calculating permitted load
Table 210.21(B)(2)
$^1/_{1.25} = 80\%$
15 A OCPD x 80% = 12 A

Step 4: Verifying loading
430.53(B); and **Table 210.21(B)(2)**
11.6 A is less than 12 A

Step 5: Protecting smaller motor
430.53(B) and **Table 430.52**
2.2 A x 250% = 5.5 A

Step 6: Selecting OCPD
430.53(B), 240.4(B), and **240.6(A)**
5.5 A allows 15 A CB

Solution: **Section 430.53(B) allows the next-size circuit breaker which is 15 amps.**

Note: Most inspectors allow this concept since the next size circuit breaker is 15 amps.

**SMALLEST RATED MOTOR PROTECTED
NEC 430.53(B)**

Figure 18-29. Determining the number of motors allowed on a 15 amp branch circuit.

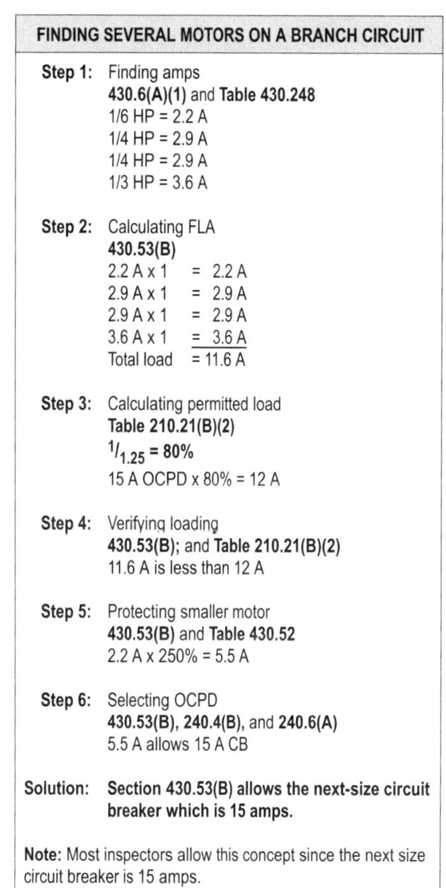

Figure 18-28. Determining the number of motors allowed on a 15 amp branch circuit.

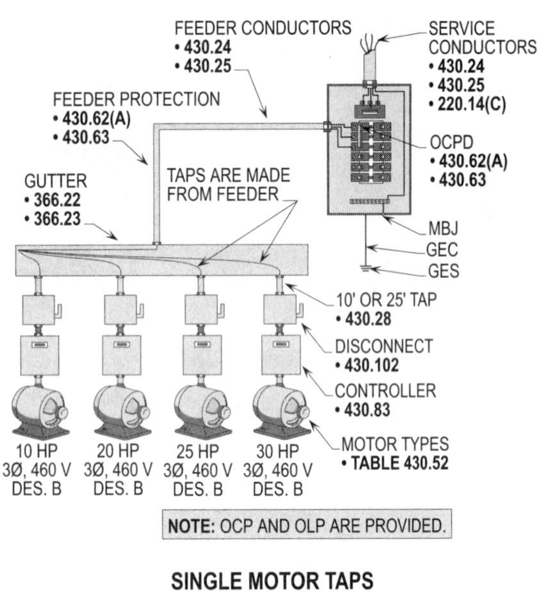

Figure 18-30. Taps can be made from a feeder with the proper conductor size and overcurrent protection device with each tap.

MAXIMUM SIZE OVERLOAD PROTECTION DEVICE 430.32(C)

The selection of the running overload protection (overload relay) shall be permitted to be selected at higher percentages if the percentages of **430.32(A)(1)** are not sufficient. The running overload protection device shall be selected to trip or shall be rated no larger than the following percentages of the motor's (nameplate) full-load current rating:

- Motors, with marked service factor not less than 1.15, use 140% x FLA

- Motors, with a marked temperature rise not over 40°C, use 140% x FLA

- All other motors, 130% x FLA

(See Figure 18-32)

SIZING THE CONTROLLER TO START AND STOP THE MOTOR 430.81 AND 430.83

The sizes and types of motor controllers shall be installed with a horsepower rating at least equal to the motor to be controlled. However, there is an Exception to this rule for motors rated at and below a certain horsepower rating.

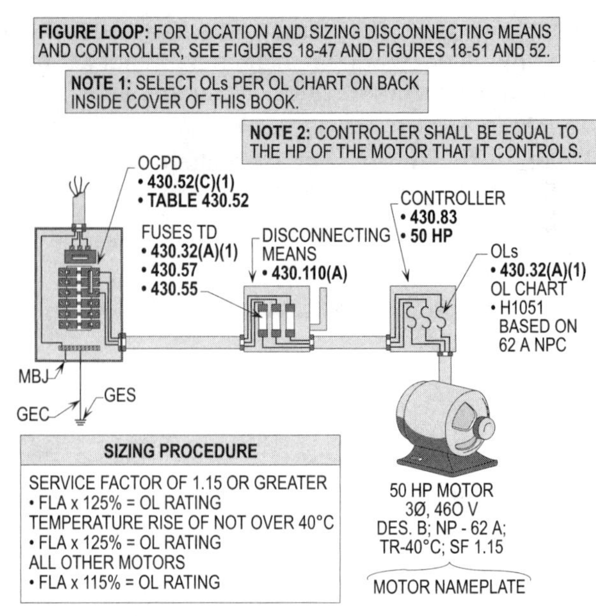

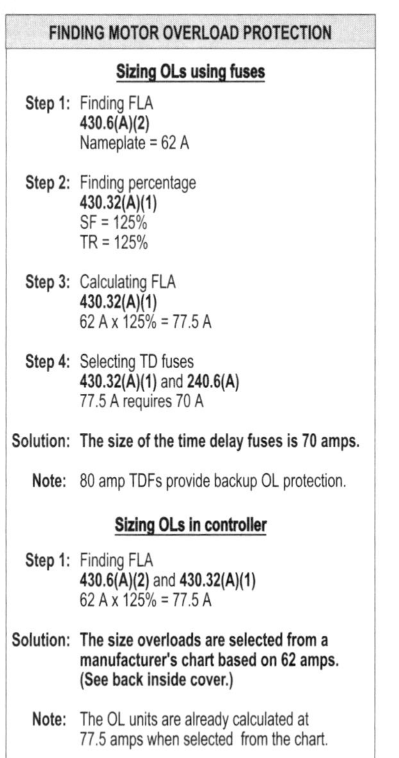

FINDING MOTOR OVERLOAD PROTECTION

Sizing OLs using fuses

Step 1: Finding FLA
430.6(A)(2)
Nameplate = 62 A

Step 2: Finding percentage
430.32(A)(1)
SF = 125%
TR = 125%

Step 3: Calculating FLA
430.32(A)(1)
62 A x 125% = 77.5 A

Step 4: Selecting TD fuses
430.32(A)(1) and 240.6(A)
77.5 A requires 70 A

Solution: The size of the time delay fuses is 70 amps.

Note: 80 amp TDFs provide backup OL protection.

Sizing OLs in controller

Step 1: Finding FLA
430.6(A)(2) and 430.32(A)(1)
62 A x 125% = 77.5 A

Solution: The size overloads are selected from a manufacturer's chart based on 62 amps. (See back inside cover.)

Note: The OL units are already calculated at 77.5 amps when selected from the chart.

MINIMUM SIZE OVERLOAD PROTECTION DEVICE
NEC 430.32(A)(1)

Figure 18-31. Determining the minimum size overloads based upon service factor and temperature rise.

STATIONARY MOTOR OF 1/8 HORSEPOWER OR LESS 430.81(A)

The branch circuit protective device shall be permitted to serve as the controller where the motor is rated 1/8 horsepower or less. **(See Figure 18-33)**

For example, for motors less than 1/8 horsepower, where they are mounted stationary or permanent and the construction is such that one or more might fail during operation, the branch circuit elements plus the motor(s) won't be damaged. In other words, the components of the circuit won't be burned out, etc.

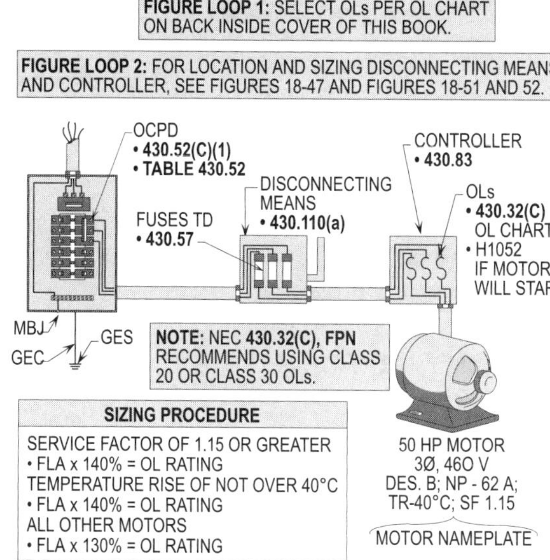

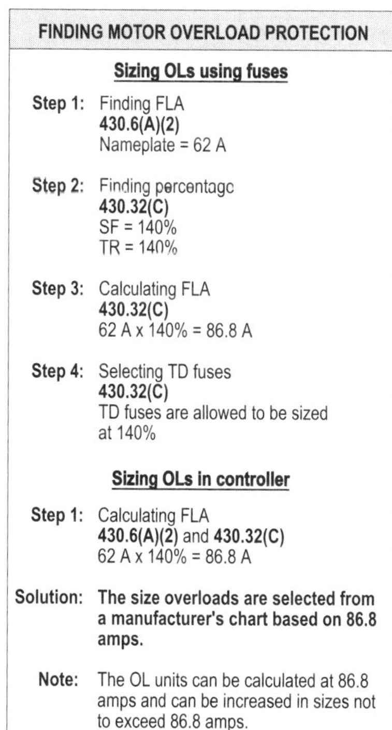

MAXIMUM SIZE OVERLOAD PROTECTION DEVICE
NEC 430.32(C)

Figure 18-32. Determining the maximum size overloads based upon service factor and temperature rise.

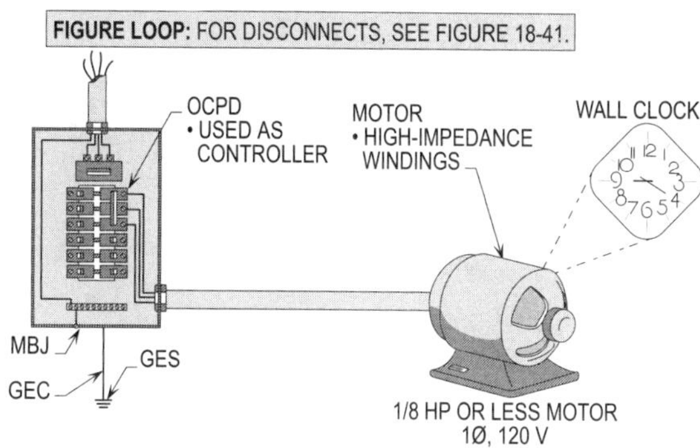

STATIONARY MOTOR
1/8 HORSEPOWER OR LESS
NEC 430.81(A)

Figure 18-33. The branch-circuit overcurrent protection device shall be permitted to serve as a controller for 1/8 HP or less motor.

PORTABLE MOTOR OF 1/3 HORSEPOWER OR LESS
430.81(B)

The controller shall be permitted to be an attachment plug and receptacle that is acceptable for use with portable motors rated 1/3 horsepower or less. **(See Figure 18-34)**

GENERAL REQUIREMENTS
430.83

The controller shall have a rating as specified in **430.83(A)**, unless otherwise permitted in **430.83(B)** or **(C)**, or as outlined in **(D)**, under the conditions specified in these sections.

HORSEPOWER RATINGS
430.83(A)(1)

Controllers, other than inverse time circuit breakers and molded case switches, shall have horsepower ratings at the application voltage not lower than the horsepower rating of the motor. **(See Figure 18-35)**

CIRCUIT BREAKERS
430.83(A)(2)

Inverse-time circuit breakers shall only be permitted to be installed as a controller where rated in amps. If such a circuit breaker is also used for motor overload protection, it shall be sized at 125 percent or less of the motor's nameplate current rating per **430.6(A)(2)** and **430.32(A)(1)**. **(See Figure 18-36)**

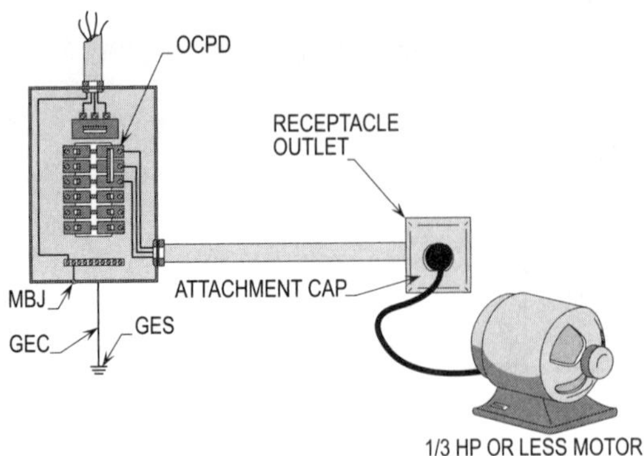

Figure 18-34. The controller for a motor of 1/3 HP motor or less shall be permitted to be an attachment cap and receptacle.

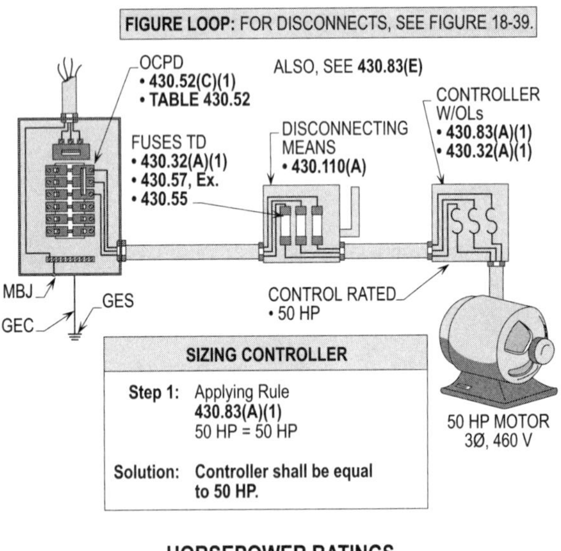

Figure 18-35. Controllers shall have a horsepower rating at least equal to the horsepower of the motor.

When used as a disconnecting means for a motor, it shall be sized with a interrupting rating of at least 115 percent of the motor's FLC rating (in amps) per **430.110(A)**.

SMALL MOTORS
430.83(B)

Stationary motors rated 1/8 horsepower or less and portable motors rated 1/3 horsepower or less shall be permitted to serve as controllers and shall not be required to be horsepower rated. These horsepower-rated motors, because of their smaller locked-rotor currents, can be disconnected by cord-and-plug connections.

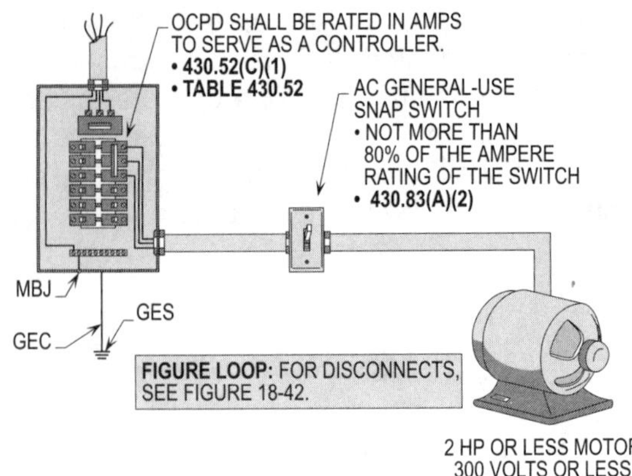

Figure 18-36. A circuit breaker rated at 125 percent of the motor's FLA can be used as a controller for the motor and also provide overload protection.

STATIONARY MOTORS OF
2 HORSEPOWER OR LESS
430.83(C)

For a stationary motor rated 2 horsepower or less, the controller shall be permitted to be a general-use switch rated for at least twice the motor's full-load current. An AC general-use snap switch shall be permitted to be installed as the controller where the full-load current rating of the switch does not exceed 80 percent (1÷1.25 = 80%) of the branch-circuit rating. **(See Figure 18-37) Note,** review **430.83(E)**.

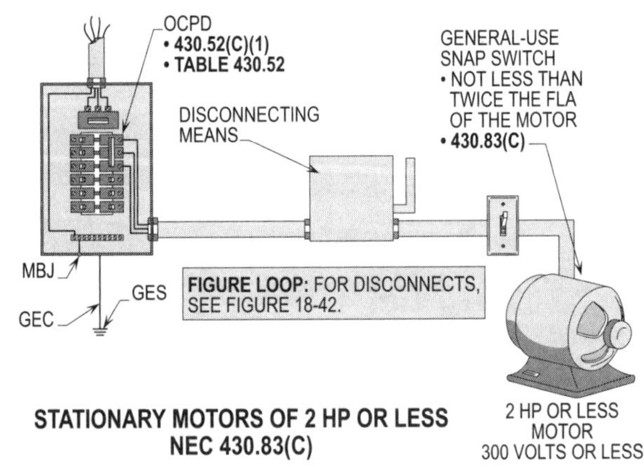

Figure 18-37. For stationary motors rated 2 HP or less, a general-use snap switch shall be permitted to be used if sized not less than twice the motor's full-load current in amps.

TORQUE MOTORS
430.83(D)

The motor controller for a torque motor shall have a continuous duty, full-load current rating not less than the nameplate current rating of the motor. **(See Figure 18-38)**

> **Design Tip:** If the motor controller is rated in horsepower and not marked or rated as above, to determine the amperage or horsepower rating, use **Tables 430.247 through 430.250**.

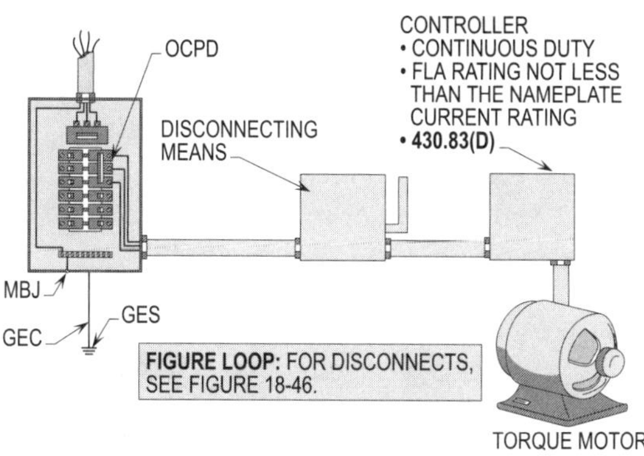

TORQUE MOTORS
NEC 430.83(D)

Figure 18-38. The controller for a torque motor shall be capable of holding the amps indefinitely.

SIZING THE DISCONNECTING MEANS TO DISCONNECT BOTH THE CONTROLLER AND MOTOR 430.109(A) AND 430.110(A)

The disconnecting means for motor circuits shall have an ampere rating of at least 115 percent of the full-load current rating of the motor per **430.110(A)**. The disconnecting means shall be horsepower rated and capable of deenergizing locked-rotor currents per **Tables 430.251(A)** and **(B)** of the NEC.

Figure Loop: For sizing controller, see **Figure 18-35**. For sizing control circuits, see **Figures 18-51** and **18-52**.

OTHER THAN HORSEPOWER RATED 430.109(B) THRU (G)

Sections **430.109(B) through (G)** permit other than a horsepower rated disconnecting means to be used to deenergize the power circuit to certain types of motors:

- Stationary motors rated 1/8 horsepower or less
- Stationary motors rated 2 horsepower or less (300 volts or less)
- Autotransformer-type controlled motors
- Torque motors

GENERAL REQUIREMENTS 430.109(A)

The disconnecting means shall be permitted to be one of the following, as specified in this section:

- A listed motor-circuit switch rated horsepower,
- A listed molded case circuit breaker,
- A listed molded case switch,
- An instantaneous trip circuit breaker that is part of a listed combination motor controller, or
- Listed self-protected combination controller.

See **Figure 18-39** and **Figure 18-40** for permitted disconnecting means to deenergize the power circuit.

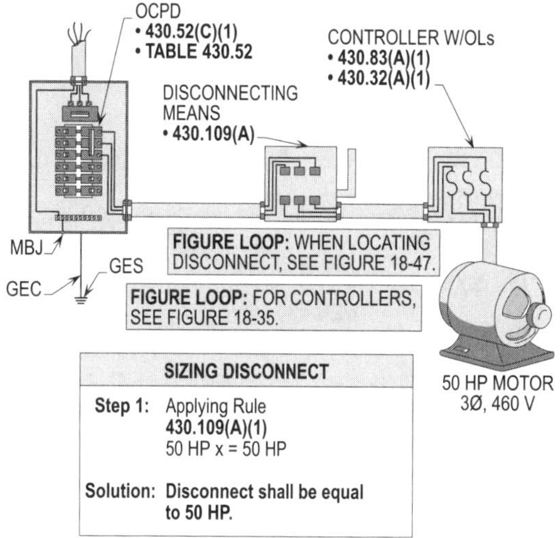

GENERAL RULE
NEC 430.109(A)(1)

Figure 18-39. Disconnecting means shall be at least equal to the horsepower of the motor.

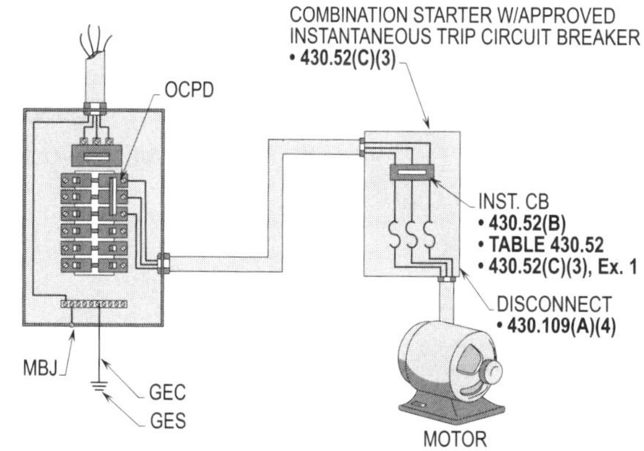

INSTANTANEOUS TRIP CIRCUIT BREAKER
NEC 430.109(A)(4)

Figure 18-40. The disconnecting means for a motor shall be permitted to be an approved instantaneous trip circuit breaker.

STATIONARY MOTORS OF 1/8 HORSEPOWER OR LESS 430.109(B)

For a stationary motor rated 1/8 horsepower or less, the branch-circuit overcurrent protective device shall be permitted to serve as the disconnecting means. This rule is permitted because the windings of such motors do not produce locked-rotor currents high enough to damage such motors, circuit conductors, or elements. **(See Figure 18-41)**

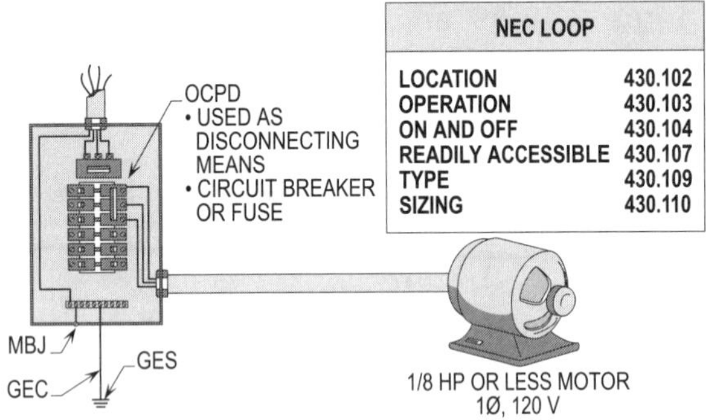

Figure 18-41. Motors rated 1/8 horsepower or less shall be permitted to be disconnected by the overcurrent protection device located in the panelboard that is used to supply the circuit.

STATIONARY MOTORS OF 2 HORSEPOWER OR LESS 430.109(C)

For a stationary motor rated 2 horsepower or less, the controller shall be permitted to be a general-use switch rated for at least twice the motor's full-load current. An AC general-use snap switch may be installed as the controller, where the full-load current rating, in amps, of the switch, does not exceed 80 percent (1÷1.25 = 80%) of the branch circuit rating. **(See Figure 18-42)**

AUTOTRANSFORMER-TYPE CONTROLLED MOTORS 430.109(D)

Motors rated over 2 horsepower through 100 horsepower shall be permitted to be installed with a separate disconnecting means (general-use switch) if the motor is equipped with an autotransformer-type controller and complies with all the following conditions:

- The motor drives a generator that is provided with overload protection.

- The controller is capable of interrupting the locked-rotor current of the motor.

- The controller is provided with a no-voltage release.

- The controller is provided with running overload protection not exceeding 125 percent of the motor's full-load current rating, in amps.

- Separate fuses or an inverse-time circuit breaker is rated at 150 percent or more of the motor's full-load current, in amps. **(See Figure 18-43)**

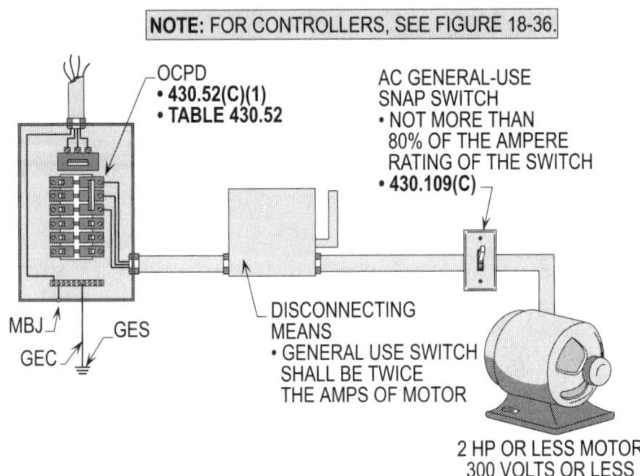

Figure 18-42. The above lists the rules pertaining to the disconnecting means for motors rated 2 horsepower or less.

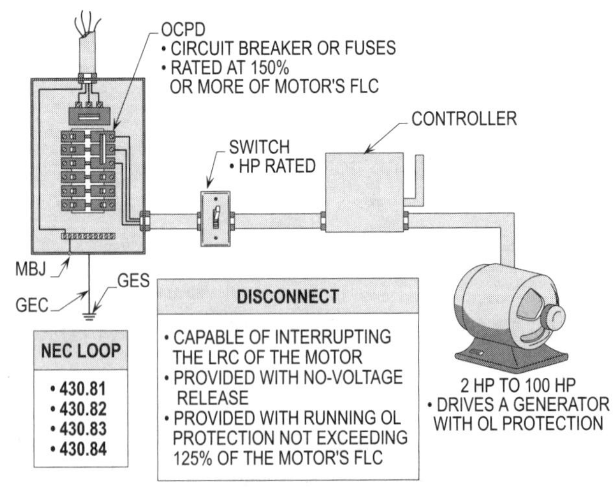

Figure 18-43. The above lists the rules for a disconnecting means and controller used to disconnect and control motors rated 2 horsepower to 100 horsepower.

ISOLATING SWITCHES
430.109(E)

The disconnecting means shall be permitted to be a general-use or isolating switch for DC stationary motors rated at 40 horsepower or greater and AC motors rated 100 horsepower or greater. However, such disconnects shall be plainly marked, "Do not operate under load." **(See Figure 18-44)**

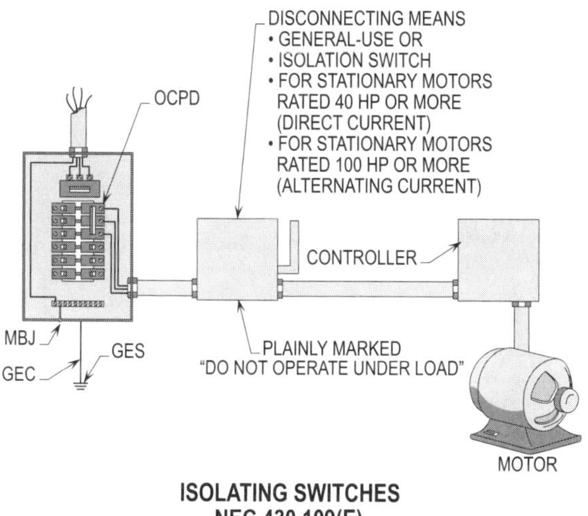

Figure 18-44. The above lists the rules for disconnecting means used to disconnect motors rated at 40 horsepower or more.

CORD-AND-PLUG CONNECTED MOTORS
430.109(F)

For a cord-and-plug connected motor, a horsepower-rated attachment plug and receptacle having ratings no less than the motor ratings shall be permitted to serve as the disconnecting means. A horsepower-rated attachment plug and receptacle shall not be required for a cord-and-plug connected appliance in accordance with **422.33**, a room air conditioner in accordance with **440.63**, or a portable motor rated 1/3 horsepower or less. **(See Figure 18-45)**

TORQUE MOTOR
430.109(G)

The disconnecting means for a torque motor shall be permitted to be installed as a general-use switch. Such switch shall be capable of handling the locked-rotor current, in amps, of the motor indefinitely. **(See Figure 18-46)**

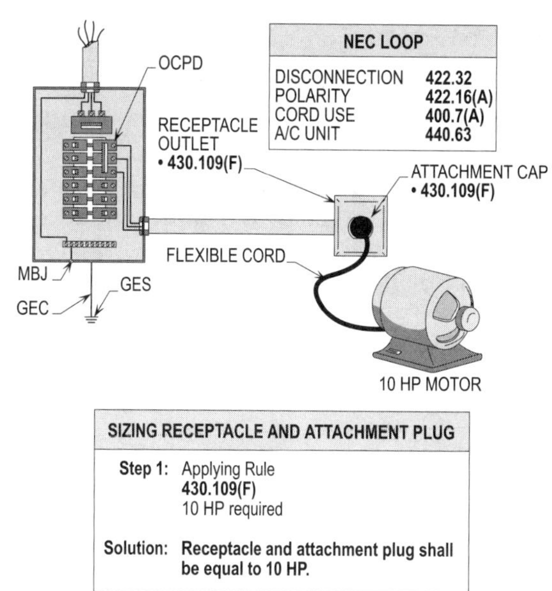

Figure 18-45. A receptacle and attachment cap used as a disconnecting means for motors shall be at least equal to motor's horsepower rating.

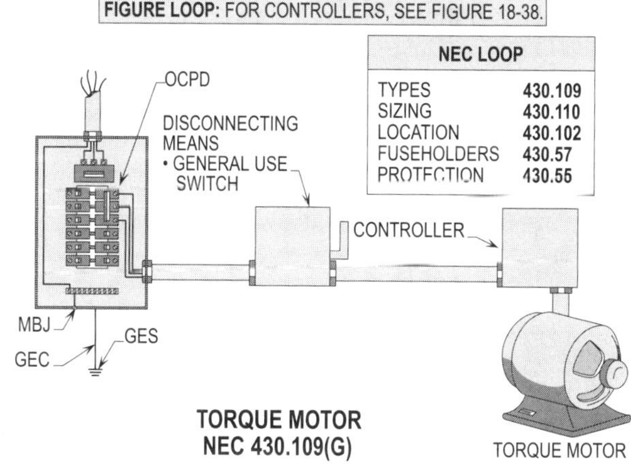

Figure 18-46. The disconnecting means for a torque motor shall be permitted to be a general-use switch.

LOCATION OF THE DISCONNECTING MEANS FOR THE CONTROLLER AND MOTOR
430.102 AND 430.107

A motor and its driven machinery or load shall be installed within sight of the controller for the motor. This rule provides safety for electricians and maintenance personnel while servicing such machinery and circuit elements.

WITHIN SIGHT
ARTICLE 100, 430.102(A), AND 430.102(B)(2)

The disconnecting means shall be installed within sight of the motor controller. All of the ungrounded (phase) conductors shall be disconnected from both the motor and controller supplying the motor circuit. The disconnecting means shall be installed within sight of the motor and not more than 50 ft (15 m) from the motor. If such disconnecting means is not installed within 50 ft (15 m) of the controller, motor, and driven equipment, other provisions for disconnecting the motor shall be made. The controller has a direct relationship to the disconnecting means and shall be installed within sight and within 50 ft (15 m) of the disconnecting means. The motor does not have a direct relationship with the controller. **(See Figure 18-47)**

Note that a disconnecting means shall always be required to be located in sight from the controller location. A single disconnecting means shall be permitted to be located adjacent to a group of coordinated controllers mounted adjacent one to another, such as on a multi-motor continuous process machine. For further information pertaining to sizing, selecting, and locating such controllers and disconnecting means, review **430.83, 430.102, 430.103, 430.107,** and **430.109** very carefully.

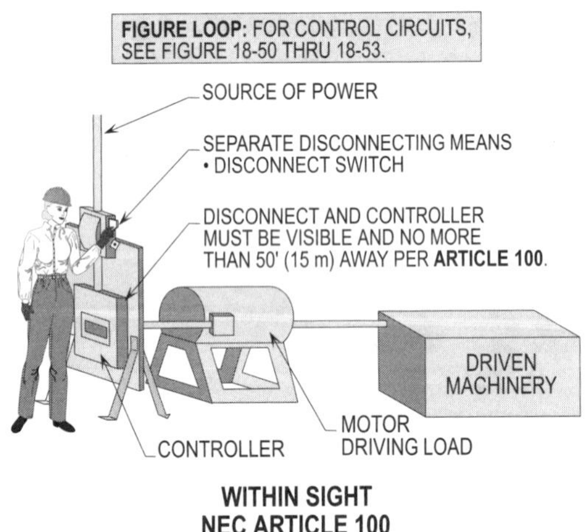

Figure 18-47. The disconnecting means shall be within sight and within 50 ft (15 m) of the controller, motor, and driven machinery.

LOCKED IN THE OPEN POSITION
430.102(A) AND (B)

Section **430.102(A)** and **(B)** permits the disconnect on the line side of the controller, if within sight and within 50 ft (15 m), and capable of being individually locked open, to serve as the disconnecting means for both the controller and motor. In this case, note that the motor shall be installed within sight and within 50 ft (15 m) of the disconnecting means of the controller. **[See Figures 18-48(a) and (b)]**

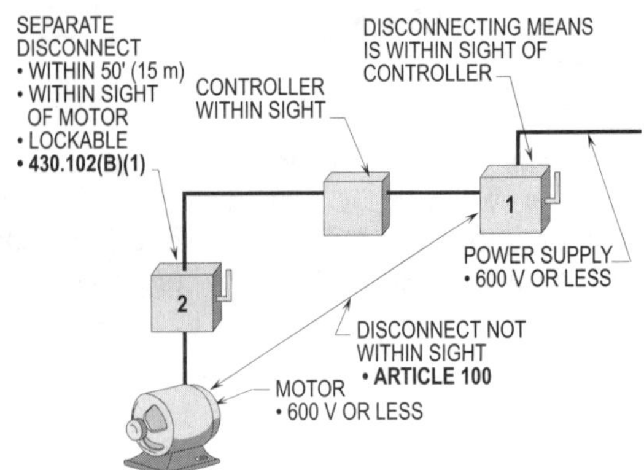

(a) DISCONNECTING MEANS 1 IS NOT WITHIN SIGHT OF MOTOR

LOCKED IN THE OPEN POSITION
NEC 430.102(A) AND (B)(1)

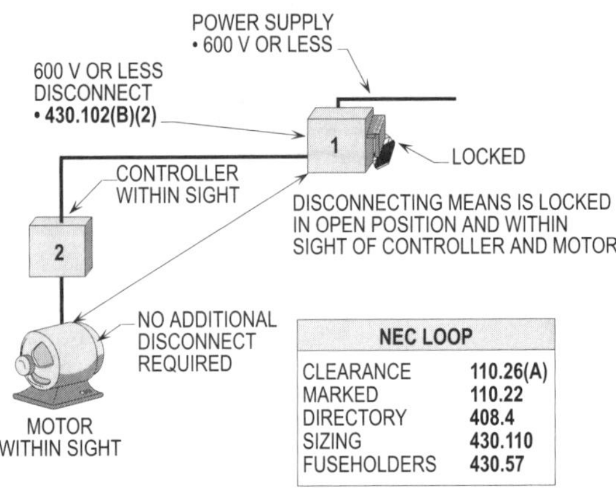

NEC LOOP	
CLEARANCE	110.26(A)
MARKED	110.22
DIRECTORY	408.4
SIZING	430.110
FUSEHOLDERS	430.57

(b) DISCONNECTING MEANS 2 IS WITHIN SIGHT OF MOTOR AND CAN BE LOCKED IN THE OPEN POSITION.

LOCKED IN THE OPEN POSITION
430.102(B)(2)

Figure 18-48(a). Locating the disconnecting means to disconnect power conductors to motors rated 600 volts or less.

CANNOT BE LOCKED IN THE OPEN POSITION
430.102(A) AND 430.102(B)

For motors rated 600 volts or less, an additional disconnecting means shall be mounted by the motor and within sight where the disconnecting means installed by the controller cannot be locked in the open position. **[See Figure 18-48(a)]** The controller disconnecting means for a motor branch circuit over 600 volts shall be permitted to be located out of sight of the motor branch circuit controller and motor. However, the controller shall have a warning label that marks and lists the location and identification of the disconnecting means. To completely satisfy this rule, such disconnecting means shall be capable of being locked in the open position. **[See Figure 18-49(a)]**

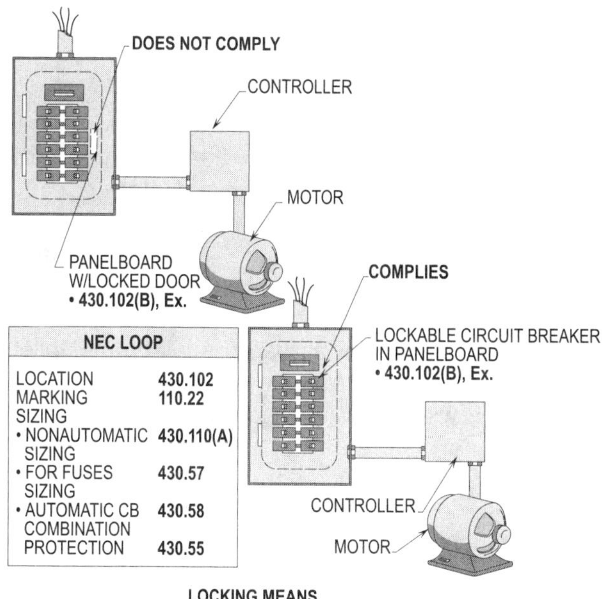

NEC LOOP	
LOCATION	430.102
MARKING	110.22
SIZING	
• NONAUTOMATIC SIZING	430.110(A)
• FOR FUSES SIZING	430.57
• AUTOMATIC CB	430.58
COMBINATION PROTECTION	430.55

Figure 18-48(b). The locked door of a panelboard shall not be permitted to serve as the required disconnecting means for a motor. However, an individual locked circuit breaker shall be permitted to serve as the disconnecting means.

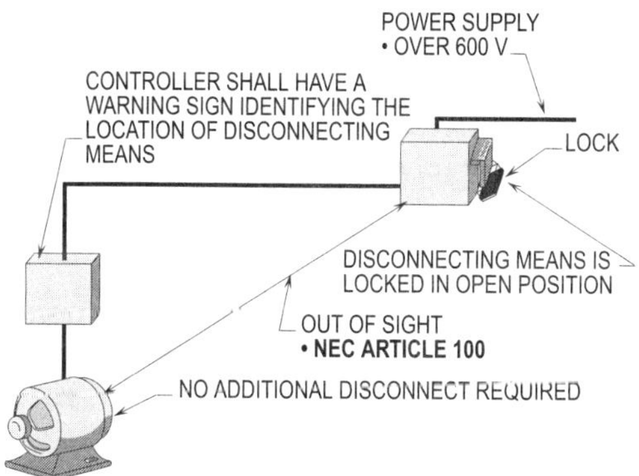

Figure 18-49(a). If the disconnecting means is within sight and 50 ft (15 m) of controller and can't be locked in the open position, an additional disconnecting means shall be installed by the motor.

APPLYING EXCEPTION 430.102(B)

Ex. (a) and **(b)** to **430.102(B)** do not require an additional disconnect to be installed within sight of the motor where the disconnecting means would be impractical or increase hazards. An additional disconnecting means is not required

where it is located in an industrial installation that has written safety procedures and only qualified employees are permitted to work on the equipment involved. **[See Figure 18-49(b)]**

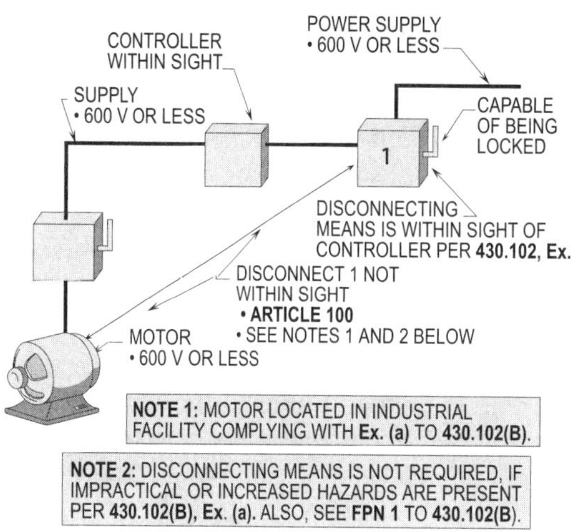

Figure 18-49(b). Under certain conditions of use, an additional disconnecting means is not required to be installed within sight of the motor. When disconnecting valve actuator motors (vams), see **430.102(A), Ex. 3**.

SIZING CONDUCTORS FOR CONTROL CIRCUIT 430.72 AND 725.43

A motor control circuit tapped on the load side of fuses and circuit breakers utilized for motor branch circuits shall protect such conductors, or supplementary protection devices shall be provided.

The size of the control circuit conductors and the rating of the motor's branch-circuit device will be determined by this method of protection. Motor control circuits are classified as remote-control circuits where such circuits derive their power from other than the motor's branch-circuit conductors. Various situations permit fuses or circuit breakers to be utilized to protect remote motor control circuits. For further information, see **725.43** and **725.45**.

Design Tip: Remote-control circuits shall have their disconnecting means located immediately adjacent to the disconnecting means used to disconnect the branch circuit conductors supplying the controller and motor. Sometimes an interlock in the disconnect for the motor controller is used for this purpose that allows the controller, motor, and remote-control circuit to be disconnected simultaneously.

CONDUCTOR PROTECTION
430.72(B)

Conductors larger than 14 AWG are selected from **Tables 310.16 through 310.19** for motor-control circuit conductors that are tapped from a motor power circuit. Overcurrent protection for conductors smaller than 14 AWG shall not exceed the values listed in **Table 430.72(B), Column A**. Conductors 18 AWG and 16 AWG shall be protected at the following amperage ratings:

- 18 AWG shall be protected at 7 amps when used for remote-control circuits

- 16 AWG shall be protected at 10 amps when used for remote-control circuits

Fuses selected at either 1 amp, 3 amp, 6 amp, or 10 amp are normally used to protect these conductors from short circuits, ground faults, and overloads. See **240.6(A)** for selection of such fuse sizes.

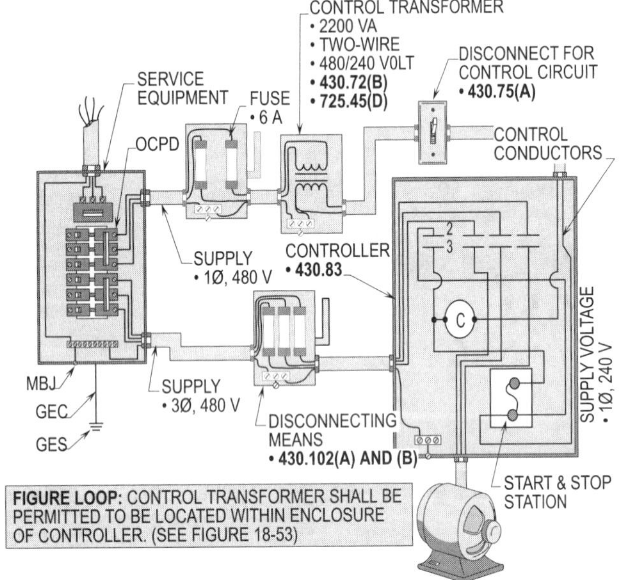

PROTECTION OF CONDUCTORS
NEC 430.72(B), Ex. 2

Figure 18-50. Control circuit conductors are supplied by a control transformer and protected by fuses in primary side.

PROTECTION OF CONDUCTORS
430.72(B), Ex. 2

The secondary conductors of the control transformer circuit shall be permitted be protected by the primary side of the transformer. The transformer shall be protected per **450.3(B)** and **Table 450.3(B)**. A two-wire secondary for a transformer installed outside or within the control starter enclosure shall be permitted to be protected per **240.4(F)** and **240.21(C)(1)**.

The secondary conductor ampacity shall be multiplied by the secondary-to-primary voltage ratio to provide protection in accordance with **450.3(B)** and **Table 450.3(B)**. Where the rated primary current is 9 amps or greater and 125 percent of this current does not correspond to a standard rating of a fuse or circuit breaker, the next higher standard size can be selected. Where the rated primary current is less than 9 amps, but is 2 amps or greater, an overcurrent protection device rated or set at not more than 167 percent of the primary current can be used. Where the rated primary current is less than 2 amps, an overcurrent protection device rated or set not greater than 300 percent to 500 percent shall be permitted to be used. **(See Figure 18-50)**

For example, if the primary full-load current of a motor control transformer is less than 2 amps, the overcurrent protection device shall be permitted to be calculated and sized at 500 percent times such full-load current in amps per **430.72(C)(4)**.

PROTECTION OF CONDUCTORS
430.72(B)(2)

Motor-control circuit conductors that do not extend beyond the control equipment enclosure shall be permitted to be protected by the motor branch circuit fuses or circuit breakers. **Table 430.72(B), Column B** permits this type of installation where the devices do not exceed 400 percent of the ampacity rating of sizes 14 AWG and larger conductors. Overcurrent protection for conductors smaller than 14 AWG shall not exceed the values listed in **Table 430.72(B), Column B**. Conductors rated 18 AWG through 10 AWG can be protected with the following sized overcurrent protection devices:

- 18 AWG = 25 amps (7 A x 400% = 28 A and requires 25 A OCPD)

- 16 AWG = 40 amps (10 A x 400% = 40 A and requires 40 A OCPD)

- 14 AWG = 100 amps (25 A x 400% = 100 A and requires 100 A OCPD)

- 12 AWG = 120 amps (30 A x 400% = 120 A and requires 110 A OCPD)

- 10 AWG = 160 amps (40 A x 400% = 160 A and requires 150 A OCPD)

- 8 AWG and larger = 400 percent

> **Design Tip:** The free air ampacities of **Table 310.17** for 60°C wire are used to determine the ampacity ratings for the control circuit conductors. This type of installation has more free space to dissipate the heat where control conductors are installed in the open air space of enclosures instead of enclosed raceways.

See Figure 18-51 for selecting such conductors based upon the overcurrent protection device rating.

PROTECTION OF CONDUCTORS
430.72(B)(2)

Motor-control circuit conductors that extend beyond the control equipment enclosure shall be permitted to be protected by the motor's branch-circuit fuse or circuit breaker. **Table 430.72(B), Column C** allows this type of installation where the devices do not exceed 300 percent of the ampacity rating of sizes 14 AWG and larger conductors. Overcurrent protection for conductors smaller than 14 AWG shall not exceed the values listed in **Table 430.72(B), Column C**. Conductors rated 18 AWG through 10 AWG can be protected with the following sized overcurrent protection devices:

- 18 AWG = 7 amps and requires 7 A OCPD
- 16 AWG = 10 amps and requires 10 A OCPD
- 14 AWG = 45 amps (15 A OCPD x 300% = 45 A and requires 45 A OCPD)
- 12 AWG = 60 amps (20 A OCPD x 300% = 60 A and requires 60 A OCPD)
- 10 AWG = 90 amps (30 A OCPD x 300% = 90 A and requires 90 A OCPD)
- 8 AWG and larger = 300 percent

Note that the above protection shall be required anytime the control circuit is used for the remote control of a coil in the motor controller enclosure.

See Figure 18-52 for selecting such conductors based upon the overcurrent protection device rating.

CONTROL CIRCUIT TRANSFORMER
430.72(C)(1) THRU (C)(5)

When a motor-control circuit transformer is provided, the transformer shall be protected by the rules and regulations of **Article 450**. A fuse or circuit breaker shall be permitted to be installed in the secondary circuit of the transformer per **430.72(C)(1) through (C)(5)**. The primary overcurrent protection device shall be permitted to be used to provide protection for the conductors tapped from the secondary side of a control transformer, if its rating does not exceed the secondary-to-primary ratio of the transformer. (**See Figure 18-53**)

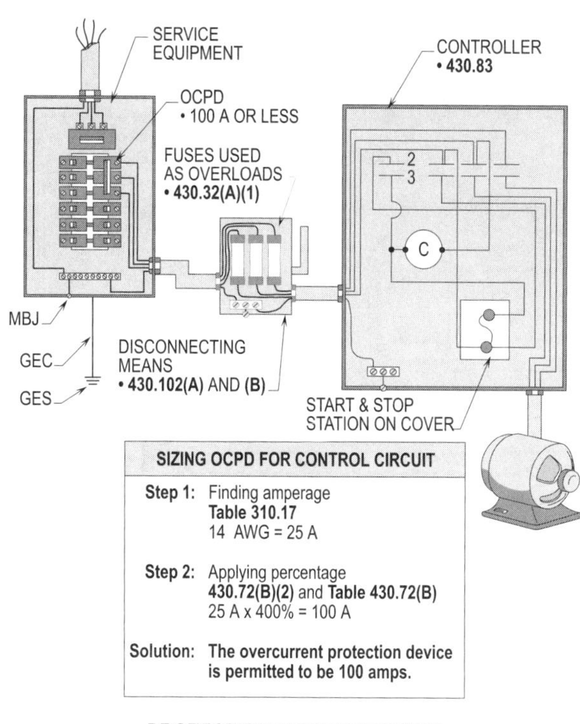

Figure 18-51. Control circuit conductors located in controller and protected by the branch-circuit overcurrent protection device.

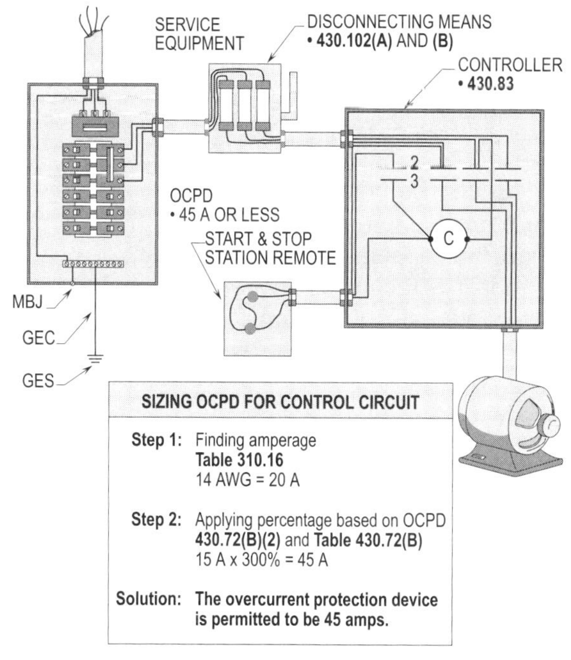

Figure 18-52. Control circuit conductors are run remote and protected by the branch circuit's overcurrent protection device in the panelboard.

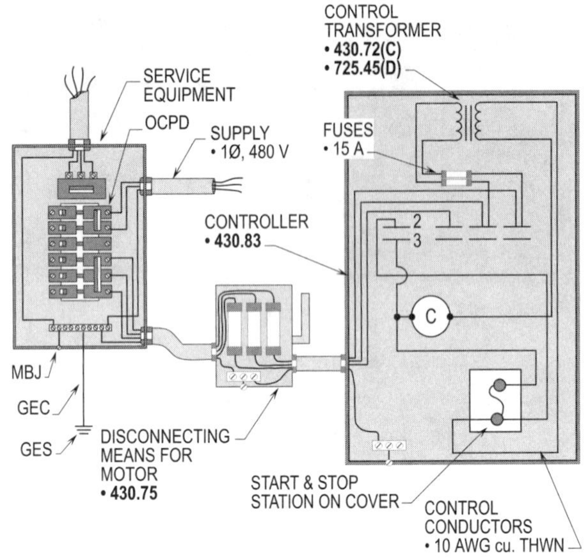

SIZING OCPD FOR THE CONTROL CIRCUIT	
Step 1:	Finding ratio of transformer Pri. and Sec. **725.45(D)** Ratio = (240 V Sec. ÷ 480 V Pri.) x 35 A Ratio = 17.5 A
Step 2:	Selecting OCPD for control circuit **725.45(D)** and **240.6(A)** 17.5 A requires 15 A OCPD
Step 3:	Checking ratio **725.45(D), 240.6(A),** and **430.72(C)(5)** Ratio = 480 V Pri. ÷ 240 V Sec. = 2 to 1 OCPD = 35 A (10 AWG cu.) ÷ 2 (ratio) OCPD = 17.5 A
Solution:	**The size OCPD for the control circuit is 15 amps, which complies with the NEC.**

CONTROL CIRCUIT TRANSFORMER
NEC 430.72(C)(1) THRU (C)(5)

Figure 18-53. The primary overcurrent protection device shall be permitted to be used to provide the protection for the conductors tapped from the secondary side of a control transformer, if its rating does not exceed the secondary-to-primary voltage ratio of the transformer.

MOTOR CONTROL AND MOTOR POWER CIRCUIT CONDUCTORS
430.75

Motor control circuits shall be disconnected from all sources of supply when the disconnecting means is in the open position. The disconnecting means for the starter may be installed to serve as the disconnecting means for the motor circuit conductors if the control circuit conductors are tapped from the line terminal of the magnetic starter. An auxiliary contact shall be installed in the disconnecting means of the controller, or an additional disconnecting means shall be mounted adjacent to the controller, to disconnect the motor

control circuit conductors if they are fed from another source and not tapped from the starter conductors. **(See Figure 18-54)**

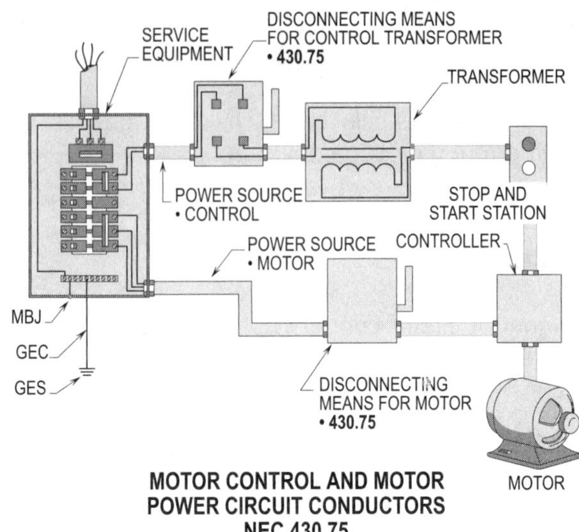

MOTOR CONTROL AND MOTOR POWER CIRCUIT CONDUCTORS
NEC 430.75

Figure 18-54. One disconnecting means or a number of disconnects may be required to disconnect a control circuit and power supply to a motor.

CAPACITOR
460.8(A)

The ampacity of capacitor circuit conductors shall not be less than 135 percent of the rated current of the capacitor. The leads for a capacitor that supplies a motor shall not be less than one-third the ampacity of the motor circuit conductors. The larger of the above conductors shall be used for the capacitor supply conductors. **(See Figure 18-55)**

ROUTING CONTROL CIRCUIT CONDUCTORS
300.3(C)(1) AND 725.48

Conductors of different systems shall be permitted to occupy the same raceway without regard to the use of AC or DC per **300.3(C)(1)**. Conductors shall be insulated for the maximum voltage of any one conductor when occupying the same raceway. These conductors shall not exceed 600 volts. Class 1 conductors shall be permitted to occupy the same raceway when installed with the power conductors supplying the magnetic starter and motor per **725.48(B)(1)**. For control circuit conductors to occupy the same raceway as the motor circuit conductors, they shall be functionally associated with the motor system. **(See Figure 18-56)**

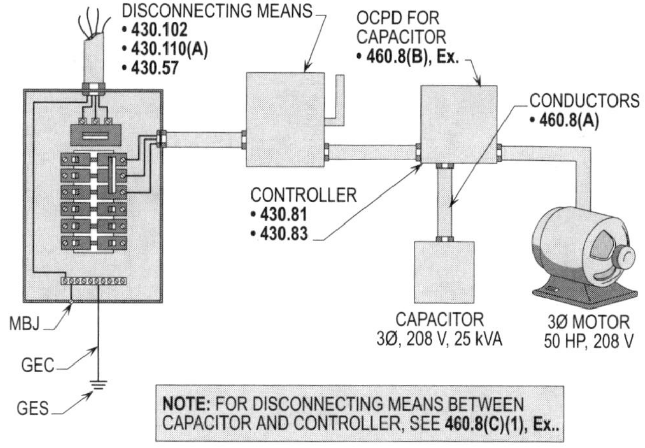

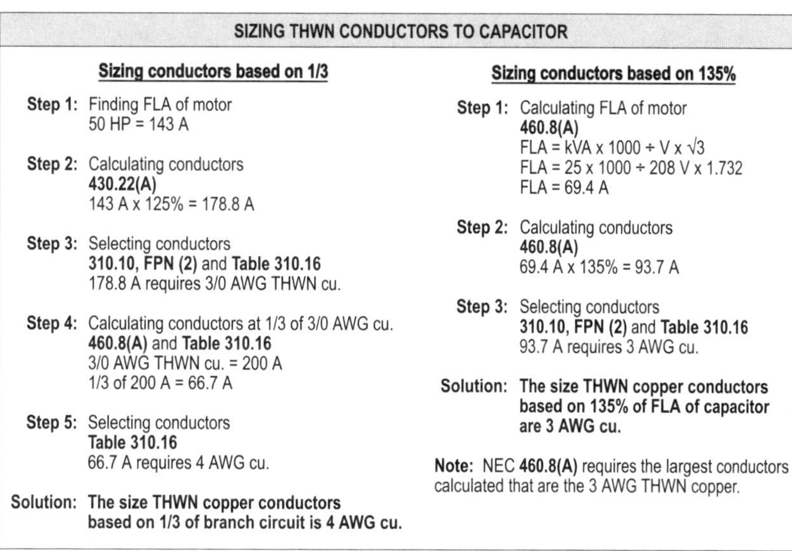

SIZING THWN CONDUCTORS TO CAPACITOR

Sizing conductors based on 1/3

Step 1: Finding FLA of motor
50 HP = 143 A

Step 2: Calculating conductors
430.22(A)
143 A x 125% = 178.8 A

Step 3: Selecting conductors
310.10, FPN (2) and **Table 310.16**
178.8 A requires 3/0 AWG THWN cu.

Step 4: Calculating conductors at 1/3 of 3/0 AWG cu.
460.8(A) and **Table 310.16**
3/0 AWG THWN cu. = 200 A
1/3 of 200 A = 66.7 A

Step 5: Selecting conductors
Table 310.16
66.7 A requires 4 AWG cu.

Solution: The size THWN copper conductors
based on 1/3 of branch circuit is 4 AWG cu.

Sizing conductors based on 135%

Step 1: Calculating FLA of motor
460.8(A)
FLA = kVA x 1000 ÷ V x √3
FLA = 25 x 1000 ÷ 208 V x 1.732
FLA = 69.4 A

Step 2: Calculating conductors
460.8(A)
69.4 A x 135% = 93.7 A

Step 3: Selecting conductors
310.10, FPN (2) and **Table 310.16**
93.7 A requires 3 AWG cu.

Solution: The size THWN copper conductors
based on 135% of FLA of capacitor
are 3 AWG cu.

Note: NEC **460.8(A)** requires the largest conductors
calculated that are the 3 AWG THWN copper.

CAPACITOR
NEC 460.8(A)

Figure 18-55. There are two calculations to be performed, and one of them shall be selected to size the capacitor circuit conductors. Note that the greater of the 1/3 calculation or 135 percent calculation shall be used.

OCCUPYING THE SAME ENCLOSURE
300.3(C)(2)(d) AND 300.32

Motor excitation, magnetic starter, control relay, or ammeter conductors shall be permitted to occupy the same enclosure if rated at 600 volts and less or over 600 volts. These conductors shall not be permitted to occupy the same raceway where installed as a combination of conductors of 600 volts or less with conductors of over 600 volts. The motor enclosure and the starter enclosure can contain motor excitation, control, relay, and ammeter conductors of 600 volts or less, together with power conductors of over 600 volts. **(See Figure 18-57)**

MAGNETIC STARTER CONTACTOR AND
ENCLOSURE
725.51(A) AND (B)

When installing four or more current-carrying conductors that are continuously operated in a raceway and used for Class 1 remote-control, signal, and power-limited circuit conductors, they may have to be derated by the derating factors of **310.15(B)(2)(a)** per **725.51(B)(1)** or **(B)(2)**. Conductors that are noncontinuously operated shall not be required to be derated by the derating factors of **310.15(B)(2)(a)**. When installing power and control conductors in the same raceway, **310.15(B)(2)(a)** shall be applied to all current-carrying conductors operating for three hours or more. **(See Figure 18-58)**

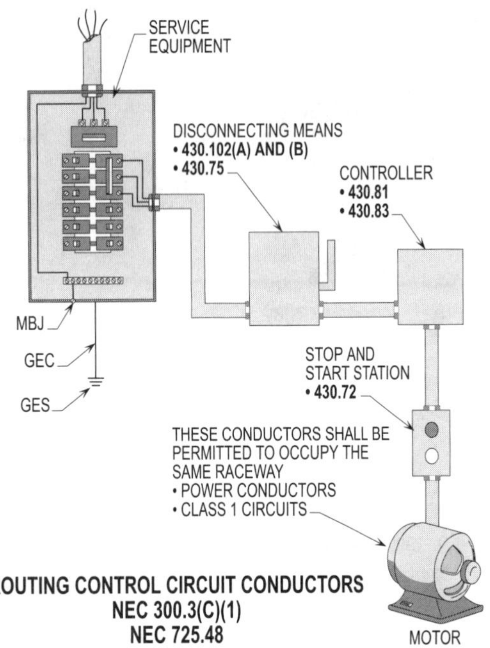

Figure 18-56. When functionally associated with motor operation, a Class 1 control circuit shall be permitted to occupy the same raceway as the motor's power circuit conductors.

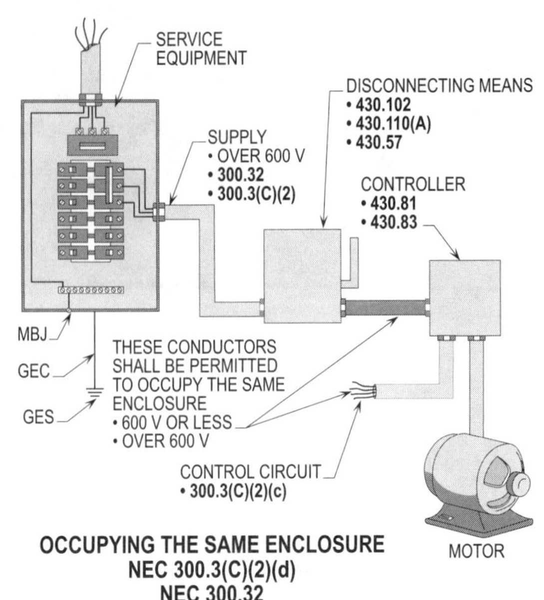

Figure 18-57. Conductors over 600 volts and those 600 volts or less shall be permitted to occupy the same enclosure under certain conditions of use.

Design Tip: Control circuit conductors operating for three hours or more shall not be required to be derated per **310.15(B)(2)(a)** if their ampacity does not exceed 10 percent of the control conductor's ampacity rating.

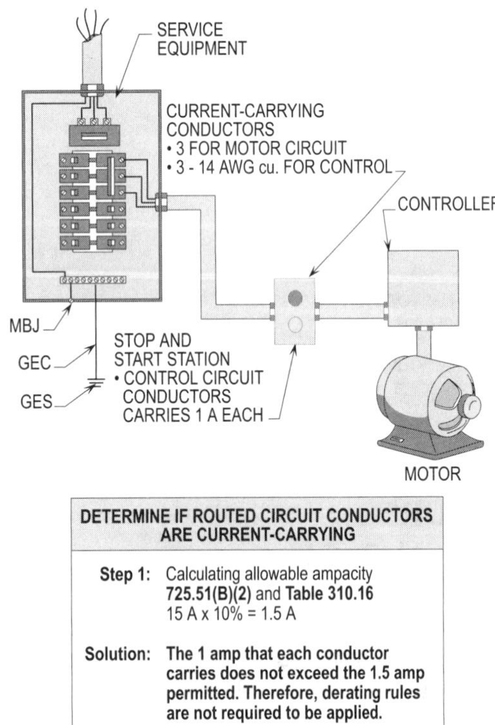

DETERMINE IF ROUTED CIRCUIT CONDUCTORS ARE CURRENT-CARRYING	
Step 1:	Calculating allowable ampacity **725.51(B)(2)** and **Table 310.16** 15 A x 10% = 1.5 A
Solution:	The 1 amp that each conductor carries does not exceed the 1.5 amp permitted. Therefore, derating rules are not required to be applied.

Figure 18-58. Circuit conductors that do not carry more than 10 percent of their ampacity shall not be considered current-carrying conductors.

TABLE CURRENT IN AMPS
TABLES 430.247 THRU 430.250

The following methods can be used to determine the full-load current rating in amps for motors that are not listed in **Tables 430.247 through 430.250**:

- The horsepower rating of a listed motor shall be selected to be below that of the unlisted motor.

- The motor's full-load current rating shall be divided by its horsepower rating to obtain the multiplier.

- The multiplier times the horsepower of the unlisted motor derives FLC in amps for the unlisted motor.

Design Tip: The full-load current rating of the motor is determined by multiplying these values by the horsepower rating of the unlisted motor. **(See Figure 18-59)**

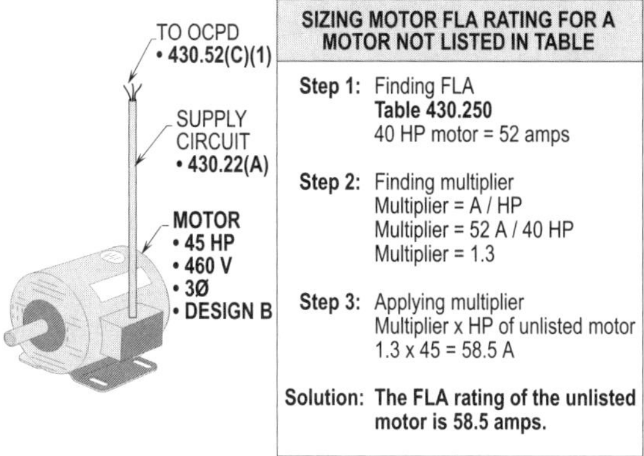

TABLE CURRENT IN AMPS
NEC TABLES 430.247 THRU 430.250

Figure 18-59. The above illustrates the procedure for calculating the FLA of a motor not listed in **Tables 430.247 through 430.250**.

RULE-OF-THUMB – AMPS

The full-load current of a motor may be found by using the rule-of-thumb method in **Table 430.248** for single-phase and **Table 430.250** for three-phase. The Table current will not always be exactly the same as the rule-of-thumb amps.

Overcurrent protection devices, conductors, and other elements can be sized with the full-load current ratings obtained by the rule-of-thumb method. The full-load current ratings are within a usable range when applying the rule-of-thumb method to determine the full-load current rating in amps. These amperage ratings will provide values to calculate elements for a complete and safe electrical motor system.

The following percentages can be applied when using a rule-of-thumb method to derive full-load amps for a particular size motor:

- When installing 550, 575, or 600 volt, three-phase motors, the horsepower rating of the motor shall be multiplied by 1.00 to obtain full-load current in amps.

- When installing 440, 460, or 480 volt, three-phase motors, the horsepower rating of the motor shall be multiplied by 1.25 to obtain full-load current in amps.

- When installing 220, 230, or 240 volt, three-phase motors, the horsepower rating of the motor shall be multiplied by 2.50 to obtain full-load current in amps.

- When installing 220, 230, or 240 volt, single-phase motors, the horsepower rating of the motor shall be multiplied by 5.00 to obtain full-load current in amps.

- When installing 110, 115, or 120 volt, single-phase motors, the horsepower rating of the motor shall be multiplied by 10.00 to obtain full-load current in amps.

See Figure 18-60 for a detailed illustration of calculating full-load currents in amps for motors using the rule-of-thumb method.

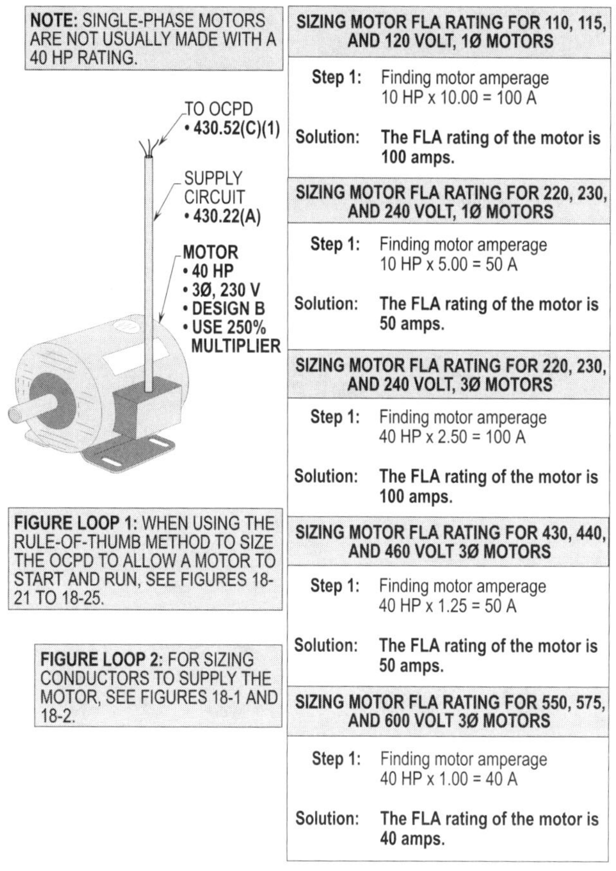

RULE-OF-THUMB – AMPS

Figure 18-60. The above is a rule-of-thumb method used by the electrical industry to determine the FLA of motors whether found or not found in **Tables 430.248 through 430.250**. Note that this method is used in the field or office when a code book is not available.

NUMBER OF MOTORS SERVED BY EACH DISCONNECT AND CONTROLLER
430.87, Ex. 1 AND 430.112

Each controller shall be installed with a disconnecting means located within sight and within 50 ft. Unauthorized energizing of the electrical power circuit shall be prevented during maintenance procedures. A single disconnect and controller shall be permitted to be installed to control and disconnect any number of motors of 600 volts or less per **430.87, Ex. 1** and **430.112, Ex.** Motors used to drive different parts of the same machine is an example of a number of motors that can be protected by a single disconnect and controller. **(See Figure 18-61)**

A single overcurrent protection device shall be permitted to be installed to protect a group of motors only if they are rated in the fractional-horsepower range and installed on a general purpose circuit of 20 amps or less per **430.87, Ex. 2** and **430.53(A)**.

Motors installed to be served by the same disconnect and controller shall be within sight and located within one room. Unauthorized energizing of the electrical power circuit can be prevented during maintenance procedures if the above rules are complied with. For lockout and tagout procedures, see **OSHA 1910, Subpart S** and **NFPA 70E, Chapter 1**.

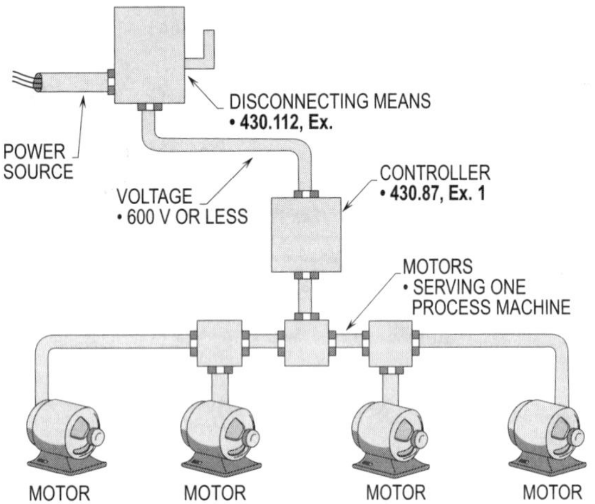

Figure 18-61. Under certain conditions of use, one disconnect and controller shall be permitted to be used for several motors serving a machine.

REDUCED STARTING METHODS

As to methods of starting a motor, there are seven methods used in the electrical industry, and they are as follows:

- Full voltage (not reduced at starting)
- Resistor starting
- Reactor starting
- Autotransformer starting
- Solid state starting
- Part winding
- Start wye and run delta

AC motors up to about 10 or 15 horsepower are usually started on full voltage. For larger motors, a resistance or reactance is inserted in the motor circuit at starting to reduce the starting current. The voltage at starting can be reduced by the use of an autotransformer. A solid-state starter can also be used to reduce starting currents.

ADJUSTABLE-SPEED DRIVE SYSTEMS
PART X TO ARTICLE 430

When designing and installing electrical systems for adjustable-speed drives, the installation provisions of Part I through Part IX are applicable unless modified or supplemented by Part X.

BRANCH/FEEDER-CIRCUIT CONDUCTORS
430.122(A)

Circuit conductors supplying power conversion equipment included as part of an adjustable-speed drive system shall have an ampacity not less than 125 percent of the rated input to the power conversion equipment. **(See Figure 18-62)**

Note, electrical resonance can result from the interaction of the nonsinusoidal currents from this type of load with power factor correction capacitors.

BYPASS DEVICE
430.122(B)

For an adjustable speed drive system that utilizes a bypass device, the conductor ampacity shall not be less than required by **430.6**. The ampacity of circuit conductors supplying power conversion equipment included as part of an adjustable speed drive system that utilizes a bypass device shall be the larger of either of the following:

(1) 125 percent of the rated input to the power conversion equipment

(2) 125 percent of the motor full-load current rating as determined by **430.6**

For an illustrated description, see **Figure 18-63**.

OVERLOAD PROTECTION
430.124

Overload protection of the motor shall be provided.

INCLUDED IN POWER CONVERSION EQUIPMENT
430.124(A)

Where the power conversion equipment is marked to indicate that motor overload protection is included, additional overload protection shall not be required.

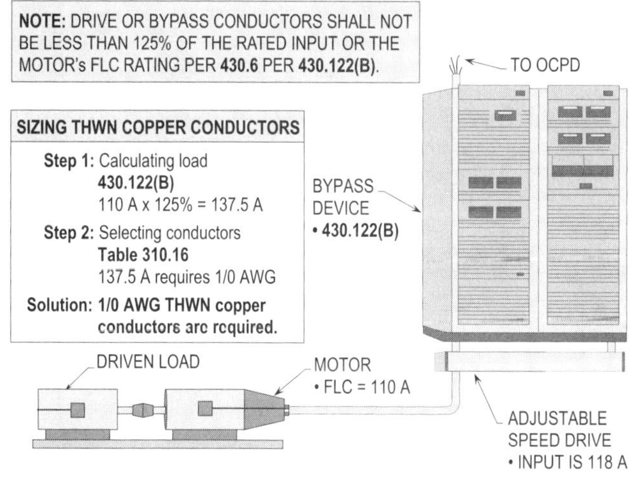

BRANCH/FEEDER-CIRCUIT CONDUCTORS
NEC 430.122(A)

Figure 18-62. The above illustrates the procedure for calculating the load in amps to size the conductors supplying the power conversion equipment.

BYPASS DEVICE
NEC 430.122(B)

Figure 18-63. The above illustrates the procedures for calculating the load in amps to size the conductors for a bypass drive system.

BYPASS CIRCUITS
430.124(B)

For adjustable speed drive systems that utilize a bypass device to allow motor operation at rated full load speed, motor overload protection as described in **Article 430, Part III**, shall be provided in the bypass circuit.

MULTIPLE MOTOR APPLICATIONS
430.124(C)

For multiple motor application, individual motor overload protection shall be provided in accordance with **Article 430, Part III**. **(See Figure 18-64)**

MOTOR OVERTEMPERATURE PROTECTION – GENERAL
430.126(A)

Adjustable speed drive systems shall protect against motor overtemperature conditions where the motor is not rated to operate at the nameplate rated current over the speed range required by the application. This protection shall be provided in addition to the conductor protection required in **430.32**. Protection shall be provided by one of the following means:

(1) Motor thermal protector in accordance with **430.32**

(2) Adjustable speed drive system with load- and speed-sensitive overload protection and thermal memory retention upon shutdown or power loss

Thermal memory retention upon shutdown or power loss is not required for continuous duty loads per **430.126(A)(2), Ex**.

(3) Overtemperature protection relay utilizing thermal sensors embedded in the motor and meeting the requirements of **430.32(A)(2)** or **(B)(2)**

(4) Thermal sensor embedded in the motor whose communications are received and acted upon by an adjustable speed drive system

For a detailed description of these requirements, see **Figure 18-65** and **FPN** to **430.126(A)**.

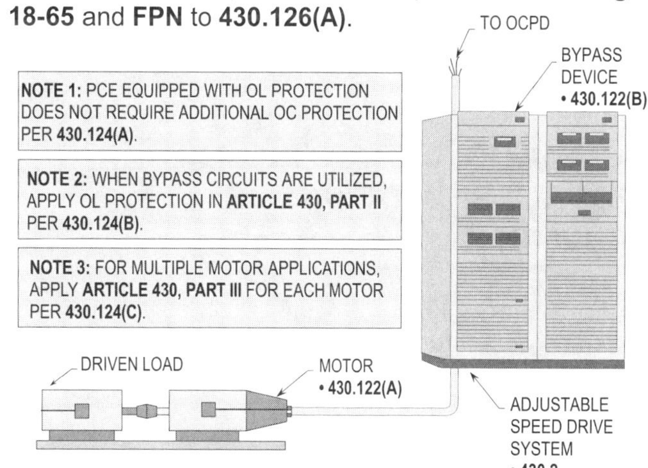

BYPASS DEVICE
NEC 430.124(A) THRU (C)

Figure 18-64. The above illustrates the procedure for protecting the motor from overload when using a bypass device.

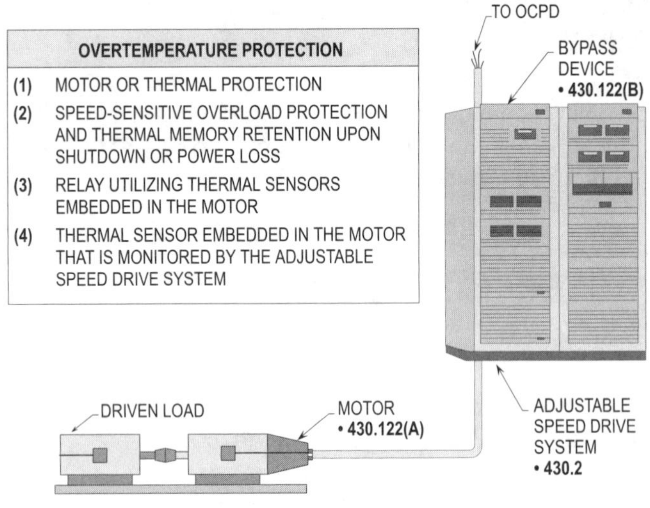

**MOTOR OVERTEMPERATURE PROTECTION – GENERAL
NEC 430.126(A)(1) THRU (A)(4)**

Figure 18-65. The above illustrates the procedure for determining protection for motor overtemperature problems.

MULTIPLE MOTOR APPLICATIONS
430.126(B)

For multiple motor application, individual motor overtemperature protection shall be provided per **430.126(A)**.

Note, the relationship between motor current and motor temperature changes when the motor is operated by an adjustable speed drive. When operated at reduced speed, overheating of motors may occur at current levels less than or equal to a motor's rated full load current. This is the result of reduced motor cooling when its shaft-mounted fan is operating at less than rated nameplate RPM.

AUTOMATIC RESTARTING AND
ORDERLY SHUTDOWN
430.126(C)

The provisions of **430.43** and **430.44** shall apply to the motor overtemperature protection means.

DISCONNECTING MEANS
430.128

The disconnecting means shall be permitted to be in the incoming line to the conversion equipment and shall have a rating not less than 115 percent of the rated input current of the conversion unit.

For example, if the rated input current of the conversion unit is 173.9 amps, the disconnecting means shall be rated at least 200 amps (173.9 A x 115% = 199.9 A) per **430.128**.

INDUSTRIAL CONTROL PANELS
ARTICLE 409

Article 409 covers the installation of industrial control panels intended for general use and operating procedures at 600 volts or less. UL 508A governs the procedures for safety when installing components in industrial control panels.

CONDUCTOR – MINIMUM SIZE
AND AMPACITY
409.20

The size of the industrial control panel supply conductor shall have an ampacity not less than 125 percent of the full-load current rating of all resistance heating loads, plus 125 percent of the full-load current rating of the highest rated motor plus the sum of the full-load current ratings of all other connected motors and apparatus based on their duty cycle that may be in operation at the same time. **(See Figure 18-66)**

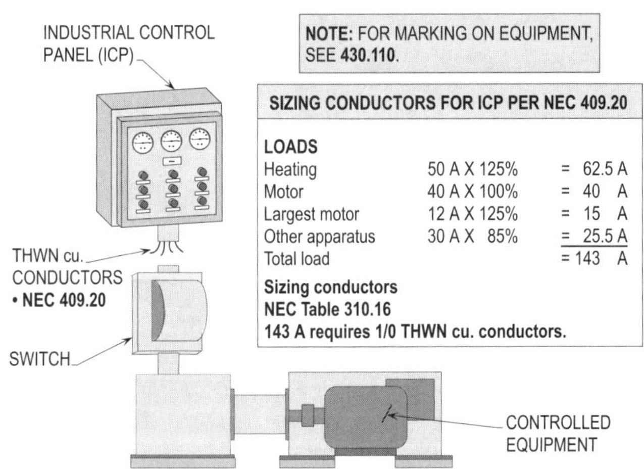

**CONDUCTOR – MINIMUM SIZE AND AMPACITY
NEC 409.20**

Figure 18-66. The above illustrates the procedure for calculating the load in amps to size the conductors supplying the industrial control panel.

OVERCURRENT PROTECTION –
GENERAL
409.21(A)

Industrial control panels shall be provided with overcurrent protection in accordance with **Parts I, II** and **IX** of **Article 240**.

LOCATION
409.21(B)(1) AND (B)(2)

Overcurrent protection for each incoming supply circuit shall be provided by either of the following:

(1) An overcurrent protective device located ahead of the industrial control panel

(2) A single main overcurrent protective device located within the industrial control

Where overcurrent protection is provided as part of the industrial control panel, the supply conductors shall be considered as either feeders or taps as covered by **240.21**.

RATING
409.21(C) AND Ex.

The rating or setting of the overcurrent protective device for the circuit supplying the industrial control panel shall not be greater than the sum of the largest rating or setting of the branch-circuit short circuit, and ground fault protective device provided with the industrial control panel, plus 125 percent of the full-load current rating of all resistance heating loads, plus the sum of the full-load currents of all other motors and apparatus that could be in operation at the same time. **(See Figure 18-67)**

Applying Exception to **409.21(C)**: Where one or more instantaneous trip circuit breakers or motor short circuit protectors are used for motor branch-circuit, short circuit, and ground fault protection as permitted by **430.52(C)**, the procedure specified above for determining the maximum rating of the protective device for the circuit supplying the industrial control panel shall apply, with the following provision:

• For the purpose of the calculation, each instantaneous trip circuit breaker or motor short circuit shall be assumed to have a rating not exceeding the maximum percentage of motor full-load current permitted by **Table 430.52** for the type of control panel supply circuit protective device employed.

Where no branch-circuit short circuit and ground fault protective device is provided with the industrial control panel for motor or combination of motor and non-motor loads, the rating or setting of the overcurrent protective device shall be based on **430.52** and **430.53**, as applicable.

DISCONNECTING MEANS
409.30

Disconnecting means that supply motor loads shall comply with **Part IX** of **Article 430**.

For example, if the motor loads on an industrial process add up to 173 amps, the size disconnecting would be 200 amps (173 A x 115% = 198.95 A) per **430.110(A)** of the NEC. **(See Figure 18-68)**

WIRING SPACE IN INDUSTRIAL CONTROL PANELS – GENERAL
409.104(A)

Industrial control panel enclosures shall not be used as junction boxes, auxiliary gutters, or raceways for conductors feeding through or tapping off to other switches or overcurrent devices, unless adequate space for this purpose is provided. The conductors shall not fill the wiring space at any cross section to more than 40 percent of the cross-sectional area of the space, and the conductors, splices,

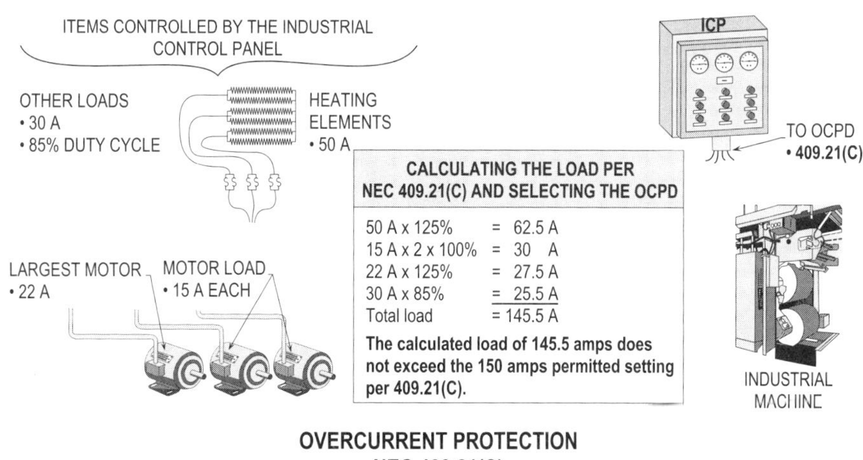

Figure 18-67. The above illustrates the procedure for calculating the load in amps to size the overcurrent protection device protecting the industrial control panel.

and taps shall not fill the wiring space at any cross section to more than 75 percent of the cross-sectional area of that space. **(See Figure 18-69)**

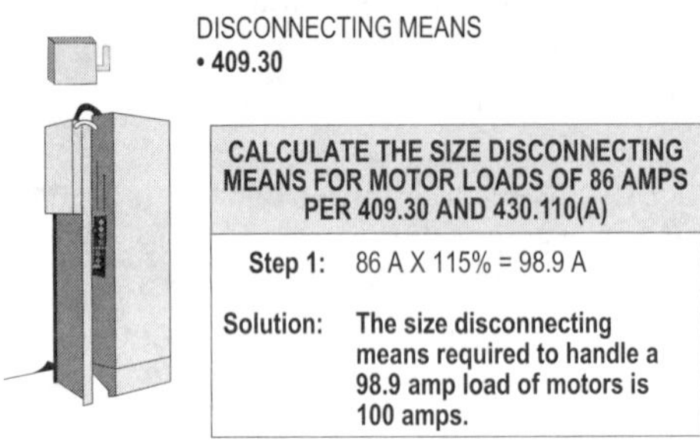

DISCONNECTING MEANS
NEC 409.30

Figure 18-68. The above illustrates the procedure for calculating motor loads in amps to size the disconnecting means.

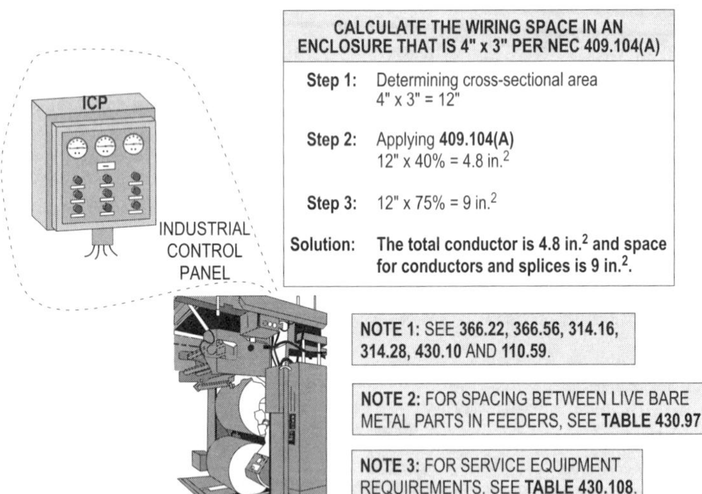

WIRING SPACE IN INDUSTRIAL CONTROL PANELS – GENERAL
NEC 409.104(A)

Figure 18-69. The above illustrates the procedure for calculating the wiring space necessary for accommodating conductors and splices in an industrial control panel.

PHASE CONVERTERS
455.6(A) AND 455.7(A)

Phase converters are used to convert single-phase power to three-phase power. The disconnecting means shall be located within 50 ft (15 m) and within sight per **455.8(A)**. Where the voltage is not the same, the output-to-input ratio shall be applied per **455.6(A)**.

Branch-circuit conductors shall be sized at 125 percent times the phase converter's nameplate single-phase input full-load current rating, in amps. The overcurrent protection device

shall be sized at 125 percent times the phase converter's nameplate single-phase input full-load amps. The overcurrent protection device shall not exceed the 125 percent but shall be equal to or lower than 125 percent. **(See Figure 18-70)** Branch-circuit elements such as overcurrent protection devices and conductors supplying specific loads shall be calculated at 250 percent of the equipment's full-load amp rating. **(See Figure 18-71)**

Feeder conductors that convert single-phase power to three-phase power to supply power to two or more phase converters shall be sized at 250 percent times the three-phase amperage of all motors and other loads served. The overcurrent protection device shall be sized at 250 percent times the full-load three-phase amps of all motors and other loads. If the percentage does not correspond to a standard size, the next size overcurrent protection device above this percentage shall be permitted to be selected per **455.7**. **(See Figure 18-72)**

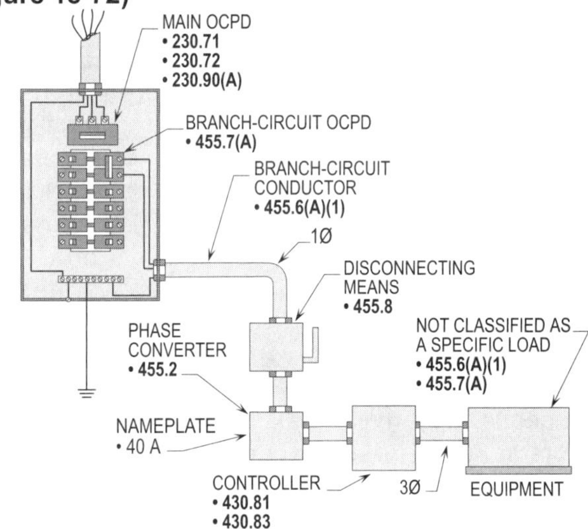

FINDING THWN CONDUCTORS AND OCPD FOR PHASE CONVERTER	
Sizing conductors	**Sizing OCPD**
Step 1: Finding amperage **455.4** Nameplate = 40 A	**Step 1:** Finding amperage **455.4** Nameplate = 40 A
Step 2: Calculating amperage **240.4(G)** and **455.6(A)(1)** 40 A x 125% = 50 A	**Step 2:** Calculating amperage **455.7(A)** 40 A x 125% = 50 A
Step 3: Selecting conductors **Table 310.16** 50 A = 8 AWG THWN cu.	**Step 3:** Selecting OCPD **455.7(A)** and **240.6(A)** 50 A = 50 A OCPD
Solution: The size THWN copper conductors are required to be 8 AWG.	**Solution:** A 50 amp OCPD is required.

PHASE CONVERTERS
NEC 455.6(A)
NEC 455.7(A)

Figure 18-70. Branch-circuit conductors shall be sized at 125 percent times the phase converter's nameplate single-phase input full-load amperage. The overcurrent protection device shall be sized at 125 percent times the phase converter's nameplate single-phase input full-load current, in amps.

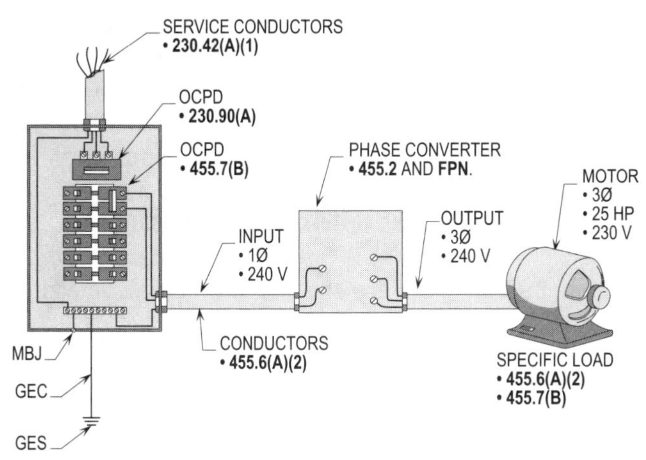

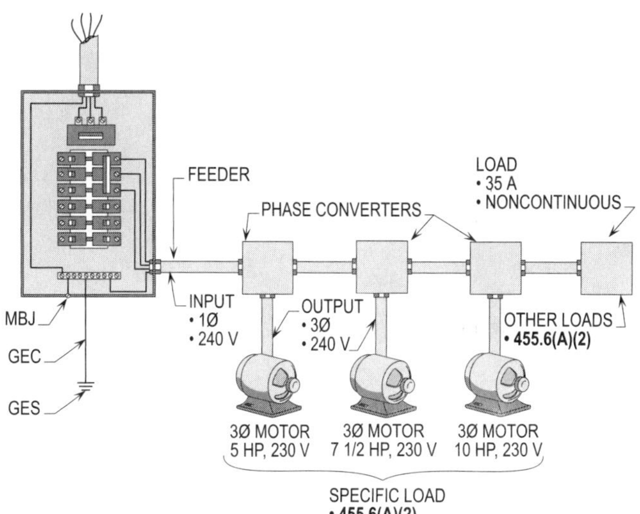

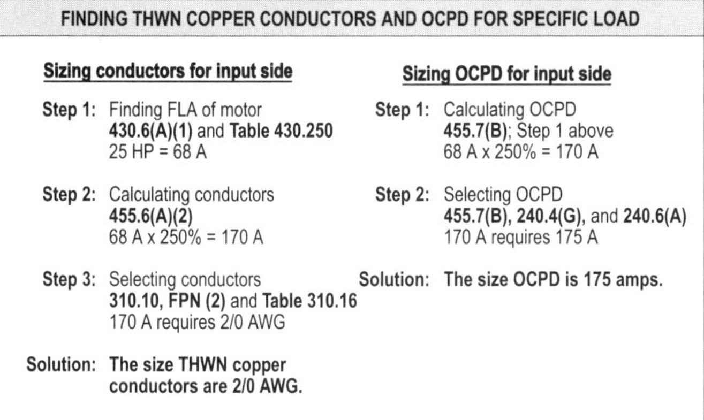

FINDING THWN COPPER CONDUCTORS AND OCPD FOR SPECIFIC LOAD

Sizing conductors for input side

Step 1: Finding FLA of motor
430.6(A)(1) and **Table 430.250**
25 HP = 68 A

Step 2: Calculating conductors
455.6(A)(2)
68 A x 250% = 170 A

Step 3: Selecting conductors
310.10, FPN (2) and **Table 310.16**
170 A requires 2/0 AWG

Solution: The size THWN copper
conductors are 2/0 AWG.

Sizing OCPD for input side

Step 1: Calculating OCPD
455.7(B); Step 1 above
68 A x 250% = 170 A

Step 2: Selecting OCPD
455.7(B), 240.4(G), and **240.6(A)**
170 A requires 175 A

Solution: The size OCPD is 175 amps.

PHASE CONVERTERS
NEC 455.6(A)(2)
NEC 455.7(B)

Figure 18-71. Branch-circuit elements such as overcurrent protection devices and conductors supplying specific loads shall be calculated at 250 percent of the equipment's full-load current rating, in amps.

FIRE PUMPS
ARTICLE 695

Article 695 covers the installation of electric power sources, interconnecting circuits, and switching and control equipment dedicated to fire pumps. Note that for more information on fire pump installations, see **NFPA 20, NFPA 37,** and for maintenance checks, **NFPA 110**.

POWER SOURCES
695.3(A)

Section **695.3(A)** covers power sources that are allowed to supply power to fire pump installations.

Power sources such as a reliable service, an on-site generator, a separately derived system, or a tap ahead of the service disconnecting means are considered dependable power supply systems when serving fire pumps and other related equipment. **(See Figure 18-73)**

NEC HELP

CONDUCTOR SIZING
• **455.6(A)**
OCPD SIZING
• **455.7**
DISCONNECTING MEANS
• **455.8**
HP RATED DISCONNECT
• **455.8(C)(2)**
CAPACITOR INSTALLATIONS
• **455.23**

FINDING OCPD AND THHN COPPER CONDUCTORS FOR FEEDER

Sizing conductors for input side

Step 1: Finding FLA for input side
430.6(A)(1) and **Table 430.250**
5 HP = 15.2 A
7 1/2 HP = 22.0 A
10 HP = 28.0 A

Step 2: Finding other loads
215.2(A)(1)
35 A x 100% = 35 A

Step 3: Calculating conductors
455.6(A)(2)
15.2 A + 22 A + 28 A + 35 A = 100.2 A
100.2 A x 250% = 250.5 A

Step 4: Selecting conductors
310.10, FPN (2) and **Table 310.16**
250.5 A requires 250 KCMIL cu.

Solution: The size THWN copper conductors
are 250 KCMIL copper.

Sizing OCPD for input side

Step 1: Calculating OCPD
455.7(B), Step 3 above
100.2 A x 250% = 250.5 A

Step 2: Selecting OCPD
455.7(B) and **240.6(A)**
250.5 A allows 250 A

Solution: The size OCPD is 250 amps.

PHASE CONVERTERS
NEC 455.6(A)(2)
NEC 455.7
NEC 455.7(B)

Figure 18-72. Feeder conductors that convert single-phase power to three-phase power for supplying power to two or more phase converters shall be sized at 250 percent times the three-phase amperage of all motors and other loads served. The overcurrent protection device shall be sized at 250 times the full-load three-phase amps of all motors and other loads.

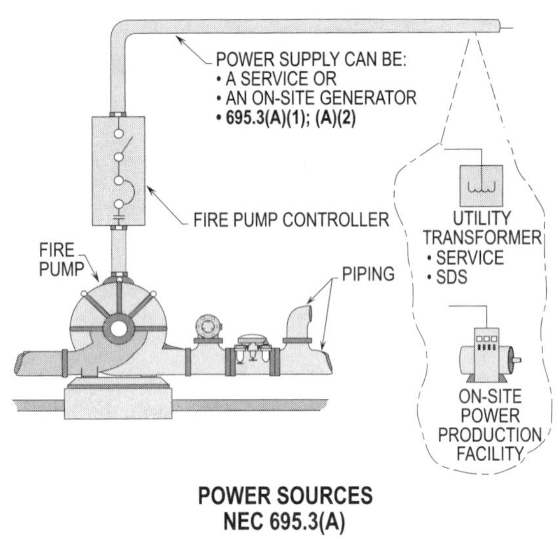

POWER SOURCES
NEC 695.3(A)

Figure 18-73. The above shows power sources that are permitted to supply fire pump installations.

SIZING CONDUCTORS
695.6(C)(1) AND 430.22(A)

Conductors shall be sized with enough capacity so that they are protected against short-circuit currents. By sizing the conductors to the fire pump motors at 125 percent of the motor's FLA, this should be accomplished. For sizing the conductors to one motor, see **430.22(A)** and for more than one motor, plus other loads, see **430.24** of the NEC. **(See Figure 18-74)**

SIZING OVERCURRENT PROTECTION DEVICE
695.5(B), (C)(2), AND 230.90(A), Ex. 4

The overcurrent protection device shall protect the conductors and fire pump motor and accessories from short circuits. **(See Figure 18-75)**

SIZING TRANSFORMER USED AS A SEPARATELY DERIVED SYSTEM
695.5(A)

Section **695.5(A)** allows a transformer dedicated to supplying a fire pump to be rated at a minimum of 125 percent of the sum of the rated full load of the fire pump motor(s), the rated full loads of pressure maintenance pump motor(s), and the full-load amps of any associated fire pump accessory equipment connected to the transformer.

Secondary overcurrent protection for the transformer shall not be permitted, and the primary overcurrent protection device shall not be set above 600 percent of the transformer's full-load current rating, in amps.

SIZING TRANSFORMER ELEMENTS

Section **695.5(A)** covers the requirements for sizing a separately derived system, **695.5(B)** deals with sizing the overcurrent protection device, and **695.5(C)(2)** outlines the rules that require the overcurrent protection device to carry the locked-rotor current of the transformer indefinitely.

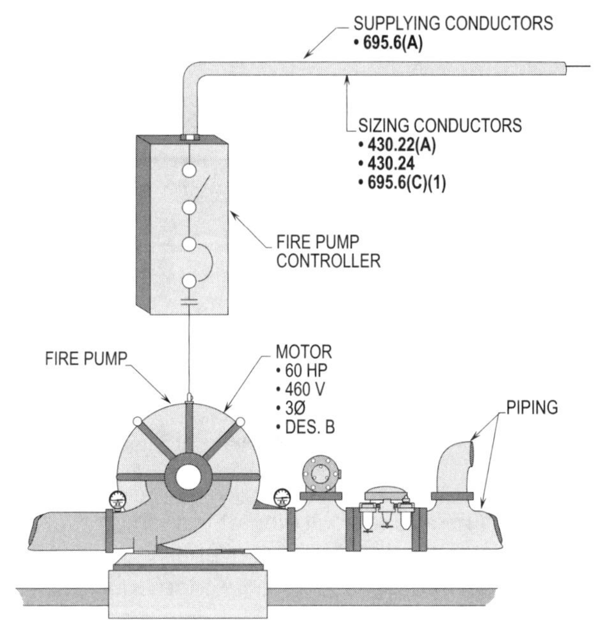

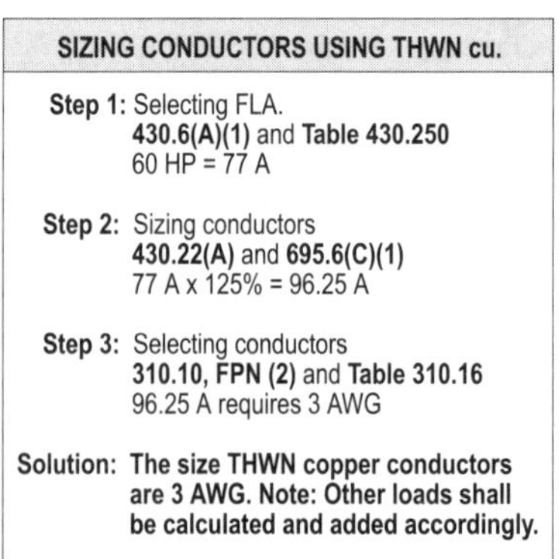

SIZING CONDUCTORS USING THWN cu.

Step 1: Selecting FLA.
430.6(A)(1) and Table 430.250
60 HP = 77 A

Step 2: Sizing conductors
430.22(A) and 695.6(C)(1)
77 A x 125% = 96.25 A

Step 3: Selecting conductors
310.10, FPN (2) and Table 310.16
96.25 A requires 3 AWG

Solution: The size THWN copper conductors are 3 AWG. Note: Other loads shall be calculated and added accordingly.

SIZING CONDUCTORS
NEC 695.6(C)(1)
NEC 430.22(A)

Figure 18-74. The above shows the procedure for sizing the conductors to supply a fire pump.

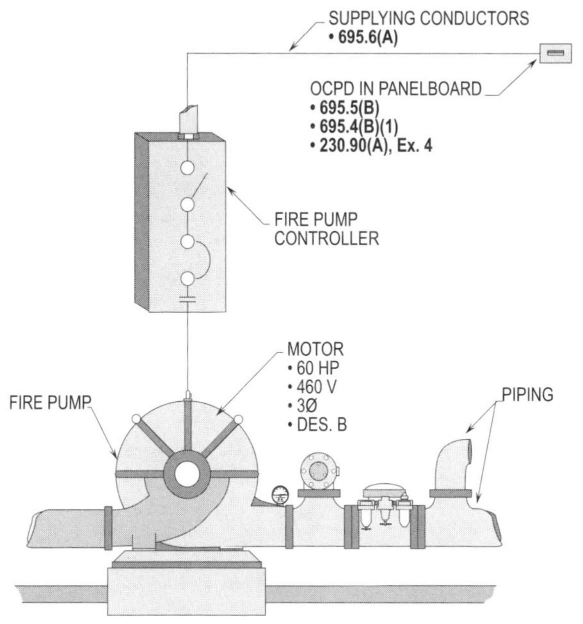

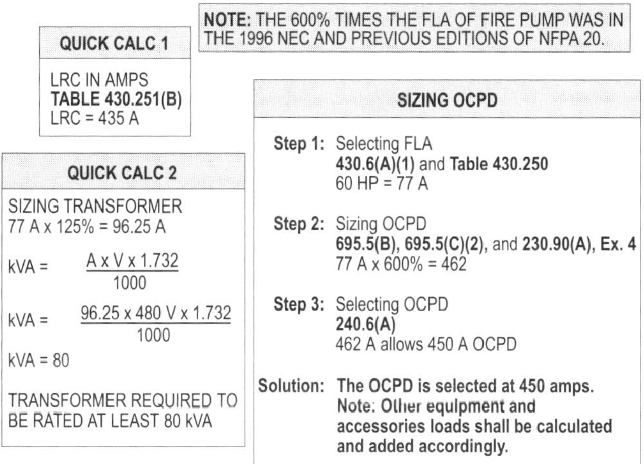

Figure 18-75. The above shows the procedure for sizing the overcurrent protection device to protect a fire pump.

SIZING OVERCURRENT PROTECTION DEVICE FOR A SEPARATELY DERIVED SYSTEM
695.5(B)

Section **696.5(B)** requires the overcurrent protection device on the primary side of a separately derived system, supplying power to a fire pump installation, to carry the secondary circuit indefinitely.

Note that such secondary currents includes both normal full-load operating current, in amps, as well as the locked-rotor current, in amps, of the motor. **(See Figure 18-76)**

When separately derived systems are used to supply power to fire pumps and accessories, they are usually installed in the fire pump room with the fire pump controller. Note that the transformer shall supply power until pump motor failure. This requirement allows the motor to pump water to fight the fire for as long as possible.

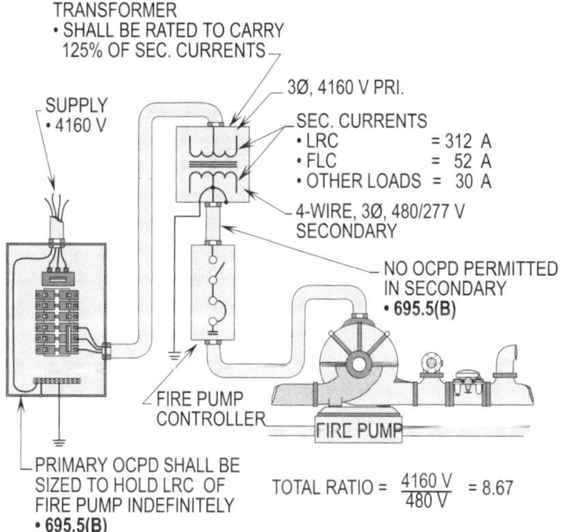

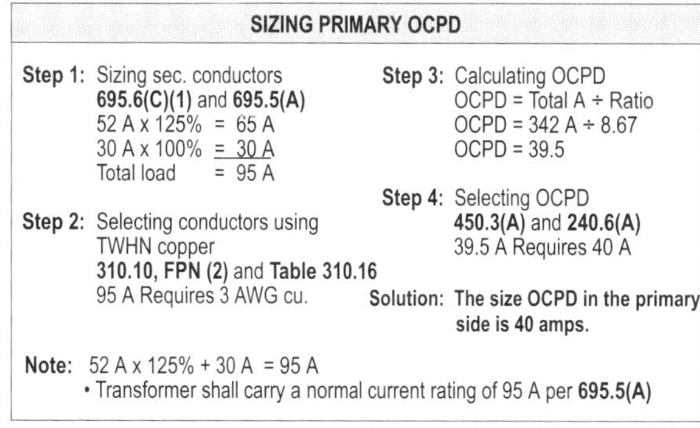

Figure 18-76. The above shows the procedure for sizing the transformer and primary overcurrent protection device for a fire pump installation.

Chapter 18. Motors

Section   Answer

_________   _________   1. Conductors for a motor used for periodic duty and rated for 5 minutes shall
                            be sized with a current-carrying capacity of ______ percent of the motor's
                            full-load amps.
                            (a) 85   (b) 100
                            (c) 110   (d) 125

_________   _________   2. Wound-rotor motors are three-phase motors that are installed with ______
                            sets of leads.
                            (a) 1   (b) 2
                            (c) 3   (d) 4

_________   _________   3. Time-delay fuses that are sized at ______ percent or less of a motor's
                            FLA rating can provide overload protection for the motor.
                            (a) 100   (b) 110
                            (c) 115   (d) 125

_________   _________   4. The maximum percentage that shall be permitted to be applied to a time-
                            delay fuse is ______ percent of a motor's FLA.
                            (a) 175   (b) 200
                            (c) 225   (d) 300

_________   _________   5. Two or motors shall be permitted to be installed without individual
                            overcurrent protection devices if rated less than ______ horsepower each
                            and if the full-load current rating, in amps, of each motor does not exceed
                            6 amps.
                            (a) 1   (b) 2
                            (c) 3   (d) 5

_________   _________   6. Thermal protectors and thermal relays shall be permitted to be installed
                            to provide running overload protection for motors rated more than ______
                            horsepower.
                            (a) 1   (b) 2
                            (c) 3   (d) 4

_________   _________   7. The controller shall be permitted to be an attachment plug and receptacle
                            that is acceptable for use with portable motors rated ______ horsepower or
                            less.
                            (a) 1/8   (b) 1/4
                            (c) 1/3   (d) 1/2

_________   _________   8. The controller shall be permitted to be a general-use switch rated for at
                            least twice the motor's full-load current, in amps, for stationary motors rated
                            ______ horsepower or less.
                            (a) 1/8   (b) 1/2
                            (c) 1   (d) 2

9. For a stationary motor rated _____ horsepower or less, the branch-circuit overcurrent protection device shall be permitted to serve as the disconnecting means.
    (a) 1/16   (b) 1/8
    (c) 1/4   (d) 1/2

10. The general rule requires the disconnecting means for a motor to be installed within sight of the motor and not more than _____ ft from the motor.
    (a) 20   (b) 25
    (c) 50   (d) 75

11. Branch-circuit conductors shall be sized at _____ percent times the phase converter's (variable loads) nameplate single-phase input FLA rating.
    (a) 125   (b) 175
    (c) 200   (d) 250

12. Feeder conductors supplying power to two or more phase converters shall be sized at _____ percent times the three-phase amperage of all motors and other loads served.
    (a) 125   (b) 175
    (c) 200   (d) 250

13. Branch-circuit conductors supply a single motor shall have an ampacity not less than _____ percent of the motor's FLC, in amps.
    (a) 100   (b) 125
    (c) 135   (d) 150

14. The selection of conductors for wye start and delta run motors between the controller and motor shall be based on _____ percent of the motor's full load current, in amps, times 125 percent for continuous use.
    (a) 58   (b) 67
    (c) 75   (d) 80

15. The disconnecting means for power conversion equipment shall not be less than _____ percent of the input FLA rating of the conversion unit.
    (a) 100   (b) 110
    (c) 115   (d) 125

16. The full-load current rating, in amps, of the largest motor shall be multiplied by _____ percent to select the size of conductors for a feeder supplying a group of two or more motors.
    (a) 100   (b) 110
    (c) 115   (d) 125

17. The motor branch-circuit overcurrent device shall be capable of carrying the _____ current, in amps, of the motor.
    (a) starting   (b) varying
    (c) torque   (d) continuous

18. When a motor won't start and run, a nontime-delay fuse not exceeding 600 amperes in rating shall be permitted to be increased up to _____ percent of the full-load current, in amps, of the motor.
    (a) 250   (b) 300
    (c) 400   (d) 600

19. When a motor won't start and run, a inverse-time circuit breaker greater than 100 amps shall be permitted to be increased up to _____ percent of the motor's full-load current, in amps.
    (a) 250   (b) 300
    (c) 400   (d) 600

20. Three-phase squirrel-cage induction motors have three separate windings per pole on the stator that generate magnetic fields that are _____ degrees out of phase with each other.
    (a) 120   (b) 125
    (c) 150   (d) 180

21. Time-delay fuses will hold _____ times their rating, which will permit most induction motors to start and accelerate their driven loads.
    (a) 1   (b) 2
    (c) 3   (d) 5

22. Inverse-time circuit breakers will hold about _____ times their rating for different periods of time based upon their frame size.
    (a) 1   (b) 2
    (c) 3   (d) 5

23. An overcurrent protection device rated at _____ amps or less can protect a 120 volt or less branch circuit supplying motors rated less than 1 horsepower.
    (a) 10   (b) 15
    (c) 20   (d) 30

24. The branch-circuit protective device shall be permitted to serve as the controller for stationary motors where the motor is rated _____ horsepower or less.
    (a) 1/16   (b) 1/8
    (c) 1/4   (d) 1/2

25. A branch-circuit inverse-time circuit breaker rated in amperes shall bc permitted as a _____ disconnecting means.
    (a) motor   (b) horsepower
    (c) controller   (d) stationary

26. The motor controller for a torque motor shall have a continuous duty, FLC rating, in amps, of not less than the _____ current rating of the motor.
    (a) labeling   (b) nameplate
    (c) listing   (d) table

27. A listed motor circuit switch rated in _____ shall be permitted as a disconnecting means.
    (a) horsepower   (b) amperage
    (c) voltage   (d) resistance

28. The disconnecting means shall be permitted to be a general-use or isolating switch for AC motors rated _____ horsepower or greater.
    (a) 40   (b) 60
    (c) 75   (d) 100

29. A motor and its driven machinery or load shall be installed within sight and within _______ of the controller for the motor.
    (a) 25   (b) 50
    (c) 75   (d) 100

30. The ampacity of capacitor circuit conductors shall not be less than _______ percent of the rated current, in amps, of the capacitor.
    (a) 100   (b) 125
    (c) 135   (d) 150

31. When applying the rule-of-thumb method, the horsepower rating of the motor shall be multiplied by _______ to obtain full-load current in amps for 440, 460, and 480 volt, three-phase motors.
    (a) 1.00   (b) 1.10
    (c) 1.25   (d) 5.0

32. When applying the rule-of-thumb method, the horsepower rating of the motor shall be multiplied by _______ to obtain full-load current in amps for 220, 230, and 240 volt, single-phase motors.
    (a) 1.00   (b) 1.10
    (c) 1.25   (d) 5.0

33. Power conversion equipment requires the conductors to be sized at _______ percent of the rated input, in amps, of such equipment.
    (a) 100   (b) 115
    (c) 125   (d) 150

34. For Design B motors, the setting of an instantaneous circuit breaker shall be permitted to be adjusted up to _______ percent to allow the motor to start and run.
    (a) 1000   (b) 1200
    (c) 1300   (d) 1700

35. A nontime-delay fuse will hold _______ times its rating for approximately 1/4 to 2 seconds based upon type used.
    (a) 2   (b) 3
    (c) 5   (d) 10

36. Motors with a marked service factor not less than 1.15 shall have the minimum running overload protection sized at _______ percent.
    (a) 115   (b) 125
    (c) 130   (d) 140

37. Motors with a marked temperature rise not over 40°C shall have the minimum running overload protection sized at _______ percent.
    (a) 115   (b) 125
    (c) 130   (d) 140

38. Motors with a marked service factor not less than 1.15 shall have the maximum running overload protection sized at _______ percent.
    (a) 115   (b) 125
    (c) 130   (d) 140

39. Motors with a marked temperature rise not over 40°C shall have the maximum running overload protection sized at _______ percent.
    (a) 115   (b) 125
    (c) 130   (d) 140

40. An AC general-use snap switch shall be permitted to be installed as the controller for a stationary motor where the full-load current of the switch does not exceed _____ percent of the branch-circuit rating.
    (a) 50   (b) 75
    (c) 80   (d) 90

41. When applying the rule-of-thumb method, the horsepower rating of the motor shall be multiplied by _____ to obtain full-load current in amps for 220, 230, and 240 volt, three-phase motors.
    (a) 100   (b) 125
    (c) 250   (d) 500

42. The overcurrent protection device (variable loads) shall be sized at _____ percent times the phase converter's nameplate single-phase input FLA rating.
    (a) 100   (b) 125
    (c) 250   (d) 500

43. Circuit conductors supplying power conversion equipment included as part of an adjustable-speed drive system shall have an ampacity not less than _____ percent of the rated input to the power conversion equipment.
    (a) 100   (b) 125
    (c) 150   (d) 250

44. The disconnecting means shall be permitted to be in the incoming line to the conversion equipment and shall have a rating not less than _____ percent of the rated input current of the conversion unit.
    (a) 115   (b) 125
    (c) 135   (d) 150

45. The conductors in industrial control panels shall not fill the wiring space at any cross section to more than _____ percent of the cross-sectional area of the space.
    (a) 20   (b) 30
    (c) 40   (d) 75

46. The conductors, splices, and taps in industrial control panels shall not fill the wiring space at any cross section to more than _____ percent of the cross-sectional area of that space.
    (a) 20   (b) 30
    (c) 40   (d) 75

47. Motors rated over 2 horsepower through _____ horsepower shall be permitted to be installed with a separate disconnecting means (general-use switch) if the motor is equipped with an autotransformer-type controller and certain conditions are complied with.
    (a) 50   (b) 60
    (c) 100   (d) 150

48. The disconnecting means shall be permitted to be an isolating switch for DC stationary motors rated at _____ horsepower or greater.
    (a) 40   (b) 50
    (c) 100   (d) 150

49. Conductors supplying a fire pump motor(s), pressure maintenance pumps, and associated fire pump accessory equipment shall have a rating not less than ______ percent of the sum of the fire pump motor(s) and pressure maintenance motor(s) full load currents.
   (a) 100   (b) 125
   (c) 150   (d) 250

50. Conductors that are 16 AWG in size shall be protected at ______ amps when used for remote-control circuits.
   (a) 3   (b) 6
   (c) 7   (d) 10

51. What size THWN branch-circuit copper conductors are required for a 3 HP, 208 volt, single-phase, Design B motor?

52. What size THWN branch-circuit copper conductors are required for a 20 HP, 230 volt, three-phase, Design B motor?

53. What size THWN branch-circuit copper conductors are required for a 75 HP, 460 volt, three-phase, 15 minute-rated intermittent duty cycle motor?

54. What size THWN copper conductors are required to supply power conversion equipment with a rated input of 112 amps?

55. What size THWN branch-circuit copper conductors are required to supply a 50 HP, 208 volt, three-phase, Design B part-winding motor?

56. What size THWN branch-circuit copper conductors are required to supply a 30 HP, 40 HP, and 50 HP, 460 volt, three-phase, Design B motor?

57. What size THWN branch-circuit copper conductors are required to supply an 10 HP, 208 volt, three-phase, 5-minute rated intermittent duty cycle, an 15 HP, 208 volt, three-phase, 15-minute rated intermittent duty cycle motor, and a 20 HP, 208 volt, three-phase motor?

58. What is the LRC (maximum) for a 40 HP, 230 volt, three-phase, code letter G motor?

59. What is the LRC for a 40 HP, 230 volt, three-phase, Design letter B motor?

60. What is the rounded-down and rounded-up size nontime-delay fuse for a 50 HP, 230 volt, three-phase, Design letter B motor?

61. What is the rounded-down and rounded-up size time-delay fuse for a 50 HP, 230 volt, three-phase, Design letter B motor?

62. What is the minimum and maximum setting for an instantaneous trip circuit breaker for a 50 HP, 230 volt, three-phase, Design letter B motor?

63. What is the rounded-down and rounded-up size inverse time circuit breaker for a 50 HP, 230 volt, three-phase, Design letter B motor?

64. What is the maximum size nontime-delay fuse for a 50 HP, 230 volt, three-phase, Design letter B motor?

65. What is the maximum size time-delay fuse for a 50 HP, 230 volt, three-phase, Design letter B motor?

66. What is the maximum size inverse time circuit breaker for a 50 HP, 230 volt, three-phase, Design letter B motor?

67. What size overcurrent protection device (CB) is required for a feeder that supplies a 10 HP, 15 HP, 20 HP, and 25 HP, 460 volt, three-phase, Design letter B group of motors?

68. What size overload protection (minimum) is required for a 20 HP, 460 volt, three-phase, Design letter B motor with a nameplate rating of 48 amps, temperature rise of 40°C, and a service factor of 1.15?

69. What size overload protection (maximum) is required for a 20 HP, 460 volt, three-phase, Design letter B motor with a nameplate rating of 48 amps, temperature rise of 40°C, and a service factor of 1.15?

70. What size nonfused disconnect is required for a 50 HP, 460 volt, three-phase, Design letter B motor?

71. What size overcurrent protection device (maximum) is required for motor control circuit conductors located in the controller and supplied by a 12 AWG conductor?

72. What size overcurrent protection device (maximum) is required for motor control circuit conductors that are run remote and supplied by a 12 AWG conductor?

73. What size overcurrent protection device is required for motor control circuit conductors that are supplied by a 2400 VA, 480 volt, two-wire control transformer?

74. What size THWN copper conductors are required to supply a 20 kVA, 208 volt, three-phase capacitor with a 40 HP, 208 volt, three-phase, Design letter B motor?

75. What size FLA rating is required for a 30 HP, 115 volt, three-phase, Design letter B motor using the rule-of-thumb method?

76. What size FLA rating is required for a 30 HP, 220 volt, three-phase, Design letter B motor using the rule-of-thumb method?

77. What size FLA rating is required for a 30 HP, 440 volt, three-phase, Design letter B motor using the rule-of-thumb method?

78. What size FLA rating is required for a 30 HP, 575 volt, three-phase, Design letter B motor using the rule-of-thumb method?

79. What size overcurrent protection device and THWN copper conductors are required for a phase converter with a nameplate rating of 35 amps supplying a piece of equipment classified as a variable load?

80. What size overcurrent protection device and THWN copper conductors are required for a phase converter supplying a 20 HP, 230 volt, three-phase, Design letter B motor classified as a specific load?

Compressor Motors

Article 440 deals with individual or group installations having hermetically sealed motor-compressors. The techniques for designing the proper size conductors, disconnecting means, and controllers are discussed.

The conductors supplying power to heating, air conditioning, and refrigeration (HACR) equipment are sized from the full-load amp (FLA) ratings of the compressor and condenser motor. These FLA ratings are increased by 125 percent per **440.32** to compensate for the starting periods and overload conditions.

The overcurrent protection devices protecting the branch circuits from short-circuit and ground fault currents are sized from the provisions listed in **440.22(A)**, which require the FLA ratings to be increased from 175 percent up to 225 percent to allow the HACR equipment to start and run without tripping the overcurrent protection device ahead of the circuit.

Note that the elements used to supply the branch circuits to HACR equipment may be required to be selected by the branch circuit selection currents listed on the nameplate of such equipment per **440.4(C)** and **110.3(B)**.

NAMEPLATE LISTING
440.1

The overcurrent protection devices, running overload protection devices, conductors, disconnecting means, and controllers shall be sized and selected by the information provided on the nameplate listing for air conditioning and refrigeration equipment. The information on the nameplate is very important to installers and service personnel; therefore, the nameplate shall never be removed from the air conditioner or refrigeration equipment.

MARKING ON HERMETIC REFRIGERANT MOTOR-COMPRESSORS AND EQUIPMENT
440.4

Hermetic refrigerant motor-compressors shall be provided with a marking on the nameplate giving the manufacturer's name, trademark, or symbol and designating the identification, number of phases, voltage, and frequency. The information provided on the nameplate of the hermetic refrigerant motor-compressor is used to determine the ratings of branch-circuit conductors, ground fault protection, short circuits, disconnecting means, controllers, and other elements of the electrical system.

MARKING ON CONTROLLERS
440.5

Controllers shall be marked with information that lists the manufacturer's name, trademark or symbol, identifying voltage, phases, full-load current, locked-rotor current rating, or horsepower.

AMPACITY AND RATING
440.6

The full-load current rating listed on the nameplate of the motor-compressor shall be used to determine the branch-circuit conductor rating, short circuit protection rating, motor overload protection rating, controller rating, or disconnecting means rating. The branch-circuit selection current (if greater) shall be applied if shown instead of the full-load current rating. The full-load current rating, in amps, shall be used to determine the motor's overload protection rating. The full-load current rating listed on the compressor nameplate shall be used when the nameplate for the equipment does not list a full-load current rating based upon the branch-circuit selection current. (**See Figure 19-1**)

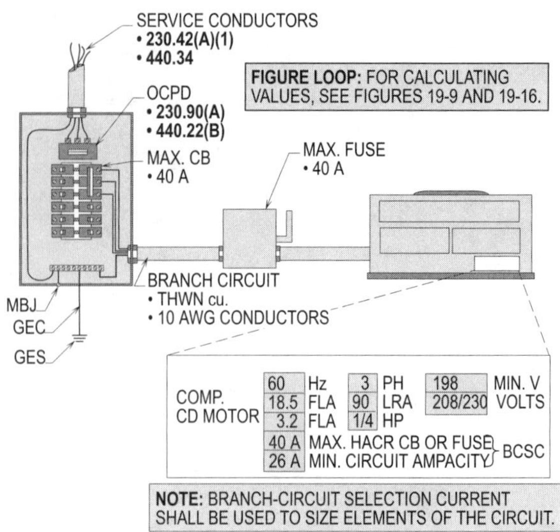

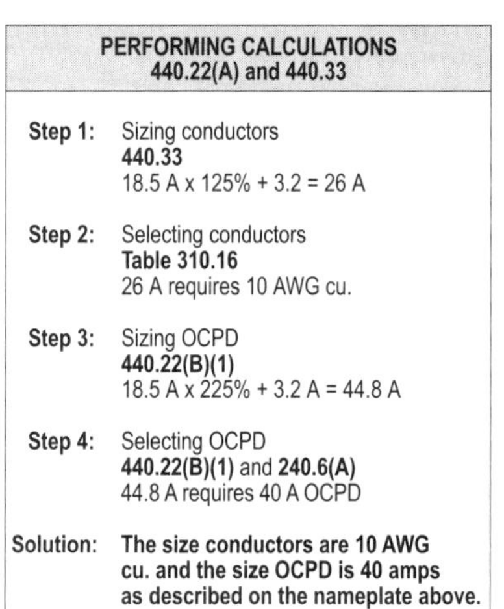

AMPACITY AND RATING
NEC 440.6

Figure 19-1. If the branch-circuit selection current (BCSC) on the nameplate calls for a certain circuit size and overcurrent protection device size, this rating shall be used instead of actually calculating such values and sizes per **440.22(A)** and **440.32**.

HIGHEST RATED (LARGEST) MOTOR
440.7

When sizing the conductors for a feeder supplying A/C units and motors per **430.24**, the full-load current in amps of the largest motor is multiplied by 125 percent. The full-load current ratings of the remaining motors are added to this total to derive the FLA. **See Figures 18-9** and **18-27** for a specific illustration pertaining to this rule.

When sizing the overcurrent protection device for two or motors per **430.62(A)**, the full-load current (in amps) of the largest motor is multiplied by the percentages listed in **Table 430.52**. The full-load current ratings (in amps) of the remaining motors are added to this total to derive the FLA. (Also, see **430.63.**)

The full-load current ratings listed on the nameplate of the motor-compressor shall be used to determine the size conductors and overcurrent protection device using the same procedure. The larger of the two is used.

See Figures 19-2(a) and (b) for a feeder supplying motors and A/C units. Note that the A/C unit is the largest motor and not one of the motors in the group.

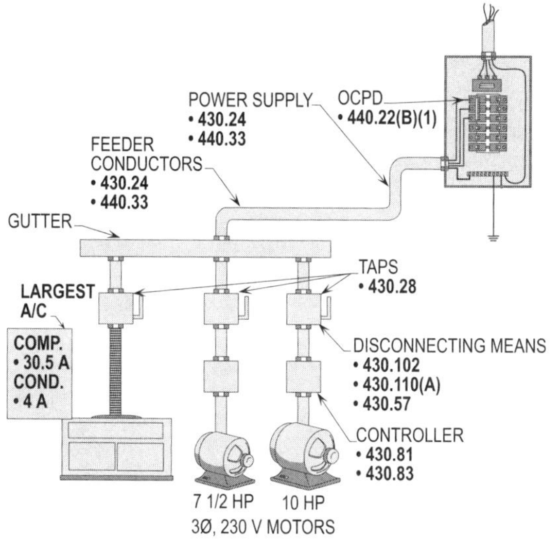

**HIGHEST RATED (LARGEST) MOTOR
NEC 440.7**

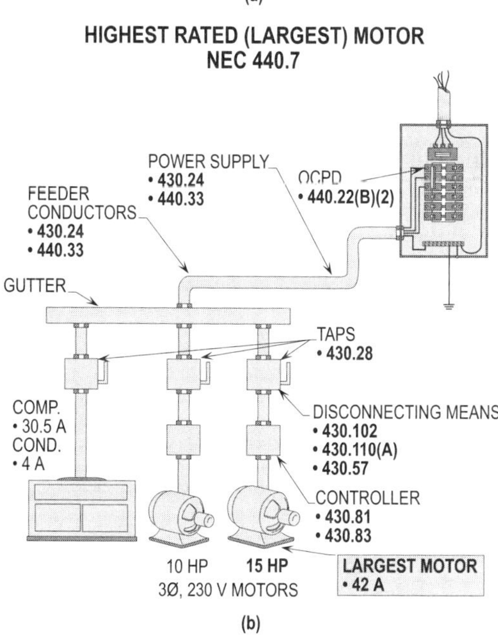

**HIGHEST RATED (LARGEST) MOTOR
NEC 440.7**

Figure 19-2(a) and (b). The above shows the calculation procedure for sizing conductors and overcurrent protection device where the A/C unit or motor is the largest in the group of A/C units and motors.

SINGLE MACHINE
440.8

Each motor controller shall be provided with an disconnecting means. Air conditioning and refrigeration systems are considered a single machine even through they consist of any number of motors. The number of disconnecting means to be provided are determined by applying **430.87, Ex. 1** and **430.112, Ex**.

DISCONNECTING MEANS
440.11

The full-load current rating of the nameplate or the nameplate branch-circuit selection current of the compressor, whichever is greater, shall be used to size the branch-circuit conductors and the disconnecting means to disconnect air conditioner and refrigeration equipment.

RATING AND INTERRUPTING CAPACITY
440.12

The full-load current rating of the nameplate or the nameplate branch-circuit selection current of the compressor, whichever is greater, shall be sized at 115 percent to size the disconnecting means. A horsepower rated switch, circuit breaker, or other switches shall be permitted to be used as the disconnecting means per **430.109** and **430.110(A)**. **(See Figure 19-3)**

> **Design Tip:** A minimum load is derived when applying 115 percent for sizing the disconnecting means. Therefore, on larger units the 115 percent may not be of sufficient ampacity for opening the circuit under load.

The horsepower amperage rating shall be selected from **Tables 430.247 through Table 430.250** when corresponding to the nameplate rating or branch circuit selection current of the motor-compressor or equipment when listed in amperage and not horsepower. The horsepower amperage rating for locked-rotor current shall be selected from **Tables 430.251(A)** and **(B)** when the nameplate fails to list the locked-rotor current. Note that the disconnecting means shall be sized with enough capacity in horsepower that is capable of disconnecting the total locked-rotor current. **(See Figure 19-4)**

The full-load current rating (in amps) of the nameplate shall be permitted to be used to size a circuit breaker at 115 percent or more to disconnect a hermetically sealed motor from the power circuit. **(See Figure 19-5)**

Design Tip: The circuit breaker shall be sized at 115 percent or more of the branch-circuit selection current if it is greater in rating, so as to be capable of disconnecting the circuit safely.

Two or more hermetic motors or combination loads such as hermetic motor loads, standard motor loads, and other loads shall have their separate values totaled to determine the rating of a single disconnecting means. This total rating shall be sized at 115 percent to determine the size disconnecting means required to disconnect the circuits and elements in a safe and reliable manner. (**See Figure 19-6**)

CORD-CONNECTED EQUIPMENT 440.13

Cord-and-plug connected equipment such as room air conditioners, home refrigerators and freezers, drinking water coolers, and beverage dispensers, a separable connector or an attachment plug and receptacle shall be permitted to be used to serve as a disconnecting means. (**See Figure 19-7**)

Design Tip: In some cases, room air conditioners shall not be permitted to have a cord-and-plug connection to serve as their disconnecting means, as when unit switches for manual control are installed in air conditioners mounted over 6 ft (1.8 m) above finished grade.

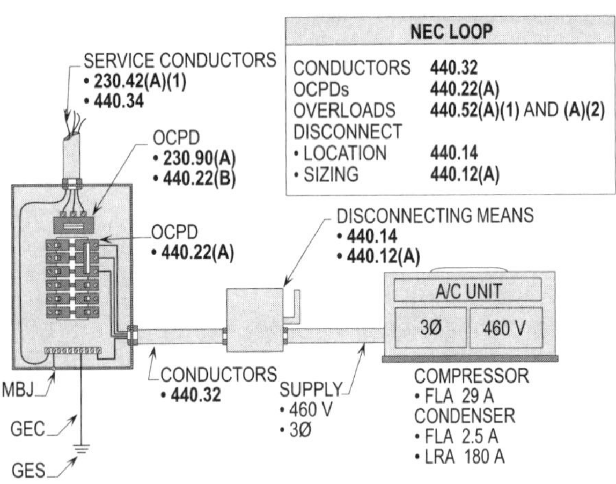

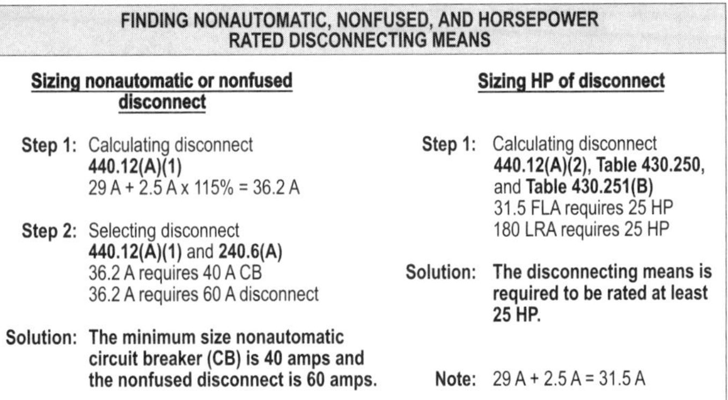

FINDING NONAUTOMATIC, NONFUSED, AND HORSEPOWER RATED DISCONNECTING MEANS	
Sizing nonautomatic or nonfused disconnect	**Sizing HP of disconnect**
Step 1: Calculating disconnect **440.12(A)(1)** 29 A + 2.5 A x 115% = 36.2 A	**Step 1:** Calculating disconnect **440.12(A)(2), Table 430.250,** and **Table 430.251(B)** 31.5 FLA requires 25 HP 180 LRA requires 25 HP
Step 2: Selecting disconnect **440.12(A)(1)** and **240.6(A)** 36.2 A requires 40 A CB 36.2 A requires 60 A disconnect	**Solution:** **The disconnecting means is required to be rated at least 25 HP.**
Solution: **The minimum size nonautomatic circuit breaker (CB) is 40 amps and the nonfused disconnect is 60 amps.**	**Note:** 29 A + 2.5 A = 31.5 A

RATING AND INTERRUPTING CAPACITY
NEC 440.12(A)

Figure 19-3. The full-load current rating (in amps) of the nameplate or the nameplate's branch-circuit selection current of the compressor, whichever is greater, shall be sized at 115 percent to size the disconnecting means.

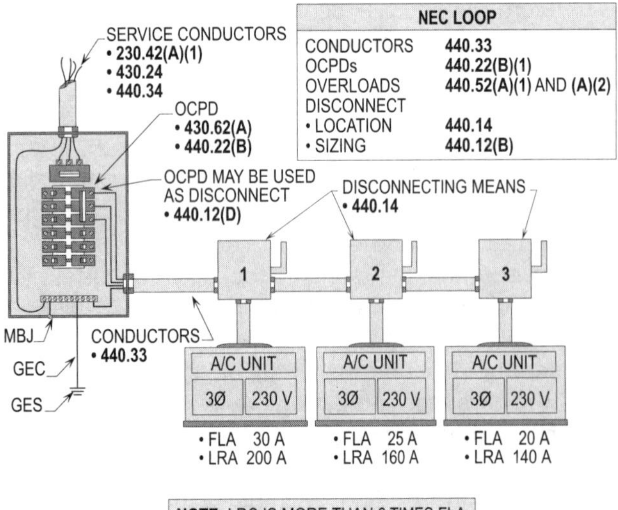

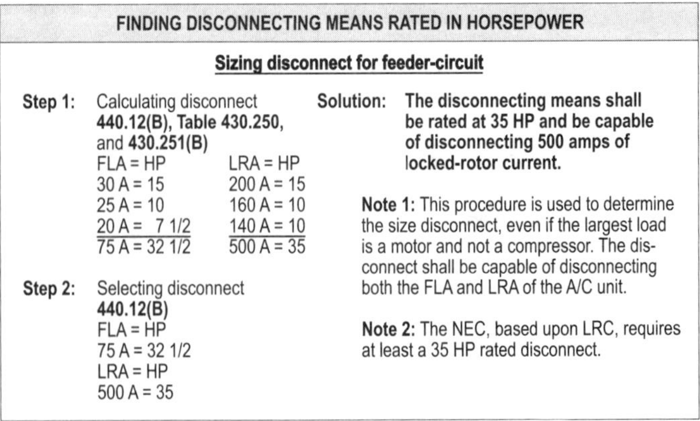

FINDING DISCONNECTING MEANS RATED IN HORSEPOWER	
Sizing disconnect for feeder-circuit	
Step 1: Calculating disconnect **440.12(B), Table 430.250,** and **430.251(B)** FLA = HP LRA = HP 30 A = 15 200 A = 15 25 A = 10 160 A = 10 20 A = 7 1/2 140 A = 10 75 A = 32 1/2 500 A = 35	**Solution:** **The disconnecting means shall be rated at 35 HP and be capable of disconnecting 500 amps of locked-rotor current.**
Step 2: Selecting disconnect **440.12(B)** FLA = HP 75 A = 32 1/2 LRA = HP 500 A = 35	**Note 1:** This procedure is used to determine the size disconnect, even if the largest load is a motor and not a compressor. The disconnect shall be capable of disconnecting both the FLA and LRA of the A/C unit. **Note 2:** The NEC, based upon LRC, requires at least a 35 HP rated disconnect.

RATING AND INTERRUPTING CAPACITY
NEC 440.12(B)

Figure 19-4. Sizing horsepower rating to select disconnecting means based upon locked-rotor current.

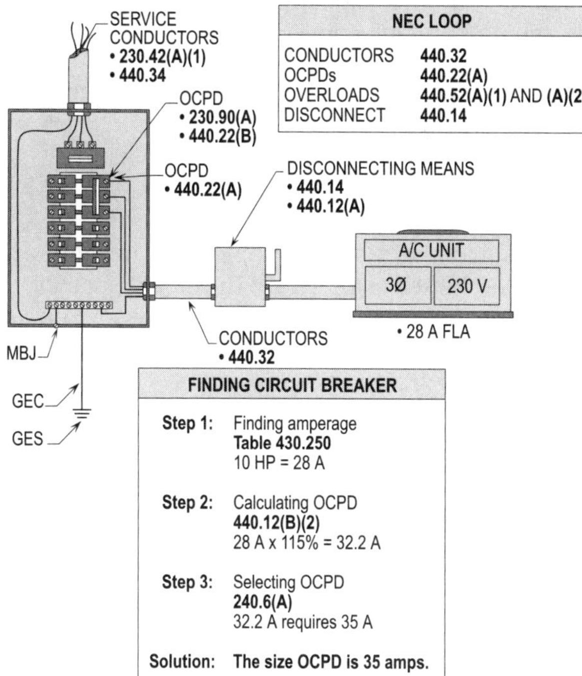

RATING AND INTERRUPTING CAPACITY
NEC 440.12(B)(1)
NEC TABLE 430.250

Figure 19-5. The full-load current rating of the nameplate shall be permitted to be used to size a circuit breaker at 115 percent or more to disconnect a hermetically sealed motor from the power circuit.

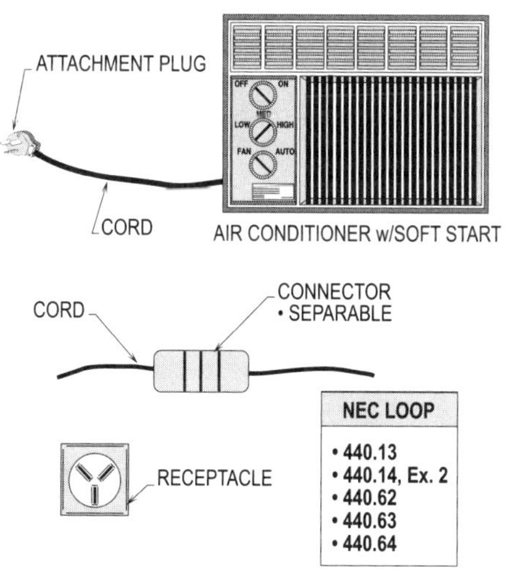

CORD-CONNECTED EQUIPMENT
NEC 440.13

Figure 19-7. Cord-and-plug connected equipment such as room air conditioners, home refrigerators and freezers, drinking water coolers, and beverage dispensers shall be permitted to be disconnected by a cord and receptacle. A separable connector or an attachment plug and receptacle shall be permitted to be used to serve as such disconnecting means.

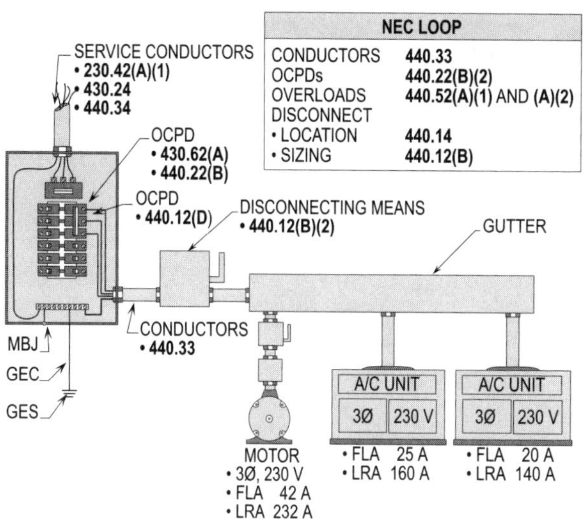

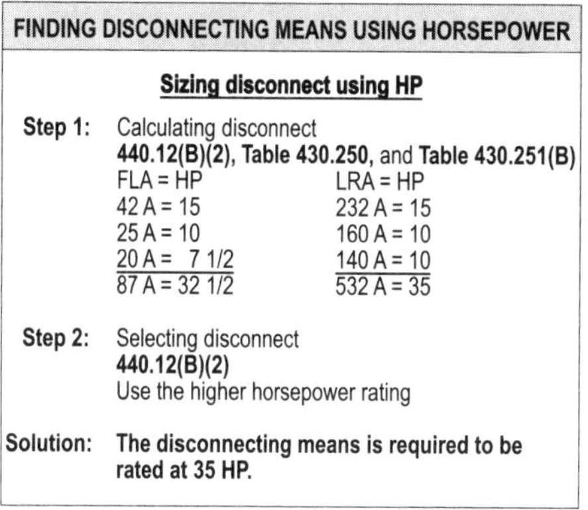

RATING AND INTERRUPTING CAPACITY
NEC 440.12(B)(2)

Figure 19-6. Two or more hermetic motors or combination loads, such as hermetic motor loads, standard motor loads, and other loads, shall have their separate values totaled to determine the rating of a single disconnecting means.

LOCATION
440.14

The disconnecting means for air conditioning or refrigeration equipment shall be located within sight and within 50 ft (15 m) and shall be readily accessible to the user. An additional circuit breaker or disconnecting switch shall be provided at the equipment if the air conditioning or refrigeration equipment is not within sight or within 50 ft (15 m). The disconnecting means shall be permitted to be installed within or on the air conditioning or refrigeration equipment. For the use of unit switches located in air conditioning units, review **422.34** per the AHJ. **(See Figure 19-8)**

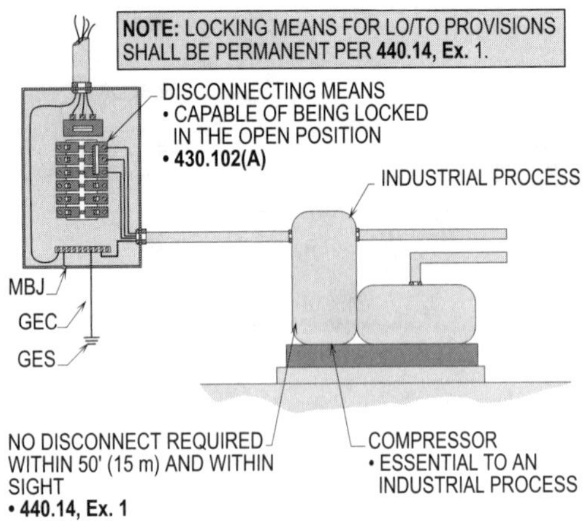

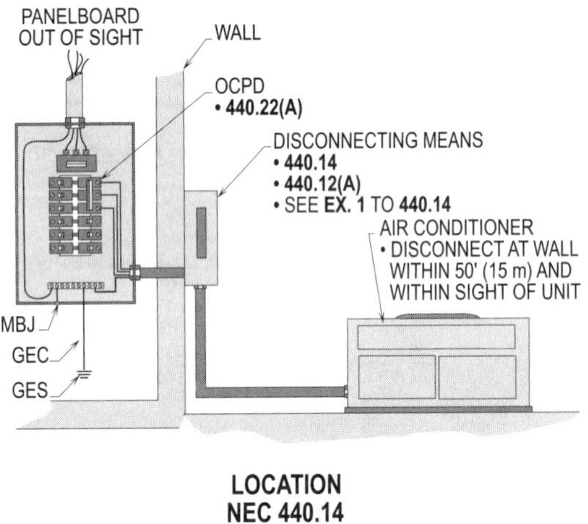

Figure 19-8. The disconnecting means for air conditioning or refrigeration equipment shall be located within sight and within 50 ft (15 m) and shall be readily accessible to the user. An additional circuit breaker or disconnecting switch shall be provided at the equipment if the air conditioning or refrigeration equipment is not within sight or within 50 ft (15 m).

APPLICATION AND SELECTION
440.22

The branch-circuit fuse or circuit breaker ratings for hermetically sealed motors shall be sized with enough capacity to allow the motor to start and develop speed without tripping open the overcurrent protection device due to the momentary inrush current of the compressor and other elements. Maximum protection is always provided by the ratings and settings of the overcurrent protection device being sized with values as low as possible. Hermetic refrigerant motor-compressors shall be protected by properly sizing and selecting the ratings and settings of the overcurrent protection devices to protect the branch-circuit

conductors and other elements in the circuit from short-circuit and ground-fault conditions.

RATING AND SETTING FOR INDIVIDUAL MOTOR-COMPRESSORS
440.22(A)

The overcurrent protection device for hermetic seal compressors shall be selected at 175 percent (for minimum) or 225 percent (for maximum) of the compressor FLA rating or the branch-circuit selection circuit current, whichever is greater. **(See Figure 19-9)**

Overcurrent protection devices for hermetically sealed compressors shall be permitted to be selected up to 225 percent to permit the motor to start if the compressor will not start and develop speed when the rating is 175 percent or less.

> **Design Tip:** A normal circuit breaker shall not be installed when the equipment is marked for a particular fuse size or HACR circuit breaker rating. The branch-circuit conductors shall be protected only by that specified fuse size or HACR circuit breaker rating.

RATING OR SETTING FOR EQUIPMENT
440.22(B)

When sizing the overcurrent protection device, the rating or setting shall be selected and comply with the number of hermetic motors, or combination of hermetic motors, and standard motors installed on a circuit.

SIZING OVERCURRENT PROTECTION DEVICE FOR TWO OR MORE HERMETIC MOTORS
440.22(B)(1)

The overcurrent protection device for a feeder supplying two or more air conditioning or refrigerating units shall be sized to allow the largest unit to start and allow the other units to start at different intervals of time. The full-load current rating, in amps, of the nameplate or the branch-circuit selection current rating of the largest motor, whichever is greater, shall be sized at 175 percent if there are two or more hermetically sealed motors installed on the same feeder. **(See Figure 19-10)**

Overcurrent protection devices for hermetically sealed motors shall be permitted to be selected up to 225 percent to allow the motor to start if the motor will not start and develop speed when the rating is selected at 175 percent or less. **(See Figure 19-11)**

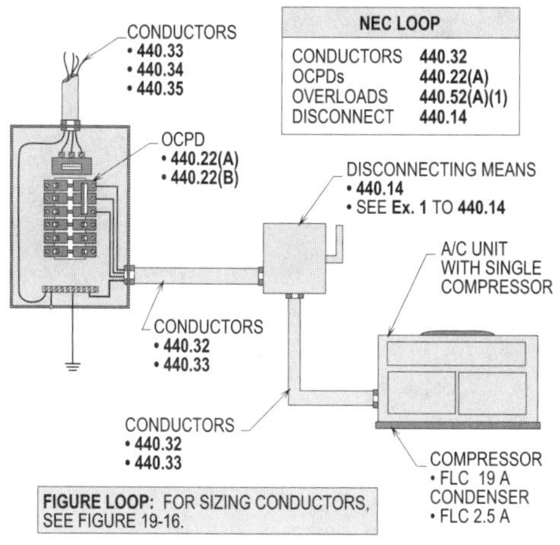

FINDING MINIMUM AND MAXIMUM SIZE OCPD OF THE A/C UNIT

Sizing minimum OCPD	Sizing maximum size OCPD
Step 1: Calculating OCPD **440.22(A)** 19 A x 175% + 2.5 A = 35.75 A	**Step 1:** Calculating OCPD **440.22(A)** 19 A x 225% + 2.5 A = 45.25 A
Step 2: Selecting OCPD **240.4(G) and 240.6(A)** 35.75 A requires 35 A	**Step 2:** Selecting OCPD **240.4(G) and 240.6(A)** 45.25 A requires 45 A
Solution: The minimum size OCPD is 35 amps.	**Solution:** The maximum size OCPD is 45 amps.

RATING AND SETTING FOR INDIVIDUAL MOTOR-COMPRESSORS
NEC 440.22(A)

Figure 19-9. The overcurrent protection device for hermetically sealed compressors shall be selected at 175 percent (for minimum) or 225 percent (for maximum) of the compressor's FLA rating or the branch-circuit selection circuit current, whichever is greater. Note that a smaller size overcurrent protection device shall be permitted to be used than selected per solution, if A/C unit will start and run.

SIZING OVERCURRENT PROTECTION DEVICE FOR HERMETIC MOTOR AND OTHER LOADS WHEN A HERMETICALLY SEALED MOTOR IS THE LARGEST
440.22(B)(1)

When installing hermetically sealed motors and other loads such as motors on the same circuit, and the largest motor of the group is hermetic, the same procedure used for two or more hermetic motors on a feeder shall be used to size the overcurrent protection device. The full-load current rating of the nameplate or the branch-circuit selection current rating (in amps) of the largest hermetic motor, whichever is greater, shall be sized at 175 percent, and the sum of the full-load current ratings, in amps, of the other motors added to this largest hermetic motor load.

SIZING OVERCURRENT PROTECTION DEVICE FOR HERMETIC MOTORS AND OTHER LOADS WHEN A MOTOR IS THE LARGEST
440.22(B)(2)

When installing hermetically sealed motors and other loads such as motors on the same circuit, and the largest in the group is a motor, the overcurrent protection device is sized and selected based on the percentages from **Table 430.52**. The maximum branch-circuit overcurrent protection device shall be used when the standard motor is the largest of the group, and the sum of the full-load current ratings of the remaining hermetically sealed motor and other motors of the group added to the largest motor. The next lower standard size overcurrent protection device below this total sum shall be installed per **240.6(A)**. (**See Figure 19-12**)

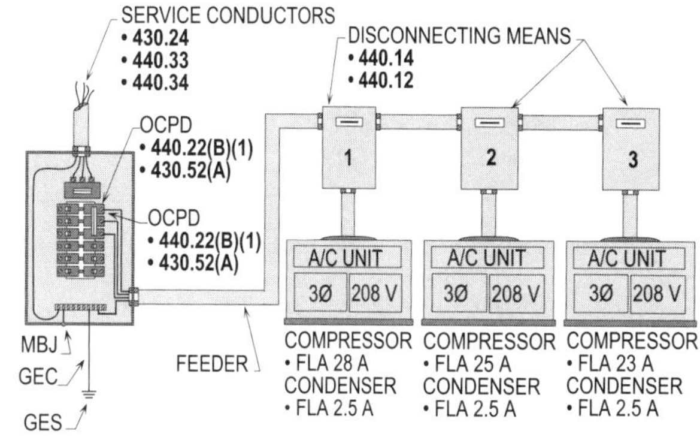

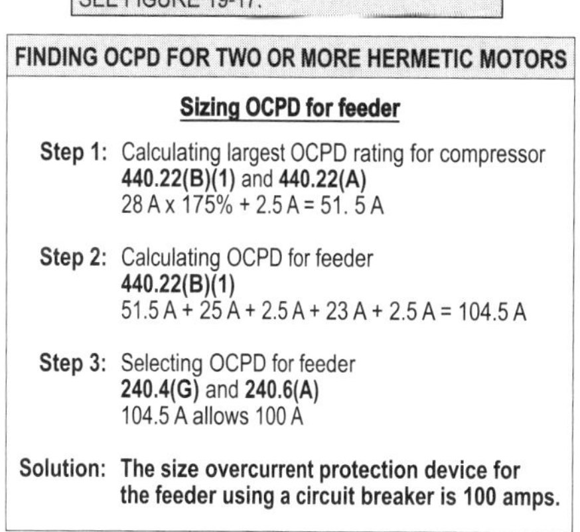

FINDING OCPD FOR TWO OR MORE HERMETIC MOTORS

Sizing OCPD for feeder

Step 1: Calculating largest OCPD rating for compressor
440.22(B)(1) and 440.22(A)
28 A x 175% + 2.5 A = 51. 5 A

Step 2: Calculating OCPD for feeder
440.22(B)(1)
51.5 A + 25 A + 2.5 A + 23 A + 2.5 A = 104.5 A

Step 3: Selecting OCPD for feeder
240.4(G) and 240.6(A)
104.5 A allows 100 A

Solution: The size overcurrent protection device for the feeder using a circuit breaker is 100 amps.

SIZING OVERCURRENT PROTECTION DEVICE FOR TWO OR MORE HERMETIC MOTORS
NEC 440.22(B)(1)

Figure 19-10. The full-load current rating (in amps) of the nameplate or the branch-circuit selection current rating of the largest motor, whichever is greater, shall be sized at 175 percent if there are two or more hermetically sealed motors installed on the same feeder.

Design Tip: The next larger standard size is not permitted to be installed, for there is not an exception to allow the next higher size per **440.22(B)(2)** or **430.62(A)**.

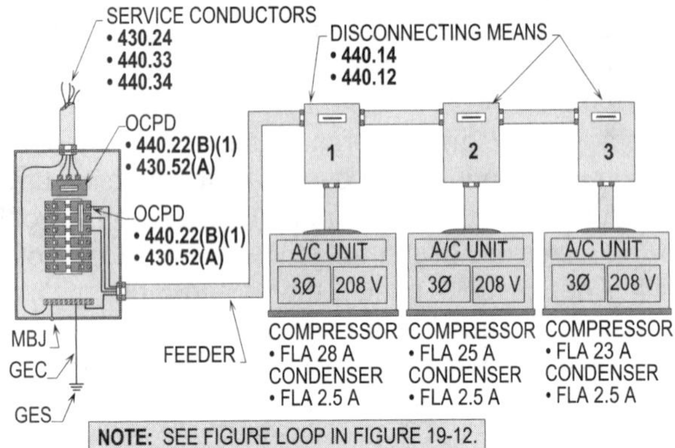

FINDING OCPD FOR TWO OR MORE HERMETIC MOTORS

Sizing OCPD for feeder

Step 1: Calculating largest OCPD rating for compressor
440.22(B)(1) and **440.22(A)**
28 A x 225% + 2.5 A = 65.5 A

Step 2: Calculating OCPD for feeder
440.22(B)(1)
65.5 A + 25 A + 2.5 A + 23 A + 2.5 A = 118.5 A

Step 3: Selecting OCPD for feeder
240.4(G) and **240.6(A)**
118.5 A allows 110 A

Solution: The size overcurrent protection device for the feeder using a circuit breaker is **110 amps.**

SIZING OVERCURRENT PROTECTION DEVICE FOR
TWO OR MORE HERMETIC MOTORS
NEC 440.22(B)(1)

Figure 19-11. Overcurrent protection devices for hermetically sealed motors shall be permitted to be selected up to 225 percent to allow the motor to start if the motor will not start and develop speed.

USING A 15 OR 20 AMP OVERCURRENT PROTECTION DEVICE
440.22(B)(2), Ex. 1

Where the equipment will start, run, and operate on a 15 or 20 amp, 120 volt, single-phase branch circuit, or a 15 amp, 208 volt or 240 volt, a single-phase branch circuit, with a 15 or 20 amp overcurrent protection device, such device shall be permitted to be used to protect the branch circuit. However, the values of the overcurrent protection device in the branch circuit shall not exceed the values marked on the nameplate of the equipment. **(See Figure 19-13)**

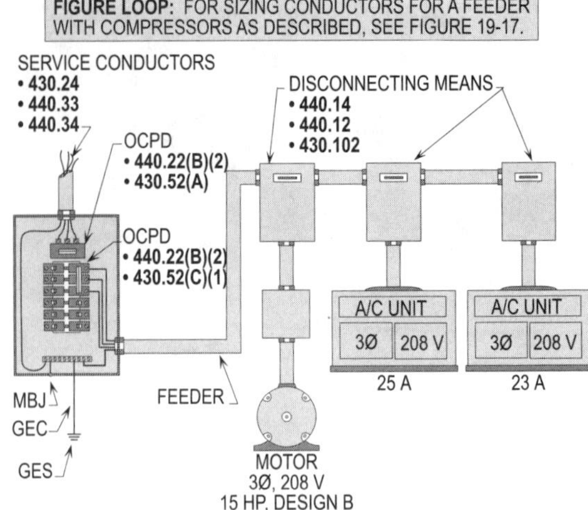

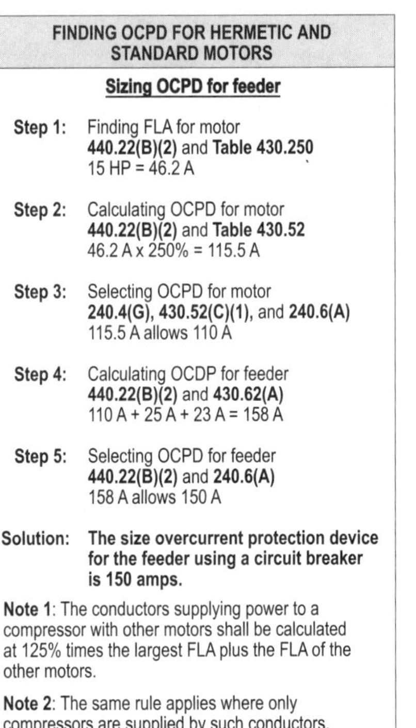

FINDING OCPD FOR HERMETIC AND STANDARD MOTORS

Sizing OCPD for feeder

Step 1: Finding FLA for motor
440.22(B)(2) and **Table 430.250**
15 HP = 46.2 A

Step 2: Calculating OCPD for motor
440.22(B)(2) and **Table 430.52**
46.2 A x 250% = 115.5 A

Step 3: Selecting OCPD for motor
240.4(G), **430.52(C)(1)**, and **240.6(A)**
115.5 A allows 110 A

Step 4: Calculating OCDP for feeder
440.22(B)(2) and **430.62(A)**
110 A + 25 A + 23 A = 158 A

Step 5: Selecting OCPD for feeder
440.22(B)(2) and **240.6(A)**
158 A allows 150 A

Solution: The size overcurrent protection device for the feeder using a circuit breaker is 150 amps.

Note 1: The conductors supplying power to a compressor with other motors shall be calculated at 125% times the largest FLA plus the FLA of the other motors.

Note 2: The same rule applies where only compressors are supplied by such conductors.

SIZING OVERCURRENT PROTECTION DEVICE FOR
HERMETIC MOTORS AND OTHER LOADS WHEN A
MOTOR IS THE LARGEST
NEC 440.22(B)(2)

Figure 19-12. When hermetically sealed motors and other loads, such as motors, are being installed on the same circuit, and the largest in the group is a motor, the overcurrent protection device shall be sized and selected based on the percentages from **Table 430.52**.

USING A CORD-AND-PLUG CONNECTION NOT OVER 250 VOLTS
440.22(B)(2), Ex. 2

The rating of the overcurrent protection device shall be determined by using the rating of the nameplate of the cord-and-plug connected equipment serving single-phase, 250 volts or less, hermetically sealed motors. **(See Figure 19-14)**

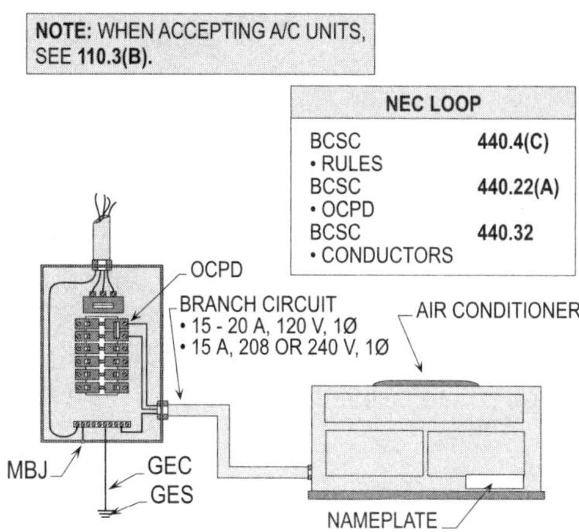

USING A 15 OR 20 AMP OVERCURRENT PROTECTION DEVICE
NEC 440.22(B)(2), Ex. 1

Figure 19-13. Where the equipment will start, run, and operate on a 15 or 20 amp, 120 volt, single-phase branch circuit, or a 15 amp, 208 volt or 240 volt, a single-phase branch circuit, a 15 or 20 amp overcurrent protection device shall be permitted to be used to protect the branch circuit.

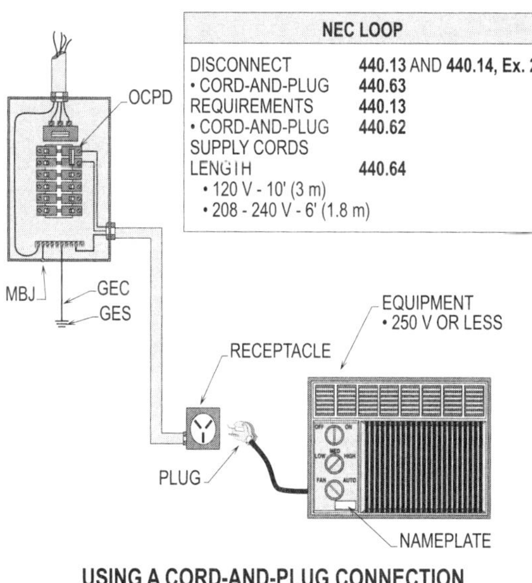

USING A CORD-AND-PLUG CONNECTION NOT OVER 250 VOLTS
NEC 440.22(B)(2), Ex. 2

Figure 19-14. The rating of the overcurrent protection device shall be determined by using the rating of the nameplate of the cord-and-plug connected equipment having single-phase, 250 volt or less, hermetically sealed motors.

PROTECTIVE DEVICE RATING NOT TO EXCEED THE MANUFACTURER'S VALUES
440.22(C)

The manufacturer's values marked on the equipment shall not be exceeded by the overcurrent protection device rating, where the maximum overcurrent protective device ratings on the manufacturer's heater table for use with a motor controller are less than the rating or setting per **440.22(A)** and **(B)**. **(See Figure 19-15)**

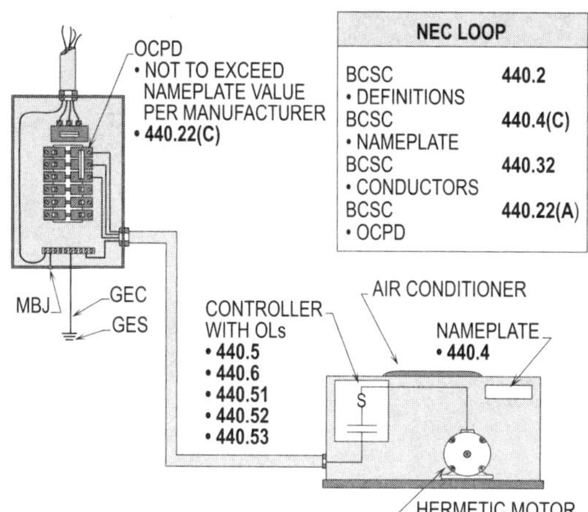

PROTECTIVE DEVICE RATING NOT TO EXCEED THE MANUFACTURER'S VALUES
NEC 440.22(C)

Figure 19-15. The manufacturer's values marked on the equipment shall not be exceeded by the overcurrent protection device rating where the maximum overcurrent protective device ratings on the manufacturer's heater table for use with a motor controller are less than the rating or setting per **440.22(A)** and **(B)**.

BRANCH-CIRCUIT CONDUCTORS
440.31

In general, to prevent conductors and motor elements of the branch circuit from overheating, the conductors shall be sized with enough capacity to allow a hermetic motor to start and run. To ensure adequate sizing, a derating factor of 80 percent shall be applied to the branch-circuit conductors or such conductors shall be sized at 125 percent of the load.

SINGLE MOTOR-COMPRESSORS 440.32

The conductors supplying power to an air conditioning or refrigerating unit shall be sized to carry the load of the unit plus an overload for a period of time that won't damage the elements. The full-load current rating of the nameplate or branch-circuit selection current, whichever is greater, shall be sized at 125 percent to size and select the conductors supplying hermetically sealed motors. **(See Figure 19-16)**

TWO OR MORE MOTOR-COMPRESSORS 440.33

Two or more compressors plus other motor loads can be connected to a feeder. The largest compressor shall be calculated at 125 percent of its FLA, and the remaining compressor loads are added to this total at 100 percent of their FLA ratings. For units with a branch-circuit selection current, the circuit conductors shall be selected and based upon the nameplate values. **(See Figure 19-17)**

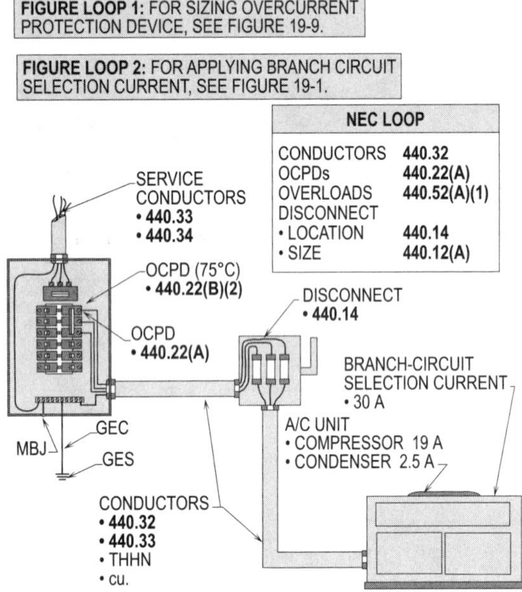

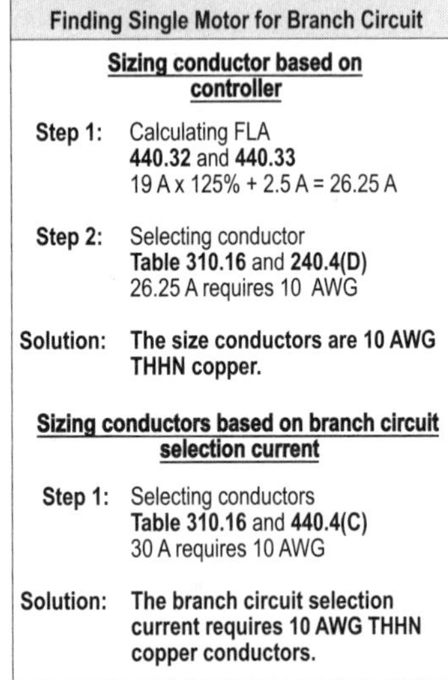

SINGLE MOTOR-COMPRESSORS NEC 440.32

Figure 19-16. The full-load current rating of the nameplate or branch-circuit selection current, whichever is greater, shall be sized at 125 percent to size and select the conductors supplying hermetically sealed motors.

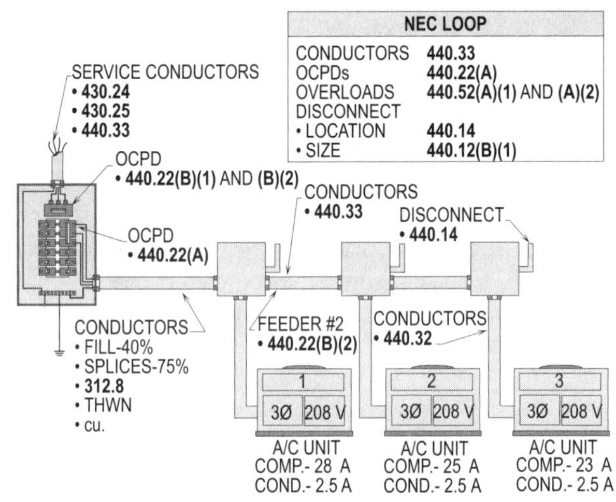

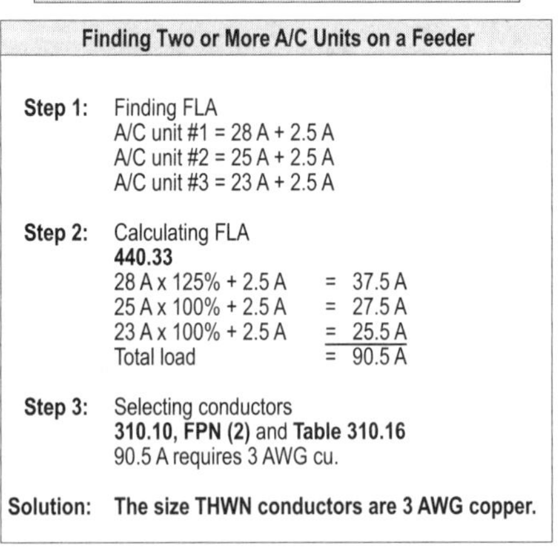

TWO OR MORE MOTOR-COMPRESSORS NEC 440.33

Figure 19-17. Two or more compressors plus other motor loads can be connected to a feeder. The largest compressor shall be calculated at 125 percent of its FLA and the remaining compressor loads are added to this total at 100 percent of their FLA ratings.

COMBINATION LOAD
440.34

Two or more motor-compressors with motor loads plus other loads may be connected to a feeder or service conductors. The largest compressor or motor load shall be calculated at 125 percent plus 100 percent of the remaining compressors and motors. The other loads shall be calculated at 125 percent for continuous and 100 percent for noncontinuous operation per **215.2(A)(1)**, and these total values used to select conductors. **(See Figure 19-18)**

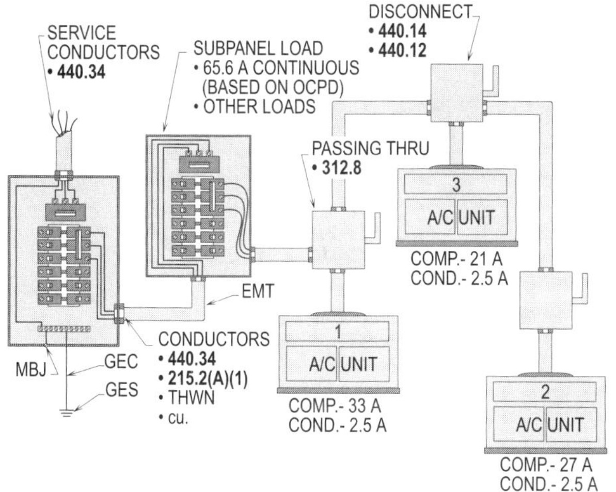

Finding Two or More A/C Units with Other Loads	
Step 1:	Finding FLA **440.34** A/C unit #1 = 33 A + 2.5 A A/C unit #2 = 27 A + 2.5 A A/C unit #3 = 21 A + 2.5 A Other loads = 65.6 A
Step 2:	Calculating FLA **440.34** and **215.2(A)(1)** 33 A x 125% + 2.5 A = 43.75 A 27 A x 100% + 2.5 A = 29.5 A 21 A x 100% + 2.5 A = 23.5 A 65.6 A x 125% = 82 A Total load = 178.75 A
Step 3:	Selecting conductors **310.10, FPN (2)** and **Table 310.16** 178.75 A requires 3/0 AWG cu.
Solution:	**The size THWN conductors for the feeder are 3/0 AWG copper.**
Note:	The load of 178.75 amps requires the conductors and overcurrent protection device to be rated at 200 amps, respectively

COMBINATION LOAD
NEC 440.34

Figure 19-18. Two or more motor-compressors with motor loads plus other loads may be connected to a feeder or service conductors. The largest compressor or motor load shall be calculated at 125 percent, plus 100 percent of the remaining compressors and motors, plus the other loads.

MULTIMOTOR AND COMBINATION LOAD EQUIPMENT
440.35

The marking on the nameplate shall be used when sizing the branch circuit conductors for multimotor and combination load equipment. The conductors shall be installed to have a rating equal to the nameplate rating. Each individual motor or load contained in the unit shall not be required to be calculated individually to size and select the conductors.

CONTROLLERS FOR MOTOR-COMPRESSORS
440.41

When installing the wiring for a motor controller, the circuit supply conductors are run from a motor controller and connected to the terminals of the compressor. The full-load current rating and the locked-rotor current rating of the compressor motor shall be sized for continuous operation.

MOTOR-COMPRESSOR CONTROLLER RATING
440.41(A)

The full-load current rating of the nameplate or the branch circuit selection current ratings, whichever is greater, shall be used to size and select the motor controller. If necessary, the locked-rotor current rating of the motor shall be permitted to be used to size and select the motor controller. **(See Figure 19-19)**

> **Design Tip:** The motor controller shall be sized and selected using the same procedure as used for the sizing of the disconnecting means.

MOTOR-COMPRESSOR AND BRANCH-CIRCUIT OVERLOAD PROTECTION
440.51

The overload (OL) protection for compressors may be accomplished by using overcurrent protection devices in separate enclosures, separate overload relays, or thermal protectors that are an integral part of the compressor.

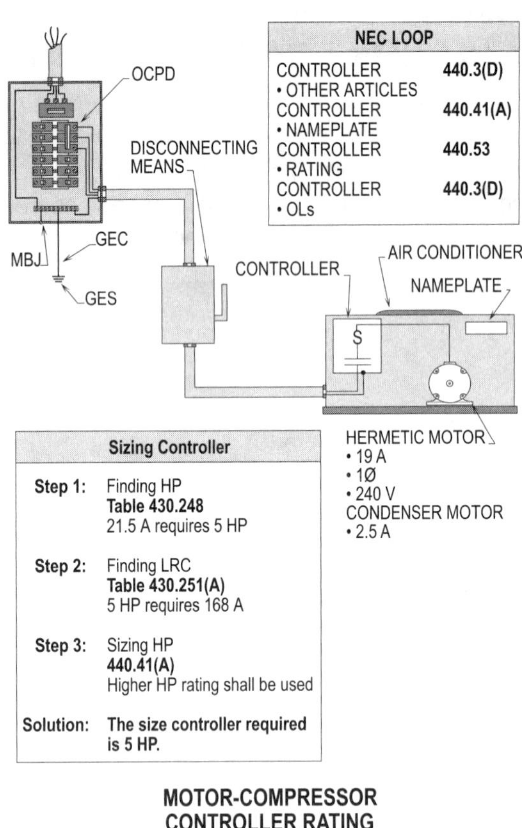

Figure 19-19. The full-load current rating of the nameplate or the branch-circuit selection current ratings, whichever is greater, shall be used to size and select the motor controller.

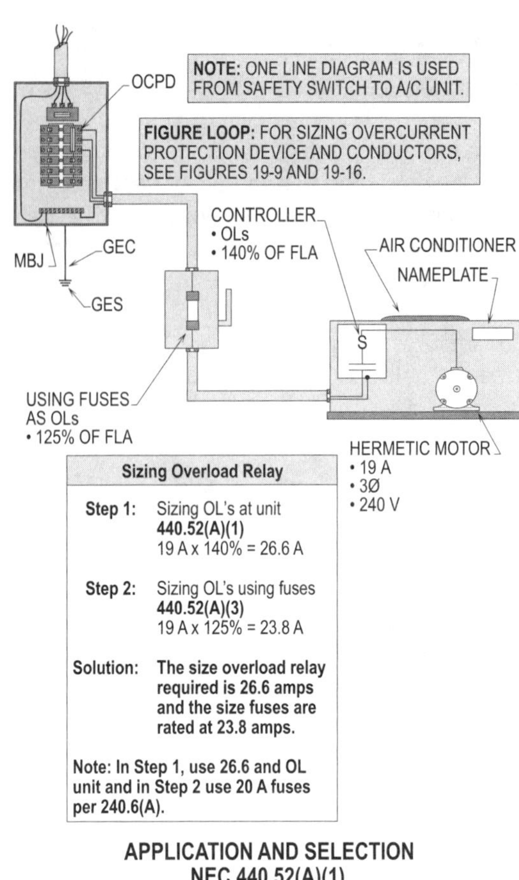

Figure 19-20. The overload relay for the motor-compressor shall trip at not more than 140 percent of the full-load current rating. If a fuse or circuit breaker is used for the protection of the motor-compressor, it shall trip at not more than 125 percent of the full-load current rating.

APPLICATION AND SELECTION 440.52

The overload relay for the motor-compressor shall trip at not more than 140 percent of the full-load current rating. If a fuse or circuit breaker is used for the protection of the motor-compressor, it shall trip at not more than 125 percent of the full-load current rating. (**See Figure 19-20**)

OVERLOAD RELAYS 440.53

Short circuit and ground fault protection is not provided by overload relays and thermal protectors. Overload relays and thermal protectors respond to any type of heat buildup and open with a delay action that will not operate instantly, even on short circuits or ground faults. The branch circuit overcurrent protection device for the circuit shall operate and clear the circuit under short circuit and ground fault conditions.

MOTOR-COMPRESSORS AND EQUIPMENT ON A 15 OR 20 AMP BRANCH CIRCUIT NOT CORD-AND-PLUG CONNECTED 440.54

Overload protection shall be provided for direct or fixed-wired motor-compressors and equipment that is connected to 15 or 20 amp, 120 volt, single-phase branch circuits. Note that 15 amps are used for 240 volt, single-phase branch circuits.

The full-load current rating of the hermetically sealed motor shall be selected at 140 percent when sizing separate overload relays. Hermetic motors shall be provided with fuses or circuit breakers that provide sufficient time delay to allow the motor to come up to running speed without tripping open the circuit due to the high inrush current. (**See Figure 19-21**)

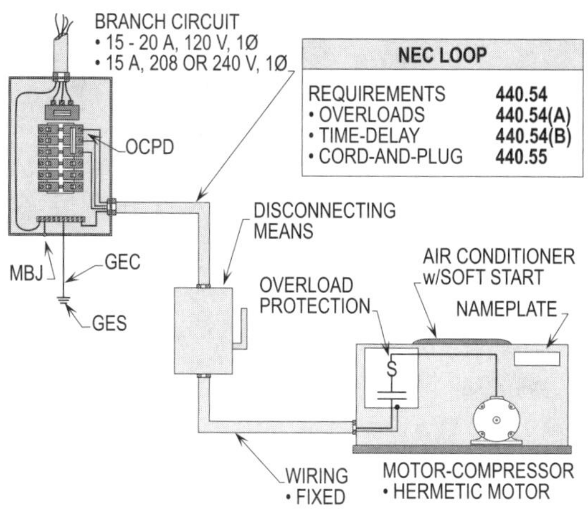

Figure 19-21. Overload protection shall be provided for direct or fixed-wired motor-compressors and equipment that is connected to 15 or 20 amp, 120 volt, single-phase branch circuits.

CORD-AND-ATTACHMENT PLUG CONNECTED MOTOR-COMPRESSORS AND EQUIPMENT ON 15 OR 20 AMP BRANCH CIRCUITS 440.55

When attachment plugs and receptacles are used for circuit connection they shall be rated no higher than 15 or 20 amps for 120 volt, single-phase circuit, or 15 amps, for 208 or 240 volt, single-phase branch circuits. (**See Figure 19-22**)

ROOM AIR CONDITIONERS 440.60

Room air conditioners are usually cord-and-plug connected when installed on 120/240 volt, single-phase systems. However, they may be hard-wired. Air conditioners are always hard-wired when installed on three-phase systems, or on electrical supply systems over 250 volts.

GROUNDING 440.61

The following wiring methods, when utilized to wire in room air conditioners, shall be connected to an equipment grounding conductor:
- Cord-and-plug connected
- Hard-wired (if within reach of the ground or grounded object)

- If in contact with metal
- Operating over 150 volts-to-ground
- Wired with metal-clad wiring
- Located in a hazardous location
- Installed in damp location (within reach of the user)

See Figure 19-23 for wiring methods that shall be permitted to be used to ground room air conditioners.

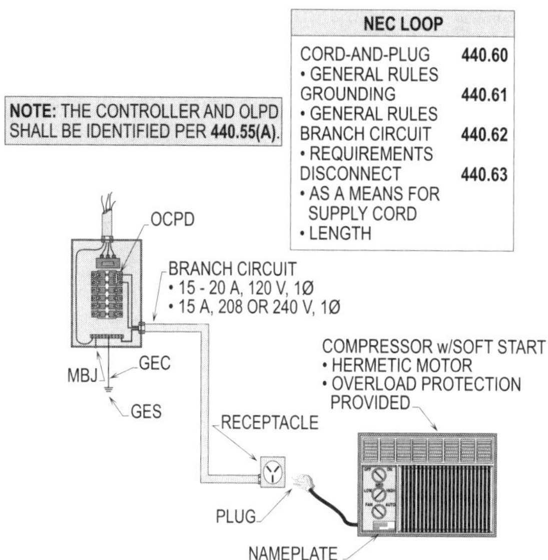

Figure 19-22. When attachment plugs and receptacles are used for circuit connection, they shall be rated no higher than 15 or 20 amps, for 120 volt, single-phase circuit, or 15 amps, for 208 or 240 volt, single-phase branch circuits.

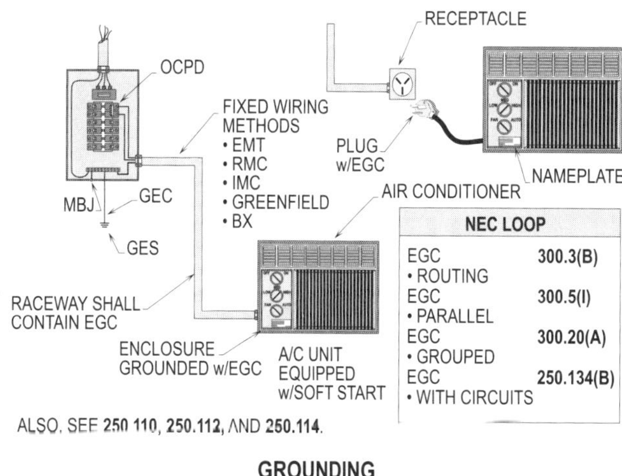

Figure 19-23. The above are wiring methods that shall be permitted to be used to ground room air conditioners.

BRANCH CIRCUIT REQUIREMENTS
440.62

The full-load current rating of the room air conditioner shall be marked on the nameplate and shall not operate at more than 40 amps on 250 volts. The branch-circuit overcurrent protection device shall be installed with a rating no greater than the circuit conductor's ampacity or the rating of the receptacle serving the unit, whichever is less. The ampacity of a cord-and-plug connected air conditioning window unit shall not exceed 80 percent of the branch circuit where no other loads are served. If other loads are served by the branch circuit, the cord-and-plug connected air conditioner unit shall not exceed 50 percent of the branch circuit. **[See Figures 19-24(a) and (b)]**

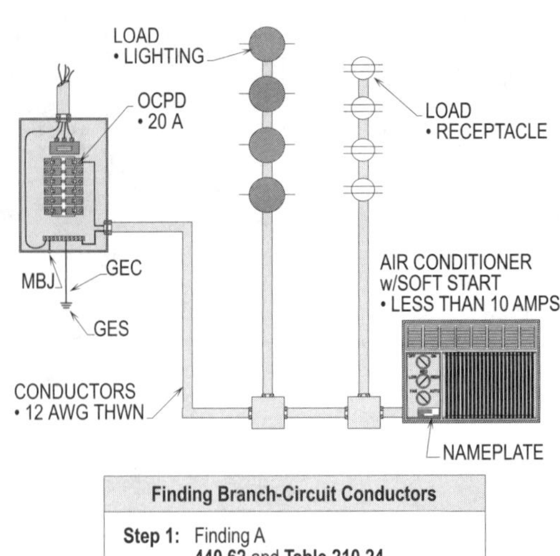

Finding Branch-Circuit Conductors

Step 1: Finding A
440.62 and **Table 210.24**
12 AWG requires 20 A

Step 2: Calculating A
440.62(C) and **210.23(A)**
20 A x 50% = 10 A

Solution: The maximum conductor ampacity permitted is 10 amps.

BRANCH CIRCUIT REQUIREMENTS
NEC 440.62(C)

Figure 19-24(b). If other loads are served by the branch circuit, the cord-and-plug connected air conditioner unit shall not exceed 50 percent of the branch circuit.

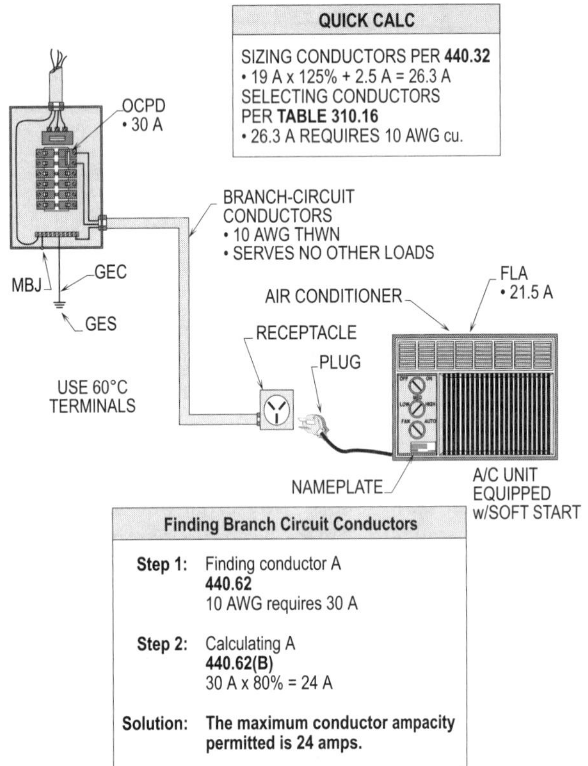

Finding Branch Circuit Conductors

Step 1: Finding conductor A
440.62
10 AWG requires 30 A

Step 2: Calculating A
440.62(B)
30 A x 80% = 24 A

Solution: The maximum conductor ampacity permitted is 24 amps.

BRANCH CIRCUIT REQUIREMENTS
NEC 440.62(B)

Figure 19-24(a). The ampacity of a cord-and-plug connected air conditioning window unit shall not exceed 80 percent of the branch circuit where no other loads are served.

DISCONNECTING MEANS
440.63

A cord-and-plug shall be permitted to serve as the disconnecting means for the room air conditioner if all the following conditions are complied with:

- Operates at 250 volts or less
- Controls are manually operated
- Controls are within 6 ft (1.8 m) of the floor
- Controls are readily accessible to user

A room air conditioner may be hard-wired and located within sight of the service equipment, or it may be wired so that it is readily accessible to an disconnecting switch for the user. However, such switch shall be located within sight and within the unit. **[See Figure 19-25(a) and (b)]**

Design Tip: The rules for three-phase room air conditioners shall not be used for these types of units. Three-phase room air conditioners shall be hard-wired and shall be installed with a disconnecting means that is readily accessible to the user.

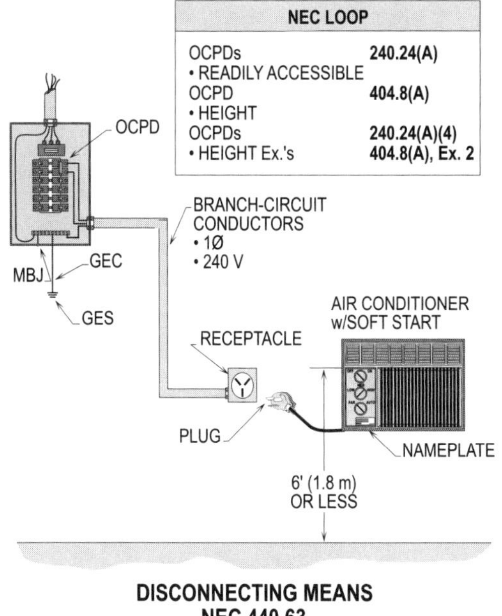

Figure 19-25(a). A cord-and-plug shall be permitted to serve as the disconnecting means for a room air conditioner if it operates at 250 volts or less and its controls are manually operated, within 6 ft (1.8 m) of the floor, and readily accessible to the user.

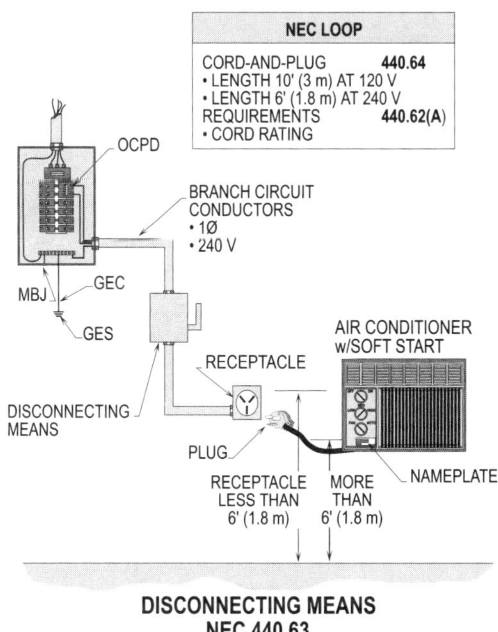

Figure 19-25(b). A cord-and-plug shall be permitted to serve as the disconnecting means even if the room air conditioner's manual controls are located above 6 ft (1.8 m) from the finished grade.

SUPPLY CORDS
440.64

Room air conditioners installed with flexible cords shall be a length that is limited to 10 ft (3 m) for 120 volt circuits and 6 ft (1.8 m) for 208 or 240 volt circuits. Long cords shall not be used for they are dangerous. Long cords can also be a shock or fire hazard. **(See Figure 19-26)**

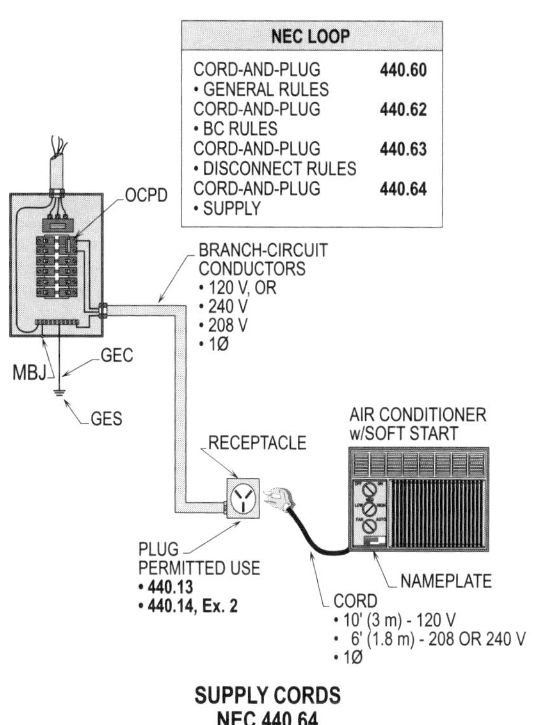

Figure 19-26. Room air conditioners installed with flexible cords shall have a length that is limited to 10 ft (3 m) for 120 volt circuits and 6 ft (1.8 m) for 208 or 240 volt circuits.

Chapter 19. Compressor Motors

Section   Answer

_______ _______   1. An additional circuit breaker or disconnecting means shall be provided at the equipment if an A/C unit is not within sight or within ______ ft of the disconnect.
   (a) 25   (b) 40
   (c) 50   (d) 75

_______ _______   2. An overcurrent protection device for hermetic-seal compressors shall be permitted to be selected up to ______ percent of its FLA to allow the motor to start and run.
   (a) 175   (b) 200
   (c) 225   (d) 250

_______ _______   3. If a fuse or circuit breaker is used for the protection of the motor compressor, it shall trip at not more than ______ percent of the full-load current rating.
   (a) 100   (b) 110
   (c) 115   (d) 125

_______ _______   4. The ampacity of a cord-and-plug connected air conditioning (window) unit shall not exceed ______ percent of the branch circuit where no other loads are served.
   (a) 80   (b) 100
   (c) 115   (d) 150

_______ _______   5. Room air conditioners installed with flexible cords shall be of a length that is limited to ______ ft for 120 volt circuits.
   (a) 6   (b) 10
   (c) 12   (d) 15

_______ _______   6. When sizing the conductors for a feeder supplying A/C units and motors, the full-load current in amps of the largest motor shall be multiplied by ______ percent.
   (a) 100   (b) 110
   (c) 115   (d) 125

_______ _______   7. The full-load current rating of the nameplate or the nameplate branch-circuit selection current of the compressor, whichever is greater, shall be sized at ______ percent of the disconnecting means.
   (a) 100   (b) 110
   (c) 115   (d) 125

_______ _______   8. The disconnecting means for air conditioning or refrigeration equipment shall be located within sight and within ______ ft and shall be readily accessible to the user.
   (a) 20   (b) 50
   (c) 75   (d) 100

9. The full-load current rating on the nameplate or the branch-circuit selection
    current rating of the largest motor, whichever is greater, shall be sized at _____
    percent if there are two or more hermetically sealed motors installed on the
    same feeder.
    (a) 125   (b) 135
    (c) 150   (d) 175

10. Two or more motor-compressors plus other motor loads can be connected to a
    feeder, with the largest (motor) compressor calculated at 125 percent of its
    FLA and the remaining compressor loads added to this total at _____ percent
    of their FLA ratings.
    (a) 100   (b) 110
    (c) 115   (d) 125

11. The overload relay for the motor-compressor shall trip at not more than _____
    percent of the full-load current rating.
    (a) 110   (b) 115
    (c) 135   (d) 140

12. When attachment plugs and receptacles are used for circuit connection, they
    shall be rated no higher than _____ amps for 208 or 240 volt, single-phase
    branch circuits.
    (a) 15   (b) 20
    (c) 30   (d) 50

13. The full-load current rating of the room air conditioner shall be marked on the
    nameplate and shall not operate at more than _____ amps on 250 volts, single-
    phase.
    (a) 20   (b) 30
    (c) 40   (d) 60

14. A cord-and-plug shall be permitted to serve as the disconnecting means even
    if the room air conditioner manual controller is above _____ ft.
    (a) 3   (b) 6
    (c) 10   (d) 12

15. Room air conditioners installed with flexible cord are to be a length that is
    limited to _____ ft for 208 or 240 volt circuits.
    (a) 3   (b) 5
    (c) 6   (d) 10

16. The overcurrent protection device for a hermetically sealed compressor shall
    be selected at _____ percent (for minimum) of the compressor's FLA rating or
    the branch-circuit selection current, whichever is greater.
    (a) 125   (b) 150
    (c) 175   (d) 225

17. The rating of the overcurrent protection device shall be determined by using
    the rating on the nameplate of the cord-and-plug connected equipment having
    single-phase, _____ volt or less, hermetically sealed motors.
    (a) 120   (b) 250
    (c) 480   (d) 600

18. Overload protection shall be provided for direct or fixed-wired motor-compressors and equipment that is connected to _____ or _____ amp, 120 volt, single-phase branch circuits.
    (a) 15, 20   (b) 20, 30
    (c) 30, 40   (d) 40, 50

19. Room air conditioners shall be grounded for the protection of _____ .
    (a) equipment   (b) personnel
    (c) all of the above   (d) none of the above

20. A cord-and-plug shall be permitted to serve as the disconnecting means for a room air conditioner if it operates at _____ volts or less.
    (a) 120   (b) 150
    (c) 250   (d) 300

21. What size nonautomatic circuit breaker is required to disconnect an A/C unit with a compressor rated at 29 amps and a condenser rated at 2.5 amps?

22. What size nonfused disconnect is required for an A/C unit with a compressor rated at 29 amps and a condenser rated at 2.5 amps?

23. What size horsepower rated disconnect is required for an A/C unit with a compressor rated at 29 amps and a condenser rated at 2.5 amps with a 180 amp locked-rotor current rating? (Note that the supply is 480 volts, three-phase.)

24. What size horsepower rated disconnect is required for the following loads on a three-phase, 230 volt system?

• A/C unit with a compressor rated at 29 amps, with a 200 amp locked-rotor current rating
• A/C unit with a compressor rating 24 amps, with a locked-rotor current of 150 amps
• A/C unit with a compressor rating of 20 amps, with a locked-rotor current of 140 amps

25. What size circuit breaker is required to disconnect a hermetically sealed motor for an A/C unit rated at 29 amps? (Note that the supply is 208 volts, three-phase.)

26. What size disconnecting means using horsepower is required to disconnect the following loads on a three-phase, 230 volt system?

• Motor rated at 38 amps, with a 212 amp locked-rotor current rating
• A/C unit with a compressor rated at 28 amps, with a 160 amp locked-rotor current rating
• A/C unit with a compressor rated at 24, amps with a 160 amp locked-rotor current rating

27. What is the minimum size overcurrent protection device required for an A/C unit with a compressor rated at 20 amps and a condenser rated at 2.5 amps?

28. What is the maximum size overcurrent protection device required for an A/C unit with a compressor rated at 20 amps and a condenser rated at 2.5 amps?

_______________ _______________

29. What is the minimum size overcurrent protection device required for a feeder with the following loads on a 230 volt, three-phase system?

• A/C unit with a compressor rated at 29 amps and a condenser rated at 2.5 amps
• A/C unit with a compressor rated at 26 amps and a condenser rated at 2.5 amps
• A/C unit with a compressor rated at 22 amps and a condenser rated at 2.5 amps

_______________ _______________

30. What is the maximum size overcurrent protection device required for a feeder with the following loads on a 230 volt, three-phase system?

• A/C unit with a compressor rated at 29 amps and a condenser rated at 2.5 amps
• A/C unit with a compressor rated at 26 amps and a condenser rated at 2.5 amps
• A/C unit with a compressor rated at 22 amps and a condenser rated at 2.5 amps

_______________ _______________

31. What size overcurrent protection device (time delay fuse) is required for a feeder with the following loads on a 230 volt, three-phase system?

• A/C unit with a compressor rated at 26 amps and a condenser rated at 2.5 amps
• A/C unit with a compressor rated at 24 amps and a condenser rated at 2.5 amps
• 10 HP, 230 volt, three-phase, Design letter B motor

_______________ _______________

32. What size THHN copper conductors are required to supply an A/C unit with a compressor rated at 20 amps and a condenser rated at 2.5 amps?

_______________ _______________

33. What size THHN copper conductors are required to supply an A/C unit with a branch-circuit selection current of 30 amps?

_______________ _______________

34. What size THWN copper conductors are required for a feeder with the following loads on a 208 volt, three-phase system?

• A/C unit with a compressor rated at 30 amps and a condenser rated at 3 amps
• A/C unit with a compressor rated at 28 amps and a condenser rated at 2.5 amps
• A/C unit with a compressor rated at 24 amps and a condenser rated at 2.5 amps

_______________ _______________

35. What size THWN copper conductors are required for a feeder with the following loads on a 208 volt, three-phase system?

• A/C unit with a compressor rated at 30 amps and a condenser rated at 3 amps
• A/C unit with a compressor rated at 28 amps and a condenser rated at 2.5 amps
• A/C unit with a compressor rated at 24 amps and a condenser rated at 2.5 amps
• Other loads of 80 amps (continuous operation)

_______________ _______________

36. What size controller is required for an A/C unit with a compressor rated at 20 amps and a condenser rated at 2.5 amps on a 230 volt, single-phase system?

37. What size overload relay and fuses is required for an A/C unit with a compressor
rated at 20 amps on a 240 volt, three-phase system?

38. What is the allowable ampacity for a 10 AWG THWN copper conductor supplying
an air conditioning window unit?

19-21

39. What is the allowable ampacity available for an air conditioning window unit
with other loads such as lighting and receptacle loads that are supplied by size
10 AWG THWN copper conductor?

20

Transformers

Transformers must be sized with enough capacity to supply power to loads and allow loads with high inrush currents to start and run. In addition, they must be protected by properly sized overcurrent protection devices and be equipped with conductors having allowable ampacity ratings to supply the loads. Overcurrent protection devices and conductors must be designed and installed in such a manner to safely protect the windings of such power sources from dangerous short circuits, ground faults, and overloads.

The overcurrent protection devices and conductors are sometimes required to be adjusted in size in order to protect the transformer windings or the conductors from overload conditions.

Transformers may be utility or customer owned and fall under the rules of the *National Electrical Safety Code*. If customer owned, they are still required to follow the *National Electrical Safety Code* as well as the *National Electrical Code*. And these requirements must be complied with.

Note, When selecting the actual size circuit breaker or fuse for the protection of electrical systems rated over 600 volts, see one of the *ANSI C Standards*. For example, for fuses rated at 100 amps or less, see ANSI C 37.46, and for over 100 amps see ANSI 37.46 and ANSI 37.40. When circuit breakers are used to protect high-voltage systems, see ANSI C 37.06. However, there may be protection designs that require reference to other ANSI C standards, and the designer must be prepared to refer to such standards.

SIZING TRANSFORMERS

The total volt-amps of all loads in a building shall be used to size a transformer. Depending on the load requirements of a building, single-phase and three-phase currents may be used to supply the building. The windings are connected in configurations necessary to supply voltage and load requirements of the facility.

WYE-CONNECTED SECONDARIES

The size transformers required to supply a wye-connected secondary system can be found by applying the following:

- Adding the total single-phase and three-phase loads together for an individual transformer.

- Dividing the load in VA by 1/3 (.33) to derive three transformers.

The kVA rating of three transformers, if they are separately connected together, will add up to one individual transformer. Using this method, a single transformer rating can be sized and selected from the total volt-amps. One transformer with three windings is sized by adding the total VA of all the loads together and selecting the transformers kVA rating based upon this value. **(See Figure 20-1)**

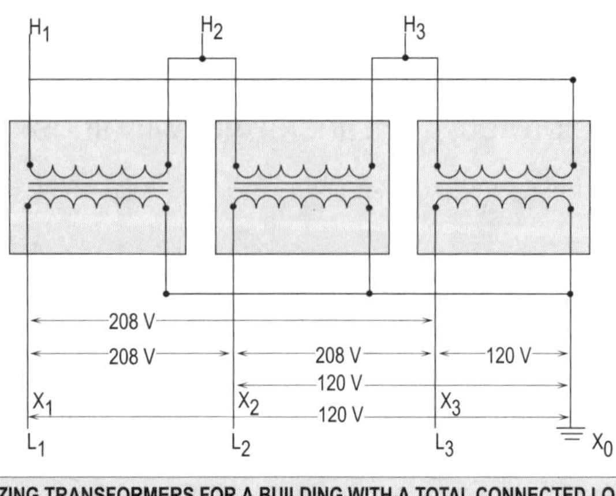

SIZING TRANSFORMERS FOR A BUILDING WITH A TOTAL CONNECTED LOAD OF 20 kVA FOR SINGLE-PHASE LOADS AND 30 kVA FOR THREE-PHASE LOADS

Step 1: Finding kVA load for three transformers
Single-phase: 20 kVA x .33 = 6.6 kVA
Three-phase: 30 kVA x .33 = 9.9 kVA
Total load = 16.5 kVA

Step 2: Sizing individual transformers
Three 20 kVA lighting and power transformers of the same size must be installed (20 kVA x 3 = 60 kVA)

Step 3: Sizing kVA for one transformer
20 kVA + 30 kVA = 50 kVA

Solution: Three transformers of 20 kVA each and one transformer of at least one 60 kVA are recommended. Wye connected secondaries.

WYE-CONNECTED SECONDARIES

Figure 20-1. Sizing wye-connected transformers with single-phase and three-phase loads.

CLOSED DELTA-CONNECTED SECONDARIES

The size transformer required to supply a closed delta-connected secondary system can be found by multiplying the following:

- Multiply the single-phase load in VA by 67 percent.

- Multiply the three-phase load in VA by 33 percent.

- Add the kVA load of (1) and (2) together to derive the two lighting and power transformers.

- Multiply the single-phase and three-phase loads in VA by 33 percent.

- Use this total load in kVA to derive the two power transformer.

See Figure 20-2(a) for the rules and regulations for sizing the transformer used in a closed delta-connected system.

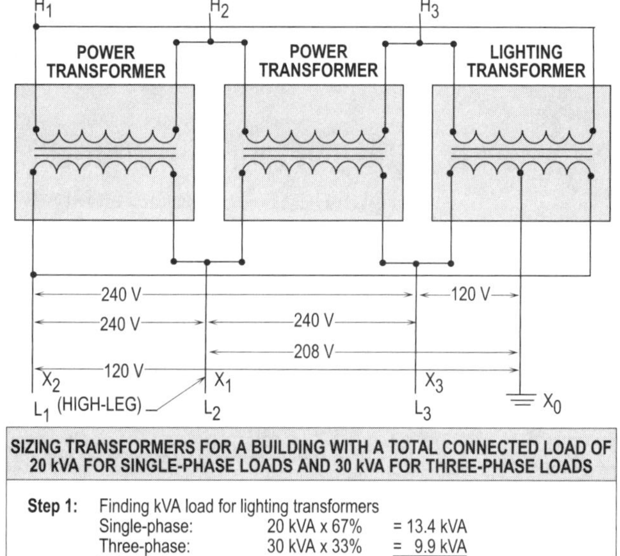

SIZING TRANSFORMERS FOR A BUILDING WITH A TOTAL CONNECTED LOAD OF 20 kVA FOR SINGLE-PHASE LOADS AND 30 kVA FOR THREE-PHASE LOADS

Step 1: Finding kVA load for lighting transformers
Single-phase: 20 kVA x 67% = 13.4 kVA
Three-phase: 30 kVA x 33% = 9.9 kVA
Total load = 23.3 kVA

Solution: Lighting transformer is 25 kVA.

Step 1: Finding kVA load for power transformers
Single-phase: 20 kVA x 33% = 6.6 kVA
Three-phase: 30 kVA x 33% = 9.9 kVA
Total load = 16.5 kVA

Solution: Two power transformers of 20 kVA each are required.

Note: Lighting transformer is 25 kVA and the two power transformers are 20 kVA each.

CLOSED DELTA-CONNECTED SECONDARIES

Figure 20-2(a). Sizing closed delta-connected transformers with single-phase and three-phase loads.

OPEN DELTA-CONNECTED SECONDARIES

Open delta-connected secondary systems can be determined by calculating the single-phase load at 100 percent and the three-phase load at 58 percent and using this total value to size the transformer. By adding these two loads together, the size of a mid-tap transformer can be determined. A power transformer can be sized by calculating the three-phase load at 58 percent, which is the reciprocal of the square root of 3 (1 ÷ 1.732 = 58%). This reduced total is then used to size the power transformer, which will be smaller in rating than the lighting and power transformer. **[See Figure 20-2(b)]**

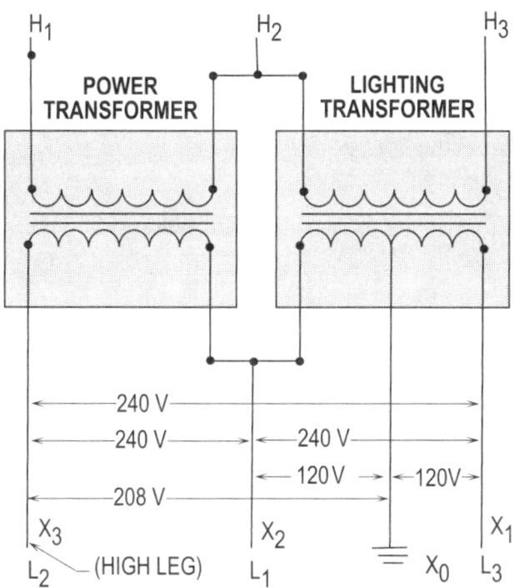

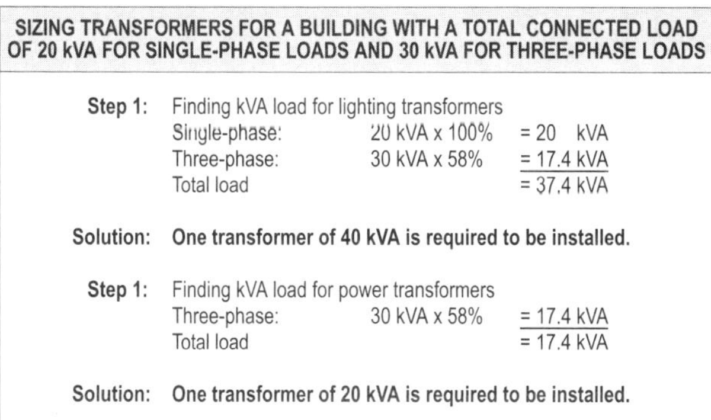

SIZING TRANSFORMERS FOR A BUILDING WITH A TOTAL CONNECTED LOAD OF 20 kVA FOR SINGLE-PHASE LOADS AND 30 kVA FOR THREE-PHASE LOADS

Step 1:	Finding kVA load for lighting transformers		
	Single-phase:	20 kVA x 100%	= 20 kVA
	Three-phase:	30 kVA x 58%	= 17.4 kVA
	Total load		= 37.4 kVA
Solution:	One transformer of 40 kVA is required to be installed.		
Step 1:	Finding kVA load for power transformers		
	Three-phase:	30 kVA x 58%	= 17.4 kVA
	Total load		= 17.4 kVA
Solution:	One transformer of 20 kVA is required to be installed.		

OPEN DELTA-CONNECTED SECONDARIES

Figure 20-2(b). Sizing open delta-connected transformers with single-phase and three-phase loads.

CALCULATING PRIMARY AND SECONDARY CURRENTS

The transformer's primary amp rating shall be equivalent to the amps of the connected load when installing a feeder to supply the primary of a transformer to step up or step down the voltage. To determine the FLA of a transformer, the kVA of the transformer must be divided by the voltage x 1.732 if the supply is three-phase. **[See Figures 20-3(a) and (b)]**

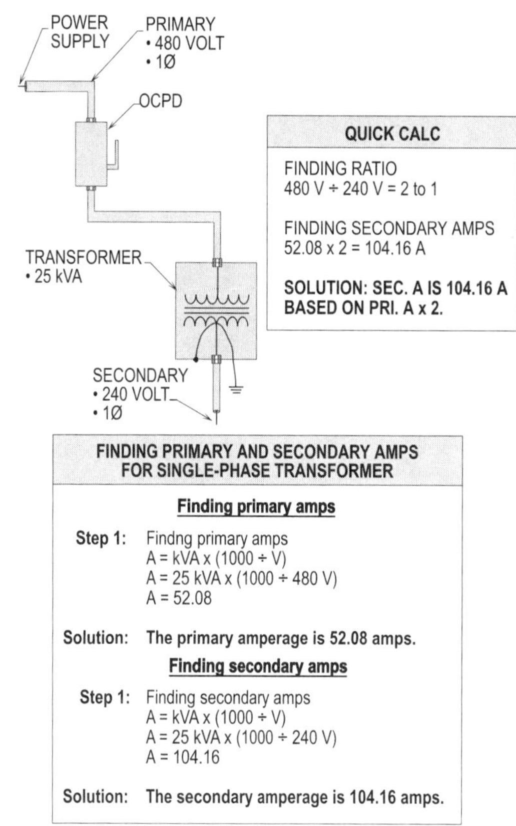

FINDING PRIMARY AND SECONDARY AMPS FOR SINGLE-PHASE TRANSFORMER

Finding primary amps

Step 1:	Finding primary amps A = kVA x (1000 ÷ V) A = 25 kVA x (1000 ÷ 480 V) A = 52.08
Solution:	The primary amperage is 52.08 amps.

Finding secondary amps

Step 1:	Finding secondary amps A = kVA x (1000 ÷ V) A = 25 kVA x (1000 ÷ 240 V) A = 104.16
Solution:	The secondary amperage is 104.16 amps.

CALCULATING PRIMARY AND SECONDARY CURRENTS

Figure 20-3(a). Finding amps of a single-phase transformer.

FINDING AMPERAGE

The kVA or amp rating for the primary or secondary of a transformer can be determined for a single-phase system by applying the following formula:

kVA = volts x amps ÷ 1000
amps = kVA x 1000 ÷ volts

The following formula shall be applied to determine the ratio of a transformer having a 480 volt primary and 240 volt secondary:

primary ÷ secondary
480 V ÷ 240 V
2:1 ratio

The amp rating for the primary or secondary can be determined for a three-phase system by applying the following formula:

$$kVA = volts \times 1.732 \times amps \div 1000$$
$$amps = (kVA \times 1000) \div (volts \times 1.732)$$

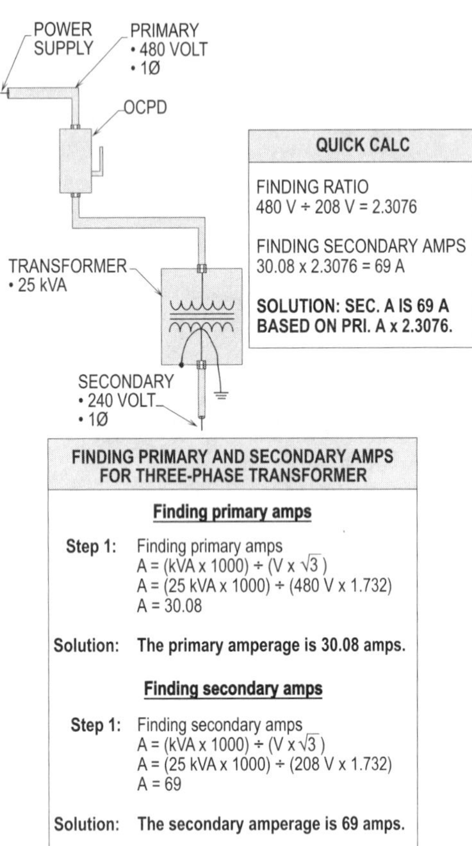

Figure 20-3(b). Sizing amps of a three-phase transformer.

INSTALLING TRANSFORMERS
ARTICLE 450

Transformers shall be installed and protected per **Article 450**. Transformers shall be permitted to be installed inside or outside buildings based upon their design and type.

LOCATION
450.13

Transformers shall be located where readily accessible to qualified personnel for inspection and maintenance. Where it is necessary to use a ladder, lift, or bucket truck to get to a transformer, it shall not be considered readily accessible. See definition of *readily accessible*, in **Article 100** of the NEC. **(See Figure 20-4)**

There are two exceptions to the general rule to the accessibility rules, and they are explained under the next two headings.

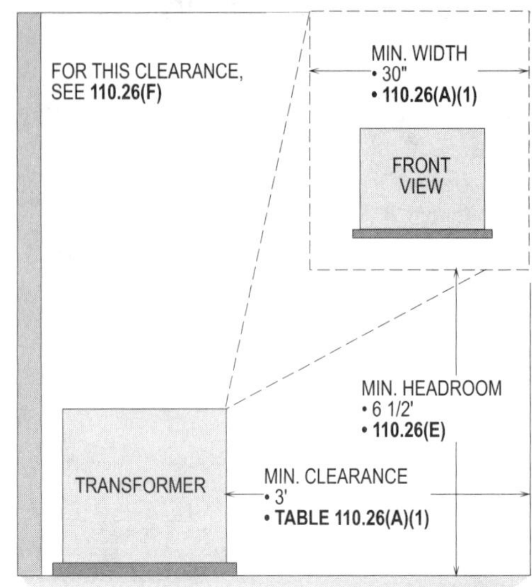

Figure 20-4. The general rule of **450.13** requires transformers to be readily accessible for maintenance, repair, and service. See AHJ for this requirement.

HUNG FROM WALL OR CEILING
450.13(A)

- Dry-type transformers not over 600 volts and located on open walls or steel columns shall not be required to be readily accessible. It is permissible to gain access to this type of installation using a portable ladder or bucket lift. **[See Figure 20-5(a)]**

MOUNTED IN CEILING
450.13(B)

- Dry-type transformers not over 600 volts and 50 kVA shall be permitted to be installed in hollow spaces of buildings. The transformers cannot be permanently closed in and there shall be some access to the transformers, but they do not have to be readily accessible per **Article 100** of the NEC. It was not clear in the 1993 or previous editions of the NEC whether dry-type transformers not exceeding 600 volts, nominal, and rated 50 kVA or less were permitted to be installed in the space above suspended ceilings with removable panels, even if the transformer was accessible and provided with proper working clearances. Note that the space where the transformer is installed shall comply with the ventilation requirements of **450.9** and be designed by the rules of **450.21(A)** and **(B)**. If such ceiling space is used as a return air space for air conditioning, **300.22(C)** shall be reviewed and the provisions of this section shall be complied with. **[See Figure 20-5(b)]**

Design Tip: The two exceptions to the general rule are for dry-type transformers. The exceptions do not apply for oil or askarel-filled transformers due to the damage of a possible oil spillage or the threat of fire because of a rupture occurring in the case.

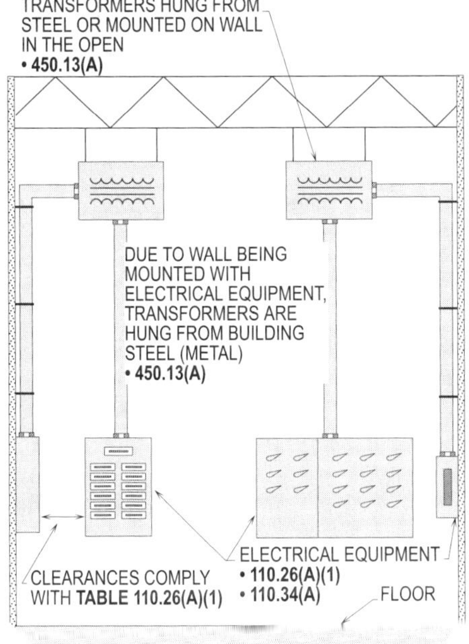

Figure 20-5(a). Transformers hung from the wall or ceiling shall not be required to be readily accessible.

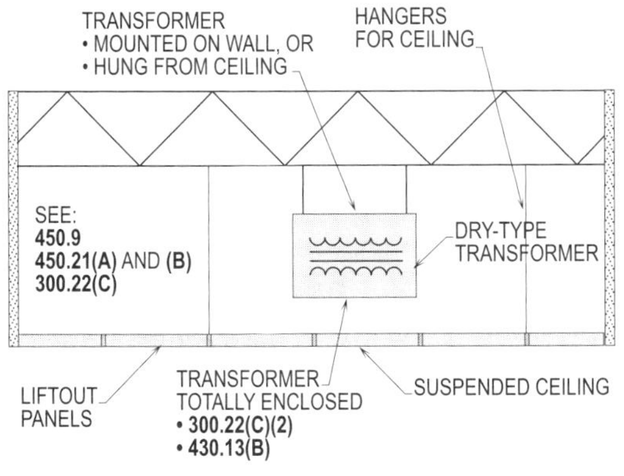

Figure 20-5(b). Transformers mounted in ceiling shall not be required to be readily accessible.

OVERCURRENT PROTECTION
450.3(A) AND (B)

There are two sets of rules when providing overcurrent protection of transformers: rules for transformers rated over 600 volts and transformers of 600 volts or less. The overcurrent protection device may be placed in the primary only or in the primary and secondary side of the transformer.

PRIMARY ONLY – OVER 600 VOLTS
450.3(A) AND TABLE 450.3(A)

The term *primary* is often inferred in the field as being the high side, and the term *secondary* as the low side of the transformer. This is really not the proper terminology. The primary is the input side of the transformer and the secondary is the output side. Thus, voltage has nothing to do with "high" or "low."

Each transformer shall be protected by an overcurrent device in the primary side. If the overcurrent protection is fuses, they shall be rated not greater than 250 percent (2.5 times) of the rated primary current of the transformer. When circuit breakers are used, they shall be set not greater than 300 percent (3 times) of the rated primary current. **(See Figure 20-6)**

This overcurrent protection device shall be permitted to be mounted in the vault or at the transformer, if approved for such purpose. It shall also be permitted to be mounted in the panelboard and be designed to protect the windings and circuit conductors supplying the transformer.

If not installed in a vault, the overcurrent protection device shall be permitted to be installed outdoors on a pole, with a disconnecting means installed in the vault to disconnect supply conductors.

APPLYING NOTE 1
TABLE 450.3(A), Note 1

Where 250 percent (2.5 times) of the rated primary current of the transformer does not correspond to a standard rating of a fuse, the next higher standard rating shall be permitted, per **240.6(A)**.

PRIMARY AND SECONDARY
OVER 600 VOLTS
450.3(A) AND TABLE 450.3(A)

A transformer over 600 volts, nominal, having an overcurrent protection device on the secondary side rated to open not greater than the values listed in **Table 450.3(A)**, or a transformer equipped with a coordinated thermal overload protection by the manufacturer, shall not be required to have individual protection in the primary. However, a feeder overcurrent protection device rated or set to open at not greater than the values listed in **Table 450.3(A)** shall be provided.

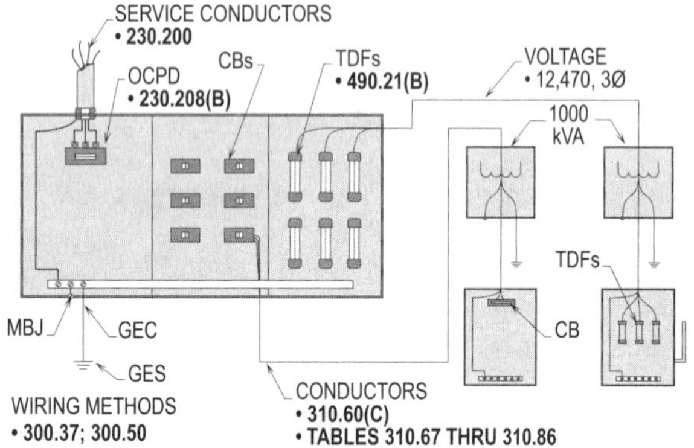

FINDING INDIVIDUAL OCPD FOR THE PRIMARY SIDE OF THE TRANSFORMER

Sizing OCPD using CBs

Step 1: Finding FLA of primary
450.3(A)
FLA = (kVA x 1000) ÷ (V x $\sqrt{3}$)
FLA = (1000 x 1000) ÷ (12,470 V x 1.732)
FLA = 46.3 A

Step 2: Calculating FLA for OCPD
450.3(A) and Table 450.3(A)
46.3 A x 300% = 138.9 A

Step 3: Selecting OCP
Table 450.3(A), Note 1 and 240.6(A)
138.9 A requires 150 A

Solution: The size circuit breaker is 150 amps.

Sizing OCPD using TDFs

Step 1: Calculating FLA for TDFs
450.3(A) and Table 450.3(A)
46.3 A x 250% = 115.8 A

Step 2: Selecting TDFs
Table 450.3(A), Note 1 and 240.6(A)
115.8 A allows 125 A

Solution: The size time delay fuses are 125 amps.

Note: Before actually selecting the overcurrent protection device, see ANSI C 37.46 and C 37.06.

INDIVIDUAL PROTECTION
NEC 450.3(A) AND TABLE 450.3(A)

Figure 20-6. If the overcurrent protection is fuses, they shall be rated not greater than 250 percent (2.5 times) of the rated primary current of the transformer. When circuit breakers are used, they shall be set at not greater than 300 percent (3 times) of the rated primary current.

NONSUPERVISED LOCATIONS
450.3(A) AND TABLE 450.3(A)

Overcurrent protection for a nonsupervised location shall be permitted to be placed in the primary and secondary side of high-voltage transformers if the overcurrent protection devices are designed and installed according to the provisions listed in **Table 450.3(A)**.

If the secondary voltage is 600 volts or less, the overcurrent protection device and conductors on the secondary side shall be sized at 125 percent of the FLC rating. Overcurrent protection devices sized at 125 percent of the FLC in amps protect the conductors and windings of the transformer from dangerous overload conditions. With higher voltage on the secondary side of the transformer, the percentages for sizing the overcurrent protection devices shall be selected from **Table 450.3(A)** (any location) based upon the particular voltage level. **[See Figure 20-7(a)]**

SUPERVISED LOCATIONS
450.3(A) AND TABLE 450.3(A)

Overcurrent protection shall be permitted to be placed in the primary and secondary side of high-voltage transformers if the overcurrent protection devices are designed and installed according to the provisions listed in **Table 450.3(A)**.

Where the facility has trained engineers and maintenance personnel, the overcurrent protection device for the secondary shall be sized at not more than 250 percent of the FLC for voltage 600 volts or less. With higher voltage on the secondary side of the transformer, the percentages for sizing the overcurrent protection devices shall be selected from **Table 450.3(A)** based upon the particular voltage level. **[See Figures 20-7(b) and (c)]**

Note: See Figure 20-8 for certain design conditions that permit the primary overcurrent protection device to be used to protect the primary and secondary sides of two-wire to two-wire connected transformers and three-wire to three-wire delta connected transformers per **240.4(F)** and **240.21(C)(1)**.

PRIMARY ONLY – 600 VOLTS OR LESS
450.3(B) AND TABLE 450.3(B)

A transformer 600 volts or less, nominal, having an individual overcurrent protection device on the primary side shall be sized at no more than 125 percent of the transformer's full-load current rating. Note, with the overcurrent protection device and conductors sized at 125 percent or less of the transformer's FLC, the supply conductors and transformer windings shall be considered protected from overload conditions. It appears that individual protection in the primary is not recognized per **450.3(B)** and **Table 450.3(B)**. **(See Figure 20-9)**

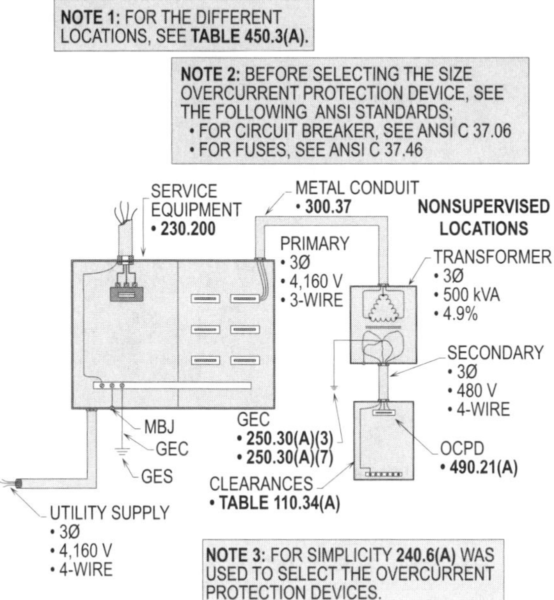

SIZING OCPD FOR PRIMARY SIDE	
Step 1:	Finding FLA of transformer FLA = (kVA x 1000) ÷ (V x √3) FLA = (500 x 1000) ÷ (4160 V x 1.732) FLA = 69.4 A
Step 2:	Calculating FLA for OCPD **450.3(A)** and **Table 450.3(A)** FLA = 69.4 A x 600% FLA = 416.4 A
Step 3:	Selecting OCPD **Table 450.3(A), Note 1** and **240.6(A)** 416.4 A allows 450 A
Solution:	**The size OCPD for primary side is 450 amps.**

Note 1: Higher percentages than 125 percent per **Table 450.3(A)** shall be used to size the overcurrent protection device in the secondary side where the secondary voltage is greater than 600 volts.

SIZING OCPD FOR SECONDARY SIDE	
Step 1:	Finding FLA of transformer FLA = (kVA x 1000) ÷ (V x √3) FLA = (500 x 1000) ÷ (480 V x 1.732) FLA = 601.7 A
Step 2:	Calculating FLA for OCPD **450.3(A)** and **Table 450.3(A)** FLA = 601.7 A x 125% FLA = 752 A
Step 3:	Selecting OCPD **Table 450.3(A), Note 1** 752 A allows 800 A
Solution:	**The size OCPD for secondary side is 800 amps.**

Note 2: If the secondary voltage is 4160, the overcurrent protection device using a circuit breaker shall be sized at 300 percent and a fuse shall be sized at 250 percent of transformer's FLC.

**NONSUPERVISED LOCATIONS
(ANY LOCATION)
NEC 450.3(A) AND TABLE 450.3(A)**

Figure 20-7(a). Sizing the primary and secondary side of a transformer in a nonsupervised (any) location.

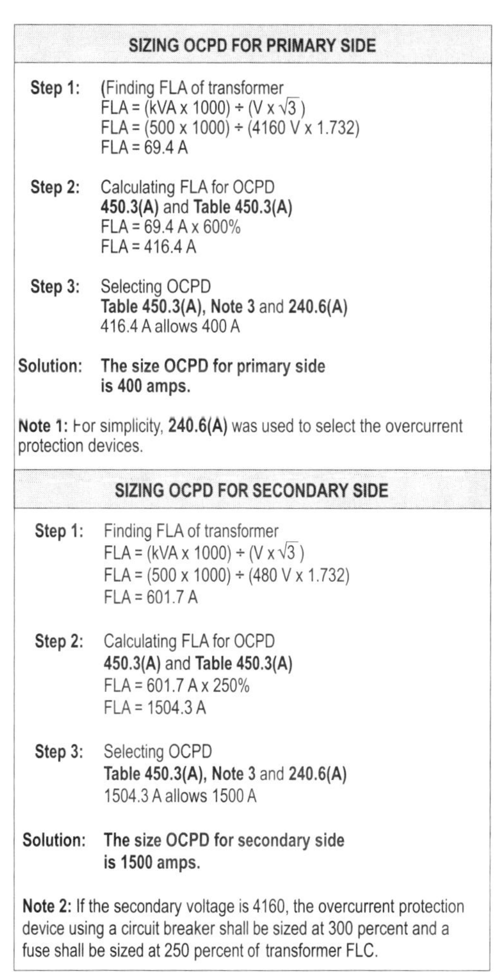

SIZING OCPD FOR PRIMARY SIDE	
Step 1:	(Finding FLA of transformer FLA = (kVA x 1000) ÷ (V x √3) FLA = (500 x 1000) ÷ (4160 V x 1.732) FLA = 69.4 A
Step 2:	Calculating FLA for OCPD **450.3(A)** and **Table 450.3(A)** FLA = 69.4 A x 600% FLA = 416.4 A
Step 3:	Selecting OCPD **Table 450.3(A), Note 3** and **240.6(A)** 416.4 A allows 400 A
Solution:	**The size OCPD for primary side is 400 amps.**

Note 1: For simplicity, **240.6(A)** was used to select the overcurrent protection devices.

SIZING OCPD FOR SECONDARY SIDE	
Step 1:	Finding FLA of transformer FLA = (kVA x 1000) ÷ (V x √3) FLA = (500 x 1000) ÷ (480 V x 1.732) FLA = 601.7 A
Step 2:	Calculating FLA for OCPD **450.3(A)** and **Table 450.3(A)** FLA = 601.7 A x 250% FLA = 1504.3 A
Step 3:	Selecting OCPD **Table 450.3(A), Note 3** and **240.6(A)** 1504.3 A allows 1500 A
Solution:	**The size OCPD for secondary side is 1500 amps.**

Note 2: If the secondary voltage is 4160, the overcurrent protection device using a circuit breaker shall be sized at 300 percent and a fuse shall be sized at 250 percent of transformer FLC.

**SUPERVISED LOCATIONS
NEC 450.3(A) AND TABLE 450.3(A)**

Figure 20-7(b). Sizing the primary and secondary side of a transformer in a supervised location.

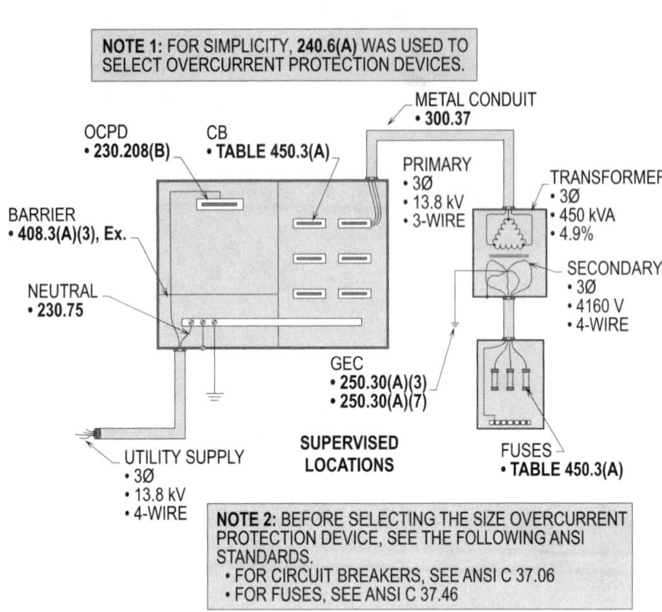

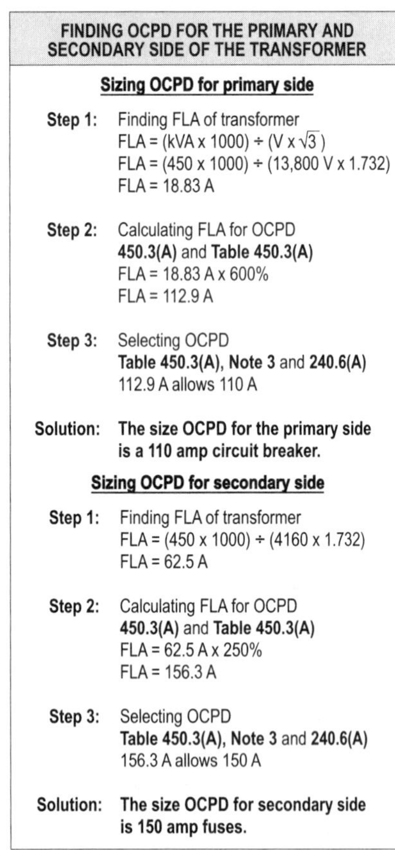

SUPERVISED LOCATIONS
NEC 450.3(A) AND TABLE 450.3(A)

Figure 20-7(c). Sizing the primary and secondary side of a transformer in a supervised location.

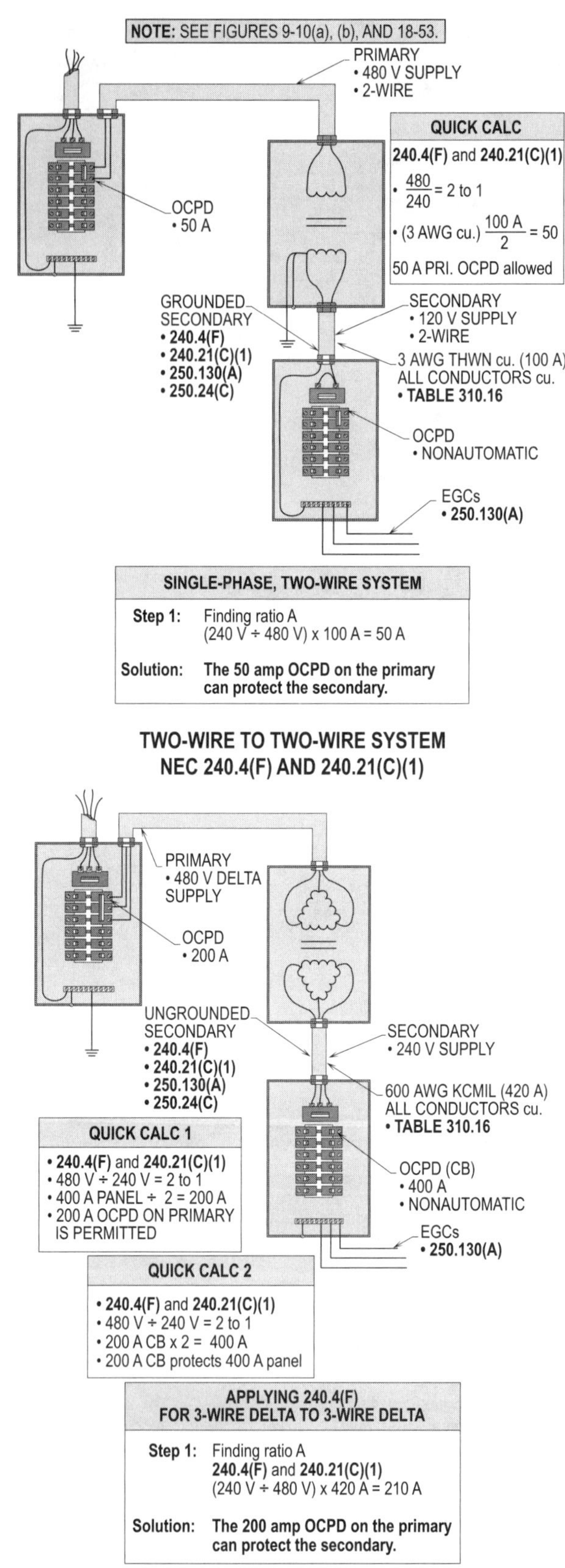

Figure 20-8. Sizing the overcurrent protection device for a single-phase, two-wire system and three-phase, three-wire system.

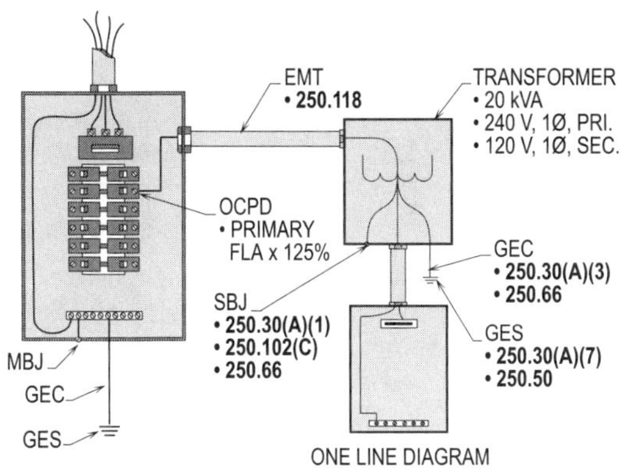

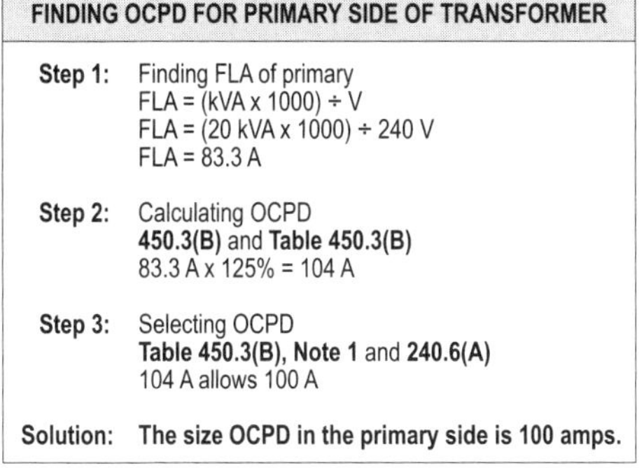

FINDING OCPD FOR PRIMARY SIDE OF TRANSFORMER

Step 1: Finding FLA of primary
FLA = (kVA x 1000) ÷ V
FLA = (20 kVA x 1000) ÷ 240 V
FLA = 83.3 A

Step 2: Calculating OCPD
450.3(B) and **Table 450.3(B)**
83.3 A x 125% = 104 A

Step 3: Selecting OCPD
Table 450.3(B), Note 1 and **240.6(A)**
104 A allows 100 A

Solution: The size OCPD in the primary side is 100 amps.

PRIMARY ONLY
600 VOLTS OR LESS
NEC 450.3(B) AND TABLE 450.3(B)

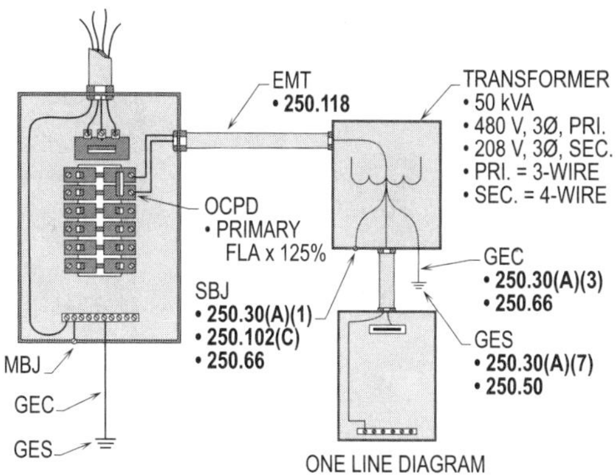

FINDING OCPD FOR PRIMARY SIDE OF TRANSFORMER

Sizing primary OCPD (First Level)

Step 1: Finding FLA of primary
FLA = (kVA x 1000) ÷ (V x √3)
FLA = (50 kVA x 1000) ÷ (480 V x 1.732)
FLA = 60.2 A

Step 2: Calculating OCPD
450.3(B) and **Table 450.3(B)**
60.2 A x 125% = 75.3 A

Step 3: Selecting OCPD
Table 450.3(B), Note 1 and **240.6(A)**
75.3 A allows 80 A

Solution: The size OCPD in the primary side is 80 amps.

PRIMARY 9 AMPS OR MORE
NEC 450.3(B) AND TABLE 450.3(B)

Figure 20-9. A transformer of 600 volts or less, nominal, having an individual overcurrent protection device on the primary side shall be sized at no more than 125 percent of the transformer's full-load current rating, in amps.

Figure 20-10(a). Where the rated primary current of a transformer is 9 amps or more and 125 percent of this current does not correspond to a standard rating of a fuse or circuit breaker, the next size shall be permitted to be used per **240.6(A)** and **Note 1** to **Table 450.3(B)**.

PRIMARY 9 AMPS OR MORE
450.3(B) AND TABLE 450.3(B)

Where the rated primary current of a transformer is 9 amps or more and 125 percent of this current does not correspond to a standard rating of a fuse or circuit breaker, the next size shall be permitted to be used per **240.6(A)**. Where the rated primary current of a transformer is less than 9 amps but more than 2 amps, an overcurrent device rated or set at no more than 167 percent of primary current shall be used. When the rated primary current of a transformer is less than 2 amps, an overcurrent protection device rated or set at not more than 300 percent shall be used. **[See Figures 20-10(a), (b),** and **(c)]**

PRIMARY AND SECONDARY –
600 VOLTS OR LESS
450.3(B) AND TABLE 450.3(B)

Combination protection shall be permitted to be provided for both the primary and secondary sides of a transformer. A current value of 250 percent of the rated primary current of the transformer shall be used if 125 percent of the rated primary current of the transformer is not sufficient to allow loads with high inrush currents to start and operate. However, the secondary overcurrent protection device shall be sized at 125 percent of the rated secondary full-load current of the transformer. Where the rated secondary current of a transformer is less than 9 amps, an overcurrent device rated or set at no more than 167 percent of secondary current shall be used. **(See Figure 20-11)**

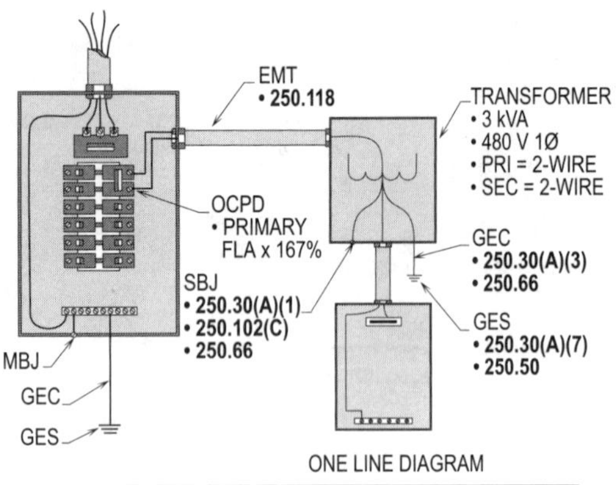

FINDING OCPD FOR PRIMARY SIDE OF TRANSFORMER

Sizing primary OCPD (Second Level)

Step 1: Finding FLA of primary
FLA = kVA x (1000 ÷ V)
FLA = 3 kVA x (1000 ÷ 480 V)
FLA = 6.25 A

Step 2: Calculating OCPD
450.3(B) and **Table 450.3(B)**
6.25 A x 167% = 10.4 A

Step 3: Selecting OCPD
Table 450.3(B) and **240.6(A)**
10.4 A allows 10 A

Solution: The size OCPD in the primary side is 10 amps.

**PRIMARY 2 AMPS OR MORE
BUT LESS THAN 9 AMPS
NEC 450.3(B) AND TABLE 450.3(B)**

Figure 20-10(b). Where the rated primary current of a transformer is less than 9 amps but 2 amps or more, an overcurrent device rated or set at no more than 167 percent of primary current shall be used.

9 AMPS OR MORE
TABLE 430.3(B), NOTE 1

Where the rated secondary current of a transformer is 9 amps or more and 125 percent of this current does not correspond to a standard rating of a fuse or circuit breaker, the next size shall be permitted to be used per **240.6(A)**.

GROUNDING AUTOTRANSFORMERS
450.5

Autotransformers are connected to three-phase, three-wire ungrounded systems to derive a three-phase, four-wire grounded system. Three autotransformers connected in a star (wye) configuration to the three-phase ungrounded system converts to a three-phase, four-wire grounded system.

Autotransformers are installed today because many electrical systems are not grounded. Existing ungrounded delta systems are grounded with autotransformers to derive

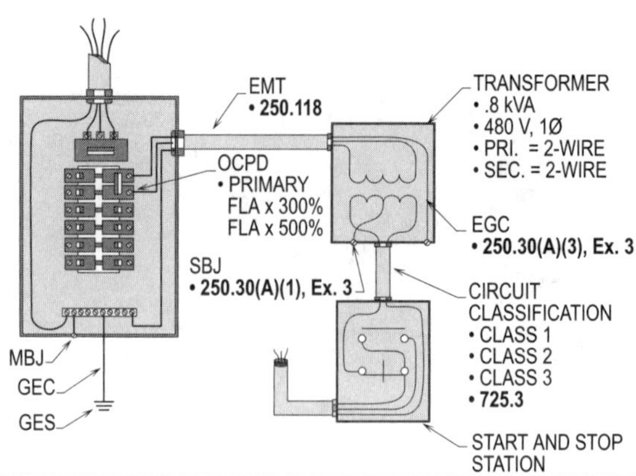

FINDING OCPD FOR PRIMARY SIDE OF TRANSFORMER

Sizing min. primary OCPD (Third Level)

Step 1: Finding FLA of primary
FLA = kVA x (1000 ÷ V)
FLA = .8 kVA x (1000 ÷ 480 V)
FLA = 1.7 A

Step 2: Calculating OCPD
450.3(B) and **Table 450.3(B)**
1.7 A x 300% = 5.1 A

Step 3: Selecting OCPD
Table 450.3(B) and **240.6(A)**
5.1 A allows 3 A

Solution: The minimum size OCPD in the primary side is 3 amps.

Sizing max. primary OCPD (Third Level)

Step 1: Calculating OCPD
430.72(C)(4) and **Step 1**
1.7 A x 500% = 8.5 A

Step 2: Selecting OCPD
430.72(C)(4) and **240.6(A)**
8.5 A allows 6 A

Solution: The maximum size OCPD in the primary side is 6 amps.

Note: The secondary side is a motor controller circuit.

**PRIMARY LESS THAN 2 AMPS
NEC 450.3(B), NEC 430.72(C)(4), AND TABLE 450.3(B)**

Figure 20-10(c). When the rated primary current of a transformer is less than 2 amps, an overcurrent protection device rated or set at not more than 300 percent shall be used, unless **430.72(C)(4)** is applied.

a neutral. Three-phase zigzag transformers are generally installed for this purpose.

THREE-WIRE CIRCUIT TO THREE-PHASE, FOUR-WIRE CIRCUIT
450.5(A)

Grounding autotransformers are connected to derive a neutral from a three-phase, three-wire ungrounded system to a three-phase, four-wire grounded system, the following conditions shall apply:

• Proper connections shall be made.

• Overcurrent protection shall be provided.

• Transformer fault sensing shall be installed.

• Rating shall be adequately sized.

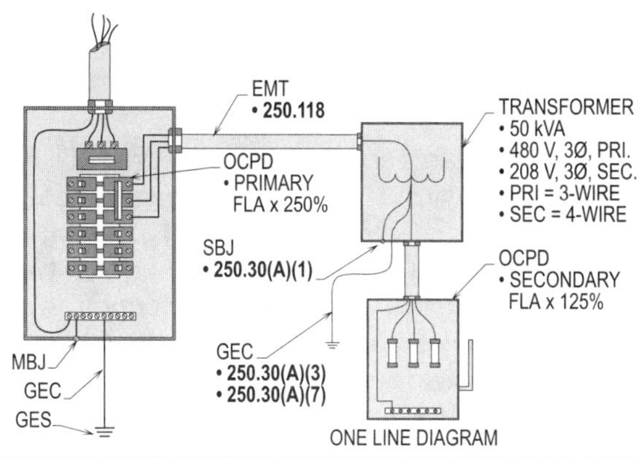

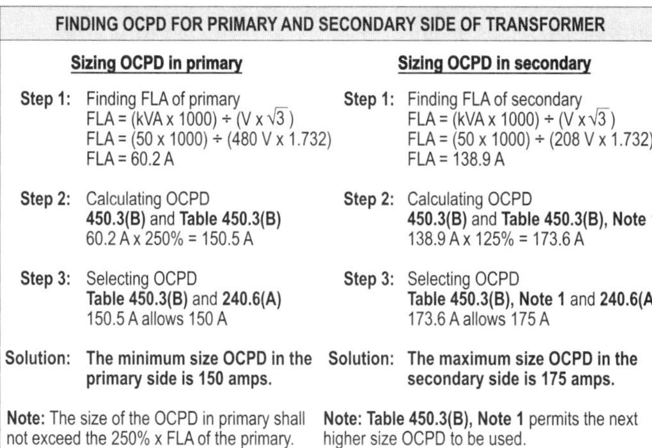

FINDING OCPD FOR PRIMARY AND SECONDARY SIDE OF TRANSFORMER

Sizing OCPD in primary	Sizing OCPD in secondary
Step 1: Finding FLA of primary FLA = (kVA x 1000) ÷ (V x $\sqrt{3}$) FLA = (50 x 1000) ÷ (480 V x 1.732) FLA = 60.2 A	**Step 1:** Finding FLA of secondary FLA = (kVA x 1000) ÷ (V x $\sqrt{3}$) FLA = (50 x 1000) ÷ (208 V x 1.732) FLA = 138.9 A
Step 2: Calculating OCPD **450.3(B)** and **Table 450.3(B)** 60.2 A x 250% = 150.5 A	**Step 2:** Calculating OCPD **450.3(B)** and **Table 450.3(B), Note 1** 138.9 A x 125% = 173.6 A
Step 3: Selecting OCPD **Table 450.3(B)** and **240.6(A)** 150.5 A allows 150 A	**Step 3:** Selecting OCPD **Table 450.3(B), Note 1** and **240.6(A)** 173.6 A allows 175 A
Solution: **The minimum size OCPD in the primary side is 150 amps.**	**Solution:** **The maximum size OCPD in the secondary side is 175 amps.**
Note: The size of the OCPD in primary shall not exceed the 250% x FLA of the primary.	**Note: Table 450.3(B), Note 1** permits the next higher size OCPD to be used.

**PRIMARY AND SECONDARY
600 VOLTS OR LESS
NEC 450.3(B) AND TABLE 450.3(B)**

Figure 20-11. Sizing overcurrent protection device for the primary and secondary side of a transformer rated 600 volts or less.

CONNECTIONS
450.5(A)(1)

Transformers shall be directly connected to the ungrounded (phase) conductors with no switches or overcurrent protection devices installed between the connection and the autotransformer.

OVERCURRENT PROTECTION
450.5(A)(2)

An overcurrent protection sensing device shall be designed to trip at 125 percent of its continuous current per phase or neutral rating. The next higher standard rating shall be permitted to be installed where the input current is 9 amps or more and calculated at 125 percent. Input current of 2 amps or less shall not exceed 167 percent. **(See Figure 20-12)**

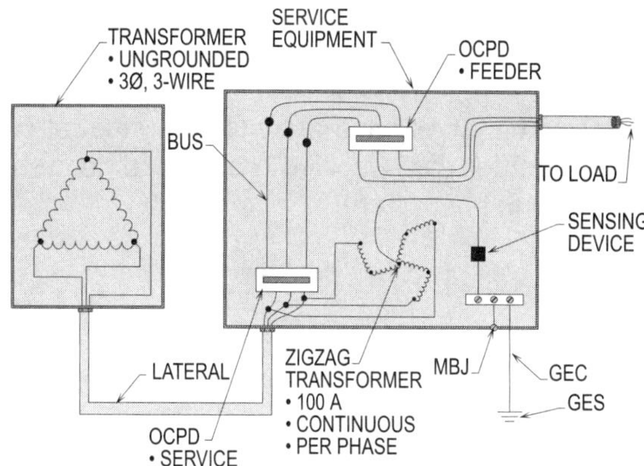

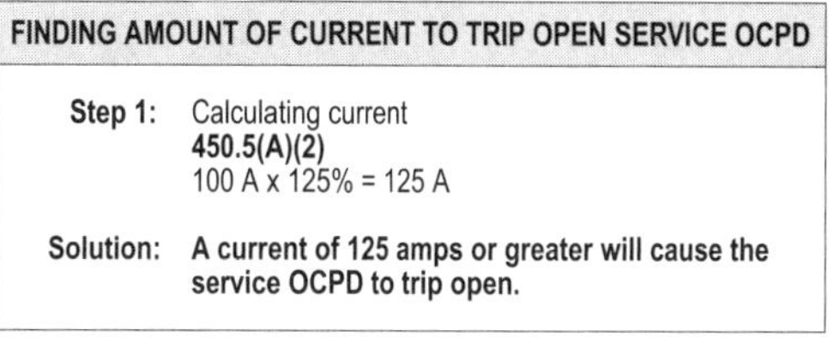

FINDING AMOUNT OF CURRENT TO TRIP OPEN SERVICE OCPD

Step 1: Calculating current
450.5(A)(2)
100 A x 125% = 125 A

Solution: A current of 125 amps or greater will cause the service OCPD to trip open.

**OVERCURRENT PROTECTION
NEC 450.5(A)(2)**

Figure 20-12. An overcurrent protection sensing device shall be designed to trip at 125 percent of its continuous current per phase or neutral rating.

TRANSFORMER FAULT SENSING
450.5(A)(3)

A main switch or common-trip overcurrent protection device for the three-phase, four-wire system shall be provided with fault sensing systems to guard against single-phasing or internal faults.

RATING
450.5(A)(4)

Autotransformers shall be designed with a continuous neutral current rating sufficient to handle the maximum possible unbalanced neutral load current that could flow in the four-wire system.

DETECTING GROUNDS ON THREE-PHASE, THREE-WIRE SYSTEMS
450.5(B)

The following conditions shall apply when autotransformers are used to detect grounds on three-phase, three-wire systems:

- Proper rating
- Overcurrent protection sized adequately
- Ground reference for damping transitory overvoltages

RATING
450.5(B)(1)

Autotransformers shall have a continuous neutral current rating sufficient for the specified ground fault current that could develop in the system.

OVERCURRENT PROTECTION
450.5(B)(2)

The overcurrent protection device shall open simultaneously with a common trip all ungrounded conductors and be set to trip at not more than 125 percent of the rated phase current of the transformer. Note that 42 percent of the overcurrent protection device's rating may be used if connected in the autotransformer's neutral connection. When dealing with high impedance grounded systems per **250.36**, review the **Ex.** to **450.5(B)(2)** and **110.9** of the NEC. **(See Figure 20-13)**

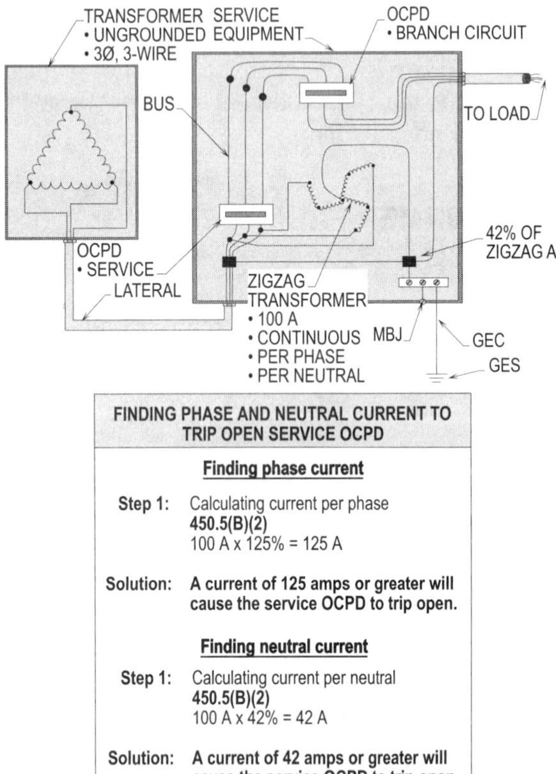

FINDING PHASE AND NEUTRAL CURRENT TO TRIP OPEN SERVICE OCPD

Finding phase current

Step 1: Calculating current per phase
450.5(B)(2)
100 A x 125% = 125 A

Solution: **A current of 125 amps or greater will cause the service OCPD to trip open.**

Finding neutral current

Step 1: Calculating current per neutral
450.5(B)(2)
100 A x 42% = 42 A

Solution: **A current of 42 amps or greater will cause the service OCPD to trip open.**

OVERCURRENT PROTECTION
NEC 450.5(B)(2)

Figure 20-13. The overcurrent protection device shall open simultaneously with a common trip all ungrounded conductors and be set to trip at not more than 125 percent of the rated phase current of the transformer. Note that 42 percent of the overcurrent protection device's rating shall be used if connected in the autotransformer's neutral connection. Review Sections **250.24(B), 250.30(A)(1),** and **250.32(B), Ex**.

SECONDARY TIES
450.6

In large industrial plants and facilities, a "network" distribution system is usually utilized for supplying power loads. Three-phase banks of transformers are located at various points throughout the plant or facility. There are normally two high-tension primary circuits feeding such transformers. A double-throw switch that is located at each transformer bank allows either primary circuit to serve any bank of transformers. The primary circuit conductors are sized with enough capacity so that either circuit is capable of carrying the entire load if a fault develops in the other circuit. Secondary voltage is usually three-phase systems rated 600 volts or less. The transformer secondaries are connected together in a network system, and all transformers are used to feed all the loads involved that can be all at once or as necessary.

Secondary ties shall be protected at both ends, and such protection shall be permitted to be fuses based upon the current-carrying capacity of the conductors per **450.6(A)** or the ties protected by a limiter installed at each end per **450.6(A)(3)**. A limiter protects the elements against short circuit; however, it does not provide overload protection. Usually, limiters, rather than fuses, are used for protection of the ties due to the fact they are very current limiting and will protect the circuit elements from damage during short-circuit conditions.

There is normally a load center connected to the tie at the points where a transformer bank connects to the tie. The transformer is protected by a circuit breaker in the secondary leads between the transformer and the load center. Circuit breaker setting shall be permitted to be up to 250 percent (2.5 times) of transformer's secondary current rating per **450.6(B)**. A reverse power relay shall be provided per **450.6(B)** that opens the circuit in case the transformer should fail for any reason. A reverse power relay is provided to prevent current from being fed to an out-of-service transformer from the other transformers of the network. Where the secondary voltage is greater than 150 volts to ground, to ensure adequate protection, ties shall be provided with a switch at each end per **450.6(A)(5)**.

The rules for designing and installing secondary ties can be summed up as follows:

• Where transformers are tied together in parallel and connected by tie conductors that do not have overcurrent protection as per **Article 240** of the NEC, the ampacity of the ties connecting conductors shall not be less than 67 percent of the rated secondary current of the largest transformer in the tie circuit. **(See Figure 20-14)**

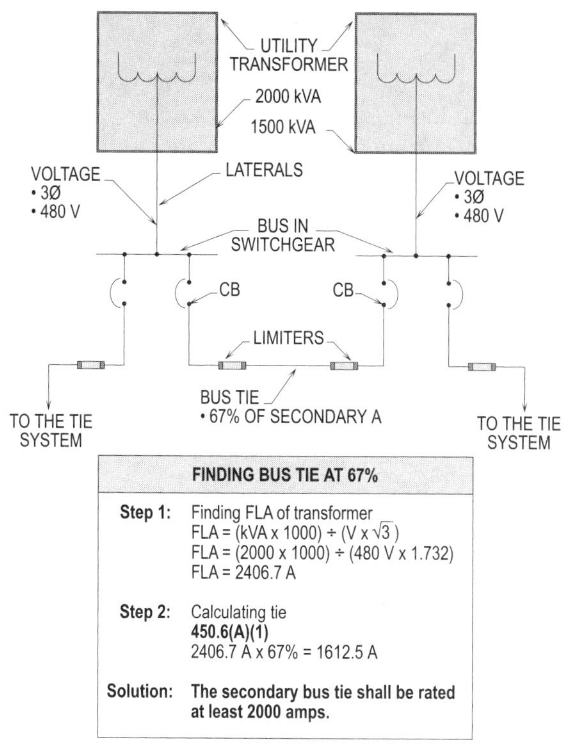

LOADS AT TRANSFORMER SUPPLY POINTS ONLY
NEC 450.6(A)(1)

Figure 20-14. The ampacity of the ties connecting conductors shall not be less than 67 percent of the rated secondary current of the largest transformer in the tie circuit.

The paralleling of transformers is common, but great care should be exercised to ensure that the transformers are similar in all conditions of use. If they are not, one transformer will try to carry more of the load than the other transformer load. If the transformers are of the same capacity and similar characteristics, they each, in theory, will carry 50 percent of the total load. The 67 percent allows for differences in transformer sizes and thus allows for adjusting solutions.

- Where the load is connected to the tie at any point between the transformer supply points, and overcurrent protection is not provided by the provisions listed in **Article 240** of the NEC, the rated ampacity of the tie shall not be less than 100 percent of the rated secondary current of the largest transformer connected to the secondary tie system except as provided in **450.6(A)(4). (See Figure 20-15)**

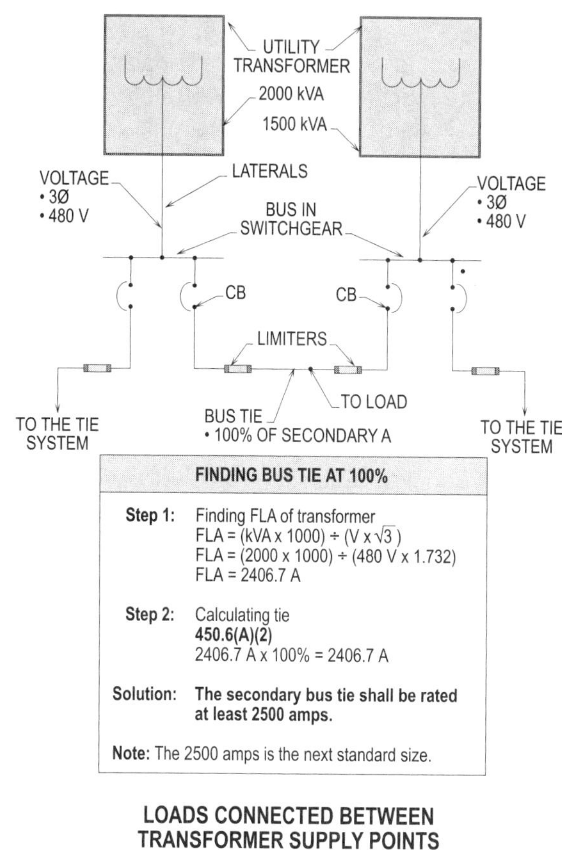

LOADS CONNECTED BETWEEN
TRANSFORMER SUPPLY POINTS
NEC 450.6(A)(2)

Figure 20-15. The rated ampacity of the tie shall not be less than 100 percent of the rated secondary current of the largest transformer connected to the secondary tie system except as provided in **450.6(A)(4)**.

- Sections **450.6(A)(1)** and **(A)(2)** state that both ends of each tie connection shall be provided with a protective device that opens at a certain temperature of the tie conductor. This prevents damage to the tie conductor and its insulation, and such installations shall consist of:

(a) A limiter is a fusible link cable connector. The limiter is selected and designed for the insulation, conductor material, etc. on the tie conductors.

(b) A circuit breaker, actuated by devices having characteristics that are comparable to the above, can be used if designed and sized properly.

- Where the tie consists of more than one conductor per phase, the conductors of each phase shall be interconnected in order to create a load supply point. The protection required in **450.6(A)(3)** is to be provided in each tie conductor at this point, except as follows:

Section **450.6(A)(4)(b)** permits the loads to be connected to the individual conductor(s) of each phase and without the protection listed in **450.6(A)(3)** if, at load connection points, the tie conductors of each phase have a combined capacity of not less than 133 percent of the rated secondary current of the largest transformer connected to the secondary tie system. The total load of such taps shall not exceed the rated secondary current of the largest transformer, and the loads shall be equally divided on each phase and on the individual conductors of each phase as closely as possible. **(See Figure 20-16)**

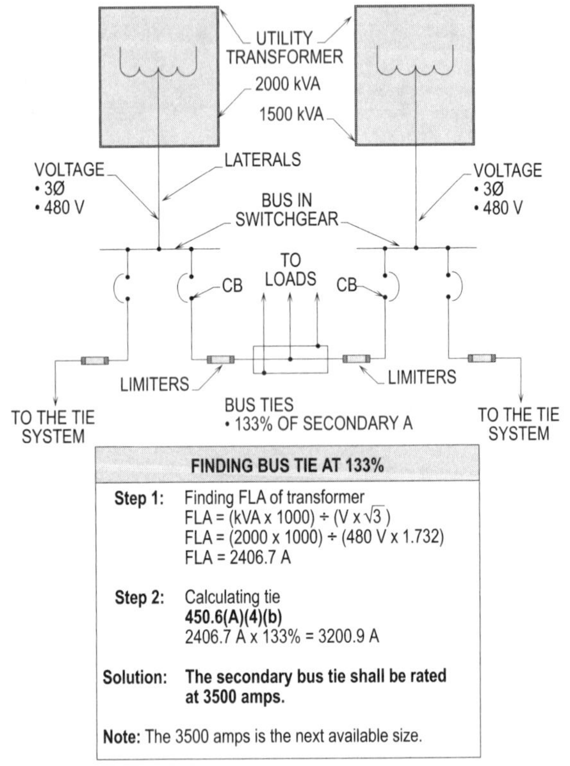

FINDING BUS TIE AT 133%

Step 1: Finding FLA of transformer
$FLA = (kVA \times 1000) \div (V \times \sqrt{3})$
$FLA = (2000 \times 1000) \div (480\ V \times 1.732)$
$FLA = 2406.7\ A$

Step 2: Calculating tie
450.6(A)(4)(b)
$2406.7\ A \times 133\% = 3200.9\ A$

Solution: **The secondary bus tie shall be rated at 3500 amps.**

Note: The 3500 amps is the next available size.

**INTERCONNECTION OF PHASE CONDUCTORS
BETWEEN TRANSFORMER SUPPLY POINTS
NEC 450.6(A)(4)(b)**

Figure 20-16. Section **450.6(A)(4)(b)** permits the loads to be connected to the individual conductor(s) of each phase and without the protection listed in **450.6(A)(3)** if, at load connection points, the tie conductors of each phase have a combined capacity of not less than 133 percent of the rated secondary current of the largest transformer connected to the secondary tie system.

The use of multiple conductors on each phase and the requirement that loads do not have to tap the multiple conductors of the same phase might possibly set up unbalanced current flow in the multiple conductors on the same phase.

The requirement that the combined capacity of the multiple conductors on the same phase be rated at 133 percent of the secondary current of the largest transformer is satisfied. Limiters are necessary at the tap or connections to the transformers that are tied together to properly protect the elements of the circuit.

Design Tip: The above applies where the interconnection of phase conductors between transformers supply points occurs, per **450.6(A)(4)(b)**.

- If the operating voltage of secondary ties exceeds 150 volts to ground, there shall be a switch ahead of the limiters and tie conductors that is capable of deenergizing the tie conductors and the limiters. This switch shall comply with the following:

    (a) The current rating of the switch shall not be less than the current rating of the conductors connected to such switch.

    (b) The switch shall be capable of opening its rated current.

    (c) The switch shall not open under the magnetic forces caused by short-circuit currents.

Design Tip: The above applies where the tie circuit control is located as mentioned in **450.6(A)(5)**.

- When secondary ties from transformers are used, an overcurrent device in the secondary of each transformer that is rated or set at not greater than 250 percent (2.5 times) of the rated secondary current of the transformer shall be provided. In addition, there shall be a circuit breaker actuated by a reverse-current relay; the breaker is to be set at not greater than the rated secondary current of the transformer. Such overcurrent protection protects against overloads and short-circuit conditions, and the reverse-current relay and circuit breaker shall be designed to handle any reversal of current flow into the transformer. **(See Figure 20-17)**

Design Tip: The above applies where overcurrent protection for secondary connections is installed to protect the system as required by **450.6(B)**.

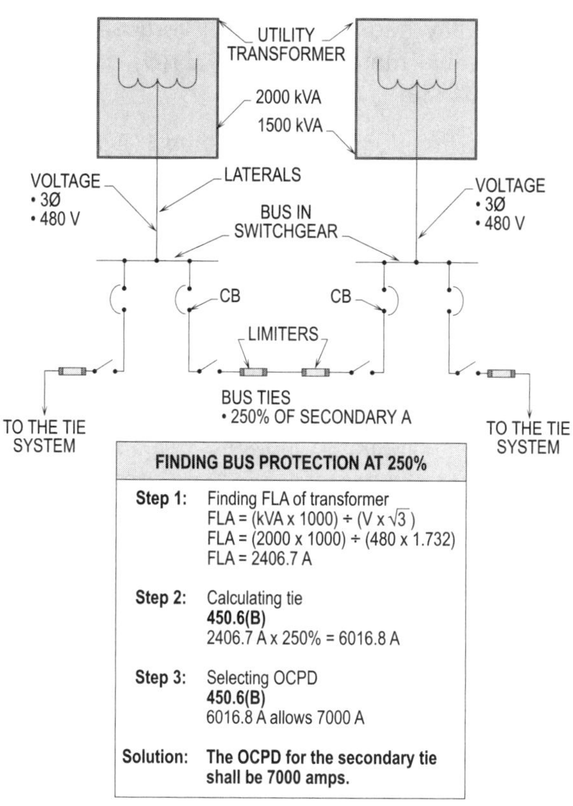

Figure 20-17. When secondary ties from transformers are used, an overcurrent device in the secondary of each transformer that is rated or set at not greater than 250 percent (2.5 times) of the rated secondary current of the transformer shall be provided.

PARALLEL OPERATION
450.7

Transformers shall be permitted to be connected in parallel and switched as a unit, provided each transformer has overcurrent protection that is properly sized and complies with **450.3(A)** and **450.3(B)**. **(See Figure 20-18)**

Persons working with paralleled transformers or transformer tie circuits should be extremely careful that there are no feedbacks or other conditions that would affect safety. In order to secure a balance of current between paralleled transformers, all transformers should have characteristics that are very much alike, such as voltage, impedance, and other such pertinent elements.

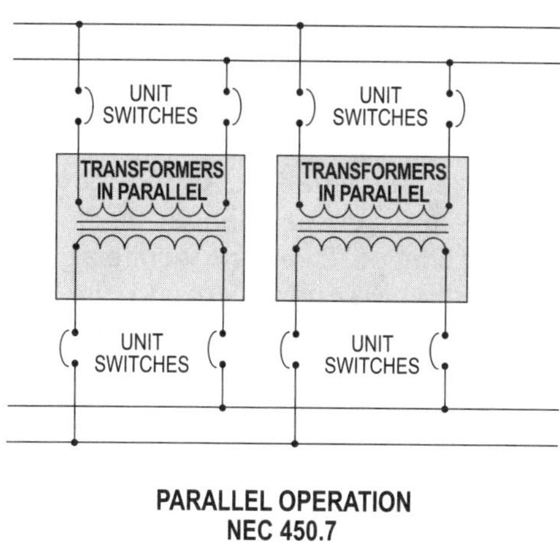

Figure 20-18. Transformers shall be permitted to be connected in parallel and switched as a unit, provided each transformer has overcurrent protection that is properly sized.

GUARDING
450.8

Transformers shall be permitted to be isolated in a room or accessible only to qualified personnel to prevent accidental contact with live parts. To safeguard live parts from possible damage, the transformer shall be elevated. The following are acceptable means of safeguarding live parts as required in **110.27(A)** and **110.34(E)**.

- Transformers shall be permitted to be isolated in a room or accessible only to qualified personnel.

- Permanent partitions or screens shall be permitted to be installed.

- Transformers shall be elevated at least 8 ft (2.5 m) above the floor to prevent unauthorized personnel from contact.

Design Tip: Signs indicating the voltage of live exposed parts of transformers, or other suitable markings, shall be used in areas where transformers are located.

VENTILATION
450.9

Transformers shall be located and installed in rooms or areas that are not subject to exceedingly high temperatures to prevent overheating and possible damage to windings. Transformers with ventilating openings shall be installed so that the ventilating openings are not blocked by walls or other obstructions that could block air flow.

GROUNDING
450.10

Transformer cases shall be grounded per **250.110**. Transformers enclosed by fences or guards shall be grounded and bonded. Live parts require a guard such as a fence to guard against unauthorized entry and to protect the general public and unqualified persons from dangerous electrical parts. Such fences shall be grounded and bonded to prevent metal elements from being accidentally energized with voltage. **(See Figure 20-19)**

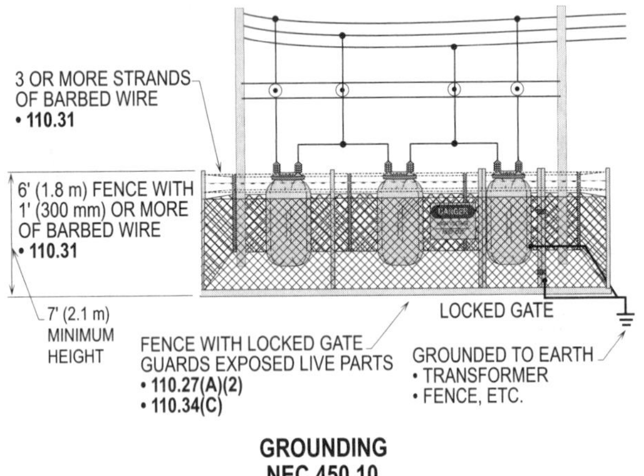

Figure 20-19. Transformers enclosed by fences or guards shall be grounded and bonded. Live parts require a guard such as a fence to guard against unauthorized entry and to protect the general public and unqualified persons from dangerous electrical parts.

TYPES OF TRANSFORMERS
PART II TO ARTICLE 450

Transformers are either used indoors or outdoors based upon their type and condition of use. The type of transformer determines where it is permitted to be installed inside or outside of the building. Sometimes it becomes necessary to build rooms of certain fire-rated material or confinement areas to house transformers due to their design and installation.

DRY-TYPE TRANSFORMERS
INSTALLED INDOORS
450.21

The rules for installing dry-type transformers indoors can be summed up as follows:

- Dry-type transformers greater than 112-1/2 kVA, having Class 155 or higher insulation systems, shall have a fire-resistant, heat-insulating barrier placed between transformers and combustible material, or,

if no barrier, shall be separated at least 6 ft (1.83 m) horizontally and 12 ft (3.7 m) vertically from the combustible material per **450.21(B), Ex. 1**. **(See Figure 20-20)**

- Dry-type transformers greater than 112-1/2 kVA and having Class 155 or higher of insulation systems shall be installed in a fire-resistant transformer room per **450.21(B), Ex. 2**. **(See Figure 20-21)**

- Dry-type transformers rated at 112-1/2 kVA or less and 600 volts or less shall have a fire-resistant, heat-insulating barrier between transformers and combustible material, or, without a barrier, shall be separated at least 12 in. (300 mm) from the combustible material where the voltage is 600 volts or less per **450.21(A)**. **(See Figure 20-22)**

- Dry-type transformers rated 112-1/2 kVA or less and 600 volts or less shall not be required to have a 12 in. (300 mm) separation or barrier if they are completely enclosed, except for vent openings. **(See Figure 20-23)**

- All indoor dry-type transformers of over 35,000 volts shall be installed in a vault. Vault requirements shall fully comply with **Part III** of **Article 450** of the NEC. **(See Figure 20-24)**

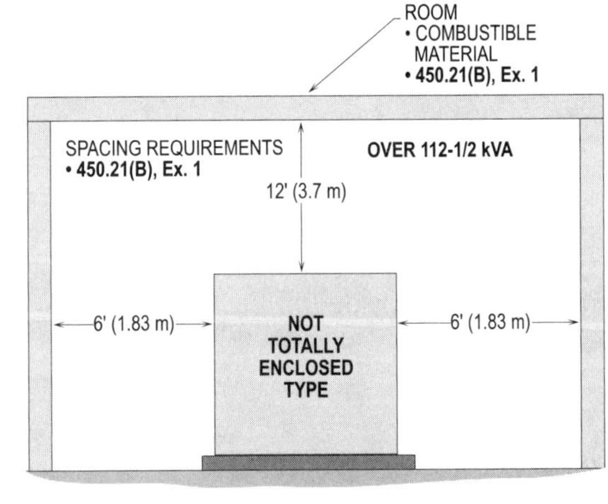

Figure 20-20. Dry-type transformers greater than 112-1/2 kVA and having Class 155 or higher insulation systems shall have a fire-resistant, heat-insulating barrier placed between transformers and combustible material, or, if no barrier, shall be separated at least 6 ft (1.83 m) horizontally and 12 ft (3.7 m) vertically from the combustible material.

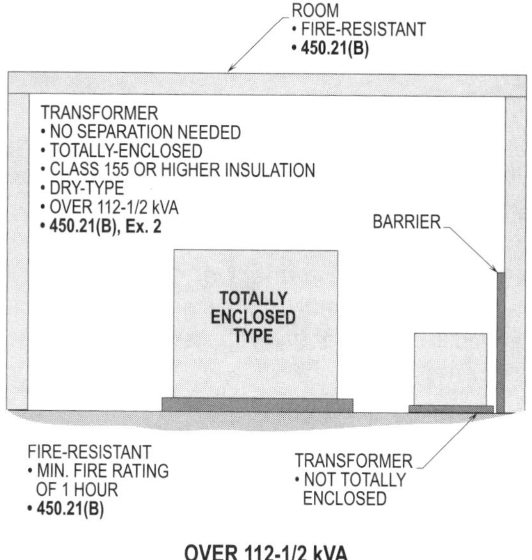

Figure 20-21. Dry-type transformers greater than 112-1/2 kVA and having Class 155 or higher insulation systems shall be installed in a fire-resistant transformer room.

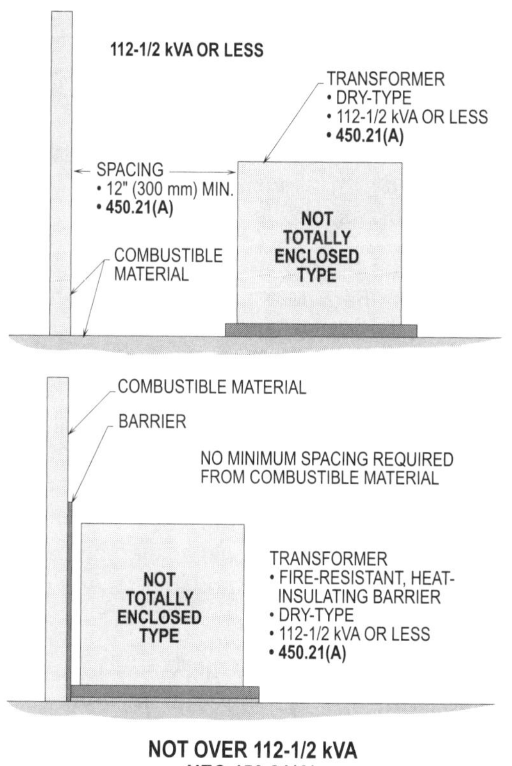

Figure 20-22. Dry-type transformers rated 112-1/2 kVA or less and 600 volts or less shall have a fire-resistant, heat-insulating barrier between transformers and combustible material, or, without a barrier, shall be separated at least 12 in. (300 mm) from the combustible material where the voltage is over 600 volts.

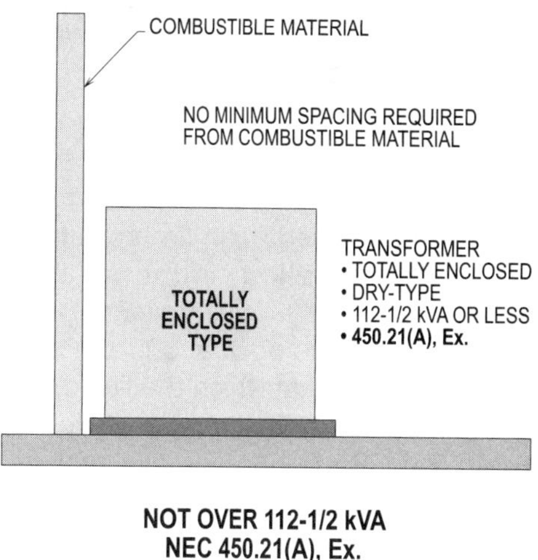

Figure 20-23. Dry-type transformers rated 112-1/2 kVA or less, and 600 volts or less shall not be required to have a 12 in. (300 mm) separation or barrier if they are completely enclosed except for vent openings.

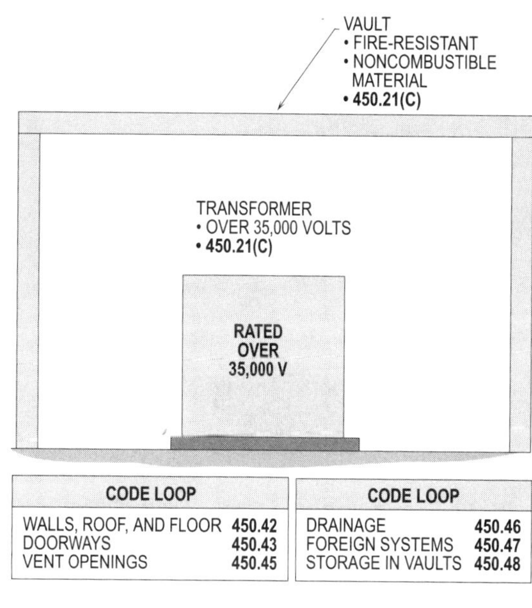

CODE LOOP		CODE LOOP	
WALLS, ROOF, AND FLOOR	450.42	DRAINAGE	450.46
DOORWAYS	450.43	FOREIGN SYSTEMS	450.47
VENT OPENINGS	450.45	STORAGE IN VAULTS	450.48

Figure 20-24. All indoor dry-type transformers of over 35,000 volts shall be installed in a vault.

DRY-TYPE TRANSFORMERS INSTALLED OUTDOORS
450.22

Dry-type transformers installed outdoors shall have weatherproof enclosures. See the definition in **Article 100** of the NEC for the difference between "weatherproof" and "watertight."

LESS FLAMMABLE LIQUID-INSULATED TRANSFORMERS
450.23

Transformers using a "listed" high fire point liquid shall be permitted to be installed indoors, but only in "noncombustible" areas of "noncombustible" buildings. The NEC sets the minimum fire point at 300°C (572°F). This is the minimum temperature at which the liquid ignites. Such transformers shall be permitted to be installed indoors, for voltages up to 35,000. Higher voltages require a vault if they are installed indoors. This is due to the safety required because of the higher voltage and associated equipment. **(See Figure 20-25)**

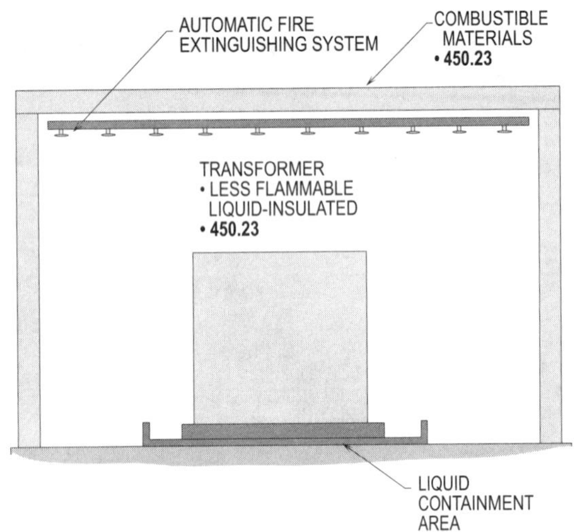

Figure 20-25. Transformers using a "listed" high fire point liquid shall be permitted to be installed indoors, but only in "noncombustible" areas of "noncombustible" buildings. The NEC sets the minimum fire point at 300°C (572°F).

NONFLAMMABLE FLUID-INSULATED TRANSFORMERS
450.24

Transformers using a "dielectric" nonflammable liquid shall be permitted to be installed indoors in any location, for voltages up to 35,000. Higher voltages require a vault, when installed indoors due to the safety required for the higher voltage and associated equipment.

For the purpose of this section, a nonflammable dielectric fluid is one that does not have a flash point or fire point, and is not flammable in air.

ASKAREL-INSULATED TRANSFORMERS INSTALLED INDOORS
450.25

Askarel is a liquid that does not burn; therefore, it is safer than oil for use as a transformer liquid. However, arcing in askarel produces greater gases that are nonexplosive.

Askarel-insulated transformers of over 25 kVA shall be furnished with a relief vent such as a chimney to relieve the pressure built up by gases that may be generated within the transformer.

In rooms that are well ventilated, the vent may be discharged directly to the room. In rooms that are poorly ventilated, the vent shall be piped to a flue or chimney that is capable of carrying the gases out of the room. Or, as an alternative to such ventilating, the transformer can be fitted with a gas absorber placed inside the case. When there is a gas absorber, the vent may also be discharged to the room.

Askarel transformers of more than 35,000 volts shall be installed in a vault because the oil and higher voltage are a hazard to unqualified personnel. **(See Figure 20-26)**

OIL-INSULATED TRANSFORMERS INSTALLED INDOORS
450.26

The rules for installing oil-insulated transformers indoors can be summed up as follows:

- Indoor oil-filled transformers greater than 600 volts shall be installed in a vault, with the following exceptions, where, regardless of voltage, a vault is not required:
  (a) Electric furnace transformers with a total rating of 75 kVA or less shall be permitted to be located in a fire-resistant room.
  (b) Oil-filled transformers may be installed in a building without a vault, provided the building is accessible to qualified personnel only and is used solely for providing electric service to other buildings.

- If suitable provisions are provided to prevent a possible oil fire from igniting other materials, oil-filled transformers of 600 volts or less shall be permitted to be installed without a vault. When installed without a vault, the total kVA rating of all transformers allowed in a room or section of a building is limited to 10 kVA for nonfire-resistant buildings and to 75 kVA for fire-resistant buildings.

See Figure 20-27 for installing rules when applying the **Ex. 1** to **450.26**.

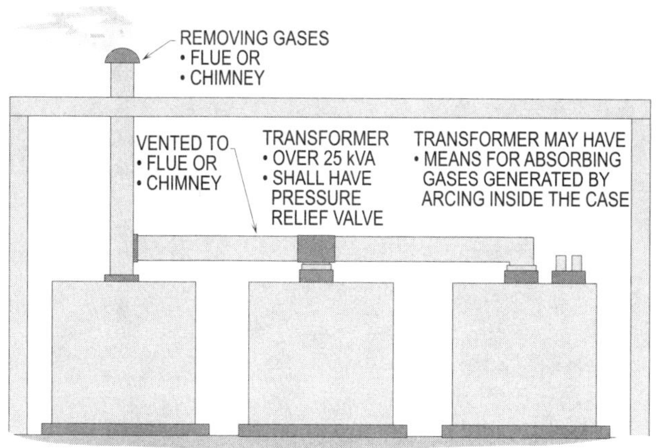

Figure 20-26. Askarel-insulated transformers of over 25 kVA shall be furnished with a relief vent such as a chimney to relieve the pressure built up by gases that may be generated within the transformer.

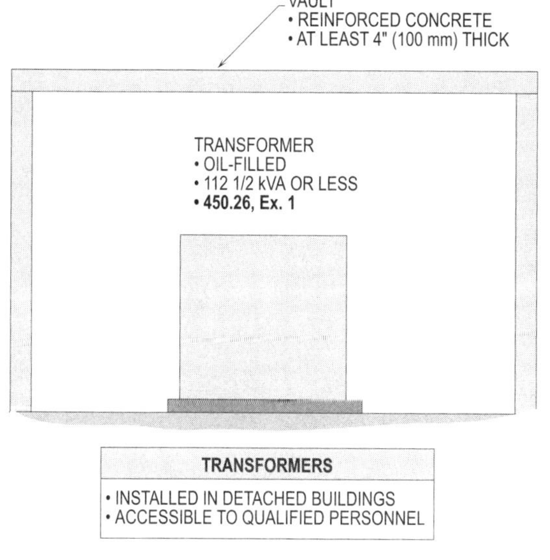

Figure 20-27. Oil-filled transformers rated 112-1/2 kVA or less that are installed in detached buildings and accessible to only qualified personnel shall be installed in a vault with reinforced concrete at least 4 in. (100 mm) thick.

OIL-INSULATED TRANSFORMERS INSTALLED OUTDOORS
450.27

When oil-filled transformers are installed on or adjacent to combustible buildings or material, the building or material shall be safeguarded from possible fire originating in a transformer. Fire-resistant barriers, water-spray systems, and enclosures for the transformers are approved safeguards if, where used, they are installed by the rules of the NEC. **[See Figures 20-28(a) and (b)]**

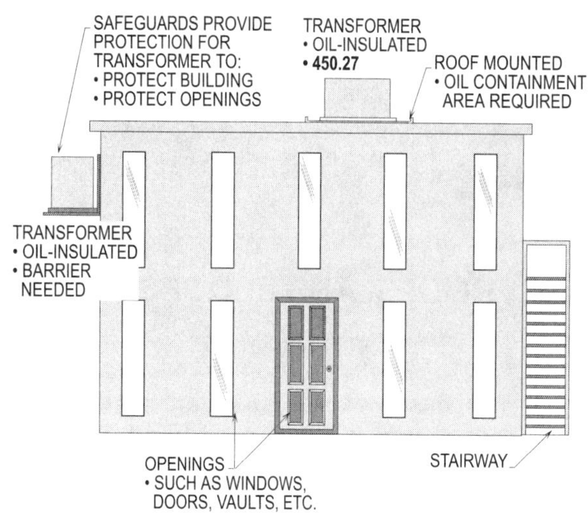

Figure 20-28(a). When oil-filled transformers are installed on or adjacent to combustible buildings or material, the building or material shall be safeguarded from possible fire originating in a transformer.

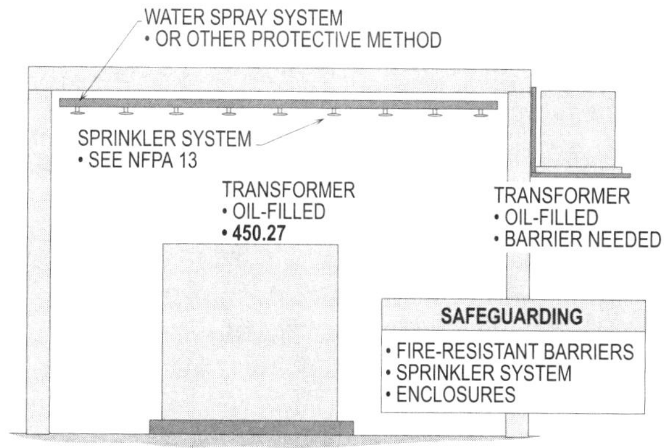

Figure 20-28(b). Fire-resistant barriers, water-spray systems, and enclosures for the transformers are approved safeguards if, where used, they are installed by the rules of the NEC.

LOCATION OF TRANSFORMER VAULTS
450.41

Vaults are used to house dry-type transformers that are rated over 35 kV or transformers filled with combustible material used as an aid in cooling their windings.

Vaults shall be designed and built with specific rules and regulations.

Wherever possible, transformer vaults shall be located at an outside wall of the building. This rule is intended to allow ventilation direct to the outside without using ducts, flues, etc. per **450.45**.

WALLS, ROOFS, AND FLOORS
450.42

The rules for construction of vaults are set forth in this section. Floor, walls, and roof shall be of fire-resistant material such as concrete and be capable of withstanding heat from a fire within for at least three hours. A 6 in. (150 mm) thickness is specified for the walls and roof. The floor, when laid and in contact with the earth, shall be at least 4 in. (100 mm) thick. Walls, roof, and floor shall have at least have a 3 hour fire rating. **(See Figure 20-29)**

DOORWAYS
450.43

The door to a transformer vault shall be built according to the standards of the National Fire Protection Association, which requires a 3 hour fire rating. The door sill shall be at least 4 in. (100 mm) high. This is to prevent any oil that may accumulate on the floor from running out of the transformer room and moving to other areas. Doors shall be kept locked at all times to prevent access of unqualified persons to the vault.

> **Design Tip:** Personnel doors shall swing out and be equipped with panic bars, pressure plates, or other devices that open under simple pressure per **450.43(C)**.

VENTILATION OPENINGS
450.45

Where ventilation is direct to the outside, without the use of ducts or flues, the vent opening shall have an area of at least 3 sq. in. (1900 mm²) for each kVA of transformer capacity, but never less than 1 sq. ft (0.1 m²) in area. The vent opening shall be fitted with a screen or grating and an automatic closing damper. If ducts are used in the vent

system, the ducts shall have sufficient capacity to maintain a suitable vault temperature. **(See Figure 20-30)**

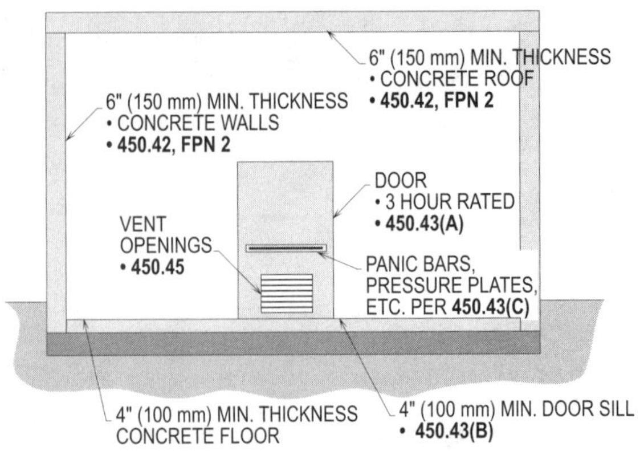

WALLS, ROOFS, AND FLOORS
NEC 450.42

Figure 20-29. Floor, walls, and roof shall be of fire-resistant material such as concrete and be capable of withstanding heat from a fire within for at least three hours. A 6 in. (150 mm) thickness is specified for the walls and roof. The floor, when laid and in contact with the earth, shall be at least 4 in. (100 mm) thick.

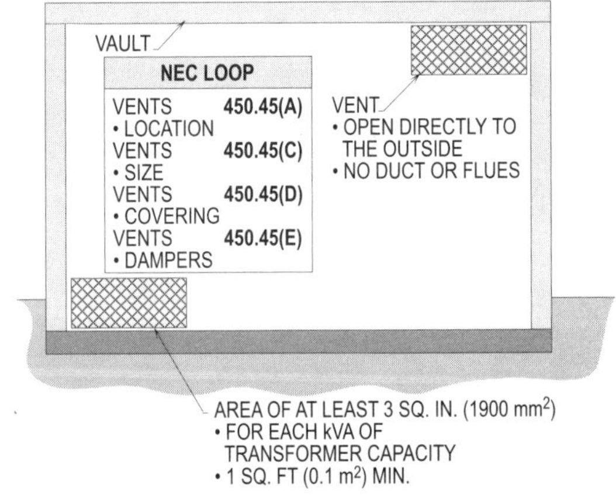

VENTILATION OPENINGS
NEC 450.45

Figure 20-30. Where ventilation is direct to the outside, without the use of ducts or flues, the vent opening shall have an area of at least 3 sq. in. (1900 mm²) for each kVA of transformer capacity, but never less than 1 sq. ft (0.1 m²) in area.

DRAINAGE
450.46

Drains shall be provided to drain off oil that might accumulate on the floor due to a leak in a transformer caused by an accident. This rule is designed to prevent a fire hazard from occurring.

WATER PIPES AND ACCESSORIES
450.47

Piping for fire protection within the vault or piping to water-cooled transformers shall be permitted to be present in a vault. No other piping or duct system shall enter or pass through. Valves or other fittings of a foreign piping or duct system shall not be permitted in a vault containing transformers. **(See Figure 20-31)**

STORAGE IN VAULTS
450.48

No storage of any kind shall be permitted in a vault other than the transformers and equipment necessary for their operation. This typically means that transformer vaults are not to be used for warehouses or storage areas but to contain transformers and accessories only. The reasons the vault is to be kept clear are the high voltage and safety measures needed for personnel servicing such equipment. Also, consideration shall be given to foreign material being a threat of fire under certain conditions. **(See Figure 20-32)**

CONNECTIONS FROM THE SECONDARY OF TRANSFORMERS
240.21(B) AND (C)

Overcurrent protection devices of circuits shall be located at the point where the service to those circuits originates. However, it shall be permitted to make connections from the secondary side of transformers. Such conductors shall be designed and installed by the rules and regulations of **240.21(B)** and **(C)** of the NEC. Sizing connections, not over 25 ft (7.5 m) long, shall be designed and installed per **240.21(B)(3)** and **(C)(5)**. Transformer secondary conductors of separately derived systems for industrial locations are sized per **240.21(C)(2)**, **(C)(3),** and **(C)(6)**. Outside transformer connections are sized per **240.21(C)(4)**. Overcurrent protection shall be provided by **450.3(B)** and **Table 450.3(B)**. (See **240.92(B)** and **(D)** and **Figure 20-33** for illustrated diagrams)

NOT OVER 10 FT (3 m) LONG
240.21(C)(2)

Conductors shall be permitted to be connected, without overcurrent protection at the connection, to a feeder or transformer secondary where all the following conditions are met:

- Connecting conductors do not exceed 10 ft (3 m) in length.

- Connecting conductors shall have a current rating

not less than the combined calculated loads of the circuits supplied by connecting conductors. Their ampacity shall not be less than the rating of the overcurrent protection device at the termination of the connecting conductors.

- The connecting conductors shall not extend beyond the switchboard, panelboard, disconnecting means, or control devices they supply.

- Connecting conductors shall be enclosed in a raceway that will extend from the connection to the enclosure of an enclosed switchboard, panelboard, or control devices, or to the back of an open switchboard.

- The rating of the overcurrent device protecting the primary of the transformer, multiplied by the primary to secondary voltage ratio, shall not exceed 10 times the ampacity of the secondary conductor for field installations where the secondary conductors leave the enclosure or vault.

Overcurrent protection for panelboards shall comply with the provisions outlined in **408.36**, including **Ex. 1, Ex. 2**, and **Ex. 3**, whichever applies. The maximum number of overcurrent devices shall be permitted to be determined per **408.54** and **408.55, Ex. 1**.

See Figure 20-34 for the proper procedure for making a connection using the 10 ft rule.

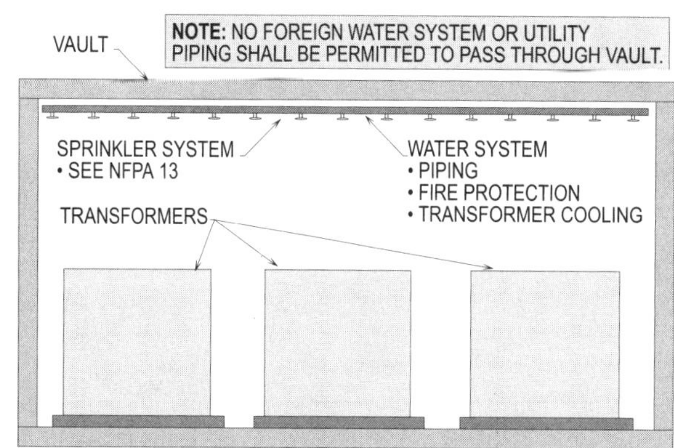

WATER PIPES AND ACCESSORIES
NEC 450.47

Figure 20-31. Piping for fire protection within the vault or piping to water-cooled transformers shall be permitted to be present in a vault. No other piping or duct system shall enter or pass through.

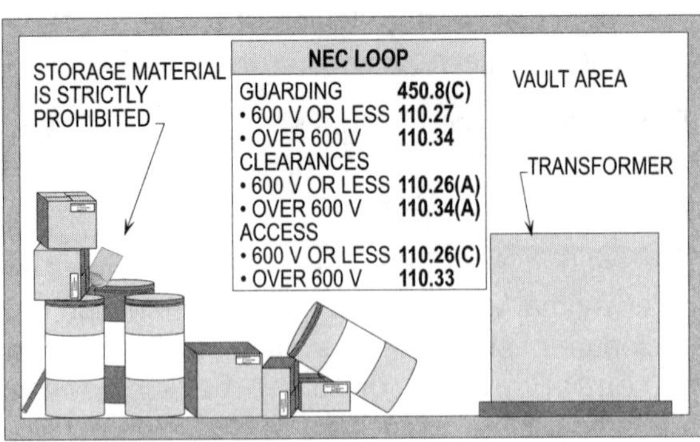

STORAGE IN VAULTS
NEC 450.48

Figure 20-32. No storage material of any kind shall be placed in a vault other than the transformers and equipment necessary for their operation.

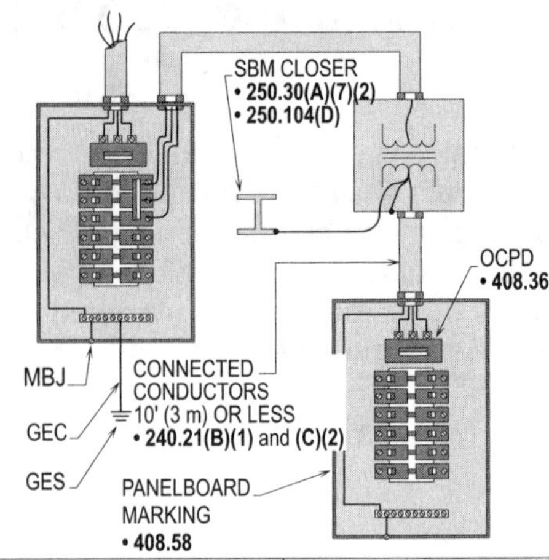

SIZING THWN CU. CONDUCTORS	SIZING OCPD
Step 1: Calculating min. size connection **240.21(C)(2)(1)** Calculated load is 148 A	**Step 1:** Calculating OCPD **240.4(E), 240.21(C)(2)(1) and 240.4(B)** 1/0 AWG cu. = 150 A OCPD rated at 150 A protects conductors from overload
Step 2: Sizing conductors **Table 310.16** 1/0 AWG THWN cu. = 150 A	**Solution:** **The size OCPD is permitted to be 150 amps.**
Step 3: Verifying size **240.21(C)(2)(1)** 150 A is greater than 148 A	
Solution: **The size THWN copper conductors are 1/0 AWG rated at 150 amps.**	

NOT OVER 10 FT (3 m) LONG
NEC 240.21(C)(2)

Figure 20-34. The above illustration shows the procedure for sizing a 10 ft (3 m) connection from the secondary of a transformer.

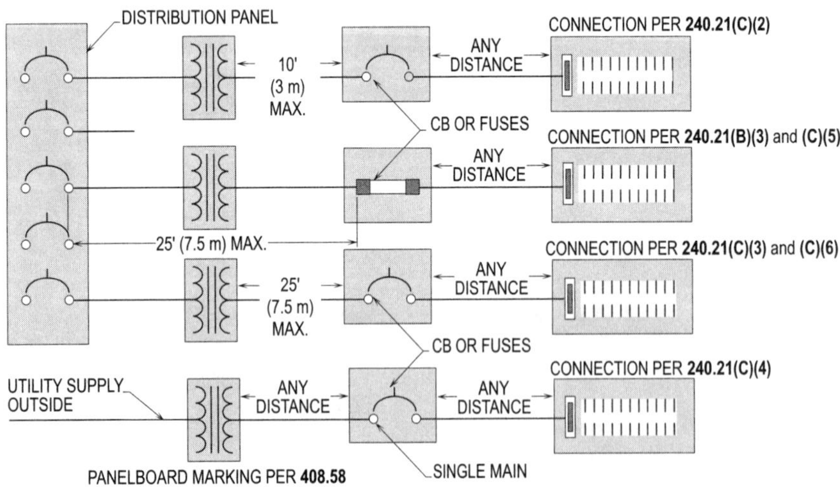

CONNECTIONS FROM THE SECONDARY OF TRANSFORMERS
NEC 240.21(B) AND (C)

Figure 20-33. The illustration above shows the four most used transformer secondary connections that are utilized to supply electrical systems with transformer secondary conductors.

NOT OVER 25 FT (7.5 m) LONG
240.21(B)(3) AND (C)(5)

Conductors supplying a transformer shall be permitted to be tapped, without overcurrent protection at the tap, from a feeder where all the following conditions are met:

- Tapped conductors supplying the primary shall have an ampacity at least 1/3 of the rating of the feeder being tapped.

- Connecting conductors supplying the secondary shall have an ampacity at least 1/3 of the rating of the feeder being connected, based on the primary-to-secondary voltage ratio.

- The total length of one primary plus one secondary conductor shall not be over 25 ft (7.5 m).

- The primary and secondary conductors shall be protected from physical damage.

- Secondary conductors shall terminate in a single circuit breaker or set of fuses, sized to protect the secondary.

See Figure 20-35 for the proper procedure for making a tap and connection using the 25 ft (7.5 m) rule.

INDUSTRIAL INSTALLATION
SECONDARY CONDUCTORS
NOT OVER 25 FT (7.5 m) LONG
240.21(C)(3)

Conductors shall be permitted to be connected to a transformer secondary of a separately derived system for industrial locations, without overcurrent protection at the connection, where all the following conditions are met:

- Secondary conductors shall not exceed 25 ft (7.5 m) in length.

- Ampacity of connected conductors shall be equivalent to current rating of the transformer, and the overcurrent protection devices shall not exceed the ampacity of the connected conductors.

- All overcurrent devices are grouped.

- Connected conductors shall be protected from physical damage.

See Figure 20-36 for the procedure to be applied when a 25 ft (7.5 m) connection rule is installed from the secondary side of a transformer.

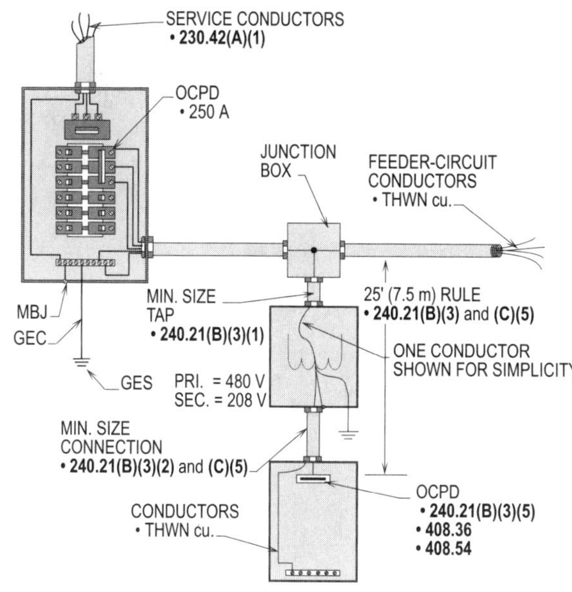

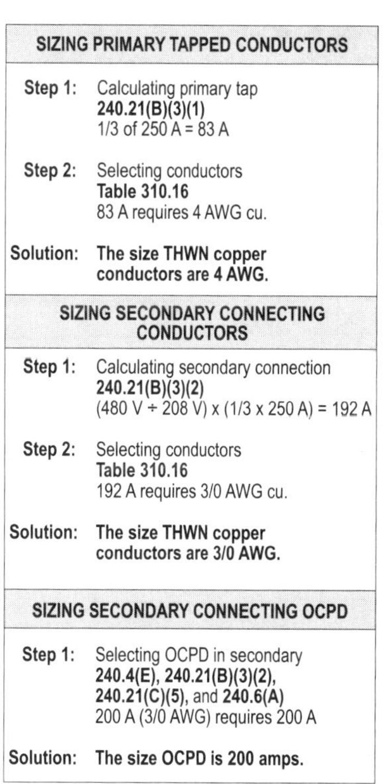

SIZING PRIMARY TAPPED CONDUCTORS	
Step 1:	Calculating primary tap **240.21(B)(3)(1)** 1/3 of 250 A = 83 A
Step 2:	Selecting conductors **Table 310.16** 83 A requires 4 AWG cu.
Solution:	**The size THWN copper conductors are 4 AWG.**

SIZING SECONDARY CONNECTING CONDUCTORS	
Step 1:	Calculating secondary connection **240.21(B)(3)(2)** (480 V ÷ 208 V) x (1/3 x 250 A) = 192 A
Step 2:	Selecting conductors **Table 310.16** 192 A requires 3/0 AWG cu.
Solution:	**The size THWN copper conductors are 3/0 AWG.**

SIZING SECONDARY CONNECTING OCPD	
Step 1:	Selecting OCPD in secondary **240.4(E), 240.21(B)(3)(2),** **240.21(C)(5), and 240.6(A)** 200 A (3/0 AWG) requires 200 A
Solution:	**The size OCPD is 200 amps.**

NOT OVER 25 FT (7.5 m) LONG
NEC 240.21(B)(3) AND (C)(5)

Figure 20-35. The primary tap for this connection rule shall be at least 1/3 of the overcurrent protection device protecting the larger feeder conductors. The secondary connecting conductors shall be at least 1/3 of the overcurrent protection device protecting the feeder conductors based on the primary-secondary transformer ratio.

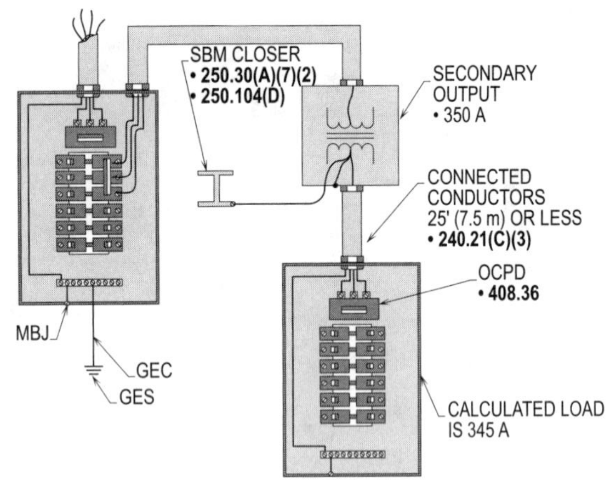

SIZING THWN cu. CONDUCTORS	SIZING OCPD
Step 1: Calculating min. size connection **240.21(C)(3)(2)** Calculated load is 345 A	**Step 1:** Calculating OCPD **240.4(E), 240.21(C)(3)(1), and 240.4(B)** 500 KCMIL cu. = 380 A OCPD rated at 350 A protects conductors from overload
Step 2: Sizing conductors **Table 310.16** 500 THWN cu. = 380 A	**Solution: The size OCPD is permitted to be 350 amps.**
Step 3: Verifying size **240.21(C)(3)(1)** 380 A is greater than 345 A	**Note:** If 6 CBs are used, they must equal to one main.
Solution: The size THWN copper conductors are 500 KCMIL rated at 380 amps.	

INDUSTRIAL INSTALLATION SECONDARY CONDUCTORS NOT OVER 25 FT (7.5 m) LONG NEC 240.21(C)(3)

Figure 20-36. The above illustration shows the procedure for sizing a 25 ft (7.5 m) connection from the secondary of a transformer.

OUTSIDE SECONDARY CONDUCTORS 240.21(C)(4)

Outside conductors shall be permitted to be connected to a feeder or be connected at the transformer secondary without overcurrent protection at the connection. However, all of the following conditions shall be complied with:

- The connected conductors are suitably protected from physical damage.

- The conductors terminate at a single circuit breaker or a single set of fuses that will limit the load to the ampacity of the conductors. This single overcurrent protection device can supply any number of additional overcurrent devices of its load side.

- The overcurrent protection device for the conductors is an integral part of a disconnecting means or shall be located immediately adjacent thereto.

- The disconnecting means for the conductors are installed at a readily accessible location either outside of a building or structure or inside, nearest the point of entrance of the conductors.

See Figure 20-37 for the rule pertaining to outside transformer connections from the secondary side of transformers.

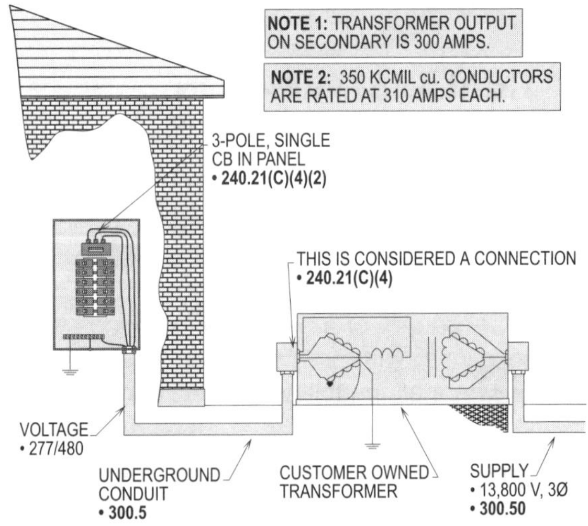

SIZING CONDUCTOR USING THWN cu.	SIZING OCPD BASED UPON SEC. OUTPUT
Step 1: Sizing conductors 300 A requires 350 KCMIL	**Step 1:** Sizing OCPD 300 A output requires 300 A OCPD
Solution: The size THWN copper conductors are 350 KCMIL.	**Solution: The size OCPD required is 300 amps.**

OUTSIDE SECONDARY CONDUCTORS NEC 240.21(C)(4)

Figure 20-37. The above illustration shows the rules for sizing the conductors and overcurrent protection device for a feeder connection from a transformer located outside.

TRANSFORMER SECONDARY CONDUCTORS IN LENGTHS OF 10 FT (3 m) TO 25 FT (7.5 m) 240.21(C)(6)

Conductors over 10 ft (3 m) and up to 25 ft (7.5 m) in length shall be permitted to be connected to the secondary size of a transformer. When applying this section of the code, the 25 ft (7.5 m) secondary connection shall be terminated in a single overcurrent protection device (circuit breaker or fuses) to limit the load and to also comply with the 1/3 rule when multiplied by the secondary-to-primary voltage ratio. The secondary conductors shall be protected from physical damage and abuse. **(See Figure 20-38)**

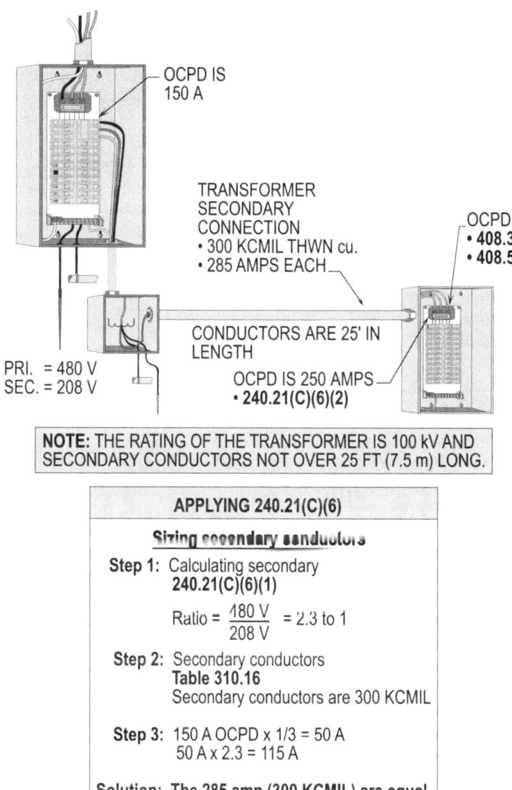

Figure 20-38. The above shows the rules for making a 25 ft (7.5 m) secondary conductor connection in other than industrial locations.

SUPERVISED INDUSTRIAL INSTALLATIONS – FEEDER AND BRANCH-CIRCUIT CONDUCTORS 240.92(A)

Feeder and branch-circuit conductors shall be protected at the point where the conductors receive their supply. However, this permits a variation of requirements for transformer secondary conductors taken from separately derived systems and outside feeder taps. **(See Figure 20-39)**

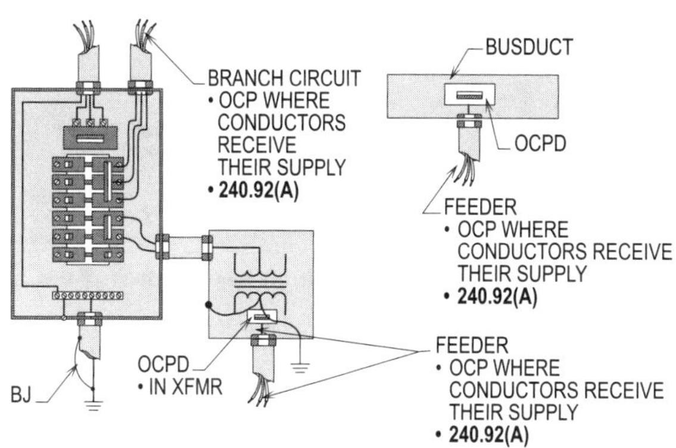

Figure 20-39. The above illustration shows feeder and branch-circuit conductors protected at the point where the conductors receive their supply. (Also, see **240.21(C)(4)** for a similar rule.)

SUPERVISED INDUSTRIAL INSTALLATIONS AND CONNECTIONS UP TO 100 FT (30 m) 240.92(C)(1)(1) AND (2)

Unprotected lengths of secondary conductors shall be permitted at up to 100 ft (30 m) if the transformer primary overcurrent is sized at a value (reflected to the secondary by the transformer phase voltage ratio) of not more than 150 percent of the secondary conductor ampacity.

Additionally, the conductors shall be protected by a differential relay with a trip setting equal to or less than the conductor ampacity. Note that a differential relay provides superior short-circuit protection at a trip open value that is almost always well below the conductor ampacity. **(See Figure 20-40)**

SHORT-CIRCUIT AND GROUND FAULT PROTECTION
240.92(C)(1)(3)

Conductors up to 100 ft (30 m) in length shall be permitted if calculations are made under engineering supervision and it is determined that the secondary conductors will be protected within recognized times versus current limits for all short-circuits and ground fault conditions that could occur. **(See Figure 20-40)**

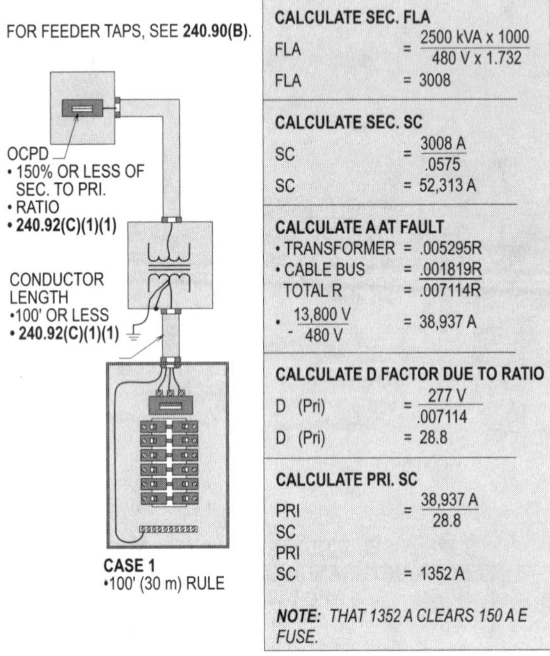

DESIGN TIP: Consider a 2500 kVA, 5.75% z transformer with a 13.8 kV to 480/277 V ratio and 500 mVA available short-circuit current on the primary for 100 circuit feet of 4000 amp cable bus, and a three-phase bolted fault at the end of the bus (worst case), about 38,937 amps will flow from the system, or 1352 amps on the primary, which will clear a typical 150E fuse (which meets the maximum 150 percent requirement) within .42 seconds. This time vs. current value is well within the rating of the secondary conductors. (Cable bus has a resistance of .001819 and the transformer has a resistance of .005295.)

**SHORT-CIRCUIT AND GROUND FAULT PROTECTION
NEC 240.92(C)(1)(1), (2), AND (3)**

Figure 20-40. The above illustration shows methods of providing short-circuit and ground fault protection for transformers and conductors.

OVERLOAD PROTECTION
240.92(C)(2)

To provide overload protection, the secondary conductors shall be permitted to be terminated in a single overcurrent protection device or in lugs of the bus, if not more than six overcurrent protection devices with a combined rating are installed that do not exceed the ampacity of the conductors. Another method of protection is to provide overload current relaying with the ability (design into) to trip either the primary overcurrent protection devices or the downstream overcurrent protection devices so that the load current doesn't exceed the conductor's ampacity. **(See Figure 20-41)**

> **Transformer Tip:** In some cases, the short-circuit and ground fault protective arrangements may provide overload protection. Note that if engineering calculations prove this to be the case, separate overload protection isn't really needed.

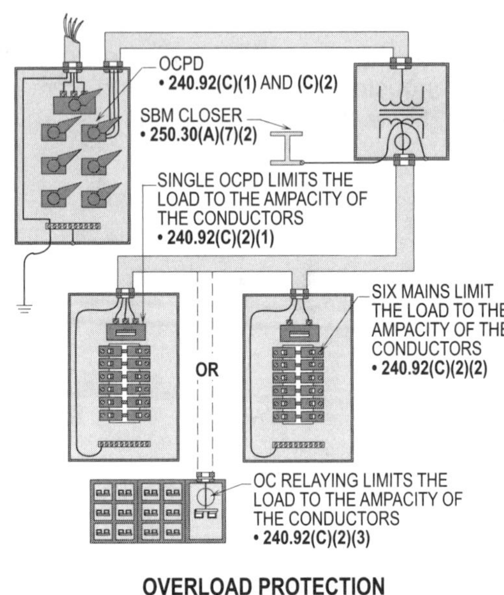

Figure 20-41. The above illustration shows methods of providing overload protection.

SUPERVISED INDUSTRIAL INSTALLATIONS – OUTSIDE FEEDER TAPS
240.92(D)

Section **240.92(D)** permits alternate means of protecting transformer secondary conductors in supervised industrial installations where the transformer is located outside. The secondary conductors shall be protected against (1) overloads, with the additional stipulation that (2) they are suitably protected against physical damage. **[See Figures 20-42(a) and (b)]**

> **Transformer Tip:** Such protection shall be permitted to be provided by six or less overcurrent protection devices where the total rating does not exceed the ampacity of the conductors routed per **240.92(D)**. Note that up to six overcurrent protection devices can be used instead of one overcurrent protection device at the feeder termination.

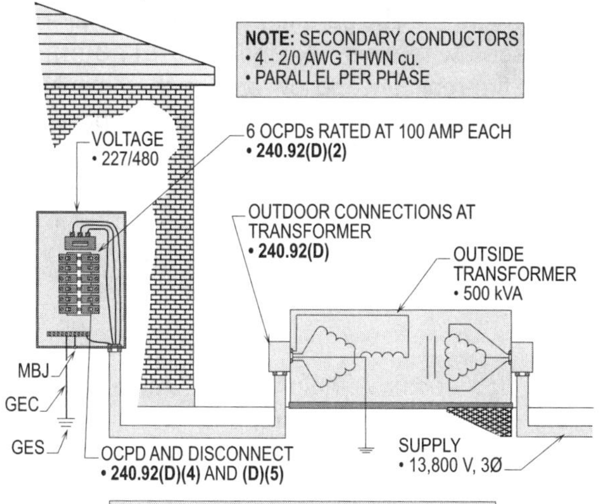

APPLYING 240.92(C)(2)
Step 1: Calculating FLA (Sec.) FLA = (500 kVA x 1000) ÷ (480 V x 1.732) FLA = 601.7 A
Step 2: Sizing secondary OCPD **240.92(D)(2)** 600 A secondary OCPD is less than the 601.7 A output (100 A CB x 6 = 600 A)
Solution: Six OCPDs rated 100 amps each are equal to a 600 A main (100 A CB x 6 = 600 A) and protects 4 - 2/0 AWG cu. conductors in parallel (175 A x 4 = 700 A).

SUPERVISED INDUSTRIAL INSTALLATIONS
OUTSIDE FEEDER TAPS
NEC 240.92(D)

Figure 20-42(a). The above illustration shows alternate means allowed for protecting conductors tapped to a transformer located outside.

PROTECTION BY PRIMARY OVERCURRENT DEVICE 240.92(E)

Conductors supplied by the secondary side of a transformer shall be permitted to be protected by overcurrent protection, provided on the primary (supply) side of the transformer, provided the primary device time-current protection characteristic, multiplied by the maximum effective primary-to-secondary transformer voltage ratio, effectively protects the secondary conductors. **(See Figure 20-43)**

Design Tip: This section recognizes overcurrent protection installations where a device in series, with a transformer primary, is used to protect secondary conductors. The ratios given in **240.92(E)** are recognized in numerous industry standards and references, including IEEE Standard C37.91.

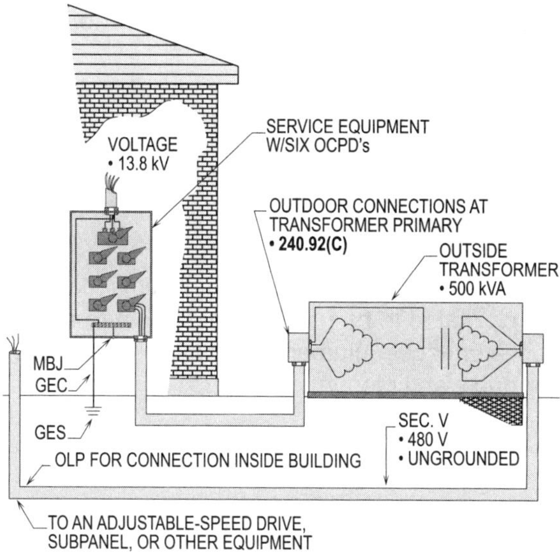

SUPERVISED INDUSTRIAL INSTALLATIONS
OUTSIDE FEEDER TAPS
NEC 240.92(D)

Figure 20-42(b). The above illustration shows alternate means allowed for protecting conductors connected to a transformer located outside.

EXCEPTION TO 725.3(C)

Note that Type CL2P or CL3P cables and plenum signaling raceways shall be permitted for Class 2 and Class 2 circuits installed in other spaces used for environmental air, if in accordance with **725.154(A)**.

WIRING AND EQUIPMENT IN SPACES USED FOR ENVIRONMENTAL AIR 300.22(C)(1)

Only wiring methods of a specific type shall be permitted to be installed in ducts used to transport dust, loose stock, or flammable vapors. Wiring methods of any type shall not be permitted to be installed in any duct, or shaft containing only such ducts, used for vapor removal or ventilation of commercial type cooking equipment.

The following wiring methods are only permitted to be installed in ducts or plenums used for environmental air:

- MI cable
- MC cable
- AC cable
- Factory-assembled multiconductor control or power cable listed for the use

- Listed prefabricated cable assemblies of metallic manufactured wiring systems without nonmetallic sheath

- Electrical metallic tubing

- Flexible metallic tubing

- Intermediate metal conduit

- Rigid metal conduit

Flexible metal conduit and liquidtight flexible metal conduit shall be permitted to be used in lengths not to exceed 4 ft (1.2 m) to connect equipment and devices permitted to be in these ducts and plenum chambers.

Flexible metal conduit and liquidtight flexible metal conduit in single lengths shall be permitted to be used in air-handling ceiling spaces where not exceeding 6 ft (1.8 m). Cables that have a fire-resistant and low-smoke characteristics shall be permitted to be used in air-handling ceiling spaces without conduit per **725.3(C), 760.3(B)** and **800.3(B)**.

See Figure 20-44 for other types of wiring methods permitted in other types of spaces used for environmental air.

> **Design Tip:** Electrical equipment with metal enclosures or nonmetallic enclosures that are listed shall be permitted to be installed in other spaces used for environmental air per **300.22(C)(2)**. A transformer that is totally enclosed may be installed to meet this requirement per **450.13(B)** if approved for such use.

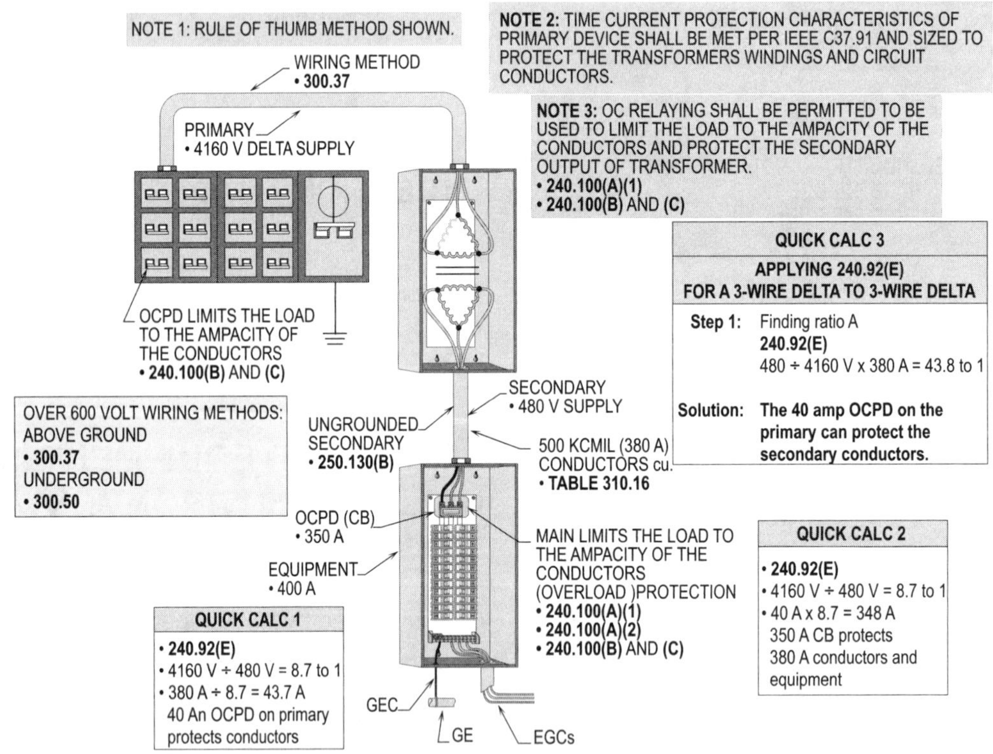

Figure 20-43. The primary protection of a transformer shall be permitted to be used to protect the secondary conductors, provided the primary device time-current protection characteristic, multiplied by the maximum effective primary-to-secondary voltage ratio will effectively protect the secondary conductors.

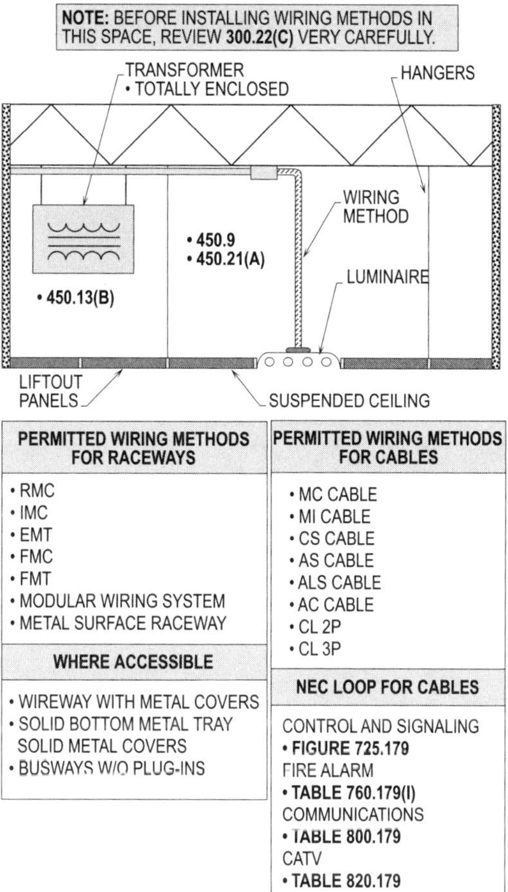

Figure 20-44. The above illustration shows the permitted wiring methods in spaces used for environmental air.

Chapter 20. Transformers

Section	Answer

1. Lighting transformers for a closed delta-connected secondary system can be found by multiplying the three-phase load by _____ percent.
    (a) 33   (b) 67
    (c) 75   (d) 100

2. The lighting transformer for an open delta-connected secondary system can be found by multiplying the three-phase load by _____ percent and adding this value to the single-phase load.
    (a) 33   (b) 58
    (c) 67   (d) 75

3. A transformer 600 volts or less, nominal, having an individual overcurrent protection device on the primary side shall be sized at not more than _____ percent of the transformer's full-load current rating, in amps. (General rule)
    (a) 100   (b) 110
    (c) 115   (d) 125

4. An overcurrent protection sensing device shall be designed to trip at _____ percent of its continuous current per phase or neutral rating.
    (a) 100   (b) 125
    (c) 150   (d) 250

5. Askarel transformers of more than _____ volts shall be installed in a vault.
    (a) 10,000   (b) 25,000
    (c) 35,000   (d) 50,000

6. Walls, roof, and floor for a transformer vault shall at least have a _____ hour fire resistance rating.
    (a) 1   (b) 2
    (c) 3   (d) 5

7. Flexible metal conduit in single lengths shall be permitted to be used in air-handling spaces where not exceeding _____ ft.
    (a) 3   (b) 4
    (c) 5   (d) 6

8. A closed delta-connected secondary system can be found by multiplying the single-phase load by _____ percent.
    (a) 67   (b) 89
    (c) 100   (d) 125

9. Dry-type transformers not over 600 volts that are located on open walls or steel columns do not have to be _____ accessible.
    (a) easily   (b) readily
    (c) employee   (d) permanently

10. If installing fuses, the individual overcurrent protection device for the primary side of a transformer rated over 600 volts shall be rated not greater than _____ percent of the rated primary current of the transformer. (Supervised location)
    (a) 125   (b) 150
    (c) 200   (d) 250

11. For a transformer in a nonsupervised location with a secondary voltage rated 600 volts or less, the overcurrent protection device and conductors on the secondary side shall be sized at _____ percent of the FLC rating. (Any location)
    (a) 125   (b) 150
    (c) 200   (d) 250

12. For a transformer 600 volts or less, nominal, where the rated secondary current of a transformer is less than 9 amps, an overcurrent protection device rated or set at no more than _____ percent of secondary current shall be permitted to be used.
    (a) 133   (b) 150
    (c) 167   (d) 225

13. Autotransformers shall have a _____ neutral current rating sufficient for the specified ground fault current that could develop in the system.
    (a) noncontinuous   (b) varying
    (c) periodic   (d) continuous

14. Where transformers are tied together in parallel and connected by tie conductors that do not have overcurrent protection, the ampacity of the ties connecting conductors shall not be less than _____ percent of the rated secondary current of the largest transformer in the tie circuit.
    (a) 33   (b) 67
    (c) 85   (d) 100

15. When secondary ties from transformers are used, an overcurrent protection device in the secondary of each transformer that is rated or set at not greater than _____ percent of the rated secondary current of the transformer shall be provided.
    (a) 125   (b) 150
    (c) 200   (d) 250

16. Transformers shall be elevated at least _____ ft above the floor or working space to prevent unauthorized personnel from contact.
    (a) 6   (b) 8
    (c) 10   (d) 15

17. Dry-type transformers greater than 112-1/2 kVA having Class 155 or higher insulation shall be separated at least 6 ft horizontally and _____ ft vertically from the combustible material if no fire-resistant, heat-insulating barrier is provided.
    (a) 8   (b) 10
    (c) 12   (d) 15

18. All indoor dry-type transformers of over _____ volts shall be installed in a vault.
    (a) 15,000   (b) 25,000
    (c) 30,000   (d) 35,000

19. Dry-type transformers installed outdoors shall have _____ enclosures.
    (a) weatherproof   (b) weathertight
    (c) rainproof   (d) raintight

20. The walls and roof for a transformer vault shall have a _____ in. thickness.
    (a) 2   (b) 3
    (c) 4   (d) 6

21. The door sills for a transformer vault shall be at least _____ in. high.
    (a) 2   (b) 3
    (c) 4   (d) 6

22. Connecting primary and secondary conductors not over 25 ft long (taps supplying transformer) shall have an ampacity of at least _____ of the rating of the feeder's overcurrent protection device, based on the primary-to-secondary voltage ratio.
    (a) 1/4   (b) 1/3
    (c) 1/2   (d) 3/4

23. The lighting transformer for an open delta-connected secondary system can be determined by multiplying the single-phase load by _____ percent plus 58 percent of the three-phase load.
    (a) 58   (b) 67
    (c) 100   (d) 125

24. For a transformer in a supervised location with 6 percent impedance and a secondary voltage rated 600 volts or less, the overcurrent protection device on the secondary side shall be sized at not more than _____ percent of the FLC rating.
    (a) 100   (b) 125
    (c) 225   (d) 250

25. Dry-type transformers rated 112-1/2 kVA or less shall be separated at least _____ in. from the combustible material where the voltage is 600 volts or less.
    (a) 6   (b) 12
    (c) 18   (d) 24

26. Askarel-insulated transformers of over _____ kVA shall be furnished with a relief vent such as chimney.
    (a) 25   (b) 35
    (c) 50   (d) 75

27. The floor for a transformer vault shall be at least _____ in. thick.
    (a) 2   (b) 4
    (c) 6   (d) 12

28. Dry-type transformers exceeding 112 kVA shall not be located within _____ in. of combustible materials of building, unless the transformer has Class 155 insulation systems or higher and is completely enclosed except for ventilation openings.
    (a) 3   (b) 6
    (c) 10   (d) 12

29. Askarel-insulated transformers installed indoors and rated over _____ kVA shall be furnished with a pressure-relief vent.
    (a) 25,000   (b) 30,000
    (c) 35,000   (d) 50,000

30. Where practicable, vaults containing more than _______ kVA transformer capacity shall be provided with a drain or other means.

    (a) 50   (b) 75
    (c) 100   (d) 150

31. What is the primary and secondary amperage for a 20 kVA, 480/240 volt, single-phase transformer?

32. What is the primary and secondary amperage for a 20 kVA, 480/240 volt, three-phase transformer?

33. What is the individual overcurrent protection device rating (using circuit breakers and time-delay fuses) for the primary side of a transformer with the following?
- supervised location
- 1500 kVA transformer
- 12,470 volts
- three-phase

34. What is the overcurrent protection device rating for the primary and secondary side of a transformer with the following?
- 400 kVA transformer
- 4160/480 volts
- three-wire to four-wire
- three-phase
- nonsupervised location (any location)

35. What is the overcurrent protection device rating for the primary and secondary side of transformer with the following?
- 400 kVA transformer
- 4160/480 volts
- three-wire to four-wire
- three-phase
- supervised location

36. What is the overcurrent protection device rating for the primary and secondary side of transformer with the following?
- 400 kVA transformer
- 13,800/4160 volts
- three-wire to four-wire
- three-phase
- supervised location

37. What is the individual overcurrent protection device rating for the primary side of a transformer with the following?
- 25 kVA transformer
- 240/120 volts
- single-phase

38. What is the individual overcurrent protection device rating for the primary side of a transformer with the following?
- 2 kVA transformer
- 480 volts
- two-wire to two-wire
- single-phase

39. What is the minimum and maximum individual overcurrent protection device rating for the primary side of a transformer with the following?
- .7 kVA transformer
- 480 volts
- two-wire to two-wire
- single-phase

40. What is the overcurrent protection device rating for the primary and secondary side of a transformer with the following?
- 40 kVA transformer
- 480/208 volts
- three-wire to four-wire
- three-phase

41. What size overcurrent protection device and THWN copper conductors are required for a 10 ft tap with a calculated load of 142 amps?

42. What size overcurrent protection device (secondary) and THWN copper conductors (primary and secondary) are required for a 25 ft tap with the following?
- 200 amp overcurrent protection device (primary)
- 480 volt primary
- 240 volt secondary

Hazardous (Classified) Locations

The intent of this chapter is to assist in the classification of areas or locations with respect to hazardous conditions. Such areas or locations are hazardous due to atmospheric concentrations of hazardous gases, vapors, deposits, or accumulations of materials that may be readily ignited.

The requirements are covered for the installation of electrical equipment and wiring in locations that are classified depending on the properties of the flammable vapors, liquids or gases, or combustible dusts that may be present and the likelihood that a flammable or combustible concentration of quantity is present. The hazardous (classified) locations to be covered are assigned the designations as follows:

(1) Class I, Division 1
(2) Class I, Division 2
(3) Class II, Division 1
(4) Class II, Division 2
(5) Class III, Division 1
(6) Class III, Division 2

(For the metric system dimensions, see those listed in the *National Electrical Code*)

GENERAL OVERVIEW

Hazardous areas and locations are classified by group, class, and division. They are determined by the atmospheric mixtures of various gases, vapors, dust, and other materials. The intensity of the explosion that can occur depends upon the concentrations, temperatures, and many other factors, which are listed in the codes from the National Fire Protection Association (NFPA).

Note that all areas designated as hazardous (classified) locations shall be properly documented. Such documentation shall be available to those authorized to design, install, inspect, maintain, or operate electrical equipment in the location involved per **500.4(A)**.

DIVISIONS (GASES)
500.5(B)

Class I locations are identified in the NEC as those in which flammable gases, flammable liquid-produced vapors, or combustible liquid-produced vapors are or may be present in the air in quantities sufficient to produce explosive or ignitible mixtures. The amount of vapor varies all the way from being continuously present to never present at all. Naturally, if vapors are not present, the area is not a classified (hazardous) location. However, where vapors are present, it is very likely that a flammable mixture may or may not be present.

From a designing standpoint, greater care must be exercised if a particular condition is likely to occur, such as the presence of a flammable mixture of vapor and air within the explosive range, than is needed if a flammable mixture within the explosive range is unlikely to occur. Therefore, this is the reason why it is necessary to divide hazardous locations into two divisions.

DIVISION 1 (GASES)
500.5(B)(1)

Division 1 hazardous locations are defined in the NEC as those locations (a) in which ignitible concentrations of flammable gases, flammable liquid-produced vapors, or combustible liquid-produced vapors can exist under normal operating conditions, or (b) in which ignitible concentrations of such flammable gases, flammable liquid-produced vapors, or combustible liquid-produced vapors above their flash points may exist frequently because of repair or maintenance operations or because of leakage, or (c) in which breakdown or faulty operation of equipment or processes might release ignitible concentrations of flammable gases, flammable liquid-produced vapors, or

combustible liquid-produced vapors and might also cause simultaneous failure of electric equipment in such a way as to directly cause the electrical equipment to become a source of ignition. Note that in each case, ignitible concentrations are mentioned. This means concentrations between the lower and upper flammable or explosive limits.

There are minimum and maximum concentrations of flammable gases and vapors that are above and below the mixture and it will not burn or explode. The lower limit is where the substance is too lean to burn or explode. The upper limit is where the mixture has become too rich to burn or explode.

FPN 1 to **500.5(B)(1)** of the NEC describes a number of areas and occupancies that are normally classified as Division 1 locations. Such locations are as follows:

(1) Petroleum refining facilities

(2) Dip tanks containing flammable or combustible liquids

(3) Dry cleaning plants

(4) Plants manufacturing organic coatings

(5) Spray finishing areas
(residue must be considered)

(6) Petroleum dispensing areas

(7) Solvent extraction plants

(8) Plants manufacturing or using pyroxylin (nitro-cellulose) type and other plastics

(9) Locations where inhalation anesthetics are utilized

(10) Utility gas plants and operations involving storage and handling of liquefied petroleum and natural gas plants, and

(11) Aircraft hangars and fuel servicing areas

See Figure 21-1 for a detailed illustration of Class I, Division 1 locations.

> **Design Tip: Article 100** defines a flammable liquid as a liquid having a closed-cup flash point below 38°C (100°F), as determined by the test procedures and apparatus set forth in NFPA 30. Flash point is the minimum temperature at which a liquid gives off vapor in sufficient concentraion to form an ignitible mixture with air near the surface of the liquid, as specified by test. (See Sec. 3.3.6 of NFPA 497 for more details concerning flammable liquids.)

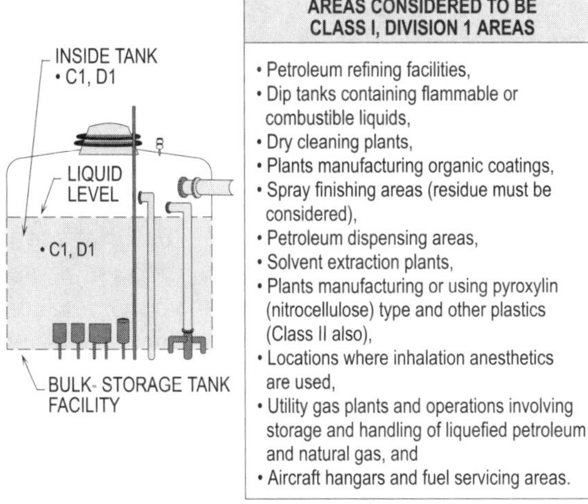

Figure 21-1. The facilities that are listed in the above illustration are considered, under certain conditions, to be Class I, Division 1.

DIVISION 2 (GASES)
500.5(B)(2)

The NEC defines Division 2 locations as those locations (a) in which volatile flammable gases, flammable liquid-produced vapors, or combustible liquid-produced vapors are handled, processed, or used, but in which the liquids, vapors, or gases will normally be confined within closed containers or closed systems from which they can escape only in case of accidental rupture or breakdown of such containers or systems or in case of abnormal operation of equipment, or (b) in which ignitible concentrations of flammable gases, flammable liquid-produced vapors, or combustible liquid-produced vapors are normally prevented by positive mechanical ventilation and which might become hazardous through failure or abnormal operation of the - ventilating equipment or (c) that are adjacent to a Class I, Division 1 location, and to which ignitible concentrations of flammable gases, flammable liquid-produced vapors, or combustible liquid-produced vapors above their flash points might occasionally be communicated unless such communication is prevented by adequate positive-pressure ventilation from a source of clean air and effective safeguards against ventilation failure are provided.

FPNs 1 and **2** to **500.5(B)(2)** of the NEC describe a number of areas and occupancies that are normally classified as Division 2 locations. For example, piping without valves, checks, meters, and devices does not usually introduce a hazardous condition even though such piping is utilized for flammable liquids. This type of piping is considered a contained system. Therefore, the area around shall be

classified as a Division 2 location by the AHJ. The following areas are considered to be Division 2 locations:

(1) Locations where volatile flammable liquids or flammable gases or vapors are used.

(2) Locations that, in the judgment of the AHJ, would become hazardous only in case of an accident or some unusual operating condition.

(3) Locations where the quantity of flammable material that might escape in case of accident, the adequacy of ventilating equipment, the total area involved, and the record of the industry or business with respect to explosions or fires are all factors that merit consideration in determining the classification and extent of each location.

(4) Locations where piping without valves, checks, meter, and similar devices would not ordinarily introduce a hazardous condition even though used for flammable liquids or gases.

(5) Locations used for the storage of flammable liquids or of liquefied or compressed gases in sealed containers would not normally be considered hazardous unless subject to other hazardous conditions also.

(6) Electrical conduits and their associated enclosures separated from process fluids by a single seal or barrier shall be classified as a Division 2 location if the outside of the conduit and enclosure is an unclassified location.

See Figure 21-2 for an illustration of Class I, Division 2 locations.

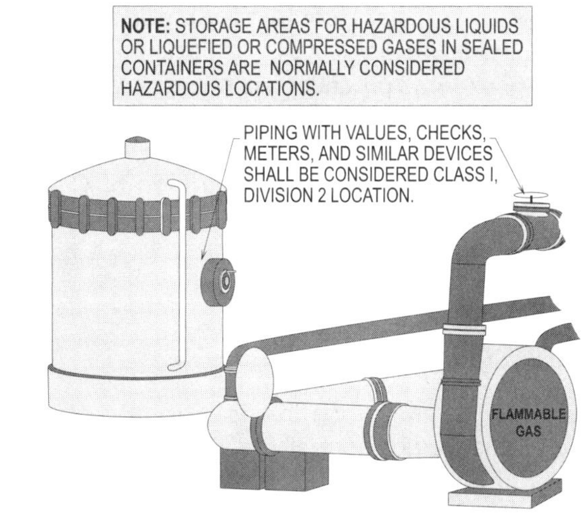

Figure 21-2. Areas in refineries where vessels and pipelines are equipped with valves, meters, and similar devices shall be considered Class I, Division 2 locations.

DIVISION 1 (DUST)
500.5(C)(1)

A Class II, Division 1 location is a location (1) in which combustible dust is in the air under normal operating conditions in quantities sufficient to produce explosive or ignitible mixtures, or (2) where mechanical failure or abnormal operation of machinery or equipment might cause such explosive or ignitible mixtures to be produced, and might also provide a source of ignition through simultaneous failure of electric equipment, through operation of protection devices, or from other causes, or (3) in which Group E combustible dusts may be present in quantities sufficient to be hazardous.

Areas that are considered Class II, Division 1 locations are based upon certain designing techniques and conditions, as follows:

(1) Combustible dusts that are electrically nonconductive include dusts produced in the handling and processing of:

    (a) grain and grain products,

    (b) pulverized sugar and cocoa,

    (c) dried egg and milk powders,

    (d) pulverized spices, starch and pastes,

    (e) potato and wood flour, oil meal from beans, and

    (f) seed, dried hay, and other organic materials that may produce combustible dusts when processed or handled.

(2) Only Group E dusts are considered to be electrically conductive for classification purposes. Dusts containing magnesium or aluminum are particularly hazardous, and the use of extreme precaution is necessary to avoid ignition and explosion.

See Figure 21-3 for a detailed illustration of Class II, Division 1 locations.

DIVISION 2 (DUST)
500.5(C)(2)

A Class II, Division 2 location is a location where combustible dust is not normally in the air in quantities sufficient to produce explosive or ignitible mixtures and where dust accumulations are normally insufficient to interfere with the normal operation of electrical equipment or other apparatus, but combustible dust may be in suspension in the air as a result of infrequent malfunctioning of handling or processing equipment and where combustible dust accumulations on, in, or in the vicinity of the electrical equipment may be sufficient to interfere with the safe dissipation of heat from electrical equipment or may be ignitible by abnormal operation or failure of electrical equipment.

Areas that are considered Class II, Division 2 locations are based upon certain designing techniques and conditions, as follows:

(1) The quantity of combustible dust that may be present and the adequacy of dust removal systems are factors that merit consideration in determining the classification and may result in an unclassified area.

(2) Where products such as seed are handled in a manner that produces low quantities of dust, the amount of dust deposited may not warrant classification.

See Figure 21-4 for a detailed illustration of Class II, Division 2 locations.

DIVISION 1 (FIBERS/FLYINGS)
500.5(D)(1)

A Class III, Division 1 location is a location in which easily ignitible fibers/flyings are handled, manufactured, or used.

(1) Such locations usually include some parts of:

    (a) rayon, cotton, and other textile mills,

    (b) combustible fiber manufacturing and processing plants,

    (c) cotton gins and cotton-seed mills,

    (d) flax-processing plants,

    (e) clothing manufacturing plants,

    (f) woodworking plants, and

    (g) establishments and industries involving similar hazardous processes or conditions.

(2) Easily ignitible fibers and flyings include:

    (a) rayon,

    (b) cotton (including cotton liners and cotton waste),

    (c) sisal or henequen,

    (d) istle,

    (e) jute, hemp, tow, cocoa fiber, oakum,

    (f) baled waste kapok,

    (g) spanish moss, excelsior, and

    (h) other materials of similar nature.

See Figure 21-5(a) for a detailed illustration of Class III, Division 1 locations.

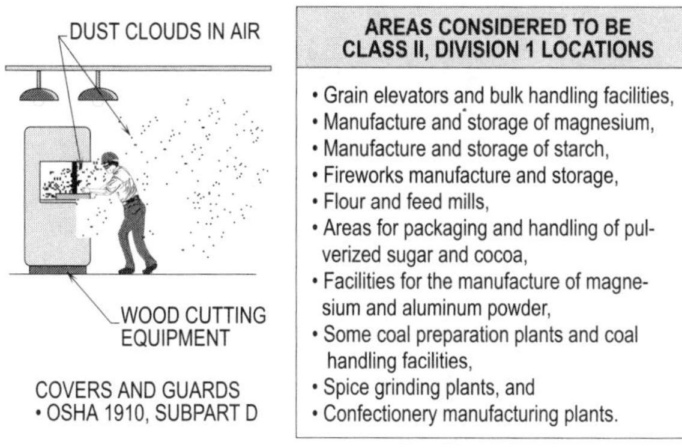

AREAS CONSIDERED TO BE
CLASS II, DIVISION 1 LOCATIONS

• Grain elevators and bulk handling facilities,
• Manufacture and storage of magnesium,
• Manufacture and storage of starch,
• Fireworks manufacture and storage,
• Flour and feed mills,
• Areas for packaging and handling of pulverized sugar and cocoa,
• Facilities for the manufacture of magnesium and aluminum powder,
• Some coal preparation plants and coal handling facilities,
• Spice grinding plants, and
• Confectionery manufacturing plants.

DIVISION 1 (DUST)
NEC 500.5(C)(1)

Figure 21-3. The above are facilities that are considered under certain conditions to be Class II, Division 1 locations.

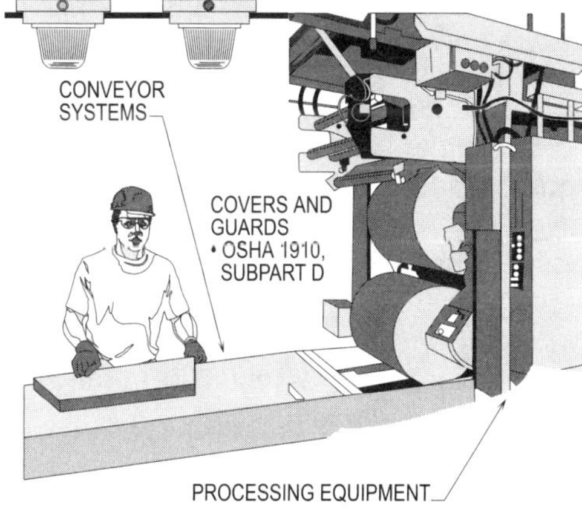

AREAS CONSIDERED TO BE
CLASS II, DIVISION 2 LOCATIONS

• Rooms containing closed spouts and conveyors, closed bins, or hoppers.
• Machines and equipment from which quantities of dust could escape under abnormal operating conditions,
• Rooms adjacent to Class II areas,
• Rooms in which control over suspension of dust in explosive or ignitible quantities is present,
• Warehouses and shipping rooms where dust-producing materials are stored, and
• Any other similar locations.

DIVISION 2 (DUST)
NEC 500.5(C)(2)

Figure 21-4. The above are facilities that are considered under certain conditions to be Class II, Division 2 locations.

DIVISION 2 (FIBERS/FLYINGS) 500.5(D)(2)

A Class III, Division 2 location is a location in which easily ignitible fibers/flyings are stored or handled other than in the process of manufacture. **[See Figure 21-5(b)]**

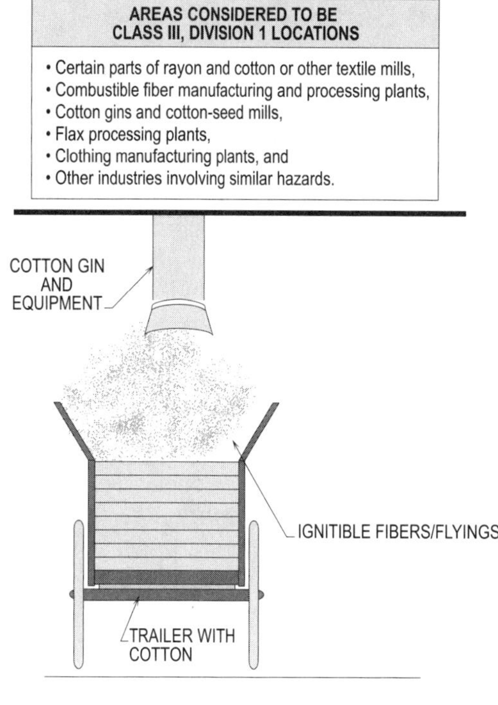

AREAS CONSIDERED TO BE
CLASS III, DIVISION 1 LOCATIONS

• Certain parts of rayon and cotton or other textile mills,
• Combustible fiber manufacturing and processing plants,
• Cotton gins and cotton-seed mills,
• Flax processing plants,
• Clothing manufacturing plants, and
• Other industries involving similar hazards.

DIVISION 1 (FIBERS/FLYINGS)
NEC 500.5(D)(1)

Figure 21-5(a). Listed above are facilities that are considered, under certain conditions, to be Class III, Division 1 locations.

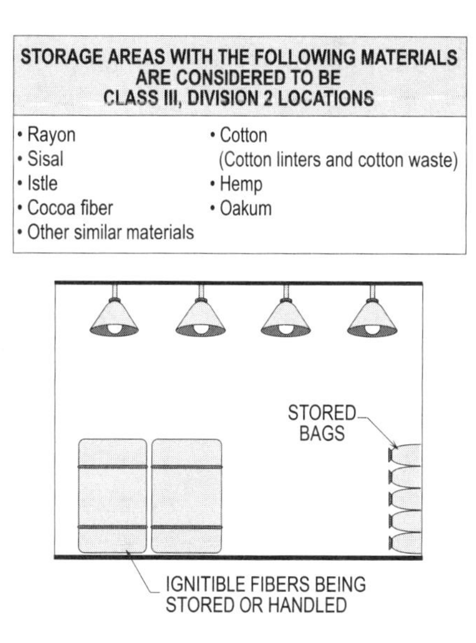

STORAGE AREAS WITH THE FOLLOWING MATERIALS
ARE CONSIDERED TO BE
CLASS III, DIVISION 2 LOCATIONS

• Rayon
• Sisal
• Istle
• Cocoa fiber
• Other similar materials
• Cotton (Cotton linters and cotton waste)
• Hemp
• Oakum

DIVISION 2 (FIBERS/FLYINGS)
NEC 500.5(D)(2)

Figure 21-5(b). Listed above are facilities that are considered under certain conditions to be Class III, Division 2 locations.

MATERIAL GROUPS (CLASS I)
500.6(A)

Until publication of the 1937 edition of the NEC, Class I hazardous locations were not divided into groups. All flammable gases and vapors were classified as a single degree of hazard. It was recognized, however, that the degrees of hazard varied and that equipment suitable only for use where gasoline was handled was not necessarily suitable for use where hydrogen or acetylene was handled. It was also recognized that manufacturing equipment and enclosures for use in hydrogen atmospheres was very difficult and that the equipment, even if built, was expensive. It was not logical, from a designing standpoint, to require explosion-proof equipment in gasoline filling stations that was also suitable for use in hydrogen atmospheres. Not only would this unnecessarily increase the cost of the electrical installation in one of the most common types of hazardous locations, but it would make some types of equipment unavailable.

By placing flammable materials into groups and classifying them based upon the explosive characteristics of such gases and vapors, electrical equipment and wiring methods could be selected.

GROUP A (GASES)
500.6(A)(1)

As mentioned above, combustible and flammable gases and vapors are divided into four groups; the classification involving determinations of maximum explosion pressures, and maximum safe clearance between parts of a clamped or threaded joint in an enclosure.

Really, there is no consistent relationship between Groups A, B, C, or D classification and flash point/ignition temperature/explosive limits. Instead, the groups are classified by chemical families. Certain chemicals produce higher explosive pressures and heat when ignited. Group A gas produces the greater pressures during an explosion, and therefore is the most difficult to handle and control. Group B is next highest in pressure, then Group C, and lastly Group D.

> **Design Tip:** This is the reason why a Group A or B listing is more difficult to get than a Group C or D listing for electrical equipment. **(See Figure 21-6)**

For example, Group A is an atmosphere containing acetylene, that is capable of producing a very dangerous explosive mixture. Note that equipment falling in this group is not available. Therefore, special equipment such as purged and pressurized shall be designed and used.

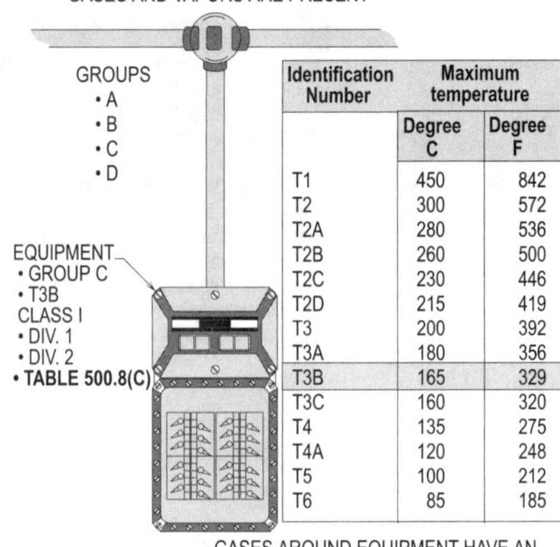

Identification Number	Maximum temperature	
	Degree C	Degree F
T1	450	842
T2	300	572
T2A	280	536
T2B	260	500
T2C	230	446
T2D	215	419
T3	200	392
T3A	180	356
T3B	165	329
T3C	160	320
T4	135	275
T4A	120	248
T5	100	212
T6	85	185

Figure 21-6. The above equipment is required to have a T3B (T-Code) identification number when installed in a Class I, Division 1 location.

GROUP B (GASES)
500.6(A)(2)

Group B is a flammable gas, flammable liquid-produced vapor, or combustible liquid-produced vapor mixed with air that may burn or explode, having either a maximum experimental safe gap (MESG) value less than or equal to 0.45 mm or a minimum igniting current ratio (mic ratio) less than or equal to 0.40. (See NFPA 497, Sec. 3.3.5.1.2)

Group B is an atmosphere containing hydrogen, fuel, and combustible process gases containing more than 30 percent hydrogen by volume, or it contains gases or vapors of equivalent hazard such as butadiene, ethylene oxide, propylene oxide, and acrolein.

GROUP C (GASES)
500.6(A)(3)

Group C is a flammable gas, flammable liquid-produced vapor, or combustible liquid-produced vapor mixed with air that may burn or explode, having either a maximum environmental safe gas (mesg) value greater than 0.45 mm and less than or equal to 0.75 mm, or a minimum igniting current ratio (MIC ratio) greater than 0.40 mm and less than or equal to 0.80. (See NFPA 497, Sec. 3.3.5.1.3)

Group C is an atmosphere such as ethyl ether, ethylene, or gases or vapors of equivalent hazard. Manufacturers make all types of equipment that fall in this group of explosive equipment. [See **500.7(A)**]

GROUP D (GASES)
500.6(A)(4)

Group D is a flammable gas, flammable liquid-produced vapor, or combustible liquid-produced vapor mixed with air that may burn or explode, having either a maximum experimental safe gas (MESG) value greater than 0.75 mm or a minimum igniting current ratio (MIC ratio) greater than 0.80. (See NFPA 497, Sec. 3.3.5.1.4)

Group D is an atmosphere such as acetone, ammonia, benzene, butane, cyclopropane, ethanol, gasoline, hexane, methanol, methane, natural gas, naphtha, propane, or gases or vapors of equivalent hazard. There are many types of equipment available in this group.

MATERIAL GROUPS (CLASS II)
500.6(B)

Combustible dusts are divided into three groups; the classification involving the tightness of the joints of assembly and shaft openings to prevent entrance of dust in the dust-ignition proof enclosure, the blanketing effect of layers of dust on the equipment that may cause overheating, and the ignition temperature of the dust are the main concerns.

GROUP E (DUST)
500.6(B)(1)

Group E atmospheres contain combustible metal dusts, including aluminum, magnesium, and their commercial alloys, or other combustible dusts of similar hazard.

Design Tip 1: "Only Group E dusts are considered to be electrically conductive." These dusts are metal dusts, such as aluminum, magnesium, and their commercial alloys or other dusts of small particle size, abrasiveness and/or electrical conductivity that presents a similar hazard.

Design Tip 2: When the dust is electrically conductive, care must be taken, for these dusts may ignite from bridging the gap between energized terminals, from arcs or from failure of equipment. Where Group E dusts are encountered in hazardous quantities, only Class II, Division 1 electrical equipment can be utilized. There is no such classification as Class II, Group E, Division 2. Either the location has enough electrically conductive dusts to make it a Division 1 location, or there is not enough dust present to even classify it as a hazardous area.

GROUP F (DUST)
500.6(B)(2)

Group F atmospheres contain combustible carbonaceous dusts, including carbon black, charcoal, coal, or coke dusts, that has more than 8 percent total entrapped volatiles or dusts that have been sensitized by other materials.

GROUP G (DUST)
500.6(B)(3)

Group G atmospheres contain combustible dusts not included in Group E or F, including flour, grain, wood, plastic and chemicals.

Equipment to be used in these atmospheres shall not only be approved for Class I, II, or III, but also for the specific group [Class I (A, B, C, or D) and II (E, F, or G)].

Design Tip: The explosion characteristics of air mixtures of dust vary with the materials involved. For Class II locations, Groups E, F, and G, the classification involves the tightness of the joints of assembly and shaft openings to prevent the entrance of dust in the dust-ignition proof enclosure. The blanketing effect of layers of dust on the equipment that may cause overheating and the ignition temperature of the dust are also considered. It is necessary, therefore, that equipment be approved not only for the class, but also for the specific group of dust that will be present.

See **Figure 21-7** for a detailed illustration of Groups E, F, and G equipment.

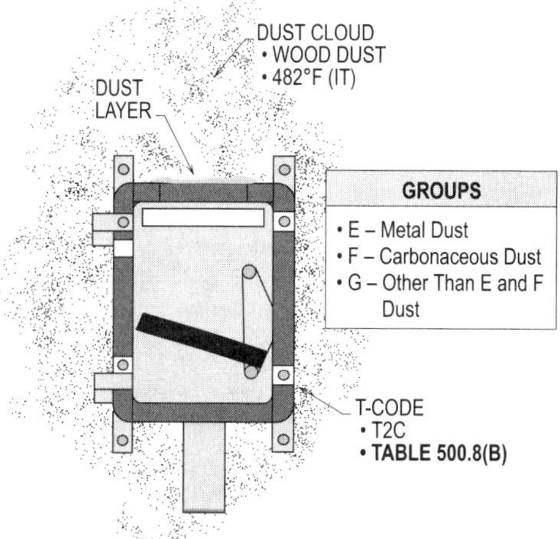

Figure 21-7. The above equipment is required to have a T2C identification number per Figure 21-6.

CLASSES AND DIVISIONS
500.5

Classes are utilized to identify hazardous locations that are subjected to materials that can be explosive if mixed with air to create an ignitible mixture.

Hazardous locations must be well understood by anyone designing, installing, working on, or inspecting electrical equipment and wiring methods located in such areas. These locations are dangerous due to the threat of flammable or combustible gases, vapors, or dusts being present some of the time or all of the time.

CLASS I, DIVISION 1
500.5(B)(1)

A Class I, Division 1 location is where (a) ignitible concentrations of flammable gases, flammable liquid-produced vapors, or combustible liquid produced vapors can exist under normal operating conditions, or (b) ignitible concentrations of such flammable gases, flammable liquid-produced vapors, or combustible liquids above their flash points may exist frequently because of repair or maintenance operations or because of leakage, or (c) breakdown or faulty operation of equipment or processes might release ignitible concentrations of flammable gases, flammable liquid-produced vapors, or combustible liquid-produced and might also cause simultaneous failure of electrical equipment in such a way as to directly cause the electrical equipment to become a source of ignition. **[See Figure 21-8(a)]**

CLASS I, DIVISION 2
500.5(B)(2)

A Class I, Divison 2 location is where (a) volatile flammable gases, flammable liquid-produced vapors or combustible liquid-produced vapors are handled, processed, or used, but in which the liquids, vapors, or gases will normally be confined within closed containers or closed systems from which they can escape only in case of accidental rupture or breakdown of such containers or systems or in case of abnormal operation of equipment, or (b) where ignitible concentrations of flammable gases, flammable liquid-produced vapors, or combustible liquid-produced vapors are normally prevented by positive mechanical ventilation and which might become hazardous through failure or abnormal operation of the ventilating equipment, or (c) is adjacent to a Class I, Division 1 location and to which ignitible concentrations of flammable gases, flammable liquid-produced vapors, or combustible liquid-produced vapors above their flash points might occasionally be communicated unless such communication is prevented by adequate positive-pressure ventilation from a source of clean air and effective safeguards against ventilation failure are provided. **[See Figure 21-8(b)]**

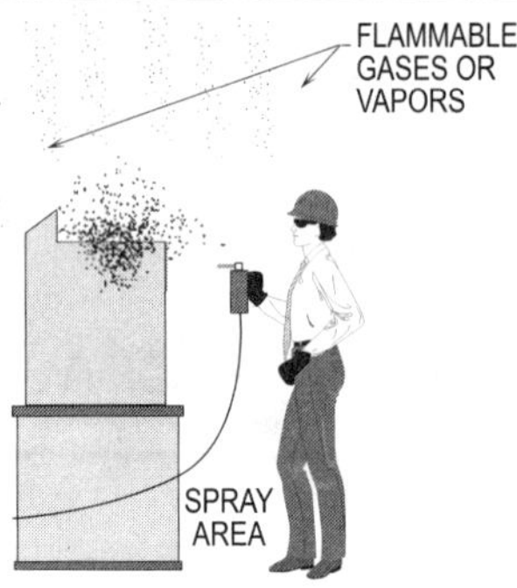

Figure 21-8(a). The above illustration is a Class I, Division 1 location, due to the presence of gases or vapors during operation, maintenance, or repair.

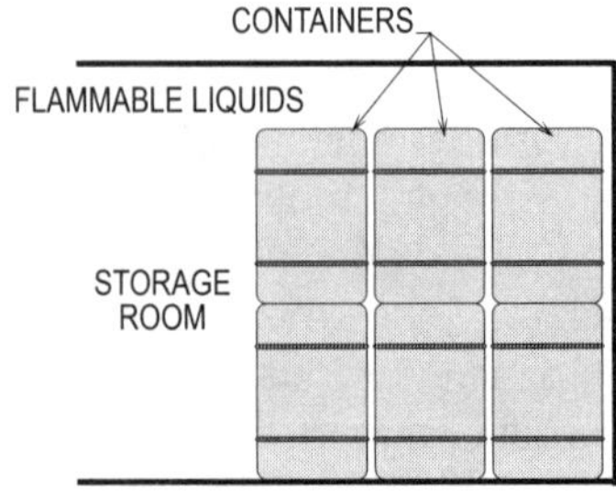

Figure 21-8(b). The above is a Class I, Division 2 location, due to being confined in containers or closed systems.

CLASS II, DIVISION 1
500.5(C)(1)

A Class II, Division 1 location is a location where (a) combustible dust is in the air under normal operating conditions in quantities sufficient to produce explosive or ignitible mixtures, or (b) where mechanical failure or abnormal operation of machinery or equipment might cause such explosive or ignitible mixtures to be produced and might also provide a source of ignition through simultaneous failure of electrical equipment, through operation of protective devices, or from other causes, or (c) where Group E combustible dusts may be present in quantities sufficient to be hazardous. **[See Figure 21-9(a)]**

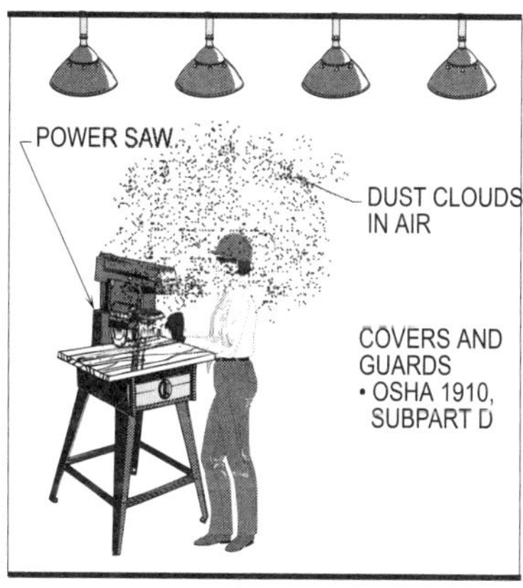

Figure 21-9(a). The above is a Class II, Division 1 location, due to the presence of dust during operation or mechanical failure of equipment.

CLASS II, DIVISION 2
500.5(C)(2)

A Class II, Division 2 location is a location where (a) combustible dust, due to abnormal operations, may be present in the air in quantities sufficient to produce explosive or ignitible mixtures, or (b) dust accumulations are present but are normally insufficient to interfere with the normal operation of electrical equipment or other apparatus but

could as a result of infrequent malfunctioning of handling or processing equipment become suspended in the air, or (c) combustible dust accumulations on, in, or in the vicinity of the electrical equipment could be sufficient to interfere with the safe dissipation of heat from electrical equipment or could be ignitable by abnormal operation or failure of electrical equipment. **[See Figure 21-9(b)]**

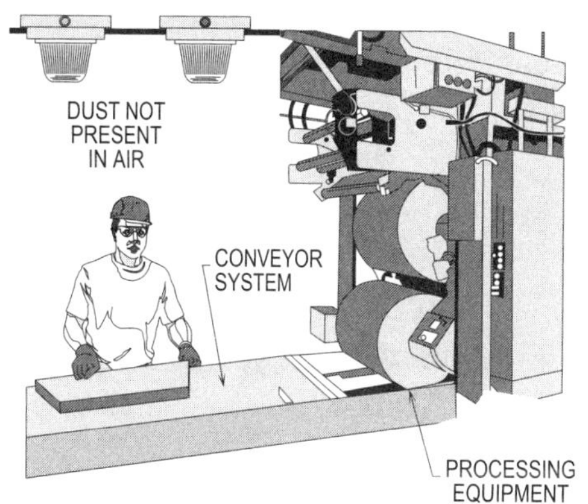

Figure 21-9(b). The above is a Class II, Division 2 location, due to dust being limited during operation of equipment.

CLASS III, DIVISION 1
500.5(D)(1)

A Class III, Division 1 location is a location in which easily ignitible fibers/flyings are handled, manufactured, or used. **[See Figure 21-10(a)]**

CLASS III, DIVISION 2
500.5(D)(2)

A Class III, Division 2 location is a location in which easily ignitible fibers/flyings are stored or handled other than in the process of manufacture. **[See Figure 21-10(b)]**

Figure 21-10(a). The above illustration is a Class III, Division 1 location, due to the presence of easily ignitible fibers/flyings.

Figure 21-10(b). The above is a Class III, Division 2 location, due to easily ignitible fibers being stored or handled.

EXPLOSIVE PROPERTIES OF GASES AND VAPORS
NFPA 497 – CHAPTER 3

To select the type of equipment allowed to be installed in Class I, Division 1 and 2 locations, it is necessary to understand the meaning of certain terms and how they are to be applied.

FLASH POINT
NFPA 497, SECTION 3.3.7

Flash point of a liquid is the minimum temperature at which the liquid gives off vapor in sufficient concentration to form an ignitible mixture with the air near the surface of the liquid, as specified by test. An ignitible mixture is within the flammable range (between upper and lower limits) that is capable of propagating flame away from the source of ignition when ignited. Some evaporation takes place below the flash point but not in sufficient quantities to form an ignitible mixture.

Design Tip: The flash point applies mostly to flammable and combustible liquids, although there are certain solids, such as camphor and naphthalene, that slowly evaporate or volatilize at ordinary room temperature, and liquids, such as benzene, that freeze at relatively high temperatures and therefore have flash points while in the solid state. **(See Figure 21-11)**

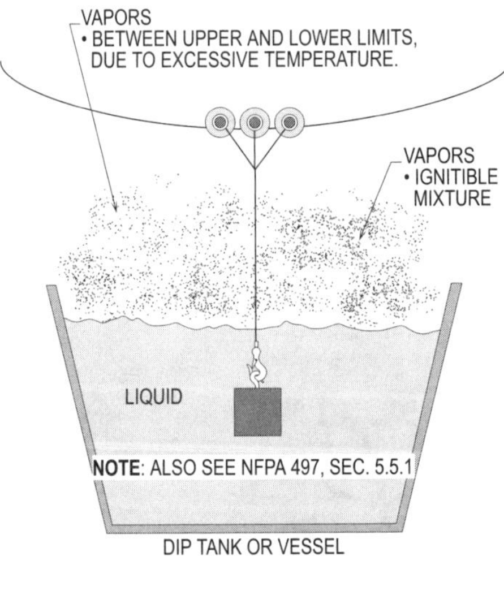

Figure 21-11. The above shows a situation where sufficient vapors are driven off by heat due to the surrounding temperature creating an explosive mixture of vapors and air.

IGNITION TEMPERATURE
NFPA 497 – SECTION 4.2

Ignition temperature of a substance, whether solid, liquid, or gaseous, is the minimum temperature required to initiate or cause self-sustained combustion independently of the heating or heated element.

Ignition temperatures observed under one set of conditions may be changed substantially by a change of conditions. For this reason, ignition temperatures should be looked upon only as approximations. Some of the variables known to affect ignition temperatures are the percentage composition of the vapor or gas-air mixture, shape and size of the space where the ignition occurs, rate and duration of heating, kind and temperature of the ignition source, catalytic or other effect of materials that may be present, and oxygen concentration. As there are many differences in ignition temperature test methods, such as size and shape of containers, method of heating and ignition source, ignition temperatures are affected by the test method. **(See Figure 21-12)** Also, see NFPA 497, 5.5.1 through 5.5.6.

> **Design Tip:** Ignition temperature of a substance, whether solid, liquid, or gaseous, is the minimum temperature required to initiate or cause self-sustained combustion in the absence of any source of ignition.

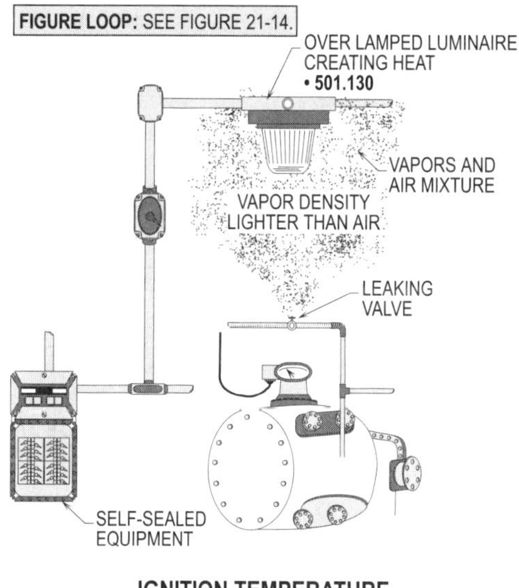

Figure 21-12. The above illustration shows a mixture of vapors and air that could explode due to the excessive heat from the surface temperature of the luminaire.

FLAMMABLE (EXPLOSIVE) LIMITS API 500 – SECTION 3.2.19

In the case of gases or vapors that form flammable mixtures with air or oxygen, there is a minimum concentration of vapor in air or oxygen below which propagation of flame does not occur on contact with a source of ignition. There is also a maximum proportion of vapor or gas in the air above in which propagation of flame does not occur. These boundary line mixtures of vapor or gas with air, which if ignited will just propagate flame, are known as the "lower and upper flammable or explosive limits" and are usually expressed in terms of percentage by volume of gas or vapor in air. **(See Figure 21-13)**

> **Design Tip:** In popular terms, a mixture below the lower flammable limit is too lean to burn or explode and a mixture above the upper flammable limit is too rich to burn or explode.

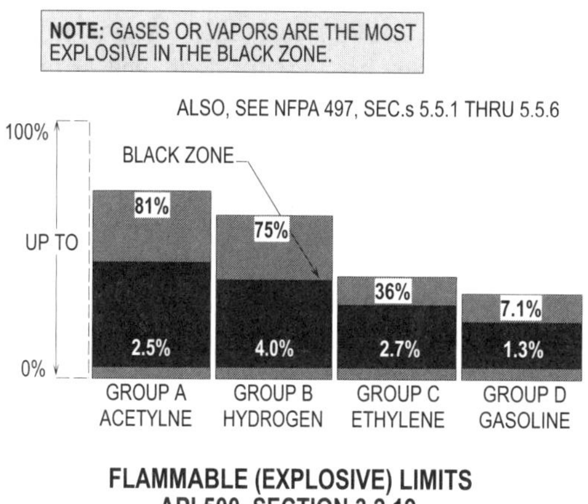

Figure 21-13. The above shows gases or vapors in their lower and upper limits.

VAPOR DENSITY NFPA 497– SECTION 4.2

Vapor densities are used mainly to determine the settling or rising tendency of a mixture. A figure of less than 1.0 indicates that the vapor is lighter than air and will tend to rise. A figure greater than 1.0 indicates the vapor will settle and move along the grade.

> **Design Tip:** An example of a lighter-than-air gas is methane natural gas, that has a density of 0.6 and 0.1, respectively. This type of gas in indoor locations will tend to concentrate near the ceiling where luminaires are usually mounted. On the other hand, heavier-than-air mixtures, like ethyl ether vapors, have the ability to travel at low levels for a considerable distance to locations where sources of ignition might be available.

See Figure 21-14 for an illustration showing the different vapor densities of flammable materials.

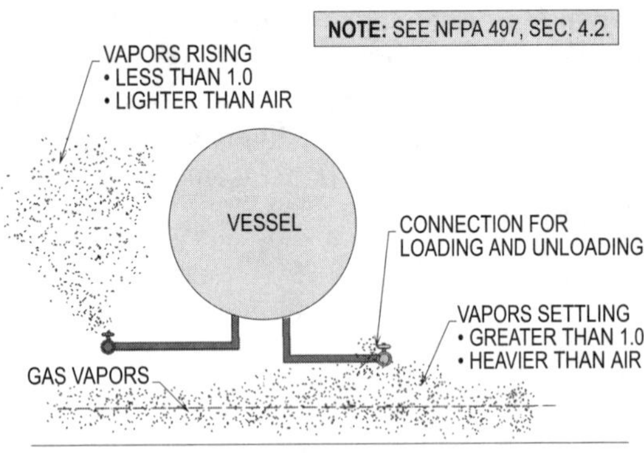

VAPOR DENSITY
NFPA 497, SECTION 4.2

Figure 21-14. The above shows that vapors lighter than air rise and that vapors heavier than air will settle near the grade.

COMBUSTION PRINCIPLES
NFPA 497 – SECTION 4.2

The following three basic conditions shall be satisfied for a fire or explosion to occur:

(1) A flammable liquid, vapor, or combustible shall be present in sufficient quantity.

(2) The flammable liquid, vapor, or combustible dust shall be mixed with air or oxygen in the proportions required to produce an explosive mixture.

(3) A source of energy shall be applied to the explosive mixture.

In applying these principles, the quantity of the flammable liquid or vapor that may be liberated and its physical characteristics shall be recognized. Vapors from flammable liquids have a natural tendency to disperse into the atmosphere and rapidly become diluted to concentrations below the lower explosion limit, particularly when there is natural or mechanical ventilation. Finally, the possibility that the gas concentration may be above the upper explosion limit does not ensure any degree of safety, as the concentration shall first pass through the explosive range to reach the upper explosion limit. (See NFPA 497, Sec. 4.2.6 and 4.2.7)

SOURCES OF IGNITION
NFPA 497 – SECTION 4.3

A source of energy is all that is needed to touch off an explosion where flammable gases or combustible dust are mixed in the proper proportion with air. One prime source of energy is electricity.

Equipment such as switches, circuit breakers, motor starters, push-button stations, or plugs and receptacles can produce arcs or sparks in normal operation when contacts are opened and closed, which could easily cause ignition.

Other hazards are devices that produce heat, such as luminaires and motors. Here, surface temperatures may exceed the safe limits of many flammable atmospheres. Finally, many parts of the electrical system can become potential sources of ignition in the event of insulation failure. This group would include wiring (particularly splices in the wiring), transformers, impedance coils, solenoids, and other low-temperature devices without make-or-break contacts. Nonelectrical hazards such as sparking metal can also easily cause ignition. Thus a hammer, a file, or another tool that is dropped on masonry or on a ferrous surface is a hazard unless the tool is made of non-sparking material. For this reason, portable electrical equipment is usually made from aluminum or other material that will not produce sparks if the equipment is dropped. Therefore, electrical safety is of crucial importance.

The electrical installation must prevent accidental ignition of flammable liquids, vapors, and dusts released into the atmosphere. In addition, since much of this equipment is used outdoors or in corrosive atmospheres, the material and finish shall be such that maintenance costs and shutdowns are minimized. **(See Figure 21-15)**

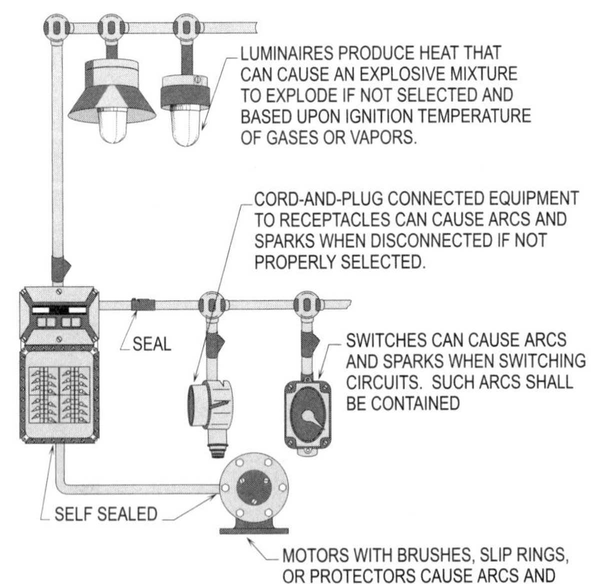

SOURCES OF IGNITION
NFPA 497, SECTION 4.3

Figure 21-15. Ground joint construction type enclosures shown above will withstand an internal explosion and cool the hot escaping gases through the ground surface of the joint.

ENCLOSURES
NEMA 250

Each area that contains gases or dusts that are considered hazardous must be carefully evaluated to make certain the correct electrical equipment is selected. To conform with the NEC, the use of fittings and enclosures that are identified or listed for the specific hazardous gas or dust involved is required.

In Class I, Division 1 and 2 locations, conventional relays, contactors, and switches that have arcing contacts shall be enclosed in explosionproof housings, except for those few cases where general purpose enclosures are permitted by the NEC. By definition, enclosures for these locations shall prevent the ignition of an explosive gas or vapor that may surround it. In other words, an explosion inside the enclosure shall be prevented from starting a larger explosion on the outside. Adequate strength is one requirement for such an enclosure.

For explosionproof equipment, a safety factor of 4 is used. That is, the enclosure shall withstand a hydrostatic pressure test of four times the maximum pressure from an explosion within the enclosure. In addition to being strong, the enclosure shall be flametight. This term does not imply that the enclosure is hermetically sealed but rather that the joints cool the hot gases resulting from an internal explosion so that by the time they reach the outside hazardous atmosphere, they are too cool to effect ignition.

TYPES OF ENCLOSURES
NEMA 250

Type 7 and 10 enclosures, when properly installed and maintained, are designed to contain an internal explosion without causing an external hazard. Type 8 enclosures are designed to prevent combustion through the use of oil-immersed equipment. Type 9 enclosures are designed to prevent the ignition of combustible dust. Descriptions and tests in this standard publication cover equipment that is suitable for installation in locations classified as Division 1 or Division 2. In Division 2 locations, other types of protection and enclosures for nonhazardous locations may be installed if the equipment does not constitute a source of ignition under normal operating conditions.

Intrinsically safe and nonincendive equipment is not capable of releasing sufficient electrical or thermal energy under normal or abnormal conditions to cause ignition of specific hazardous atmospheres and may be installed in any type of enclosure otherwise suitable for the environmental conditions expected.

Equipment installed in enclosures that are suitable for nonhazardous locations, including the hazardous area, may be reduced or eliminated by adequate positive pressure ventilation from a source of clean air in conjunction with effective safeguards against ventilation failure.

TYPE 7 ENCLOSURES
NEMA 250

Type 7 enclosures are for indoor use in locations classified as Class I, Groups A, B, C, or D, as defined in the NEC.

Type 7 enclosures shall be capable of withstanding the pressures resulting from an internal explosion of specified gases and containing such an explosion sufficiently that an explosive gas-air mixture existing in the atmosphere surrounding the enclosure will not be ignited. Enclosed heat generating devices shall not cause external surfaces to reach temperatures capable of igniting explosive gas-air mixtures in the surrounding atmosphere. Enclosures shall meet explosion, hydrostatic, and temperature design tests.

When Type 7 enclosures are completely and properly installed, they will:

(1) Provide a degree of protection to a hazardous gas environment from an internal explosion or from operation of internal equipment,

(2) Not develop surface temperatures that exceed prescribed limits for the specific gas corresponding to the atmospheres for which the enclosure is intended, when internal equipment is operated at rated load,

(3) Withstand a series of internal explosion design tests that determine:

    (a) The maximum pressure effects of the gas mixture

    (b) Propagation effects of the gas mixtures, and

(4) Withstand, without rupture or permanent distortion, an internal hydrostatic design test based on the maximum internal pressure obtained during explosion tests and on a specified safety factor be marked with the appropriate class and group(s) for which they have been qualified.

TYPE 8 ENCLOSURES
NEMA 250

Type 8 enclosures are for indoor or outdoor use in locations classified as Class I, Groups A, B, C, and D, as defined in the NEC.

Type 8 enclosures and enclosed devices are arranged so that all arcing contacts, connections, and any other parts that could cause arcing are immersed in oil. Arcing is confined under the oil so that it will not ignite an explosive mixture of the specified gases in internal spaces above the oil or in the atmosphere surrounding the enclosure. Enclosed heat generating devices shall not cause external surfaces to reach temperatures capable of igniting explosive gas-air mixtures in the surrounding atmosphere. Enclosures shall meet operation and temperature design tests. Enclosures intended for outdoor use shall also meet the rain test.

When Type 8 enclosures are completely and properly installed, they will:

(1) Provide, by oil immersion, a degree of protection to a hazardous gas environment from operation of internal equipment,

(2) Not develop surface temperatures that exceed prescribed limits for the specific gas corresponding to the atmospheres for which the enclosure is intended when internal equipment is at rated load,

(3) Withstand a series of operation design tests with oil levels arbitrarily reduced and with flammable gas-air mixtures introduced above the oil,

(4) Exclude water when subjected to a water spray design test simulating a beating rain, if installed outdoors, and

(5) Be marked with the appropriate Class and Group(s) for which they have been qualified.

TYPE 9 ENCLOSURES
NEMA 250

Type 9 enclosures are intended for indoor use in locations classified as Class II, Group E or G, as defined in the NEC.

Type 9 enclosures shall be capable of preventing the entrance of dust. Enclosed heat generating devices shall not cause external surfaces to reach temperatures capable of igniting or discoloring dust on the enclosure or igniting dust-air mixtures in the surrounding atmosphere. Enclosures shall meet dust penetration and temperature design tests and prevent aging of gaskets (if used).

When Type 9 enclosures are completely and properly installed, they will:

(1) Provide a degree of protection to a hazardous dust environment from operation of internal equipment,

(2) Not develop surface temperatures that exceed prescribed limits for the Group corresponding to the atmospheres for which the enclosure is intended when internal equipment is operated at rated load,

(3) Withstand a series of operation design tests while exposed to a circulating dust-air mixture to determine that dust does not enter the enclosure and that operation of devices does not cause ignition of surrounding atmosphere, and

(4) Be marked with the appropriate Class and Group(s) for which they have been qualified.

Note: For more information on enclosure types, see www.neca.org.

PREVENTION OF EXTERNAL IGNITION AND EXPLOSION

Explosives are hazardous enough by themselves, but around electricity there are even more hazards present. For this reason, special enclosures are necessary to house electrical elements that are capable of producing arcs and sparks. An arc, spark, or hot surface can easily initiate an explosion. Therefore, these ignition sources shall be contained or the equipment housing such apparatus shall be installed outside the area.

Each area that contains gases or dusts that are considered hazardous must be carefully evaluated to make certain the correct electrical equipment is selected. Many hazardous atmospheres are Class I, Group D, or Class II, Group G. However, certain areas may involve other groups, particularly Class I, Groups B and C. Conformity with the NEC requires the use of fittings and enclosures that are identified or listed for the specific hazardous gas or dust involved. (See NEC **Articles 500** and **501** for more information.)

CLASS I, DIVISION 1 EQUIPMENT
500.2, 500.7, AND 500.8(A)(1)

Class I, Division 1 equipment shall provide a form of construction that will ensure safe performance under conditions of proper use and maintenance and will operate in such a manner to prevent igniting an explosive mixture of gases and air. Note that Class I, Division 1 equipment shall be permitted to be used in Class II, Division 2 locations. When approving equipment installed in classified areas, see **500.8(A)(1)** and Figure 21-60 in this chapter.

EXPLOSIONPROOF EQUIPMENT (CLASS I, DIVISION 1) 500.2 AND 500.7(A)

The enclosure shall withstand an internal explosion of gases or vapors and prevent those gases or vapors from igniting gases and vapors in the surrounding atmosphere outside of the enclosure.

These enclosures are not intended to exclude flammable or combustible gases or vapors, but rather to withstand an internal explosion and prevent the ignition of external gases and vapors. The internal gases are dissipated or cooled before they are released from the enclosure, thereby preventing ignition or an explosion of gases or vapors in the surrounding atmosphere. **(See Figure 21-16)**

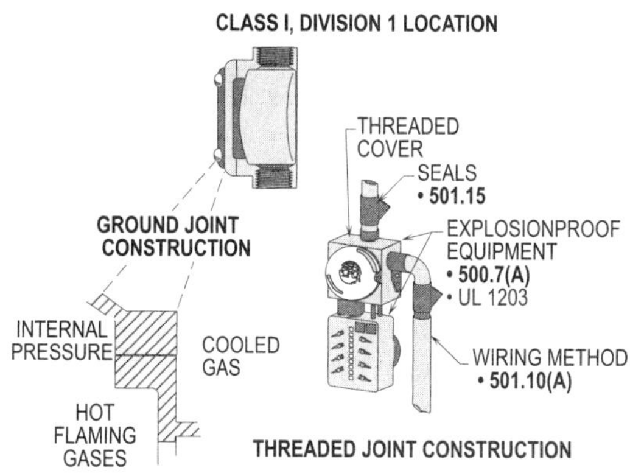

Figure 21-16. The above illustration shows an enclosure that is capable of withstanding an internal explosion of gases or vapors and prevent those gases or vapors from igniting gases and vapors in the surrounding atmosphere outside of the enclosure.

PURGED AND PRESSURIZED SYSTEMS (CLASS I, DIVISION 1) 500.2 AND 500.7(D)

Purged and pressurized systems permit the safe operation of electrical equipment under conditions of hazard for which approved equipment may not be commercially available. For instance, most switchgear units and many large-size motors do not come in designs listed for Class I, Groups A and B. Whether cast metal enclosures for hazardous locations or sheet metal enclosures with pressurization should be used is mainly a question of economics if both types are available.

As a typical example, if an installation had many electronic instruments that could be enclosed in a single sheet metal enclosure, the installation lends itself to the purging/ pressurization system. However, if the instruments, due to their nature, had to be installed in separate enclosures, the cast metal in hazardous location housing would almost invariably prove more economical. Pressurized enclosures require:

(1) A source of clean air or inert gas,

(2) A compressor to maintain the required pressure on the system, and

(3) Pressure control valves, to prevent the power from being applied before the enclosure have been purged and to deenergize the system should pressure fall below a safe value.

In addition, door interlock switches shall prevent access to the equipment while the circuits are energized. It can readily be seen that all of these accessories can add up to a considerable expenditure.

Design Tip: For a detailed description of purged and pressurized systems, see NFPA 496-2003, "*Standard for Purged and Pressurized Enclosures for Electrical Equipment.*" **(See Figure 21-17)**

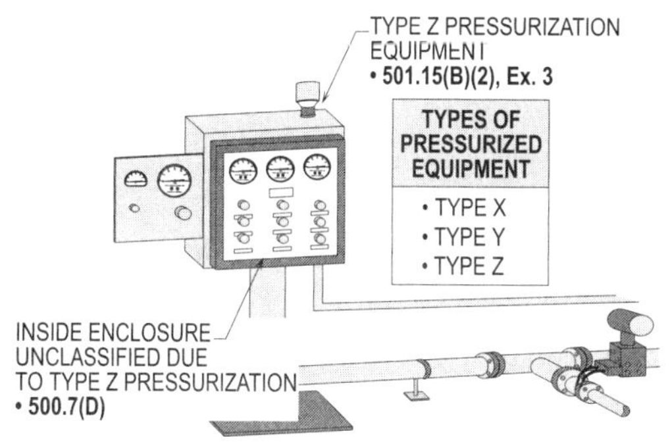

Figure 21-17. The above illustration shows the inside of an enclosure being an unclassified area, based upon a Type Z pressurized piece of equipment. Purged or pressurized equipment shall be permitted to be used in Class I, Division 1 or 2 locations.

INTRINSICALLY SAFE EQUIPMENT (CLASS I, DIVISION 1) 504.2 AND 500.7(E)

The use of intrinsically safe equipment is primarily limited to process control instrumentation, since these electrical systems lend themselves to the low energy requirements. ANSI/UL 913-1997 and ANSI/ISA-RP 12.06.01-2003 provide information on the design test and evaluation. The installation rules are covered in **Article 504** of the NEC. The definition of intrinsically safe circuit is: "A circuit in which any spark or thermal effect is incapable of causing ignition of a mixture of flammable or combustible material in air under prescribed test conditions." UL and Factory Mutual list several devices in this category.

Design Tip: The equipment and its associated wiring shall be installed so they are positively separated from the nonintrinsically safe circuits. Induced voltages could defeat the concept of intrinsically safe circuits. **(See Figure 21-18)**

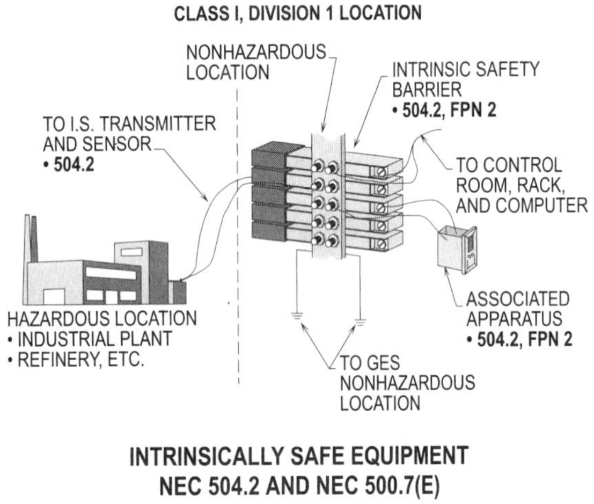

Figure 21-18. Intrinsically safe equipment shall be permitted to be utilized to supply low-voltage instruments, etc. in Class I, Division 1 and 2 locations. See **501.105(B)** and **501.150(B)** of the NEC and Figure 4-20 in Chapter 4.

NONINCENDIVE EQUIPMENT (CLASS I, DIVISION 2) 500.2, 500.7(F), (G), AND (H)

Nonincendive equipment is a form of intrinsic safety designed for use in Class I, Division 2 locations. Such equipment is defined as components having contacts that make or break an incendive circuit and elements. The enclosures in which the contacts are enclosed are so constructed that the components are not capable of igniting the surrounding flammable gases and air mixtures.

General purpose enclosures shall be permitted to be used to house such circuits and components that under normal operating conditions will not release enough energy to ignite a specific explosive mixture.

Design Tip: The housing of nonincendive components is not intended to exclude a flammable atmosphere or contain an explosion. **(See Figure 21-19)**

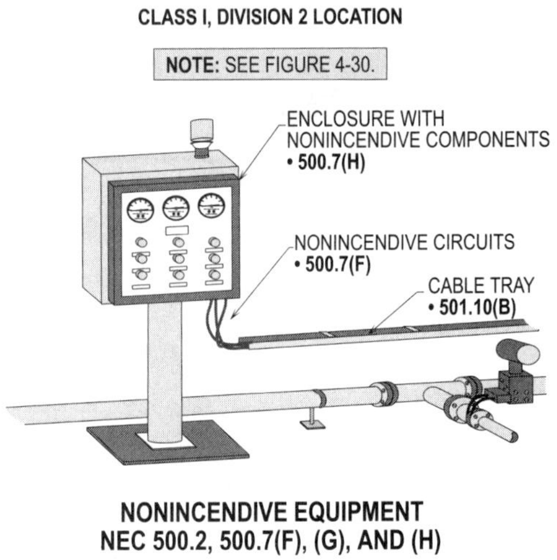

Figure 21-19. The above shows nonincendive components and circuits used in Class I, Division 2 locations.

OIL IMMERSION AND HERMETICALLY SEALED CONTACTS 500.2, 500.7(I), AND (J)

The general rule requires switches, circuit breakers, and make-and-break contacts of push buttons, relays, alarm bells, and horns to be installed in enclosures that are identified for Class I, Division 1 locations. However, in Class I, Division 2 locations, general purpose enclosures shall be permitted to be used if the current-interrupting contacts are:

(1) Immersed in oil, or

(2) Enclosed within a chamber hermetically sealed against the entrance of gases or vapors.

See Figure 21-20 for contacts immersed in oil and sealed within an enclosed hermetically sealed chamber.

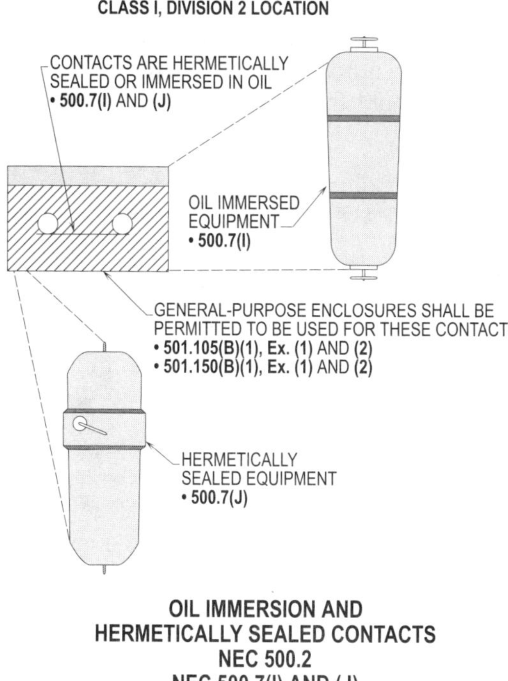

Figure 21-20. The above shows contacts installed in general purpose enclosures that are immersed in oil and hermetically sealed chambers. Such enclosures and controls shall be permitted to be installed in Class I, Division 2 locations.

CLASS I, DIVISION 1 WIRING METHODS 501.10(A)(1)

In Class I, Division 1 locations, threaded rigid metal conduit (RMC), threaded steel intermediate metal conduit (IMC), or Type MI cable with termination fittings that are identified for the location shall be the wiring method employed. All boxes, fittings, and joints shall be threaded for connection to conduit or cable terminations and shall be of explosionproof type per **510.10(A)(3)**. Threaded joints shall be made up with at least five threads fully engaged per **500.8(E)**. Type MI cable shall be installed and supported in such a manner to avoid tensile stress at the termination fittings. Where it is necessary to utilize flexible connections, as at motor terminals, flexible fittings that are identified for Class I locations shall be used per **501.10(A)(2)** and **501.140**.

TYPE PVC OR RTRC (CLASS I, DIVISION 1) 501.10(A)(1)(a), Ex.

The **Ex.** to **501.10(A)(1)(a)** allows type PVC or RTRC conduit to be installed in Class I, Division 1 areas, if encased in a concrete envelope of at least 2 in. (50 mm) and buried below the surface in not less than 24 in. (600 mm) of earth.

MC CABLE AND ITC CABLE (CLASS I, DIVISION 1) 501.10(A)(1)(c) AND (d)

Section **501.10(A)(1)(c)** and **(d)** permits Type MC and ITC cable that is listed for such use and equipped with a gas/vaportight continuous corrugated aluminum sheath with an overall jacket of suitable polymeric material to be installed in Class I, Division 1 locations. **(See Figure 21-21)**

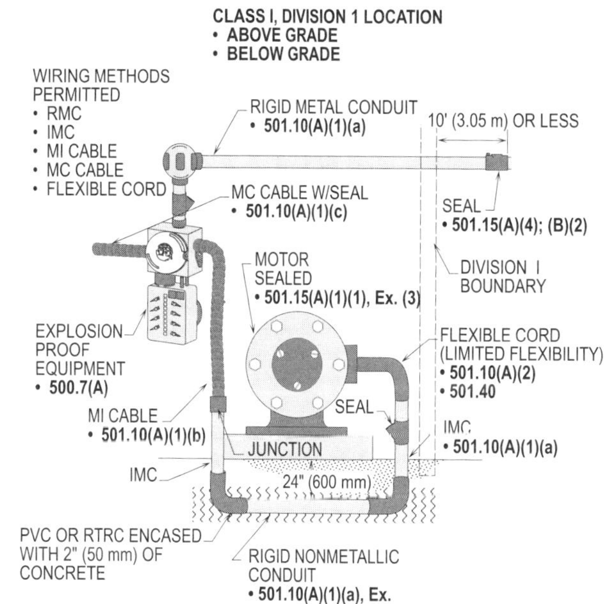

Figure 21-21. As shown above, rigid metal conduit, IMC, type PVC or RTRC, MI and MC cable, and flexible cord under certain conditions of use shall be permitted to be utilized in Class I, Division 1 locations.

WIRING METHODS USED FOR LIMITED FLEXIBILITY (CLASS I, DIVISION 1) 501.10(A)(2)

Flexible fittings that are listed for use in Class I locations are required where it is necessary to employ flexible connections such as at motor terminals, etc.

Section **501.10(A)(2)** refers to **501.140,** that allows flexible cord to be used for that portion of the circuit where the fixed wiring method does not provide the necessary movement for fixed and mobile electrical utilization equipment in an industrial establishment. However, for this rule to be applied, proper maintenance and supervision shall be available. **(See Figure 21-21)**

CLASS I, DIVISION 2 EQUIPMENT
500.2 AND 500.7

Class I, Division 2 equipment shall provide a form of construction that will operate in such a manner as to prevent igniting flammable mixtures of gases and air.

TYPES OF EQUIPMENT
(CLASS I, DIVISION 2)
500.7(A), (D), AND (E)

Equipment installed in Class I, Division 1 areas can be installed in Class I, Division 2 locations. Such equipment is the following:

(1) Explosionproof

(2) Purged and pressurized

(3) Intrinsically safe

> **Design Tip:** See and review the equipment that is either identified or listed and illustrated in this chapter under Class I, Division 1 Equipment.

COMBUSTIBLE GAS DETECTION
SYSTEM
500.2 AND 500.7(K)

Under certain conditions of use, a combustible gas detection system shall be permitted to be used. For example, Class I, Division 2 equipment shall be permitted to be installed in Class I, Division 1 locations and electrical equipment for Class I, Division 2 locations shall be permitted to be installed in unclassified locations. However, note that the following requirements shall be complied with when using a combustible gas detection system:

- In a Class I, Division 1 location that is so classified due to inadequate ventilation, electrical equipment suitable for Class I, Division 2 locations shall be permitted.

- In a building located in, or with an opening into, a Class I, Division 2 location where the interior does not contain a source of flammable gas or vapor, electrical equipment for unclassified locations shall be permitted.

- In the interior of a control panel containing instrumentation utilizing or measuring flammable liquids, gases or vapors, electrical equipment suitable for Class I, Division 2 locations shall be permitted.

- Gas detection equipment shall be listed for detection of the specific gas or vapor that is to be encountered.

- System shall be installed in industrial establishments, with restricted public access and where the conditions of supervision and maintenance ensure that only qualified persons service the installation.

- The types of detection equipment, installation location(s), alarm and shutdown criteria, and calibration frequency shall be documented when combustible gas detectors are used as a protection technique.

See Figure 21-22 for an illustration of a typical combustible gas detection system.

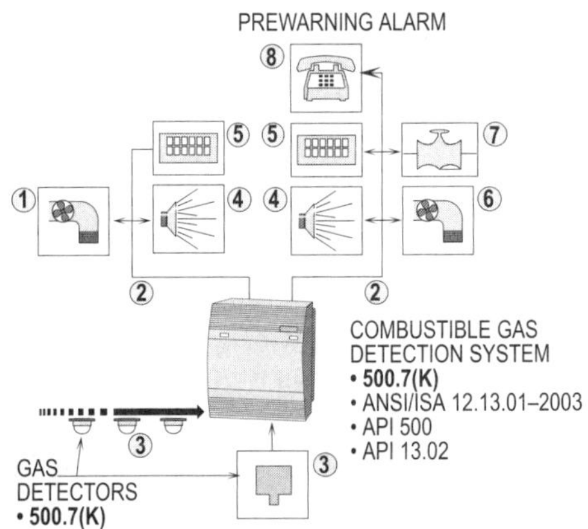

**COMBUSTIBLE GAS DETECTION SYSTEM
NEC 500.2 AND 500.7(K)**

Figure 21-22. The above shows a typical combustible gas detection system.

CLASS I, DIVISION 2 WIRING METHODS 501.10(B)(1)(1) THRU (7)

In Class I, Division 2 locations, threaded rigid metal conduit, threaded IMC, enclosed gasketed busways, enclosed gasketed wireways, or Type PLTC cable in accordance with the provisions of **Article 725**, Type MI, MC, MV, or TC cable with listed termination fittings shall be the wiring method utilized. Type ITC, PLTC, MI, MC, MV, or TC cable shall be permitted to be installed in cable tray systems where installed in a manner to avoid tensile stress at the termination fittings.

> **Design Tip:** Boxes, fittings, and joints shall not be required to be explosionproof except when they are used in areas requiring Class I, Division 1 enclosures.

WIRING METHODS USED FOR LIMITED FLEXIBILITY (CLASS I, DIVISION 2) 501.4(B)(2)(1) THRU (5)

If provision must be made for limited flexibility, as at motor terminals, flexible metal fittings, flexible metal conduit, liquidtight flexible metal conduit, or liquidtight flexible nonmetallic conduit shall be permitted to be used.

> **Design Tip:** The above wiring methods shall be provided with listed fittings that provide a proper connection for bonding per **501.30(B)**.

Flexible cord that is listed for extra-hard usage and provided with approved bushed fittings shall be permitted to be used. However, an additional conductor for equipment grounding conductor shall be included in the flexible cord. **(See Figure 21-23)**

REQUIREMENTS FOR SWITCHES, CIRCUIT BREAKERS, MOTOR CONTROLLERS, AND FUSES 501.115

Enclosures used to house switches, circuit breakers, motor controllers, and fuses shall be rated for either Class I, Division 1 or Division 2. The Division determines the type of enclosure, based upon the amount and the time the gas or vapor is present.

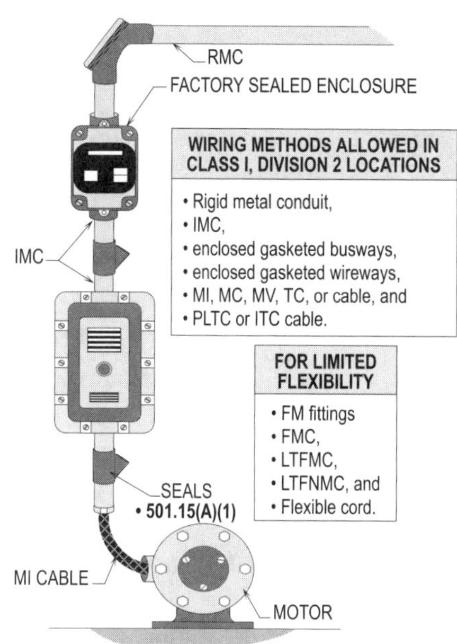

**WIRING METHODS USED FOR LIMITED FLEXIBILITY
(CLASS I, DIVISION 2)
NEC 501.10(B)(2)(1) THRU (5)**

Figure 21-23. As shown above, there are more wiring methods that are permitted in Class I, Division 2 than Class I, Division 1 locations. **See Figure 21-21** for wiring methods permitted in Class I, Division 1 locations.

CLASS I, DIVISION 1 501.115(A)

Identified Class I, Division 1 enclosures shall be used for switches, circuit breakers, motor controllers and fuses, including push buttons, relays and similar devices, in order to be installed in Class I, Division 1 locations. **(See Figure 21-24)**

CLASS I, DIVISION 2 501.115(B)

Identified Class I, Division 1 enclosures shall be used for switches, circuit breakers, motor controllers and fuses, including push buttons, relays and similar devices, in order to be installed in Class I, Division 2 locations.

However, general purpose enclosures shall be permitted to be used if the interruption of current occurs in **(1)** a hermetically sealed chamber, or **(2)** the current make-and-break contacts are oil-immersed.

The use of general purpose enclosures shall be permitted if the interruption of current occurs within a factory-sealed explosionproof chamber that is identified for the location. **(See Figure 21-24)**

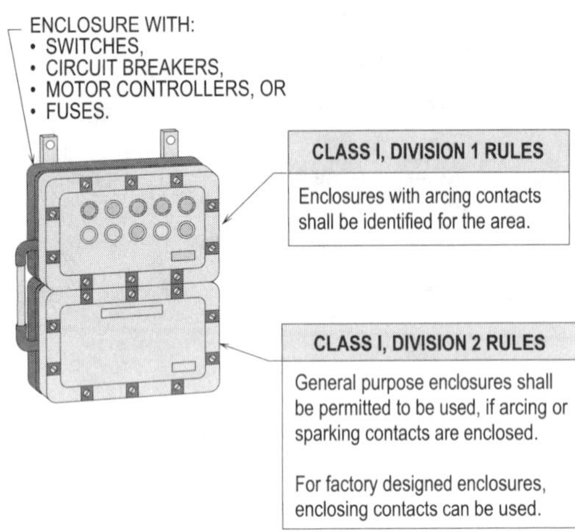

Figure 21-24. Enclosures containing arcing contacts shall be identified for Class I, Division 1 locations. In Class I, Division 2 locations, general-purpose and factory enclosures shall be permitted to be used if the contacts are enclosed in such a manner to handle arcing and sparking conditions.

Design Tip: The use of disconnects and isolating switches in general purpose enclosures in Class I, Division 2 locations shall be permitted if neither the switch nor fuse operates as a normal current interrupting device. In such cases, fuses can only be used for short-circuit protection.

REQUIREMENTS FOR MOTORS AND GENERATORS
501.125

Electric motors are needed to drive pumps, compressors, fans, blowers, and conveyors, so their presence in hazardous atmospheres is sometimes unavoidable.

The types of hazardous atmospheres and corrosive conditions are major factors in motor selection. The hazardous area dictates the type of motor needed to avoid excessive maintenance and expensive shutdowns. The types available vary all the way from "drip-proof" to "totally enclosed" and "fan cooled" motors.

CLASS I, DIVISION 1
501.125(A)

In Class I, Division 1 locations, only motors that are the explosionproof, totally enclosed, and pressurized with clean air, totally enclosed inert gas filled, and special submerged type shall be permitted to be used.

The NEC makes it clear that Class I, Division 1 totally enclosed motors shall have no external surface operating temperature in excess of 80 percent of the ignition temperature of the gas or vapor involved.

Also required are devices to detect and automatically deenergize the motor (or provide an effective alarm) in case of overheating.

In addition, auxiliary equipment shall be of a type that is identified for the location in which it is installed. **(See Figure 21-25)**

CLASS I, DIVISION 2
501.125(B)

Motors for use in Class I, Division 2 locations in which sliding contacts, switching mechanisms, or integral resistance devices are employed shall be explosionproof (Class I, Division 1) or purged and pressurized.

However, open type motors such as squirrel-cage induction motors without any arcing devices shall be permitted to be used in Class I, Division 2.

Design Tip: UL provides a procedure in which listed explosionproof motors can be repaired. Motor personnel must consult as to which repair shops have been authorized to make such repairs. Unauthorized maintenance of an explosionproof motor can result in voiding the manufacturer's listing.

To enable the AHJ to determine whether or not the maximum temperature exceeds 80 percent of the ignition temperature of the gas or vapor involved, space heaters used in a motor shall have a permanent, visible nameplate marking on the motor, indicating the maximum surface temperature (based on 40°C ambient).

If the temperature of the heater does not exceed 80 percent of the ignition temperature, the motor shall be considered suitable for use in that atmosphere. **(See Figure 21-25)**

REQUIREMENTS FOR LUMINAIRES
501.130

In locations where explosive gases or vapors exist, bare lamps or non-explosionproof enclosed luminaires can create extreme hazards. Bare lamps may be broken, and the arc or spark can cause explosions. For this reason, lamps shall be enclosed in luminaires that are specifically designed for Class I, Divisions 1 or 2 locations, whichever applies.

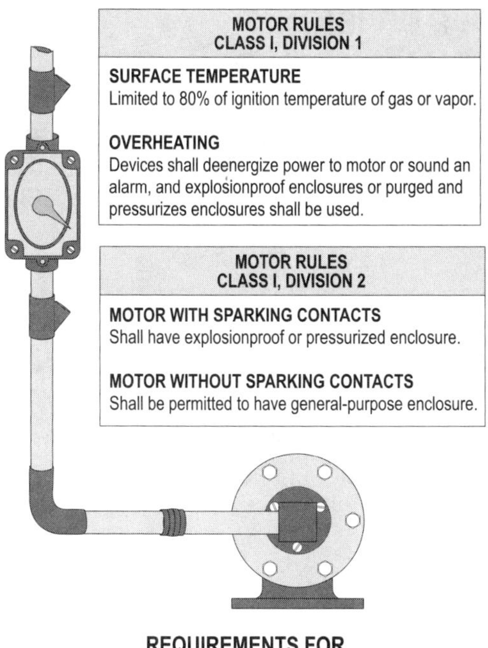

Figure 21-25. The above rules apply when installing motors in Class I, Division 1 and 2 locations.

CLASS I, DIVISION 1
501.130(A)(1)

Surface temperature of Class I, Division 1 luminaires shall not exceed the ignition temperature of the gases or vapors surrounding such luminaires.

Each luminaire shall be identified as a complete assembly for the Class I, Division 1 location. In addition, it shall be clearly marked to indicate the maximum wattage of lamps for which it is identified.

PENDANT LUMINAIRES
501.130(A)(3)

Pendant luminaires for Class I, Division 1 shall be suspended and supplied by threaded metal or IMC conduit stems. Threaded joints of Class I, Division 1 luminaires shall have set-screws to prevent loosening on rigid stems. Where rigid stems (conduit portion from mounting box to luminaire) are longer than 12 in. (300 mm), they shall be braced against movement or provided with flexible fittings. **(See Figure 21-26)**

PORTABLE LIGHTING EQUIPMENT
501.130(A)(1)

Luminaires intended for portable use shall be specifically identified as a complete assembly for that use. Personnel shall not take portable lighting equipment into such classified areas if it is not identified for such area and use. **(See Figure 21-26)**

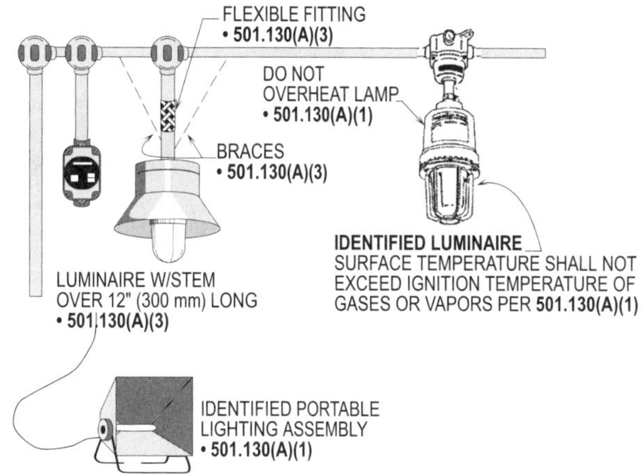

Figure 21-26. The above rules and regulations apply when installing luminaires in Class I, Division 1 locations.

CLASS I, DIVISION 2
501.130(B)(1)

For Class I, Division 2 luminaires, it is the lamp surface temperature, and not the luminaire surface temperature, that shall not exceed the ignition temperature of the gases or vapors present.

Basically, in locations that are classified Class I, Division 2, fixed luminaires that have been tested shall be permitted to operate up to the ignition temperature of the gas or vapor present. However, if the luminaires have not been tested for such use, the maximum temperature shall not exceed 80 percent of the ignition temperature of the gas or vapor present. **(See Figure 21-27)**

PENDANT LUMINAIRES
501.130(B)(3)

In Class I, Division 2 locations, pendant luminaires shall be suspended by flexible hangers unless rigid stems not over 12 in. (300 mm) long are used, or longer stems are permanently braced within 12 in. (300 mm) of the luminaire.

Pendant luminaires shall be suspended and supplied by threaded metal or IMC conduit stems that are screwed into the threaded joints of the luminaires. Set-screws, to prevent loosening on rigid stems, shall also be provided. **See Figure 21-26** for rules pertaining to pendant-hung luminaires.

PORTABLE LIGHTING
501.130(B)(4)

Portable lighting used in Class I, Division 2 locations shall not be required to be identified for Class I, Division 1 if it is mounted on a movable stand and connected by an approved flexible cord. The luminaire only needs to be identified for Class I, Division 2 and be specified as follows:

(1) The luminaire be protected by a suitable guard or location,

(2) The luminaire has a suitable enclosure to prevent sparks or hot metal from a lamp from causing ignition of the surrounding atmosphere, and

(3) The luminaire shall not exceed temperature limitations.

See Figure 21-27 for rules concerning luminaires installed and used in Class I, Division 2 locations.

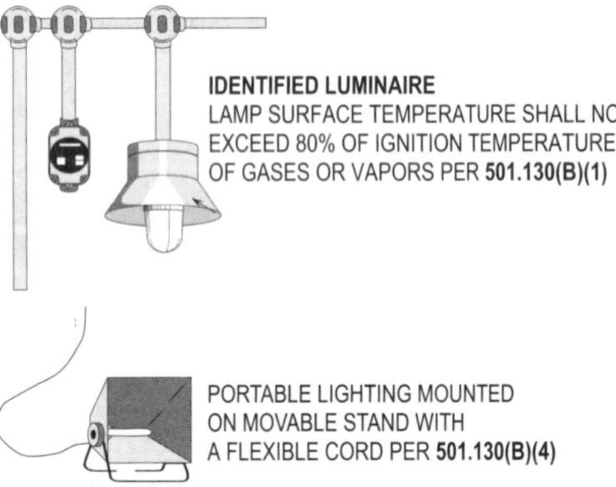

Figure 21-27. The above rules and regulations apply when installing luminaires in Class I, Division 2 locations.

REQUIREMENTS FOR UTILIZATION EQUIPMENT
501.135

Utilization equipment installed in hazardous (classified) areas shall be considered to be appliances specifically designed for such areas.

CLASS I, DIVISION 1
501.135(A)

Utilization equipment, including electrically heated and motor-driven equipment, that is used in Class I, Division 1 locations shall be identified for such location. (**See Figure 21-28**)

CLASS I, DIVISION 2
501.135(B)

Utilization equipment installed in Class I, Division 2 locations shall comply with certain design criteria.

HEATERS
501.135(B)(1)

Heating appliances shall be identified for Class I, Division 1 locations. Heating appliances would include water heaters and room heaters. There are exceptions that permit heaters to be the general purpose type, and they are as follows:

(1) If the maximum operating temperature of any exposed surface does not exceed 80 percent of the ignition temperature of the gas or vapor surrounding the heater, when operating continuously at 120 percent rated voltage, the heater shall be permitted to be the general-purpose type.

(2) If the maximum operating temperature of any exposed surface does not exceed 80 percent of the ignition temperature of the gas or vapor surrounding the heater, when operating continuously at 100 percent rated voltage, the heater shall be permitted to be the general-purpose type, provided it is supplied with a temperature controller.

To meet the requirements of Class I, Division 1 equipment, operating temperature at rated voltage would have to be no greater than about 55 percent of the ignition temperature of the gas or vapor and all heaters shall be completely enclosed.

See Figure 21-29 for rules for installing heaters in Class I, Division 2 locations.

APPLIANCES WITH MOTORS
501.135(B)(2)

Appliances with motors shall not be required to be identified for the location unless the motor has brushes, a centrifugal switch, or a built-in thermal protector, in which case the appliances shall be identified for Class I, Division 1 locations. (See **501.115(B)(1)** for further information concerning motors.) **(See Figure 21-29)**

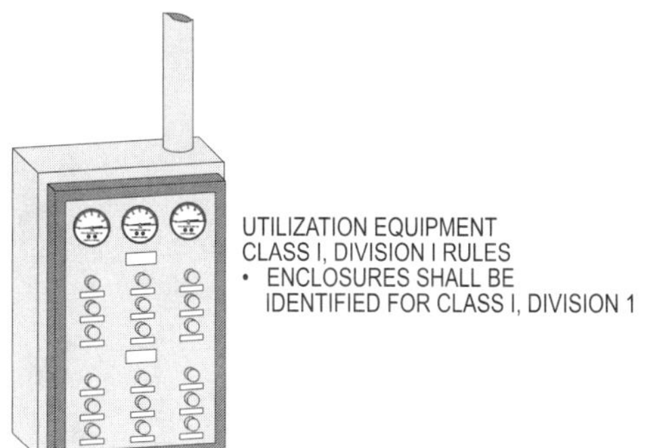

Figure 21-28. Enclosures, used for the components of utilization equipment, shall be identified for Class I, Division 1 locations.

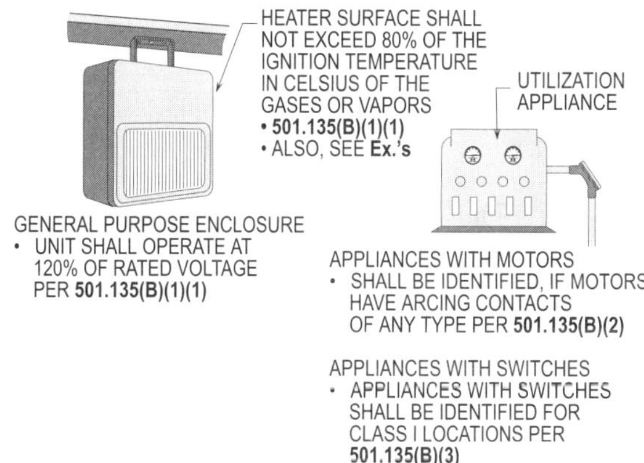

Figure 21-29. The above rules and regulations apply when installing utilization equipment (appliances) in Class I, Division 2 locations.

APPLIANCES WITH SWITCHES
501.135(B)(3)

If an appliance has a switch, the switch shall be identified for Class I locations. The rest of the appliance may in some cases have a general purpose enclosure, but the switch shall always be identified for Class I locations.

> **Design Tip:** Except as required by **501.135(B)(1), (B)(2),** and **(B)(3)** above, appliances in Class I, Division 2 locations shall be permitted to be the general-purpose type. **(See Figure 21-29)**

USING FLEXIBLE CORD IN CLASS I, DIVISION 1 AND 2 AREAS
501.140

A flexible cord shall be permitted to be used for connection between a portable lamp or other portable utilization equipment. The fixed portion of its supply circuit and where such connections are used shall:

(1) Be of a type identified for extra-hard usage,

(2) Contain, in addition to the conductors of the circuit, an equipment grounding conductor conforming to **250.118**, that is, of green color and used for no other purpose than for grounding,

(3) Be connected to terminals or to supply conductors in an approved manner,

(4) Be supported by clamps or other suitable means in such a manner that there will be no tension on the terminal connections, and

(5) Have suitable seals provided where the flexible cord enters boxes, fittings, or enclosures of the explosionproof type.

In Class I, Division 1 locations, flexible cord shall be permitted to be used for that portion of the circuit where the fixed wiring method of **501.10(A)(1)** cannot provide the necessary degree of movement for fixed and mobile electrical utilization equipment. However, proper maintenance and supervision shall be provided, and such equipment shall be located in an industrial establishment.

CORD-AND-PLUG CONNECTED SUBMERSIBLE PUMPS
501.140(A)(3)

Electric submersible pumps with means for removal without entering the wet pit shall be considered portable utilization equipment. The extension of the flexible cord between the wet pit and power source shall be permitted to be enclosed in a suitable raceway.

CORD-AND-PLUG CONNECTED ELECTRIC MIXERS
501.140(A)(4)

Electric mixers traveling in and out of open-type mixing tanks or vats shall be permitted to be classified as portable utilization equipment so that wiring procedures will not have to be too restrictive.

See Figure 21-30 for a detailed illustration of flexible cord being used for such installations.

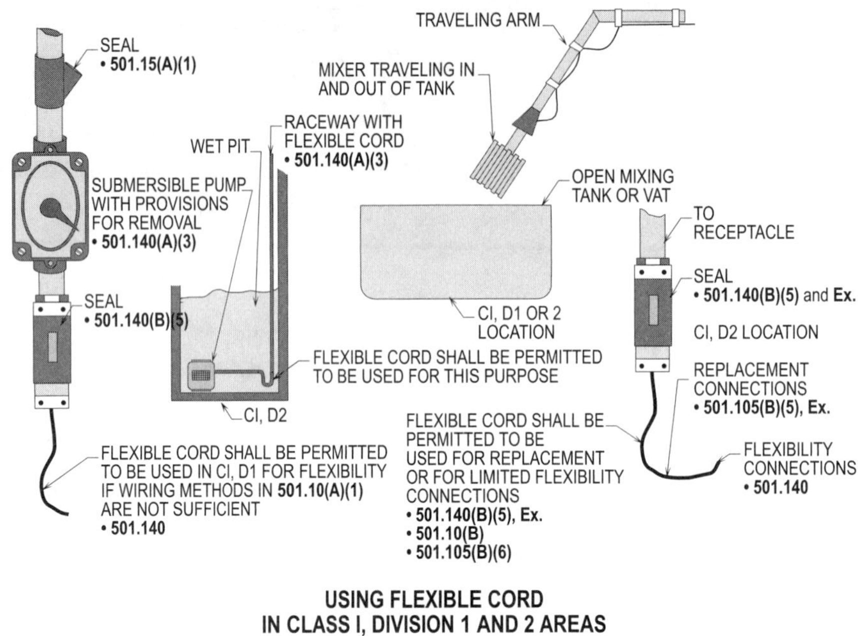

Figure 21-30. The above rules and regulations apply when using flexible cord in Class I, Division 1 and 2 locations.

USING FLEXIBLE CORD TO FACILITATE REPLACEMENTS 501.140(B)(5), Ex. AND 501.105(B)(6)

To facilitate replacements, process control instruments shall be permitted to be connected through flexible cord, attachment plug, and receptacle. However, the following shall be provided:

(1) A switch complying with the rules of **501.105(B)(1)** above is provided so that the attachment plug is not depended on to interrupt current,

(2) The current does not exceed 3 amps at 120 volts,

(3) The power supply cord does not exceed 3 ft (900 mm) and is listed for extra-hard usage or for hard usage if protected by location and is supplied through an attachment plug or receptacle of the locking and grounding type,

(4) Only necessary if receptacles are provided, and

(5) The receptacle carries a label warning against unplugging under load. **(See Figure 21-30)**

USING FLEXIBLE CORD FOR FLEXIBILITY CONNECTIONS 501.140(B)(5), Ex. AND 501.10(B)

Where provisions must be made for a flexible connection, as at motors, flexible cord shall be permitted to be used but additional grounding shall be provided around these flexible connections to ensure an effective ground path for clearing short circuits and ground faults.

Extra-hard usage flexible cord with listed bushed fittings and an extra equipment grounding conductor shall be permitted to be used for this purpose and fully complies with such requirements as above. **(See Figure 21-30)**

REQUIREMENTS FOR RECEPTACLES AND ATTACHMENT CAPS 501.145

Arcing at exposed contacts shall be prevented in Class I, Division 1 or 2 locations. To accomplish this, receptacles are designed so that plug contacts are safely within an explosionproof enclosure when they are electrically engaged, confining arcing, if any, to the receptacle interior.

RECEPTACLES WITH SWITCHES 501.145 AND UL 498

With this configuration, the plug cannot be inserted unless the switch is in the OFF position and cannot be withdrawn with the receptacle in the ON position. The reason for this is the receptacle contacts are interlocked with a switch located in an explosionproof enclosure. Therefore, arcing does not occur outside the enclosure, due to mated parts being deenergized during plug insertion and removal. **(See Figure 21-31)**

RECEPTACLES WITHOUT SWITCHES 501.145, AND UL 498

Receptacles without switches rely on a mechanical means that provides a delayed action to confine arcing to the receptacle interior during plug insertion and withdrawal. The design used in these receptacles prevents removal of the plug until any flame, spark, or hot metal from an arc has cooled sufficiently to prevent ignition of the surrounding explosive atmosphere. **(See Figure 21-31)**

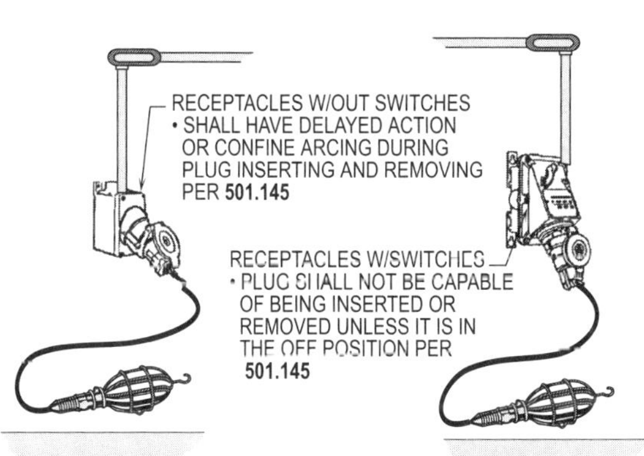

Figure 21-31. The above rules and regulations apply when using cord-and-plug connected equipment in Class I, Division 1 and 2 locations.

WIRING METHODS IN CLASS II, DIVISION 1 AND 2 LOCATIONS 502.10

In Class II, Division 1 and 2 locations, certain types of wiring methods and enclosures shall be utilized to properly contain and handle dust-related areas.

WIRING METHODS (CLASS II, DIVISION 1) 502.10(A)(1)(1), (2), AND (3)

In Class II, Division 1 locations, the wiring method shall be either metal conduit or Type MI cable (mineral-insulated cable).

Section **502.10(A)(1)(3)** permits Type MC cable, listed for such use and equipped with a gas/vaportight continuous corrugated aluminum sheath with an overall jacket of suitable polymeric material, to be installed in Class II, Division 1 locations.

> **Design Tip:** To utilize this wiring method, a separate equipment grounding conductor, in compliance with **250.122** and with termination fittings listed for the application, shall be used. **(See Figure 21-32)**

FITTINGS AND BOXES (CLASS II, DIVISION 1) 502.10(A)(1)(4)

There are two main parts to this rule that shall be applied, and they are as follows:

(1) For locations where the dusts is hazardous but not of a combustible electrically conductive nature. This includes grain, cocoa, dried egg and milk dust, starch dust, and hay dust locations.

Boxes and fittings that contain splices or taps shall be identified for Class II locations when used in such locations.

Boxes and fittings without splices or taps shall not be required to be identified, but shall have dusttight enclosures with no openings such as screw holes.

(2) For locations where dusts of a combustible electrically conductive nature are present. This includes coal dust, coke, carbon black, charcoal dust, magnesium, and aluminum dust locations.

All fittings and boxes, with or without splices or taps, shall be identified for Class II locations when installed in these locations. **(See Figure 21-32)**

FLEXIBLE CONNECTIONS
502.10(A)(2)

In Class II, Division 1 locations, liquidtight flexible metal conduit and liquidtight flexible nonmetallic conduit with listed fittings or flexible cord, listed for extra-hard usage and equipped with bushed fittings, shall be permitted to be used. Unless other means are provided for grounding, cords shall have an equipment grounding conductor. Where dusts of an electrically conducting nature are present, cords shall have dusttight seals at both ends. **(See Figure 21-32)**

WIRING METHODS
(CLASS II, DIVISION 2)
502.10(B)(1)(1) THRU (6)

In Class II, Division 2 locations, the following wiring methods shall be permitted:

(1) Rigid metal conduit

(2) Intermediate metal conduit (IMC)

(3) Electrical metallic tubing (EMT)

(4) Dusttight wireways

(5) Type MI cable (mineral-insulated cable)

(6) Type MC cable (metal-clad power cable)

(7) Type ITC in cable trays

(8) Type PLTC in cable trays

(9) Type MC, MI, or TC cable in cable trays

(10) Nonincendive circuits in any suitable wiring method

FLEXIBLE CONNECTIONS
502.10(B)(2)

In Class II, Division 2 locations, liquidtight flexible metal conduit and liquidtight flexible nonmetallic conduit with approved fittings or flexible cord, approved for extra-hard usage and equipped with bushed fittings, shall be permitted to be used. Unless other means are provided for grounding, cords shall have an equipment grounding conductor. Where dusts of an electrically conducting nature are present, cords shall have dusttight seals at both ends. **(See Figure 21-33)**

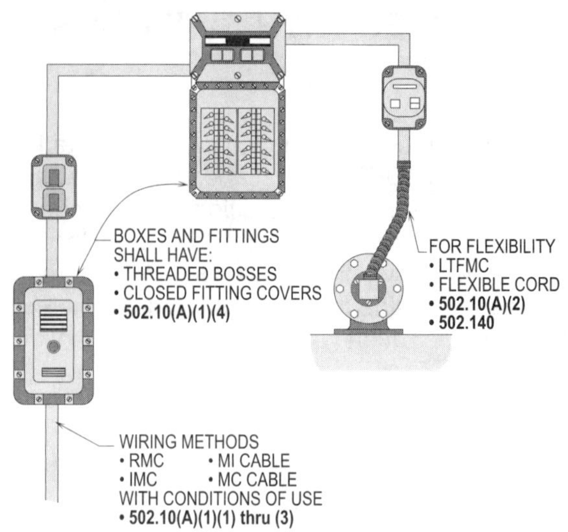

Figure 21-32. The above rules and regulations apply to wiring methods and equipment installed in Class II, Division 1 locations.

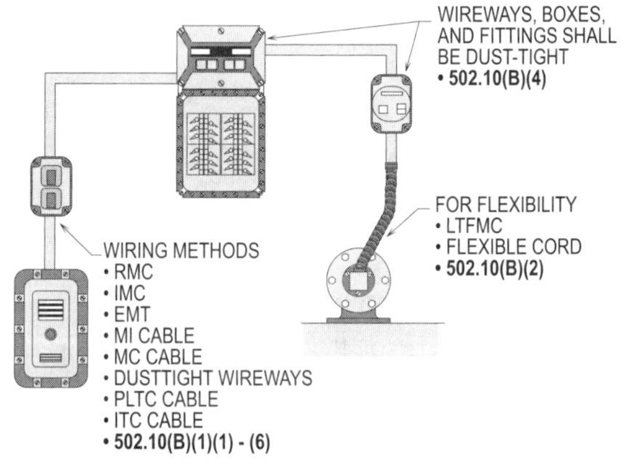

Figure 21-33. The above rules and regulations apply for wiring methods and equipment installed in Class II, Division 2 locations.

WIREWAYS, FITTINGS, AND BOXES
502.10(B)(4)

Wireways, fittings, and boxes in Class II, Division 2 locations are not necessarily required to be approved for Class II locations. However, they shall have close-fitting enclosures with no openings, such as screw holes, that might transmit sparks to the outside of the wireway, fitting, or box. Sparking or burning material shall not escape, because it could ignite adjacent combustible material. **(See Figure 21-33)**

REQUIREMENTS FOR SEALS AND DRAINS (CLASS I, DIVISIONS 1 AND 2) 501.15

Seals shall be provided in conduit and cable systems to minimize the passage of gases or vapors from one portion of the system to another portion. Another purpose of the seals is to prevent transmitting an explosion or to keep ignition from traveling between sections of the system.

NEED FOR SEALS 501.15(A)(1) THRU (A)(4)

Seals and drains are recommended to be installed as follows:

(1) To restrict the passage of gases, vapors, or flames from one portion of the electrical installation to another at atmospheric pressure and normal ambient temperatures,

(2) To limit explosions to the sealed-off enclosure and prevent compression or "pressure piling" in conduit systems,

(3) While not a code requirement, many engineers consider it good practice to sectionalize long conduit runs by inserting seals not more than 50 ft (15 m) to 100 ft (30 m) apart, depending on the conduit size, to minimize the effects of "pressure piling." Sealing fittings are required, and

(4) At each entrance to an enclosure housing having an arcing or sparking device and used in Class I, Division 1 and 2 hazardous locations. To be located as close as practical and in no case more than 18 in. (450 mm) from such enclosures. **(See Figure 21-34)**

(5) At each entrance 2 in. (53 mm) size or larger to an enclosure or fitting, housing terminals, splices, or taps when used in Class II, Division 1 location. To be located as close as practical and in no case more than 18 in. (450 mm) from such enclosures. **(See Figure 21-35)**

(6) In conduit systems when leaving a Class I, Division 1 or Division 2 location. **(See Figure 21-36)**, and

(7) In cable systems when the cables either do not have a gas/vaportight continuous sheath or are capable of transmitting gases or vapors through the cable core where these cables leave a Class I, Division 1 or Division 2 location. **(See Figure 21-37)**

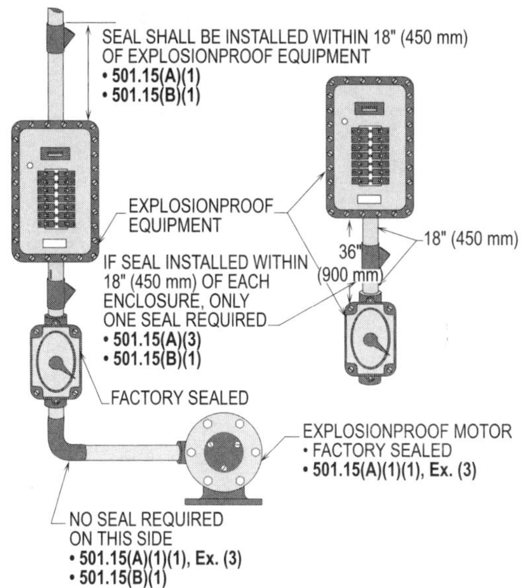

Figure 21-34. Explosionproof equipment or equipment with arcing or sparking devices shall have a seal placed within 18 in. (450 mm) of such equipment.

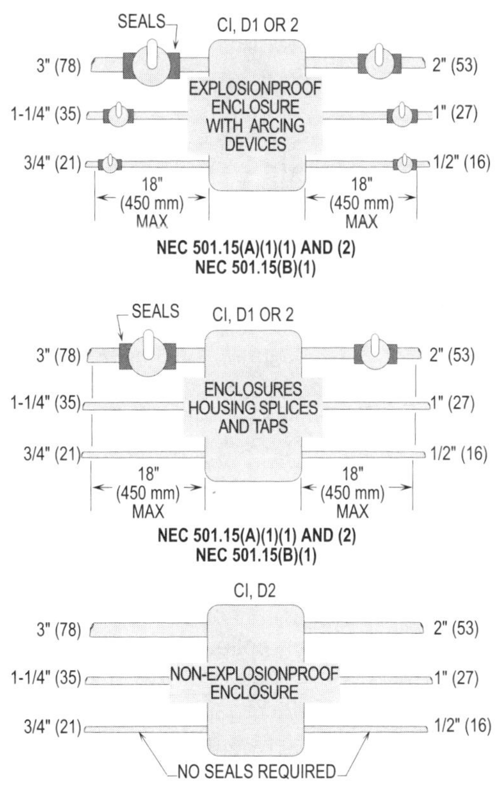

Figure 21-35. The above rules apply to the enclosures of splices, taps, and arcing or sparking devices if they are installed in Class I, Divisions 1 or 2 locations.

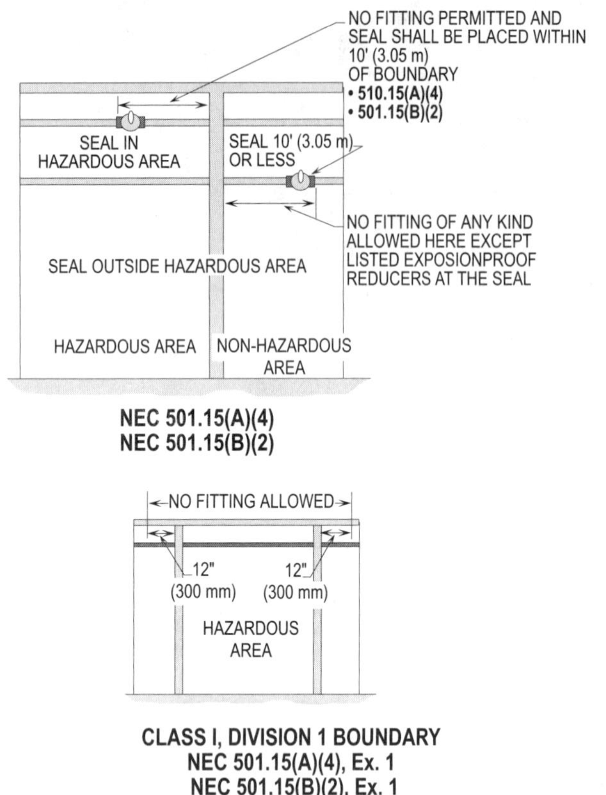

Figure 21-36. The above rules apply when installing seals in conduit runs leaving or passing through a Class I, Divisions 1 and 2 location.

NEED FOR DRAINS
501.15(F)

In humid atmospheres or in wet locations, where it is likely that water can gain entrance to the interiors of enclosures or raceways, the raceways should be inclined so that water will not collect in enclosures or in seals but will be led to low points where it may pass out through ECD drains. Frequently the arrangement of raceway runs makes this method impractical if not impossible. In such instances, Type EZD drain seal fittings shall be used. These fittings prevent accumulations of water above the seal.

In locations that usually are considered dry, surprising amounts of water frequently collect in conduit systems. No conduit system is airtight; therefore, it may breathe. Alternate increases and decreases in temperature and/or barometric pressure due to weather changes or due to the nature of the process carried on in the location where the conduit is installed will cause breathing. Outside air is drawn into the conduit system when it breathes in. If this air carries sufficient moisture, it will be condensed within the system when the temperature decreases and chills this air. Due to the internal conditions being unfavorable to evaporation, the resultant water accumulation will remain

and be added to by repetitions of the breathing cycle. In view of this likelihood, it is good practice to ensure against such water accumulations and probable subsequent insulation failures by installing EZD with drain cover or inspection cover even though conditions prevailing at the time of planning or installing may not indicate their need. **(See Figure 21-38)**

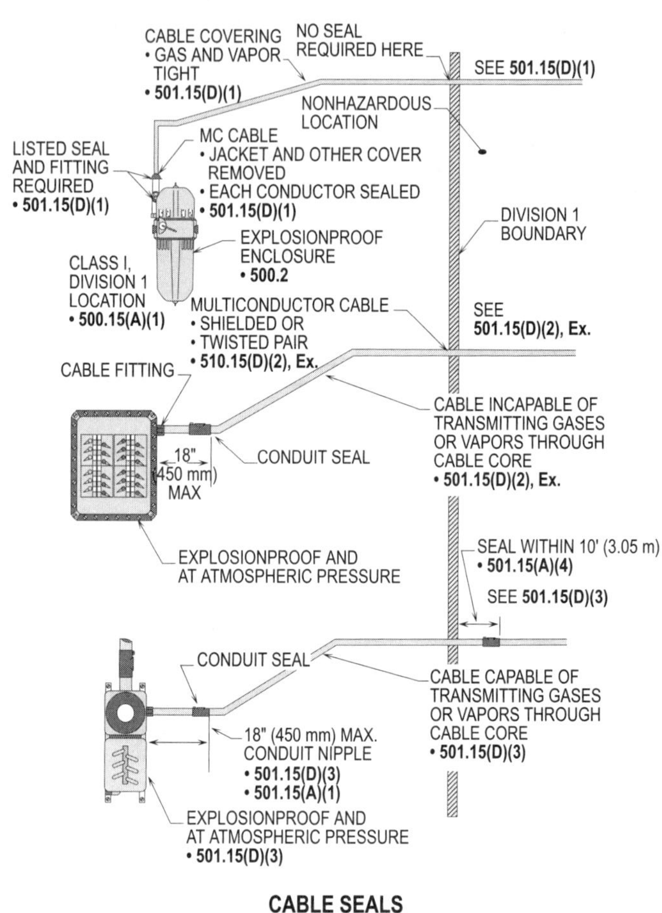

Figure 21-37. The above rules apply when sealing multiconductor cables leaving a Class I, Division 1 location.

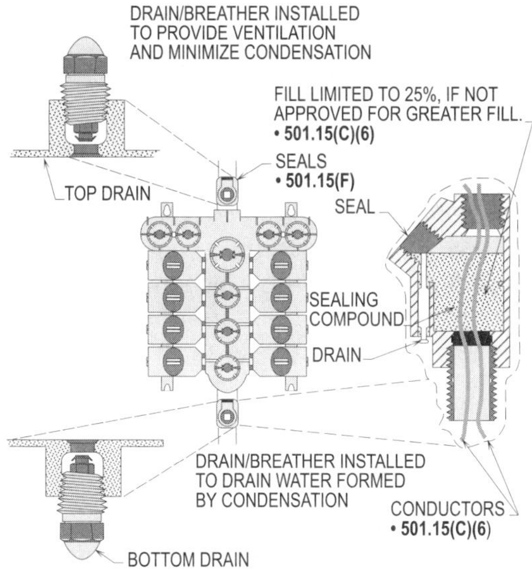

Figure 21-38. Drains and seals installed as shown above will minimize water conditions formed by condensation and help prevent mildew from forming.

SELECTION OF SEALS AND DRAINS

There are different types of seals and drains that are made to be utilized for vertical or horizontal installations and are to be used only for the purpose for which they are designed. Care shall be taken when selecting and installing such fittings.

The following primary considerations shall be used when selecting seals and drains:

(1) Select the proper sealing fitting for the hazardous vapor involved, such as Class I, Groups A, B, C, or D.

(2) Select a sealing fitting for the proper use in respect to mounting position. This is particularly critical when the conduit runs between hazardous and nonhazardous areas. Improper positioning of a seal may permit hazardous gases or vapors to enter the system beyond the seal and permit them to escape into another portion of the hazardous area or to enter a nonhazardous area. Some seals are designed to be mounted in any position; others are restricted to horizontal or vertical mounting.

(3) Install the seals on the proper side of the partition or wall as recommended by the manufacturer.

(4) Only trained personnel can install seals, which shall be installed in compliance with the instruction sheets furnished with the seals and sealing compound. Precautionary notes shall be included on installation diagrams to stress the importance of following manufacturer's instructions.

(5) Splices or taps in sealing fittings are strictly prohibited.

(6) Sealing fittings are listed by UL for use in Class I hazardous locations with CHICO A compound only. This compound, when properly mixed and poured, hardens into a dense, strong mass, that is insoluble in water, is not attacked by chemicals, and is not softened by heat. It will withstand ample safety factor pressure of the exploding trapped gases or vapor.

> **Design Tip:** Seals shall be poured with the compound recommended by its manufacturer.

(7) Conductors sealed in the compound shall be permitted to be approved thermoplastic or rubber insulated types. Either may or may not be lead covered (the lead need not be removed).

See Figure 21-39 for installing seals and drains.

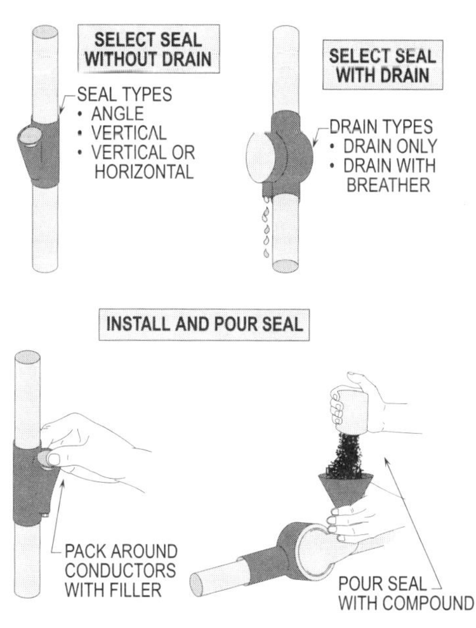

Figure 21-39. The above rules shall be applied when selecting and installing seals/drains in Class I, Division 1 locations.

TYPES OF SEALING FITTINGS
501.15(C)(1)

The following sealing fittings meet the requirements of the NEC when they are properly installed:

(1) EYS Sealing Fittings—A certain style of EYS sealing fittings are for use with vertical or nearly vertical conduit in sizes from 1/2 (16) through 1 in. (27) Other styles are available in sizes 1/2 (16) through 6 in. (155) for use in vertical or horizontal conduits. In horizontal runs, these are limited to face-up openings. Sizes from 1-1/4 (155) through 6 in. (155) have extra large work openings and separate filling holes, so that CHICO X fiber dams are easy to make. Overall diameter of sizes 1-1/4 (35) through 6 in. (155) is scarcely greater than that of unions of corresponding sizes, permitting close conduit spacings.

Note: Check with the manufacturer for the type of compound to be used with a particular seal.

(2) EZS Sealing Fittings—EZS seals are for use with conduit running at any angle, from vertical through horizontal.

See Figure 21-40 for the different types of sealing fittings.

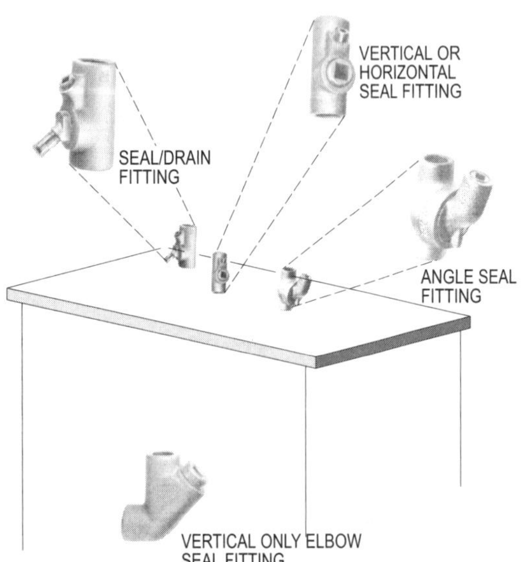

TYPES OF SEALING FITTINGS
NEC 501.15(C)(1)

Figure 21-40. As shown above, it is most important to select and install the proper seal or seal with drain based upon the type of installation.

REQUIREMENT FOR SEALS (CLASS II, DIVISIONS 1 AND 2)
502.15(1) THRU (4)

When a dust-ignitionproof enclosure and one that is not are installed, there can be communication of dust between the two enclosures under certain conditions of use. The entrance of dust into the dust-ignitionproof enclosure shall be prevented to avoid explosions. This can be accomplished by applying one of the following sealing methods:

(1) By use of a permanent and effective seal, as used in Class I locations,

(2) By connection to a horizontal raceway that is not less than 10 ft (3.05 m) in length, and

(3) By means of a vertical conduit not less than 5 ft (1.5 m) in length, that is installed downward from the dust-ignitionproof enclosure.

> **Design Tip:** Where a conduit provides communication between an enclosure that is required to be dust-ignitionproof and an enclosure in an unclassified location, seals are not required.

COMMERCIAL GARAGES, REPAIR, AND STORAGE
ARTICLE 511

Article 511 of the NEC applies to commercial garages in which repair and service work is done on automobiles, trucks, buses, tractors, etc.

Garages used only for parking or storage are not hazardous areas. The requirements of **Article 511** do not apply to garages in which no maintenance repair or service work is being performed. However, the NEC does require that indoor storage and parking spaces be adequately ventilated to remove exhaust fumes from engines.

EIGHTEEN INCHES ABOVE THE FLOOR AND BELOW CEILING AREA
511.3(C)(1)(b) AND (D)(1)(b)

For each floor, the entire area up to a level of 18 in. (450 mm) above the floor or 18 in. (450 mm) from the ceiling is considered to be a Class I, Division 2 location, except where the AHJ determines that there is sufficient mechanical ventilation to provide a minimum of four air changes per hour.

ANY PIT OR DEPRESSION BELOW FLOOR LEVEL
511.3(C)(3)(b) AND (D)(3)(b)

Pits or depressions below floor level that extend up to the floor level shall be considered as Class I, Division 1 locations. However, if the pit or depression is ventilated, the AHJ may classify and judge the area to be a Class I, Division 2 or unclassified location.

> **Design Tip:** Any pit or depression in which six air changes per hour are exhausted at the floor level of the pit can be judged by the AHJ to be a Class I, Division 2 location per **511.3(C)(3)(a)** and **(D)(3)(a).**

ADJACENT AREAS
511.3(E)(1)

Areas adjacent to Class I, Division 2 locations in which there is no likelihood of hazardous vapors being released, such as stock rooms, offices, etc. are still classified as Class I, Division 2 locations. If the floor of the adjacent area is elevated 18 in. (450 mm) above the floor of the hazardous area or a separation between the two areas is provided by either an 18 in. (450 mm) tight curb or partition, the area can be classified as nonhazardous.

> **Design Tip:** Gasoline vapors are heavier than air and settle to and move along the floor area. Since these vapors settle to the floor, they may be transmitted into other rooms that may seem otherwise to be nonhazardous areas. Transmittal of such hazardous vapors can be stopped by an elevated floor, a tight curb, or a properly built partition, each of which shall be at least 18 in. (450 mm) high, between the two rooms.

SPECIAL PERMISSION
ARTICLE 100 AND 511.3(E)(1)

Adjacent areas that by reason of adequate ventilation, air pressure differentials, or physical spacing are such that, in the opinion of the AHJ, no ignition hazardous condition exists can be classified as nonhazardous.

See Figure 21-41 for rules and regulations pertaining to garages.

FUEL DISPENSING UNITS
511.4(B)(1)

When fuel dispensing units (other than liquid petroleum gas, which is prohibited), are located within buildings, the requirements of **Article 514** shall be applied.

If adequate mechanical ventilation is provided in the dispensing area, the controls shall be interlocked so that the dispenser cannot operate without ventilation as required by **500.5(B)(2).**

PORTABLE LIGHTING EQUIPMENT
511.4(B)(2)

If fuel pumps are located inside, the requirements of **Article 514** pertaining to service stations are required to be applied to the pumps and the area around the pumps.

Portable lighting equipment used in garages shall:

(1) Be of hard rubber or other nonmetallic material,

(2) Be equipped with a handle,

(3) Have a guard and hook,

(4) Be the unswitched type, and

(5) Have no plug-in.

> **Design Tip:** If portable lighting equipment is to be used in a hazardous area near the floor, for instance, It shall be a type that is identified for hazardous areas.

If the cord is connected in such a manner that the lamp cannot reach a hazardous level when in use, a portable lamp can be the general purpose type.

WIRING AND EQUIPMENT IN CLASS I LOCATIONS
511.4(A) AND 511.3

Hazardous vapors in a garage are concentrated in low areas within a short distance above the floor and within the ceiling area. The code sets the distance as 18 in. (450 mm), and the area below an imaginary plane of 18 in. (450 mm) above the floor is classified as a hazardous area.

Raceways beneath the floor or embedded in a masonry shall be considered to be within the hazardous area if any connections or extensions lead into the hazardous area per **511.3.**

Hazardous vapors seek either a low or high level. Therefore, any ceiling or pit in a garage floor shall be considered to be hazardous down to the bottom of the pit or up to the ceiling area.

All wiring and equipment located in the space below the hazardous level shall be suited for a Division or Zone location if adequate ventilation is not utilized.

Design Tip: If all of the wiring and equipment is routed and installed above the 18 in. (450 mm) height from the floor or below the 18 in. (450 mm) ceiling area, EMT and general purpose equipment and devices shall be permitted to be used. If the wiring is installed in conduit below the 18 in. (450 mm) boundary in the floor slab, this part of the wiring system shall be a Class I location and the wiring method would have to comply with the rules of this classification. To pass from the hazardous area and be routed into the nonhazardous area, seals would have to be installed at the 18 in. (450 mm) boundary.

SEALING REQUIREMENTS 511.9

Seals shall be provided as follows for wiring methods entering or leaving the Class I, Division 2 area:

(1) If conduit is used, a vertical run, unless unbroken, shall be sealed at the hazardous boundary level, which is 18 in. (450 mm) above the floor.

(2) For a horizontal run, if the run leaves the area at a height below the hazardous boundary level, that is less than 18 in. (450 mm) from the floor, the wiring method shall be sealed at the point where it leaves the hazardous area.

(3) Seals shall be placed in each conduit run entering the enclosures for switches, circuit breakers, fuses, relays, resistors, or any device that produces arcing, sparking, or high temperature, and only if the enclosure is located beneath the hazardous boundary.

(4) In each conduit run of 2 in. (53) or larger, seals shall be installed at junction boxes and fittings if the box or fitting contains splices, taps, or connections and is located beneath the hazardous boundary.

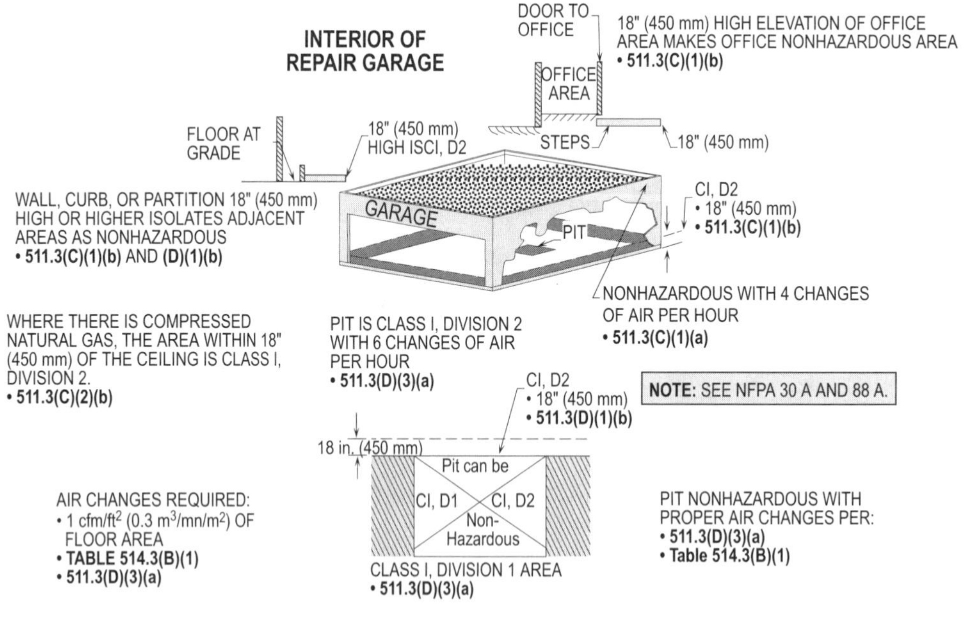

Figure 21-41. In classification of a floor in a garage where vehicles are repaired and volatile flammable fluids are present, the floor area, up to 18 in. (450 mm), is Class I, Division 2. However, if adequate ventilation is present, the 18 in. (450 mm) space above the floor shall be considered to be nonhazardous. Likewise, if the pit is adequately ventilated, it can be judged to be a Class I, Division 2 location or a nonhazardous location.

> **Design Tip:** The seal shall be installed within 18 in. (450 mm) of the box or fitting to comply with NEC rules and regulations. [See **501.15** and **501.15(B)(1)**]

WIRING ABOVE CLASS I LOCATIONS 511.7(A)(1)

Above the 18 in. (450 mm) level, the following wiring methods shall be permitted:

(1) Metal raceways

(2) Rigid nonmetallic conduit

(3) Electrical nonmetallic tubing

(4) EMT

(5) Intermediate metal conduit

(6) Type MC cable (metal clad)

(7) Type MI cable (mineral-insulated)

(8) Type TC cable (power and control tray)

(9) Type PLTC

(10) ITC

(11) Flexible metal conduit

(12) Liquidtight flexible metal conduit

(13) Liquidtight nonmetallic conduit

> **Design Tip:** Cellular metal floor raceways shall be permitted to be used only to supply ceiling outlets or extensions to areas located below the floor. However, outlets shall not be connected above the floor. No electrical conductor shall be permitted to be installed in any cell, header, or duct that contains pipe for any service except electrical or compressed air.

For example, cellular metal floor raceways are a part of the building structure, and if the floor in the garage has cellular metal floor raceways installed, the raceway shall be permitted to be used to supply equipment in the room below the garage. However, it shall not be permitted to be used to supply equipment in the garage. Where the cellular raceway is in the floor above, it shall be permitted to be used for ceiling outlets in the garage.

GROUNDING AND BONDING REQUIREMENTS 511.16

When a circuit that supplies portables or pendants includes an identified grounded conductor (white or gray) as provided in **Article 200**, receptacles, attachment plugs, connectors,

and similar devices shall be of the polarized type and the grounded conductor of the flexible cord shall be connected to the screw shell of any lampholder or to the grounded terminal of any utilization equipment supplied.

> **Design Tip:** Grounding continuity of the equipment grounding conductor to noncurrent-carrying parts shall be maintained.

WRING AND EQUIPMENT INSTALLED ABOVE CLASS I LOCATIONS 511.7(A) AND (B)

Any equipment that might produce an arc or spark, and that is less than 12 ft (3.7 m) above floor level, shall have a totally enclosed or tight enclosure that will prevent the escape of any arc or spark. Charging panels, motors, generators, and switches, if installed less than 12 ft (3.7 m) above the floor, should be the totally enclosed type, per **511.7(B)(1)(a)**.

Luminaires shall be the totally enclosed type or of a construction that does not allow the escape of arcs or sparks, if located less than 12 ft (3.7 m) above lanes where vehicles are commonly driven or if located where exposed to physical damage, per **511.7(B)(1)(b)**. (**See Figure 21-42**)

> **Design Tip:** Fluorescent and incandescent luminaires in Class I locations shall be fitted with a lens or globe. Exposed tubes and light bulbs shall not be permitted.

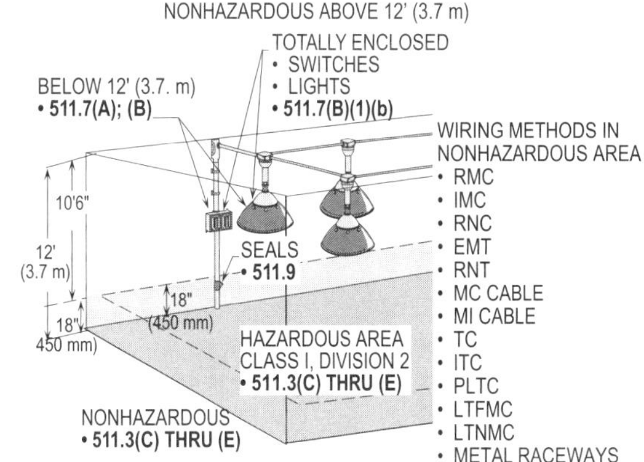

Figure 21-42. The above are rules and regulations pertaining to wiring methods and equipment installed in and above the hazardous area in garages.

AIRCRAFT HANGARS
ARTICLE 513

Article 513 applies to hangars that are used for servicing or storing aircraft containing gasoline or other hazardous liquids or gases. Certain areas in aircraft hangars are a Class I location, except as follows:

(1) Locations used exclusively for aircraft that has never contained such volatile flammable gases or liquids.

(2) Aircraft that has been drained or properly purged.

CLASSIFICATION OF LOCATIONS
513.3(A) THRU (D)

Pits or depressions located below the hangar floor level shall be considered as Class I, Division 1 or Zone 1 locations, and this classification extends to the floor level per **513.3(A)**.

The entire area of the hangar shall be considered a Class I, Division 2 or Zone 2 location to a height of 18 in. (450 mm) above the floor and includes adjacent areas into which the hazards may be communicated unless suitably cut off from the hangar per **513.3(B)** and **(D)**.

The area immediately adjacent to the aircraft shall be classified as a Class I, Division 2 or Zone 2 location. Such an area is defined as being within 5 ft (1.5 m) horizontally from aircraft power plants, aircraft fuel tanks, or aircraft structures containing fuel. These locations shall extend from the floor level vertically to a level of 5 ft (1.5 m) above the wings and above the engine enclosures per **513.3(C)(1)**.

Aircraft painting hangars shall be classified as Class I, Division 1 or Zone 1 where the area within 10 ft (3 m) horizontally from aircraft surfaces from the floor to 10 ft (3 m) above the aircraft. The area horizontally from aircraft surfaces between 10 ft (3 m) and 30 ft (19 m) from the floor to 30 ft (9 m) above the aircraft surface shall be classified as Class I, Division 2 or Zone 2.

Adjacent areas into which hazardous vapors are likely to be communicated or released, such as stockrooms and electrical control rooms, shall be considered a Class I, Division 2 location up to a height of 18 in. (450 mm). These adjacent areas need not be classified as hazardous if they are effectively cut off from the hangar by walls or partitions and are adequately ventilated per **513.3(D)**. (See **513.3(B)** above) **(See Figure 21-43)**

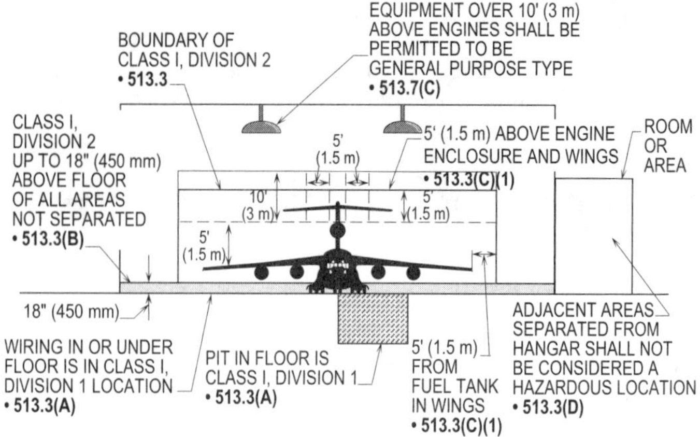

Figure 21-43. The above are the rules and regulations for equipment and wiring methods installed in aircraft hangars.

WIRING AND EQUIPMENT IN
CLASS I LOCATIONS
513.4

Section **513.3** presents the classification of areas within a hangar as either a Class I, Division 1 or Zone 1 or Class I, Division 2 or Zone 2 location. All wiring and equipment within a hazardous area in a hangar must comply with the requirements for either a Class I, Division 1, or Zone 1 or Class I, Division 2 or Zone 2 location, based upon the classification of the area in which the wiring and equipment is installed per **513.4**. Note that wiring in or under the hangar floor shall be classified as Class I, Division 1 per **513.8**.

Attachment plugs and receptacles in hazardous areas are subject to the approval of the AHJ. Such devices shall be identified for Class I locations or, as an alternative, shall be of a type so designed that they cannot be plugged or unplugged unless the circuit is deenergized.

WIRING NOT INSTALLED IN
CLASS I LOCATIONS
513.7(A)

Wiring that is outside the hazardous areas in a hangar shall be as follows:

(1) Metal raceways,

(2) Type MI cable (mineral-insulated),

(3) Type TC cable (power and control tray cable), and

(4) Type MC cable (metal-clad)

This requirement is for nonhazardous areas in the room where the aircraft is stored, such as areas above the 18 in. (450 mm) level. In adjoining nonhazardous rooms, the wiring method shall be permitted to be any approved type.

GROUNDING AND BONDING REQUIREMENTS 513.16

When circuits supplying portable equipment and pendants include an identified grounded (neutral) conductor (white or gray) per **Article 200**, they shall have the following devices of the polarized type, and the grounded (neutral) conductor of the flexible cord shall be connected to the screw shell of any lampholder or to the identified terminal of utilization of equipment such as:

(1) Receptacles,

(2) Attachment plugs,

(3) Connectors, and

(4) Similar devices.

Grounding continuity shall be provided between fixed raceways and the noncurrent-carrying metal parts of:

(1) Pendant luminaires,

(2) Portable lamps, and

(3) Portable utilization equipment.

This may be accomplished by properly bonding the equipment grounding conductor to the metal raceway system. The insulation of equipment grounding conductors shall be green or green with one or more yellow stripes.

EQUIPMENT NOT INSTALLED IN CLASS I LOCATIONS 513.7(C) AND (D)

In locations other than those described in **513.3**, equipment capable of producing arcs, sparks, or particles of hot metals, such as lamps and lampholders for fixed lighting, cutouts, switches, receptacles, charging panels, generators, motors, or other equipment having make-and-break or sliding contacts, shall be of such a type that the escape of sparks or hot metal particles will be prevented. Equipment of the totally enclosed type, or so constructed as to prevent the escape of sparks or hot metal particles, shall comply with this rule. However, in areas described in **513.3(D)**, equipment shall be permitted to be of the general purpose type.

> **Design Tip:** Equipment that might produce an arc or spark and that is less than 10 ft (3 m) above wings and engine of aircraft shall have a tight enclosure that will prevent the escape of hot particles from an arc or spark.

STANCHIONS, ROSTRUMS, AND DOCKS 513.7(E)

Where these items are located, or likely to be located within 5 ft (1.5 m) of aircraft fuel tanks or engines outlined as in **513.3(C)(1)**, the entire stanchion, rostrum, or dock shall be considered to be a hazardous location. All wiring and equipment shall comply with the rules and regulations for Class I, Division 2 or Zone 2 locations.

Where they are not located, or not likely to be located, as described in **513.7**, they shall be considered hazardous only up to a height of 18 in. (450 mm) above the floor. Above the 18 in. (450 mm) level, wiring and equipment shall be permitted to be installed by the provision of **513.4(B)**. Wiring and equipment below the 18 in. (450 mm) level shall comply with the requirements for Class I, Division 2 or Zone 2 locations. **(See Figure 21-44)**

> **Design Tip:** Mobile stanchions containing electrical equipment that conforms to **513.3(C), 513.7**, and **513.4(B)** shall carry at least one sign, permanently affixed to the stanchion, that must read as follows:
>
> ## WARNING
> ## KEEP 5 FT (1.5 m) CLEAR OF AIRCRAFT ENGINES AND FUEL TANK AREAS

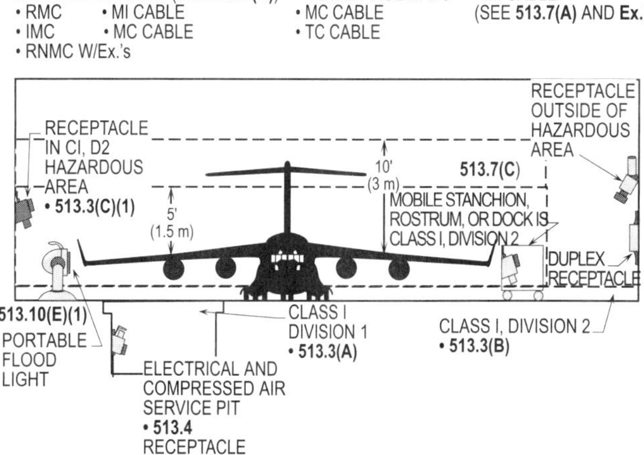

Figure 21-44. The above shows wiring methods, equipment, and portable equipment installed in hazardous and nonhazardous areas in aircraft hangars.

SEALING
513.9

A seal shall be provided at every point where a conduit is routed from a hazardous area to a nonhazardous area or from a Division 1 or Zone 1 to a Division 2 of Zone 2 location. Note that conduits beneath the hangar floor shall be considered to be in a Class I, Division 1 of Zone 1 location per **501.15** and **505.16.**

Conduits that are embedded in or under concrete shall be classified as being in the same hazardous area as that installed above the floor, when any connections lead into or through such areas.

Conduits emerging from such areas shall be considered as passing from a Class I, Division 1 or Zone 1 area to a Class I, Division 2 of Zone 2 area.

> **Design Tip:** Seals shall be provided when connecting conduits to an enclosure where there is a possibility of arcs or sparks or when passing from a hazardous to a nonhazardous area.

SPECIAL EQUIPMENT
513.10(A) THRU (E)

When an aircraft is stored in a hangar, the battery shall be disconnected. It should also be disconnected, if possible, when the aircraft is brought into the hangar for repairs or maintenance.

Aircraft batteries shall not be charged when the aircraft is in a hangar. Such charging shall be done when the aircraft is entirely outside the hangar. Battery chargers should be kept in a separate building when they are not in use.

Electrical equipment on energizers shall be kept at least 5 ft (1.5 m) clear of aircraft engines and fuel tank areas and in addition shall be at least 18 in. (450 mm) above the floor.

> **Design Tip:** Power systems shall have a permanently affixed sign that reads as follows:
>
> ## WARNING
> ## KEEP 5 FT (1.5 m) CLEAR
> ## OF AIRCRAFT ENGINES
> ## AND FUEL TANK AREAS

MOBILE SERVICING EQUIPMENT WITH ELECTRICAL COMPONENTS
513.10(D)

Mobile servicing equipment that is not suitable for Class I, Division 2 or Zone 2 locations shall be so designed and mounted that all fixed wiring and equipment will be at least 18 in. (450 mm) above the floor. This equipment includes vacuum cleaners, air compressors, air movers, and similar type equipment.

> **Design Tip:** Unless such equipment is approved by the AHJ for Class I, Division 2, or Zone 2 locations, the electrical equipment shall be kept at least 5 ft (1.5 m) clear of aircraft engines and fuel tank areas.

Flexible cords for mobile equipment shall be suitable for the type of service, be listed for extra-hard usage, and include an equipment grounding conductor. Receptacles and attachment plugs shall be suitable for the type of service and approved by the AHJ for the location in which they are installed.

> **Design Tip:** Equipment that is not of a type suitable for Class I, Division 2 or Zone 2 locations shall not be operated in areas where maintenance operations are likely to release hazardous vapors in progress of repair. Such equipment shall carry a warning sign, to read as follows:
>
> ## WARNING
> ## KEEP 5 FT (1.5 m) CLEAR
> ## OF AIRCRAFT ENGINES
> ## AND FUEL TANK AREAS

GROUNDING AND BONDING REQUIREMENTS
513.16

All metallic raceways and all noncurrent-carrying metallic portions of fixed or portable equipment, regardless of voltage, shall be grounded and bonded according to the provision of **Article 250**. The insulation of the equipment grounding conductors shall be green or green with one or more yellow stripes.

GASOLINE DISPENSING AND SERVICE STATIONS
ARTICLE 514

One of the most common applications of the requirements in **Article 514** would be gasoline service stations. Basically, hazardous locations are areas where fuel (gasoline) is transferred to the fuel tanks of self-propelled vehicles or to auxiliary tanks. **Articles 510** and **511** appy to locations in filling stations such as lubritoriums, service rooms, repair rooms, offices, salesrooms, compressor rooms, storage rooms, and rest rooms, as well as other associated areas.

Gasoline service stations are not the only areas subjected to the requirements of **Article 514**. Any location having a gasoline dispensing pump, whether in a service station or elsewhere, would fall under the rules of this Article.

CLASS I LOCATIONS
TABLE 514.3(B)(1)

Table 514.3(B)(1) shall be applied where Class I liquids are stored, handled, or dispensed and shall be used to delineate and classify service stations. A Class I location does not extend beyond an unpierced wall, roof, or other solid partition. Locations where the AHJ can satisfactorily determine that flammable liquids such as gasoline having a flash point below 38°C (100°F) will not be present shall be permitted to be classified as nonhazardous. (See **Table 514.3(B)(1)** for the dimensions that are used in the following text up to **514.4.**)

THE SPACE WITHIN
A GASOLINE PUMP

The space within a gasoline pump shall be classified as a Class I, Division 1 location up to a height of 4 ft (1.2 m) above the base and immediately underneath the base. Normally, the space surrounding a gasoline pump, 18 in. (450 mm) out from the pump and 4 ft (1.2 m) up from its base, is Class I, Division 2. (See ANSI 87 and **Table 514.3(B)(1)** for these dimensions)

OUTSIDE AREA OF PUMP

In addition to the above, the circular outside area of 20 ft (6 m) horizontally surrounding the gasoline pump and out from the 18 in. (450 mm) circle shall be classified as a Class I, Division 2 location up to a height of 18 in. (450 mm) above ground. A building or any part of a building falling within this circle shall be a Class I, Division 2 location, if not suitably cut off by a barrier. In most cases, a solid wall or an 18 in. (450 mm) elevation above grade provides an adequate barrier.

OUTSIDE AREA OF FILL PIPES

The circular outside area surrounding a fill pipe, 10 ft (3 m) out from the pipe, shall be considered a Class I, Division 2 location up to a height of 18 in. (450 mm) above ground. The space within a building that is within the 10 ft (3 m) circle shall be classified as a Class I, Division 2 if not suitably cut off.

> **Design Tip:** A wall or an 18 in. (450 mm) elevation above grade would in most cases provide a suitable barrier.

SUSPENDED PUMP

The area up to 18 in. (450 mm) above grade level within 20 ft (6 m), horizontally measured from a point vertically below the edge of any dispenser enclosure, shall be classified as a Class I, Division 2 location.

SURROUNDING SPACE
FROM DISPENSER

The surrounding space out to a distance of 20 ft (6 m), measured from a point vertically below the edge of the dispenser enclosure, shall be classified as a Class I, Division 2 location up to a height of 18 in. (450 mm) above grade. If a building or part of a building is constructed within this space, it is considered nonhazardous if suitably cut off by a ceiling or wall that acts as a barrier.

OUTSIDE AREA OF VENT PIPES

The area 3 ft (900 mm) in all directions from the discharge ends of a vent pipe shall be classified as Class I, Division 1. The space between 3 ft (900 mm) and 5 ft (1.5 m) of open end of vent, extending in all directions shall be classified a Class I, Division 2 location.

AREA AROUND A
LUBRICATION ROOM

A lubrication room shall be classified as a Class I, Division 2 location up to a height of 18 in. (450 mm) above the floor. A pit in a lubrication room shall be classified as a Class I, Division 1 location from the floor down to the bottom of the pit. **See Figure 21-45** for a detailed illustration of classifying area around dispensers and gas stations.

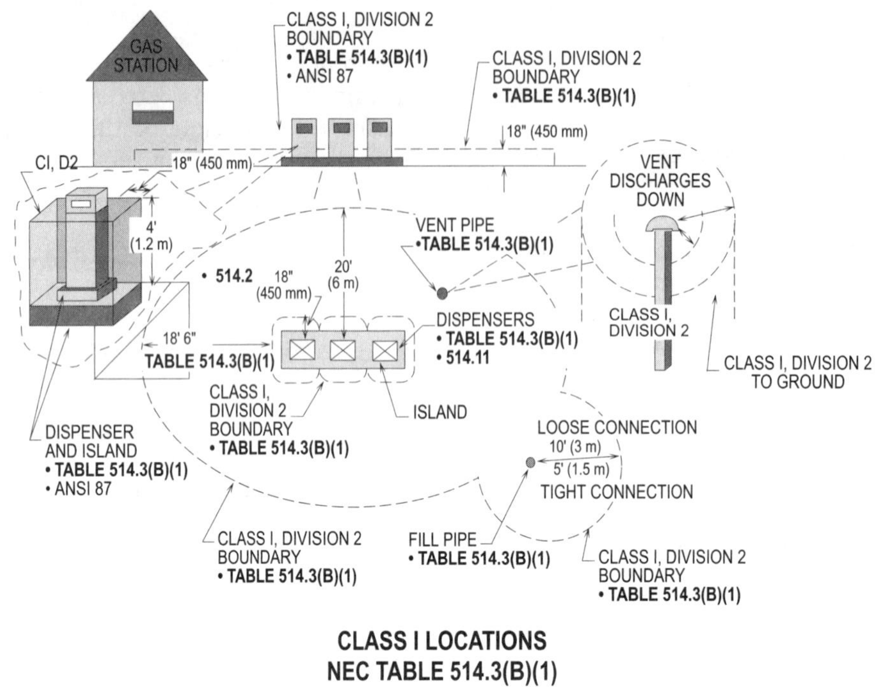

Figure 21-45. The above illustrates the classified area around dispensers and gas stations.

WIRING AND EQUIPMENT WITHIN CLASS I LOCATIONS 514.4

Electrical equipment and wiring installed in the Class I, Division 1 and Class I, Division 2 locations, shall be approved and suitable for such locations by the AHJ. These wiring methods shall be permitted in these areas:

(1) Rigid metal conduit and

(2) Type MI cable.

Note: See **Article 514.3** and **Article 501**.

> **Design Tip:** Underwriters Laboratories has a directory with listings on types of insulations on conductors that are approved for these locations.

WIRING AND EQUIPMENT ABOVE CLASS I LOCATIONS 514.7

The rules for wiring and equipment above hazardous areas (above the 18 in. (450 mm) level) are the same as for commercial garages per **514.3** and **511.7**.

These wiring methods shall be permitted in these locations:

(1) Metal raceways and

(2) Type MI cable.

CIRCUIT DISCONNECTS 514.11 AND 514.13

Each circuit leading to or through a dispensing pump shall be provided with a switch or other acceptable means to disconnect simultaneously from the source of supply all conductors of the circuit, including the grounded (neutral) conductors, if used per **514.11** and **514.13**.

The intent of this requirement is to ensure that the supply is disconnected from the source and there are no conductors connected that lead to or through a dispensing pump.

> **Design Tip:** There are special breakers available that have a pigtail that ties to the neutral bus and ensures that the ungrounded (phase) and grounded (neutral) conductor(s) are disconnected when the breaker is in the OFF position.

Therefore, the threat of an arc or spark is limited, should the grounded (neutral) conductor be disconnected with another circuit still energized that is sharing the neutral with other circuits. **(See Figure 21-46)**

> **Design Tip:** Each dispensing device shall be provided with a means to remove all external voltage sources, including feedback, during periods of maintenance and service of the dispensing equipment per **514.13**.

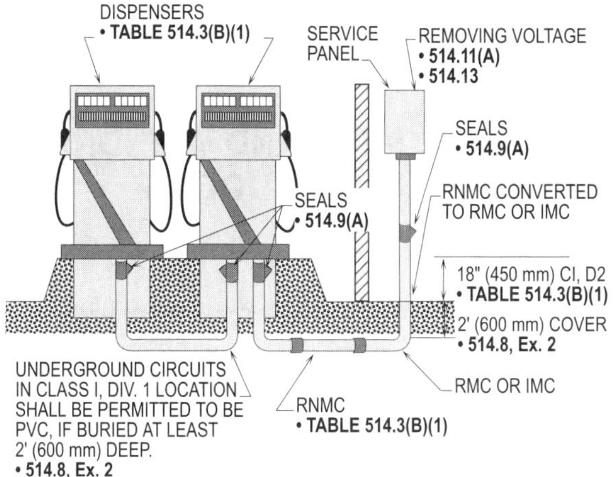

Figure 21-46. The above illustrates the rules and regulations for wiring methods and seals used for dispensers and boundaries to gas stations.

ATTENDED SELF-SERVICE STATIONS 514.11(B)

For attended self-service stations, the emergency controls specified in **514.11(A)** and **(B)** and **514.13** shall be installed at a location acceptable to the AHJ. Such emergency controls shall not be located more than 100 ft (30 m) from the dispensers.

UNATTENDED SELF-SERVICE STATIONS 514.11(C)

For unattended self-service stations, the emergency controls specified in **514.11(A)** and **(C)** shall be installed at a location acceptable to the AHJ. Such emergency controls shall be more than 20 ft (6 m) but less that 100 ft (30 m) from the dispensers.

Additional emergency controls shall be installed on each group of dispensers or the indoor equipment shall be used to control the dispenser.

The main emergency disconnect usually is an across-the-line device. However, a remote control device shall be permitted for the additional controls required, but they shall be capable of being manually reset.

> **Design Tip:** Emergency controls shall shut off all power to all dispensing equipment at the station. Controls shall be manually reset only in a manner approved by the AHJ. (Review **514.11** and **514.13** of the NEC.)

SEALING AT DISPENSER 514.9(A)

An approved seal shall be provided in each conduit run entering or leaving a dispenser or any cavities or enclosures in direct communications therewith. The sealing fitting shall be the first fitting after the conduit emerges from the earth or concrete.

> **Design Tip:** All seals shall be readily accessible and no seal shall be buried or installed in an inaccessible wall. This rule makes it very clear that no fitting of any kind shall be installed in the conduit between the floor and seal. For example, if a conduit run is below a panelboard and there is a hazardous location below, a seal shall be installed at 18 in. (450 mm) or more above the floor without a fitting between the seal and floor. **(See location of seals in Figure 21-46)**

SEALING AT BOUNDARY 514.9(B)

Additional seals shall be provided as required in **501.15**, **501.15(A)(4)** and **501.15(B)(2)**. Such sealing rules shall apply to both horizontal and vertical boundaries between the hazardous and nonhazardous areas. **(See Figure 21-47)**

GROUNDING AND BONDING 514.16

Metallic portions of dispensing pumps, metallic raceways, and all noncurrent-carrying metal parts of electric equipment, regardless of voltage, shall be grounded and bonded as required in **Article 250**. Such grounding shall extend back to the service equipment and service ground.

In case of an accidental fault, a poor grounding connection can create an arc and cause an explosion to occur.

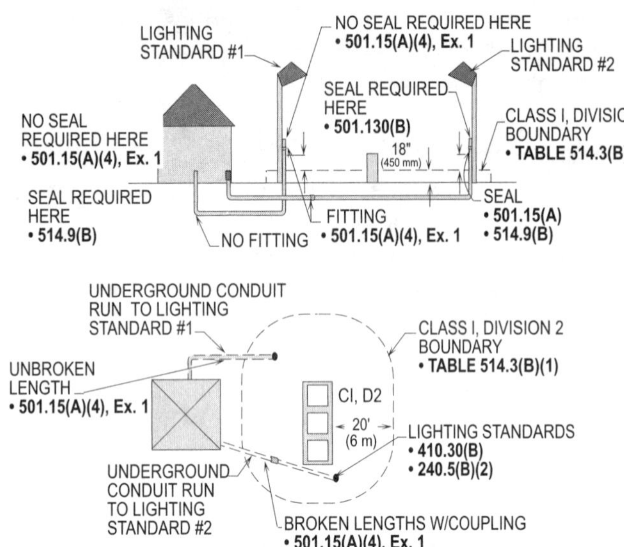

Figure 21-47. The above illustrates additional seals that shall be provided.

UNDERGROUND WIRING
514.8 and Ex. 2

Underground wiring shall be installed in rigid metal conduit or threaded steel intermediate metal conduit, or, where buried under not less than 2 ft (600 mm) of earth, it shall be permitted to be installed PVC if it complies with **Article 352**. Where PVC is used, threaded rigid metal conduit or threaded steel intermediate metal conduit shall be used for the last 2 ft (600 mm) of the underground run to emergence or to the connection to the aboveground raceway; an equipment grounding conductor shall be included to provide electrical continuity of the raceway system and for grounding noncurrent-carrying metal parts. **(See Figures 21-46 and 47)**

> **Design Tip:** Nonmetallic conduit was added to the NEC to take the place of metallic conduit and ease the problem of corrosion, which might let fumes or gasoline into the conduit system. The use of nonmetallic conduit does not change the rules for sealing. To protect from physical damage, metallic conduit shall be routed up into the dispenser or the nonhazardous location.

BULK STORAGE PLANTS
ARTICLE 515

A bulk storage plant is that portion of a property where flammable liquids are received by tank vessel, pipelines, tank car, or tank vehicle and are stored or blended in bulk for the purpose of distributing such liquids by tank vessel, pipeline, tank car, tank vehicle, portable tank, or container. For example, such designation includes locations where gasoline or other volatile flammable liquids are stored in tanks having an aggregate capacity of one carload or more and from which these products are distributed normally by tank truck, rail tank, etc.

CLASS I LOCATIONS
515.3 AND TABLE 515.3

Table 515.3 shall be applied where Class I liquids are stored, handled, or dispensed and shall be used to delineate and classify bulk storage plants. The Class I location shall not be required to extend beyond an unpierced wall, roof, or other solid partition.

PUMPS, BLEEDERS, AND WITHDRAWAL FITTINGS LOCATED INDOORS
TABLE 515.3

Such devices, when installed in a gasoline pipeline, can be subject to leakage and therefore present a hazardous condition. Where such devices are located indoors, the hazardous area around the devices extends as follows:

(1) 5 ft (1.5 m) in all directions, to which this boundary includes an imaginary sphere around such device, and

(2) 25 ft (7.5 m) out horizontally and 3 ft (900 mm) up from the floor grade, to which this boundary includes an imaginary circular area around such device.

The above areas are considered Class I, Division 2 areas and all wiring and equipment installed in these areas shall comply with the requirements for a Class I, Division 2 location.

PUMPS, BLEEDERS, AND WITHDRAWAL FITTINGS LOCATED OUTDOORS
TABLE 515.3

When such devices are located outdoors, the Class I, Division 2 hazardous area around these devices extends as follows:

(1) 3 ft (900 mm) in all directions, forming an imaginary sphere, and

(2) 10 ft (3 m) out along the ground and 18 in. (450 mm) up from the ground, forming an imaginary circular area.

See Figure 21-48 for a detailed illustration of such devices installed outdoors.

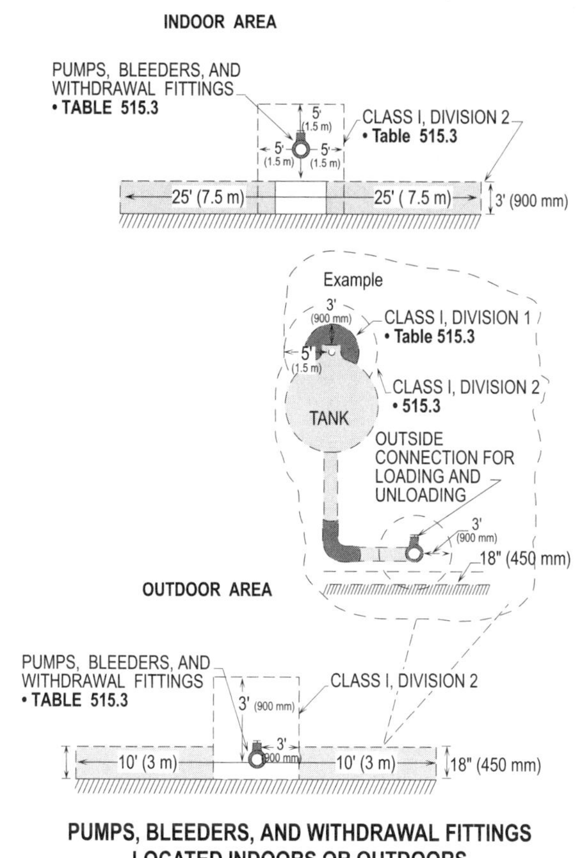

Figure 21-48. Classifying the hazardous area around a device or fitting located indoors or outdoors.

TANK VEHICLES AND TANK CARS IN OUTSIDE LOCATIONS
TABLE 515.3

The hazardous area extends 15 ft (4.5 m) in all directions around an open dome or vent, forming a 15 ft (4.5 m) sphere. The first 3 ft (900 mm) shall be classified as a Class I, Division 1 location, and the remaining 12 ft (3.7 m) shall be classified as a Class I, Division 2 location.

For bottom loading or unloading, the hazardous area extends 3 ft (900 mm) in all directions from the fixed connection and within a 10 ft (3 m) radius and 18 in. (450 mm) up from the ground. These hazardous areas shall be classified as a Class I, Division 2 location, but only during loading and unloading operations. **(See Figure 21-49)**

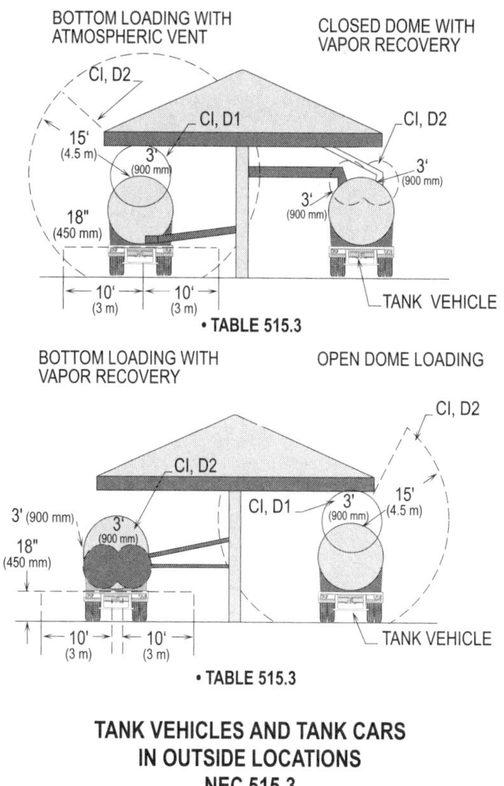

Figure 21-49. The above illustrates the hazardous areas, under certain conditions, around tank and car vehicles, based on top or bottom loading or unloading procedures.

ABOVEGROUND TANKS
TABLE 515.3

The space above the roof and within the shell of a floating roof type tank shall be classified as a Class I, Division 1 location.

For other tanks of this type, the area within 10 ft (3 m) in all directions shall be classified as a Class I, Division 2 location, except the area within 5 ft (1.5 m) in all directions shall be classified as a Class I, Division 1 location. The area between 5 ft (1.5 m) and 10 ft (3 m) out shall be classified a Class I, Division 2 location. **(See Figure 21-50)**

PITS
TABLE 515.3

A pit that is located in the 10 ft (3 m) or 25 ft (7.5 m) sphere around pumps, bleeders, meters, etc. shall be classified as a Class I, Division 1 location.

A pit located in the 10 ft (3 m) sphere around a fill pipe opening shall be classified a Class I, Division 1 location.

A pit located in a nonhazardous area shall be considered as nonhazardous where there are no piping, valves, or fittings in the pit. If the pit has piping, valves or fittings, it shall be classified as a Class I, Division 2 area.

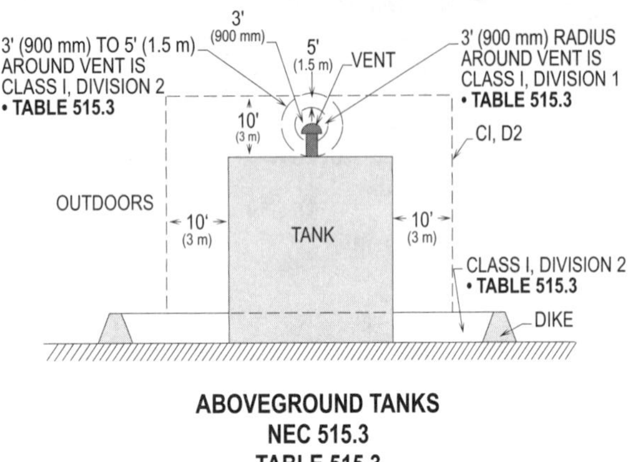

Figure 21-50. The above illustrates the hazardous area around an outdoor bulk storage tank.

GARAGES FOR TANK VEHICLES
TABLE 515.3

Storage and repair garages for tank vehicles shall be classified a Class I, Division 2 location up to a height of 18 in. (450 mm) above the floor and shall be wired with wiring methods and equipment rated for such locations.

ADJACENT LOCATIONS
TABLE 515.3

Office buildings, boiler rooms, etc. shall not be considered hazardous locations if they are not used for handling and storage of gasoline or other flammable material.

WIRING AND EQUIPMENT
WITHIN IN CLASS I LOCATIONS
515.3

All wiring and equipment in a hazardous area shall comply with the requirements of a Class I, Division 1 or 2 location or **Article 505** for the Zones, whichever applies, based upon the flammable material.

WIRING AND EQUIPMENT
ABOVE HAZARDOUS LOCATIONS
515.7

The requirements of this Section are basically the same as those for commercial garages. The main concern is sparks, arcs, or hot metal particles that might drop into the hazardous area and create an explosion if there was a proper mixture of gases and air.

Electrical equipment that might produce arcs, sparks, or particles of hot metal, such as lamps and lampholders for fixed lighting, cutouts, switches, receptacles, motors, or other such equipment having make-and-break or sliding contacts, shall be totally enclosed or constructed in such a manner as to prevent the escape of sparks or hot metal particles.

> **Design Tip:** Portable lamps or utilization equipment shall be suitable and shall comply with the provisions of **Article 501** for the hazardous location in which they are to be used.

The wiring methods used above the 3 ft (900 mm) and 18 in. (450 mm) levels to connect the electrical equipment listed above shall be permitted to be one of the following or any combination of such:

(1) Metal raceways,

(2) Type MI cable,

(3) Type TC cable,

(4) Type MC cable,

(5) PVC Sch. 80, and

(6) Rigid nonmetallic conduit

UNDERGROUND WIRING
515.8(A) THRU (C)

For underground wiring routed to and around aboveground storage tanks shall be permitted to be one of the following:

(1) Where buried less than 2 ft (600 mm) in the earth, rigid or IMC conduit shall be used.

(2) Where buried 2 ft (600 mm) or more, nonmetallic conduit shall be permitted to be used. However, a change shall be made to metal conduit for the last 2 ft (600 mm) before emergence from the ground. Cable requires 2 ft (600 mm) of cover.

(3) Where nonmetallic conduit is used, an equipment grounding conductor shall be routed with the circuit conductors to ground all noncurrent-carrying parts.

(4) Where underground cable without a metal sheath is used, the cable shall have an equipment grounding conductor to ensure proper grounding of equipment with metal parts.

(5) Where cable is used, it shall be protected by rigid metal conduit or IMC at points where it emerges from the earth.

(6) Conductor insulation shall be a type approved for the location, to protect from gasoline in case of leakage or an accidental rupture of equipment.

> **Design Tip:** Nonmetallic conduit and cables shall be buried at least 2 ft (600 mm) in the earth or have a 2 ft (600 mm) cover.

SEALING REQUIREMENTS
515.9

Seals shall be provided in accordance with **501.15**. Sealing requirements in **501.15(A)(4)** and **501.15(B)(2)** apply to horizontal as well as to vertical boundaries of the defined Class I locations. Buried raceways under defined hazardous locations shall be considered to be within such locations.

Seals, when required, shall be installed as follows:

(1) Seals shall be placed in each conduit run entering the enclosure for switches, circuit breakers, fuses, relays, resistors, or any other device that is capable of producing arcing, sparking, or high temperature and that is located in a hazardous area. These seals shall be placed no further than 18 in. (450 mm) from such enclosures.

(2) A seal shall be required at the point where the conduit leaves a hazardous area and passes into a nonhazardous area. The seal shall not be required to be installed at the exact boundary of the hazardous area. It shall be permitted to be installed at a convenient point in the run, on either side of the boundary. However, it is recommended to place the seal as near to the boundary as possible.

A vertical run shall be sealed where it enters or leaves the hazardous level either at 18 in. (450 mm) or 36 in. (900 mm) above the floor or ground. For a horizontal run, if the run leaves the area at a height below the hazardous level, it is mandatory that the conduit be sealed where it leaves the hazardous area.

An exception allows a conduit that passes through a hazardous area, with no fitting within or 12 in. (300 mm) beyond the hazardous boundary, to be installed without seals.

(3) For 2 in. (53) and larger conduit runs, seals shall be required at the entry to an enclosure, if the box or fitting contains a splice, tap, or connection and is located within a hazardous area. Note that such seal shall be installed within 18 in. (450 mm) of the enclosure.

> **Design Tip:** Conduit installed underneath a hazardous area shall be considered to be in the hazardous area beneath that it is buried and shall be treated as such.

GASOLINE DISPENSING PUMPS
515.10

When gasoline is dispensed from a bulk storage plant, the rules of **Article 514**, that covers the dispensing of gasoline, shall be applied.

In other words, if there are gasoline pumps on the premises, the pump area shall comply with the rules and regulations for service stations.

GROUNDING AND BONDING
515.16

All metal raceways, metal-jacketed cables, and all non-current-carrying metal parts of fixed or portable electrical equipment, regardless of voltage, shall be grounded and bonded by the provisions in **Article 250**. Grounding in Class I locations shall comply with the rules of **501.30**.

> **Design Tip:** This grounding technique shall be required for all wiring in order to provide proper grounding for raceways and equipment where there is no raceway continuity.

The grounding continuity shall be complete back to the service equipment and the service ground for non-hazardous as well as hazardous areas.

FINISHING PROCESSES
ARTICLE 516

Article 516 applies for locations where paint, lacquer, or powdered finishing materials are frequently applied by brushing, spraying, or dipping. Conditions of process shall always be evaluated based upon use and the effects of the material.

For example, the vapors from finishing processes and the storage of paints, lacquers, or other flammable finishes shall be considered.

Another problem is the residue that is the result of the finishing process. Vapors can be easier to control than the residue because residue can be ignited by heat and cause fires to occur. The right mixture of vapors and air can create a dangerous explosion when ignited by an arc or a spark.

> **Design Tip:** NFPA 33 states that water-based paints, when mixed, create no problem. However, when sprayed, a residue condition exists, and this could cause a problem.

CLASS I OR II LOCATIONS
516.3

Classification is in respect to the effects of and exposure to flammable gases and vapors, and in some cases deposits of paint spray.

> **Design Tip:** For deposits and residues requirements, review **516.3(B)** and **(C)**.

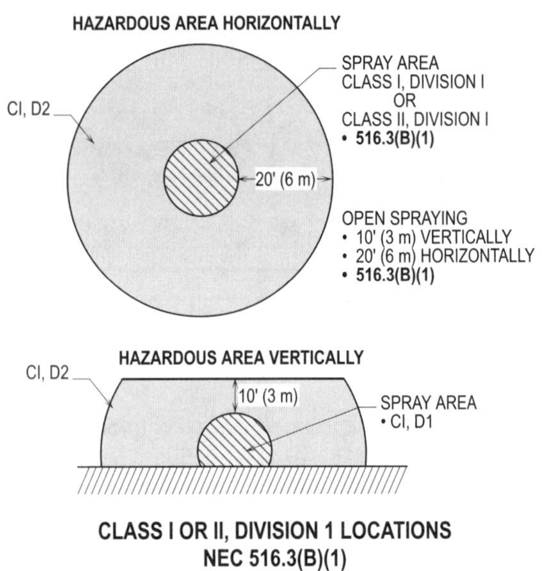

Figure 21-51. The above illustrates the hazardous area for an unenclosed area used for spraying.

CLASS I OR II, DIVISION 1 LOCATIONS
516.3(B)(1) THRU (B)(6)

Where brushing, spraying, or dipping is not performed in a spray booth, the hazardous areas are to be considered as follows:

(1) The space around a spraying operation, which contains a dangerous quantity of spray vapor, shall be classified a Class I, Division 1 location. The NEC does not define a definite boundary for such space. The extent of the hazardous boundary is always subject to the judgment of the AHJ.

(2) The space beyond the Class I, Division 1 space shall be classified a Class I, Division 2 location up to 20 ft (6 m) beyond, and up from the floor to a height of 10 ft (3 m) above the spraying area.

(3) For dipping operations, all space within 5 ft (1.5 m) in any direction from the liquid, whether the liquid is in the tank or as a wet coating on the drain board or on the dipped object, shall be classified a Class I, Division 1 location.

> **Design Tip:** Where the brushing, spraying, or dipping is performed in a spray booth, the entire spray booth shall be classified as Class I, Division 1. Exhaust ducts for the spray booth shall also be classified as a Class I, Division 1 location. **(See Figure 21-51)**

CLASS I OR II, DIVISION 2 LOCATIONS
516.3(C)(2)(a) AND (b)

Where the spray booth is equipped with an open front, the space outside the booth shall be classified as a Class I, Division 2 location and rules are applied as follows:

(1) For spray booths with an interlocked vent system, so designed and arranged that spraying cannot be performed except when the vent system is operating under these conditions, the space outside the booth shall be classified as a Class I, Division 2 location.

(2) For spray booths without an interlocked vent system, the space outside the spray booth shall be classified as a Class I, Division 2 location, due to the possibility of spraying without proper ventilation.

(3) For open-top spray booths, the Class I, Division 2 hazardous area shall be considered to extend to 3 ft (900 mm) above the booth and 3 ft (900 mm) from all openings.

See Figures 21-52(a), (b), and (c) for a detailed illustration of hazardous area around spray booths.

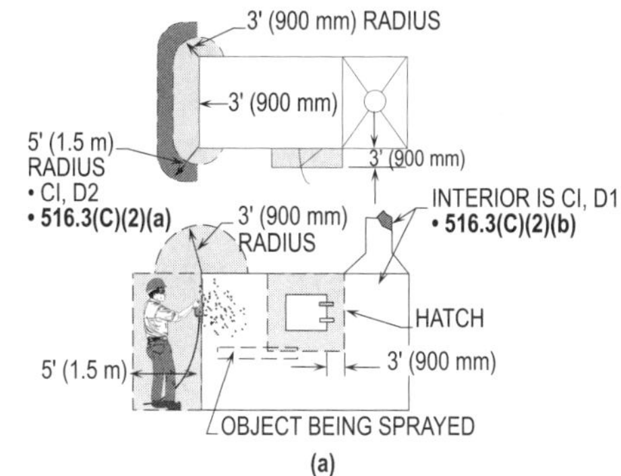

CLASS I OR II, DIVISION 2 LOCATIONS
NEC 516.3(C)(2)(a)

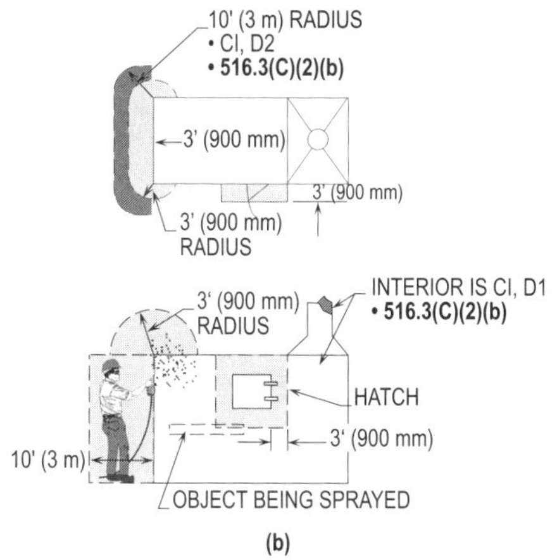

CLASS I OR II, DIVISION 2 LOCATIONS
NEC 516.3(C)(2)(b)

Figure 21-52(a). The above illustrates the hazardous areas around spray booths, based upon the ventilation being interlocked or not interlocked with spray operation.

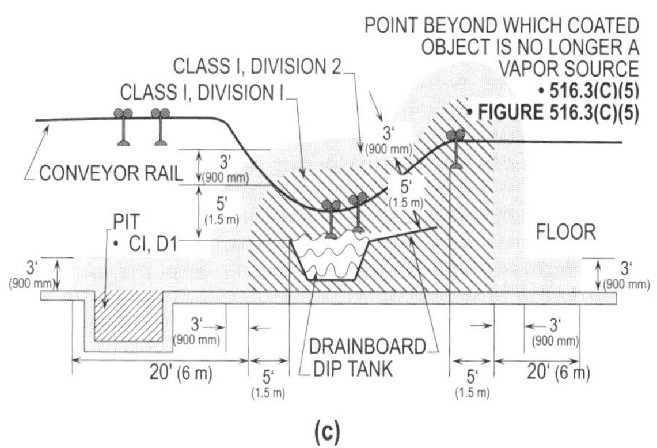

CLASS I OR II, DIVISION 1 LOCATIONS
NEC 516.3(C)(4) AND (C)(5)

Figure 21-52(c). The above illustrates the hazardous area in and around dip tanks and drain boards.

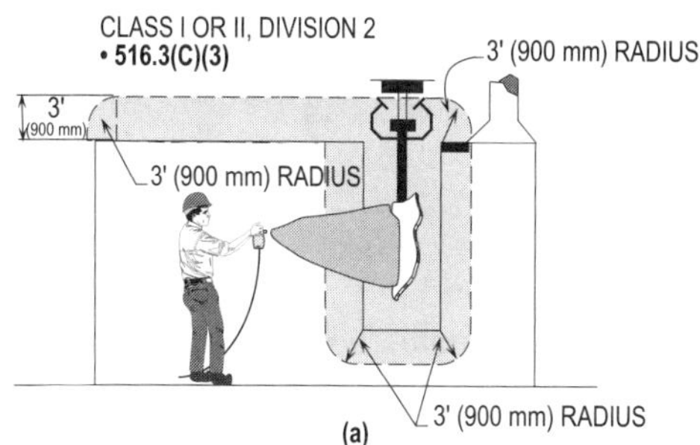

CLASS I OR II, DIVISION 2 LOCATIONS
NEC 516.3(C)(3)

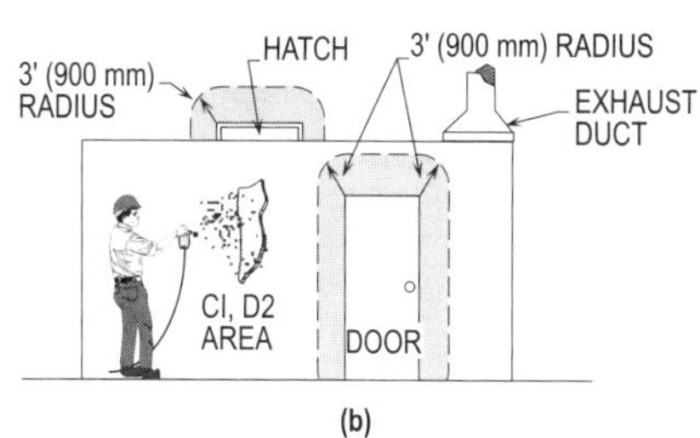

CLASS I OR II, DIVISION 2 LOCATIONS
NEC 516.3(C)(4)

Figure 21-52(b). The above illustrates the hazardous areas in and around open and enclosed spray booths.

WIRING AND EQUIPMENT IN CLASS I LOCATIONS 516.4

All electric wiring and equipment installed within the Class I location containing vapor only and not residue shall comply with the following rules:

(1) All wiring and equipment in the hazardous areas shall comply with the NEC requirements for a Class I, Division 1 or Class I, Division 2 location, whichever applies.

(2) Any equipment installed in a location where paint or lacquer could accumulate on the equipment during operation, to such an extent that a hazard might exist, shall be approved for such condition of use.

(3) Luminaires within a spray booth shall be approved for Class I, Division 1 locations and are usually of an explosionproof type. To avoid use of such expensive luminaires, lighting can be accomplished through glass panels in the wall or ceiling of the spray booth. If the luminaire is installed outside of the Class I, Division 1 location, a less expensive luminaire shall be permitted to be used. **(See Figure 21-53)**

(4) Portable luminaires shall be of a type that are approved for Class I, Division 1 locations. However, portable appliances shall be permitted to be used in a spray booth only when no spraying, brushing, or dipping operation is being performed.

Where deposits of residue present a heating problem, illumination shall be permitted to be provided through panels of glass or other approved transparent or translucent material if the following rules are followed:

(a) Fixed lighting units are utilized.

(b) The panel is of a material unlikely to break during conditions of use.

(c) Panels properly isolate the hazardous location from the area in which the lighting units are installed.

(d) The accumulation of hazardous residues on the surface of the panel will not rise to a dangerous temperature during spray operations.

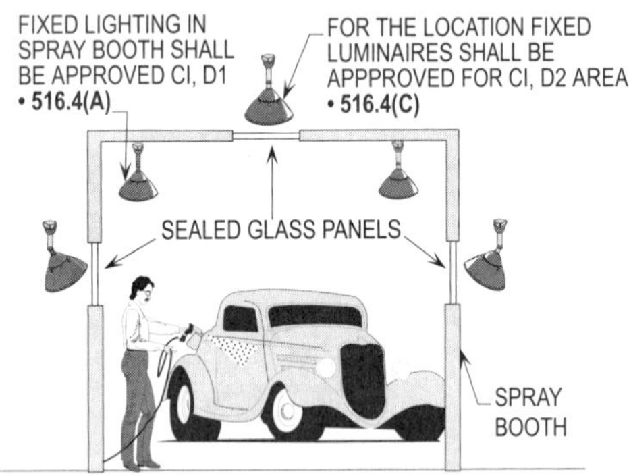

Figure 21-53. The above illustrates the rules for installing luminaires inside or outside of spray booths.

WIRING ABOVE CLASS I LOCATIONS 516.7(A)

Above hazardous areas, the following wiring methods shall be permitted as follows:

(1) Metal raceways,

(2) Type MI cable,

(3) Type TC cable,

(4) Type MC cable,

(5) Electrical nonmetallic tubing, and

(6) Rigid nonmetallic conduit

Design Tip: Leads without seals may be routed out of a cellular metal raceway in the ceiling. However, such leads can only be routed out of a cellular metal raceway in the floor if seals are properly installed.

CLASSIFICATION OF AREA

There are basically seven steps to be applied for classifying an area, and they are as follows:

(1) Determining the governing codes, standards, and authorities,

(2) Determining flammable materials,

(3) Determining group,

(4) Determining class,

(5) Determining division,

(6) Determining extent (boundaries) of classified areas, and

(7) Preparing plans and specifications

DETERMINING THE GOVERNING CODES, STANDARDS, AND AUTHORITIES

The first step in area classification consists of establishing the relevant codes, standards, and the authority having jurisdiction (AHJ), and obtaining all permits and other licenses. There are many U.S. companies that have their own operating standards, and these should also be consulted.

The AHJ is the organization, office, or individual responsible for approving equipment, installation, etc. The AHJ can be federal, state, or local, a fire chief or marshal, fire preventive bureau, labor department, a rating bureau, etc. The following are some of the relevant codes, standards, and testing laboratories:

(1) *Underwriters Laboratories* (UL)

(2) *Factory Mutual Research Corporation* (FM)

(3) *Electrical Testing Laboratories* (ETL)

(4) *American Gas Association* (AGA)

(5) *National Electrical Manufacturers Association* (NEMA)

(6) *American Petroleum Institute* (API)

(7) *Occupational Safety and Health Administration* (OSHA)

(8) *Mine Safety and Health Administration* (MSHA)

(9) *United States Coast Guard* (USCG)

(10) *Instrument Society of America* (ISOA)

(11) *Institute of Electrical and Electronic Engineers* (IEEE)

(12) *International Electrotechnical Commission* (IEC)

(13) Local building codes

(14) Manufacturer's data

(15) *National Fire Protection Association* (NFPA)

 • *National Electrical Code* (NEC®)

DETERMINING FLAMMABLE MATERIALS

Flammable liquids (Class I) are defined as being any liquid that has a closed-cup flash point below 100°F (37.8°C) and a reid vapor pressure not exceeding 40 ps/a (2068.6 mm Hg) absolute at 100°F (37.8°C). Flammable and combustible liquids are subdivided into Classes I, II, and III. Classes are used to identify flammable and combustible liquids and should not be confused with class as defined in the NEC.

Flammable liquids (Class IA) are those liquids that have flash points below 73°F (22.8°C) and boiling points below 100°F (37.8°C). Class IB liquids are those liquids that have flash points below 73°F (22.8°C) and boiling points at or above 100°F (37.8°C). Class IC liquids are those liquids that have flash point at or above 73°F (22.8°C) and below 100°F (37.8·C).

Combustible liquids (Class II) are those liquids that have flash points at or above 100°F (37.8°C) and below 140°F (60°C). Class III A liquids are those having flash points at or above 140°F (60°C) and below 200°F (93°C). Class IIIB liquids are those having flash points at or above 200°F (93°C).

DETERMINING GROUP

Equipment shall be selected, tested, and approved for the type of flammable material involved. Maximum explosive pressures, MESG, safe operating temperatures (ignition temperature), and ignition energy shall be considered in the design and testing of safe electrical equipment for installation in classified areas. The NEC has a partial list of the most commonly encountered materials that have been tested. They are grouped on the basis of their flammability characteristics:

Group A	acetylene
Group B	hydrogen
Group C	carbon monoxide, ethylene
Group D	gasoline, benzine, propane, ethyl alcohol, methane
Group E – G	dusts

DETERMINING CLASS

The NEC refers to three classes:

Class I	flammable gases and vapors
Class II	combustible dusts
Class III	easily ignitible fibers or flyings

DETERMINING DIVISION

The NEC recognizes two divisions, and they are as follows:

Division 1: A location that is likely to have flammable gases or vapors present under normal conditions.

Division 2: A location that is likely to have flammable gases or vapors present only under abnormal conditions.

DETERMINING EXTENT (BOUNDARIES) OF CLASSIFIED AREAS

It is important to exercise sound engineering experience and judgment in determining the extent (boundaries) of a classified area. Having established the presence of flammable substances and determined the perimeter of the classified area, the class, group, and division is selected. The next step is to select the equipment and installation techniques. The following steps will aid in determining the extent of the boundaries of the classified area:

(1) The flammable materials involved,

(2) The type of installation,

(3) The Class (I, II, or III),

(4) The Group,

(5) The Division (1 or 2),

(6) Flash point of material,

(7) Ignition of sources,

(8) Volume and pressure release,

(9) Ventilation,

(10) Fire walls and barriers, and

(11) Purging and pressurization

PREPARING PLANS AND SPECIFICATIONS

Plans and specifications should be prepared so as to outline the hazardous area and the extent of the classified boundaries. In addition, plans that clearly illustrate the divisions and the extent of the areas should be prepared. Tables should also be developed with properties of the flammable materials and a written description based on the classification that explains the methods and procedures of each area being classified.

QUESTIONS AND ANSWERS FOR CLASSIFYING DIVISIONS

The following questions can be used to determine the Division of a classified location.

ASSIGNMENT OF DIVISION 1 LOCATIONS

(1) Is an ignitible atmospheric mixture likely to exist under "normal" operating conditions?

(2) Is an ignitible atmospheric mixture likely to occur frequently due to repair, maintenance, or leakage?

(3) Would failure of process equipment, piping, or vessels be likely to cause a failure of the electrical system simultaneous with the release of the combustible material?

(4) Is a piping system containing combustible material in an inadequately ventilated space, and is the system likely to leak?

(5) Is the space or area in question below grade level such that heavier-than-air vapors may accumulate there?

(6) Is the space or area in question above grade level such that lighter-than-air vapors may accumulate there?

An answer of yes to any of these questions would require a Division 1 classification.

ASSIGNMENT OF DIVISION 2 LOCATIONS

(1) Is a piping system containing combustible material in an inadequately ventilated space, and is the system not likely to leak?

(2) Is a process equipment system containing a combustible material in an inadequately ventilated area and can the material escape only during abnormal situations such as failure of gaskets or packing?

(3) In the location adjacent and open to a Division 1 location by trenches, pipes, or ducts?

(4) If mechanical ventilation is used, can failure or abnormal operation of the ventilation equipment permit an ignitable atmospheric mixture?

An answer of yes to any of these questions would require a Division 2 classification.

GENERAL REQUIREMENTS 505.6(A) THRU (C)

Article 505 in Chapter 5 of the NEC covers the general requirements and how such rules apply to the electrical wiring and equipment installed in locations classified as Class I, Zone 0, Zone 1, or Zone 2. The concept of using Zones, per the IEC, instead of Divisions, per the NEC, has been used internationally for many years. In the past, the International Electrotechnical Commission (IEC) has regulated the requirements for such a system.

The IEC system uses two groups to identify the hazards involved. Group I is used for mining, while Group II is used for surface industries and offshore installations.

Group II consists of three subgroups A, B, and C. Subgroups A, B, and C are designed to represent categories of flammable gases or vapors that are based upon the minimum ignition energy of the hazard. Subgroup A represents the most difficult flammable gas or vapor to ignite, and Subgroup C is the easiest to ignite. **(See Figure 21-54)**

> **Design Tip:** Check with local NEC and OSHA inspectors to obtain their permission to use the Zone concept when designing and installing electrical equipment and wiring methods in hazardous locations.

ZONE CLASSIFICATION 505.5(A) AND (B)

The IEC system uses three Zones to represent the different levels of risk. Zone 0 represents areas in which an explosive gas-air mixture is continuously present or present for long periods of time. Zone 1 represents areas in which an explosive gas-air mixture is likely to occur in normal operation. Zone 2 represents areas in which gas-air mixtures are not likely to occur, and should they occur they exist only for a short period of time. See **500.5(B)(1)** and **(B)(2)** to define Divisions 1 and 2 for similar situations, where gas/air mixtures present risk levels based upon conditions. **(See Figure 21-55)**

CLASS I, ZONE 0 LOCATIONS 505.5(B)(1)

A Class I, Zone 0 location is a location in which an explosive gas atmosphere is present continuously or present for long periods of time.

Note, this usually applies to the inside of containers or apparatus such as vaporizers, reactors, storage tanks, etc.

	NEC GROUP	REPRESENTATIVE GAS	IEC GROUP	
GROUP A AND B IS MORE IGNITIBLE • 500.6(A)(1); (A)(2)	A	ACETYLENE	IIC	GROUP IIC IS MORE IGNITIBLE • 505.6(A), FPN
	B	HYDROGEN		
GROUP C • 500.6(A)(3)	C	ETHYLENE	IIB	GROUP IIB • 505.6(B), FPN
GROUP D • 500.6(A)(4)	D	PROPANE	IIA	GROUP IIA • 505.6(C), FPN

**A COMPARISON OF NEC AND IEC GROUPING AND CLASSIFICATION SCHEME
NEC 505.6(A) THRU (C)**

Figure 21-54. The Table above shows a comparison of the NEC and IEC grouping and classification scheme.

Long periods of time could be considered as the time needed for gases and vapors to mix with air and create an explosive mixture. **(See Figure 21-56)**

CLASS I, ZONE 1 LOCATIONS
NEC 505.5(B)(2)

A Class I, Zone 1 location is a location:

(1) in which ignitible concentrations of flammable gases or vapors are likely to exist under normal operating conditions.

(2) in which ignitible concentrations of flammable gases or vapors may exist frequently because of repair or maintenance operations or because of leakage.

(3) in which equipment is operated or processes are carried on, of such a nature that equipment breakdown or faulty operations could result in the release of ignitible concentrations of flammable gases or vapors and also cause simultaneous failure of electrical equipment in a mode to cause the electrical equipment to become a source of ignition.

(4) that is adjacent to a Class I, Zone 0 location from which ignitible concentrations of vapor could be communicated, unless communication is prevented by adequate positive pressure ventilation from a source of clean air and effective safeguards against ventilation failure are provided. **(See Figure 21-57)**

TABLE 1		
IEC ZONE CHART		
ZONES	HAZARDOUS RISK LEVELS	
ZONE 0	IN WHICH AN EXPLOSIVE GAS-AIR MIXTURE IS CONTINUOUSLY PRESENT. • 505.5(B)(1)	
ZONE 1	IN WHICH AN EXPLOSIVE GAS-AIR MIXTURE IS LIKELY TO OCCUR IN NORMAL OPERATION. • 505.5(B)(2)	
ZONE 2	IN WHICH AN EXPLOSIVE GAS-AIR MIXTURE IS NOT LIKELY TO OCCUR IN NORMAL OPERATION AND IF IT OCCURS WILL ONLY LAST FOR A SHORT PERIOD. • 505.5(B)(3)	
TABLE 2		
COMPARISON OF IEC ZONES AND NEC DIVISIONS		
IEC ZONES	0	1	2
NEC DIVISIONS		1	2

**ZONE CLASSIFICATION
NEC 505.5(B)(1) THRU (B)(3)**

Figure 21-55. The Table above can be used to define the different Zones, based upon how often a hazardous gas or vapor is present, using the IEC system concept.

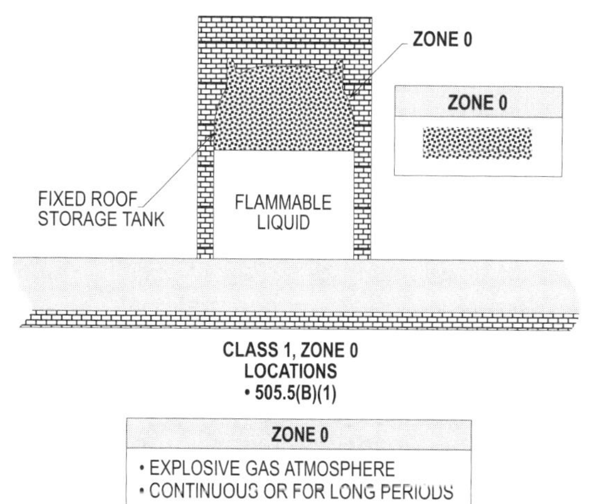

**CLASS I, ZONE 0 LOCATIONS
NEC. 505.5(B)(1)**

Figure 21-56. The above illustration shows a Class I, Zone 0 location.

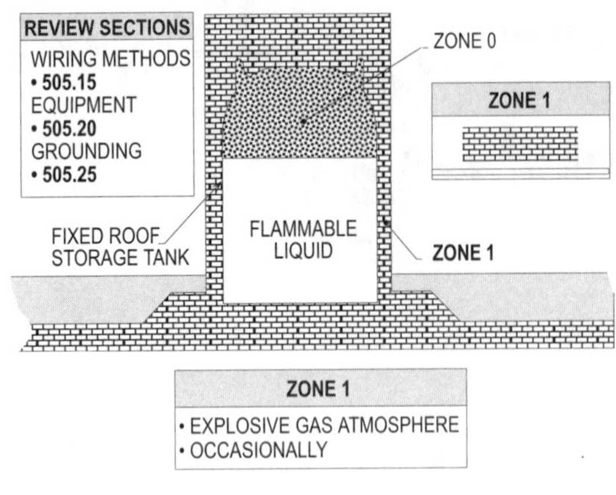

Figure 21-57. The above illustration shows a Class I, Zone 1 location.

CLASS I, ZONE 2 LOCATIONS
505.5(B)(3)

A Class I, Zone 2 location is a location:

(1) in which ignitible concentrations of flammable gases or vapors are not likely to occur in normal operation, and if they do occur, will exist only for a short period.

(2) in which volatile flammable liquids, flammable gases, or flammable vapors are handled, processed, or used but in which the liquids, gases, or vapors normally are confined within closed containers of closed systems from which they can escape, only as a result of accidental rupture or breakdown of the containers or system, or as a result of the abnormal operation of the equipment with which the liquids or gases are handled, processed, or used.

(3) in which ignitible concentrations of flammable gases or vapors normally are prevented by positive mechanical ventilation, but may become hazardous as the result of failure or abnormal operation of the ventilation equipment.

(4) that is adjacent to a Class I, Zone 1 location, from which ignitible concentrations of flammable gases or vapors could be communicated, unless such communication is prevented by adequate positive-pressure ventilation from a source of clean air, and effective safeguards against ventilation failure are provided. **(See Figure 21-58)**

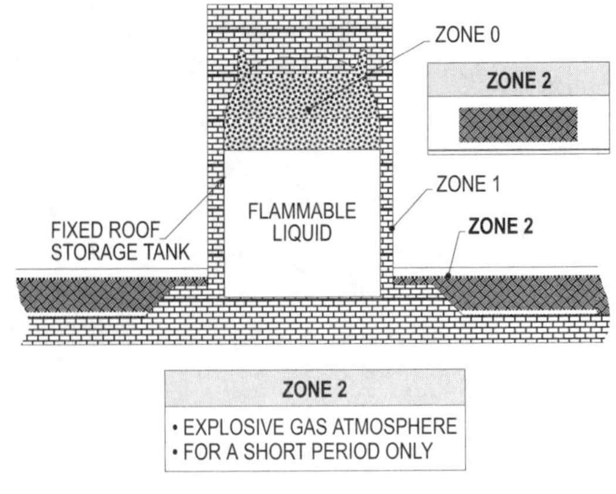

Figure 21-58. The above illustration sows a Class I, Zone 2 location.

CLASS I, ZONE 0, 1, and 2
EQUIPMENT
505.8(A) THRU (K)

Equipment installed in the zone locations shall be provided in a form of construction that will ensure safe performance under conditions of use. Proper maintenance of the equipment is also necessary to prevent the possible igniting of an explosive mixture of gases and air. **(See Figure 21-59)**

SUITABILITY OF EQUIPMENT
505.9(A) AND (B)

Suitability of identified equipment shall be determined by one of the following:

(1) Equipment listing or labeling

(2) Evidence of equipment evaluation from a qualified testing laboratory or inspection agency concerned with product evaluation

(3) Evidence acceptable to the AHJ, such as a manufacturer's self-evaluation or an owner's engineering judgment. **(See Figure 21-60)**

LISTING
505.9(B)(1) AND (B)(2)

Section **505.9(B)(1)** permits equipment that is listed for a Zone 0 location to be used in Zone 1 or Zone 2 locations of the same gas group. Likewise, equipment listed or otherwise acceptable for a Zone 1 location shall be permitted to be used in a Zone 2 location of the same gas group. **(See Figure 21-61)**

Equipment shall be permitted to be listed for a specific gas or vapor, specific mixtures of gases or vapors, or any combination of gases or vapors per **505.9(B)(2)**.

FPN: One common example is equipment marked for "IIB + H2."

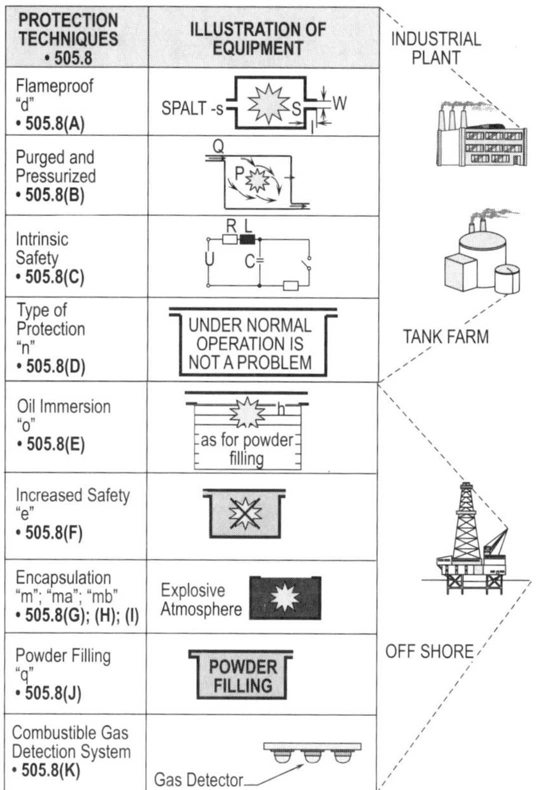

Figure 21-59. The illustration above shows equipment that is permitted in the zone locations.

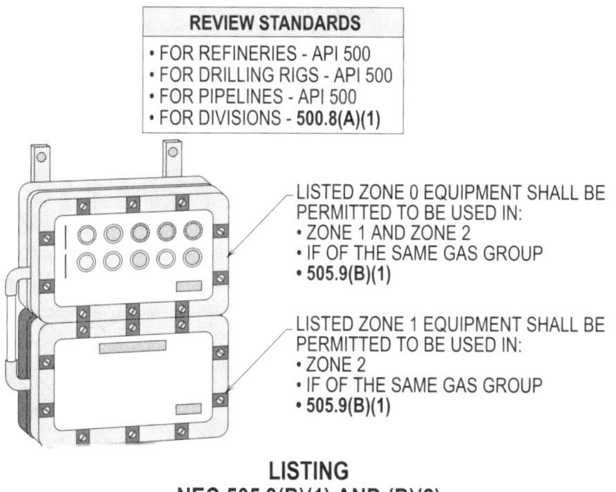

Figure 21-61. Equipment, under certain conditions, shall be permitted to be used from one zone to another.

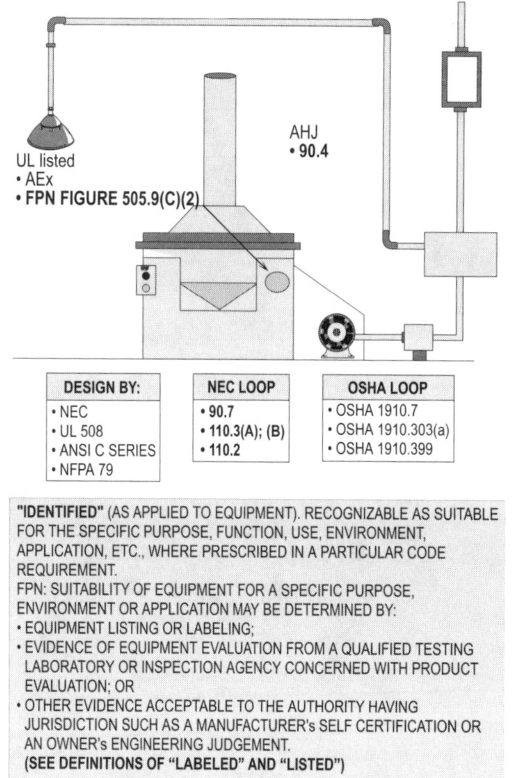

Figure 21-60. The above illustration shows how equipment shall be accepted in Division and Zone locations.

MARKING
505.9(C)

Section **505.9(C)(1)** or **(C)(2)** requires electrical equipment, approved for operation at ambient temperatures exceeding 45°C, to be marked with the maximum ambient temperature for which the equipment is approved and the operating temperature range at that specific ambient temperature.

Design Tip: Electrical equipment shall be selected by a T-code number so that the enclosure won't overheat and cause an explosion to occur. The outer surface shall not exceed the ignition temperature of the specific gas or vapor if the proper T-code number is selected and used. **(See Figure 21-62)**

WIRING METHODS FOR CLASS I, ZONE 0, 1, AND 2 LOCATIONS
505.15(A) THRU (C)

Wiring methods that are permitted to be used in a Class I, Zones 0, 1, or 2 locations are listed in **Figure 21-63** and are based upon area classification.

IEC SYSTEM		NEC SYSTEM	
TEMPERATURE CLASS	MAX. SURFACE TEMPERATURE °C	TEMPERATURE IDENTIFICATION NUMBER	IGNITION TEMPERATURE OF GASES OR VAPORS (°C)
T1	450	T1	450
T2	300	T2	300
		> T2A	> 280
		> T2B	> 260
		> T2C	> 230
		> T2D	> 215
T3	200	T3	200
		> T3A	> 180
		> T3B	> 165
		> T3C	> 160
T4	135	T4	135
		> T4A	> 120
T5	100	T5	100
T6	85	T6	85

>: NEC T-CODE NUMBERS FOR TEMPERATURE IDENTIFICATION; THEY ARE NOT LISTED BY THE IEC SYSTEM BUT ARE LISTED IN THE NEC PER TABLE 500.8(C).

WHAT "T" NUMBER IS REQUIRED FOR IGNITION
TEMPERATURE OF A GAS RATED AT 300°C?

STEP 1: FINDING T NUMBER
TABLE 505.9(D)(1)
300°C REQUIRES T2

SOLUTION: THE T-CODE NUMBER IS
REQUIRED TO BE T2.

MARKING
NEC 505.9(C)

Figure 21-62. The above illustration shows the procedure for selecting the temperature identification number for electrical equipment.

WIRING METHODS PERMITTED

ZONE 0

• INTRINSICALLY SAFE WIRING
  PER **505.15(A)**
• RMC
• IMC
• MI CABLE PER **505.15(B)(1)(c)**
• NONINCENDIVE
• OPTICAL FIBER CABLE PER

ZONE 1

WIRING METHODS PERMITTED
ARE THOSE IN:
• CLASS I, DIVISION 1
• CLASS I, ZONE 0
• PLUS SEALING PER **505.16(A)**

ZONE 2

WIRING METHODS PERMITTED
ARE THOSE IN:
• CLASS I, DIVISION 2
• CLASS I, DIVISION 1
• CLASS I, ZONE 0
• CLASS I, ZONE 1
• PLUS SEALING PER **505.16**

WIRING METHODS FOR
CLASS I, ZONE 0, 1, AND 2 LOCATIONS
NEC 505.15(A) THRU (C)

Figure 21-63. The above chart lists the equipment allowed in Class I, Zones 0, 1, and 2 locations.

EQUIPMENT
505.20(A) THRU (D)

Equipment shall be listed and marked specifically for the Zone in which it is going to be installed. **Ex.s** to **505.20(A), (B),** and **(C)** permit equipment, under certain conditions, to be used from one Zone to another Zone.

Design Tip: Electrical equipment installed using the Zone concept shall comply with **Article 505** and be listed for use by a testing laboratory such as UL, utilizing the symbol AEx. **(See Figure 21-64)**

COMPARISON BETWEEN EQUIPMENT USING THE IEC ZONE CONCEPT OR THE 1996 NEC DIVISION CONCEPT			
EQUIPMENT PERMITTED BY THE IEC		EQUIPMENT PERMITTED BY THE NEC	
ZONE 0	ONLY INTRINSICALLY SAFE EQUIPMENT IS PERMITTED	CLASS I, DIVISION 1	THE FOLLOWING EQUIPMENT SHALL BE PERMITTED TO BE USED: • EXPLOSIONPROOF ENCLOSURES • PURGING • INTRINSIC SAFETY
ZONE 1	EQUIPMENT PERMITTED TO BE USED: "d" FLAMEPROOF ENCLOSURE "p" PRESSURIZED APPARATUS "i" INTRINSIC SAFETY "o" OIL IMMERSION "e" INCREASED SAFETY "q" POWDER FILLING		
ZONE 2	ALL EQUIPMENT CERTIFIED FOR ZONE 0 OR 1 REQUIREMENTS SHALL BE PERMITTED TO BE USED IN ZONE 2	CLASS I, DIVISION 2	ALL EQUIPMENT CERTIFIED FOR DIVISION 1 SHALL BE PERMITTED TO BE USED PLUS: • HERMETICALLY SEALED CONTACTS • OIL IMMERSION OF CONTACTS • NONINCENDIVE CIRCUITS

EQUIPMENT
NEC 505.20(A) THRU (D)

Figure 21-64. The above chart lists the equipment permitted in Class I, Zones 0, 1, and 2.

CLASSIFICATION OF AREA

When applying the zone concept, the protection methods required to ensure safety are related to the likelihood of the presence of an explosive gas/air mixture. The higher the probability, the greater the care needed to ensure safety.

Installations where flammable gases, vapors, or liquids are present can be categorized into areas where the probability of the presence of explosive gas/air mixtures is high or low, and the electrical apparatus shall be selected and installed accordingly. It is clear that the overall risk of exposure to an explosive atmosphere is assessed with respect to both the anticipated frequency of release of flammable material and the probable duration of its presence.

See Figure 21-65 for an illustrated comparison between the IEC and NEC classifications, based upon the presence of gas/air mixtures and the conditions of exposure to personnel and equipment.

LISTING, MARKING, AND DOCUMENTATION 505.9(A), (B), (C), AND 505.4(A)

Zone equipment, if used instead of Division equipment, shall be marked with Class, Zone, and the symbol AEx to verify to the user that such equipment complies with American Standards as well as IEC standards. **(See Figures 21-65 and 21-66** for a detailed illustration of such markings**)**

IEC (Publication 79-10)		NEC (Section **500.5(B)(1) and (B)(2)**)	
Zone 0	Area in which an explosive gas-air mixture is continuously present or present for long periods. (Example: vapor space of a process vessel or storage tank.)	Class I, Division 1	A Class I, Division 1 location • in which ignitible concentrations of flammable gases of vapors exist under normal operating conditions, or • in which ignitible concentrations of such gases or vapors may exist frequently because of repair or maintenance operations or because of leakage, or
Zone 1	Area in which an explosive gas-air mixture is likely to occur in normal operation.		•in which breakdown or faulty operation of equipment or processes might release ignitible concentrations of flammable gases or vapors, and might also cause simultaneous failure of electric equipment.
Zone 2	Area in which an explosive gas-air mixture is not likely to occur, and if it occurs it will only exist for a short time. NOTE: ALSO, SEE **505.5(B)(1) THRU (B)(3)**	Class I, Division 2	A Class I, Division 2 location •in which volatile flammable liquids or flammable gases are handled, processed or used but in which the liquids, vapors, or gases will normally be confined within closed containers or closed systems from which they can escape only in case of accidental rupture or breakdown of such containers or systems, or in case of abnormal operation of equipment, or •in which ignitible concentrations of gases or vapors are normally prevented by positive mechanical ventilation, and which might become hazardous through failure or abnormal operation of the ventilating equipment, or •that is adjacent to a Class I, Division 1 location, and to which ignitible concentrations of gases or vapors might occasionally be communicated unless such communication is prevented by adequate positive-pressure ventilation from a source of clean air, and effective safeguards against ventilation failure are provided.

CLASSIFICATION OF AREA

Figure 21-65. The above table illustrates a comparison between IEC and NEC area classifications.

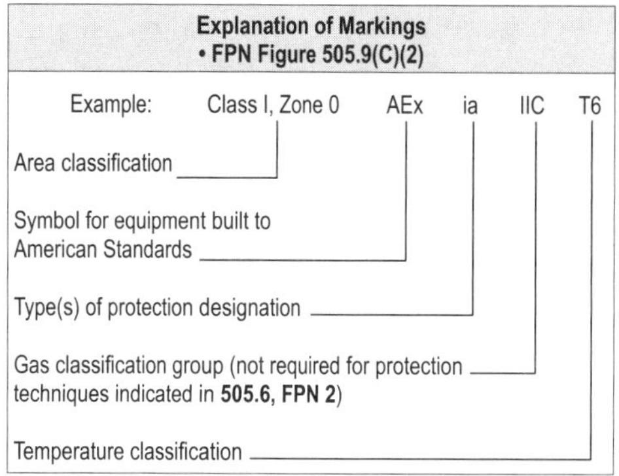

LISTING, MARKING, AND DOCUMENTATION NEC 505.9(A), (B), (C), AND 505.4(A)

Figure 21-66. The above table illustrates an explanation of markings that are used on Zone equipment to identify certain characteristics.

Chapter 21. Hazardous (Classified) Locations

Section   Answer

1. Class I, Division 2 locations are those locations in which volatile _____ gases are handled, processed, or used.
   (a) flammable   (b) combustible
   (c) ignitible   (d) explosive

2. Class II, Division 1 locations are those locations in which _____ dust is in the air under normal operation conditions in quantities sufficient to produce explosive or ignitable mixtures.
   (a) flammable   (b) combustible
   (c) ignitible   (d) explosive

3. Class III, Division 2 locations are those locations in which easily _____ fibers/flyings are stored or handled except in process of manufacturing.
   (a) flammable   (b) combustible
   (c) ignitible   (d) explosive

4. Equipment used in the Divisions or Zones shall be _____ and suitable for the location.
   (a) marked   (b) listed
   (c) labeled   (d) identified

5. Group _____ atmospheres contain combustible metal dusts.
   (a) A   (b) C
   (c) E   (d) G

6. Vapor _____ are used mainly to determine the settling or rising tendency of a mixture.
   (a) percentages   (b) densities
   (c) quantities   (d) temperatures

7. Type _____ enclosures are used only for indoor locations classified as Class I, Groups A, B, C, and D.
   (a) 7   (b) 8
   (c) 9   (d) 10

8. Type _____ enclosures are intended for indoor use in locations classified as Class II, Groups E or G.
   (a) 7   (b) 8
   (c) 9   (d) 10

9. Flexible cord that is listed for _____ usage and provided with listed bushed fittings shall be permitted to be used in Class I, Division 1 locations.
   (a) hard   (b) dry
   (c) wet   (d) extra-hard

10. Surface temperature of Class I, Division 1 luminaires shall not exceed the _____ temperature of the gases or vapors surrounding such units.
    (a) ignition
    (b) flammable
    (c) combustible
    (d) normal

11. In commercial garages, EMT shall be permitted to be installed above the _____ in. level.
    (a) 6
    (b) 12
    (c) 18
    (d) 24

12. A lubrication room shall be considered Class I, Division 2 up to a height of _____ in. above the floor.
    (a) 12
    (b) 18
    (c) 24
    (d) 30

13. For open-top spray booths, the Class I, Division 2 hazardous area shall be considered to extend to _____ ft above the booth and 3 ft from all openings.
    (a) 3
    (b) 4
    (c) 5
    (d) 6

14. A Class II, Division 2 location is a location where combustible _____ is not normally in the air in quantities sufficient to produce explosive or ignitible mixtures.
    (a) liquids
    (b) dusts
    (c) vapors
    (d) fibers/flyings

15. A Class III, Division 1 location is a location in which easily ignitible _____ or materials are handled, manufactured, or used.
    (a) liquids
    (b) dusts
    (c) vapors
    (d) fibers/flyings

16. Group B is an atmosphere containing flammable gases or _____ liquids.
    (a) flammable
    (b) volatile
    (c) combustible
    (d) ignitible

17. Group F atmospheres contain combustible _____ dusts.
    (a) flammable
    (b) metal
    (c) charcoal
    (d) carbonaceous

18. Flash point of a liquid is the minimum _____ at which the liquid gives off sufficient vapor to form an ignitable mixture with the air near the surface of the liquid or within the vessel.
    (a) temperature
    (b) ignition
    (c) density
    (d) level

19. Explosionproof enclosures will withstand an _____ explosion of gases or vapor and prevent those gases or vapors from igniting gases and vapors in the surrounding atmosphere outside of the enclosure.
    (a) volatile
    (b) internal
    (c) external
    (d) flammable

20. Threaded joints of rigid metal conduit shall be made up with at least _____ threads fully engaged.
    (a) 3
    (b) 4
    (c) 5
    (d) 6

21. Rigid nonmetallic conduit shall be permitted to be installed in Class I, Division 1 locations if encased in a concrete envelope of at least 2 in. and buried below the surface in not less than _______ ft of earth.
  (a) 1   (b) 2
  (c) 3   (d) 5

22. In Class I, Division 2 locations, pendant luminaires shall be suspended by a flexible hanger unless rigid stems not over _______ in. long are used.
  (a) 6   (b) 8
  (c) 10   (d) 12

23. Explosionproof equipment or equipment with arcing or sparking devices shall have a seal placed within _______ in. of such equipment.
  (a) 6   (b) 12
  (c) 18   (d) 24

24. In classification of a floor of a garage where vehicles are repaired and volatile flammable fluids are present, the floor area up to _______ in. is Class I, Division 2.
  (a) 12   (b) 18
  (c) 24   (d) 30

25. Any equipment in commercial garages that might produce an arc or spark, and that is less than _______ ft above the floor level, shall have a tight enclosure that will prevent the escape of any arc or spark.
  (a) 12   (b) 15
  (c) 18   (d) 20

26. Any equipment installed in aircraft hangers that is over _______ ft above engines shall be permitted to be the general purpose type.
  (a) 3   (b) 5
  (c) 6   (d) 10

27. The surrounding space out to a distance of _______ ft, measured from a point vertically below the edge of the dispenser enclosure, is Class I, Division 2 up to a height of 18 in. above grade.
  (a) 12   (b) 15
  (c) 18   (d) 20

28. Emergency controls for unattended self-service stations shall be more than 20 ft but less than _______ ft from the dispensers.
  (a) 20   (b) 50
  (c) 75   (d) 100

29. A pit for a bulk-storage tank that is located in the 10 ft or _______ ft sphere (horizontally) around pumps, bleeders, meters, etc. shall be considered as a Class I, Division 1 location.
  (a) 15   (b) 20
  (c) 25   (d) 40

30. Storage and repair garages for tank vehicles shall be considered as Class I, Division 2 up to a height of _______ in. above the floor.
  (a) 18   (b) 24
  (c) 30   (d) 36

___________ ___________ **31.** Underground wiring routed to and around aboveground storage tanks shall be permitted to be installed with rigid metal conduit where buried less than _____ ft in the earth.

    (a) 2   (b) 3
    (c) 5   (d) 6

___________ ___________ **32.** Portable lamps within a spray booth shall be of a type that are _____ for Class I, Division 1 locations.

    (a) listed   (b) identified
    (c) labeled   (d) approved

___________ ___________ **33.** Receptacles used to cord-and-plug connect equipment in Class I, Division 1 or 2 locations shall be _____ for such use.

    (a) listed   (b) labeled
    (c) identified   (d) approved

___________ ___________ **34.** Group _____ is an atmosphere such as ethyl ether, ethylene, or gases or vapors of equivalent hazard.

    (a) A   (b) B
    (c) C   (d) D

___________ ___________ **35.** Group _____ is an atmosphere that contains combustible dusts such as flour, grain, wood, plastic, and chemicals.

    (a) D   (b) E
    (c) F   (d) G

Residential Calculations

Residential calculations are the most difficult to perform because the rules and regulations of the *National Electrical Code* are more restrictive than those for commercial and industrial facilities. The standard or optional calculation shall be permitted to be used to calculate the loads to size and select the elements for the feeder or service. The optional calculation seems to be the favorite method used by designers and electricians. This is true because once loads are calculated, it produces smaller VA or amps than the more complicated standard calculation. Therefore, smaller components are required in the electrical system and greater savings in wiring methods are achieved. The procedure for laying out residential calculations will be different in some ways from those used for commercial and industrial.

APPLYING THE STANDARD CALCULATION
ARTICLE 220, PARTS II AND III

Residential occupancies are known in the industry as dwelling units or single-family dwellings. This chapter mainly deals with one- and two-family dwelling units.

When using the standard calculation for calculating loads for a residential occupancy, all loads are divided into three groups and four columns. The groups are as follows:

Group 1
General lighting and receptacle loads
Group 2
Small appliance loads
Group 3
Special appliance loads

See Figure 22-1 for a detailed illustration of the three groups of loads.

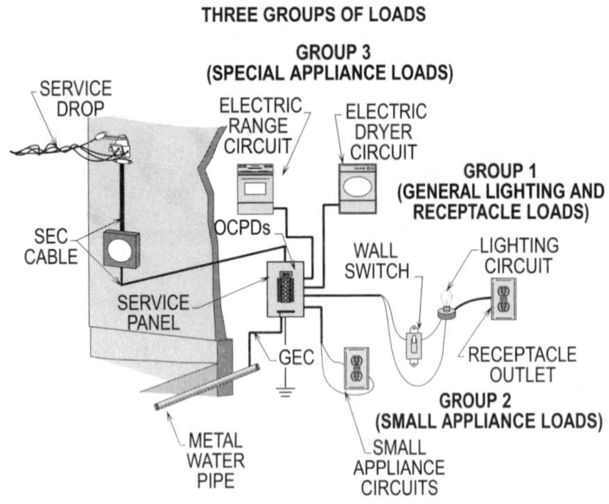

Figure 22-1. The loads in a dwelling unit are divided into three groups and four columns in which demand factors shall be permitted to be applied. The demand factors reduce the loads by a percentage based on the NEC.

GENERAL LIGHTING LOAD
220.12 AND TABLE 220.12

The general lighting load for a dwelling unit shall be determined by multiplying the square footage by 3 VA per sq. ft per **Table 220.12**. The required square footage per unit load VA (volt-amps) is found in **Table 220.12**. All lighting loads and general purpose receptacle loads in a dwelling shall be determined by the general lighting load. All lighting loads per **210.70(A)**, and general purpose receptacle loads per **Article 100** and **210.52(A) through (H)** shall be calculated per **Table 220.12**. The locations in which general-purpose receptacle outlets are installed are bedrooms, bathrooms, dens, living rooms, halls, garages, and outside areas.

See Figure 22-2 for calculating the general purpose lighting and receptacle loads in a dwelling unit.

For example: Calculate the general lighting and receptacle load for an 2400 sq. ft dwelling unit.

Step 1:   Calculating the load in VA
Table 220.12
2400 sq. ft x 3 VA = 7200 VA

Solution: **Table 220.12 requires 7200 VA for the general lighting and receptacle loads.**

For example: Calculate the general lighting and receptacle load for a 60 ft x 55 ft dwelling unit.

Step 1:   Calculating the load in VA
220.12
60' x 55' = 3300 sq. ft

Step 2:   Calculating VA
Table 220.12
3300 sq. ft x 3 VA = 9900 VA

Solution: **Table 220.12 requires 9900 VA for the general lighting and receptacle loads.**

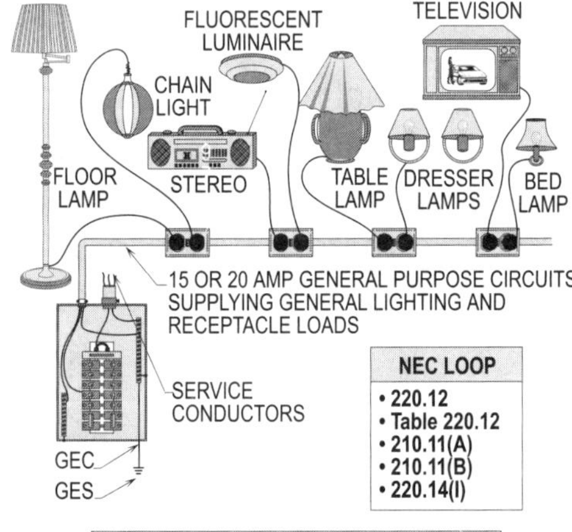

Figure 22-2. The general lighting load for a dwelling unit shall be calculated by multiplying the square footage by 3 VA per sq. ft per **Table 220.12**. All lighting loads and general-purpose receptacle loads in a dwelling shall be calculated by the general lighting load calculation. See the asterisk by the dwelling units in **Table 310.16** and the **Note** below the Table.

SMALL APPLIANCE AND LAUNDRY LOADS
220.52(A) AND (B)

All small appliance circuits shall be located in a dwelling unit per **210.52(B)(1)** and **210.11(C)(1)**. At least two small appliance circuits shall be required to supply receptacle outlets located in the kitchen, breakfast room, pantry, and dining room per **220.52(A)**. A laundry room receptacle outlet shall be required per **220.52(B)**. All small appliance and laundry loads shall be calculated at 1500 VA to determine the size feeder conductors and elements to size the service.

See Figure 22-3 for calculating the small appliance and laundry loads in a dwelling unit.

For example: Calculate the minimum VA rating for small appliance and laundry loads required in a dwelling unit.

Step 1: Calculating the load in VA
220.52(A)
1500 VA x 2 = 3000 VA

Step 2: Calculating the load in VA
220.52(B)
1500 VA x 1 = 1500 VA

Solution: Sections 220.52(A) and (B) require 4500 VA for the small appliance and laundry circuits.

SMALL APPLIANCE CIRCUITS AND LOADS
COLUMN 1

NEC LOOP	
Number Required	210.52(B) and 210.11(C)(1)
Countertops	210.52(C)
VA Rating	220.52(A)
Laundry	220.52(B) and 210.11(C)(2)
Use	210.50(C)
Installation	210.52(B) and 210.52(C)
GFCI	210.8(A)(6)

Calculate the minimum VA rating for the small appliance and laundry loads required in a dwelling unit.

Step 1: Calculating small appliance load
220.52(A)
1500 VA x 2 = 3000 VA

Step 2: Calculating laundry load
220.52(B)
1500 VA x 1 = 1500 VA

Solution: The load is 4500 VA.

SMALL APPLIANCE AND LAUNDRY LOADS
NEC 220.52(A) AND (B)

Figure 22-3. All small appliance and laundry loads shall be calculated at 1500 VA to determine the feeder conductors and elements to size the service. At least three small appliance circuits shall be required. However, more circuits are permitted.

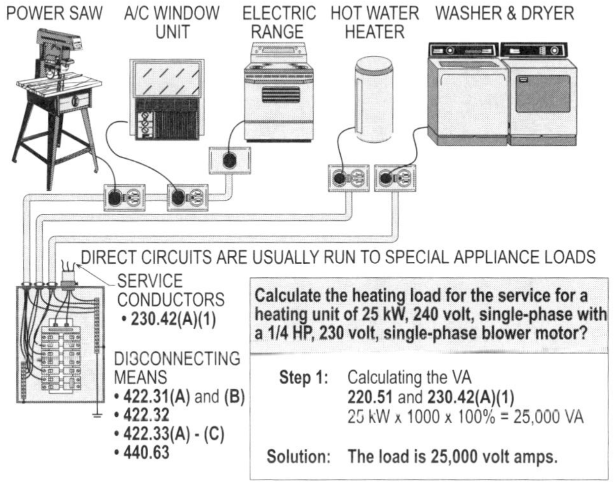

Calculate the heating load for the service for a heating unit of 25 kW, 240 volt, single-phase with a 1/4 HP, 230 volt, single-phase blower motor?

Step 1: Calculating the VA
220.51 and 230.42(A)(1)
25 kW x 1000 x 100% = 25,000 VA

Solution: The load is 25,000 volt amps.

SPECIAL APPLIANCE LOADS
NEC 230.42(A) AND (B)

Figure 22-4. All direct circuit loads shall not be permitted to be connected to the general lighting load. The kilowatt rating listed on the nameplate for special appliance loads shall be calculated at 125 percent for continuous operation and 100 percent noncontinuous operation.

SPECIAL APPLIANCE LOADS
220.14(A) THRU (C), 230.42(A), AND (B)

Special appliance loads are usually supplied by direct circuits. These loads are water heaters, heating units, ranges, air conditioners, dishwashers, motors, etc. Direct circuit loads shall not be connected to the general lighting load. The VA rating for the overcurrent protection device for special appliance loads shall be calculated at 125 percent for continuous operation and 100 percent for noncontinuous operation, with demand loads being determined by the kilowatt rating listed on the nameplate times a percentage.

See Figure 22-4 for calculating the special appliance loads.

For example: Calculating the special appliance load for a 8 kW water heater.

Step 1: Calculating the VA
210.19(A)(1), 230.42(A)(1), and 422.13
8 kW water heater = 8000 watts (volt-amps)

Solution: Sections 230.42(A)(1), 210.19(A)(1) and 220.14(A) requires 8000 VA for the water heater load.

For example: Calculate the heating load for the service and branch circuit load for a heating unit of 20 kW, 240 volt, single-phase with a 1/4 HP, 230 volt, single-phase blower motor.

Step 1: Calculating the VA
220.51
20 kW x 1000 x 100% = 20,000 VA

Solution: Section 220.51 requires 20,000 VA for the service load.

Step 1: Calculating the branch circuit load
424.3(B) and **Table 430.248**
Heating load   = 20,000 VA
Motor load (2.9 A x 240 V) =   696 VA
Total load   = 20,696 VA

Step 2: Calculating the total VA
424.3(B)
20,696 VA x 125% = 25,870 VA

Solution: Section 424.3(B) requires 25,870 VA for the branch circuit load.

Design Tip: Some designers and inspectors include the blower motor load with the kW rating of heating unit. Verify with local codes for interpretation.

DEMAND FACTORS
ARTICLE 220, PART III

The following four loads are separated into two columns of loads and demand factors are applied:

Column 1:
General lighting and receptacle loads
and small appliance loads   **Table 220.42**

Column 2:
Cooking equipment loads   **220.55** and **Table 220.55**
Fixed appliance loads   **220.53**
Dryer loads   **220.54** and **Table 220.54**

General lighting and receptacle loads and the small appliance circuits plus laundry circuit shall be permitted to have demand factors applied.

The total number of fixed appliance loads shall be permitted to have demand factors applied. The total number of ranges and dryers in a dwelling unit shall be permitted to be reduced by a percentage.

See Figure 22-5 for the four loads that shall be permitted to have demand factors applied.

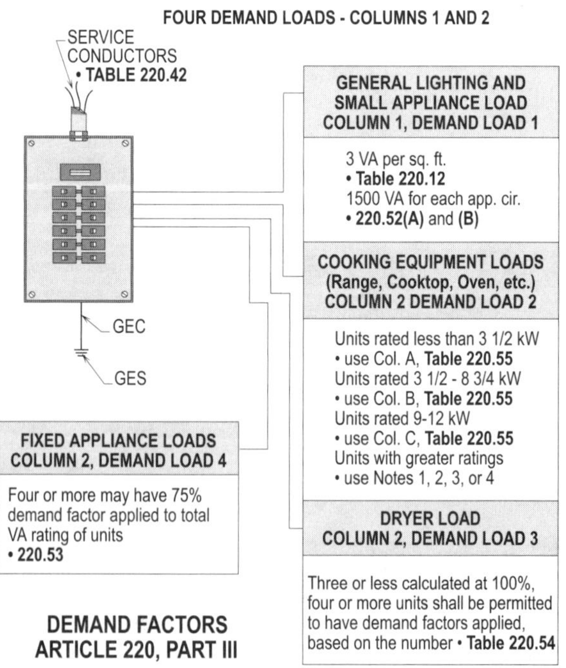

Figure 22-5. General lighting and receptacle loads and small appliance loads, cooking equipment loads, fixed appliance loads, and dryer loads are four loads in Columns 1 and 2 to which demand factors shall be permitted to be applied.

GENERAL LIGHTING AND RECEPTACLE LOADS AND SMALL APPLIANCE AND LAUNDRY LOADS
COLUMN 1 – TABLE 220.42, 220.52(A) AND (B)

The general lighting load for a dwelling unit shall be calculated by multiplying the square footage by 3 VA per sq. ft per **Table 220.12**. The required square footage per unit load (volt-amps) is found in **Table 220.12**. All small appliance and laundry loads shall be calculated at 1500 VA per **220.52(A)** and **(B)**. A demand factor shall be permitted per **Table 220.42**. The demand factors for the general lighting and receptacle loads per **Table 220.42** for dwelling units are as follows:

	Volt-amps	
0 -	3,000	at 100%
3,001 -	120,000	at 35%
120,001 and up		at 25%

See Figure 22-6 for calculating the general purpose lighting and receptacle loads, including the small appliance and laundry loads.

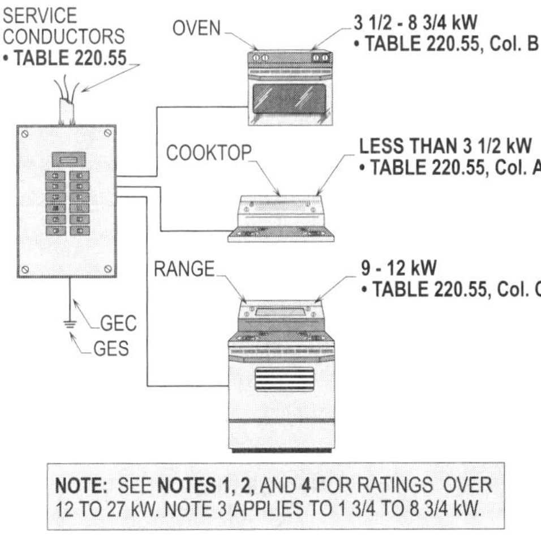

Figure 22-7. The demand factors listed in **Table 220.55** shall be used to apply the demand loads for cooking equipment. The demand factors of **Table 220.55, Columns A, B,** or **C** or based on the kW rating of the cooking equipment. If the kW rating is greater than the kW rating of the equipment listed in **Columns A, B,** or **C**, **Notes 1, 2, 3,** or **4** to **Table 220.55** shall be applied.

For example: Calculate the demand load for the general lighting load for a 2800 sq. ft dwelling unit.

Column 1 and Demand load 1

Step 1: General lighting load
Table 220.12
2800 sq. ft x 3 VA = 8400 VA

Step 2: Small appliance load
220.52(A)
1500 VA x 2 = 3000 VA

Step 3: Laundry load
220.52(B)
1500 VA x 1 = 1500 VA

Step 4: Total load
General lighting load = 8,400 VA
Small appliance load = 3,000 VA
Laundry load = 1,500 VA
Total load = 12,900 VA

Step 5: Applying demand factors
Table 220.42
First 3000 VA x 100% = 3000 VA
Next 9900 VA x 35% = 3465 VA
Total load = 6465 VA

Solution: Demand load 1 requires 6465 VA for the general lighting and receptacle loads, including the small appliance and laundry loads.

COOKING EQUIPMENT LOADS
COLUMN 2 – 220.55 AND TABLE 220.55

The cooking equipment loads and Demand load 2 are separated from Group 3 and placed in Column 2, and demand factors applied accordingly.

The demand factors listed in **Table 220.55** apply to the demand loads for cooking equipment. The **Footnotes** are based on the kW rating and number of units. Ranges, wall-mounted ovens, and counter-mounted cooktops are units of cooking equipment per **Table 220.55**.

See Figure 22-7 for the demand factors listed in **Table 220.55, Columns A, B,** or **C**.

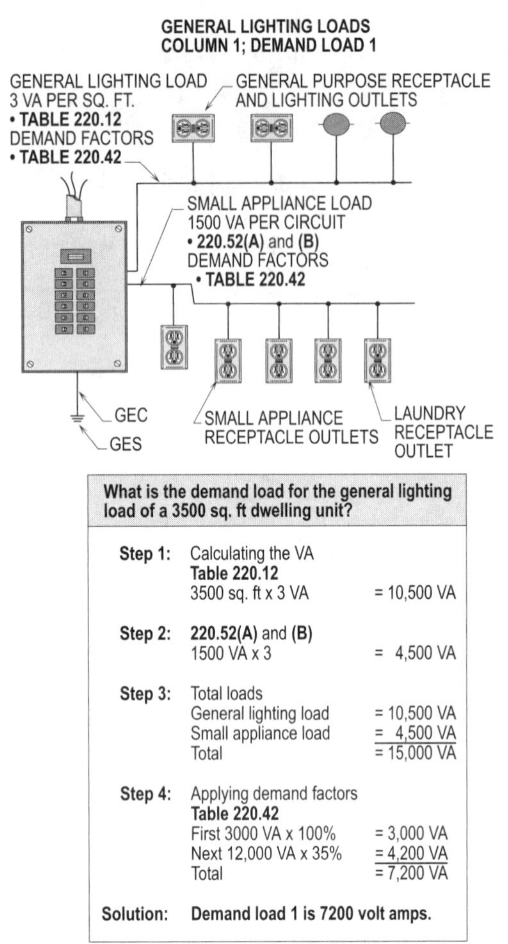

What is the demand load for the general lighting load of a 3500 sq. ft dwelling unit?

Step 1:	Calculating the VA Table 220.12 3500 sq. ft x 3 VA	= 10,500 VA
Step 2:	220.52(A) and (B) 1500 VA x 3	= 4,500 VA
Step 3:	Total loads General lighting load Small appliance load Total	= 10,500 VA = 4,500 VA = 15,000 VA
Step 4:	Applying demand factors Table 220.42 First 3000 VA x 100% Next 12,000 VA x 35% Total	= 3,000 VA = 4,200 VA = 7,200 VA
Solution:	Demand load 1 is 7200 volt amps.	

**GENERAL LIGHTING AND RECEPTACLE
LOADS AND SMALL APPLIANCE AND LAUNDRY LOADS
TABLE 220.42
NEC 220.52(A) AND (B)**

Figure 22-6. The general lighting and receptacle loads, including the small appliance loads, are calculated in VA when determining the size service-entrance equipment and conductors. This is Demand load 1 in the standard calculation.

DEMAND LOAD 2
TABLE 220.55, COLUMN C

The demand load of cooking equipment is calculated in kW. The maximum demand for the size and number of ranges is already calculated for selecting the elements of the feeder or service. Therefore, cooking equipment does not need to be multiplied by the kW rating of the unit by a percentage until **Columns A** or **B** are utilized.

See Figure 22-8 for demand loads that shall be permitted to be applied in **Table 220.55, Column C**.

For example: Calculating the demand load for a 11.5 kW range.

Step 1:	Calculating the load in VA **Table 220.55, Column C** 11.5 kW = 8 kVA	
Solution:	**Column C to Table 220.55 allows 8 kVA for the range.**	

For example: Calculating the demand load for a 12 kW and 10 kW cooking unit.

Step 1:	Calculating VA **Table 220.55, Column C** 12 kW and 10 kW = 11 kVA	
Solution:	**Column C to Table 220.55 allows 11 kVA for the two cooking units.**	

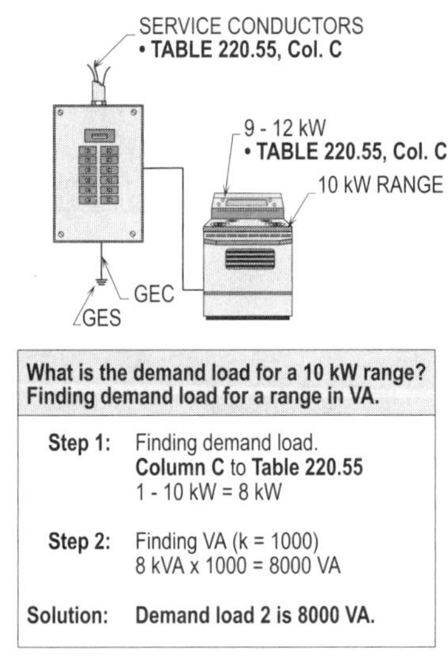

What is the demand load for a 10 kW range? Finding demand load for a range in VA.

Step 1:	Finding demand load. Column C to Table 220.55 1 - 10 kW = 8 kW	
Step 2:	Finding VA (k = 1000) 8 kVA x 1000 = 8000 VA	
Solution:	Demand load 2 is 8000 VA.	

**COOKING EQUIPMENT LOADS
TABLE 220.55, Col. C**

Figure 22-8. The demand load in kW for cooking equipment in **Column C** is already calculated per **Table 220.55**.

DEMAND LOAD 2
TABLE 220.55, COLUMN A

The nameplate (kW) wattage rating of a range shall be calculated by the number of cooking units times the percentage factor found in **Table 220.55**. This calculation is used to obtain the maximum demand load.

See **Figure 22-9** for demand loads to be applied in **Table 220.55, Column A**.

For example, what is the percentage allowed for three cooking units rated at 3 kW, 2 1/2 kW, and 2 kW respectively? The first step is to select the percentage to be applied, based upon the number of cooking units. Three cooking units permit a demand factor of 70 percent per **Table 220.55, Column A**.

For example: Calculating the demand load for a piece of cooking equipment with a 2.5 kW rating.

Step 1: Calculating the VA
Table 220.55, Column A
2.5 kW x 80% = 2 kVA

Solution: Column A to Table 220.55 allows 2 kVA for the cooking equipment.

DEMAND LOAD 2
TABLE 220.55, COLUMN B

The nameplate (kW) wattage rating of a range shall be calculated by the number of cooking units times the percentage factor applied in **Table 220.55**. This calculation derives the maximum demand load.

See **Figure 22-10** for demand loads to be applied from **Table 220.55, Column B**.

For example: Calculating the demand load for a piece of cooking equipment with a 8.5 kW rating.

Step 1: Calculating the VA
Table 220.55, Column B
8.5 kW x 80% = 6.8 kVA

Solution: Column B to Table 220.55 allows 6.8 kVA for the cooking equipment.

DEMAND LOAD 2
TABLE 220.55, NOTE 1

For cooking equipment rated over 12 kW to 27 kW in **Column C**, each kW over 12 kW shall be increased 5 percent to calculate the demand load per **Note 1**.

See **Figure 22-11** for demand factors to be applied from **Table 220.55, Column C, Note 1**.

For example: Calculating the demand load for a range with a 25 kW rating.

Step 1: Calculating the percentage
Table 220.55, Note 1
25 kW - 12 kW = 13 kW
13 kW x 5% = 65%

Step 2: Calculating the VA
Table 220.55, Column C
8 kW x 165% = 13.2 kVA

Solution: Note 1 to Table 220.55 allows 13.2 kVA for the range.

DEMAND LOAD 2
TABLE 220.55, NOTE 2

Cooking equipment of unequal values rated over 12 kW to 27 kW in **Column C, Note 2** shall be calculated by adding the kW ratings of all units and dividing by the number of units; all ranges below 12 kW shall be calculated at 12 kW. When an average rating is found, the number of units shall be increased by 5 percent for each kW exceeding 12 kW, to derive the allowable kW.

See **Figure 22-12** for demand factors that are applied per **Table 220.55, Column C, Note 2**.

Note, the demand factor selected from **Table 220.55, Column C** shall be based upon the number of units.

For example: Determine the demand load for three pieces of cooking equipment with a 12 kW, 14 kW, and 20 kW rating respectively.

Step 1: Calculating the percentage
Table 220.55, Note 2
Total kW rating
12 kW + 14 kW + 20 kW = 46 kW
Average rating
46 kW ÷ 3 = 15.3 (round up)
16 - 12 = (4 x 5%) = 20%

Step 2: Calculating the VA
Table 220.55, Column C
14 kW x 120% = 16.8 kVA

Solution: Note 2 to Table 220.55 allows 16.8 kVA demand load for three pieces of cooking equipment.

Design Tip: It is permissible per **Note 3** to **Table 220.55** to add all pieces of cooking equipment with ratings over 1-3/4 kW through 8-3/4 kW together and multiply by the percentage of **Columns A** or **B**, whichever produces the smaller kW rating. This calculation shall be permitted to be applied if it provides the smaller kW rating of all the methods available in **Table 220.55** and **Notes**.

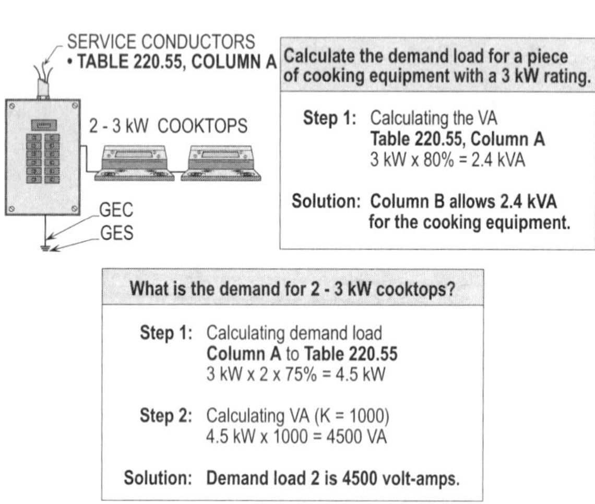

COOKING EQUIPMENT LOADS
TABLE 220.55, COLUMN A

Figure 22-9. The maximum demand load shall be calculated by the number of cooking units times the percentage factor from **Table 220.55, Column A**.

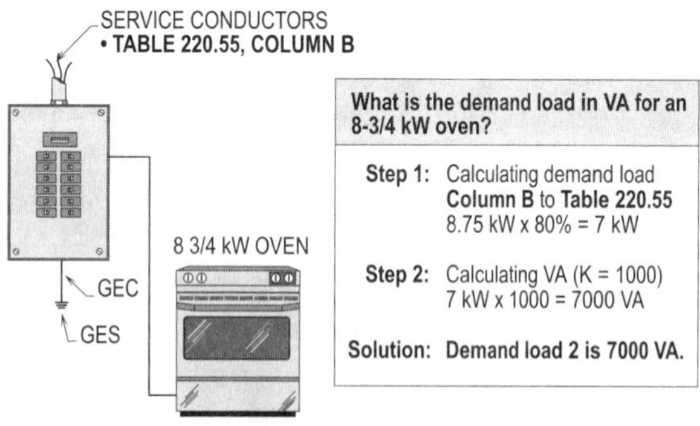

COOKING EQUIPMENT LOADS
TABLE 220.55, COLUMN B

Figure 22-10. The maximum demand load shall be calculated by the number of cooking units times the percentage factors from **Table 220.55, Column B**. This calculation produces smaller volt-amp ratings and allows smaller elements.

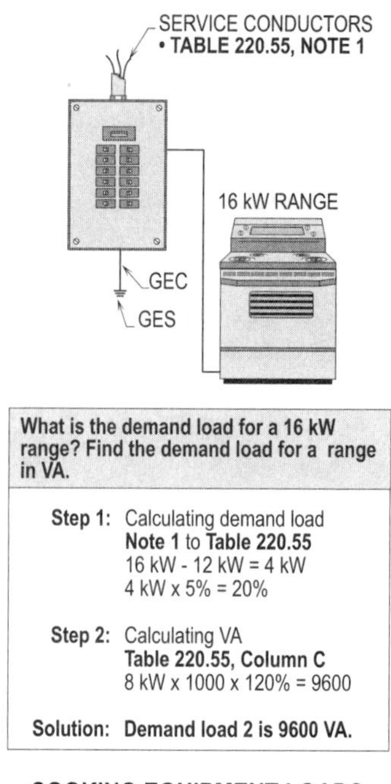

COOKING EQUIPMENT LOADS
TABLE 220.55, NOTE 1

Figure 22-11. For cooking equipment rated over 12 kW to 27 kW in **Column C**, each kW over 12 kW shall be increased 5 percent to calculate the demand load per **Note 1**. This percentage times 8 kW will calculate the demand load to be used to size the elements.

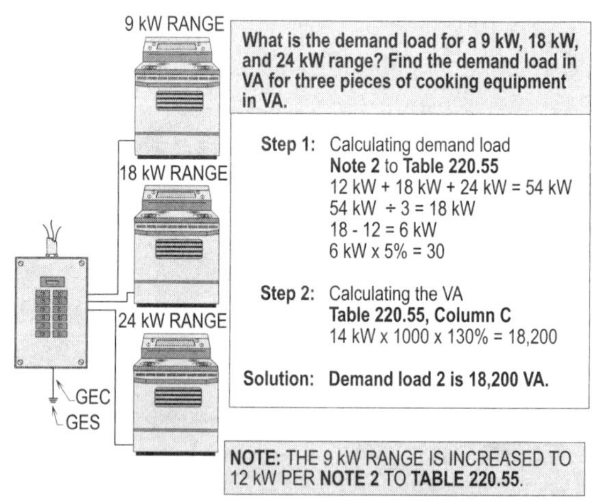

COOKING EQUIPMENT LOADS
TABLE 220.55, NOTE 2

Figure 22-12. For cooking equipment of unequal values rated over 12 kW to 27 kW in **Column C, Note 2** shall be calculated by adding the kW ratings of all units and dividing by the number of units; all ranges below 12 kW shall be calculated at 12 kW. When an average rating is determined, the number of units shall be increased 5 percent for each kW exceeding 12 kW.

DEMAND LOAD 2
TABLE 220.55, NOTE 4

The demand load for a cooktop and two or less wall-mounted ovens shall be calculated by finding the average rating of each unit in kW. Each kilowatt that exceeds 12 kW shall be increased by 5 percent. The nameplate rating of the appliance shall be calculated per **Table 220.55, Column C**. Sections **240.21(A)** and **210.19(A)(3), Ex. 1** list requirements for each piece of cooking equipment to be supplied by a tap. Each piece of cooking equipment shall be installed in the same room in order to comply with **Note 4** to **Table 220.55**. The demand load for each kW exceeding 12 kW is found by multiplying the kW rating by 5 percent to determine the average rating per **Table 220.55, Column C**.

See **Figure 22-13** for demand factors to be applied from **Table 220.55, Column C, Note 4**.

For example: Calculating the demand load for a 12 kW cooktop and 2 - 6 kW ovens for a dwelling unit.

Step 1: Calculating the percentage
Table 220.55, Note 4
Total kW rating
12 kW + 6 kW + 6 kW = 24 kW
24 kW - 12 kW = 12 kW
12 kW x 5% = 60%

Step 2: Calculating the VA
Table 220.55, Column C
8 kW x 160% = 12.8 kVA

Solution: **Note 4 to Table 220.55 allows 12.8 kVA demand load for one cooktop and two ovens.**

NEC LOOP	
DEMAND FACTOR	TABLE 220.55, NOTE 4
TAPS	210.19(A)(3), Ex. 1
RECEPTACLE OUTLET	220.14(B)

What is the demand load for an 8 kW, 10 kW and 12 kW oven in VA?

Step 1: Calculating demand load
Note 4 to **220.55**
Total kW rating - 8 kW + 10 kW + 12 kW = 30 kW
30 kW - 12 kW = 18 kW
18 kW x 5% = 90%

Step 2: Calculating the VA
Table 220.55, Column C
8 kW x 1000 x 190% = 15,200

Solution: Demand load 2 is 15,200 VA.

**COOKING EQUIPMENT LOADS
TABLE 220.55, NOTE 4**

See **Figure 22-13.** The demand load for a cooktop and two or less wall-mounted ovens shall be calculated by finding the amperage rating of each unit in kW. Each kilowatt that exceeds 12 kW shall be increased by 5 percent to obtain the multiplier. The multiplier times the demand for one unit in **Column C** to **Table 220.55** derives the demand load for the elements of the circuit.

FIXED APPLIANCE LOAD
COLUMN 2 – 220.53

The fixed appliance load for cooking equipment loads, dryer equipment loads, air conditioning loads, and heating equipment loads for three or less fixed appliances is determined by adding wattage (volt-amps) values by the nameplate ratings for each appliance. A 75 percent demand factor shall not be applied for these loads per **220.53**. However, these fixed appliance loads shall be permitted to have demand factors applied if there are four or more. All other fixed appliances of four or more grouped into a special appliance load shall be found by adding wattage ratings from appliance nameplates and multiplying the total wattage (volt-amps) by 75 percent to obtain demand load.

See **Figure 22-14** for demand factors to be applied per **220.53**. Note that this is demand load 4.

Three or less fixed appliance loads shall be calculated at 100 percent. The 75 percent per **220.55** shall not be applied unless there are four or more present.

Fixed appliance loads include dishwashers, compactors, disposals, water heaters, attic fans, water circulating pumps, etc. It shall be permitted to apply a demand factor to these loads to calculate the elements of the service or feeder equipment. However, the branch circuit elements shall be calculated and sized according to other Sections in the NEC.

For example, the elements of a water heater shall be calculated at 125 percent times the amperage of the total elements load per **422.13**. A dishwasher shall be calculated at 100 percent of its nameplate rating per **422.10(A)** and **422.62**.

An attic fan shall be calculated per **430.22(A)** based upon the FLA selected from **Table 430.248** for single-phase motors used for such purposes.

> **Design Tip:** The elements for branch circuits shall be calculated at 125 percent for continuous operation and 100 percent for noncontinuous operation. This includes the conductors and overcurrent protection devices feeding and protecting branch circuits.

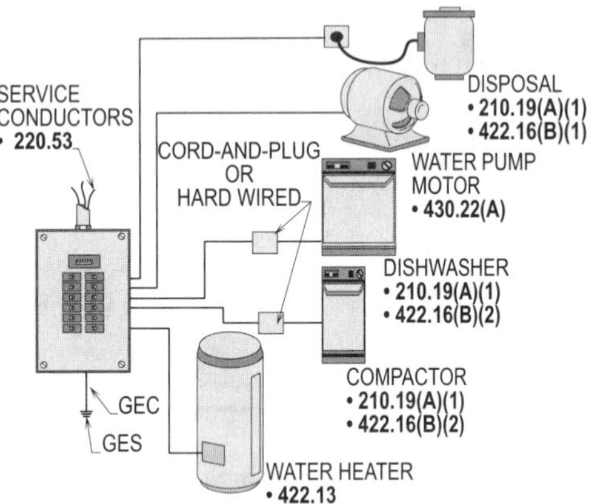

What is the demand load for the fixed appliance load in VA?

5000 VA A/C	240 V, single-phase
15,000 VA heating unit	240 V, single-phase
6000 VA water heater	240 V, single-phase
9000 VA oven	240 V, single-phase
7500 VA cooktop	240 V, single-phase
1600 VA water pump	240 V, single-phase
900 VA disposal	120 V, single-phase
1200 VA compactor	120 V, single-phase
1000 VA microwave	120 V, single-phase
5000 VA dryer	240 V, single-phase
700 VA blower motor	240 V, single-phase

Step 1: Finding the demand load
220.53
Remove the following fixed appliances

Heating load	15,000 VA heating unit
Air-conditioning load	5000 VA A/C unit
Dryer load	5000 VA dryer
Cooking equipment load	9000 VA oven and 7500 VA cooktop

Step 2: Calculate the VA
220.53

water heater load	6,000 VA
water pump load	1,600 VA
disposal load	900 VA
compactor load	1,200 VA
microwave load	1,000 VA
blower motor load	700 VA
Total load	11,400 VA

Step 3: Applying demand factors
220.53
11,400 VA x 75% = 8550 VA

Solution: Demand load 4 is 8550 VA.

FIXED APPLIANCE LOAD
NEC 220.53

For example: Calculating the demand load for the following fixed appliances.

6000 VA A/C	240 V, single-phase
10,000 VA heating unit	240 V, single-phase
5000 VA water heater	240 V, single-phase
8000 VA oven	240 V, single-phase
8500 VA cooktop	240 V, single-phase
2600 VA water pump	240 V, single-phase
1000 VA disposal	120 V, single-phase
1200 VA compactor	120 V, single-phase
1600 VA dishwasher	120 V, single-phase
1000 VA microwave	120 V, single-phase
5000 VA dryer	240 V, single-phase
800 VA blower motor	240 V, single-phase

Step 1: Special appliance loads, removing the following loads:
220.53
Heating load
• 10,000 VA heating unit
Air-conditioning load
• 6000 VA A/C unit
Dryer load
• 5000 VA dryer
Cooking equipment load
• 8000 VA oven
• 8500 VA cooktop

Step 2: Calculate the VA
220.53

Water heater load	=	5,000 VA
Water pump load	=	2,600 VA
Disposal load	=	1,000 VA
Compactor load	=	1,200 VA
Dishwasher load	=	1,600 VA
Microwave load	=	1,000 VA
Blower motor load	=	800 VA
Total load	=	13,200 VA

Step 3: Applying demand factors
220.53
13,200 VA x 75% = 9900 VA

Solution: **Section 220.53 allows 9900 VA demand load for the fixed appliance.**

Figure 22-14. All fixed appliances of four or more grouped in a special appliance load shall be found by adding wattage ratings from appliance nameplates and multiplying the total wattage (volt-amps) by 75 percent. Calculation derives demand load.

> **Design Tip:** After the heating load, A/C load, cooking equipment load, and dryer load have been removed from the fixed appliance load, all other appliances shall be considered fixed appliances per **220.53**.

DRYER LOAD
COLUMN 2 – 220.54 AND TABLE 220.54

The demand load for household dryers shall be calculated at 5 kVA or the nameplate rating, whichever is greater. Dryer equipment of four or fewer dryers shall be calculated at 100 percent of the nameplate rating. Dryer equipment of five or more dryers shall be permitted to have a percentage applied based on the number of units per **Table 220.54**.

See Figure 22-15 for demand factors to be applied in **220.54** and **Table 220.54**.

For example: Calculate the demand load for a 4 kW dryer.

Step 1: Calculating the kW
220.54
4 kW = 5 kW

Step 2: Applying demand factors
Table 220.54
Four or fewer dryers = 100%
5 kW x 100% = 5 kVA

Solution: Table 220.54 allows 5 kVA demand load for the dryer.

For example: Calculate the demand load for 5 dryers rated at 8000 VA each.

Step 1: Selecting percentage
220.54; Table 220.54
5 dryers = 85%

Step 2: Applying demand factors
Table 220.54
8000 VA x 5 x 85% = 34,000 VA

Solution: Table 220.54 allows 34,000 VA demand load for 5 dryers.

Note, a dryer rated at 4500 VA shall be calculated at 5000 VA for determining the components for sizing a branch circuit, feeder, or service. Where applying the optional calculation, the volt-amp rating shall be calculated at the nameplate values only.

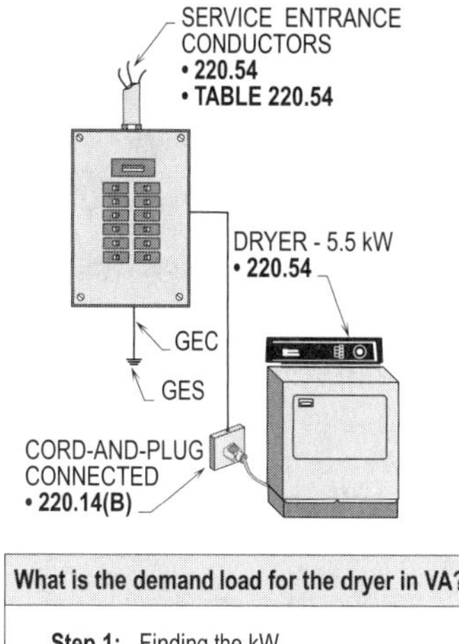

Figure 22-15. The demand load for household dryers shall be calculated at 5 kVA or the nameplate rating, whichever produces the greater rating. Four or fewer dryers shall be calculated at 100 percent of the nameplate rating. Five or more dryers shall be permitted to have a percentage applied based on the number of units per **Table 220.54**. The Table is based on five dryers being used at different times, and the load is limited to about 85 percent of total. **Table 220.54** is used for dwelling units in apartment complexes.

LARGEST LOAD BETWEEN HEATING AND A/C
COLUMN 3 – 220.60

Section **220.60** shall be applied when determining the largest load between heating and A/C. The heating and A/C loads shall be calculated at 100 percent, and the smaller of the two loads is dropped.

See Figure 22-16 for demand factors to be applied per **220.60**.

Design Tip: Some designers recognize the A/C unit as still being eligible for the largest motor per **220.50**, even if it has been dropped per **220.60**.

For example: Calculating the load for a 10 kW heating unit and a 5.5 kW A/C unit load in a dwelling unit.

Step 1:   Calculating the VA
220.60
Heating load
10 kW x 100% = 10 kVA
A/C load
5.5 kW x 100% = 5.5 kVA

Solution: **Section 220.60 requires 10 kVA load for the largest load between the heating and A/C load.**

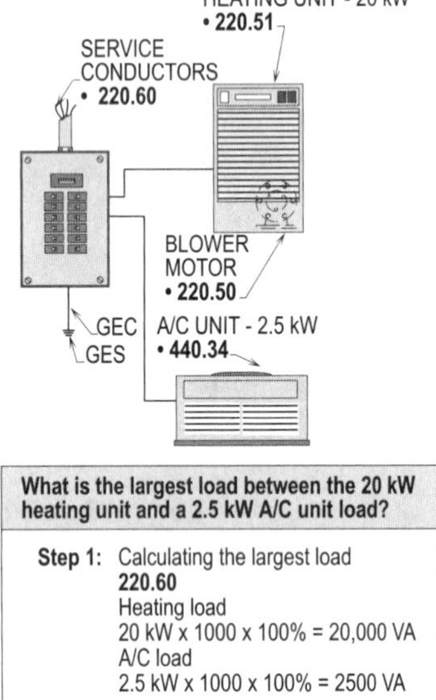

Figure 22-16. The heating and A/C loads shall be calculated at 100 percent and the smaller of the two dropped and not used again until the branch circuit for each is calculated. **Note:** Heat pumps are usually calculated with the heating load. (Their total VA ratings are combined.)

LARGEST MOTOR LOAD
COLUMN 4 – 220.50

The motor's total full-load current rating, in amps, shall be calculated at 25 percent per **220.50**, which is required per **430.24** and **430.25** for calculating the load of one or more motors with other loads.

See Figure 22-17 for demand factors to be applied in **220.50**.

For example: Calculating the VA load for a 3 HP, 230 volt, single-phase motor, which is to be used for the largest load.

Step 1:   Finding FLA
Table 430.248
3 HP = 17 A

Step 2:   Calculating A
220.50; 430.22(A), 430.24
17 A x 25% = 4.25 A

Step 3:   Calculating VA
Text
4.25 A x 240 V = 1020 VA

Solution: **Section 220.50 requires 1020 VA load to supply the largest motor load.**

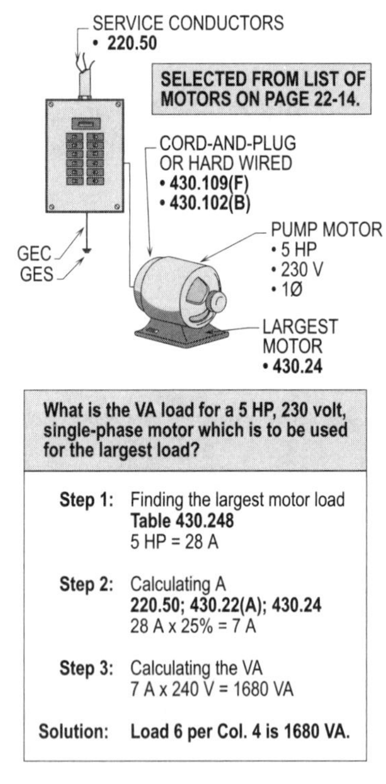

Figure 22-17. The motor's total full-load current rating, in amps, shall be calculated at 25 percent per **220.50**, which refers to **430.24** for calculating the load of one or more motors with other loads.

For example: Calculating the load for a 7.5 HP, 230 volt, single-phase motor with 1400 VA of other loads.

Step 1: Finding FLA
430.6(A)(1) and **Table 430.248**
Motor load
7.5 HP = 40 A
40 A x 240 V   = 9,600 VA
Other loads
1400 VA   = 1,400 VA
Total loads   = 11,000 VA

Step 2: Calculating the VA
220.50, 430.24, and **430.22(A)**
Total loads   = 11,000 VA
Largest motor load
9600 VA x 25%   = 2,400 VA
Total load   = 13,400 VA

Solution: **Section 220.50 requires 13,400 VA demand load to supply the largest motor load.**

Select largest motor from fixed appliance load

20 kW heating unit	240 V, single-phase
5.5 kW A/C unit	240 V, single-phase
6 kW water heater	240 V, single-phase
1.2 kW compactor	120 V, single-phase
1.8 kW dishwasher	120 V, single-phase
1.7 kW disposal	120 V, single-phase
1.6 kW attic fan	120 V, single-phase
1.3 kW microwave	120 V, single-phase
5 HP pump motor	240 V, single-phase
12 kW range	240 V, single-phase

The first step is to select the largest motor from the list of loads that are eligible to be considered the largest motor. The 20 kW heating unit is larger than the 5.5 kW A/C unit, which eliminates the A/C unit per **220.60**. After scanning the other motors to find the largest, the **5 HP motor** becomes the largest per **220.50**.

The largest motor load shall always be added into the calculation, and it really doesn't matter what size the largest motor is. The NEC does not distinguish between the size of the motor, but requires one to be selected and calculated based upon HP and voltage. **(See Design Tip on Page 22-10)**

ONE-FAMILY DWELLING STANDARD CALCULATION

For example, calculating the following loads of a 2800 sq. ft dwelling unit and size the service-entrance conductors required for Phases A and B and the Neutral.

Loads grouped

Group 1:
General lighting and receptacle load
Group 2:
Small appliance and laundry load
Group 3:
Special appliance load

6000 VA A/C	240 V, single-phase
10,000 VA heating unit	240 V, single-phase
5000 VA water heater	240 V, single-phase
8000 VA oven	240 V, single-phase
8500 VA cooktop	240 V, single-phase
2600 VA water pump	120 V, single-phase
1000 VA disposal	120 V, single-phase
1200 VA compactor	120 V, single-phase
1600 VA dishwasher	120 V, single-phase
1000 VA microwave	120 V, single-phase
5000 VA dryer	240 V, single-phase
800 VA blower motor	240 V, single-phase

Column 1
General lighting and receptacle load

Table 220.12
2800 sq. ft x 3 VA = 8,400 VA

Small appliance and laundry loads
220.52(A) and **(B)**
1500 VA x 2 = 3,000 VA
1500 VA x 1 = 1,500 VA
Total load = 4,500 VA

Applying demand factors for demand load 1
Table 220.42
General lighting load = 8,400 VA
Small appliance and
laundry load = 4,500 VA
Total load = 12,900 VA

First 300 VA x 100% = 3,000 VA
Next 9900 VA x 35% = 3,465 VA
Total load = **6,465 VA** • √

Column 2
Special appliance loads

Cooking equipment loads and demand load 2
Table 220.55, Col. B
Total kW rating
8 kW + 8.5 kW = 16.5 kW
16.5 kW x 65% = 10.725 kVA •
10.725 kVA x 100% x 1000 = **10,725 kVA** •

Fixed appliance load and demand load 3
220.53
Water heater = 5,000 VA
Water pump = 2,600 VA √
Disposal = 1,000 VA √
Compactor = 1,200 VA √
Dishwasher = 1,600 VA √
Microwave = 1,000 VA √
Blower motor = 800 VA
Total load = 13,200 VA
13,200 VA x 75% = **9,900 VA** •

Dryer load and demand load 4
220.54; Table 220.54
5000 VA x 100% = **5,000 VA** •

Column 3
Largest between heating or A/C load

220.60
Heating load
10,000 VA x 100% = **10,000 VA** •

Column 4
Largest motor load

220.50
Water pump
2600 VA x 25% = **650 VA** •

Finding total VA for Phases A and B

General lighting load = 6,465 VA •
Cooking equipment = 10,725 VA •
Fixed appliance load = 9,900 VA •
Dryer load = 5,000 VA •
Heating load = 10,000 VA •
Largest motor load = 650 VA •
Total load = 42,740 VA

Finding amps for Phases A and B
220.5(B)

$I = VA \div V$
$I = 42{,}740 \text{ VA} \div 240 \text{ V}$
$I = 178 \text{ A}$

Total VA loads (Neutral)

Column 1
General lighting load and demand load 1

220.61(A)
6465 VA = **6,465 VA** √

Column 2
Cooking equipment load and demand load 2

(Use 70% of cooking load)
220.61(B)(1)
10,725 VA x 70% = **7,508 VA** √

Column 2 - Use 120 V loads (√)
Fixed appliance load and demand load 3

(Use 75% of 120 V load)
7400 VA x 75% = 5550 VA
220.53
5550 VA = **5,550 VA** √

Column 2
Dryer load and demand load 4

(Use 70% of dryer load)
220.61(B)(1)
5000 VA x 70% = **3,500 VA** √

Column 4
Largest motor load

(Use 25% of largest motor load)
2600 VA x 25% = **650 VA** √

Finding total VA for neutral

General lighting load	=	6,465 VA √
Cooking equipment	=	7,508 VA √
Dryer load	=	3,500 VA √
Fixed appliance load	=	5,550 VA √
Largest motor load	=	650 VA √
Total load	=	23,673 VA

Finding amps for neutral
I = VA ÷ V
I = 23,673 ÷ 240
I = 99 A

Table 310.16 and **Table 310.15(B)(6)**
Phases A and B

Table 310.16 permits
3/0 AWG THWN cu. conductors

Table 310.15(B)(6) permits
2/0 AWG THWN cu. conductors

Neutral is 3 AWG THWN cu. conductor

See Design Problem 22-1 on page 22-19 for a detailed illustration of calculating the loads for a dwelling unit using the standard calculation.

APPLYING THE OPTIONAL CALCULATION
220.82

The optional calculation for dwelling units provides a easier method for calculating the load per **220.82(B)** and **(C)**. The loads are separated into two columns. The first column of loads consist of all loads except heating and A/C loads, which are described as general loads. Heating and A/C loads are the second column of loads. The elements of the service shall be determined by using the total loads of these two columns.

The optional calculation has percentages applied that are derived from the demand factors. The total kVA of a dwelling unit shall be used in determining the demand factors to be applied. The percentages in **220.82(B)** and **(C)** shall be used to calculate the load in dwelling units. Separate units located in a multifamily dwelling shall be permitted to have the optional calculation applied per **220.82(B)** and **(C)**. The optional calculation method shall only be applied when an ampacity of at least 100 amps is applied to the service conductors. This method shall be permitted to be applied to newly constructed or older existing dwelling units. **(See Design Problem 22-2 on page 22-20)**

GENERAL LOADS
220.82(B)

General loads are the first columns of loads to be calculated. General lighting and general purpose receptacle loads shall be calculated at 3 VA per sq. ft per **220.82(B)(1)**. Small appliance and laundry loads shall be calculated at 1500 VA per **220.82(B)(2)**, as covered in **210.11(C)(1)** and **(C)(2)**. These loads are then applied to the other loads in the column. Heating and A/C loads shall not be applied to the other loads in the column. Heating and A/C loads shall not be applied to other loads per **220.82(B)(3)**. Special appliance loads shall be added to general loads per **220.82(B)(3)**. Section **220.82(B)(4)** includes motor loads in the general loads column. Section **220.50** shall not require the largest motor load at 25 percent to be added to general loads. The nameplate ratings in the general columns of loads shall be calculated at 100 percent. The first 10,000 VA of the general loads shall be calculated at 100 percent of the demand factor, and the remaining VA shall be calculated at 40 percent of the demand factor allowed per **220.82(B)**.

HEAT OR A/C LOADS
220.82(C)(1) THRU (C)(6)

The second column of loads consist of heating, A/C. or heating or heat pumps. The loads shall be determined by the following steps:

- Three or fewer units shall be calculated at 65 percent of the total kW rating of the heating load.

- Four or more units shall be calculated at 40 percent of the total kW rating of the heating load.

- The A/C load shall be calculated at 100 percent of its kVA rating.

- The total load VA rating of the heating unit shall be compared to the A/C load, and the smaller load is dropped per **220.82(C)**.

- Heat pumps that operate with the heating unit shall be calculated at 100 percent and added to the heating load. **(See Design Problem 22-3 on page 22-21)**

APPLYING THE OPTIONAL CALCULATION FOR EXISTING UNITS
220.83

Existing dwelling units per **220.83** shall be permitted to use the optional calculation to size the components for the service equipment to determine if additional loads can be added to the service. A 120/240 or 120/208 volt, three-wire, single-phase system shall be required to supply the service. Two columns are used in separating the loads in the dwelling unit. General lighting and general purpose receptacle loads shall be calculated at 3 VA per sq. ft per **220.82(A)(1)**. All other loads shall be calculated at 100 percent of their nameplate ratings. Heating and A/C loads shall be calculated at 100 percent, and the smaller load is dropper per **220.83(B)**.

OTHER LOADS
220.83

Existing loads are the first column to be selected. This column shall consist of adding 3 VA per sq. ft to the small appliance loads plus the nameplate ratings of each special appliance load to determine the existing load in a dwelling unit. The demand factors in **220.83** shall be used in reducing the VA rating to the existing load, and this value is then added to the second column, which contains the appliance load to be added.

ADDED APPLIANCE LOAD
220.83

Added appliance loads are the second column to be applied. This column shall be determined by adding the total VA rating of the load at 100 percent. The existing loads shall be added to the added loads to determine the total VA rating of the dwelling unit. The added load usually, consist of a heating unit, air conditioner, dryer, or some other type of special appliance load.

From the following loads in an existing dwelling unit, it will be determined if a 5040 VA A/C unit load can be added to the service conductors. (Service is rated 100 amps.)

- 1800 sq. ft dwelling unit
- 2 small appliance circuits
- 1 laundry circuit

• 12 kW range	240 V, single-phase
• 5 kW dryer	240 V, single-phase
• 1000 VA compactor	120 V, single-phase
• 900 VA disposal	120 V, single-phase
• 1200 VA dishwasher	120 V, single-phase

See Design Problem 22-4 on page 22-22 to verify if the A/C load can be added without upgrading the size of the service conductors.

It is a different story if a heat pump is involved with the heating unit load. There are two methods in which a heat pump shall be calculated in relation to the heating unit. The first method is where the heat pump operates independently and does not operate with the heating unit. In this case, it is not added to the heating unit load at 100 percent to derive the total load. If it does operate with the heating load, it is added to the heating load at 100 percent to derive the total load.

Heat pumps are usually not effective when the temperature is below freezing. Heating elements are normally installed to melt the frozen ice from the coil. Section **220.82(B)(3)** addresses the supplemental heating required to provide this service. However, the supplementary heating elements performing this duty shall be permitted to be disconnected when the temperature reaches a certain temperature and the heating unit can operate independently.

ONE-FAMILY DWELLING OPTIONAL CALCULATION

For example, calculating the following loads of a 2800 sq. ft dwelling unit and size the service-entrance conductors required for Phases A and B and the neutral.

Loads grouped

Group 1:
- General lighting and receptacle loads
- Small appliance and laundry loads
- Special appliance loads

5000 VA water heater	240 V, single-phase
8000 VA oven	240 V, single-phase
8500 VA cooktop	240 V, single-phase
2600 VA water pump	120 V, single-phase
1000 VA disposal	120 V, single-phase
1200 VA compactor	120 V, single-phase
1600 VA dishwasher	120 V, single-phase
1000 VA microwave	120 V, single-phase
5000 VA dryer	240 V, single-phase
800 VA blower motor	240 V, single-phase

Group 2:

6000 VA A/C	240 V, single-phase
10,000 VA heating unit	240 V, single-phase

Column 1
General (other) loads

General lighting and receptacle load
220.82(B)(1) and **Table 220.12**
2800 sq. ft x 3 VA   = 8,400 VA •

Small appliance and laundry loads
220.82(B)(2), 210.11(C)(1), and **(C)(2)**
1500 VA x 2   = 3,000 VA •
1500 VA x 1   = 1,500 VA •

Special appliance loads
Table 220.82(B)(3) and **(B)(4)**

Oven	= 8,000 VA •
Cooktop	= 8,500 VA •
Water heater	= 5,000 VA •
Water pump	= 2,600 VA •
Disposal	= 1,000 VA •
Compactor	= 1,200 VA •
Dishwasher	= 1,600 VA •
Microwave	= 1,000 VA •
Dryer	= 5,000 VA •
Blower motor	= 800 VA •
Total load	= 47,600 VA

Column 2
Largest load between heating and A/C load

220.82(C)(1), (C)(2), and **(C)(4)**

A/C unit (6000 VA x 100%)	= 6,000 VA
Heating unit (10,000 VA x 65%)	= 6,500 VA
Total load (heating selected)	= 6,500 VA

Calculating general loads and largest load per 220.82(B)

First 10,000 VA x 100%	= 10,000 VA
Next 37,600 VA x 40%	= 15,040 VA
Largest load (Between heat and A/C)	= 6,500 VA

Calculating the VA

General loads	= 10,000 VA
	= 15,040 VA
Largest load (Between heating and A/C)	= 6,500 VA
Total load	= 31,540 VA

Calculating the amps

I = VA ÷ V
I = 31,540 VA ÷ 240 V
I = 131 A

Table 310.16 and **Table 310.15(B)(6)**
Phases A and B

Table 310.16 permits
1/0 AWG THWN cu. conductors

310.15(B)(6) permits
1 AWG THWN cu. conductors

Neutral is 3 AWG THWN cu. conductor

Note, see the calculation for the neutral in the previous explanation of the standard calculation on **pages 22-14** and **15**.

MULTIFAMILY STANDARD CALCULATION
PARTS III AND IV TO ARTICLE 220

When applying the standard calculation for multifamily dwelling units, the loads shall be calculated the same as using the standard calculation for one-family dwellings. The only difference is to calculate the loads of each unit and multiply the total number of dwelling units and pieces of electrical equipment together to derive the total VA or amps to size the elements of the service or feeder. **(See Design Problem 22-5 on page 22-23)**

MULTIFAMILY OPTIONAL CALCULATION
PARTS III AND IV TO ARTICLE 220

When applying the optional calculation for multifamily dwelling units, the loads shall be calculated the same as using the optional calculation for one-family dwellings. The only difference is to calculate the total number of dwelling units and pieces of electrical equipment together and apply a percentage based upon the number to derive the total VA or amps to size the elements of the service or feeder. **(See Design Problem 22-6 on page 22-24)**

FEEDER TO MOBILE HOME STANDARD CALCULATION
550.18

Service calculations for mobile homes are performed at the factory. However, it is the responsibility of the designer or electrician to size and select the proper size feeder to supply power to the mobile home using the standard calculation. Such feeder shall be a four-wire circuit with all conductors insulated. The size of the conductors shall be permitted to be selected from **Table 310.16** or **310.15(B)(6)**. The service equipment on a pole or pedestal shall be rated at least 100 amps. **(See Design Problem 22-7 on page 22-25)**

MOBILE HOME PARK SERVICE AND FEEDERS - OPTIONAL CALCULATION
550.31

The elements of the service shall be permitted to be calculated using the optional calculation. All loads are added together based upon the total number of mobile homes and multiplied by a percentage to derive the total load. Use 16,000 VA for each mobile home if the calculated load per **550.31** produces a lower VA rating. **(See Design Problem 22-8 on page 22-26)**

CALCULATION PROBLEMS
PARTS III AND IV TO ARTICLE 220

The elements of electrical systems shall be permitted to be calculated by using the standard or optional calculation. The size of these elements are determined by whichever method the designer chooses to calculate these loads.

The following calculations are typical examples of how these loads are calculated, sized, and selected. The step-by-step procedures are easy to follow and have condensed the more complicated rules pertaining to calculating loads into a compact listing, that provides easier understanding of how to perform calculations according to the provisions of the NEC.

A broad assortment of basic NEC calculations have been selected to represent the main principles of designing and installing electrical systems in residential occupancies.

MINIMUM RATING OF SERVICE EQUIPMENT FOR A MOBILE HOME
550.32(A)

Generally, the mobile home service equipment shall be located adjacent to the mobile home and not mounted in or on the mobile home. The service equipment shall be located in sight from and not more than 30 ft (9 m) from the exterior wall of the mobile home it serves. Mobile home service equipment shall be rated at not less than 100 amps at 120/240 volts, and provisions shall be made for connecting a mobile home feeder assembly by a permanent wiring method. Note that 50 amp power outlets shall also be permitted to be used as service equipment.

DESIGN PROBLEM 22-1. What is the load in VA and amps for a residential dwelling unit with the following loads?

General lighting and receptacle load

- 2500 sq. ft dwelling unit
- 2 small appliance circuits
- 1 laundry circuit

120 V, single-phase loads

- 2600 VA water pump
- 1000 VA disposal
- 1200 VA compactor
- 1600 VA dishwasher

240 V, single-phase loads

- 5000 VA dryer
- 6000 VA A/C unit
- 20,000 VA heating unit
- 6000 VA water heater
- 10,000 VA oven
- 9000 VA cooktop
- 800 VA blower motor
- 1000 VA pool pump

Sizing phases = •
Sizing neutral = √

COLUMN 1
CALCULATING GENERAL LIGHTING AND RECEPTACLE LOAD

Step 1: General lighting and receptacle load
Table 220.12
2500 sq. ft x 3 VA = 7,500 VA

Step 2: Small appliance and laundry load
220.52(A) and **(B)**
1500 VA x 2 = 3,000 VA
1500 VA x 1 = 1,500 VA
Total load = 12,000 VA

Step 3: Applying demand factors
Demand load 1; **Table 220.42**
First 3000 VA x 100% = 3,000 VA
Next 9000 VA x 35% = 3,150 VA
Total load = **6,150 VA** • √

COLUMN 2
CALCULATING COOKING EQUIPMENT LOAD

Step 1: Applying demand factors for phases A and B
Demand load 2; **Table 220.55, Col. C**
9 kW and 10 kW = **11,000 VA** •

Step 2: Applying demand factors for neutral
220.61(B)(1)
11,000 VA x 70% = **7,700 VA** √

COLUMN 2
CALCULATING DRYER LOAD

Step 1: Applying demand factors for phases A and B
Demand load 3; **Table 220.54**
5000 VA x 100% = **5,000 VA** •

Step 2: Applying demand factors for neutral
220.61(B)(1)
5000 VA x 70% = **3,500 VA** • √

COLUMN 2
CALCULATING FIXED APPLIANCE LOAD

Step 1: Applying demand factors for phases A and B
Demand load 4; **220.53**
2600 VA x 75% = 1,950 VA √
1000 VA x 75% = 750 VA √
1200 VA x 75% = 900 VA √
1600 VA x 75% = 1,200 VA √
800 VA x 75% = 600 VA
1000 VA x 75% = 750 VA
6000 VA x 75% = 4,500 VA
Total load = **10,650 VA** •

Step 2: Applying demand factors for neutral
Demand load 4; **220.53** and **220.61(A)**
2600 VA x 75% = 1,950 VA
1000 VA x 75% = 750 VA
1200 VA x 75% = 900 VA
1600 VA x 75% = 1,200 VA
Total load = **4,800 VA** √

COLUMN 3
LARGEST LOAD BETWEEN HEATING AND A/C LOAD

Step 1: Selecting largest load
Demand load 5; **220.51** and **220.60**
Heating unit
20,000 VA x 100% = **20,000 VA** •

COLUMN 4
CALCULATING LARGEST MOTOR LOAD

Step 1: Selecting largest motor load for phases A and B
220.50 and **430.24**
2600 VA x 25% = **650 VA** •

Step 2: Selecting largest motor load for neutral
220.50 and **430.24**
2600 VA x 25% = **650 VA** √

CALCULATING PHASES (ADD ALL •)

- General lighting load = 6,150 VA •
- Cooking load = 11,000 VA •
- Dryer load = 5,000 VA •
- Appliance load = 10,650 VA •
- Heating load = 20,000 VA •
- Largest motor load = 650 VA •
Total load = **53,450 VA**

CALCULATING NEUTRAL (ADD ALL √)

- General lighting load = 6,150 VA √
- Cooking load = 7,700 VA √
- Dryer load = 3,500 VA √
- Appliance load = 4,800 VA √
- Largest motor load = 650 VA √
Total load = **22,800 VA**

FINDING AMPS FOR PHASES A AND B

$I = VA \div V$
$I = 53,450\ VA \div 240\ V$
$I = 223\ A$

FINDING AMPS FOR NEUTRAL

$I = VA \div V$

DESIGN PROBLEM 22-2. What is the load in VA and amps for a residential dwelling unit with the following loads? (See standard calculation in Design Problem 22-1 for sizing the neutral)

General lighting and receptacle load

- 2500 sq. ft dwelling unit
- 2 small appliance circuits
- 1 laundry circuit

120 V, single-phase loads

- 2600 VA water pump
- 1000 VA disposal
- 1200 VA compactor
- 1600 VA dishwasher

240 V, single-phase loads

- 5000 VA dryer
- 6000 VA A/C unit
- 20,000 VA heating unit
- 6000 VA water heater
- 10,000 VA oven
- 9000 VA cooktop
- 800 VA blower motor
- 1000 VA pool pump

Sizing phases = •

I = 22,800 VA ÷ 240 V
I = 95 A

COLUMN 1
GENERAL LOAD

Step 1: General lighting load
220.82(B)(1)
2500 sq. ft x 3 VA = 7,500 VA

Step 2: Small appliance and laundry load
220.82(B)(2); 210.11(C)(1) and (C)(2)
1500 VA x 2 = 3,000 VA
1500 VA x 1 = 1,500 VA

Step 3: Appliance load
220.82(B)(3) and (B)(4)
Cooktop load = 9,000 VA
Oven load = 10,000 VA
Dryer load = 5,000 VA
Water heater load = 6,000 VA
Disposal load = 1,000 VA
Compactor load = 1,200 VA
Dishwasher load = 1,600 VA
Pool pump load = 1,000 VA
Blower motor load = 800 VA
Water pump load = 2,600 VA
Total load = 50,200 VA

Step 4: Applying demand load
220.82(B)
First 10,000 VA x 100% = 10,000 VA
Next 40,200 VA x 40% = 16,080 VA
Total load = 26,080 VA •

COLUMN 2
LARGEST LOAD BETWEEN HEATING AND A/C LOAD

Step 5: Selecting largest load
220.82(C)(1), (C)(2), and (C)(4)
Heating load
20,000 VA x 1 x 65% = 13,000 VA •
A/C load
6000 VA x 1 x 100% = 6,000 VA
Total load = 13,000 VA •

TOTALING COLUMNS 1 AND 2
220.82(B); 220.82(B) AND (C)

Col. 1 ld. = 26,080 VA •
Col. 2 ld. = 13,000 VA •
Total load = 39,080 VA

FINDING AMPS FOR PHASES A AND B

I = VA ÷ V
I = 39,080 VA ÷ 240 V
I = 163 A

FINDING AMPS FOR NEUTRAL

I = VA ÷ V
I = 22,800 VA ÷ 240 V
I = 95 A

Note: See "Calculating neutral" and "Finding amps for neutral" in Design Problem 22-1 on page 22-19.

DESIGN PROBLEM 22-3. What is the load in VA and amps for a residential dwelling unit with the following loads? (See standard calculation in Design Problem 22-1 for sizing the neutral)

General lighting and receptacle load

- 2500 sq. ft dwelling unit
- 2 small appliance circuits
- 1 laundry circuit

120 V, single-phase loads

- 2600 VA water pump
- 1000 VA disposal
- 1200 VA compactor
- 1600 VA dishwasher

240 V, single-phase loads

- 5000 VA dryer
- 6000 VA heat pump
- 20,000 VA heating unit
- 6000 VA water heater
- 10,000 VA oven
- 9000 VA cooktop
- 800 VA blower motor
- 1000 VA pool pump

Sizing phases = •

Note: Heat pump may be used with heating unit.

COLUMN 1
GENERAL LOADS

Step 1:   General lighting load
220.82(B)(1)
2500 sq. ft x 3 VA   = 7,500 VA

Step 2:   Small appliance and laundry load
220.82(B)(2); 210.11(C)(1) and (C)(2)
1500 VA x 2   = 3,000 VA
1500 VA x 1   = 1,500 VA

Step 3:   Appliance load
220.82(B)(3) and (B)(4)

Cooktop load	= 9,000 VA
Oven load	= 10,000 VA
Dryer load	= 5,000 VA
Water heater load	= 6,000 VA
Disposal load	= 1,000 VA
Compactor load	= 1,200 VA
Dishwasher load	= 1,600 VA
Pool pump load	= 1,000 VA
Blower motor load	= 800 VA
Water pump load	= 2,600 VA
Total load	= 50,200 VA

Step 4:   Applying demand load
220.82(B)
First 10,000 VA x 100%   = 10,000 VA
Next 40,200 VA x 40%   = 16,080 VA
Total load   = 26,080 VA •

COLUMN 2
LARGEST LOAD BETWEEN HEATING AND A/C LOAD

Step 5:   Selecting largest load
220.82(C)(1), (C)(2), and (C)(4)
Heating load
20,000 VA x 1 x 65%   = 13,000 VA
Heat pump
6000 VA x 1 x 100%   = 6,000 VA
Total load   = 19,000 VA •

TOTALING COLUMNS 1 AND 2
220.82(B) AND (C)

Col. 1 ld.   = 26,080 VA •
Col. 2 ld.   = 19,000 VA •
Total load   = 45,080 VA

FINDING AMPS FOR PHASES A AND B

$I = VA \div V$
$I = 45{,}080 \text{ VA} \div 240 \text{ V}$
$I = 188 \text{ A}$

FINDING AMPS FOR NEUTRAL

$I = VA \div V$
$I = 22{,}800 \text{ VA} \div 240 \text{ V}$
$I = 95 \text{ A}$

Note: The neutral load in VA and amps is calculated as shown in Design Problem 22-1 on page 22-19. See "Calculating neutral" and "Finding amps for neutral."

DESIGN PROBLEM 22-4: Can a 5040 VA combination A/C unit heat pump be added to the existing dwelling unit without upgrading the service elements? (Service is rated 100 amps)

General lighting and receptacle load

• 1800 sq. ft dwelling unit
• 2 small appliance circuits
• 1 laundry circuit

120 V, single-phase loads

•   900 VA disposal
• 1000 VA compactor
• 1200 VA dishwasher

240 V, single-phase loads

• 5000 VA dryer
• 12,000 VA range
• 5040 VA A/C unit and heat pump to be added

COLUMN 1
CALCULATING EXISTING LOADS

Step 1: General lighting load
220.83(B)(1)
1800 sq. ft x 3 VA = 5,400 VA

Step 2: Small appliance and laundry load
220.83(B)(2); 210.11(C)(1) and (C)(2)
1500 VA 2 = 3,000 VA
1500 VA x 1 = 1,500 VA

Step 3: Existing load
220.83(B)(3)
Range load = 12,000 VA
Dryer load = 5,000 VA
Disposal load = 900 VA
Compactor load = 1,000 VA
Dishwasher load = 1,200 VA
Total load = 30,000 VA

Step 4: Applying demand load
220.83(B)
First   8000 VA x 100% = 8,000 VA
Next 22,000 VA x 40% = 8,800 VA
Total load = 16,800 VA •

COLUMN 2
ADDED LOAD

Step 5: Calculating added load
220.83(B)
Fixed appliance load
5040 VA x 100% = 5,040 VA •

TOTALING COLUMNS 1 AND 2
220.83(B)(1) THRU (B)(3)

Col. 1 ld.   = 16,800 VA •
Col 2 ld.   = 5,040 VA •
Total load   = 21,840 VA

FINDING AMPS FOR PHASES A AND B

I = VA ÷ V
I = 21,840 VA ÷ 240 V
I = 91 A
Existing service   = 100 A
New calculated   = 91 A

Note: If calculated load is less than service load, the new load can be added. Since 91 amps is less than the 100 amp service listed in Design Problem 22-4, the A/C unit heat pump load of 5040 VA may be added to the existing service.

DESIGN PROBLEM 22-5. What is the load in VA and amps for 25 multifamily dwelling units with the following loads? **Note:** Parallel service conductors, 6 times per phase.

General lighting and receptacle load	120 V, single-phase loads	240 V, single-phase loads	
• 25 - 1000 sq. ft dwelling unit	• 25 - 1000 VA dishwashers	• 25 - 12,000 VA ranges	Sizing phases = •
• 2 small appliance circuits per unit	• 25 - 1200 VA disposals	• 25 - 6000 VA water heaters	Sizing neutral = √
• 1 laundry circuit per unit		• 25 - 20,000 VA heating units	

COLUMN 1
CALCULATING GENERAL LIGHTING AND RECEPTACLE LOAD

Step 1: General lighting and receptacle load
Table 220.12
1000 sq. ft x 3 VA x 25 = 75,000 VA

Step 2: Small appliance and laundry load
220.52(A) and **(B)**
1500 VA x 2 x 25 = 75,000 VA
1500 VA x 1 x 25 = 37,500 VA
Total load = 187,500 VA

Step 3: Applying demand factors
Demand load 1; **Table 220.42**
First 3000 VA x 100% = 3,000 VA
Next 117,000 VA x 35% = 40,950 VA
Remaining 67,500 VA x 25% = 16,875 VA
Total load = **60,825 VA • √**

COLUMN 2
CALCULATING COOKING EQUIPMENT LOAD

Step 1: Applying demand factors for phases A and B
Demand load 2; **Table 220.55, Col. C**
25 - 12,000 VA ranges = **40,000 VA •**

Step 2: Applying demand factors for neutral
220.61(B)(1)
40,000 VA x 70% = **28,000 VA √**

COLUMN 2
CALCULATING FIXED APPLIANCE LOAD

Step 1: Applying demand factors for phases A and B
Demand load 4; **220.53**
1000 VA x 25 x 75% = 18,750 VA
1200 VA x 25 x 75% = 22,500 VA
6000 VA x 25 x 75% = 112,500 VA
Total load = **153,750 VA •**

Step 2: Applying demand factors for neutral
Demand load 4; **220.61(A)**
1000 VA x 25 x 75% = 18,750 VA
1200 VA x 25 x 75% = 22,500 VA
Total load = **41,250 VA √**

COLUMN 3
LARGEST LOAD BETWEEN HEATING AND A/C LOAD

Step 1: Selecting largest load
Demand load 5; **220.60**
Heating unit
20,000 VA x 25 x 100% = **500,000 VA •**

COLUMN 4
CALCULATING LARGEST MOTOR LOAD

Step 1: Selecting largest motor load for phases A and B
220.50 and **430.24**
1200 VA x 25% = **300 VA •**

Step 2: Selecting largest motor load for neutral
220.50 and **430.24**
1200 VA x 25% = **300 VA √**

CALCULATING PHASES A AND B (add all •)

• General lighting load	= 60,825 VA •
• Cooking load	= 40,000 VA •
• Appliance load	= 153,750 VA •
• Heating load	= 500,000 VA •
• Largest motor load	= 300 VA •
Total load	= 754,875 VA

CALCULATING NEUTRAL (add all √)

• General lighting load	= 60,825 VA √
• Cooking load	= 28,000 VA √
• Appliance load	= 41,250 VA √
• Largest motor load	= 300 VA √
Total load	= **130,375 VA**

FINDING AMPS FOR PHASES A AND B

I = VA ÷ V
I = 754,875 VA ÷ 240 V
I = 3145 A

FINDING AMPS FOR NEUTRAL

I = VA ÷ V
I = 130,375 VA ÷ 240 V
I = 543 A

FINDING SIZE CONDUCTORS IN PARALLEL FOR PHASES A AND B

Phases A and B
310.4
I = 3145 ÷ 6 (No. runs per phase)
I = 524 A

Neutral (applying demand factors)
220.61(B)(2)

543 A
First 200 A x 100% = 200 A
Next 343 A x 70% = 240 A
Total load = **440 A**
310.4 and **250.24(C)(2)**

I = 440 A ÷ 6 (No. runs per phase)
I = 73 A

Table 310.16

Phases A and B
6 - 1000 KCMIL THWN copper conductors per phase
Neutral
6 - 1/0 AWG THWN copper conductors per phase

Note: The neutral conductor shall be 2/0 AWG per **250.24(C)(2)**.
Column 1.

DESIGN PROBLEM 22-6. What is the load in VA and amps for 25 multifamily dwelling units with the following loads? (See standard calculation in Design Problem 22-5 for sizing the neutral) **Note:** Parallel service conductors, 6 times per phase.

General lighting and receptacle load

• 25 - 1000 sq. ft dwelling unit
• 2 small appliance circuits per unit
• 1 laundry circuit per unit

120 V, single-phase loads

• 25 - 1000 VA dishwashers
• 25 - 1200 VA disposals

240 V, single-phase loads

• 25 - 12,000 VA ranges
• 25 -   6000 VA water heaters
• 25 - 20,000 VA heating units

Sizing phases = •

COLUMN 1
CALCULATING GENERAL LIGHTING AND RECEPTACLE LOAD

Step 1: General lighting and receptacle load
220.84(C)(1)
1000 sq. ft x 3 VA x 25 = 75,000 VA

Step 2: Small appliance and laundry load
220.84(C)(2)
1500 VA x 2 x 25 = 75,000 VA
1500 VA x 1 x 25 = 37,500 VA
Total load = **187,500 VA** •

COLUMN 2
CALCULATING COOKING EQUIPMENT LOAD

Step 1: Applying demand factors for phases A and B
Demand load 2; **220.82(C)(3)**
12,000 VA x 25 = **300,000 VA** •

COLUMN 2
CALCULATING FIXED APPLIANCE LOAD

Step 1: Applying demand factors for phases A and B
Demand load 4; **220.82(C)(3)**
1000 VA x 25 = 25,000 VA
1200 VA x 25 = 30,000 VA
6000 VA x 25 = 150,000 VA
Total load = **205,000 VA** •

COLUMN 3
LARGEST LOAD BETWEEN HEATING AND A/C LOAD

Step 1: Selecting largest load
Demand load 5; **220.82(C)(5)**
Heating unit
20,000 VA x 25 x 100% = **500,000 VA** •

CALCULATING PHASES

• General lighting load = 187,500 VA •
• Cooking load = 300,000 VA •
• Appliance load = 205,000 VA •
• Heating load = 500,000 VA •
Total load = **1,192,500 VA**

APPLYING DEMAND FACTORS
TABLE 220.84

$I = VA \div V$
$I = 1{,}192{,}500 \text{ VA} \div 240$
I = 4969

FINDING AMPS FOR PHASES A AND B
TABLE 220.84

$I = A \times \%$
$I = 4969 \times 35\%$
I = 1739

FINDING SIZE CONDUCTORS FOR PHASES A AND B

Phases A and B
310.4
$I = 1739 \div 6$ (No. runs per phase)
I = 290 A

Table 310.16
Phases A and B
6 - 350 KCMIL THWN copper conductors

Note: The neutral in VA and amps is calculated as shown in Design Problem 22-5 on page 22-23. See "Calculating neutral," "Finding amps for neutral," and "Sizing conductors for neutral."

DESIGN PROBLEM 22-7: What is the load in VA and amps for a mobile home with the following loads?

General lighting and receptacle load

- 800 sq. ft dwelling unit
- 2 small appliance circuits
- 1 laundry circuit
- 8500 VA range
- 6000 VA water heater
- 540 VA disposal
- 800 VA dishwasher
- 5500 VA heating

Sizing phases = •
Sizing neutral = √

CALCULATING GENERAL LIGHTING AND RECEPTACLE LOAD

Step 1:   General lighting and receptacle load
550.18(A)(1)
800 sq. ft x 3 VA   = 2,400 VA

Step 2:   Small appliance and laundry load
550.18(A)(2) and (A)(3)
1500 VA x 2   = 3,000 VA
1500 VA x 1   = 1,500 VA
Total load   = 6,900 VA

Step 3:   Applying demand factor
550.18(A)(5)
First 3000 VA x 100%   = 3,000 VA
Next 3900 VA x 35%   = 1,365 VA
Total load   **= 4,365 VA • √**

CALCULATING SPECIAL APPLIANCE LOAD

Step 1:   Applying demand factors for phases A and B
550.18(B)(2), (B)(3), and (B)(4)
Water heater   = 6,000 VA
Dishwasher   = 800 VA √
Disposal   = 540 VA √
Heating   = 5,500 VA
Total load   **= 12,840 VA •**

CALCULATING RANGE LOAD

Step 1:   Applying demand factor for phases A and B
550.18(B)(5)
8500 VA x 80%   = 6,800 VA •
6800 VA x 70%   = 4,760 VA √

CALCULATING LARGEST MOTOR LOAD

Step 1:   Selecting largest motor load for phases A and B
550.18(B)(3)
540 VA x 25%   = **135 VA • √**

CALCULATING LOAD FOR PHASES A AND B (add all •)

- General lighting load   = 4,365 VA •
- Special appliance load   = 12,840 VA •
- Range load   = 6,800 VA •
- Largest motor load   = 135 VA •
Total load   **= 24,140 VA**

FINDING AMPS FOR PHASES A AND B

$I = VA \div V$
$I = 24{,}140 \text{ VA} \div 240 \text{ V}$
I = 101 A

CALCULATING NEUTRAL LOAD (add all √)

- General lighting load   = 4,365 VA √
- Dishwasher load   = 800 VA √
- Disposal load   = 540 VA √
- Range   = 4,760 VA √
- Largest motor   = 135 VA √
Total load   **= 10,600 VA**

FINDING AMPS FOR NEUTRAL

$I = VA \div V$
$I = 10{,}600 \text{ VA} \div 240 \text{ V}$
I = 44 A

Finding size conductors for phases A and B
Table 310.16
Phases A and B
101 A requires 2 AWG THWN cu.
Neutral
44 A requires 8 AWG THWN cu.

310.15(B)(6)
Phases A and B
101 A requires 3 AWG THWN cu.
Neutral
44 A requires 8 AWG THWN cu.

DESIGN PROBLEM 22-8: What is the load in VA and amps for 28 mobile homes with the following loads?

Mobile homes
- 28 units
- 17,000 VA each based upon calculated load

FINDING TOTAL VA LOAD

Step 1: Finding VA load
550.31
Mobile home = 17,000 VA

Step 2: Calculating VA load
550.31(2) and **Table 550.31**
17,000 VA x 28 x 24% **= 114,240 VA**

SIZING CONDUCTORS

Step 1: Finding VA for phases A and B
550.31
Total VA = 114,240 VA

Step 2: Calculating A for phases A and B
550.31
A = 114,240 VA ÷ 240 V
A = 476

Step 3: Selecting size conductors for phases A and B
Table 310.16 and **310.4**
Phases A and B (Parallel)
A = 476 ÷ 3 (No. runs per phase)
A = 158.7 (round up)
159 A requires 2/0 AWG THWN cu.

SIZING NEUTRAL BASED UPON CALCULATED LOAD

Step 1: Finding VA for neutral
550.31
Total VA = 114,240 VA

Step 2: Calculating A for neutral
550.31
A = 114,240 VA ÷ 240 V
A = 476

Step 3: Applying demand factors
220.61(B)
First 200 A x 100% = 200 A
Next 276 A x 70% = 193 A
Total load = 393 A

Step 4: Selecting conductors for phases A and B and neutral
Table 310.16 and **310.4**
Neutral (parallel)
A = 393 A ÷ 3 (No. runs per phase)
A = 131
131 A requires 1/0 AWG THWN cu.

SIZING ELEMENTS
DESIGN PROBLEM 22-1

Using the service amps of Design Problem 22-1 on page 22-19 of this Chapter, size the elements of the following:

- Size THWN copper conductors per **Table 310.16**
- Size overcurrent protection device
- Size panelboard
- Size conduit using RMC
- Size grounding electrode conductor (cu.)
- Size supplementary ground
- Size grounded (neutral) conductor

SIZING THE CONDUCTORS FOR THE SERVICE PER TABLE 310.16 AND TABLE 310.15(B)(6)

Step 1: Calculated loads
Phases = 223 A
Neutral = 95 A

Step 2: Selecting conductors
Table 310.16
Phases = 223 A requires 4/0 AWG THWN cu.
Neutral = 95 A requires 3 AWG THWN cu.

Step 3: Applying **Table 310.15(B)(6)**
Phases = 223 A requires 3/0 AWG THWN cu.
Neutral = 95 A requires 3 AWG THWN cu.

Solution: The ungrounded (phase) conductors are 4/0 AWG THWN copper conductors and the grounded (neutral) conductor is a 2 AWG copper conductor, per 250.24(C)(1).

Note that 4/0 AWG THWN copper conductors were selected based upon Step 2 above and not Step 3.

SIZING THE OVERCURRENT PROTECTION DEVICE FOR THE SERVICE CONDUCTORS PER TABLE 310.16 AND 110.14(C)(1) AND (C)(2)

Step 1: Amperage of load or conductor
240.4(A) - (G) and **240.6(A)**
223 A requires 225 A OCPD

Solution: The size OCPD required is 225 amps.

SIZING THE PANELBOARD FOR SERVICE PER 240.4(A) - (G); 408.36

Step 1: Amperage of load or conductor
240.4(A) - (G) and **240.6(A)**
223 A requires 225 A panelboard

Solution: The size panelboard required is 225 amps.

SIZING THE CONDUIT FOR THE SERVICE CONDUCTORS USING RIGID METAL CONDUIT PER TABLES 5 AND 4 TO CHAPTER 9

Step 1: Different size conductors (max. size)
Table 5 and Table 4 to Chapter 9
4/0 AWG THWN
.3237 sq. in. x 2 = .6474
2 AWG THWN
.1158 sq. in. x 1 = .1158
Total sq. in. area = .7632

Step 2: Selecting size
Table 4 to Chapter 9
.7632 sq. in requires 1-1/2" (41) conduit

Solution: The size rigid metal conduit required is 1-1/2 in. (41).

SIZING THE GEC TO GROUND THE SERVICE TO A METAL WATER PIPE PER 250.66 AND TABLE 250.66

Step 1: Size of ungrounded (phase) conductors (max. size)
250.66 and **Table 250.66**
4/0 AWG THWN requires 2 AWG cu.

Solution The size GEC required is 2 AWG cu.

SIZING THE GEC REQUIRED TO GROUND THE SERVICE TO A DRIVEN ROD TO SUPPLEMENT THE METAL WATER PIPE PER 250.53(D)(2), 250.104(A), AND 250.52(A)(1)

Step 1: Size of ungrounded (phase) conductors (max. size)
250.66(A)
4/0 AWG THWN requires 6 AWG cu.

Solution: The size of the grounding electrode conductor is 6 AWG cu.

**SIZING THE GROUNDED (NEUTRAL) CONDUC-
TOR REQUIRED TO CLEAR A GROUND FAULT
PER 220.61 AND 250.24(B)(1)**

Step 1: Size of ungrounded (phase)
conductors
250.24(C)(1), 220.61(B)(2), and
Table 250.66
4/0 AWG requires 2 AWG cu.

**Solution: The size grounded service
conductor is required to be 2 AWG
copper.**

Note 1: The calculated grounded (neutral) per **220.61(B)(2)**
produces the smaller grounded (neutral) conductor.
However, such grounded (neutral) conductors shall be sized
per Table **250.66**, because it is used as a grounded (neutral)
conductor and equipment grounding conductor. See the
neutral calculation of Design Problem 22-1 on Page 22-19.

Note 2: For selecting the proper size grounded (neutral)
conductor to carry the amount of fault current that it might
be called on to carry, see pages 11-15 through 11-18 of
Chapter 11.

Chapter 22. Residential Calculations

Section   Answer

___________ ___________   1. The general lighting load for a dwelling unit shall be determined by multiplying the square footage by _____ VA.
        (a) 1   (b) 2
        (c) 3   (c) 3 1/2

___________ ___________   2. All small appliance and laundry loads shall be calculated at _____ VA to determine the size feeder conductors and elements to size the service.
        (a) 1000   (b) 1500
        (c) 1800   (c) 2000

___________ ___________   3. The VA rating for the overcurrent protection device for special appliance loads shall be calculated at 100 percent for noncontinuous duty and _____ percent for continuous duty, with demand loads determined by the kilowatt rating listed on the nameplate times a percentage.
        (a) 100   (b) 110
        (c) 115   (c) 125

___________ ___________   4. For four or more fixed appliances that are grouped into a special appliance load can be found by adding wattage ratings from appliance nameplates and multiplying the total wattage (volt-amps) by _____ percent to obtain the demand load.
        (a) 75   (b) 80
        (c) 100   (c) 125

___________ ___________   5. The demand load for household dryers shall be calculated at _____ kVA or the nameplate rating, whichever is greater.
        (a) 2   (b) 3
        (c) 4   (c) 5

___________ ___________   6. The heating and A/C loads shall be calculated at _____ percent and the smaller of the two is dropped.
        (a) 75   (b) 100
        (c) 115   (c) 125

___________ ___________   7. The optional calculation method shall only be permitted to be applied when an ampacity of at least _____ amps is applied to the service conductors.
        (a) 50   (b) 75
        (c) 100   (c) 175

___________ ___________   8. When adding appliance loads for the optional calculation method, the total VA rating of the load shall be determined at _____ percent.
        (a) 50   (b) 70
        (c) 80   (c) 100

___________ ___________   9. The VA rating for the overcurrent protection device for special appliance loads shall be calculated at 125 percent for continuous duty and _____ percent for noncontinuous duty, if a demand factor shall not be applied based on a number.
        (a) 100   (b) 115
        (c) 125   (c) 150

_______ _______ **10.** The fixed appliance load for three or less fixed appliances shall be determined by adding wattage (volt-amps) values by ______ percent.
(a) 75 (b) 100
(c) 125 (c) 150

_______ _______ **11.** Dryer equipment of four or fewer units shall be calculated at ______ percent of the nameplate rating or 5000 VA, whichever is greater.
(a) 70 (b) 75
(c) 80 (c) 100

_______ _______ **12.** The largest motor load shall be calculated at ______ percent, for calculating the load of one or more motors with other loads.
(a) 50 (b) 100
(c) 125 (c) 150

_______ _______ **13.** The general lighting load for a dwelling unit shall be determined by multiplying the square footage by ______ VA.
(a) 1 (b) 2
(c) 3 (c) 4

_______ _______ **14.** At least two small appliance circuits shall be required to supply receptacle outlets located in the:
(a) kitchen (b) pantry
(c) dining room (c) all of the above

_______ _______ **15.** The demand load for four or more fixed appliances grouped into a special appliance load is found by adding wattage ratings from the appliance nameplates and multiplying the total wattage (volt-amps) by:
(a) 65 percent (b) 75 percent
(c) 80 percent (c) 100 percent

_______ _______ **16.** The demand load for a household dryer shall be calculated by the nameplate rating or ______ kVA, whichever is greater.
(a) 2 (b) 3
(c) 4 (c) 5

_______ _______ **17.** The heating and A/C load shall be calculated at ______ percent and the smaller of the two dropped.
(a) 80 (b) 100
(c) 125 (c) 150

_______ _______ **18.** Dryer equipment shall be permitted to have a demand factor of ______ percent for 8 dryers.
(a) 60 (b) 65
(c) 70 (c) 75

_______ _______ **19.** A service supplying cooking equipment shall be permitted to have an additional demand factor of ______ percent.
(a) 60 (b) 65
(c) 70 (c) 75

_______ _______ **20.** Fixed electric space-heating loads shall be calculated at ______ percent of the total connected load.
(a) 70 (b) 100
(c) 115 (c) 125

21. What is the general lighting and receptacle load in VA for a 50 ft x 50 ft dwelling unit?

22. What is the small appliance load in VA for 2 small appliance circuits and 1 laundry circuit?

23. What is the largest load between the heating and A/C load in VA of a 20 kW heating unit and a 6000 VA A/C unit?

24. What is the demand load in VA for a 9 kW range?

25. What is the demand load in VA for a 3-1/2 kW cooktop?

26. What is the demand load in VA for two 3-1/2 kW cooktops?

27. What is the demand load in VA for a 8-3/4 kW oven?

28. What is the demand load in VA for a 18 kW range? (Use **Note 1** to **Table 220.19**)

29. What is the demand load in VA for a 8 kW, 16 kW, and 24 kW range? (Use **Note 2** to **Table 220.19**)

30. What is the demand load in VA for a 8 kW and 10 kW oven and a 10 kW cooktop? (Use **Note 4** to **Table 220.19**)

31. What is the demand load for the fixed (special) appliance load in VA for a 6000 VA A/C unit, 20,000 VA heating unit, 5000 VA water heater, 10,000 VA oven, 8000 VA range, 1400 VA water pump, 1000 VA compactor, 800 VA blower motor, 1200 VA microwave, 900 VA disposal, and a 4500 VA dryer?

32. What is the demand load in VA for a 4500 VA dryer?

33. What is the demand load in VA for a 5500 VA dryer?

34. What is the demand load in VA for 6 dryers rated at 6000 VA?

35. What is the largest motor load in VA for a 5 HP, 230 volt, single-phase motor?

36. What is the amp load for a 3000 sq. ft dwelling unit with a 240 volt, single-phase service with the following loads? (use the standard calculation)

- 2 small appliance circuits
- 1 laundry circuit
-   6,000 VA A/C unit   240 volt, single-phase
- 35,000 VA heating unit   240 volt, single-phase
-   5000 VA water heater   240 volt, single-phase
- 12,000 VA oven   240 volt, single-phase
-   9000 VA cooktop   240 volt, single-phase
-   5000 VA dryer   240 volt, single-phase
-   2400 VA sump pump   240 volt, single-phase
-   1000 VA pool pump   240 volt, single-phase
-     900 VA microwave   120 volt, single-phase
-   1600 VA dishwasher   120 volt, single-phase
-   1050 VA disposal   120 volt, single-phase
-     900 VA compactor   120 volt, single-phase

_________ _________   **(a)** What size (minimum and maximum) THWN copper conductors are required.

_________ _________   **(b)** What size overcurrent protection device based on calculated load is required.

_________ _________   **(c)** What size panelboard is required.

_________ _________   **(d)** What size RMC conduit is required.

_________ _________   **(e)** What size cu. grounding electrode conductor is required.

_________ _________   **37.** What is the amp load for a 3000 sq. ft dwelling unit with a 240 volt, single-phase service with the following loads? (use the optional calculation)

     • 2 small appliance circuits
     • 1 laundry circuit

• 6 000 VA A/C unit	240 volt, single-phase
• 35,000 VA heating unit	240 volt, single-phase
• 5000 VA water heater	240 volt, single-phase
• 12,000 VA oven	240 volt, single-phase
• 9000 VA cooktop	240 volt, single-phase
• 5000 VA dryer	240 volt, single-phase
• 2400 VA sump pump	240 volt, single-phase
• 1000 VA pool pump	240 volt, single-phase
• 900 VA microwave	120 volt, single-phase
• 1600 VA dishwasher	120 volt, single-phase
• 1050 VA disposal	120 volt, single-phase
• 900 VA compactor	120 volt, single-phase

_________ _________   **(a)** What size (max.) THWN copper conductors are required?

_________ _________   **(b)** What size overcurrent protection device based on calculated load is required?

_________ _________   **(c)** What size panelboard is required?

_________ _________   **(d)** What size RMC conduit is required?

_________ _________   **(e)** What size cu. grounding electrode conductor is required?

Note: Use the standard calculation to size the grounded (neutral) conductor. (See problem 36)

_________ _________   **38.** From the following loads in an existing dwelling unit, determine if a 25 amp, A/C unit load (calculated load) can be added to the service conductors. The existing service conductors are 3 AWG THWN copper conductors. (use the existing optional calculation)

     • 2000 sq. ft dwelling unit
     • 2 small appliance circuits
     • 1 laundry circuit

• 11,000 VA range	240 volt, single-phase
• 5000 VA dryer	240 volt, single-phase
• 1200 VA compactor	120 volt, single-phase
• 800 VA disposal	120 volt, single-phase
• 1400 VA dishwasher	120 volt, single-phase

39. What is the amp load for 20-1000 sq. ft multifamily dwelling units with   _________ _________
a 240 volt, single-phase service with the following loads? (Use the standard
calculation – parallel service conductors, six times per phase)

    22-33

* 2 small appliance circuits
* 1 laundry circuit
* 20 - 11,000 VA ranges   240 volt, single-phase
* 20 -   5000 VA water heaters   240 volt, single-phase
* 20 - 25,000 VA heating units   240 volt, single-phase
* 20 -   1000 VA dishwashers   120 volt, single-phase
* 20 -   1200 VA disposals   120 volt, single-phase

40. What is the amp load for 20-1000 sq. ft multifamily dwelling units with a 240   _________ _________
volt, single-phase service with the following loads? (Use the optional calculation
– parallel service conductors, six times per phase)

* 2 small appliance circuits
* 1 laundry circuit
* 20 - 11,000 VA ranges   240 volt, single-phase
* 20 -   5000 VA water heaters   240 volt, single-phase
* 20 - 25,000 VA heating units   240 volt, single-phase
* 20 -   1000 VA dishwashers   120 volt, single-phase
* 20 -   1200 VA disposals   120 volt, single-phase

41. What is the amp load for a 1000 sq. ft mobile home with a 240 volt, single-   _________ _________
phase service with the following loads?

* 2 small appliance circuits
* 1 laundry circuit
*   8500 VA range   240 volt, single-phase
*   5500 VA water heater 240 volt, single-phase
*   600 VA disposal   120 volt, single-phase
*   800 VA dishwasher   120 volt, single-phase
*   6000 VA heating unit   240 volt, single-phase

42. What is the amp load for 26 mobile homes with following loads?   _________ _________
* 26 units
* 15,000 VA each based on calculated load
* Per **550.18** in the NEC

23

Commercial Calculations

Commercial facilities such as offices, banks, stores, and restaurants have diverse loads. These loads are classified as continuous or noncontinuous, or such loads may cycle on and off, allowing demand factors to be applied.

The procedure and manner in which the lighting, receptacle, and equipment loads are used in the electrical system determines how they are classified.

Loads shall be calculated based upon the type of occupancy and the requirements of the equipment supplied. Either the standard or optional calculation is utilized to calculate the loads in volt-amps or amps to size the service equipment and associated elements.

APPLYING THE STANDARD CALCULATION
ARTICLE 220, PART III

The standard calculation may be used to calculate the VA or amp rating in order to size and select the elements of the service equipment and associated components. The selection of loads for applying the standard calculation is arranged in a different manner for commercial loads than it is for loads used in residential occupancies. There are seven loads utilized to determine the service load. Based upon conditions of use, demand factors shall be permitted to be applied to certain loads, which reduces the VA or amps.

The loads are grouped into seven individual loads, and the proper *National Electrical Code* rule is applied to each of these loads based upon use. The loads are grouped and classified as follows:

(1) Lighting loads
  • General lighting load per **220.12**
  • Show window load per **220.43(A)**
  • Lighting track load per **220.43(B)**
  • Low voltage lighting load per **Article 411**
  • Outside lighting load per **230.42(A)(1)** and **(A)(2)**
  • Outside sign lighting load per **220.14(F)**

(2) Receptacle loads
  • General purpose receptacle load
    [Noncontinuous per **220.14(I)**, **Table 220.44**,
    and **230.42(A)(1)**; **(A)(2)**]
  • General purpose receptacle load
    [Calculated per **220.14(I)**, **230.42(A)(1)**, **(A)(2)**
    and **Table 220.44**]
  • Multioutlet assembly load
    [Used simultaneously per **220.14(H)(2)**]
    [Not used simultaneously per **220.14(H)(1)**]

(3) Special appliance loads
  • Noncontinuous load per **230.42(A)(1)** and
    220.14(A)
  • Continuous load per **230.42(A)(1)**
  • Demand factor per various Sections of the NEC

(4) Compressor loads
  • Refrigeration per **440.34** and **230.42(A)(1)**
  • Cooling per **440.34** and **230.42(A)(1)**

(5) Motor loads
  • Single-phase per **430.25**
  • Three-phase per **430.25**

(6) Heat or A/C loads
  • Heating per **220.51**
  • A/C per **440.34**
  • Heat pump per **440.34**

(7) Largest motor loads
  • Taken from loads (4), (5), or (6) per **220.50**

The seven loads shall be calculated at continuous operation or noncontinuous operation. Demand factors shall be applied to the noncontinuous operated loads by specific Sections of the NEC.

Generally, no demand factors shall be permitted to be applied to the loads for commercial occupancies, as the loads are usually used at continuous operation. However, **Table 220.42** permits the general lighting loads in hospitals, hotels, motels, and warehouses to have demand factors applied because of load diversity.

LIGHTING LOADS
ARTICLES 220, 410, AND 411

Lighting loads are the first of the loads to be calculated. Six lighting loads shall be calculated to derive the total lighting load in commercial facilities. These six loads are as follows:

  • General lighting loads per **Table 220.12**
  • Show window loads per **220.43(A)**
  • Lighting track loads per **220.43(B)**
  • Low voltage lighting per **Article 411**
  • Outside lighting load per **230.42(A)(1)** and **(A)(2)**
  • Sign lighting load per **220.14(F)**

Each lighting load shall be calculated by the operation in which it is used. Loads shall be calculated at continuous or noncontinuous operation, and other lighting loads shall be permitted to have demand factors applied, where permitted by the NEC.

GENERAL LIGHTING LOADS
220.12 AND TABLE 220.12

General lighting loads consist of lighting units installed inside the facility. This load shall be calculated by the VA rating times sq. ft from **Table 220.12** based on the type of commercial occupancy.

The general lighting load found in listed occupancies is calculated according to the number of VA per sq. ft and not based upon the number of outlets served. When fluorescent lighting is used, either the area load of VA per sq. ft of the facility or the total connected load of each ballast is used, whichever is greater in rating.

Lighting loads – either the VA per sq. ft or individual units shall be calculated at 100 percent for noncontinuous operation or at 125 percent for continuous operation.

For example: A noncontinuous lighting load of 60 amps and a continuous load of 100 amps is calculated as follows:

Step 1: Calculating load
230.42(A)(1)
60 A x 100% = 60 A
100 A x 125% = 125 A
Total load   = 185 A

Solution: The calculated load for the lighting is 185 amps.

NONCONTINUOUS AND CONTINUOUS OPERATION
230.208(B), 230.42(A)(1), AND (A)(2)

The requirements for derating the overcurrent protection device by 80 percent for continuous operation is no longer found in 384-16(d), as it was in the 1999 NEC. For overcurrent protection devices rated over 600 volts, the rules are listed in **230.208(B)**, and under certain conditions the 80 percent derating rule does not apply. For feeders rated over 600 volts, see **215.2(B)**.

To apply the 80 percent derating rule for loads rated at 600 volts or less, **230.42(A)(1)** requires such loads to be calculated at 125 percent (1 ÷ 80% = 1.25) if they operate for three hours or more. Loads operating for a period of less than three hours shall be calculated at 100 percent per **230.42(A)(1)**. The ampacity of the conductors as well as the overcurrent protection device shall be calculated and sized at 125 percent and 100 percent or a combination of both based on the operation of such loads. [For feeders, see **215.2(A)(1)**.]

Section **230.90(A)** with **Ex.s** refers to various Sections of the NEC to calculate the size of overcurrent protection devices. Loads that are not motor related shall be calculated based upon their operation procedures, that are listed in **230.42(A)(1)** and **(A)(2)**. **(See Figure 23-1)**

LISTED OCCUPANCIES
220.12 AND TABLE 220.12

Table **220.12** is used to select the VA rating for listed occupancies. For example, a store building shall be calculated at 3 VA per sq. ft and an office building shall be calculated at 3.5 VA per sq. ft Other types of occupancies have different VA ratings per sq. ft based on the type of occupancy involved.

For example: What is the load in VA for the general-purpose lighting load in an 8000 sq. ft office per **Table 220.12**?

Step 1: Calculating load
Table 220.12
8000 sq. ft x 3.5 = 28,000 VA

Solution: The general purpose lighting load is 28,000 VA. Note: If the number of receptacle outlets is not known, an extra 1 VA per sq. ft shall be added to the above.

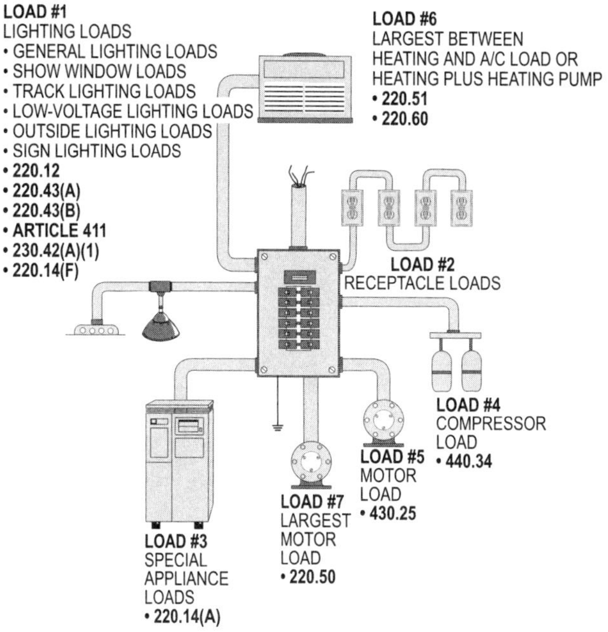

Figure 23-1. As outlined above, there are seven loads to be calculated when applying the standard calculation to determine the service load in VA or amps.

See Figure 23-2 for a detailed illustration of calculating the lighting load for a listed occupancy.

UNLISTED OCCUPANCY
220.18, 220.14(D), AND (L)

If an occupancy is not listed in **Table 220.12**, the general-purpose lighting load shall be calculated in the following manner.

- Lamps for incandescent lighting per **220.14(L)**
- Lamps for recessed luminaires per **220.14(D)**
- Ballasts for electric discharge lighting per **220.18(B)**
- Track lighting unit per **220.43(B)**
- Low voltage systems per **Article 411**

An unlisted VA rating shall be calculated at 125 percent for sizing the overcurrent protection device and conductors for continuous operation and at 100 percent for noncontinuous operated loads per **230.42(A)(1)**.

For example: What is the load in VA for 60, 120 volt, lighting ballasts rated at 1.5 amps each and used for 12 hours a day?

Step 1: Calculating load in A
220.18(B)
60 x 1.5 A = 90 A

Step 2: Calculating continuous load
230.42(A)(1)
90 A x 125% = 112.5 A

Step 3: Calculated VA
112.5 A x 120 V = 13,500 VA

Solution: The lighting load for the unlisted occupancy is 13,500 VA.

See Figure 23-3 for a detailed illustration of calculating the lighting load for an unlisted occupancy.

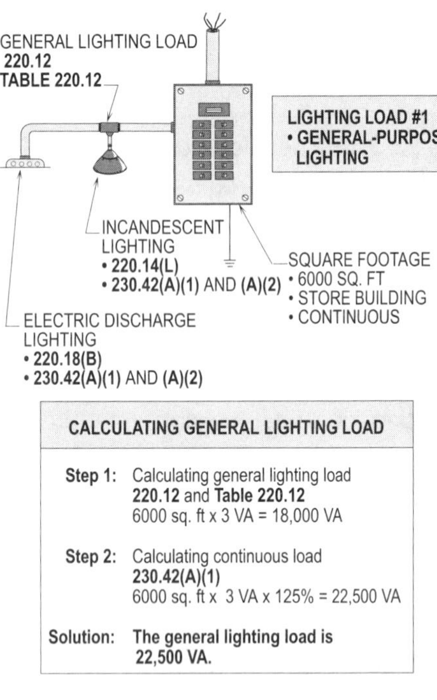

Figure 23-2. The general lighting load (listed occupancy) shall be calculated by multiplying the sq. ft of the facility by the VA per sq. ft per **Table 220.12**.

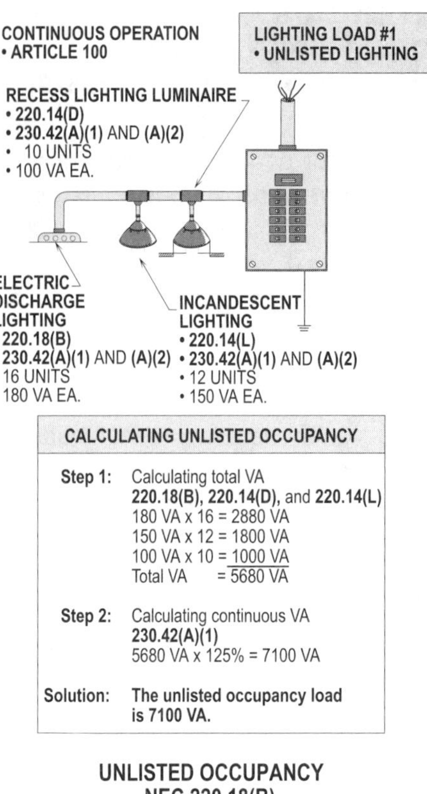

Figure 23-3. The unlisted lighting load shall be permitted to be calculated by multiplying the VA of each lighting load by the number and increasing this value by 125 percent if used for three hours or more.

SHOW WINDOW LIGHTING LOAD 220.43(A)

The lighting load in VA for the show window shall be calculated by multiplying the linear feet of the show window by 200 VA per foot. Such lighting load shall be calculated at 100 percent for noncontinuous operation and at 125 percent for continuous operation per **230.42(A)(1)**. Conductors and overcurrent protection devices shall be increased to comply with **240.3** and **240.4** in the NEC.

For example: What is the lighting load in VA for a 80 ft show window used at noncontinuous or continuous operation?

Step 1: Calculating noncontinuous load
220.43(A) and **230.42(A)(1)**
80' x 200 VA x 100% = 16,000 VA

Step 2: Calculating continuous load
220.43(A) and **230.42(A)(1)**
80' x 200 VA x 125% = 20,000 VA

Solution: The noncontinuous load is 16,000 VA and the continuous load is 20,000 VA.

If the number of lighting outlets is known, the VA rating of each luminaire shall be multiplied by 125 percent for sizing the show window load. If the VA is not known, each outlet shall be calculated at 180 VA times 125 percent to obtain the lighting load in VA for the show window. Note that the greater of either the 200 VA per linear foot or each individual calculation shall be used.

See Figure 23-4 for a detailed illustration of calculating the show window lighting load.

TRACK LIGHTING LOAD
220.43(B)

The lighting load in VA for lighting track shall be calculated by multiplying the lighting track by 150 VA and dividing by 2. Such VA rating shall be multiplied by 100 percent or 125 percent based upon noncontinuous or continuous operation.

For example: What is the load in VA for 80 ft of lighting track used at noncontinuous or continuous operation?

Step 1: Calculating noncontinuous load
220.43(B) and **230.42(A)(1)**
80' ÷ 2' x 150 VA x 100% = 6000 VA

Step 2: Calculating continuous load
220.43(B) and **230.42(A)(1)**
80' ÷ 2' x 150 VA x 125% = 7500 VA

Solution: The noncontinuous load is 6000 VA and the continuous load is 7500 VA.

See Figure 23-5 for a detailed illustration of calculating the track lighting load.

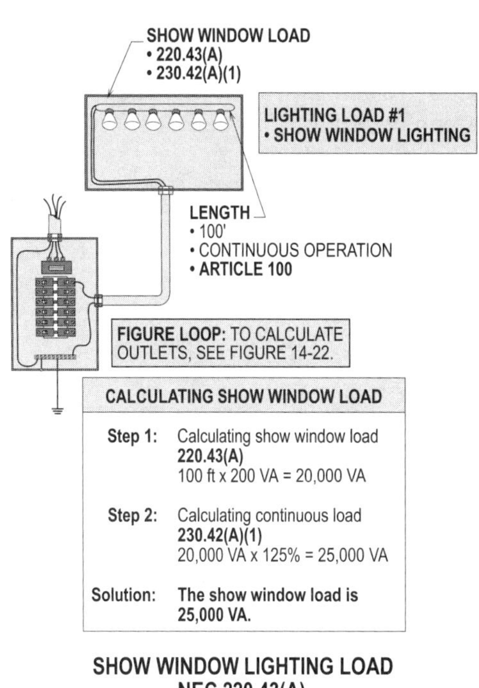

Figure 23-4. The show window load shall be calculated by multiplying the linear foot by 200 VA. This value shall be increased by 125 percent since such load is continuous.

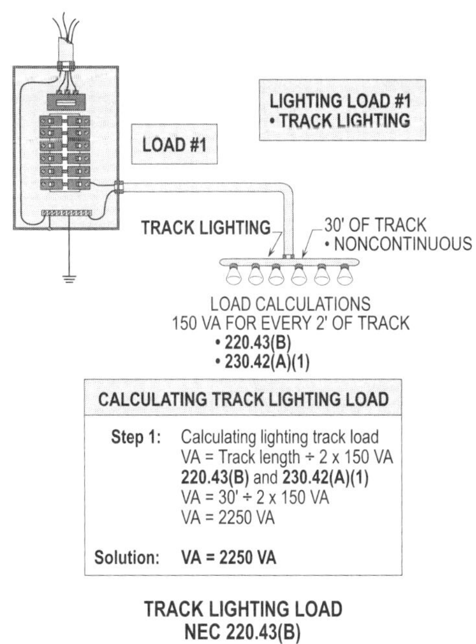

Figure 23-5. The lighting track load shall be calculated by dividing the length of the track by 2 and multiplying by 150 VA. This value shall be increased by 125 percent if it is used for three hours or more.

LOW VOLTAGE LIGHTING LOAD
ARTICLE 411, 230.42(A)(1), AND (A)(2)

The lighting load in VA for low voltage lighting systems shall be calculated by multiplying the FLA of the isolation transformer by 100 percent for noncontinuous operation and by 125 percent for continuous operation.

For example: What is the load in VA for a low voltage lighting system supplied by an isolation transformer with a FLA of 50 amps used at noncontinuous or continuous operation?

Step 1: Calculating noncontinuous load
Article 411 and **230.42(A)(1)**
50 A x 100% = 50 A

Step 2: Calculating continuous load
Article 411 and **230.42(A)(1)**
50 A x 125% = 62.5 A

Solution: The noncontinuous load is 50 amps and the continuous load is 62.5 amps.

See Figure 23-6 for a detailed illustration of calculating the low voltage lighting load.

OUTSIDE LIGHTING LOAD
220.18(B), 230.42(A)(1), AND (A)(2)

The lighting load in VA for outside lighting loads shall be calculated by multiplying the VA rating of each lighting unit by 100 percent for noncontinuous operation and by 125 percent for continuous operation.

For example: What is the lighting load in VA for 30 continuous operated luminaires with a 75 VA ballast in each unit and for 10 noncontinuous operated luminaires with each ballast having a rating of 75 VA?

Step 1: Calculating load
220.18(B) and **230.42(A)(1)**
75 x 30 x 125%   = 2812.5 VA
75 x 10 x 100%   =   750   VA
Total load   = 3562.5 VA

Solution: The total outside lighting load is 3562.5 VA.

See Figure 23-7 for a detailed illustration for calculating the outside lighting load at continuous operation.

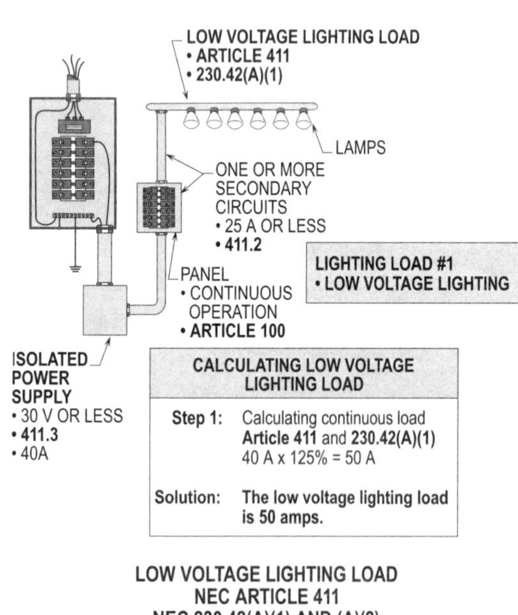

Figure 23-6. The low voltage lighting load shall be calculated by multiplying the FLA of the isolation transformer by 125 percent if used for three hours or more.

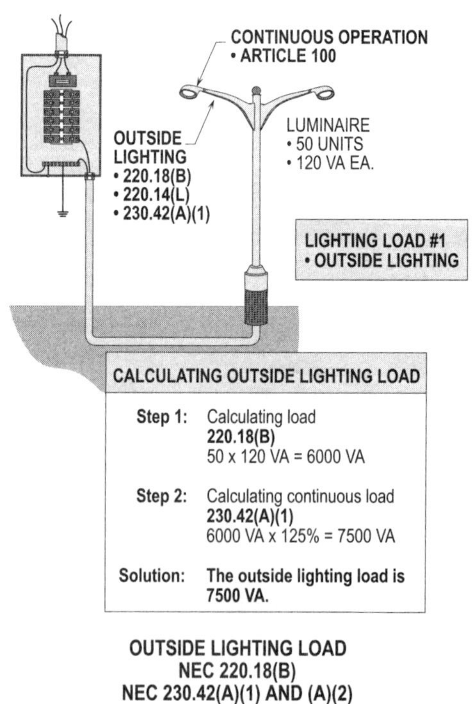

Figure 23-7. The outside lighting load shall calculated by multiplying the number of ballasts by 125 percent if the load operates for three hours or more.

OUTSIDE SIGN LIGHTING LOAD
220.14(F), 230.42(A)(1), AND (A)(2)

The lighting loads for signs shall be calculated according to the ground floor footage accessible to pedestrians in a commercial occupancy or facility. Occupancies with grade level access for pedestrians shall have a minimum of 1200 VA provided for a sign lighting load. This VA rating shall be multiplied by 125 percent for signs operating for three hours or more and by 100 percent for those operating less than three hours.

See Figure 23-8 for a detailed illustration for calculating the outside sign lighting load.

For example: What is the load in VA for 48 general purpose receptacles used to serve noncontinuous and continuous related loads?

Step 1: Calculating noncontinuous load
220.14(I) and **230.42(A)(1)**
180 VA x 48 x 100% = 8640 VA

Step 2: Calculating continuous load
220.14(I) and **230.42(A)(1)**
180 VA x 48 x 125% = 10,800 VA

Solution: **The noncontinuous load is 8640 VA and the continuous load is 10,800 VA.**

See Figure 23-9 for a detailed illustration for calculating the general purpose receptacle load.

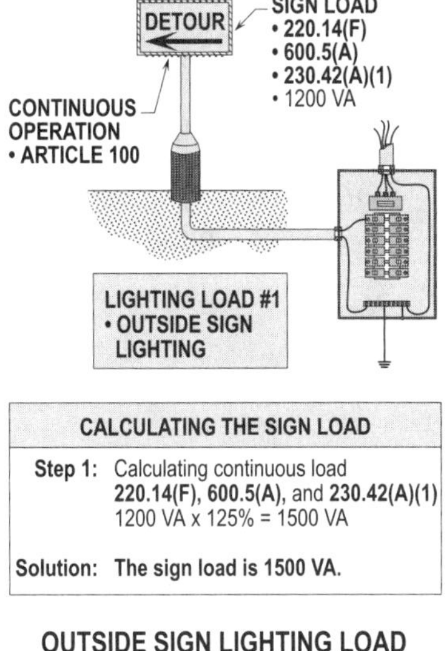

Figure 23-8. The outside sign lighting load shall be calculated by multiplying the VA of the sum by 125 percent if it operates for three hours or more.

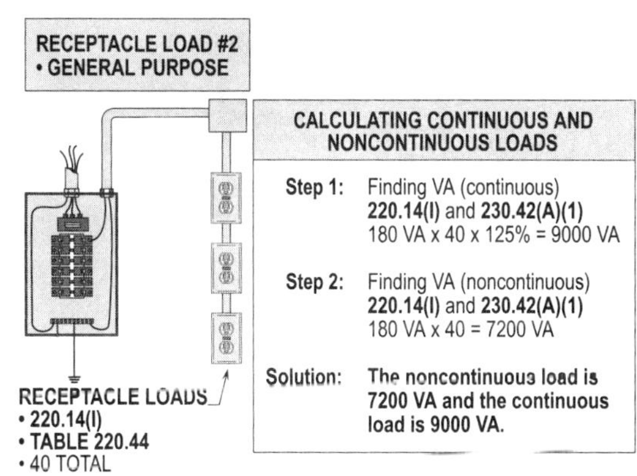

Figure 23-9. The general purpose receptacle load shall be calculated by multiplying the VA rating of each receptacle by the number and increasing this value by 125 percent if used for three hours or more.

RECEPTACLE LOADS
220.14(I) AND TABLE 220.44

Receptacle loads are the second group of loads to be calculated. Such loads are divided into two subgroups as follows:

- General purpose receptacle outlets
- Multioutlet assemblies

Each load in the subgroup shall be calculated differently to derive total VA. The load in VA for the general purpose receptacle load shall be calculated by multiplying the number of outlets times 180 VA each, times 100 percent for noncontinuous operation and times 125 percent for continuous operation.

APPLYING DEMAND FACTORS
220.14(I) AND TABLE 220.44

General purpose receptacle outlets for cord-and-plug connected loads used at noncontinuous operation are calculated per **220.14(I)** and **Table 220.44**. Noncontinuous operated receptacles with a VA rating of 10,000 VA or less shall be calculated at 100 percent. If the VA rating of the receptacle load exceeds 10,000 VA, a demand factor of 50 percent shall be permitted to be applied to all VA exceeding 10,000 VA per **Table 220.44**.

For example: What is the VA rating for 125 general purpose receptacle outlets to cord-and-plug connect loads used at noncontinuous operation?

Step 1: Calculating load
220.14(I) and 230.42(A)(1)
125 x 180 VA = 22,500 VA

Step 2: Applying demand factors
Table 220.44
First 10,000 VA x 100% = 10,000 VA
Next 12,500 VA x 50% = 6,250 VA
Total load = 16,250 VA

Solution: The demand load is 16,250 VA.

See Figure 23-10(a) for a detailed illustration for calculating the (noncontinuous) receptacle demand load.

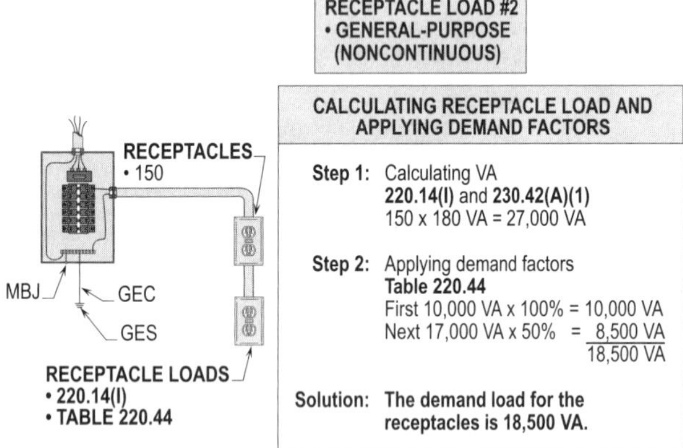

Figure 23-10(a). The demand load for general purpose receptacles (noncontinuous) shall be calculated by taking the first 10,000 VA at 100 percent and all remaining VA above 10,000 VA at 50 percent.

MULTIOUTLET ASSEMBLIES
220.14(H)(1) AND (H)(2)

For connected loads not operating simultaneously, the VA rating shall be calculated by dividing the length of the assembly by 5 ft and multiplying by 180 VA. For connected loads operating simultaneously, each foot of multioutlet assembly is multiplied by 180 VA. The fixed multioutlet assembly load shall be permitted to be added to the noncontinuous receptacle load and demand factors applied per **Table 220.44**.

For example: What is the load in VA for 100 ft of multioutlet assembly used to cord-and-plug connect loads that are not used simultaneously and used simultaneously? **[220.14(H)(1)]**

Step 1: Connecting load for nonsimultaneous use
220.14(A)(1)
VA = length ÷ 5' x 180 VA
VA = 100' ÷ 5' x 180 VA
VA = 3600

Step 2: Calculating load for simultaneous use
220.14(A)(2)
VA = length x 1' x 180 VA
VA = 100' x 1' x 180 VA
VA = 18,000

Solution: The load in VA for the nonsimultaneous load is 3600 VA and for the simultaneous load is 18,000 VA.

See Figures 23-10(b) and (c) for detailed illustrations for calculating the multioutlet assembly load.

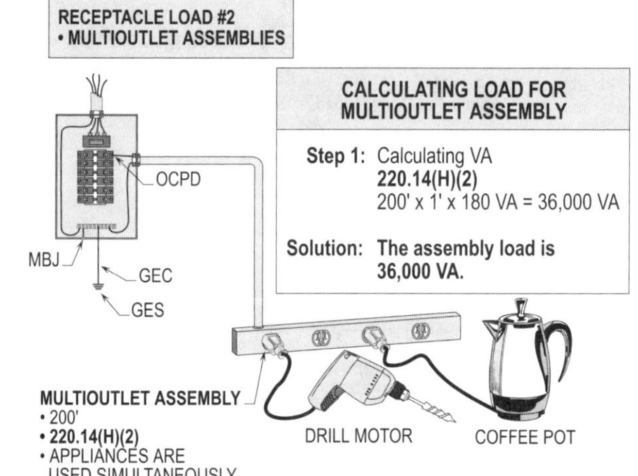

Figure 23-10(b). The multioutlet assembly load shall be calculated by multiplying the total length of the assembly (each 1 ft) by 180 VA where appliances are likely to be used simultaneously.

SPECIAL APPLIANCE LOADS
220.14(A), 230.42(A)(1), AND (A)(2)

Special appliance loads are the third group of loads to be calculated. These loads, which include calculators, processing machines, etc., are usually served by individual circuits.

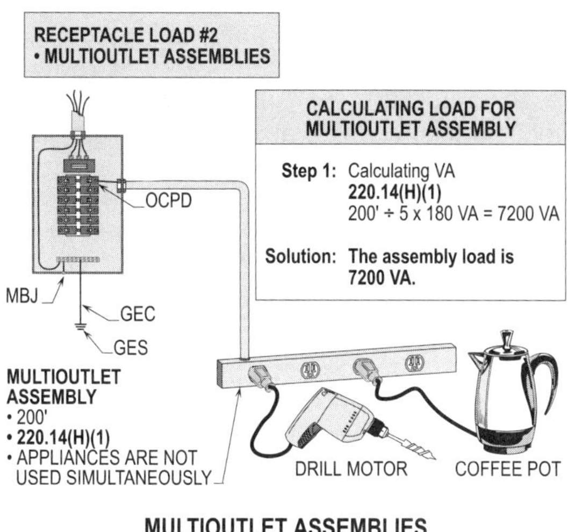

Figure 23-10(c). The multioutlet assembly load shall be calculated by multiplying the total length of the assembly by 180 VA and dividing by 5 where appliances are not likely to be used simultaneously.

CONTINUOUS AND NONCONTINUOUS OPERATION
230.42(A)(1) AND (A)(2)

The load in VA for special appliance loads shall be calculated by multiplying the VA rating of each load by 100 percent for noncontinuous operation and by 125 percent for continuous operation. To determine classification, special appliance loads operating for less than three hours shall be classified as a noncontinuous operated load. However, a special appliance load operating for three hours or more shall be classified as continuous operated load.

For example: What is the VA rating for a 208 volt, three-phase, 65 amp special appliance load operating for ten hours and supplied by an individual branch circuit?

Step 1: Calculating VA
220.5(A)
VA = V x 1.732 (360 V) x I
VA = (208 V x 1.732) x 65
VA = 23,400

Step 2: Calculating continuous load
230.42(A)(1)
23,400 VA x 125% = 29,250 VA

Solution: **The load at continuous operation is 29,250 VA.**

For example: Consider and calculate the VA rating for a special appliance load of 52 amps operating at 480 volts, three-phase, for a period of 2-1/2 hours every four hours.

Step 1: Calculating VA
220.5(A)
VA = V x 1.732 (831 V) x I
VA = (480 V x 1.732) x 52 A
VA = 43,212

Step 2: Calculating noncontinuous load
230.42(A)(1)
43,212 VA x 100% = 43,212 VA

Solution: **The load in VA for the noncontinuous load is 43,212 VA.**

See Figure 23-11 for a detailed illustration for calculating the VA or a continuous special appliance load.

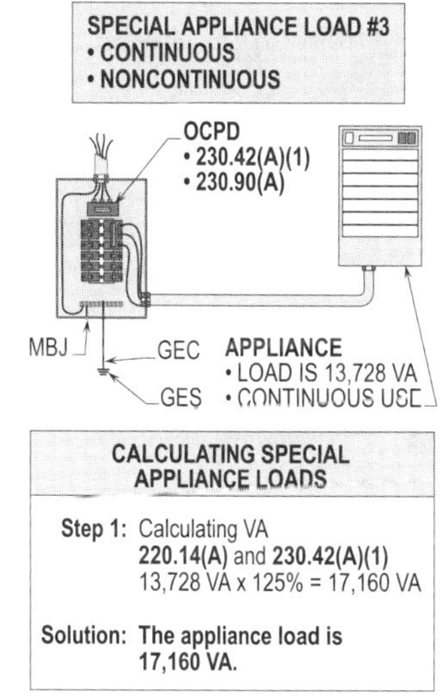

Figure 23-11. The appliance load shall be calculated by multiplying the VA of the appliance by 125 percent if used for three hours or more.

APPLYING DEMAND FACTORS
TABLE 220.56

Demand factors shall be permitted to be applied to cooking equipment in restaurants with three or more cooking units. The load in VA for cooking equipment shall be permitted to be calculated by applying the optional calculation listed in **Table 220.56**.

For example: What is the demand load in amps for 10 - 208 volt, three-phase cooking units rated 8 kW each?

Step 1: Calculating VA
220.56
8 kW x 10 = 80 kW

Step 2: Calculating amps
220.5(A)
I = (kW x 1000) ÷ (V x 1.732)
I = (80 kW x 1000) ÷ (208 V x 1.732)
I = 222 A

Step 3: Applying demand factors
Table 220.56
222 A x 65% = 144.3 A

Solution: The demand load is 144.3 A.

See **Figure 23-12** for a detailed illustration for calculating demand loads for certain types of equipment.

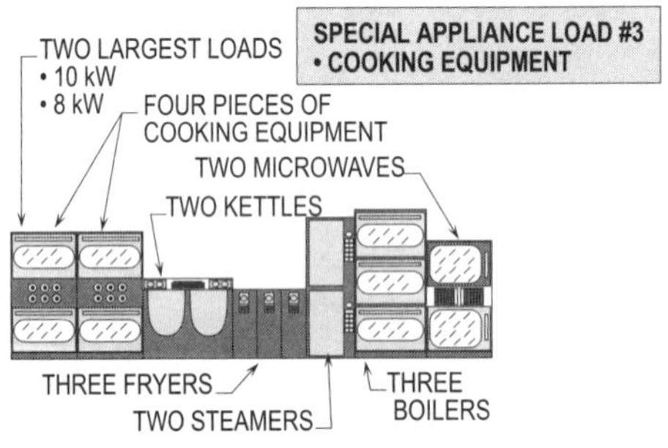

CALCULATING LOAD FOR COOKING EQUIPMENT

Step 1: Calculating percentage
Table 220.56
16 pieces allowed 65%

Step 2: Applying demand factors
Table 220.56 and 220.56
82 kW x 65% = 53.3 kVA

Solution: The demand load of 53.3 kVA is greater than the two largest loads of 18 kVA.

APPLYING DEMAND FACTORS
NEC 220.56
NEC TABLE 220.56

Figure 23-12. The demand load shall be calculated based upon the number and the percentage per **Table 220.56**.

COMPRESSOR LOADS
440.34

Compressor loads are the fourth group of loads to be calculated. Special considerations shall be applied when calculating loads for hermetic sealed compressors supplying refrigerant and cooling related equipment.

CONTINUOUS OR NONCONTINUOUS OPERATION
440.34, 230.42(A)(1), AND (A)(2)

Compressor-related equipment shall be calculated at 100 percent of their VA or amps. If one of such is the largest motor per load #7, it is added to the total calculation of all loads at 125 percent of its FLA rating.

For example: What is the load in VA for 5 compressors rated at 26.5 amps each and supplied by a 480 volt, three-phase supply?

Step 1: Calculating VA
220.5(A) and **440.34**
26.5 A x 480 V x 1.732 = 22,031 VA
22,031 VA x 5 = 110,155 VA

Step 2: Calculating continuous load
230.42(A)(1) and **440.34**
Largest motor
26.5 A x 480 V x 1.732 = 22,031 VA
22,031 Va x 25% = 5508 VA
22,031 + 5508 VA = 27,539 VA

Solution: The continuous load rating is 27,539 VA.

See **Figure 23-13** for a detailed illustration for calculating compressor related loads.

MOTOR LOADS
220.50 AND 430.24

Motor loads are the fifth group of loads to be calculated. The VA rating of motors is converted from FLA to VA by multiplying the FLA from **Table 430.248** for single-phase or **Table 430.250** for three-phase by the supply voltage.

Design Tip: Motors can be used as a single unit to drive a piece of equipment. Motors used in an approved assembly such as a processing machine are not usually considered individual motor loads.

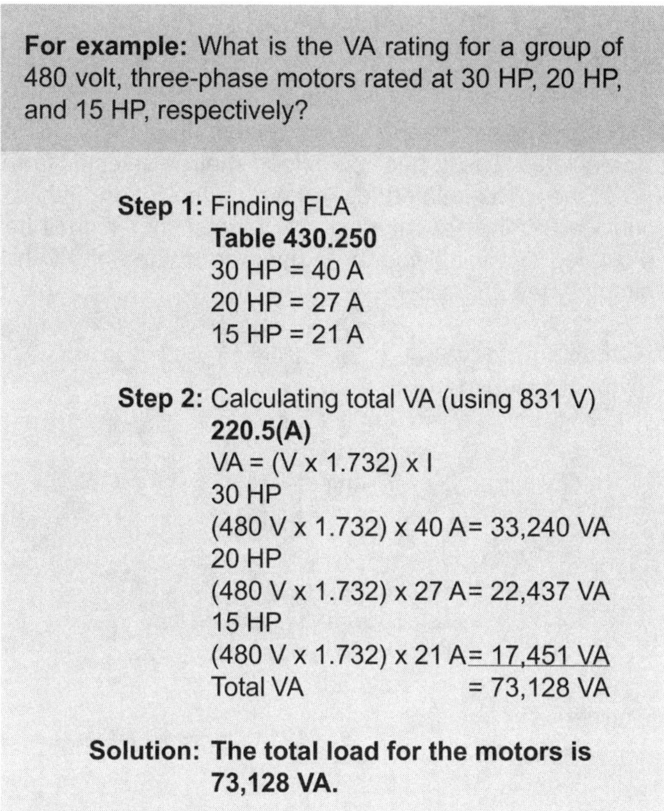

For example: What is the VA rating for a group of 480 volt, three-phase motors rated at 30 HP, 20 HP, and 15 HP, respectively?

    Step 1: Finding FLA
        Table 430.250
        30 HP = 40 A
        20 HP = 27 A
        15 HP = 21 A

    Step 2: Calculating total VA (using 831 V)
        220.5(A)
        VA = (V x 1.732) x I
        30 HP
        (480 V x 1.732) x 40 A = 33,240 VA
        20 HP
        (480 V x 1.732) x 27 A = 22,437 VA
        15 HP
        (480 V x 1.732) x 21 A = 17,451 VA
        Total VA   = 73,128 VA

    Solution: The total load for the motors is 73,128 VA.

See **Figure 23-14** for a detailed illustration for calculating individual motor loads.

HEAT OR A/C LOADS
220.60

Heating or A/C loads is the sixth group of loads to be calculated. The largest VA rating between the heating or A/C load shall be selected, and the smaller of the two loads is dropped. To determine the largest of the two loads, the VA rating of each load shall be calculated at 100 percent and the largest load of the two is selected. The load dropped is not used again in the calculation. **(See Figure 23-15)**

Design Tip: There is no need to calculate both loads to select the elements of the service equipment, for the loads are never used simultaneously in the electrical system.

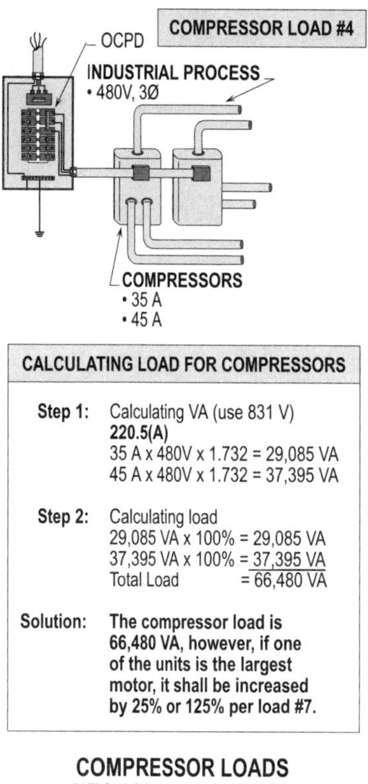

Figure 23-13. The compressor load shall be calculated at 100 percent for each compressor, and if one of such is the largest motor, it shall be increased by 25 percent (or 125 percent) and added to the other loads.

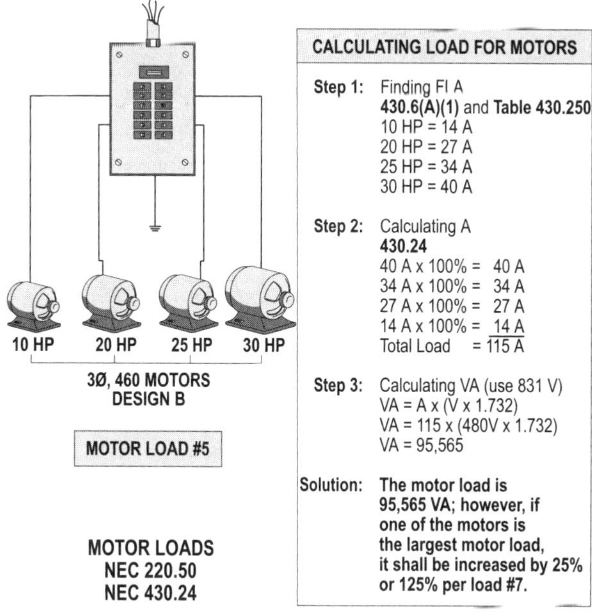

Figure 23-14. The motor load shall be calculated at 100 percent for each motor, and if one of such is the largest motor load, it shall be increased by 25 percent (or 125 percent) and added to the other loads.

For example: What is the largest load between a 30 kW heating unit and a 32.5 amp A/C unit? The voltage is supplied by 208 volts, three-phase system.

Step 1: Selecting largest load (using 360 V)
220.60
Heating load
30 kW x 1000 x 100% = **30,000 VA**
A/C load
(208 V x 1.732) x 32.5 A = **11,700 VA**

Solution: The 30,000 VA heating unit is the largest load.

See Figure 23-15 for a detailed illustration for calculating the largest VA rating between the heat and A/C load.

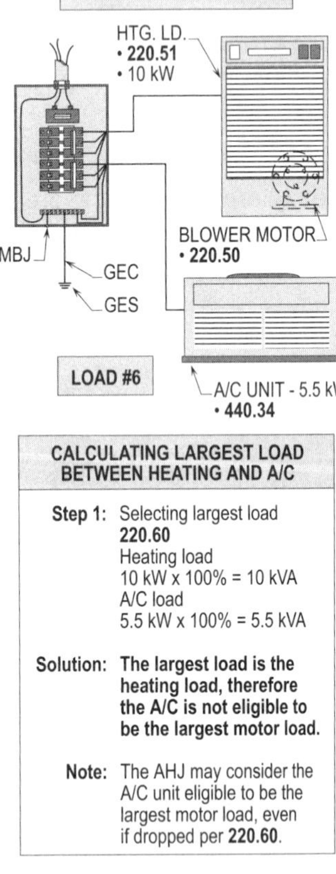

Figure 23-15. The largest load between the heating and A/C unit in VA shall be selected, and the smaller load dropped.

LARGEST MOTOR LOAD
220.50 AND 430.24

The largest motor load in VA is the seventh of the loads to be calculated. The largest motor load shall be selected from one of the motor related loads listed in the fourth, fifth, or sixth loads. The VA rating of the largest motor shall be calculated by multiplying the amperage of the unit by the voltage times 25 percent.

For example: What is the largest motor from the following loads?
Fourth load = compressor of 35 A
Fifth load = motor of **40 A**
Sixth load = A/C unit = 30 A
(larger than the gas heating)

Step 1: Selecting largest load
220.50, 440.34, and **430.24**
The motor load of 40 A is the largest load

Solution: The largest motor load is 40 amps.

See Figure 23-16 for a detailed illustration of calculating the largest motor load in amps.

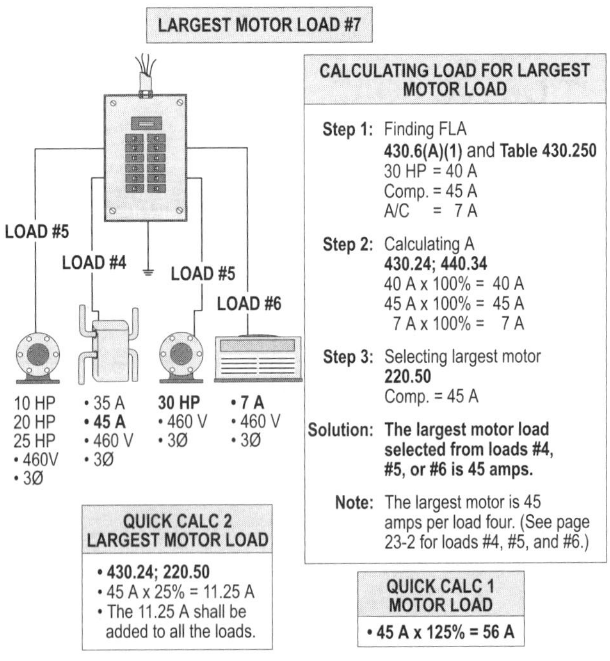

Figure 23-16. The largest motor load is the largest between the compressor loads, A/C unit loads, where eligible, or the motor loads, whichever is greater.

APPLYING THE OPTIONAL CALCULATION ARTICLE 220, PART IV

The load in VA and amps may be calculated by the optional calculation instead of the standard calculation. The optional calculation is based upon specific use of the electrical system or type of occupancy and its use and operation.

KITCHEN EQUIPMENT
220.56 AND TABLE 220.56

Table 220.56 shall be permitted to be used for load calculation for commercial electrical cooking equipment, such as dishwashers, booster heaters, water heaters, and other kitchen equipment. The demand factors shown in the Table are applicable to all equipment that is thermostatically controlled or is only intermittently used as part of the kitchen equipment. In no way do the demand factors apply to the electric heating, ventilating, or air conditioning equipment. In calculating the demand, the demand load shall not be permitted to be less than the sum of the two largest kitchen equipment loads. **(See Figure 23-12)**

SCHOOLS
220.86 AND TABLE 220.86

Table 220.86 shall be permitted to be used to calculate the service or feeder loads for schools if they are equipped with electric space heating, air conditioning, or both. The demand factors in **Table 220.86** apply to both interior and exterior lighting, power, water heating, cooking, or other loads, and the larger of the space-heating load or the air conditioning load.

When using this optional calculation, the grounded (neutral) conductor of the service or feeder loads shall be permitted to be calculated as required in **220.61**. Feeders within the building or structure where the load is calculated by this optional method may use the reduced ampacity as connected, but the ampacity of any feeder need not be larger than the individual ampacity for the entire building. Portable classrooms or buildings are not included in this Section. **(See Figure 23-17)**

NEW RESTAURANTS
220.88 AND TABLE 220.88

When calculating the service or feeder load for a new restaurant, and the feeder carries the entire load, **Table 220.88** shall be permitted to be used to size the elements necessary to supply the load. Overload protection shall be in accordance with **230.90, 215.3,** and **240.4**. Also, feeder or subfeeder conductors do not have to be larger than service conductors, regardless of calculations. **(See Figure 23-18)**

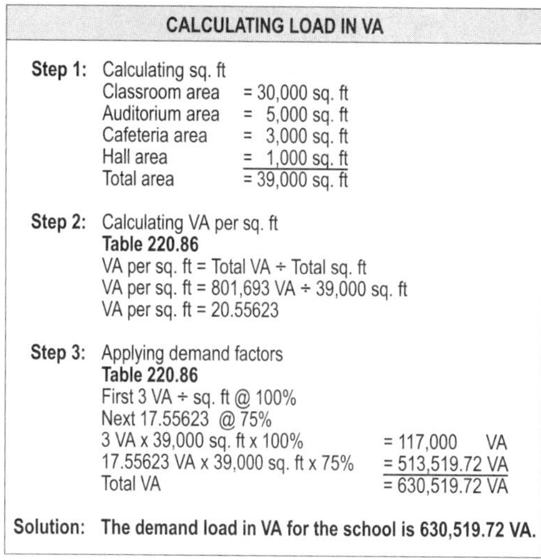

CALCULATING LOAD IN VA

Step 1: Calculating sq. ft
Classroom area = 30,000 sq. ft
Auditorium area = 5,000 sq. ft
Cafeteria area = 3,000 sq. ft
Hall area = 1,000 sq. ft
Total area = 39,000 sq. ft

Step 2: Calculating VA per sq. ft
Table 220.86
VA per sq. ft = Total VA ÷ Total sq. ft
VA per sq. ft = 801,693 VA ÷ 39,000 sq. ft
VA per sq. ft = 20.55623

Step 3: Applying demand factors
Table 220.86
First 3 VA ÷ sq. ft @ 100%
Next 17.55623 @ 75%
3 VA x 39,000 sq. ft x 100% = 117,000 VA
17.55623 VA x 39,000 sq. ft x 75% = 513,519.72 VA
Total VA = 630,519.72 VA

Solution: The demand load in VA for the school is 630,519.72 VA.

TOTAL SQUARE FOOTAGE OF SCHOOL

Classroom area	= 30,000 sq. ft
Auditorium area	= 5,000 sq. ft
Cafeteria area	= 3,000 sq. ft
Hall area	= 1,000 sq. ft

SCHOOLS
NEC 220.86
NEC TABLE 220.86

Figure 23-17. The above calculation shows the optional calculation being applied for a school.

OPTIONAL CALCULATIONS FOR ADDITIONAL LOADS TO EXISTING INSTALLATIONS
220.87

When additional loads are added to existing facilities having feeders and service as originally calculated, the maximum kVA calculations in determining the load on the existing feeders and service shall be permitted to be used if the following conditions are complied with:

- If the maximum data of the demand in kVA is available for a minimum of one year, such as demand meter ratings.

- If the demand ratings for that period of one year at 125 percent and the addition of the new load does not exceed the rating of the service. Where demand meters are used, in most cases the load as calculated will probably be less than the demand meter indications.

- If the overcurrent protection meets **230.90, 215.3,** and **240.4** for the feeder or service.

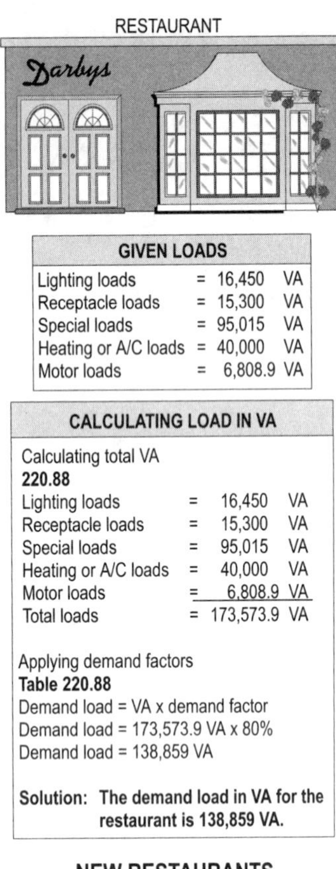

Figure 23-18. The above calculation shows the optional calculation being applied for a restaurant.

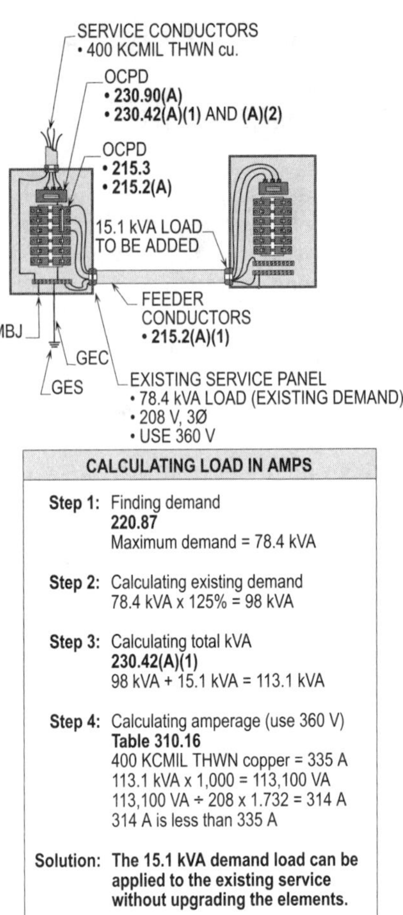

Figure 23-19. The above illustration is the calculation for a service or feeder.

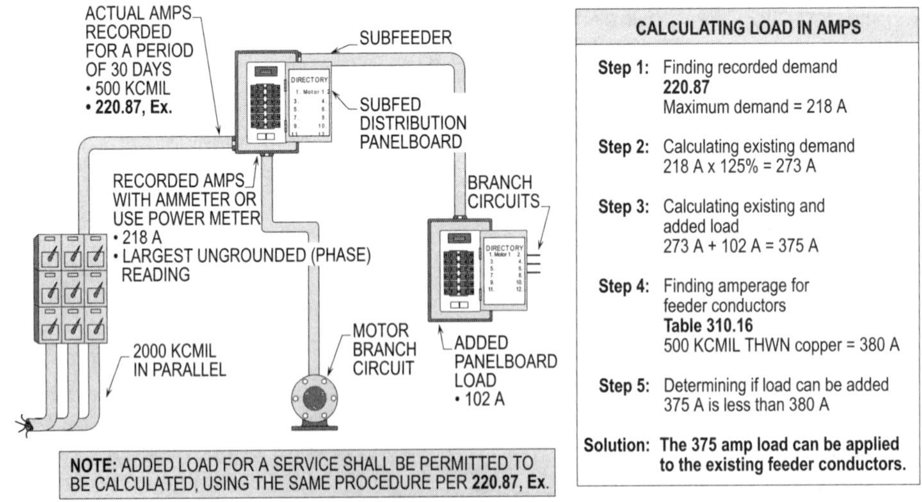

Figure 23-20. The above calculation shows the optional calculation being applied for adding a load to an existing feeder.

APPLYING Ex. TO 220.87
220.87, Ex.

If the maximum demand data for a one year period is not available, the calculated load shall be permitted to be based on the maximum demand (measure of average power demand over a 15-minute period) continuously recorded over a minimum 30 day period using a recording ammeter or power meter connected to the highest loaded ungrounded (phase) of the feeder or service, based on the initial loading at the start of the recording. **(See Figure 23-19)**

> **Design Tip:** By measurement or calculation, the larger of the heating or cooling equipment load shall be included in the load to be added. **(See Figure 23-20)**

CALCULATING THE GROUNDED (NEUTRAL) CONDUCTOR
220.61 AND 310.15(B)(4)(c)

For a service or feeder, the maximum unbalanced load controls the ampacity selected for the grounded (neutral) conductor. Grounded (neutral) feeder load shall be considered wherever a grounded (neutral) conductor is used in conjunction with one or more ungrounded (phase) conductors. On a single-phase feeder using one ungrounded (phase) conductor and a grounded (neutral) conductor, the grounded (neutral) conductor will carry the same amount of current as the ungrounded (phase) conductor. A two-wire feeder is seldom used, so in considering the grounded (neutral) feeder current, always assume that there is a grounded (neutral) conductor and two or more ungrounded (phase) conductors. If there are two ungrounded (phase) conductors that are connected to the same phase, and a grounded (neutral) conductor, the grounded (neutral) conductor would be required to carry the total current from both ungrounded (phase) conductors, which would not be an accepted practice.

For three-wire DC or single-phase AC; four-wire, three-phase; three-wire, two-phase; and five-wire, two-phase systems, a further demand factor of 70 percent shall be permitted to be applied to that portion of the unbalanced load in excess of 200 amperes. There shall be no reduction of the grounded (neutral) conductor capacity for that portion of the load that consists of electric-discharge lighting, electronic calculator/data processing, or similar equipment, when supplied by four-wire, wye-connected, three-phase systems. **(See Figure 23-21)**

   For example, on a four-wire, three-phase wye circuit where the major portion (over 50 percent) of the load consists of nonlinear loads, there are harmonic

currents present in the grounded (neutral) conductor, and the grounded (neutral) conductor shall be considered to be a current-carrying conductor. In other words, the ampacity of the conductor shall be derated per **310.15(B)(2)(a)**.

Review the rules and examples of **Chapters 14** and **15** for the sizing and use of the grounded (neutral) conductor in service, feeders and branch circuit installations.

As an example, the grounded (neutral) conductor served by a 277/480 volt supply would be calculated per **Figure 23-21,** where there is electrical discharge lighting (inductive), incandescent lighting (resistive), and other resistive related loads.

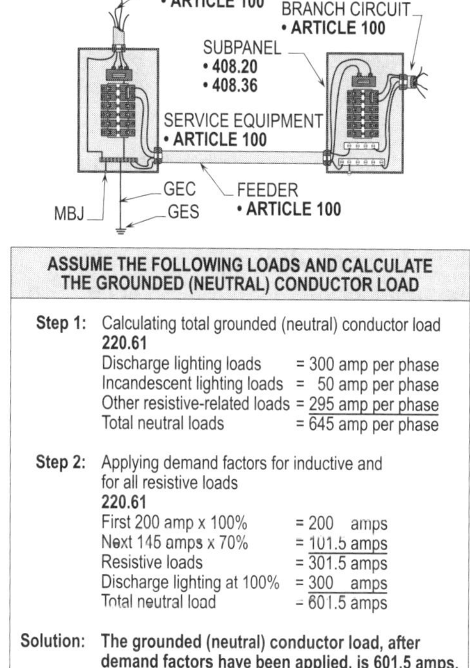

ASSUME THE FOLLOWING LOADS AND CALCULATE THE GROUNDED (NEUTRAL) CONDUCTOR LOAD

Step 1: Calculating total grounded (neutral) conductor load
220.61

Discharge lighting loads	= 300 amp per phase
Incandescent lighting loads	= 50 amp per phase
Other resistive-related loads	= 295 amp per phase
Total neutral loads	= 645 amp per phase

Step 2: Applying demand factors for inductive and for all resistive loads
220.61

First 200 amp x 100%	= 200 amps
Next 145 amps x 70%	= 101.5 amps
Resistive loads	= 301.5 amps
Discharge lighting at 100%	= 300 amps
Total neutral load	= 601.5 amps

Solution: The grounded (neutral) conductor load, after demand factors have been applied, is 601.5 amps.

CALCULATING THE
GROUNDED (NEUTRAL) CONDUCTOR
NEC 220.61
NEC 310.15(B)(4)(c)

Figure 23-21. The above calculation shows the procedure for calculating the grounded (neutral) conductor for a service or feeder and applying demand factors, where permitted.

CALCULATION PROBLEMS
ARTICLE 220, PARTS III AND IV

The elements of electrical systems shall be permitted to be calculated by using the standard or optional calculation. The size of these elements are determined by which method the designer chooses to calculate these loads. The following calculations are typical examples of how these loads are calculated, sized, and selected. The step-by-step procedures are easy to follow and have condensed the more complicated rules pertaining to calculating loads into a compact listing, that provides easier understanding of how to perform calculations according to the provisions of the NEC. A broad assortment of basic code calculations have been selected to represent the main principles of designing and installing electrical systems per NEC rules.

PROCEDURE FOR CALCULATING THE LOADS IN VA AND AMPS ARTICLE 220, PARTS III AND IV

Commercial loads shall be calculated by applying the following steps:

- The first step is to calculate the general purpose lighting load by the VA per sq. ft according to the provisions of **Table 220.12** and **230.42(A)(1)** and **(A)(2)** if the occupancy is listed.

- The second step is to calculate the general purpose receptacle load by multiplying the total number by 180 VA by the provisions of **220.14(I)** and **230.42(A)(1)** and **(A)(2)**. If the total exceeds 10 kVA, a demand factor of 50 percent shall be permitted to be applied to all VA ratings exceeding 10 kVA. **Tables 220.42** and **220.44** permit the reduction of such loads.

- The third step is to calculate the total load for all special appliances available in the facility. These appliances shall be calculated at nameplate rating based upon noncontinuous and continuous use or a combination of both.

- The fourth step is to calculate the compressor load at noncontinuous use and set aside as the largest in case it is the largest motor load.

- The fifth step is to calculate the motor loads. Each motor in VA or amps shall be added together at 100 percent value to obtain the total and set aside as the largest in case it is the largest motor load.

- The sixth step is to calculate the largest of the air conditioning (A/C) or electric heating load, and drop the smaller of the two.

- The seventh step is to calculate the largest motor load by 25 percent (or 125 percent) and add to the other loads in the steps above. The largest motor load is selected from one of the loads in steps four, five and six, respectively. (Apply 25 percent rule)

The final step is to total the calculated loads in all the steps and divide by the voltage if in VA and to size the electrical elements for a service or feeder.

STORE BUILDING SUPPLIED BY 120/240 VOLT POWER SOURCE TABLE 220.12 – (LISTED OCCUPANCY)

It is the loads in a facility that determines the voltage for designing and sizing the service or feeder elements.

For example, 120/240 volt supply is usually used to supply a facility having an equal number of three-phase and single-phase related loads.

The disadvantage of using a 120/240 volt supply is it produces larger ratings for selecting the elements for service equipment and feeders. **(See Design Problem 23-1 on page 23-25)**

STORE BUILDING SUPPLIED BY 120/208 VOLT POWER SOURCE TABLE 220.12 – (LISTED OCCUPANCY)

A facility equipped with a large number of 120 volt and 208 volt three-phase loads is normally supplied by a four-wire, three-phase wye-connected system.

On a three-phase wye-connected system, three 120 volt ungrounded (phase) conductors are obtained from ground; therefore, three ungrounded (phase) conductors may be routed with each grounded (neutral) conductor. This type of installation requires fewer grounded (neutral) conductors and therefore saves on the amount of copper or aluminum that would otherwise be needed if one grounded (neutral) conductor was pulled with each ungrounded (phase) conductor.

See Design Problem 23-2 on page 23-26 for calculating the total load for sizing the elements to be used with 120/208 volt, three-phase, four-wire systems.

STORE BUILDING SUPPLIED BY 277/480 VOLT POWER SOURCE TABLE 220.12 – (LISTED OCCUPANCY)

Smaller store buildings are usually supplied by 120/208 volt power systems. Loads such as lighting, receptacles, and other related loads are served by the lower voltage of 120 volts. Equipment is normally served by the higher voltage rated at 208 volts phase-to-phase.

Larger office buildings utilize 277/480 volt, three-phase, four-wire systems. The higher 480 volt, three-phase voltage supplies the heavier equipment and the 277 volts, single-phase voltages supplies lighting. Transformers are used to step down the 480 volts to 120/208 volts or 120/240 volts to supply the lower voltage loads and equipment. **(See Design Problem 23-3 on page 23-27)**

STORE BUILDING SUPPLIED BY 120/240 VOLT, THREE-PHASE POWER SOURCE TABLE 220.12 – (LISTED OCCUPANCY)

Store buildings with a greater number of three-phase loads than single-phase loads may be supplied by a three-phase, four-wire, 120/240 volt service. The 120 volts may be obtained from two of the ungrounded (phase) conductors to ground. This phase-to-ground voltage is taken from the lighting and power transformer.

Transformers on a delta-connected system are connected in an open or closed delta configuration.

Closed delta systems require three transformers, while open delta systems require only two transformers. One of the advantages of closed delta systems is that if one transformer fails, the remaining two may be connected in an open delta configuration and continue to supply the load until a replacement transformer is purchased and installed. **(See Design Problem 23-4 on page 23-28)**

OFFICE BUILDING SUPPLIED BY 277/480 VOLT POWER SOURCE TABLE 220.12 – (LISTED OCCUPANCY)

Office buildings consist mostly of lighting loads and receptacle loads supplying small sensitive electronic machines used for office-related work. Heavier loads such as equipment, heating, A/C, and other three-phase loads are supplied by the higher voltage.

The advantage of 277/480 volt systems is that they permit overcurrent protection devices, conductors, and other electrical elements to be smaller in size. Naturally, these smaller elements reduce the cost of installation. **(See Design Problem 23-5 on page 23-30)**

SCHOOL BUILDING SUPPLIED BY 277/480 VOLT POWER SOURCE TABLE 220.12 – (LISTED OCCUPANCY)

School buildings supplied by 277/480 volt services are calculated with the basic steps used to calculate the load for any other commercial occupancy. **(See Design Problem 23-6 on page 23-31)**

RESTAURANT SUPPLIED BY 120/208 VOLT POWER SOURCE TABLE 220.12 – (LISTED OCCUPANCY)

Small restaurants consist of lighting and receptacle loads, with the larger loads being the cooking equipment and other such pertinent apparatus.

Based upon the number of cooking units, demand factors shall be permitted to be applied to the total load per **220.56** and **Table 220.56**. Larger restaurants are supplied with 277/480 volt services, with step-down transformers being utilized to serve smaller 120 volt loads. **(See Design Problem 23-7 on page 23-32)**

HOSPITAL BUILDING SUPPLIED BY 277/480 VOLT POWER SOURCE TABLE 220.12 – (LISTED OCCUPANCY)

The loads in hospitals are calculated by their conditions of use. They are either calculated at continuous operation or noncontinuous operation, or demand factors shall be permitted to be applied for certain loads. The service voltage is determined by the size of the facility and related equipment. The procedure for calculating the load is to use the seven steps listed in this chapter for calculating the total load for a premises. **(See Design Problem 23-8 on page 23-33)**

WELDING SHOPS SUPPLIED BY 120/208 VOLT POWER SOURCE ARTICLE 220, PARTS II AND III – (UNLISTED OCCUPANCY)

Unlisted occupancies are those not appearing in **Table 220.12**, therefore, their VA ratings shall be calculated from other sections than those listed in **Table 220.12**. The load for unlisted occupancies shall be calculated by applying the same step procedure listed in this chapter. The only difference is the lighting is not obtained per **Table 220.12**. **(See Design Problem 23-9 on page 23-34)**

SIZING ELEMENTS

Design problems 23-1 through 23-9 cover specific questions on problems pertaining to sizing elements that are found in commercial facilities. Questions are based on the calculations of design problems 23-1 through 23-9. Elements such as overcurrent protection devices, conductors, conduits, etc. are determined using step-by-step procedures.

<table><tr><td valign="top">

**SIZING ELEMENTS
DESIGN PROBLEM 23-1**
(See page 23-25)

(1) What size THWN copper conductors are required for the service when they are paralleled 5 times per phase?

Step 1: Paralleling conductors
310.4
Amps of conductors = service A ÷ No. in parallel
A = 1274 A ÷ 5
A = 254.8

Step 2: Selecting conductors size
Table 310.16 and **220.5(B)**
254.8 A requires 250 KCMIL
250 KCMIL = 255 A

Solution: **It takes 5 - 250 KCMIL THWN copper conductors to supply a load of 1274 amps (255 A x 5 = 1275 A).**

(2) What size overcurrent protection device is required for the service based upon the ampacity of the conductors?

Step 1: Sizing OCPD
230.90(A), **240.4(C)**, and **240.6(A)**
1274 A requires 1250 A

Solution: **The size OCPD based upon ampacity of conductors is 1250 amps.**

</td><td valign="top">

(3) What size grounded (neutral) conductor is required for the service when they are paralleled 5 times per phase?

Step 1: Selecting conductor size
310.4
Amps of conductors = service A ÷ No. in parallel
A = 759 A ÷ 5
A = 151.8

Step 2: Selecting conductor size
Table 310.16 and **220.5(B)**
152 A requires 2/0 AWG cu.
2/0 AWG cu. = 175 A

Solution: **It takes 5 - 2/0 AWG THWN copper conductors to supply a load of 759 amps (175 A x 5 = 875 A).**

(4) What size rigid metal conduit is required to enclose the conductors for each run?

Step 1: Sizing sq. in. area
Table 5, Ch. 9
250 KCMIL cu.   = 0.397
2/0 AWG THWN cu.   = 0.2223

Step 2: Calculating total sq. in. area
Table 5, Ch. 9
0.3970 x 2   = .794   sq. in.
0.2223 x 1   = .2223 sq. in.
Total sq. in.   = 1.0163 sq. in.

Step 3: Selecting RMC
Table 4, Ch. 9
1.0163 sq. in. = 2" (53)

Solution: **The size rigid metal conduit for each is 2 in. (53).**

</td></tr></table>

SIZING ELEMENTS
DESIGN PROBLEM 23-2
(See page 23-26)

(1) What size THWN copper conductors are required for the service when they are paralleled 5 times per phase?

    Step 1: Paralleling conductors
        310.4
    Amps of conductors = service A ÷ No. in parallel
        A = 850 A ÷ 5
        A = 170

    Step 2: Selecting conductors size
        Table 310.16
        170 A requires 2/0 AWG cu.
        2/0 AWG cu. = 175 A

Solution: It takes 5 - 2/0 AWG THWN copper conductors to supply a load of 850 amps (175 A x 5 = 875 A).

(2) What size OCPD is required for the service based upon the calculated load?

    Step 1:   Sizing OCPD
        230.90(A), 240.4(C), and **240.6(A)**
        850 A requires 800 A

Solution: The size OCPD based upon calculated load is 800 amps.

(3) What size grounded (neutral) conductor is required for the service when they are paralleled 5 times per phase?

    Step 1: Selecting conductor size
        310.4
    Amps of conductors = service A ÷ No. in parallel
        A = 506 A ÷ 5
        A = 101

    Step 2: Selecting conductor size
        310.4, Table 310.16, 220.5(B), and
        250.24(C)(2)
        101 A requires 1/0 AWG cu.
        1/0 AWG cu. ⁻ 150 A

Solution: It takes 5 - 1/0 AWG THWN copper conductors to supply a load of 506 amps (150 A x 5 = 750 A).

(4) What size rigid metal conduit is required to enclose the conductors for each run?

    Step 1: Sizing sq. in. area
        Table 5, Ch. 9
        2/0 AWG THWN = 0.2223
        1/0 AWG THWN = 0.1855

    Step 2: Calculating total sq. in. area
        Table 5, Ch. 9
        0.2223 x 3   = .6669 sq. in.
        0.1855 x 1   = .1885 sq. in.
        Total sq. in.   = .8554 sq. in.

    Step 3: Selecting RMC
        Table 4, Ch. 9
        .8554 sq. in. = 2" (53)

Solution: The size rigid metal conduit for each run is 2 in. (53).

(5) What size copper grounding electrode conductor is required to ground the service to building structural steel (BSS)?

    Step 1: Sizing GEC
        Table 8, Ch. 9
        2/0 AWG cu. = 133,100 CM
        CM = CM x No. of conductors
        CM = 133,100 CM x 5
        CM = 665,500 CM
        KCMIL = CM ÷ 1,000
        KCMIL = 665,500 CM ÷ 1000
        KCMIL = 665.5

    Step 2: Selecting GEC
        Table 250.66
        665.5 KCMIL = 2/0 AWG cu.

Solution: The size of the GEC required to ground to BSS is 2/0 AWG copper.

SIZING ELEMENTS
DESIGN PROBLEM 23-3
(See page 23-27)

(1) What size THWN copper conductors are required for the service when they are paralleled 2 times per phase?

Step 1: Paralleling conductors
310.4
Amps of conductors = service A ÷ No. in parallel
A = 369 A ÷ 2
A = 184.5

Step 2: Selecting conductors size
Table 310.16 and **220.5(B)**
185 A requires 3/0 AWG THWN cu.
3/0 AWG THWN cu. = 200 A

Solution: It takes 2 - 3/0 AWG THWN copper conductors to supply a load of 369 amps (200 A x 2 = 400 A).

(2) What size OCPD is required for the service based upon the ampacity of conductors?

Step 1: Sizing OCPD
230.90(A), 240.4(B), and **240.6(A)**
369 A requires 400 A

Solution: The size OCPD based upon the ampacity of the conductors is 400 amps.

(3) What size grounded (neutral) conductor is required for the service when they are paralleled 5 times per phase?

Step 1: Selecting conductor size
310.4
Amps of conductors = service A ÷ No. in parallel
A = 150 A ÷ 2
A = 75

Step 2: Selecting conductor size
310.4, Table 310.16, and **250.24(C)(2)**
75 A requires 1/0 AWG cu.
1/0 AWG cu. = 150 A

Solution: It takes 2 - 1/0 AWG THWN copper conductors to supply a load of 150 amps (150 A x 2 = 300 A).

(4) What size rigid metal conduit is required for each run?

Step 1: Sizing sq. in. area
Table 5, Ch. 9
3/0 AWG THWN = 0.2679
1/0 AWG THWN = 0.1855

Step 2: Calculating total sq. in. area
Table 5, Ch. 9
0.2679 x 3   = .8037 sq. in.
0.1855 x 1   = .1855 sq. in.
Total sq. in.   = .9892 sq. in.

Step 3: Selecting RMC
Table 4, Ch. 9
.9892 sq. in. = 2" (53)

Solution: The size rigid metal conduit for each run is 2 in. (53).

(5) Will the 3/0 AWG THHN copper conductors supply the service load of 356 amps if four conductors in each RMC run are current-carrying?

Step 1: Applying derating factors
310.15(B)(2)(a) and **110.14(C)**
3/0 AWG THWN cu. = 225 A
225 A x 80% = 180 A

Step 2: Checking amps
310.4
180 A x 2 = 360 A

Solution: Yes, the derating of 360 amps will supply a load of 356 amps.

SIZING ELEMENTS
DESIGN PROBLEM 23-4
(See Page 23-28)

(1) What size THWN copper conductors are required to supply Phases A and C when they are paralleled 4 times per phase?

Step 1: Paralleling conductors
310.4
Amps of conductors = service A ÷ No. in parallel
A = 1119 A ÷ 4
A = 279.75

Step 2: Selecting conductors
Table 310.16 and **220.5(B)**
280 A requires 300 KCMIL
300 KCMIL THWN cu. = 285 A

Solution: It takes 4 - 300 KCMIL THWN copper conductors to supply a load of 1119 amps (285 A x 4 = 1140 A).

(2) What size THWN copper conductors are required to supply the high leg, which is Phase B? (Phase B is also paralleled 4 times per phase.)

Step 1: Paralleling conductors
310.4 and **220.5(B)**
No. of conductors = 215 A ÷ 4 = 54 A

Solution: Section 310.4 requires at least 1/0 AWG conductors to be connected in parallel. Therefore, Phase B requires 4 THWN copper conductors per phase.

(3) What size THWN copper conductors are required to supply the grounded (neutral) load?

Step 1: Paralleling conductors
310.4 and **220.5(B)**
Amps of conductors = service A ÷ No. in parallel
A = 759 A ÷ 4
A = 189.8

Step 2: Selecting conductors
Table 310.16 and **250.24(C)(2)**
190 A requires 3/0 AWG cu.
3/0 AWG cu. THWN = 200 A

Solution: It takes 4 - 3/0 AWG THWN copper conductors to supply a load of 759 amps (200 A x 4 = 800 A).

(4) What color is the high leg?

Step 1: Determining color code
110.15, 230.56, and **408.3(F)**
Orange shall be used

Solution: Orange is the color required for the high leg when used in an open or closed delta-connected system.

SIZING ELEMENTS
DESIGN PROBLEM 23-5
(See Page 23-30)

(1) What size GEC is required per **Table 250.66** when the service conductors are paralleled 6 times per phase?

Step 1: Sizing conductors in parallel
310.4, Table 310.16, and **220.5(B)**
A of conductors ÷ No. in parallel
A = 1042 A ÷ 6
A = 174 A
2/0 AWG THWN cu. = 175 A
175 A supplies 174 A

Step 2: Calculating KCMIL
Table 8, Ch. 9
2/0 AWG cu. = 133,100 CM
CM = 133,100 CM x 6
CM = 798,600
KCMIL = 798,600 ÷ 1000
KCMIL = 798.6

Step 3: Selecting GEC
Table 250.66 and **220.5(B)**
798.6 KCMIL requires 2/0 AWG cu.

Solution: The size GEC is 2/0 AWG copper.

(2) What size copper equipment bonding jumper is required to bond all of the rigid metal conduits to the grounded busbar?

Step 1: Selecting EBJ
250.102(C) and **Table 250.66**
798.6 KCMIL requires 2/0 AWG cu.

Solution: The size equipment bonding jumper is 2/0 AWG copper.

(3) What size THWN copper grounded (neutral) conductors are required per phase?

Step 1: Selecting grounded (neutral) conductor
250.24(C)(2) and **Table 250.66**
798.6 KCMIL requires 2/0 AWG cu.

Step 2: Calculating A
310.4 and **220.5(B)**
Amps of conductors ÷ No. in parallel
A = 634 A ÷ 6
A = 106 A
106 A requires 2/0 AWG cu. per
250.24(C)(2)

Solution: The size grounded (neutral) conductor per phase is 2/0 AWG THWN copper, which is the correct size per 250.24(C)(2).

SIZING ELEMENTS
DESIGN PROBLEM 23-6
(See Page 23-31)

(1) What is the allowable ampacity for 4 THHN copper current-carrying conductors routed through an ambient temperature of 120°F? (See page 23-31)

Step 1: Selecting phase conductors
310.10, FPN(2) and **Table 310.16**
335 A requires 400 KCMIL cu.

Step 2: Applying adjustment factors
Table 310.15(B)(2)(a)
380 A x 80% = 304 A

Step 3: Applying correction factors
Table 310.16 and **220.5(B)** (too small)
304 A x 82% = 249 A

Step 4: Applying adjustment and correction factors (large enough)
Table 310.16 and **Table 310.15(B)(2)(a)**
700 KCMIL cu. = 520 A
520 A x 80% x 82% = 341 A

Solution: The size THHN copper conductors are 700 KCMIL. Note: It would be more logical to parallel the number of conductors necessary to supply the load.

SIZING ELEMENTS
DESIGN PROBLEM 23-7
(See Page 23-32)

(1) What size kVA transformer is required to supply the calculated load? (Based on calculated load)

> **Step 1:** Calculating the load in VA
> Load = 137,185 VA
>
> **Step 2:** Sizing transformer
> page 23-32
> kVA = VA ÷ 1000
> kVA = 137,185 VA ÷ 1000
> kVA = 137.185
>
> **Step 3:** Selecting transformer
> Chart (inside back cover of book)
> 137.185 kVA requires 150 kVA
>
> **Solution: The size transformer to supply the load is 150 kVA.**

(2) What size panelboard is required to supply the calculated load?

> **Step 1:** Calculating the load in amps
> **220.5(A)**
> $I = VA \div (V \times \sqrt{3})$
> $I = 137,185 \text{ VA} \div (208 \text{ V} \times 1.732)$
> $I = 381 \text{ A}$
>
> **Step 2:** Selecting panelboard
> **408.36, 240.4(B),** and **220.5(B)**
> Chart (inside back cover of book)
> 381 A requires 400 A
>
> **Solution: The size panelboard to handle the calculated load is 400 amps.**

SIZING ELEMENTS
DESIGN PROBLEM 23-8
(See Page 23-33)

(1) What size busway is required to supply the calculated load?

> **Step 1:** Calculating the load in amps
> Chart
> $I = VA \div (V \times \sqrt{3})$
> $I = 1,142,563 \text{ VA} \div (480 \text{ V} \times 1.732)$
> $I = 1375 \text{ VA}$
>
> **Step 2:** Selecting busway
> Chart (inside back cover of book)
> 1375 A requires 1600 A
>
> **Solution: The size busway required to supply the calculated load is 1600 amps. Note: If available by the manufacturer, a 1500 amp busway shall be permitted to be used.**

(2) What size disconnect switch is required ahead of the distribution panelboard?

> **Step 1:** Sizing fuses
> **240.6(A)**
> Chart (inside back cover of book)
> 1375 A requires 1400 A
>
> **Solution: The size disconnect switch required to hold the fuses is 1600 amps. Note: 240.6(A) allows a listed manufacturer fuse of 1400 amps to be used.**

SIZING ELEMENTS
DESIGN PROBLEM 23-9
(See Page 23-34)

(1) What is the load, in amps, for Phases A, B, and C when adding the amps of each individual load?

Step 1: Calculating amps
220.5(A)
General lighting load (•)

$I = VA \div (V \times \sqrt{3})$
$I = 13,800 \text{ VA} \div (208 \text{ V} \times 1.732) \, 360$
$I = \textbf{38.33 A}$

Receptacle loads (•)

$I = 13,500 \text{ VA} \div 360 \text{ V}$
$I = \textbf{37.5 A}$

Special loads (•)

$I = 56,468 \text{ VA} \div 360 \text{ V}$
$I = \textbf{156.86 A}$

Compressor and motor loads (•)

$I = 13,464 \text{ VA} \div 360 \text{ V}$
$I = \textbf{37.4 A}$

Heating load (•)

$I = 20,000 \text{ VA} \div 360 \text{ V}$
$I = \textbf{55.55 A}$

Largest motor load (•)

$I = 2178 \text{ VA} \div 360 \text{ V}$
$I = \textbf{6.05 A}$
Total load = **331.69 A**

Solution: Round up to 332 amps per Tables, Ch. 9 and 220.5(B).

(2) Does the added amps per phase for each load equal the calculated amps in Design Problem 23-9 on page 23-34?

Step 1: Calculating amps
Added amps per phase equals
calculated amps per phase

Solution: Yes, the amps per phase are equal.

DESIGN PROBLEM 23-1: What is the load in VA and amps to calculate and size the elements for 120/240 volt, single-phase service supplying a 40,000 sq. ft store with 20,000 sq. ft of warehouse space?

120 V, single-phase loads

- 80 linear ft of show window (noncontinuous operation)
- 120 ft of lighting track
- 30 - 180 VA ballasts outside lighting (continuous operation)
- 3600 VA sign lighting (continuous operation)
- 65 receptacles (noncontinuous operation)
- 28 receptacles (continuous operation)
- 80 ft multioutlet assembly (heavy duty)

240 V, single-phase loads

- 7380 VA freezer
- 5580 VA ice cream boxes
- 1 - 1/2 HP exhaust fan
- 10,000 VA water heater
- 9540 VA walk-in cooler
- 50,000 VA heating unit
- 22,320 VA A/C unit
- 1 - 2 HP water pump

Sizing phases = •
Sizing neutral = √
Sizing total load = *

Note 1: For sizing elements, see pages 23-18 of this chapter.

Note 2: Add Asterisks and and checks to obtain total load.

CALCULATING LIGHTING LOAD

Step 1: General lighting load
Table 220.12 and **230.42(A)(1)**

40,000 sq. ft x 3 VA	= 120,000 VA √
120,000 VA x 125%	= 150,000 VA *
20,000 sq. ft x 1/4 VA	= 5,000 VA √
5,000 VA x 125%	= 6,250 VA *

Step 2: Show window load
220.43(A)

80 ft x 200	= 16,000 VA * √

Step 3: Track lighting load
220.43(B)

120 ft ÷ 2 x 150 VA	= 9,000 VA * √

Step 4: Outside lighting load
230.42(A)(1)

30 x 180 VA	= 5,400 VA √
5400 VA x 125%	= 6,750 VA *

Step 5: Sign lighting load
220.14(F) and **230.42(A)(1)**

3600 VA x 100%	= 3,600 VA √
3600 VA x 125%	= 4,500 VA *
Total load	= 192,500 VA •

CALCULATING RECEPTACLE LOAD

Step 1: Noncontinuous operation
220.14(H)(2), 220.14(I), and **230.42(A)(1)**

65 x 180 VA	= 11,700 VA
80' x 180 VA	= 14,400 VA
Total Load	= 26,100 VA
Table 220.44	
First 10,000 VA x 100%	= 10,000 VA
Next 16,100 VA x 50%	= 8,050 VA
Total load	= 18,050 VA * √

Step 2: Continuous operation
220.14(I) and **230.42(A)(1)**

28 x 180 VA	= 5,040 VA √
5040 VA x 125%	= 6,300 VA *
Total load	= 24,350 VA •

CALCULATING SPECIAL LOAD

Step 1: Water heater load

10,000 VA x 100%	= 10,000 VA * •

CALCULATING COMPRESSOR LOAD

Step 1: Freezer load
230.42(A)(1) and **440.34**

7380 VA x 100%	= 7,380 VA *

Step 2: Ice cream boxes

5580 VA x 100%	= 5,580 VA *

Step 3: Walk-in cooler

9540 VA x 100%	= 9,540 VA *
Total load	= 22,500 VA •

CALCULATING MOTOR LOADS

Step 1: Water pump load
430.24 and **Table 430.248**

12 A x 240 V x 100%	= 2,880 VA *

Step 2: Exhaust fan load

4.9 A x 240 V x 100%	= 1,176 VA *
Total load	= 4,056 VA •

CALCULATING HEATING OR A/C LOAD

Step 1: Heating load selected
220.60

50,000 VA x 100%	= 50,000 VA * •
23,320 VA x 100%	= 22,320 VA

CALCULATING LARGEST MOTOR LOAD

Step 1: Walk-in cooler
220.50, 430.24, and **440.34**

9540 VA x 25%	= 2,385 VA * •

CALCULATING FOR PHASES A AND B (ADD ALL •)

Lighting loads	= 192,500 VA •
Receptacle loads	= 24,350 VA •
Special loads	= 10,000 VA •
Compressor loads	= 22,500 VA •
Motor loads	= 4,056 VA •
Heating load	= 50,000 VA •
Largest motor load	= 2,385 VA •
Total load for facility	= 305,791 VA

FINDING AMPS FOR PHASES A AND B

I = VA ÷ V
I = 305,791 VA ÷ 240 V
I = 1274 A

CALCULATING NEUTRAL (ADD ALL √)

Lighting load	= 159,000 VA √
Receptacle load	= 23,090 VA √
Total load	= 182,090 VA

FINDING AMPS FOR NEUTRAL (CALCULATED AT 100% OF VA)

I = VA ÷ V
I = 182,090 VA ÷ 240 V
I = 759 A

See **220.5(B), 220.61** and Design Tip No. 5 on page 23-29.

DESIGN PROBLEM 23-2: What is the load in VA and amps to calculate and size the elements for 120/208 volt, three-phase, four-wire service supplying a 40,000 sq. ft store with 20,000 sq. ft of warehouse space?

120 V, single-phase loads

- 80 linear ft of show window (noncontinuous operation)
- 120 ft of lighting track
- 30 - 180 VA ballasts outside lighting (continuous operation)
- 3,600 VA sign lighting (continuous operation)
- 65 receptacles (noncontinuous operation)
- 28 receptacles (continuous operation)
- 80 ft multioutlet assembly (heavy duty)

208 V, three-phase loads

- 7,380 VA freezer
- 5,580 VA ice cream boxes
- 1 - 1 HP exhaust fan
- 10,000 VA water heater
- 9,540 VA walk-in cooler
- 50,000 VA heating unit
- 22,320 VA A/C unit
- 1 - 2 HP water pump

Sizing phases = •
Sizing neutral = √
Sizing total load = *

Note 1: For sizing elements, see page 23-19 of this chapter.

Note 2: Add Asterisks and checks to obtain total load.

CALCULATING LIGHTING LOAD

Step 1: General lighting load
Table 220.12 and **230.42(A)(1)**

40,000 sq. ft x 3 VA	= 120,000 VA √
120,000 VA x 125%	= 150,000 VA *
20,000 sq. ft x 1/4 VA	= 5,000 VA √
5000 VA x 125%	= 6,250 VA *

Step 2: Show window load
220.43(A)

80ft x 200	= 16,000 VA * √

Step 3: Track lighting load
220.43(B)

120 ÷ 2 x 150 VA	= 9,000 VA * √

Step 4: Outside lighting load
230.42(A)(1)

30 x 180 VA	= 5,400 VA √
5400 VA x 125%	= 6,750 VA *

Step 5: Sign lighting load
220.14(F) and **230.42(A)(1)**

3600 VA x 100%	= 3,600 VA √
3600 VA x 125%	= 4,500 VA *
Total load	**= 192,500 VA •**

CALCULATING RECEPTACLE LOAD

Step 1: Noncontinuous operation
220.14(H)(2), 220.14(I), and **230.42(A)(1)**

65 x 180 VA	= 11,700 VA
80' x 180 VA	= 14,400 VA
Total Load	= 26,100 VA
Table 220.44	
First 10,000 VA x 100%	= 10,000 VA
Next 16,100 VA x 50%	= 8,050 VA
Total load	= 18,050 VA * √

Step 2: Continuous operation
220.14(H) and **230.42(A)(1)**

28 x 180 VA	= 5,040 VA √
5040 VA x 125%	= 6,300 VA *
Total load	**= 24,350 VA •**

CALCULATING SPECIAL LOAD

Step 1: Water heater load

10,000 VA x 100%	**= 10,000 VA * •**

CALCULATING COMPRESSOR LOAD

Step 1: Freezer load
230.42(A)(1) and **440.34**

7380 VA x 100%	= 7380 VA *

Step 2: Ice cream boxes

5580 VA x 100%	= 5,580 VA *

Step 3: Walk-in cooler

9540 VA x 100%	= 9,540 VA *
Total load	**= 22,500 VA •**

CALCULATING MOTOR LOADS

Step 1: Water pump load
430.24 and **Table 430.250**

7.5 A x 360 V x 100%	= 2,700 VA *

Step 2: Exhaust fan load

4.6 A x 360 V x 100%	= 1,656 VA *
Total load	**= 4,356 VA •**

CALCULATING HEATING OR A/C LOAD

Step 1: Heating load selected
220.60

50,000 VA x 100%	**= 50,000 VA * •**
23,320 VA x 100%	**= 22,320 VA**

CALCULATING LARGEST MOTOR LOAD

Step 1: Walk-in cooler
220.50, 430.24, and **440.34**

9540 VA x 25%	= 2,385 VA * •

CALCULATING FOR PHASES A, B, AND C (ADD ALL •)

Lighting loads	= 192,500 VA •
Receptacle loads	= 24,350 VA •
Special loads	= 10,000 VA •
Compressor loads	= 22,500 VA •
Motor loads	= 4,356 VA •
Heating load	= 50,000 VA •
Largest motor load	= 2,385 VA •
Total load for facility	**= 306,091 VA**

FINDING AMPS FOR PHASES A, B, AND C

$I = VA ÷ V$
$I = 306{,}091\ VA ÷ (208\ V \times 1.732)\ 360\ V$
I = 850 A

CALCULATING NEUTRAL (ADD ALL √)

Lighting load	= 159,000 VA √
Receptacle load	= 23,090 VA √
Total load	**= 182,090 VA**

FINDING AMPS FOR NEUTRAL

$I = VA ÷ V$
$I = 182{,}090\ VA ÷ (208\ V \times 1.732)\ 360\ V$
I = 506 A

See Design Tip 5 on page 23-29.

DESIGN PROBLEM 23-3: What is the load in VA and amps to calculate and size the elements for 277/480 volt, three-phase, four-wire service supplying a 40,000 sq. ft store with 20,000 sq. ft of warehouse space with 277 volt lighting?

120 V, single-phase loads

- 80 linear ft of show window (noncontinuous operation)
- 120 ft of lighting track
- 30 - 180 VA ballasts outside lighting (continuous operation)
- 3,600 VA sign lighting (continuous operation)
- 65 receptacles (noncontinuous operation)
- 28 receptacles (continuous operation)
- 80 ft multioutlet assembly (heavy duty)

480 V, three-phase loads

- 7,380 VA freezer
- 5,580 VA ice cream boxes
- 1 - 1 HP exhaust fan
- 10,000 VA water heater
- 9,540 VA walk-in cooler
- 50,000 VA heating unit
- 22,320 VA A/C unit
- 1 - 2 HP water pump

Sizing phases = •
Sizing neutral = √
Sizing total load = *

Note 1: For sizing elements, see page 23-20 of this chapter.

Note 2: Add Asterisks and checks to obtain total load.

CALCULATING LIGHTING LOAD

Step 1: General lighting load
Table 220.12 and **230.42(A)(1)**

40,000 sq. ft x 3 VA	=	120,000 VA √
120,000 VA x 125%	=	150,000 VA *
20,000 sq. ft x 1/4 VA	=	5,000 VA √
5000 VA x 125%	=	6,250 VA *

Step 2: Show window load
220.43(A)
80 ft x 200 = 16,000 VA *

Step 3: Track lighting load
220.43(B)
120 ÷ 2 x 150 VA = 9,000 VA *

Step 4: Outside lighting load
230.42(A)(1)
30 x 180 VA = 5,400 VA
5400 VA x 125% = 6,750 VA *

Step 5: Sign lighting load
220.14(F) and **230.42(A)(1)**
3600 VA x 100% = 3,600 VA
3600 VA x 125% = 4,500 VA *
Total load = 192,500 VA •

CALCULATING RECEPTACLE LOAD

Step 1: Noncontinuous operation
220.14(H)(2), 220.14(I), and **230.42(A)(1)**

65 x 180 VA	=	11,700 VA
80' x 180 VA	=	14,400 VA
Total Load	=	26,100 VA

Table 220.44

First 10,000 VA x 100%	=	10,000 VA
Next 16,100 VA x 50%	=	8,050 VA
Total load	=	18,050 VA *

Step 2: Continuous operation
220.14(I) and **230.42(A)(1)**

28 x 180 VA	=	5,040 VA
5040 VA x 125%	=	6,300 VA *
Total load	=	**24,350 VA** •

CALCULATING SPECIAL LOAD

Step 1: Water heater load
10,000 VA x 100% = **10,000 VA** * •

CALCULATING COMPRESSOR LOAD

Step 1: Freezer load
230.42(A)(1) and **440.34**
7380 VA x 100% = 7,380 VA *

Step 2: Ice cream boxes
5580 VA x 100% = 5,580 VA *

Step 3: Walk-in cooler
9540 VA x 100% = 9,540 VA *
Total load = **22,500 VA** •

CALCULATING MOTOR LOAD

Step 1: Water pump load
430.24 and **Table 430.250**
3.4 A x 831 V x 100 = 2,825 VA *

Step 2: Exhaust fan load
2.1 A x 831 V x 100 = 1,745 VA *
Total load = **4,570 VA** •

CALCULATING HEATING OR A/C LOAD

Step 1: Heating load selected
220.60
50,000 VA x 100% = **50,000 VA** * •
23,320 VA x 100% = **22,320 VA**

CALCULATING LARGEST MOTOR LOAD

Step 1: Walk-in cooler
220.50 and **430.24**
9540 VA x 25% = **2,385 VA** * •

CALCULATING FOR PHASES A, B, AND C (ADD ALL •)

Lighting loads	=	192,500 VA•
Receptacle loads	=	24,350 VA•
Special loads	=	10,000 VA•
Compressor loads	=	22,500 VA•
Motor loads	=	4,570 VA•
Heating load	=	50,000 VA•
Largest motor load	=	2,385 VA•
Total load for facility		**= 306,305 VA**

FINDING AMPS FOR PHASES A, B, AND C

I = VA ÷ V
I = 306,305 VA ÷ (480 V x 1.732) 831 V
I = 369 A

CALCULATING NEUTRAL (ADD ALL √)

Lighting load	= 125,000 VA √
Total load	**= 125,000 VA**

FINDING AMPS FOR NEUTRAL

I = VA ÷ V
I = 125,000 VA ÷ (480 V x 1.732) 831 V
I = 150 A

DESIGN PROBLEM 23-4: What is the load in VA and amps to calculate and size the elements for 120/240 volt, three-phase, four-wire service supplying a 40,000 sq. ft store with 20,000 sq. ft of warehouse space?

120 V, single-phase loads

- 80 linear ft of show window (noncontinuous operation)
- 120 ft of lighting track
- 30 - 180 VA ballasts outside lighting (continuous operation)
- 3,600 VA sign lighting (continuous operation)
- 65 receptacles (noncontinuous operation)
- 28 receptacles (continuous operation)
- 80 ft multioutlet assembly (heavy duty)

240 V, three-phase loads

- 7,380 VA freezer
- 5,580 VA ice cream boxes
- 1 - 1 HP exhaust fan
- 10,000 VA water heater
- 9,540 VA walk-in cooler
- 50,000 VA heating unit
- 22,320 VA A/C unit
- 1 - 2 HP water pump

Sizing phases = •
Sizing neutral = √
Sizing total load = *

Note 1: For sizing elements, see pages 23-21 of this chapter.

Note 2: Add Asterisks and checks to obtain total load.

CALCULATING LIGHTING LOAD

Step 1: General lighting load
Table 220.12 and **230.42(A)(1)**

40,000 sq. ft x 3 VA	= 120,000 VA √
120,000 VA x 125%	= 150,000 VA *
20,000 sq. ft x 1/4 VA	= 5,000 VA √
5,000 VA x 125%	= 6,250 VA *

Step 2: Show window load
220.43(A)
80 ft x 200 = 16,000 VA * √

Step 3: Track lighting load
220.43(B)
120 ÷ 2 x 150 VA = 9,000 VA * √

Step 4: Outside lighting load
230.42(A)(1)

30 x 180 VA	= 5,400 VA √
5400 VA x 125%	= 6,750 VA *

Step 5: Sign lighting load
220.14(F) and **230.42(A)(1)**

3600 VA x 100%	= 3,600 VA √
3600 VA x 125%	= 4,500 VA *
Total load	**= 192,500 VA** •

CALCULATING RECEPTACLE LOAD

Step 1: Noncontinuous operation
220.14(H)(2), **220.14(I)**, and **230.42(A)(1)**

65 x 180 VA	= 11,700 VA
80' x 180 VA	= 14,400 VA
Total load	= 26,100 VA
Table 220.44	
First 10,000 VA x 100%	= 10,000 VA
Next 16,100 VA x 50%	= 8,050 VA
Total load	= 18,050 VA * √

Step 2: Continuous operation
220.14(I) and **230.42(A)(1)**

28 x 180 VA	= 5,040 VA √
5040 VA x 125%	= 6,300 VA *
Total load	**= 24,350 VA** •

CALCULATING SPECIAL LOAD

Step 1: Water heater load
10,000 VA x 100% **= 10,000 VA** * •

CALCULATING COMPRESSOR LOAD

Step 1: Freezer load
230.42(A)(1) and **440.34**
7380 VA x 100% = 7,380 VA *

Step 2: Ice cream boxes
5580 VA x 100% = 5,580 VA

Step 3: Walk-in cooler

9540 VA x 100%	= 9,540 VA *
Total load	**= 22,500 VA** •

CALCULATING MOTOR LOAD

Step 1: Water pump load
430.24 and **Table 430.250**
6.8 A x 416 V x 100% = 2,829 VA *

Step 2: Exhaust fan load

4.2 A x 416 V x 1.732 x 100%	= 1,747 VA *
Total load	**= 4,576 VA** •

CALCULATING HEATING OR A/C LOAD

Step 1: Heating load selected
220.60

50,000 VA x 100%	**= 50,000 VA** * •
23,320 VA x 100%	**= 22,320 VA**

CALCULATING LARGEST MOTOR LOAD

Step 1: Walk-in cooler
220.50 and **430.24**
9540 VA x 25% = 2,385 VA * •

SINGLE-PHASE LOAD

Lighting loads	= 192,500 VA •
Receptacle loads	= 24,350 VA •
Total load	**= 216,850 VA**

THREE-PHASE LOAD (SPECIAL LOAD)

Water heater load	= 10,000 VA •
Compressor load	= 22,500 VA •
Motor load	= 4,576 VA •
Heating load	= 50,000 VA •
Largest motor load	= 2,385 VA •
Total load	**= 89,461 VA**

SINGLE-PHASE NEUTRAL LOAD

Lighting loads	= 159,000 VA √
Receptacle loads	= 23,090 VA √
Total load	**= 182,090 VA**

CALCULATING SINGLE-PHASE LOAD

I = 216,850 VA ÷ 240 V
I = 904 A

CALCULATING THREE-PHASE LOAD (HIGH LEG)

I = 89,461 VA ÷ (240 V x 1.732) 416 V
I = 215 A

CALCULATING NEUTRAL LOAD

I = 182,090 VA ÷ 240 V
I = 759 A

CALCULATING PHASES A AND C

Single-phase load	=	904 A
Three-phase load	=	215 A
Total load		**= 1,119 A**

CALCULATING PHASE B (HIGH LEG)

Three-phase load   **= 215 A**

(1)
Design Tip (OPEN DELTA SYSTEM): The power and lighting transformer will consist of 120/240 volt single-phase loads plus the three-phase loads at 240 volts, respectively. The power transformer will consist of the three-phase loads only. Note that there will be one larger transformer (power plus lighting loads) and one smaller transformer (power three-phase loads only).

(2)
Design Tip (ADDING LOADS TOGETHER ON AN OPEN DELTA SYSTEM): The larger transformer is determined by adding the single-phase and three-phase loads together. The smaller transformer is determined by adding the three-phase loads together, which will not include the single-phase loads.

(3)
Design Tip (CLOSED DELTA SYSTEM): There will be three transformers on a closed delta system that are connected at each corner of the windings to form a closed delta system. A closed delta connected system should be used when the greater of the loads are three-phase motors, compressors, etc.

(4)
Design Tip (ADDING LOADS TOGETHER ON A CLOSED DELTA SYSTEM): The two larger transformers supplying power to the single-phase and three-phase loads is determined by adding the single-phase and three-phase loads together.

The smaller transformer(s) is determined by adding the three-phase loads together. However, in some installations, it is possible to add the 240 volt single-phase loads of phases A and C to phase B, which is the high leg. When this is done, the high leg load will be greater in size. Note that phases A and C in a delta system will usually always have a greater calculated load than phase B.

(5)
Design Tip (CALCULATING LOAD FOR THE GROUNDED (NEUTRAL) CONDUCTOR): There is really no reason for calculating the load in VA or amps at 125 percent to size the grounded (neutral) conductor. Remember that the grounded (neutral) conductor connects to the lugs of the busbar and not to the terminals of an overcurrent protection device. However, there are designers who will calculate such load at 125 percent. It is your choice whether to calculate the grounded (neutral) conductor at 125 percent or not. The grounded (neutral) conductor in this book is calculated at 100 percent of the VA or amps. Section **366.23(A)** in the NEC does not require a busbar (bare copper) to be derated 80 percent of its rating. Therefore, the 125 percent rule in **210.19(A)(1)** and **210.20** or **215.2(A)(1)** and **215.3** does not necessarily have to be applied.

DESIGN PROBLEM 23-5: What is the load in VA and amps to calculate and size the elements for 277/480 volt, three-phase, four-wire service supplying a 150,000 sq. ft office facility with 3000 sq. ft hall area equipped with 277 volt lighting units?

120 V, single-phase loads

• 60 ft of lighting track
• 20 - 180 VA ballasts outside lighting (continuous operation)
• 4800 VA sign lighting (continuous operation)
• 182 receptacles (noncontinuous operation)
• 121 receptacles (continuous operation)
• 6000 VA isolation transformer for LVLS continuous operation

Sizing phases = •
Sizing neutral = √
Sizing total load = *

208 V, three-phase loads

• 200 ft multioutlet assembly (heavy duty)
• 5 - 1450 VA copying machines
• 8500 VA water heater
• 25 - 225 VA data processors
• 10 - 175 VA word processors
• 4 - 1200 VA printers

480 V, three-phase loads

• 40 HP elevator (15 minute intermittent duty)
• 40 kW heating unit
• 12,000 VA A/C unit

Note 1: For sizing elements, see page 23-22 of this chapter

Note 2: Add Asterisks and checks to obtain total load.

CALCULATING LIGHTING LOAD

Step 1: General lighting load
Table 220.12 and **230.42(A)(1)**

150,000 sq. ft x 3.5 VA	= 525,000 VA √	
525,000 VA x 125%	= 656,250 VA *	
3,000 sq. ft x 1/2 VA	= 1,500 VA √	
1,500 VA x 125%	= 1,875 VA *	

Step 2: Track lighting load
220.43(B)
60 ft ÷ 2 x 150 VA = 4,500 VA *

Step 3: Low-voltage lighting load
Art. 411 and **230.42(A)(1)**
6000 VA x 100% = 6,000 VA
6000 VA x 125% = 7,500 VA *

Step 4: Outside lighting load
230.42(A)(1)
20 x 180 VA = 3,600 VA
3600 VA x 125% = 4,500 VA *

Step 5: Sign lighting load
220.14(F) and **230.42(A)(1)**
4800 VA x 100% = 4,800 VA
4800 VA x 125% = 6,000 VA *
Total load = 680,625 VA •

CALCULATING RECEPTACLE LOAD

Step 1: Noncontinuous operation
220.14(I) and **230.42(A)(1)**
182 x 180 VA = 32,760 VA
200' x 180 VA = 36,000 VA
Total load = 68,760 VA
Table 220.44
First 10,000 VA x 100% = 10,000 VA
Next 58,760 VA x 50% = 29,380 VA
Total load = 39,380 VA *

Step 2: Continuous operation
220.14(I) and **230.42(A)(1)**
121 x 180 VA = 21,780 VA
21,780 VA x 125% = 27,225 VA *
Total load = 66,605 VA •

CALCULATING SPECIAL LOAD

Step 1: Copying machine load
230.42(A)(1)
1450 VA x 5 = 7,250 VA
7250 VA x 125% = 9,063 VA *

Step 2: Water heater load
422.13 and **230.42(A)(1)**
8500 VA x 100% = 8,500 VA *

Step 3: Data processor load
225 VA x 25 = 5,625 VA
5625 VA x 125% = 7,031 VA *

Step 4: Word processor load
230.42(A)(1)
175 VA x 10 = 1,750 VA
1750 VA x 125% = 2,188 VA *

Step 5: Printer load
230.42(A)(1)
1200 VA x 4 = 4,800 VA
4800 VA x 125% = 6,000 VA *
Total load = 32,782 VA •

CALCULATING MOTOR LOAD

Step 1: 40 HP elevator (using 831 V)
430.24;, 430.22(E), and **Table 430.22(E)**
52 A x (480 V x 1.732) x 85% = 36,730 VA * •

CALCULATING HEATING OR A/C LOAD

Step 1: Heating load selected
220.60 and **220.51**
40,000 VA x 100% = 40,000 VA * •

CALCULATING LARGEST MOTOR LOAD

Step 1: 40 HP elevator
220.50 and **430.24**
36,730 VA x 25% = 9,183 VA *•
Total load for facility = 865,925 VA

FINDING AMPS FOR PHASES A, B AND C

I = VA ÷ (V x √3)
I = 865,925 VA ÷ (480 V x 1.732) 831 V
I = 1,042 A

CALCULATING NEUTRAL (ADD ALL √)

General lighting load
(office building) = 525,000 VA √
(halls) = 1,500 VA √
Total load = 526,500 VA

FINDING AMPS FOR NEUTRAL

I = VA ÷ (V x √3)
I = 526,500 VA ÷ (480 V x 1.732) 831 V
I = 634 A

DESIGN PROBLEM 23-6: What is the load in VA and amps to calculate and size the elements for 277/480 volt, three-phase, four-wire service supplying a 30,000 sq. ft classroom area, 5000 sq. ft auditorium area, and 1000 sq. ft assembly hall area? (School building general lighting load is supplied by 277 volt luminaires.)

120 V, single-phase loads

- 200 receptacles (noncontinuous duty)
- 50 receptacles (continuous duty)
- 200 ft multioutlet assembly (heavy duty)

Single-phase and three-phase motor loads

- 4 - 1 HP hood fans
  208 V, single-phase
- 3 - 3/4 HP grill vent fans
  208 V, single-phase
- 20 - 3/4 HP exhaust fans
  480 V, three-phase

Cooking equipment

• 2 - 1 kW toasters 120 V, single-phase	• 4 - 12 kW ranges 208 V, single-phase
• 4 - 1.5 kW refrigerators 120 V, single-phase	• 3 - 9 kW ovens 208 V, single-phase
• 3 - 1.5 kW freezers 120 V, single-phase	• 4 - 4 kW fryers 208 V, single-phase

Note: All loads are continuous

Sizing phases = •
Sizing neutral = √
Sizing total load = *

Note 1: For sizing elements, see page 23-22 of this chapter.

Note 2: Add Asterisks and checks to obtain total load.

CALCULATING LIGHTING LOAD

Step 1: General lighting load
Table 220.12 and **230.42(A)(1)**

30,000 sq. ft x 3 VA	= 90,000 VA √
90,000 VA x 125%	= 112,500 VA *
5,000 sq. ft x 1	= 5,000 VA √
5,000 VA x 125%	= 6,250 VA *
1,000 sq. ft x 1	= 1,000 VA √
1,000 VA x 125%	= 1,250 VA *
Total load	**= 120,000 VA •**

CALCULATING RECEPTACLE LOAD

Step 1: Noncontinuous operation
220.14(I) and **230.42(A)(1)**

200 x 180 VA	= 36,000 VA
200 ft x 180 VA	= 36,000 VA
Total load	= 72,000 VA

Table 220.44

First 10,000 VA x 100%	= 10,000 VA
Next 62,000 VA x 50%	= 31,000 VA
Total load	= 41,000 VA *

Step 2: Continuous operation
220.14(I) and **230.42(A)(1)**

50 x 180 VA	= 9,000 VA
9000 VA x 125%	= 11,250 VA *
Total load	**= 52,250 VA •**

CALCULATING SPECIAL LOAD

Step 1: Kitchen equipment
220.56

Toasters	
2 x 1 kW x 1000	= 2,000 VA
Refrigerators	
4 x 1.5 kW x 1000	= 6,000 VA
Freezers	
3 x 1.5 kW x 1000	= 4,500 VA
Ranges	
4 x 12 kW x 1000	= 48,000 VA
Ovens	
3 x 9 kW x 1000	= 27,000 VA
Fryers	
4 x 4 kW x 1000	= 16,000 VA
Total load	= 103,500 VA

Step 2: Applying demand factors

103,500 VA x 65%	= 67,275 VA * •

CALCULATING MOTOR LOAD
TABLES 430.248 AND 430.250

Step 1: Exhaust fans

32 A x 100% x 480 V	= 15,360 VA
15,360 VA x 1.732	= 26,604 VA *
(1.6 A x 20 = 32 A)	

Step 2: Hood fans

35.2 A x 100% x 208 V	= 7,322 VA *
(8.8 A x 4 = 35.2 A)	

Step 3: Grill vent fans

22.8 A x 100% x 208 V	= 4,742 VA *
(7.6 A x 3 = 22.8)	
Total load	**= 38,668 VA •**

CALCULATING LARGEST MOTOR LOAD

Step 1: Hood fan

8.8 A x 100% x 208 V	= 1,830 VA
1830 VA x 25%	= 458 VA * •
Total load for facility	**= 278,651 VA**

FINDING AMPS FOR PHASES A, B, AND C

$I = VA \div (V \times \sqrt{3})$
$I = 278{,}651 \text{ VA} \div (480 \text{ V} \times 1.732)\ 831 \text{ V}$
$I = 335 \text{ A}$

CALCULATING NEUTRAL (ADD ALL √)

General lighting load	= 96,000 VA √
Largest motor load	= 375 VA √
Total load	**= 96,375 VA**

FINDING AMPS FOR NEUTRAL

$I = VA \div (V \times \sqrt{3})$
$I = 96{,}375 \text{ VA} \div (480 \text{ V} \times 1.732)\ 831 \text{ V}$
I – 116 A

Note: Largest 120 volt motor is calculated by taking 1.5 kW x 1000 x .25% = 375 VA.

DESIGN PROBLEM 23-7: What is the load in VA and amps to calculate and size the elements for 120/208 volt, three-phase, four-wire service supplying a restaurant with an area of 5600 sq. ft?

120 V, single-phase loads

Lighting load

30 ft lighting track (continuous)
10 - 180 VA outside lighting (continuous)
1200 VA sign lighting (continuous)

Receptacle load

35 receptacles (noncontinuous)
25 receptacles (continuous)
20 ft multioutlet assembly (heavy duty)

208 V, three-phase loads

Special loads

2 - 20 kW heating units
208 V, three-phase
2 - 8650 VA A/C units
208 V, three-phase

Motor loads

7322 VA hood fans
208 V, single-phase
4742 VA grill vent fans
208 V, single-phase

208 V, three-phase loads

Kitchen equipment

3800 VA boiler
2 - 2700 deep fat fryers
20 A walk-in cooler
6000 VA water heater

208 V, single-phase loads

13 A freezer
11,000 VA cooktop
2 - 9000 VA ovens
12,000 VA range
14 A refrigerator
3650 VA ice cream box

Note 1: For sizing elements, see page 23-23 of this Chapter.

Note 2: Add Asterisks and checks to obtain total load.

CALCULATING LIGHTING LOAD

Step 1: General lighting load
Table 220.12 and **230.42(A)(1)**
5600 sq. ft x 2 VA = 11,200 VA √
11,200 VA x 125% = 14,000 VA *

Step 2: Track lighting load
220.43(B) and **230.42(A)(1)**
30 ft ÷ 2 x 150 VA = 2,250 VA √
2250 VA x 125% = 2,813 VA *

Step 3: Outside lighting load
220.14(L) and **230.42(A)(1)**
180 VA x 10 = 1,800 VA √
1800 VA x 125% = 2,250 VA *

Step 4: Sign lighting load
220.14(F), 230.42(A)(1), and **(A)(2)**
1200 VA x 100% = 1,200 VA √
1200 VA x 125% = 1,500 VA *
Total load = **20,563 VA** •

CALCULATING RECEPTACLE LOAD

Step 1: Receptacle load (noncontinuous)
220.14(I) and **230.42(A)(2)**
35 x 180 VA = 6,300 VA * √

Step 2: Receptacle load (continuous)
220.14(I) and **230.42(A)(1)**
25 x 180 VA = 4,500 VA √
4500 VA x 125% = 5,625 VA *

Step 3: Multioutlet assembly
220.14(H)(2) and **230.42(A)(1)**
20 ft x 180 VA = 3,600 VA * √
Total load = **15,525 VA** •

CALCULATING SPECIAL LOAD

Step 1: Kitchen equipment
220.56
Boiler = 3,800 VA
Deep fat fryer = 5,400 VA
Walk-in cooler = 7,200 VA
Water heater = 6,000 VA
Ice cream box = 3,650 VA
Freezer = 2,704 VA
Cooktop = 11,000 VA
Ovens = 18,000 VA

Range = 12,000 VA
Refrigerator = 2,912 VA
Total load = 72,666 VA

Applying demand factor
Table 220.56
72,666 VA x 65% = **47,233 VA** * •

CALCULATING MOTOR LOAD

Step 1: Hood fans
430.22(A), 430.24, and **430.25**
7322 VA x 100% = 7,322 VA *

Step 2: Grill vent fans
4742 VA x 100% = 4,742 VA *
Total load = **12,064 VA** •

CALCULATING HEATING OR A/C LOAD

Step 1: Heating load
220.60
20 kW x 2 x 1000 = **40,000 VA** * •

CALCULATING LARGEST MOTOR LOAD

Step 1: Walk-in cooler (using 360 V)
220.50, 430.22(A), and **440.34**
20 A x 100% x (208 V x 1.732) = 7,200 VA
7200 VA x 25% = 1,800 VA * •
Total load = **137,185 VA**

CALCULATING VA LOAD (NEUTRAL)
220.61, 230.42(A)(1), AND (A)(2)

Lighting load = 16,450 VA √
Receptacle load = 14,400 VA √
Total load = **30,850 VA**

FINDING AMPS FOR PHASES A, B, AND C

I = VA ÷ (V x $\sqrt{3}$)
I = 137,185 VA ÷ (208 V x 1.732) 360 V
I = 381 A

FINDING AMPS FOR NEUTRAL

I = VA ÷ (V x $\sqrt{3}$)
I = 30,850 VA ÷ (208 V x 1.732) 360 V
I = 86 A

DESIGN PROBLEM 23-8: What is the load in VA and amps to calculate and size the elements for 277/480 volt, three-phase service supplying a hospital with an office area of 150,000 sq. ft illuminated by 277 volt lighting units?

277 V, single-phase loads

• 150,000 sq. ft office
• 3000 sq. ft halls
• 800 sq. ft of closets
• 1000 sq. ft of hallways
• 4000 sq. ft of storage space

Note 1: For sizing elements, see page 23-23 of this chapter.

Note 2: Add Asterisks and checks to obtain total load.

480 V, three-phase motor loads

• 6 - 40 HP elevators (15 minute intermittent duty)
• 12,000 VA A/C unit

480 V, three-phase loads

Emergency system loads

• 40 kW heating unit
• 60,000 VA life safety branch
• 35,000 VA critical branch
• 30,000 VA life support equipment
• 45,000 VA essential system loads
(All loads continuous)

Sizing phases = •
Sizing neutral = √
Sizing total load = *

CALCULATING LIGHTING LOAD

Step 1: General lighting load
Table 220.12 and **230.42(A)(1)**
(office)

150,000 sq. ft x 3.5 VA	=	525,000 VA √
525,000 VA x 125%	=	656,250 VA *
(halls)		
3000 sq. ft x 1/2 VA	=	1,500 VA √
1500 VA x 125%	=	1,875 VA *
(closets)		
800 sq. ft x 1/2 VA	=	400 VA √
400 VA x 125%	=	500 VA *
(stairways)		
1000 VA x 1/2 VA	=	500 VA √
500 VA x 125%	=	625 VA *
(storage space)		
4000 VA x 1/4 VA	=	1,000 VA √
1000 VA x 125%	=	1,250 VA *
Total load	=	**660,500 VA** •

CALCULATING SPECIAL LOAD

Step 1: Emergency system loads
230.42(A)(1)

(Life safety branch)		
60,000 VA x 125%	=	75,000 VA *
(Critical branch)		
35,000 VA x 125%	=	43,750 VA *
(Life support equipment)		
30,000 VA x 125%	=	37,500 VA *
(Essential system)		
45,000 VA x 125%	=	56,250 VA *
Total load	=	**212,500 VA** •

CALCULATING MOTOR LOAD

Step 1: 40 HP elevators (using 831 V)
430.24, 430.22(E), and **Table 430.22(E)**

52 A x (480 V x 1.732) x 85%	=	36,730 VA
36,730 VA x 6	=	220,380 VA * •

CALCULATING HEATING OR A/C LOAD

Step 1: Heating load
220.60 and **220.51**

40,000 VA x 100%	=	40,000 VA * •

CALCULATING LARGEST MOTOR LOAD

Step 1: 40 HP elevator
220.50 and **430.24**

36,730 VA x 25%	=	9,183 VA * •

CALCULATING FOR PHASES A, B AND C

General lighting load	=	660,500 VA •
Special loads	=	212,500 VA •
Motor loads	=	220,380 VA •
Heating loads	=	40,000 VA •
Largest motor load	=	9,183 VA •
Total load	=	**1,142,563 VA**

CALCULATING NEUTRAL (ADD ALL √)

General lighting load		
(Office)	=	525,000 VA √
(Halls)	=	1,500 VA √
(Closets)	=	400 VA √
(Stairways)	=	500 VA √
(Storage)	=	1,000 VA √
Total load	=	**528,400 VA**

FINDING AMPS FOR PHASES A, B, AND C

$I = VA \div (V \times \sqrt{3})$
$I = 1,142,563 \text{ VA} \div (480 \text{ V} \times 1.732)$ 831
I = 1375 A

FINDING AMPS FOR NEUTRAL

$I = VA \div (V \times \sqrt{3})$
$I = 528,400 \text{ VA} \div (480 \text{ V} \times 1.732)$ 831
I = 636 A

DESIGN PROBLEM 23-9: What is the load in VA and amps to calculate and size the elements for 120/208 volt, three-phase service supplying a welding shop?

120 V, single-phase loads

• 9000 VA inside lighting loads (continuous operation)
• 6 - 180 VA outside lighting loads (continuous operation)
• 1200 VA sign lighting loads (noncontinuous operation)
• 60 receptacles (continuous operation)

Sizing phase = •
Sizing neutral = √
Sizing total load = 13,800 VA
Sizing total load = *

Note 1: A welding shop is not a listed occupancy per **Table 220.12**.
Note 2: For sizing elements, see page 23-24 of this chapter.
Note 3: Add Asterisks and checks to obtain total load

208 V, three-phase loads

• 2 - 10 kW heating units
• 5400 VA A/C units
• 7.5 HP air compressor
• 2 - 1 1/2 HP grinders
• Welders - resistance (50% duty cycle)
• 12 kW
• 8 kW
• Welders - motor-generator arc (90% duty cycle)
• 14 kW
• 12 kW
• Welders - nonmotor-generator arc (80% duty cycle)
• 13 kW
• 9 kW

CALCULATING LIGHTING LOAD

Step 1: Inside lighting load
230.42(A)(1)
9000 VA x 100%   = 9,000 VA √
9000 VA x 125%   = 11,250 VA *

Step 2: Outside lighting load
230.42(A)(1)
6 x 180 VA   = 1,080 VA √
1080 VA x 125%   = 1,350 VA *

Step 3: Sign lighting load
220.14(F) and **230.42(A)(1)**
1200 VA x 100%   = 1,200 VA * √
Total load   = **13,800 VA** •

CALCULATING RECEPTACLE LOAD

Step 1: Continuous duty
220.14(I) and **230.42(A)(1)**
60 x 180 VA   = 10,800 VA √
10,800 VA x 125%   = **13,500 VA** * •

CALCULATING SPECIAL LOAD

Step 1: Welders - resistance
630.31(A) and **(B)**
12,000 VA x 71%   = 8,520 VA *
8000 VA x 71% x 60%   = 3,408 VA *

Step 2: Welders - motor-generator arc
630.11(A) and **(B)**
14,000 VA x 96%   = 13,440 VA *
12,000 VA x 96%   = 11,520 VA *

Step 3: Welders - nonmotor-generator arc
630.11(A) and **(B)**
13,000 VA x 89%   = 11,570 VA *
9000 VA x 89%   = 8,010 VA *
Total load   = **56,468 VA** •

CALCULATING MOTOR LOAD

Step 1: Air compressor (using 360 V)
430.24, 430.22(E), and **Table 430.22(E)**
24.2 A x 100% x (208 V x 1.732) = 8,712 VA *

Step 2: Grinders
(6.6 A x 2) x 100% x (208 V x 1.732) = 4,752 VA *
Total load   = **13,464 VA** •

CALCULATING HEATING OR A/C LOAD

Step 1: Heating load
220.60 and **220.51**
20,000 VA x 100%   = **20,000 VA** * •

CALCULATING LARGEST MOTOR LOAD

Step 1: Air compressor (using 360 V)
24.2 A x (208 V x 1.732) x 25% = **2,178 VA** * •

CALCULATING TOTAL LOAD

Lighting loads   = 13,800 VA •
Receptacle loads   = 13,500 VA •
Special loads   = 56,468 VA •
Compressor and motor load   = 13,464 VA •
Heating loads   = 20,000 VA •
Largest motor load   = 2,178 VA •
Total load   = 119,410 VA

FINDING AMPS FOR PHASES A, B, AND C

$I = VA \div (V \times \sqrt{3})$
I = 119,410 VA ÷ (208 V x 1.732) 360 V
I = 332 A

CALCULATING NEUTRAL (ADD ALL √)

220.61
Lighting loads (9000 + 1080 + 1200 VA)   = 11,280 VA √
Receptacle loads   = 10,800 VA √
Total load   = **22,080 VA**

FINDING AMPS FOR NEUTRAL

$I = VA \div (V \times \sqrt{3})$
I = 22,080 VA ÷ (208 V x 1.732) 360 V
I = 61 A

Chapter 23. Commercial Calculations

Section   Answer

1. The lighting load in VA for lighting track shall be calculated by multiplying the lighting track by _____ VA and dividing by 2.
   (a) 100   (b) 150
   (c) 180   (d) 200

2. Occupancies with grade level access for pedestrians shall have a minimum of _____ VA provided for a sign lighting load.
   (a) 1200   (b) 1500
   (c) 1800   (d) 2400

3. There are _____ loads to be calculated when using the standard calculation to determine the load in VA or amps for the receptacle load.
   (a) 2   (b) 3
   (c) 4   (d) 5

4. Multioutlet assemblies with cord-and-plug connected appliances that operate simultaneously shall be calculated at _____ VA per foot.
   (a) 120   (b) 150
   (c) 180   (d) 200

5. In a commercial building, receptacles shall be calculated at _____ VA for each outlet. (General rule)
   (a) 150   (b) 180
   (c) 200   (d) 225

6. The total VA rating for a 8000 sq. ft office is _____ per **Table 220.12**.
   (a) 16,000   (b) 20,000
   (c) 24,000   (d) 28,000

7. The total rating for 60 ft of show window area is _____ VA if used noncontinuous.
   (a) 9000   (b) 10,800
   (c) 12,000   (d) 15,000

8. The total rating for 70 ft of multioutlet assembly having connected loads that are not simultaneously is _____ VA.
   (a) 2520   (b) 2530
   (c) 2560   (d) 2620

9. The total VA rating for 40 ft of lighting track is _____ VA when used at continuous operation.
   (a) 2000   (b) 2500
   (c) 3000   (d) 3750

10. The demand load for 130 general purpose receptacles that are used noncontinuous in an office is _____ VA.
    (a) 16,600   (b) 16,650
    (c) 16,700   (d) 23,400

_________ _________   **11.** The total load for 130 receptacles used at continuous operation in a bank is ______ VA.
      (a) 16,000   (b) 16,700
      (c) 23,400   (d) 29,250

_________ _________   **12.** If a 460 volt, three-phase compressor (largest motor load) is rated at 42 amps, the total rating for the largest motor load is ______ VA.
      (a) 8640   (b) 8726
      (c) 34,680   (d) 34,902

_________ _________   **13.** The total rating for a continuous 120 volt, single-phase, low voltage isolation transformer with a nameplate rating of 20 amps is ______ VA.
      (a) 2200   (b) 2400
      (c) 2500   (d) 3000

_________ _________   **14.** The total rating for a service load having a three-phase 35,000 VA heating unit and a three-phase 8280 VA heat pump is ______ amps. (277 / 480 volts) **Note,** the units can operate together.
      (a) 10   (b) 42
      (c) 52   (d) 65

_________ _________   **15.** The total rating for 50 ft of lighting track operating for 15 hours a day is ______ VA.
      (a) 3550.5   (b) 3660.8
      (c) 3750   (d) 4687.5

_________ _________   **16.** The demand for one of the following receptacle load is ______ VA.
    • 150 ft of multioutlet assembly appliances used simultaneously
    • 100 ft of multioutlet assembly appliances not used simultaneously
    • 70 general purpose receptacles operated at noncontinuous use
      (a) 1300   (b) 26,600
      (c) 39,600

_________ _________   **17.** The total rating for 45 receptacles used at continuous operation in a bank is ______ VA.
      (a) 8000   (b) 8100
      (c) 10,125

_________ _________   **18.** The total rating for the largest motor load based on the following motors is ______ amps. (Apply 25 percent rule)
    • 20 HP, three-phase, 460 volt motor
    • 15 HP, three-phase, 208 volt motor
    • 10 HP, three-phase, 230 volt motor
      (a) 6.75   (b) 7
      (c) 11.5

_________ _________   **19.** The total rating for the above motor loads is ______ amps for a 480 volt, three-phase supply. (Round up amps.)
      (a) 35   (b) 61
      (c) 68

_________ _________   **20.** The total load for a 40 ft x 200 ft bank with 60 general purpose receptacles is ______ VA. (Calculate VA without applying 125 percent rule.)
      (a) 35,600   (b) 38,400
      (c) 38,800

21. The total load for a 10,000 sq. ft office with 50 general purpose receptacles is
______ VA. (Calculate VA without applying 125 percent rule.)
    (a) 44,000   (b) 52,750
    (c) 55,000

22. The total load for a 400 ft hallway, 600 ft of stairway areas, and 200 ft of
storage space is ______ VA. (Calculate VA without applying 125 percent rule.)
    (a) 550   (b) 600
    (c) 750

23. The total load for 2000 sq. ft church with an 8000 sq. ft auditorium is ______ VA.
(Calculate for continuous load)
    (a) 9500   (b) 10,000
    (c) 12,500

24. The total load for 40 ballasts rated at .86 amp each and used for ten hours a
day is ______ amps. (Calculated applying 125 percent rule)
    (a) 33   (b) 34
    (c) 43

25. The total load for a small 20,000 sq. ft school with 2400 sq. ft of assembly hall
area is ______ VA. (Calculate VA without applying 125 percent rule.)
    (a) 62,400   (b) 75,000
    (c) 76,200

26. What is the amp load for a 50,000 sq. ft store including a 30,000 sq. ft warehouse
space with a 120/240 volt, single-phase service with the following loads:

120 V, single-phase loads
- 100 linear feet of show window (noncontinuous operation)
- 120 ft of lighting track (noncontinuous operation)
- 40 - 180 VA ballasts outside lighting (continuous operation)
- 4200 VA sign lighting (continuous operation)
- 74 receptacles (noncontinuous duty)
- 24 receptacles (continuous duty)
- 100 ft multioutlet assembly (simultaneously operated)

240 V, single-phase loads
- 12,000 VA water heater
- 60,000 VA heating unit
- 24,800 VA A/C unit
- 7240 VA freezer
- 6480 VA ice cream box
- 9560 VA walk-in cooler
- 1 - 1/2 HP exhaust fan
- 1 - 2 HP water pump

_______________ _______________

27. What is the amp load for a 50,000 sq. ft store including a 30,000 sq. ft warehouse space with a 120/208 volt, three-phase service with the following loads:

120 V, single-phase loads
- 100 linear feet of show window (noncontinuous operation)
- 120 ft of lighting track
- 40 - 180 VA ballasts outside lighting (continuous operation)
- 4200 VA sign lighting (continuous operation)
- 74 receptacles (noncontinuous duty)
- 24 receptacles (continuous duty)
- 100 ft multioutlet assembly (simultaneously operated)

208 V, three-phase loads (Use 208 V x 1.732 = 360 V)
- 12,000 VA water heater
- 60,000 VA heating unit
- 24,800 VA A/C unit
- 7240 VA freezer
- 6480 VA ice cream box
- 9560 VA walk-in cooler
- 1 - 1/2 HP exhaust fan
- 1 - 2 HP water pump

_______________ _______________

28. What is the amp load for a 50,000 sq. ft store including a 30,000 sq. ft warehouse space with a 277 volt lighting system and a 277/480 volt, three-phase service with the following loads:

120 V, single-phase loads
- 100 linear feet of show window (noncontinuous operation)
- 120 ft of lighting track
- 40 - 180 VA ballasts outside lighting (continuous operation)
- 4200 VA sign lighting (continuous operation)
- 74 receptacles (noncontinuous duty)
- 24 receptacles (continuous duty)
- 100 ft multioutlet assembly (simultaneously operated)

480 V, three-phase loads (Use 480 V x 1.732 = 831 V)
- 12,000 VA water heater
- 60,000 VA heating unit
- 24,800 VA A/C unit
- 7240 VA freezer
- 6480 VA ice cream box
- 9560 VA walk-in cooler
- 1 - 1/2 HP exhaust fan
- 1 - 2 HP water pump

29. What is the amp load for a 50,000 sq. ft store including a 30,000 sq. ft warehouse space with a 120/240 volt, three-phase service with the following loads:

120 V, single-phase loads
- 100 linear feet of show window (noncontinuous operation)
- 120 ft of lighting track
- 40 - 180 VA ballasts outside lighting (continuous operation)
- 4200 VA sign lighting (continuous operation)
- 74 receptacles (noncontinuous duty)
- 24 receptacles (continuous duty)
- 100 ft multioutlet assembly (simultaneously operated)

240 V, three-phase loads (Use 240 V x 1.732 = 416 V)
- 12,000 VA water heater
- 60,000 VA heating unit
- 24,800 VA A/C unit
- 7240 VA freezer
- 6480 VA ice cream box
- 9560 VA walk-in cooler
- 1 - 1/2 HP exhaust fan
- 1 - 2 HP water pump

30. What is the amp load for a 150,000 sq. ft office facility with a 2500 sq. ft of halls, equipped with a 277 volt lighting system supplied with a 277/480 volt, three-phase service with the following loads:

120 V, single-phase loads
- 50 ft of lighting track (noncontinuous operation)
- 20 - 180 VA ballasts outside lighting (continuous operation)
- 4200 VA sign lighting (continuous operation)
- 174 receptacles (noncontinuous duty)
- 114 receptacles (continuous duty)
- 6000 VA isolation transformer for LVLS (continuous operation)
- 160 ft multioutlet assembly (simultaneously operated)

208 V, three-phase loads
- 4 - 1275 VA copying machines (noncontinuous load)
- 1 - 8000 VA water heater (noncontinuous load)
- 22 - 225 VA data processors (continuous load)
- 8 - 175 VA work processors (continuous load)
- 2 - 1000 VA printers (continuous load)

480 V, three-phase loads
- 40 HP elevator (15 minute intermittent duty)
- 40 kW heating unit

_________ _________

31. What is the amp load for a 20,000 sq. ft classroom area, a 4000 sq. ft auditorium area, and 1000 sq. ft assembly hall area with a 277/480 volt, three-phase service with the following loads: (lighting supplied by 277 volts)

120 V, single-phase loads
- 170 receptacles (noncontinuous duty)
- 40 receptacles (continuous duty)
- 160 ft multioutlet assembly (simultaneously operated)

120 V, single-phase cooking equipment
- 2 - 1   kW toasters
- 4 - 1.5 kW refrigerators
- 2 - 1.5 kW freezers

208 V, single-phase cooking equipment
- 4 -   9 kW ranges
- 3 - 10 kW ovens
- 4 -   3 kW fryers

208 V, single-phase motor loads
- 3 -   1 HP vent-hood fans
- 3 - 3/4 HP grill-vent fans

480 V, three-phase motor loads
- 18 - 3/4 HP exhaust fans

_________ _________

32. What is the amp load for a 6000 sq. ft restaurant with a 120/208 volt (use 360 volt), three-phase service that is equipped with the following loads:

120 V, single-phase loads
- 35 ft of lighting track (continuous duty)
- 10 - 180 VA outside lighting (continuous duty)
- 1200 VA sign lighting (continuous duty)
- 35 receptacles (noncontinuous duty)
- 30 receptacles (continuous duty)
- 20 ft multioutlet assembly (simultaneously operated)

208 V, single-phase loads
- 6950 VA vent-hood fans
- 4758 VA grill-vent fans
- 13 A   freezer
- 8000 VA cooktop
- 2 - 10,000 VA range
- 11,000 VA range
- 14 A refrigerator
- 3750 VA ice cream box

208 V, three-phase loads
- 2 - 25 kW heating units
- 2 - 7850 VA A/C units
- 3600 VA boiler
- 2 - 2600 VA deep fat fryers
- 20 A walk-in cooler
- 6500 VA water heater

33. What is the amp load for a hospital having an office area of 150,000 sq. ft that
is illuminated by a 277 volt lighting system with a 277/480 volt (use 831 volt),
three-phase service with the following loads:

277 V, single-phase loads
- 150,000 sq. ft office space
- 2500 sq. ft hall space
- 600 sq. ft closet space
- 1200 sq. ft hallway
- 3800 sq. ft of storage space

480 V, three-phase motor loads
- 4 - 40 HP elevators (15 minute intermittent duty)
- 11,000 VA A/C unit

480 V, three-phase loads (emergency system loads)
- 50 kW heating unit (continuous duty)
- 50,000 VA life safety branch (continuous duty)
- 35,000 VA critical branch (continuous duty)
- 40,000 VA life support equipment (continuous duty)
- 40,000 VA essential system load (continuous duty)

34. What is the amp load for a welding shop with a 120/208 volt, three-phase
service with the following loads:

120 V, single-phase loads
- 8500 VA inside lighting loads (continuous duty)
- 6 - 180 VA outside lighting loads (continuous duty)
- 1200 VA sign lighting load (noncontinuous duty)
- 50 receptacles (continuous duty)

208 V, three-phase loads
- 2 - 12 kW heating units
- 6,000 VA A/C unit
- 7.5 HP air-compressor
- 2 - 1 1/2 HP grinders
- 2 - welders - resistance (30% duty cycle)
    - 11 kW
    - 9 kW
- 2 - welders - motor generator arc (80% duty cycle)
    - 12 kW
    - 10 kW
- 2 - welders - nonmotor-generator arc (90% duty cycle)
    - 12 kW
    - 9 kW

24

Industrial Calculations

The *National Electrical Code* recognizes certain rules for calculating loads for sizing and selecting elements of electrical systems used to supply power to industrial occupancies. According to NEC requirements, each service and feeder shall be calculated and sized with enough capacity to carry a load current that is not less than the sum of all branch circuits it supplies in the electrical system. These calculations vary, depending upon the type of facility, and the size and nature of the total load served.

In any electrical system, the distribution system consists of the equipment and wiring methods used to carry power from the supply transformer to the service equipment's overcurrent devices.

Distribution systems are used to carry power to lighting panelboards, power panelboards, switchboards, and motor control centers that house feeders and branch circuit protective devices for supplying individual and multiple power loads. Adequate calculations will ensure that the elements of the system will provide the right amount of power at the right voltage to each distribution point.

LAYING OUT THE LOADS
ARTICLE 220, PART III

Since industrial occupancies are not a listed occupancy, lighting loads shall be calculated without the use of **Table 220.12**. However, the other loads shall be calculated in the same manner as commercial facilities. They are laid out using the standard calculation as follows:

- Lighting loads
- Receptacle loads
- Special loads
- Compressor loads
- Motor loads
- Largest between heat and A/C loads
- The largest motor load

Note: See Chapter 23 for a detailed description of these loads and the number and types of loads that are associated with each specific load.

UNLISTED OCCUPANCY
220.14(A) THRU (L)

If an occupancy is not listed in **Table 220.12**, the general-purpose lighting load shall be calculated in the following manner per **220.14**.

- Lamps for incandescent lighting per **220.14(L)**
- Lamps for recessed lighting per **220.14(D)**
- Ballasts for electric discharge lighting per **220.18(B)**
- Show windows per **220.43(A)**
- Track lighting units per **220.43(B)**
- Low voltage systems per **Article 411**

An unlisted VA rating shall be calculated at 125 percent for sizing the overcurrent protection device and conductors for continuous operation and at 100 percent for noncontinuous operated loads per **215.2(A)(1)** and **230.42(A)(1)**. **(See Figure 23-3)**

For example: What is the load in VA for 160, 120 volt, lighting ballasts rated at 1.5 amps each and used for 12 hours a day?

Step 1: Calculating load in A
220.18(B)
160 x 1.5 A = 240 A

Step 2: Calculating continuous load
215.2(A)(1) and **230.42(A)(1)**
240 A x 125% = 300 A

Step 3: Calculated VA
300 A x 120 V = 36,000 VA

Solution: **The lighting load for the unlisted occupancy is 36,000 VA.**

SHOW WINDOW LIGHTING LOAD
220.43(A)

The lighting load in VA for a show window shall be calculated by multiplying the linear feet of the show window by 200 VA per foot. Such lighting load shall calculated at 100 percent for noncontinuous operation and at 125 percent for continuous operation per **215.2(A)(1)** and **230.42(A)(1)**. Conductors and overcurrent protection devices shall be increased to comply with **240.3** and **240.4**.

For example: What is the lighting load in VA for a 40 ft show window used at noncontinuous or continuous operation?

Step 1: Calculating noncontinuous load
220.43(A), **215.2(A)(1)**, and
230.42(A)(1)
40' x 200 VA x 100% = 8000 VA

Step 2: Calculating continuous load
220.43(A), **215.2(A)(1)**, and
230.42(A)(1)
40' x 200 VA x 125% = 10,000 VA

Solution: **The noncontinuous load is 8000 VA and the continuous load is 10,000 VA.**

If the number of lighting outlets is known, the VA rating of each luminaire shall be multiplied by 125 percent for sizing the show window load. If the VA is not known, each outlet shall be calculated at 180 VA times 125 percent to obtain the lighting load in VA for the show window. Note that the greater of either the 200 VA per linear foot or each individual calculation shall be used.

See Figure 23-4 for a detailed illustration of calculating the show window lighting load.

TRACK LIGHTING LOAD
220.43(B)

The lighting load in VA for lighting track shall be calculated by multiplying the lighting track by 150 VA and dividing by 2. Such VA rating shall be multiplied by 100 percent or by 125 percent based upon noncontinuous or continuous operation per **90.7** and **110.3(B)**. **(See Figure 23-5)**

For example: What is the load in VA for 180 ft of lighting track used at noncontinuous or continuous operation?

Step 1: Calculating noncontinuous load
220.43(B), 215.2(A)(1), and
230.42(A)(1)
180' ÷ 2' x 150 VA x 100% = 13,500 VA

Step 2: Calculating continuous load
220.43(B), 215.2(A)(1), and
230.42(A)(1)
180' ÷ 2' x 150 VA x 125% = 16,875 VA

Solution: The noncontinuous load is 13,500 VA and the continuous load is 16,875 VA.

LOW VOLTAGE LIGHTING LOAD
ARTICLE 411, 215.2(A)(1), AND 230.42(A)(1)

The lighting load in VA for low voltage lighting systems shall be calculated by multiplying the full-load amps of the isolation transformer by 100 percent for noncontinuous operation and by 125 percent for continuous operation.

For example: What is the load in VA for a low voltage lighting system supplied by an isolation transformer with a FLA of 150 amps used at noncontinuous or continuous operation?

Step 1: Calculating noncontinuous load
Article 411, 215.2(A)(1), and
230.42(A)(1)
150 A x 100% = 150 A

Step 2: Calculating continuous load
Art. 411, 215.2(A)(1), and
230.42(A)(1)
150 A x 125% = 187.5 A

Solution: The noncontinuous load is 150 amps and the continuous load is 187.5 amps.

OUTSIDE LIGHTING LOAD
220.18(B), 215.2(A)(1), AND 230.42(A)(1)

The lighting load in VA for outside lighting loads shall be calculated by multiplying the VA rating of each lighting unit by 100 percent for noncontinuous operation and by 125 percent for continuous operation. **(See Figure 23-7)**

For example: What is the lighting load in VA for 130 continuously operated luminaires with a 175 VA ballast in each unit and 110 noncontinuously operated units with each ballast having a rating of 175 VA?

Step 1: Calculating load
220.18(B), 215.2(A)(1), and
230.42(A)(1)
175 x 130 x 125%   = 28,437.5 VA
175 x 110 x 100%   = 19,250   VA
Total load   = 47,687.5 VA

Solution: The total outside lighting load is 47,687.5 VA.

OUTSIDE SIGN LIGHTING LOAD
220.14(F), 215.2(A)(1), AND 230.42(A)(1)

The lighting loads for signs shall be calculated based on the commercial occupancy or facility having ground floor footage accessible to pedestrians. Occupancies with grade level access for pedestrians shall have a minimum of 1200 VA provided for a sign lighting load. This VA rating is multiplied by 125 percent for signs operating for three hours or more and by 100 percent for those operating less than three hours. **(See Figure 23-8)**

RECEPTACLE LOADS
220.14(I) AND TABLE 220.44

Receptacle loads are the second group of loads to be calculated. Such loads are divided into two subgroups as follows:

- General purpose receptacle outlets
- Multioutlet assemblies

Each load in the subgroup shall be calculated differently to derive total VA.

GENERAL PURPOSE RECEPTACLE LOADS
220.14(I), 215.2(A)(1), AND 230.42(A)(1)

The load in VA for the general purpose receptacle load shall be calculated by multiplying the number of outlets times 180 VA each, times 100 percent for noncontinuous operation and 125 percent for continuous operation. **(See Figure 23-9)**

For example: What is the load in VA for 55 general purpose receptacles used to serve noncontinuous and continuous related loads?

Step 1: Calculating noncontinuous load
220.14(I), 215.2(A)(1), and 230.42(A)(1)
180 VA x 55 x 100% = 9900 VA

Step 2: Calculating continuous load
220.14(I), 215.2(A)(1), and 230.42(A)(1)
180 VA x 55 x 125% = 12,375 VA

Solution: The noncontinuous load is 9900 VA and the continuous load is 12,375 VA.

Note that a demand factor as listed in **Table 220.44** shall not be permitted to be applied to the continuous load of 12,375 VA even if this value exceeds 10,000 VA.

APPLYING DEMAND FACTORS
220.14(I) AND TABLE 220.44

General-purpose receptacle outlets for cord-and-plug connected loads used at noncontinuous operation shall be calculated per **220.14(I)** and **Table 220.44**. Noncontinuous operated receptacles with a VA rating of 10,000 VA or less shall be calculated at 100 percent. If the VA rating of the receptacle load exceeds 10,000 VA, a demand factor of 50 percent shall be permitted to be applied to all VA exceeding 10,000 VA per **Table 220.44**. **[See Figure 23-10(a)]**

For example: What is the VA rating for 225 general purpose receptacle outlets to cord-and-plug connect loads used at noncontinuous operation?

Step 1: Calculating load
220.14(I)
225 x 180 VA = 40,500 VA

Step 2: Applying demand factors
Table 220.44
First 10,000 VA x 100% = 10,000 VA •
Next 30,500 VA x 50% = 15,250 VA •
Total load   = 25,250 VA

Solution: The demand load is 25,250 VA.

MULTIOUTLET ASSEMBLIES
220.14(H)(1) AND (H)(2)

For connected loads not operating simultaneously, the VA rating shall be calculated by dividing the length of the assembly by 5 ft and multiplying by 180 VA. For connected loads operating simultaneously, each foot of multioutlet assembly shall be multiplied by 180 VA. The fixed multioutlet assembly load shall be permitted to be added to the noncontinuous receptacle load and a demand factor applied per **Table 220.44**. **[See Figures 23-10(b) and (c)]**

For example: What is the load in VA for 200 ft of multioutlet assembly used to cord-and-plug connect loads that are not used simultaneously and used simultaneously? **[220.14(A)(1)]**

Step 1: Calculating load for
nonsimultaneous use
VA = length ÷ 5 ft x 180 VA
VA = 200 ft ÷ 5 ft x 180 VA
VA = 7200

Step 2: Calculating load for simultaneous use
VA = length x 180 VA
VA = 200 ft x 180 VA
VA = 36,000

Solution: The load in VA for the nonsimultaneous load is 7200 VA and for the simultaneous load is 36,000 VA.

SPECIAL LOADS
215.2(A)(1) AND 230.42(A)(1)

Special appliance loads are the third group of loads to be calculated. These loads, which include computers, processing machines, etc., are usually served by individual circuits.

CONTINUOUS AND NONCONTINUOUS OPERATION
215.2(A)(1) AND 230.42(A)(1)

The load in VA for special appliance loads shall be calculated by multiplying the VA rating of each load by 100 percent for noncontinuous operation and by 125 percent for continuous operation. To determine the classification, special appliance loads operating for less than three hours shall be classified as a noncontinuous operated load. However, a special appliance load operating for three hours or more shall be classified as a continuous operated load. **(See Figure 23-11)**

For example: What is the VA rating for a 208 volt, three-phase, 165 amp special appliance load operating for t10 hours and supplied by an individual branch circuit? [Use 360 V (208 x 1.732)]

Step 1: Calculating VA (using 360 V)
220.5(A)
VA = (V x 1.732) x I
VA = (208 x 1.732) x 165
VA = 59,400

Step 2: Calculating continuous load
215.2(A)(1) and 230.42(A)(1)
59,400 VA x 125% = 74,250 VA

Solution: The load at continuous operation is 74,250 VA.

For example: Consider and calculate the VA rating for a special appliance load of 82 amps operating at 480 volts, three-phase, for a period of 2-1/2 hours every four hours. [Use 831 V (480 V x 1.732)]

Step 1: Calculating VA (using 831 V)
220.5(A)
VA = (V x 1.732) x I
VA = (480 V x 1.732) x 82 A
VA = 68,142

Step 2: Calculating noncontinuous load
215.2(A)(1) and 230.42(A)(1)
68,142 VA x 100% = 68,142 VA

Solution: The load in VA for the noncontinuous load is 68,142 VA.

COMPRESSOR LOADS
440.34

Compressor loads are the fourth group of loads to be calculated. Special considerations shall be applied when calculating loads for hermetic sealed compressors supplying refrigerant and cooling related equipment.

CONTINUOUS OR NONCONTINUOUS OPERATION
440.34, 215.2(A)(1), AND 230.42(A)(1)

Compressor-related equipment shall be calculated at 100 percent of their VA or amps. If one of such is the largest motor per load #7, it is added to the total calculation of all loads at 125 percent of its nameplate rating. **(See Figure 23-13)**

For example: What is the load in VA for 6 compressors rated at 42 amps each and supplied by a 480 volt, three-phase supply? (All used in one process at the same time)

Step 1: Calculating VA (using 831 V)
220.5(A)
(42 A x 6) x (480 V x 1.732) =
209,412 VA

Step 2: Calculating continuous load
440.34, 215.2(A)(1), and **230.42(A)(1)**
209,412 VA x 125% = 261,765 VA

Solution: **The load in VA for the 6 compressors used in an industrial process at the same time is 261,765 VA.**

MOTOR LOADS
220.50 AND 430.24

Motor loads are the fifth group of loads to be calculated. The VA rating of motors is converted from amperage to VA by multiplying the amperage from **Table 430.248** for single-phase, or **Table 430.250** for three-phase, by the supply voltage. **(See Figure 23-14)**

For example: What is the VA rating for a group of 480 volt, three-phase motors rated at 125 HP, 40 HP and 30 HP?

Step 1: Finding FLA
Table 430.250
125 HP = 156 A
40 HP = 52 A
30 HP = 40 A

Step 2: Calculating total VA (using 831 V)
220.5(A)
VA = (V x 1.732) x I
125 HP
(480 V x 1.732) x 156 A = 129,636 VA •
40 HP
(480 V x 1.732) x 52 A = 43,212 VA •
30 HP
(480 V x 1.732) x 40 A = 33,240 VA •
Total VA = 206,088 VA

Solution: **The total load for the motors is 206,088 VA.**

Design Tip: Motors can be used as a single unit to drive a piece of equipment. Motors used in an approved assembly such as a processing machine are not usually considered individual motor loads.

HEAT OR A/C LOADS
220.60

Heating or A/C loads is the sixth group of loads to be calculated. The largest VA rating between the heating or A/C load is selected, and the smaller of the two loads is dropped. To determine the largest of the two loads, the VA rating of each load shall be calculated at 100 percent and the largest load of the two is selected. The load dropped is not used again in the calculation. **(See Figure 23-15)**

Design Tip: There is no need to calculate both loads to select the elements of the service equipment, for the loads are never used simultaneously in the electrical system. For heat pump rules, **see Figure 23-16.**

For example: What is the largest load between a 240 kW heating unit and a 97 amp A/C unit? The voltage is supplied by 480 volts, three-phase system.

Step 1: Selecting largest load (using 831 V)
220.60
Heating load
240 kW x 1000 x 100% = 240,000 VA
A/C load
97 A x (480 V x 1.732) = 80,607 VA

Solution: The 240,000 VA heating unit is the largest load.

LARGEST MOTOR LOAD
220.50 AND 430.24

The largest motor load in VA is the seventh of the loads to be calculated. The largest motor load is selected from one of the motor related loads listed in the fourth, fifth, or sixth loads. The VA rating of the largest motor shall be calculated by multiplying the amperage of the unit by the voltage, times 25 percent. **(See Figure 23-16)**

For example: What is the largest motor from the following loads?

Fourth load   = compressor of 52 A
Fifth load   = motor of 65 A
Sixth load   = A/C unit = 23 A

Step 1: Selecting largest load
220.50, 440.34, and **430.24**
The motor load of 65 A is the largest load

Solution: The largest motor load is 65 amps.

OPTIONAL CALCULATIONS FOR ADDITIONAL LOADS TO EXISTING INSTALLATIONS
220.87

When additional loads are added to existing facilities having feeders and service as originally calculated, the maximum kVA calculations in determining the load on the existing feeders and service shall be permitted to be used if the following conditions are complied with:

- If the maximum data of the demand in kVA is available for a minimum of one year, such as demand meter ratings.

- If the demand ratings for that period of one year at 125 percent and the addition of the new load does not exceed the rating of the service. Where demand meters are used, in most cases the load as calculated will probably be less than the demand meter indications.

- If the overcurrent protection device meets **230.90, 215.3,** and **240.4** for feeders or a service. **(See Figure 23-19)**

APPLYING Ex. TO 220.87
220.87, Ex.

If the maximum demand data for a one year period is not available, the calculated load shall be based on the maximum demand (measure of average power demand over a 15 minute period) continuously recorded over a minimum 30 day period using a recording ammeter or power meter connected to the highest loaded phase of the feeder or service, based on the initial loading at the start of the recording. **(See Figure 23-20)**

Design Tip: By measurement or calculation the larger of the heating or cooling equipment load shall be included if it is not in the demand data.

CALCULATING THE NEUTRAL
220.61 AND 310.15(B)(4)(c) TO TABLE 310.16

For a service or feeder, the maximum unbalanced load controls the ampacity of the grounded (neutral) conductor. The grounded (neutral) conductor feeder load shall be considered wherever a grounded (neutral) conductor is used in conjunction with one or more ungrounded (phase) conductors. On a single-phase feeder using one ungrounded (phase) conductor and a grounded (neutral) conductor, the grounded (neutral) conductor will carry the same amount of current as the ungrounded (phase) conductor. A two-wire feeder is seldom used, so in considering the grounded (neutral) conductor feeder current, always assume that there is a grounded (neutral) conductor and two or more ungrounded (phase) conductors. If there are two ungrounded (phase) conductors that are connected to the same phase and a neutral, the grounded (neutral) conductor would be required to carry the total current from both ungrounded (phase) conductors, which would not be an accepted practice.

For three-wire DC or single-phase AC; four-wire, three-phase; three-wire, two-phase; and five-wire, two-phase systems, a further demand factor of 70 percent shall be permitted to be applied to that portion of the unbalanced load in excess of 200 amperes. There shall be no reduction of the grounded (neutral) conductor capacity for that portion of the load that consists of electric-discharge lighting, electronic computer/data processing, or similar equipment, when supplied by four-wire, wye-connected, three-phase systems.

For example, on a four-wire, three-phase wye circuit where the major portion (over 50 percent) of the load consists of nonlinear loads, there are harmonic currents present in the grounded (neutral) conductor, and the grounded (neutral) conductor is considered to be a current-carrying conductor. In other words, the ampacity of the conductor shall be derated per **310.15(B)(2)(a)**.

Review the rules and examples of Chapters 14 and 15 for the sizing and use of the grounded (neutral) conductor in service, feeder, and branch circuit installations.

USING ONE-LINE DIAGRAM 90.8

When calculating the load for an existing electrical system or to design an entirely new system, an important tool is a good, up-to-date line diagram of the system. It indicates by single lines and standard symbols the routed course and component parts of an electric circuit or system or circuits.

A schedule of loads and values shall be developed, and such load values calculated to size and select components and wiring methods. **(See Figure 24-1)**

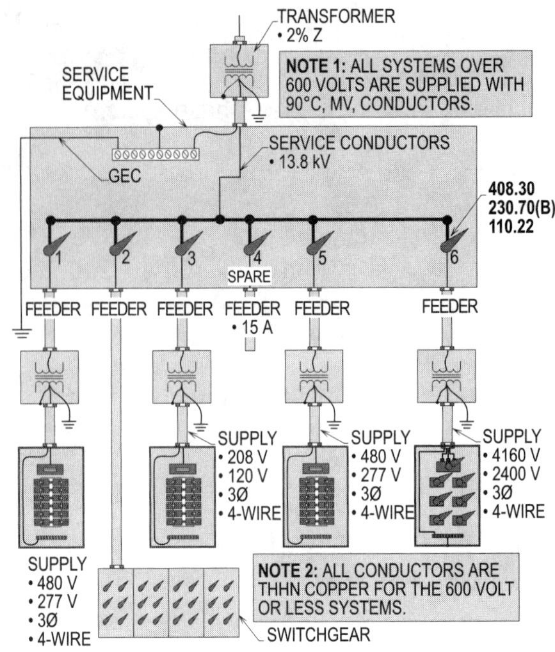

Figure 24-1. The above is a one-line diagram depicting the electrical equipment in an industrial plant.

CALCULATING FEEDER NO. 1

(1) Lighting loads

Number	Type load	Phases	Volts	Amps
• 60	2 ballast	1Ø	277 V	.38 A ea.
• 42	HID 1000 W	1Ø	480 V	.86 A ea.
• 6	Large equip.	3Ø	480 V	21 A ea.
• 10	process units	3Ø	480 V	34 A ea.

Note: All loads are used at continuous operation.

(2) Calculating total amps for feeder No. 1
    220.18(B); 215.2(A)(1); 440.34
        (2 ballast at .38 A ea.)
            • 60 x 2 x .38 x 125%   = 57 A •
        (HID 1000 W at .86 A ea.)
            • 42 x .86 A x 125%   = 45.15 A •
            • 6 x 21 A x 125%   = 157.5 A •
            • 10 x 34 A x 125%   = 425   A •
            Total load   = 684.65 A •

Rounded up to 685 amps per **220.5(B)**

(3) Sizing conductors for feeder No. 1
    Table 310.16

    Step 1: Paralleling three per phase
           310.4
           A of conductors = A ÷ No. of conductors
           A = 685 A ÷ 3
           A = 228

    Step 2: Selecting conductors for feeder No. 1
           Table 310.16
           229 A requires 4/0 AWG cu.
           4/0 AWG THHN cu. = 230 A

    Step 3: Total amps of conductors
           310.4
           4/0 AWG cu. = 230 A x 3 = 690 A
           690 A supplies 685 A

    Solution: The size THHN copper conductors are 4/0 AWG paralleled 3 times per phase.

(4) Sizing OCPD based upon load
    240.4(B); 240.6(A)

685 A or 690 A requires 700 A OCPD

    Solution: The size OCPD for 3 - 4/0 paralleled THHN copper conductors is 700 amps.

CALCULATING FEEDER NO. 2

(1) What size 90ºC, MV, copper conductors and OCPD is required to supply power to the switchgear based on the following loads per **Part III** to **Article 220**?

Number	Type load	Phases	Volts	Amps
• 2	Heavy equip.	3Ø	13.8 kV	2 A ea.
• 2	Large motors	3Ø	13.8 kV	1.5 ea.
• 1	Distribution center	3Ø	13.8 kV	15 A

Note: All loads are used at continuous operation.

    (2) Calculating total amps for feeder No. 2
        230.202; 230.208(B); 215.2(A)(1); 430.24
        (2 heavy equipment at 2 A ea.)
            • 2 x 2 A x 125%   = 5   A
        (2 large motors at 1.5 ea.)
            • 2 x 1.5 A x 125%   = 3.75 A
        (1 distribution center at 15 A)
            • 1 x 15 A x 125%   = 18.75 A
        (largest motor load)
            • 1.5 A x 25%   =   .375 A
            Total loads   = 27.875 A

(2) What size 90°C, MV copper conductors are required for an underground run in conduit to supply the distribution equipment?

Number	Type load	Phases	Volts	Loads
• 1	distribution run	3Ø	13.8 kV	27.875 A

(3) Calculating load for underground feeder
    Figure 310.60, Detail 3; Table 310.77

        27.875 A requires 6 AWG cu.

    Solution: The size 90°C, MV conductors for the underground run is 6 AWG copper.

(4) How many mains are allowed for disconnecting the feeder if it were run to a separate building?

    Step 1: Finding number of mains
           225.33(A); 225.34(A)
           The number of mains is 6

    Solution: Six mains are allowed to disconnect the feeder.

CALCULATING FEEDER NO. 3
PART II TO ARTICLE 220

Number	Type load	Phases	Volts	Amps
• 150	4 - 34 W fluorescent	1Ø	120 V	.86 A ea.
• 50	recessed luminaires	1Ø	120 V	1.25 A ea.
• 40	150 W bulbs	1Ø	120 V	1.25 A ea.
• 200	receptacles (180 W ea.)	1Ø	120 V	1.5 A ea.
• 10	100 ft multi-outlet assembly	1Ø	120 V	
• 10	isolation receptacles	1Ø	120 V	1.5 A ea.

Note: All the loads except the receptacles and bullet four are used at continuous operation.

(1) Calculating total amps for feeder No. 3
220.14(I); 220.14(H)(2); 215.2(A)(1); Table 220.44
(150 fluorescent at .86 A ea.)
    • 150 x .86 A x 125%   = 161.25 A
(50 recessed luminaires at 1.25 A ea.)
    • 50 x 1.25 A x 125%   = 78.125 A
(40 - 150 W bulbs at 1.25 A ea.)
    • 40 x 1.25 A x 125%   = 62.5   A
(200 receptacle at 1.5 A ea.)
    • 200 x 180 W   = 36,000 W
    First 10,000 W x 100%   = 10,000 VA
    • 10 x 180 VA   = 1,800 VA
    Total load   = 37,800 VA
    Next 27,800 W x 50%   = 13,900 VA
    Total load   = 23,900 VA

    I = 23,900 VA ÷ (208 x 1.732) = 66.38 A
(10 isolation receptacles at 1.5 A ea.)
    • 10 x 1.5 x 125%   = 18.75 A
    Total load   = 387.005 A

(2) What is the minimum size OCPD required on the primary side of the transformer for feeder No. 3?

Step 1: Sizing the transformer based on secondary
GE manual (based on 125% of LD.)
kVA = I x (V x √3) ÷ 1000
kVA =
387.005 A x (208 V x 1.732) ÷ 1000
kVA = 139.322

Step 2: Selecting size of the transformer
ACME chart (based on 125% of LD.)
139.322 kVA requires 225 kVA

Solution: **The size transformer to supply the load is 225 kVA.**

Step 3: Finding FLA of transformer
I = (kVA x 1000) ÷ (V x √3)
I = (225 kVA x 1000) ÷ (13,800 V x 1.732)
I = 9.4 A

Step 4: Finding size of OCPD for primary
Table 450.3(A); 240.6(A)
9.4 A x 600% = 56.4 A
56.4 A requires 50 A

Solution: **Under certain conditions, 230.208(B) allows the load for over 600 volt systems to be calculated at 100 percent. Size OCPD is 50 amp.**

(3) What size 90°C, MV copper conductors, based on the OCPD, are required to supply the primary of the transformer for Feeder No. 3?

Step 1: Selecting conductors based on OCPD
Table 310.77; 110.40
50 A OCPD requires 6 AWG cu.

Solution: **The size of the conductors for feeder No. 3 are 6 AWG copper based on the OCPD.**

CALCULATING FEEDER NO. 4
PART II TO ARTICLE 220

Number	Type load	Phases	Volts	Amps
• 1	Spare	3Ø	13.8 kVA	15 A

(1) Providing load capacity for Feeder No. 4

A load capacity of 15 amps is provided for future use

CALCULATING FEEDER NO. 5
PART III TO ARTICLE 220

Number	Type load	Phases	Volts	Amps
• 10	10 HP motors	3Ø	460 V	14 A ea.
• 12	15 HP motors	3Ø	460 V	21 A ea.
• 15	5 HP motors	3Ø	460 V	7.6 A ea.
• 16	3 HP motors	3Ø	460 V	4.8 A ea.

(1) What size conductors using the 10 ft (3 m) connection rule are required to supply the motors on the secondary side of the transformer?

Step 1: Calculating motor load
430.24

10 x 14 A	= 140 A
12 x 21 A	= 252 A
15 x 7.6 A	= 114 A
16 x 4.8 A	= 76.8 A
21 A x 25%	= 5.3 A
Total load	= 588.1 A

(2) Using the 10 ft (3 m) connection rule, how many 3/0 AWG, THHN copper conductors are required for a parallel hook-up between the transformer and panel?

Step 1: Calculating the number of conductors
310.4; Table 310.16
(3/0 AWG = 200 A)
No. of conductors = total A ÷ A of conductors
No. of conductors = 588.1 A ÷ 200 A
No. of conductors = 2.94 A

Step 2: Finding the number of conductors
310.4; Table 310.16
2.94 requires 3

Solution: The parallel hook-up requires 3 - 3/0 AWG per phase.

(3) What size OCPD (CB) is required for the connected secondary conductors?

Step 1: Sizing OCPD for largest motor load
430.62(A); 430.52(C)(1), Table 430.52
21 A x 250% = 52.5 A

Step 2: Selecting size OCPD
430.52(C)(1); Table 430.52
52.5 A requires 50 A

Step 3: Sizing OCPD for tapped conductors
430.28(1); 430.62(A)

Largest OCPD	= 50	A
Other motors	= 140	A
(252 A - 21 A = 231 A)	= 231	A
	= 114	A
	= 76.8 A	
Total load	= 611.8 A	

Step 4: Selecting OCPD for the secondary conductors
430.62(A); 240.6(A)
611.8 A requires 600 A

Solution: The size OCPD for the feeder using a circuit breaker is 600 amps. Note: There is no Ex. to 430.62(A) that will allow the next size OCPD above 611.8 amps to be used.

(4) What size copper grounding electrode conductor, bonding jumper, and grounded (neutral) conductor are required to ground the secondary to building steel?

Step 1: Sizing the GEC
250.30(A)(3); Table 250.66;
Table 8, Ch. 9
3/0 AWG = 167,800 CM x 3 in
parallel = 503,400 CM
KCMIL = 503,400 CM ÷ 1000 = 503.4 KCMIL
503.4 KCMIL requires 1/0 AWG cu.

Solution: The size of the GEC is 1/0 AWG copper.

Step 1: Sizing the BJ
250.30(A)(1); 250.102(C);
Table 250.66
503.4 KCMIL requires 1/0 AWG cu.

Solution: The size of the bonding jumper shall be at least 1/0 AWG copper.

Step 1: Sizing the grounded (neutral) conductor
250.24(C)(2); Table 250.66; 310.4
503.4 KCMIL requires 1/0 AWG copper

Solution: The size of the grounded (neutral) conductor shall be 1/0 AWG copper in each conduit run.

CALCULATING ELEMENTS FOR FEEDER NO. 6

Number	Type load	Phases	Volts	Amps
2	Heavy load (machine)	3Ø	4160 V	12 A
4	Large motors (in a process)	3Ø	4160 V	
1	motor			20 A
1	motor			26 A
1	motor			15 A
1	motor			25 A
3	motors			50 A
1	distribution panel	3Ø	4160 V	50 A
1 motor control center				
	• 1 - 200 HP	3Ø	4160 V	27.9 A
	• 1 - 150 HP	3Ø	4160 V	20.9 A
	• 1 - 125 HP	3Ø	4160 V	18.5 A
	• 2 - 50 HP	3Ø	4160 V	7.6 A

(1) What is the total load for Feeder No. 6?

    Step 1: Calculating total load
        215.2(B); 430.24; 430.25

(2 heavy loads)		
2 x 12 A	=	24 A
(7 large motors)		
1 x 20 A	=	20 A
1 x 26 A	=	26 A
1 x 15 A	=	15 A
1 x 25 A	=	25 A
3 x 50 A	=	150 A
(1 distribution panel at 50 A)		
1 x 50 A	=	50 A
(motor control center - 5 motors)		
1 x 27.9 A	=	27.9 A
1 x 20.9 A	=	20.9 A
1 x 18.5 A	=	18.5 A
2 x 7.6 A	=	15.2 A
(largest motor load)		
50 A x 25%	=	12.5 A
Total load	=	405 A

CALCULATING THE SERVICE LOAD IN AMPS FOR EACH VOLTAGE
230.42(A)(1); 430.24; 440.34

    Step 1: Calculating amps of each feeder
        220.5(A)

Feeder No. 1 (277 ÷ 480 V)	=	684.65 A
Feeder No. 2 (13.8 kV)	=	27.875 A
Feeder No. 3 (120 ÷ 208 V)	=	387.005 A
Feeder No. 4 (277 / 480 V)	=	15 A
Feeder No. 5 (13.8 kV)	=	588.1 A
Feeder No. 6	=	405 A
Total load	=	2,107.63 A

Step 2: Calculating kVA of feeders 1, 3, 5 and 6 based on supply voltage and load
    220.5(A)
    Feeder No. 1 on page 24-8:
      kVA = 684.65 A x 480 V x 1.732 ÷ 1000
      kVA = 569.19
    Feeder No. 3 on page 24-9:
      kVA = 387.005 A x 208 V x 1.732 ÷ 1000
      kVA = 139.42
    Feeder No. 5 on page 24-10:
      kVA = 588.1 x 480 V x 1.732 ÷ 1000
      kVA = 488.92
    Feeder No. 6 on page 24-11:
      kVA = 405 x 4160 V x 1.732 ÷ 1000
      kVA = 2918.07

Step 3: Calculating amps of each feeder based on supply voltage and load
    220.5(A)
    Feeder No. 1:
      $I = (kVA \times 1000) \div (V \times \sqrt{3})$
      $I = (569.19 \times 1000) \div (13,800 \times 1.732)$
      I = 23.81 A
    Feeder No. 2 on page 24-8:
      Amps at 13,800 V = 27.85 A
    Feeder No. 3:
      $I = (kVA \times 1000) \div (V \times \sqrt{3})$
      $I = (139.42 \times 1000) \div (13,800 \times 1.732)$
      I = 5.83 A
    Feeder No. 4 on page 24-9:
      Amps at 13,800 V = 15 A
    Feeder No. 5:
      $I = (kVA \times 1000) \div (V \times \sqrt{3})$
      $I = (488.92 \times 1000) \div (13,800 \times 1.732)$
      I = 20.46 A
    Feeder No. 6:
      $I = (kVA \times 1000) \div (V \times \sqrt{3})$
      $I = (2918.07 \times 1000) \div (13,800 \times 1.732)$
      I = 122.08 A
    Total amps = 214.93 A

Solution: **The service load in amps to size the elements of the service is 214.93 amps.**

Step 4: Calculating the 90°C, MV, conductors of the service
    Table 310.77, Col. 4
    214.93 A at 13.8 kV requires 4 AWG

Solution: **The size conductors for the service are 4 AWG copper.**

WHAT SIZE TRANSFORMER IS REQUIRED TO SUPPLY THE SERVICE EQUIPMENT LOADS?
220.5(A); PAGE 24-11

    Step 1: kVA = I x (V x √3)
            kVA = 214.93 A x (13,800 V x 1.732)
            kVA = 5137.17

    Step 2: Selecting transformers
            ACME chart
            Use a transformer bank

  Solution: Use a transformer bank that can handle 5194.06 at 13,800 volts.

WHAT IS THE FLC IN AMPS OF THE TRANSFORMER?

    Step 1: FLC = (kVA x 1000) ÷ (V x √3)
           FLC = (5137.17 kVA x 1000) ÷ (13,800 V x 1.732)
           FLC = 214.92 A

  Solution: The FLC of the transformer is 214.92 amps.

WHAT IS THE AVAILABLE FAULT-CURRENT AT THE TERMINALS OF THE TRANSFORMER?

    Step 1: AFC = FLA of transformer ÷ Z
           AFC = 214.92 A ÷ .02
           AFC = 10,746 A

  Solution: The AFC at the terminals of the transformer is 10,746 amps.

SEE FIGURE 24-2 FOR THE SIZE ELEMENTS OF EACH FEEDER
220.2

Note 1: Figure 24-2 has the size elements of each feeder shown for easy identification after such elements have been calculated and selected.

Note 2: Figure 24-2 has a summary of the elements in Figure 24-1 after they were sized from the load calculations.

Note 3: The NEC defines low voltage systems as 600 volts or less, and high voltage systems are those operating over 600 volts.

Note 4: IEEE-141 defines low voltage systems as those that are 600 volts or less and medium voltage as systems over 600 volts up to 69,000 volts. High voltage systems are greater than 69,000 volts.

Feeder No. 1 on page 24-8

Size OCPD for the panelboard
  • 700 A
Size conductors for panelboard
  • 3 - 4/0 AWG THWN cu. conductors per phase

Feeder No. 2 on page 24-8

Size OCPD for the switchgear
  • 30 A or 50 A based on conductors
Size conductors for the switchgear
  • 3 - 6 MV cu. conductors

Feeder No. 3 on page 24-9

Size OCPD for the primary of transformer
  • 50 A
Size conductors for the primary of transformer
  • 6 MV cu. conductors

Feeder No. 4 on page 24-9

Size OCPD for the future load
  • 15 A
Size conductors for the future load
  • 6 MV cu. based on NEC

Feeder No. 5 on page 24-10

Size OCPD for the connection
  • 600 A
Size conductors for the connection
  • 3 - 3/0 THWN cu. conductors per phase

Feeder No. 6 on page 24-11

Total load for feeder in amps
  • 405 A

Service elements on pages 24-11 and 12

Size conductor for the service
  • 4 MV cu. conductors per phase
Size transformer for the service
  • transformer bank

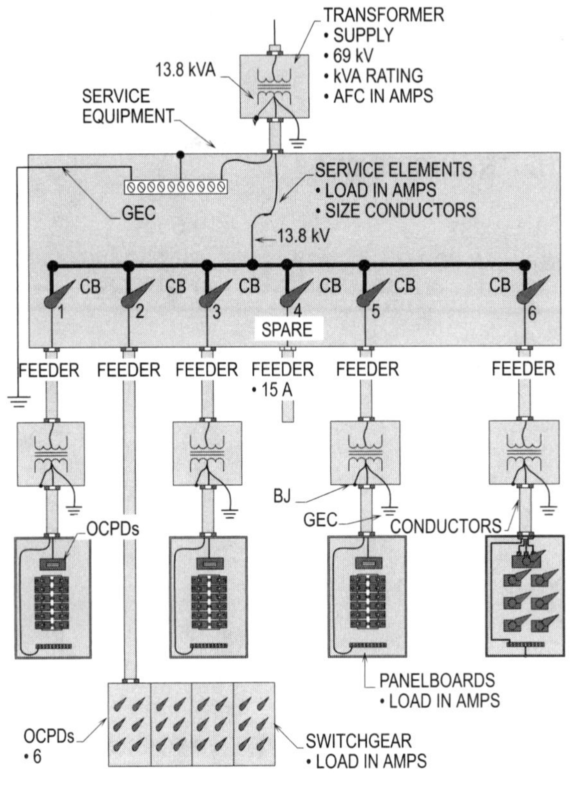

Figure 24-2. One-line diagram showing electrical equipment and wiring methods, with the size of individual elements of service and feeders after they were sized from load calculations. For calculation procedure for each element, refer to the page number listed below the sizes shown.

Chapter 24. Industrial Calculations

Section   **Answer**

1. What is the load in VA for an unlisted occupancy feeder having 140 - 120 volt lighting ballasts rated at 1.5 amps each and used for 6 hours a day?
   - (a) 25,200 VA
   - (b) 30,000 VA
   - (c) 31,500 VA
   - (d) 33,500 VA

2. What is the lighting load in VA for an unlisted occupancy feeder having 50 ft show window used at continuous operation?
   - (a) 9,000 VA
   - (b) 10,000 VA
   - (c) 10,500 VA
   - (d) 12,500 VA

3. What is the load in VA for 200 ft of lighting track on a service used at noncontinuous operation?
   - (a) 15,000 VA
   - (b) 18,000 VA
   - (c) 18,750 VA
   - (d) 22,500 VA

4. What is the load in VA for 50 general purpose receptacles on a service used to serve continuous related loads?
   - (a) 9,000 VA
   - (b) 10,000 VA
   - (c) 11,250 VA
   - (d) 12,500 VA

5. What is the load in VA for 200 general purpose receptacle outlets to cord-and-plug connect loads used at noncontinuous operation?
   - (a) 23,000 VA
   - (b) 28,750 VA
   - (c) 36,000 VA
   - (d) 45,000 VA

6. What is the load in VA for 150 ft of multioutlet assembly used to cord-and-plug connect loads that are used simultaneously?
   - (a) 18,000 VA
   - (b) 27,000 VA
   - (c) 30,000 VA
   - (d) 33,750 VA

7. What is the VA rating for a 208 volt, three-phase, 155 amp special appliance load operating for eight hours and supplied by an individual branch circuit?
   - (a) 55,800 VA
   - (b) 64,480 VA
   - (c) 69,750 VA
   - (d) 80,600 VA

8. What is the VA rating for a 208 volt, three-phase, 155 amp special appliance load operating for two hours and supplied by an individual branch circuit?
   - (a) 32,240 VA
   - (b) 40,300 VA
   - (c) 55,000 VA
   - (d) 69,750 VA

9. What is the load in amps for 4 compressors rated at 38 amps each and supplied by a 480 volt, three-phase supply? (All used in one process at the same time)
   - (a) 152
   - (b) 162
   - (c) 168
   - (d) 174

10. What is the VA rating for a group of 480 volt, three-phase motors rated at 100 HP, 50 HP, and 25 HP? (Calculate for a feeder.)
    - (a) 213
    - (b) 254
    - (c) 262
    - (d) 266

Appendix

Table A

Size AWG/ KCMIL	Ohms to Neutral per 1000 feet					
	AC Resistance for Uncoated Copper Wires			AC Resistance for Aluminum Wires		
	PVC Conduit	Alu. Conduit	Steel Conduit	PVC Conduit	Alu. Conduit	Steel Conduit
14	3.1	3.1	3.1	—	—	—
12	2.0	2.0	2.0	3.2	3.2	3.2
10	1.2	1.2	1.2	2.0	2.0	2.0
8	0.78	0.78	0.78	1.3	1.3	1.3
6	0.49	0.49	0.49	0.81	0.81	0.81
4	0.31	0.31	0.31	0.51	0.51	0.51
3	0.25	0.25	0.25	0.40	0.41	0.40
2	0.19	0.20	0.20	0.32	0.32	0.32
1	0.15	0.16	0.16	0.25	0.26	0.25
1/0	0.12	0.13	0.12	0.20	0.21	0.20
2/0	0.10	0.10	0.10	0.16	0.16	0.16
3/0	0.077	0.082	0.079	0.13	0.13	0.13
4/0	0.062	0.067	0.063	0.10	0.11	0.10
250	0.052	0.057	0.054	0.085	0.090	0.086
300	0.044	0.049	0.045	0.071	0.076	0.072
350	0.038	0.043	0.039	0.061	0.066	0.063
400	0.033	0.038	0.035	0.054	0.059	0.055
500	0.027	0.032	0.029	0.043	0.048	0.045
600	0.023	0.028	0.025	0.036	0.041	0.038
750	0.019	0.024	0.021	0.029	0.034	0.031
1000	0.015	0.019	0.018	0.023	0.027	0.025

Note, See **Table 9, Ch. 9** of the NEC.

Table B

(Three Single Conductors) "C" Value for Conductors and Busways					
Copper AWG or kcmil	**Three Single Conductor Conduit Steel**			**Nonmagnetic**		
	600V	**5KV**	**15KV**	**600V**	**5KV**	**15KV**
14	389	389	389	389	389	389
12	617	617	617	617	617	617
10	981	981	981	981	981	981
8	1557	1551	1557	1558	1555	1558
6	2425	2406	2389	2430	2417	2406
4	3806	3750	3695	3825	3789	3752
3	4760	4760	4760	4802	4802	4802
2	5906	5736	5574	6044	5926	5809
1	7292	7029	6758	7493	7306	7108
1/0	8924	8543	7973	9317	9033	8590
2/0	10755	10061	9389	11423	10877	10318
3/0	12843	11804	11021	13923	13048	12360
4/0	15082	13605	12542	16673	15351	14347
250	16483	14924	13643	18593	17120	15865
300	18176	16292	14768	20867	18975	17408
350	19703	17385	15678	22736	20526	18672
400	20565	18235	16365	24296	21786	19731
500	22185	19172	17492	26706	23277	21329
600	22965	20567	17962	28033	25430	22690
750	24136	21386	18888	28303	25430	22690
1000	25278	22539	19923	31490	28083	24887
Aluminum						
14	236	236	236	236	236	236
12	375	375	375	375	375	375
10	598	598	598	598	598	598
8	950	950	951	951	950	951
6	1480	1476	1472	1481	1478	1476
4	2345	2332	2319	2350	2341	2333
3	2948	2948	2948	2958	2958	2958
2	3713	3669	3626	3729	3701	3672
1	4645	4574	4497	4678	4631	4580
1/0	5777	5669	5493	5838	5766	5645
2/0	7186	6968	6733	7301	7152	6986
3/0	8826	8466	8163	9110	8851	8627
4/0	10740	10167	9700	11174	10749	10386
250	12122	11460	10848	12862	12343	11847
300	13909	13009	12192	14922	14182	13491
350	15484	14280	13288	16812	15857	14954
400	16670	15355	14188	18505	17321	16233
500	18755	16827	15657	21390	19503	18314
600	20093	18427	16484	23451	21718	19635
750	21766	19685	17686	23491	21769	19976
1000	23477	21235	19005	28778	26109	23482

Table B

(Three Conductor Cable) "C" Value for Conductors and Busways					
Copper AWG or kcmil	**Three Single Conductor Conduit Steel**			**Nonmagnetic**		
	600V	**5KV**	**15KV**	**600V**	**5KV**	**15KV**
14	389	389	389	389	389	389
12	617	617	617	617	617	617
10	981	981	981	981	981	981
8	1559	1557	1559	1559	1558	1559
6	2431	2424	2414	2433	2428	2420
4	3830	3811	3778	3837	3823	3798
3	4760	4790	4760	4802	4802	4802
2	5989	5729	5827	6087	6022	5957
1	7454	7364	7188	7579	7507	7364
1/0	9209	9086	8707	9472	9372	9052
2/0	11244	11045	10500	11703	11528	11052
3/0	13656	13333	12613	14410	14118	13461
4/0	16391	15890	14813	17482	17019	16012
250	18310	17850	16465	19779	19352	18001
300	20617	20051	18318	22524	11938	20163
350	22646	21914	19821	24904	24126	21982
400	24253	23371	21042	26915	26044	23517
500	26980	25449	23125	30028	28712	25916
600	28752	27974	24896	32236	31258	27766
750	31050	20024	26932	32404	31338	28303
1000	33864	32688	29320	37197	35748	31959
Aluminum						
14	236	236	236	236	236	236
12	375	375	375	375	375	375
10	598	598	598	598	598	598
8	951	951	951	951	951	951
6	1481	1480	1478	1482	1481	1479
4	2351	2347	2339	2353	2349	2344
3	2948	2956	2948	2958	2958	2958
2	3733	3719	3693	3739	3724	3709
1	4686	4663	4617	4699	4681	4646
1/0	5852	5820	5717	5875	5851	5771
2/0	7327	6968	6733	7301	7152	6986
3/0	9077	8980	8750	9242	9164	8977
4/0	11184	10121	10642	11408	11277	10968
250	12796	12636	12115	13236	13105	12661
300	14916	14698	13973	15494	15299	14658
350	15413	16490	15540	17635	17351	16500
400	18461	18063	16921	19587	17321	16233
500	18755	16827	15657	21390	19243	18154
600	23633	23195	21348	25750	25243	23294
750	26431	25789	23750	25682	25141	23491
1000	29864	29049	26608	32938	31919	29135

Note, These values are equal to one over the impedence per foot for impedences found in IEEE. std. 241-1990. IEEE Recommended Practice for Commercial Building Power System.

Table B

AWG or MCM	Copper Three Single-conductors				Copper Three-Conductor Cable	
	Steel-conduit		Nonmagnetic Conduit		Steel Conduit	Nonmagnetic Conduit
	600 V and 5 kV Nonshielded	5 kV Shielded and 15 kV	600 V and 5 kV Nonshielded	5 kV Shielded and 15 kV	600 V and 5 kV Nonshielded	600 V and 5 kV Nonshielded
12	588		588			
10	909		1449			
8	1429	1230	1449	1230	1230	1230
6	2222	1940	2273	1940	1950	1950
4	3333	3040	3448	3070	3080	3090
3	4167	3830	4348	3870	3880	3900
2	5000	4670	5263	4780	4830	4850
1	6250	5750	6250	5920	6020	6100
1/0	7692	6990	7692	7250	7410	7580
2/0	9091	8260	9091	8770	9090	9350
3/0	10638	9900	11364	10700	11100	11900
4/0	12500	10800	13514	12600	13400	14000
250	13699	12500	17857	14000	14900	15800
300	15385	13600	16949	15500	16700	17900
350	16667	14700	18868	17000	18600	20300
400	17857	15200	20408	17900	19500	21100
500	20000	16500	23256	19700	21900	24000
600	21277	17200	25000	20900	23300	25700
750	23256	18300	27778	22500	25600	28200
1000	25000		31250			

Note, These tables were used for fault calculations before Tables B on page III and IV were available.

AWG or MCM	Aluminum Three Single Conductors Or Three Conductor Cables	
	Steel Conduit	Nonmagnetic Conduit
	600 V and 5 kV Nonshielded	600 V and 5 kV Nonshielded
12	357	357
10	555	556
8	909	909
6	1388	1408
4	2173	2173
3	2702	2702
2	3333	3333
1	4000	4166
1/0	5000	5000
2/0	6250	6250
3/0	7142	7692
4/0	9090	9090
250	10000	10638
300	11363	12195
350	12500	13698
400	13698	15384
500	15625	17543
600	17241	19607
750	19230	22222
1000	21739	25641

Table C

Size AWG/ KCMIL	Area Cir. Mills	Conductors				DC Resistance at 75°C (167°F)		
		Standing		Overall		Copper		Aluminum
		Quantity	Diam. In.	Diam. In.	Area Sq. In.^{2}	Uncoated ohm/kFT	Coated ohm/kFT	ohm/kFT
18	1620	1	——	0.040	0.001	7.77	8.08	12.8
18	1620	7	0.015	0.046	0.002	7.95	8.45	13.1
16	2580	1	——	0.051	0.002	4.89	5.08	8.05
16	2580	7	0.019	0.058	0.003	4.99	5.29	8.21
14	4110	1	——	0.064	0.003	3.07	3.19	5.06
14	4110	7	0.024	0.073	0.004	3.14	3.26	5.17
12	6530	1	——	0.081	0.005	1.93	2.01	3.18
12	6530	7	0.030	0.092	0.006	1.98	2.05	3.25
10	10380	1	——	0.102	0.008	1.21	1.26	2.00
10	10380	7	0.038	0.116	0.011	1.24	1.29	2.04
8	16510	1	——	0.128	0.013	0.764	0.786	1.26
8	16510	7	0.049	0.146	0.017	0.778	0.809	1.28
6	26240	7	0.061	0.184	0.027	0.491	0.510	0.808
4	41740	7	0.077	0.232	0.042	0.308	0.321	0.508
3	52620	7	0.087	0.260	0.053	0.245	0.254	0.403
2	66360	7	0.097	0.292	0.067	0.194	0.201	0.319
1	83690	19	0.066	0.332	0.087	0.154	0.160	0.253
1/0	105600	19	0.074	0.373	0.109	0.122	0.127	0.201
2/0	133100	19	0.084	0.419	0.138	0.0967	0.101	0.159
3/0	167800	19	0.094	0.470	0.173	0.0766	0.0797	0.126
4/0	211600	19	0.106	0.528	0.219	0.0608	0.0626	0.100
250	——	37	0.082	0.575	0.260	0.0515	0.0535	0.0847
300	——	37	0.090	0.630	0.312	0.0429	0.0446	0.0707
350	——	37	0.097	0.681	0.364	0.0367	0.0382	0.0605
400	——	37	0.104	0.728	0.416	0.0321	0.0331	0.0529
500	——	37	0.116	0.813	0.519	0.0258	0.0265	0.0424
600	——	61	0.099	0.893	0.626	0.0214	0.0223	0.0353
700	——	61	0.107	0.964	0.730	0.0184	0.0189	0.0303
750	——	61	0.111	0.998	0.782	0.0171	0.0176	0.0282
800	——	61	0.114	1.03	0.834	0.0161	0.0166	0.0265
900	——	61	0.122	1.09	0.940	0.0143	0.0147	0.0235
1000	——	61	0.128	1.15	1.04	0.0129	0.0132	0.0212
1250	——	91	0.117	1.29	1.30	0.0103	0.0106	0.0169
1500	——	91	0.128	1.41	1.57	0.00858	0.00883	0.0141
1750	——	127	0.117	1.52	1.83	0.00735	0.00756	0.0121
2000	——	127	0.126	1.63	2.09	0.00643	0.00662	0.0106

Note, See **Table 8, Ch. 9** of the NEC.

Abbreviations

A

A - amps
AC - alternating current
A/C - air conditioning
AEGCP - assured equipment grounding conductor program
AHJ - authority having jurisdiction
Alu. - aluminum
ASCC - available short-circuit current
AWG - American Wire Gauge

B

BC - branch circuit
BCSC - branch-circuit selection current
BJ - bonding jumper
BK - black
BL - blue
BR - brown

C

°C - Celsius
CB - circuit breaker
CEE - concrete-encased electrode
CL - code letter
CM - circular mills
CMP - Code-making panel
Comp. - compressor
Cond. - condenser
Cont. - continuous
cu. - copper
cu. in. - cubic inches

D

DC - direct-current
dia. - diameter
DPCB - double pole circuit breaker

E

EBJ - equipment bonding jumper
EExde - increase safety
Eexe - flameproof/increased safety components
Eff. - efficiency
EGB - equipment grounding bar
EGC - equipment grounding conductor
EMT - electrical metallic tubing
ENT or ENMT - electrical nonmetallic tubing
Epf - explosionproof
Ex. - Exception

F

°F - Fahrenheit
FLA - full-load amperage
FLC - full-load current
FMC - flexible metal conduit
FPN - fine print note
Ft - foot

G

G - ground
GE - grounding electrode
GEC - grounding electrode conductor
GES - grounding electrode system
GFCI - ground fault circuit interrupter
GFL - ground fault limiter
GFPE - ground fault protection of equipment
GR - green
GRY - gray
GSC - Grounded service conductor

H

H - hot conductor
HACR - heating, air conditioning, cooling, and refrigeration
HP - horsepower
Htg. - heating
Hz - hertz

I

I - amperage or current
IEC - International Electrotechnical Commission
IG - Isolated ground
in. - inches
INST. CB - instantaneous trip circuit breaker
INVT - inverse-time circuit breaker
IRA - inrush amps
ITSC - intrinsically safe circuits

K

kFT - 1,000 ft.
kV - kilo volts
kVA - kilo volt-amps
kW - kilo watts
kWH - kilowatt-hour

L

L - length of conductor
Ld. - load
LFMC - liquidtight flexible metal conduit
LFNC - liquidtight flexible nonmetallic conduit
LPB - lighting panelboard
LRA - locked rotor amps
LRC - locked-rotor current
LTR - long time rated

M

MA - milli-amps
max. - maximum
MEL - maximum energy level
mf - microfarads
MGFA - maximum ground fault available
min. - minimum
min. - minute
MR - momentary rated
Mt. - motor

N

N - neutral
NACB - nonautomatic circuit breaker
NB - neutral bar
NEC - National Electrical Code
NEMA - National Electrical Manufacturers Association
NFD - nonfused disconnect
NFPA - National Fire Protection Association
NLTFMC - nonmetallic liquidtight flexible metal conduit
NPC - nameplate current of motor
NTDF - nontime-delay fuse

O

OCP - overcurrent protection
OCPD - overcurrent protection device
OL - overload
OLP - overload protection
OLs - overloads
OR - orange
OSHA - Occupational Safety and Health Administration

P

PF - power factor
Ph. - phases
pri. - primary
PSA - power supply assembly
PU - purple

R

R - ohms or resistance
RD - red
RMC - rigid metal conduit

S

SBS - structural building steel
SCC - short-circuit current
SDS - separately derived system
sec. - secondary
SF - service factor
SIA - seal-in amps
SP - single-pole
SPCB - single-pole circuit breaker
sq. ft - square foot (feet)
sq. in. - square inches
STR - short-time rated
SWD - switched disconnect
SWG - switchgear

T

TDC - time-delay cycle
TDF - time-delay fuse
TDL - time-delay limiter
TP - thermal protector
TR - temperature rise
TS - trip setting
TV - touch voltage

U

UF - underground feeder

V

V - volts
VA - volt-amps
VD - voltage drop

W

W - watts
WT - white
WP - weatherproof

X

XFMR - transformer

Y

YEL - yellow

Glossary

A

Across-the-line starter is a device consisting of contactor and overload relay that is used to start an electric motor by connecting it directly to the supply line.

Active power is the true electrical power or real power supplying the load.

Air gap is the air space between two electrically related parts, such as the space between poles of a magnet or poles in an electric motor.

Alternating current (AC) is the current in an electrical circuit that alternates in flowing first with a positive polarity and then with a negative polarity.

Alternator is a rotating machine whose output is AC.

Ambient conditions are the conditions of the atmosphere adjacent to electrical equipment.

Ambient temperature is the temperature of the surrounding atmosphere cooling medium, which comes into contact with the heated parts of equipment.

Ambient temperature compensated is a device, such as an overload relay, that is not affected by the temperature surrounding it.

Ampacity is the current, in amperes, that a conductor can carry continuously under the conditions of use without exceeding its temperature rating.

Ampere is a unit of intensity of electrical current produced in a conductor by an applied voltage.

Apparatus is a set of control devices used to help perform the intended control functions.

Apparent power is the sum of active power and reactive power. It is determined by multiplying voltage times current.

Arc-chute is a cover around contacts, designed to protect surrounding parts from arcing effects.

Armature is a specially designed rotor.

Armature reaction is the reaction of the magnetic field produced by the current on the magnetic lines of force, which are produced by the field coil of an electric motor or generator.

Automatic is a means of self acting; operating by its own mechanism when actuated by some impersonal influence such as, for example, a change in current, pressure, temperature, or mechanical configuration.

Automatic controller is a motor or other that mechanism which uses automatic pilot devices as activating devices. These devices may be pressure switches, level switches or thermostats.

Autotransformer starter is equipped with an autotransformer that is designed to reduce the voltage to the motor terminals and reduce the starting current, and still start the motor, etc.

Auxiliary contacts are contacts in addition to the main circuit contacts and function with the movement of the latter.

Auxiliary device is any device other than a motor or motor starter necessary to fully operate the machine or equipment.

Auxiliary interlock:

• Mechanical - A physical device or arm so arranged that it cannot close both starter circuits at the same time.

• Electrical - An additional contact mounted on the side of a magnetic starter.

B

Bearings are devices used to support the motor shaft and allow it to rotate smoothly.

Bimetallic disc is a disc made up of two strips of dissimilar metals combined to form a single strip.

Branch circuits are the circuit conductors between the final overcurrent device protecting the circuit and the outlet(s).

Breakdown torque is the maximum torque that a motor develops under increasing load conditions at rated voltage and frequency without an sudden drop in rotating speed.

Brushes are sliding contacts, usually made of carbon, which are located between a commutator and the outside circuit in a generator or motor.

C

Capacitance is the ability to store electricity in an electrostatic field.

Capacitor is a device that is designed to introduce capacitance into an electric circuit.

Capacitor-start motor is an AC split-phase induction motor that has a capacitor connected in series with an auxiliary winding, which provides a way for it to start. This auxiliary circuit is designed to disconnect to the motor when it reaches its speed.

Circuit is an electrical network of conductors that provides one or more paths for current.

Circuit breaker is a device designed to open and close a circuit by a nonautomatic means and to open the circuit automatically on a predetermined overcurrent, without damage to itself, when properly applied within its rating.

Combination starter is a magnetic starter having a manually operated disconnecting means built into the enclosure that houses the magnetic contactor or starter.

Commutator is a device that reverses the connections to the revolving loops on the armature.

Compensating windings are the windings embedded in the main pole pieces of a compound DC motor.

Component is the smallest element of a circuit.

Contactor is an electro-mechanical device for connecting and disconnecting an electric power circuit.

Contacts are connecting parts which co-act with other parts to connect or disconnect a circuit.

Control is a device or group of devices that is used in some predetermined manner to govern the electric power delivered to an apparatus.

Control circuit is the control apparatus that carries the signals directing the performance of the controller.

Control, three-wire is a control function that utilizes a momentary contact pilot device and a holding circuit contact to provide voltage to the coil of a controller. The holding circuit maintains the control circuit voltage.

Control, two-wire is a control function which utilizes a maintained contact-type pilot device to provide undervoltage release.

Control circuit transformer is utilized to supply a reduced voltage suitable for the operation of control devices.

Control circuit voltage is the voltage providing the operation to the coils of magnetic devices.

Controller is a device or group of devices that serves to govern, in some predetermined manner, the electric power delivered to the apparatus to which it is connected.

Copper loss is the electrical power lost through the resistance of the coils due to the current flowing through the wire of the coils.

Core is the magnetic path through the center of coil or transformer.

Core losses are the loss of power in the coil (core) due to eddy currents and hysteresis.

Core transformer is an electrical transformer which has the core inside of the coils.

Counter torque is a repulsion force between two magnetic fields.

Counterelectromotive force is the voltage induced in the armature coil of an electric motor.

D

Delta-delta connected is a coil connection in which the primary and secondary coils are delta connected.

Delta-wye connected is a connection in which the primary winding is delta connected while the secondary windings are wye connected.

Diagram is a connection diagram showing the electrical connection between the parts of the control and the external connections.

Direct current (DC) is a current that always flows in only one direction.

Disconnecting means is a motor circuit switch which is intended to connect and disconnect a circuit to a motor. It must be rated in horsepower and capable of interrupting the maximum current.

E

Eddy currents are the electrical currents circulating in the core of a transformer as the result of induction.

Efficiency is the ratio of output power (watts) to input power (watts).

Electric motor is a machine that converts electrical energy to mechanical energy.

Electricity is electrical charges in motion. Such movement is called current and is measured in amps.

Electrolytic capacitor is a capacitor that uses a liquid or paste as one of its electrical storage plates.

Electromagnet is a magnet comprised of a coil of wire wound around a soft iron core. When current is passed through the wire, a magnetic field is produced.

Electromagnetism is a magnetic field that exists around a wire or other conductor if a current is passing through it.

Electromechanical is a device that uses electrical energy to create mechanical motion of force.

Electromotive force is a voltage or force which causes free electrons to move in a conductor.

Electron is a negative electric charge.

Electron flow is the flow of electrons from a negative point to a positive point in a conductor.

Electrostatic charge is the electrical charge stored by a capacitor.

Electrostatic field is the stored electrical charges on the surface of an insulator.

Excitation is creating a magnetic field to be used to create electromagnetics when an electric current is passed through a coil.

F

Feeders are all circuit conductors between the service equipment, the source of a separately derived system, or other power supply source and the final branch-circuit overcurrent device.

Float switch is a switch that is operated by a float and is responsive to the level of liquid.

Foot switch is a switch that is suitable for operation by an operator's foot.

Frequency is the rate at which AC changes its direction of flow; it is normally expressed in terms of hertz (cycles) per second.

Fuse is an overcurrent protection device with a circuit opening fuseable member that opens when overheated by current passing through it.

G

Gate is one of the leads on a thyristor. This lead is the one that normally controls output when it is correctly biased.

Generator is a rotating machine that changes mechanical energy into DC.

Generator action is inducing voltage into a wire that is cutting a magnetic field.

Ground is the earth.

Guarded is covered, shielded, fenced, enclosed, or otherwise protected by means of suitable covers or casings, barriers, rails or screens, mats or platforms to remove the likelihood of dangerous contact or approach by persons or objects to a point of danger.

H

Hermetic refrigerant motor compressor is a combination consisting of a compressor and motor, both of which are enclosed in the same housing, with no external shaft or shaft seals, and the motor operating in the refrigerant.

Hertz is a measurement of frequency; it actually means "cycles per second" of AC.

High side is a transformer marking that indicates the high voltage winding.

Horsepower is a unit of measure for power that represents the force times distance times time. For example, one horsepower (HP) equals 746 watts, or 33,000 ft. lb. per minute, or 550 ft. lb. per second.

Hysteresis is the property of a magnetic substance that causes the magnetization to lag behind the magnetizing force.

I

Impedance is the total opposition to current flow in a circuit and is measured in ohms.

Induced current is the current that flows in a conductor because of a changing magnetic field.

Inductance is electromotive force resulting from a change in magnetic flux surrounding a circuit or conductor.

Induction is the generation of electricity by magnetism.

Inductive reactance is the opposition in ohms to an AC as a result of induction. This is voltage resulting from cutting lines of magnetic force.

Interlock is an electrical or mechanical device actuated by the operation of a different device to which it is directly related.

Intermittent duty is a requirement or service that demands operations for alternate intervals of (1) load and noload; (2) load and rest; or (3) load, noload, and rest such alternate intervals being definitely specified.

Isolating transformer is transformer used to electrically isolate one circuit from another.

Inverter is a circuit capable of receiving a positive signal and sending out a negative one, or vice versa. It is a device which changes AC to DC or vice versa.

J

Jogging is the rapid and repeated opening and closing of the circuit to start a motor from rest for the purpose of creating small movements of the motor or driven load.

K

kVA is the term used to rate transformers.

KVAR is the reactive power in a circuit.

kW is used to rate the load of certain types of equipment, etc.

L

Lamination consists of sheet material that is sandwiched together to construct a stator or rotor of a rotating machine.

Limit switch is a switch that is operated by a part or motion of a power-driven piece of equipment. Such operation alters the electric or electronic circuits related to the equipment.

Locked-rotor current of a motor is the current taken from the line when the motor starts or the rotor becomes locked in place.

Low side is a transformer marking that indicates which is the low voltage winding.

M

Magnetic starter is a starter that is actuated by an electro-magnetic means.

Manual controller is a device that manually closed or opened .

Manual reset is a device which requires manual action to re-engage the contacts after an overload.

Magnetic drive (magnetic clutch) is an electromagnetic device that is connected between a three-phase motor and its load. Its main purpose is regulating the speed at load rotated speed.

Magnetic field is the invisible lines of force found between the north and south poles of a magnet.

Magnetic lines of force in a magnetic field are imaginary lines which show the direction of the magnetic flux.

Maintained contacts close the circuit when the push button is pressed and will open the circuit when the push button is pressed again.

Megaohm is one million ohms.

Motor action is the mechanical force that exists between magnets. Two magnets approaching each other will either pull toward or push away from the other. In other words, there is a pull and push action between the rotor and field poles of the motor.

Multi-speed motor is a motor that is capable of operating at two or more fixed speeds.

N

NEMA (National Electrical Manufacturers Association) is an organization which establishes certain voluntary standards relating to motors such as operating characteristics, terminology, basic dimensions, ratings, and testing.

No-load speed is the speed reached by the rotor or armature when it rotates.

Nonautomatic is action requiring personal intervention for its control as applied to an electric controller; nonautomatic control does not necessarily imply a manual controller, but only that personal intervention is necessary.

Nonreversing is a control function that, provides for operation in one direction only.

Normally open and normally closed is a term when applied to a magnetically operated switching device, signifies the position that the contacts are in.

Normally closed contacts are motor control contacts (set) that are open when the push button is depressed.

Normally open contacts are contacts (set) which are closed when the push button is depressed.

O

Ohm is a unit of electrical resistance of a conductor.

Out-of-phase is a condition where two or more phases of AC are changing direction at different intervals of time.

Overcurrent protection device (OCPD) is a device that operates on excessive current, which causes the interruption of power to the circuit if necessary.

Over-excited is a condition where a synchronous motor is equipped with a DC field that supplies more magnetization than is needed.

Overload protector is a device affected by an abnormal operating condition, which causes the interruption of current flow to the device governed.

Overload relay is a device that provides overload protection for conductors and electrical equipment.

P

Parallel circuit is a circuit in which all positive terminals are connected at a common point and all negative terminals are connected at another point.

Periodic duty is a type of intermittent duty in which the load conditions are regularly recurrent.

Permanent-capacitor motor is a single-phase electric motor which uses a phase winding and capacitor in conjunction with the main winding. The phase winding is controlled by the capacitor, which remains in the circuit at all times.

Permanent magnetism is a condition in which a magnet keeps its magnetic properties indefinitely.

Permeability is a condition in which domains in a magnetic core can be made to line up to create magnetism.

Phase is the relationship of two wave forms to have the same frequency.

Phase angle is the difference in angle between two sine wave vectors.

Phase shift is the creation of a lag or advance in voltage or current in relation to another voltage or current in the same electrical circuit.

Phase voltage is the voltage across a coil.

Polarity is a condition where a magnet has north and south poles are positive and negative charge.

Polyphase is more than one phase, usually three-phase, that when related to generators, transformers, and motors.

Pounds force is an English unit of conventional measurement for force.

Power factor is the figure that indicates what portion of the current delivered to the motor is used to do work.

Primary coil is one of two coils in a transformer.

Prime mover is the primary power source that can be used to drive a generator.

Pull-in to torque is the maximum torque at which an induction motor will pull into step.

Pull-out torque is the maximum torque developed by a motor for one minute before it pulls out of step due to an overload.

Pull-up torque is the minimum torque developed by an induction motor during the period of acceleration from rest to full speed.

Push button control is the control and operation of equipment through push buttons used to activate relays.

Push button switch is a switch utilizing a button for activating a coil and contact to open or close a circuit.

R

Rainproof is an enclosure constructed, protected, or treated so as to prevent rain from interfering with the successful operation of the apparatus under specified test conditions.

Raintight is an enclosure constructed or protected so that exposure to a beating rain will not result in the extrance of water under specified test conditions.

Rated-load current is the current for a hermetic refrigerant motor-compressor is the current resulting when the motor-compressor is operated at the rated load, rated voltage, and rated frequency of the equipment it serves.

Rating is a designated limit of operating characteristics based on conditions of use such as load, voltage, frequency, etc.

Rating, continuous is the rating which defines the substantially constant load that can be carried for an indefinitely long time.

Reactive power is the reactive voltage times the current, or voltage times the reactive current, in an AC circuit.

Rectifier is an electrical device which converts AC to DC by allowing the current to move in only one direction.

Relay is a device that operates by a variation of a condition that affects the operation of other devices in an electric circuit.

Relay contacts are the contacts that are closed or opened by movement of a relay armature.

Reluctance is the ratio between the magnetomotive force and the resulting flux.

Reset is to restore a mechanism or device to a prescribed state.

Reset, automatic is a function that operates automatically to re-establish certain circuit conditions.

Reset, manual is a function that requires a manual operation to re-establish certain circuit conditions.

Residual magnetism is the magnetism remaining in the core of a coil or an electromagnet after the current flow has been removed.

Resistance is a property of conductors that makes them resist the movement of current flow.

Resistance starting is a reduced-voltage starting method employing resistances which are short circuited in one or more steps to complete the starting cycle of a motor, etc.

Resistors are electrical-electronic devices that are attached to a circuit to produce resistance to current flow.

Rheostat is a variable resistor with a fixed terminal and a movable contact.

Rotor is the rotating section that rotates within the stator of a motor.

Rotor impedance is the phasor sum of resistance and inductive reactance.

RPM is the number of revolutions per minute.

Running torque is the torque or turning effort determined by the horsepower and speed of a motor at any given point of operation.

S

Saturated is the point at which an electrical or magnetic component can not receive any more electrical current or magnetism.

Saturation is a point at which a magnet will receive any more flux density.

Sealing, voltage or current is the voltage or current required to seat the armature of a magnetic circuit closing device to the make position.

Secondary coil is the coil that is connected to the load in the electrical circuit.

Self excitation is a condition of supplying excitation voltages by a device on the generator rather than from an outside source.

Self induction is a counterelectromotive force produced in a conductor when the magnetic field produced by the conductor collapses or expands after a change in current flow.

Separate excitation is a condition of producing generator field current from an independent source.

Series circuit is a circuit in which all resistances and other components are connected so that the same current flows from point to point.

Series field is the total magnetic flux caused by the action of the series winding in a rotating piece of machine.

Series motor is a motor in which the field and armature circuits are connected in series.

Service factor is the number by which the horsepower rating is multiplied to determine the maximum safe load that a motor can carry continuously at its rated voltage and frequency.

Shaded-pole motor is a single-phase squirrel cage induction motor with stator poles slotted and used to create two sections in each pole.

Shading coil is a copper ring or coil that is set into a section of the pole piece; its function is to produce the lagging part of a rotating magnetic field for starting torque.

Short is any two points of a motor where there is zero, or extremely low, resistance between them or between two motor components.

Short-time rating is referring to the motor load that can be carried for a short and definitely specified time.

Shunt field is a type of field coil designed for a DC motor that is connected in parallel with the armature.

Silicon controlled rectifier (SCR) is a semiconductor device that has the ability to block a voltage that is applied in either direction. On a signal applied to its gate, it is capable of conducting current even when the signal has been removed.

Single-phase is having only one AC or voltage in a circuit.

Slip is the difference between the synchronous speed of a motor and the speed at which it operates.

Slip ring motor is a motor that has a rotor with the same number of magnetic poles at the stator.

Slip rings are equipped with circular bands on a rotor, which are used to transmit current from rotor coils to brushes.

Slip speed is the difference between the rotor speed and synchronous speed in an induction motor.

Soft neutral position is a condition where the brushes of a repulsion electric motor are aligned with the stator field.

Solid state controls are devices that control current to motors through semiconductors.

Solid state devices contain circuits and components using semiconductors.

Solid state relay is a relay that uses semiconductor devices.

Split-phase (resistance-start) motor is a single-phase induction motor equipped with an auxiliary winding that is connected in parallel with the main winding.

Squirrel-cage rotor is designed with a rotor which is made up of metal bars that are short-circuited at each end.

Stall torque is the amount of torque that the rotor of an energized motor produces when the rotor is not rotating.

Starter is a controller for accelerating a motor from rest to its running speed.

Starter, automatic is a starter that automatically controls the starting of a motor.

Starter, autotransformer is a starter that is provided with an autotransformer that provides a reduced voltage for starting.

Starter, part-winding is a starter that provides applied voltage to partial sections of the primary winding of an AC motor.

Starter, reactor is a starter that includes a resistor connected in series with the primary winding of an induction motor to provide reduced voltage for starting.

Starter, wye-delta is a starter that connects the motor leads in a wye configuration for reduced voltage in starting and reconnects the leads in a delta configuration for the run position.

Starting torque is the amount of torque produced by a motor as it breaks the motor shaft from standstill and accelerates to its running speed.

Static electricity is electricity at rest. It is also known in the industry as a static charge.

Stator is the portion of motor that contains the stationary parts of the magnetic circuit with their associated windings.

Stator field contains a magnetic field that is set up in the electric motor when the motor is energized and electric current is flowing.

Stator poles are the shoes on an electric motor stator that hold the windings and the magnetic poles of the stator.

Switch is a device for making, breaking, or changing the connections in an electric circuit.

Switch, float is a switch that is responsive to the level of a liquid.

Switch, foot is a switch that is operated by an operator's foot.

Switch, general-use is a general-use type non-horse rated switch that is capable of interrupting the rated current at the rated voltage.

Switch, limit is operated by some part or motion that alters the electrical circuit associated with the equipment.

Switch, master controls the operation of contactors, relays, or other similar operated devices.

Switch, motor circuit is rated in horsepower and is capable of interrupting the maximum operating current of the motor.

Switch, pressure is operated by fluid pressure, etc.

Switch, selector is a manually operated multiposition switch that is used for selecting an alternative control circuit.

Synchronous is a condition where the currents and voltages are in-step or in-phase.

Synchronous motor is an induction motor that runs at synchronous speed.

Synchronous speed is the constant speed to which an AC motor adjusts itself, depending on the frequency of the power source and the number of poles in the motor.

T

Tachometer is a device that is capable of measuring the rotational speed of rotating machines.

Tap changer is a mechanical device that has the ability to change the voltage output of a transformer.

Taps are fixed electrical connections that are located at specific positions on a transformer's coil.

Temperature, ambient is the temperature of the medium, such as air, oil, etc., into which the heat of the equipment is dissipated.

Terminal is a point at which an electrical element may be connected to another electrical element.

Terminal board is an insulating base equipped with one or more terminal connectors used for making electrical connections.

Thermal, cutout is an overcurrent protection device that has a heater element that affects a fusible member that open the circuits due to an overload.

Thermal protector is a protective device that is an integral part of the motor; that is designed to protect the motor windings from dangerous overloads.

Thermocouple is a device that consists of two unlike metals that are joined together; when heat is applied, a current will flow.

Thermostat is an instrument that responds to changes in temperature to affect control over an operating condition.

Three-phase alternator is a rotating machine that generates three separate phases of AC.

Three-phase electric motor is a motor that operates from a three-phase power supply.

Timer is a device that is designed to delay the closing or opening of a circuit for a specific period of time.

Torque is a force that produces a rotating or twisting action.

Torque, breakdown is the maximum torque that a motor develops with rated voltage when applied at rated frequency.

Torque, locked rotor is the minimum torque that a motor develops at standstill when rated voltage is applied at rated frequency.

Transformer is an device designed to change the voltage in an AC electrical circuit. Step-up transformers increase the voltage and lower the current. Step-down transformers decrease the voltage and raise the current.

Transformer efficiency is the ratio of input power to output power.

Turns ratio is the ratio of the number of turns in the primary winding of a transformer to the number of turns in the secondary winding of a transformer.

Two-capacitor motor is an induction motor which uses one capacitor for starting and one for running.

U

Under-excited is a term used to describe the magnetizing power of a synchronous motor.

Unity power factor is a power factor of 1; this is the best PF that can be obtained in an electrical system.

Undervoltage protection is a device that operates on the reduction or failure of voltage and has the ability to maintain the interruption of power.

Undervoltage release is a device that operates on the reduction or failure of voltage and has the ability to interrupt the power, but not to prevent the re-establishment of the circuit.

V

Vector is an in-phasor diagram having lines with a specific length and direction.

Voltage is a force that, when applied to a conductor, produces a current in the conductor.

W

Watt is a unit of electrical power; it is the product of voltage and amperage.

Wattmeter is an instrument used for measuring electrical power.

Wye or star connection is an electrical connection in that all terminals are joined at the neutral junction, resembling a wye connection.

Wye-wye connection is the coil arrangement in which both the primary and the secondary coils are wye-connected.

C

cabinets 6-18, 7-1, 7-10, 11-54, 11-56
cable 6-6, 6-7, 6-12, 6-14, 6-16, 6-17, 6-18, 6-19, 6-24, 6-32, 7-2, 7-10, 8-4, 8-6, 8-9, 8-11, 8-12, 8-13, 8-15, 9-18, 11-44, 11-54, 11-56, 11-60, 11-61, 11-67, 11-68, 11-69, 11-73, 12-3, 12-4, 12-8, 12-10, 12-11, 12-12, 12-23, 12-24, 12-25, 12-26, 13-1, 13-3, 13-4, 13-5, 13-6, 13-7, 13-13, 13-15, 13-17, 13-18, 13-19, 14-10, 14-11, 14-13, 15-4, 15-9, 15-10, 16-3, 16-8, 16-11, 17-2, 17-17, 17-18, 20-13, 20-27, 21-17

F

G

H